Properties of

SILICON

The Institution of Electrical Engineers

EMIS Datareviews Series No. 4

Published by: INSPEC, The Institution of Electrical Engineers, London and New York

© 1988   The Institution of Electrical Engineers

ISBN 0 85296 475 7

D
620·193

PRO

Printed in the United Kingdom by Unwin Brothers, The Gresham Press, Old Woking, Surrey, England.

Properties of
SILICON

EMIS Datareviews Series

The series is bringing together materials property data, obtained in the course of many different and varied research programmes, into collections comprising short articles which have been called 'Datareviews'.

Each Datareview normally focuses on one property of a narrowly-defined material system. A Datareview contains both a summary of data extracted from selected articles which have been published in scientific journals or conference proceedings and bibliographic references to these articles and any other key sources. The Datareview is distinguished from the conventional review by its shorter length, high degree of specialisation and by its emphasis on hard data. However, like any scientific review it seeks to discuss, compare and illuminate research results, thus enabling the reader to gain an insight and up-to-date awareness of its subject matter as well as to conduct further study by means of the references provided.

Datareviews are compiled by researchers from universities and industrial research laboratories and by the EMIS Group at INSPEC (the Information Services Division of the Institution of Electrical Engineers). The selection of articles for the compilation of Datareviews, and the completed Datareviews themselves, are refereed by prominent researchers in the appropriate field of study. INSPEC is thus indebted to a large number of people who give of their time in the preparation of Datareviews for these volumes. They are gratefully acknowledged within the publication.

In addition to appearing in this and other publications of the series, Datareviews are published electronically in the EMIS (Electronic Materials Information Service) knowledge bank, a publicly available online service produced by INSPEC.

The EMIS knowledge bank was created and is being developed by INSPEC to satisfy the need for the latest and most authoritative data on the properties of electronic materials. Experts in the field, who pointed to this need, assist INSPEC staff in collecting, selecting and analysing the most recently published data.

This knowledge bank provides a collecting point for Datareviews prior to publication in the EMIS Datareviews Series, and also a facility for updating existing Datareviews as new data become available.

Individual publications in the EMIS Datareviews Series will be reprinted as and when new data added to the EMIS knowledge bank represent a significant advance in the state-of-the-art.

Foreword

Silicon is undoubtably the most widely researched and applied semiconductor.
There are more than ten well-established annual international conferences
devoted to silicon-related topics, ranging from the very fundamental,
such as defects and the silicon/silicon dioxide interface, to the very
applied, such as devices, circuits, VLSI logic and VLSI memory technology.
The evolution of silicon technology has been extremely rapid on all of
the established fronts, while at the same time new fronts are continually
being established. For example, in the technology mainstream, the focus
was metal-gate NMOS and bipolar in the 1960s, and silicon gate NMOS and
bipolar in the 1970s. Since the early 1980s the focus has evolved into
CMOS, bipolar, bipolar/CMOS combination (BiCMOS), sensors, and monolithic
integration of digital circuits with special purpose silicon and/or
non-silicon devices.

Active researchers of today and tomorrow in silicon technology are likely to
be specialists in selected subfields, but must be able to draw upon the vast
data and literature in other related subfields. The requirement of breadth
comes from the vastly increased complexity and sophistication of the silicon
device and chip fabrication processes. I believe it is in satisfying this
requirement of breadth that active researchers will find this book in the
EMIS Datareviews Series very timely and valuable. It is virtually
impossible even for the well-informed researcher to keep up with the rapid
progress in the many subfields of silicon technology development. This book
should make referencing and information searching much easier.

The approximately 250 specialized reviews by 70 internationally-known
researchers around the world make this book the most ambitious and
complete volume on the properties of silicon. With periodic update and
revision, made possible by the computer medium used in the data storage,
this EMIS book is expected to be useful to research scientists and
technologists in silicon for years to come.

Tak H. Ning
IBM Thomas J Watson Research Center
Yorktown Heights
August, 1987

Properties of Silicon

Introduction

Silicon is the most important material exploited by mankind. Life as we know it in the modern world is built around the integrated circuit, in our work, our communications, our leisure, our entertainment, our shopping. In less than 30 years, the move in electronics from the vacuum tube to the transistor has transformed the equipment that we use daily. Silicon is everywhere.

That being so, it is difficult for the non-specialist to understand why this book should be the fourth in a series, rather than the first. To explain this, we have to probe into the motivations of scientists, and to briefly recall the early history of semiconductor physics. In the beginning there was only germanium, which died early, to be replaced by silicon. There was much research on silicon in the 1950s but by 1962 most scientists thought that all was now known, and the future of semiconductor physics lay in documentation of the variation of properties with temperature and magnetic field. The escape from such a boring prospect could come only by research into new materials, e.g. compound semiconductors like gallium arsenide and cadmium mercury telluride. There was some truth in this view - the excitement of the semiconductor laser, of electron transfer, of two-dimensional structures and superlattices, certainly passed silicon by. For silicon is not a remarkable semiconductor. Its mobility is low, its band-gap indirect, its electrons heavy. It is a workhorse rather than a racing stallion. But its uses are truly remarkable, and often represent the triumph of attention to fine detail over these limitations in physical properties. Those who have chosen the excitement of pushing back frontiers of performance with exotic semiconductors are always conscious of the silicon practitioners breathing down their necks, gradually overcoming the hurdles with greater circuit complexity and exquisitely toleranced dimensions. Almost nothing seems impossible now that the design and testing of the new generation of chips is largely in the hands of the last generation, the human setting only the strategy.

So the pragmatist beats the romantic, the tortoise overtakes the hare.

These refinements in device design and the wider spread of applications become possible only if we have much knowledge about silicon as a material. We need to know not just how pure silicon behaves, but also what impurities we can introduce with advantage. We need to know how silicon interacts with the metals and alloys we want to exploit as contacts. Above all, we need to know a great deal about the insulating layers that are deposited or created on the silicon surface, for it is in that interface that the seeds of success or failure as a device often lie.

This knowledge has been garnered over many years by scientists around the world, but it has not been previously available as a coherent body of information in a convenient form, distilled by experts who can distinguish the sloppy attempt to publish a quick measurement from the result of painstaking care and scholarship. The section headings which follow this introduction indicate the breadth of our coverage, the authors' names speak to the quality of the data and comments.

Though the greatest care has been taken in establishing the most accurate information for inclusion, in this field there is constant revision and addition to the data. The format of the EMIS books allows the reader to be brought up to date.

All the reviews which appear in this book are included in a central knowledge bank which can be accessed via a terminal, and they can be revised at any time.

This is a valuable addition to the literature on a material of critical importance to all of us. It is a bold enterprise which deserves success.

Cyril Hilsum
GEC Hirst Research Centre
Wembley
England
September, 1987

Contributing Authors

H Ahmed	Microelectronics Research Laboratory, UK	18.4,24.4
L N Aleksandrov	Academy of Sciences, USSR	11.6
P J Ashby	Mullard (Southampton), UK	12.3,12.6 12.9,12.12
D E Aspnes	Bell Communications Research, USA	2.1-2.7
K G Barraclough	Royal Signals and Radar Establishment, UK	12.2,12.4, 12.5,12.8
H S Bennett	Semiconductor Electronics Division, NBS, USA	7.3
I W Boyd	University College, London, UK	16.2,16.4, 24.1
J C Brice	Copthorne Bank, UK	1.18-1.20
A A Brown	STC Technology Ltd, UK	13.1,13.7
P J Burnett	Shell Research Ltd, UK	1.6, 1.9-1.11
L T Canham	Royal Signals and Radar Establishment, UK	12.10,12.11 15.1-15.6
A Chantre	CNET, France	18.1
C Y-P Chao	University of Illinois, USA	17.4,17.14 17.7-17.15, 17.17,17.21
A J Chen	University of Illinois, USA	17.4,17.14, 17.17,17.18 17.19
L J Chen	National Tsing Hua University	19.5,19.6, 19.13,19.15, 19.16,19.20
J P R David	University of Sheffield, UK	4.1-4.3
D de Cogan	University of Nottingham, UK	13.2-13.5, 13.10-13.13
J A del Alamo	NTT, Japan	5.5 5.7-5.9, 6.3-6.6

M Di Marco	UMIST, UK	10.4
P J Dobson	Philips Research Laboratories, UK	22.1
D J Dunstan	University of Surrey, UK	7.1,7.2, 7.4,7.5
J P Ellul	Northern Telecom Electronics Ltd, Canada	16.3
A G R Evans	University of Southampton, UK	3.4
Wu Fengmei	Nanjing University, China	11.5
P J French	University of Southampton, UK	3.4
H Fritzsche	University of Chicago, USA	27.5
G A Garcia	Naval Oceans Systems Center, USA	18.5,18.6
D J Godfrey	GEC Hirst Research Centre, UK	13.17,13.18
S M Goodnick	Oregon State University, USA	17.23
H G Grimmeiss	University of Lund, Sweden	10.6
V Grivickas	Vilnius V Kapsukas State University, Lithuanian SSR	8.1-8.7, 8.4-8.7
C R M Grovenor	University of Oxford, UK	26.1,26.2
S Guimaries	UMIST, UK	10.2
B Hamilton	UMIST, UK	9.1-9.3
S C Hardy	Metallurgy Division, NBS, USA	1.21
R Harper	UK Atomic Energy Authority, UK	24.3, 30.4-30.7
M Hart	University of Manchester, UK	1.2
N Hecking	Universitat Dortmund, German Federal Republic	11.4
P L F Hemment	University of Surrey, UK	18.10,18.11
C Hill	Plessey Research (Caswell) Ltd, UK	26.3-26.5
A Hodge	Royal Signals and Radar Establishment, UK	18.7

K K Ng	AT&T Bell Laboratories, USA	20.1-20.14, 21.1-21.11
K H Nicholas	Philips Research Laboratories, UK	3.3
O H Nielsen	NORDITA, Copenhagen, Denmark	1.4,1.5,1.7, 1.8
T Nishida	University of Illinois, USA	17.4,17.14, 17.18-17.21
D Nobili	LAMEL, Bolognia, Italy	13.9, 13.14-13.16
A O Ogura	NEC Corporation, Japan	18.9
H Ohsaki	Shinshu University, Japan	1.1
P Pan	IBM, Essex Junction, USA	17.22,17.25, 17.26
M Pawlik	GEC Hirst Research Centre, UK	3.1
A R Peaker	UMIST, UK	10.1-10.3
H R Philipp	General Electric Research and Development Center, USA	28.1-28.4, 29.1,29.2
C Pickering	Royal Signals and Radar Establishment, UK	18.7
E H Poindexter	US Army Electronics Technology and Devices Laboratory, USA	17.6
M R Polcari	IBM Thomas J. Watson Research Center, USA	23.1-23.5
K V Ravi	Crystallume, USA	11.1-11.3, 25.1-25.3 31.1-31.3
R E Reedy	Naval Oceans Systems Center, USA	18.5,18.6
J M Rorison	Royal Signals and Radar Establishment, UK	5.3,5.4
P J Rosser	STC Technology Ltd, UK	13.1,13.6, 13.7,13.8, 16.1,19.19, 30.2,30,3
G Rossi	Laboratoire pour l'Utilisation Rayonnement Electromagnetique, France	19.4,19.8, 19.11,19.17, 19.22

Acknowledgements

It is a pleasure to acknowledge the work both of the contributing authors named on the previous page and of the following experts in the field who also have participated in the production of this book:

J Albers	NBS, USA
V A Arora	Solar Energy Research Institute, USA
R L Baker	STC Technolgoy Ltd, UK
P Balk	Aachen Technical University, German Federal Republic
J C Brice	Copthorne Bank, UK
T Carlberg	Royal Institute of Technology, Sweden
A G Cullis	Royal Signals and Radar Establishment, UK
D F Edwards	Lawrence Livermore National Laboratory, USA
R B Fair	Microelectronics Center of North Carolina, USA
J M Gibson	AT&T Bell Laboratories, USA
A Glaccum	British Telecom Research Laboratories, UK
L A Grant	STC Technology Ltd, UK
R Heckingbottom	British Telecom Research Laboratories, UK
C Heslop	British Telecom Research Laboratories, UK
A Hobbs	University of Dublin, Ireland
S Horita	Kanazawa University, Japan
T Inoue	Iwaki-Meisei University, Japan
C Jacoboni	University of Modena, Italy

L Kirkup	Paisley College of Technology, UK
H K urz	Aachen Technical University, German Federal Republic
B Leroy	IBM, France
E C Lightowlers	King's College, London, UK
E Luder	University of Stuttgart, German Federal Republic
J F Marchiando	NBS, USA
W P Maszara	Allied Bendix Aerospace, USA
D Mathiot	Centre National d'Etudes des Telecommun, France
C J McHargue	Oak Ridge National Laboratory, USA
J McLaughlin	University of Southampton, UK
Sir Nevill Mott, FRS	Milton Keynes, UK
A Neugroschel	Florida University, USA
R C Newman	University of Reading, UK
D C Northrop	UMIST, UK
S L Partridge	GEC Hirst Research Centre, UK
M Pepper	Cavendish Laboratory, Cambridge, UK
M Prutton	University of York, UK
P Scovell	STC Technology Ltd, UK
R Sinclair	Stanford University, USA
E H Smith	GEC Hirst Research Centre, UK
H J Stein	Sandia National Laboratories, USA
C W Turner	King's College, London, UK
P E van Camp	University of Antwerp, Belgium
M D Vaudin	NBS, USA

L Vescan	Aachen Technical University, German Federal Republic
Y K Vohra	Cornell University, USA
D P Vu	Centre National d'Etudes des Telecommun, France
B L Weiss	University of Surrey, UK
R H Williams	University of College, Cardiff, UK
A F W Willoughby	University of Southampton, UK
C Wood	GEC Hirst Research Centre, UK
K Yamazaki	Matsushita Electric Industrial Co Ltd, Japan
S Zaima	Nagoya University, Japan

Abbreviations and Commonly Used Symbols

AEI	avalanche electron injection
AES	Auger electron spectroscopy
AHI	avalanche hole injection
APCVD	atmospheric pressure chemical vapour deposition
ARUPS	angle-resolved ultraviolet photoelectron spectroscopy
bcc	body centred cubic
BH	barrier height
CMOS	complementary metal-oxide-semicondcutor
CPU	central processing unit
CRN	continuous random network
CVD	chemical vapour deposition
CW	continuous wave
CZ	Czochralski
DC	direct current
DLTS	deep level transient spectroscopy
E_c	conduction band edge
E_f	Fermi level
E_g	indirect band gap
E_v	valence band edge
EBIC	electron beam induced current
EELS	electron energy loss spectroscopy
ENDOR	electron nuclear double resonance
EPR	electron paramagnetic resonance
ESCA	electron spectroscopy for chemical analysis
ESR	electron spin resonance
EXAFS	extended X-ray absorption fine structure
exp	exponential
fcc	face centred cubic
FIM	field ion microscopy
FZ	float zone
hcp	hexagonal close packed
IR	infrared
LEED	low energy electron diffraction
LOCOS	local oxidation of silicon
LPCVD	low pressure chemical vapour deposition
LPVPE	low pressure vapour phase epitaxy
LVM	localised vibrational mode
MBE	molecular beam epitaxy
ML	monolayer
MODFET	modulation doped field effect transient
MOS	metal-oxide-semiconductor
MOSC	metal-oxide-semiconductor capacitor
MOSFET	metal-oxide-semiconductor field effect transistor

PBT	permeable base transistor
PC	photoconductivity
PECVD	plasma enhanced chemical vapour deposition
PSG	phosphosilicate glass
RBS	Rutherford backscattering
RF	radio frequency
RHEED	reflection high energy electron diffraction
RT	room temperature
RTA	rapid thermal annealing
RTP	rapid thermal processing
SEM	scanning electron microscopy
SEXAFS	surface extended X-ray absorption fine structure
SIMOX	separation by implanted oxygen
SIMS	secondary ion mass spectroscopy
SOI	silicon on insulator
SOS	silicon on sapphire
SPE	solid phase epitaxy
SQRT	square root
STEM	scanning transmission electron microscopy
TCP	temperature coefficient of piezoresistance
TCR	temperature coefficient of resistance
TEM	transmission electron microscopy
TSC	thermally stimulated current or
	thermally stimulated capacitance
UHV	ultrahigh vacuum
ULSI	ultra large scale integration
UPS	ultraviolet photoelectron spectroscopy
UV	ultraviolet
VHN	Vickers hardness number
VLSI	very large scale integration
VPE	vapour phase epitaxy
VUV	vacuum ultraviolet
XPS	X-ray photoelectron spectroscopy
YAG	yttrium aluminium garnet

CONTENTS

1.   MECHANICAL AND THERMAL PROPERTIES   1

      O H Nielsen, P J Burnett, M N Wybourne, J C Brice,
      Y Tatsumi, M Hart, J Z Hu, T Soma, H Matsuo Kagaya,
      J A van Vechten, S C Hardy, H Ohsaki

Information on hole transport both here and
elsewhere in the book is comprehensively
indicated in the subject index.

CHAPTER 1

MECHANICAL AND THERMAL PROPERTIES

1.1 DENSITY OF CRYSTALLINE AND AMORPHOUS Si

by Y.Tatsumi, H.Ohsaki

June 1987
EMIS Datareview RN=15747

A. CRYSTALLINE Si

Density of silicon single crystals (c-Si) has been accurately measured by researchers from the U.S. National Bureau of Standards (NBS), for the purpose of obtaining a solid density standard to take the place of water which is the commonly adopted density standard [1-7]. The density is given to be 2.3290028 g/cm*3 at 25 deg C. They have measured the densities of several c-Si specimens of lightly doped transistor grades and have reported it to be 2.328993 - 2.329037 (note) g/cm*3 at 25 deg C (note: this value is corrected using the thermal expansion coefficient given by Lyon [8]). Petley discusses a relation between the atomic weight and the density, and points out that the total range of variation of the atomic weight for silicon is +- 0.001 (i.e +- 36 ppm) [9]. Accordingly, we deduce that the density of c-Si is 2.3290 +- 0.00008 g/cm*3 at 25 deg C.

Because impurities which add weight to the lattice also cause it to expand, to a first order, impurities do not affect the density. At normal doping levels (< 10*18/cm*3) density changes should be negligible compared with the uncertainties given above. At very high doping levels (> 10*19/cm*3) corrections may be necessary.

Density of c-Si at various temperature can be evaluated from the thermal expansion coefficients given both at low temperature (6-340 deg K) by Lyon [8] and at high temperature (350-1500 deg K) by Okada [10]. It is then found that the density d (g/cm*3) at T (deg K) has values:

TABLE 1

T	< 55	60-85	90-150	160-170	200	250	300
d	2.3306	2.3307	2.3308	2.3307	2.3304	2.3298	2.3290

T	350	400	500	600	700	800	900
d	2.3280	2.3269	2.3245	2.3220	2.3192	2.3164	2.3136

T	1000	1100	1200	1300	1400	1500
d	2.3106	2.3077	2.3047	2.3016	2.2986	2.2955

The apparent uncertainties are less than +- 0.0001 g/cm*3.

B. AMORPHOUS Si

Amorphous Si (a-Si) is prepared by vacuum evaporation, sputtering, glow-discharge decomposition and so forth. The physical properties depend on preparation conditions. The densities spread in the region of 70-100% of that of c-Si [11-21].

The a-Si films are classified into a-Si films deposited by vacuum evaporation and hydrogenated a-Si films (a-Si:H) deposited by various preparation methods. The densities are given separately.

TABLE 2: densities of the a-Si films deposited by vacuum evaporation
(with deposition conditions)

Density (g/cm*3)	Density Ratio of a-Si to c-Si (%)	Substrate Temp.(deg C)	Deposition Rate (nm/s)	Pressure (Pa)	Ref
1.71+-0.09	73	RT	0.05-0.19	6 X 10*-4 6 X 10*-6	[11]
2.16+-0.10	93	RT	1.5-3	–	[12]
2.2	94	RT	–	10*-5	[13]
	89-97	25-400	0.1-1.6	7 X 10*-5	[14]
	75-80	RT	0.2-1	10*-4	[15]
	> 90	275	0.2-1	10*-4	[15]

The a-Si:H films have become a major interest in the field of electronics and their physical properties have been studied by many researchers. Density of the a-Si:H films depends on preparation methods and deposition conditions.

TABLE 3: densities of sputtered a-Si:H films [16]

Density (g/cm*3)	Substrate Temp. (deg C)	Deposition Rate (nm/s)	H2 Partial Pressure (microns)
1.96	200	0.1	1.5
2.10	200	0.1	0.4
2.29	250	0.1	0

Densities of the a-Si:H films deposited at a substrate temperature of 77 deg K decrease from 1.71 to 1.30 g/cm*3 with increasing H2 partial pressure from 2.5 to 30% [17].

Density of the a-Si:H films deposited by thermal decomposition of silane (CVD: Chemical Vapour Deposition) at temperature around 600 deg C is 0.97 of that of c-Si [18].

TABLE 4: densities of plasma deposited a-Si:H films obtained as a function of hydrogen content, substrate temperature and DC bias [19]

Density (g/cm*3)	H content Relative to Si	Substrate Temp. (deg K)	DC bias
2.10	0.25	300	cathode
2.15	0.16	540	cathode
2.27	0.20	720	cathode
1.92	0.35	300	anode
2.17	0.13	540	anode
2.21	0.12	720	anode

Density of the a-Si:H films deposited by glow-discharge decomposition is 1.9 +- 0.15 g/cm*3 for a deposition rate in the region of 0.2 - 1.2 nm/s at room temperature [20]. Furthermore, the densities are given as a function of substrate temperature [21]. Typical values (from graph) are:

Substrate Temp. (deg C)	100	150	200	250	300	350
Density (g/cm*3)	2.07	2.10	2.18	2.28	2.35	2.38

Uncertainties are less than +-0.11 g/cm*3.

Densities of a-Si or a-Si:H films change with preparation conditions. So far no definite relation between the density and the preparation conditions has been established. Therefore, on referring to the density of a-Si or a-Si:H, it is necessary to study full preparation conditions in the original paper.

REFERENCES
----

[1]   R.D.Deslattes et al [Phys. Rev. Lett. (USA) vol.33 no.8 (1974)
      p.463-6]
[2]   H.A.Bowman, R.M.Schoonover, C.L.Carroll [J. Res. Natl. Bur. .
      Stand. (USA) vol.78A no.1 (1974) p.13-40]
[3]   H.A.Bowman, R.M.Schoonover, C.L.Carroll [Metrologia (Germany)
      vol.10 (1974) p.117-21]
[4]   R.S.Davis [Metrologia (Germany) vol.18 (1982) p.193-201]
[5]   R.D.Deslattes [Proc. of Course LXVIII Metrology and Fundamental
      Constants, Enrico Fermi Summer School of Physics, Varenna, Italy,
      1976 Soc. Italiana di Fisica, Bologna, p.38-113 (1980)]
[6]   A.Peuto, R.S.Davis [J. Res. Natl. Bur. Stand. (USA) vol.90 no.3
      (1985) p.217-27]
[7]   R.D.Deslattes, A.Henins, R.M.Schoonover, C.L.Carroll, H.A.Bowman
      [Phys. Rev. Lett. (USA) vol.36 no.15 (1976) p.898-900]
[8]   K.G.Lyon, G.L.Salinger, C.A.Swenson, G.K.White [J. Appl. Phys.
      (USA) vol.48 no.3 (1977) p.865-8]
[9]   B.W.Petley [The Fundamental Physical Constants and the Frontier
      of Measurement (Bristol, Hilger 1985) Sections 3.4.1 nd 3.5]
[10] Y.Okada, Y.Tokumaru [J. Appl. Phys. (USA) vol.56 no.2 (1984)
      p.314-20]
[11] Y.Tatsumi et al [Jpn. J. Appl. Phys. (Japan) vol.25 no.8 (1986)
      p.1152-5]
[12] D.Beaglehole, M.Zavetova [J. Non-Cryst. Solids (Netherlands)
      vol.4 (1970) p.272-8]
[13] R.Mosseri, C.Sella, J.Dixmier [Phys. Status Solidi a (Germany)
      vol.52 (1979) p.457-9]
[14] M.H.Brodsky, D.Kaplan, J.F.Ziegler [Appl. Phys. Lett. (USA)
      vol.21 no.7 (1972) p.305-7]
[15] G.K.M.Thutupalli, S.G.Tomlin [J. Phys. C (GB) vol.10 (1977)
      p.467-77]
[16] E.C.Freeman, W.Paul [Phys. Rev. B (USA) vol.20 no.2 (1979)
      p.716-28]
[17] J.J.Hauser, G.A.Pasteur, A.Staudinger [Phys. Rev. B (USA) vol.24
      no.10 (1981) p.5844-51]
[18] M.Hirose, M.Taniguchi, Y.Osaka [Proc. 7th Int. Conf. Amorphous
      and Liquid Semiconductors (Edinburgh, 1977) Ed. W.E.Spear p.352-6]
[19] C.C.Tsai, H.Fritzsche, M.H.Tanielian, P.J.Gaczi, P.D.Persans,
      M.A.Vesaghi [Proc. 7th Int. Conf. Amorphous and Liquid Semicon.,
      (Edinburgh, 1977) Ed. W.E.Spear p.339-42]
[20] R.C.Chittick [J. Non-Cryst. Solids (Netherlands) vol.3 (1970)
      p.255-70]
[21] I.Weitzel, R.Primig, K.Kempter [Thin Solid Films (Switzerland)
      vol.75 (1981) p.143-50]

1.2 LATTICE PARAMETER OF Si

by M. Hart

August 1987
EMIS Datareview RN=3491

PERFX floating zone dislocation free silicon, free of oxygen and carbon with total impurity content less than 10*16 atoms per cm*3, with a range of lattice parameter DELTA d / d of 8 (+- 3) X 10*-8 in six crystals [1], has according to direct comparison of (220) Bragg reflection with a He-Ne laser stabilised to the B peak of I-129 by simultaneous X-ray and optical interferometry, a lattice parameter of

> 5.4310626 +- 0.0000008 A at 1 atm
> 5.4310644 +- 0.0000008 A under vacuum

at 25 deg C [2-4]. The units are absolute (1 A = 10*-10 m) and d(220) = 1.9201706 +- 0.0000003 A at 1 atm, 25 deg C.

Dislocation free float zone Si containing carbon has typical contractions DELTA d / d in the range -0.3 to -1.03 ppm, and

> DELTA d / d = -6.5 X 10*-24 X Nc

where Nc is the carbon atom concentration per cm*3.

Oxygen causes a lattice expansion typically amounting to +3 X 10*-6 in the case of dislocation free crystals grown from quartz crucibles by the Czochralski method.

REFERENCES
----

[1] J.A.Baker et al [J. Appl. Phys. (USA) vol.39 (1968) p.4365]
[2] E.G.Kessler, Jr, R.D.Deslattes, A.Henins [Phys. Rev. A (USA) vol.19 no.1 (Jan 1979) p.215-18]
[3] R.D.Deslattes, A.Henins, R.M.Schoonover, C.L.Carrol, H.A.Bowman [Phys. Rev. Lett. (USA) vol.36 no.15 (12 Apr 1976) p.898-900]
[4] R.D.Deslattes, A.Henins [Phys. Rev. Lett. (USA) vol.31 no.16 (15 Oct 1973) p.972-5]
[5] M.Ando et al [Acta Crystallogr. Sect. A (Denmark) vol.34 (1978) p.484]

1.3 LATTICE PARAMETERS OF Si AT HIGH PRESSURES

by J.Z.Hu

July 1987
EMIS Datareview RN=15791

Until now, seven high pressure phases of Si have been discovered at room temperature. The seven phases exist in different pressure regions. The labelling, structure and pressure region of each phase is given in TABLE 1.

TABLE 1: designation of high pressure phases of Si

Designation	Structure	Pressure region (GPa)	Refs
I	Cubic (diamond)	0-12.5	[1-6]
II	Body centered tetragonal (beta-Sn)	8.8-16	[2-10]
III	Body centered cubic (bcc)	10-0	[2,3,6,11]
V	Primitive hexagonal (ph)	14-35	[2,3,4,9,12]
VI	(Undetermined)	34-40	[4,13]
VII	Hexagonal close packed (hcp)	40-78.3	[3,4,8,14,17]
VIII	Cubic (fcc)	78.3	[9,14,17]

Si (IV) is a high temperature phase after pressure treatment [15] and not included in this Datareview. Si (III) is a phase on decompression from Si (V) and (II) and metastable at atmospheric pressure [2,11]. For Si (VI) it was assumed that the structure corresponded to a different stacking of hexagonal close pack layers in 34-40 GPa by Olijnyk et al [4] and it was concluded that it was a four-layer dhcp structure by Vohra et al [13]. The very recent results obtained by Vohra, Duclos and Ruoff showed that Si (VI) phase is a complicated structure phase rather than dhcp [18].

The lattice parameters of Si at high pressures for these phases are respectively given in the following tables. For comparison, some theoretical data are also shown in these tables. The zero pressure means atmospheric pressure here.

TABLE 2: lattice parameters of Si (I) (diamond phase)

	P(GPa)	a(A)	Refs
Exp.	0	5.435	[1]
	11.2 +-0.2	5.268 +- 0.010	[2]
	12.5	5.283	[5,7]
Theo.	12.4-15.2	5.214-5.226	[8]
	9.3	5.301 (note)	[9]

TABLE 3: lattice parameters of Si (II) (beta-Sn phase)

	P(GPa)	a(A)	c(A)	c/a	Refs
Exp.	11.2 +- 0.2	4.690 +- 0.006	2.578 +- 0.005	0.550	[2]
	12.5	4.686	2.585	0.554	[5,7]
Theo.	12.4-15.2	4.705-4.720	2.499-2.503	0.536-0.53	[8]
	9.3	4.709 (note)	2.595 (note)	0.551	[9]
	12	4.682 (note)	2.580 (note)	0.551	[9]
	14	4.692	2.463	0.525	[12]

TABLE 4: lattice parameters of Si (V) (ph phase)

	P(GPa)	a(A)	c(A)	c/a	Refs
Exp.	16	2.551 +- 0.006	2.387 +- 0.007	0.936	[2,3]
	25			0.941	[2,3]
	16	2.527	2.378	0.937	[4]
	34			0.947	[4]
Theo.	12	2.558 (note)	2.443 (note)	0.955	[9]
	12	2.400	2.297	0.957	[12]

TABLE 5: lattice parameters of Si (VII) (hcp phase)

	P(GPa)	a(A)	c(A)	c/a	Refs
Exp.	41-42	2.524 +- 0.009	4.142 +- 0.050	1.64 +- 0.02	[3]
	43	2.444	4.152	1.698	[4]
Theo.	41	2.475 (note)	4.194 (note)	1.695	[9]
	116	2.331 (note)	3.952 (note)	1.695	[9]

TABLE 6: lattice parameters of Si (VIII) (fcc phase)

	P(GPa)	a(A)	Refs
Exp.	87 +- 7	3.34 +- 0.010	[17]
Theo.	116	3.32	[9]

TABLE 7: lattice parameters of Si (III) (bcc phase)

	P(GPa)	a(A)	Refs
Exp.	10	6.405 +- 0.005	[3]
	0	6.62 +- 0.01	[2,3]
	0	6.636 +- 0.005	[11]
Theo.		6.67	[16]

note: Calculated according to normalized volume Vt and Vo = 20.002
A*3/atom in ref. [9].

REFERENCES
----

[1]   J.F.Cannon [J. Phys. & Chem. Ref. Data (USA) vol.3 (1974) p.781]
[2]   J.Z.Hu, I.L.Spain [Solid State Commun. (USA) vol.51 (1984)
        p.263]
[3]   J.Z.Hu, L.D.Merkle, C.S.Menoni, I.L.Spain [Phys. Rev. B (USA)
        vol.34 (1986) p.4679]
[4]   H.Olijnyk, S.K.Sikkle, W.B.Holzapfel [Phys. Lett. A
        (Netherlands) vol.103 (1984) p.137]
[5]   J.C.Jamieson [Science (USA) vol.139 (1963) p.762]
[6]   T.I.Dyuzheva, S.S.Kabalkina, V.Novichkov [Sov. Phys. - JETP
        (USA) vol.47 (1978) p.931]
[7]   G.J.Piermerini, S.Block [Rev. Sci. Instrum. (USA) vol.46 (1975)
        p.973]
[8]   T.Soma [Phys. Status Solidi b (Germany) vol.92 (1979) p.K51]
[9]   K.J.Chang, M.L.Cohen [Phys. Rev. B (USA) vol.31 (1985) p.7819]
[10]   M.T.Yin, M.L.Cohen [Phys. Rev. Lett. (USA) vol.45 (1980)
        p.1004]; M.T.Yin, M.L.Cohen [Phys. Rev. B (USA) vol.26 (1982)
        p.5668]
[11]   S.Kasper, S.M.Richards [Acta Crystallogr. (Denmark) vol.77
        (1964) p.752]
[12]   R.J.Needs, R.M.Martin [Phys. Rev. B (USA) vol.30 (1984)
        p.5390]
[13]   Y.K.Vohra, K.E.Brister, S.Desgreniers, A.L.Ruoff, K.J.Chang,
        M.L.Cohen [Phys. Rev. Lett. (USA) vol.56 (1986) p.1944]
[14]   A.K.McMahan, J.A.Moriarty [Phys. Rev. B (USA) vol.27 (1983)
        p.3235]
[15]   I.L.Aplekar, E.Y.Tonkov [Sov. Phys.-Solid State (USA) vol.17
        (1975) p.966]
[16]   M.T.Yin [Phys. Rev. B (USA) vol.30 (1984) p.1773]
[17]   S.J.Duclos, Y.K.Vohra, A.L.Ruoff [Phys. Rev. Lett. (USA) vol.58
        (1987) p.775]
[18]   Y.K.Vohra, S.J.Duclos, A.L.Ruoff [to be published]

1.4 BULK MODULUS OF Si

by O.H.Nielsen

June 1987
EMIS Datareview RN=15735

The bulk modulus B is the volume-derivative of the hydrostatic pressure
P, B = - V dP/dV, and consequently has the same unit as the pressure.
An analogous quantity is the compressibility K defined as K = 1/B. In
practice B is not measured directly, but is given as the following
linear combination of elastic constants:

$$B \; = \; \frac{C11 \, + \, 2C12}{3}$$

valid for all cubic crystals (see, e.g. Kittel [1]).

Beyond pressures of a few kilobar the nonlinearity in B (i.e., non-
constancy) becomes noticeable and is described as

$$B(P) = B(P=0) + B' \; P$$

where B' is the pressure-derivative of B: B' = dB/dP. B' is
dimensionless, and like B it is given in terms of elastic constants as:

$$B' \; = \; \frac{- \, (C111 \, + \, 6C112 \, + \, 2C123)}{9B}$$

This equation for B' and consequently the numerical values for B'
depend upon the type of nonlinear elastic theory used in the data
analysis. The above definition results from using Lagrangian elastic
theory. For a summary of various nonlinear theories see, e.g. Nielsen
[2].

The numerical value of B is obtained with the elastic constants
discussed in ref [4]. Using the adiabatic Cij and Cijk of Hall [3,4]
we get

$$B = 0.9784 \; X \; 10*12 \; dyne/cm*2 \quad and \quad B' = 4.11$$

at ambient pressure and T = 25 deg C. The error in B is 0.02%, and the
error in B' is about 2%.

(Unit conversion: 1 Mbar = 100 GPa = 10*11 N/m*2 = 10*12 dyne/cm*2).

REFERENCES
----

[1] C.Kittel [Introduction to Solid State Physics, 4th ed. (Wiley,
     1971) ch.4]
[2] O.H.Nielsen [Phys. Rev. B (USA) vol.34 (1986) p.5808]
[3] J.J.Hall [Phys. Rev. (USA) vol.161 (1967) p.756]
[4] O.H.Nielsen [EMIS Datareview RN=15736 (June 1987) 'Stiffness of
     Si']

1.5 STIFFNESS OF Si

by O.H.Nielsen

June 1987
EMIS Datareview RN=15736

A. INTRODUCTION

The stiffness or elastic constants C11, C12 and C44 for the cubic Si
crystal are defined as in Kittel [1]. Bulk moduli and compliances
are expressed in terms of the Cij [15,16].

The Cij are usually measured by ultrasonic methods. Early data for Si
were reported by McSkimin [2], and a proper theory involving Cij as
well as higher order constants was presented by Thurston and Brugger
[3].

The Cij for pure Si obtained by a number of groups are summarised by
Metzger and Kessler [4]. The most accurate data seem to be those of
Hall [5]:

$$C11 = 1.6564 \times 10*12 \text{ dyne/cm*2}$$

$$C12 = 0.6394 \times 10*12 \text{ dyne/cm*2}$$

$$C44 = 0.7951 \times 10*12 \text{ dyne/cm*2}$$

at ambient pressure and T = 25 deg C. The error in Cij is 0.02%. The
values of Cij found by other workers [4] are generally within 1% of
Hall's values.

(Unit conversion: 1 Mbar = 100 GPa = 10*11 N/m*2 = 10*12 dyne/cm*2).

B. PRESSURE DEPENDENCE OF Cij

Beyond pressures of a few kilobar the nonlinearity in the Cij's (i.e.,
non-constancy) becomes noticeable and is described in terms of higher
order elastic constants. The conventional description is due to
Thurston and Brugger [3], who use the Lagrangian form of elastic
tensors. The alternative Eulerian form has been developed by Birch [6].
A broad theoretical treatment is given by Wallace [7].

A cubic crystal has 6 third order elastic constants Cijk which are
measured by monitoring the ultrasonic velocities as a function of
external pressure and strain upon the Si crystal. The procedure of
Thurston and Brugger [3] has been used by Hall [5] and others [8]
to yield for pure Si:

C111 = - 7.95 X 10*12 dyne/cm*2

C112 = - 4.45 X 10*12 dyne/cm*2

C123 = - 0.75 X 10*12 dyne/cm*2

C144 = + 0.15 X 10*12 dyne/cm*2

C166 = - 3.10 X 10*12 dyne/cm*2

C456 = - 0.86 X 10*12 dyne/cm*2

at ambient pressure and T = 25 deg C. The error is 0.10 X 10*12 dyne/cm*2 for C111 and C112, and 0.05 X 10*12 dyne/cm*2 for the remaining Cijk. The external pressures were typically up to 10*9 dyne/cm*2.

A study by Goncharova et al [9] applied hydrostatic pressure up to 8 GPa (= 80 kbar or 8 X 10*10 dyne/cm*2) and determined the C11, C12 and C44 as functions of pressure P. The linear terms in Cij(P) yield 3 linear combinations of the Cijk which can be derived using TABLE I of ref [3].

C. TEMPERATURE DEPENDENCE OF Cij

Close to absolute zero the Cij are given by Hall [5] as:

C11 = 1.6754 X 10*12 dyne/cm*2

C12 = 0.6492 X 10*12 dyne/cm*2

C44 = 0.8024 X 10*12 dyne/cm*2

at ambient pressure and T=4.2 deg K. The error in Cij is estimated as 0.02%. Hall [5] also presents curves of C11, C12 and C44 as functions of temperature in the range 4-300 deg K for pure as well as doped Si. Below 100 deg K the Cij of pure Si change very little. Between approx. 300-1000 deg K the Cij decrease fairly linearly with temperature [10] at the rates:

Cij	(1/Cij) X dCij/dT (units K*-1)
C11	-9.4 X 10*-5
C12	-9.8 X 10*-5
C44	-8.3 X 10*-5

D. DOPING DEPENDENCES OF Cij

The elastic properties of semiconductors depend upon the carrier
concentration, as pointed out by Keyes [11]. The heavily doped n-type
Si was studied by Hall [5] and by Beilin et al [12] in terms of the
elastic constants as functions of the temperature for concentrations
of n = 2.0 and 4.8 X 10*19/cm*3, respectively. Heavily doped p-type
Si was studied by Mason and Bateman [13]. The effect of heavy doping is
typically a decrease of the Cij's by roughly 1-5% [5], with some
temperature dependence of the decrease. For theoretical approaches see
Keyes [11] and Kim [14] and references cited therein.

REFERENCES
----

[1]   C.Kittel [Introduction to Solid State Physics, 4th Edn (Wiley,
      1971) ch.4]
[2]   H.J.McSkimin [J. Appl. Phys. (USA) vol.24 (1953) p.988]
[3]   R.N.Thurston, K.Brugger [Phys. Rev. (USA) vol.133 (1964)
      p.A1604; K.Brugger [ibid. p.A1611]
[4]   H.Metzger, F.R.Kessler [Z. Naturforsch. a (Germany) vol.25a
      (1970) p.904]
[5]   J.J.Hall [Phys. Rev. (USA) vol.161 (1967) p.756]
[6]   F.Birch [J. Geophys. Res. (USA) vol.57 (1952) p.227; see also
      F.Birch [Phys. Rev. (USA) vol.71 (1947) p.809] but note the
      error in eqn (33) & (35)]
[7]   D.C.Wallace [in Solid State Physics vol.25 Eds. H.Ehrenreich,
      F.Seitz, D.Turnbull (Academic, 1970) p.301]
[8]   H.J.McSkimin, P.Andreatch,Jr [J. Appl. Phys. (USA) vol.35
      (1964) p.3312]
[9]   V.A.Goncharova, E.V.Chernysheva, F.F.Voronov [Sov. Phys.-Solid
      State (USA) vol.25 (1983) p.2118]
[10] [Landolt-Bornstein Crystal and Solid State Physics Ed.
      K.-H.Hellwege vol.11 (Springer 1979) p.116]
[11] R.W.Keyes [IBM J. Res. & Dev. (USA) vol.5 (1961) p.266; and in:
      Solid State Physics, vol.20 Eds. H.Ehrenreich, F.Seitz,
      D.Turnbull, (Academic, 1976) p.37]
[12] V.M.Beilin, Yu.Kh.Vekilov, O.M.Krasil'nikov [Sov. Phys.-Solid
      State (USA) vol.12 (1970) p.531]
[13] W.P.Mason, T.B.Bateman [Phys. Rev. (USA) vol.134 (1964)
      p.A1387]
[14] C.K.Kim [J. Appl. Phys. (USA) vol.52 (1981) p.3693]
[15] O.H.Nielsen [EMIS Datareview RN=15735 (June 1987) 'Bulk Modulus
      of Si']
[16] O.H.Nielsen [EMIS Datareview RN=15737 (June 1987) 'Compliance of
      of Si']

1.6 STIFFNESS OF ION-IMPLANTED Si

by P.J.Burnett

May 1987
EMIS Datareview RN=15713

A. INTRODUCTION

Ion implantation into silicon results in both structural, chemical and
electronic changes in the surface to a depth (typically) of < 1 micron.
These changes are expected to be manifest in the modifications to
mechanical properties such as stiffness. In particular, radiation damage
(i.e. the bond-breaking processes associated with displacements of host
atoms by collisions with the incident ions) is expected to exert a
profound influence on the elastic properties of crystalline silicon by
reducing the total bonding within the crystal. In addition, implantation
of electrically active dopant species (e.g. As+, B+) to sufficiently
high doses may also affect the elastic constants of crystalline silicon
[1].

B. RADIATION DAMAGE EFFECTS

Burnett and Briggs [2] have used the line focus acoustic microscope [3]
to investigate the variation of room temperature surface acoustic wave
velocity for (111) plane-family Si implanted with a range of doses of
both electrically active (As+) and iso-valent (Si+) ion species. Using
the approximation that silicon is elastically isotropic together with
isotropic multi-layer wave propagation theory [4] they were able to
infer that the elastic constants C11 and C44 decreased approximately
linearly until the cumulative effects of the radiation damage rendered
the material amorphous (this corresponding to a displacement level of
approx. 0.15 displacements per host atom (dpa) for implantation at
ambient temperatures [5]) by which point the elastic constants appeared
to have decreased by approx. 30%. Thus, the variation of elastic
constants may be approximated by:

C11 (implanted) = C11 (unimplanted) X (1- (2 X dpa))

C44 (implanted) = C44 (unimplanted) X (1- (2 X dpa))

The actual figures determined are given in TABLE 1.

TABLE 1: the elastic constants C11 and C44 of unimplanted and ion
implantation amorphised Si

	C11 (dyne/cm*2)	(GPa)	C44 (dyne/cm*2)	(GPa)
Unimplanted Si	1.865 X 10*12	(186.5)	0.665 X 10*12	(66.5) from [6]
Amorphous Si	1.305 X 10*12	(130.5)	0.466 X 10*12	(46.6)

The assumption of elastic isotropy used in the calculations means that
these values may be in error by as much as 10%. Also, no information on
the possible variations of C12 was obtained.

C. DOPANT EFFECTS

In bulk silicon Keyes and others [1,7-9] have reported the variation of
the shear elastic constant (C11-C12) by a few percent when carrier
concentrations exceeded typically 10*18/cm*3 (i.e. concentration levels
of approx. 0.02% or greater). These and higher dopant concentrations
are easily achievable using ion implantation and so similar effects upon
elastic constants might be expected when implantations of electrically
active species are performed. However, for the as-implanted silicon
studied by Burnett and Briggs [2] no significant differences between
the behaviour of the As+ implants and the Si+ implants were found,
presumably the large changes in elastic properties resulting from
radiation damage effects masking the smaller dopant effects.

D. ANNEALED ION-IMPLANTED SI

No data is yet available on the elastic properties of annealed ion-
implanted silicon, however, since annealing removes radiation damage
and increases the number of implanted aliovalent atoms that are
electrically active, changes in elastic constants of the order suggested
by Keyes [1] would be expected.

REFERENCES

[1] R.W.Keyes [IEEE Trans. Sonics & Ultrason. (USA) vol.SU-29 (1982)
     p.99]
[2] P.J.Burnett, G.A.D.Briggs [J. Mater. Sci. (GB) vol.21 (1986)
     p.1828]
[3] J.Kushibiki, N.Chubachi, [IEEE Trans. Sonics & Ultrason. (USA)
     vol.SU-32 (1985) p.189]
[4] L.M.Brekhovskikh [Waves in layered media - 2nd Edition, (Academic
     Press, New York, 1980)]
[5] L.A.Christel, J.F.Gibbons, T.W.Sigmon [J. Appl. Phys. (USA) vol.52
     (1981) p.7143]
[6] O.L.Anderson [in 'Physical acoustics' vol.3 part.B, Ed. W.P.Mason
     (Academic Press, New York, 1965) p.77]
[7] R.W.Keyes [Solid State Phys. (USA) vol.20 (1967) p.37]
[8] J.J.Hall [Phys. Rev. (USA) vol.161 (1967) p.756]
[9] N.G.Einspruch, P.Csavinsky [Appl. Phys. Lett. (USA) vol.2 (1963)
     p.1]

1.7 COMPLIANCE OF Si

by O.H.Nielsen

June 1987
EMIS Datareview RN=15737

The compliance (Sij) and stiffness (Cij) constants are inverse to each
other (see Kittel [1]) as given by the relations:

$$C11 + 2C12 = \frac{1}{S11 + 2S12}$$

$$C11 - C12 = \frac{1}{S11 - S12}$$

$$C44 = \frac{1}{S44}$$

Using the adiabatic Cij of Hall [2] we get

S11 =   0.7691 X 10*-12 cm*2/dyne

S12 = - 0.2142 X 10*-12 cm*2/dyne

S44 =   1.2577 X 10*-12 cm*2/dyne

at ambient pressure and T = 25 deg C. For more information on the Cij,
their accuracy and dependence upon various parameters see [3].

REFERENCES
----

[1] C.Kittel [Introduction to Solid State Physics, 4th ed. (Wiley,
     1971) ch.4]
[2] J.J.Hall [Phys. Rev. (USA) vol.161 (1967) p.756]
[3] O.H.Nielsen [EMIS Datareview RN=15736 (June 1987) 'Stiffness of
     Si']

1.8 INTERNAL STRAIN EFFECTS IN Si

by O.H.Nielsen

June 1987
EMIS Datareview RN=15738

Crystals with more than one atom per unit cell (such as Si with 2
atoms per unit cell) may undergo 'internal displacements' (also called
'internal strains') when exposed to external non-hydrostatic strain.
This is the case for the Si crystal when strained along, e.g., the
[110] or [111] crystal axes. This effect was described by Kleinman [1]
who parametrised the displacement away from the perfectly strained Si
lattice in terms of a dimensionless 'zeta' parameter. Normally, zeta
would be expected to be in the range of 0-1 depending upon the ratio of
bond-bending to bond-stretching force constants. This zeta is intimately
related to the magnitude of the C44 elastic constant as:

$$C44 = C44(0) - Const. \; X \; zeta*2 \; X \; OMEGA(k=0)*2$$

where C44(0) is the C44 value that would appear in the absence of
internal strain, and OMEGA(k=0) is the optical k=0 phonon frequency,
as described by Nielsen and Martin [2]. The general theory of internal
strain has been explored and reviewed by Cousins [3].

Recent experiments by d'Amour et al [4] and Cousins et al [5]
monitoring the 006 forbidden X-ray reflection established values of
zeta for Si as 0.74 and 0.72, respectively. Very recent measurements by
Cousins et al [6] have shown that previous experimental values have to
be revised due to crystal surface effects in the X-ray experiments. The
experimental value of zeta = 0.54 with an error of +/- 0.04 is now
found, in agreement with theoretical calculations by Nielsen and Martin
[2] giving zeta = 0.53.

REFERENCES

[1] L.Kleinman [Phys. Rev. (USA) vol.128 (1962) p.2614]
[2] O.H.Nielsen, R.M.Martin [Phys. Rev. Lett. (USA) vol.50 (1983)
     p.697; Phys. Rev. B (USA) vol.32 (1985) p.3792]
[3] C.S.G.Cousins [J. Phys. C (GB) vol.15 (1982) p.1857; J. Phys.
     C (GB) vol.11 (1978) p.4867]
[4] H.d'Amour, W.Denner, H.Schulz, M.Cardona [J. Appl. Crystallogr.
     (Denmark) vol.15 (1982) p.148]
[5] C.S.G.Cousins, L.Gerward, J.Staun Olsen, B.Selsmark, B.J.Sheldon
     [J. Appl. Crystallogr. (Denmark) vol.15 (1982) p.154]
[6] C.S.G.Cousins, L.Gerward, J.Staun Olsen, B.Selsmark, B.J.Sheldon
     [J. Phys. C (GB) vol.20 (1987) p.29]

1.9 HARDNESS OF Si

by P.J.Burnett

March 1987
EMIS Datareview RN=15710

A. INTRODUCTION

Hardness and indentation techniques are often the only way of
introducing plastic deformation into a brittle material (such as
silicon), the high hydrostatic stress state generated beneath an
indentor inhibiting catastrophic fracture. Indentation with diamond
pyramids is commonly used for such hardness measurements, the Vickers,
Knoop and Berkovitch profile indentors being those most often found.
Whatever the profile diamond indentor the hardness is defined as the
applied load divided by the (projected) area of the impression; the
units are thus those of stress. Indeed, hardness reflects the ease of
dislocation motion in crystalline materials (such as silicon) and,
even in silicon which is usually regarded as a brittle material,
dislocations have been observed underneath hardness impressions by
several workers, e.g. [1-3].

Many factors affect the measured hardness of single crystal silicon,
but the most important of these is temperature.

B. LOW TEMPERATURE HARDNESS BEHAVIOUR

The majority of hardness values for silicon found in the literature lie
in the relatively athermal low temperature regime. The room temperature
hardness of silicon is usually quoted as being around 1000 kg/mm*2
(9.8 GPa) for all types of indentor profile. TABLE 1 lists some
representative values.

TABLE 1: typical room temperature Vickers hardness of silicon

Hardness (GPa)	7-20	10.3	6.9	7.5-10.9	10.2
Reference	[4]	[5]	[6]	[7]	[8]

The hardness of silicon is found to vary little with temperature up to
about 400 deg C. In this athermal region dislocation motion has been
modelled by Gilman [9,10] in terms of double kink motion by phonon
assisted quantum mechanical tunnelling across a potential barrier. The
small temperature dependence of plastic flow is a consequence of the
effect of temperature on phonon energy. The hardness variation in the
athermal region is given approximately by

$$H/Ho = 1 - 2 k \, theta/Uo \, [\, coth \, (theta/T)-1 \,]$$

where Ho = hardness at 0 deg K, k = Boltzmann's constant, theta is the phonon cut-off temperature and Uo is an energy barrier at 0 deg K. Gilman [10] determined values of theta and 2k theta/Uo as 750 deg K and 1.4, whilst Naylor and Page [8] determined Ho as 10.4 GPa.

It has been suggested that other mechanisms may be responsible for the athermal nature of the low temperature hardness behaviour; these include the postulation of reversible stress-induced phase transformations [11,12].

C. HIGH TEMPERATURE HARDNESS

Above approximately 400 deg C single crystal silicon softens appreciably, this being associated with thermal activation of dislocation motion. TABLE 2 gives examples of the data to be found in the literature.

TABLE 2: hot hardness values for silicon

Temperature	From [8]	From [10]
24	10.2	7.9
100	9.8	7.4
200	9.3	7.1
400	5.6	6.2
600	2.6	2.3
800	1.0	0.8

Activation energy plots reveal a temperature/hardness relationship of the form

H is proportional to exp (-E/kT)

where E is an activation energy for dislocation motion evaluated as 35 kJ/mol for 100 gram Vickers indentation [8].

D. INDENTATION SIZE EFFECTS

It is often found that hardness is dependent upon test load (and hence indentation size). This indentation size effect is often described using a power law expression [13] such that

H = a d*(n-2)

where the exponent n is usually between 1.5 and 2.0. For n = 2 the hardness is independent of load, but for n < 2 the hardness is observed to increase as the load decreases. Much of the scatter in the hardness values of TABLE 1 may be due to differing test loads/indentation sizes. Brookes [13] has determined the variation of n for Knoop indentation on (111) silicon along the [110] direction at a range of temperatures (given in TABLE 3).

TABLE 3: Knoop indentation size effect exponent for (111) Si

Temperature (deg C)	25	100	200	300	400
n	1.54	1.73	1.8	1.82	1.83

The indentation size effect indices for various silicon surfaces have also been derived at room temperature using Vickers indentation techniques and typical values are given in TABLE 4.

TABLE 4: Vickers indentation size effect exponents

Surface	n	Ref
(111)	1.92	[8]
(111)	1.7	[14]
(100)	1.85	[15]

E. HARDNESS ANISOTROPY

Since the hardness of single crystal silicon is dominated by the crystallographic nature of the plastic flow (i.e. slip on (111) plane-family and <110>) it is reasonable to expect some variation of hardness with indentation for non-axisymmetric indentors. The 2-fold rotational symmetry of the Knoop indentor renders it the most sensitive to the effects of hardness anisotropy [13]. Silicon has the diamond cubic structure and is expected to display hardness anisotropy similar to that of diamond itself with the 'hard' direction for (001) surface occurring with the Knoop indentation parallel to [100] and the 'soft' direction parallel to [110]. Diamond itself shows a difference in hardness of about 11% between these two directions [14]. No data were found in the literature for silicon itself. However, work by Zhitaru et al [16] showed the presence of hardness anisotropy on (111) surfaces using Vickers indentation of two orientations 45 deg C apart. They obtained hardness of 11.7 GPa and 12.5 GPa in these orientations, but omitted to give the absolute orientation of the indentation.

F. INDENTATION CREEP

In normal microhardness testing, indenter dwell times of between 10
and 30 seconds are used. However, if significantly longer dwell times
are utilised indentation creep may become significant, i.e. the longer
the dwell time the softer the material appears. For instance, TABLE 5
lists the dwell time and measured hardness ((111) surface - [1 1 bar2]
direction) for silicon indented at 300 deg C and 450 deg C (from
[17]).

TABLE 5: indentation creep of silicon

Dwell Time	Hardness (GPa)	
(secs)	300 deg C	450 deg C
10	8.8	5.4
100	8.0	3.5
1000	7.0	2.4
10000	6.1	1.6
50000	5.4	1.1

G. DOPANT EFFECTS

Dislocation velocities in silicon are known to be strongly dependent
upon doping levels [18-20] and therefore it might be expected that
hardness might also show some doping dependence. Roberts et al [20]
have used indentation techniques (at 400 deg C) combined with a
dislocation etch-pit technique to determine the distance to which the
dislocations propagate and hence a critical shear stress for dislocation
motion in silicon doped in the range $10*12$ to 2 X $10*18$ atoms/cm*3
(both p- and n-type dopants being used). No observable differences in
microhardness were found between the different dopant levels. However,
significant differences were found in the calculated dislocation shear
stresses; low dopant levels yielded critical shear stress levels of
about 103 MPa whilst the higher dopant levels gave values of as low
as 73.1 MPa.

In contrast to these results, Maugis et al [21] found that increasing
dopant levels increased the room temperature hardness of single crystal
silicon. This may be due to the impurity atoms acting as dislocation
pinning centres at room temperature, i.e. conventional solid-solution
hardening mechanisms may be operating. Danyluk et al [22] have
observed that doping with isovalent carbon above a concentration of
3.5 X $10*23/m*3$ results in a hardening of silicon, an increase in
hardness of about 8% being found at the highest C concentration used,
7.9 X $10*23/m*3$.

H. ENVIRONMENTAL AND SURFACE EFFECTS

The hardness of many non-metallic materials may be affected by the test environment (e.g. dry air, moist air) and these effects are known as chemomechanical or Rehbinder effects. Hanneman and Westbrook [23] showed that the presence of adsorbed water could cause a 26% softening of the surface of silicon. Certain other polar solvents may also produce similar softening. Lee et al [24] found that maximum softening occurred when tests were performed under liquids whose dielectric constant was close to 28; methanol, ethanol and glycerol all producing significant effects (softening of up to 30% being reported).

In addition to the chemomechanical effects, the surface preparation route may also affect the apparent hardness of silicon. For instance Zhitaru et al [16] found that silicon held in a 4% solution of GdCl3 for up to 1 hr prior to testing showed increased hardness (by up to 15%), whilst those held in ErCl3 for similar times showed slight softening (by up to 7%). However, these results may not represent the true silicon hardness since a contaminant film was reported to be present on the GdCl3 treated specimens. Leighly and Oglesbee [7] have also reported a dependence of hardness on surface treatment - here it was found that oxidising surface treatments resulted in a higher hardness than those that produced a 'clean' surface (Knoop hardnesses of 10.8 and 7.6 - 8.8 GPa respectively). Clearly, the presence of an oxide film may produce effects analogous to those observed by Zhitaru et al [16].

I. PHOTOPLASTIC EFFECTS

There is evidence that light with a wavelength of 2.5-4 microns (photons of 0.5-0.1 eV) may affect both dislocation velocity [25] and hardness [7] by providing the activation energy required for the bond-breaking processes required in silicon for dislocation motion. Leighly and Oglesbee [7] found that reductions in hardness of up to 22.6% were possible in the presence of light but that the degree of photoplastic effect depended critically on whether or not an oxide layer was present. This layer acted as a filter removing photons of the required energies and thus reduced the level of photoplastic effect to approx 10-13%. Temperature dependence of the photoplastic effect was found to be of the form:

delta H/H (dark) is proportional to 1/T (absolute)

REFERENCES

[1]   M.J.Hill, D.J.Rowcliffe [J. Mater. Sci. (GB) vol.9 no.10 (Oct 1974) p.1569-76]
[2]   B.R.Lawn, B.J.Hockey, S.M.Wiederborn [J. Mater. Sci. (GB) vol.15 no.5 (May 1980) p.1207]
[3]   J.Samuels, P.Pirouz, S.G.Roberts, P.D.Warren, P.B.Hirsch [Inst. Phys. Conf. Ser. (GB) no.76 (1985) p.49]
[4]   J.B.Pethica, R.Hutchings, W.C.Oliver [Philos. Mag. A (GB) vol.48 (1983) p.593]
[5]   J.J.Gilman [in 'The Science of Hardness Testing and its Research Application' Eds J.W.Westbrook, H.Conrad, (ASM, Ohio, USA, 1973) p.51]

[6]   A.Sher, A-B.Chen, W.E.Spicer [Appl. Phys. Lett. (USA) vol.46
        no.1 (1 Jan 1985) p.54-6]

[7]   H.P.Leighly, R.M.Oglesbee [in 'The Science of Hardness Testing
        and its Research Application' Eds J.W.Westbrook, H.Conrad (ASM,
        Ohio, USA, 1973) p.445]

[8]   M.G.S.Naylor, T.F.Page [Third Annual Technical Report, Grant
        No. DA-ERO-78-G-010, European Research Office, United States
        Army, London, England]

[9]   J.J.Gilman [J. Appl. Phys. (USA) vol.39 (1968) p.6086]

[10]   J.J.Gilman [J. Appl. Phys. (USA) vol.46 no.12 (Dec 1975)
        p.5110-13]

[11]   V.G.Ermenko, V.I.Nikitenko [Phys. Status Solidi a (Germany)
        vol.14 (1972) p.317]

[12]   I.V.Gridneva, Yu.V.Milman, V.I.Trefilov [Phys. Status Solidi a
        (Germany) vol.14 (1972) p.177]

[13]   C.A.Brookes [in 'The Science of Hard Materials' Eds
        R.K.Viswanadham, D.Rowcliffe, J.Gurland (Plenum Press, New York,
        1983) p.181]

[14]   P.J.Burnett, T.F.Page [J. Mater. Sci. (GB) vol.19 no.3 (Mar 1984)
        p.845-60]

[15]   S.G.Roberts, T.F.Page [in 'Ion Implantation into Metals'
        (GB, 1982) p.135]

[16]   R.P.Zhitaru, S.V.Donn, E.I.Purlich, M.D.Mazus, T.I.Malinovskii
        [Sov. Phys.-Dokl. (USA) vol.26 (1981) p.1166]

[17]   G.Morgan [in 'The Science of Hard Materials' Eds R.K.Viswanadham,
        D.Rowcliffe, J.Gurland (Plenum Press, New York, 1983) p.181]

[18]   H.Alexander, P.Haasen [Solid State Phys. (USA) vol.22 (1968)
        p.27]

[19]   A.George, G.Champier [Phys. Status Solidi a (Germany) vol.53
        (1979) p.529]

[20]   S.G.Roberts, P.Pirouz, P.B.Hirsch [J. Phys. Colloq. (France)
        vol.44 no.C-4 (1983) p.75]

[21]   D.Maugis, J.Guyonnet, R.Courtel [C.R. Acad. Sci. Paris (France)
        vol.272A (1971) p.971]

[22]   S.Danyluk, D.S.Lim, J.Kalejs [J. Mater. Sci. Lett. (GB) vol.4
        no.9 (Sept 1985) p.1135-7]

[23]   R.E.Hanneman, J.E.Westbrook [Philos. Mag. (GB) vol.18 (1968)
        p.73]

[24]   S.W.Lee, D.S.Lim, S.Danyluk [J. Mater. Sci. Lett. (GB) vol.3
        no.7 (Jul 1984) p.651-3]

[25]   K.H.Kusters, H.Alexander [Physica B & C (Netherlands) vol.116B
        (1983) p.594]

1.10 HARDNESS OF ION-IMPLANTED Si

by P.J.Burnett

March 1987
EMIS Datareview RN=15709

A. INTRODUCTION

Hardness measurements represent the resistance of a material to plastic
deformation under point or concentrated surface contacts (hardness
values being also related to uniaxial yield strengths). In crystalline
materials such as metals, ceramics and semiconductors, this plastic
deformation occurs principally by propagation of dislocations through
the crystal structure (this even being found in brittle materials
where plastic flow is not normally observed [1]). Since ion
implantation results in a considerable disruption of the crystal
structure by displacement-collision radiation damage processes,
dislocation motion in an ion implanted surface is expected to become
more difficult resulting in an increased hardness. Indeed, radiation
hardening of metals by neutron irradiation is well documented.

B. HARDENING OF Si BY ION IMPLANTATION

The effect of ion implantation upon the ease of dislocation motion has
been studied in two ways. Firstly, Hu and Schwenker [2] indented (at
about 600 deg C) the (100) plane-family surfaces of unimplanted
silicon and silicon implanted at 50 kV to doses of up to 10*16 ion/cm*2.
The ion species used were Ar, As, B, N and O. Using an etching
technique they subsequently showed that dislocations generated by the
indentation process travelled further from the indentation centre on
the unimplanted silicon than those on the implanted silicon. Typically
the distance of dislocation travel was reduced three-fold indicating
an increase in dislocation flow stress of about 800%. These authors
concluded that the observed 'hardening' was principally due to the
presence of radiation damage induced defects.

Burnett and Briggs [3] used direct microhardness measurements made
using 25 gf (0.098 N) and 50 gf (0.295 N) Knoop indentations of
(111) plane-family surfaces to monitor changes in hardness with dose
for Si+ and As+ implanted silicon. TABLE 1 lists the Knoop
microhardness values obtained.

TABLE 1: Knoop microhardness of ion implanted silicon

Dose (ions/cm*2)	Knoop Hardness (GPa)	
	Load = 0.098 N	Load = 0.245 N
Unimplanted	7.49	8.41
10*14 As+	8.67	9.04
2 X 10*14 As+ (see note)	6.36	8.17
10*15 As+ (see note)	7.22	7.36
10*14 Si+	8.55	8.66
2 X 10*14 Si+	8.20	8.32
10*15 Si+ (see note)	7.25	8.26

note: Amorphous material present.

Here, as in [2], an increase in hardness was observed at doses below
those required to produce amorphous silicon layers (as a result of
cumulative displacement damage). However, when amorphous material
was present as a sub-surface layer softening was apparent (see next
section).

C. SOFTENING OF Si BY ION IMPLANTATION

Roberts and Page [4] and Burnett and Page [5,6] have studied the
softening of silicon by ion implantation. Softening has been clearly
linked to the formation of an amorphous layer by ion implantation. By
analogy with the implantation hardening/softening behaviour of other
ceramic materials of both covalent and ionic bond types (e.g. SiC,
MgO, Al2O3 [6-9]) the ion implanted surface is expected to reach a
peak hardness just before the onset of amorphisation (the dose at
which this occurs is critically dependent upon ion species, ion energy
and implant temperature). Thereafter, the microhardness test (which
typically penetrates the surface by 1-5 micron) samples the hardness
of a sandwich of crystalline radiation-hardened surface layer
overlaying amorphous sub-surface layer which in turn overlays the
crystalline substrate containing the 'tail' of the radiation damage
profile. The microhardness-dose data of TABLE 1 shows this trend for
both the As+ and Si+ implants.

At sufficiently high doses a surface amorphous layer is formed.
Although this layer is too thin to determine the microhardness directly
(any hardness impressions penetrating the layer and hence sampling the
hardness of the substrate and coating) without the use of specialised
ultra microhardness testing techniques [10] the indentation process may
be modelled to yield a 'true' layer hardness. Using Vickers indentation,
Burnett and Page [5] evaluated amorphous layer hardness as approximately
600 VHN (5.9 GPa). A later refinement of this analysis yielded a
figure of 500 VHN (4.9 GPa). Thus the amorphous material is some 50%
softer than the crystalline parent material [11].

D. OTHER POSSIBLE EFFECTS

The experimentally derived low-load hardness of ion-implanted silicon may also be affected by the chemical nature of the implant species. For instance it is known that electrically active implant species may increase dislocation velocity of bulk silicon [12,13,15]. However, both Hu and Schwenker [2] and Burnett and Page [5] failed to find any dopant sensitivity in their hardness results.

Annealing is expected to remove the radiation damage in the non-amorphised material and to recrystallise the amorphised silicon. Both would result in the hardness returning to close to that of bulk silicon. If any electrically active species were implanted their effects might now be expected to become apparent since the dominant radiation damage effect will have been removed. There is, however, no data on this.

It is expected that changes in the surface structure brought about by ion implantation will also affect near-surface hardness phenomena such as chemomechanical and photomechanical effects [15]. Indeed, Skupov [14] has reported an increase in photomechanical softening (in the presence of light of wavelength about 600-700 nm) from less than 10% for unimplanted (111)Si to 40-45% at doses of > $10*14$ Ar+/cm*2. The saturation of the photomechanical effect at > $10*4$ Ar+/cm*2 was attributed to amorphisation of the silicon.

REFERENCES

[1]   B.R.Lawn, B.J.Hockey, S.M.Wiederhorn [J. Mater. Sci. (GB) vol.15
        (1980) p.1207]
[2]   S.M.Hu, R.O.Schwenker [J. Appl. Phys. (USA) vol.49 no.6 (Jun
        1978) p.3259-65]
[3]   P.J.Burnett, G.A.D.Briggs [J. Mater. Sci. (GB) vol.21 no.5 (May
        1986) p.1828]
[4]   S.G.Roberts, T.F.Page [in 'Ion Implantation into Metals', Eds
        V.Ashworth, W.A.Grant, R.P.M.Proctor (Pergamon, Oxford, GB, 1982)
        p.135]
[5]   P.J.Burnett, T.F.Page [J. Mater. Sci. (GB) vol.19 no.3 (Mar 1984)
        p.845-60]
[6]   P.J.Burnett, T.F.Page [J. Mater. Sci. (GB) vol.19 no.11 (Nov 1984)
        p.3524-45]
[7]   P.J.Burnett, T.F.Page [Mater. Res. Soc. Symp. Proc. (USA) vol.27
        (1984) p.401]
[8]   P.J.Burnett, T.F.Page [Proc. Int. Conf. Science of Hard
        Materials II, Rhodes, Greece, 23-28 Sept 1984, Eds C.A.Almond,
        C.A.Brookes, R.Warren (Adam Hilger, Bristol, England, 1986) p.789]
[9]   C.J.McHargue, C.S.Yust [J. Am. Ceram. Soc. (USA) vol.67 (1984)
        p.117]
[10]   J.B.Pethica, R.Hutchings, W.C.Oliver [Philos. Mag. A (GB) vol.48
        (1983) p.593]
[11]   P.J.Burnett, D.S.Rickerby [Surf. Eng. (GB) vol.3 no.1 (1987)
        p.69]
[12]   A.George, G.Champier [Phys. Status. Solidi a (Germany) vol.53a
        (1979) p.529]
[13]   S.G.Roberts, P.Pirouz, P.B.Hirsch [J. Phys. Colloq. (France)
        vol.44 no.C-4 (1983) p.75-83]
[14]   V.D.Skupov [Inorg. Mater. (USA) vol.19 (1983) p.910]
[15]   P.J.Burnett [EMIS Datareview RN=15710 (Mar 1987) 'Hardness of
        Si']

1.11 TOUGHNESS OF UNIMPLANTED AND ION-IMPLANTED Si

by P.J.Burnett

April 1987
EMIS Datareview RN=15708

A. TOUGHNESS OF Si

The toughness of brittle materials such as silicon is difficult to
determine experimentally since machining of conventional toughness test
specimens (such as single edge notched beams) usually results in the
introduction of unwanted flaws that cause premature failure.
Consequently, the toughness of these materials is frequently measured
using the indentation bend strength method [1] or indentation fracture
mechanics method [2,3].

A selection of room temperature fracture toughness (K1C) values for
(111) plane-family orientation Si found in the literature is given in
TABLE 1.

TABLE 1: room temperature K1C values for (111) plane-family single
         crystal silicon

Reference	[1]	[4]	[5]	[6]	[2]	[7]	[8]
Fracture Toughness (MPa m*1/2)	0.82	0.6	0.93	1.24-2.85	0.7	0.6	0.79-0.89

Reference [1] used the indentation bend strength technique, [4]
and [5] used a double cantilever technique, [6] used a notched beam
technique (and found a dependence of toughness on notch tip radius),
[2] used a double torsion technique and [7] and [8] used indentation
fracture techniques. Chen and Leipold [1] and Yasutake et al [9] also
determined the variation of K1C with crystal orientation, the results
being given in TABLE 2.

TABLE 2: variation of K1C with crystal orientation

Orientation plane-family	(111)	(110)	(100)	Reference
K1C (MPa m*1/2)	0.82	0.9	0.95	[1]
	-	0.9	0.95	[9]

In addition, these workers determined the toughness of polysilicon as
0.75 MPa m*1/2. This value was determined for a specimen containing
precipitated copper; considerable internal stresses were present that
might have significantly modified the toughness value when compared to
the toughness of pure polycrystalline Si.

B. TOUGHNESS OF ION IMPLANTED Si

The indentation fracture technique lends itself to the testing of ion
implanted silicon surfaces. Here the toughness is evaluated from the
ratio of the indentation diagonals to the lengths of radial cracks
found around indentations. Using this method Burnett and Page [8]
determined the toughness of (111) plane-family silicon surfaces
implanted with 2 X 10*17 (N2)+/cm*2 at 90 keV and 5 X 10*17 Al+/cm*2
at 300 keV (TABLE 3).

TABLE 3: indentation fracture toughness of ion implanted Si

Dose	Indentation Load (g)	Indentation Fracture Toughness (MPa m*1/2)
Unimplanted	200	0.74
	500	0.89
2 X 10*17 (N2)+/cm*2	200	0.96
	500	1.04
5 X 10*17 Al+/cm*2	200	1.29
	500	1.19

Both of the implanted surfaces studied possessed an implantation
induced surface amorphous layer and both showed an increased toughness
when compared to the unimplanted surface. Further, microscopical
examination of both these (111) surfaces and the (100) implanted
surface studied by Roberts and Page [10] showed an increased resistance
to the chipping (or lateral) mode of indentation fracture. Studies
of the toughness of other ion-implanted ceramic surfaces have shown
that an amorphous layer is not required to produce toughening, rather
it is the generation of an intense compressive surface stress that
seems responsible for the improved resistance to fracture, peak
toughness being coincident with peak hardness [14] at the dose just
prior to the onset of amorphisation [11,12]. It must be emphasised
that these toughness values are characteristic of the implanted
layer-substrate system rather than being a property solely of the
layer.

Finally, Hu [13] attempted to study the effect of ion implantation
upon the critical fracture stress (a quantity related to toughness)
for Si loaded using a concentric ring apparatus (so avoiding
specimen edge effects) and found that Ar implantation followed by a
one-step anneal (900 deg C, 1 hour) could significantly increase the
fracture strength whilst the as-implanted specimen showed no
significant increase. Whilst this behaviour apparently contradicts the
indentation fracture results, it should be recognised that in Hu's
tests brittle failure will propagate from surface flaws, i.e. the
fracture stress is critically determined by both K1C and the surface
flaw distribution, whilst in the former case fracture is initiated in
the bulk of the specimen and propagates to the surface. Consequently,
the improvement in critical fracture stress observed after a one-step
anneal probably represents a change in the surface flaw characteristics.

The toughness fell after the second stage anneal presumably due to a
further change in flaw distribution and possibly related to the
recrystallisation process producing interfaces that encourage fracture.

REFERENCES
----

[1]   C.P.Chen, M.H.Leipold [Am. Ceram. Soc. Bull. (USA) vol.59 (1980)
        p.469]
[2]   B.R.Lawn, A.G.Evans, D.B.Marshall [J. Am. Ceram. Soc. (USA)
        vol.63 (1980) p.574]
[3]   G.R.Anstis, P.Chantikul, B.R.Lawn, D.B.Marshall [J. Am. Ceram.
        Soc. (USA) vol.64 (1981) p.533]
[4]   Inferred by [1] from the data in R.J.Jaccodine [J. Electrochem.
        Soc. (USA) vol.10 no.6 (1963) p.524]
[5]   C.St.John [Philos. Mag. (GB) vol.32 (1975) p.1193]
[6]   R.J.Myers, B.M.Hillberry [Proc. 4th Int. Conf. on Fracture
        vol.3B, Ed. D.M.R.Taplin (Pergamon Press, New York, USA 1977) p.1001]
[7]   B.R.Lawn, D.B.Marshall [J. Am. Ceram. Soc. (USA) vol.62 (1979)
        p.347]
[8]   P.J.Burnett, T.F.Page [J. Mater. Sci. (GB) vol.19 no.11 (Nov 1984)
        p.3524-45]
[9]   K.Yasutake, M.Iwata, K.Yoshi, M.Umeno, H.Kawabe [J. Mater. Sci.
        (GB) vol.21 no.6 (Jun 1986) p.2185-92]
[10]   S.G.Roberts, T.F.Page [in 'Ion Implantation into Metals' Eds.
        V.Ashworth, W.A.Grant, R.P.M.Procter (Pergamon Press, Oxford, GB
        1982) p.135]
[11]   P.J.Burnett, T.F.Page [J. Mater. Sci. (GB) vol.20 (1985) p.4624]
[12]   C.J.McHargue, C.S.Yust [J. Am. Ceram. Soc. (USA) vol.67 (1984)
        p.117]
[13]   S.M.Hu [J. Appl. Phys. (USA) vol.53 no.8 (May 1982) p.3576-80]
[14]   P.J.Burnett [EMIS Datareview RN=15709 (Mar 1987) 'Hardness of
        ion-implanted Si']

1.12 THERMAL EXPANSION COEFFICIENT OF Si

by T.Soma, H.-Matsuo Kagaya

March 1987
EMIS Datareview RN=15706

Data relevant to the linear thermal expansion coefficient alpha (/deg K) of Si are given in refs [1-14]. Recommended data are summarized in TABLE 1.

TABLE 1

T (deg K)	alpha (10*-6/deg K)	T (deg K)	alpha (10*-6/deg K)	T (deg K)	alpha (10*-6/deg K)
5	0.6 X 10*-4	180	1.061	700	4.016
10	0.48 X 10*-3	200	1.406	800	4.151
20	-0.29 X 10*-2	220	1.715	900	4.185
40	-0.164	240	1.986	1000	4.258
60	-0.400	260	2.223	1100	4.323
80	-0.472	280	2.432	1200	4.384
100	-0.339	300	2.616	1300	4.442
120	-0.057	400	3.253	1400	4.500
140	0.306	500	3.614	1500	4.556
160	0.689	600	3.842	1600	4.612

NOTES (O:Observed, T:Theoretical and E:Extrapolated)
-----

1. Extrapolation to the melting point (1687 deg K) gives
    4.661 X 10*-6/deg K. (E)

2. The thermal expansion coefficient has a minimum at 80 deg K. (O)

3. Below 14 deg K, the tabulated data is given by:-

    alpha = 4.8 X 10*-13 T*3 within +- 0.01 X 10*-8 /deg K [12]. (T)

4. From 120 deg K to 1500 deg K, a smooth curve of the linear thermal
    expansion coefficient given by:-

    alpha = 3.725 X 10*-6 [1 - exp (-5.88 X 10*-3 (T-124))]
            + 5.548 X 10*-10T within +- 2 X 10*-7 /deg K

    is proposed [14]. (T)

5. Above 1000 deg K the tabulated data is given by:-

    alpha = (3.684 + 0.00058T) X 10*-6 within +- 0.004 X 10*-6/deg K. (T)

6. Above 600 deg K, the tabulated values obey the Gruneisen relation
    in which the ratio of thermal expansion coefficient to specific heat
    is (1.101 +- 0.022) X 10*-6 g/cal. (T)

7. The fact that there is a systematic difference of lattice parameters
    between powder and single-crystal samples is experimentally reported
    [14]. (O)

REFERENCES
----

[1]   D.F.Gibbons [Phys. Rev. (USA) vol.112 (1958) p.136]
[2]   S.I.Novikova, P.G.Strelkov [Sov. Phys.-Solid State (USA) vol.1
      (1960) p.1687]
[3]   S.I.Novikova [Sov. Phys.-Solid State (USA) vol.6 (1964) p.269]
[4]   D.N.Batchelder, R.O.Simmons [J. Chem. Phys. (USA) vol.41 (1964)
      p.2324]
[5]   R.H.Carr, R.D.McCammon, G.K.White [Philos. Mag. (GB) vol.12 (1965)
      p.157]
[6]   P.W.Sparks, C.A.Swenson [Phys. Rev. (USA) vol.163 (1967) p.779]
[7]   V.V.Zhdanova, M.G.Kekua, T.Z.Samaclashvill [Inorg. Mater. (USA)
      vol.3 (1967) p.1112]
[8]   A.P.Nasekovs'kii [Ukr. Fiz. Zh. (USSR) vol.12 (1967) p.1352]
[9]   H.Ibach [Phys. Status Solidi b (Germany) vol.31 (1969) p.625]
[10] G.K.White [J. Phys. D (GB) vol.6 (1973) p.2070]
[11] M.A.Norton, J.W.Berthold, III, S.F.Jacobs, W.A.Plummer [J. Appl.
      Phys. (USA) vol.47 no.4 (Apr 1976) p.1683-5]
[12] K.G.Lyon, G.L.Salinger, C.A.Swenson, G.K.White [J. Appl. Phys.
      (USA) vol.48 no.3 (Mar 1977) p.865-8]
[13] R.B.Roberts [J. Phys. D (GB) vol.14 (1981) p.L163]
[14] Y.Okada, Y.Tokumaru [J. Appl. Phys. (USA) vol.56 no.2 (15 Jul 1984)
      p.314-20]

1.13 SPECIFIC HEAT OF Si

by H.-Matsuo Kagaya and T.Soma

March 1987
EMIS Datareview RN=15705

Data relevant to the specific heat at constant pressure (C) of Si
are given in refs [1-6]. Recommended data are summarized in TABLE 1
together with the Debye temperature (theta).

TABLE 1

T (deg K)	C (J/g/K)	theta (deg K)	T (deg K)	C (J/g/K)	theta (deg K)
5	0.340×10^{-4}	641	280	0.691	645
10	0.276×10^{-3}	631	300	0.713	643
20	0.341×10^{-2}	545	400	0.785	650
40	0.441×10^{-1}	463	500	0.832	653
60	0.115	492	600	0.849	654
80	0.188	535	700	0.866	
100	0.259	572	800	0.883	655
120	0.328	599	900	0.899	
140	0.395	617	1000	0.916	656
160	0.456	629	1100	0.933	
180	0.511	637	1200	0.950	
200	0.557	643	1300	0.967	
220	0.597	647	1400	0.983	
240	0.632	648	1500	1.000	
260	0.665	647	1600	1.017	

NOTES   (O:Observed, T:Theoretical and E:Extrapolated)
-----

1. Extrapolation of the specific heat to the melting point (1687 deg
    K) gives 1.032 J/g/K. (E)

2. The Debye temperature extrapolated to the absolute zero gives 645
    deg K and reaches to about 657 deg K at the melting point. (E)

3. The Debye temperature has a minimum 463 deg K at T = 39 deg K. (O)

4. At low temperatures below 5 deg K [2], specific heat is given by:-

    C = (0.716T + 0.243T*3) X 10*-6 within +- 0.002 X 10*-6 J/g/K. (T)

5. Above 5 deg K the ratio C/T*3 has a maximum at 39 deg K [2,4] and
    decreases rapidly with the approximate dependence of T*-2.22 from
    100 deg K to 500 deg K. (T)

6. Above 500 deg K the tabulated data is given by:-

    C = 0.748 + 0.000168T within +- 0.004. (T)

7. Anharmonic contributions to the specific heat and the difference
    between the specific heat at constant pressure and that at
    constant volume of Si are appreciable at high temperatures (see
    refs [7-10]). (T)

REFERENCES
----

[1]   C.T.Anderson [J. Am. Chem. Soc. (USA) vol.52 (1930) p.2301]
[2]   N.Pearlmon, P.H.Keesom [Phys. Rev. (USA) vol.88 (1952) p.398,
      vol.85 (1952) p.730]
[3]   P.V.Gul'tyaev, A.V.Petrov [Sov. Phys. Solid State (USA) vol.1
      (1959) p.330]
[4]   P.Flubacher, A.I.Leadbetter, J.A.Morrison [Philos. Mag. (GB)
      vol.4 (1959) p.273]
[5]   P.B.Kantor, O.M.Kisel, E.M.Fomichev [Ukr. Fiz. Zh. (USSR) vol.5
      (1960) p.358]
[6]   H.R.Shanks, P.D.Maycock, P.H.Sidles, G.C.Danielson [Phys. Rev.
      (USA) vol.130 (1963) p.1743]
[7]   M.E.Fine [J. Chem. Phys. (USA) vol.21 (1953) p.1427]
[8]   D.M.T.Newsham [Phys. Rev. (USA) vol.152 (1966) p.841]
[9]   P.C.Trivedi, H.O.Sharma, L.S.Kothari [J. Phys. C (GB) vol.10 no.18
      (28 Sep 1977) p.3487-97]
[10] P.C.Trivedi [J. Phys. C (GB) vol.18 no.5 (20 Feb 1985) p.983]

1.14 THERMAL CONDUCTIVITY OF Si

by M.N.Wybourne

May 1987
EMIS Datareview RN=15711

The thermal conductivity of single crystal silicon has been measured
over a wide temperature range and on a variety of samples by many
workers [1-9]. One of the earliest measurements was by Koenigsberger
and Weiss [10] who found a value of 0.84 W/cm/deg K at 290 deg K for
an impure sample. Recently, a number of measurements have been
presented and compared in graphical form by Beadle, Tsai and Plummer
[11]. Rowe and Bhandari [12] have reviewed the techniques used to
measure the thermal conductivity of semiconductors and have also
discussed the data for doped siicon reported by Steigmeier [13],
Holland and Neuringer [8] and Slack [9].

In general, the thermal conductivity of single crystal silicon
exhibits a temperature dependence characteristic of single crystal
dielectric materials. For the purest silicon samples, the thermal
conductivity goes through a maximum at a temperature of about
22 deg K. Above this temperature the conductivity falls as the phonon
mean-free-path is reduced by phonon-phonon scattering processes. At
temperatures below the maximum the absolute value of the conductivity
is highly dependent on the sample but, in broad terms, it falls with
a temperature dependence close to that of the specific heat of the
lattice (T*3).

For the purpose of this Datareview it is convenient to consider two
temperature regions; one from 200 deg K to the melting point of
silicon and the other up to 200 deg K. Above 200 deg K the thermal
conductivity is largely independent of the particular sample
specifications and the various reported data coalesce to be in
agreement to within a factor of two. Representative data in this
temperature range are taken from the measurements by Glassbrenner and
Slack [7] and are presented in TABLE 1.

TABLE 1

T(deg K)	K(W/cm/deg K)	T(deg K)	K(W/cm/deg K)
200	2.66	1000	0.31
300	1.56	1100	0.28
400	1.05	1200	0.26
500	0.80	1300	0.25
600	0.64	1400	0.24
700	0.52	1500	0.23
800	0.43	1600	0.22
900	0.36	1681	0.22

Note: These measurements were made on a [111] axis cylindrical sample
130 mm long and 26 mm in diameter. The sample was grown by the
float-zone process in a argon atmosphere. The sample was n-type
with a room-temperature resistivity of 440 ohm cm and a carrier
concentration of 1.27 X 10*13/cm*3.

In the temperature range up to 200 deg K details of the temperature dependence of the thermal conductivity together with position and magnitude of the conductivity maximum, are sensitive to the size, perfection, doping level and type, oxygen concentration [14], orientation [15] and surface quality [16] of the sample. It is only possible, therefore, to provide illustrative data in this temperature range, TABLE 2 [8].

TABLE 2

T(deg K)	K(W/cm/deg K)	T(deg K)	K(W/cm/deg K)
150	4.10	20	47.7
100	9.13	10	24.0
50	28.0	8	16.4
40	36.6	6	8.99
30	44.2	4	3.11
		2	0.44

Note: These measurements were made on a bar sample 31.75 mm long and 6.30 X 6.07 mm*2 in cross-section. The sample was grown by the float-zone process and contained 1.0 X 10*13/cm*3 boron.

In TABLE 3 we present data from the measurements of Holland and Neuringer [8] as a guide to the dependence of the thermal conductivity maximum on the concentration of boron impurities. It should be noted that the presence of oxygen impurities causes resonant phonon scattering at about 40 deg K [17] which suppresses both the magnitude and the temperature of the maximum for a given boron concentration.

TABLE 3

Boron Concentration (deg K)	Thermal conductivity Maximum (W/cm/deg K)	Temperature (deg K)	Sample Cross-section (mm*2)
1.0 X 10*13	48	22	6.30 X 6.07
4.2 X 10*14	43	25	6.17 X 6.32
4.2 X 10*14	38	26	3.45 X 3.50
1.0 X 10*15	43	25	6.19 X 6.32
4.0 X 10*15	33	27	5.86 X 5.86
4.0 X 10*16	18	37	6.24 X 6.27

Finally, thermal conductivity measurements of doped silicon samples have been used as a technique to study the nature of the impurities themselves [18-21]. Many of these studies have been reviewed elsewhere [22].

REFERENCES
----

[1]   B.B.Kuprovski, P.V.Geld [Fiz. Met. & Metalloved. (USSR) vol.3 (1956) p.182]
[2]   W.D.Kingery [J. Am. Ceram. Soc. (USA) vol.42 (1959) p.617]
[3]   H.Mette, W.W.Gartner, C.Loscoe [Phys. Rev. (USA) vol.117 (1960) p.1491]
[4]   A.D.Stuckes [Philos. Mag. (GB) vol.5 (1960) p.84]
[5]   R.G.Morris, J.G.Hust [Phys. Rev. (USA) vol.124 (1961) p.1426]
[6]   R.G.Morris, J.J.Martin [J. Appl. Phys. (USA) vol.34 (1963) p.2388]

[7]   C.J.Glassbrenner, G.A.Slack [Phys. Rev. (USA) vol.134 (1964)
       p.A1058]

[8]   M.G.Holland, L.G.Neuringer [Proc. Int. Conf. Physics of
       Semiconductors, Exeter, England, 1962 (Inst. Phys., Bristol)
       p.474]

[9]   G.A.Slack [J. Appl. Phys. (USA) vol.35 (1964) p.3460]

[10]   J.Koenigsberger, J.Weiss [Ann. Phys. vol.35 (1911) p.1]

[11]   W.E.Beadle, J.C.C.Tsai, R.D.Plummer [Quick Reference Manual
       for Silicon Integrated Circuit Technology (John Wiley & Sons,
       1985) p.2-64 - 2-66]

[12]   D.M.Rowe, C.M.Bhandari [Prog. Cryst. Growth & Charact. (GB)
       vol.13 (1986) p.233]

[13]   E.F.Steigmeier [in 'Thermal Conductivity' vol.2 Ed. R.P.Tye
       (Academic Press, 1969)]

[14]   D.Fortier, H.Djersassi, K.Suzuki, H.J.Albany [Phys. Rev. B (USA)
       vol.9 (1974) p.4340]

[15]   A.K.McKurdy, H.J.Maris, C.Elbaum [Phys. Rev. B (USA) vol.2
       (1970) p.4077]

[16]   W.S.Hurst, D.R.Frankl [Phys. Rev. (USA) vol.186 (1969) p.801]

[17]   A.M.de Goer, M.Locatelli, K.Lassmann [J. Phys. Colloq. (France)
       vol.42 (1981) p.235]

[18]   D.Fortier, K.Suzuki [J. Phys. (France) vol.37 (1976) p.143]

[19]   D.Fortier, K.Suzuki [Phys. Rev. B (USA) vol.9 (1974) p.2530]

[20]   A.Adolf, D.Fortier, J.H.Albany, K.Suzuki, M.Locatelli [in 'Phonon
       Scattering in Condensed Matter' Ed. H.J.Marris (Plenum Press,
       1980) p.421]

[21]   K.A.Adilov et al [Sov. Phys.- Solid State (USA) vol.28 (1986)
       p.1070]

[22]   L.J.Challis, A.M.de Goer [in 'The Dynamical Jahn-Teller effect
       in Localized Systems' Eds Y.E.Perlin, M.Wagner (Elsevier, 1984)
       p.533]

1.15 PHONON SPECTRUM OF Si

by M.N.Wybourne

June 1987
EMIS Datareview RN=15728

Details of the phonon spectrum of silicon have been studied primarily
by the techniques of inelastic neutron scattering, Raman spectroscopy,
and infrared spectroscopy. Some of the earliest inelastic neutron
scattering measurements of the phonon dispersion were reported by
Brockhouse [1]. Later, with improved resolution and calibration,
Dolling [2] presented dispersion curve data at 300 deg K for the three
directions [zeta 0 0], [zeta zeta zeta], and [0 zeta zeta]. For the two
high symmetry directions [zeta 0 0] and [zeta zeta zeta], the modes are
strictly longitudinal (L) or transverse (T), while in the [0 zeta zeta]
direction symmetry restrictions only require that the polarisation
vector of the modes is in the [0 1 1] mirror plane. Some of the phonon
frequencies, with polarisation type, are given in TABLE 1

TABLE 1

Reduced wavevector coordinates			Acoustic (THz) T	L	Optical (THz) T	L
0.0	0.0	0.0	--------		15.53+-0.23	
0.2	0.0	0.0	2.00+-0.08	3.20+-0.12	15.40+-0.36	15.55+-0.38
0.4	0.0	0.0	3.45+-0.06	6.10+-0.15	14.65+-0.33	15.15+-0.35
0.6	0.0	0.0	4.27+-0.10	8.50+-0.20	14.18+-0.35	14.60+-0.35
0.8	0.0	0.0	4.46+-0.07	10.60+-0.20	13.95+-0.34	13.60+-0.27
1.0	0.0	0.0	4.49+-0.06	12.32+-0.20	13.90+-0.30	12.32+-0.20
0.1	0.1	0.1	1.53+-0.03	3.00+-0.13	15.40+-0.37	15.45+-0.35
0.2	0.2	0.2	2.62+-0.05	5.60+-0.15	15.10+-0.30	14.78+-0.30
0.3	0.3	0.3	3.22+-0.05	8.03+-0.15	14.83+-0.30	14.13+-0.30
0.4	0.4	0.4	3.41+-0.05	10.12+-0.20	13.78+-0.27	13.30+-0.22
0.5	0.5	0.5	3.43+-0.05	11.35+-0.30	14.68+-0.30	12.60+-0.32
0.0	0.2	0.2	2.91+-0.07	4.55+-0.10	15.10+-0.35	
0.0	0.4	0.4	5.11+-0.09	7.60+-0.20	14.80+-0.40	13.87+-0.28
0.0	0.5	0.5	6.06+-0.12	8.60+-0.18	14.55+-0.38	12.83+-0.27
0.0	0.75	0.75	6.56+-0.14	10.85+-0.30	14.40+-0.38	11.15+-0.30
0.0	0.85	0.85	5.60+-0.11	11.70+-0.30	14.20+-0.34	11.70+-0.30

Inelastic neutron scattering measurements have also been made by Nilsson and Nelin [3] to study the homology of the phonon dispersion relations of silicon and germanium. These authors presented data at 300 and 80 deg K with less uncertainty than the results of Dolling [2]; however, there are no significant differences between these sets of data.

Many different models have been used to describe the neutron scattering data, the earliest being the shell model of Cochran [4]. One of the best fits to the measured phonon frequencies (approx. 2%) is given by the Weber adiabatic bond-charge model [5]. Other calculations using the electron-ion Hamiltonian of the system have also been developed to model the phonon dispersion. These calculations treat the Hamiltonian in two ways; one in which the phonons are considered as small perturbations of the system [6] and the other in which the total energy of the lattice containing 'frozen-phonons' of arbitrary displacements is calculated [7]. Results of such calculations model the phonon frequencies to within approx. 5%, but are only tractable at the Brillouin zone centre and at points of high symmetry. A new ab initio approach to calculate the phonon characteristics was presented by Kunc and Martin [8] in which the force constants in real space are determined directly. Fleszar and Resta [9] have used this technique for silicon and have reported convergence to better than 3% for both the optical and acoustic phonon frequencies.

The Raman spectrum of silicon has been thoroughly studied by many authors and, with the help of the inelastic neutron scattering data, the interpretation is well understood. The diamond structure of silicon allows only one first-order Raman active phonon at the Brillouin zone centre. The first-order Raman spectra has been investigated by Russell [10], Parker et al [11], Hart et al [12], and Chang et al [13] and the zone centre optical phonon frequency has been determined to be 15.56 +- 0.03 THz. The second order Raman spectrum has been studied by Weinstein and Cardona [14] and Temple and Hathaway [15]. Using different polarisations, these workers were able to separate the independent components of the Raman tensor so that a comparison with the theoretical Raman spectrum [16], together with the unambiguous assignment of the spectral features, was possible. Weinstein and Cardona, and Temple and Hathaway have also discussed the comparison of the two phonon Raman spectrum intensity with the calculated phonon density of states for silicon [17].

The infrared absorption spectrum of silicon was first studied by Johnson [18] before neutron-scattering data were available. Further measurements and assignments of the spectral features were made later by Johnson and Loudin [19] and Balkanski and Nusimovici [20]. A comparison of the phonon frequencies obtained at high symmetry points in the Brillouin zone by neutron-scattering [2], Raman spectroscopy [15] and infrared spectroscopy [20] are given in TABLE 2.

TABLE 2

Critical points		Phonon Frequencies (THz)		
	Mode	Infrared [20]	Raman [15]	Neutron [2]
GAMMA		15.50 +- 0.06	15.56 +- 0.03	15.53 +- 0.09
X	TO	13.46 +- 0.09	13.79 +- 0.06	13.91 +- 0.30
	TA	4.65 +- 0.15	4.53 +- 0.06	4.50 +- 0.06
L	TO	14.78 +- 0.06	14.69 +- 0.06	14.69 +- 0.30
	TA	3.39 +- 0.06	3.39 +- 0.06	3.42 +- 0.06
W	TO	14.33 +- 0.39	14.09 +- 0.06	14.42 +- 0.39

Stress-induced changes in the phonon spectrum offer a convenient method by which to study anharmonic effects and to study the lattice dynamical models of silicon. Weinstein and Piermarini [21] have studied the first and second order Raman spectra of silicon under large applied hydrostatic pressures. The general effect of the pressure on the phonon dispersion near the zone edge was found to be that the transverse acoustic modes softened while the optical modes shifted to higher energies. The best fit to the pressure dependence of the first order Raman peak is;

omega = 15.56 +- 0.03 THz + (15.6 +- 0.9 GHz/kbar)P

+ (-21 +- 6 MHz/kbar*2)(P*2)

where P is the pressure. Estel, Kalus, and Pintschovius [22] have reported phonon frequency changes in the [zeta zeta zeta] transverse acoustic branch, with applied uniaxial stress, studied by inelastic neutron scattering. With a stress of 1.01 GPa in the [1 1 1] direction, the energy shifts reported are from +0.5% near the zone centre to -1.2% at the zone edge.

The internal stress of silicon deposited epitaxially onto sapphire has been studied using Raman spectroscopy to measure the shift in the zone-centre optical phonon frequency. In this way, Englert, Abstreiter, and Pontcharra [23] were able to determine an internal stress of about 7.0 +- 0.3 kbar for [001] silicon layers on [1012] sapphire substrates. Subsequently, this technique has been used to study the stress relief of deposited films that have been annealed [24] or silicon implanted and then annealed [25].

REFERENCES

[1]   B.N.Brockhouse [Phys. Rev. Lett. (USA) vol.2 (1959) p.256]
[2]   G.Dolling [Symp. Inelastic Scattering of Neutrons in Solids
       and Liquids (IAEA, Vienna, 1963) vol.2 p.37]
[3]   G.Nilsson, G.Nelin [Phys. Rev. B (USA) vol.6 (1972) p.3777]
[4]   W.Cochran [Proc. R. Soc. London Ser A. (GB) vol.253 (1959)
       p.260]
[5]   W.Weber [Phys. Rev. B (USA) vol.15 (1977) p.4789]
[6]   R.M.Pick, M.H.Cohen, R.M.Martin [Phys. Rev. B (USA) vol.1 (1970)
       p.910]
[7]   J.Ihm, M.T.Yin, M.L.Cohen [Solid State Commun. (USA) vol.37
       (1981) p.491]
[8]   K.Kunc, R.M.Martin [Phys. Rev. Lett. (USA) vol.48 (1982) p.406]
[9]   A.Fleszar, R.Resta [Phys. Rev. B (USA) vol.34 (1986) p.7140]
[10]   J.P.Russell [J. Phys. Radium vol.26 (1965) p.620]
[11]   J.H.Parker,Jr, D.W.Feldman, M.Ashkin [Phys. Rev. (USA) vol.155
       (1967) p.712]
[12]   T.R.Hart, R.L.Aggarwal, B.Lax [Phys. Rev. B (USA) vol.1 (1970)
       p.638]
[13]   R.K.Chang, J.M.Ralston, D.E.Keating [in 'Light Scattering Spectra
       of Solids, Ed. G.B.Wright (Springer-Verlag, Solid State 1969)
       p.369]
[14]   B.A.Weinstein, M.Cardona [Solid State Commun. (USA) vol.10 (1972)
       p.961]
[15]   P.A.Temple, C.E.Hathaway [Phys. Rev. B (USA) vol.7 (1973)
       p.3685]
[16]   R.A.Cowley [J. Phys. (France) vol.26 (1965) p.659]
[17]   G.Dolling, R.A.Cowley [Proc. Phys. Soc. (GB) vol.88 (1966)
       p.463]
[18]   F.A.Johnson [Proc. Phys. Soc. (GB) vol.73 (1959) p.265]
[19]   F.A.Johnson, R.Loudon [Proc. R. Soc London Ser. A (GB) vol.281
       (1964) p.274]
[20]   M.Balkanski, M.Nusimovici [Phys. Status Solidi (Germany) vol.5
       (1964) p.635]
[21]   B.A.Weinstein, G.J.Piermarini [Phys. Rev. B (USA) vol.12 (1975)
       p.1172]
[22]   J.Estel, J.Kalus, L.Pintschovius [Phys. Status Solidi b
       (Germany) vol.126 (1984) p.121]
[23]   Th.Englert, G.Abstreiter, J.Pontcharra [Solid-State Electron.
       (GB) vol.23 (1980) p.31]
[24]   G.A.Sai-Halasz, F.F.Fang, T.O.Sedgwick, A.Segmuller [Appl. Phys.
       Lett. (USA) vol.36 (1980) p.419]
[25]   Y.Ohmura, T.Inoue, T.Yoshii [J. Appl. Phys. (USA) vol.54 (1983)
       p.6779]

1.16 THERMAL DIFFUSIVITY OF Si

by M.N.Wybourne

August 1987
EMIS Datareview RN=16129

Some of the earliest measurements of the thermal diffusivity of
single-crystal silicon were reported by Shanks et al [1]. Using the
method of Kennedy et al [2] these workers measured the thermal
diffusivity of a (111) orientation 107 ohm cm p-type sample over the
temperature range 300-1400 deg K (TABLE 1). Shanks et al also
presented data for two n-type samples of resistivity 33 and 1010 ohm cm
with the orientations (111) and (100) respectively. The data from all
three samples agreed to within the accuracies of the experiment. In the
temperature range 300-900 deg K, the values of the thermal conductivity
derived from these data are approximately 10% below those obtained from
direct measurement [3]. Above 1000 deg K, direct measurements show that
the thermal conductivity continues to decrease with temperature up to
almost the melting point [3]. This is inconsistent with the thermal
diffusivity measurements which suggest that the thermal conductivity
should be temperature independent in this temperature range. The
discrepancy is about 20% at 1400 deg K and is of unknown origin.

TABLE 1

T deg K	D cm*2/s	T deg K	D cm*2/s
300	0.86	800	0.19
400	0.52	900	0.16
500	0.37	1000	0.14
600	0.29	1200	0.12
700	0.24	1400	0.12

The values of thermal conductivity obtained from thermal diffusivity
measurements made by Abeles et al [4] on a 100 ohm cm silicon sample
over the temperature range 300-1200 deg K, are between those of
Glassbrenner et al [3] and Shanks et al [1].

More recent measurements of the thermal diffusivity of silicon have
been made by Turkes [5] using an ion-implanted resistor as the thermal
transient sensor. This technique removed the problems associated with
the thermal response time of sensors used in transient methods of
determining the diffusivity. Turkes studied two samples, a 6 ohm cm
phosphorus doped sample and a 1 X 10*17/cm*3 aluminium doped sample.

Data were taken in the range 20-250 deg K. Above 50 deg K, the
diffusivity was sample independent and had a temperature dependence
governed by the phonon-phonon scattering rate (TABLE 2). Below
50 deg K, the phonon-phonon interaction is not predominant and the
temperature dependence of the diffusivity is determined by a
combination of isotope, impurity and boundary scattering rates. Note
that at the lowest temperatures where the phonon scattering rate is
solely dominated by the boundaries of the sample, the diffusivity will
be dependent on the sample geometry and will be temperature
independent.

TABLE 2

T deg K	D cm*2/s	T deg K	D cm*2/s
50	63.5	100	11.3
60	37.6	150	4.24
70	24.8	200	2.23
80	17.1	250	1.23
90	14.4		

Ebrahimi [6] has developed a technique to measure the thermal
diffusivity of small silicon chips. Two bipolar transistors on the
silicon chip are used as a thermal source and a thermal detector with
the spacing between the devices acting as a thermal delay. Using a
simplification of Angstrom's method [7], the diffusivity can be
determined from the phase difference between the devices. Measurements
were made over a limited temperature range, 220-380 deg K, and the
values of the diffusivity obtained were approx. 15% higher than those
obtained by Shanks et al [1].

The data given here refer to relatively lightly doped samples. Thermal
conductivity data show that the diffusivity of highly doped silicon
will be lower by a significant amount. For example, at 100 deg K a
sample doped to over 10*20/cm*3 would have a diffusivity that is an
order of magnitude below the value for a lightly doped sample (TABLE 2).
At 300 deg K, this reduction has fallen to a factor of two.

Rowe and Bhandari [8] reviewed thermal conductivity data and the
methods for measuring both the conductivity and the diffusivity of
doped semiconductors. They point out that both random and systematic
errors occur in these measurements. Most authors quote only two
significant figures implying an error between 0.5 and 5%. For the data
given here, errors of the order of 5% should be assumed.

REFERENCES
----

[1] H.R.Shanks, P.D.Maydock, P.H.Sidles, G.C.Danielson [Phys. Rev. (USA) vol.130 (1963) p.1743]

[2] W.L.Kennedy, P.H.Sidles, G.C.Danielson [Adv. Energy Convers. (GB) vol.2 (1962) p.53]

[3] C.J.Glassbrenner, G.A.Slack [Phys. Rev. (USA) vol.134 (1964) p.A1058]

[4] B.Abeles, D.S.Beers, G.D.Cody, J.P.Dismukes [Phys. Rev. (USA) vol.125 (1962) p.44]

[5] P.Turkes [Phys. Status Solidi a (Germany) vol.75 no.2 (16 Feb 1983) p.519-23]

[6] J.Ebrahimi [J. Phys. D (GB) vol.3 (1970) p.236]

[7] H.S.Carslaw, J.C.Jaeger [Conduction of Heat in Solids (Clarendon Press, Oxford, 1948) ch.4]

[8] D.M.Rowe, C.M.Bhandari [Prog. Cryst. Growth & Charact. (GB) vol.13 (1986) p.233]

1.17 SELF-DIFFUSIVITY OF Si

by J.A.Van Vechten

August 1987
EMIS Datareview RN=16118

At first view the literature on bulk self-diffusion in Si is a
bewildering array of apparently contradictory reports; activation
enthalpies ranging from 3.6 eV [1] to 5.13 eV [2] and pre-exponential
factors ranging from 0.3 cm*2/s [3] to 9000 cm*2/s [2] are derived from
evidently well-designed experiments by reputable scientists.
Particularly at lower temperatures, the absolute rate of self-diffusion
is reported at values ranging over two orders of magnitude [3]. The
situation is no better with surface self-diffusion where reports of the
activation enthalpy on the (111) surface range from 1.1 eV [4] to 3.6 eV
[5] with pre-exponential factors varying from 9.5 [6] to 9.4 X 10*6
cm*2/s [7]. As one might expect, there is controversy regarding the
mechanisms of diffusion; the only broad agreement seems to be that
there is more than one mode and their relative importance varies with
temperature, T. (Ref [3] is convincing on this point.) Arguments have
been made for the importance of simple interstitials [8,9], simple
single vacancies [2,10,11], extended interstitials (dense amorphous
zones) [12], extended vacancies (open amorphous zones) [13],
divacancies [14,15], and for ring-exchange mechanisms involving no
ambient point defect [16]. While recent positron annihilation
measurements [17] provide direct evidence for the presence of single
vacancies in concentrations exceeding 10*16/cm*3 in high temperature
Si, which (in this author's opinion) makes it very difficult to
justify an assumption that substantial numbers of simple interstitials
are also present, this review will try to avoid the controversy as to
the mechanisms, but instead show how the absolute rate of self-
diffusion may be predicted for a particular case in the face of the
seemingly contradictory literature.

Let us first consider the surface self-diffusion. In this case almost
all measurements have been taken near 1000 deg C so there can be no
evidence for different mechanisms at different temperatures. The lowest
reported activation enthalpy on a (111) surface, 1.1 eV, was obtained
from observations of surface reconstruction by LEED on surfaces that
had been roughened by laser irradiation [4]. The highest activation
enthalpy was obtained also from LEED observations of surface
reconstruction but on surfaces that had not been roughened [5]. Of
course, the activation enthalpy for diffusion can generally be
separated into an enthalpy of formation of the mobile species and an
enthalpy of migration of that species between equivalent sites:

$$Hd = Hf + Hm.$$

One supposes that the difference between reports [4] and [5] is that the
Hf does not appear in the value of [4] because the mobile species,
presumably surface adatoms, were produced by the irradiation, but this
term does appear with [5]. This explanation is supported in [6]. This
implies that Hm(111) = 1.1 eV. One can verify that this is a value
that could have been anticipated by making a ballistic model [18,19]
estimate:

$$Hm = 0.5\ Mv*2$$

where M is the mass of the mobile species, the Si atom, and v is its initial velocity. It has been found that in most cases v can be approximated by assuming the atom hops the distance between sites in one zone boundary phonon period. In bulk Si this gives Hm = 1.2 eV. On a (111) surface, the distance between adatom sites (on the ideal, unreconstructed surface) is (8/3)*0.5 times greater than the bulk nearest neighbour sites but the phonon modes of these adatoms, which have only 1 of 4 bonds complete, are softened by a factor close to 3*0.5. These two factors almost cancel so the (111) surface v should be about the same as the bulk v and so should Hm(111) be about the same as the bulk ballistic value; Hm(111) = approx. 1.1 eV-1.2 eV = Hm(bulk).

Now the Hf for an adatom on a surface will depend on the nature of the sites from which the adatom must be obtained. The simplest treatment of the Hf follows from consideration of the surface energy increase caused by the process that removes the atom from this site to an isolated adatom position. It has been shown [20,21] that macroscopic surface tension values can be used to approximate the surface energies of cavities even on an atomic scale with accuracy comparable to experimental uncertainties. Using the empirical surface tension of Si, one calculates the surface energy of a bulk Si vacancy, which is an estimate of its formation enthalpy, to be Hf(V bulk) = 2.8 eV. On a flat (111) surface, the same assumptions lead to an estimate of the surface energy of a surface vacancy that is 3/4 the bulk value, because 3/4 as much surface area is present, but there is also an increase of surface energy at the adatom which would be estimated at the same value. Thus, the total Hf(flat 111) = 4.2 eV and the sum, Hm(flat 111) = 5.3 eV is larger than any reported value. (Note that there is no surface adatom contribution to the enthalpy of formation of a bulk vacancy because the number of surface atoms does not change when a bulk vacancy is created [20].) However, it is generally agreed that surface adatoms are not created by extracting atoms out of flat sections of perfect surfaces but from kink sites on steps on the surface, so long as kink sites are available, or lacking these, from straight sections of steps. The removal of an atom from a straight section of a step generates two new kink sites but can be the rate limiting process and therefore may establish the activation enthalpy. When an atom is removed from a kink site, the kink simply moves along the step and the energy of the step does not change. However, the adatom again presents a surface 3/4 the size of a bulk vacancy for which one may estimate an energy of 2.1 eV. This leads to Hm(kink 111) = 3.2 eV. This is slightly larger than the value 3.09 eV obtained [7] from observation of grain boundary grooving behaviour on polycrystalline samples (where there should have been enough steps and kinks that the steps would not be exhausted in the experiment). It is somewhat less than the value 3.6 eV from LEED observation of reconstruction on a non-irradiated surface [5] where there were few steps and the initial kinks may have been exhausted. Using again the surface tension times increase in surface area approximation, one estimates the increase in energy at the step when an atom is removed to be about 1/3 that to form a vacancy or 0.9 eV. However, because two kink sites are created in this process, the effect on Hm should be just 1/2 of that or 0.5 eV. This then leads to the simple estimate Hm(step 111) = 3.7 eV, which is slightly larger than the empirical value 3.6 eV for the experiment expected to be influenced by kink site nucleation [6].

Turning to the bulk self-diffusion, we may first note that the simple estimates of Hf and Hm for single bulk vacancies imply Hd(Bulk V) = 2.8 + 1.2 = 4.0 eV, which value is typical of those obtained below 1000 deg C. The lowest of these, 3.6 eV, was obtained at 850 deg C from analysis of strain hardening in terms of diffusion controlled motion of jogged screw dislocations [1]. A similar analysis of the annealing of dislocation loops in intrinsic Si between 970 and 1070 deg C [22] led to an activation enthalpy of 4.1 eV with a pre-exponential factor, Do, of 5.8 cm*2/s. Probably the most definitive low temperature experiment on intrinsic Si was done using ion implantation of the stable isotope Si-30 followed by profiling using a nuclear transmutation technique [3]. This allows for full annealing of ion implantation damage which may have clouded earlier reports. For 800 < T < 1100 deg C, they find 3.9 < Hm < 4.4 eV with 0.3 < Do < 30 cm*2/s. Ref [22] obtained Do = 5.8 cm*2/s, while similar treatment of other data in the same range of T and also using Si-30 [23] yield 3.95 < Hm < 4.45 eV with 0.6 < Do < 16 cm*2/s. It should be noted that if the 1150-1200 deg C data points from either [3] or [22] are combined with the lower T values and fitted with a single straight line, then the least squares value of Hm rises to about 4.4 eV for [3] or 4.65 eV for [22] with a Do of 20 or 155 cm*2/s and a substantially poorer fit. The latter values are sometimes attributed to the low T regime but in the author's view this is an error. Values of Hm = 4.1 eV and Do = 8 cm*2/s for 900 < T < 1100 deg C were also obtained by Hirvonen and Anttila [24] by entirely different techniques.

As are the values of Hm for bulk diffusion in intrinsic Si at low T, the above quoted values of Do are of the order of magnitude expected for neutral single vacancy diffusion [13].

For intrinsic Si with 1100 < T < 1412 deg C, i.e., the melting point, both Hm and Do are found to be much larger than the above. As noted initially the reasons for this are controversial but the values are relatively fixed. The classic paper of Fairfield and Masters [2], which used Si-31 radio isotopes, found Hm = 5.13 eV with Do = 9000 cm*2/s. More recent work with similar methods [25] finds Hm = 5.02 eV and Do = 1460 cm*2/s. These values were first reported by Peart [14], who attributed them to divacancy mediated diffusion. (See [15] for a recent defence of that theory.)

The reader is reminded that the above discussion applies only to intrinsic Si. In both the high T and the low T regimes, the rate of self-diffusion is increased by doping and in both regimes n-type doping has more effect than does p-type doping. Counter doping experiments find that the effect follows the carrier concentration, and thus the Fermi level, rather than the net impurity concentration so the effect is not due either to any metallurgical weakening of the lattice with impurities or to the formation of stable complexes of charged point defects with dopants that might migrate as a complex. Indeed, [22] found Do varied in proportion to [n], the free electron concentration, in n-type Si for 970 < T < 1070 deg C. At high T, the effect is less than proportional; [2] shows that, e.g., a 39% difference in [n] due to As doping at 1088 deg C causes a 10% difference in D, and presumably in Do. [2] also finds that a 2.75 times difference in hole concentration, [p], due to heavy doping with B produces a 46% difference in D (and thus Do). For 8 X 10*19 doping, [2] reports 55% greater D (and Do) for n-type doping (P) than for p-type doping (B) near 1090 deg C.

This increase in D and Do with doping is indicative of diffusion mediated by ambient point defects that are amphoteric, i.e., that have both donor and acceptor ionisation levels. The greater response to n-type doping than to p-type implies the acceptor levels are deeper within the fundamental gap than are the donor levels. While the equilibrium concentration of neutral point defects is independent of Fermi level, Ef, the concentrations of all ionised defects vary in response to Ef because part of the enthalpy of formation of the neutral defect is regained when a carrier is allowed to fall from Ef to an acceptor level below or to a donor level above Ef [18,21]. One presumes that the ionised point defects mediate self-diffusion as effectively as the neutral defects so Do is proportional to the sum over all ionisation states of the defects that mediate the diffusion. The linear response at low T and less than linear response at high T implies that the low T acceptors are much shallower than the low T donors while the high T donors are at depth comparable to the acceptors.

Under the tentative assumption that only two modes contribute to Si self-diffusion and labelling these 1 and 2 for the low and the high T regimes, we are led to

$$D(Ef, T) = Do1 \ ([1(Ef)]/[1(o)]) \ exp(-Hm1/kT)$$
$$+ Do2 \ ([2(Ef)]/[2(o)]) \ exp(-Hm2/kT)$$

where Do1 and Do2 are the intrinsic values of Do for the two modes, approximately 10 and 2000 cm*2/s as discussed above, with Hm1 = 4.1 eV at T with Ef, [1(o)] is the concentration of 1 for T and Ef at the intrinsic level, and the corresponding for defect 2. Of course, a few more terms could be added, and would be justified by the various theories proposing different mechanisms. If the parameters thus introduced are regarded as free, one could certainly improve the fit to any set of data one wished to accept. However, the existent data are not yet adequate to uniquely fix several parameters. The calculation of the ratios [j(Ef)]/[j(o)] is straightforward once the ionisation levels of the assumed point defect j are established (which is a controversial issue) for the T of the experiment. However, it would take too much space to describe here. It is illustrated in [15] for the assumption that 1 is a single vacancy and 2 is a divacancy. It has been established by exhaustive cryogenic experiments that single Si vacancies have two acceptor levels and two donor levels, with the donors much deeper than the acceptors, while divacancies have two acceptor levels and one donor level that is comparably deep. Thus, with plausible extrapolation of these levels to the T's in question one can fit the observed data well. Most calculations find simple Si interstitials (which have never been observed directly) to have several donor but no acceptor levels [9], which argues against them in either regime. No estimates regarding the ionisation levels of extended interstitials, extended vacancies, or any effect of Ef upon ring mechanisms are available.

REFERENCES
----

[1]   H.Siethoff, W.Schroeter [Philos. Mag. A (GB) vol.37 (1978)
       p.711]
[2]   J.M.Fairfield, B.J.Masters [J. Appl. Phys. (USA) vol.38 (1967)
       p.3148]
[3]   F.J.Demond, S.Kalbitzer, H.Mannsperger, H.Damjantschitsch [Phys.
       Lett. A (Netherlands) vol.93 (1983) p.503]
[4]   S.M.Bedair [Surf. Sci. (Netherlands) vol.42 (1974) p.595]
[5]   B.Z.Olshanetski, S.M.Repinski, A.A.Shklyaev [JETP Lett. (USA)
       vol.27 (1978) p.378]
[6]   Y.L.Gavrilyuk, Y.S.Kaganovskii, V.G.Lifshifts [Kristallografiya
       (USSR) vol.26 (1981) p.561]
[7]   W.M.Robertson [J. Am. Ceram. Soc. (USA) vol.64 (1981) p.9]
[8]   T.H.Han, U.Goesele [Appl. Phys. Lett. (USA) vol.40 (1982) p.616]
[9]   R.Car, P.J.Kelly, A.Oshiyama, S.T.Pantelides [Phys. Rev. Lett.
       (USA) vol.54 no.4 (28 Jan 1985) p.360-3]
[10]   J.A.Van Vechten [Phys. Rev. B (USA) vol.10 (1974) p.1482]
[11]   M.Lannoo, G.Allen [Phys. Rev. B (USA) vol.25 (1982) p.4089]
[12]   A.Seeger, K.P.Chik [Phys. Status. Solidi b (Germany) vol.29
       (1968) p.455]
[13]   J.A.Van Vechten [J. Electron. Mater. (USA) vol.14a (1985)
       p.293]
[14]   R.F.Peart [Phys. Status Solidi b (Germany) vol.15 (1966) p.K119]
[15]   J.A.Van Vechten [Phys. Rev. B (USA) vol.33 no.4 (15 Feb 1986)
       p.2674-89]
[16]   K.C.Pandey [Phys. Rev. Lett. (USA) vol.57 no.18 (3 Nov 1986)
       p.2287-90]
[17]   S.Dannefaer, P.Mascher, D.Kerr [Phys. Rev. Lett. (USA) vol.56
       no.20 (19 May 1986) p.2195-8]
[18]   J.A.Van Vechten [Phys. Rev. B (USA) vol.12 (1975) p.1247]
[19]   J.A.Van Vechten, J.F.Wager [Phys. Rev. B (USA) vol.32 no.8
       (15 Oct 1985) p.5259-64]
[20]   J.A. Van Vechten [J. Electrochem. Soc. (USA) vol.122 (1975)
       p.405]
[21]   F.A.Kroger [Annu. Rev. Mater. Sci. (USA) vol.7 (1977) p.449]
[22]   I.R.Sanders, P.S.Dobson [J. Mater. Sci. (GB) vol.9 (1974)
       p.1987]
[23]   L.Kalinowski, R.Seguin [Appl. Phys. Lett. (USA) vol.36 (1980)
       p.171]
[24]   J.Hirvonen, A.Anttila [Appl. Phys. Lett. (USA) vol.35 (1979)
       p.703]
[25]   G.Hettich, H.Mehrer, K.Maier [Inst. Phys. Conf. Ser. (GB) no.46
       (1979) p.500]

1.18 MELTING POINT OF Si

by J.C.Brice

June 1987
EMIS Datareview RN=15743

A. INTRODUCTION

The accurate measurement of temperatures at about 1400 deg C is far
from easy (see refs [1-3]). Temperature transducers require careful
calibration and have poor ageing characteristics. In this range the
transducers are usually thermocouples (comprising platinum group
metals). The majority show negative ageing: the output falls by of
the order of the equivalent of 3 degrees after about 1000 hours at
temperature and temperature cycling which is an essential part of the
usual method (differential thermal analysis) of measuring melting
points, increases the ageing rate. Thus thermocouples need to be
calibrated regularly. Early workers in the field (prior to 1948) had
the added problem that there was no universally accepted temperature
scale in this range. This may account for some of the scatter of
results from early work which gave a scatter of recommended melting
point values over the range 1400 to 1420 deg C. Individual results
spread over a bigger range (1396 to 1444 deg C). See the early
results reported in refs [4-6]. These early results will be ignored
here. The later data given in [4-6] and in [7] and [8] show a much
smaller scatter. See TABLE 1 and the other data given below.

Measuring the melting point of silicon presents two particular problems:

(a) Silicon, particularly when molten, is very reactive and most, if
     not all, impurities depress the melting point. The usual choice of
     material to contain the melt and protect the thermocouple is
     vitreous silica but this dissolves slowly in silicon leading to
     contamination by oxygen which lowers the melting point by about 1
     degree for each atom per cent in the melt. (See the phase diagram
     in [9]).

(b) Spontaneously nucleated silicon is bounded by (111) faces and to
     achieve a reasonable growth rate a measurable supercooling is
     needed (a growth rate of 2 mm/min requires a supercooling of
     about 2 degrees - Takao Abe, private communication). Thus thermal
     arrests on cooling curves will occur at temperatures below the
     melting point and the average of arrests on heating and cooling
     will be depressed.

In 1948 the International Practical Temperature Scale (IPTS-48) was
published. This scale was revised in 1968 leading to the current scale
(IPTS-68). At 1400 deg C, these scales differ by 2.0 degrees. See ref
[10 p.4-6]. Adding 2.0 degrees to an IPTS-48 value yields an IPTS-68
value. In what follows, this has been done when necessary so that from
here on all temperatures are on the 1968 scale.

. T 0,3
B. PREVIOUS RECOMMENDATIONS

In 1950 [11] a value of 1412 deg C was recommended in a US Atomic
Energy Commission report. In 1952 NBS published its Circular 500 [12]
which also gave 1412 deg C. In 1968 Hultgren et al [14] gave 1414 deg C
and this value was given again in 1973 [15].

C. SOME DATA AND DISCUSSION

Post 1948 data from [4-8] is summarised in TABLE 1 which gives the
number of times a value has been reported.

TABLE 1: melting temperatures

Temperature (deg C)	1408	1410	1412	1414	1416
Number of Reports	1	1	7	8	3

The mean of these data is 1413.1 deg C with a standard error of 0.31
degrees. However, as discussed in the Introduction, all the likely
errors are negative so that the mean is likely to be below the true
value. Thus a better choice would be the median (i.e. 1414 deg C).
This has a standard error of about 0.4 degrees. Note that 1414 deg C
is also the value most frequently observed (the mode).

D. RECOMMENDATION

A safe choice (half way between the previous recommendations) would
be 1413 +- 2 deg C. However, on the basis of the evidence, a reasonable
recommendation is 1414.0 deg C with an uncertainty corresponding to a
90% confidence level of 0.9 degrees.

REFERENCES
----

[1]   ASTM [Manual on the Use of Thermocouples in Temperature
       Measurement (ASTM, Philadelphia, 1981)]
[2]   J.V.Nicholas, D.R.White [Traceable Temperatures (DSIR,
       Wellington, New Zealand, 1982)]
[3]   T.J.Quinn [Temperature (Academic Press, London, 1983)]
[4]   M.Hansen [Constitution of Binary Alloys (McGraw-Hill, New York,
       1958)]
[5]   R.P.Elliott [Constitution of Binary Alloys: First Supplement
       (McGraw-Hill, New York, 1959)]
[6]   F.A.Shunk [Constitution of Binary Alloys: Second Supplement
       (McGraw-Hill, New York, 1969)]
[7]   E.A.Brandes [Smithells Metals Reference Book (Butterworth,
       London, 1983)]
[8]   W.G.Moffatt [The Handbook of Binary Phase Diagrams (Genium
       Publishing Corp., Schenectady, continuing)]

[9]   J.C.Brice [Curr. Top. Mater. Sci. (Netherlands) vol.4 Ed.
      E.Kaldis (North Holland, Amsterdam, 1980) ch.5]
[10] D.W.Gray [American Institute of Physics Handbook (McGraw-Hill,
      New York, 1972)]
[11] US Atomic Energy Commission [Report ANL-5750], which is cited
      by Hodgman [Handbook of Chemistry and Physics (Chemical Rubber
      Publishing, Cleveland, 1960) p.1870-1]
[12] F.D.Rossini, D.D.Wagman, W.H.Evans, S.Levine, I.Jaffe [Selected
      Values of Chemical Thermodynamics Properties (NBS Circular 500,
      Washington, 1952)]
[13] H.L.Anderson [Physics Vade Mecum (AIP, New York, 1981)]
[14] R.R.Hultgren, R.L.Orr, P.D.Anderson, K.K.Kelley [Selected Values
      of Thermodynamic Properties of Metals and Alloys (Wiley, New York,
      1968)]
[15] R.R.Hultgren, P.O.Desai, D.T.Hawkins, K.K.Kelley, D.Wagman
      [Selected Values of Thermodynamic Properties of Elements (Wiley,
      New York, 1973)]

1.19 MELTING POINT OF Si, PRESSURE DEPENDENCE

by J.C.Brice

June 1987
EMIS Datareview RN=15744

Under a pressure of about one atmosphere, silicon melts at 1414.0 deg C (1687.1 deg K) [1]. The effect of pressure on the melting point is easily calculated using the Clausius-Clapeyron equation which shows that

$$dTm/dP = Tm[V(1) - V(s)]/H \qquad (1)$$

where V(s) and V(1) are given by M/p(s) and M/p(1) and P is the pressure. The other symbols are explained in the TABLE 1 which gives the values found in [2].

TABLE 1

Symbol	Parameter	Value	Units
H	Latent heat of fusion	50660	J/mole
Tm	Melting point	1687.0	deg K
M	Atomic weight	28.086	g/mole
p(s)	Density of solid at Tm*	2.30	g/cm*3
p(1)	Density of liquid at Tm	2.51	g/cm*3
V(s)	Molar volume of solid	0.0000122	m*3/mole
V(1)	Molar volume of liquid	0.0000112	m*3/mole

* Calculated from the room temperature density (2.34 g/cm*3) and the thermal expansion coefficient which was assumed to be 8 X 10*-6/deg K on average.

From equation (1), the rate of fall of Tm is

-dTm/dP = 3.3 X 10*-8 degrees/Pascal

or since 1 atmosphere = 101325 Pascals,

dTm/dP = -0.0034 degrees/atmosphere.

Note that the accuracy of this result depends critically on the accuracy of the density values. With the values used here, the result is very uncertain: errors of 10 to 50% are probable. It must be emphasised that this estimate applies at low pressures. At high pressures, it is necessary to take account of the compressibilities of the solid and the liquid when estimating V(s) and V(1).

According to Landolt-Bornstein [3], Tm = 1190 deg C at P = 40 kB. This data gives an average dTm/dP of -0.0057 deg K/atmosphere in the range 1 to 40 kB and suggests that dTm/dP is a decreasing function of pressure.

REFERENCES
----

[1]   J.C.Brice [EMIS Datareview RN=15743 (June 1987) 'Melting point of Si']
[2]   E.A.Brandes [Smithells Metals Reference Book (Butterworth, London, 1983) p.3-1, 8-3, 14-2 and 14-8]
[3]   [Landolt-Bornstein 'Numerical Data and Functional Relationships in Science and Technology (New Series)' vol.III/17a (Springer-Verlag, Berlin, 1982) section B1]

1.20 HEAT OF FUSION OF Si

by J.C.Brice

June 1987
EMIS Datareview RN=15745

Little work on the heat of fusion of silicon has been published.
Reports in 1950 [1] and 1952 [2] review early work in the field and
suggest a 'doubtful' value of 11.1 kcal/mole (46.5 kJ/mole). Hultgren
in 1968 [3] and again in 1973 [4] gives a recommended value of
12.1 kcal/mole (50.6 kJ/mole). This value should be reliable to of the
order of 3%.

REFERENCES
----

[1]   US Atomic Energy Commission [Report ANL-5750] which is cited
       by Hodgman [Handbook of Chemistry and Physics (Chemical Rubber
       Publishing, Cleveland, 1960) p.1870-1]
[2]   F.D.Rossini et al [Selected Values of Chemical Thermodynamic
       Properties (NBS Circular 500, Washington, 1952)]
[3]   R.R.Hultgren et al [Selected Values of Thermodynamic Properties
       of Metals and Alloys (Wiley, New York, 1968)]
[4]   R.R.Hultgren et al [Selected Values of the Chemical Thermodynamic
       Properties of Elements (Wiley, New York, 1973)]

1.21 SURFACE TENSION OF LIQUID Si

by S.C. Hardy

September 1986
EMIS Datareview RN=15701

The surface tension of liquid silicon at the melting point (1683 deg K)
is 885 mJ/m*2 and decreases linearly with slope -0.28 mJ/m*2 deg K
as the temperature increases [1]. These values for the surface tension
and its temperature coefficient were determined using the sessile drop
technique in purified flowing argon. The silicon samples were supported
in shallow boron nitride cups. Auger analysis of the solidified silicon
drops found no evidence of contamination by the cup material. The
oxygen partial pressure in these measurements is thought to be very low
because the surface tension values obtained are in good agreement with
an earlier measurement at 1723 deg K in pure hydrogen [2]. Surface
tension values found in the only other systematic measurement were
significantly lower than the above result, probably due to the presence
of surface active impurities [3]. Other measurements of the surface
tension of liquid silicon are described in refs [4-8].

References
----

[1]   S.C.Hardy [J. Cryst. Growth (Netherlands) vol.69 nos.2/3 (1934)
      p.456-60]
[2]   W.D.Kingery, M.Humenik [J. Phys. Chem. (USA) vol.57 (1953)
      p.359]
[3]   S.V.Lukin, V.I.Zhuchkov, N.A.Vatolin [J. Less-Common Met.
      (Switzerland) vol.67 (1979) p.399]
[4]   P.H.Keck, W.Van Horn [Phys. Rev. (USA) vol.91 (1953) p.512]
[5]   P.V.Geld, M.S.Petrushevski [Izv. Akad. Nauk SSSR Otd. Tekhn.
      Nauk (USSR) vol.3 (1961) p.160]
[6]   Yu.M.Shashkov, T.P.Kolesnikova [Russ. J. Phys. Chem. (USSR)
      vol.37 (1963) p.747]
[7]   N.K.Dshemilev, S.I.Popel, B.V.Zarevski [Fiz. Metal. Metalloved.
      (USSR) vol.18 (1964) p.83]
[8]   V.P.Eljutin, V.I.Kostikov, V.Ya.Levin [Izv. Vysshikh Uchebn.
      Zaved. Tsvetn. Met. (USSR) vol.2 (1970) p.131]

CHAPTER 2

OPTICAL FUNCTIONS (Complex Refractive Index and
Absorption Coefficient)

2.1 OPTICAL FUNCTIONS OF INTRINSIC Si: GENERAL REMARKS

by D.E.Aspnes

November 1987
EMIS Datareview RN=17801

The optical functions of Si have been discussed in three recent reviews
[1-3]. Li [1] provides refractive index values from 1.2 to 14 microns
(0.088 to 1.033 eV) with comprehensive descriptions of samples and
measurement conditions for all 55 different works reported from 1949 to
1978. The review of Edwards [2] includes 33 references and provides
refractive index and extinction coefficient values from 0.62 to
333 microns (0.00372 to 2000 eV). Aspnes and Studna [3] assess 38
optical investigations in the visible/deep UV from 206.6 to
826.4 nm (1.500 to 6.000 eV). The reader interested in more detail is
encouraged to consult these works.

The standard optical functions are the complex dielectric function e =
e1 + ie2 (where e denotes permittivity), the complex refractive index
n' = n + ik = e*0.5 where n and k are the ordinary refractive index
and the extinction coefficient, respectively, and the absorption
coefficient a = 4 pi k / Y, where Y is the wavelength of light. The
extinction coefficient is occasionally defined as n' = n (1 + ik). The
sign of the imaginary part of the above expressions follows the physics
convention; the optics/engineering literature follows the opposite sign
convention, with n, k, and e2 remaining positive definite in both
representations.

The common terminology 'optical constants' is misleading because e1,
e2, n, k, and a are all strong functions of Y. The use of 'constant'
should be restricted to the infrared and static dielectric constants
e(inf) and e(o), which are low-frequency limits of e1 above and below,
respectively, the lattice absorption bands. As the Si lattice has
inversion symmetry, it is infrared-inactive and e(inf) and e(o) are
identical.

In principle only one set of optical functions is needed because they
are interrelated. However, the different representations are useful
for different purposes. Wave propagation is described by the
propagation vector k = 2 pi n' / Y, which is proportional to n'. The
fundamental response of a solid to an electric field is described by
e according to D = eE = E + 4 pi P, where D and E are the macroscopic
(observable) displacement and electric fields, respectively, and P
is the dipole moment per unit volume. As macroscopic dipole moments are
approximately additive, the e representation provides a convenient means
of dealing with extrinsic modifications to the optical response caused
by the presence of, e.g., point defects and free carriers. Also, simple
analytic representations for e, such as the Drude expression for free
carriers or the Sellmeier expression for dispersion in regions of
transparency, permit efficient calculation of optical functions to
supplement tabulated values.

Energy (E) to wavelength (Y) conversion is done according to E(Y) =
hc/n(air) = 1.2395086 eV micron, where h = 6.626176(36) X 10*-34 J s, c =
2.99792458(1.2) X 10*10 cm/s, 1 eV = 1.6021892(46) X 10*-19 J, and n(air)
(ordinary refractive index of dry air) = 1.000277 at 15 deg C, 760 mm
Hg, Y = 550 nm, where all data are from the CRC Handbook [4].

REFERENCES
----

[1]   H.H.Li [J. Phys. & Chem. Ref. Data (USA) vol.9 (1980) p.561-658]
[2]   D.F.Edwards [in 'Handbook of Optical Constants of Solids' Ed.
      E.Palik (Academic, New York, 1985) p.547-69]
[3]   D.E.Aspnes, A.A.Studna [Phys. Rev. B (USA) vol.27 no.2 (15 Jan
      1983) p.985-1009]
[4]   [CRC Handbook of Chemistry and Physics, 64th Edn., Ed. R.C.Weast
      (CRC Press, Boca Raton, 1983). The fundamental constants are on
      p.F198; the datum for n(air) is on p.E364]

2.2 OPTICAL FUNCTIONS OF INTRINSIC Si: DIELECTRIC CONSTANT

by D.E.Aspnes

November 1987
EMIS Datareview RN=17802

Silicon has inversion symmetry, so the lattice is infrared-inactive and the static and infrared dielectric constants e(o) and e(inf) are identical. Values of e(o) have been surveyed by Young and Fredrikse [1]. Reported values range from 11.7 to 12.1.

Using the equality between e(o) and e(inf), an accurate value of e(o) can be obtained from the zero-frequency limit of n. The data of Loewenstein et al [2] summarized in our listing [3] allow such an extrapolation to be made, provided that the apparent Drude free carrier contribution for energies less than about 0.0250 eV (wavelength greater than about 50 microns) is neglected (see [4]). We calculate that e(o) = 11.695 +- 0.03, where we have assigned as the uncertainty the 0.005 discrepancy in n between the data sets of Edwards and Ochoa [5] and Icenogle et al [6]. The value of n(o) follows as 3.420 +- 0.005.

The temperature dependence of e(o) is nonlinear due to the temperature nonlinearities of phonon densities, absorption thresholds, and other contributing physical processes. The absence of significant oscillator strength for energies less than 1.120 eV (wavelengths greater than 1.107 microns) suggests that de(o)/dT can also be estimated from the long-wavelength limit of n. The data of Icenogle et al [6] at both 5.19 and 10.27 microns (0.239 and 0.121 eV) yield de/dT values of 1.011 X 10*-3/deg K and 1.052 X 10*-3/deg K, respectively, supporting common values of de(o)/dT = approx. 1.0 X 10*-3/deg K and dn(o)/dT = approx. 1.5 X 10*-4/deg K at room temperature.

REFERENCES
----

[1] K.F.Young, H.P.R.Fredrikse [J. Phys. & Chem. Ref. Data (USA) vol.2 (1973) p.313-409]

[2] E.V.Loewenstein, D.R.Smith, R.L.Morgan [Appl. Opt. (USA) vol.12 (1973) p.398-406]

[3] D.E.Aspnes [EMIS Datareview RN=17807 (Nov 1987) 'Optical functions of intrinsic Si: table of refractive index, extinction coefficient and absorption coefficient vs energy (0 to 400 eV)']

[4] D.E.Aspnes [EMIS Datareview RN=17803 (Nov 1987) 'Optical functions of intrinsic Si: region of near transparency']

[5] D.F.Edwards, E.Ochoa [Appl. Opt. (USA) vol.19 (1980) p.4130-1]

[6] H.W.Icenogle, B.C.Platt, W.L.Wolfe [Appl. Opt. (USA) vol.15 (1976) p.2348-51]

2.3 OPTICAL FUNCTIONS OF INTRINSIC Si: REGION OF NEAR TRANSPARENCY

by D.E.Aspnes

November 1987
EMIS Datareview RN=17803

Si is transparent up to the indirect absorption edge [1.107 microns
(1.120 eV) at room temperatures] except for weak two-phonon absorption
processes in the 0.050-0.250 eV (5-25 microns) spectral range. These
two-phonon processes are too weak to influence the refractive index
noticeably, although they are important in transmission studies of
impurity absorption [1].

Many groups have reported refractive index data in this spectral range,
as summarized by Li [2]. In his review, Edwards [3] selected only those
data from groups who obtained results on samples that were
single-crystal of good or optical quality, who reported their data in
tabular form or in a figure that could be read to a few places in the
third decimal place, and whose data agreed to within 5 in the third
decimal place in regions of overlap. The six sets of results that meet
these criteria are: Primak [4], 1.12 to 2.16 microns (0.574 to 1.107
eV); Salzberg and Villa [5], 1.357 to 11.04 microns (0.112 to 0.913 eV);
Edwards and Ochoa [6], 2.437 to 25 microns (0.0496 to 0.509 eV); Icenogle
et al [7], 2.554 to 10.27 microns (0.121 to 0.486 eV); Loewenstein et al
[8], 28.57 to 333 microns (0.0037 to 0.0434 eV); and Randall and Rawcliffe
[9], 67.8 to 588 microns (0.0021 to 0.0183 eV).

The data that Edwards [3] actually listed were those of Primak [4],
Edwards and Ochoa [6], and Loewenstein et al [8], and we follow the
same selection here [10]. The original Primak entry of 3.501 for 0.903
eV does not fit the general trend and is apparently in error. The
Loewenstein et al data [8] deviate from a smooth extrapolation to zero
energy for energies less than about 0.0250 eV (wavelengths greater than
50 microns) by what appears to be a free-carrier contribution, although
Loewenstein et al [8] point out that the deviation is about 4 times
larger than that expected from the nominal resistivity. Whatever the
explanation, it is clearly not an intrinsic property of the material.

Edwards [3] determined an approximate Sellmeier dispersion relation to
represent these data analytically:

$$n*2 = 11.6858 + 0.939816/(Y/micron)*2 + \frac{0.009934}{(Y/micron)*2 - 1.22567}$$

where Y denotes the wavelength.

The maximum deviations between values calculated from this expression
and our tabulated refractive index data [10] for all wavelengths longer
than 1.2 microns (energies less than 1.03 eV) are +0.004 and -0.002
at 2 and 5 microns (0.620 and 0.248 eV), respectively. Edwards's
expression [3] better represents the physics than the simpler expression
of Li [2], which does not include a separate term for the contribution of
the indirect absorption threshold.

Edwards [3] also determined polynomial representations of the temperature derivatives dn/dT at selected wavelengths from the data of Icenogle et al [7]. Removing unnecessary significant figures from Edwards's expressions yields:

TABLE 1

wavelength (microns)	dn/dT (10*-4 per deg K)
2.55	-1.11 + 6.22t - 3.60t*2
2.75	-0.10 + 2.90t - 1.09t*2
5.19	-0.41 + 4.01t - 2.11t*2
10.27	-0.65 + 4.91t - 2.92t*2

where t = T/300 deg K

Intrinsic optical absorption below 1.107 eV (1.12 microns) is difficult to measure accurately due to extrinsic contributions from surface damage, free carriers, shallow donors and acceptors, and lattice defects. Theoretically, the only intrinsic absorption mechanism should be the two-phonon process that extends from 5 to 25 microns (0.050 to 0.250 eV). Absorption coefficients of the order of 1 /cm in the far IR, as reported by Loewenstein et al [8] and Randall and Rawcliffe [9] are probably extrinsic in origin. Photoacoustic measurements by Hordvik and Skolnik [11] at the 3.8 and 2.7 micron (0.33 and 0.46 eV) DF and HF laser lines indeed show very low absorption coefficients of 4.2 and 1.2 X 10*-4/cm, respectively, and these authors suggest that the absorption coefficient at 0.46 eV may already be determined by impurity effects. Values remain virtually negligible to within 0.05 eV of the indirect threshold, as shown by Cody and Brooks [12].

Accordingly, we list [10] the absorption coefficient values of Johnson [13] and Bendow et al [14] in the two-phonon band, the HF/DF data of Hordvik and Skolnik [11], and the near-absorption-edge data of Cody and Brooks [12]. The absorption coefficient values of Brooks are about 20-25% larger at any given energy than the earlier values of Macfarlane et al [15] in the common energy range of 1.0 to 1.2 eV (1.033 to 1.240 microns), but improvements in techniques and materials suggest the later data are more reliable.

REFERENCES
----

[1]   L.Canham, P.J.Ashby [EMIS Datareviews entered in 1987:
       RN=16162 'IR absorption due to O in Si'
       RN=16163 'IR absorption due to C in Si'
       RN=15789 'IR absorption due to N in Si']
[2]   H.H.Li [J. Phys. & Chem. Ref. Data (USA) vol.9 (1980)
       p.561-658]
[3]   D.F.Edwards [in 'Handbook of Optical Constants of Solids' Ed.
       E.Palik (Academic, New York, 1985) p.547-69]
[4]   W.Primak [Appl. Opt. (USA) vol.10 (1971) p.759-63]
[5]   C.D.Salzberg, J.J.Villa [J. Opt. Soc. Am. (USA) vol.47 (1957)
       p.244-6]

[6]   D.F.Edwards, E.Ochoa [Appl. Opt. (USA) vol.19 (1980) p.4130-1]

[7]   H.W.Icenogle, B.C.Platt, W.L.Wolfe [Appl. Opt. (USA) vol.15
       (1976) p.2348-51]

[8]   E.V.Loewenstein, D.R.Smith, R.L.Morgan [Appl. Opt. (USA) vol.12
       (1973) p.398-406]

[9]   C.M.Randall, R.D.Rawcliffe [Appl. Opt. (USA) vol.6 (1967)
       p.1889-95]

[10]   D.E.Aspnes [EMIS Datareview RN=17807 (Nov 1987) 'Optical
       functions of intrinsic Si: table of refractive index, extinction
       coefficient and absorption coefficient vs energy (0 to 400 eV)']

[11]   A.Hordvik, L.Skolnik [Appl. Opt. (USA) vol.16 (1977) p.2919-24]

[12]   G.D.Cody, B.G.Brooks [private communication (1986)]

[13]   F.A.Johnson [Proc. Phys. Soc. London, (GB) vol.73 (1959) p.265-72]

[14]   B.Bendow, H.G.Lipson, S.P.Yukon [Appl. Opt. (USA) vol.16 (1977)
       p.2909-13]

[15]   G.G.Macfarlane, T.P.McLean, J.E.Quarrington, V.Roberts [Phys.
       Rev. (USA) vol.111 (1958) p.1245-54]

2.4 OPTICAL FUNCTIONS OF INTRINSIC Si: INDIRECT ABSORPTION EDGE TO THE VACUUM UV

by D.E.Aspnes

November 1987
EMIS Datareview RN=17804

As indirect absorption is relatively weak, the absorption coefficient, a, and the extinction coefficient k=aY/4pi (Y is wavelength) can be determined by transmission nearly up to the first direct optical transition at 3.35 eV (370 nm). However, the transparency is not sufficient for minimum-beam-deviation or channel-spectrum measurements, so n must be determined by less accurate reflectance or ellipsometric methods. Above 3.4 eV (356 nm) all optical data must be obtained in reflection geometries.

From 1 to 3 eV (1.24 microns to 413 nm), the absorption coefficient increases by 7 orders of magnitude. Most transmission data agree to within a factor of 2 in this spectral range. The 1.0-1.2 eV (1.032 to 1.240 micron) data of Cody and Brooks [1] and the 1.1-3.1 eV (1.127 micron to 400 nm) data of Dash and Newman [2] are probably the most accurate and are listed in the table [3].

The value of a measured by Dash and Newman [2] is 60% larger than that determined by Cody and Brooks [1] at 1.2 eV, even though the values near 1.15 eV (1.078 microns) are in agreement. The Dash-Newman absorption coefficient values at energies above 1.2 eV (wavelengths less than 1.032 microns) have been substantiated by a number of independent measurements over the years, specifically by Runyon [4], Kazmerski [5], and Taft [6]. Selected data from the last two sets are also listed in the table [3] for comparison.

Aspnes and Studna [7] critically reviewed reflectance-type measurements from 1.5 to 6.0 eV (wavelengths from 206.6 to 826.3 nm). There is no shortage of reported data because Si is technologically important and because the ready availability of high-quality material made it the natural substance on which to try different experimental methods. Results varied far more than intrinsic instrumental inaccuracies because surface preparation, already important in the infrared, becomes crucial as light penetration depths are as low as 5 nm and a 1 nm residual oxide or contamination layer can have a 20% effect. From this perspective the more recent ellipsometric measurements have an advantage, because an ellipsometer can be used to assess surface quality while the surface is being prepared, and e1 and e2 can be determined directly without requiring a Kramers-Kronig transformation involving extrapolations of reflectance data into experimentally inaccessible wavelength regions.

From 1.5 to 6.0 eV (206.6 to 826.3 nm), e is determined most accurately from the pseudodielectric function data of Aspnes and Studna [7], and with the exception of the k values for energies less than 2.8 eV (see end of paragraph), these are the data given in the table [3]. Rather than attempt to correct their data for the presence of undetermined overlayers, these workers eliminated surface overlayers as far as possible using the ellipsometer to assess surface quality during sample preparation. Estimated accuracy was about 2% in mod e and 1% in mod n. However, values of e2 less than 1 (energies less than about 2.7 eV) could not be determined to this accuracy due to instrumental limitations, so they used the absorption data of Dash and Newman [2] to improve accuracy in this spectral range. Rather than use a weighted average as in [7], we use the Dash-Newman a values directly to calculate k below 2.8 eV and show both sets of results from 2.8 to 3.0 eV. Refractive index and extinction coefficient data determined by Philipp [8] from a Kramers-Kronig transformation of reflectance data, discussed more fully in the next section, are shown at every 0.5 eV for comparison.

To our knowledge, no refractive index data are available between the 1.107 eV (1.12 micron) Primak [9] datum and the 1.500 eV (826.3 nm) lower-energy limit of ellipsometric results [7]. We fill this gap by extrapolating the ellipsometric data and list the values so obtained with an asterisk to indicate that these values are estimates. The extrapolation joins smoothly to the transmission results at 1.1 eV, but a small change in slope indicates that the ellipsometric values may be about 1% too large at 1.5 eV, which is about their estimated accuracy.

Temperature effects in this spectral range originate from the shift and broadening of the two major critical point structures, E1 at 3.4 eV (365 nm) and E2 at 4.2 eV (295 nm). Systematic investigations from 1.64 to 4.73 eV (262 to 756 nm) have been reported by Jellison and Modine [10] for temperatures of 10, 297, 465, 676, 874 and 972 deg K from 1.7 to 5.7 eV by Lautenschlager et al [11] for temperatures from 30 to 820 deg K. The temperature dependences of the E1 and E2 transitions were fit in the latter work to an equation of Varshni form:

$$E(T) = E(0) - pT*2/(q + T)$$

where p and q are 4.7 X 10*-4 eV/deg K and 350 deg K for E1, and 2.9 X 10*-4 eV/deg K and 124 deg K for E2. The normally sharp features in the optical spectrum are severely degraded by 800 deg K.

REFERENCES
----

[1]   G.D.Cody, B.G.Brooks [private communication(1986)]
[2]   W.C.Dash, R.Newman [Phys. Rev. (USA) vol.99 (1955) p.1151-5]
[3]   D.E.Aspnes [EMIS Datareview RN=17807 (Nov 1987) 'Optical
       functons of intrinsic Si: table of refractive index, extinction
       coefficient and absorption coefficient vs energy (0-400 eV)']
[4]   W.R.Runyon [Final Report, NASA Grant NGR 44-007-016 (1963)
       (SMU 83-13)]
[5]   L.Kazmerski [in 'Solar Materials Science', Ed. L.E.Muir
       (Academic, New York, 1980) ch.15]
[6]   E.A.Taft [J. Electrochem. Soc. (USA) vol.125 no.6 (1978)
       p.968-71]
[7]   D.E.Aspnes, A.A.Studna [Phys. Rev. B (USA) vol.27 no.2 (15 Jan
       1983) p.985-1009]

[8]   H.R.Philipp [J. Appl. Phys. (USA) vol.43 (1972) p.2835-9]

[9]   W.Primak [Appl. Opt. (USA) vol.10 (1971) p.759-63]

[10]   G.E.Jellison, F.A.Modine [Phys. Rev. B (USA) vol.27 no.12 (15
       Jun 1983) p.7466-72]

[11]   P.Lautenchlager, M.Garriga, L.Vina, M.Cardona [Phys. Rev. B
       (USA) vol.36 no.9 (15 Sep 1987) p.4821-30]

2.5 OPTICAL FUNCTIONS OF INTRINSIC Si: VACUUM UV

by D.E.Aspnes

November 1987
EMIS Datareview RN=17805

Optical functions from O to 25 eV (wavelengths longer than 49.6 nm)
were originally determined by Philipp and Ehrenreich [1] by a
Kramers-Kronig transformation of normal-incidence reflectance data
taken over the same energy range. These data were later corrected for
overlayer artifacts by Philipp [2]. The corrected data and the
spectroellipsometric data of Aspnes and Studna [3] agree to within
about 10% in n and 5% in k within their region of overlap of 1.5 to 6.0
eV; the Philipp data shown in the table [4] in increments of 0.5 eV from
2.0 to 6.0 eV allow an assessment of the discrepancies.

Data from 20 to 95 eV (13 to 62 nm) were obtained by Hunter on
evaporated films, as described and tabulated by Edwards [5]. The
absence of absorption thresholds in this spectral range results in a
smooth dependence of the optical functions on energy.

Optical functions in the 70-400 eV (3.1 to 17.7 nm) spectral range
have been determined by Filatova et al [6]. This range includes the
L(2,3) soft-X-ray threshold near 100 eV (12.4 nm), which leads to
fine structure in the 99-106 eV (11.7-12.5 nm) spectral range. The
Filatova optical-function values were obtained by a Kramers-Kronig
transformation of reflectance data taken as a continous function of
wavelength from 95 to 200 eV (6.2 to 13.0 nm) supplemented by
discrete-wavelength data at the 72.4, 192.6, 277 and 392.4 eV (17.1,
6.44, 4.47, and 3.16 nm) characteristic X-ray lines of Al, Mo, C and
N, respectively. The values listed in the table [3] are those obtained
from the 4 deg angle-of-incidence curves in the figures in this
reference.

Absorption coefficient (a) data were taken in the 99-200 eV (6.2 to
12.5 nm) spectral range by Brown et al [7]. These a data agree well
with the Filatova et al results [6] above 120 eV (wavelengths shorter
than 10 nm), but are about a factor of 2 lower at the L(2,3)
soft-X-ray threshold itself. This difference may be due to the apparent
enhancement of oscillator strength of this transition in reflectance
due to SiO2 on the surface, but further measurements are needed to
resolve the discrepancy. We list the Filatova et al [6] data in the
table [4] because they give both n and k.

REFERENCES
----

[1] H.R.Philipp, H.Ehrenreich [Phys. Rev. (USA) vol.129 (1963) p.1550-60]

[2] H.R.Philipp [J. Appl. Phys. (USA) vol.43 (1972) p.2835-9]

[3] D.E.Aspnes, A.A.Studna [Phys. Rev. B (USA) vol.27 no.2 (15 Jan 1983) p.985-1009]

[4] D.E.Aspnes [EMIS Datareview RN=17807 (Nov 1987) 'Optical functions of intrinsic Si: table of refractive index, extinction coefficient an absorption coefficient vs energy (0 to 400 eV)']

[5] W.R.Hunter [as listed by D.F.Edwards in 'Handbook of Optical Constants of Solids' Ed. E.Palik (Academic, New York, 1985) p.547-69]

[6] E.O.Filatova, A.S.Vinogradov, I.A.Sorokin, T.M.Zimkina [Sov. Phys.-Solid State (USA) vol.25 (1983) p.736-9]

[7] F.C.Brown, R.Z.Bachrach, M.Skibowski [Phys. Rev. B (USA) vol.15 no.10 (15 May 1977) p.4781-8]

2.6 OPTICAL FUNCTIONS OF INTRINSIC Si: TABLE OF REFRACTIVE INDEX, EXTINCTION
COEFFICIENT AND ABSORPTION COEFFICIENT VS ENERGY (O TO 400 EV)

by D.E.Aspnes

November 1987
EMIS Datareview RN=17807

All data in TABLE 1 are for nominally room-temperature samples. The
complex dielectric function e = e1 + e2 is related to the index of
refraction and the extinction coefficient by e = (n + ik)*2. The
extinction coefficient k and absorption coefficient a are related by

$$a = \frac{4 \pi k}{wavelength}$$

Energy is converted into wavelength by the expression

$$wavelength \ in \ microns = \frac{1.2395086}{energy \ in \ eV}$$

Original references are indicated in brackets. All entries in a given
column appearing between the same reference numbers have been taken
from that reference.

Extinction coefficient values are calculated from transmission data
(absorption coefficients) for all references except [12,13,14 and 16].
For these four references absorption coefficients have been calculated
from extinction coefficients.

TABLE 1: optical functions of single crystal Si

Energy (eV)	n		k	a (cm*-1)
0.000	3.420	[1]	0	0
0.00372	3.4155	[2]	0	0
0.00496	3.4160			
0.00620	3.4165			
0.00744	3.4170			
0.00868	3.4175			
0.00992	3.4180			
0.01116	3.4182			
0.0124	3.4185			
0.0136	3.4188			
0.0149	3.4190			
0.0161	3.4191			
0.0174	3.4192			
0.0186	3.4194			
0.0198	3.4195			
0.0211	3.4195			
0.0223	3.4196			
0.0236	3.4197			
0.0248	3.4197			
0.0260	3.4198			
0.0273	3.4198			
0.0285	3.4198			
0.0297	3.4198			

Energy (eV)	n	k	a (cm*-1)
0.0310	3.4199		
0.0322	3.4199		
0.0335	3.4200		
0.0347	3.4200		
0.0359	3.4200		
0.0372	3.4200		
0.0384	3.4200		
0.0397	3.4200		
0.0409	3.4200		
0.0421	3.4200		
0.0434	3.4200 [2]		
0.0496	3.4201 [3]	9. X 10*-5	0.5
0.0516	3.4202	1.1 X 10*-4	0.6
0.0539	3.4203	1.3	0.7
0.0563	3.4203	1.6	0.9
0.0590	3.4204	2.3	1.4
0.0620	3.4204	2.9	1.8
0.0652	3.4205	3.2	2.1
0.0689	3.4205	3.7	2.6
0.0707		4.2	3.0
0.0729	3.4206	3.8	2.8
0.07513		9.9 X 10*-4	7.5
0.0762		1.22 X 10*-3	9.5
0.0775	3.4206	7.6 X 10*-4	6.0
0.0787		2.0	1.6
0.0800	3.4207	1.6	1.3
0.0826	3.4207	1.3	1.1
0.0855	3.4208	1.2	1.0
0.0868		1.0 X 10*-4	0.9
0.0885	3.4208	1.7	1.5
0.0918	3.4209	3.1	2.9
0.0921		3.1	3.1
0.0954	3.4209	2.3	2.3
0.0992	3.4210	1.9	1.9
0.0998		1.7	1.7
0.1016		2.0	2.1
0.1033	3.4211	1.6	1.7
0.1042		1.4	1.5
0.1078	3.4212	2.0	2.2
0.1103		2.1	2.3
0.1127	3.4213	1.8	2.1
0.1181	3.4214	1.3 X 10*-4	1.5
0.124	3.4215	7. X 10*-5	0.9
0.130	3.4217	4.	0.5
0.135		4.9	0.67 [5]
0.138	3.4219		
0.140		7.	1.0 [4]
0.140		6.1	0.87 [5]
0.145		4.	0.6 [4]
0.145		4.5	0.67 [5]
0.146	3.4221		
0.150		2.	0.3 [4]
0.150		2.2	0.33 [5]
0.155	3.4224	1.5	0.24
0.160		2.3	0.36
0.165	3.4227	1.7	0.28

Energy (eV)	n	k	a (cm*-1)
0.170		1.6	0.28
0.175		1.9	0.33
0.177	3.4231		
0.180		2.2 X 10*-5	0.40
0.185		5.3 X 10*-6	0.10
0.190		1.1 X 10*-6	0.021
0.191	3.4236		
0.195		9.0 X 10*-7	0.018
0.200		7.4	0.015
0.205		6.1	0.018
0.207	3.4242		
0.210		4.2	0.0089
0.215		4.1	0.0089
0.220		4.0	0.0089
0.225	3.4250	3.3	0.0075
0.230		3.0	0.0069
0.235		2.9	0.0069
0.240		2.8	0.0069
0.245		2.3	0.0058
0.248	3.4261		
0.250		1.9	0.0049
0.255		2.0	0.0052
0.260		2.0	0.0053
0.265		2.2	0.0058
0.270		2.3 X 10*-7	0.0063 [3]
0.275	3.4275		
0.291	3.4283		
0.310	3.4294		
0.326		1.3 X 10*-8	0.00042 [6]
0.354	3.4321		
0.363	3.4327		
0.375	3.4335		
0.413	3.4361		
0.457	3.4393		
0.459		2.6 X 10*-9	0.00012 [6]
0.496	3.4424		
0.509	3.4434 [3]		
0.574	3.443 [7]		
0.620	3.449		
0.689	3.458		
0.731	3.464		
0.775	3.471		
0.809	3.478		
0.885	3.489		
0.903	3.501 (?)		
1.015		1.6 X 10*-8	0.0017 [8]
1.020		4.1	0.0042
1.025		5.6 X 10*-8	0.0058
1.030		1.1 X 10*-7	0.011
1.033	3.519		
1.035		2.1	0.022
1.040		3.1	0.032
1.045		3.9	0.041
1.050		6.4 X 10*-7	0.064
1.055		1.4 X 10*-6	0.145
1.060		2.2	0.24
1.065		3.1	0.34

Energy (eV)	n	k	a (cm*-1)
1.070		4.2	0.46
1.075		5.5	0.60
1.080		7.0	0.70
1.084	3.533		
1.085		8.8 X 10*-6	0.97
1.090		1.10 X 10*-5	1.26
1.095		1.33	1.48
1.100		1.54	1.72 [8]
1.100		0.85	0.95 [9]
1.105		1.77	1.99 [8]
1.107	3.536 [7]		
1.110		2.1	2.4
1.120		2.7	3.1
1.130		3.6	4.1
1.140		4.6	5.3
1.150		5.9	6.9 [8]
1.150	3.550* [10]	8.1	9.4 [9]
1.160		7.6 X 10*-5	8.9 X 10*0 [8]
1.170		1.00 X 10*-4	1.2 X 10*1
1.180		1.36	1.6
1.190		1.8	2.2
1.200		2.4	2.9 [8]
1.200	3.565*	4.5	5.4 X 10*1 [9]
1.250	3.582*	8.8 X 10*-4	1.12 X 10*2 [9]
1.250		1.2 X 10*-3	1.5 [11]
1.300	3.599*	1.5 X 10*-3	1.9 [9]
1.350	3.617*	2.2	2.9
1.400	3.636*	2.9	4.1
1.450	3.655* [10]	3.7	5.4
1.500	3.673 [12]	4.5	6.8 [9]
1.500		5.	7. [11]
1.550	3.693	5.4	8.5 X 10*2 [9]
1.600	3.714	6.7	1.09 X 10*3
1.650	3.732	8.1	1.4
1.700	3.752	9.7 X 10*-3	1.7
1.750	3.773	0.0112	2.0 [9]
1.750		0.01	2. [11]
1.800	3.796	0.0134	2.4 [9]
1.850	3.820	0.015	2.8
1.900	3.847	0.017	3.2
1.950	3.876	0.019	3.7
2.000		0.020	4.1 [9]
2.000		0.02	4. [11]
2.000	3.906 [12]	0.022 [12]	4.5 [9]
2.000	3.886 [13]	0.019 [13]	3.9
2.050	3.937 [12]	0.023	4.7
2.100	3.969	0.024	5.1
2.150	4.004	0.026	5.6
2.200	4.042	0.028	6.1
2.250	4.081	0.030	6.9 [9]
2.250		0.03	7. [11]
2.270	4.086 [14]	0.031 [14]	7.1
2.300	4.123	0.033	7.7 [9]
2.350	4.168	0.036	8.6
2.400	4.215	0.039	9.4 X 10*3
2.450	4.266	0.042	1.05 X 10*4

Energy (eV)	n	k	a (cm*-1)
2.500		0.048	1.21 [9]
2.500		0.06	1.5 [11]
2.500	4.320 [12]	0.073 [12]	1.9
2.500	4.291 [13]	0.051 [13]	1.3
2.550	4.379 [12]	0.052	1.35 [9]
2.600	4.442	0.057	1.5
2.650	4.508	0.064	1.7
2.700	4.583	0.074	2.0
2.750	4.665	0.083	2.3
2.800		0.095	2.7 [9]
2.800	4.753	0.163 [12]	4.6
2.850		0.110	3.2 [9]
2.850	4.851	0.183 [12]	5.3
2.900		0.124	3.6 [9]
2.900	4.961	0.203 [12]	6.0
2.950		0.142	4.3 [9]
2.950	5.083	0.237 [12]	7.1
3.000		0.160	4.9 [9]
3.000	5.222 [12]	0.269 [12]	8.2
3.000	5.150 [13]	0.198 [13]	6.0
3.050		0.184	5.7 [9]
3.050	5.385 [12]	0.320 [12]	9.9 X 10*4
3.100	5.570	0.387	1.21 X 10*5
3.150	5.792	0.479	1.53
3.200	6.062	0.630	2.0
3.250	6.384	0.876	2.9
3.300	6.709	1.321	4.4
3.350	6.820	2.018	6.9
3.400	6.522	2.705	9.3 X 10*5
3.450	5.987	3.007	1.05 X 10*6
3.500	5.610 [12]	3.014 [12]	1.07
3.500	5.949 [13]	2.952 [13]	1.05
3.550	5.411 [12]	2.986 [12]	1.07
3.600	5.296	2.987	1.09
3.650	5.216	3.013	1.11
3.700	5.156	3.058	1.15
3.750	5.105	3.115	1.18
3.800	5.065	3.182	1.23
3.850	5.034	3.258	1.27
3.900	5.016	3.346	1.32
3.950	5.010	3.453	1.38
4.000	5.010 [12]	3.587 [12]	1.45
4.000	5.280 [13]	3.580 [13]	1.45
4.050	5.019 [12]	3.758 [12]	1.54
4.100	5.020	3.979	1.65
4.150	4.990	4.267	1.80
4.200	4.888	4.639	1.98
4.250	4.612	5.074	2.19
4.300	4.086	5.395	2.35
4.350	3.536	5.426	2.39
4.400	3.120	5.344	2.38
4.450	2.766	5.232	2.36
4.500	2.451 [12]	5.082 [12]	2.32
4.500	2.579 [13]	5.075 [13]	2.32
4.550	2.186 [12]	4.891 [12]	2.26

Energy (eV)	n		k		a (cm*-1)
4.600	1.988		4.678		2.18
4.650	1.851		4.466		2.10
4.700	1.764		4.278		2.04
4.750	1.703		4.118		1.98
4.800	1.658		3.979		1.94
4.850	1.623		3.857		1.90
4.900	1.597		3.749		1.86
4.950	1.582		3.650		1.83
5.000	1.570	[12]	3.565	[12]	1.81
5.000	1.684	[13]	3.577	[13]	1.81
5.100	1.571	[12]	3.429	[12]	1.77
5.200	1.589		3.354		1.77
5.300	1.579		3.353		1.80
5.400	1.471		3.366		1.84
5.500	1.340	[12]	3.302	[12]	1.84
5.500	1.508	[13]	3.401	[13]	1.90
5.600	1.247	[12]	3.206	[12]	1.82
5.700	1.186		3.119		1.80
5.800	1.133		3.045		1.79
5.900	1.083		2.982		1.78
6.000	1.010	[12]	2.909	[12]	1.77
6.000	1.108	[13]	3.045	[13]	1.85
6.250	0.968		2.89		1.83
6.500	0.847		2.73		1.80
6.750	0.756		2.58		1.77
7.000	0.682		2.45		1.74
7.250	0.618		2.32		1.71
7.500	0.563		2.21		1.68
7.750	0.517		2.10		1.65
8.000	0.478		2.00		1.62
8.250	0.444		1.90		1.59
8.500	0.414		1.82		1.57
8.750	0.389		1.73		1.53
9.000	0.367		1.66		1.51
9.250	0.348		1.58		1.48
9.500	0.332		1.51		1.45
9.750	0.318		1.45		1.43
10.00	0.306		1.38		1.40
10.50	0.286		1.26		1.34
11.00	0.272		1.16		1.29
11.50	0.263		1.06		1.24
12.00	0.257		0.963		1.17
12.50	0.255		0.875		1.11
13.00	0.258		0.792		1.04 X 10*6
13.50	0.265		0.714		9.7 X 10*5
14.00	0.275		0.641		9.1
14.50	0.288		0.573		8.4
15.00	0.313		0.479		7.3
15.50	0.323		0.450		7.1
16.00	0.345		0.394		6.4
16.50	0.369		0.342		5.7
17.00	0.397		0.296		5.1
17.50	0.426		0.255		4.5
18.00	0.455		0.219		4.0
18.50	0.485		0.189		3.5
19.00	0.514		0.163		3.1
19.50	0.542		0.140		2.8

Energy (eV)	n	k	a (cm*-1)
20.00	0.569 [13]	0.122 [13]	2.5
20.00	0.567 [15]	0.0835	1.7 [15]
21.01	0.627	0.058	1.2 X 10*5
22.14	0.675	0.0405	9.1 X 10*4
23.39	0.722	0.0292	6.9
24.31	0.752	0.0243	6.0
25.30	0.778	0.0205	5.3
26.38	0.803	0.0178	4.8
27.55	0.824	0.0158	4.4
28.83	0.843	0.0147	4.3
30.24	0.860	0.0138	4.2
32.63	0.885	0.0128	4.2
34.44	0.899	0.0121	4.2
36.47	0.918	0.0113	4.2
38.75	0.925	0.0104	4.1
40.00	0.930	0.010	4.1
42.75	0.942	0.0090	3.9
45.92	0.952	0.00785	3.7
47.69	0.956	0.0073	3.5
51.66	0.964	0.0061	3.2
56.36	0.972	0.0050	2.9
61.99	0.978	0.00393	2.5
68.88	0.987	0.00297	2.1
77.49	0.991	0.00215	1.7
88.56	1.000 [15]	0.00143	1.3 [15]
98.5	1.003 [16]	0.011 [16]	1.1 X 10*5
99.0	1.006	0.011	1.1
99.5	1.014	0.014	1.4
100.0	1.016	0.022	2.2
100.3	1.030		
100.5	1.012	0.043	4.4
101.0	1.000	0.040	4.0
101.5	0.996	0.028	2.8
102.0	0.996	0.027	2.7
103.0	0.996	0.026	2.7
104.0	0.996	0.026	2.7
106.0	0.997	0.026	2.8
110	0.995	0.025	2.8
115	0.994	0.025	2.9
120	0.990	0.025	3.0
125	0.988	0.024	3.0
130	0.986	0.021	2.8
135	0.986	0.020	2.7
140	0.987	0.018	2.5
145	0.988	0.017	2.5
150	0.988	0.016	2.4
155	0.988	0.014	2.2
160	0.989	0.014	2.2
165	0.989	0.013	2.2
170	0.988	0.012	2.1
180	0.989	0.009	1.6
200	0.990	0.007	1.5
250	0.993	0.004	0.9
300	0.995	0.003	0.8
350	0.997	0.002	0.7
400	0.998 [16]	0.002 [16]	0.7 X 10*5

REFERENCES

[1]   D.E.Aspnes [EMIS Datareview RN=17802 (Nov 1987) 'Optical
        functions of intrinsic Si: dielectric constant']
[2]   E.V.Loewenstein, D.R.Smith, R.L.Morgan [Appl. Opt. (USA) vol.12
        (1973) p.398-406; temperature T = 300 deg K]
[3]   D.F.Edwards, E.Ochoa [Appl. Opt. (USA) vol.19 (1980) p.4130-1]
[4]   F.A.Johnson [Proc. Phys. Soc. London (GB) vol.73 (1959) p.265-72]
[5]   B.Bendow, H.G.Lipson, S.P.Yukon [Appl. Opt. (USA) vol.16 (1977)
        p.2909-13]
[6]   A.Hordvik, L.Skolnik [Appl. Opt. (USA) vol.16 (1977) p.2919-24]
[7]   W.Primak [Appl. Opt. (USA) vol.10 (1971) p.759-63]
[8]   G.D.Cody, B.G.Brooks [private communication (1986)]
[9]   W.C.Dash, R.Newman [Phys. Rev. (USA) vol.99 (1955) p.1151-5]
[10]   Interpolation between n data of Primak [7] and of Aspnes and
        Studna [11]
[11]   L.Kazmerski [in 'Solar Materials Science' Ed. L.E.Muir
        (Academic, New York, 1980) ch.15]
[12]   D.E.Aspnes, A.A.Studna [Phys. Rev B (USA) vol.27 no.2 (15 Jun
        983) p.985-1009]
[13]   H.R.Philipp [J. Appl. Phys. (USA) vol.43 (1972) p.2835-9]
[14]   E.A.Taft [J. Electrochem. Soc. (USA) vol.125 (1978) p.968-71]
[15]   W.R.Hunter [as listed by by D.F.Edwards in 'Handbook of Optical
        Constants of Solids' Ed. E.Palik (Academic, New York, 1985)
        p.547-69]
[16]   E.O.Filatova, A.S.Vinogradov, I.A.Sorokin, T.M.Zimkina [Sov.
        Phys.-Solid State vol.25 (1983) p.736-9]

2.7 OPTICAL FUNCTIONS OF LIQUID Si

by D.E.Aspnes

November 1987
EMIS Datareview RN=17806

The metallic character of liquid Si makes its optical response
substantially different from that of the solid at temperatures above its
1410 deg K melting point, a property that is particularly useful for
in-situ optical studies of laser annealing. The optical functions of
liquid Si produced by pulsed laser melting of a wafer surface were
recently determined at 632.8 nm (1.959 eV) and the six wavelengths of
the Ar+ ion laser by Jellison and Lowndes [1]. An ellipsometer with a
nanosecond response time was used for the measurements, and estimated
confidence limits were 0.05 on n and k. The data, given in tabular form
below, were fit to the Drude expression

$$e = 1 - \frac{(35.1 \text{ eV})*2}{E(E + i8.2 \text{ eV})}$$

which yields n and k values within 15% of the data and also gives an
estimated DC conductivity of 50 micro-ohm cm, in reasonable agreement
with a measured value of 80 micro-ohm cm.

The results are also consistent with previous measurements of n and k
by Li and Fauchet [2] and n by Shvarev et al [3]. The k values of
Shvarev et al appear to be low by about 1.

TABLE 1: optical functions of liquid Si according to ref [1]

Energy eV	n	k	a 10*6/cm
1.959	3.80	5.20	1.03
2.409	3.11	4.89	1.19
2.540	2.94	4.99	1.29
2.707	2.74	4.96	1.36
3.407	1.93	4.32	1.49
3.531	1.80	4.26	1.53
3.711	1.74	4.09	1.54

REFERENCES
----

[1]   G.E.Jellison Jr, D.H.Lowndes [Appl. Phys. Lett. (USA) vol.51 no.5
      (3 Aug 1987) p.352-4]
[2]   K.D.Li, P.M.Fauchet [Solid State Commun. (USA) vol.61 no.3 (1987)
      p.207-9]
[3]   K.M.Shvarev, B.A.Baum, P.V.Gel'd [Sov. Phys.-Solid State (USA)
      vol.16 (1975) p.2111-2]

CHAPTER 3

RESISTIVITY AND PIEZORESISTANCE

3.1 RESISTIVITY OF n- AND p-TYPE Si, DOPING DEPENDENCE

by M.Pawlik

October 1987
EMIS Datareview RN=14808

The relationship between resistivity and dopant density has been
systematically measured in boron and phosphorus doped silicon
[1]. Bulk silicon slices were used having dopant densities in the
range 3 X 10*13 to 1 X 10*20/cm*3 for n-type and 1 X 10*14 to
1 X 10*20/cm*3 for p-type. The techniques of capacitance voltage
(CV) for densities less than 1 X 10*18/cm*3 and Hall effect for
densities greater than 1 X 10*18/cm*3 were used in conjunction with
four point probe measurements of resistivity. Additional data points
from other published sources have been added to extend the range of
validity. Estimates of the self consistency of the conversion between
resistivity and dopant concentration and vice versa are given and
vary over the range 1 X 10*12 to 1 X 10*21 for boron and 1 X 10*12
to 5 X 10*20 for phosphorus, being at worst 4.5%.

It is emphasized that there is conversion between dopant density and
resistivity and not carrier concentration and resistivity. No account
has been taken of possible incomplete ionisation although in the
range of direct measurement deviations are expected to be small or
nonexistent.

These data have been tabulated and presented in graphical form [2],
together with estimates of the deviation from the other much quoted
work of Irvin [3]. The deviation is found to be as large as 60% over
a part of the range. TABLE 1 shows data at convenient values of
resistivity and TABLE 2 shows data at convenient values of dopant
concentration.

TABLE 1

Resistivity (ohm cm)	Dopant density (/cm*3)	
	n-type	p-type
10*-4	1.6 X 10*21	1.2 X 10*21
10*-3	7.38 X 10*19	1.17 X 10*20
10*-2	4.53 X 10*18	8.49 X 10*18
10*-1	7.84 X 10*16	2.77 X 10*17
1	4.86 X 10*15	1.46 X 10*16
10*1	4.45 X 10*14	1.34 X 10*15
10*2	4.27 X 10*13	1.33 X 10*14
10*3	4.2 X 10*12	1.3 X 10*13
10*4	4.0 X 10*11	1.3 X 10*12

TABLE 2

Dopant density (/cm*3)	Resistivity (ohm cm)	
	n-type	p-type
10*12	4.3 X 10*3	1.3 X 10*4
10*13	4.3 X 10*2	1.3 X 10*3
10*14	4.29 X 10*1	1.31 X 10*2
10*15	4.48	1.33 X 10*1
10*16	5.23 X 10*-1	1.44
10*17	8.38 X 10*-2	2.02 X 10*-1
10*18	2.36 X 10*-2	4.35 X 10*-2
10*19	5.78 X 10*-3	8.87 X 10*-3
10*20	7.70 X 10*-4	1.17 X 10*-3
10*21	1.6 X 10*-4	1.3 X 10*-4

A number of possible interferences may arise which can affect these
conversions. Primary amongst these is the compensation of silicon
through the presence of impurities other than the majority dopant
density impurity. Normally these are unintentionally introduced
dopants of opposite polarity and similar effects due to the presence
of deep level impurities, crystalline defects, interstitial and
lattice damage by radiation are observed.

At high doping levels, the possibility of the formation of compounds
and complexes arises and such effects prevent some dopant atoms from
being electrically active. This is particularly likely in phosphorus
or arsenic doped silicon above the limit of solid solubility. Since
bulk silicon material with dopant densities greater than 1 X 10*20/cm*2
is not available, measurements in this range are made on ion implanted
or diffused material. Differences are observed depending on the method
of preparation of the samples, in particular the method of annealing,
and no universal agreement has been reached about the conversion or
mobility changes. The paper of Masetti [4] describes the latest and most
comprehensive attempts to determine mobility and carrier concentration
relationships.

REFERENCES
----

[1] W.R.Thurber, R.L.Mattis, Y.M.Liu, J.J.Filliber [National
     Bureau of Standards, Special Publication SP400-64 (May 1981)
     Washington (USA)]
[2] [ASTM Standard Practice, Designation F723-82, Annual Book of
     Standards vol.10.05 (1984, Philadelphia, USA) p.598-614]
[3] J.C.Irvin [Bell Syst. Tech. J. (USA) BSTA vol.41 (1962)
     p.387-410]
[4] G.Masetti, M.Severi, S.Solmi [IEEE Trans. Electron Devices (USA)
     vol.T-ED-30 no.7 (Jul 1983) p.764-9]

3.2 CARRIER CONCENTRATIONS IN LIGHTLY- AND MODERATELY-DOPED Si

by D.Schechter

July 1987
EMIS Datareview RN=15775

A. SYMBOLS

Due to the limited character set available the following symbols are
used in this Datareview:

Ea = acceptor binding energy
Ec = energy of conduction band edge
Ed = donor binding energy
Ef = Fermi energy
Ev = energy of valence band edge
g1 = acceptor degeneracy
g2 = donor degeneracy
k = Boltzmann constant
n = density of electrons in the conduction band
n2 = density of electrons bound to donors
Na = density of acceptors
Nc = effective density of states in conduction band
Nd = density of donors
Nv = effective density of states in valence band
p = density of holes in the valence band
p1 = density of holes bound to acceptors
T = temperature in deg K

B. LIGHTLY DOPED SILICON (< 1 X 10*16/cm*3)

Carrier concentrations in lightly-doped material at thermal equilibrium
and in the absence of applied fields are calculated by assuming charge
neutrality [1].

$$p + p1 + Nd = n + n2 + Na$$

For parabolic bands [2]

$$p = Nv \exp[(Ev - Ef)/kT]$$

$$n = Nc \exp[-(Ec-Ef)/kT]$$

$$n2 = Nd/[1 + (1/g2) \exp(Ed-Ef)/kT]$$

$$p1 = Na/[1 + (1/g1) \exp(Ef-Ea)/kT]$$

The effective densities of states in silicon are

$$Nv = 1.04 \ X \ 10*19 \ X \ (T/300)*(3/2)/cm*3$$

$$Nc = 2.8 \ X \ 10*19 \ X \ (T/300)*(3/2)/cm*3$$

In practice, the Fermi energy is initially unknown. However, one finds
it by iteratively trying various values until the charge neutrality
relation is satisfied.

Further discussion will be confined to n-type silicon. In n-type
material the Fermi level is in the vicinity of the conduction band
edge (at least at low temperatures) and hence one may, to a good
approximation, take p = p1 = 0. Hence for lightly doped material the
charge neutrality equation becomes

    Nd = n + n2 + Na

C. MODERATELY DOPED SILICON (< 1 X 10*18/cm*3)

All of the above relations hold reasonably well up to dopings of about
1 X 10*16/cm*3. Many of these relationships must be modified in more
heavily doped materials. A practical model for treating heavier donor
dopings, up to about 1 X 10*18/cm*3, has been devised by Lee and
McGill [3,4]. The random electric fields of the ionised impurities (in
n-type materials all of the acceptors are ionised) cause various
effects - a shift in the conduction band edge, conduction band tailing
(so that the conduction band edge is no longer sharply defined), and
a random shift in the binding energy of the various donor impurity
atoms. In addition, the overlap of the impurity electron wavefunctions
of adjacent donor impurities produces a further random shift in donor
binding energies. These random shifts in donor binding energies result
in the donor energy lying within a range, rather than having a sharply
defined value, and hence one speaks of an impurity band.

In this model the number of electrons in the conduction band is given
by

$$n = \int_{-inf}^{+inf} \frac{Nc(E)\ dE}{1 + \exp[(E-Ef)/kT]}$$

where Nc(E) is suitably modelled average density of conduction band
states that results from the conduction band tailing.

The density of ionised donors in this model is given by

$$Nd+ = Nd - n2$$

$$= \int_{-inf}^{+inf} \frac{rho(i)\ (E-Ed')dE}{1 + 2\exp[(Ef-E)/kT]}$$

where rho(i) (E-Ed') is a suitably modelled impurity band density of
states and Ed' is the average shifted donor binding energy.

The procedure for calculating the carrier concentration using this
model is analogous to that for lightly doped material. The Fermi energy
is iterated until charge neutrality is satisfied, the difference being
that the charge neutrality computations are much more complicated
since the conduction band density of states and the impurity band
density of states depend on the carrier concentration and the extent
of ionisation of the donors, which in turn depends on the Fermi level.
Hence these computations must be done self-consistently.

Based on the available experimental data, the Lee-McGill model appears
to be quite successful at describing n-type silicon up to a maximum
donor concentration of about 1 X 10*18/cm*3. This author has published
detailed computations of activation energy and carrier concentration
[4]. TABLES 1 and 2 give carrier concentration as a function of donor
and compensating acceptor doping level in As doped Si at temperatures
of 80 deg K and 300 deg K, respectively. Carrier concentrations were
also computed at 10 deg K, but the number of carriers was so small that
the results were not tabulated.

Computations were also carried out in the doping range 1 X 10*15/cm*3
< Nd < 1 X 10*16/cm*3 to compare the dilute and Lee-McGill models. At
a temperature of 300 deg K the two models agreed within 1% for all
three shallow group V impurities (P, As, and Sb) for Nd = 1 X
10*16/cm*3. This agreement held for P and Sb doped material down to Nd
= 2 X 10*15/cm*3. For As doped Si the agreement steadily decreased as
Nd decreased below 1 X 10*16/cm*3. For all dopants, the Lee-McGill
model ceased to give reasonable results for Nd < 2 X 10*15/cm*3.

As the temperature is decreased, the agreement between the two models
also decreases. At 80 deg K, the two models agree to about 20% at Nd =
1 X 10*16/cm*3. As Nd is decreased, the percent difference decreased
steadily to about 5% at Nd = 2 X 10*15/cm*3 (12% for As doped
material). At this temperature, the Lee-McGill model again fails for
Nd < 2 X 10*15/cm*3. For all dopants the carrier concentration
calculated by the Lee-McGill model at low temperatures exceeded that
calculated by the dilute model.

TABLE 1: carrier concentration in As doped Si at 80 deg K

| Nd (/cm*3) | ----------------------Na (/cm*3)---------------------- | | | | |
	1 X 10*12	1 X 10*13	1 X 10*14	1 X 10*15	1 X 10*16
1 10*14	8.9 10*13	8.1 10*13	-	-	-
2 10*14	1.6 10*14	1.6 10*14	8.1 10*13	-	-
5 10*14	3.5 10*14	3.4 10*14	2.7 10*14	-	-
1 10*15	5.8 10*14	5.7 10*14	5.1 10*14	-	-
2 10*15	9.2 10*14	9.2 10*14	8.6 10*14	3.7 10*14	-
5 10*15	1.6 10*15	1.6 10*15	1.6 10*15	1.1 10*15	-

1 10*16	3.0 10*15	3.0 10*15	2.9 10*15	2.6 10*15	-
2 10*16	4.6 10*15	4.6 10*15	4.6 10*15	4.3 10*15	2.0 10*15
5 10*16	8.1 10*15	8.1 10*15	8.0 10*15	7.8 10*15	6.3 10*15
1 10*17	1.2 10*16	1.2 10*16	1.2 10*16	1.2 10*16	1.1 10*16
2 10*17	1.9 10*16	1.9 10*16	1.9 10*16	1.9 10*16	1.9 10*16
5 10*17	3.8 10*16	3.8 10*16	3.8 10*16	3.8 10*16	3.9 10*16
1 10*18	7.1 10*16	7.1 10*16	7.1 10*16	7.2 10*16	7.5 10*16

Note - carrier concentrations are expressed in units of /cm*3.

TABLE 2: carrier concentration in As doped Si at 300 deg K.

Nd (/cm*3)	Na (/cm*3)				
	1 X 10*12	1 X 10*13	1 X 10*14	1 X 10*15	1 X10*16
1 10*14	9.9 10*13	9.0 10*13	-	-	-
2 10*14	2.0 10*14	1.9 10*14	1.0 10*14	-	-
5 10*14	5.0 10*14	4.9 10*14	4.0 10*14	-	-
1 10*15	1.0 10*15	9.9 10*14	9.0 10*14	-	-
2 10*15	2.0 10*15	2.0 10*15	1.9 10*15	1.0 10*15	-
5 10*15	5.0 10*15	5.0 10*15	4.9 10*15	4.0 10*15	-
1 10*16	9.9 10*15	9.9 10*15	9.8 10*15	8.9 10*15	-
2 10*16	2.0 10*16	2.0 10*16	2.0 10*16	1.9 10*16	9.9 10*15
5 10*16	4.9 10*16	4.9 10*16	4.9 10*16	4.8 10*16	3.9 10*16
1 10*17	9.6 10*16	9.6 10*16	9.6 10*16	9.5 10*16	8.7 10*16
2 10*17	1.9 10*17	1.9 10*17	1.9 10*17	1.9 10*17	1.8 10*17
5 10*17	4.4 10*17	4.4 10*17	4.4 10*17	4.4 10*17	4.3 10*17
1 10*18	8.1 10*17	8.1 10*17	8.1 10*17	8.1 10*17	8.0 10*17

Note - carrier concentrations are expressed in units of /cm*3

REFERENCES
----

[1]   W.Shockley [in 'Electrons and holes in Semiconductors' (D. Van
      Nostrand, 1950) p.465-71]
[2]   K.Seeger [Semiconductor Physics (Springer-Verlag, 1973)
      p.35-45]
[3]   T.F.Lee, T.C.McGill [J. Appl. Phys. (USA) vol.46 no.1 (Jan 1975)
      p.373-80]
[4]   D.Schechter [J. Appl. Phys. (USA) vol.61 no.2 (15 January 1987)
      p.591-8]

3.3   SHEET RESISTANCE OF ION-IMPLANTED BULK Si

by K.H.Nicholas

July 1987
EMIS Datareview RN=15768

The accurate control of the dose of dopant ions and the ability to
implant through a passivating layer make ion implantation an attractive
technique for forming doped layers for use as resistors or as layers in
active devices. Special electrical properties can also be achieved by
ion implantation.

The sheet resistance of a uniformly doped fully annealed ion implanted
layer is identical with such a layer produced by diffusion or growth.
The dopant species used for fully annealed layers are generally those
used in other doping methods (elements of groups III and V). There
are however some features specific to implanted layers and some species
not generally considered for diffusion (In, N, etc.) can be
incorporated by ion implantation.

The concentration profile of an implanted layer is well defined [1] and
controlled and full activation can be achieved without seriously
degrading the profile. The sheet resistance can then be calculated from
standard bulk concentration versus resistivity data [2]. Often a
Gaussian profile can be assumed as the resistance is not very sensitive
to small amounts of channelling. Standard computer models, e.g. SUPREM
(Stanford, USA) and ICECREM (IFT Munich, West Germany) can be used
to obtain accurate values. An approximate value can be obtained from the
simple formula $R = K D*-0.7 s*-0.3$ [3] where R is the sheet resistance,
D the implant dose, s the standard deviation of the implanted profile
and K a constant. For a p-type dopant such as boron K has the value 1 X
$10*11$ ohm/cm*1.1 while for n-type dopants it is 5.8 X $10*10$ ohm/cm*1.1.
The formulae give a first approximation (+- 10% B, +- 20% P&As)
between doses of 3 X $10*12$/cm*2 and $10*15$/cm*2. If the anneal does not
cause significant diffusion the value of s can be related to the
implant energy. For implant energies of 30 to 200 keV the following
formulae can be used:

Boron   s = 2.64 X $10*-6$ (ln E) - 4.6 X $10*-6$

Phosphorus s = 1.56 X $10*-6$ (ln E)*2 - 9.5 X $10*-6$ (ln E) + 1.66 X $10*-5$

Arsenic s = 8.5 X $10*-7$ (ln E)*2 - 5.7 X $10*-6$ (ln E) + 1.07 X $10*-5$

where E is the implant energy in keV. Agreement with standard tables
[1] is to better than 5%. Resistors with sheet resistance greater than
about 50 kohm/sq tend to be non-linear with voltage and sensitive to
surface related charges and substrate doping. The special resistors
described later can ease the problem. The lower limit is set by the
solid solubility of impurities and acceptable layer depth. It can be
difficult to reduce resistance much below 10 ohm/sq without the special
annealing techniques described later. There is some evidence that at
very high concentrations mobilities in p and n type layers converge [4]
but at fairly high concentrations those for different n-type species
diverge [2,16].

The temperature necessary for complete annealing (removal of damage effects on mobility and reaching activation levels as in thermal equilibrium) depends on dose, species and other factors. Some general trends can however be given provided the silicon has not been rendered amorphous by the implantation damage; (a) the higher the dose the higher the temperature of anneal required (typically 500 deg C to 950 deg C), (b) the higher the temperature of implant the less the damage and minimum anneal temperature (provided it does not prevent amorphisation) and (c) oxidising atmospheres during annealing can cause defects to grow rather than anneal out [5].

If the layer has become amorphous regrowth can take place at temperatures as low as 650 deg C [6]. Redistribution of dopant can occur and so the sheet resistance estimates are not necessarily valid. For high doses of certain dopants, loss of dopant to the surface can be significant particularly with (111) material which does not regrow so rapidly as (100) material.

Resistors with markedly different properties can be made by incomplete or rapid annealing. 500 deg C is a high enough anneal temperature to stabilise the layer and the sheet resistance is not generally very sensitive to anneal time and temperature [7]. There is often a minimum in sheet resistance against anneal temperature graphs at 500 deg C particularly for p-type layers. Anneals at 500 deg C therefore tend to give stable resistors of reproducible value while being compatible with aluminium and silicide layers.

The temperature coefficients of resistance of fully annealed implanted layers are similar to those of diffused layers. They are always positive. Residual damage tends to lead to effective dopant level depths deeper in the band gap than without damage. This leads to a more negative temperature coefficient of resistance [8]. By adjusting the amount of damage temperature coefficients near zero can be obtained as are desired for resistors in linear integrated circuits [9]. The damage can be adjusted by the choice of annealing temperature and extra damage can be produced by implanting a species neutral in silicon [10].

To make reproducible high value resistors the sensitivity to oxide charge and to depletion at the junction with the substrate must be reduced. This can be achieved by reducing carrier mobility or deepening the dopant energy depth.

The smaller the mobility the smaller the change in resistance for a given change in oxide or depleted charges. The mobility is lower for p-type layers and can be reduced further by compensation [11] or damage [10]. Ion implantation is ideal for making compensated resistors because of the good control of dose. Even greater improvements can be achieved by implantation damage.

Only a small fraction of dopant atoms are ionised at room temperature for deep level dopants but all are ionised in depletion layers. Thus for a given change in depletion layer charge the number of ionised dopants, and so carriers, changes by a much smaller amount. There are p-type dopants with deeper energy levels (In, Tl [12]) and ion implantation helps in avoiding any problems due to their diffusion in oxides. Implantation damage also causes the effective energy depth of p and n-type dopants to be deeper.

High value resistors can therefore be made by damaging implants into
boron doped layers or just implants of neutral ions (such as Ne) [13]
which give p-type activity with an acceptor depth of 0.25 eV. In this
way resistors with a sheet resistance of 250 kohm/sq can be made. The
temperature coefficient of such layers is strongly negative because of
the dopant energy depth. Linearity of resistors of a given value is
also improved by damage because of the reduced sensitivity to depletion
at the junction with the substrate.

Rapid annealing (1000 deg C for less than a minute) can lead to
supersaturation of dopant atoms for high doses. Concentrations well
above solid solubility can be achieved. Mobilities tend to be low
because of ionised impurity scattering but sheet resistances much lower
than with thermal equilibrium anneals can be achieved [14]. Rapid
annealing techniques can be divided into those in which the slice as a
whole reaches a uniform temperature and those so rapid that heating is
localised laterally and in depth. Strip heaters [15], optical heating
[16] and electron beam [17] generally give uniform slice temperatures.
The sheet resistances for implant doses below those causing
supersaturation are then in good agreement with the calculations
mentioned earlier because diffusion is negligible in terms of its
effect on sheet resistance. The advantage of these forms of rapid
annealing is the full activation with limited diffusion at a
relatively low cost compared to the localised annealing techniques.
Pulsed laser [18] or electron beam annealing can involve local melting
of the crystal surface. The implant profile is then altered to provide
a rectangular profile for high frequency emitters [19] etc. The sheet
resistance can also be slightly lower than with other rapid annealing
techniques.

REFERENCES
----

[1]   B.J.Smith [Ion implantation range data for silicon and germanium
       technologies (GB Learned Information Ltd, Europe, 1977)]
[2]   G.Masetti, M.Severi, S.Solmi [Extended Abstr. Electrochem. Soc.
       (All Div.) (USA) vol.83-1 (1983) p.639]
[3]   K.H.Nicholas [Radiat. Eff. (GB) vol.28 (1976) p.177]
[4]   C.Hill [Inst. Phys. Conf. Ser. (GB) no.69 Solid State Devices
       (1983) p.172]
[5]   I.R.Sanders, P.S.Dobson [Philos. Mag. (GB) vol.20 (1969) p.881];
       P.Ashburn, C.Bull, K.H.Nicholas, G.R.Booker [Solid-State
       Electron. (GB) vol.9 (1977) p.731]
[6]   L.Csepregi, W.K.Chu, H.Muller, J.W.Mayer, T.W.Sigmon [Radiat.
       Eff. (GB) vol.28 (1976) p.277]
[7]   J.M.Shannon, R.A.Ford, R.A.Gard [Radiat. Eff. (GB) vol.6 (1970)
       p.217]
[8]   D.P.Oostkoek, J.A.den Boer, W.K.Hofker [European Conf. on Ion
       Implantation (Reading, GB, Stevenage) (1970) p.88]
[9]   G.Kelson, H.H.Stellrecht, D.S.Perloff [IEEE J. Solid-State
       Circuits (USA) vol.SC-8 (1973) p.336]
[10]   B.A.Maciver [Electron. Lett. (GB) vol.11 (1975) p.484]
[11]   J.D.MacDougal, K.E.Manchester, P.E.Roughan [Proc. IEEE (USA)
       vol.57 (1969) p.1538]

[12] K.R.Whight, P.Blood, K.H.Nicholas [Solid-State Electron. (GB) vol.19 (1976) p.1021]

[13] K.H.Nicholas, R.A.Ford [Proc. 2nd Int. Conf. Ion Implantation in Semiconductors (Springer-Verlag, Berlin, 1971) p.357-61]

[14] W.K.Hokfer, G.E.J.Eggermont, K.Tamminga, D.P.Oosthoek [Laser-Solid Interactions and Laser Processing (USA, New York 1979) p.425-8]

[15] B.-Y.Tsaur, J.P.Donnelly, J.C.C.Fan, M.W.Geis [Appl. Phys. Lett. (USA) vol.39 (1981) p.93]

[16] G.K.McGinty, K.H.Nicholas [ESSDERC 1983 Europhysics Conf. Abs. (EPS Switzerland) vol.F7 (1983) p.159-61]

[17] H.J.Smith, E.Ligeon, A.Bontemps [Appl. Phys. Lett. (USA) vol.39 (1981) p.93]

[18] D.H.Austin, C.M.Surka, T.N.C.Venkatesen, R.E.Slushier, J.A.Golovchenko [Appl. Phys. Lett. (USA) vol.33 (1978) p.437]

[19] C.Hill [Laser Solid Interactions and Transient Thermal Processing of Materials (North Holland, NY, USA,1983) p.381]

3.4 PIEZORESISTANCE IN SINGLE CRYSTAL AND POLYCRYSTALLINE Si

by P.J.French and A.G.R.Evans

July 1987
EMIS Datareview RN=15724

A. BACKGROUND

When a crystal is subjected to a stress the resulting change in
resistance is derived from two effects. The first, the volumetric
effect, is due to the change in dimensions of the resistor and the
second, the piezoresistive effect, is due to the change in the
resistivity of the material itself.

The piezoresistive effect was first discovered in 1857 by Lord Kelvin
when investigating stress effects in metals. It was not, however, until
the 1930s that any further research was carried out and the potential
use of the piezoresistive effect realised. With the rise of silicon
technology in the 1950's and 1960's, this research was extended.

Several different terms are used in this work and these should be
clarified.

1) Piezoresistance coefficient (Pi) - Fractional change in resistivity
   per unit stress (units - cm*2/10*12 dyne or m*2/newton)

2) Elastoresistance coefficient (m) - Fractional change in resistivity
   per unit strain (dimensionless)

3) Gauge factor (G) (also K-factor) - Fractional change in resistance
   per unit strain (dimensionless). A positive gauge factor means
   that a positive change in resistance will result from a tensile
   strain and is defined by:

   $$dR/R = dl/l - dx/x - dy/y + dr/r$$

   $$G = 1 + 2v + Pi\ Y$$

   where l, x and y are the resistor dimensions, R the resistance,
   r the resistivity, Y Young's modulus and v Poisson's ratio.

4) Longitudinal/transverse gauge factor - the gauge factor for a
   strain parallel to and perpendicular to the current flow,
   respectively.

The piezoresistive coefficients are generally more widely used than the
elastoresistance coefficients, but they are related by Young's modulus
which is well defined for silicon by the equation Pi = m/Y [1].

Part of the volumetric effect is due to Poisson's ratio, which defines
the strain developed perpendicular to the applied strain. The total
contribution to the gauge factor from the volumetric effect is between
+1 and +2 for all materials. The insignificance of the volumetric effect
can be seen for silicon in contrast to other strain gauge materials in
TABLE 1 [2].

TABLE 1: gauge factor of common strain gauge materials

Material	Gauge Factor
Nichrome V	2.1
Isoelastic	3.6
Platinum/tungsten	4.0
Silicon	
Single crystal	-102 to +135
Polycrystalline	-30 to +40

The principal application of piezoresistance is for strain gauges which can be discrete or integrated devices with on-chip signal processing.

This survey is on the piezoresistive effect in both single crystal silicon and polycrystalline silicon and comments on the application of piezoresistance for strain gauges.

B. SINGLE CRYSTAL SILICON

The piezoresistive effect in single crystal silicon was first discovered by Smith [3] and Adams [4] in 1954. With the increased use of silicon in the 1960's, this study was greatly expanded [5-9], and the potential as a strain gauge material realised.

B1 Coefficients

The experimental values for the piezoresistive and elastoresistive coefficients from Smith are still widely used today and are given in TABLE 2. The elastoresistance coefficients were calculated from Smith [3] using Young's modulus [1].

TABLE 2: piezoresistive and elastoresistive coefficients for single crystal silicon

	Piezoresistive coefficients (cm*2/10*12 dynes)			Elastoresistive coefficients (dimensionless)		
	Pi11	Pi12	Pi44	m11	m12	m44
n-type	-102.2	+53.4	-13.6	-132.9	+69.4	-10.8
p-type	+6.6	-1.1	+138.1	+8.6	-1.4	+110.0

For an anisotropic material such as silicon both sets of coefficients shown in TABLE 2 are fourth rank tensors and are fully explained in the literature [2]. The notation in TABLE 2 is as follows: Pi11 and Pi12 are the fractional changes in resistivity for a stress parallel to and perpendicular to the current flow, respectively. The term Pi44 represents a voltage measured perpendicular to the current flow for a shear stress. All coefficients refer to currents measured along the principal (<100>) axes.

A similar set of coefficients can be found for the elastic constants, Sii (units - cm*2/10*12 dyne) or Cii (units - 10*12 dynes/cm*2). Young's modulus is given by $Y = 1/S11$ and Poisson's ratio by $v = S12/S11$. The values of the coefficients are given in TABLE 3 [50].

TABLE 3: the compliance coefficient and stiffness constant related to the principal axes

Compliance coefficient (cm*2/10*12 dyne)			Stiffness constant (10*12 dynes/cm*2)		
S11	S12	S44	C11	C12	C44
0.768	−0.214	1.26	1.657	0.639	0.796

TABLES 2 and 3 have shown the various coefficients related to the principal axes. If, however, an alternative crystallographic direction is required, an axis rotation must be used [10]. Such a rotation will yield the high anisotropy found with silicon and this is shown in the calculated values in TABLE 4 for the (100) plane.

TABLE 4: maximum and minimum values of piezoresistive and elastic constants in the (100) plane.

	n-type				p-type			
	max		min		max		min	
	mag	dir	mag	dir	mag	dir	mag	dir
Pil	−102.2	<100>	−31.2	<110>	+71.8	<110>	+6.6	<100>
Pit	+53.4	<100>	0.0	< * >	−66.3	<110>	−1.1	<100>
GF l*	−133.1	<100>	−52.7	<110>	+121.3	<110>	+9.7	<100>
GF t*	+69.5	<100>	0.0	< * >	−112.0	<110>	−0.3	<100>
1/Sii	+1.69	<110>	+1.30	<100>	+1.69	<110>	+1.30	<100>
−Sij/Sii	+0.28	<100>	+0.06	<110>	+0.28	<100>	+0.06	<110>

[* Gauge factor calculations include the volumetric effect 1+2v]

The '< * >' represents a direction 15 degrees from the <110> direction. The transverse gauge factor for n-type material will yield a negative sign for the <110> direction. Thus a sensor relying on the difference between Pil and Pit will have a greatly reduced sensitivity.

Furthermore, the piezoresistive effect is dopant dependent resulting in a reduced sensitivity for high doping concentrations. This is illustrated by the theoretical values in TABLE 5 [10,11].

TABLE 5: longitudinal piezoresistance as a function of doping
         concentration as a percentage of the low doped value

Doping concn (/cm*3)	1 X 10*18	5 X 10*18	1 X 10*19	5 X 10*19	1 X 10*20
Pil (n-type)	98	92	85	52	34
Pil (p-type)	95	80	66	28	20

(Note: with p-type material Kanda [10] quoted a deviation of calculated values (above) from experimental values [12] of +13% and -21% at doping concentrations of 5 X 10*18/cm*3 and 3 X 10*19/cm*3, respectively)

B2 Temperature Coefficients

The temperature coefficients of resistance and piezoresistance are of great importance for sensor design, as will become apparent in a later section. Both coefficients as a function of doping concentration are shown in TABLE 6 [10,13].

TABLE 6: temperature coefficients of resistance (TCR) and
         piezoresistance (TCP) as a function of doping concentration

doping concentration (/cm*3)	n		p	
	TCR (%/deg C)	TCP (%/deg C)	TCR (%/deg C)	TCP (%/deg C)
5 X 10*18	0.01	-0.28	0.0	-0.27
1 X 10*19	0.05	-0.27	0.01	-0.27
3 X 10*19	0.09	-0.19	0.06	-0.18
1 X 10*20	0.19	-0.12	0.17	-0.16

The small temperature coefficients of elastic constants have also been measured and are as follows [14]:

C11 -0.0075%/deg C; C12 -0.0025%/deg C; and C44 -0.0056%/deg C.

C. THEORETICAL MODEL

The theory of piezoresistance, in both n- and p-type material, is based on the shifting of bands relative to each other when the crystal is subjected to stress. Silicon is an indirect band gap material with six conduction band minima, all located on the principal axes. The constant energy surfaces (in k-space) are ellipsoidal with anisotropic mobility and effective mass. Smith [2] proposed that the piezoresistive effect in n-type material was due to the shifting of these bands relative to each other and the subsequent transfer of electrons. The result would be to offset the balance of high and low mobility electrons in the field direction. Several theoretical models have since been proposed based on the same basic principle [7,10,15].

In p-type silicon the light and heavy hole bands are degenerate at k=0 and are distorted from the simple parabolic shape. When the system is strained, this degeneracy is lifted and the distortion is reduced. The piezoresistive effect is caused by the transfer of holes between the two bands and the change in effective mass due to the change in the shape of the bands. Several theoretical studies of piezoresistance in p-type material have been made mainly based on these principles [7,16,17].

D. POLYCRYSTALLINE SILICON

The structure of polysilicon consists of small single crystal grains separated by a thin amorphous layer (grain boundary). The texture of the material has been found to be dominated by a small number of grain orientations [18].

The piezoresistive effect in polysilicon was first investigated by Onuma et al in 1974 [19] and has been further investigated by several researchers [20-27]. The inhomogeneous structure of a polysilicon film has a significant effect on both its electrical and piezoresistive properties. The gauge factor of polysilicon was shown in TABLE 1 to be significantly lower than that of single crystal silicon. Polysilicon has the advantage that it can be easily and cheaply deposited on a wide range of substrates including insulators. This avoids the p-n junction isolation necessary for single devices. The drop in gauge factor is due to a) the grain boundary effects and b) the averaging effect over high and low sensitive orientations.

The theory of piezoresistance has been subjected to several investigations [21,23,25-27]. As polysilicon films are inhomogeneous with grain and grain barrier regions care must be taken when modelling both electrical and piezoresistive properties. The two regions have been modelled respectively as a single crystal and a Schottky barrier (see [28,29] for conduction model).

The barrier piezoresistive effect can be modelled from the contribution to thermionic emission from each minimum or band edge and the subsequent change due to the shift in the bands. The gauge factor equation for a single polysilicon grain considering both grain and barrier is as follows [25]:

$$G = V + \frac{(L1-L2) \; rg \; Pig}{L1 \; r \; Sii} + \frac{L2 \; rb \; Pib}{L1 \; r \; Sii}$$

where V is the volumetric effect, L1 and L2 the grain size and barrier width, r,rg,rb the resistivities of the film, undepleted grain and barrier and Pig and Pib the piezoresistive coefficients of the grain and grain boundary, respectively. The second reducing factor is due to the averaging of high and low sensitivities from different orientations. The significance of this effect is shown by the values, calculated by averaging [25], in TABLE 7.

TABLE 7: averaging of piezoresistive coefficients within 5 planes

		plane					
		(100)	(111)	(311)	(110)	(331)	random
Pil	(n-type)	-68.0	-31.6	-56.9	-40.9	-38.2	-55.8
G1*	(n-type)	-97.2	-53.4	-85.7	-61.4	-59.4	-82.1
Pil	(p-type)	+39.4	+72.8	+49.5	+64.2	+66.7	+53.1
G1*	(p-type)	+61.5	+122.9	+77.7	+111.7	+114.5	+87.5
Pit	(n-type)	+18.5	+30.1	+24.1	+27.4	+28.3	+37.0
Gt*	(n-type)	+24.0	+50.9	+35.0	+43.1	+45.3	+55.5
Pit	(p-type)	-33.8	-23.1	-28.7	-25.6	-24.9	-18.1
Gt*	(p-type)	-53.3	-39.1	-45.9	-44.4	-42.8	-30.3
Y	(1/Sii)	+1.50	+1.71	+1.56	+1.67	+1.68	+1.60
v	(-Sij/Sii)	+0.25	+0.21	+0.23	+0.22	+0.22	+0.27

[* G1, etc. represent the contribution to gauge factor from each orientation (e.g. Pil/Sii) and do not include the volumetric effect]

The electrical and piezoresistive properties of polysilicon are highly dependent upon the structure of the films and thus on the processing. For example, theoretical estimates [25] show that a 50% increase in grain size will yield a 7% increase in gauge factor, whereas a 50% increase in grain barrier height will yield a 9% drop in gauge factor. A maximum gauge factor is usually achieved for p-type material at doping concentrations of the order of 1 X 10*19/cm*3. The maximum values quoted in the literature are of the order of 40 [26]. Higher doping concentrations will yield the same high doping effect found in single crystal. However, sensor design requires a consideration of temperature coefficients and these are shown in TABLE 8 [26].

TABLE 8: temperature coefficients for polysilicon films

| doping | n | | p | |
| concentration | TCR | TCP | TCR | TCP |
(/cm*3)	(%/deg C)	(%/deg C)	(%/deg C)	(%/deg C)
5 X 10*18	-0.6	-0.20	-0.41	-0.22
1 X 10*19	-0.30	-0.20	-0.17	-0.23
3 X 10*19	-0.06	-0.21	+0.03	-0.19
1 X 10*20	+0.12	-0.15	+0.15	-0.16

(Note - due to the nature of polysilicon these figures are merely
examples for a given structure and are not applicable to all
polysilicon films)

In an attempt to improve the performance of polysilicon sensors some
workers have moved to laser annealing [30,31]. This has been used to
achieve an increased grain size and gauge factors of about 50 have
been reached.

E. RECENT DEVELOPMENTS IN THE APPLICATION OF PIEZORESISTANCE

The piezoresistive effects in single crystal and polycrystalline
silicon are well documented in the literature. More recent work has
concentrated on the improvement of performance or alternative
structures.

The main problems with all sensors are those of non-linearity/
hysteresis, temperature coefficients etc.. There are two reasons for the
non-linearity derived from the packaging and the piezoresistive effect
itself [33]. The first is determined by the structure and shape of the
diaphragm [33] and can cause the ballooning effect [34]. The second
effect has been shown to increase with increasing strain [35] and to be
greater for n-type material than for p-type material [36]. Further
problems can develop from the different thermal expansion coefficients
[35] or from the creep of aluminium [37]. Problems of packaging are
reviewed by Poppinger [33]. With the increased use of microprocessor
systems, however, many of these effects can be compensated. The
same is true of temperature coefficients although other techniques are
available. The simplest of these is to choose a doping concentration to
match TCR and TCP [38], a technique used for the current fed Wheatstone
bridge devices. Alternatively, a feed-back system or on-chip
temperature sensor can be used to control the bridge current supply
[39,40].

New structures have included digital or frequency output sensors,
micromechanical structures and four terminal resistors. Digital or
frequency output sensors, for example, can use either active or passive
devices as the sensing element. Both MOS and bipolar transistors are

strain sensitive [41-43]. Alternatively, a piezoresistor can be used to control the frequency of a ring oscillator [44] or the balance in a flip-flop [45]. Along with more sophisticated etching techniques has come a wide range of new sensor possibilities. Micromachined bridges of polysilicon [46] or single crystal [47,48] have been presented. An alternative structure uses the shear piezoresistive effect which was first discussed in detail by Pfann and Thurston [6], showing that a significant voltage perpendicular to current flow can be measured when a shear stress is applied. This technique has since been put to use in sensor applications [49,56]. A new structure which has high potential in the pressure sensor field is the Silicon-on-Sapphire structure [51-55]. The sapphire substrate has good measurement properties and is corrosion and radiation resistant. Sensors have been produced with a pressure range of 10 bar-200 bar with a linearity of < 0.5% FSO and a hysteresis of < 0.05% FSO [52,53]. Furthermore a frequency response of 140 kHz was measured and radiation resistance of up to 1 meV electron beam at 5 X 10^{14}/cm^2 [53]. Measurements of the piezoresistive coefficients within an inversion layer of a SOS-MOS device showed a marked difference from that of standard devices [54]. For example the ratio $Pi11/Pi12$ was found to be unity as opposed to 2 as found for silicon on silicon devices. This is thought to be due to the stresses at the silicon/sapphire interface (measured at 6-13 X 10^9 dynes/cm^2) and stresses at the free silicon surface (measured at 5.5-7.5 X 10^9 dynes/cm^2) [55]. The use of an isolator as the insulating layer avoids the use of a pn junction isolation and thus a wider operating temperature range is possible (up to 425 deg C [52]). These results show silicon on sapphire has great promise for pressure sensors which have to operate in hostile environments.

REFERENCES
----

[1]   J.J.Wortman, R.A.Evans [J. Appl. Phys. (USA) vol.36 (1965) p.153]

[2]   J.W.Dally, W.F.Riley [Experimental Stress Analysis (McGraw-Hill Kogakusha Ltd, 1978)]

[3]   C.S.Smith [Phys. Rev. (USA) vol.94 (1954) p.42]

[4]   E.N.Adams [Phys. Rev. (USA) vol.96 (1954) p.803]

[5]   W.P.Mason, R.N.Thurston [J. Acoust. Soc. Am. (USA) vol.29 (1957) p.1906]

[6]   W.G.Pfann, R.N.Thurston [J. Appl. Phys. (USA) vol.32 (1961) p.2008]

[7]   C.Herring, E.Vogt [Phys. Rev. (USA) vol.101 (1956) p.944]

[8]   R.W.Keyes [Solid State Phys. (USA) vol.11 (1960) p.149]

[9]   O.N.Tufte, E.L.Stelzer [J. Appl. Phys. (USA) vol.34 (1963) p.313]

[10]   Y.Kanda [IEEE Trans. Electron Devices (USA) vol.ED-29 no.1 (Jan 1982) p.64-70]

[11]   O.N.Tufte, E.L.Stelzer [Phys. Rev. (USA) vol.133 (1964) p.A1705]

[12]   W.P.Mason, J.J.Forst, L.M.Tornillo [Instrum. Soc. Am. Conf. Prepr. vol.15 (1960) p.NY-60]

[13]   W.M.Bullis, F.H.Brewer, C.D.Kolstad, L.J.Swartzendruber [Solid-State Electron. (GB) vol.11 no.7 (Jul 1968) p.639-46]

[14]   H.J.McSkimin, W.L.Bond, E.Buckler, G.K.Teal [Phys. Rev. (USA) vol.83 (1951) p.1080]

[15] P.I.Baranskii, V.V.Kolomoets, Yu.A.Okhrimenko [Sov. Phys. Semicond. (USA) vol.20 no.4 (Apr 1986) p.487-8]

[16] M.Sweid, K.Hess, K.Seeger [J. Phys. & Chem. Solids (GB) vol.39 no.4 (1978) p.393-402]

[17] J.A.Chroboczek, F.H.Pollak, H.F.Staunton [Philos. Mag. B (GB) vol.50 (1984) p.113]

[18] T.I.Kamins, M.Mandurah, K.C.Saraswat [J. Electrochem. Soc. (USA) vol.125 no.6 (June 1978) p.927-32]

[19] Y.Onuma, K.Sekiya [Jpn. J. Appl. Phys. (Japan) vol.11 (1972) p.420]

[20] J.Y.W.Seto [J. Appl. Phys. (USA) vol.47 no.11 (1976) p.4780-3]

[21] H.Mikoshiba [Solid-State Electron. (GB) vol.24 (1981) p.221]

[22] W.Germer, W.Todt [Sens. & Actuators (Switzerland) vol.4 (1983) p.183]

[23] J.C.Erskine [IEEE Trans. Electron. Devices vol.ED-30 (1983) p.796]

[24] S.Nishida, M.Konagai, K.Takahashi [Thin Solid Films (Switzerland) vol.112 no.1 (3 Feb 1984) p.7-16]

[25] P.J.French, A.G.R.Evans [Sens. & Actuators (Switzerland) vol.8 no.3 (Nov 1985) p.219-26]

[26] P.J.French, A.G.R.Evans [J. Phys. E (GB) vol.19 (1986) p.1055]

[27] D.Schubert, W.Jenschke, T.Uhlig, F.D.Schmidt [Sens. & Actuators (Switzerland) vol.11 no.2 (Mar 1987) p.145-56]

[28] M.M.Mandurah, K.C.Saraswat, T.I.Kamins [IEEE Trans. Electron Devices (USA) vol.ED-28 no.10 (Oct 1981) p.1163-71]

[29] S.N.Singh, R.Kishore, P.K.Singh [J. Appl. Phys. (USA) vol.57 no.8 pt 1 (15 Apr 1985) p.2793-80]

[30] J.Binder, W.Henning, E.Obermeier, H.Schaber, D.Cutter [Sens. & Actuators (Switzerland) vol.4 (1983) p.527]

[31] E.Luder [Sens. & Actuators (Switzerland) vol.10 nos 1&2 (Sep/Oct 1986) p.9-24]

[32] J.Detry, D.Koneval, S.Blakewell [Proc. Transducers '85 Conf. Philadelphia, PA, USA, 11-14 June 1985 (IEEE, New York, USA, 1985) p.278]

[33] M.Poppinger [Solid State Devices, 1985 Eds P.Balk, O.G.Folberth (Elsevier, Amsterdam, 1986) p.53]

[34] S.Way [Trans. Am. Mech. Eng. vol.APM-56 (1934) p.627]

[35] K.Yamada, M.Nishihara, S.Shimada, M.Tanabe, M.Shimazoe, Y.Matsuoko [IEEE Trans. Electron Devices (USA) vol.ED-29 no.1 (Jan 1982) p.71-7]

[36] J.Bretschi [IEEE Trans. Electron Devices (USA) vol.ED-23 (1976) p.59]

[37] A.K.Sinha, T.T.Sheng [Thin Solid Films (Switzerland) vol.48 no.1 (2 Jan 1978) p.117-26]

[38] J.H.Greenwood [Electron. & Power (GB) vol.29 (1983) p.170]

[39] K.Yamada, M.Hishihara, R.Kanzawa [Sens. & Actuators (Switzerland) vol.4 (1983) p.63]

[40] T.Ishihara, K.Suzuki, S.Suwazono, M.Hirata [Proc. 6th Sensor Symp. (1986) p.17]

[41] W.Sansen, P.Vandeloo, B.Puers [Sens. & Actuators (Switzerland) vol.3 no.4 (Jul 1983) p.343-54]

[42] A.P.Dorey, T.S.Maddern [Solid-State Electron. (GB) vol.12 (1969) p.185]

[43] J.Neumeister, G.Schuster, W.von Munch [Sens. & Actuators (Switzerland) vol.7 no.3 (Jul 1985) p.167-76]

[44] P.J.French, A.P.Dorey [Sens. & Actuators (Switzerland) vol.3 (1983) p.77]

[45] W.Lian, S.Middelhoek [Sens. & Actuators (Switzerland) vol.9
      no.3 (May 1986) p.259-68]
[46] R.T.Howe, R.S.Muller [Sens. & Actuators (Switzerland) vol.4
      (1983) p.447]
[47] W.Benecke, L.Csepregi, A.Heuberger, K.Kuhl, H.Seidl
      [Proc. Transducers '85 Conf., Philadelphia, PA, USA, 11-14 June
      1985 (IEEE, New York, USA, 1985) p.105]
[48] S.Sugiyama, T.Suzuki, K.Kawahata, M.Takigawa, I.Igarashi [Proc.
      6th Sensor Symp. (1986) p.23]
[49] Y.Kanda, A.Yasukawa [Sens. & Actuators (Switzerland) vol.2 no.3
      (Jul 1982) p.283-96]
[50] W.P.Mason [Physical Acoustics and Properties of Solids (D. van
      Nostrand Co., New York, (1955)]
[51] R.Allen [Electronics (USA) vol.52 (1979) p.42]
[52] H.W.Keller, K.Anagnostopoulos [Proc. Transducers '87 Conf.,
      Tokyo, Japan, 2-5 June 1987 (IEE of Japan, Tokyo, Japan, 1987)
      p.316]
[53] C.Qinggui, Z.Jinghuan, T.Zhenglie [Proc. Transducers '87 Conf.,
      Tokyo, Japan, 2-5 June 1987 (IEE of Japan, Tokyo, Japan, 1987)
      p.320]
[54] S.Zaima, Y.Yasudu, S.Kawaguchi, M.Tsuneyoshi, T.Nakamura,
      A.Yoshida [J. Appl. Phys. (USA) vol.60 no.11 (1 Dec 1986)
      p.3959-66]
[55] K.Yamazaki, M.Yamada, K.Yamamoto, K.Abe [Jpn. J. Appl. Phys.
      Part 1 (Japan) vol.23 no.6 (Jun 1984) p.681-6]
[56] M.Bao, Y.Wang [Sens. & Actuators (Switzerland) vol.12 no.1 (Jul
      1987) p.49-56]

CHAPTER 4

CARRIER IONISATION RATES

4. 1   Carrier ionisation coefficients in Si,
            temperature dependence

4. 2   Carrier ionisation coefficients in Si,
            orientation dependence

4. 3   Carrier ionisation coefficients in Si,
            electric field dependence

4.1 CARRIER IONISATION COEFFICIENTS IN Si, TEMPERATURE DEPENDENCE

by J.P.R.David

January 1987
EMIS Datareview RN=15702

Unlike the case of ionisation coefficient variation with electric field in Si there has been relatively little work done on the temperature dependence of a, the electron ionisation coefficient, and b, the hole ionisation coefficient.

Crowell and Sze [1] fitted the experimental electron ionisation coefficient a of Lee et al [2] who were the first to measure a and b on the same junction. The experimental results were obtained from microplasma free junctions and relatively pure electron and hole injection was used ensuring reasonably accurate data. The results were fitted over an electric field range of 2-5 X 10*5 V/cm and temperature range 100-300 deg K and are given in TABLE 1:

TABLE 1

E-field (10*5 V/cm)	a (/cm) - electron ionisation coefficient		
	100 deg K	213 deg K	300 deg K
4	6.80 X 10*4	6.0 X 10*4	5.0 X 10*4
3.33	3.95 X 10*4	3.0 X 10*4	2.2 X 10*4
2.85	1.80 X 10*4	1.2 X 10*4	8.6 X 10*3
2.5	7.60 X 10*3	4.7 X 10*3	2.9 X 10*3
2.22	2.70 X 10*3	1.6 X 10*3	8.0 X 10*2

Schroeder and Haddad [3] refer to a private communication by Udelson and Ward who have measured a at T=200 deg C. For E > 3 X 10*5 V/cm they found that:

a(E) = 1.8 X 10*6 exp (-1.64 X 10*6/E)   /cm

For E < 3 X 10*5 V/cm this expression gave a slight over-estimation of a; and b was estimated at 200 deg C by assuming that K (defined as the a/b ratio) would be the same as at room temperature and hence they down shifted the room temperature data of Lee et al [2] to give:

b(E) = 1 X 10*7 exp (-3.2 X 10*6/E)   /cm

Grant [4] carried out measurements on diffused p+/n mesa diodes using a technique similar to [2] to inject pure carriers. A numerical technique for obtaining a and b without either assuming a simplified field profile or assuming a relationship between a and b was used. The electric field is obtained from the doping profile via C-V measurements and a computer is used to iteratively solve the relationship between a, b and multiplication which is represented in a matrix form. The effect of the threshold energy which gives rise to a dead space in the depletion region in which no ionisation occurs and the effective lowering of the maximum field in heavily doped structures due to this dead space was included. Using a heated sample stage the photo-multiplication was measured over the temperature range 22 - 150 deg C. If the ionisation curves are interpreted in terms of the empirical form:

$$a,b = A \exp{(-B/E)}$$

then A is relatively constant, with the major variation with temperature appearing in the exponent. This is consistent with Crowell and Sze's [1] observation that at high electric fields a and b are less temperature sensitive.

Over the electric field range 2.4 - 5.3 X 10*5 V/cm the temperature dependent ionisation coefficients were found to be:

$$a(E) = 6.2 \; X \; 10*5 \exp{[-(1.05 \; X \; 10*6 + 1.3 \; X \; 10*3 \; T)/E]} \; /cm$$

$$b(E) = 2.0 \; X \; 10*6 \exp{[-(1.95 \; X \; 10*6 + 1.1 \; X \; 10*3 \; T)/E]} \; /cm$$

where T is in degrees centigrade (T > 20 deg C) and E is in volts/cm.

Decker and Dunn [5] used a different technique to obtain the temperature dependence of a and b. They used diffused epitaxial silicon avalanche diodes on which avalanche voltages and broadband microwave small signal admittance measurements were carried out over the range 280 - 590 deg K. An accurate theoretical model of the diode was used to obtain the junction voltage and admittance characteristics. The temperature dependent theory of impact ionisation developed by Crowell and Sze [1] was modified by taking into account a more realistic ionisation cross-section. Subsequently values of ionisation coefficients at an electric field of 4 X 10*5 V/cm were determined at each temperature by a numerical minimisation routine to obtain the best fit between the calculated and measured data. Their ionisation coefficient parameters for the expression:

$$a(E) = A \exp{(-B/E)}$$

are given in TABLE 2:

TABLE 2

T deg K	Electrons		Holes	
	A(10*6/cm)	B(10*6 V/cm)	A(10*6/cm)	B(10*6 V/cm)
280	1.89	1.66	1.08	1.57
317	2.05	1.72	1.00	1.48
363	2.83	1.86	1.75	1.73
385	2.60	1.84	1.94	1.75
395	2.86	1.88	1.51	1.70
434	3.72	2.00	1.87	1.82
437	2.12	1.80	2.62	2.02
482	4.05	2.07	2.65	1.97
544	4.26	2.15	2.26	1.99
586	3.93	2.12	3.68	2.25

More recently Rang [6] has determined more accurate impact-ionisation
coefficients and their temperature dependence over a wide range of
electric field strengths using experimental data from the literature.
Little information is given on how this is done other than to state
that a numerical method was used on Chynoweth's formula [7]. The
results for a and b over the range 5 X 10*4 - 1 X 10*6 V/cm have the
form:

$$a(E) = 1.286 \times 10^6 \; [\; 1 + 4.95 \times 10^{-4} \; (T-300) \;]$$

$$\exp [\; -1.4 \times 10^6 \; (\; 1 + 6.43 \times 10^{-4} \; (T-300)/E \;) \;] \; /cm$$

$$b(E) = 1.438 \times 10^6 \; [\; 1 + 5.05 \times 10^{-4} \; (T-300) \;]$$

$$\exp [\; -2.02 \times 10^6 \; (\; 1 + 6.21 \times 10^{-4} \; (T-300)/E \;] \; /cm$$

where T is the temperature in Kelvin and E is the electric field in V/cm.

REFERENCES
----

[1] C.R.Crowell, S.M.Sze [Appl. Phys. Lett. (USA) vol.9 no.6 (Sept
     1966) p.242-4]
[2] C.A.Lee, R.A.Logan, R.L.Bardorf, J.J.Kleimack, W.Wiegmann [Phys.
     Rev. (USA) vol.134 no.3A (May 1964) p.761-73]
[3] W.E.Schroeder, G.I.Haddad [Proc. IEEE (USA) vol.61 no.2 (Feb
     1973) p.153-82]
[4] W.N.Grant [Solid-State Electron. (GB) vol.16 no.10 (Oct 1973)
     p.1189-203]
[5] D.R.Decker, C.N.Dunn [J. Electron. Mater. (USA) vol.4 no.3
     (Jun 1975) p.527-47]
[6] T.Rang [Radioelectron. & Commun. Syst. (USA) vol.28 no.5 (1985)
     p.91-3]
[7] A.G.Chynoweth [Phys. Rev. (USA) vol.109 no.5 (Mar 1958)
     p.1537-40]

4.2 CARRIER IONISATION COEFFICIENTS IN Si, ORIENTATION DEPENDENCE

by J.P.R.David

January 1987
EMIS Datareview RN=15703

The early workers on silicon did not pay much attention to the
orientation dependence of a, the electron ionisation coefficient and
b, the hole ionisation coefficient, sometimes neglecting to mention
the orientation of the layers on which the measurements were made.

Chynoweth and McKay [1] measured the multiplication threshold energies
on low resistivity diffused junctions in (100), (110) and (111) planes
and found that there was a slight suggestion that the (111) direction
possessed the highest threshold energy and the (100) direction the
lowest. Since the ionisation process depends critically on threshold
energies, this would imply that the (100) direction would have the
largest ionisation coefficients with the (111) direction having the
lowest.

Lee et al [2] carried out measurements on (111) orientation layers and
were the first to measure a and b on the same junction. They ensured
that the junctions were free of microplasmas and used relatively pure
electron and hole injection to give accurate results. Their results
covered an electric field range of 2 - 4 X 10*5 V/cm for a and slightly
less for b as shown in TABLE 1.

TABLE 1

E-field (10*5 V/cm)	a (/cm)	b (/cm)
3.63	3.0 X 10*4	2.9 X 10*3
3.08	1.5 X 10*4	6.0 X 10*2
2.66	6.0 X 10*3	1.2 X 10*2
2.22	2.0 X 10*3	-
2.0	6.5 X 10*2	-
1.82	2.1 X 10*2	-

Grant [3] carried out measurements on (100) orientation p+/n mesa
diodes. A numerical technique for obtaining a and b without either
assuming a simplified field profile or assuming a relationship between
a and b was used. The electric field is obtained from the doping profile
via C-V measurements and a computer is used to iteratively solve the
relationship between a, b and multiplication which is represented in a
matrix form. The effect of the threshold energy which gives rise to a

dead space in the depletion region in which no ionisation occurs and the effective lowering of the maximum field in heavily doped structures due to this dead space was included. The results obtained were:

Electrons ($2.4 \times 10^5 < E < 3 \times 10^5$ V/cm)

$a(E) = 6.2 \times 10^5 \exp (-1.08 \times 10^6/E)$ /cm

Holes ($2 \times 10^5 < E < 5.3 \times 10^5$ V/cm)

$b(E) = 2 \times 10^6 \exp (-1.97 \times 10^6/E)$ /cm

For $E < 2.4 \times 10^5$ V/cm a decreases more rapidly with inverse field and is given by:

$a(E) = 2.6 \times 10^6 \exp (-1.43 \times 10^6/E)$ /cm

For $E > 5.3 \times 10^5$ V/cm a and b tend to saturate and can be approximated by:

$a(E) = 5 \times 10^5 \exp (-9.9 \times 10^5/E)$ /cm

$b(E) = 5.6 \times 10^5 \exp (-1.32 \times 10^6/E)$ /cm

These results show some difference from those of Lee et al [2]. However, it would be wrong to attribute this solely to orientation differences as different experimental techniques were used which could easily account for the differences.

Lee [4] was the first to make a comparison of ionization coefficient measurements on (100) and (110) orientation samples while using the previously measured results [2] for information on the (111) orientation. The measurement technique used was similar to that used previously [2] with slight differences in the case of all three orientations and this may account for the considerable orientation dependence observed. The highest ionisation coefficient and smallest K (defined as a/b) was found in the (100) direction while the multiplication was slightly less and K slightly larger in the (110) direction. In the (111) direction [2], a was slightly lower than a (110) but b was much lower than either b (100) or b (110) giving it the largest K value.

The electric field range considered was 2 - 4 X 10*5 V/cm and the results for the two orientations are given in TABLE 2:

TABLE 2

	(100)	
E-field (10*5 V/cm)	a (/cm)	b (/cm)
3.67	2.4 X 10*4	1.1 X 10*4
3.3	2.2 X 10*4	9.6 X 10*3
2.86	1.7 X 10*4	7.0 X 10*3
2.5	1.2 X 10*4	5.0 X 10*3
2.22	8.6 X 10*3	3.9 X 10*3
2.05	7.0 X 10*3	3.1 X 10*3

	(110)	
E-field (10*5 V/cm)	a (/cm)	b (/cm)
3.67	2.1 X 10*4	8.6 X 10*3
3.33	1.8 X 10*4	7.0 X 10*3
2.86	1.0 X 10*4	3.9 X 10*3
2.5	6.0 X 10*3	2.2 X 10*3
2.2	3.5 X 10*3	1.3 X 10*3
2.05	2.35 X 10*3	8.2 X 10*2

Due to several possible sources of inaccuracies in the experiment and the fact that the (111) direction was not remeasured, there is some doubt as to the correctness of this result.

Recently Robbins et al [5] have examined a and b in (100) and (111) orientations with an independent check made by noise measurements. Reach through APD's were used and great care was taken to produce pure carriers by shining different wavelength light on each side of the diode. Baseline corrections to the photocurrent curve were made to get the correct primary photocurrent and Grant's numerical technique [3] was used to calculate a and b. No difference in a and b was found in the two orientations. K varied from 70 at 2.4 X 10*5 V/cm to 22 at 2.9 X 10*5 V/cm. They also obtained good agreement between calculated and

measured excess noise factors for electron injection. Measured excess
noise factors for hole injection were larger than would be expected and
this was attributed to inaccuracies in field determination. Breakdown
voltages measured showed no anisotropy which supports the case for no
orientation dependence of a and b. Their graphical data is given in
TABLE 3 over the field range 1.5 - 3 X 10*5 V/cm.

TABLE 3

E-field (10*5 V/cm)	a (/cm)	b (/cm)
2.86	1.7 X 10*4	7.5 X 10*2
2.63	1.3 X 10*4	3.0 X 10*2
2.38	6.5 X 10*3	1.2 X 10*2
2.17	3.0 X 10*3	-
1.85	5.7 X 10*2	-
1.60	1.2 X 10*2	-

REFERENCES
----

[1] A.E.Chynoweth, K.G.McKay [Phys. Rev. (USA) vol.108 no.1
     (Oct 1957) p.29-34]
[2] C.A.Lee, R.A.Logan, R.L.Bardorf, J.J.Kleimack, W.Wiegmann [Phys.
     Rev. (USA) vol.134 no.3A (May 1964) p.761-73]
[3] W.N.Grant [Solid-State Electron. (GB) vol.16 no.10 (Oct 1973) p.1189]
     p.1189-203]
[4] C.A.Lee [RADC Technical report, RADC-TR-93-182 (1983)]
[5] V.M.Robbins, T.Wang, K.F.Brennan, K.Hess, G.E.Stillman [J. Appl.
     Phys. (USA) vol.58 no.12 (15 Dec 1985) p.4614-17]

4.3 CARRIER IONISATION COEFFICIENTS IN Si, ELECTRIC FIELD DEPENDENCE

by J.P.R.David

July 1986
EMIS Datareview RN=15700

The study of ionisation coefficients in Si began in the early 1950's but not until Miller's [1] work in 1957 was there a proper study of the effects of separate hole and electron injection across the reverse biassed junction enabling a, the electron ionisation coefficient, and b, the hole ionisation coefficient to be obtained. Electron and hole multiplication data were obtained from complementary p+ n and n+ p Si step junctions respectively. By fitting empirical relations to the multiplication versus bias measurements, the analytical expressions for multiplication when a and b are different were solved. The electric field range covered was 3.5-6.5 X 10*5 V/cm and the graphical results are tabulated below:-

E-field (10*5 V/cm)	3.5	4.5	5.5	6.5
a(E) (/cm)	1.0 X 10*4	3.3 X 10*4	1.0 X 10*5	2.3 X 10*5
b(E) (/cm)	4.0 X 10*3	1.3 X 10*4	2.8 X 10*4	4.4 X 10*4

Clearly a and b are not equal and a is greater than b over the entire field range.

Chynoweth [2] used a more refined method for measuring multiplication versus bias and determined a and b over a wider electric field range. Again complementary junctions were used and for each junction a = b was assumed in working out a and b, the only error being a slight decrease in a at high fields. The field range covered was 2.5-6 X 10*5 V/cm for holes and 2-4 X 10*5 V/cm for electrons and the results are tabulated below:-

E-field (10*5 V/cm)	3.64	3.08	2.66	2.35	2.10
a(E) (/cm)	1.8 X 10*4	8.9 X 10*3	4.2 X 10*3	1.8 X 10*3	8.0 X 10*2

E-field (10*5 V/cm)	5.70	4.44	3.64	3.08	2.66
b(E) (/cm)	6.3 X 10*4	2.0 X 10*4	7.1 X 10*3	2.5 X 10*3	7.4 X 10*2

The trend in a agrees with Miller's data [1] but b here is higher.

Moll and Van Overstraeten [3] again used complementary p-n junctions to obtain their data. The field profile was obtained by C-V profiling which removes any errors due to approximations being used. A constant K (defined as the ratio of a/b) was assumed at all electric fields and the effects of increased collection efficiency due to junction widening were also included for the first time. a is found to be almost an extension of Chynoweth's data [2] and over the field range 2-6.7 X 10*5 V/cm is given by:-

a(E) = 1.6 X 10*6 exp(-1.65 X 10*6/E) /cm

b over the field range 5-8 X 10*5 V/cm is given by:-

b(E) = 5.5 X 10*5 exp(-1.65 X 10*6/E) /cm

The discrepancy in b between this and [2] is attributed to the fact that no correction for electrons in [2] was made in working out b (and vice-versa) but since K is greater than 1 the error would be larger in b than a. The b here agrees well with [1].

Lee et al [4] found that existing a and b could not be fitted to Baraff's [5] theory with reasonable parameters. They were the first to measure a and b on the same junction and ensured that junctions were free of any microplasmas. By shining penetrating and non-penetrating light on the device pure injection of electrons and holes on either side of the junction could be ensured. A new and simplified approach was used to obtain a and b from the multiplication data. The electric field was derived from C-V measurements and the range covered was approximately 2-4 X 10*5 V/cm.

E-field (10*5 V/cm)	3.63	3.08	2.66	2.22	2	1.82
a(E) (/cm)	3.0 X 10*4	1.5 X 10*4	6.0 X 10*3	2.0 X 10*3	6.5 X 10*2	2.1 X 10*2
b(E) (/cm)	2.9 X 10*3	6.0 X 10*2	1.2 X 10*2	-	-	-

a shows a steeper slope with inverse field than previous results and is slightly larger. Also b is an order of magnitude smaller at E less than 3 X 10*5 V/cm than previous results giving K approximately equal to 50. These results are thought to be due to the pure hole injection now possible.

Baertsch [6] carried out measurements over a similar field range as Lee et al using a similar technique. The electron data agrees well, but the hole results are larger than theirs by a factor of 2-10 depending on field. The tabulated form of b is:-

E-field (10*5 V/cm)	3.33	2.86	2.5	2.2
b(E) (/cm)	2.8 X 10*3	9 X 10*2	3 X 10*2	1 X 10*2

Noise measurements were also carried out on these devices and the noise due to hole and electron primary photocurrents agreed well with theoretical predictions for K is approximately 5-10. This gives a degree of confirmation to the ionisation measurements.

Van Overstraeten and De Man [7] also measured a and b on the same diffused p-n Si junctions. Single carrier type injection was achieved by a technique similar to that of Lee et al [4]. A non-constant K with electric field was used and the effect of threshold energy on charge multiplication was also taken into account. Their results are:-

Electrons   (E ranging from 1.75 X 10*5 to 6 X 10*5 V/cm)

a(E) = 7.03 X 10*5 exp(-1.231 X 10*6/E) /cm

Holes   (E ranging from 1.75 X 10*5 to 4 X 10*5 V/cm)

b(E) = 1.582 X 10*6 exp(-2.036 X 10*6/E) /cm

Holes   (E ranging from 4 X 10*5 to 6 X 10*5 V/cm)

b(E) = 6.71 X 10*5 exp(-1.693 X 10*6/E) /cm

Breakdown voltages computed from these data gave good agreement with experimental data. The differences between these results and those of Moll and Van Overstraeten [3] are thought to be due to the neglect of the electron threshold energy. This is also thought to account for the discrepancies with the data of Lee et al [4] though numerical inaccuracies might also be to blame as their a and b are incompatible with their multiplication data.

Grant [8] carried out measurements on p+ -n diffused mesa diodes using a similar technique to [4] to inject pure carriers. A numerical technique for obtaining a and b without either assuming a simplified field profile or assuming a relationship between a and b was used. The electric field is obtained from the doping profile via C-V measurements and a computer is used to iteratively solve the relationship between a, b and multiplication which is represented in a matrix form. The effect of the threshold energy which gives rise to a dead space in the depletion region in which no ionisation occurs and the effective lowering of the maximum field in heavily doped structures due to this dead space was included. The results are:-

Electrons   (E ranging from 2.4 X 10*5 to 3 X 10*5 V/cm)

a(E) = 6.2 X 10*5 exp(-1.08 X 10*6/E) /cm

Holes   (E ranging from 2 X 10*5 to 5.3 X 10*5 V/cm)

b(E) = 2 X 10*6 exp (-1.97 X 10*6/E) /cm

For E greater than 5.3 X 10*5 V/cm a and b tend to saturate and can be approximated by:-

a(E) =   5 X 10*5 exp(-9.9 X 10*5/E) /cm

b(E) = 5.6 X 10*5 exp(-1.32 X 10*6/E) /cm

For E less than 2.4 X 10*5 V/cm a decreases more rapidly with inverse field and is given by:-

a(E) = 2.6 X 10*6 exp(-1.43 X 10*6/E) /cm

These results agree more closely with Van Overstraeten and De Man [7] than those of Lee et al [4] though they are about 1.4 times larger. Breakdown voltages were obtained by solving the diode continuity equations and Poisson's equations and this independent check on the accuracy of a and b gave good agreement with experimental results.

Ogawa [9] measured a and b over a lower field range (E less than 2.5 X 10*5 V/cm) than had been considered before by using high voltage p-i-n junctions. Correction for collector efficiency was allowed for when the depletion region extended into the heavily doped regions and devices were screened for anomalous photoresponse. The results in a tabulated form are:-

E-field (10*5 V/cm)	1.66	1.43	1.25	1.11
a(E) (/cm)	2 X 10*2	50	10	2.2
b(E) (/cm)	2	0.6	-	-

a appears to follow on from Chynoweth's data [2] quite well but b was obtained over a limited field range only and appears to be lower with a smaller slope than would be expected.

Kuzmin et al [10] also carried out measurements over the lower field range with APD's and using a He-Ne laser to create electrons and holes. Over the field range 0.9-3.5 X 10*5 V/cm considered the results are given by:-

a(E) = 7.4 X 10*5 exp(-1.16 X 10*6/E) /cm

b(E) = 7.25 X 10*5 exp(-2.20 X 10*6/E) /cm

a agrees with an extrapolation of Van Overstraeten and De Man's data [7] to lower fields and so does b within the limits of experimental error. However a is greater than Ogawa's [9] by a factor of 3-5 though there is closer agreement in b. Kuzmin's data also shows that K is exponentially dependent on E and was 200 at 2 X 10*5 V/cm.

Lee [11] used linearly graded junctions formed by diffusion and the standard technique of Lee et al [4] was used for pure carrier injection across a thinned wafer. [100] and [110] orientation crystals were used and significant differences were noted in a and b. Data for the [111] orientation were taken from a previous publication [4]. Highest multiplication and smallest K were found in the [100] orientation. The field range considered was 2-4 X 10*5 V/cm and the results for the two orientations are tabulated below:-

[100]

E-field (10*5 V/cm)	3.67	3.3	2.86	2.5	2.22	2.05
a(E) (/cm)	2.4 X 10*4	2.2 X 10*4	1.7 X 10*4	1.2 X 10*4	8.6 X 10*3	7.0 X 10*3
b(E) (/cm)	1.1 X 10*4	9.6 X 10*3	7.0 X 10*3	5.0 X 10*3	3.9 X 10*3	3.1 X 10*3

[110]

E-field (10*5 V/cm)	3.67	3.3	2.86	2.5	2.2	2.05
a(E) (/cm)	2.1 X 10*4	1.8 X 10*4	1.0 X 10*4	6.0 X 10*3	3.5 X 10*3	2.35 X 10*3
b(E) (/cm)	8.6 X 10*3	7.0 X 10*3	3.9 X 10*3	2.2 X 10*3	1.3 X 10*3	8.20 X 10*2

Recently Robbins et al [12] have examined a and b in [100] and [111] orientations with an independent check made by noise measurements. Reach through APD's were used and great care was taken to produce pure carriers by shining different wavelength light on each side of the diode. Baseline corrections to the photocurrent curve were made to get the correct primary photocurrent and Grant's numerical technique [8] was used to calculate a and b. No difference in a and b was found in the two orientations. K varied from 70 at 2.4 X 10*5 V/cm to 22 at 2.9 X 10*5 V/cm. They also obtained good agreement between calculated and measured excess noise factors for electron injection. Measured excess noise factors for hole injection was larger than would be expected and this was attributed to inaccuracies in field determination. Breakdown voltages measured showed no anisotropy which supports the case for no orientation dependence of a and b. Their graphical data is tabulated below over the field range 1.5-3 X 10*5 V/cm.

E-field (10*5 V/cm)	2.86	2.63	2.38	2.17	1.85	1.6
a(E) (/cm)	1.7 X 10*4	1.3 X 10*4	6.5 X 10*3	3 X 10*3	5.7 X 10*2	1.2 X 10*2
b(E) (/cm)	7.5 X 10*2	3.0 X 10*2	1.2 X 10*2	-	-	-

REFERENCES
----

[1]   S.L.Miller [Phys. Rev. (USA) vol.105 no.4 (Feb 1957) p.1246-9]
[2]   A.G.Chynoweth [Phys. Rev. (USA) vol.109 no.5 (Mar 1958) p.1537-40]
[3]   J.L.Moll, R.van Overstraeten [Solid-State Electron. (GB) vol.6 (1963) p.147-57]
[4]   C.A.Lee, R.A.Logan, R.L.Bardorf, J.J.Kleimack, W.Wiegmann [Phys. Rev. (USA) vol.134 no.3A (May 1964) p.A761-73]
[5]   G.A.Baraff [Phys. Rev. (USA) vol.128 (1962) p.2507]
[6]   R.D.Baertsch [IEEE Trans. Electron Devices. (USA) vol.ED-13 (1966) p.987]
[7]   R.van Overstraeten, H.de Man [Solid-State Electron. (GB) vol.13 (1970) p.583-608]
[8]   W.N.Grant [Solid-State Electron. (GB) vol.16 no.10 (Oct 1973) p.1189-203]
[9]   T.Ogawa [Jpn. J. Appl. Phys. (Japan) vol.4 no.7 (Jul 1965) p.473-84]
[10]   V.A.Kuz'min, N.N.Kryukova, A.S.Kyuregyan, T.T.Mnatsakanov, V.B.Shuman [Sov. Phys-Semicond. (USA) vol.9 no.4 (1975) p.481-2]
[11]   C.A.Lee [RADC Technical report, RADC-TR-93-182 (1983)]
[12]   V.M.Robbins, T.Wang, K.F.Brennan, K.Hess, G.E.Stillman [J. Appl. Phys. (USA) vol.58 no.12 (15 Dec 1985) p.4614-7]

CHAPTER 5

ELECTRON MOBILITY, DIFFUSION AND LIFETIME

5.1 ELECTRON MOBILITY IN BULK Si

by P.J.Mole

June 1987
EMIS Datareview RN=15739

A. INTRODUCTION

In this review the mobility of majority carrier electrons in bulk
silicon is considered over a temperature range of 200 to 400 deg K,
where the models presented for the mobility are claimed to be accurate
to about 5%. This temperature range includes that over which silicon
devices are used. The mobility outside this range is also given in
some of the references cited. In addition, the mobility of electrons
when they are minority carriers (typical of high injection
conditions) and their mobility at high electric fields are considered
in separate reviews in this book.

B. INTRINSIC MOBILITY, TEMPERATURE DEPENDENCE

The intrinsic mobility of silicon is limited by scattering of the
electrons by lattice vibrations (phonons). This gives rise to a
mobility which falls rapidly as the temperature increases (as this
increases the density of phonons present). Historically, the model
for the mobility is based on experimental measurement of the Hall
mobility [1-2]; subsequently [3-4], the measurements have been
corrected for the Hall coefficient and the model presented here is
for the conductivity mobility. The simplest model for the intrinsic
mobility is:

$$mu(L) = mu(o) (T/To)^{-alpha}$$

In [5,6], the parameters for the model are calculated to be
mu(o) = 1430-1448 cm*2/V.s at To = 300 deg K and alpha = 2.20-2.33.

The simplicity of this model belies the complexity of the phonon
scattering processes which contribute to the carrier scattering. In
effect there will be contributions from acoustic and optic phonons via
inter and intra valley scattering processes. This added complexity is
rarely necessary for the analysis of silicon devices; however, details
can be found in [7,8].

C. THE EFFECT OF LATTICE STRAIN ON MOBILITY

The effect of strain on mobility is covered in more detail in the section on piezoresistance of silicon. However its effect on the mobility is notable in devices where the silicon is processed or packaged in a manner which introduces strain, for example hetero-epitaxy as is commonly used in silicon-on-sapphire devices. This is elegantly illustrated in [9] where the Hall mobility in a silicon-on-sapphire film is measured as a function of the lattice stress (determined by Raman spectroscopy). A compressive stress of 5 X 10*9 dyne/cm*2 reduces the mobility by about 30%. This figure can only be used as a guide since the mobility in silicon-on-sapphire is also reduced by film defects and so cannot be compared directly to intrinsic silicon.

D. THE EFFECT OF ELECTRICALLY ACTIVE IMPURITIES ON MOBILITY

The deliberate introduction of electrically active impurities into silicon allows the control of silicon devices. Over the temperature range of interest, it will be assumed that these impurities are fully active and therefore neutral impurity scattering will be ignored [10]. Donors (usually phosphorus or arsenic) give up a valence electron and introduce a fixed positive electronic charge density, Nd (/cm*3) into the silicon. Acceptors (usually boron) accept an electron and introduce a fixed negative electronic charge density, Na (/cm*3), into the silicon. These charge centres will introduce coulombic scattering which in the presence of a high mobile carrier density will be reduced by screening.

In [11] the following model for the doping dependence of mobility at 300 deg K was introduced:

$$
mu = mu(min) + \frac{mu(o) - mu(min)}{1 + [(Nd + Na)/Nref]*alpha}
$$

[12] gives mu(min) = 92 cm*2/V.s, mu(o) = 1360 cm*2/V.s, alpha = 0.91 Nref = 1.3 X 10*17/cm*3 with a mean square error of 6%.

These values, however, do not allow simple prediction of mobility over a range of temperatures. In [13] an attempt is made to introduce a temperature dependence of the mobility parameters above. However an alternative approach is to derive an expression based on the Brooks-Herring formula for impurity scattering as in [5,6]. Firstly a contribution to the scattering due to impurity scattering is introduced:

$$
mu(I) = \frac{AT*3/2}{Nscat} [\ln (1 + \frac{BT*2}{Ncar}) - \frac{BT*2}{Ncar + BT*2}] *-1
$$

The parameters are given by A = 4.61 X 10*17/cm/V/s/K*3/2

B = 1.52 X 10*15/cm*3/K*2

and the distinction introduced in [6] between the number of scattering
sites introduced by the impurities (Nscat = Nd+Na) and the number of
free carriers available for screening (Ncar = Nd-Na) is made.

The contribution to the mobility from impurity scattering can then be
combined with that from phonons using the approach of [14] and the
approximation in [5]:

mu = mu(L) f(x)

where f(x) can be approximated by:-

$$\frac{1.025}{[~~1 + (x/1.68)*1.43~~]} - 0.025$$

and x = SQRT [6 mu(L)/mu(I)]

At high doping levels (> 10*19/cm*3) the above approaches still cannot
account for all the effects of impurities. The presence of such a high
concentration of impurity introduces several effects. Not all the
electrically active impurity atoms surrender their electrons to the
conduction band. This gives rise to a reduced number of charged impurity
scattering sites and correspondingly increases the number of neutral
scattering centres [10]. This is most significant at the lower
temperatures. The conduction band and the band of states generated by
the impurities merge complicating the band structure and leading to the
dependence of the scattering on donor type. The electron gas will become
degenerate and the form of the scattering will change accordingly e.g.
[17]. At high concentrations the impurities precipitate becoming
electrically inactive and provide additional scattering centres; the
degree of precipitation will depend on the dopant type and the
processing history [see section 13 of reference [17]]. In practice
there is no tendency for the mobility to reach a minimum at high
concentrations and a distinct dependence on dopant type is reported
[15]. This can be modelled at 300 deg K by [15]:

$$mu ~=~ mu(o) + \frac{mu(max) - mu(o)}{1+(p/Cr)*alpha} - \frac{mu(1)}{1+(Cs/p)*beta}$$

p = carrier concentration

TABLE 1: the parameters calculated for the data measured on both arsenic
and phosphorus doped samples [15]

Parameter	Phosphorus	Arsenic	Units
mu(o)	68.5	52.2	cm*2/V.s
mu(max)	1414	1417	cm*2/V.s
mu(1)	56.1	43.4	cm*2/V.s
Cr	9.20 X 10*16	9.68 X 10*16	/cm*3
Cs	3.41 X 10*20	3.42 X 10*20	/cm*3
alpha	0.711	0.680	
beta	1.98	2.00	

E. THE EFFECT OF MATERIAL QUALITY ON MOBILITY

There is no detailed report of variation of mobility with material
preparation. It therefore seems reasonable to assume that current day
crystal growth techniques and epitaxial layer growth techniques do not
introduce significant densities of carrier scattering defects. However
current hetero-epitaxy techniques e.g. silicon-on-sapphire [9] and
other new 'silicon-on-alternative substrate' growth techniques cannot
be guaranteed to reduce the carrier scattering defects to such levels.
Thus the application of all the data presented in this review to
silicon films with abnormally high defect levels must be made with care.

F. A MORE ACCURATE DETERMINATION OF MOBILITY

For certain applications a more accurate knowledge of the mobility is
required over a wider temperature range. The models described in [16]
are extensions to the models above which give a greater accuracy except
at high doping levels where the corrections discussed in [15] are not
included. The reader is referred directly to them.

REFERENCES
----

[1]   G.L.Pearson, J.Bardeen [Phys. Rev. (USA) vol.75 (1948) p.865]
[2]   F.J.Morin, J.P.Maita [Phys. Rev. (USA) vol.96 (1954) p.28]
[3]   D.Long [Phys. Rev. (USA) vol.120 (1960) p.2024]
[4]   P.Norton, T.Bragginns, H.Levinstein [Phys. Rev. B (USA) vol.8
      (1973) p.5632]
[5]   J.M.Dorkel, Ph.Letureq [Solid-State Electron. (GB) vol.24
      (1981) p.821]

[6]   N.D.Arora, J.R.Hauser, D.J.Roulston [IEEE Trans. Electron
        Devices (USA) vol.ED-29 (1982) p.292]
[7]   K.Seeger [Semiconductor Physics (Springer, 1973) ISBN
        3-211-81186-9]
[8]   C.T.Sah, P.C.H.Chan, C.K.Wang, R.L.Sah, K.A.Yamakawa, R.Lutwack
        [IEEE Trans. Electron Devices (USA) vol.ED-28 (1981) p.304]
[9]   Y.Kobayashi, M.Nakamura, T.Suzuki [Appl. Phys. Lett. (USA)
        vol.40 no.12 (15 Jun 1982) p.1040-2]
[10]   S.S.Li, W.R.Thurber [Solid-State Electron. (GB) vol.20 no.7
        (Jul 1977) p.609-16]
[11]   D.M.Caughey, R.E.Thomas [Proc. IEEE (USA) vol.55 (1967) p.2192]
[12]   G.Baccarani, P.Ostoja [Solid-State Electron. (GB) vol.18 (1975)
        p.579]
[13]   S.Selberherr [Analysis and Simulation of Semiconductor Devices
        (Springer 1984) ISBN 3-211-81800-6]
[14]   P.P.Debye, E.M.Conwell [Phys. Rev. (USA) vol.93 (1954) p.693]
[15]   G.Masetti, M.Severi, S. Solmi [IEEE Trans. Electron Devices
        (USA) vol.ED-30 no.7 (Jul 1983) p.764-9]
[16]   R.D.Larrabee, W.R.Thurber, W.M.Bullis [NBS Special publ.
        400-63 'A FORTRAN program for calculation the electrical
        parameters of extrinsic silicon' (1980)]
[17]   C.Herring, E.Vogt [Phys. Rev. (USA) vol.101 (1986) p.944]

5.2   ELECTRON MOBILITY AT A (100) Si SURFACE

by P.J.Mole

June 1987
EMIS Datareview RN=15740

A. INTRODUCTION

The silicon MOSFET is now one of the most important devices made in
silicon because of its central role in VLSI technology. Conduction
occurs in an inversion layer most commonly at a (100) surface. The
degree of inversion is controlled both by the doping within the device
and the electric field in the insulator above the surface. The
knowledge of the factors influencing this surface mobility (as
distinct from the mobility in the bulk of the material) including the
effect of the controlling electric field, is therefore of considerable
importance. In this review values are reported for the effective
carrier mobility as opposed to the field effect mobility or the Hall
mobility, the distinction between these terms is explained in [1].

B. THE INFLUENCE OF THE CONTROLLING FIELD

The electric field which induces the inversion layer is due to the gate
voltage; increasing this electric field forces the electrons closer to
the interface and decreases the width of the potential well in which
the carriers reside. The classic theory of influence of normal field on
mobility [2], relies on the increased scattering as the electrons are
confined closer to a rough or heavily charged surface. In [3] it was
shown that if the mobility is considered as a function of effective
confining electric field (E(eff) = (QB+Qn/2)/e where QB is the charge
density in the surface depletion layer, Qn is the electron charge
density in the inversion layer and e is the dielectric constant of
silicon), then the dependence of mobility on substrate bias, surface
doping and gate voltage can be expressed by one 'universal function'.
The data in [1,4-6] are expressed in a variety of ways by the authors;
but the data can be modelled by the expression:

$$mu = \frac{mu(o)}{1 + theta\ E(eff)}$$

where at 300 deg K mu(o) = 640-880 cm*2/V.s and theta = 1.15-1.08 X
10*-6 cm/V.

The relatively broad range of values observed in the literature (the
discrepancies being largest at low effective fields) probably reflects
the variation in surface quality between manufacturers. This view is
supported by the influence of oxide growth conditions on the mobility
as reported in [1]. The measured data are taken over a range of fields
up to about 5 X 10*5 V/cm. However it should be noted that when the
electric field drops below that level required for strong inversion,
the model no longer applies [7].

There have been several different models proposed for this normal
field dependence. In [11] the classic theory in [2] is modified by
acknowledging the carrier density to be governed by Fermi-Dirac
statistics (rather than Boltzmann statistics) at the carrier
densities typical of strongly inverted MOSFET's. This also offers an
explanation for the temperature dependence discussed below. Alternative
theories based on strong quantisation of the inversion layer have been
proposed e.g. [12] which also explain the form of the electric field
variation. There is not yet one accepted explanation for the observed
behaviour.

C. TEMPERATURE VARIATION

The variation of mobility with temperature is dominated by the
scattering of the electrons by phonons. As for bulk silicon this
variation is most simply modelled by a variation of the form:

$$mu(L) = mu(o) \ (T/To)^{-alpha}$$

For To=300 deg K, mu(o) is given in the previous section; the value
for alpha is considerably less than the bulk value and [5] report a
value close to 1.5 over the temperature range 200-400 deg K.

D. THE EFFECT OF CHARGED IMPURITIES

Charged impurities at the interface or due to bulk doping can affect
the mobility. The latter affect the mobility through their influence
on the surface field (see above) as well as through coulombic
scattering which is not well documented, but in [8] a mobility model
is introduced which draws, without justification, on the impurity
scattering parameters from bulk silicon. However the scattering of
carriers by surface charges has been examined in [9] and it is shown
that the scattering by surface charges depends on the interface charge
density and the inversion layer charge density. No model is given but
for an inversion carrier density of 2 X 10*12/cm*2 and a surface
electronic charge density of 10*11/cm*2 the mobility can be given by:

$$\frac{1}{mu} = \frac{1}{mu(o)} + \frac{1}{mu(cl)}$$

where mu(o) is the mobility in the absence of interface charge and
mu(cl) is the interface contribution (6 X 10*3 cm*2/V.s). This
contribution varies inversely with surface charge.

E. THE EFFECT OF HIGH DRIFT ELECTRIC FIELDS

Under the influence of high drift electric fields (E), the electrons
become heated and as a result their mobility is affected as new
scattering mechanisms become available to them. The effect has been
measured by careful interpretation of the drain current in a short
channel MOSFET [5,10] and by time of flight measurements [4]. The
references cited in this review use a model for the mobility of the
form:

$$mu = \frac{mu(o)}{[\ 1 + (E\ mu(o)/v(sat))*alpha\ ]*(1/alpha)}$$

The data are summarised below

ref	[4]	[5]	[10]	
alpha	1.92	2.9	2	
v(sat)	9.23	7.4 10*6	about 8.5	10*6 cm/s

There is some debate amongst the authors as to whether the maximum
velocity v(sat) should be dependent on the controlling field; the
close agreement between the two recent measurements [4,10] lends
weight to their values.

REFERENCES
----

[1]   S.C.Sun, J.D.Plummer [IEEE Trans. Electron. Devices (USA)
        vol.ED-27 (1980) p.1497]
[2]   J.R.Schrieffer [Phys. Rev. (USA) vol.97 (1955) p.641]
[3]   A.G.Sabnis, J.T.Clemens [IEEE International Electron Devices
        Meeting Tech. Digest (IEEE,1979) p.18]
[4]   J.A.Cooper, D.F.Nelson [J. Appl. Phys. (USA) vol.54 (1983)
        p.1445]
[5]   R.W.Coen, R.S.Muller [Solid-State Electron. (GB) vol.23 no.1
        (Jan 1980) p.35-40]
[6]   S.J.Murray, P.J.Mole [Proc. Simulation of semiconductor devices
        and processes (Pineridge Press 1986) ISBN 0-906674-59-x]
[7]   J.T.C.Chen, R.S.Muller [J. Appl. Phys. (USA) vol.45 no.2
        (Feb 1974) p.828-34]
[8]   K.Yamaguchi [IEEE Trans. Electron Devices (USA) vol.ED-30
        (1983) p.658]
[9]   S.Manzini [J. Appl. Phys. (USA) vol.57 no.2 (15 Jan 1985)
        p.411-14]
[10]   W.Mueller, I.Eisele [Solid-State Commun. (USA) vol.34 (1980)
        p.447]
[11]   G.Baccarani, Z.Mazzone, C.Morandi [Solid State Electron. (GB)
        vol.17 (1974) p.785]
[12]   A.Rothwarf [IEEE Electron Device Lett. (USA) vol.EDL-8 (1987)
        p.499]

5.3 ELECTRON MOBILITY IN 2D n-TYPE Si: THEORETICAL ASPECTS

by J.M.Rorison

October 1987
EMIS Datareview RN=16173

A. INTRODUCTION

When silicon is fabricated into devices boundaries are introduced:
either edges marking the extent of the homogeneous silicon structure
or interfaces marking the boundaries between different component
materials of a silicon heterostructure. These boundaries produce
additional physics which becomes increasingly important in understanding
electron transport as the device scale decreases. These effects can be
classified into 2 types: i) those arising from the new physics, i.e. new
scattering processes, sub-band structures etc.; and ii) those arising out
of non-equilibrium effects in small structures.

B. NEW PHYSICS IN THE n-INVERSION LAYER OF SiO2/Si

B1 Sub-band Structure

Most devices of current interest which exhibit 2D behaviour are those
involving heterostructures, the most well studied being the n-inversion
layer formed at the SiO2/Si interface in MOS structures. The well known
behaviour of the MOSFET is discussed by Sze [1]. The electrons in the
n-channel reside in a confining potential perpendicular to the
interface (z), created by space-charge effects, which can be solved
with a solution of the Schrodinger wave equation coupled with Poisson's
equation. The solution of these equations [2] results in a sub-band
structure for the energy levels of the form

$$E = E_n + \frac{h^2}{8\pi^2 m^*} [k(x)^2 + k(y)^2] \qquad (1)$$

where $k(x), k(y)$ are wavevector components for motion parallel to the
surface, m^* is the electron effective mass and E_n are the electronic
quantum levels arising from confinement in the narrow potential well.
These types of systems are said to have 'dynamical' 2D character [2].
The band structure of silicon is also affected by confinement with the
6-fold degeneracy of the conduction band being lifted at the <100> and
<110> silicon surfaces. On the <100> surface the 4 valleys (for which
the effective mass for motion perpendicular to the surface is the light
mass m_t^*) are expected to have a higher E_n than the 2 valleys whose
long axis (corresponding to the heavy effective mass m_l^*) is
perpendicular to the surface. It is possible to define a density of
states effective mass $(m_d)^* = [m(x) m(y)]^{*(1/2)}$ and a conductivity or
optical mass

$$m(op)^* = \frac{2}{1/m(x) + 1/m(y)}$$

These values are given by Ando et al [2] for the different orientations
of the silicon surface and for the different subband series (for
silicon <100> the lowest subband series has $m(op)^* = m(d)^* = 0.19$). If
the electrons are only in the first subband (low n, low temperature
(T) and large DELTA En) they behave 2 dimensionally while if there is
a significant population in several subbands (high n, high T or small
DELTA En) the behaviour of the electrons will approach that of the bulk
system.

B2 Mobility in the n-Channel

The simplest measurable quantity is the sheet conductivity defined as

$$sigma = e\ n\ mu = gd\ L/W \qquad\qquad (2)$$

where e is the electronic charge, n is the areal free charge density,
mu is the mobility, gd is the source-drain conductance, L is the length
of the MOSFET structure and W its width (the mobility mu is defined as
$mu = e<tau>/mt^*$ where tau is the scattering rate or mean time between
collisions).

Experimental measurements of sigma are difficult because of non-ohmic
effects, contact resistances comparable with channel resistance,
leakage currents or poorly defined channel lengths. If the measurement
can be made the next step is to determine n and mu independently. The n
is usually estimated from the gate voltage dependence of the capacitance
and the conductance. Several mobilities are used to characterise
MOSFETS:

i) the effective mobility [2]

$$mu(eff) = \frac{sigma}{Cox\ (Vg-Vt)} \qquad\qquad (3)$$

where sigma is the small signal conductivity, Cox is the capacitance
per unit across the oxide, Vg is the gate voltage and Vt is the
threshold gate voltage. It can be related to the mobility mu according
to $mu(eff) = f\ mu$ where f is the fraction of 'free' electrons.

ii) the field effect mobility [2]

$$mu(FE) = \frac{gm}{Cox\ Vd} = \frac{dId/dVg}{Cox\ Vd} \qquad\qquad (4)$$

where gm is the transconductance, Id is the drain current and Vd the
drain voltage. dId/dVg in this case refers to the partial differential
of Id with respect to Vg at constant Vd.

iii) the Hall mobility [2]

$$mu(H) = r\ mu = R(H)\ c\ sigma \qquad\qquad (5)$$

where r is the Hall ratio, R(H) the Hall constant and c is the velocity
of light.

The experimental and technical difficulties in measuring these different
mobilities and their inter-relationships are discussed in detail in Ando
et al [2]. The first comprehensive transport experiments were carried
out in 1968 by Fang and Fowler [3] and since then many experimental
groups have made detailed studies [2]. This large mass of data on
transport in n-Si inversion layers is summarised by Ando et al [2]. In
general at room temperature mu is less than the bulk mobility mu(B) due
to increased scattering. At low temperatures it is higher partly because
the electrons are in the lowest subband so that the effective mass
parallel to the surface is light and partly because they are separated
from their compensating positive charge in the gate which reduces
Coulomb scattering.

B3 Modelling the Mobility

Unfortunately mu is dependent upon many parameters which are very sample
dependent such as the thickness of the oxide, the number of charges per
unit area Nox and the concentration of fixed charges in the depletion
layer N(depl) in addition to the dependence upon n, T and the source
drain field Ed.

The important scattering processes determining mu are: coulomb
scattering (oxide charge), interface roughness, optical and acoustic
phonon for the intrasubband and intersubband-intravalley scattering.

At low temperatures coulomb scattering and interface roughness
scattering are the most important processes which shows up in the
experimental dependence of mu on n and Nox [4]. Coulomb scattering
involves scattering of the electrons from charged centres in the oxide
and the silicon. It is often called oxide charge scattering due to the
dominance of charges in the oxide. As n increases the scattering length
decreases and the electron-electron screening, epsilon(ee), increases,
reducing the number of contributing scattering centres to those
located very near the interface. Therefore it decreases with increasing
n and increases with increasing Nox. Hartstein et al [4] have
extensively studied the behaviour of mu limited by coulomb scattering.
A comparison of their experiments with theoretical models [5] is given
in Ando et al [2]. Surface roughness scattering from interface roughness
is important at high n. The scattering of electrons from roughness in
quantum subbands was first theoretically studied by Prange and Nee [6]
and has been updated for the inversion layer problem most recently by
Ando [7]. It is found to be dependent on [n + 2N(depl)]*2 and
epsilon(ee). A comparison of theory [7] and experiment [5] is given in
[2].

These two scattering mechanisms are sufficient to explain qualitatively
the mu dependence at low T. The mu increases first, peaks to a maximum
value at n of the order of 10*12/cm*2 and then decreases with
increasing n. (The values of mu at low n are very dependent upon
Nox). A simple empirical formula to describe this behaviour can be
written as

$$\frac{1}{mu(eff)} = C + a\ Nox + b\ Nox*2 \qquad (6)$$

where the first term is related to surface roughness, the second to
single coulomb scattering and the third to multiple coulomb scattering
[2].

At higher T phonon effects become important and at room T they dominate. As T is raised intravalley acoustic phonon scattering becomes important involving phonons with low energy which is almost elastic. As T is raised still higher intervalley scattering can be induced by the emission or absorption of high energy, high momentum phonons, either optical or acoustic phonons. In addition as T is raised the electron population moves into higher subbands making analysis difficult. Experimentally it is found that mu is flattish for n < 2 X 10*11/cm*2 with mu > 1000 cm*2/V.s for good samples [2]. The mu then decreases with increasing n and for 10*12 < n < 5 X 10*12/cm*2 mu decreases as approximately [n + N(depl)]*-1/3. The experimental T dependence of mu is proportional to T*-1.5. Using a 2D model for the electrons and a 3D phonon model to model intravalley acoustic phonon scattering Kawaji [8] found mu is proportional to n*(-1/3)/T. The calculated mu was found to be much larger than the experimental mu. Inclusion of interface modes was found to have little effect [9] while inclusion of intersubband scattering was found to reduce mu substantially [10]. While this improved the agreement between theory and experiment for the T dependence it removed the dependence of mu on n which is still not understood. Intervalley phonon scattering is important in inversion layers at high T, as in the bulk. Unfortunately its analysis depends on the details of the subband structure and so this is mainly qualitative.

The dependence of mu on the source drain field Ed is discussed in reviews by Ando et al [2], Hess [12] and Ferry [13]. It can be viewed as heating of the electrons as they travel down the channel leading to hot electron effects. It has been used extensively to study intervalley phonon scattering. If mu is T dependent then it is possible to measure a change in conductance with Ed and relate it to a change of electron temperature (DELTA T = approx. 0.3 (Ed)*(3/2)) [14]. If mu is only weakly T dependent it is possible to use the T dependence of the magneto-oscillations as a thermometer [14]. When the conductance is measured as a function of Vd at constant T (lattice) for low Ed the data can be fitted to an expression similar to that used for the bulk mobility in the 'warm field' regime.

$$mu(Ed) = mu(0) [1 + beta (Ed)*2)] \qquad\qquad (7)$$

where beta is determined by acoustic and elastic scattering [15]. At high fields even at 300 deg K, most of the electron energy is transferred to the lattice through optical phonons resulting in velocity saturation, as in the bulk [14]. The saturation velocity is approximately 6 X 10*6 cm/s for the Si<100> surface at 300 deg K [14], slightly lower than the bulk value of about 10*7 cm/s [16]. Observations of negative differential resistance have been reported and studied recently [2].

C. NON-EQUILIBRIUM EFFECTS

In bulk materials the spatial and temporal variation of an electric field E is weak so that at each point a steady state of equilibrium is attained. This arises from an equilibrium between perturbations due to E and scattering mechanisms. For large scale devices bulk treatments (i.e. Boltzmann transport equation) are valid. For medium scale devices with active lengths < 0.5 micron bulk models can be applied with interface scattering and quantum effects included in the inversion layers (i.e. SiO2/Si interfaces in MOSFETs). For small devices with active lengths < 0.1 micron the Boltzmann transport equation is no longer valid [7]. In addition effective mass models and band structure approximations are starting to break down. Also because of the very high fields usually found in this type of structure collisions cannot be assumed to occur instantaneously in space and time (intra-collisional field effect [18]). Therefore a quantum theory of transport is needed. This is currently an active area of work.

Even for the medium scale devices new phenomena arise such as ballistic motion and overshoot velocity. Overshoot velocity can occur due to fast temporal variation of the applied field producing a drift velocity Vd of 2.5 X 10*7 cm/s for n-Si at 293 deg K [19] which is much larger than the equilibrium value. Similar effects can arise from short spatial scales. These effects are currently being studied for understanding of the basic physics and exploitation in device fabrication.

REFERENCES
----

[1]   S.M.Sze [Physics of Semiconductor Devices (Wiley, New York, 2nd Ed. 1981)]
[2]   T.Ando, A.B.Fowler, F.Stern [Rev. Mod. Phys. (USA) vol.54 no.2 (Apr 1982) p.437-672]
[3]   F.F.Fang, A.B.Fowler [Phys. Rev. (USA) vol.169 (1968) p.619]
[4]   A.Hartstein, T.H.Ning, A.B.Fowler [Surf. Sci. (Netherlands) vol.58 (1976) p.178]; A.Hartstein, A.B.Fowler, M.Albert [Surf. Sci. (Netherlands) vol.98 (1980) p.181]
[5]   F.Stern, W.E.Howard [Phys. Rev. (USA) vol.163 (1967) p.816]; F.Stern [Surf. Sci. (Netherlands) vol.73 (1978) p.193]
[6]   R.E.Prange, T.W.Nee [Phys. Rev. (USA) vol.168 (1968) p.779]
[7]   T.Ando [J. Phys. Soc. Jpn. (Japan) vol.43 (1977) p.1616]
[8]   S.Kawaji [J. Phys. Soc. Jpn. (Japan) vol.27 (1969) p.906]
[9]   H.Ezawa, S.Kawaji, K.Nakamura [Jpn. J. Appl. Phys. (Japan) vol.13 no.1 (Jan 1974) p.126-55 and vol.14 (1975) p.921-2 (E)]
[10]   H.Ezawa [Surf. Sci. (Netherlands) vol.58 no.1 (Aug 1976) p.25-32]
[11]   D.K.Ferry [Surf. Sci. (Netherlands) vol.57 (1976) p.218]; D.Roychoudhury, P.K.Basu [Solid-State Electron. (GB) vol.20 (1977) p.1023]
[12]   K.Hess [Solid-State Electron. (GB) vol.21 no.1 (Jan 1978) p.123-32]
[13]   D.K.Ferry [Solid-State Electron. (GB) vol.21 no.1 (Jan 1978) p.115-21]
[14]   F.F.Fang, A.B.Fowler [J. Appl. Phys. (USA) vol.41 (1970) p.1825]
[15]   K.Hess, C.T.Sah [Phys. Rev. B (USA) vol.10 (1974) p.3375]
[16]   C.Canali, G.Manni, R.Minder, G.Ottaviani [IEEE Trans. Electron Devices (USA) vol.ED-22 (1975) p.1045]
[17]   E.Constant [Inst. Phys. Conf. Ser. (GB) no.57 (1981) p.141-68]
[18]   J.R.Barker, D.K.Ferry [Phys. Rev. Lett. (USA) vol.42 (1979) p.1779]
[19]   J.G.Ruch [IEEE Trans. Electron Devices (USA) vol.ED-19 (1972) p.652]; J.P.Nougier, J.C.Vaissiere, D.Gasquet, J.Zimmerman, E.Constant [J. Appl. Phys. (USA) vol.52 no.2 (Feb 1981) p.825-32]

5.4   ELECTRON MOBILITY IN n-TYPE BULK Si, ELECTRIC FIELD DEPENDENCE

by J.M.Rorison

October 1987
EMIS Datareview RN=16136

A. INTRODUCTION: THE BASIC MODEL

For extrinsic semiconductors the conduction is determined by the
acceleration of the excess carriers, in the case of n-type materials
electrons, and the scattering processes which arrest their motion. For
n less than about 10*18/cm*3 at 300 deg K, Maxwell-Boltzmann (non-
degenerate) statistics are appropriate to describe the electrons'
behaviour, while for higher carrier densities Fermi-Dirac (degenerate)
statistics are needed. This reflects the transition from semiconductor
to metallic-like behaviour as the doping level is increased. In the
absence of an external field, either an electric field E or a
concentration gradient Vn, the electrons move with a random thermal
velocity v(th) which depends on their thermal energy epsilon(th)
according to:-

$$epsilon(th) = \frac{m* \; [v(th)]*2}{2} = \frac{3}{2} \, kT \qquad (1)$$

where m* is the electron effective mass of the conduction band, k is
Boltzmann's constant and T is the temperature. The effect of
introducing an electric field E is to distort this velocity
distribution, imparting a drift velocity vd to this 'gas' of electrons.
For linear response, valid at low field, the mobility mu of the
electrons is defined as

$$mu = vd/E \qquad (2)$$

The mobility mu can be related to the scattering processes through the
relation:-

$$mu = \frac{e(tau)'}{m*} = \frac{el'}{m*v'} \qquad (3)$$

where e is the electron charge, tau' and l' are the mean time and
length between collisions respectively and v' is the effective velocity
including field effects (heating). Eqn (2) is written in the most
general form where mu is in general a tensor but for cubic

semiconductors it is scalar in the linear regime and is usually
generalised to be a scalar outside this regime as well. This approach
leads to a current defined for the electrons of:-

$$je = e \, n \, mu \, E \qquad (4)$$

where n is the carrier concentration.

The response of electrons to a concentration gradient is defined [1] in
terms of a diffusion coefficient D as

$$jD = -e \, D \, Vn \qquad (5)$$

where D is a scalar at low E field.

At equilibrium mu and D are connected by the Einstein relation:-

$$D = \frac{mu \; kT}{e} \qquad (6)$$

where T is the thermal temperature.

The equation for the total current j is the sum of jD and je which is
known as the drift diffusion equation and is valid for low electric
field with mu and D taken as scalars. In general both mu and D are
field dependent tensors outside the low field region but we will only
focus on the electric field dependence of mu and vd.

A1 General Results

The band structure of the material can be seen to enter into the
mobility in m* using effective mass theory which is suitable for low
field. The band structure details will also be important in the
evaluation of the individual scattering processes. Certain general
features have been found for covalent semiconductors independent of
the details of the material. These are [1]:

   (i)   The mobility goes down as the electric field increases.

   (ii)   The average energy <epsilon> increases slowly at low field
          and more quickly as the field increases.

   (iii) D increases slowly with increasing field.

These results are linked to the general results that for covalent
semiconductors the scattering mechanisms become more effective as the
electron energy increases.

The resulting behaviour for drift velocity and mobility for silicon and n-silicon at room temperature as a function of field is [1]:

     (i)   low field: vd is linear and mu is a constant mu(o) (ohmic behaviour);

    (ii)   warm field (E < 10*3 V/cm): vd is sub-linear due to acoustic and intervalley scattering and mu is well described by the expression mu(E) = mu(o) [1 + beta E*2] where beta is termed the warm electron coefficient;

   (iii)   intermediate field (10*3 V/cm < E): vd is increasingly sub-linear due to the increasing role of longitudinal optical phonons resulting in a decreasing of mu(E);

    (vi)   high field (E > 10*4 V/cm): vd saturates to vs through the dominance of longitudinal optical phonon scattering

          vs = approx. SQRT [8 Eo/3 pi m*]

         where Eo is the energy of the longitudinal optical phonon (hwo/2 pi);

     (v)   very high field (E > 10*5 V/cm): is characterised by rapid carrier heating and breakdown.

This gradual increase to velocity saturation for silicon is in sharp contrast to gallium arsenide where intervalley effects produce negative differential mobility (NDM). In addition the lower m* of gallium arsenide results in a higher velocity for low field than in silicon. For high n-type silicon vd is always lower for the same electric field compared with pure silicon. It saturates at a higher field and to a lower vs which decreases with n as shown in Jacoboni et al [2].

B. MODELLING ELECTRON TRANSPORT

Simple empirical or semi-empirical treatments for electron velocity or mobility as a function of E, T and n in bulk semiconductors are useful for device simulation. For n-type silicon simple modelling of mu(E) by parametrising or analysing beta [3-8] are reasonably successful.

A very commonly used empirical expression for vd as a function of E and n for T = 300 deg K is the Scharfetter and Gummel expression [9]

$$vd = \frac{mu(o)\ E}{\left[\ 1\ +\ \dfrac{N}{N(ref)/S\ +\ N(ref)}\ +\ \dfrac{(E/A)*2}{(E/A\ +\ F)}\ +\ (\ \dfrac{E}{B}\ )*2\ \right]*0.5}$$

where mu(o) = 1400 cm*2/V.s, N(ref) = 3 X 10*16/cm*3, S = 350, A = 3.5 X 10*3 V/cm, F = 8.8 and B = 7.4 X 10*3 V/cm.

To attain more understanding of the physics of electron transport it is usual to go to a semi-classical approach in which the electron is treated classically within a Boltzmann equation formalism while the scattering events are treated quantum mechanically. This approach only yields analytical results usually employing a heated Maxwellian distribution [1] for very simple models. Numerical methods, chiefly Monte Carlo simulation techniques [1,2,10,11] have been used to derive results. This approach simulates the path of an electron through many scattering events in a probabilistic approach and results in a mean energy <epsilon> and mean velocity <vd>. It is also possible to derive a distribution function from this approach.

B1 Scattering Processes

Detailed understanding of the scattering processes involved are needed for this approach. Scattering can occur from phonon deformation of the lattice (deformation potential interaction) or from the electrostatic field set up by these phonons (piezoelectric interaction). It is important to distinguish the difference between these two types of scattering as the latter type will be screened very efficiently by electrons for n-type systems reducing their effect. The screening due to electrons denoted as epsilon(e-e)(q,w) is both wavevector (q) and frequency (w) dependent [12] although a static approximation w=0 or a long wavelength approach q=0 is often made. The role of electrons in screening scattering processes and also in relaxing the distribution function towards a Maxwellian form is an area under current investigation [13].

The scattering processes are derived independently of the electric field and are usually divided into intravalley and intervalley scattering. The important scattering processes for n-silicon are: intravalley acoustic deformation potential scattering, intervalley phonon scattering, ionised impurity scattering*, neutral impurity scattering* and electron-electron scattering*. (Those marked with * will be screened efficiently by the electrons.) These processes are dealt with in detail in [1] in terms of the physics and derivation of the scattering matrix elements. Values of these parameters are given in [2]. Good agreement of theory and experiment can be derived through analysis of the details of the scattering processes when the anisotropy and anharmonicity of the band structure of silicon is included [2].

At room temperature and low electric field the acoustic deformation potential and intervalley phonon scattering are the most important scattering mechanisms. While acoustic phonon scattering is well understood the identification of the type of phonons involved in intervalley phonon scattering has been the subject of much work [14,15] and is still unresolved.

At low temperatures and low fields ionised impurity scattering becomes an important scattering mechanism and reduces mu. Impurity scattering can be treated within the Brooks-Herring or Conwell-Weiskoff formalisms which have been reconciled by Ridley [16]. The effect of n on mu as a function of T for low T is given in Jacoboni et al [2]. At room temperature if n > 10*16/cm*3 impurity scattering again becomes significant [2]. The temperature dependence of the mobility at low T (T < 50 deg K) is due to acoustic phonons while at higher T (T > 50 deg K) intervalley phonon scattering becomes important. A room

temperature mu(T) behaviour of T*-2.42 is found empirically which
deviates from the T*-1.50 behaviour expected for purely acoustic phonon
scattering [2].

For higher fields there is a deviation from Ohm's law which occurs at
lower fields as T decreases which makes vd(E) become sub-linear. In
addition for higher fields (E is > 10 V/cm at 8 deg K, E > 10*2 V/cm
at 77 deg K, E > 2 X 10*3 V/cm at 300 deg K [2]) vd exhibits anisotropy
behaviour with respect to the orientation of the electric field on the
crystal. The anisotropy is due to a repopulation of the valleys [17].
When E is parallel to <111> the six valleys are equally orientated with
respect to E but for E parallel to <100> two valleys exhibit ml* in the
direction of the field while the remaining four exhibit mt* where mt*
< ml*. Electrons in transverse valleys therefore have a higher mobility and
are heated to a greater extent by the field and transfer electrons to the
colder and slower valleys. The net result is a lower vd for E parallel to
<100>. The effect becomes stronger as the temperature falls since
relaxation effects are less effective at low T. For T < 45 deg K negative
differential mobility (NDM) occurs for 20 V/cm < E < 60 V/cm. For the high
field limit the anisotropy disappears. Experimental results of vd(E,T) are
given in [2]. The role of optical phonon (deformation potential)
scattering becomes important as the field increases and results in the
saturation of the velocity of the electrons. Detailed theories show
that the other scattering processes also contribute to vs and that
band non-parabolicity is also important. For T < 45 deg K vs = 1.3-1.4
X 10*7 cm/s independent of T [2]. For higher T vs(T) decreases steadily
as T increases and at room temperature vs is approx. 1 X 10*7 cm/s [2].
In general the mobility of electrons in bulk silicon is well understood
as a function of electric field, temperature and impurity content.

C. EXPERIMENTAL COMPARISONS

Experimental measurements of vd and D result from conductivity
measurements, microwave techniques and time of flight techniques [1,2].
Conductivity methods have been used since the 1950's and are based on
measurement of the current density which flows through a sample of
extrinsic semiconductor when the current is as in Eqn (3). By knowing
the carrier concentration (Hall effect) vd can be determined.
Microwave techniques can be applied in two ways; either the electrons
are heated by a pulsed E field and their mobility is determined by
measurement of the attenuation of a low microwave field or the charge
carriers are heated up by a large pulsed microwave field and their
mobility is determined with the aid of a low DC electric field. The
former method is useful in the determination of the energy relaxation
time. Time of flight techniques involve measurement of the drift
velocity which is a variation of the Hayes and Shockley experiment. The
measurement of vs can be achieved by analysis of one carrier injection
current under space charge limited conditions at high field. A detailed
comparison of experimental results for vd as a function of E for electrons
at 77 deg K and 300 deg K using the different experimental techniques
is given in Jacoboni et al [2].

REFERENCES

----

[1]   C.Jacoboni, L.Reggiani [Adv. Phys. (GB) vol.28 no.4 (1979)
      p.493-553]
[2]   C.Jacoboni, C.Canali, G.Ottaviani, A.Alberigi Quaranta
      [Solid-State Electron. (GB) vol.20 (1977) p.77]
[3]   C.Canali, G.Majni, R.Minder, G.Ottaviani [Electron. Lett. (GB)
      vol.10 no.24 (1974) p.523-4]
[4]   C.Canali, G.Majni, R.Minder, G.Ottaviani [IEEE Trans. Electron
      Devices (USA) vol.ED-22 no.11 (1975) p.1045-47]
[5]   G.Bosman, R.J.J.Zijlstra, F.Nava [Solid-State Electron. (GB)
      vol.24 no.1 (Jan 1981) p.5-9]
[6]   N.Ahmad, V.K.Arora [IEEE Trans. Electron Devices (USA) vol.ED-33
      no.7 (Jul 1986) p.1075-7]
[7]   S.A.Schwarz, S.E.Russek [IEEE Trans. Electron Devices (USA)
      vol.ED-30 no.12 (Dec 1983) p.1629-33]
[8]   S.A.Schwarz, S.E.Russek [IEEE Trans. Electron Devices (USA)
      vol.ED-30 no.12 (Dec 1983) p.1634-9]
[9]   D.L.Scharfetter, H.K.Gummel [IEEE Trans. Electron Devices (USA)
      vol.ED-16 (1969) p.64]
[10] P.J.Price [Semicond. & Semimet. (USA) vol.14 'Lasers,
      Junctions, Transport' (Academic Press, New York, 1979) ch.4
      p.249-308]
[11] C.Jacoboni, L.Reggiani [Rev. Mod. Phys. (USA) vol.55 (1983)
      p.645]
[12] J.M.Rorison, D.C.Herbert [J. Phys. C (GB) vol.19 no.21 (30 Jul
      1986) p.3991-4010]
[13] P.Lugli, D.K.Ferry [Physica B&C (Netherlands) vol.129 no.1-3
      (Mar 1985) p.532-6]
[14] M.H.Jorgensen, N.O.Gram, N.I.Meyer [Solid State Commun. (USA)
      vol.10 (1972) p.337]
[15] J.G.Nash, J.W.Holm-Kennedy [Appl. Phys. Lett. (USA) vol.24 no.3
      (1974) p.139-41]
[16] B.K.Ridley [J. Phys. C (GB) vol.10 (1977) p.1589]
[17] W.Sasaki, M.Shibuya, K.Mizougchi, G.M.Hatoyama [J. Phys. &
      Chem. Solids (GB) vol.8 (1959) p.250]

5.5   ELECTRON MOBILITY IN n-TYPE EPITAXIAL Si

by J.A. del Alamo

July 1987
EMIS Datareview RN=15731

Electron mobilities in epitaxially grown n-type Si have to be compared
with mobilities measured in bulk doped as-grown Si [1]. These bulk
mobilities are, within experimental error, identical to those obtained
in ion-implanted or diffused layers up to solid solubility values [2].
A small dependence on the dopant species is observed for doping levels
higher than about 1 X 10*19/cm*3.

A word of caution is required when comparing reported mobility data.
Mobility is usually measured using the Hall technique. The result of
this measurement is the Hall mobility which is related to the
conductivity mobility (the parameter of most interest in device
applications) through the Hall factor. Systematic measurements of the
Hall factor have been carried out for phosphorus-doped Si [3]. In spite
of it not actually being the case [3], most authors assume a Hall factor
of unity or, in other words, they assume that the Hall and conductivity
mobilities are identical. This review will therefore address Hall
mobilities measured in epitaxial layers grown by various techniques.

a)   Vapour Phase Epitaxy (VPE). Systematic measurements of electron
     mobility in P-doped Si grown in a H2/SiH4/PH3 gas system at
     1050 deg C have been reported in the doping range of 2.5 X 10*17 to
     1.6 X 10*20/cm*3 [4,5]. At all doping levels electron mobilities
     within 4% of the Si bulk values were obtained.

b)   Low Pressure Vapour Phase Epitaxy (LPVPE). LPVPE-growth of P-doped
     Si using the H2/SiCl2H2/PH3 gas system in the temperature range
     785-933 deg C have been reported [6,7]. The maximum doping level is
     5 X 10*18/cm*3. Mobilities identical to bulk Si values have been
     measured down to 823 deg C. Unintentionally doped n-type layers in
     the 1 X 10*16/cm*3 level show mobilities about 90% of the bulk
     values.

c)   Molecular Beam Epitaxy (MBE). MBE growth of P-, As-, and Sb-doped
     Si has been extensively studied. In P-doped Si, doping levels
     up to 1.5 X 10*19/cm*3 have been obtained [8]. Within the
     experimental error, the mobility is identical to that of bulk Si
     at 300 deg K and 77 deg K. In As-doped Si, mobility values have
     been reported for doping levels up to 8 X 10*19/cm*3 [9]. The best
     values in [9] and those of other authors at lower doping levels
     [8] are essentially identical to bulk values. Sb-doped Si layers
     have been grown with doping levels up to 1 X 10*20/cm*3 [10]. Again
     identical results to bulk Si were obtained [8,10,11].

REFERENCES

[1]   W.R.Thurber, R.L.Mattis, Y.M.Liu, J.J.Filliben [J. Electrochem.
       Soc. (USA) vol.127 no.8 (Aug 1980) p.1807-12]
[2]   G.Masetti, M.Severi, S.Solmi [IEEE Trans. Electron Devices
       (USA) vol.ED-30 no.7 (July 1983) p.764-9]
[3]   J.A.del Alamo, R.M.Swanson [J. Appl. Phys. (USA) vol.57 no.6
       (15 Mar 1985) p.2314-7]
[4]   J.A.del Alamo, R.M.Swanson [J. Appl. Phys. (USA) vol.56 no.8
       (15 Oct 1984) p.2250-2]; Erratum in [J. Appl. Phys. (USA)
       vol.57 no.6 (15 Mar 1985) p.2346]
[5]   J.A.del Alamo, R.M.Swanson [J. Electrochem. Soc. (USA) vol.132
       no.12 (Dec 1985) p.3011-6]
[6]   L.Vescan, H.Beneking, O.Meyer [Mater. Res. Soc. Symp. Proc. (USA)
       vol.71 (1986) p.133-8]
[7]   L.Vescan, U.Breuer, Ch.Werres, H.Beneking [Proc. Int. Symp. on
       Trends and New Applic. in Thin Films, Strasbourg, France,
       10-13 Mar 1987 p.217-21]
[8]   R.A.A.Kubiak, S.M.Newstead, W.Y.Leong, R.Houghton, E.H.C.Parker,
       T.E.Whall [Appl. Phys. A (Germany) vol.42 no.3 (Mar 1987)
       p.197-200]
[9]   T.Sugiyama, T.Itoh [J. Electrochem. Soc. (USA) vol.133 no.3
       (March 1986) p.604-9]
[10]   Y.Shiraki [Prog. Cryst. Growth & Charact. (GB) vol.12 nos 1-4
       (1986) p.45-66]
[11]   R.A.A.Kubiak, W.Y.Leong, E.H.C.Parker [J. Electrochem. Soc.
       (USA) vol.132 no.11 (Nov 1985) p.2738-42]

5.6 ELECTRON MOBILITY IN Si, PRESSURE DEPENDENCE

by D.Lancefield

November 1987
EMIS Datareview RN=16195

The mobility of electrons in silicon is found to increase with
increasing hydrostatic pressure at room temperature. For a sample with
n = 10*16/cm*3; mu = 1000 cm*2/V.s, d(ln mu)/dP = 2.5 X 10*-3/kbar [1].
This is in contrast to Ge and the direct gap III-V semiconductors where
the mobility generally decreases with increasing pressure because of
the dominant effect of the pressure dependence of the electron effective
mass [2].

From a consideration of the band structure of silicon and its pressure
dependence the pressure dependence of the density of states effective
mass is predicted to be small, i.e. d(ln m*)/dP up to 3 X 10*-4/kbar,
compared to Ge where it is predicted to be almost an order of magnitude
larger. Consequently at room temperature where intervalley phonon
scattering is dominant it is the pressure induced increase in the
phonon energies, causing a reduction in the phonon population, that
causes the mobility to increase.

REFERENCES
----

[1] D.Lancefield [PhD Thesis, 1985, University of Surrey, Guildford,
     Surrey, UK]
[2] D.Lancefield, A.R.Adams, M.A.Fisher [J. Appl. Phys. (USA) vol.62
     (1987) p.2342]

5.7 MINORITY CARRIER ELECTRON MOBILITY IN p-TYPE Si

by J.A. del Alamo

August 1987
EMIS Datareview RN=16111

Measurements of the minority carrier electron mobility, mu(e), in
p-type Si are still very scarce in comparison with its value as a
majority carrier, MU(e), in n-type Si.

In high purity p-type Si, several measurements [1-4] indicate that
mu(e) is independent of the acceptor concentration, Na, and is
essentially identical to MU(e) [5]. As the acceptor concentration
increases above about 1 X 10*16/cm*3, mu(e) and MU(e) start to
decrease. Up to a doping level of about 5 X 10*17/cm*3, mu(e) and
MU(e) coincide within the scattering of the available experimental
data [4,6,7]. From that doping level on, mu(e) decreases less strongly
than MU(e) and in the highest doping regime [8] saturates to a value
about 2.5 times larger than the saturated MU(e) [5].

A fit to all the data was given in [8]:

$$mu(e) = 232 \; + \; \frac{1180}{1 + [Na/(8 \; X \; 10*16)]*0.9}$$

in cm*2/V.s, with Na in cm*-3.

Limited temperature dependence data have been reported in the high
purity regime [1]. In the temperature range from 77 deg K to 300 deg K
mu(e) appears to behave as T*-1.8.

Recently, time-of-flight measurements of the mobility-electric field
relationship for minority carrier electrons have been performed [7,9].
The results indicate that mu(e) abruptly decreases with the electric
field to about half of its zero field value in the range from 0 to 200
V/cm. This is attributed to the drag effect of electron-hole scattering
in a drift field [7,9]. At higher electric fields (E > 200 V/cm), the
mobility decreases in a manner reminiscent of the usual hot-electron
effects, i.e., approximately as E*-0.5 [7,9].

A recent report indicates that in the proximity of a n+/p junction,
mu(e) increases by as much as four times its bulk value [10]. Lattice
stress induced by impurity diffusion appears to be at the origin of
this anomalous behaviour.

REFERENCES
----

[1]   M.Morohashi, N.Sawaki, I.Akasaki [Jpn. J. Appl. Phys. Part 1
      (Japan) vol.24 no.6 (Jun 1985) p.661-5]
[2]   A.Skumanich, D.Fournier, A.C.Boccara, N.M.Amer [Appl. Phys.
      Lett. (USA) vol.47 no.4 (15 Aug 1985) p.402-4]

[3]   A.Neugroschel [IEEE Electron. Device Lett. (USA) vol.EDL-6 no.8
        (Aug 1985) p.425-7]
[4]   M.C.Carotta, M.Merli, L.Passari, E.Susi [Appl. Phys. Lett. (USA)
        vol.49 no.1 (7 July 1986) p.44-5]
[5]   W.R.Thurber, R.L.Mattis, Y.M.Liu, J.J.Filliben [J. Electrochem.
        Soc. (USA) vol.127 no.10 (Oct 1980) p.2291-4]
[6]   J.Dziewior, D.Silber [Appl. Phys. Lett. (USA) vol.35 no.2 (15
        Jul 1979) p.170-2]
[7]   D.D.Tang, F.F.Fang, M.Scheuermann, T.C.Chen, G.Sai-Halasz [Proc.
        Int. Electron. Device Meeting, Los Angeles, CA, USA, Dec 1986 (IEEE,
        USA, 1986) p.20-3]
[8]   S.E.Swirhun, Y.-H.Kwark, R.M.Swanson [Proc. Int. Electron Device
        Meeting, Los Angeles, CA, USA, 1986 (IEEE, USA, 1986) p.24-7]
[9]   D.D.Tang, F.F.Fang, M.Scheuermann, T.C.Chen [Appl. Phys. Lett.
        (USA) vol.49 no.22 (1 Dec 1986) p.1540-1]
[10]   V.G.Weizer, R.DeLombard [Appl. Phys. Lett. (USA) vol.49 no.4
        (28 Jul 1986) p.201-3]

5.8 MINORITY CARRIER ELECTRON DIFFUSION LENGTH IN p-TYPE Si

by J.A. del Alamo

August 1987
EMIS Datareview RN=16115

The electron diffusion length in p-type silicon, Le, is related to the electron lifetime, tau(e) and the electron mobility, mu(e) through the equation:

$$Le = SQRT [(kT/q) \, mu(e) \, tau(e)] \qquad (1)$$

where k is the Boltzmann constant, T is the absolute temperature, and q is the electron charge (SQRT denotes the square root of the bracketed expression). The electron diffusion length can therefore be computed at any doping level using the available experimental data for mu(e) and tau(e) given in other chapters of this book [1,2].

In lightly doped unprocessed p-type Si grown by the float-zone technique, electron diffusion lengths as high as 3000 microns have been measured [3]. In Czochralski-grown Si and after processing Le is usually much smaller than the above value due to the sensitivity of tau(e) to contaminants and crystal defects.

As the acceptor concentration increases beyond about 1 X 10*16/cm*3, Le starts to decrease. In comparison to n-type silicon, systematic measurements of Le as a function of Na in p-type silicon are very scarce [4]. An approximate fit to the measurements of [4] for doping levels over 1 X 10*18/cm*3 is:

$$Le = 7.13 \ X \ 10*14 \quad Na*(-0.923) \ cm \qquad (2)$$

with Na in cm*-3. Because the material used in [4] was unprocessed, to compare Eqn (2) with the data collected in [1,2], the linear term of Eqn (1) in [2] should be dropped. Even in this case the agreement is no better than about a factor of 3 in the 10*18 - 10*20/cm*3 regime. More measurements are certainly needed in this range.

No temperature dependent measurements of Le are known to this author.

REFERENCES
----

[1] J.A. del Alamo [EMIS Datareview RN=16111 (August 1987) 'Minority carrier electron mobility in p-type Si']
[2] J.A. del Alamo [EMIS Datareview RN=16112 (July 1987) 'Minority carrier electron lifetime in p-type Si']
[3] D.Huber, A.Bachmeier, R.Wahlich, H.Herzer [Proc. Conf. Semiconductor Silicon 1986, Eds. H.R.Huff, T.Abe, B.Kobessen (Electrochemical Society, USA, 1986) p.1022-32]
[4] L.Passari, E.Susi [J. Appl. Phys. (USA) vol.54 no.7 (July 1983) p.3935-7]

5.9   MINORITY CARRIER ELECTRON LIFETIME IN p-TYPE Si

by J.A. del Alamo

July 1987
EMIS Datareview RN=16112

In high purity p-type silicon, the electron lifetime, tau(e), is
independent of the acceptor concentration but is sensitive to other
impurities (like O, Fe, or C), the degree of crystal perfection
(stacking faults, dislocation loops, point defects), and the
cleanliness and thermal history of the fabrication procedures. The
highest reported values of electron lifetime in unprocessed p-type
silicon fabricated by the float-zone technique are of the order of
20 ms [1]. These high lifetimes were also found to be insensitive to
gettering techniques that usually improve the lifetime of lower quality
material. The data presented in [1] may then constitute a fundamental
upper limit for electrons, restricted by an intrinsic defect like a
vacancy.

Beyond an acceptor impurity concentration of about 1 X 10*17/cm*3 the
electron lifetime begins to decrease as the acceptor density, Na,
increases. The rate of decrease also increases with doping level. Time
decay measurements of tau(e) in as-grown Czochralski [2,3] and
epitaxially grown and processed silicon [4] have yielded different
results in the moderate doping regime. In the latter, a fit to the
obtained data valid for Na > 1 X 10*17/cm*3 is [4]:

   1/tau(e) = [(3.45 X 10*-12 Na) + (9.5 X 10*-32 Na*2)] s*-1   (1)

with Na in cm*-3.

In unprocessed Czochralski-grown silicon only the quadratic term in
Eqn (1) is required to acceptably fit the experimental data in this
doping regime. The value of the constant is, however, essentially
identical to the one given in Eqn (1). The Auger coefficient in heavily
doped p-type silicon appears then to be Cae = 9.5 X 10*-32 cm*6/s.

The temperature dependence of the electron lifetime in lightly doped
p-type silicon around room temperature has been explained by invoking
a Shockley-Read-Hall recombination mechanism. The energy position above
the valence band of the trap responsible appears to change with doping
from 45 meV [5] to 180 meV [6]. In intentionally Au-contaminated
moderately doped (Na = 2 X 10*17/cm*3) Si, tau(e) was observed to
depend quadratically on the absolute temperature [7]. In the
heavily-doped regime, the Auger coefficient Cae increases with
temperature from a value of 7.8 X 10*-32 cm*6/s at 77 deg K to 1.2 X
10*-31 cm*6/s at 400 deg K [3].

In highly excited Si (when both the hole and electron concentrations
are equal and exceed the background doping level), lifetimes of 40 ms
have recently been measured [8]. Measurements of the sum of the hole
and electron Auger coefficients, Cah + Cae, have also been carried out
in this regime. By comparing data (at room temperature) from a number

of authors (see TABLE 1), Cah + Cae is found to decrease as the
carrier concentration increases. The value at the highest carrier
concentrations [9,11] is close to the sum of Cah and Cae of heavily
doped n- and p-type silicon (2.75 X 10*-31 cm*6/s).

TABLE 1: sum of Auger coefficients in highly injected Si

References	n(/cm*3)	Cae + Cah (cm*6/s)
Yablonovitch & Gmitter [8]	4 X 10*16	2 X 10*-30
Sinton & Swanson [10]	6 X 10*16	1.7 X 10*-30
Svantesson & Nilsson [11]	2.5 X 10*18	3.4 X 10*-31
Fauchet & Nighan [9]	3 X 10*20	2 X 10*-31

This highly excited Auger coefficient was found by Svantesson and
Nilsson [12] to be approximately proportional to T*0.6 in the
temperature interval 195-372 deg K.

REFERENCES
----

[1]   D.Huber, A.Bachmeier, R.Wahlich, H.Herzer [Proc. Conf.
       Semiconductor Silicon 1986, Eds. H.R.Huff, T.Abe, B.Kobessen
       (Electrochemical Society, USA, 1986) p.1022-32]
[2]   J.D.Beck, R.Conradt [Solid-State Commun. (USA) vol.13 no.1
       (1 Jul 1973) p.93-5]
[3]   J.Dziewior, W.Schmid [Appl. Phys. Lett. (USA) vol.31 no.5 (1 Sep
       1977) p.346-8]
[4]   S.E.Swirhun, Y.-H.Kwark, R.M.Swanson [Proc. Int. Electron Device
       Meeting, Los Angeles, CA, USA, Dec 1986 (IEEE, USA, 1986) p.24-7]
[5]   A.Ohsawa, K.Honda, R.Takizawa, N.Toyokura [Rev. Sci. Instrum.
       (USA) vol.54 no.2 (Feb 1983) p.210-2]
[6]   P.C.Mathur, R.P.Sharma, P.Saxena, J.D.Arora [J. Appl. Phys.
       (USA) vol.52 no.5 (May 1981) p.3651-4]
[7]   W.Schmid, J.Reiner [J. Appl. Phys. (USA) vol.53 no.9 (Sep 1982)
       p.6250-2]
[8]   E.Yablonovitch, T.Gmitter [Appl. Phys. Lett. (USA) vol.49 no.10
       (8 Sep 1986) p.587-9]
[9]   P.M.Fauchet, W.L.Nighan,Jr [Appl. Phys. Lett. (USA) vol.48
       no.11 (17 Mar 1986) p.721-3]
[10]   R.Sinton, R.M.Swanson [IEEE Trans. Electron Devices (USA)
       vol.ED-34 no.6 (June 1987) p.1380-9]
[11]   K.G.Svantesson, N.G.Nilsson [Solid-State Electron. (GB) vol.21
       no.11-12 (Nov-Dec 1978) p.1603-8]
[12]   K.G.Svantesson, N.G.Nilsson [J. Phys. C (GB) vol.12 no.18
       (28 Sep 1979) p.5111-20]

CHAPTER 6

HOLE MOBILITY, DIFFUSION AND LIFETIME

6.1 HOLE MOBILITY IN BULK Si

by P.J.Mole

July 1987
EMIS Datareview RN=15742

A. INTRODUCTION

In this review the mobility of majority carrier holes in bulk silicon
is considered over a temperature range of 200 to 400 deg K, where the
models presented for the mobility are claimed to be accurate to about
5%. This temperature range includes that over which silicon devices
are used. The mobility outside this range is also given in some of
the references cited. In addition, the mobility of holes when they
are minority carriers (typical of high injection conditions) is
considered in ref. [16].

B. INTRINSIC MOBILITY, TEMPERATURE DEPENDENCE

The intrinsic mobility of silicon is limited by scattering of the holes
by lattice vibrations (phonons). This gives rise to a mobility which
falls rapidly as the temperature increases (as this increases the
density of phonons present). Historically, the model for the mobility
is based on experimental measurement of the Hall mobility [1-2];
subsequently [3], the mobility has been calculated from the resistivity
by obtaining the doping density of the sample from MOS C-V measurements
on the same sample; this ensures the model is for the conductivity
mobility. The simplest model for the intrinsic mobility is:

$$mu(L) = mu(o) \ (T/To)^{-alpha}$$

In [4,5], the parameters for the model are calculated to be
$mu(o)$ = 473-495 cm*2/V.s at To = 300 deg K and alpha = 2.23-2.20.

The simplicity of this model belies the complexity of the phonon
scattering processes which contribute to the carrier scattering. In
effect there will be contributions from acoustic and optic phonons
scattering holes in the light, heavy or spin-orbit band. This added
complexity is rarely necessary for the analysis of silicon devices;
however further details of such approaches to calculating the
mobility can be found in [6-8].

C. THE EFFECT OF ELECTRICALLY ACTIVE IMPURITIES ON MOBILITY

The deliberate introduction of electrically active impurities into silicon permits the control of silicon devices. Over the temperature range of interest, it will be assumed that these impurities are fully active and therefore neutral impurity scattering will be ignored [9]. Donors (usually phosphorus or arsenic) give up a valence electron and introduce a fixed positive electronic charge density, Nd (/cm*3), into the silicon. Acceptors (usually boron) accept an electron and introduce a fixed negative electronic charge density, Na (/cm*3), into the silicon. These charge centres will introduce coulombic scattering which in the presence of high mobility carrier density will be reduced by screening.

In [10] the following model for the doping dependence of mobility at 300 deg K was introduced:

$$
mu = mu(min) + \frac{mu(o) - mu(min)}{1 + [(Nd + Na/Nref)*alpha]}
$$

[11] gives mu(min) = 47.7 cm*2/V.s, mu(o) = 495 cm*2/V.s, alpha = 0.76 and Nref = 6.3 X 10*16/cm*3.

These values, however, do not allow simple prediction of mobility over a range of temperatures. In [12] an attempt is made to introduce a temperature dependence of the mobility parameters above. However an alternative approach is to derive a formula based on the Brooks-Herring formula for impurity scattering as in [4,5]. Firstly a contribution to the scattering due to impurity scattering is introduced:

$$
mu(I) = \frac{AT*3/2}{Nscat} \left[\ln \left(1 + \frac{BT*2}{Ncar} \right) - \frac{BT*2}{Ncar + BT*2} \right] *-1
$$

The parameters are given by A = 1.0 X 10*17/cm/V/s/K*3/2

$$
B = 6.25 \text{ X } 10*14/cm*3/K*2
$$

and the distinction introduced in [5] between the number of scattering sites introduced by the impurities (Nscat = Nd+Na) and the number of free carriers available for screening (Ncar = Na-Nd) is made.

The contribution to the mobility from impurity scattering can then be combined with that from phonons using the approach of [13] and the approximation in [4]:

$$mu = mu(L) \ f(x)$$

where f(x) can be approximated by:-

$$\frac{1.025}{[\ 1 + (x/1.68)*1.43\ ]} - .025$$

and $x = SQRT \ [\ 6 \ mu(L)/mu(I)\ ]$

At high doping levels (> 10*19/cm*3) the above approaches still cannot account for all the effects of impurities. The presence of such a high concentration of impurity introduces several effects. Not all the electrically active impurity atoms accept electrons from the valence band. This gives rise to a reduced number of charged impurity scattering sites and correspondingly increases the number of neutral scattering centres [9]. This is most significant at the lower temperatures. The valence band and the band of states generated by the impurities merge complicating the band structure and leading to the dependence of the scattering on acceptor type. The hole gas will become degenerate and the form of the scattering will change accordingly [e.g. 16]. At high concentrations the impurities precipitate becoming electrically inactive and provide additional scattering centres, the degree of precipitation will depend on the dopant type and the processing history [see section 13 of ref 16]. The data in the literature is limited to boron doped silicon where in practice there is no tendency for the mobility to reach a minimum at high concentrations [14]. This can be modelled at 300 deg K by [14]:

$$mu = mu(o)exp\ (-pc/p) + \frac{mu(max) - mu(o)}{1+(p/Cr)*alpha} - \frac{mu(1)}{1+(Cs/p)*beta}$$

p = carrier concentration

TABLE 1: the parameters calculated for the data measured on boron doped
samples [14]

parameter	value	units
mu(o)	44.9	cm*2/V.s
mu(max)	470.5	cm*2/V.s
mu(1)	29.0	cm*2/V.s
Cr	2.23 X 10*17	/cm*3
Cs	6.10 X 10*20	/cm*3
alpha	0.719	
beta	2.00	
pc	9.23 X 10*16	/cm*3

D. THE EFFECT OF MATERIAL QUALITY ON MOBILITY

There is no detailed report of variation of mobility with material preparation. It therefore seems reasonable to assume that current day crystal growth techniques and epitaxial layer growth techniques do not introduce significant densities of defects that scatter carriers. However, current hetero-epitaxy techniques e.g. silicon-on-sapphire and other new silicon-on-alternative substrate growth techniques cannot be guaranteed to reduce the carrier scattering defects. Thus the application of all the data presented in this review to silicon films with abnormally high defect levels must be made with care.

E. A MORE ACCURATE DETERMINATION OF MOBILITY

For certain applications a more accurate knowledge of the mobility is required over a wider temperature range. The models described in [15] are extensions to the models which give a greater accuracy except at high doping levels where the corrections discussed in [14] are not included. The reader is referred directly to them.

REFERENCES

[1]   G.L.Pearson, J.Bardeen [Phys. Rev. (USA) vol.75 (1948) p.865]
[2]   F.J.Morin, J.P.Maita [Phys. Rev. (USA) vol.96 (1954) p.28]
[3]   S.S.Li [Solid-State Electron. (GB) vol.21 no.9 (Sep 1978)
        p.1109-17]
[4]   J.M.Dorkel, Ph.Leturcq [Solid-State Electron. (GB) vol.24
        (1981) p.821]
[5]   N.D.Arora, J.R.Hauser, D.J.Roulston [IEEE Trans.Electron
        Devices (USA) vol.ED-29 (1982) p.292]
[6]   F.Szmulowicz [Phys. Rev. B (USA) vol.28 (1983) p.5943]
[7]   K.Seeger [Semiconductor Physics (Springer 1973)
        ISBN 3-211-81186-9]
[8]   C.T.Sah, P.C.H.Chan, C.K.Wang, R.L.Sah, K.A.Yamakawa, R.Lutwack
        [IEEE Trans. Electron Devices (USA) vol.ED-28 (1981) p.304]
[9]   S.S.Li, W.R.Thurber [Solid-State Electron. (GB) vol.20 no.7
        (Jul 1977) p.609-16]
[10]   D.M.Caughey, R.E.Thomas [Proc. IEEE vol.55 (1967) p.2192]
[11]   C.Jacobini, C.Canali, G.Ottaviani, A.Quaranta [Solid-State
        Electron. (GB) vol.20 no.2 (Feb 1977) p.77-89]
[12]   S.Selberherr ['Analysis and Simulation of Semiconductor Devices'
        (Springer, 1984) ISBN 3-211-81800-6]
[13]   P.P.Debye, E.M.Conwell [Phys. Rev. (USA) vol.93 (1954) p.693]
[14]   G.Masetti, M.Severi, S.Solmi [IEEE Trans. Electron Devices
        (USA) vol.ED-30 no.7 (Jul 1983) p.764-9]
[15]   R.D.Larrabee, W.R.Thurber, W.M.Bullis [NBS Spec. Publ. 400-63
        'A FORTRAN program for calculating the electrical
        parameters of extrinsic silicon' (1980)]
[16]   J.A.del Alamo [EMIS Datareview RN=16113 (Aug 1987) 'Minority
        carrier hole mobility in n-type Si']
[17]   C.Herring, E.Vogt [Phys. Rev. (USA) vol.101 (1956) p.994]

6.2 HOLE MOBILITY AT A (100) Si SURFACE

by P.J.Mole

June 1987
EMIS Datareview RN=15741

A. INTRODUCTION

The silicon MOSFET is now one of the most important devices made in
silicon because of its central role in VLSI technology. Conduction
occurs in an inversion layer most commonly at a (100) surface. The
degree of inversion is controlled both by the doping within the device
and the electric field in the insulator above the surface. The knowledge
of the factors influencing this surface mobility (as distinct from
the mobility in the bulk of the material), including the effect of the
controlling electric field, is therefore of considerable importance. In
this review values are reported for the effective carrier mobility as
opposed to the field effect mobility or the Hall mobility; the
distinction between these terms is explained in [1].

B. THE INFLUENCE OF THE CONTROLLING FIELD

The electric field which induces the inversion layer is due to the gate
voltage; increasing this electric field forces the holes closer to
the interface and decreases the width of the potential well in which the
carriers reside. The classic theory of influence of normal field on
mobility [2] relied on the increased scattering as the carriers
are confined closer to a rough or heavily charged surface. In [3] it was
shown that (for electrons) if the mobility is considered as a function
of effective confining electric field ($E(eff) = (QB+Qn/2)/e$ where QB
is the charge density in the surface inversion layer, Qn is the electron
charge density in the inversion layer and e is the dielectric constant
of silicon), then the dependence of mobility on substrate bias, surface
doping and gate voltage can be expressed by one 'universal function'.
There is considerably less data in the literature for hole mobility.
The data in [4,5] can be modelled by the expression:

$$mu = \frac{mu(o)}{1 + theta\ E(eff)}$$

where at 300 deg K, $mu(o) = 220-320$ cm*2/V.s and theta = 2.5-4.7 X
10*-6 cm/V

The relatively broad range of values observed in the literature (the
discrepancies being largest at low effective fields) probably reflects
the variation in surface quality between manufacturers. This view is
supported by the influence of oxide growth conditions on the mobility
as reported in [1]. The measured data are taken over a range of effective
fields up to about 5 X 10*5 cm/V. However it should be noted that when
the electric field drops below that level required for strong inversion,
the model no longer applies [6].

There have been several different models proposed for this normal
field dependence. In [8] the classic theory in [2] is modified by
acknowledging the carrier density to be governed by Fermi-Dirac
statistics (rather than Boltzmann statistics) at the carrier densities
typical of strongly inverted MOSFET's. This also offers an explanation
for the temperature dependence discussed below. Alternative theories
based on strong quantisation of the inversion layer have been proposed
e.g. [9] which also explain the form of the electric field variation.
There is not yet one accepted explanation for the observed behaviour.

C. TEMPERATURE VARIATION

The variation of mobility with temperature is dominated by the scattering
of the electrons by phonons. As for bulk silicon this variation is most
simply modelled by a variation of the form:

$$mu(L) = mu(o) \ (T/To)^{-alpha}$$

For To = 300 deg K, mu(o) is given in the previous section; the value
for alpha is considerably less than the bulk value and [4] report a
value close to 1.5 over the temperature range 200-400 deg K.

D. THE EFFECT OF CHARGED IMPURITIES

Charged impurities at the interface or due to bulk doping can affect
the mobility. The latter affect the mobility through their influence
on the surface field (see above) as well as through coulombic
scattering. In [5] it is shown that at surface concentrations of about
3 X 10*16/cm*3 the mobility starts to fall through the influence of
impurity scattering. The scattering of carriers by surface charges has
not been examined, but it is expected that the effects will be similar
to those reported for electrons [7].

E. THE EFFECT OF HIGH DRIFT ELECTRIC FIELDS

Under the influence of high drift electric fields (E), the holes become
heated and as a result their mobility is affected as new scattering
mechanisms become available to them. The effect has been measured by
careful interpretation of the drain current in a short channel MOSFET
[4]. A model for the mobility of the form:

$$mu = \frac{mu(o)}{[\ 1 + (E \ mu(o)/v(sat))*alpha \ ]*(1/alpha)}$$

has been proposed with alpha = 2.6 and v(sat) = 2.5 X 10*7 cm/s.
However because of the high fields that must be obtained to measure
these parameters (2 X higher than for electrons) these values are
difficult to measure.

REFERENCES
----

[1] S.C.Sun, J.D.Plummer [IEEE Trans Electron Devices (USA) vol.ED-27
     (1980) p.1497]
[2] J.R.Schrieffer [Phys. Rev. (USA) vol.97 (1955) p.641]
[3] A.G.Sabnis, J.T.Clemens [IEEE Int. Electron Devices Meeting
     Tech Digest (1979) p.18]
[4] R.W.Coen, R.S.Muller [Solid-State Electron. (GB) vol.23 no.1
     (Jun 1980) p.35-40]
[5] S.J.Murray, P.J.Mole [Proc. Simulation of Semiconductor Devices
     and Processes (Pineridge Press, 1986) ISBN 0-906674-59-x]
[6] J.T.C.Chen, R.S.Muller [J. Appl. Phys. (USA) vol.45 no.2
     (Feb 1974) p.828-34]
[7] S.Manzini [J. Appl. Phys. (USA) vol.57 no.2 (15 Jun 1985) p.411-14]
[8] G.Baccarani, A.Mazzone, C.Morandi [Solid State Electron. (GB)
     vol.17 (1974) p.785]
[9] A.Rothwarf [IEEE Electron Device Lett. (USA) vol.EDL-8 (1987)
     p.499]

6.3   HOLE MOBILITY IN p-TYPE EPITAXIAL Si

by J.A.del Alamo

July 1987
EMIS Datareview RN=15732

Mobility measurements in p-type epitaxial Si layers are to be compared
to those reported in bulk doped as-grown Si [1]. These results agree
with measurements carried out in B ion-implanted Si up to solid
solubility values [2]. The authors in [1,2] have assumed a Hall
scattering factor of 0.8 in B-doped Si (see discussion in [3]). No
systematic measurements of this parameter are known to this author.

The following summarises results obtained in epitaxial p-type Si layers
grown by various techniques:

a)   Low Pressure Vapour Phase Epitaxy (LPVPE). B doping levels in the
     range of 3 X 10*18 to 5 X 10*19/cm*3 have been reported [4,5].
     Mobilities are found to be 90% of those of bulk Si.

b)   Molecular Beam Epitaxy (MBE). B doped MBE-grown Si layers have
     been reported up to doping levels of 1 X 10*20/cm*3 [6]. Identical
     mobilities to bulk Si have been obtained throughout the whole
     doping range at both 300 deg K and 77 deg K. The work on Ga-doped
     Si is more difficult to appraise because of the lack of systematic
     measurements in bulk Si. For in-situ solid phase epitaxy (SPE) of
     MBE deposited layers doping levels as high as 2 X 10*20/cm*3
     have been reported [7,8]. The data reported by several authors at
     300 deg K agree with each other in their common doping range, i.e.
     1 X 10*16 to 1 X 10*18/cm*3 [6,7,9], and are almost identical to
     bulk B-doped Si values. In the 1 X 10*19 to 1 X 10*20/cm*3 range,
     the reported mobilities at 300 deg K are much lower than in bulk
     B-doped Si [7,8]. At 77 deg K lower values are also obtained in the
     1 X 10*15 to 1 X 10*16/cm*3 range [6].

c)   Limited Reaction Processing (LRP). A single data point has been
     reported for this new rapid epitaxial technique [10]. In Si doped
     with B to a concentration of 6 X 10*19/cm*3 a hole mobility of
     42 cm*2/V.s has been reported. This value is identical to bulk Si of
     the same doping level.

REFERENCES
----

[1]   W.R.Thurber, R.L.Mattis, Y.M.Liu, J.J.Filliben [J. Electrochem.
      Soc. (USA) vol.127 no.10 (Oct 1980) p.2291-4]
[2]   G.Masetti, M.Severi, S.Solmi [IEEE Trans. Electron. Devices (USA)
      vol.ED-30 no.7 (July 1983) p.764-9]
[3]   J.A.del Alamo [EMIS Datareview RN=15731 (July 1987) 'Electron
      mobility in n-type epitaxial Si']
[4]   L.Vescan, H.Beneking, O.Meyer [Mater. Res. Soc. Symp. Proc. (USA)
      vol.71 (1986) p.133-8]
[5]   L.Vescan, H.Beneking, O.Meyer [J.Cryst. Growth (Netherlands)
      vol.76 no.1 (July 1986) p.63-8]

[6] R.A.A.Kubiak, S.M.Newstead, W.Y.Leong, R.Houghton, E.H.C.Parker,
     T.E.Whall [Appl. Phys. A (Germany) vol.42 no.3 (Mar 1987) p.197-
     200]

[7] A.Casel, H.Jorke, E.Kasper, H.Kibbel [Appl. Phys. Lett. (USA)
     vol.48 no.14 (7 April 1986) p.922-4]

[8] L.Vescan, E.Kasper, O.Meyer, M.Maier [J. Cryst. Growth
     (Netherlands) vol.73 no.3 (Dec 1985) p.482-6]

[9] Y.Shiraki [Prog. Cryst. Growth & Charact. (GB) vol.12 nos.1-4
     (1986) p.45-66]

[10] C.M.Gronet, J.C.Sturm, K.E.Williams, J.F.Gibbons, S.D.Wilson
     [Appl. Phys. Lett. (USA) vol.48 no.15 (14 April 1986) p.1012-4]

6.4   MINORITY CARRIER HOLE MOBILITY IN n-TYPE Si

by J.A. del Alamo

August 1987
EMIS Datareview RN=16113

In spite of the importance that an accurate knowledge of the minority
carrier mobilities represents in injection-type devices, like solar
cells and bipolar transistors, very few measurements have been reported
in the literature. This, no doubt, is due to the difficulty of the
measurement techniques.

In high purity n-type Si, the mobility of minority carrier holes, mu(h),
is experimentally found to be independent of donor doping, Nd, and the
material preparation technique. Measurements of this low doping hole
mobility put its value at around 500 cm*2/V.s, within about 20% [1-4].
This value is essentially identical to that of the majority carrier
hole mobility in p-type silicon, MU(h) for the same doping level [5].

As the donor doping level increases, mu(h) starts to decrease just as
MU(h) does. However, the onset of decay of mu(h) does not occur up to
a doping level of about 1 X 10*17/cm*3 [6,7], while the drop of MU(h)
is already significant at an acceptor concentration of 1 X 10*16/cm*3
[5]. In consequence, in this moderately doped regime, mu(h) is larger
than MU(h) at the same doping level.

At high doping levels, much fewer experimental data are available. Only
in [8,9] have measurements been reported in the 1 X 10*19/cm*3 decade
(up to 6.5 X 10*19/cm*3). In this regime, as is also the case for
MU(h), mu(h) saturates with doping. The saturation value is however
about 130 cm*2/V.s, over two times larger than MU(h) [5].

A fit to all the available experimental data was given in [8,9]:

$$mu(h) = 130 \ + \ \frac{370}{1 + [Nd/(8 \ X \ 10*17)]*1.25}$$

in cm*2/V.s, with Nd in cm*-3.

A report on the temperature dependence of mu(h) exists in the lightly
doped regime [4]. In the temperature range from 77 to 290 deg K, mu(h)
appears to behave as T*-1.9.

REFERENCES
----

[1]   J.Muller, H.Bernt, H.Reichl [Solid-State Electron. (GB) vol.21
      no.8 (Aug 1978) p.999-1003]
[2]   D.Gasquet, J.P.Nougier, G.Gineste [Physica B&C (Netherlands)
      vol.129 no.1-3 (Mar 1985) p.524-6]

[3]   A.Neugroschel [IEEE Electron Device Lett. (USA) vol.EDL-6 no.8
       (Aug 1985) p.425-7]
[4]   M.Morohashi, N.Sawaki, T.Somatani, I.Akasaki [Jpn. J. Appl. Phys.
       Part 1 (Japan) vol.22 no.2 (Feb 1983) p.276-80]
[5]   W.R.Thurber, R.L.Mattis, Y.M.Liu, J.J.Filliben [J. Electrochem.
       Soc. (USA) vol.127 no.8 (Aug 1980) p.1807-12]
[6]   J.Dziewior, D.Silber [Appl. Phys. Lett. (USA) vol.35 no.2 (15
       Jul 1979) p.170-2]
[7]   D.E.Burk, V. de la Torre [IEEE Electron. Device Lett. (USA)
       vol.EDL-5 no.7 (July 1984) p.231-3]
[8]   J.A del Alamo, S.Swirhun, R.M.Swanson [Proc. Int. Electron. Device
       Meeting, Washington DC, Dec 1985 (IEEE, USA, 1985) p.290-3]
[9]   S.E.Swirhun, J.A. del Alamo, R.M.Swanson [IEEE Electron Device
       Lett. (USA) vol.EDL-7 no.3 (Mar 1986) p.168-71]

6.5 MINORITY CARRIER HOLE DIFFUSION LENGTH IN n-TYPE Si

by J.A. del Alamo

August 1987
EMIS Datareview RN=16110

The hole diffusion length is the average distance that an excess hole
travels in a field-free n-type semiconductor before recombining. By
definition, therefore, the hole diffusion length, Lh, is related to the
hole lifetime, tau(h), and the hole mobility, mu(h), through the
following equation:

$$Lh = SQRT [(kT/q) mu(h) tau(h)] \qquad\qquad (1)$$

where k is the Boltzmann constant, T the absolute temperature, q the
electron charge. One can therefore compute the diffusion length at any
donor doping level by using the available experimental mu(h) and tau(h)
data given in other chapters of this book [1,2].

In practice, however, it is much easier to measure a diffusion length
than both the lifetime and a minority carrier mobility. The measurement
of the diffusion length can be strictly carried out in the steady
state, while lifetime and minority carrier mobility measurements
require time-dependent techniques which become very complex in the high
doping regime.

In lightly doped unprocessed n-type Si grown by the float-zone technique
maximum hole diffusion lengths in excess of 2000 microns have recently
been measured [3]. As the donor doping level increases beyond 1 X 10*16
/cm*3 Lh starts to drop. At a doping level of 1 X 10*17/cm*3 a value of
200 microns has been measured [3]. These values are considerably higher
than those found in Czochralski or processed Si wafers, since Lh is
very sensitive to contamination and crystal defects.

As the doping level increases, the diffusion length decreases very fast.
A large amount of experimental Lh data has been reported in the highly
doped regime, obtained by a variety of techniques: lateral bipolar
transistors [4,5], electron-beam-induced current (EBIC) measurements
[6,7], and optical techniques [8,9]. A fit to the data above a donor
doping level, Nd, of 1 X 10*18/cm*3 was given in [5]:

$$Lh = 2.77 X 10*14 Nd*(-0.941) cm \qquad\qquad (2)$$

with Nd in cm*-3. All the data agree with each other within approximately
a factor of two. No definite trends of Lh are observed with varying the
dopant species or the doping techniques.

No reports are known to this author on the temperature dependence of Lh.

REFERENCES

[1]   J.A. del Alamo [EMIS Datareview RN=16113 (August 1987) 'Minority
      carrier hole mobility in n-type Si']
[2]   J.A. del Alamo [EMIS Datareview RN=16114 (August 1987) 'Minority
      carrier hole lifetime in n-type Si']
[3]   D.Huber, A.Bachmeier, R.Wahlich, H.Herzer [Proc. Conf.
      Semiconductor Silicon 1986, Eds. H.R.Huff, T.Abe, B.Kobessen
      (Electrochemical Society, USA, 1986) p.1022-32]
[4]   A.W.Wieder [IEEE Trans. Electron Devices (USA) vol.ED-27 no.8
      (Aug 1980) p.1402-8]
[5]   J. del Alamo, S.Swirhun, R.M.Swanson [Proc. Int. Electron. Device
      Meeting, Washington DC, Dec 1985 (IEEE, USA, 1985) p.290-3]
[6]   D.E.Burk, V. de la Torre [IEEE Electron. Device Lett. (USA)
      vol.EDL-5 no.7 (Jul 1984) p.231-3]
[7]   G.E.Possin, M.S.Adler, B.J.Baliga [IEEE Trans. Electron Devices
      (USA) vol.ED-31 no.1 (Jan 1984) p.3-17]
[8]   R.P.Mertens, J.L.van Meerbergen, J.F.Nijs, R.J.van Overstraeten
      [IEEE Trans. ELectron. Devices (USA) vol.ED-27 no.5 (May 1980)
      p.949-55]
[9]   L.Passari, E.Susi [J. Appl. Phys. (USA) vol.54 no.7 (Jul 1983)
      p.3935-7]

6.6 MINORITY CARRIER HOLE LIFETIME IN n-TYPE Si

by J.A. del Alamo

August 1987
EMIS Datareview RN=16114

In lightly doped n-type silicon, the hole lifetime, tau(h), is experimentally found to be independent of the donor concentration and very sensitive to the quality of the crystal and the cleanliness of the device fabrication process. The highest reported values of hole lifetime in unprocessed n-type silicon grown by the float-zone technique are of the order of 30 ms [1].

As the donor concentration increases beyond about 1 X 10*16/cm*3, the hole lifetime is observed to start to decrease. The drop of tau(h) becomes more pronounced as the doping level increases. Reliable measurements of tau(h) at the highest doping level have not been possible until recently. Direct measurements of hole lifetime have been carried out by observing the temporal decay of the photo-luminescence radiation emitted by the band-to-band recombination of electron-hole pairs after excitation by a laser pulse [2-4]. Other time-transient techniques have also been used [5,6]. A fit to the available experimental data for donor concentrations, Nd, higher than 1 X 10*17/cm*3 was given in [7]:

$$1/tau(h) = [(7.8 \ X \ 10*-13 \ Nd) + (1.8 \ X \ 10*-31 \ Nd*2)] \ s*-1 \qquad (1)$$

with Nd in cm*-3. To the first order, in this high doping regime, the hole lifetime is independent of the donor species and the material preparation technique (doped as-grown, solid-state diffusion, ion implantation, epitaxy). Extension of the fit to donor concentrations below 1 X 10*17/cm*3 was not carried out because tau(h) varies widely from author to author.

As Eqn (1) shows, in the very heavily doped regime, there is a quadratic dependence of the inverse lifetime on doping level. This exponent, however, cannot be determined accurately due to the factor of two to three scattering in the available experimental data. A quadratic dependence is theoretically predicted to occur if the main recombination process is of the Auger type in which the energy released in the recombination process is given to a neighboring electron. The constant of the quadratic term of Eqn (1) is therefore denoted as Auger coefficient (Cah). In heavily doped n-type silicon, the value which better fits all the experimental data is Cah = 1.8 X 10*-31 cm*6/s [7], although the value 2.8 X 10*-31 cm*6/s, given by Dziewior and Schmid [3], is very commonly used.

The temperature dependence of the hole lifetime in n-type silicon has received little attention experimentally. For moderately doped samples (Nd = 1.1 X 10*17/cm*3) intentionally contaminated with Au, a square-root temperature dependence of the hole lifetime has been reported [8]. In more heavily doped samples, measurements have showed that the Auger coefficient in n-type silicon appears to vary little with temperature [4,9]. The results of Dziewior and Schmid [4] give a maximum variation of Cah of less than 20% from 77 deg K to 400 deg K (Cah increases with temperature).

In highly excited Si, a lifetime (sum of hole plus electron lifetimes) of 40 ms has been recently measured [10]. Additionally, measurements of the sum of the hole and electron Auger coefficients have been carried out in highly injected Si. A review of the main results is given in a different section of this book [11]. It appears that the sum of the Auger coefficients increases by as much as one order of magnitude as carrier concentration is decreased from the 1 X 10*20/cm*3 injection level to the 1 X 10*16/cm*3 level.

REFERENCES
----

[1]   D.Huber, A.Bachmeier, R.Wahlich, H.Herzer [Proc. Conf. Semiconductor Silicon 1986 Eds. H.R.Huff, T.Abe, B.Kobessen (Electrochemical Society, USA, 1986) p.1022-32]
[2]   J.D.Beck, R.Conradt [Solid State. Commun. (USA) vol.13 no.1 (1 Jul 1973) p.93-5]
[3]   J.Dziewior, W.Schmid [Appl. Phys. Lett. (USA) vol.31 no.5 (1 Sep 1977) p.346-8]
[4]   S.E.Swirhun, J.A. del Alamo, R.M.Swanson [IEEE Electron Device Lett. (USA) vol.EDL-7 no.3 (Mar 1986) p.168-71]
[5]   Y.Vaitkus, V.Grivitskas [Sov. Phys.-Semicond. (USA) vol.15 no.10 (Oct 1981) p.1102-8]
[6]   D.E.Burk, V.de la Torre [IEEE Electron Device Lett. (USA) vol.EDL-5 no.7 (July 1984) p.231-3]
[7]   J. del Alamo, S.Swirhun, R.M.Swanson [Proc. Int. Electron. Device Meeting, Washington DC, USA, December 1985 (IEEE USA 1985) p.290-3]
[8]   W.Schmid, J.Reiner [J. Appl. Phys. (USA) vol.53 no.9 (Sep 1982) p.6250-2]
[9]   L.A.Delimova [Sov. Phys.-Semicond. (USA) vol.15 no.7 (Jul 1981) p.778-80]
[10]   E.Yablonovitch, T.Gmitter [Appl. Phys. Lett. (USA) vol.49 no.10 (8 Sep 1986) p.587-9]
[11]   J.A. del Alamo [EMIS Datareview RN=16112 (July 1987) 'Minority carrier electron lifetime in p-type Si']

CHAPTER 7

BAND STRUCTURE

7.1 INDIRECT ENERGY GAP OF Si, TEMPERATURE DEPENDENCE

by D.J.Dunstan

August 1987
EMIS Datareview RN=16116

As in any semiconductor, the silicon band-gap Eg(T) is expected to
depend on temperature T, through two effects, lattice dilation and
electron-phonon interaction. Lattice dilation gives a linearly
decreasing Eg(T) at high temperature [1] and at low temperature its
effect is small. In any case, dilation accounts for only 20-25% of the
total dependence. The major contribution, electron-phonon interaction,
is expected to give a reduction in Eg(T), with a shift proportional to
T for T << TD and T*2 for T >> TD, where TD is the Debye temperature.

The most reliable data covering a wide temperature range (4 deg K to
400 deg K) is that of MacFarlane et al [2] and of Bludau et al [3]
obtained by optical absorption. The latter authors fitted their data
with the following empirical formulae, accurate to 0.2 meV:

  Eg(T) = 1.170 + (1.059 X 10*-5)T - (6.05 X 10*-7)T*2
                                        for 0 < T < 190 deg K,

and

  Eg(T) = 1.1785 - (9.025 X 10*-5)T - (3.05 X 10*-7)T*2
                                        for 150 < T < 300 deg K

Varshni [4] proposed a semi-empirical formula for Eg(T) which is now
widely used:

  Eg(T) = Eo - AT*2/(T+B)

where Eo is the energy gap at 0 deg K. A and B are constants and
B is expected to approximate to the Debye temperature. The original
fit to the data of MacFarlane et al used the parameters

  Eo = 1.1557   A = 7.021 X 10*-4   B = 1108

Thurmond [5] showed that the fitted values of A and B are strongly
interdependent, and, including the data of Bludau et al, proposed the
fit

  Eo = 1.170 +- 0.001 eV

  A = (4.73 +- 0.25) X 10*-4 eV/deg K

  B = 636 +- 50 deg K

and these remain the best values to date. Thus the room temperature
band gap may be taken to be Eg(293 deg K) = 1.126 eV.

REFERENCES
----

[1] J.Bardeen, W.Shockley [Phys. Rev. (USA) vol.80 (1950) p.72]

[2] G.G.MacFarlane, T.P.MacLean, J.E.Quarrington, V.Roberts [Phys.
     Rev. (USA) vol.111 (1958) p.1245; J. Phys. & Chem. Solids (GB)
     vol.8 (1959) p.388]

[3] W.Bludau, A.Orton, W.Heinke [J. Appl. Phys. (USA) vol.45 (1974)
     p.1846]

[4] Y.P.Varshni [Physica A (Netherlands) vol.34 (1967) p.149]

[5] C.D.Thurmond [J. Electrochem. Soc. (USA) vol.122 no.8 (Aug 1975)
     p.1133-41]

7.2 INDIRECT ENERGY GAP OF Si, PRESSURE DEPENDENCE

by D.J.Dunstan

September 1987
EMIS Datareview RN=16167

Silicon is typical of the III-V and IV crystalline semiconductors, which all have pressure shifts of the conduction band minima with respect to the valence band maximum of about +10 eV/Mbar (gamma-point), +5 eV/Mbar (L-point) and -2 eV/Mbar (X-point) until the first phase transition [1].

Early determinations of the pressure coefficient for the silicon conduction band X-minimum valence band gamma-point indirect gap used electrical transport measurements from which the band-gap could be deduced (for a review, see [1]). More reliable data was obtained by Slykhouse and Drickamer [2] by optical transmission; they obtained a pressure coefficient of -2.0 eV/Mbar, from data at pressures up to 140 kbar. It is now thought that the first phase transition in Si occurs at 120 +- 5 kbar [3,4] so this value may be a little high.

Welber et al [4] observed optical absorption in a diamond anvil cell at pressures up to the phase transition, and obtained a coefficient of -1.41 +- 0.06 eV/Mbar. This remains the best value. Theoretical calculations [5] give a value of -1.6 eV/Mbar [5] and, using a more accurate band structure, -1.432 eV/Mbar [6] in excellent agreement with the experimental value.

REFERENCES

[1] W.Paul, D.M.Warschauer [in 'Solids under Pressure' Eds W.Paul, D.M.Warschauer (McGraw-Hill, 1963) ch.8 p.226]
[2] T.E.Slykhouse, H.G.Drickamer [J. Phys. & Chem. Solids (GB) vol.7 (1958) p.210]
[3] B.A.Weinstein, G.J.Piermarini [Phys. Rev. B (USA) vol.12 (1975) p.1172]
[4] B.Welber, C.K.Kim, M.Cardona, S.Rodriguez [Solid State Commun. (USA) vol.17 (1975) p.1021]
[5] K.Chang, S.Froyen, M.L.Cohen [Solid State Commun. (USA) vol.50 (1984) p.105]
[6] P.E.Van Camp, V.E.Van Doren, J.T.Devreese [Phys. Rev. B (USA) vol.34 no.2 (15 Jul 1986) p.1314-16]

7.3 INDIRECT ENERGY GAP OF Si, DOPING DEPENDENCE

by H.S.Bennett

March 1987
EMIS Datareview RN=15707

A. INTRODUCTION

The doping dependence of the indirect energy gap of silicon has received
considerable experimental and theoretical attention in recent years.
Many of these efforts are motivated by the need to describe the effect
of high concentrations on the performance of electronic devices. Even
with all this activity, a consensus for the doping dependence in n-type
and p-type silicon does not exist.

This Datareview is a guide with commentary to assist readers in
selecting which values are best for their applications. Unfortunately,
our knowledge in this area is such that the final application for the
data on bandgap changes may determine which values to use.

The dependence of the indirect energy gap of silicon on doping density
is usually expressed in terms of the bandgap change,

$$delta\ Eg(N) = Eg(N) - Eg(N=0;\ intrinsic)$$

where N is the doping density for uncompensated material. Several
determinations of the dependence of the bandgap change for n-type and
p-type silicon appear in the literature. These determinations may be
grouped into four categories: 1) electrical measurements at room
temperature [1-5]; 2) photoluminescence measurements at low
temperatures and at 300 deg K [6-11]; 3) optical or infrared absorption
measurements at both low temperatures and at room temperature [12-14];
and 4) theoretical calculations at 0 deg K and at 300 deg K [15-19].

Statistical analyses of these data [20] show that bandgap changes
determined by: 1) optical absorption measurements [12-14] are
consistent at both 300 deg K and at temperatures below 35 deg K with
theoretical calculations [15-19]; 2) electrical measurements [1-5] and
optical measurements [6-14] are not consistent with each other; and 3)
low temperature optical absorption measurements [12-14] and
photoluminescence measurements [6-11] are consistent with each other
when the physically correct density of states is used to interpret the
experimental data. Photoluminescence data show more apparent bandgap
narrowing than the low temperature optical absorption data. However,
this difference is reconciled by the fact that these two methods are
sensitive to different regions of the distorted densities of states.

B. ELECTRICAL MEASUREMENTS AT ROOM TEMPERATURE

There are two common methods [1-3] based on compact-analytic device
models which use electrical measurements to obtain an 'effective
bandgap' change as a function of the donor or acceptor density. Both
methods involve building either a diode structure or a bipolar
transistor. The latter may be either lateral or vertical.

The first method [1] is based on measuring the minority carrier current injected from a lightly doped emitter into a uniformly heavily doped substrate. This measurement has many uncertainties since the minority carrier current must be isolated from other current components. Also, since the compact-analytic models used to interpret the measurements are usually one dimensional, the effects due to surface and sidewall currents must be known. Additional measurements are then required to obtain values for the minority carrier diffusion coefficient and diffusion length. The measurements are interpreted in terms of a compact-analytic device model that contains the effective intrinsic carrier concentration. An 'effective bandgap' change that is, in fact, simply a fitting parameter which has the units of energy is then inferred from this effective intrinsic carrier concentration. This fitting parameter, which frequently is confused with the physical bandgap change, contains the effects of everything not considered elsewhere in the device model. These effects include carrier-dopant-ion interactions, carrier-carrier interactions, degeneracy, minority mobilities which are not equal to majority mobilities, and physical lifetimes which are different from the empirical ones assumed in device models.

The second method [2,3] is based upon measuring the collector current of a bipolar transistor. Ancillary measurements then include the 'average' minority carrier diffusion coefficient or mobility and the area of the emitter. As in the first method, these measurements are then interpreted in terms of a compact-analytic device model which contains the effective intrinsic carrier concentration.

The interpretations of most electrical measurements contain the assumption that minority and majority mobilities are equal at the same doping density. Recent work [5,21] shows that this assumption is not correct, particularly in silicon doped above 10*19/cm*3. Many interpretations also assume uniform doping and this may not be valid for the devices on which the measurements were made.

C. PHOTOLUMINESCENCE MEASUREMENTS AT LOW TEMPERATURES AND AT 300 DEG K

Photoluminescence experiments usually involve immersing the doped silicon samples in liquid helium between 2.4 deg K and 4 deg K and exciting them with light, for example, with 514.5 nm (2.409 eV) photons from an Ar-ion laser. Even though the quantum efficiencies are small for indirect bandgap semiconductors, photoluminescence spectra from the recombination of the electrons and holes can be obtained. A few measurements have been done at 300 deg K [8].

Four main uncertainties exist in the interpretation of photoluminescence spectra from indirect bandgap semiconductors. These are associated with: 1) the initial and final states of the carriers; 2) the appropriate admixture of phonon assisted transitions included in the calculated photoluminescence spectra for which the bandgap and Fermi energy are parameters; 3) the Fermi energy and distorted densities of states used in the calculated photoluminescence spectra; and 4) the extrapolation of low temperature photoluminescence results to room temperature. Very few room temperature photoluminescence measurements have been reported [8]. A consensus does not exist for the temperature dependence of the bandgap change at constant doping density. Understanding the room temperature behaviour of electron devices is one reason why the extrapolation procedure of the earlier low temperature photoluminescence data [6,7] has been attempted. The more recent photoluminescence measurements [8] at room temperature give conflicting results. Theory [20] gives a temperature dependence for the bandgap change. The resolution of these conflicting results may lie in the fact that the interpretations of experiments depend on whether intrinsic or distorted densities of states are used.

D. OPTICAL ABSORPTION MEASUREMENTS AT 300 DEG K AND AT LOW TEMPERATURES

The optical or infrared absorption coefficients, alpha, are obtained usually from measurements of the transmission coefficient T by inverting the relation

$$T = \frac{(1-R)*2 \ exp(-alpha \ d)}{1 - [R*2 \ exp(-2 \ alpha \ d)]}$$

where R is the reflection coefficient and d is the sample thickness [12-14]. Separate measurements of d and R then give experimental curves for the absorption coefficients as a function of the absorbed photon energies. Such data have been obtained for a range of temperatures from 4 deg K to 300 deg K in heavily doped silicon. The interpretation of the absorption coefficients to yield the bandgap values and the bandgap changes involve the first three uncertainties given above for the photoluminescence data. Optical absorption measurements at 300 deg K have the two advantages that 1) no questionable extrapolation procedures are needed to obtain room temperature values for bandgap changes and 2) they are easier to perform.

E. THEORY AT 300 DEG K AND AT 0 DEG K

Most theories treat the dopant ions as point charges and thereby only distinguish between the sign of the dopant ion. They do not distinguish between ion type, for example, As and P. Recent theoretical work on bandgap variations with doping density has reduced the confusion noted by Mahan [22]. The recent 0 deg K theories of Ghazali and Serre [23], Berggren and Sernelius [15], and Abram, Childs, and Saunderson [24] are in general agreement with the low temperature photoluminescence measurements for bandgap narrowing in n-type silicon. However, they are not adequate to compute effective intrinsic carrier concentrations at room temperature. Computing the latter requires integrating perturbed densities of states over the distorted band structure to give the electron and hole densities. The doping dependence of the effective intrinsic carrier concentration is an essential input for numerical simulations of electronic devices.

The 300 deg K theories for bandgap changes given in refs [16-19] are compromises between two conflicting goals: 1) calculating the band structure changes as rigorously as possible and 2) having the 300 deg K results incorporated readily and economically into device models. The 0 deg K theories [15,22,23] do not meet the second goal because they cannot be easily extended to 300 deg K. The 0 deg K theory of ref [24] can be extended to finite temperatures and, according to its authors, this extension will be demonstrated in a future paper. The bandgap narrowing predictions in refs [16-19] are in general agreement with the low temperature and room temperature absorption measurements.

The methods of refs [16,18,21] are based on quantum mechanical concepts and give self-consistent values for the effective intrinsic carrier concentration, band edge changes, perturbed or distorted densities of states, minority carrier mobilities, and minority carrier lifetimes. All these quantities occur in device models.

F. GUIDE TO SELECTED BANDGAP DATA

A. Room Temperature - 300 deg K

   1.   Electrical Measurements

      a)   Tang [2], As dopant ion

      b)   Slotboom and DeGraaff [3], empirical formula given by Eqn (18) for B dopant ion. This formula is strictly valid for the doping range of $4 \times 10^{15}/cm^3$ to $2.5 \times 10^{19}/cm^3$ in p-type silicon. However, many device models for n-type silicon use the same empirical formula.

      c)   Wieder [4], As and B dopant ions
One of the conclusions several authors have reached from Wieder's measurements [4] on As doped silicon is that similar, but not exactly equal, values for the bandgap changes in As and B doped silicon are obtained. Theory [16] gives larger differences between n-type and p-type silicon at the same doping density.

      d)   del Alamo, Swirhun and Swanson [5], P dopant ions

   2.   Optical Absorption Measurements

      a)   Schmid [12], As dopant ion

      b)   Vol'fson and Subashiev [13], As dopant ion

      c)   Balkanski et al [14], P dopant ion

   3.   Photoluminescence Measurements

      a)   Wagner [8], P and B dopant ions

   4.   Theory

      a)   Bennett [17], n-type

      b)   Bennett [18], n-type and p-type

      c)   Lowney [19], n-type and p-type

B. Low Temperature

    1.   Optical Measurements

        a)   Schmid [12], 4 deg K, As dopant ion

        b)   Balkanski et al [14], 35 deg K, P dopant ion

    2.   Photoluminescence Measurements

        a)   Dumke [6,7], 2.4 deg K, As dopant ion

        b)   Wagner [8], 5 deg K and 20 deg K, P and B dopant ions

    3.   Theory

        a)   Lowney and Bennett [16], 0 deg K, n-type

G. SOME APPLICATIONS OF BANDGAP CHANGES

In conventional device physics, the parameters used to describe high concentration effects on the band structure are determined from interpretations of electrical measurements on the devices being modelled. Ambiguous results may occur when extracting such parameters from electrical measurements on devices. Such extractions for model parameters usually are based on a lower level device model and hence become dependent on that lower level device model itself. Conventional procedures give acceptable predictions for the effects of small variations in processing and in device geometry and dimensions. However, models built in this way may not lead to a fundamental understanding of the physical mechanisms responsible for the performance of the device, particularly when dimensions are less than a micrometer.

When possible it is preferable to obtain parameters for high concentration effects by calculations based on quantum mechanics or by alternative measurements such as optical absorption or photoluminescence. The latter measurements do not require device models for their interpretation. First-principles procedures give acceptable results for larger changes in fabrication processes and in device geometry and dimensions than do empirical procedures. They also have a much higher probability of leading to a fundamental understanding of device behaviour. However, it is important to remember that the physics may be different in graded structures than in uniform material.

REFERENCES
----

[1]   R.P.Mertens, J.L.Van Meerbergen, J.F.Nijs, R.J.Van Overstraeten [IEEE Trans. Electron Devices (USA) vol.ED-27 (1980) p.949]

[2]   D.D.Tang [IEEE Trans. Electron Devices (USA) vol.ED-27 (1980) p.563]

[3]   W.Slotboom, H.C.deGraaff [Solid-State Electron. (GB) vol.19 (1976) p.586]

[4]   A.W.Wieder [Proc. Int. Electron Devices Meeting (USA) (1978) p.460 and IEEE Trans. Electron Devices (USA) vol.ED-27 (1980) p.1402]

[5]   J. del Alamo, S.Swirhun, R.M.Swanson [Proc. Int. Electron
       Devices Meeting (USA) (1985) p.290 and Solid-State Electron.
       (GB) vol.28 (1985) p.47]
[6]   W.P.Dumke [Appl. Phys. Lett. (USA) vol.42 (1983) p.196] In
       table I, 'EG + EF' should read 'EG + EF - hbar omega p'.
[7]   W.P.Dumke [J. Appl. Phys. (USA) vol.54 (1983) p.3200]
[8]   J.Wagner [Solid-State Electron. (GB) vol.28 (1985) p.25 and
       Phys. Rev. B (USA) vol.32 (1985) p.1323]
[9]   P.E.Schmid, M.L.W.Thewalt, W.P.Dumke [Solid State Commun. (USA)
       vol.38 (1981) p.1091]
[10]   R.R.Parson [Solid State Commun. (USA) vol.29 (1979) p.763]
[11]   A.Selloni, S.T.Pantelides [Phys. Rev. Lett. (USA) vol.49 (1982)
       p.586]
[12]   P.E.Schmid [Phys. Rev. B (USA) vol.23 (1981) p.5531]
[13]   A.A.Vol'fson, V.K.Subashiev [Sov. Phys.-Semicond. (USA) vol.1
       (1967) p.327]
[14]   M.Balkanski, A.Aziza, E.Amzallag [Phys. Status Solidi (Germany)
       vol.31 (1969) p.323]
[15]   K.F.Berggren, B.E.Sernelius [Phys Rev. B (USA) vol.24 (1981)
       p.1971]
[16]   J.R.Lowney, H.S.Bennett [J. Appl. Phys. (USA) vol.53 (1982)
       p.433]
[17]   H.S.Bennett [IEEE Trans. Electron Devices (USA) vol.ED-30 (1983)
       p.920]
[18]   H.S.Bennett [J. Appl. Phys. (USA) vol.59 no.8 (15 Apr 1986)
       p.2837-44]
[19]   J.R.Lowney [J. Appl. Phys. (USA) vol.59 no.16 (15 Mar 1986)
       p.2048-53]
[20]   H.S.Bennett, C.L.Wilson [J. Appl. Phys. (USA) vol.55 (May 1984)
       p.3582]
[21]   H.S.Bennett [Solid-State Electron. (GB) vol.26 (1983) p.1157]
[22]   G.D.Mahan [J. Appl. Phys. (USA) vol.51 (1980) p.2634]
[23]   A.Ghazali, J.Serre [Phys. Rev. Lett. (USA) vol.48 (1982) p.886]
[24]   R.A.Abram, G.N.Childs, P.A.Saunderson [J. Phys. C (GB) vol.17
       (1984) p.6105]

7.4   EFFECTIVE MASS OF ELECTRONS IN Si

by D.J.Dunstan

September 1987
EMIS Datareview RN=16168

The lowest conduction band minima in crystalline silicon are along the
delta directions (k,0,0), approximately 85% of the way to the
Brillouin zone edge from the zone centre. Because of the lowered
symmetry, the surfaces of constant energy are ellipsoidal and the
effective mass is anisotropic.

Early measurements by Dresselhaus et al [1] and Dexter et al [2] by
cyclotron resonance gave the longitudinal effective mass ml = 0.97 +-
0.02 and 0.98 +- 0.04 respectively and the transverse value was mt =
0.19 +- 0.01. These values correspond to a density of states effective
mass of 0.33. Subsequently, Hensel et al [3] measured cyclotron
resonance with uniaxial strain (to lift the valley degeneracy) and,
with an omega-tau product of 160 obtained values of ml = 0.09163 +-
0.0004 and mt = 0.1905 +- 0.0001. Ning and Sah [4] used these values to
obtain an equivalent m* = 0.2982 for calculating binding energies in a
single spherical valley approximation and this value agrees with the m*
= 0.23 obtained by Lawaetz [5] on theoretical grounds.

Recent authors have used the value ml = 0.98 (see, e.g. [6,7])
usually without references or discussion. However, the accepted
best values are those of Hensel et al given above (see, e.g. [8]).

Away from the conduction band minima, the effective mass increases
(non-parabolicity). This effect can be parametrised using

$$m(e) = mo (1 + 2 A E)$$

where mo is the value at the bottom of the band. For the silicon band
edge, A has the value 0.5/eV [7]. All the effective masses also
increase with temperature and with doping density, for details see [8].

REFERENCES
----

[1]   G.Dresselhaus, A.F.Kip, C.Kittel [Phys. Rev. (USA) vol.98 (1955)
      p.368]
[2]   R.N.Dexter, H.J.Zeiger, B.Lax [Phys. Rev. (USA) vol.104 (1956)
      p.637]
[3]   J.C.Hensel, H.Hasegawa, M.Nakayama [Phys. Rev. (USA) vol.138
      (1965) p.A225]
[4]   T.H.Ning, C.T.Sah [Phys. Rev. B (USA) vol.4 (1971) p.3468]
[5]   P.Lawaetz [Phys. Rev. B (USA) vol.4 (1971) p.3460]
[6]   S.M.Sze [Physics of Semiconductor Devices (Wiley, New York, USA,
      1981)]
[7]   C.Jacobini, L.Reggiani [Rev. Mod. Phys. (USA) vol.55 (1983)
      p.645]; and Adv. Phys. (GB) vol.28 (1979) p.493]
[8]   [Landolt-Bornstein, Zahlenwerte und Funktionen, New Series III-17a
      Section 1.2.1]

7.5 EFFECTIVE MASS OF HOLES IN Si

by D.J.Dunstan

August 1987
EMIS Datareview RN=16117

The top of the valence band in silicon is typical of tetrahedral
semiconductors. The $J = 3/2$ states are degenerate at the Brillouin zone
centre, and are spin-orbit split from the lower-lying $J = 1/2$ states.
Away from the zone centre, the fourfold $J = 3/2$ degeneracy is not
allowed by symmetry and the splitting can be described in terms of a
heavy hole and a light hole band [1,2]. The bands are not spherical, and
so effective masses are approximations appropriate for some purposes
only. From cyclotron resonance, parameters describing the valence band
precisely may be determined, from which the effective masses may be
derived approximately. Ref [3] gives $A = 4.0$, $B = 1.1$ and $C = 4.0$, from
which Cardona [4] derives the zone centre heavy hole mass $mh = 0.52$,
the light hole mass $ml = 0.16$; Dresselhaus et al [3] gave the spin-orbit
split-off mass $mso = 0.25$.

Effective masses may also be derived from k.p theory [5] from which
$mh = 0.49$ and $ml = 0.16$. Accepted values for the effective masses are:

$mh = 0.49$ [6]

$ml = 0.16$ [5,6]

$mso = 0.24$ [5]

Going away from the zone centre, the heavy hole band is most
non-parabolic in the (110) directions and the light hole band in the
(111) directions. In the region of the spin-orbit splitting energy,
44 meV, the bands are not parabolic and an effective mass is not an
appropriate approximation. At energies much higher than the spin-orbit
splitting, the bands are again parabolic, with effective masses [7]

$mh = 1.26$

$ml = 0.36$

REFERENCES
----

[1]   G.F.Koster, J.O.Dimmock, R.G.Wheeler, H.Statz [Properties of the
       32 Point Groups (MIT Press, Ma, USA, 1963)]
[2]   R.A.Smith [Semiconductors (Cambridge University Press, 1978)]
[3]   G.Dresselhaus, A.F.Kip, C.Kittel [Phys. Rev. (USA) vol.98 (1955)
       p.368]
[4]   M.Cardona [J. Phys. & Chem. Solids (GB) vol.24 (1963) p.1543]
[5]   P.Lawaetz [Phys. Rev. B (USA) vol.4 (1971) p.3460]
[6]   S.M.Sze [Physics of Semiconductor Devices (Wiley, New York, USA,
       1981)]
[7]   C.Jacoboni, L.Reggiani [Rev. Mod. Phys. (USA) vol.44 (1983)
       p.645]

CHAPTER 8

PHOTOCONDUCTIVITY AND PHOTOGENERATED CARRIER LIFETIMES

8.1 PHOTOCONDUCTIVITY OF Si: GENERAL REMARKS

August 1987
EMIS Datareview RN=15793

In Si excess carriers can be created by light absorption in the
fundamental band, photon energy greater than or equal to 1.1 eV, or by
light absorption from bound levels in the forbidden gap, photon energy
< 1.1 eV. Also creation of excess carriers can be performed by e.g.
electron beam and injection from a rectifying contact. Photoconductivity
(PC) measurements can be divided into two types: steady state PC and
non-steady state PC; both are developed for uniform and non-uniform
samples. Some of the methods are published by Bube [1] Ryvkin [2],
Milnes [3] and Marfaing [4]. The steady state PC, based on the spectral
absorption cross section dependence, is an excellent tool for
monitoring deep energy levels in Si (Grimmeiss et al [5]). A typical
example of monitoring shallow impurities at low temperatures in the case
of Si:In is given by Peschel et al [6]. The steady state methods are
very effective for high resistivity, highly compensated Si material and
for the structure with pn-junction. Recombination lifetime, tau,
diffusion coefficient, D, and carrier mobility, mu, are semiconductor
parameters that characterise the time-resolved PC. The recombination
carrier lifetime is defined as the time required for the concentration
of excess carriers in the material to decay by recombination to 1/e of
its initial value. The recombination lifetime depends to some extent on
the excess carrier concentration, n. For both the Shockley-Read-Hall
(SRH) process and for the Auger type process, this dependence saturates
for high and low concentrations of the excess carriers and
correspondingly the high-level (HL) and the low-level (LL) carrier
lifetimes are defined.

The diffusion length L of excess carriers is related to the carrier
lifetime tau according to the equation $L*2 = D \times tau$, where D is the
relevant diffusion coefficient corresponding to minority carriers in
the LL case (n << no) and ambipolar for the HL case (n >> no). The
diffusion coefficient is related to the mobility of carriers by the
Einstein relation (Meyer [7]). In general, for the HL case, tau, D and
mu at intermediate or low doping level depend to a great extent on the
Si band structure and on the carrier effective mass (Auger type band-
band recombination, multicarrier scattering, band-filling and
band-narrowing). Accurate measurements of the tau, D and mu parameters
in the LL case and at high doping levels are important for material
and processing specialists as a tool for monitoring lattice defects
caused by contamination or improper thermal treatment.

The classical PC decay method for measuring recombination lifetime was
improved by using a pulsed laser source (Blair and Seiler [8]).
Material characterisation by contactless PC measurement can be carried
out by light induced absorption (Warabisako et al [9]), light-induced

reflection (Bergner and Bruckner [10]), time-resolved microwave conductivity (Kunst and Werner [11]) and other methods. In this way, the exponential decay of the signal time can be easily controlled. Deviations from exponential decay are usually not observed except for:

1) small samples with high tau values, wherein the carriers mainly recombine at the surface (Luke and Cheng [12], Grivitskas et al [13])

2) crystals with highly inhomogeneous carrier lifetimes (Grivitskas et al [14])

3) sweep out of carriers at one of the contacts because of high electric fields

4) samples containing traps.

Finally carrier parameters tau, D and mu of silicon crystals after heat treatment are influenced by the following conditions (Greff and Fischer [15]):

1) atmosphere during heat treatment
2) cleanliness of the furnace and substrate
3) rate of cooling the sample to room temperature
4) type of crystal conductivity
5) purity of the crystal
6) cleanliness of the sample surface.

Some time elapsed effects of the parameters are observed in the cases of sawing crystals, etching crystals, illumination by light and quenching after heating [15], as well as after ion implantation (Giedrys et al [16]), and after long heating at T > 1200 deg C or irradiation by gamma rays (Bolotov et al [17]).

REFERENCES
----

[1]   R.H.Bube [Photoconductivity of Solids (Wiley, NY, USA, 1960)]
[2]   S.M.Ryvkin [Photoelectric Effects in Semiconductors (Cons. Bureau, NY, USA, 1964)]
[3]   A.G.Milnes [Deep Impurities in Semiconductors (J. Wiley and Sons, NY, USA, 1973)]
[4]   Y.Marfaing [in 'Handbook of Semiconductors' vol.2 Ed. M.Balkanski (North Holland, 1980) ch.7]
[5]   H.G.Grimmeiss [Annu. Rev. Mater. Sci. (USA) vol.7 (1977) p.341]
[6]   W.Peschel, R.Kuhnert, M.Schulz [Appl. Phys. A (Germany) vol.A30 (1983) p.59-62]
[7]   J.R.Meyer [Phys. Rev. B (USA) vol.21 (1980) p.1554-8]
[8]   P.D.Blair, C.F.Seiler [Lifetime Factors in Silicon Ed. J.B.Wheeler et al, ASTM STP 712 (ASTM, Baltimore, 1980) p.148-58]
[9]   T.Warabisako, T.Saitoh, T.Motooko, T.Tokuyama [Jpn. J. Appl. Phys. Suppl. (Japan) vol.22 (1983) no.22-1 p.577-60]
[10] H.Bergner, V.Bruckner [Opt. & Quantum. Electron. (GB) vol.15 (1983) p.477-85]
[11] M.Kunst, A.Werner [Mater. Res. Soc. Symp. Proc. (USA) vol.69 (1986) p.361-6]

[12] Keung L.Luke, L.-J.Cheng [J. Appl. Phys. (USA) vol.61 no.6
      (15 Mar 1987) p.2282-93]
[13] V.Grivitskas, M.Willander, J.A.Tellefsen [11th Nordic
      Semiconduct. Meeting, ESPOO, Finland (1984) p.123-6]
[14] V.Grivitskas, D.Noreika, V.Amstibovskij [Sov. Phys.-Collect.
      (USA) vol.28 no.1 (1988)]
[15] K.Graff, H.Fischer [Top. Appl. Phys. (Germany) vol.31 (1979)
      p.173-211]
[16] T.Giedrys, V.Grivickas, L.Pranevicius, A.Ragauskas, J.Vaitkus
      [Nucl. Instrum. & Methods Phys. Res. Sect. B (Netherlands) vol.B6
      no.1/2 (Jan 1985) p.427-9]
[17] V.V.Bolotov, V.A.Korotchenko, A.P.Mamontov, A.B.Rzchanov,
      L.S.Smirnov, S.S.Shaimeev [Sov. Phys.-Semicond. (USA) vol.14
      (1980) p.2257-60]

8.2 PHOTOCONDUCTIVITY SPECTRA OF ION-IMPLANTED Si

by M.Willander and B.G.Svensson

July 1987
EMIS Datareview RN=15794

Only a few reports exist in the literature on PC-spectra from ion
implanted silicon and furthermore, PC-measurements are not a common
topic at international conferences related to ion implantation and
atomic collision effects in solids. A possible explanation may be
that energetic atomic particles create large defect clusters due to
the high density of energy deposited into atomic displacements, and
as a result, well-defined energy levels are hard to observe. Similar
ideas were discussed by Kalma and Corelli [1] who found a broad
'energy band' over-riding single-defect levels in samples irradiated
by 45 MeV electrons, which create a substantial amount of complex
defects. For 1.5 MeV electrons, which predominantly generate 'simple'
defects, very little of the broad 'energy band' was observed and the
single levels dominated.

Netange et al [2] implanted high-resistivity n- and p-type silicon
with 100 keV B+ ions at room temperature (RT) to doses of about
10*14/cm*2 and obtained spectra similar to those for electron-
irradiated samples, i.e., a major level at 0.32 eV and an important
increase of the photoconductivity at about 0.4 eV. The level at 0.32 eV
was attributed to a transition of the divacancy from a singly positive
to a neutral charge state, in accordance with results for neutron-
irradiated p-type silicon [3]. Cheng [3] based his identification
on similarities of band shape and stress-induced dichroism of the
0.32 eV PC-band with the corresponding properties of the divacancy
measured by other experimental techniques (infrared absorption and
electron paramagnetic resonance (ESR)). According to ESR and deep
level transient spectroscopy (DLTS) measurements the transition from
neutral to singly positive charge state of the divacancy occurs
0.25 eV above the valence band which, however, may not be directly
comparable to the 0.32 eV level due to the different excitation
processes involved [3].

Netange et al [2] associated the sharp increase near 0.4 eV in the
PC-spectra with the so-called Di-defects [4] (in this case
interstitial boron, B(i)). However, its temperature stability (about
300 deg C) does not agree with that for B(i) since interstitial boron
atoms are mobile at RT, and Watkins [5] showed that B(i) anneals out
below RT. Consequently, we conclude that the 0.4 eV transition is not
due to B(i), but some other irradiation induced defect. According to
results from other experimental techniques, e.g. DLTS [6], several
energy levels assigned to different defects appear in the region of
0.4 eV above the valence band.

PC-spectra of silicon samples (p-type, 80-2000 ohm cm) implanted with 10*15/cm*2 selenium ions and subsequently annealed at 1200 deg C showed two of three well-known levels found in selenium diffused samples [7], namely the levels at Ec-0.19 eV (attributed to Se2-centres) and Ec-0.27 eV. Astrova et al [7] performed experiments with two different cooling procedures (rapid cooling in air and quenching in cooled water) and found a strong influence on the PC-spectra. A considerable reduction of the concentration of the levels was observed in the quenched samples but no conclusive explanation was presented. Furthermore, a numerical value for the photo-ionisation cross section of the level Ec-0.19 eV was extracted and found to be about 100 times smaller than that of the level Ec-0.27 eV.

The level positions of indium and thallium in silicon, introduced by ion implantation and a subsequent anneal, have been determined by Peschel et al [8] using PC-measurements. In order to increase the sensitivity of the measurements, a pn-junction was formed between the implanted thin film and the silicon substrate by an additional implant. During the PC-measurements this pn-junction was reverse biased by applying a voltage to a back-side contact, and as a result an improvement of the sensitivity by 2-3 orders of magnitude was achieved. Good agreement was found between the implanted films and bulk wafers with respect to spectral sensitivity and positions of the levels. The ionisation energies were extracted by a novel method which separates the contribution of photothermal excitation to photoconductivity via excited bound states from direct photoexcitation into the band continuum, utilising the fact that the two contributions have different temperature dependence.

Finally, PC-spectra of silicon implanted with high doses of carbon (5.5 X 10*17 ions/cm*2) contain a dominant peak at 0.37 eV and two additional transitions at 0.73 and 1.0 eV [9]. After a 30 min anneal at 1100 deg C the peaks disappear. Akimchenko et al [9] assigned the level at 0.37 eV to the singly negatively charged divacancy (V2)-. Provided that the interaction takes place with the conduction band, this may be consistent with results from DLTS where the transition from (V2)- to V2(0) has been found to occur at Ec-0.41 eV [6]. However, it must be emphasised that several defect levels appear close to Ec-0.4 eV and the assignment made by Akimchenko et al [9] must be regarded as tentative, especially since the material is reported to be hole conductive. The transition at 1.0 eV was attributed to interstitial carbon which is in fair agreement with an acceptor level about 0.12 eV below the conduction band, as found by other experimental techniques, if the interaction takes place with the valence band. On the other hand interstitial carbon atoms are mobile at RT and form more stable complexes by pairing and/or reactions with other impurities or defects. Thus, isolated interstitial carbon atoms are not expected to exist after implantation at RT (or elevated temperatures) and therefore, the 1.0 eV transition is believed to be due to some other irradiation induced defect which may be carbon related.

REFERENCES
----

[1]   A.H.Kalma, J.C.Corelli [in 'Radiation Effects in Semiconductors'
      Ed. F.L.Vook (Plenum Press, New York, 1968) p.153]

[2]   B.Netange, M.Cherki, P.Baruch [Appl. Phys. Lett. (USA) vol.20 no.9
      (1 May 1972) p.349-51]

[3]   L.J.Cheng [in 'Radiation Effects in Semiconductors' Ed. F.L.Vook
      (Plenum Press, New York, 1968) p.143]

[4]   M.Cherki, A.H.Kalma [Phys. Rev. B (USA) vol.1 no.2 (Jan 1970)
      p.647-57]

[5]   G.D.Watkins [Phys. Rev. B (USA) vol.12 no.12 (Dec 1975)
      p.5824-39]

[6]   L.C.Kimerling [Inst. Phys. Conf. Ser. (GB) no.31, Radiation
      Effects in Semiconductors, Dubrovnik, Sept 1976, Eds N.B.Urli,
      J.W.Corbett (Inst. Phys., Bristol, 1977) p.221-30]

[7]   E.V.Astrova, I.B.Bol'shakov, A.A.Lebedev, O.A.Mikhno [Sov. Phys.
      Semicond. (USA) vol.19 no.5 (May 1985) p.565-7]

[8]   W.Peschel, R.Kuhnert, M.Schulz [Appl. Phys. A (Germany) vol.A30
      (1983) p.59-62]

[9]   I.P.Akimchenko, K.V.Kisseleva, V.V.Krasnopevtsev, A.G.Touryanski,
      V.S.Vavilov [Radiat. Eff. (GB) vol.48 (1980) p.7-11]

8.3 PHOTOCONDUCTIVITY SPECTRA OF ELECTRON IRRADIATED Si

by M.Willander and J.L.Lindstrom

July 1987
EMIS Datareview RN=15795

Many of the known defects in silicon introduced by electron irradiation
have been investigated by photoconductivity (PC) measurements. The
method is direct and sensitive. As these defects often are anisotropic
and can be preferentially aligned by uniaxial stress it is often
possible to get more detailed information about a defect using stress
and polarised light in connection with PC [1]. When doing the
measurements one has usually to vary the frequency of the chopped light
to optimise the PC signal [2]. A possible influence from defects at
the surface should also be considered. When interpreting the results
there can be a problem in determining the background signal for a
particular defect level since the PC signal is additive. The early work
using PC on e-irradiated silicon has been summarised by Fan and Ramdas
[3]. Later reviews have been done by Vavilov [4] and Curtis [5].
TABLE 1 is a summary of the reported energy levels from PC on electron
irradiated silicon.

One of the most studied defects in silicon, the vacancy-oxygen centre
(A-centre), appears in two charge states VO(0) and VO(-). It acts as
an acceptor and gives rise to a level 0.16-0.18 eV below the
conduction band. There is a spread in values reported from different
experimental techniques (resistivity, Hall effect and PC). From PC
the value Ec-0.16 eV is found, see TABLE 1. Since the formation of
A-centres requires the presence of vacancies and oxygen atoms, a great
variation in concentration can occur for a certain electron fluence
depending upon the concentration of oxygen and dopants. Impurities
trapping vacancies, like P and Sn will lower the A-centre
concentration, while the presence of traps for self-interstitials will
suppress the vacancy-self-interstitial annihilation and therefore will
give a higher concentration. Thus the conditions for observing the
A-centre with PC are that the irradiation has to be performed on a
suitable material and to an electron fluence so that the Fermi level
is above Ec-0.16 eV.

Another important defect studied in PC is the divacancy. This defect
appears in four different charge states. Levels at Ec-0.36, 0.39, 0.54
and Ev+0.25 eV have been associated with it [6,7]. With the Fermi level
below Ec-0.22 eV only the 0.54 eV level could be observed [6]. Kalma
and Corelli [6] used a technique reported by Cheng [8] to study the
Ec-0.39 and Ec-0.54 eV levels using polarised light and stress
induced dichroism.

They found excellent agreement with results from EPR and IR-absorption measurements and concluded that the two levels are related to the same defect, the divacancy. Several defects formed by interaction between vacancies, interstitials and dopants have been reported.

In phosphorus doped material the strength of the level at E(c)-0.43 eV increases with increasing phosphorus doping at constant electron fluence. The level has been assigned to the vacancy-phosphorus centre (E-centre) [7].

Since the mobility of the self-interstitial in silicon has a different temperature dependence than the mobility of the vacancy, experiments have been performed where the temperature of the sample during irradiation has been varied.

Vavilov et al [9] observed a continous distribution of allowed levels introduced by electron irradiation at 80 deg K. When raising the temperature to 300 deg K they disappeared and new levels appeared at Ev+0.45 eV in boron-doped FZ material and at Ec-0.16, Ev+0.30 and Ev+0.45 eV in high oxygen material.

Matsui and Baruch [7] found energy levels at 0.29, 0.445, 0.525 and 1.0 eV above the valence band in p-type material irradiated at 20 deg K by 1.5 MeV electrons. In n-type material levels at 0.16, 0.43 or 0.48 eV below the conduction band were observed.

Cherki and Kalma [10] investigated electron-irradiated boron and aluminium doped silicon irradiated at 4.2, 20.4, 77 and 300 deg K. They observed at all temperatures the same defect levels Ev+0.43 eV (+- 0.005 eV) in boron doped and Ev+0.39 eV (+- 0.010 eV) in aluminium doped silicon. Their conclusion was that, as only the self-interstitial is mobile at the lowest of these temperatures, the observed defects are formed when a self-interstitial changes place with a substitutional boron or aluminium atom. They also did stress-induced dichroism measurements and found a C3v symmetry of these defects. Both defects annealed in the temperature range of 250 to 300 deg C.

Li-doped silicon irradiated with 1.5 MeV electrons at 77 deg K was studied by Caillot, Cherki and Kalma [11]. In FZ material they observed at least four kinds of complexes. They discuss different formation and annealing mechanisms and found a Li-concentration dependence of the annealing. In pulled crystals they observed two different complexes which are stable up to > 730 deg K. These were interpreted as Li-O (or vacancy) complexes.

TABLE 1: energy levels in electron-irradiated silicon observed by photoconductivity measurements

Position of level	Defect assignment	References
Ec-0.16 eV	vacancy-oxygen (A-centre)	[9,3,7]
Ec-0.36 eV	divacancy	[6]
Ec-0.39 eV	divacancy	[6,11]
Ec-0.43 eV	phosphorus-related	[7]
Ec-0.48 eV	arsenic-related	[7]
Ec-0.54 eV	divacancy	[6]
Ev+1.0 eV	unknown	[7]
Ev+0.53 eV	unknown	[7]
Ev+0.525 eV	unknown	[7]
Ev+0.445 eV	unknown	[7]
Ev+0.45 eV	unknown	[9]
Ev+0.430 eV	boron-related	[10]
Ev+0.39 eV	aluminium-related	[10]
Ev+0.35 eV	unknown	[10]
Ev+0.30 eV	interstitial-oxygen complex?	[9,7]
Ev+0.25 eV	divacancy	[7]

REFERENCES
----

[1]   J.W.Corbett [Electron Radiation Damage in Semiconductors and
       Metals, Solid State Physics (USA) . Supp.7 (Academic Press, New
       York, 1966)]
[2]   P.Vajda, L.J.Cheng [in 'Radiation Effects in Semiconductors',
       Ed. J.W.Corbett, G.D.Watkins (Gordon & Breach, New York, 1971)
       p.245]
[3]   Y.H.Fan, A.K.Ramdas [J. Appl. Phys. (USA) vol.30 (1959) p.1127]

[4]   V.S.Vavilov [Radiation Damage in Semiconductors (Dunod & Cie.,
        Paris, 1965) p.115]
[5]   O.L.Curtis,Jr [in 'Point Defects in Solids' vol.2 Eds
        J.H.Crawford,Jr, L.M.Slifkin (Plenum Press, New York-London,
        1975)]
[6]   A.H.Kalma, J.C.Corelli [in 'Radiation Effects in Semiconductors'
        Ed. F.L.Vook (Plenum, New York, 1968) p.153]
[7]   K.Matsui, P.Baruch [Lattice Defects in Semiconductors, (Univ. of
        Tokyo Press, Tokyo, 1968) p.282]
[8]   L.J.Cheng [Phys. Lett. A (Netherlands) vol.24 A (1967) p.729]
[9]   V.S.Vavilov, S.I.Vintovkin, A.S.Lyutovich, A.F.Plotnikov,
        A.A.Sokolova [Sov. Phys.-Solid State (USA) vol.7 (1965) p.399]
[10]   M.Cherki, A.H.Kalma [IEEE Trans. Nucl. Sci. (USA) vol.NS-16
        (1969) p.24]
[11]   M.Calliot, M.Cherki, A.H.Kalma [Phys. Status Solidi a (Germany)
        vol.4 (1971) p.121]

8.4 LIFETIMES OF PHOTOGENERATED CARRIERS IN UNDOPED Si

by M.Willander and V.Grivickas

July 1987
EMIS Datareview RN=15796

We define undoped Si as a material which is grown without intentional doping or intentional compensation of impurities and with resistivity 50 ohm cm or above at 300 deg K. In most cases, the carrier lifetime in undoped Si can be expressed as the single constant, tau, for photo-excitation of the carriers in the fundamental absorption band.

For injected carrier density below 10*17/cm*3 at 300 deg K, the carrier lifetime is well described by the capture-emission process at a single deep recombination trap, the SRH model [1,2]. The density of the recombination defects usually is in the 10*11 - 10*14/cm*3 range. It has been observed that the upper limit of tau is approximately 10*-3 s for non-heated FZ Si samples [3-5]. Fossum and Lee [6] have shown that this fundamental (unavoidable) defect is of the acceptor-type, possibly a vacancy or a divacancy complex. The activation energy of the defect, /Ed - Ei/, is approximately 0.2 eV at 300 deg K, and the capture parameters for electrons and holes are Cno = Cp- = approx. 10*-8/cm*3 s.

To create the defect in the Si lattice an activation energy of 1.4 eV is required and the solubility temperature for the defect is 300-400 deg C. The density of the fundamental defects depends also on the doping level. However, the defects are more soluble for n-type Si than for p-type Si, which suggests that the fundamental limit for the electron lifetime is longer than that for the hole at a given unavoidable doping density.

Passari and Susi have shown [7] that the carrier lifetime in CZ Si varies between 10*-5 and 10*-4 s both for electrons and holes. FZ Si has a longer carrier lifetime than CZ Si which can be attributed to the unavoidable presence of oxygen and carbon in CZ crystals. Although these impurities are not electrically active, they create recombination trap complexes with lattice defects [8].

In silicon on sapphire (SOS), the carrier lifetime, as shown by Grivitskas et al [9], is strongly dependent on the distance of the carrier from the sapphire substrate. This carrier lifetime is mainly influenced by lattice distortion, internal stress and autodoping from the substrate [10].

For FZ, CZ and SOS, the photogenerated carrier lifetime depends weakly on the temperature in the range 100-600 deg K. If the samples are annealed to high temperatures, the carrier lifetime will be changed. At 450 deg C and annealing time t = 10-300 h, the lifetime increases, typically up to ten times due to oxygen related thermal donor formation [11,12]. When heating to 450-650 deg C the lifetime decreases down to 10*-6 s. After annealing to temperatures higher than 650 deg C the carrier lifetime depends strongly on the preheating procedure and the oxygen and carbon diffusion, precipitated complex formation, dislocation generation and wafer contamination. Usually the Si samples have a nonuniformity of the carrier lifetime distribution, depending on the thickness and radius of the wafer [8]. The temperature induced recombination traps can be soluble in the matrix only after heating to 1300 deg C and to higher temperatures.

If the photogenerated carrier density is higher than 10*17/cm*3, then the lifetime in undoped Si decreases approximately with the inverse square of the injected carrier density. This dependence is caused by the fundamental band-band type recombination. The radiative recombination rate coefficient has been determined as 10*14 cm*3/s by Gerlach et al [13]. Thus, it is possible that the radiative recombination rate can come up to roughly 10% of the SRH and the Auger type recombination rate, when the injection density is in the range 10*17 - 10*18/cm*3. It has been shown that Auger recombination in Si can be of direct n-n-p type [14] with an activation energy of En less than or equal to 50 meV, direct p-p-n type [15] with Ep = 25-125 meV and phonon-assisted type (PAA) [16,17] with Ep = approx. 0. There is a big experimental uncertainty in the carrier lifetime at high carrier density mainly due to photo-induced heating, absorption, surface recombination and many-body effects. The total Auger coefficient, gamma n + gamma p, has been determined to be in the range 10*-30 - 10*-32 cm*6/s. In the degenerated carrier plasma the phonon-assisted Auger recombination dominates (gamma n + gamma p = approx. 10*-31 cm*6/s for PAA recombination). This conclusion can be drawn from the investigations of the temperature dependence of gamma n + gamma p [18], the dependence on the occupation of the conduction band valley [19] and the ratio in e-h drops at low temperature [20]. For carrier densities in the intermediate range i.e. around 10*18/cm*3 and for T = 200-400 deg K the experimental results of Grivitskas and Vaitkus [21,22] can be interpreted with the direct Auger recombination rate coefficient being of the order of 10*-30 cm*6/s.

REFERENCES
----

[1]   W.Shockley, W.T.Read [Phys. Rev. (USA) vol.87 (1952) p.835]
[2]   N.Hall [Phys. Rev. (USA) vol.87 (1952) p.387]
[3]   D.L.Kendall [Solar Cells: Outlook for improved efficiency
       (Washington, NRC, 1972)]
[4]   K.Graff, H.Pieper [Lifetime factors in Si (ASTM STP 712)
       Philadelphia (1980)]
[5]   R.D.Nasby, C.M.Garner, H.T.Weaver, F.W.Sexton, J.L.Rodriguez
       [15th IEEE Photovoltaic Specialists Conf., Kissimee, FL, USA
       12-15 May 1981 (IEEE, New York, USA, 1981)]
[6]   J.G.Fossum, D.S.Lee [Solid-State Electron. (GB) vol.25 (1982)
       p.741]
[7]   L.Passari, E.Susi [J. Appl. Phys. (USA) vol.54 no.7 (Jul 1983)
       p.3935-7]
[8]   K.V.Ravi [Imperfections and Impurities in Semiconductors (John
       Wiley and Sons, NY, USA, 1981)]
[9]   V.Grivitskas, M.Willander, J.A.Tellefsen [J. Appl. Phys. (USA)
       vol.55 no.8 (15 Apr 1984) p.3169-72]
[10]   S.Cristoloveanu, G.Gribando, G.Kamarinos [Rev. Phys. Appl.
       (France) vol.19 (1984) p.161]

[11]   K.D.Glinchuk, N.M.Litovchenko [Phys. Status Solidi a (Germany)
       vol.58 (1980) p.549]

[12]   L.F.Makarenko, L.I.Murin [Sov. Phys.-Semicond. vol.20 no.8 (Aug
       1986) p.961-2]

[13]   W.Gerlach, H.Schlangenotto, H.Maeder [Phys. Status Solidi a
       (Germany) vol.13 (1972) p.277]

[14]   L.Huldt [Phys. Status Solidi a (Germany) vol.8 (1971) p.173]

[15]   V.Grivitskas, Yu.Vaitkus [Sov. Phys.-Semicond. (USA) vol.13 no.11
       Nov 1979) p.1226-8]

[16]   A.Haug [Solid State Commun. (USA) vol.28 no.3 (Oct 1978)
       p.291-3]

[17]   W.Lochmann [Phys. Status Solidi a (Germany) vol.45 no.2 (16 Feb
       1978) p.423-32]

[18]   L.Huldt, N.G.Nilsson, K.G.Svantesson [Appl. Phys. Lett. (USA)
       vol.35 no.10 (15 Nov 1979) p.776-7]

[19]   L.A.Delimova [Sov. Phys.-Semicond. (USA) vol.15 (1981) p.778]

[20]   A.Haug [Solid State Commun. (USA) vol.25 no.7 (Feb 1978)
       p.477-9]

[21]   Yu.Vaitkus, V.Grivitskas [Sov. Phys.-Semicond. (USA) vol.15 (1981)
       p.1102]

[22]   V.Grivickas [Sov. Phys.-Collect. (USA) vol.22 (1982) p.83]

8.5 LIFETIMES OF PHOTOGENERATED CARRIERS IN DOPED Si

by M.Willander and V.Grivickas

August 1987
EMIS Datareview RN=15797

We define doped Si as a material with intentional doping during crystal growth or by diffusion, ion implantation or neutron transmutation.

In shallow impurity doped Si the minority carrier lifetime was measured by Passari et al [1]. It was shown that for a majority carrier concentration less than 10*17/cm*3, the minority carrier lifetime is approximately equal to that of undoped Si. For concentrations higher than 10*17/cm*3, the carrier lifetime decreases approximately by the inverse quadrate of the increasing majority carrier concentration. Analyses carried out by Fossum et al [2] and Tyagi and Van Overstraeten [3] show that the recombination in the high doping region can be explained by a superposition of the SHR-type mechanism through the single deep level and by the phonon-assisted band-band Auger recombination with the recombination rate coefficient approximately equal to 10*-31 cm*6/s. The scattering of the lifetime data which is within one order of magnitude at the intermediate doping region, depends on the technological processing and on the uncertainty of the measuring technique.

In Si doped with a deep impurity, the carrier lifetime decreases very significantly which is an important device implication.

The decrease in carrier lifetime is inversely proportional to the deep level concentration and the carrier capture cross section sigma i, where i = n,p, as well as for the trap occupancies, which may vary during the transient. For attractive defects in Si the carrier cross sections can be expressed according to the Lax cascade capture theory [4]. Abakumov et al [5] have compared the temperature dependence of sigma for most impurities in Si. At 300 deg K, the values for sigma are usually in the range 10*-15 - 10*-16 cm*2. The temperature dependence can be approximated to sigma proportional to T*-3 for kT much larger than m X s*2, and to sigma proportional to T*-1 for kT much less than m X s*2, where m=carrier effective mass and s=sound velocity. For neutral defects, sigma is essentially independent of the temperature. Typical values for the capture cross sections of the gold and acceptor states were determined by Wu and Peaker [6]. Most deep impurities have the soluble concentration in Si in the interval 10*16 - 10*17/cm*3 [7] and therefore the lowest values of carrier lifetimes are in the 10*-9 - 10*-7 s range at T = 300 deg K. Some high values of sigma (10*-15 - 10*-16 cm*2) obtained for neutral deep impurities in Si were interpreted as a trap-assisted Auger recombination [7]. At low temperatures the lifetimes of the excitons bound to different shallow impurities (N = 10*17 - 3 X 10*18/cm*3) have been measured by Schmid [8] and Osbourn et al [9]. The recombination is an Auger type localised recombination. There is a strong dependence of the carrier lifetime, tau, on the impurity binding energy, the tau function for acceptors is proportional to Ea*-4.5 and the tau function for donors is proportional to Ed*-3.9. The lifetime values vary from 10*-6 s for Si:B to 2.7 X 10*-9 s for Si:In and they are independent of N.

For ion-implanted, electron, proton, neutron or Co-60 gamma irradiated
Si, the carrier lifetime is strongly dependent on the introduced
radiation defects. This is usually described by the lifetime
degradation factor K defined by: 1/tau = 1/tau(c) + K X Fi where tau is
the lifetime after irradiation, tau(c) is the lifetime before
irradiation and Fi is the irradiation fluency of particles/cm*2.
Typical values of K for undoped materials vary between 10*-7 to 10*-9
cm*2/s for Fi = 10*13 - 10*15/cm*2 and K has a dose [10], energy [11]
and doping dependence [12]. Among recombination centres, the A centre
(Ec - 0.18 eV) with sigma An = 10*-14 cm*2 and the divacancy centre
(Ec - 0.23 eV) with sigma Fn = 2 X 10*-15 cm*2 at 300 deg K are very
active recombination centres which are responsible for the lifetime
control [10]. At low doses also the phenomenon of lifetime increase with
increasing Fi has been reported by Bolotov et al [13]. As was shown in
FZ Si by Willander [14] and in SOS by Smith et al [15] radiation damage
caused by ion-implantation changes the carrier lifetime by several
orders of magnitude and can be as short as 10*-11 - 10*-12 s at the
amorphisation dose.

The influence on carrier lifetime due to radiation damage of pure FZ-Si
from neutron transmutation was measured by Maekawa et al [16]. They
showed that a trapping effect of minority carriers (holes) will
occur. Annealing above 700 deg C will result in the lifetime beginning
to increase and around 1000 deg C a sharp increase in lifetime appears.

REFERENCES

[1]   L.Passari, E.Susi [J. Appl. Phys. (USA) vol.54 no.7 (Jul 1983)
       p.3935-7]
[2]   J.G.Fossum, R.P.Mertens, D.S.Lee, J.F.Nijs [Solid-State
       Electron. (GB) vol.26 (1983) p.569]
[3]   H.S.Tyagi, R.Van Overstraeten [Solid-State Electron. (GB) vol.26
       no.6 (Jun 1983) p.577-97]
[4]   M.Lax [Phys. Rev. (USA) vol.119 (1960) p.1502]
[5]   V.N.Abakumov, V.I.Perel, I.N.Jassievic [Sov. Phys.-Semicond.
       (USA) vol.12 (1978) p.1]
[6]   R.H.Wu, A.R.Peaker [Solid-State Electron. (GB) vol.25 (1982)
       p.643]
[7]   A.G.Milnes [Deep Impurities in Semiconductors (J. Wiley and
       Sons, 1973)]
[8]   W.Schmid [Solid-State Electron. (GB) vol.21 no.11-12 (Nov-Dec
       1978) p.1285-7]
[9]   G.C.Osbourn, S.A.Lyon, K.R.Elliott [Solid-State Electron. (GB)
       vol.21 no.11-12 (Nov-Dec 1978) p.1339-42]
[10]   S.D.Brotherton, P.Bradley [J. Appl. Phys. (USA) vol.53 (1982)
       p.5720]
[11]   N.I.Maslov, G.D.Pugachev, N.I.Kheitets [Sov. Phys.-Semicond.
       (USA) vol.16 (1983) p.328]
[12]   V.V.Mikhinovich [Sov. Phys. Semicond. (USA) vol.16 (1982)
       p.364]
[13]   V.V.Bolotov, V.A.Korotchenko, A.P.Mamontov, A.V.Rzhanov,
       L.S.Smirnov, S.S.Shalmeev [Sov. Phys.-Semicond. (USA) vol.14
       no.11 (Nov 1980) p.2257-60]
[14]   M.Willander [J. Appl. Phys. (USA) vol.56 (1984) p.3006]
[15]   P.R.Smith, D.H.Auston, A.M.Johnson, W.M.Augustyniak [Appl. Phys.
       Lett. (USA) vol.38 no.1 (1 Jan 1981) p.47-50]
[16]   T.Maekawa, S.Inove, A.Aiura, A.Usami [Semicond. Sci. & Technol.
       (GB) vol.1 no.5 (1986) p.305-12]

8.6 DIFFUSION AND MOBILITY OF PHOTOGENERATED CARRIERS IN UNDOPED Si

by M.Willander and V.Grivickas

August 1987
EMIS Datareview RN=15798

The ambipolar diffusion coefficient, Da, and the ambipolar mobility,
mu(e) + mu(h), are determined by different methods such as
time-resolved photoconductivity, free carrier absorption, free carrier
reflection, transient grating, transient photo-Hall or time-of-flight
techniques.

When the excited carrier density is lower than a critical concentration,
nc, the mobility and diffusion are controlled by the lattice scattering.
At 290 deg K the values of Da can be as high as 18.5 cm*2/s, mu(e) +
mu(h) as high as 1870 cm*2/V.s and the ambipolar Hall mobility
mu(H) = (re[mu(e)*2] - rh[mu(h)*2])/[mu(e)*2 + mu(h)*2] = 1200 cm*2/V.s.
These values agree well with known drift values for electrons (1450
cm*2/V.s), and holes (400 cm*2/V.s), and for the expression of Da
according to the Einstein relation [1].

If the excited carrier density, n, is higher than nc, then the
ambipolar mobility will decrease due to electron-hole (e-h)
scattering, which is in agreement with the Debye-Conwell mixed
scattering formula [2]. The decrease in mobility is more pronounced at
low temperatures. As was shown by Grivitskas et al [3] the empirical
temperature dependence for nc (/cm*3) is given by

nc = 10*16 X (T/300 deg K)*1.4 X 1800/[mu(e) + mu(h)]

(with mobilities in units of cm*2/V.s). In the high carrier
density region (n > 10*17/cm*3, 290 deg K) the additional effect of
electron-electron (e-e) and hole-hole (h-h) scattering takes place
as well. The intercarrier scattering has a dramatic influence on device
functions [4]. At 290 deg K the relation between the carrier
density and the conductivity-mobility, mu(i), can be expressed by the
empirical Caughey-Thomas formula [3]:

mu(i) = (mu+ - mu-)/(1 + (n/nref)*a) + mu-, where i = e,h.

The constants are listed in TABLE 1

TABLE 1

	mu+ (cm*2/V.s)	mu- (cm*2/V.s)	a	nref (/cm*3)
electrons:	1550	80	0.7	5.8 X 10*16
holes:	470	60	0.64	5.8 X 10*16

An increase of mu(e) + mu(h) in the degenerated region
(n > 10*19/cm*3) at 80 deg K has been reported by Auston et al [5].

In contrast to mobility, the ambipolar diffusion coefficient Da is not affected by e-h scattering, but many-body effects become important at high carrier density [6]. Bergner et al [7] have measured Da at 300 deg K and found a slow decrease of Da down to 15 cm*2/s for n =10*18 - 10*19/cm*3, and an increase in Da up to 45 - 60 cm*2/s in the degenerated region (n > 10*20/cm*3).

The influence of the electric field, E, on the ambipolar diffusion coefficient has been experimentally investigated by Vaitkus et al [8,9]. They measured a decrease of Da when E is parallel to the diffusion and an increase of Da when E is transverse to the diffusion for E less than or equal to 2 kV/cm, n = approx. 10*17/cm*3 and at room temperature.

REFERENCES
----

[1] C.Jacoboni, L.Reggiani [Adv. Phys. (GB) vol.28 no.4 (1979) p.493-553]

[2] P.Debye, E.Conwell [Phys. Rev. (USA) vol.93 (1954) p.693]

[3] V.Grivitskas, M.Willander, J.Vaitkus [Solid-State Electron. (GB) vol.27 no.6 (Jun 1984) p.565-72]

[4] F.Dannhauser [Solid-State Electron. (GB) vol.15 no.12 (Dec 1972) p.1371-5]; J.Krausse [Solid-State Electron. (GB) vol.15 no.12 (1972) p.1377-81]

[5] D.H.Auston, A.M.Johnson [Ultra-Short Light Pulses vol.18 (Springer-Verlag, NY, USA 1977) p.245-7]

[6] J.F.Young, H.M.van Driel [Phys. Rev. B (USA) vol.26 (1982) p.2147]

[7] H.Bergner, V.Bruckner, M.Schubert [J. Lumin. (Netherlands) vol.30 (1985) p.114]

[8] J.Vaitkus, L.Subacius, K.Jarasiunas [Sov. Phys.-Collect. (USA) vol.25 no.4 (1985) p.75-80]

[9] J.Vaitkus, K.Jarasiunas, E.Gaubas, L.Jonikas, R.Pranaitis, L.Subacius [IEEE J. Quantum ELectron. (USA) vol.QE-22 no.8 (Aug 1986) p.1298-305]

8.7 DIFFUSION AND MOBILITY OF PHOTOGENERATED CARRIERS IN DOPED Si

by M.Willander and V.Grivickas

September 1987
EMIS Datareview RN=15799

If the doping density, N, is larger than 10*15/cm*3, then the majority
carrier mobility and the diffusion coefficient will begin to be
dependent on N at 300 deg K. This dependence is mainly caused by ionised
impurity scattering. At low temperatures, ionised impurity scattering is
more important, since the relaxation time is proportional to T*-1.5
which is in contrast to the lattice scattering relaxation time, which
is proportional to T*-2.6 for n-Si and T*-2.3 for p-Si in the
intermediate range of T. Li and Thurber have shown [1,2] that the
influence of intercarrier scattering must be considered both for the
lattice and for the ionised scattering mobility, for doping densities
greater than approximately 10*17/cm*3.

At high doping density, N > 3 X 10*18/cm*3, the shallow levels in the
forbidden gap begin to merge with the free band edge and the doping
atoms are always ionised. For these doping levels the screening radius
of ionised impurities decreases considerably. The carrier mobility is
approximately independent of doping density and temperature. In the
doping range 2 X 10*19 - 2 X 10*20/cm*3, several authors [3-6] have
measured the mobility values as 80 +- 20 cm*2/V.s for electrons and 50
+- 20 cm*2/V.s for holes in as-grown material, diffused and ion-
implanted layers. An empirical mobility expression related to
temperature, doping and injection level has been derived by Dorkel and
Leturcq [7] and Arora et al [8]. As was shown by Boltaks et al [9]
doping with deep impurities such as Fe, Co, Ni, Mg and Au will give
lower mobility values than doping with the same amount of shallow
impurities. This effect is related to the multi-ionised states of the
deep impurities. For high compensation of Si by deep impurities, the
carrier mobility decreases due to an increasing screening factor
[9,10].

The minority carrier diffusion and mobility were measured by Dziewior
and Silber [11] and by Burk and de la Torre [12]. In the intermediate
doping density range, 10*15 - 10*18/cm*3, the minority carrier
mobilities and the diffusion coefficients are somewhat higher than
those for the majority carriers in the opposite type of material but
with the same doping level. This is explained by screening of majority
carriers and differences in the coloumbic forces. In the high doping
range, when N is approximately equal to 10*20/cm*3, the mobilities of
the minority carriers are considerably lower than those of the
majority carriers. An interpretation based on the minority shallow
bound state or on minority carrier trapping in the band tail has been
made by Fossum et al [13] and also by Neugroschel and Lindholm [14].
Hansch and Mahan [15] have used thermodynamic arguments to show that
minority carrier drag upon the majority carriers can take place.

REFERENCES
----

[1]   S.S.Li, W.R.Thurber [Solid-State Electron. (GB) vol.20 no.7
        (Jul 1977) p.609-16]
[2]   S.S.Li [Solid-State Electron. (GB) vol.21 no.9 (Sep 1978)
        p.1109-17]
[3]   J.A.del Alamo, R.M.Swanson [J. Appl. Phys. (USA) vol.56 no.8
        (15 Oct 1984) p.2250-2]
[4]   G.Masetti, S.Solmi [IEE J. Solid-State & Electron. Devices (GB)
        vol.3 (1979) p.65]
[5]   G.Masetti, M.Severi, S.Solmi [IEEE Trans. Electron. Devices
        (USA) vol.ED-30 no.7 (Jul 1983) p.764-9]
[6]   V.I.Fistul [Heavily Doped Semiconductors (Plenum Press, NY, USA
        1969) p.77]
[7]   J.M.Dorkel, Ph.Leturcq [Solid-State Electron. (GB) vol.24 (1981)
        p.821]
[8]   N.D.Arora, J.R.Hauser, D.J.Roulston [IEEE Trans. Electron.
        Devices (USA) vol.ED-29 (1982) p.292]
[9]   B.I.Boltaks (Ed) [Compensated Silicon (in Russian), (Leningrad,
        Nauka, 1972) p.292-5]
[10]   B.K.Ridley [J. Phys. C (GB) vol.10 no.10 (28 May 1972)
        p.1589-93]
[11]   J.Dziewior, D.Silber [Appl. Phys. Lett. (USA) vol.35 no.2 (15 Jul
        1987) p.170-2]
[12]   D.E.Burk, V.de la Torre [IEEE Electron Device Lett. (USA)
        vol.EDL-5 no.7 (1984) p.231-3]
[13]   J.G.Fossum, D.E.Burk, S.Y.Yung [IEEE Trans. Electron Devices
        (USA) vol.ED-32 no.9 (Sep 1985) p.1874-7]
[14]   A.Neugroschel, F.A.Lindholm, C.T. Sah [Solar Cells (Switzerland)
        vol.14 no.3 (Jul 1985) p.211-18]
[15]   W.Hansch, G.D.Mahan [J. Phys. & Chem. Solids (GB) vol.44 (1983)
        p.663]

CHAPTER 9

PHOTOLUMINESCENCE

9. 1   Photoluminescence of heavily-doped Si

9. 2   Photoluminescence of Si: intrinsic and
               shallow dopant emission

9. 3   Photoluminescence of Si due to C and O impurities

9.1 PHOTOLUMINESCENCE OF HEAVILY DOPED Si

by B.Hamilton

November 1987
EMIS Datareview RN=17830

At high doping levels important changes take place in the characteristics
of PL spectra. These effects are distinct from, say, concentration
broadening of excitonic lines. Many effects can influence the
luminescence but the following are known to be important in the case of
Si:

(i)   Impurity band tails effectively reducing the gap [1]

(ii)   Gap shrinkage to particle interaction [2-6]

(iii)   The role of metal semiconductor transition when carrier density
         exceeds the Mott density [7]

(iv)   Defects produced by heavy doping [8]

It has been shown by Parsons that the luminescence observed is
characteristic of the excitation level [9], low level excitation producing
additional and shifted bands compared to high level excitation. In general
luminescence features are broad in heavily-doped material. The current
view on these bands is that the low level emission results from
recombination at rather deep defects, produced by the heavy doping; and
that the high level band results from band-to-band recombination. TABLE
1 gives examples of the approximate positions of these bands as the
doping level is varied. The data are for B and As doping. The exact
peak positions are not precise and the numbers quoted represent averages
taken from several data sources.

TABLE 1: spectral positions (eV) of the high level (HL) and low level
         (LL) bands in B and As doped Si (T is about 4.2 deg K)

Approx. Nd-Na (/cm*3)	Boron HL	LL	Approx. Nd-Na (/cm*3)	Arsenic HL only
5 X 10*17	1.08	1.08(5)	5 X 10*18	1.064
3 X 10*18	1.077	1.076	2 X 10*19	1.055
5 X 10*18	1.069	1.052	3 X 10*19	1.053
10*19	1.068	1.051	5 X 10*19	1.051
2 X 10*19	1.066	1.050		

Dumke [10] has analysed heavy doping photoluminescence and compared
the data with optical absorbed and transport measurements. The author
concluded that a non-rigid gap shift takes place in n-type silicon.
The Fermi level and gap data from the analysis of ref [10] are given
in TABLE 2

TABLE 2: Fermi level and gap data for heavily doped n-type Si
         at 4.2 deg K

Nd (10*18/cm*3)	Eg+Ef (eV)	Ef -- Es	Ef fitted meV	Ef calc. meV	Ego (eV)	Ego* (eV)
6	1.064	8	17	11	1.147	1.149
24	1.057	8	28	28	1.123	1.131
32	1.058	6	33	34	1.118	1.126
40	1.057	6	36	39	1.112	1.120
100	1.057	10	40	73	1.070	1.10

Here Ef is the energy resolution of the PL data, Ef (fitted) is the
fitting to the optical spectrum, Ego is the value of Eg extrapolated to
the measurement temperature. Ego* is the effective energy gap defined
through

$$Ego* + Efb = Ego + Ef$$

where $Efb = \dfrac{kT}{q} \ln \dfrac{Nd}{Nc}$

Wagner [7] has used PL and PL excitation spectroscopy to study heavily
doped n- and p- type silicon, and concludes that the gap shrinkage
follows an n*(1/3) law below the Mott transition and is linear with n
above the transition. The PLE data allow independent measurements of
the optical gap.

REFERENCES
----

[1]   R.A.Abram, G.J.Rees, B.L.H.Wilson [Adv. Phys. (GB) vol.27 no.6
       (Nov 1978) p.799-892]
[2]   P.E.Schmid [Phys. Rev. B (USA) vol.23 (1981) p.5531]
[3]   R.R.Parsons [Can. J. Phys. (Canada) vol.56 (1978) p.814]
[4]   P.E.Schmid, M.L.W.Thewalt, W.P.Dumke [Solid State Commun. (USA)
       vol.38 (1981) p.1091]
[5]   J.W.Slotboom, H.C.de Graf [Solid-State Electron (GB) vol.19
       (1976) p.857]
[6]   R.W.Keyes [Comments Solid State Phys. (GB) vol.7 (1977) p.149]
[7]   J.Wagner [Phys. Rev. B (USA) vol.29 no.4 (15 Feb 1984) p.2002-9]
[8]   R.R.Parsons, J.A.Rostworowski, B.Bergersen [Inst. Phys Conf.
       Ser. (GB) no.43 (1979) p.1267]
[9]   R.R.Parsons [Solid State Commun. (USA) vol.29 (1979) p.763]
[10]   W.P.Dumke [Appl. Phys. Lett. (USA) vol.42 no.2 (1983) p.196-8]

9.2 PHOTOLUMINESCENCE OF Si: INTRINSIC AND SHALLOW DOPANT EMISSION

by B.Hamilton

November 1987
EMIS Datareview RN=17829

A. GENERAL COMMENTS

Silicon is an indirect gap material and so optical transitions require
some form of momentum conservation. Most photoluminescence data are
taken at cryogenic temperatures (below about 20 deg K). Excitonic
transitions usually dominate, though free to bound (F-B) and donor
acceptor pair (DAP) spectra have been observed in Si. Excitonic features
are complex: momentum can be conserved with lattice (TO TA LO) phonons;
in addition a second, almost zero crystal momentum phonon (such as a
He zone centre or phonon) may be involved. Thus spectra may be rich in
phonon replicas. In spite of the indirect gap, no-phonon (NP) lines
are observed for bound exciton transitions; in simple terms these
transitions result from the (variable) carrier localisation at the
impurity which spreads the particle wave functions across the zone. The
non-zero component at the zone centre results in NP emission. Impurities
in Si may bind more than one exciton and the decay of multiple bound
excitons results in additional spectral lines.

Discrimination of chemical species using photoluminescence (PL) is possible
but concentration broadening for impurity densities above about $10*16/cm*3$
can begin to make this difficult. Stress broadening or splitting can
also cause problems. Nevertheless, PL remains a very powerful technique
for defect analysis in Si.

B. INTRINSIC LUMINESCENCE

Intrinsic luminescence in the form of free exciton (FE) emission and
also a condensed phase of FE, the electron hole droplet (EHD)
emission is well documented [1-4]. EHD luminescence is produced by high
carrier injection levels.

TABLE 1 is a summary of data on FE and EHD emission. EHD luminescence is
characteristically broad and FE luminescence is also broader than bound
exciton lines by the particle kinetic energy terms which reflect carrier
temperatures. These effects reduce the precision of the data.

TABLE 1: intrinsic Si luminescence

Feature	Energy (eV)	Temperature approx.	Comments
FE(NP)	1.156	10 deg K	These data are spectroscopic energies and are not calculated using values of phonon energy. TO and LO features often difficult to resolve
FE(TA)	1.136	"	
FE(LO)	1.098 (5)	"	
FE(TO)	1.096 (8)	"	
FE((TO)+ O(GAMMA))	1.032	"	
EHD(TO+LO)	1.0803 (typically 12 meV width)	2.1 deg K	Line shape fitted to mixture of LO and TO assisted transitions

Other replicas of the EHD band are seen and are reported in [3,4] and references contained therein. These references also report values of condensate densities, work function, electrochemical potential etc. based on Fermi liquid theory.

C. SHALLOW DOPANTS: SPECTRAL FEATURES

In this section the standard dopants P, As, Sb, B, Ga and Al are reviewed.

Detailed spectroscopy in lightly doped crystals shows that these impurities exhibit multi-exciton capture and recombination often referred to as bound multiexciton complex luminescence (BMEC). A number of excitons (index m) can be bound. Such complexes have been successfully modelled using a shell model [5,6] which has been extensively reviewed by Thewalt [7]. A very large number of BMEC lines exist and the reader is referred to [7] for details. The spectroscopic positions of the m=1 and m=2 alpha transitons for the above dopants are given in TABLE 2. Other BMEC transitions involving excited initial and final states exist.

TABLE 2: spectral energies (eV) of m=1 and m=2 BMEC emission at 4.2 deg K

Impurity	NO phonon m=1	m=2	TO replica m=1	m=2
P	1.1500(1)	1.1464(7)	1.0921	1.0884
As	1.1492	1.1457	1.0909	1.0877
Sb	1.1501	1.1467	1.0926	1.0886
B	1.1507(2)	1.485(3)	1.0930	1.0909
Al	1.1495(6)	1.1423(7)	1.0918	1.0903
Ga	1.1490(3)	1.1483(5)	1.0913	1.0887

High resolution spectroscopy shows that the deeper acceptors exhibit
triplet structure due to the splitting of the exciton ground state.
This can be clearly seen for example in the detailed PL spectrum due
to Al [8]. The m=2 luminescence mirrors the bound exciton final state
and three components of m=2 luminescence due to Ga have been observed
[9]. BMEC data show very characteristic ratios of the m=1,2,3...
components. For example, B, which shows relatively simple BMEC
luminescence due to the negligible splitting of the exciton ground
state, shows strong m=1,2 and 3 components. The first hole shell closes
at m=3, leading to very weak higher m components.

TABLE 3 shows the spectroscopy determined binding energies of some BMEC
systems using the shell model; precision of these data is around +- 0.5
meV.

TABLE 3: spectroscopic binding energies of BMEC's in meV [7]

Element	m=1	2	3	4
B	3.9	6.1	8.3	
Al	5.1	5.2	8.5	
P	4.0	3.8	6.7	9.0

Limited agreement between these spectroscopic data and the thermalization
related binding energies exists [7].

D. DOPANT CONCENTRATION

Photoluminescence is not in essence a quantitative technique for
measuring impurity state concentrations. The main reason for this is the
variability of competition for the injected carriers by competing
(radiative or non-radiative) recombination processes.

Two approaches have been made to produce calibration curves linking PL
data to electrically measured dopant concentration. The first is to use
very low excitation levels making EHD and BMEC features weak. Bound and
free exciton features both increase lineraly with excitation level, and
the ratio of extrinsic features to intrinsic features can be used to
generation calibration data [10,12]. The extinction of dopant
concentration relies on an analysis of rate equations for the various
decay routes.

The second approach is to use higher excitation densities since it has
been shown experimentally that the ratio of bound to free exciton
luminescence tends towards saturation with increasing power density.
Tajima and co-workers have carried out measurements in this regime
[13,14]. Calibraton curves for B and P based on the ratio of BE and FE,
TO replicas obtained by these authors show linear behaviour at or below
impurity concentratons of 10*13/cm. Above this concentration sub-linear
variation is seen.

Dopant calibration is reviewed by McL.Collney and Lightowlers [8] who have made accurate calibration of the dopants B, P and Al using Fourier transform photoluminescence measurements. These authors use the ratio of the impurity NP-lines to the FE TO replica to make the calibration. Both heights and areas were used. They fix excitation conditions of 500 meV (514 nM line of an argon laser) unfocused laser power, attenuated by mirrors, filters, windows etc. The samples were held in bubbling liquid He. Their curves show constant slope of nearly unity on log-log plots and cover the concentration range. Fitting was made to the expression

$$\log C = m (\log R - \log \langle R \rangle) + \log \langle C \rangle$$

where C is the impurity concentration, m is the slope, R and $\langle R \rangle$ are the measured and mean ratios, and $\langle C \rangle$ is the mean concentration (/cm*3). The fitting parameters for peak height ratios only are given in TABLE 4.

TABLE 4: fitting parameter for B, P and Al calibration from [8]

Element	$\langle R \rangle$	$\langle C \rangle$	m
B	8.58 X 10*-2	1.94 X 10*13	0.987 +- 0.025
P	2.98	4.29 X 10*-14	1.036 +- 0.015
Al	47.3	8.89 X 10*-13	1.010 +- 0.071

E. BOUND EXCITON LIFETIMES

Impurity exciton lifetimes are found to be much shorter than the theoretical radiative lifetime and have been measured by Schmid [15]. TABLE 5 lists experimental values.

TABLE 5: bound exciton lifetimes at 4.2 deg K

Element	Lifetime ns
P	272
As	183
B	1055
Al	76
Ga	77

The short lifetimes (indicative lifetimes are calculated to be about 1 ms) are explained by assuming a localised phononless Auger process at the neutral defect.

REFERENCES

(The literature on these dopants is very large indeed. The papers listed here are considered to be representative and sufficient for a useful literature search)

[1]   J.R.Haynes [Phys. Rev. Lett. (USA) vol.17 (1966) p.860]
[2]   M.A.Vouk, E.C.Lightowlers [J. Lumin. (Netherlands) vol.15 (1977) p.357]
[3]   R.B.Hammon, T.C.McGill, J.W.Meyer [Phys. Rev. B (USA) vol.13 no.8 (1975) p.3566]

[4]   M.A.Vouk, E.C.Lightowlers [J. Phys. C (GB) vol.8 no.21
        (7 Nov 1975) p.3695-702]

[5]   P.J.Dean, D.C.Herbert, D.Bimberg, W.J.Choyke [Phys. Rev. Lett.
        (USA) vol.37 no.24 (13 Dec 1976) p.1635-8]

[6]   A.S.Kaminskii, Ya.E.Pokrovskii [Sov. Phys.-JETP (USA) vol.40
        (1978) p.523]

[7]   M.L.W.Thewalt [Excitons, Ed. Rashiba and Sturge (North
        Holland, 1982)]

[8]   P.McL.Colley, E.C.Lightowlers [Semicond. Sci. & Technol. (GB)
        vol.2 no.3 (Mar 1987) p.157-66]

[9]   S.A.Lyon, D.I.Smith, T.C.McGill [Solid State Commun. (USA) vol.28
        (1978) p.317]

[10]   A.S.Kaminskii, L.I.Kolesink, B.M.Leiferou, Ya.E.Prokovskii [J.
        Appl. Spectrosco. (USA) vol.36 (1982) p.516]

[11]   R.B.Hammond, J.M.Mees [in Proc. 3rd Int. Conf. Neutron
        Doping of Silicon, Copenhagen, Denmark, 27-29 Aug 1980, Ed.
        J.Guldberg (Plenum, New York, 1981) p.417]

[12]   H.Nakayama, T.Nisho, Y.Hamakawa [Jpn. J. Appl. Phys. (Japan)
        vol.19 no.3 (Mar 1980) p.501-11]

[13]   M.Tajima [Appl. Phys. Lett. (USA) vol.32 no.11 (1 Jun 1978)
        p.719-21]

[14]   M.Tajima, A.Yusa [in Proc. 3rd Int. Conf. Neutron Transmutation
        Doping of Silicon, Copenhagen, Denmark 27-29 Aug 1980, Ed.
        J.Guldberg (Plenum, New York, 1981)]

[15]   W.Schmid [Phys. Status Solidi (Germany) b vol.84 no.2 (Dec 1977)
        p.529-40]

9.3 PHOTOLUMINESCENCE OF Si DUE TO C AND O IMPURITIES

by B.Hamilton

November 1987
EMIS Datareview RN=17852

A. INTRODUCTION

Carbon and oxygen occur as process contaminants in Si, oxygen being a
major impurity in CZ material. They combine to form many defects in
silicon. In most cases the defect atomic configuration is not known.
Oxygen is a key component in the formation of thermal donor states in
CZ material, which show strong variation in properties, including PL
signatures, depending on the precise processing of the wafer. Carbon
and oxygen are also strongly involved in many radiation damage states.

Of the very many PL lines associated with carbon and oxygen a small
number are relatively well understood. These are now discussed.

B. THE G LINE

The G line has been observed strongly in electron irradiation float
zone material and its association with carbon has long been suspected.
Recent detailed work confines this assignment and has provided some
detailed models of possible microsocpic configurations. The main
spectral features of the G line are given in TABLE 1.

TABLE 1: G line spectroscopic features at 5 deg K

Component	Energy (mev)	Ref
No phonon line	970	[1-4]
local mode replica (E line)	71.9	[5]
local mode replica (E' line)	71.9+71.5 displacement	[5]
local mode replica	6735 displacement	[6]

The G line defect is considered to involve both carbon atom(s) and
interstial silicon [6,7]. The strength of the G line in irradiated FZ
material suggests the possibility of using PL, or more quantitatively
optical absorption to deduce carbon concentrations [8].

C. THE C AND P LINES

These lines are known to be definitely associated with oxygen though
they also involve carbon. The association with oxygen as made clear in
refs [1,2,4,9,10]. Wagner and co-workers [11]used a combination of PL
and PL excitation spectroscopy to elucidate many of the C line-
associated components. TABLE 2 summarises the data.

TABLE 2: C line components at 20 deg K

Componenet	Energy (meV)	
no phonon	790	
excited state	795	
local mode replica	724.5	
local mode replica	717.5	
excited states seen in	800	[12]
absorption	800.3	[12]

The C line spectrum is also strong in TO, LA and TA replicas.

Isotope shifts implicate O directly in the formation of the C line
defect [13]. The involvement of an interstitial Si atom is deduced from
deconvolution of 3 components (reflecting the natural abundance of
Si-28, Si-29 and Si-30) of one of the local mode replicas [12]. Isotope
shift of the no phonon line [14] proves the involvement of carbon in
this defect.

The P line has remarkably similar properties to the C line. Both lines
have similar vibrational properties and PLE work [14,15] has revealed
excited state distributions consistent with effective mass donor theory.
The binding energies of the donor states predicted by this theory are
[14,15]

'C line donor' 38.3 meV

'P line donor' 34.7 meV

The PL spectral characteristics of the P line are given in TABLE 3.

There is much similarity between the annealing characteristics of the C
line and thermal donors. The P line is most effective when generated by
annealing at about 450 deg C and becomes quenched at around 600 deg C
[16]. Thermal donors produce complex and sometimes broad PL spectra
[17,18].

TABLE 3: P line components at 20 deg K

Component	Energy (meV)
C line (no phonon component)	767.5
local mode replica	702.0
local mode replica	696.5
excited state	774.2
	774.8
	776.0
	780.5

[Additional excited states fitting to EMT in ref [15]]

D. OTHER LINES

Many other PL lines which can be associated with either carbon or oxygen (or both) have been reported. TABLE 4 shows a selection of these.

TABLE 4: other C- or O-related optical components

energy meV	main association	popular designation	ref
951	C similar vibrations		[6]
953			[6]
954			[6]
957			[6]
949.9			[6]
929.1			[6]
928.6			[6]
925.5			[6]
919.8			[6]
875.4	C and Ga		[19]
768.6	C		[6]
1082.0	C and Li		[20]
925.5	O	H	[21]
965.2	O	I	[21]
935.5	O	T	[21]

REFERENCES

[1] R.J.Spry, W.D.Compton [Phys. Rev. (USA) vol.138 (1965) p.A543]

[2] A.V.Yukhnevich [Sov. Phys.-Solid State (USA) vol.7 (1965) p.257]

[3] V.D.Tkachev, A.V.Mudryi [Inst. Phys. Conf. Ser. (GB) no.31 (1977) p.231-43]

[4] C.G.Kirkpatrick, D.R.Myers, B.G.Streetman [Radiat. Eff. (GB) vol.31 (1977) p.175]

[5] K.Thonke, H.Klemisch, J.Weber, R.Sauer [Phys. Rev. B (USA) vol.24 (1981) p.5874]

[6] G.E.Davies, E.C.Lightowlers, M.do Carmo [J. Phys. C (GB) vol.16 (1983) p.5503]

[7] K.P.O'Donnell, K.M.Lee, D.G.Watkins [Physica B&C (Netherlands) vol.116 (1983) p.258]

[8] G.E.Davies, E.C.Lightowlers, M. do Carmo, J.G.Wilkes, G.R.Wolstenholme [Solid State Commun. (USA) vol.50 no.12 (Jun 1984) p.1057-61]

[9] A.V.Yukhnevich, V.D.Tkachev, M.V.Bortnik [Sov. Phys.-Solid State (USA) vol.8 (1967) p.2571]

[10] C.E.Jones, W.D.Compton [Radiat. Eff. (GB) vol.9 (1971) p.183]

[11] J.Wagner, A.Doernen, R.Sauer [Phys. Rev. B (USA9 vol.29 (1984) p.705]

[12] C.P.Foy [Physica B&C (Netherlands) vol.116 (1983) p.276]

[13] G.Davies, E.C.Lightowlers, R.Woolley, R.C.Newman, A.S.Oates [J. Phys. C (GB) vol.17 (1984) p.L499]

[14] K.Thanke,, G.D.Watkins R.Sauer [Solid State Commun. (USA) vol.51 (1984) p.127]

[15] J.Wagner, A.Doernen, R.Sauer [Phys. Rev. B (USA) vol.31 no.8 (15 Apr 1985) p.5561-4]

[16] A.Dornen, R.Sauer, J.Weber [J. Electron. Mater. (USA) vol.49 (1985) p.653]

[17] M.Tajima, A.Kanomori, S.Kishino, T.Iizuka [Jpn. J. Appl. Phys. (Japan) vol.19 (1980) p.L755]

[18] M.Tajima, P.Stallhofer, D.Huber [Jpn. J. Appl. Phys. Part.2 (Japan) vol.2 (1983) p.L586]

[19] J.R.Norman, C.G.Kirkpatrick, B.G.Streetman [Solid State Commun. (USA) vol.15 (1974) p.1055]

[20] E.C.Lightowlers, L.J.Canham, G.Davies, M.L.W.Thewalt, S.P.Watkins [Phys. Rev. B (USA) vol.29 no.8 (15 Apr 1984) p.4517-23]

[21] M.S.Minaev, A.V.Mudryi [Phys. Status Solidi a (Germany) vol.68 (1981) p.501]

CHAPTER 10

DEEP LEVELS

10.1 DISLOCATION-RELATED DEEP LEVELS IN Si

by A.R.Peaker and E.C.Sidebotham

November 1987
EMIS Datareview RN=17836

A. INTRODUCTION

Extensive work has been carried out in characterising deep defect levels
in bulk silicon associated with dislocations introduced by plastic
deformation. Deep level transient spectroscopy (DLTS) studies of the
deep levels in n-type [1-7] and in p-type [2,3,5,8] silicon have been
undertaken. On introducing a high density of dislocations by plastic
deformation, typically 10*9/cm*2, a complex spectrum of energy levels
results, which simplifies on annealing.

More work has been done on traps in n-type silicon than in p-type. The
results for n-type are more consistent than for p-type, although there
is still differing opinion between authors as to the nature of the
defects giving rise to the deep levels in n- as well as p-type silicon.

The deep levels introduced by deformation depend on the temperature,
stress and strain conditions under which the dislocations were produced.
The effect of background impurities has not yet been investigated.
Kimerling suggests that the variety of deep levels introduced by
deformation are probably related to point defect aggregates left behind
as debris by the deformation process. References [2 and 3] propose that
the change in the DLTS spectra on annealing is due to reconstruction of
dangling bands in the dislocation core, although this is disputed by Ono
[8].

A DLTS study of hydrogen passivation of deep levels associated with
dislocations is reported by [9]. Two studies [10,11] of deep levels
associated with heavily oxygen precipitated silicon report levels
similar to those seen in heavily dislocated material. Chan [10]
suggests that this might be because severe precipitation causes lattice
deformation resulting in dislocations and other related defects.

B. ELECTRON TRAPS

Omling [1] reviews the DLTS results obtained by [2-7] and concludes that
the same 5 deep levels are observed by each author in n-type silicon.
These traps are listed in the TABLE 1 [1-4]. All five deep levels were
found to have a logarithmic capture process for longer filling pulses,
which is characteristic of states associated with dislocations [7]. All
the authors report similar results for annealing experiments and [1,3,4]
compare DLTS and EPR results in attempts to identify the deep states.
Drawing on the results of [1-5] the characteristics and likely identity
of each of the five frequently observed deep levels are, using Omling's
notation, as follows:

D and B are removed by annealing at 800 deg C but C2, C1 and A are stable at this temperature. The initial capture cross section of D which, from correlation with EPR spectra, is a dislocation dangling bond state, is 6 X 10*-16 cm*2. C2 is either a singly occupied dislocation dangling bond or is related to kinks in the dislocations. C1 may also be related to dislocation kinks. B has an initial capture cross section of 7 X 10*-16 cm*2 which is typical of a neutral centre and comparison with EPR spectra leads [4] to suggest that it is a shallow dopant located in or around dislocations. The thermally stable deep level A is also believed to be related to dopants in or around the dislocation core.

C. HOLE TRAPS

Kimerling [2] reports eight different hole traps in heavily dislocated p-type silicon, one of which at Ev+0.35 eV remains after a 900 deg C anneal. Kveder [3] reports similar results. A five defect spectrum simplifies to a broad peak centred at Ev+0.35 eV after a short anneal at only 780 deg C. Kveder suggests that DH2, DH3 and DH5 are associated with dislocation dangling bonds and that DH1 is due to point defects localised less than 100 A from the dislocation core. Although there is general agreement between the two groups, the defect energy levels are different and there is a large discrepancy in the annealing behaviours. The differences may be due, at least in part, to the different deformation temperatures used. Both believe that the change in the defect spectrum on annealing is due to reconstruction of dangling bonds at the dislocation core.

Ono [8] reports three main hole traps in dislocated p-type silicon. Although DH(0.33) seems to correspond to the Ev+0.35 eV states reported by both Kveder and Kimerling, this state as well as the other two, almost completely vanishes on annealing at 900 deg C. Ono suggests that DH(0.33) which largely controls the temperature dependence of the carrier concentration is due to irregularities such as kinks or jogs along the dislocations. DH(0.24) and DH(0.56) are thought to be due to agglomerations of point defects which were the debris resulting from dislocation formations.

Ono [8] reports TEM observations that the 900 deg C anneal eliminates jogs, kinks and agglomerates, supporting his defect models. He also suggests that if the reduction in the spectra is due to reconstruction of dangling bonds, as proposed by Kimerling and Kveder, then it would be likely to start at a temperature lower than those reported by these authors.

TABLE 1: electron traps in silicon associated with dislocations

Name	Activation Energy (meV)	Temperature (deg K) e=100/s	Reference
D	540	262	[1]
E(0.54)	540	262	[4]
DE4	600	259	[3]
E(0.63)	530	–	[2]
C2	510	235	[1]
DE3	510	232	[3]
E(0.48)	480	–	[2]
E(0.40)	400	–	[4]
C1	520	218	[1]
DE2	370	205	[3]
E(0.41)	410	–	[2]
E(0.29)	290	157	[4]
B	290	157	[1]
DE1	270	146	[3]
E(0.28)	280	–	[2]
A	190	107	[1]
E(0.19)	190	107	[4]
E(0.18)	180	–	[2]

TABLE 2: hole traps in silicon associated with dislocations

Name	Activation Energy (meV)	Temperature (deg K) e=100/s	Reference
H(0.63)	630	–	[2]
H(0.40)	400	–	[2]
H(0.36)	360	–	[2]
H(0.35)	350	–	[2]
H(0.29)	290	–	[2]
H(0.26)	260	–	[2]
H(0.23)	230	–	[2]
H(0.09)	90	–	[2]
DH(0.56)	560	234	[8]
DH(0.33)	330	175	[8]
DH(0.24)	240	105	[8]
DH6 (note 1)	350	214	[3]
DH5 "	670	263	[3]
DH4 "	470	243	[3]
DH3 "	400	203	[3]
DH2 "	390	178	[3]
DH1 "	250	110	[3]

note 1: Temperatures listed are those for which emission rate is 80/s
        not 100/s

REFERENCES
----

[1]   P.Omling, E.R.Weber, L.Montelious, H.M.Alexander [J. Phys. Rev.
       B (USA) vol.32 (1985) p.6571]

[2]   L.C.Kimerling, J.R.Patel [Appl. Phys. Lett. (USA) vol.34 no.1
       (1 Jun 1979) p.73-5]

[3]   V.V.Kveder, Yu.A.Osyipan, W.Schroter [Phys. Status Solidi a
       (Germany) vol.72 (1982) p.701]

[4]   E.R.Weber, H.Alexander [J. Phys. Colloq. (France) vol.44 no.C-4
       (Sep 1983) p.319]

[5]   J.R.Patel, L.C.Kimerling [Cryst. Res. & Technol. (Germany) vol.16
       (1981) p.187]

[6]   W.Szkielko, O.Breitenstein, R.Pickenhein [Cryst. Res. & Technol.
       (Germany) vol.16 (1981) p.197]

[7]   W.Schroter, M.Seibt [J. Phys Colloq. (France) vol.44 no.C-4
       (Sep 1983) p.329-37]

[8]   H.Ono, K.Sumino [J. Appl. Phys. (USA) vol.57 no.2 (1985) p.287]

[9]   I.T.Belash, V.V.Kveder [Sov. Phys.-Solid State (USA) vol.28
       (1986) p.776]

[10]   S.S.Chan, C.J.Varker, J.D.Whitfield, R.W.Carpenter [Mater. Res.
       Soc. Symp. Proc. (USA) vol.46 (1985) p.281]

[11]   F.D.Whitwer, H.Haddad, L.Forbes [Mater. Res. Soc. Symp. Proc.
       (USA) vol.71 (1986) p.53]

10.2 VACANCY-RELATED DEEP LEVELS IN Si

by A.R.Peaker and S.Guimaries

November 1987
EMIS Datareview RN=17835

In many materials vacancies have been studied by rapid quenching so
freezing in an excess concentration. In silicon this is not feasible as
the isolated vacancy is extremely mobile and, in practice, the defect
reacts with other impurities generating vacancy related complexes.
Similarly vacancies diffusing from the surface or strained regions
create complexes. The only successful method of isolated vacancy
creation is by electron irradiation at cryogenic temperatures. However,
even in this case, the rate of vacancy production is low unless the
vacancy and interstitial atom can be separated rapidly. This occurs in
p-type material where the group III atom traps the interstitial [1].
Hence the vacancy production rate using 1.5 MeV electrons is much higher
in p-type material than in n-type. An alternative method of separation
appropriate for n-type material is to use high electron energies in the
range 5 - 50 MeV [2,3] or alternatively to counterdope the n-type
material with group III element [1].

The high mobility of the primary defects makes studies of vacancy
related states extremely complex and some indication of this is given
in a review by Watkins [4]. However, in a number of cases studies of
the electrical activity of defects generated reproducibly have been made
and combined with techniques capable of giving structural information
such as electron paramagnetic resonance (EPR) and electron-nuclear
double resonance (ENDOR).

A list of states observed reproducibly and specifically associated with
vacancies is given in TABLE 1 (electron traps) and TABLE 2 (hole
traps).

The characteristics of the primary defect, the isolated vacancy, are
particularly interesting. The DLTS activation energy of 130 meV is
interpreted as the energy needed to emit a hole from the double charged
state (V++). However, the removal of a second hole is accompanied by a
large Jahn Teller relaxation. This was predicted by Baraff et al [5]
and the consequence is that the emission of a second hole (V+ to V
neutral) requires less energy than the emission of the first hole. This
phenomenon of negative U (the Hubbard correlation energy) means that
the DLTS activation energy will correspond to the V++ to V+ process, but
two holes will be lost (the V++ to V neutral transition) giving a DLTS
signal of twice the expected amplitude. This is indeed the case [6-8]
and EPR experiments support this, indicating that the energy for the V+
to V neutral transition is 50 meV [1].

It is expected that V- and V-- energy levels of the isolated vacancy lie
in the band gap, V- is apparent in EPR studies and V-- is inferred as V-
is created optically [1]. However, the relevant transitions have not
been observed in deep level measurements.

The isolated vacancy is quite mobile at 250 deg K and starts to react with impurities to create vacancy complexes as detailed above. The divacancy is relatively stable remaining in the material at temperatures up to 550 deg K and the literature on this defect is very comprehensive. The state at 210 meV [10] has been unambiguously linked to the emission of a hole from the positively charged state of the divacancy. Polarised excitation photocapacitance [11] has been used to support the observed agreement with EPR measurements and has been linked to introduction rate, anneal recovery kinetics and the local atomic and electronic structure.

Two electron traps are also associated with the divacancy. The activation energy of around 400 meV is believed to characterise the emission of an electron from the single negatively charged state. Unfortunately, the DLTS peak coincides in temperature with the E centre (phosphorus vacancy complex) and so care must be taken to distinguish between them by choice of material or by annealing studies. The results of Kimerling [10] and Brotherton [12] agree extremely well in all respects. Evwaraye's [15] value of electron cross section is smaller than those of the other two authors. The state at around 230 meV is probably the emission of an electron from the double negatively charged state.

Annealing characteristics are consistent with this and the cross section for electron capture is smaller than for the neutral di-V centre, quite consistent with it being more negatively charged. In addition, in contrast to the case of the neutral centre for which the cross section is independent of temperature Brotherton [12] observes the electron capture into the negatively charged state to be thermally activated and equal to:

$$4 \times 10*-16 \exp(-0.017/kT)$$

This value agrees well with the single temperature measurements of Kimerling [10] but again is greater than that of Evwaraye [15].

The electron state with an activation energy of 170-180 meV has the same energy position, introduction behaviour, annealing behaviour and correlation to oxygen content as the A centre (oxygen-vacancy pair) [20,21], the complex anneals out at around 620 deg K.

The 440 meV level is assigned to the phosphorus vacancy pair sometimes referred to as the E centre. It is the state present in highest concentration in irradiated n-type float zone silicon and its energy and annealing behaviour is in agreement with the EPR results [22]. Work prior to 1986 on the di-vacancy, the A centre and the E centre have been reviewed by Indusekhar [17]. However, recent work on this defect has shown that the situation is rather more complicated in that the DLTS peak may well have two components, one of these being the 440 meV state of a multistable complex with electron activation energies of 440, 340, 260 and 230 meV [18,19]. The situation is made even more complicated by the unfortunate coincidence that the DLTS peak associated with the divacancy is indistinguishable from that of the E centre. Additional data are also included in the table for the Al-V and Sn-V pairs, although their assignment is less definite than the other defects discussed.

TABLE 1: electron traps

Identity		Energy (meV)	Temperature (deg K) e=100/s	Measured cross section (cm*2)	Ref
di-V	(-/0)	390	-	4 X 10*-15	[10]
di-V	(-/0)	413	227	2 X 10*-15	[12]
di-V	(-/0)	410	-	1.6 X 10*-16	[15]
di-V	(--/-)	230	-	2 X 10*-16	[10]
di-V	(--/-)	246	132	See text	[12]
di-V	(--/-)	230	-	6 X 10*-17	[15]
O-V	(A centre)	169	93	10*-14	[12]
O-V		180	-	10*-14	[10]
P-V	(E centre)	456	230	3.7 X 10*-15	[12]
P-V		440	-	-	[10,14]
Al-V		480	-	-	[10]

TABLE 2: hole traps

Identity		Energy (meV)	Temperature (deg K) e=100/s	Measured cross section (cm*2)	Ref
V	(++/+)	130	72.7	-	[9]
V	(++/+)	110	-	5 X 10*-17	[10]
di-V	(+/0)	204	117	4 X 10*-16	[12,13]
di-V	(+/0)	210	-	2 X 10*-16	[10]
di-V	?	180	107	-	[16]
Sn-V		70	-	-	[7]
Sn-V		320	-	-	[7]

REFERENCES
----

[1]   G.D.Watkins [Inst. Phys. Conf. Ser. (GB) no.23 Lattice Defects in
      Semiconductors (London, 1975) p.1]
[2]   E.G.Wickner, D.P.Snowden [Bull. Am. Phys. Soc. (USA) vol.9
      (1964) p.706]
[3]   C.J.Cheng, J.C.Corelli [Phys. Rev. (USA) vol.140 (1965) p.A2130]
[4]   G.D.Watkins [in 'Deep Centres in Semiconductors' Ed S.T.Pantelides,
      (Gordon & Breach, New York, 1986) p.147-83]
[5]   G.A.Baraff, E.O.Kane, S.Schluter [Phys. Rev. Lett. (USA) vol.43
      (1979) p.956; Phys. Rev. B (USA) vol.21 (1980) p.3563; Phys. Rev.
      B (USA) vol.22 (1980) p.5662]
[6]   G.D.Watkins, J.R.Troxell [Phys. Rev. Lett. (USA) vol.44 (1980)
      p.593]

[7]   G.D.Watkins, A.P.Chatterjee, R.D.Harris [Inst. Phys. Conf. Ser. (GB) no.59 Defects and Radiation Effects in Semiconductors (London, 1981) p.199]

[8]   G.D.Watkins [Proc. Conf. Defects in Semiconductors Eds J.Narayan, T.Y.Tan (North-Holland, Amsterdam, Netherlands 1981) p.21]

[9]   G.D.Watkins, J.R.Troxell, A.P.Chatterjee [Inst. Phys. Conf. Ser. (GB) no.48 Defects and Radiation Effects in Semiconductors (London, 1976) p.16]

[10]   L.C.Kimerling [Inst. Phys. Conf. Ser. (GB) no.31 Eds N.B.Urli, J.W.Corbett (London 1977) p.221]

[11]   M.Stavola , L.C.Kimerling [J. Appl. Phys. (USA) vol.54 (1983) p.3897]

[12]   S.D.Brotherton, P.Bradley [J. Appl. Phys. (USA) vol.3 no.8 (1982) p.5720-32]

[13]   S.D.Brotherton, G.J.Parker, A.Gill [J. Appl. Phys. (USA) vol.54 no.9 (1983) p.5112-6]

[14]   A.V.Vasilev, S.A.Smagulova, S.S.Shaymeev [Sov. Phys.-Semicond. (USA) vol.16 no.11 (1982) p.1229-81]

[15]   A.O.Evwaraye, E.Sun [J. Appl. Phys. (USA) vol.47 (1976) p.3776-80]

[16]   B.N.Mukashev, N.Fukuoka, H.Saito [Radia. Eff. (GB) vol.61 (1982) p.159-63]

[17]   H.Indusekhor, V.Jumar, D.Sengupta [Phys. Status Solidi a (Germany) vol.93 (1986) p.645-53]

[18]   L.U.Song, B.W.Benson, G.D.Watkins [Phys. Rev. B (USA) vol.33 (1986) p.1452-5]

[19]   A.Chantre, L.C.Kimerling [Appl. Phys. Lett (USA) vol.48 (1980) p.1000]

[20]   G.D.Watkins, J.W.Corbett [Phys. Rev. (USA) vol.121 (1981) p.1001]

[21]   J.W.Corbett, G.D.Watkins, R.M.Chienko, R.S.McDonald [Phys. Rev. (USA) vol.121 (1961) p.1015]

[22]   G.D.Watkins, J.W.Corbett [Phys. Rev. (USA) vol.A134 (1964) p.1359]

10.3 DEEP LEVELS IN MBE Si

by A.R.Peaker

December 1987
EMIS Datareview RN=17844

A. INTRODUCTION

Silicon grown by molecular beam epitaxy differs significantly from
Czochralski material or conventional epitaxy in that the material is
dislocated with densities typically in the 10*3 to 10*4/cm*2 range. This
is significant in that a number of the electron traps observed have been
related to dislocations.

B. ELECTRON TRAPS

A recent review [1] of the deep state population of layers grown in
several different laboratories indicates a large range of concentrations
(10*12 to 10*16/cm*3). The concentration of defects appears to decrease
with increasing growth temperature [2], to be dependent on substrate
cleaning methods [3] and vary spatialy in the layer; the highest
concentrations being found near the substrate-epi interface [2,4].

TABLE 1: electron traps

Name	Energy meV	Temperature (deg K) at e = 100/s	Notes	Ref
T1	580	264	Dislocation	[1]
	580	–	Broad Peak Carbon related?	[3]
Q1	544	282	Vacancy related	[1,2]
	530	–		[4]
T3	500	212	Dislocation	[1]
Ta	500	246	Tantalum	[1,2]
T2	420	240		[1]
T4	350	178		[1]
	320	161	Dislocation	[1,2]
Q2	300	187	Vacancy-related	[1,2]
T5	140	139		[1]

The electron capture cross-sections have been measured directly for a
number of the states and are quoted in [1].

The states at 580, 500(T3) and 320 meV are all related to dislocations.
This assignment is based on the similarity of states (Arrhenius plot and
filling behaviour) observed in plastically deformed silicon [5].

In addition it appears that two other states Q1 and Q2 are related to
vacancies [6]. These states are very frequently observed in processed
silicon and are powerful recombination centres; they are difficult to
anneal out but their detailed structure is unknown.

The remaining centre which is commonly seen is related to tantalum
contamination originating from the substrate heater in particular systems
[1].

In general the deep state concentration decreases with increasing growth
temperature. In the case of the vacancy-related states this is a gradual
change. In the case of the 500 (Ta) and 320 meV states there appears to
be a sudden transition at a substrate temperature of 600 deg C. The 500
meV state is associated with tantalum whilst the 320 meV state is
associated with decorated dislocations. Hence it appears that at above
600 deg C the electrically active deep states decrease. However, there
is a simultaneous rise in the concentration of S-pits suggesting the
change is in the distribution of metallic contaminants between
precipitates and uniformly distributed atoms, rather than a change in
the total concentration of extrinsic impurity.

Xie et al [3] have studied the effect of different substrate cleaning
procedures on the deep state population. They correlate a state with an
activation energy of 580 meV to the pre-growth thermal clean cycle.
'Flashing' at 700 deg C reduced the concentration by over an order of
magnitude and a 950 deg C treatment reduced the concentration to below
the detection limit. Xie noted a parallel reduction in carbon content
at the surface. It is noteworthy that the DLTS peak observed is much
broader than the expected linewidth for a simple point defect.

In cases where thick layers (greater than one micron) have been grown,
it is noted that the concentration of deep states increases towards the
substrate-epitaxial interface [2-4].

C. HOLE TRAPS

Very little work has been published on hole traps in MBE silicon although
some data appear in [1]. The states observed are listed below:

TABLE 2: hole traps

Energy meV	Temperature (deg K) at e = 100/s	Notes	Ref
670	261	Measured in n-type	[1]
430	274	Measured in p-type	[1]
350	297	Measured in p-type	[1]

D. RECOMBINATION

Schockley Hall Read calculations [1] indicate that both the generation
lifetime and minority carrier lifetime are expected to be considerably
lower in present day MBE silicon than in conventional material. 'Typical'
values estimated are 1 microsec for generation lifetime and 10 ns for
recombination lifetime in MBE silicon. Actual generation lifetime
measurements on MOS capacitors give even lower values (30 ns) although
no gettering was used during the fabrication process [1].

REFERENCES
----

[1] A.R.Peaker, E.C.Sidebotham, B.Hamilton, M.Pawlik [Recombination-
     Generation Kinetics in MBE Silicon, Proc. 2nd Int. Symp. MBE
     Silicon, Hawaii 1987 (to be published by Electrochem. Soc.,
     1988)]
[2] E.C.Sidebotham et al [The effects of substrate temperature on
     deep states in MBE silicon, Proc. 2nd Int. Symp. MBE Silicon,
     Hawaii, 1987 (to be published by Electrochem. Soc., 1988)]
[3] Y.H.Xie, Y.Y.Wu, K.L.Wang [Proc. 1st Int. Symp. Silicon MBE,
     Toronto, Ont., Canada, May 1985 (Electrochem. Soc., Pennington,
     NJ, USA, 1985) p.93-101]
[4] A.Sandu, B.Hamilton, A.R.Peaker, R.A.Kubiak, W.Y.Leong,
     E.H.C.Parker [Proc. 1st Int. Symp. Silicon MBE, Toronto, Ont.,
     Canada, May 1985 (Electrochem. Soc., Pennington, NJ, USA, 1985)
     p.78-85]
[5] P.Omling, E.R.Weber, L.Montelius, H.Alexander, J.Michel [Phys.
     Rev. B (USA) vol.32 no.10 (15 Nov 1985) p.6571-81]
[6] W.Yau, C.T.Sah [Solid State Electron. (GB) vol.14 (1971) p.193]

10.4 DEEP LEVELS IN LAMP ANNEALED Si

by A.R.Peaker and M.Di Marco

January 1988
EMIS Datareview RN=17885

Rapid thermal annealing is a technology which is used to activate
implanted impurities and remove implantation damage with as little
redistribution of the dopants as possible. This is essential for very
small devices as used in ULSI and is usually achieved by rapidly
increasing the slice temperature, holding at the anneal temperature for
a short time, and rapidly cooling. In general temperatures greater than
1000 deg C are used for less than 100 s. A convenient and controllable
source of heat is an array of tungsten-halogen lamps although graphite
heaters are also in widespread use. The process is often referred to as
Rapid Thermal Annealing (RTA) to distinguish it from Conventional
Thermal Annealing (CTA) in furnaces where heating and cooling rates are
usually slower and anneal times longer.

In general it is believed that the opportunity for contamination from
the annealing environment is less in RTA than in CTA. However, all types
of 'as grown' silicon contain significant concentrations of unwanted
impurities and in addition high concentrations of intrinsic defects
(vacancies and interstitials) are introduced during implantation.
These defects are redistributed during RTA and frozen in during cooling.
The end result is that the defect population is very different to that
observed after CTA.

Some workers have investigated deep states in RTA silicon which has not
been implanted. In general it has been found that the concentration of
deep states in as grown silicon (unimplanted) subjected to RTA cycles
is low (10*12/cm*3) and often less than in unannealed material
provided the anneal time is longer than 30 s [1,2] or less than 5 s [2]
in n-type material. However, for intermediate anneal times (20 s) higher
concentrations of deep states are observed (10*14/cm*3) [2-5]. The deep
states reported in unimplanted material are summarised in TABLE 1
(Electron traps) and TABLE 2 (Hole traps).

TABLE 1: electron traps in unimplanted RTA silicon

Energy meV	Temperature (deg K) e = 100/s	Notes	Ref
170	–	CZ 1100 deg C 10 s	[3]
270	–	CZ 1100 deg C 10 s	[3]
280	152	CZ and FZ 1100 deg C 100 s	[1]
580	–	CZ 1100 deg C 10 s	[3]

The state at 280 meV was present only in low concentrations (10*12/cm*3)
and had identical emission characterisation to a vacancy related state
observed by Yau and Sah [6] in quenched silicon.

TABLE 2: hole traps in unimplanted RTA silicon

Name	Energy meV	Temperature (deg K) e = 100/s	Notes (concentrations in /cm*3)					Ref
H2(B87)	290	163	CZ	1000 deg C	5 s	Nt = 10*13		[2]
H3(B87)	300	210	CZ	1000 deg C	5 s	Nt = 10*13		[2]
P	300	-	epi	900 deg C	20 s	Nt = 10*15		[3]
H1(B)	320	217	CZ	900 deg C	10 s	Nt = 10*13		[4]
H1(D)	450	238	FZ	1100 deg C	100 s	Nt = 10*12		[1]
H1(B87)	450	270	CZ	1000 deg C	5 s	Nt = 10*14		[2]
H2(B)	450	310	CZ	900 deg C	10 s	Nt = 10*13		[4]

The capture cross section for holes has been measured directly for H1(B) and H2(B). The values are 9 X 10*-18 cm*2 (200 deg K) and 5 X 10*-19 cm*2 (280 deg K) respectively. The cross section of H2(B) increases with increasing temperature.

H1(B87), H2(B) and H1(D) are all believed to be related to interstitial iron. Iron is a common contaminant in silicon but its concentration is very variable even in slices from the same manufacturer. In float zone (FZ) and Czochralski (CZ) material almost all the iron present is in the form of precipitates. This dissolves during the anneal cycle and is quenched into interstitial sites during cool down. The interstitial centre is not stable at room temperature and its electrical concentration is seen to decay due to the formation of iron-boron pairs. Borenstein [4] has performed annealing experiments which support this assignment. The formation of the 450 meV states depends on two stages (dissolution and quenching). Inevitably the dependence of its concentration is a function of both the anneal time/temperature cycle and the cooling rate. Barbier [2] has undertaken a systematic investigation of some of these parameters.

The origins of P, B2(B87), H1(B) and H3(B87) are uncertain but Barbier [2] suggests that H2(B87) could be the chromium interstitial-boron pair or a boron-vacancy complex. The silicon di-interstitial and interstitial vanadium are proposed as possible candidates for H3(B87). P is thought to be a boron-vacancy complex.

After implantation the situation is rather more complex. In the case of a simple dopant implant, apart from the dopant species, vacancy interstitial pairs are created from the displacement of silicon atoms. In the case where a heavy dose of silicon or germanium is implanted to amorphise a region, and so prevent channelling, an excess of Si or Ge interstitials is present. TABLES 3 and 4 summarise the published information on electron and hole traps respectively.

TABLE 3: electron traps in implanted RTA silicon

Name	Energy meV	Notes	Ref
	170	B or BF2 into n-type	[3]
E1	210	B into P doped CZ 700 deg C 7 s	[7]
E2	230	B into P doped CZ 700 deg C 7 s	[7]
	270	B or BF2 into n-type	[3]
E3	290	B into P doped CZ 700 deg C 7 s	[7]
E4	330	B into P doped CZ 1000 deg C 7 s	[7]
	370	1100 deg C 10 s + 550 deg C furnace	[3]
	430	B or BF2 into n-type	[3]
E5	480	B into P doped CZ 100 deg C 7 s	[7]
	570	B or BF2 into n-type	[3]
	600	As into B doped CZ 1100 deg C 10 s	[8]

In the case of the samples discussed in [3] the B implant was annealed at 1280 deg C for 8.3 s and the BF2 implant at 1187 deg C for 2.9 s. The concentration of deep states was about an order less for the B implant with the higher temperature long anneal than the BF2 implant which had a lower temperature shorter anneal.

TABLE 4: hole traps in RTA implanted Si

Energy meV	Implant	Ref
240	Ar into Al doped FZ	[5]
290	PF5 or As into B doped CZ	[8]
300	(note A)	[5]
340	Si into Al doped FZ	[5]
400	PF5 or As into B doped CZ	[8]
430	Si into B doped epi	[5]
450	PF5 or As (note B)	[8]

Anneal conditions were in the range 900-1100 deg C for 5-20 s.

Note A   The 300 meV trap was observed for Ar or B implants into B doped epi and for B implants into Al doped FZ. It was not seen for Ar implants into Al doped substrates.

Note B   The 450 meV trap is ascribed to interstitial iron. Its concentration in the RTA implanted layers was less than in as-grown material subjected to RTA.

Insufficient work has been done to be able to generalise about the extremely complex reactions which ensue from RTA of implanted layers. A significant defect from the point of view of its concentration and stability is the 300 meV hole trap which appears to be a boron-vacancy complex. In a 20 s space anneal the concentration of this defect reaches a maximum value of 10*15/cm*3 after annealing at 1000 deg C. The state can be annealed out by a post RTA furnace treatment at temperatures greater than 400 deg C. The concentration is reduced to 6 X 10*13/cm*3 after 1000 deg C for 1800 s.

The hole traps at 430 and 340 meV are possibly due to silicon interstitials complexing with the shallow dopants [5].

An added complication is that DLTS studies of implant damage, particularly that resulting from boron or phosphorus pentafluoride, often reveal a broad emission spectrum which defies conventional analysis. This is ascribed to the presence of large complexes or extended defects as shown, for example, in [7].

The effect of these deep states on device performance is unclear. Usami et al [7] see no change in reverse leakage of diodes produced by a boron implant into phosphorus doped CZ material over the RTA range 700-1100 deg C. However, Remram [8] observes high leakage currents with worst results at anneal temperatures of 1000 deg C, the temperature at which maximum concentrations of the deep levels are observed.

REFERENCES
----

[1] M. Di Marco, A.R.Peaker, C.Hill, M.Hart, A.E.Glaccum [Proc. European Solid State Device Research Conf., Bologna, 1987]
[2] D.Barbier, M.Remram, J.F.Joly, A.Laugier [J. Appl. Phys. (USA) vol.61 no.1 (1 Jan 1987) p.156-60]
[3] C.M.Ransom, T.O.Sedgwick, S.A.Cohen [Mater. Res. Soc. Symp. Proc. (USA) vol.52 (1986) p.153-6]
[4] J.T.Borenstein, J.T.Jones, J.W.Corbett, G.S.Oehplein, R.L.Kleinkenz, [Appl. Phys. Lett. (USA) vol.49 no.4 (28 Jul 1986) p.199-200]
[5] G.Pensl, M.Schulz, P.Stolz, N.M.Johnson, J.F.Gibbons, J.Hoyt [Mater. Res. Soc. Symp. Proc. (USA) vol.23 (1984) p.347-58]
[6] L.O.Yau, C.T.Sah [Solid-State Electron. (GB) vol.17 (1974) p.193]
[7] A.Usami, M.Katayama, Y.Tokuda, T.Wada [Semicond. Sci. & Technol. (GB) vol.2 no.2 (Feb 1987) p.83-7]
[8] M.Remram, D.Barbier, J.F.Joly, A.Laugier [Mater. Res. Soc. Symp. Proc. (USA) vol.74 (1987) p.705-10]

10.5 DEEP LEVELS DUE TO TRANSITION METALS IN Si

by E.R.Weber

January 1988
EMIS Datareview RN=17890

Transition impurities in silicon are chemically very different from the
host crystal. Therefore it is not surprising that isolated interstitial
or substitutional transition metals and small, well-defined complexes
give rise to energy levels deep in the band gap. Recent self-consistent
calculations of these energy levels are discussed in detail in the
review by Zunger [1]. These calculations show that the metal-induced
energy levels are generally host-like electron states shifted into the
gap by the impurity atom. The valence electrons of large transition
metal clusters or precipitates are too dissimilar from the host states
so that no deep level defects are found from such clusters.

The extrapolated solubility of interstitial as well as substitutional
transition metals at room temperature is always negligible. However,
most experiments investigating the electrical activity of transition
metals are performed at or below room temperature. Therefore,
isolated transition metals are generally observed in a supersaturated
condition, in which they are only stable if the diffusion coefficient
at room temperature is low enough. This condition is easily fulfilled
for substitutional transition metals like the heavy metals Au and Pt
which have predominant substitutional solubility. Transition metals
of the 3d series, on the other hand, prefer interstitial lattice sites
and the fastest diffusing interstitial transition metals Co, Ni and Cu
cannot be retained in the interstitial site at room temperature, so that
no energy levels can be determined for the dominant interstitial atoms.

In the past, diffusion experiments with transition metals in silicon
were frequently flawed by uncontrolled contamination, which could be
detected as deep level defects, but which was not related to the
specific metal under investigation. Additional confusion in the
literature stems from the fact that the mobile interstitial transition
metals can form electrically active pairs and other simple complexes.
All transition metals analysed so far have donor levels, so that they
experience Coulomb attraction from acceptors in p-Si resulting in the
formation of donor-acceptor complexes. Energy levels of such
complexes were frequently mixed up with levels of isolated species.
This situation was improved by combination experiments, which allowed
determination of the transition metal species present in a certain
crystal. The most powerful technique for these experiments is electron
paramagnetic resonance (EPR), which allows determimation of the
character of a specific impurity in great detail. With such combination
experiments it was possible to identify many deep level defects of
well-defined transition metal species as discussed in detail in ref [2].
TABLE 1 lists those deep level defects of transition metals which can
be ascribed to specific species.

TABLE 1: deep energy levels of isolated interstitial (i) and
substitutional (s) transition metals in silicon. The symbol -/o
denotes an acceptor level, o/+ a donor level and +/++
a double donor level. Energy level positions described with '-'
refer to electron traps measured with respect to the conduction
band, those with '+' to hole traps measured with respect to
the valence band. DLTS parameters refer to Eqn (2). If several
references are given, the table gives average values.

DLTS   PARAMETERS

Metal	level type	H: energy level eV	A: prefactor /s/K*2	E: activation energy	T: eg K (e=100/s)	Ref.
Ti(i)	-/o	-0.08		0.08		[4]
	o/+	-0.26	2.3 X 10*6	0.26	152	[5]
	+/++	-0.255	3.4 X 10*5	0.290	182	[5]
V(i)	-/o	-0.16		0.16	109	[4,6,7]
	o/+	-0.45		0.45	148	[4,6,7]
	+/++	+0.32		0.47	246	[4,6,7]
Cr(i)	o/+	-0.22		0.22		[8]
Mn(i)	-/o	-0.11		0.11		[4,7]
	o/+	-0.42	1.2 X 10*8	0.43	201	[7]
	+/++	+0.25	2.9 X 10*5	0.32	200	[7]
Mn(s)	o/+	+0.34		0.35	202	[9,10]
Fe(i)	o/+	+0.385	1.0 X 10*6	0.425	244	[11,12]
Ag(s?)	-/o	-0.54	2.5 X 10*7	0.548	269	[14]
	o/+	+0.34	1.6 X 10*8	0.350	166	[13,14]
Au(s)	-/o	-0.55	2.3 X 10*7	0.553	272	[15]
	o/+	+0.35	2.8 X 10*8	0.345	160	[15]
Pt(s)	-/o	-0.23	2.2 X 10*7	0.231	122	[16]
	o/+	+0.31	1.2 X 10*7	0.321	170	[16]

TABLE 1 lists only those references which were directly used to
calculate the data in the table. In most cases similar energy levels
have been determined by many groups. References to those papers can
be found in refs [2,3,21]. The measurement techniques used for these
studies include Hall effect, Deep Level Transient Spectroscopy (DLTS)
and other electrical and optical junction techniques.

In a DLTS experiment the carrier emission rate out of a deep trap to the valence band (holes) or conduction band (electrons) is given by

$$e = c N v \exp\left[-\frac{S}{k} - \frac{H}{kT}\right] \qquad (1)$$

where c is the capture cross section, N is the density of states in the valence or conduction band, v is the thermal velocity of the carriers, and S, H are the entropy and enthalpy changes connected with the ionisation process [17]. The product N v is proportional to T*2. Therefore the analysis of DLTS spectra is commonly performed using the equation

$$e/T\!*\!2 = A \exp\left(-E/kT\right) \qquad (2)$$

The activation energy E is not always equal to the true level position H (as measured e.g. by Hall effect or photo-ionisation techniques). A difference between H and E indicates a temperature dependent capture cross section c, which has to be measured by direct trap filling experiments. Because of the generally unknown temperature dependence of c and entropy S it is not useful to combine Eqns (1) and (2) to calculate the capture cross section. Such calculated, not directly measured, capture cross sections are the origin of many confusing values for c found in original papers and review articles. Unfortunately, the literature on deep levels in silicon usually does not contain the value for A; the data for A and T(e=100/s) had therefore to be graphically determined from figures in the references cited.

Energy levels for Co, Ni and Cu have not been included in TABLE 1, because these fastest diffusing interstitial transition metals in silicon cannot be retained in the interstitial sites during quenching. On the other side, the substitutional fraction of these elements is typically not larger than about 10*13/cm*3 and therefore very difficult to distinguish from uncontrolled contamination during the diffusion treatment. There is some evidence that the levels Ec-0.41 eV and Ev +0.21 eV [18] might be due to substitutional Co and the levels at Ec-0.43 eV and Ev+0.16 eV [19] might be due to substitutional Ni, but no final conclusion is possible. The literature on Ag in Si is somewhat more consistent than the papers on Co, Ni and Cu, so that recent detailed results [14] were included in TABLE 1 although a final determination of the supposedly substitutional lattice site of this species has still to be made.

Most of the energy levels contained in TABLE 1 have been confirmed by recent self-consistent calculations reviewed by Zunger [1]. The agreement is quite good for interstitial atoms which reside in undistorted Td sites; energy levels of substitutional metal atoms differ by up to 0.2 eV from the calculated ones, which might indicate a lattice distortion induced by these heavy metals [20].

In addition to the deep levels of isolated transition metals, a large number of simple complexes of metal donors with shallow (B,Al,Co,In) and deep (Zn,Au) acceptors have been identified. In most cases these pairs form a donor level. Frequently an acceptor level is observed as well. Detailed data of energy levels of these complexes can be found in refs [3,21].

REFERENCES

[1] A.Zunger [Solid State Phys. (USA) vol.39 (1986) p.275]
[2] E.R.Weber [Appl. Phys. A (Germany) vol.30 no.1 (Jan 1983) p.1-22]
[3] E.R.Weber [in 'Impurity Diffusion and Gettering in Semiconductors' Eds R.B.Fair, G.B.Pearce, J.Washburn (Mater. Res. Soc., 1985)]
[4] K.Graff, H.Pieper [Semiconductor Silicon 1981, Ed. H.R.Huff (Electrochem. Soc., Pennington, USA, 1981) p.331]
[5] J.-W.Chen, A.G.Milnes, A.Rohatgi [Solid-State Electron (GB) vol.22 no.9 (Sep 1979) p.801-8]
[6] E.Ohta, M.Sakata [Solid State Electron. (GB) vol.23 no.7 (July 1980) p.759-64]
[7] H.Lemke [Phys. Status Solidi a (Germany) vol.64 (1981) p.549-56]
[8] H.Conzelmann, K.Graff, E.R.Weber [Appl. Phys. A (Germany) vol.36 no.3 (Mar 1983) p.169-75]
[9] H.Lemke [Phys. Status Solidi a (Germany) vol.83 (1984) p.637-43]
[10] R.Czaputa, H.Feichtinger, J.Oswald, H.Sitter, M.Haider [Phys. Rev. Lett. (USA) vol.55 no.7 (12 Aug 1985) p.758-60]
[11] K.Wuenstel, P.Wagner [Appl. Phys. A (Germany) vol.27 (1982) p.207-11]
[12] S.D.Brotherton, P.Bradley, A.Gill [J. Appl. Phys. (USA) vol.55 no.4 (15 Feb 1984) p.952-6]
[13] H.Lemke [Phys. Status Solidi a (Germany) vol.94 no.1 (1986) p.K55-9]
[14] N.Baber, H.G.Grimmeiss, M.Kleverman, P.Omling, M.Zafar Iqbal [J. Appl. Phys. (USA) vol.62 no.7 (1 Oct 1987) p.2853-7]
[15] D.V.Lang, H.G.Grimmeiss, E.Meijer, M.Jaros [Phys. Rev. B (USA) vol.27 no.7 (10 Oct 1980) p.3917-34]
[16] S.D.Brotherton, P.Bradley, J.Bicknell [J. Appl. Phys. (USA) vol.50 no.5 (May 1979) p.3396-403]
[17] H.G.Grimmeiss, E.Jansen, B.Skarstam [J. Appl. Phys. (USA) vol.51 no.7 (Jul 1980) p.3740-5]
[18] H.Kitagawa, H.Nakashima, K.Hashimoto [Jpn. J. Appl. Phys. Part 1 (Japan) vol.24 no.3 (Mar 1985) p.373-4]
[19] H.Kitagawa, H.Nakashima [Phys. Status Solidi a (Germany) vol.102 no.1 (1987) p.K238-9]
[20] A.Fazzio, M.J.Caldas, A.Zunger [Phys. Rev. B (USA) vol.22 (1985) p.934]
[21] K.Graff [Semiconductor Silicon 1986, Eds. H.R.Huff, T.Abe, B.Kolbesen (Electrochem. Soc., Pennington, USA, 1986) p.751-66

10.6 DEEP DEFECT STATES IN S-, Se- AND Te-DOPED Si

by H.G.Grimmeiss and M.Kleverman

December 1987
EMIS Datareview RN=17870

Doping silicon with chalcogens such as sulphur, selenium or tellurium
generally creates chalcogen-related defects which are electrically
and optically active. This behaviour was first reported by Ludwig [1]
in sulphur doped silicon and by Swartz [2] in selenium doped silicon
but is now known to occur also for tellurium [3]. The various donor
centres may involve one, two or more impurity atoms, at least one of
the impurity atoms being a chalcogen atom. The trend of forming
complexes decreases considerably when going from sulphur to tellurium.
In sulphur doped silicon at least three different double donors and
several other sulphur-related complexes have been reported [1,3,4]. In
contrast, the defect of the highest concentration in tellurium doped
silicon is the single substitutional double donor.

All sulphur, selenium and tellurium related centres which have been
studied in greater detail, exhibit excited states [3,4]. The excited
states (essentially all p-like states) are well described by
effective-mass theory (EMT) [5] which provides a general, characteristic
pattern easily applicable for the analysis of impurity spectra. The
s-states and, in particular, 1s-states, show large splittings due to
central cell effects and valley-orbit interactions [3,4]. The 1s states
are particularly sensitive to the local symmetry and their study provides
reliable information on defect geometry especially for more complex
centres. From the energy spacing of p-states and higher lying excited
states the charge state of a centre is readily evaluated [2-4]. This is
significant because the assignment of chalcogen related centres is often
facilitated by knowing the charge state of the centre.

Sulphur, selenium and tellurium in silicon have been studied optically in
great detail [2-4, 6-7 and refs therein]. The electrical dipole allowed
transitions have been measured either in absorption or photoconductivity.
The forbidden transitions have been investigated using Fano resonance
[7] or by tuning the interaction between allowed and forbidden transitions
with stress [8]. A selection of experimentally determined binding
energies for the ground state and excited states of neutral double
donors are listed in TABLE 1 together with corresponding values obtained
from EMT [3-5].

TABLE 1: binding energies of neutral isolated chalcogen centres in
        silicon compared with EMT values (in meV)

	EMT	S(o)	Se(o)	Te(o)
1s (A1)	31.26	318.32	306.63	158.8
1s (T2)		34.62	34.44	39.1
1s (E)		31.6	31.2	31.6
2s (A1)	8.86	18.4	18.0	15.2
2p (o)	11.49	11.48	11.45	11.5
2s (T2)	8.86	9.22	9.27	9.7
2p (+-)	6.40	6.39	6.39	6.3
3p (o)	5.48	5.45	5.48	5.5
3s	4.78	4.88	4.90	
3d (o)	3.75		3.80	4.0
4p (o)	3.31	3.31	3.29	
3p (+-)	3.12	3.12	3.12	3.12
4p (+-)	2.19	2.19	2.20	2.1
4f (+-)	1.89	1.91	1.90	
5p (+-)	1.45	1.46	1.46	
5f (+-)	1.26	1.28	1.26	
6p (+-)	1.07	1.08	1.08	
7p (+-)	0.82	0.82	0.85	

Tuning the interaction between spin-singlet and spin-triplet states of
double donors with stress does not only reveal the binding energy of
the levels involved but allows also the determination of important
electronic parameters such as the spin-orbit coupling strength and the
exchange splitting of the spin-singlet and triplet terms in the absence
of spin-orbit interaction [8]. The chemical trend in coupling strength
is in good agreement with the trend in atomic values and is consistent
with the expression that the heavier chalcogens should have larger
spin-orbit coupling.

TABLE 2: coupling strength and exchange splitting (cm*-1)

	S	Se	Te	Ref
Coupling strength	< 2	3.2	13	[8]
Exchange splitting	48.2	61		

Direct evidence that for a particular isolated chalcogen impurity the
neutral and charge centres are different states of the same defect
has been presented by Kleverman et al [9]. They showed that the hole
capture at neutral double donors is governed by a local Auger effect
involving both electrons of the neutral centres resulting in relatively
large hole capture cross sections for the neutral defect.

TABLE 3: capture cross sections at 250 deg K

Defect	Hole Cross section (cm*2)	Electron Cross section (cm*2)	Notes
S (o)	9 X 10*-17	9 X 10*-17	
Se (o)	7 X 10*-17	2 X 10*-16	
Te (o)	2 X 10*-17		Extrapolated value
S (+)	10*-20		
Se (+)	2 X 10*-23	1.2 X 10*-14	
Te (+)	2 X 10*-24		Extrapolated value
References	[9]	[10,11]	

The energy of p-like states relative to the ionisation limit is notably insensitive to the nature of the defect [3,4]. This implies that the binding energy of p-like states is almost identical for all chalcogen related centres. This is also true for charged defects if their energies are divided by four.

TABLE 4: binding energies of the lower lying states for charged isolated chalcogen centres (in meV)

		S(+)	Se(+)	Te(+)	Ref
1s	(A1)	614.0	593.3	411.2	[3,9]
1sT7	(T2)	184.3	160.0	177.2	[3,13]
1sT8	(T2)	183.5	163.7	171.2	[3,13]
1sT	(E)	130	130		[14]
2p	(o)	45.57	45.57	47.2	[3,4]
2p	(+-)	25.61	25.61	25.6	[3,5]

The symmetry of the neutral sulphur pair S-S(o) was shown by Krag et al [12] to be trigonal (<111> axis). Ludwig [1] found that the atoms in the S-S(+) pair are on close, equivalent sites, which is consistent with trigonal symmetry. High resolution spectroscopy indicates D(3d) [4]. Trigonal symmetry implies that an impurity pair lies along an axis through a nearest-neighbour pair in the host lattice. Since the electronic structures (absorption spectra) of the neutral and charged S-S, Se-Se and Te-Te centres are similar, it is assumed that the sites occupied by the S, Se and Te pairs must be the same. For other complexes of chalcogens in silicon see [2-4] and references therein.

TABLE 5: binding energies of the lower lying states for neutral
chalcogenic pairs (in meV)

	S-S(o)	Se-Se(o)	Te-Te(o)	Ref
1s (A1+)	187.61	206.44	158.2	[3,4]
1s (A1-)	26.5	25.72	25.8	[3,4]
1s (E-)	31.30	31.30	33.1	[3,4]
1s (E+)	34.4	33.2		[4]
2s (A1+)	15.3	15.9		[4]
2p (o)	11.49	11.58	11.5	[3,4]
2p (+-)	6.39	6.39	6.3	[3,4]

TABLE 6: binding energies of the lower lying states for ionised
chalcogen pairs (in meV)

	S-S(+)	Se-Se(+)	Te-Te(+)	Ref
1s (A+)	371	390	-	[4]
1s (A1-)	96.2	92.9	-	[4]
1s (E-)	149.5	124.7	-	[4]
1s (E+)	-	-	-	
2p (o)(E-)	46.3	46.3	-	[4]
2p (o)(A1-)	47.7	45.1	-	[4]
2p (+-)(A1-)	26.39	6.39	-	[4]
2p (+-)(2E-)	26.3	26.3	-	[4]

The above discussion shows that each of the chalcogens sulphur, selenium
and tellurium generally generates more than one chalcogen related defect
in silicon. For sulphur more than eight different energy levels have
been observed, all lying in the upper half of the band gap. While the p-
line states of all these levels are very similar, the 'fingerprint' of
the impurity i.e. the binding energy of the 1s ground state, is
determined by the central cell potential and reflects the chemical
nature of the defect.

REFERENCES
----

[1]   G.W.Ludwig [Phys. Rev. (USA) vol.137 (1965) p.A1520]
[2]   J.C.Swartz, D.H.Lemmon, R.N.Thomas [Solid State Commun. (USA)
        vol.36 (1980) p.331]
[3]   P.Wagner, C.Holm, E.Sirtl, R.Oeder, W.Zulehner [Festkorperprobleme
        (Germany) vol.XXIV, Ed. P.Grosse (Vieweg, Braunschnweig, 1984)
        p.151]
[4]   E.Janzen, R.Stedman, G.Grossman, H.G.Grimmeiss [Phys. Rev. B
        (USA) vol.29 no.4 (15 Feb 1984) p.1907-18]
[5]   R.A.Faulkner [Phys. Rev. (USA) vol.184 (1969) p.713]
[6]   W.E.Krag, H.J.Zeiger [Phys. Rev. Lett. (USA) vol.8 (1962) p.485]
[7]   E.Janzen, G.Grossmann, R.Stedman, H.G.Grimmeiss [Phys. Rev. B
        (USA) vol.31 no.12 (15 Jun 1985) p.8000-12]
[8]   K.Bergman, G.Grossmann, H.G.Grimmeiss, M.Stavola [Phys. Rev.
        Lett. (USA) vol.56 no.26 (30 Jun 1986) p.2827-3]

[9]   M.Kleverman, H.G.Grimmeiss, A.Litwin, E.Janzen [Phys. Rev. B (USA)
        vol.31 no.6 (15 Mar 1985) p.3659-66]
[10]   H.G.Grimmeiss, E.Janzen, B.Skarstam [J. Appl. Phys. (USA) vol.50
        no.8 (Aug 1980) p.4212-17] and [vol.51 no.7 (Jul 1980) p.3740-5]
[11]   L.Montelius [Thesis, Univ. Lund (Sweden) 1987)]
[12]   W.E.Krag, W.H.Kletner, H.J.Zeiger, S.Fischler [J. Phys. Soc. Jpn.
        Suppl. (Japan) vol.21 (1966) p.230]
[13]   H.G.Grimmeiss, E.Janzen, K.Larsson [Phys. Rev. B (USA) vol.25
        no.4 (15 Feb 1982) p.2627-32]
[14]   M.Altarelli [Proc. 16th Int. Conf. Phys. Semicond., Montpellier,
        1982, Ed. M.Averous (North-Holland, Amsterdam, 1983) p.122]

CHAPTER 11

DEFECT STRUCTURE

11.1 STRUCTURE OF EXTENDED DEFECTS IN Si

by K.V.Ravi

June 1987
EMIS Datareview RN=15729

A. INTRODUCTION

Extended defects include all defects with dimensions greater than zero.
The principal extended defects in silicon of interest include
dislocations, twins, stacking faults and grain boundaries. In this
review the properties of grain boundaries in silicon will not be
discussed. Extensive studies of these defects have included their
crystallographic structure, their methods of introduction into crystals
and films of silicon and their influence on various physical properties.

B. DISLOCATIONS

Dislocations are line defects created as a consequence of the presence
of a missing row or an extra row of atoms along close packed directions
in the crystal. Principally, two types of dislocations have been
identified in silicon, viz screw dislocations which have their Burgers
vector parallel to the dislocation and the 60 deg dislocation with the
Burgers vector at an angle of 60 deg to the dislocation. Hornstra [1]
has discussed in detail the various possible configurations of
dislocations in the diamond cubic structure. It has been postulated that
many types of dislocation can occur in the diamond cubic lattice most
of which are not generally mobile. The most commonly observed
dislocations in silicon single crystals are the 60 deg dislocations [2].

Using high resolution transmission electron microscopy techniques it
has been demonstrated that most dislocations in silicon are dissociated,
that is, they consist of partial dislocations separated from each other
with the dissociation distances being of the order of 50 A [3]. Hirth
and Lothe [4] have identified the presence of two sets of dislocations
in silicon which they characterise as the glide set and the shuffle set
of dislocations. The essential difference between the two sets of
dislocations is that the glide dislocations are glissile, that
is, they can glide on close packed glide planes, and they can
dissociate into a pair of Shockley partial dislocations which are also
glissile. The shuffle set of dislocations is formed by the insertion of
extra planes of atoms between layers of atoms between the two
interpenetrating face centered cubic lattices that have the same sign.
That is, in a stacking sequence AaBbCcAa, and so on, characterising the
diamond cubic structure, the shuffle set is created by the insertion of
an extra layer of atoms between, for instance, layers A and a, whereas
the extra plane of atoms representing a glide dislocation terminates

between atomic planes having different signs, that is, between A and b, for example. Among the dissociation models the most common one is the dissociation of a perfect 60 deg dislocation into a Frank partial and a Shockley partial according to the following reaction:

$$a/2 \ (110) = a/6 \ (112) + a/3 \ (111)$$

C. DISLOCATION GENERATION MECHANISMS

Dislocations can be introduced into silicon as a result of a variety of mechanisms. These include: dislocation generation due to thermomechanical stresses, point defect condensation, and chemical impurity effects.

Thermomechanical stresses can be introduced into silicon crystals during growth due to non-uniform thermal gradients extant in the crystal growth process. Due to thermal expansion, any perturbation or deviation of the local thermal gradient from the average will result in the generation of internal stresses that will be accommodated by the generation of dislocations. Inhomogeneous radial and axial thermal gradients in the growth of crystals can result from: (1) radial heat losses, (2) a non-zero growth rate of the crystal, and (3) rotation of the crystal about its axis. In addition to the above mechanisms, the cooling rate of the crystal from the melting point is also important in maintaining a dislocation-free state of the crystal. Rapid and nonlinear cooling can result in dislocation introduction if suitable dislocation sources are available. Dislocation generation due to thermomechanical stresses can also occur during the processing of silicon wafers. In particular high temperature and large nonlinear thermal gradients attending processes such as dopant diffusion, oxidation and epitaxial deposition can result in plastic deformation and the attendant generation of dislocations [5].

If the concentration of point defects in a growing crystal exceeds the equilibrium value at a given temperature, the excess point defects will agglomerate and condense in the form of dislocation loops. These loops can either be faulted or occur as perfect prismatic loops. Once these loops are formed they can grow, shrink and change shape depending upon a variety of factors such as the concentration of point defects in the vicinity, the temperature, the size of the initially nucleated loop, the cooling rate of the crystal and the availability of other sources and sinks for point defects.

Chemical effects are very important in the generation of dislocations with particular reference to various dopant and diffusional processes characterising silicon device processing. In crystal growth the presence of high concentrations of impurities precludes the growth of dislocation-free crystals. A particular case of interest is the influence of carbon on the generation of dislocations. Carbon concentrations in the range of $10*18$ to $10*19$ atoms/cm*3 cause a loss of perfection of the crystal. One of the major impurity related dislocation generation mechanisms is due to the presence of oxygen in the crystal. In crystals grown by the Czochralski process dissolved oxygen up to concentration levels of $10*18$ atoms/cm*3 can be present. When wafers containing oxygen are heat treated during device fabrication the oxygen can precipitate in the form of silicon dioxide.

The precipitation process can result in dislocation generation as a result of interfacial stress generated between the precipitate and the silicon matrix in response to the difference in the thermal expansion coefficients of the two phases [6]. The diffusion of impurities and dopants as well as the implantation of energetic ions into silicon wafers can cause dislocation generation. The diffusion of gold into silicon has been shown to result in dislocation generation particularly when the crystal is prone to contain point defect agglomerates formed during the crystal growth process. The diffusion of phosphorus in high concentrations for the formation of np junctions can result in the generation of a network of dislocations close to the surface into which the phosphorus is diffused with the dislocation network being planar and parallel to the surface and the pn junction. The mechanisms involved in these cases relate to misfit stresses between the silicon and the diffusing atoms. Other factors of importance in the dislocation generation process include the influence of point defects, electric field effects and the presence of any pre-existing defects which can influence the defect nucleation processes. The generation of defects by the implantation of ions into silicon is a direct process of primary and secondary collisions of the ions with the atoms of the host crystal. This form of radiation damage generates a high concentration of point defects which can agglomerate into dislocation loops. Implantation is typically followed by thermal annealing to activate the dopant atoms and to anneal the radiation damage. Annealing results in the retention of residual damage in the form of dislocations.

D. STACKING FAULTS

Stacking faults are two-dimensional defects that are formed by the equivalence of removing or inserting an extra plane of atoms between close packed (111) planes in the host lattice. The removal of a plane of atoms results in an intrinsic fault while the insertion of an extra plane of atoms results in an extrinsic fault. The most commonly occurring stacking faults in silicon are extrinsic stacking faults. The defects are composed of two extra layers of atoms which are bound by Frank partial dislocations with a Burgers vector of 1/3 (111). These defects are sessile in nature and they can only move under the action of chemical forces, i.e. as a consequence of changes in the local point defect concentration. The chemical driving force that causes changes in the fault dimensions is given by

$$C(F) = C(0) \exp(-\text{gamma } B*2/2kT)$$

where $C(F)$ is the vacancy concentration in equilibrium with the fault, $C(0)$ is the vacancy concentration in the defect-free crystal, $B*2$ is the area of the vacancy in (111) plane family and gamma is the stacking fault energy. The stacking fault will grow if there is vacancy undersaturation in the vicinity of the fault and it will shrink if an excess concentration of vacancies is supplied to the fault. Conversely, fault growth can be promoted by the supplying of interstitials, or extra atoms of silicon to the fault.

E. STACKING FAULT GENERATION MECHANISMS

Since the extrinsic nature of stacking faults involves the local
accumulation and condensation of excess silicon atoms, the generation
of these defects requires two events to take place:

1. An excess concentration of silicon self interstitials, or other
    impurity atoms, should be generated.

2. These excess atoms should then condense to form localised extra
    planes of atoms between close packed planes.

Stacking faults can be introduced into silicon crystals as a result of
oxidation, impurity and dopant diffusion and ion implantation. In
addition epitaxial or growth stacking faults which have a different
structure from oxidation or diffusion induced stacking faults can also
be introduced into epitaxial silicon films.

Oxidation of silicon results in the generation of an excess
concentration of silicon self interstitials at the oxidation front
[7]. These excess self interstitials can locally condense and form
stacking faults. The heterogeneous precipitation of the point defects
occurs as a result of the local availability of nucleating agents such
as point defect clusters formed in the crystals during the growth
process or the presence of local sites of mechanical damage [8]. The
presence of oxygen in the crystals can also lead to the generation of
stacking faults. The precipitation of supersaturated oxygen in the form
of the oxides of silicon leads to an injection of silicon interstitials
into the silicon matrix at the oxide/silicon interface due to the
volume change associated with the precipitation process. The excess
interstitials condense into extrinsic stacking faults.

F. TWINS

Twins are two-dimensional defects which do not occur very frequently in
silicon. They can be introduced into silicon crystals during growth if
a high concentration of carbon is present. It has been suggested that
such growth induced twins are made up of sheets of carbon atoms between
close packed silicon planes [9]. Twins can also be introduced by the low
temperature (< 600 deg C) plastic deformation of silicon [10].

REFERENCES
----

[1]   J.Hornstra [J. Phys. & Chem. Solids (GB) vol.5 (1958) p.129]
[2]   P.Haasen [Acta Metall. (USA) vol.5 (1957) p.598]
[3]   D.J.H.Cockayne, I.L.F.Ray, M.J.Whelan [Philos. Mag. (GB) vol.20
       (1969) p.1265]
[4]   J.P.Hirth, J.Lothe [Theory of Dislocations (McGraw-Hill, New
       York, USA 1968) p.356]
[5]   K.V.Ravi [Imperfections and Impurities in Semiconductor Silicon
       (John Wiley & Sons, New York, USA 1981) p.62]

[6]   T.W.Tan, W.K.Tice [Philos. Mag. (GB) vol.30 (1976) p.615]

[7]   W.Frank, U.Gosele, H.Mehrer, A.Seeger [in 'Diffusion in
        Crystalline Solids' Eds G.Murch, A.S.Nowick (Academic, New York,
        USA 1984) p.63]

[8]   K.V.Ravi, C.J.Varker [J. Appl. Phys. (USA) vol.45 (1974) p.263]

[9]   K.V.Ravi [J. Cryst. Growth (Netherlands) vol.39 no.1 (Jul 1977)
        p.1-16]

[10]   K.Yasutake, J.D.Stephenson, M.Umeno, H.Kawabe [Philos. Mag. A
        (GB) vol.53 no.3 (Mar 1986) p.L41-8]

11.2 STRUCTURE OF C- AND O-RELATED POINT DEFECTS IN Si

by K.V.Ravi

June 1987
EMIS Datareview RN=15730

A. INTRODUCTION

Carbon and oxygen are the main impurities in high purity semiconductor
grade silicon with typical carbon concentrations of 10*16 and 10*18
atoms/cm*3 for Czochralski (crucible pulled) and float-zoned crystals
and typical oxygen concentrations ranging from 2 X 10*17 to 2 X 10*18
atoms/cm*3 in Czochralski crystals and between 2 X 10*15 to 2 X 10*16
atoms/cm*3 in float-zoned crystals. These impurities have their origins
in a variety of sources with carbon being present in semiconductor
grade polysilicon as well as from sources such as vacuum pump oils and the
hardware in crystal growth machines. In addition carbon can also enter
the silicon by solid state diffusion if suitable impurity sources
(such as organic residues) are available at the wafer surface. The
primary source of oxygen in Czochralski crystals is the silica crucible
used in crystal growth. At its melting point silicon reacts with quartz
resulting in the dissolution of oxygen to saturation levels in the
crystal. The source of oxygen in float zoned silicon is not as clear,
with the polysilicon and the crystal growth ambient both contributing
to the oxygen content.

B. CARBON

Because of its very small segregation coefficient (k = 0.058 +- 0.005
[1]) carbon is typically distributed inhomogeneously along the length
and across the diameter of crystals. The carbon concentration along the
axis of the crystal can increase from the seed to the tang end by a
factor of 2-20. Macroscopic radial variation of the carbon
concentration of up to 70% has been found, this variation being dependent
on the shape of the solid liquid interface during crystal growth.
Substitutional carbon in silicon gives rise to an infrared absorption
band at a wavelength of 16.5 microns and this technique is frequently
utilized to detect the presence of carbon and oxygen [2]. Carbon has
been confirmed to occupy substitutional sites in the silicon lattice by
the observation of a decrease in the lattice parameter of silicon with
an increase in the carbon concentration [3]. There is evidence to
indicate that the concentration of carbon in silicon is enhanced by the
presence of dissolved oxygen [4]. The diffusion coefficient of carbon
in silicon has been observed to be enhanced as a result of processes
such as oxidation and phosphorus diffusion. This has been explained as
being due to the creation of silicon self interstitials by the
oxidation and diffusion processes and the trapping of the excess
interstitials by the carbon [5].

A particularly important effect of carbon, especially at high
concentrations (> 2 X 10*17 atoms/cm*3) is its effect on the formation
of microdefects during the growth of crystals. Carbon can trap silicon
self interstitials during solidification of the crystal to form a defect
which has been termed the 'B defect'.

These defects assume a spiral distribution across the diameter of the wafer and through the thickness of the crystal and an aggregation of the B defects and the larger A defects have been termed 'swirl defects' [6]. Carbon can also function as a nucleating agent for the precipitation of the oxides of silicon. Secondary defects such as dislocations, stacking faults and precipitates can be preferentially nucleated at carbon-containing sites in the wafers when the wafers are subjected to heat treatments. Such defects can have an adverse impact on the electrical characteristics of devices built in wafers prone to contain these defects. In general, however, the behaviour of carbon in silicon is insufficiently understood and more work needs to be done to fully understand its effects.

C. OXYGEN

Oxygen has several important roles to play in silicon:
a) oxygen, unlike carbon, is an electrically active impurity in silicon, functioning as a donor under certain conditions; b) oxygen in supersaturated solid solution can precipitate as an oxide of silicon when the oxygen-containing crystals are heat treated. Precipitation has important defect related implications. c) Dissolved oxygen can influence the mechanical properties of silicon, influencing behaviour such as the generation of thermomechanically induced dislocations. In general, the manipulation of the concentration, the distribution, the lattice location and the chemical state of oxygen in silicon is an important means for the utilisation of the beneficial effects and avoidance of the harmful effects of oxygen with regard to device electrical performance and yield.

D. DONOR FORMATION

Heat treating oxygen-containing crystals at temperatures in the range of 300 to 500 deg C results in the formation of donors with the maximum donor concentration occurring at 450 deg C. The change in the electrical nature of oxygen is a result of a change in its lattice location from the normally interstitial location to a substitutional location. When oxygen is in interstitial locations, that is, as Si-O-Si units, the asymmetric vibration of these units gives rise to an infrared absorption band at 9 microns. An absorption band attributed to SiO_2 occurs at 8.4 microns. The early explanation for the structure of donors contended that an aggregate of four oxygen atoms comprised the donors. The current view is that there are several donor centres and that the different centres involve different numbers of oxygen atoms in the cluster. Bourret [7] has reviewed the various models for donor formation, none of which seem to satisfactorily explain all the observed phenomena. These various models involve the formation and diffusion of di-oxygen molecules, oxygen-vacancy or oxygen-self interstitial interactions, an enhancement of the diffusion of oxygen and the creation of oxygen clusters made up of four or more atoms which function as the donors. A radical new proposal by Newman [8] suggests that thermal donors in silicon are in fact not due to oxygen clusters but are due to silicon self interstitial clusters. It is clear that the structure and the nature of the defect giving rise to donor behaviour are still imperfectly understood.

E. OXYGEN PRECIPITATION

Annealing oxygen-containing crystals results in the precipitation of
oxygen. The precipitation process is found to proceed in a complicated
fashion from the solid solution through several stages including local
oxygen clustering, the formation of amorphous and coherent crystalline
phases culminating in the formation of equilibrium phases [7]. Extended
annealing at low temperatures (about 650 deg C) results in the
formation of rod-like defects and small dot-like features, when
observed by transmission electron microscopy, which have been suggested
to be local oxygen clusters. At higher temperatures (800-1000 deg C)
the rod-like features are no longer observed and two morphologies of
precipitates are formed, a crystalline phase of silica identified to be
coesite and an amorphous precipitate which is suggested to be an
amorphous form of SiO2. At temperatures above 1000 deg C equilibrium
SiO2 precipitates are formed and secondary defects such as stacking
faults and dislocations are formed in association with the precipitates.
The interface between the precipitate and the silicon matrix is the
origin of a variety of phenomena such as the generation of high strain
due to the differences in the thermal expansion coefficients between
SiO2 and silicon, injection of silicon self interstitials into the
silicon which can then collapse into extended defects and the creation
of interface states between the oxide and the silicon which have been
implicated as the electrically active centres which give rise to a
second donor called the new donor at 650 deg C.

F. INFLUENCE ON MECHANICAL PROPERTIES

The presence of oxygen in solid solution has the effect of increasing
the mechanical strength of the wafer. Oxygen pins dislocations and
retards their motion. This can result in a reduced degree of wafer
warpage in response to thermomechanical stresses in device processing.
Wafer warpage becomes increasingly important as wafer diameters are
increased and device dimensions are shrunk. If oxygen is caused to
precipitate as a result of heat treatment the strengthening effect is
lost. Oxide precipitation can in fact result in the formation of
dislocations and other defects due to the generation of interstitials
at the oxide/silicon interface, contributing to the plasticity of the
crystal. Oxide precipitation is frequently utilised to provide internal
gettering agents for gettering unwanted impurities from the active
regions of the wafer to enhance device yields. The precipitation
process can have the undesirable consequence of generating dislocations,
weakening the wafer and promoting wafer warpage.

REFERENCES
----

[1] B.O.Kolbesen, A.Muhlbauer [Solid-State Electron. (GB) vol.25 no.8
     (1982) p.759]
[2] R.C.Newman, A.S.Oates, F.M.Livingston [J. Phys. C (GB) vol.16
     (1983) p.L667]
[3] J.A.Baker, T.N.Tucker, N.E.Moyer, R.C.Buschert [J. Appl. Phys.
     (USA) vol.39 (1968) p.4365]
[4] R.C.Newman [Fall meeting of the Materials Research Society,
     Boston (1985) unpublished]
[5] J.P.Kalejs, L.A.Ladd, U.Gosele [Appl. Phys. Lett. (USA) vol.45
     (1984) p.269]
[6] K.V.Ravi [Imperfections and Impurities in Semiconductor Silicon
     (John Wiley & Sons, New York, USA, 1981) p.62]
[7] A.Bourret [Proc. 13th Int. Conf. on Defects in Semiconductors
     Eds L.C.Kimerling, J.M.Parsey,Jr (New York: AIME 1985) p.129]
[8] R.C.Newman [J. Phys. C (GB) vol.18 no.30 (30 Oct 1985) p.L967-72]

11.3 SELF INTERSTITIALS AND VACANCIES IN Si

by K.V.Ravi

June 1987
EMIS Datareview RN=15716

A. GENERAL

In spite of the extensive investigations of point defects, self
interstitials and vacancies, in silicon there is still considerable
debate on the nature of these theoretically and technologically
important defects. Vacancies and interstitials are created thermally
and as a result of irradiation and they coexist at high temperatures.
Precise values of the energies of formation and migration of these
defects have not been obtained to date but a range of vales have been
obtained, shown in TABLE 1, by several investigators using different
techniques.

TABLE 1

	Formation Energy	Migration Energy	Reference
Vacancy	3.6 +- 0.2 eV		[1]
	3.6 to 4.6 eV		[2]
		0.33 to 0.44 eV	[3]
	3.8 eV		[4]
Interstitial	3.7 to 4.43 eV	0.4 eV	[5]
	4 to 4.4 eV	0.1 to 1.1 eV	[2]

Both vacancies and interstitials can be charged and as such will be
influenced by and will influence the behaviour of electrically active
impurities in silicon. Both these defects are negative-U centres and
the state of charge on them will be functions of position of the Fermi
level within the band gap. Both defects are mobile at high temperatures
but the interstitial is found to exhibit the remarkable property of
being mobile down to very low temperatures (about 4 deg K) in the
presence of ionising radiation [6]. This behaviour has been explained
on the basis that the lattice location of the self interstitial is a
function of its charge. In p-type material the interstitial is doubly
positively charged and occupies a tetrahedral interstitial site. In
the presence of ionising radiation the interstitial will capture
electrons and become neutral and move to another site, a hexagonal or
a body centered site where the neutral defect is more stable. The
subsequent loss of the electrons at the site will again cause the now
positively charged defect to move. Consequently in the presence of
charge carriers the interstitial can migrate athermally by the
successive capture of electrons and holes [7].

B. POINT DEFECT AGGLOMERATION

A key property of point defects that has practical implications is their
propensity to agglomerate and precipitate in the form of secondary
defects. Such agglomeration occurs readily during crystal growth and has
been the subject of intense study. When the concentration of intrinsic

point defects in the growing crystal exceeds the equilibrium value at any given temperature, the excess point defects will condense to form point defect clusters, dislocation loops and stacking faults. The behaviour of the excess point defects is governed by factors such as their concentration, the local temperature, the rate and uniformity of cooling, the availability of other point defect sources and sinks and the presence of impurities. In dislocation-free crystals point defect agglomeration leads to discrete defects which have been termed 'swirl' defects due to the swirl or spiral distribution pattern assumed by these defects in wafers cut perpendicular to the growth axis of the crystals. The defects that comprise the swirl defects are called microdefects and three types of microdefects have been identified, the A,B and D defects. The A defects are small, micron sized dislocation loops. These loops are found to assume several complex shapes and are often decorated by an impurity. The dislocations are found to be interstitial in nature indicating that their formation involves a local availability of excess interstitials. The nature of the B defects, which are considerably smaller than the A defects has not been unambiguously determined. They are postulated to be small clusters of silicon atoms heterogeneously nucleated by an impurity with carbon being the most common impurity which is likely to nucleate these defects. The A defects are considered to be the by-products of the growth and transformation of the B defects. The D defects have also not been unambiguously identified and generally occur at high crystal growth rates. They have tentatively been suggested to be vacancy agglomerates.

The major factors that contribute to the inhomogeneous distribution of grown-in point defects in silicon crystals include; a) thermal convection currents in the liquid, b) the process of automatically controlling the diameter of the crystal which can result in a continuing variation in the temperature input to the growing crystal as well as continuing changes to the rate of growth and c) thermal asymmetry at the solid liquid interface due to the non-coincidence of the thermal isotherm and the rotation axis. Of these, thermal convection currents in the liquid are considered to be the primary contributors to the localised temperature fluctuations.

The formation of swirl defects depends on the crystal growth rate v and the temperature gradient G. For a given value of G no defects form at very low growth rates, A and B defects form as the growth rate is increased and with further increase in v, A and B defects disappear and the crystal is defect-free again and with further increase in the growth rate the D defects appear. In addition to depending on v and G swirl defect formation is dependent in a complicated fashion on the presence of impurities, of which carbon is the most important.

Several models for swirl defect formation have been suggested all of which involve the association of vacancies or interstitials aided by the presence of an heterogeneous nucleating agent. The models that best explain the observed phenomena assume that both vacancies and interstitials are in dynamical equilibrium at high temperature and whether interstitial agglomeration or vacancy agglomeration dominates is dependent on their respective diffusivities as well as on extrinsic phenomena such as the presence of impurities [8].

C. OXIDATION AND DIFFUSION

Point defects play a very important role in diffusion and oxidation
processes in silicon. Surface oxidation of silicon wafers results in a
variety of phenomena including the nucleation and growth of oxidation
induced stacking faults, oxidation enhanced diffusion of B, Al, Ga, In,
P, and As and the oxidation retarded diffusion of Sb [9].
During surface oxidation self interstitials are injected into the
silicon. These excess interstitials can condense into interstitial
clusters and stacking faults. The transformation of the oxidation
generated interstitials into stacking faults is a heterogeneous process
with defect nucleation occurring either at regions of surface mechanical
damage on the wafer or at microdefect sites. Wafers containing
microdefects are found to give rise to stacking faults when subjected
to oxidation [10]. The microdefects can grow by the assistance of
oxidation generated interstitials and subsequently collapse on to
close packed planes to form stacking faults.

The oxidation enhanced diffusion of substitutional dopants such as P
and As indicates that a supersaturation of interstitials accompanying
oxidation causes diffusion enhancement and that these dopants diffuse
predominantly via the interstitialcy mechanism. On the other hand the
diffusion of Sb is retarded by oxidation indicating an undersaturation
of those defects that govern the diffusion of Sb. Since a
supersaturation of interstitials results during oxidation the oxidation
retarded diffusion of Sb suggests that Sb diffusion occurs via the
vacancy mechanism and that oxidation also results in a vacancy
undersaturation. With both oxidation retarded diffusion of Sb and the
oxidation enhanced diffusion of the other dopants occurring
simultaneously, oxidation of silicon results in the simultaneous
generation of a self interstitial supersaturation and a vacancy
undersaturation. The concentrations of the two species of point defects
however are not the same for all conditions of time and temperature.
By making detailed observations of the diffusion of Sb and the other
dopants as a function of oxidation time and temperature it has been
determined that to reach dynamical equilibrium between the interstitial
and the vacancy concentrations requires considerable time at high
temperatures (e.g. 1 hr at 1100 deg C [11]). At the high temperatures
local dynamical equilibrium between vacancies and interstitials prevails
whereas at lower temperatures (<1100 deg C) and short times of
processing self interstitial supersaturation is the norm. At low
temperatures interstitials move independently of the vacancies with a
diffusivity which is much higher than the equilibrium diffusivities of
vacancies and interstitials at higher temperatures, which is quite
contrary to the situation expected for a simple thermally activated
process. This behaviour suggests that a point perturbation resulting
from processes such as oxidation, diffusion and ion implantation would
influence silicon wafers during device processing over much larger
distances and more severely at lower temperatures than at higher
temperatures [8]. The essentially non-equilibrium nature of conditions
extant in most device processing operations complicates the control of
phenomena. In the fabrication of complex integrated circuits very high

temperatures and long processing times are unsuitable when attempting to achieve very shallow p-n junctions, steep concentration gradients, narrow channel or base widths. Low temperatures with very short and, in many cases, transient processing becomes the norm. As a result defect disequilibrium, non-random phenomena and attendant nonuniform properties are perennial problems which will become more prevalent as the search for higher degrees of integration and miniaturisation continue.

REFERENCES
----

[1]   S.Dannefaer, P.Mascher, D.Kerr [Phys. Rev. Lett. (USA) vol.56 no.20 (1986) p.2195-8]

[2]   Y.Bar-Yam, J.D.Joannaopoulos [Proc. Seventeenth Int. Conf. Physics of Semiconductors, Eds J.D.Chadi, W.A.Harrison (Springer-Verlag, Berlin, 1984) p.721-4]

[3]   G.D.Watkins, J.R.Troxell, A.P.Chatterje [Proc. Int. Conf. Radiation Effects in Semiconductors, Nice, Ed. J.H.Albany (1979) p.16]

[4]   S.T.Pantelides, R.Carr, P.J.Kelly, A.Oshiyama [Mat. Res. Soc. Symp. Proc. (USA) vol.63 (1985) p.7-11]

[5]   T.Soma, T.Matsuoka, H.Matsuo Kagaya [Phys. Status Solidi b (Germany) vol.135 (1986) p.85-90]

[6]   G.D.Watkins [Radiation Damage in Semiconductors (Dunod, Paris 1964) p.67]

[7]   J.Bourgoin, J.W.Corbett [Phys. Lett. A (Netherlands) vol.38A (1972) p.135]

[8]   T.Y.Tan, U.Gosele [Appl. Phys. A (Germany) vol.37 (1985) p.1-17]

[9]   W.Frank, U.Gosele, H.Mehrer, A.Seeger [in 'Diffusion in Crystalline Solids' Eds G.Murch, A.S.Nowick (Academic, New York, 1984) p.63]

[10]   K.V.Ravi [Imperfections and Impurities in Semiconductor Silicon (John Wiley & Sons, New York, 1981)]

[11]   D.A.Antoniadies, I.Moskowitz [J. Appl. Phys. (USA) vol.53 (1982) p.6788]

11.4 STRUCTURE OF ION-IMPLANTATION INDUCED DEFECTS IN Si

by N.Hecking

October 1987
EMIS Datareview RN=16187

A. GENERAL REMARKS

Ion implantation today is a standard doping technique in semiconductor
technology. During the implantation process, radiation damage, which
is in general harmful for applications, is produced within the surface
layer. The nature of the damage depends considerably on implantation
conditions. Of major importance are mass and energy of the incoming
ions, the implantation dose, dose rate and especially the target
temperature. In 1972 Gibbons [1] reviewed the fundamental ideas of
damage production and annealing.

B. CALCULATION OF DAMAGE DEPTH-DISTRIBUTIONS

During the slowing down of implanted ions primary damaged zones are
created as long as ions dissipate kinetic energy in elastic collisions
with the target atoms. The ion range and damage distributions can be
calculated within a Monte-Carlo simulation program. The binary
collision approximation is commonly used to construct the ion
trajectories for a large number of particles.

With the TRIM program of Ziegler, Biersack and Littmark [2], these
depth distributions can be calculated with sufficient accuracy (some
%), assuming an amorphous target. This corresponds to the implantation
in random directions. The influence of channelling effects on the damage
and range profiles can be computed using another Monte-Carlo code from
Robinson [3] called MARLOWE.

The critical dose for the amorphization of silicon, a value of
considerable interest, can be estimated directly from the simulated
depth profile of energy deposited in nuclear collisions. Following a
model from Stein et al [4] for the production of amorphous material it
has been shown by Vook [5] that the critical energy density of
amorphization is 6.0 X 10*23 eV/cm*3 or 12 eV/atom. This value is valid
in the case of low temperature implants independent of the projectile
ions. With increasing substrate temperature the critical energy density
and therefore the critical dose of amorphization increases.

C. EXPERIMENTAL TECHNIQUES

Numerous experimental techniques are used to determine the amorphization
threshold or the microscopical structure of defects. Gotz [6] has
compared results of backscattering/channelling measurements (RBS) with
those of optical and electron spin resonance (ESR) investigations. From
the energy distribution of back-scattered high energy alpha-particles
or protons the depth distribution of displaced atoms can be derived.

By means of ESR spectra, the structure of point defects can be identified, as has been done for example for divacancies by Corbett [7]. The production of amorphous silicon is accompanied by an additional line in ESR spectra [8].

The concentration of divacancies can be derived from optical absorption measurements at a wavelength of 1800 nm [9]. Different optical methods [10,11,12] have identified two states of amorphous silicon, the as-implanted amorphous state and a second stabilized state after furnace annealing (15 min; 800 deg K). The extension of amorphous layers and the microscopical structure of defects especially after ion irradiation and subsequent annealing can be successfully analysed by means of several electron microscopical techniques.

D. STRUCTURE OF AS-IMPLANTED SILICON (PRIMARY DAMAGE)

Maszara and Rozgonyi [13] have investigated 150 and 300 keV silicon self implantation in the dose range from 2 X 10*14/cm*2 to 1 X 10*16/cm*2 at substrate temperatures from 82 to 296 deg K. Cross section transmission electron microscopy (XTEM) is used to study the relation between dose and depth of amorphous/crystalline interfaces for different target temperatures. From this, the critical energy-density for amorphization is derived as a function of temperature and depth. A steep increase of threshold damage energy Ec (eV/atom) is observed approaching 300 deg K where Ec exceeds a value of about 40 eV/atom. The width of amorphous/crystalline interfaces at both sides of a buried amorphous layer is shown to decrease with increasing layer thickness. The morphology of a/c interfaces is found to be independent of temperature in the analysed range.

Similar investigations have been carried out by Prunier et al [14] after irradiation at temperatures up to 473 deg K, using Rutherford backscattering (RBS) and TEM. A sharp amorphous/crystalline interface is revealed by cross section high resolution electron microscopy. In the case of 473 deg K / 180 keV Si-implantation, below a 200 nm crystalline top layer a highly damaged layer is formed. This layer shows extended defects on (113) family planes. Extended (113) plane family defects have been found also for example after high energy silicon self implantation [15] at a substrate temperature of 450 deg K and together with (111) plane family loops after high temperature (estimated less than 573 deg K) 2 X 10*17/cm*2 nitrogen implantation at 120 keV [16]. In the case of heavy ion irradiation the XTEM technique allows the observation of individual amorphous cascades [17,18].

Wang et al [11] (infrared spectroscopy and TEM) and Prussin et al [19] have studied the formation of amorphous silicon after implantation of C, Si, Sn, As, P and B at energies below 500 keV and at temperatures up to 300 deg K.

The dose dependence of damage for different ion species is discussed [11] in terms of the well known overlap model [1] of Gibbons. In the case of heavy ion bombardment (Sn), it is shown that direct amorphization can be achieved by the incoming ions and no overlap of damaged regions is necessary. For lighter ions and at higher implantation temperatures single or multiple overlap is required for amorphization.

The experimental dose and temperature dependence of radiation damage
after high energy Si implantation up to doses of about 1 X 10*18/cm*2
at target temperatures of 125 to 475 deg K is pointed out by Hecking et
al [20] in connection with corresponding model calculations. The
influence of target temperature on radiation damage is also shown by
Elliman et al [21] for instance. They combine RBS/channelling and TEM
to study 1.5 MeV Ne irradiations for temperatures between 300 and 725
deg K in the dose range from 3 X 10*15/cm*2 to 1 X 10*17/cm*2. Small
point defect clusters (2 nm) are formed by room temperature
irradiation for a dose of 5 X 10*16/cm*2; at medium temperatures (475
deg K) point defect clusters, defect complexes and dislocation loops
(< 20 nm) are visible by TEM at all depths up to the ion projected
range. At the same dose for T > 575 deg K only a buried band at the
depth of maximum nuclear stopping contains a noticeable concentration
of those defects, with a surface layer of 1450 nm remaining defect-free.

E. RESIDUAL DAMAGE AFTER ANNEALING

For device fabrication, thermal annealing is necessary to remove
radiation damage and to achieve complete activation of the implanted
dopants.

Csepregi et al [22] have studied the solid phase epitaxial regrowth
(SPE) of amorphous silicon layers during furnace annealing. The
velocity of the amorphous/crystalline interface, evaluated from
RBS-measurements, can be described by an Arrhenius expression

$$V = V_o \exp (-E_a/kT)$$

Isothermal anneal of silicon irradiated amorphous layers at 825 deg K
for example results in a (100) interface velocity of 8 nm/min. By the
presence of impurities in the layer, the regrowth velocities are
shifted to lower (C, N, O) or higher (B, P, As) values depending on
ion species. For self ion-implanted amorphous silicon, Narayan et al
[17] determined V_o = 3.07 X 10*8 cm/s and E_a = 2.68 eV directly from
XTEM.

Prussin and Jones [23] studied the defect structure after amorphization
by 1 X 10*15/cm*2 phosphorus implantation in the energy range 25 to
180 keV. One hour 825 deg K annealing, resulting in solid phase
epitaxial regrowth of the surface amorphous layer, was followed by a
30 min 1175 deg K anneal. For lower energies, plane view and cross
section TEM show small dislocation loops (several tens of nanometres
in diameter) near the position of the original amorphous/crystalline
interface. At higher energies, where buried amorphous layers are
formed, a second band of dislocation loops is found at the position
where a/c interfaces meet after recrystallization. Another type of
defect originating from the a/c interface, the so called hairpin
dislocations, which are unacceptable for VLSI [13,17], follow the
advancing interface during SPE.

Maher et al [24] have reviewed the defect structure after rapid thermal
annealing of Si implanted layers. By high resolution electron microscopy
and RBS two well defined stages of solid-phase regrowth can be observed.
During the incubation regrowth at temperatures below 725 deg K the
implantation induced roughness of the a/c interface disappears in favour

of a very sharp interface. This interface reconstruction is followed by SPE at higher temperatures. A detailed analysis shows that the generation temperatures for buried amorphous layers influence the residual damage after SPE, especially the formation of hairpin dislocations.

Similar studies after arsenic implantation have used furnace annealing [25,26] as well as rapid thermal annealing (RTA) with incoherent light [27,28].

Seidel et al [28] have investigated the annealing conditions to remove the residual a/c dislocations. After 100 keV implantation of 5 X 10*15 As/cm*2, samples were rapid annealed for several seconds at temperatures between 1125 deg K and 1475 deg K. The dislocation removal is found to be thermally activated, and characterized by an activation energy of 5 eV.

Numerous investigations (until 1983) of damage after implantation of different ion species and subsequent annealing have been extensively reviewed by Mader [26] and Gyulai [29].

Elliman et al [30] have discussed in detail the influence of defect production and dynamic defect annealing processes on the nature of initially amorphous silicon layers. It has been shown, that depending on irradiation conditions, the surface amorphous layer can be either crystallized epitaxially above a critical temperature Tc or increased in thickness.

Ion-beam induced epitaxial crystallization (IBIEC) can take place at temperatures of about 475 deg K to 675 deg K, which are considerably lower than temperatures during thermal annealing, and is mainly due to the energy deposited into nuclear processes in the a/c interface region. The irradiation temperature dependence of IBIEC can be described by an activation energy of 0.24 eV, which is small compared to the value of thermal annealing.

In recent years some efforts have been focussed on dynamic annealing effects during high current density implantations.

Experiments of Cembali et al [31] have shown, that implantation at very high current densities can result in crystalline surface layers with electrically active dopants. Cerofolini et al [32] and Komarov et al [33] report on high current As implantations. Similar investigations have been published for P implantations [34,35].

High dose high current implantations starting at low substrate temperatures in general lead to the formation of an amorphous layer within a few seconds of ion bombardment. During further irradiation, dynamic annealing of damage takes place due to beam heating which results in the recrystallization of the amorphous layer. By means of RBS and XTEM, Berti et al [34] have measured the position of a/c interfaces as a function of implantation time implanting 100 keV phosphorus ions at a comparatively low power density of 6 W/cm*2 up to a dose of 1 X 10*15/cm*2. After about two seconds the regrowth process starts at a temperature of about 505 deg K and is completed after 6 seconds when the temperature has reached a value between 825 and 875 deg K.

Cannavo et al [35] have shown, that the regrowth is not only determined by the temperature increase, but it is enhanced due to mobile point defects created at the a/c interface.

REFERENCES
----

[1]   J.F.Gibbons [Proc. IEEE (USA) vol.60 no.9 (1972) p.1062-96]

[2]   J.F.Ziegler, J.P.Biersack, U.Littmark [in 'The Stopping and Range of Ions in Matter' vol.1 Ed. J.F.Ziegler (Pergamon Press, New York, 1985)]

[3]   M.T.Robinson [Top. Appl. Phys. (Germany) vol.47 'Sputtering by Particle Bombardment I' Ed. R.Behrisch (Springer, Berlin, Heidelberg, New York, 1981) p.73-144]

[4]   H.J.Stein, F.L.Vook, D.K.Brice, J.A.Borders, S.T.Picraux [Proc. 1st Int. Conf. Ion Implantation, Eds L.T.Chadderton, F.Eisen (Gordon & Breach, London, 1971) p.17-24]

[5]   F.L.Vook [Inst. Phys. Conf. Ser. (GB) no.16 (1973) p.60]

[6]   G.Gotz [Nucl. Instrum. & Methods Phys. Res. (Netherlands) vol.199 (1982) p.61-73]

[7]   J.W.Corbett [Proc. 1st Int. Conf. Ion Implantation, Eds L.T.Chadderton, F.Eisen (Gordon & Breach, London, 1971) p.1-8]

[8]   J.R.Dennis, E.B.Hale [J. Appl. Phys. (USA) vol.49 no.3 (1978) p.1119-27]

[9]   E.C.Baranova, V.M.Gusev, Yu.V.Martynenko, C.V.Starinin, I.B.Haibullin [in 'Ion Implantation in Semiconductors and other Materials' Ed. B.L.Crowder (New York, 1973) p.59-71]

[10] K.F.Heidemann [Philos. Mag. B (GB) vol.44 no.4 (1981) p.465-85]

[11] Kou-Wei Wang, W.G.Spitzer, G.K.Hubler, D.K.Sadana [J. Appl. Phys. (USA) vol.58 no.12 (15 Dec 1985) p.4553-64]

[12] M.Fried, T.Lohner, G.Vizkelethy, E.Jaroli, G.Mezey, J.Gyulai [Nucl. Instrum. & Methods Phys. Res. Sect. B (Netherlands) vol.15 (1986) p.422]

[13] W.P.Maszara, G.A.Rozgonyi [J. Appl. Phys. (USA) vol.60 no.7 (1 Oct 1986) p.2310-15]

[14] C.Prunier, E.Ligeon, A.Bourret, A.C.Chami, J.C.Oberlin [Nucl. Instrum. & Methods Phys. Res. Sect. B (Netherlands) vol.17 no.3 Oct 1986) p.227-33]

[15] J.Belz, K.F.Heidemann, H.F.Kappert, E.Te Kaat [Phys. Status Solidi a (Germany) vol.76 (1983) p.K81]

[16] Z.Liliental, R.W.Carpenter, J.C.Kelly [Thin Solid Films (Switzerland) vol.138 no.1 (1 Apr 1986) p.141-50]

[17] J.Narayan, O.S.Oen, S.J.Pennycook [AIP Conf. Proc. (USA) vol.138 (1986) p.122-37]

[18] O.W.Holland, D.Fathy, J.Narayan [Mater. Res. Soc. Symp. Proc. (USA) vol.41 (1985) p.307-12]

[19] S.Prussin, D.I.Margolese, R.N.Tauber [J. Appl. Phys. (USA) vol.57 no.2 (15 Jan 1985) p.180-5]

[20] N.Hecking, K.F.Heidemann, E.Te Kaat [Nucl. Instrum. & Methods Phys. Res. Sect. B (Netherlands) vol.15 no.1-6 (Apr 1986) p.760-4]

[21] R.G.Elliman, J.S.Williams, S.T.Johnson, A.P.Pogany [Nucl. Instrum. & Methods Phys. Res. Sect. B (Netherlands) vol.15 (1986) p.439-42]

[22] L.Csepregi, E.F.Kennedy, J.W.Mayer, T.Sigmon [J. Appl. Phys.
      (USA) vol.49 (1978) p.3906]
[23] S.Prussin, K.S.Jones [Mater. Res. Soc. Symp. Proc. (USA) vol.71
      (1986) p.191-5]
[24] D.M.Maher, R.V.Knoell, M.B.Ellington, D.C.Jacobson [Mater. Res.
      Soc. Symp. Proc. (USA) vol.52 (1986) p.93-105]
[25] K.S.Jones, S.Prussin [Mater Res. Soc. Symp. Proc. (USA) vol.71
      (1986) p.173-8]
[26] S.Mader ['Ion Implantation Damage in Silicon' in: 'Ion
      Implantation Science and Technology', Ed. J.F.Ziegler (Academic
      Press, USA, 1984)]
[27] D.Baither, R.Koegler, D.Panknin, E.Wieser [Phys. Status Solidi a
      (Germany) vol.94 no.2 (1986) p.767-72]
[28] T.E.Seidel, D.J.Lischner, C.S.Pai, R.V.Knoell, D.M.Maher,
      D.C.Jacobson [Nucl. Instrum. & Methods Phys. Res. Sect. B
      (Netherlands) vol.B7-8 pt 1 (Mar 1985) p.251-60]
[29] J.Gyulai ['Damage Annealing in Silicon and Electrical Activity',
      in 'Ion Implantation Science and Technology', Ed. J.F.Ziegler
      (Academic Press, USA 1984)]
[30] R.G.Elliman, J.S.Williams, W.L.Brown, A.Leiberich, D.M.Maher,
      R.V.Knoell [Nucl. Instrum. & Methods Phys. Res. Sect.B
      (Netherlands) vol.B19-20 pt 2 (Feb 1987) p.435-42]
[31] G.F.Cembali, P.G.Merli, F.Zignani [Appl. Phys. Lett. (USA) vol.38
      no.10 (1 Jun 1981) p.808-10]
[32] G.F.Cerofolini et al [Thin Solid Films (Switzerland) vol.129
      no.1/2 (12 Jul 1985) p.111-25]
[33] F.F.Komarov, A.P.Novikov, I.A.Radishevskii, T.T.Samoilyuk,
      V.P.Tolstykh [Sov. Phys.-Semicond. (USA) vol.20 no.9 (Sep 1986)
      p.1081-2]
[34] M.Berti et al [Mater. Res. Soc. Symp. Proc. (USA) vol.45 (1985)
      p.97-102]
[35] S.Cannavo, A. La Ferla, E.Rimini, G.Ferla, L.Gandolfi [J. Appl.
      Phys. (USA) vol.59 no.12 (15 Jun 1986) p.4038-42]

11.5 STRUCTURE OF ELECTRON-INDUCED DEFECTS IN Si

by Wu Fengmei and Zheng Xiangqin

September 1987
EMIS Datareview RN=16164

A. INTRODUCTION

Generally, electron-induced defects in silicon can only be observed if
they are associated with energy levels situated relatively deep in the
band gap. They can be studied using Hall-effect, minority-carrier-
lifetime, thermally stimulated capacitance (TSC), deep level transient
spectroscopy (DLTS), transient photocapacitance and IR absorption
measurement techniques. The use of electron paramagnetic resonance (EPR)
and electron-nuclear double resonance (ENDOR) in determining the
microscopic configurations of these defects has been most fruitful.

The stable defects in silicon created by electron irradiation can be
divided into three categories: i) defects created directly by collison
cascade, such as the divacancy, ii) defects created by the interaction
among radiation induced intrinsic defects, such as the di-interstitial,
and iii) defects created by the interaction between an intrinsic
defect and an imperfection originally in the crystal (usually an
impurity) such as the vacancy-oxygen complex.

The identification of electron irradiation induced defects is based
on comparison of defect introduction, stability and impurity dependence
with the corresponding properties of EPR or IR spectra. The main
structures that have been reported are listed below.

B. VACANCY (V)

The defects associated with vacancies at Ec-0.09 eV and Ev+0.11 eV are
important states introduced at low temperatures which anneal far below
room temperature [1,2]. The products of annealing can be related to the
oxygen vacancy pair (O-V) in n-type material and divacancy (V-V) in
p-type material. Recently, Johnson reported [3] ultrasonic detection
of the vacancy in boron-doped Si. A defect relaxation was observed
which, from comparison with EPR data, is identified as the positively
charged vacancy.

C. DIVACANCY (V-V)

One of the electron-induced defects revealed by EPR is the divacancy
[4,5]. It is produced either as a multiple displacement defect or as a
combination of two single vacancies. The divacancy is believed to
introduce three energy levels in the band gap at Ec-0.41 eV, Ec-0.23 eV
and Ev+0.25 eV [4,6]. The Ec-0.23 eV and Ec-0.41 eV states are introduced
and annealed at the same rate in lightly doped n-type material.
They have been identified as different charge states of the same defect.

The level Ec-0.23 eV is the double-negative charge state and the level
Ec-0.41 eV is the single-negative charge state of the divacancies. The
level Ev+0.25 eV is the single positive charge state. The 1.8-, 3.3- and
and 3.9- micron bands in the IR spectrum have been associated with the
divacancy [7]. The activation energy for thermal annealing of 1.4 eV has
been reported [6]. The binding energy of the divacancy has been roughly
estimated to be > 1.6 eV [4]. The photo-ionisation cross sections of
divacancies were measured [8].

Recently, an Ec-0.37 eV level has been observed in n-type Si. An
activation energy for thermal annealing of 1.71 eV with a frequency
factor of approx. 1.5 X 10*9/s has been reported [9]. The state is
possibly related to the defect complex associated with vacancies.

D. OXYGEN-VACANCY COMPLEXES

The Ec-0.17 eV level is most commonly observed in electron irradiated
silicon [10-13]. It has been well identified as the A-centre (oxygen
vacancy pair) [14,15].

Besides the A-centre, many other multi-vacancy-oxygen complexes were
observed by EPR [16]. They correspond to divacancy + oxygen, divacancy
+ two oxygen, trivacancy + oxygen, trivacancy + two oxygen and
trivacancy + three oxygen. The microscopic configurations of these
defects were also established.

Radiation-induced accumulation and annealing processes of radiation
defects with the level Ec-0.22 eV in silicon single crystals of various
perfection are analysed and it is concluded that they consist of oxygen
and two vacancies (oxygen-divacancy complex) [17].

Another level at Ev+0.42 eV has also been found in 10 MeV electron
irradiated n-type CZ silicon [11]. It behaves similarly to the A-centre,
and must be either an alternate charge state of the A-centre or an
interstitial related defect.

Jellison [18] has observed a new level at Ec-0.105 eV in electron
irradiated n-type silicon. This trap is related to another trap at
Ec-0.172 eV and one trap can be converted to the other. This metastable
behaviour suggests that both these traps are different configurations of
the same defect, possibly being an impurity atom trapped by an A-centre.

E. PHOSPHORUS-VACANCY COMPLEXES

In phosphorus-rich silicon material, an Ec-0.44 eV level is usually
observed after electron irradiation [10,12]. It has been characterised
as the E-centre (phosphorus-vacancy pair) [21].

Besides the E-centre, Sieverts et al [22] have reported another phosphorus
-related defect in heavily phosphorus doped silicon after electron
irradiation using EPR measurement. They were identified as two-phosphorus
defect complexes.

Recently, Chantre et al [12] have detected a new defect in n-type silicon. The energy position of this defect in its stable configuration is Ec-0.30 eV. It can exist in three other metastable configurations. They are Ec-0.21 eV, Ec-0.23 eV and Ec-0.29 eV. It is found that this defect is responsible for the first of the two recovery stages of the E-centre. Microscopic structure of this defect has not been explained, but it is tentatively identified as a phosphorus-carbon pair [23]. A defect with similar metastable configuration has also been reported by Song et al [24] in 2.5 MeV electron irradiated n-type silicon.

A number of workers have studied the rate of the defect formation [19,20]. It has been pointed out that the primary process plays a dominant role during the formation of A-centre, E-centre and divacancy in electron irradiated silicon.

F. CARBON INTERSTITIAL AND CARBON-RELATED COMPLEX

In electron-irradiated p-type silicon, the carbon interstitial defect is usually detected. The thermal stability (300 deg K) of this state and the independence of the introduction rate on doping impurity or growth technique have been reported [25,11].

Ev+0.33 eV [11], Ev+0.38 eV [26] and Ev+0.36 eV [27] levels appear upon annealing carbon interstitial and therefore should be carbon-related. The Ev+0.33 eV level and Ev+0.38 eV level have been identified as C(I)-C(s) [11,28] and C-O-V complexes, respectively [26]. The Ev+0.36 eV level is possibly C(I)-C(s) or C-O-V [27].

G. BORON-RELATED DEFECTS

Watkins was the first to identify an EPR spectrum labelled Si-G10 as arising from a lattice vacancy trapped by a substitutional boron in float-zone silicon [29], and he concluded indirectly the presence of a boron atom as the next-nearest neighbour to the lattice vacancy. ENDOR measurement proved the presence of a boron atom in the defect centre. Londos [30] reported that the boron substitutional-vacancy defect appears in two metastable configurations and exhibits at least three electrical levels in the DLTS spectrum: Ev+0.31 eV, Ev+0.37 eV and another one with energy E smaller than 0.13 eV above the valence band [31].

Oxygen is an effective trap for vacancies, and in the pulled crystal with the oxygen concentration about two orders of magnitude larger than the boron concentration most of the vacancies would be trapped by oxygen. Therefore, the B(s)-V pair does not appear in pulled Si [32]. B(I) and B-V anneal out below room temperature; the B(I)-B(s) state appears [33].

The B(s)-B(I), C(I)-C(s), V-V defects are the dominant levels in boron-doped p-type Si irradiated with 1 MeV electrons at room temperature [26].

Additionally, in electron irradiated boron doped silicon two new DLTS
peaks were reported [27]: the first possibly dependent on oxygen content
at an energy level of about Ev+0.13 eV and with a capture cross section
of about 10*-18 cm*2; the second with a charge-dependent peak amplitude
at an energy of Ev+0.34 eV, apparently a vacancy dependent complex
annealing out at around room temperature.

H. ALUMINIUM RELATED DEFECTS

Interstitial aluminium, Al(I), with an energy level of Ec+0.25 eV is
observed following low temperature irradiation (10 deg K); the
smaller capture cross section for holes is suggestive of a positive
charge state [11,34]. In addition, aluminium-vacancy pairs, Al(I)-V,
(Ev+0.48 eV) [11,35] and interstitial aluminium-substitutional
aluminium pair, Al(I)-Al(s), (Ev+0.23 eV) were also observed [11].

TABLE 1: data on the main structures of electron-induced defects

Origin	Energy level (eV)	Capture cross section sigma (cm*2)	Annealing properties	Ref
V	Ec-0.09		out 90 deg K	[11]
	Ec+0.11	(p) 5 X 10*-17	out 150 deg K	[11]
	Ev+0.13		out 150 deg K	[30]
0-V	Ec-0.18	(n) 10*-14	out 350 deg C	[11]
	Ec-0.169	(n) 10*-14	out 350 deg C	[13]
V2-0 ?	Ec-0.30		in 100 deg C, out 450 deg C	[11]
V3-0 ?	Ec-0.20		in 300 deg C, out 450 deg C	[11]
(V-V)+	Ev+0.21	(p) 2 X 10*-16	out 300 deg C	[11]
	Ev+0.23	(p) 3 X 10*-16	out 300 deg C	[26]
(V-V)-	Ec-0.39	(n) 4 X 10*-15	out 300 deg C	[11]
	Ec-0.413	(n) 2 X 10*-15	out 300 deg C	[13]
(V-V)--	Ec-0.23	(n) 0.6 X 10*-16	out 300 deg C	[11]
	Ec-0.246	(n) 4 X 10*-16	out 300 deg C	[13]
P-V	Ec-0.44	(n) > 10*-16	out 150 deg C	[11]
	Ec-0.456	(n) 3.7 X 10*-15	out 150 deg C	[13]

Origin	Energy level (eV)	Capture cross section sigma (cm*2)	Annealing properties	Ref
C(I)	Ev+0.27	(p) 7 X 10*-18	out 315 deg K	[11]
	Ev+0.29			[30]
C(I)-C(s)	Ev+0.33	(p) 8 X 10*-17	in 300 deg K, out 400 deg C	[11]
V-C-0	Ev+0.38	(p) 2 X 10*-18	in 30 deg C, out 400 deg C	[26]
C-related defect	Ev+0.36	(p) 4.4 X 1*-17	in 315 deg K, out > 470 deg K	[27]
Al(I)-V	Ev+0.48	(p) > 10*-16	out 200 deg C	[11]
Al(I)	Ev+0.25	(p) 7 X 10*-18	out 200 deg C	[11]
B(I)-B(s)?	Ec-0.26		out 150 deg C	[11]
B(s)-V	Ev+0.31	(p) 1.94 X 10*-19	in 200 deg K, out 300 deg K	[30]
	Ev+0.38	(p) 2 X 10*-17		[30]
V-0-B	Ev+0.30	(p) 2 X 10*-16	in 170 deg C out 400 deg C	[26]
O(I)-B(I)	Ec-0.27	(n) 2 X 10*-13	out 170 deg C	[26]
dependent on oxygen	Ev+0.13	(p) 10*-18	out > 470 deg K	[27]
Vacancy-related defect	Ev+0.34		in 215 deg K, out 315 deg K	[27]

It has been pointed out that impurity clusters and dislocations may act as sinks for primary defects - vacancies and self-interstitials. A study of interaction of point defects with dislocations by means of irradiation in an electron microscope has been reported [36]. The study indicated that the interaction of dislocations with intersititials is higher than that with vacancies, and the activation energy for the interstitial atoms to join into lattice position in a dislocation line is 1.3 +- 0.2 eV.

Bolotov et al [37] have studied the influence of heat pre-treatment on the accumulation of radiation defects in oxygen-rich n-type silicon. They suggested that as a result of pre-annealing the production rate of compensating acceptor defects decreases or the efficiencies of the sinks for silicon self-interstitials increase.

The influence of electron irradiation on the structure properties of thin crystalline films of silicon have been studied. The positive influence on imperfections in the grown film has been reported [38].

Recently, amorphization of crystalline silicon by electron and/or ion irradiation has been studied [39]. It has been shown that alone 1 MeV electron irradiation can not induce a crystalline (c) to amorphous (a) transition even at < 10 deg K to a fluence of 14 dpa. Whereas a dual irradiation of a region with 1 MeV electrons and 1.0 or 1.5 MeV krypton ions can strongly retard the c-to-a transition if the ratio of the displacement rates for electron to ion exceeds approx. 0.5. Atomic models for these phenomena are presented.

REFERENCES
----

[1]   G.D.Watkins [Phys. Rev. B (USA) vol.12 (1975) p.5824]
[2]   P.Vajda, L.J.Cheng [J. Appl. Phys. (USA) vol.42 (1971) p.2453]
[3]   W.L.Johnson, A.V.Granato [J. Phys. Colloq. (France) vol.46
        no.C-10 (1985) p.537-40]
[4]   G.D.Watkins, J.W.Corbett [Phys. Rev. (USA) vol.138 (1965)
        p.A543)]
[5]   J.W.Corbett, G.D.Watkins [Phys. Rev. (USA) vol.138 (1965)
        p.A555]
[6]   A.O.Evwaraye, Edmund Sun [J. Appl. Phys. (USA) vol.47 (1976)
        p.3776]
[7]   L.J.Cheng, J.C.Corelli, J.W.Corbett, G.D.Watkins [Phys. Rev.
        (USA) vol.152 (1966) p.761]
[8]   S.D.Brotherton, G.J.Parker, A.Gill [J. Appl. Phys. (USA) vol.54
        (1983) p.5112]
[9]   Wu Fengmei, Lai Qi-Ji, Shen Bo, Zhou Guo-Quan [Chin. Phys. (USA)
        vol.6 no.4 (1986) p.936-40]
[10]   L.C.Kimerling, H.M.De Angelis, J.W.Diebold [Solid-State Commun.
        (USA) vol.16 (1975) p.171]
[11]   L.C.Kimerling [Proc. Int. Conf. Radiation Effects in Semiconductors
        (1976) p.221]
[12]   A.Chantre, L.C.Kimerling [Appl. Phys. Lett. (USA) vol.48 (1986)
        p.1000]
[13]   S.D.Brotherton, P.Bradler [J. Appl. Phys. (USA) vol.58 (1982)
        p.5720-32]
[14]   G.D.Watkins, J.W.Corbett [Phys. Rev. (USA) vol.121 (1961) p.1001]
[15]   J.W.Corbett, G.D.Watkins, R.M.Chrenko, R.S.McDonald [Phys. Rev.
        (USA) vol.121 (1961) p.1015]
[16]   Y.H.Lee, J.W.Corbett [Phys. Rev. B (USA) vol.13 (1976) p.2653]
[17]   L.A.Kazakevich, V.I.Kuznetsov, P.F.Lugakov [Radiat. Eff. Lett.
        Sect. (GB) vol.87 (1986) p.147]
[18]   G.E.Jellison [J. Appl. Phys. (USA) vol.53 (1982) p.5715]
[19]   P.F.Lugakav, V.V.Lukyanitsa [Phys. Status Solidi a (Germany)
        vol.83 (1984) p.521]
[20]   A.G.Abdusattarov, V.V.Emtsev, V.N.Lomasov, T.V.Mashovets [Sov.
        Phys.-Semicond. (USA) vol.20 (1986) p.101]
[21]   G.D.Watkins, J.W.Corbett [Phys. Rev. (USA) vol.134 (1964) p.A1359]
[22]   E.G.Sieverts, C.A.J.Ammerlaan [Proc. Int. Conf. Radiation Effects
        in Semiconductors (1976) p.213]
[23]   A.Chantre, J.L.Benton, M.T.Ason, L.C.Kimerling [Proc. 14th Int.
        Conf. on Defect in Semiconductor (1986) p.111]
[24]   L.W.Song, B.W.Benson, G.D.Watkins [Phys. Rev. B (USA) vol.33
        no.2 (15 Jan 1986) p.1452-5]
[25]   G.D.Watkins, K.L.Brower [Phys. Rev. Lett. (USA) vol.36 (1976) p.1329]
[26]   P.M.Mooney, L.J.Cheng, M.Suli, J.D.Gerson, J.W.Corbett [Phys. Rev.
        B (USA) vol.15 (1977) p.3836]

[27] C.A.Londos, P.C.Banbury [J. Phys. C. (GB) vol.20 no.5 (20 Feb 1987)
      p.645-50]
[28] K.L.Brower [Phys. Rev. B (USA) vol.9 (1975) p.260]
[29] G.D.Watkins [Phys. Rev. B (USA) vol.13 (1976) p.2511]
[30] C.A.Londos [Phys. Rev. B (USA) vol.34 no.2 (15 Jul 986) p.1310-13]
[31] S.K.Bains, P.C.Banbury [J. Phys. C (GB) vol.18 no.5 (20 Feb 1985)
      p.L109-16]
[32] C.A.Londos [J. Phys. & Chem. Solids (GB) vol.47 no.12 (1986)
      p.1147-52]
[33] A.R.Bean, S.R.Morrison, R.C.Newman, R.S.Smith [J. Phys. C. (GB)
      vol.5 (1972) p.379]
[34] K.L.Brower [Phys. Rev. B (USA) vol.1 (1970) p.1908]
[35] G.D.Watkins [Phys. Rev. (USA) vol.155 (1967) p.802]
[36] L.I.Fedina, A.L.Aseev [Phys. Status Solidi a (Germany) vol.95 (1986)
      p.517-30]
[37] V.V.Bolotov, A.V.Karpov, V.A.Stuchinsky, K.Schmalz [Phys. Status
      Solidi a (Germany) vol.96 no.1 (1986) p.129-34]
[38] F.F.Komarov, I.S.Tashlykov, F.N.Korschunoy, A.S.Kamyshan,
      E.V.Kotov, G.I.Plaschinski [Radiat. Eff. (GB) vol.91 (1986)
      p.293]
[39] D.N.Seidman, R.S.Averback, P.R.Okamoto, A.C.Baily [Phys. Rev.
      Lett. (USA) vol.58 no.9 (2 Mar 1987) p.900-3]

11.6 STRUCTURE OF NEUTRON-INDUCED DEFECTS IN Si

by L.N.Aleksandrov

November 1987
EMIS Datareview RN=16194

Formation of lattice defects in silicon irradiated by thermal (with
energy less than 0.1 MeV) and fast neutrons occurs in various ways.
Thermal neutrons cannot knock out an atom from a lattice site, but
being absorbed by the nucleus, they cause beta particle emission. The
corresponding recoil energy (about 0.8 keV) is consumed during atomic
displacement [1]. Fast neutrons with energy of around 50 keV interact
with Si atoms, developing a cascade of displacements resulting in the
formation of a disordered region (DR) with a diameter of about 20 nm.
The efficiency of DR introduction has been estimated to be 0.15 F cm*5,
where F is the irradiation dose [2].

The structure and properties of DR have been the subject of intensive
study [3-6]. DR overlapping does not lead to amorphization of the
irradiated volume. However, the local mechanical stresses are high
and the level of deformation approaches the critical value, ensuring
silicon amorphization [7-9]. Relaxation of these stresses (down to
relative deformations of approx. 0.01 according to IR spectroscopy
data [7]) is accompanied by vacancy and interstitial atom diffusion,
formation of vacancy clusters and vacancy-impurity complexes [10,11].
In this case the main defects are: divacancies (V2-centres), multi-
vacancy V3, V4, V5 complexes, A-centre (vacancy-oxygen), E - (with
phosphorus), K - (with oxygen-carbon); interstitial and substitutional
carbon form pairs of the type C(I)+C(S).

By means of EPR Corbett et al [12] and Smirnov et al [13] found that
the formation of interstitial defects was connected with large atomic
shifts in the DR. It was shown that formation of 0.2Si(I)+ per single
DR with a simultaneous appearance of 0.5 V4(0) occurred when Si was
irradiated by neutrons in a nuclear reactor. A.H.Antonenko et al [14]
have managed to find a number of defects of the complex structure,
which is now under study. Note that the EPR spectrum is affected by the
charge of the defects and their energy state is determined through DLTS.

The structure study of neutron-induced defects in silicon was
stimulated by an extended application of the internal friction (IF)
method [15,16]. The IF method permits evaluation of the work required
for defect state change, for their motion, reorientation, for change
of their number, size and form. Therefore, the IF method provides
additional information on the processes in which the defects take part
and allows determination of the activation energy and frequency factors
of these processes as well as the dynamic characteristics of structural
defects [17].

The IF method is based on the measurement of attenuation decrement of
elastic acoustic vibrations over a wide temperature interval; the
low-temperature IF studies (down to 300 deg K) were mainly carried
out by Khiznichenko et al [18], the high-temperature ones by
Aleksandrov et al [19-21]. The low-temperature studies made it possible

to define the details of the dynamics of the silicon dislocation structure after neutron irradiation and to evaluate the activation energy of movement of built-in bends in dislocations (0.07 eV in n-Ge).

In neutron-irradiated silicon containing B and O atoms interstitial silicon was observed to move in the <111> and <100> directions with an activation energy of 0.29 eV, and in the <110> direction [18].

IF measurement in the temperature range 300 to 1200 deg K and at a frequency of 2.25 kHz were made on the dislocation-free, phosphorus-doped silicon (the resistivity is 100-190 ohm cm) after neutron irradiation in the reactor with doses of up to 10*18 neutrons/cm*2. The behaviour of interstitial silicon atoms Si(I)+, Si(I)o, carbon and vacancy complexes 0+Vn was observed by IF peaks which change their height upon annealing. It was shown that the kinetics of Si(I) and 0+Vn annealing were described by second-order relaxation equations with Ko = 4 X 10*11 and 3.2 X 10*10 s*-1 and E = 1.4 and 2.16 eV.

The main characteristics of defects in silicon after neutron irradiation at T > 300 deg K are summarized in TABLE 1. Judging from the IF data the migration of interstitial Si(I)+ atoms in the <110> direction and of Si(I)o atoms in the <100> direction (peaks 1,3) occur with dynamic characteristics, determined by Fan et al [22] for interstitial defects in silicon annealed after implantation with B+ ions.

TABLE 1: dynamic characteristics of neutron-induced defects in Si on the basis of internal friction data [20]

Type of defect	Temperature of peak (deg K)	Annealing start-up temperature (deg K)	Activation energy (eV)	Frequency factor (s*-1)
1. Interstitial silicon atoms, Si(I)+	402	470	0.73+-0.07	2 X 10*13
2. Interstitial carbon atoms C(I),C(I)+C(S) complexes	465	470	0.81+-0.07	1 X 10*13
3. Interstitial silicon atoms Si(I)o	550	500	0.96+-0.05	8 X 10*13
4. Many-vacancy complexes with oxygen Vn+0	610	750	1.15+-0.05	4 X 10*13
5. Interstitial atoms and their complexes, vacancy complexes	background	750	0.35+-0.05	-

The processes related to the peak 4 were observed only in the Czochralski-grown silicon and were caused by transformation in DR containing oxygen. The IF background level depends on interstitial complexes, the total number of vacancies and their complexes. The annealing process is accompanied by enlargement of vacancy complexes, and the parameters of this process are close to the data obtained by by Babitzkii et al [23] who obtained an activation energy of annealing equal to 2.8 eV. The defects observed in the EPR and IR spectra disappear during annealing.

The comparison with charge carrier behaviour during annealing shows that defects causing the IF background in the neutron-irradiated silicon provide p-type conductivity, and the increase of Young's modulus in this case testifies to their vacancy nature [19]. A wide range of natural frequencies of the background defect oscillations indicate multi-vacancy complexes. At above 400 deg K the EPR spectrum of V4(O) disappears and that of charged V3(-), V4(-), V4(O), V5(-) complexes appears [1].

According to capacitance spectroscopy data A-centres V2(-), V2(--) divacancies and V2(-)+E centres complexes arise in the Vasilyev-measured silicon after irradiation [24]. During annealing up to about 1000 deg K, both the DR nucleus, consisting mainly of the V2 and the surrounding impurity-defect shell, underwent changes. In this case a wide DLTS spectrum, diffused due to superimposition of various defects, is analogous to the IF background.

Recent development of the EPR method by Dvurechenskii et al [25] and Stelmach et al [26] provided a good possibility of new data on the behaviour in neutron-irradiated silicon of the pentavacancy complexes of the primary nonparamagnetic defects (with concentration around 3 X 10*18/cm*3 before annealing). During annealing up to 760 deg K the defect concentration decreases by more than a factor of 10 and the V5(-) (Si - P1 state) appears. In non-annealing cases the same defects as well as V4(-) were observed in p-type silicon with a high content of alloying impurity (around 10*19/cm*3) when irradiating with a large neutron dose (4 X 10*19/cm*2). The defects of a lower symmetry [25] contributing to the well-known Watkins defect model V2+Si2 [27] have been found.

In regions of large lattice deformations (10*-3 or more) caused by the DR there have been found non-reoriented divacancies (N-9 centres) and vacancy defects of N-10 type with a paramagnetic electron localized at the dangling bond of the silicon atom [28].

Wu et al [29] have found new defects of triclinic symmetry during neutron irradiation of zone-melted silicon grown in hydrogen.

Development of high resolution electron microscopy permitted the taking of contrast pictures of the DR and of the largest multi-vacancy clusters. The results of investigations carried out by Zakharov et al [30] and Krivanek et al [31] made it possible to observe the structure of vacancy chains in silicon, interstitial clusters and the structure of subgrain boundaries. Electron-microscopic pictures are interpreted by means of computer-aided model design taking into account electron scattering and optical aberration.

The data obtained by the new DLTS, EPR, IF and TEM methods for properties of neutron-induced defects in silicon are in accordance with previous results known from IR absorption and conductivity changes after annealing [32-35]. Study of IR spectrum lines in silicon after neutron irradiation shows divacancies behaviour, their annealing at 250 deg C and stability of multi-vacancy complexes up to 560-580 deg C [32]. The corresponding activation energy for annealing is 3.1 +-0.3 eV. This is in accordance with the data for stable defects clusters like DR. Their size is a function of impurity content, in particular of oxygen [34-35].

The energy spectrum levels of zone structure were given mainly by optical and electrical conductivity measurements. The method of neutron scattering for the dynamic study of defects was used [36]. But interpretation of diffraction pictures is not always simple, particularly as regards to structure defects of the topological disorder type [37].

REFERENCES
----

[1]   L.S.Smirnov (Ed.) [Voprosi Radiatsionnii Tekhnologii Poluprovodnikov (Nauka, 1980), in Russian (English title: Problems of Semiconductor Radiation Technology)]

[2]   H.J.Stein [Proc. 2nd Int. Conf. Neutron Transmutation Doping Ed. J.M.Messe (Plenum Press, 1979) p.229-48]

[3]   B.R.Gossik [J. Appl. Phys. (USA) vol.30 (1959) p.1214]

[4]   H.A.Uhin [Fiz. & Tekh. Poluprovodn. (USSR) vol.6 (1972) p.931, English transl. in Sov. Phys.-Semicond. (USA)]

[5]   A.I.Baranov, L.S.Smirnov [Sov. Phys.-Semicond. (USA) vol.7 no.11 (May 1974) p.1482-3]

[6]   V.P.Kojevnikov, V.V.Mikhnovich [Fiz. & Tekh. Poluprovodn. (USSR) vol.53 (1983) p.1361, English transl. in Sov. Phys.-Semicond. (USA)]

[7]   V.D.Akhmetov, V.V.Bolotov, L.S.Smirnov [Phys. Status Solidi a (Germany) vol.75 (1983) p.601]

[8]   A.V.Vasiliev, S.A.Smagulova, L.S.Smirnov [Sov. Phys.-Semicond. (USA) vol.20 no.3 (Mar 1986) p.354-6]

[9]   L.N.Aleksandrov [Phys. Status Solidi a (Germany) vol.89 no.2 (1985) p.443-50]

[10]   O.Gosele, W.Frank, A.Seeger [Solid State Commun. (GB) vol.45 (1983) p.31]

[11]   R.C.Newman [Rep. Prog. Phys. (GB) vol.45 (1982) p.1163]

[12]   Y.-H.Lee, N.N.Gerasimenko, J.W.Corbett [Phys. Rev. B (USA) vol.14 (1976) p.4506]

[13]   A.H.Antonenko, A.V.Dvurechenskii, L.S.Smirnov [Fiz. & Tekh. Poluprovodon. (USSR) vol.16 (1982) p.2018 English transl. in Sov. Phys.-Semicond. (USA)]

[14]   A.H.Antonenko, A.V.Dvurechenskii [Phys. Status Solidi a (Germany) vol.58 (1980) p.K63]

[15]   A.S.Nowick, B.S.Berry [Anelastic Relaxation in Crystalline Solids (Academ. Press, 1982)]

[16]   L.N.Aleksandrov, M.I.Zotov [Vnutrennii Trenie i Defekti v Poluprovodnikah (Nauka, 1979), in Russian (English title: Internal Friction and Defects in Semiconductors)]

[17] L.N.Aleksandrov, M.I.Zotov [Phys. Status Solidi a (Germany)
      vol.66 (1981) p.467]
[18] L.P.Khiznichenko, E.Oteniyazov, N.S.Tulyaganova [Phys. Status
      Solidi a (Germany) vol.46 (1978) p.K147]
[19] L.N.Aleksandrov, M.I.Zotov, V.F.Stas, B.P.Surin [Sov.
      Phys.-Semicond. (USA) vol.18 no.1 (Jan 1984) p.42-4]
[20] L.N.Aleksandrov, B.P.Surin, M.I.Zotov [Phys. Status Solidi a
      (Germany) vol.91 no.1 (1985) p.57-61]
[21] L.N.Aleksandrov, M.I.Zotov, B.P.Surin [in 'Vnutrennii Trenie i
      Tonkoe Stroenie Metallov i Neorganicheskikh Materialov' (Nauka,
      1985) p.165-7 (English title: Internal Friction and Sub-Structure
      of Metals and Inorganic Materials)]
[22] S.I.Fan, B.S.Barry, W.Frank [in 'Ion Implantation in
      Semiconductors and Other Materials' (Plenum Press, 1973) p.1-30]
[23] Yu.M.Babitskii, I.N.Voronov, P.M.Grinshtein, M.A.Morokhovets
      [Sov. Phys.-Semicond. (USA) vol.16 no.5 (May 1982) p.597-8]
[24] A.V.Vasiliev, S.A.Smagulova, S.S.Shaimeev [Fiz. & Tekh.
      Poluprovodn. (USSR) vol.16 (1982) p.1983, (English transl. in
      Sov. Phys.-Semicond. (USA)]
[25] A.V.Dvurechenskii, V.V.Suprunchik [Phys. Status Solidi a
      (Germany) vol.92 no.1 (1985) p.K53-6]
[26] V.F.Stelmach, V.P.Tolstych, L.V.Cvirko [Sov. Phys.-Semicond.
      (USA) vol.19 no.10 (Oct 1985) p.1145-6]
[27] G.D.Watkins [in 'Radiation Effects in Semiconductors' (Plenum
      Press, 1968) p.67-81]
[28] A.V.Dvurechenskii, A.A.Karanovitch [Sov. Phys.-Semicond. (USA)
      vol.19 no.11 (Nov 1985) p.1198]
[29] Wu En, Wu Shu-xiang, Mao Jin-chang, Qin Guo-gang [Chin. J.
      Semicond. (China) vol.7 no.1 (1986) p.109]
[30] N.D.Zakharov, M.Pasemann, V.N.Rozhanskii [Phys. Status Solidi a
      (Germany) vol.71 no.1 (16 May 1982) p.275-81]
[31] O.L.Krivanek, S.Isoda, K.Kabajashi [Philos. Mag. (GB) vol.36
      (1977) p.931]
[32] M.T.Lappo, V.D.Tkachev [Sov. Phys.-Semicond. (USA) vol.4 (1971)
      p.1882]
[33] Yu.P.Koval, V.N.Mordkovich, E.M.Temper, V.A.Kharchenko [Sov.
      Phys.-Semicond. (USA) vol.6 no.7 (Jan 1973) p.1152-5]
[34] V.N.Mordkovich, S.P.Solovev, E.M.Temper, V.A.Kharchenko [Sov.
      Phys.-Semicond. (USA) vol.8 no.5 (Nov 1974) p.666]
[35] N.A.Yuchin [Fiz. & Tekh. Poluprovodn. (USSR) vol.6 (1972) p.931,
      English transl. in Sov. Phys.-Semicond.]
[36] R.E.Beddoe, S.Messoloras, E.W.J.Mitchell, R.J.Stewart [Inst.
      Phys. Conf. Ser. (GB) vol.46 (1979) p.258-66]
[37] J.M.Ziman [Models of Disorder (Cambridge Univ. Press, 1979)]

CHAPTER 12

OXYGEN, CARBON AND NITROGEN IN SILICON

12.1 DIFFUSION OF O IN Si

by J.G.Wilkes

November 1987
EMIS Datareview RN=14687

At first sight of the broad compass of the literature, the diffusion
of oxygen in silicon appears to be a very complex phenomenon, involving
many contributory factors. To reconcile these it is necessary to
consider underlying mechanisms for a single diffusion jump in silicon.
At any given temperature the host crystal matrix contains some
equilibrium of both silicon self-interstitials, Si(I), and lattice site
vacancies, V. As shown by Brelot and Charlemagne [1] both are known to
be selectively trapped at sites adjacent to bonded interstitial oxygen
atoms. Both provide routes for a diffusion step [2]:

i.   [Oi-V] defects are formed first - the well known A-centre with
     an IR absorption at 830 cm*-1. Subseqently this traps a mobile
     silicon self-interstitial.

ii. [Oi-Si(I)] complexes are formed first, having a 935 cm*-1
     absorption, and later either trap mobile vacancies or
     dissociate.

Therefore in examining diffusion data it is important to distinguish
between studies where defect concentrations stay close to equilibrium
with temperature - normal intrinsic diffusion; and those cases where
defect concentrations have been radically shifted away from the thermal
equilibrium value - when enhanced diffusion is observed. Non-
equilibrium defect effects become of increasing importance at lower
temperatures.

Looking first at intrinsic diffusion a considerable range of work
employing various experimental techniques has built up in this field
[3-14].

Data on these various approaches to oxygen diffusion are collated in
TABLE 1 which requires some comments. The calibration of charge particle
reactors, measurement of proton current, its penetration and capture
cross section, and native surface oxide interface, contribute to the
experimental difficulties. Again, thermal donor formation kinetics and
surface depletion of thermal donors have unexplained aspects, and more
work is required to correlate oxygen concentration to thermal donor
profiles before this method can become reliable.

TABLE 1: summary of data relating to oxygen diffusion

Ed (eV)	Do (cm*2/s)	Temperature Range (deg C)	Methods Used	Ref
3.5	135	1250-1405	Oxygen diffusion into FZ Si; thermal donors formed at 450 deg C; profile by electrical measurements	[3]
2.55	0.21	1050-1100	Internal friction - ultrasonic absorption	[4]
3.5	83	1000-1200	O diffusion into FZ Si; charged particle activation	[5]
2.4	0.091	1100-1200	Lattice strain measurements	[6]
3.15	22.6	1000-1280	O-18 diffusion in Si; 4.6 MeV irradiation; decay measurement	[7]
2.44	0.07	700-1240	O-18 diffusion in FZ Si; SIMS profiles and IR 9 microns to check isotope ratio	[8]
2.77	1.5	1110-1300	IR absorption	[9]
2.54	0.17	270-400	IR stress dichroism relaxation	[10]
0.16	3 X 10*-10	37-84	Acceptor neutralisation: Miller depth profile	[11]
2.51	0.11	350-1240	(1) Matrix heat treatment; IR absorption at 4.2 deg K (2) Small angle neutron scattering (3) Stress dichroism low temp.	[12]
2.43	0.033	750,850,1050	Growth kinetics of square platelet ppts	[13]
2.53	0.14	700-1100	O diffusion in FZ Si; SIMS measurements	[14]

A recent review by Mikkelsen [15] thoroughly covers the background to various measurements and techniques in detail, and is recommended reading.

From this analysis he draws together both high and low temperature work from the various sources, to give a combined expression:

$$D = 0.13 \exp(-2.53/kT) \text{ cm*2/s}$$

Over the wide temperature range from 350-1250 deg C.

Turning to enhanced diffusion, this can be seen in a variety of ways: electron irradiation [2], metal contamination [16], preannealing at a high temperature before diffusion at a low temperature [10], and in thermal donor formation [17]. Vacancy assisted diffusion has also been observed in isothermal annealing [18,19].

It is possible to lower the diffusivity below the normal value. If crystals are tin-doped, vacancies are very efficiently trapped on adjacent sites, leading to a reduction in A-centre production, and a dramatic reduction in the rate of loss of stress dichroism [20].

REFERENCES
----

[1]   A.Brelot, J.Charlemagne [Radiat. Eff. (GB) vol.8 (1971)
      p.161-9]
[2]   R.C.Newman, J.H.Tucker, F.M.Livingston [J. Phys. C (GB) vol.16
      (1983) p.L151]
[3]   R.A.Logan, A.J.Peters [J. Appl. Phys. (USA) vol.30 (1959)
      p.108]
[4]   C.Hass [J. Phys. & Chem. Solids (GB) vol.15 (1960) p.108]
[5]   R.A.Kushner [Electrochem. Soc. Fall Meeting 1972, Extended
      Abstract No.260 p.643-5]
[6]   Y.Takano, M.Maki [Semiconductor Silicon 1973, Proc.
      Electrochem. Soc. (USA) vol.PV-73 (1973) p.469-81]
[7]   J.Gass, H.H.Muller, H.Stussi, S.Schweitzer [J. Appl. Phys. (USA)
      vol.51 (1980) p.2030]
[8]   J.C.Mikkelsen,Jr [Appl. Phys. Lett. (USA) vol.40 no.4 (15 Feb
      1982) p.336-7]
[9]   G.Vakilova, R.F.Vitman, A.A.Lebedev, S.Mukhammedov [Sov.
      Phys.-Semicond. (USA) vol.16 no.12 (Dec 1982) p.1426-7]
[10] M.Stavola, J.R.Patel, L.C.Kimerling, P.E.Freedland [Appl. Phys.
      Lett. (USA) vol.42 no.1 (1 Jan 1983) p.73-5]
[11] W.L.Hansen, S.J.Pearton, E.E.Haller [Appl. Phys. Lett. (USA)
      vol.44 no.9 (1 May 1984) p.889-91]
[12] F.M.Livingston et al [J. Phys. C (GB) vol.17 (1984) p.6253-76]
[13] K.Wada, N.Inoue [J. Cryst. Growth (Netherlands) vol.71 no.1
      (Jan-Feb 1985) p.111-17]
[14] S.-Tong Lee, D.Nichols [Appl. Phys. Lett. (USA) vol.47 no.9 (1
      Nov 1985) p.1001-3]
[15] J.C.Mikkelsen,Jr [Mater. Res. Soc. Symp. Proc. (USA) vol.59
      (1986) p.19-30]

[16]   R.C.Newman, A.K.Tipping, J.H.Tucker [J. Phys. C (GB) vol.18
        no.27 (30 Sep 1985) p.L861-6]

[17]   S.-Tong Lee, P.Fellinger [Appl. Phys. Lett. (USA) vol.49 no.26
        (29 Dec 1986) p.1793-5]

[18]   B.G.Svensson, J.L.Linstrom, J.W.Corbett [Appl. Phys. Lett. (USA)
        vol.47 no.8 (15 Oct 1985) p.841-2]

[19]   F.Shimura, W.Dyson, J.W.Moody, R.S.Hockett [Electrochem. Soc.
        Spring Meeting 1985, Extended Abstract No.198 p.278]

[20]   A.S.Oates, M.J.Binns, R.C.Newman, J.H.Tucker, J.G.Wilkes,
        A.Wilkinson [J. Phys. C (GB) vol.17 (1984) p.5695-705]

12.2 DIFFUSION OF C IN Si

by K.G.Barraclough

August 1987
EMIS Datareview RN=15781

Carbon is a common impurity in single crystal silicon grown from either
the melt (e.g. Czochralski and floating zone silicon) or from the
vapour phase (e.g. chemical vapour phase epitaxy and molecular beam
epitaxy). The diffusion coefficient of carbon in silicon is an
important property which influences defect formation during the high
temperature heat treatments characteristic of device processing.
Diffusion coefficient data are available over a wide temperature range
from 800 deg C to 1400 deg C, and generally independent investigations
using different techniques are in reasonable agreement. Surprisingly,
none of the published data have been obtained from dislocation-free
silicon which is currently used in most device lines. Although this is
not expected to be of major significance, it should be borne in mind,
for example, when considering the influence of point defects.

Newman and Wakefield [1] determined the diffusion coefficient in
dislocated (10/cm*2), high resistivity (100 ohm cm) Czochralski
pulled silicon using a radio tracer technique with either barium
carbonate or acetylene as a source of C-14. SiC(-14) was formed on the
sample surface as a result of the reaction in silica tubes with
acetylene or carbon dioxide. Temperatures ranged from 1070-1402 deg C.
Surface concentrations of C-14 were measured directly by a Geiger
Counter after removal of layers by grinding. Heating times were chosen
to give measurable counting rates at depths of at least 60 micron, thus
avoiding near-surface effects. Experiments with acetylene were liable
to error because of its low activity. Spurious data within 15 micron of
the original surface were observed for experiments with both sources as
a result of the small amounts of silicon carbide not removed with the
original layer.

The data fitted the expression:

$$D = 0.33 \exp \left[\frac{-2.92 +- 0.25 \text{ eV}}{kT} \right] \text{ cm*2/sec}$$

Experimental values are reproduced in TABLE 1 where it is seen that the
data using acetylene as C-14 source were in good agreement with the
data using CO2, despite the limitations described above.

Gruzin et al [2] carried out similar experiments to Newman and Wakefield
[1] using BaC(-14)03 as the source of C-14. High resistivity, dislocated
(5 X 10*4/cm*2), n-type, low-oxygen, floating zone silicon samples
were used. Only four temperatures were investigated between 950 and
1100 deg C. The data were not tabulated, although it was claimed that
they were in close agreement with [1].

At the time of the investigations by Newman and Wakefield [1] it was not certain that carbon occupied substitutional sites in the silicon lattice but this was correctly deduced from the similarity between the diffusion coefficients for carbon and the substitutional group III and V elements. It is expected, therefore, that the point defect concentration strongly influences the diffusion of carbon as with the group III and V elements. This has been considered in a recent series of papers [3-5].

Using SIMS measurements of the carbon out-diffusion profile for high carbon, dislocated (10*4-10*5/cm*2) Czochralski silicon crystals, Kalejs et al [3] observed an enhanced carbon diffusivity at 900 deg C after annealing in phosphine compared with nitrogen. This enhancement was attributed to the influence of phosphorus-induced self interstitials, and it was concluded that at 900 deg C carbon has a significant diffusion component due to self interstitials.

In a later paper Ladd et al [4] extended their studies in [3] to include carbon-rich dislocated (10*6-10*7/cm*2) silicon ribbon (low-oxygen) samples grown by the edge-defined film fed growth technique. Three different ambients at 900 deg C were also used: nitrogen, oxidising and phosphorus. Diffusion coefficient enhancement over the value for the nitrogen anneal was by a factor of 3 during oxidation and a factor of 40 during phosphorus in-diffusion.

In fitting the experimental data to theoretical out diffusion plots it was assumed that the equilibrium surface concentration was 5 X 10*14 atoms/cm*3, the equilibrium solubility at 900 deg C [6]. TABLE 1 includes the data from [4] and it is seen that an extended 9.33 hour phosphorus-anneal differs significantly from a 2.66 hour phosphorus-anneal, presumably due to the decrease in generation rate of interstitials for the longer anneal.

The studies in [3] and [4] were extended in [5] by investigating different temperatures between 800 and 1100 deg C. Anneal times were chosen at each temperature to give an out-diffusion profile of at least 1 micron, so it could be readily measured by SIMS. At 1000 deg C the nitrogen ambient produced the same results as an argon ambient. The diffusion data for the nitrogen anneals, phosphorus anneals and oxidations, (TABLE 1), converge at temperatures above about 1100 deg C, as expected if the self interstitial supersaturation ratios for oxidation and phosphorus diffusions decrease with temperature. If the original data in [1] are extrapolated below 1100 deg C then, within experimental error, they are in agreement with both the nitrogen- and oxygen-ambient data in [5]. The maximum diffusion enhancement of approximately a factor of 70 was observed for phosphorus diffusions at 800 deg C [5] (TABLE 1).

TABLE 1: diffusion coefficient data for carbon in silicon

Ref.	Temp (deg C)	Time (hrs)	D (cm*2/sec)	Anneal Conditions
[1]	1402	1	6.3 X 10*-10	C-14 in-diffusion (see note below)
[1]	1402	1	7.1 X 10*-10	"
[1]	1390	3	3.7 X 10*-10	"
[1]	1390	3	3.0 X 10*-10	"
[1]	1375	2	3.9 X 10*-10	"
[1]	1350	3.9	3.5 X 10*-10	"
[1]	1320	7.0	2.1 X 10*-10	"
[1]	+ 1314	10.2	2.2 X 10*-10	"
[1]	1314	15.8	1.7 X 10*-10	"
[1]	1264	25	1.2 X 10*-10	"
[1]	1264	24	1.3 X 10*-10	"
[1]	* 1250	43	5.0 X 10*-11	"
[1]	1231	26	5.6 X 10*-11	"
[1]	1210	64	3.0 X 10*-11	"
[1]	1210	64	2.4 X 10*-11	"
[1]	* 1200	112	2.6 X 10*-11	"
[1]	* 1200	112	2.2 X 10*-11	"
[1]	1150	168	1.5 X 10*-11	"
[1]	1150	168	1.7 X 10*-11	"
[1]	+ 1109	309	6.6 X 10*-12	"
[1]	+ 1109	309	5.2 X 10*-12	"
[1]	1109	309	4.8 X 10*-12	"
[1]	1109	309	9.1 X 10*-12	"
[1]	1070	403	3.0 X 10*-12	"
[5]	1000	4	5 X 10*-13	Nitrogen ambient

Ref.	Temp (deg C)	Time (hrs)	D (cm*2/sec)	Anneal Conditions
[5]	1000	2.5	1.2 X 10*-12	Oxygen ambient
[5]	1000	0.66	6 X 10*-12	Phosphorus ambient
[4]	900	2.66	3 X 10*-14	Nitrogen ambient
[5]	900	20	6 X 10*-14	Nitrogen ambient
[4]	900	2.66	9 X 10*-14	Oxygen ambient
[4]	900	9.33	9 X 10*-14	Oxygen ambient
[5]	900	20	1.5 X 10*-13	Oxygen ambient
[5]	900	0.66	2 X 10*-12	Phosphorus ambient
[4]	900	2.66	1.3 X 10*-12	Phosphorus ambient
[5]	900	8	0.4 X 10*-12	Phosphorus ambient
[4]	900	9.33	0.6 X 10*-12	Phosphorus ambient
[5]	800	108	6 X 10*-16	Nitrogen ambient
[5]	800	141	1.4 X 10*-15	Nitrogen ambient
[5]	800	141	7 X 10*-15	Oxygen ambient
[5]	800	7	7 X 10*-14	Phosphorus ambient
[5]	800	88	4 X 10*-15	Phosphorus ambient

Notes:

+ Plastically deformed prior to heating.

* Experiments with acetylene as C-14 source.

Note: C-14 in-diffusion was from SiC(-14) produced by decomposition of BaC(-14)03 in sealed silica tubes.

REFERENCES
----

[1] R.C.Newman, J.Wakefield [J. Phys. & Chem. Solids (GB) vol.19
     no.3/4 (1961) p.230-4]
[2] P.L.Gruzin, S.V.Zemskii, A.D.Bulkin, N.M.Makarov [Sov.
     Phys.-Semicond. (USA) vol.7 (Mar 1974) p.1241]
[3] J.P.Kalejs, L.A.Ladd, U.Gosele [Appl. Phys. Lett. (USA) vol.45
     no.3 (1 Aug 1984) p.268-9]
[4] L.A.Ladd, J.P.Kalejs, U.Gosele [Mat. Res. Soc. Symp. Proc. (USA)
     vol.36 (1985) p.89-94]
[5] L.A.Ladd, J.P.Kalejs [Mat. Res. Soc. Symp. Proc. (USA) vol.59
     (1986) p.445-50]
[6] A.R.Bean, R.C.Newman [J. Phys. & Chem. Solids (GB) vol.32
     (1971) p.1211]

12.3 DIFFUSION OF N IN Si

by P.J.Ashby

August 1987
EMIS Datareview RN=15788

Nitrogen in silicon whether introduced from the melt or by ion
implantation, exists predominantly as strongly bound atom pairs. The
detail of the nitrogen pair bond to the silicon lattice has not been
determined. Several possible schemes have been proposed. Pavlov et al
[1] suggested that the nitrogen could exist in molecular form at
interstitial sites. This concept was used by Abe et al [2] whose
spectroscopic data indicated that the Si-N interactions were very weak,
with the nitrogen pairs or molecules behaving as if they were isolated
from the lattice. Alternatively Stein [3] proposed several
possibilities for coupling between two Si-N-Si centres through a N-N
bond with the nitrogen pair sited at a vacancy or di-vacancy. This
uncertainty of the structure of nitrogen centres also makes the
possible diffusion mechanism unclear.

Measurement of nitrogen diffusion in silicon is made difficult by the
limited electrical activity of nitrogen and its low solubility. Ion
implanted nitrogen however is partially electrically active and can
behave as a donor; Pavlov et al discuss this at length in their review
paper [1]. They also describe how electrically inactive nitrogen can be
activated as a donor by irradiating with neon ions. These two phenomena
have formed the basis of diffusion coefficient measurements.

Clarke et al [4] implanted (N-14)+ ions into 100 ohm cm <110> p-type
silicon at an accelerating voltage of 40 keV with a dosage in the range
of 3 X 10*12 to 5 X 10*14 ions/cm*2. Hall effect measurements indicated
that the samples had formed a n-type surface layer. The depth of this
layer was then measured as a function of annealing time and temperature.
The diffusion coefficient of the nitrogen in the irradiated silicon
was determined as:

    D = 0.87 exp(-3.29(eV)/kT) cm*2/s

In an experiment reported by Denisova et al [5] the diffusion
coefficient has been determined for nitrogen diffused into nitrogen free
silicon during a thermal heat treatment. Samples of 300 ohm cm p-type
float zone silicon were heated in a nitrogen atmosphere at temperatures
between 700 and 1200 deg C. The depth to which the nitrogen diffused
into the sample was assessed by irradiating the sample with 50 keV neon
ions at a dose of 6 X 10*14 ions/cm*2 followed by an 800 deg C anneal for
30 minutes. This treatment resulted in the formation of an inversion
layer that was assumed to be coincident with the nitrogen diffused layer.
The inversion layer depth was measured by successive oxidation etching
and resistivity determinations. After making corrections for the 800 deg C
the calculated diffusion coefficient of nitrogen was:

    D = 0.03 exp(-2.63(eV)/kT) cm*2/s

in the temperature range 700-800 deg C. The formation of a protective
SiO2/Si3N4 film inhibited diffusion at 900 deg C and above.

The differences between these values illustrate the uncertainty in
determining the diffusion coefficient. They may reflect the differences
in the defect morphology of the N-implanted silicon used by Clarke [4]
and the float zone silicon of Denisova [5]. Also the determination of
the nitrogen diffusion depth depended on effects that could only be
produced in ion implanted material and may not relate directly to the
situation in an undamaged lattice.

REFERENCES
----

[1]   P.V.Pavlov, E.I.Zorin, D.I.Tetelbaum, A.F.Khokhlov [Phys. Status
       Solidi a (Germany) vol.35 (1976) p.11]
[2]   T.Ave, H.Harada, N.Ozawa, K.Adomi [Mater. Res. Soc. Symp. Proc.
       (USA) vol.59 (1986) Eds. J.C.Mikkelsen, S.J.Pearton, J.W.Corbett,
       S.J.Pennycook]
[3]   H.J.Stein [Mater . Res. Soc. Symp. Proc. (USA) vol.59 (1986) Eds.
       J.C.Mikkelsen, S.J.Pearton, J.W.Corbett, S.J.Pennycook]
[4]   A.Clarke, J.D.MacDougall, D.E.Manchester, P.E.Roghan, F.W.Anderson
       [Bull. Amer. Phys. Soc. (USA) vol.13 (1968) p.376]
[5]   N.V.Denisova, E.I.Zorin, P.V.Pavlov, D.J.Tetelbaum, A.F.Khokhlov
       [Izv. Akad Nauk. SSSR Neorg Mater. vol.11 (1975) p.2236]

12.4 SEGREGATION COEFFICIENT OF O IN Si

by K.G.Barraclough

August 1987
EMIS Datareview RN=16127

The value for the solid-liquid segregation coefficient of oxygen in
silicon, $k(ox)$, is of particular importance in bulk Czochralski (CZ)
and thin film silicon-on-insulator (SOI) melt growth processes which
are carried out in the presence of silica, SiO_2. Liquid silicon readily
dissolves solid silica which may be in the form of a crucible in the
Czochralski process or a thermally grown layer on a silicon wafer in SOI
recrystallisation techniques. The oxygen in the melt is transported by
convective flow to the solidification interface where it is incorporated
into the solid. In the Czochralski growth process most of the oxygen
is lost as silicon monoxide, SiO at the melt surface. At a given instant
the oxygen concentration in the melt is, therefore, a result of a dynamic
process involving, primarily, the dissolution rate of the silica, the
rate of transport through the melt and the rate of evaporation at the
melt surface. These processes can be strongly influenced by many CZ-
crystal growth parameters, and for this reason it is extremely diffcult
to carry out experiments which give unambiguous evidence for $k(ox)$. By
contrast many other dopants are more uniformly distributed in the melt
and, providing evaporation and contamination are not significant, the
solid-liquid segregation coefficient can be readily determined in
Czochralski pulling experiments using the normal freeze - and the Burton,
Prim, Slichter [1] relationship. This approach for oxygen in silicon is
fraught with difficulty, as outlined below.

TABLE 1 summarises the important data for $k(ox)$ taken from either
experimental determinations, theoretical calculations, extrapolations
or estimates over the last 30 years. The values range from
approximately 0.25 to 1.4. As will be seen below the strongest evidence
favours a value very close to 1.

A value of < 1 is suggested from the eutectic phase diagram given by
Sosman [2], but the dotted phase boundaries at the silicon-rich end
indicate that this is only a tentative diagram.

In a classic paper by Kaiser and Keck [3] on oxygen in silicon it was
noted that the 9 micron IR band due to interstitial oxygen did not
change when the growth rate was changed, and that $k(ox)$ must be close
to 1. Details were not given. Important observations were also made on
the effect of growth parameters on oxygen content. In particular Kaiser
and Keck [3] noted that a reduction in crystal diameter caused by an
increase in melt temperature resulted in a reduction in oxygen content
of the melt under the crystal and, therefore, a reduction in oxygen
content of the crystal. This is of relevance in the interpretation of
later investigations.

The value of k(ox) of approx. 0.5 reported by Trumbore [4] was determined by vacuum gas fusion analysis on quenched silicon samples melted in silica tubes and was attributed to a private communication with Thurmond. No other experimental details were reported.

The first detailed accounts of experiments specifically aimed at solubility determinations in solid and liquid silicon were reported by Yatsurugi et al [5] and Nozaki et al [6]. It was recognised that the 9 micron IR absorption technique in isolation is not suitable for these measurements since clustered, IR insensitive oxygen can be present in the samples at these high oxygen concentrations. Thus, their IR measurements were complemented by charged particle activation analysis. The liquid solubility was determined by rapidly quenching small liquid zones. Values for k(ox) > 1 were found, but the stated measurement precision indicated that these could be close to 1.

Jaccodine and Pearce [7] also reported a value for k(ox) close to 1 from studies of rapidly cooled samples taken from large Czochralski melts using the technique described in [8]. However, in the same paper Jaccodine and Pearce [7] present the case for a value of 0.65 based on solubility data normalised to take account of the different IR conversion factors. As stated above, however, 9 micron IR measurements for liquid- and solid solubility determinations are not reliable. It is also reported in [7] that values of k(ox) in the range 0.35-0.90 are calculated from various theories. In the writer's view the uncertainties in the theories combined with the lack of precision of the input data do not justify serious attempts to calculate k(ox).

The contributions by Lin and Hill [9] and Lin and Stavola [10] suggest that k(ox) < 1. They reported values of 0.25 [9] and approximately 0.3 [10] from two different CZ - growth experiments. In a study [9] of the macroscopic, axial distribution of oxygen in CZ-Si, attempts were made to provide a constant concentration growth environment by pulling small diameter, 12-14 mm crystals from large (6.5 kg) melts. The area of silica in contact with the melt (oxygen source) remained virtually constant throughout the run, as did the free surface area of the melt (oxygen evaporation). Constant rotations were used so that the forced melt convection was also constant. By comparing the axial distribution of interstitial oxygen with that of either boron or arsenic dopants after manual growth rate changes it was deduced that k(ox) < 1 and approx. 0.25. The only major criticism that can be levelled at this experiment is that in order to maintain the crystal diameter essentially constant during the growth rate changes the melt temperature had to be changed. This would have had two effects. Firstly the dissolution rate of the crucible would have changed and secondly there would have been a tendency for the diameter to change. Thus, it would have been difficult in this experiment to have reproduced the same oxygen content of the melt below the crystal at the two growth rates.

In a separate experiment Lin and Stavola [10] measured the microscopic distribution of interstitial oxygen using a 9 micron laser source for their IR measurements. Comparison of the oxygen striations with the arsenic dopant striations from spreading resistance measurements led to a value of approx. 0.3. The rotational striations originated from microscopic growth rate variations in response to the uneven temperature distribution of the melt. However, the possibility of any spatial variation in oxygen concentration of the melt as a result of this temperature variation was not considered. Indeed, in a similar study [11] in which striations in the form of thermal donor (oxygen content?) variations were compared with boron dopant striations it was concluded that k(ox) > 1! Recently [12], SIMS measurements across growth striations showed no spatial variation of oxygen compared with phosphorus (k is approx. 0.35). Since so little is known about melt flows in large symmetrical and asymmetrical CZ growth systems and since oxygen distributions are believed to be quite different from other dopants in CZ-melts, it is unwise to use microscopic oxygen distributions in CZ-crystals to reach conclusions about k(ox).

Recently, three independent investigations using different methods have shown that k(ox) is near 1. Harada et al [12] compared the change in the axial distribution of phosphorus and oxygen in a CZ-Si crystal which had been allowed to grow from a melt which was fully encapsulated by the crystal extending to the crucible wall. In this case evaporation of SiO was prevented and any effects due to the enhancement (k(ox) < 1) or depletion (k(ox) > 1) of oxygen in the melt as a result of segregation effects would have been detectable. No such changes were seen (compared with phosphorus) indicating that k(ox) is 1. In another experiment by Harada et al [12] changes in oxygen concentration could not be detected across a transverse section of a <111> crystal which contained a peripheral- and a central (111) facet. By contrast phosphorus exhibited a strong segregation effect on the (111) facets. Similarly, no orientation dependence was found for oxygen incorporation in twinned crystals compared with antimony (k is approx. 0.02). Whilst not quantitative, these observations indicate that k(ox) is close to 1.

Using well known thermodynamic data and a function for the standard free energy of solution of oxygen in liquid silicon [13] which had gone unnoticed in the literature, Carlberg [14] recently calculated that k(ox) must be very close to unity.

In the writer's laboratory an experiment similar to the one described by Lin and Hill [9] was carried out recently [15]. In this case, however, the melt temperature was held constant during the pull rate changes to avoid any possibility of a change in melt concentration from effects other than pure segregation. Conditions were established so that the latent heat released by a small diameter crystal during a pull rate change did not affect substantially the crystal diameter. Whereas a predictable Sb-concentration transient was readily measurable the change in axial oxygen content was less than 10% for a pull rate change from 2 mm/min to 0.4 mm/min. This indicated that k(ox) is 1 +- 0.1.

In summary, the value for the solid-liquid segregation coefficient of oxygen in silicon has been a subject of controversy over the last 30 years. A number of independent investigations in the last two years, however, indicate a value of 1 +- 0.1.

TABLE 1: published data for the segregation coefficient k(ox) of oxygen
in silicon

k(ox)	Method	Comments	Ref
< 1	Tentative phase diagram	No firm evidence	[2]
approx. 1	CZ-Si, response to growth rate changes	No quantitative data	[3]
0.5	Vacuum gas fusion	No experimental details	[4]
1.2 +- 0.17	Analysis of liquid and solid solubility	Direct measure using CPA and IR	[5]
1.4 +- 0.3	Analysis of liquid and solid solubility	Direct measure using CPA and IR	[6]
approx. 1	Sampling CZ-melts	See ref [8]	[7]
0.65	Normalised solubility data	IR inactive oxygen?	[7]
0.35-0.90	Theory	Precision of data?	[7]
approx.0.25	Effect of growth rate on macroscopic distribution in CZ crystals	Effect of temperature changes	[9]
approx. 0.3	Microscopic distribution of oxygen in CZ-Si	Spatial variation of oxygen in melt?	[10]
> 1	Microscopic distribution of thermal donors	Not quantitative	[11]
1	Oxygen distribution in CZ-Si with no SiO evaporation	Convincing	[12]
1	Oxygen distribution across facets	Qualitative	[12]
1	Oxygen distribution across growth striae	Compare ref [10]	[12]
1	Oxygen distribution across twin boundary	Qualitative	[12]
slightly < 1	Calculated solubilities from thermodynamic data	Used recent data in [13]	[14]
1 +- 0.1	Effect of growth rate on macroscopic distribution in CZ Si	Melt temperature constant (cf[9]) Crystal diameter is critical	[15]

REFERENCES
----

[1]   J.A.Burton, R.C.Prim, W.P.Slichter [J. Chem. Phys. (USA) vol.21
       (1953) p.1987]

[2]   R.B.Sosman [Trans. & J. Br. Ceram. Soc. (GB) vol.54 (1955)
       p.655]

[3]   W.Kaiser, P.H.Keck [J. Appl. Phys. (USA) vol.28 (1957) p.882]

[4]   F.A.Trumbore [Bell Syst. Tech. J. (USA) vol.39 (1960) p.205]

[5]   Y.Yatsurugi, N.Akiyama, Y.Endo, T.Nozaki [J. Electrochem. Soc.
       (USA) vol.120 no.7 (1973) p.975-9]

[6]   T.Nozaki, Y.Yatsurugi, N.Akiyama, Y.Endo, Y.Makide [J.
       Radioanal. Chem. (Switzerland) vol.19 (1974) p.109]

[7]   R.J.Jaccodine, C.W.Pearce ['Defects in Silicon' Eds
       W.M.Bullis, L.C.Kimerling (Electrochem. Soc., Pennington, NJ, USA,
       1983) p.115]

[8]   R.J.Lavigna, C.W.Pearce, R.E.Reusser [US patent no. 4,134,785]

[9]   Wen.Lin, D.W.Hill [J. Appl. Phys. (USA) vol.54 no.2 (Feb 1983)
       p.1082-5]

[10]   W.Lin, M.Stavola [J. Electrochem. Soc. (USA) vol.132 no.6 (June
       1985) p.1412-16]

[11]   A.Murgai, H.C.Gatos, W.A.Westdorp [J. Electrochem. Soc. (USA)
       vol.126 no.12 (Dec 1979) p.2240-5]

[12]   H.Harada, T.Itoh, N.Ozawa, T.Abe [Proc. 3rd Int. Symp. VLSI Sci.
       & Technol., Toronto, Ont., Canada 13-16 May 1985 Eds W.M.Bullis,
       S.Broyda (Electrochem. Soc., Pennington, NJ, USA 1985) p.526]

[13]   S.Otsuka, J.Kozuka [Trans. Jpn. Inst. Metals (Japan) vol.22
       (1981) p.558]

[14]   T.Carlberg [J. Electrochem. Soc. (USA) vol.133 no.9 (Sep 1986)
       p.1940-2]

[15]   R.W.Series, K.G.Barraclough [Extended Abst. Electrochem. Soc.
       Fall Meeting 1987, Honolulu, Hawai, 18-23 October, 1987
       (Electrochem. Soc., Pennington, NJ, USA)]

12.5 SEGREGATION COEFFICIENT OF C IN Si

by K.G.Barraclough

August 1987
EMIS Datareview RN=15783

This review is concerned with the value for the solid-liquid segregation
coefficient of carbon in silicon which determines the concentration and
distribution of carbon in either oxygen-rich Czochralski or low-oxygen
floating-zone crystals pulled from the melt. It is important to
distinguish between the effective segregation coefficient, ke and the
equilibrium segregation coefficient, ko. Whereas ke is dependent on the
dynamics of the pulling process, ko is a constant defined by the
equilibrium phase diagram as the ratio of the concentration of the
impurity in the solid, [C]S, to that in the liquid which is in equilibrium
with the solid, [C]L:

$$ko = \frac{[C]S}{[C]L} \qquad (1)$$

In common with other impurities in semiconductor grade silicon the
carbon concentrations are sufficiently small for any variation of ko
with concentration to be insignificant, so the liquidus and solidus
are essentially straight lines. As outlined below there is now
generally good agreement on the value for ko in oxygen-free silicon
but the presence of oxygen at concentrations of approximately 10*18
atoms/cm*3 is an additional, complicating factor in Czochralski
silicon.

In the study by Scace and Slack [1] the solubility of carbon in
oxygen-free liquid silicon at the melting point was reported as a
private communication with W.C.Dash. In this separate investigation,
pedestal-melted, oxygen-free samples were used to dissolve weighted
amounts of silicon carbide under vacuum conditions. Although it seems
that only small amounts of carbon would have been introduced from the
ambient, the carbon content of the starting material was not given,
and this may have influenced the result. (At the time of the
investigation by Scace and Slack [1] the IR absorption method for
measurement of substitutional carbon content in silicon had not been
developed.) Thus the value for the liquid solubility of 3 +- 0.3 X
10*18 atoms/cm*3 in [1] may have been underestimated. However, the
maximum solid solubility of carbon in oxygen-free silicon is less than
1 X 10*18 atoms/cm*3 and the amount of the under-estimate should not
exceed this quantity.

Much larger values for the solubility of carbon in liquid silicon
(10*19 atoms/cm*3) were reported by Hall [2] who studied melts which
were maintained in either silica or morganite alundum crucibles,
depending on temperature. Presumably the silica crucibles were used at
the lower temperatures close to the melting point. As we shall discuss
later, the presence of oxygen from the silica crucible may promote
the evaporation of carbon from the melt as carbon monoxide, and since
the solubility was evaluated from the weight loss of SiC immersed in

the melt, the values in [2] were probably overestimated. However, if the data from these early liquid solubility studies are combined with known solid solubility data, the equilibrium segregation coefficient is approx. 0.1. This is close to the value determined by Haas et al [3], k approx. 0.09, but significantly different from the value of 0.005 determined earlier by Newman and Willis [4] who analysed the distribution of C-14 in pulled crystals. When analysing the distribution of impurities in pulled crystals ke is usually derived from the normal freeze relationship:

$$[C]S = ke [C]L (1-g)*(ke-1) \qquad (2)$$

where [C]S is the concentration of solute in the crystal at a position defined by g, the melt fraction solidified, and [C]L is the initial concentration in the melt. Note, however, that the normal freeze distribution in Eqn (2) is an idealised case. If the melt concentration changes by either contamination or evaporation during the growth process Eqn (2) has to be modified as described in [5]. Also, the segregation curve is sensitive to (k-1) and if k ≪ 1 then detailed examination of the tail end of the crystals (g approaching 1) is necessary. Since it is usually extremely difficult to maintain constant growth parameters from the seed - to tail end of CZ-crystals, analysis of the normal freeze is not generally recommended for precise determination of ko for carbon in silicon. Nevertheless, by taking account of the change in normal freeze behaviour of CZ-crystals as a result of contamination, it was shown [5] that best-fit distributions of carbon corresponded to a ke of 0.1 which from the Burton Prim Slichter [6] relationship is equivalent to an equilibrium segregation coefficient, ko, of 0.06-0.07. This value was in good agreement with that (0.07) extrapolated to zero growth rate from ke values determined from analysis of oxygen-free floating zone crystals pulled under vacuum [7]. In fact a better extrapolation of the data in [7] would have yielded a somewhat lower value of 0.057 which is in excellent agreement with the value of 0.058 determined independently using oxygen-free floating zone material [8].

When silicon melts are contained in silica crucibles as in CZ-pulling, the continuous reaction of melt with the silica results in near-saturation levels of oxygen (approx. 2 X 10*18 atoms/cm*3) in the melt and continuous evaporation of SiO. In this case carbon may also be lost from the melt as carbon monoxide [9]. If the simple equilibrium is considered between the carbon in the liquid [C]l, the oxygen in the liquid [O]l and the carbon monoxide in the ambient [CO]v, then we have the reversible reaction.

$$[C]l + [O]l = [CO]v \quad (reversible) \qquad (3)$$

Then the equilibrium constant K is:

$$K = \frac{P[CO]}{[C]l \, [O]l} \qquad (4)$$

Since the oxygen level is near saturation in a Czochralski melt the carbon content will be more or less proportional to the partial pressure of carbon monoxide, P[CO]. Unfortunately, P[CO] is rarely monitored, yet this is a key parameter which determines either contamination from the ambient or purification through evaporation from the melt. When CZ crystals are pulled under typical reduced pressure conditions (approx. 20 torr) the partial pressure of carbon monoxide in high purity argon is usually low enough to prevent significant contamination of the melt during growth (but not necessarily during the initial melt down of the charge [5]). If the same Czochralski growth conditions (pull speed, rotations, argon pressure and input purity , etc.) are used in the presence of a vertical magnetic field (e.g. 2K Gauss) then the situation may change dramatically. In addition to increasing the carbon content of the melt through the effects of the reduced convective melt flow, the magnetic field increases the solid-liquid diffusion boundary layer and increases the effective segregation coefficient ke to approx. 0.7 [10]. This is an order of magnitude larger than the equilibrium value. Such an increase is normally only encountered in rapid growth techniques such as pulsed laser annealing.

REFERENCES

[1]   R.I.Scace, G.A.Slack [J. Chem. Phys. (USA) vol.30 (1959) p.1551]
[2]   R.N.Hall [J. Appl. Phys. (USA) vol.29 (1958) p.914]
[3]   E.Haas, W.Brandt, J.Martin [Solid-State Electron. (GB) vol.12 (1969) p.915]
[4]   R.C.Newman, J.B.Willis [J. Phys. & Chem. Solids (GB) vol.26 (1965) p.373]
[5]   R.W.Series, K.G.Barraclough [J. Cryst. Growth (Netherlands) vol.60 no.2 (Dec 1982) p.212-18]
[6]   J.A.Burton, R.C.Prim, W.P.Slichter [J. Chem. Phys. (USA) vol.21 (1953) p.1987]
[7]   T.Nozaki, Y.Yatsurugi, N.Akiyama [J. Electrochem. Soc. (USA) vol.117 (1970) p.1566]
[8]   B.O.Kolbesen, A.Muhlbauer [Solid-State Electron. (GB) vol.25 (1982) p.759]
[9]   T.Nozaki, Y.Makide, Y.Yatsurugi, N.Akiyama, Y.Endo [J. Appl. Radiation and Isotopes vol.22 (1971) p.607]
[10]   R.W.Series, K.G.Barraclough, D.T.J.Hurle, D.S.Kemp, G.J.Rae [Extended Abstracts, Electrochemical Soc., Spring Meeting (Toronto, Canada, 1985)]

12.6 SEGREGATION COEFFICIENT OF N IN Si

by P.J.Ashby

August 1987
EMIS Datareview RN=15787

The solubility limit of nitrogen in molten silicon has been determined
by Yatsurugi et al [1]. A saturated Si-Si3N4 solution was rapidly
crystallised from just above the melting point and the dissolved
nitrogen concentration assessed by charged particle activation analysis.
Thus the liquid solubility limit was determined to be 6 X 10*18 atoms/
cm*3. This is in fair agreement with an earlier value determined by
Kaiser and Thurmond [2] of 10*19 atoms/cm*3. It is, like those for other
light impurities such as carbon and oxygen, a considerably lower value
than many metallic elements which are soluble up to > 10*20 atoms/cm*3.
This may be because nitrogen reacts to form Si3N4 which has a very high
dissociation energy, similarly carbon and oxygen form stable silicides
[1].

The equilibrium solid solubility limit of nitrogen in silicon has been
determined to be 4.5 X 10*15 atoms/cm*3 [1]. Thus the equilibrium
segregation coefficient, which is the ratio between the equilibrium
solubilities of nitrogen in solid and liquid silicon at its melting
point, is 7 X 10*-4. This value is very much lower than that for other
light impurities; oxygen, 1.25; carbon, 0.07 and is a result of the
very low solid solubility of nitrogen which is discussed in a separate
Datareview [3].

The effective segregation coefficient is strongly dependent upon
crystallisation rate. Since the solubility in liquid is greater than
that in the solid, solute trapping in the solid solute occurs with
increasing crystallisation rate. At a growth velocity of 2.5 mm/min
during float zoning, the solid solubility had increased by a factor of
two over the equilibrium value [1]. During laser annealing
crystallisation rates of approximately 10 X 15 mm/min are reached and
increases of solid solubility of several orders of magnitude have been
recorded.

REFERENCES
----

[1] Y.Yatsurugi, N.Akiyama, Y.Endo, T.Nozaki [J. Electrochem. Soc.
     (USA) vol.120 (1973) p.975]
[2] W.Kaiser, C.D.Thurmond [J. Appl. Phys. (USA) vol.30 (1959)
     p.427]
[3] P.J.Ashby [EMIS Datareview RN=15786 (Aug 1987) 'Solubility of N
     in Si']

12.7 SOLUBILITY OF O IN Si

by J.G.Wilkes

November 1987
EMIS Datareview RN=16198

During Czochralski growth of silicon from silica crucibles, most of the
oxygen is incorporated into the crystal at bonded interstitial sites
bridging two host lattice atoms, which configuration gives rise to the
well known infrared 9 micron absorption band, comprised of a number of
closely related individual vibration-rotation spectral contributions.

The solid solubility decreases with temperature so, after cooling, all
as grown crystals are supersaturated, and subsequent intermediate
temperature annealing leads to internal solid state precipitation,
until, eventually, the equilibrium solid solubility, Cs (T), is
approached.

The temperature dependence of the solubility is expected to be of the
form:

$$Cs\ (T) = Cso \exp \frac{-Es}{kT}$$

Where Es is the heat of solution, and Cso is a constant.

Starting from the work of Hrostowski and Kaiser [1] the heat treatment
of CZ silicon, following the precipitation process as it approaches
equilibrium at a chosen temperature, has been used for many solubility
determinations.

An alternative approach, generally more applicable at high temperatures,
has been to follow the in-diffusion of oxygen into oxygen-free float
zone material, by annealing the silicon samples in oxygen or water
vapour; isotopic O-18 can also be used in such diffusion studies.
TABLE 1 summarises the major contributions and methods used, [1-9].
Examination of this table reveals values of the oxygen solubility in
silicon at 1100 deg C over an order of magnitude; extrapolation to
lower temperatures rapidly worsens this position.

TABLE 1: summary of data relating to oxygen solubility

Es(eV)	Cso (/cm*3)	Methods used	Ref.
0.94	1.2 X 10*21	Heat treatment of CZ Si. Multiple heat treatments discussed + IR 9 micron measurement at 4.2 deg K	[1]
0.99	1.9 X 10*21	Heat treatment of CZ Si + IR 9 micron measurements at 300 deg K and 77 deg K: restricted times	[2]
1.57	1.2 X 10*23	Heat treatments of CZ Si with temperature cycling. IR 9 micron measurement, 77 deg K	[3]
1.03	2.0 X 10*21	Heat treatment of CZ Si previously neutron irradiated, IR 9 micron measurements at 300 deg K	[4]
1.07	2.8 X 10*21	O-18(p,n)F-18 reaction to obtain profiles and concentration of diffused O-18 from an external water vapour source	[5]
1.20	8.2 X 10*21	X-ray measurements of lattice strain produced by oxygen diffusion into FZ silicon	[6]
2.3	1.8 X 10*23	Oxygen diffused into FZ Si. Heat at 450 deg C to form thermal donors; electrical measurements	[7]
1.02	6.4 X 10*20	O-18 diffusion into FZ Si. SIMS profiles and IR 9 micron at 300 deg K to check isotopic ratio	[8]
1.2	7.1 X 10*21	Heat treatment of CZ Si. No previous heat treatments. IR 9 micron and 8 micron measurements at 4.2 deg K	[9]
1.4	2.6 X 10*22	--	[3,8,9]

Reappraisal of the study conditions by Wilkes [10] has identified a
number of factors contributing to the spread of results; while
similar reappraisal of the measurement procedures and calibration
standards by Livingston et al [9] has led to recalculation of previous
data, which then shows remarkably good agreement at temperatures of
1100 deg C or above. Major factors identified include:-

1. Separating the interstitial oxygen signal from that of the
    growing oxide precipitates. This requires measurements being
    made at 77 deg K for anneals at 1100 deg C or higher; and
    4.2 deg K for lower temperature anneals.

2. The use of the ultraclean furnace conditions in double wall
    furnaces to sweep out contaminants, Capper et al [11].

3. Solid state internal precipitation processes require very long
    times before the system approaches equilibrium - 100s to 1000s
    hours. Most workers have limited the times of heat treatment to
    less than 200 hours which is clearly too short a duration.

4. Normalising the infrared absorption data to a common - and
    latest standard - where the numerical value of the 9 micron
    band peak absorption coefficient (/cm) measured at
    room temperature is multiplied by a factor 2.54 X 10*17 to give
    the oxygen concentration in atoms/cm*3. Further interconversion
    data are required to transform measurements made at 77 deg K
    or 4.2 deg K to this internationally agreed standard.

Overall, not separating interstitial dissolved oxygen signals from
precipitates, not rigidly excluding any oxygen ingress contaminant to
the furnace tube, and limiting the heat treatment times, all lead to
solubility predictions which are too high especially if the total
kinetics are not rigorously mathematically analysed. Proper
accounting for the measurement calibration coefficient has a
substantial impact on reducing the apparent scatter of previously
reported results.

In a recent excellent review Mikkelsen [12] has examined the various
techniques used to study the behaviour of oxygen in silicon in some
detail. This is recommended reading. In bringing together the
contributions to solubility he has composed a composite data
assessment leading to:

$$Cs \ (T) = 9.0 \ X \ 10*22 \ exp \ (-1.52/kT)$$

However it must be noted that the data has been replotted using the
very recent Japanese [13] determination of the IR - concentration
calibration constant, 3.03 instead of 2.45, which is still not
internationally accepted. This affects the Cso value. In addition,
Mikkelesen includes more charged particle analysis results, and in his
discussion refers to - 'the somewhat poorer agreement on the solubility
measurements'.

Continuing he also refers to the Livingston et al [9] data compilation
as - 'the most reliable IR measurement of interstitial oxygen
concentration remaining in solution after precipitation in CZ crystals'.

Therefore this Datareview aligns with the approach taking the composite results of Bean and Newman [3], Mikkelsen [8] and Livingston et al [9] arriving at

$$Cs\ (T) = 2.6\ X\ 10*22\ exp\ (-1.40/kT)$$

For temperatures from the melting point Tm = 1413 deg C, down to at least 850 deg C,

where Cs (1413) = 1.70 X 10*18 atom/cm*3

and Cs (850) = 1.35 X 10*16 atom/cm*3

Finally it should be noted that recent theoretical chemical thermodynamics calculations, made by Carlberg [14], of the equilibria in the silicon-oxygen systems, derives values for the solid solubility in reasonable agreement with the above: but also bridges into liquid solubility data for SiO2 and SiO above the melting point.

REFERENCES
----

[1]   H.J.Hrostowski, W.Kaiser [J. Phys. & Chem. Solids (GB) vol.9 (1959) p.214]

[2]   K.Tempelhoff, F.Spiegelberg, R.Gleichmann [Semiconductor Silicon 1977; Proc. Electrochem. Soc. (USA) vol.PV 77-2 (1977) p.585-95]

[3]   A.R.Bean, R.C.Newman [J. Phys. & Chem. Solids (GB) vol.32 (1971) p.1211]

[4]   R.A.Craven [Semiconductor Silicon 1981; Proc. Electrochem. Soc. (USA) vol.PV 81-5 (1981) p.254-71]

[5]   J.Gass, H.H.Muller, H.Stussi, S.Schweitzer [J. Appl. Phys. (USA) vol.51 (1980) p.2030]

[6]   Y.Takano, M.Maki [Semiconductor Silicon 1973; Proc. Electrochem. Soc. (USA) vol.PV-73 (1973) p.469-481]

[7]   R.A.Logan, A.J.Peters [J. Appl. Phys. (USA) vol.30 (1959) p.1627]

[8]   J.C.Mikkelsen,Jr [Appl. Phys. Lett. (USA) vol.40 no.4 (1982) p.336-7]

[9]   F.M.Livingston et al [J. Phys. C (GB) vol.17 (1984) p.6253]

[10]   J.G.Wilkes [J. Cryst. Growth (Netherlands) vol.65 no.1-3 (Dec 1983) p.214-30]

[11]   P.Capper, A.W.Jones, E.J.Wallhouse, J.G.Wilkes [J. Appl. Phys. (USA) vol.48 no.4 (Apr 1977) p.1646-55]

[12]   J.C.Mikkelsen,Jr [Mater. Res. Soc. Symp. Proc. (USA) vol.59 (1986) p.19-30]

[13]   T.Iizuka et al [J. Electrochem. Soc. (USA) vol.132 no.7 (Jul 1985) p.1707-13]

[14]   T.Carlberg [J. Electrochem. Soc. (USA) vol.133 no.9 (Sept 1986) p.1940-2]

12.8 SOLUBILITY OF C IN Si

by K.G.Barraclough

August 1987
EMIS Datareview RN=15782

The solubility of carbon in single crystal silicon is an important
property affecting the formation of silicon carbide precipitates which
can seriously impair the electrical properties of p-n junction devices.
Solubility of carbon in liquid silicon and the solid/liquid segregation
coefficient determine the concentration of carbon in the grown crystal,
and these two properties are considered together in [10]. Most
electronic grade silicon contains carbon in the range 5 X 10*15 -
10*17 atoms/cm*3 which may be in excess of the solubility limit at
processing temperatures.

An infra-red (IR) absorption band from localised vibrations of
substitutional carbon atoms in silicon has been identified at
607 cm*-1 and calibrated against impurity concentration [1]. Although
there is some slight disagreement on the exact value of the
calibration coefficient, (at 300 deg K, concentration of carbon is
approximately 1 X 10*17 times the absorption coefficient), the IR
absorption technique provides a convenient measure for solubility
of carbon, if equilibrium conditions are used, and if the range of
study is limited to concentrations above 5 X 10*15 atoms/cm*3, a
typical detection limit. Both these conditions were satisfied in the
pioneering work of Bean and Newman [2] who determined the solubility of
carbon in single crystal silicon at temperatures between approximately
1100 and 1375 deg C. The starting material was high resistivity
(50 ohm cm), p-type, Czochralski-pulled crystals containing
1.4 X 10*18 atoms/cm*3 interstitial oxygen (using the calibration
coefficient of Pajot [3]) and either 10*17 atoms/cm*3 or
2 X 10*18 atoms/cm*3 substitutional carbon. By heat treating the samples
for one hour at temperatures between 600 and 1000 deg C in 50 deg C
intervals the supersaturated interstitial oxygen precipitated as SiO2
and thus provided nucleation sites for silicon carbide precipitates for
subsequent solubility studies at higher temperatures (> 1100 deg C).
Attainment of equilibrium was checked by IR absorption measurements at
various stages of up to 100 hours for the lower temperatures and 2 to 3
hours at the higher temperatures. The measurements were repeated in a
second series of experiments in which the temperature was reduced. The
consistency of both sets of results indicated that equilibrium was
achieved. In another series of experiments, previously untreated samples
were irradiated at room temperature with 2 MeV electrons at doses of
5 X 10*18 - 10*19 atoms/cm*2 to provide nucleation sites for
precipitation at high temperature. Again, the solubility data were
consistent with those from previous experiments, and fitted an
expression of the form:

$$[C] = constant \ exp \left[\frac{-DELTA \ H}{RT} \right]$$

where DELTA H, the heat of solution was 222 +- 25 kJ/Mole

If the data in [2] were extrapolated to the melting point, the
solubility was 4 - 5 X 10*17 atoms/cm*3. This value was significantly
lower than the concentration of substitutional carbon in one of the
CZ-Si starting samples (2 X 10*18 atoms/cm*3) but very close to the
solubility of carbon in oxygen-free floating zone silicon at the
melting point (3.5 X 10*17 atoms/cm*3) determined independently by
Nozaki et al [4]. There is now ample additional evidence that the
maximum solubility of carbon in oxygen-rich CZ-silicon lies close to
10*18 atoms/cm*3. Diffusion measurement [5] gave surface concentrations
as high as 1.5 X 10*18 atoms/cm*3 after treatments at 1402 deg C.
Measurements of several CZ-crystals indicated a consistent value of 9 X
10*17 atoms/cm*3 substitutional carbon content at which crystal perfection
was lost during growth [6]. In CZ-silicon crystals highly supersaturated
with oxygen it is possible to introduce carbon at concentrations in
excess of 2 X 10*18 atoms/cm*3 without introducing dislocations [7].
These results indicate that the solubility of carbon in silicon is
enhanced by the presence of interstitial oxygen. This might be expected
from lattice strain compensation. Carbon, a small atom, contracts the
silicon lattice [8] whilst interstitial oxygen, a much larger atom,
expands the lattice [9]. Note, that the data in [2] apply to high
resistivity silicon lightly doped with substitutional electrically
active elements. Significantly different results might be expected in
crystals heavily doped with large atoms such as antimony.

Returning now to the discrepancy between the extrapolated data in [2]
and the available data on the maximum solubility of carbon in as-grown
CZ-Si, it appears that this can be explained by the difference in the
state of oxygen in the samples. In the heat treated CZ-Si samples used
in [2] the interstitial oxygen was precipitated (at approximately
800 deg C) in the nucleation heat treatments leaving a low-oxygen
matrix characteristic of FZ-Si. Hence, the carbon solubility data
extrapolated to a value closer to that determined for FZ-Si at the
melting point [4]. It is worth remarking here that Bean and Newman [2]
were unable to obtain consistent carbon solubility data in oxygen-free
FZ-Si using the nucleation techniques applied to their CZ-Si samples.
This highlights the importance of silicon oxide precipitates in CZ-Si,
not only in providing nucleation sites for SiC precipitates but also
in providing silicon interstitials which can be trapped at
substitutional carbon atoms and greatly assist diffusion. The
enhancement of the diffusivity of carbon in silicon during growth of
silicon oxide is considered in [11].

REFERENCES
----

[1]   R.C.Newman, J.B.Willis [J. Phys. & Chem. Solids (GB) vol.26
        (1965) p.373-9]
[2]   A.R.Bean, R.C.Newman [J. Phys. & Chem. Solids (GB) vol.32
        (1971) p.1211]
[3]   B.Pajot [Solid-State Electron. (GB) vol.12 (1969) p.923]
[4]   T.Nozaki, Y.Yatsurugi, N.Akiyama [J. Electrochem. Soc. (USA)
        vol.117 (1970) p.1566]
[5]   R.C.Newman, J.Wakefield [J. Phys. & Chem. Solids (GB) vol.19
        (1961) p.230]
[6]   F.W.Voltmer, F.A.Padovani [Semiconductor Silicon 1973 Eds.
        H.R.Huff, R.B.Burgess (Electrochemical Soc., Princeton, NJ,
        1973) p.75]

[7]   K.B.Barraclough, J.G.Wilkes [Proc. 5th Int. Symp. Silicon
        Materials Science and Technology, Semiconductor Silicon 1986
        Eds. H.R.Huff, B.Kolbesen, T.Abe (Electrochemical Soc., Pennington
        USA, 1986) p.889]
[8]   J.A.Baker, T.N.Tucker, N.E.Moyer, R.C.Buschert [J. Appl. Phys.
        (USA) vol.39 (1968) p.4365]
[9]   Y.Takano, M.Makei [Semiconductor Silicon 1973 Eds. H.R.Huff,
        R.B.Burgess (Electrochem. Soc., Princeton, 1973) p.469]
[10]   K.G.Barraclough [EMIS Datareview RN=15783 (Aug 1987) 'The
        Segregation Coefficient of C in Si']
[11]   K.G.Barraclough [EMIS Datareview RN=15781 (Aug 87) 'Diffusion
        of C in Si']

12.9 SOLUBILITY OF N IN Si

by P.J.Ashby

August 1987
EMIS Datareview RN=15786

The solid solubility of nitrogen in silicon has been determined by
Yatsurugi et al [1]. A silicon rod was cut into hemi-cylinders and
Si3N4 sandwiched between them. A 10 mm molten zone was then passed
through the rod at various velocities. Nitrogen concentrations,
determined by charged particle activation analysis, were measured
as a function of zone velocity over the range 0.5 to 2.5 mm/min. The
equilibrium solid solubility of nitrogen at the melting point of
silicon was obtained by extrapolating to zero zone velocity and
found to be:-

(4.5 +- 1.0) X 10*15 atoms/cm*3

At the higher zone velocity of 2.5 mm/min the observed solubility
increased by a factor of two due to solute trapping. Thus nitrogen
melt doped silicon often has nitrogen concentrations in excess of
the equilibrium solubility limit.

The solubility of nitrogen is very much lower then other light
impurities such as carbon, 3.2 X 10*17 atoms/cm*3, and oxygen,
2.8 X 10*18 atoms/cm*3.

Nitrogen in silicon below the solid solubility limit has been shown to
form strongly bonded nitrogen pairs [2,3]. Details of the N-pair bond
to the lattice have not been determined but several possible schemes
have been proposed. Stein [3] suggested several possibilities for
coupling between Si-N-Si centres through a N-N bond with the nitrogen
pair sited at a vacancy or di-vacancy. Alternatively Pavlov et al [4]
suggested that the nitrogen could exist in molecular form at interstitial
sites.

This concept was used by Abe et al [2] whose spectroscopic data indicated
that the Si-N interactions were very weak, with the nitrogen pairs or
molecules behaving as if they were isolated from the lattice. The large
size of the impurity and its lack of attractive force to the lattice
leads to a low equilibrium solubility.

REFERENCES
----

[1] Y.Yatsurugi, N.Akiyama, Y.Endo, T.Nozaki [J. Electrochem. Soc.
     (USA) vol.120 (1973) p.975]
[2] T.Abe, H.Harada, N.Ozawa, K.Adomi [Mater. Res. Soc. Symp. Proc.
     (USA) vol.59 Eds. J.C.Mikkelsen, S.J.Pearton, J.W.Corbett,
     S.J.Pennycook (Materials Research Society, 1986) p.537-44]
[3] H.J.Stein [Mat. Res. Soc. Symp. Proc. vol.59 Eds. J.C.Mikkelsen,
     J.W.Corbett, S.J.Pearton, S.J.Pennycook (Materials Research Soc.,
     1986) p.523-35]
[4] P.V.Pavlov, E.I.Zorin, D.I.Tetelbaum, A.F.Khokhlov [Phys. Status
     Solid a (Germany) vol.35 (1976) p.11]

12.10 IR ABSORPTION DUE TO O IN Si

by L.T.Canham

September 1987
EMIS Datareview RN=16162

A. INTRODUCTION

The well-studied infrared absorption that arises from the presence
of oxygen in silicon has proved extremely useful in unravelling the
behaviour of this technologically important impurity. Both sharp
absorption lines and broad bands have been ascribed to oxygen in
various states of incorporation in as-grown silicon, and silicon
subjected to irradiation or prolonged heat treatments.

B. OXYGEN IN AS-GROWN Si

Electrical and X-ray measurements of O-rich Si demonstrate that when
dissolved in solid solution, O is electrically inactive and dilates
the lattice. A well-established explanation for these observations is
that isolated O occupies interstitial sites, bonding to two
neighbouring Si atoms and replacing their original covalent bond. The
resulting Si-O-Si structures can be envisaged as bent tri-atomic
molecules, embedded in the Si lattice. Indeed, their infrared
absorption resembles the rotation-vibration band of such a molecule.

Kaiser et al [1] first reported in 1956 that in Si at room temperature,
the mid-infrared absorption band at 9 microns is due to isolated O
atoms. Hrostowski and Kaiser [2] subsequently attributed an additional
room temperature band at 19 microns to isolated O. More recent studies
[3-6] with higher resolution spectrometry have not revealed any further
strong O-related absorption bands at room temperature, but have
identified a number of much weaker features [6]. A compilation of the
interstitial O-related absorption observable at room temperature is
given in TABLE 1. Naturally occurring O contains 99.76% O-16, 0.04%
O-17 and 0.20% O-18.

TABLE 1: mid-infrared room temperature absorption due to interstitial
        oxygen [6]

Energy (cm*-1)	Wavelength label of principal bands	Relative absorption	Origin
514	'19 micron band'	STRONG	symmetric vibration of O-16
1013		weak	overtone of 514 cm*-1 band of O-16
1059		weak	antisymmetric vibration of O-18
1108	'9 micron band'	STRONG	antisymmetric vibration of O-16
1227		weak	antisymmetric vibration and rotational mode of O-16
1720		weak	antisymmetric vibration of O-16 plus two-phonon lattice vibration

As the temperature is lowered the 9 micron band shifts appreciably to
higher energies and below about 120 deg K exhibits complex fine
structure. In contrast, the 19 micron band position is relatively
temperature independent and exhibits no fine structure. Near liquid
nitrogen temperature (77 deg K) a third relatively strong absorption
band due to interstitial oxygen, the 8 micron band appears [7], whose
position and shape are also fairly independent of temperature. At liquid
helium temperature (4.2 deg K) the three principal bands lie at
517, 1136 and 1203 cm*-1. As indicated in TABLE 2, the fine structure
within the 9 micron band has been ascribed to the combined effects of
discrete rotational levels and the mixed isotopic composition of the
Si-O-Si molecules. Naturally occurring Si contains 92.2% Si-28, 4.7%
Si-29 and 3.1% Si-30.

TABLE 2: mid-infrared liquid helium temperature absorption due to
interstitial oxygen [7-9]

Energy (cm*-1)	Wavelength label of principal bands	Relative absorption	Origin
517	'19 micron band'	STRONG	symmetric vibration
1084		weak	antisymmetric vibration of (Si-28)-(O-18)-(Si-28)
1109		weak	antisymmetric vibration of (Si-28)-(O-17)-(Si-28)
1128.40		weak	antisymmetric vibration of (Si-28)-(O-16)-(Si-28) from excited rotational level
1129.09		weak	antisymmetric vibration of (Si-29)-(O-16)-(Si-30)
1130.89		weak	antisymmetric vibration of (Si-29)-(O-16)-(Si-29)
1132.72		weak	antisymmetric vibration of (Si-28)-(O-16)-(Si-30)
1134.51		weak	antisymmetric vibration of (Si-28)-(O-16)-(Si-29)
1136.43	'9 micron band'	STRONG	antisymmetric vibration of (Si-28)-(O-16)-(Si-28)
1203	'8 micron band'	STRONG	rotational and vibrational combination

Bosomworth et al [10] extended measurements into the far-infrared to
detect rotational transitions of the Si-O-Si molecules. A list of the
relatively sharp transitions observed at low temperatures is given in
TABLE 3.

TABLE 3: far-infrared low temperature absorption due to interstitial
        oxygen [10]

Energy (cm*-1)	Temperature (deg K)	Origin
27.3	1.8	pseudo-rotational transition of O-18
29.3	1.8	pseudo-rotational transition of O-16
35.3	35	pseudo-rotational transition of O-18
37.8	35	pseudo-rotational transition of O-16
43.3	35	pseudo-rotational transition of O-16
49.0	35	pseudo-rotational transition of O-16

Newman [11] has provided a detailed review of the early work on infrared
absorption due to isolated O in Si. Both experimental [12] and
theoretical [13] studies are continuing to this day, some thirty years
since the pioneering work of the 1950's.

There has also been considerable recent effort directed towards
accurately calibrating the 9 micron band strength with interstitial
oxygen content [14-18]. Although numerous conversion factors have been
reported, hopefully in the near-future a universally-accepted value will
become established. Alternatively, the 19 micron absorption band could
prove a more reliable means of determining interstitial oxygen
concentrations by room temperature spectroscopy [19], since there is no
oxygen precipitate induced absorption in that spectral range (c.f.
section D).

So far only isolated oxygen atoms have been considered. A fraction of
interstitial oxygen in as-grown Si will have complexed with other
impurities present, the most common example being that of oxygen-carbon
pairing. Newman and coworkers [20-21] have assigned a number of
mid-infrared absorption bands observed in as-grown CZ Si to
substitutional C - interstitial O close pairs. Either irradiation or
extended heat treatments of as-grown Si will promote further complexing.

C. OXYGEN IN IRRADIATED Si

When O-rich Si is irradiated, the strength of the infrared absorption
due to isolated interstitial O decreases, as a result of interactions
with both the primary and secondary products of displacement damage.

The primary defects arising from Si displacement at room temperature
are mobile vacancies and self-interstitials. Interstitial oxygen traps
migrating vacancies and forms the well-studied O-V complex often called
the 'A-centre' [22-23]. Although the oxygen atom still forms 2 bridging
bonds with neighbouring Si atoms, it now occupies a near-substitutional
site and is electrically-active. This irradiation-induced complex gives
rise to absorption bands at 830 and 877 cm*-1, when in its neutral and
singly negative charge state respectively [24]. In addition,
oxygen-vacancy interactions can create higher order complexes (e.g.
O2-V2, O-V3) in heavily irradiated and partially annealed CZ Si, some
of which have been deemed responsible for specific absorption bands
[25-28]. In contrast, oxygen-Si interstitial interactions are not
well-documented. Brelot and coworkers [29-30] have attributed absorption
bands at 935, 944 and 956 cm*-1 to interstitial O - interstitial Si
pairs of unknown configuration.

Of the numerous secondary defects produced during irradiation, mobile
carbon interstitials result from silicon interstitials ejecting the
substitutional impurity into a bonded interstitialcy configuration, and
markedly influence the infrared activity of oxygen impurity. Migrating
C interstitials are trapped by both interstitial O and O-V complexes,
giving rise to a variety of optical centres in irradiated Si that
contain both impurities [31-42].

TABLE 4 lists the more established vibrational absorption bands due to
O-containing defects in irradiated Si; TABLE 5 some electronic
absorption lines.

TABLE 4: mid-infrared absorption due to O-related defects in irradiated
silicon

Defect	Absorption line energy (cm*-1)	Reference
O-V	830, 877	[22-24]
O2-V ?	889	[25-28]
O3-V ?	904, 968, 1000	[25]
OI-Si(I) ?	935, 944, 956	[29]
O(I)-C(I)	529, 550, 742, 865, 1116	[35,41]
O(I)-C(I)-Si(I) ?	936, 1020	[38,41]

TABLE 5: near-infrared absorption due to O-related defects in irradiated
         silicon

Defect	Absorption line energy (cm*-1)	(meV)	Reference
O(I)-C(I)	6367	789.4	[41,61]
O,C,V ?	3942	488.7	[42]
O,C ?	6189	767.3	[37,40]

D. OXYGEN IN HEAT-TREATED Si

Extended heat treatments of O-rich Si also lead to a marked reduction
in the strength of the infrared absorption associated with the impurity
in its isolated interstitial form. Both complexing with other impurities
and growth of other Si:O phases can be thermally induced, the resulting
infrared activity of O depending on, amongst other factors, the heat
treatment temperature.

Heat treatments below about 500 deg C can remove a significant fraction
of the interstitial oxygen originally present without forming large
precipitates [43]. The kinetics of interstitial O removal at these
temperatures suggest that clustering is limited to O2 dimer, or at the
most O3 trimer, formation [43-44]. Such complexes appear to be both
electrically and optically inactive. Although no absorption
characteristics of Si-O bonds in SiO2 precipitates appears, heating
around 450 deg C does induce a large number of relatively sharp
absorption lines within the 200-550 cm*-1 and 700-1200 cm*-1 spectral
regions [45-53], due to formation of the so-called 'thermal donors'.
It is not yet established to what extent O atoms are actually
incorporated in thermal donors [54], despite their creation relying on
high concentrations of the impurity.

Heat treatments above about 650 deg C promote substantial loss of
interstitial oxygen due to precipitation. The technological benefits of
internal gettering of metallic impurities by oxygen precipitates have
motivated much research in this area. Precipitated oxygen gives rise to
Rayleigh scattering in the 1-5 microns spectral region [55] and to
relatively broad mid-infrared absorption near 9 micron [56-60]. Although
specific polymorphs of SiO2 have been deemed responsible, both amorphous
silicon dioxide and its various crystalline forms all display
vibrational frequencies around 1100-1200 cm*-1 [59], rendering
identification of a particular phase from the optical absorption very
difficult [60].

Complexing of interstitial O with other impurities, notably carbon, can
also be thermally induced. Many weak absorption lines observed in
heated CZ Si have been ascribed to defects containing both oxygen and
carbon [46] but further work is needed before specific assignments can
be made.

REFERENCES
----

[1] W.Kaiser, P.H.Keck, C.F.Lange [Phys. Rev. (USA) vol.112 (1956)
     p.136]
[2] H.J.Hrostowski, R.H.Kaiser [Phys. Rev. (USA) vol.107 (1957)
     p.966]
[3] Z.J.Xu et al [Acta Phys. Sin. (China) vol.29 (1980) p.867]
[4] M.Stavola [Appl. Phys. Lett. (USA) vol.44 (1984) p.514]
[5] M.A.Ilin [Opt. & Spectrosc. (USA) vol.51 (1981) p.117]
[6] B.Pajot, H.J.Stein, B.Cales, C.Naud [J. Electrochem. Soc. (USA)
     vol.132 no.12 (Dec 1985) p.3034-7]
[7] H.J.Hrostowski, B.J.Alder [J. Chem. Phys. (USA) vol.33 (1960)
     p.980]
[8] B.Pajot [J. Phys. & Chem. Solids (GB) vol.28 (1967) p.73]
[9] B.Pajot, J.P.Deltour [Infrared Phys. (GB) vol.7 (1967) p.195]
[10] D.R.Bosomworth, W.Hayes, A.R.L.Spray, G.D.Watkins [Proc. Royal
     Soc. London Ser. A (GB) vol.317 (1970) p.133]
[11] R.C.Newman [in 'Infrared Studies of Crystal Defects' (Taylor &
     Francis, London, 1973) p.88]
[12] B.Pajot, B.Cales [Mater. Res. Soc. Symp. Proc. (USA) vol.59
     (1986) p.39-44]
[13] C.S.Chen, D.K.Schroder [Appl. Phys. A (Germany) vol.42 no.4 (Apr
     1987) p.257-62]
[14] The first calibration factor reported was that of W.Kaiser,
     P.H.Keck [J. Appl. Phys. (USA) vol.28 (1957) p.882]. A list of
     references of subsequent work up to 1984 is given in ref [16].
[15] T.Iizuka et al [J. Electrochem. Soc. (USA) vol.132 no.7 (Jul
     1985) p.1707-13]
[16] K.G.Barraclough, R.W.Series, J.S.Hislop, D.A.Wood [J.
     Electrochem. Soc. (USA) vol.133 no.1 (1986) p.187-91]
[17] P.K.Chu, R.S.Hockett, R.G.Wilson [Mater. Res. Symp. Proc. (USA)
     vol.59 (1986) p.67-72]
[18] W.M.Bullis [Proc. 5th Int. Symp. Si Mater. Sci. & Tech. (1986)
     p.166]
[19] P.E.Freeland [J. Electrochem. Soc. (USA) vol.127 (1980) p.754]
[20] R.C.Newman, J.B.Willis [J. Phys. & Chem. Solids (GB) vol. 26
     (1965) p.373]
[21] R.C.Newman, R.S.Smith [J. Phys. & Chem. Solids (GB) vol.30
     (1969) p.1493]
[22] G.D.Watkins, J.W.Corbett [Phys. Rev. (USA) vol.121 (1961) p.1001]
[23] J.W.Corbett, G.D.Watkins, R.M.Chrenko, R.S.McDonald [Phys. Rev.
     (USA) vol.121 (1961) p.1015]
[24] R.C.Newman [in 'Infrared Studies of Crystal Defects' (Taylor &
     Francis, London, 1973) p.118]
[25] J.W.Corbett, G.D.Watkins, R.S.McDonald [Phys. Rev. (USA) vol.135
     (1964) p.A1381]
[26] J.L.Lindstrom, G.S.Oehrlein, J.W.Corbett [Phys. Status Solidi a
     (Germany) vol.95 no.1 (1986) p.179-84]
[27] H.J.Stein [Appl. Phys. Lett. (USA) vol.48 no.22 (2 Jun 1986)
     p.1540-1]
[28] J.L.Lindstrom, B.G.Svensson [Mater. Res. Soc. Symp. Proc. (USA)
     vol.59 (1986) p.45-58]
[29] A.Brelot, J.Charlemagne [in 'Radiation effects in semiconductors'
     Eds J.W.Corbett, G.D.Watkins (Gordon & Breach, New York, USA,
     1971) p.161]
[30] A.Brelot [in 'Rad. Damage & Defects in Semiconductors' (Inst.
     Phys. London, 1973) p.191]

[31]   A.K.Ramdas, M.G.Rao [Phys. Rev. (USA) vol.142 (1966) p.451]
[32]   F.L.Vook, H.J.Stein [Appl. Phys. Lett. (USA) vol.13 (1968)
        p.343]
[33]   H.J.Stein, F.L.Vook [Radiat. Eff. (GB) vol.1 (1969) p.41]
[34]   A.R.Bean, R.C.Newman, R.S.Smith [J. Phys. & Chem. Solids (GB)
        vol.31 (1970) p.739]
[35]   R.C.Newman, A.R.Bean [Radiat. Eff. (GB) vol.8 (1971) p.189]
[36]   F.A.Abou-el-Fotouh, R.C.Newman [Solid State Commun. (USA) vol.15
        (1974) p.1409]
[37]   G.Davies, E.C.Lightowlers, R.Woolley, R.C.Newman, A.S.Oates [J.
        Phys. C (GB) vol.17 (1984) p.L499]
[38]   A.S.Oates, M.J.Binns, R.C.Newman, J.H.Tucker, J.G.Wilkes,
        A.Wilkinson [J. Phys. C (GB) vol.17 (1984) p.5695]
[39]   N.Magnea, A.Lazrak, J.L.Pautrat [Appl. Phys. Lett. (USA) vol.45
        (1984) p.60]
[40]   B.Pajot, J.von Bardeleben [Proc. 13th ICDS, Coronado, USA
        (Metall. Soc. AIME, 1985) p.685]
[41]   G.Davies et al [J. Phys. C (GB) vol.19 (1986) p.841]
[42]   G.Davies, E.C.Lightowlers, M.Stavola, K.Bergman, B.Svensson
        [Phys. Rev. B (USA) vol.35 no.6 (15 Feb 1987) p.2755-66]
[43]   R.C.Newman, A.S.Oates, F.M.Livingston [J. Phys. C (GB) vol.16
        (1983) p.L667]
[44]   T.Y.Tan, R.Kleinhenz, C.P.Schneider [Mater. Res. Soc. Symp.
        Proc. (USA) vol.59 (1986) p.195-204]
[45]   H.J.Hrostowski, R.H.Kaiser [Phys. Rev. Lett. (USA) vol.1 (1958)
        p.199]
[46]   A.R.Bean, R.C.Newman [J. Phys. & Chem. Solids (GB) vol.33 (1972)
        p.255]
[47]   D.Helmreich, E.Sirtl [Semiconductor Silicon 1977, Eds H.R.Huff,
        E.Sirtl (Electrochem. Soc., 1977) p.626]
[48]   P.Gaworzewski, K.Shmalz [Phys. Status Solidi a (Germany) vol.55
        (1979) p.699]
[49]   R.Oeder, P.Wagner [in 'Defects in Semiconductors II', Eds
        S.Mahajan, J.W.Corbett (North Holland, 1983) p.14]
[50]   B.Pajot, H.Compain, J.Lerouille, B.Clerjand [Physica B&C
        (Netherlands) vol.117 (1983) p.110]
[51]   M.Suezawa, K.Sumino [Phys. Status Solidi a (Germany) vol.83
        (1984) p.235]
[52]   M.Stavola, K.M.Lee [Mater. Res. Soc. Symp. Proc. (USA) vol.59
        (1986) p.95-110]
[53]   P.Wagner [Mater. Res. Soc. Symp. Proc. (USA) vol.59 (1986)
        p.125-38]
[54]   R.C.Newman [J. Phys. C (GB) vol.18 no.30 (30 Oct 1985)
        p.L967-72]
[55]   W.Kaiser [Phys. Rev. (USA) vol.105 (1957) p.1751]
[56]   K.Tempelhoff, F.Spielberg, R.Gleichman [Semiconductor Silicon
        1977, Eds H.R.Huff, E.Sirtl (Electrochem. Soc., 1977) p.585]
[57]   S.M.Hu [J. Appl. Phys. (USA) vol.51 (1980) p.5945]
[58]   F.Shimura, Y.Ohnishi, H.Tsuya [Appl. Phys. Lett. (USA) vol.38
        (1981) p.867]
[59]   S.L.Olimpiev, V.A.Sutyagin [Sov. Phys. Semicond. (USA) vol.16
        (1983) p.1431]
[60]   R.C.Newman [Rep. Prog. Phys. (GB) vol.45 (1982) p.1176]
[61]   J.Trombetta, G.D.Watkins [Submitted to Appl. Phys. Lett. (USA),
        May 1987]

12.11 IR ABSORPTION DUE TO C IN Si

by L.T.Canham

September 1987
EMIS Datareview RN=16163

A. INTRODUCTION

The coexistence of oxygen and carbon in as-grown Si combined with their
reciprocal influence on each other's behaviour, has motivated research
into the infrared activity of the latter impurity. As for oxygen, both
sharp absorption lines and broad bands have been ascribed to carbon in
different states of incorporation in as-grown Si, and material subjected
to irradiation or prolonged heat treatments.

B. CARBON IN AS-GROWN Si

Electrical and X-ray measurements of C-rich Si demonstrate that like O,
dissolved C is electrically inactive, but in contrast to O it contracts
the lattice. Consistent with these observations it has been established
that isolated C, a small isovalent impurity, occupies substitutional
sites in as-grown material. The infrared absorption due to isolated
substitutional carbon is relatively simple, its smaller mass than Si
giving rise to a single localized vibrational mode (LVM).

The first assignment of infrared absorption to carbon impurity was that
of Balkanski et al [1] but their polycrystalline material exhibited
absorption due to SiC precipitates rather than uniformly dissolved C
[2]. It was not until 1965 that Newman and Willis [3] showed that the
mid-infrared absorption band at 16.5 microns is due to isolated
substitutional carbon. The C-related band is superimposed on the
strongest intrinsic two-phonon band of Si [4] which was responsible for
earlier studies [5] failing to correlate the sample-dependent
absorption band with carbon content. A compilation of the infrared
absorption related to isolated substitutional C is given in TABLE 1.
Naturally occurring C contains 98.9% C-12 and 1.1% C-13.

TABLE 1: mid-infrared absorption due to isolated substitutional carbon [6]

Energy (cm*-1)		Wavelength label	Relative	Origin
290 deg K	77 deg K	of principal bands	absorption	
586.3	589.1		weak	C-13 LVM
604.9	607.5	'16.5 micron band'	STRONG	C-12 LVM
1169.0	1174.5		weak	2nd harmonic of C-13 LVM
1205.8	1211.1		weak	2nd harmonic of C-12 LVM

The LVM frequencies for the various carbon isotopes (C-14 doping creates a 570.3 cm*-1 LVM at 290 deg K [6]) are all in fair agreement with the early predictions of models based on mass differences alone [7-8], and in excellent agreement with theoretical approaches that in addition account for accompanying lattice distortion [9-11].

Since the pioneering studies of Newman and coworkers [3,6] much research has been directed towards employing the strength of the 16.5 microns absorption band as a monitor of carbon content of Si wafers [12-22]. It is important to note that even in as-grown Si, a small fraction of the carbon may have other states of incorporation, both infrared active and inactive. Carbon-oxygen complexing in as-grown CZ Si gives rise to a number of mid-infrared absorption lines [6]. Graphitic microinclusions could be responsible for one form of infrared-inactive carbon.

C. CARBON IN IRRADIATED Si

When C-rich Si is irradiated, the strength of the 16.5 micron band of isolated substitutional carbon decreases as a result of interaction with Si interstitials; there is no evidence to date that substitutional carbon can capture the mobile vacancies that also result from displacement damage.

Trapping of a Si interstitial by substitutional carbon results in the impurity being ejected into a bonded interstitialcy configuration. Under low temperature irradiations the interstitial carbon atom is relatively immobile and gives rise to LVM absorption lines at 921 and 930 cm*-1 [23-27]. Its higher vibrational frequencies than that of substitutional carbon reflect the much reduced volume available for movement.

Isolated interstitial carbon atoms are only thermally stable up to about 300 deg K. During room temperature irradiations their migration results in the creation of a variety of optical centres containing carbon [28-39]. TABLES 2 and 3 list some of the vibrational and electronic absorption of irradiated Si that has been assigned to C-containing defects.

TABLE 2: mid-infrared absorption due to C-related defects in irradiated silicon

Defect	Absorption line energy cm*-1	Reference
C(I)	921, 930	[27]
C(I)-Si(I)	959, 966	[30,52]
C(I)-O(I)	529, 550, 742, 865, 1116	[27,37]
C(I)-O(I)-Si(I)	936, 1020	[35]

TABLE 3: near-infrared absorption due to C-related defects in irradiated
          silicon

Defect	Absorption line energy (cm*-1)	(meV)	Reference
C(I)	6904	856	[38]
C(I)-O(I)	6367	789.4	[37]
C(s)-Si(I)-C(s)	7819	969.5	[32,51]
C,O ?	6189	767.3	[36]
C,O,V ?	3942	488.7	[39]

The states of incorporation of carbon in very heavily irradiated Si are
not yet established. The infrared absorption due to substitutional
carbon disappears completely without any SiC precipitate-related
absorption (c.f. section D) appearing [40]. A similar effect
is observed in Si ion-implanted with C, prior to annealing [41-43].
Newman [44] has suggested that the irradiation-induced high
concentrations of carbon interstitials promote clustering into
graphitic inclusions that do not produce infrared absorption.

D. CARBON IN HEAT-TREATED Si

Prolonged heat treatments of C-rich Si also result in a marked decrease
in the strength of the infrared absorption associated with the impurity
in its isolated substitutional form. Both complexing with other
impurities and SiC precipitation can be thermally induced, and in this
respect, heat treatments may be conveniently divided into two
categories, depending on which process dominates.

At sufficiently high temperature (1000-1250 deg C) substitutional
carbon can migrate to form internal precipitates of cubic beta-SiC.
These give rise to broad band absorption around 12 microns, the width
and peak position depending on the size and shape of the internal
particles, and thus the details of the specific heat treatment [45].
The small size and incorporation of the SiC particles within the Si
lattice normally result in absorption bands shifted in energy from the
12.6 microns reststrahlen band of the compound [45-46]. Even at such
high temperatures the interactive behaviour of carbon and oxygen is
evident; carbon precipitation in CZ Si being much more pronounced than
that in FZ Si [47].

At lower temperatures where substitutional carbon is relatively
immobile, complexing with other impurities, notably oxygen, dominates
[48-50]. Bean and Newman [48] have assigned a large number of weak
absorption lines observed in heated CZ Si to defects containing both
impurities.

REFERENCES

[1]   M.Balkanski, W.Nazarewicz, E.Silva [C.R. Acad. Sci. Paris
        (France) vol.251 (1960) p.1277]
[2]   W.G.Spitzer, D.A.Kleinman, C.J.Frosch [Phys. Rev. (USA vol.113
        (1959) p.133]
[3]   R.C.Newman, J.B.Willis [J. Phys. & Chem. Solids (GB) vol.26
        (1965) p.373]
[4]   F.A.Johnson [Proc. Phys. Soc. London (GB) vol.73 (1959) p.265]
[5]   M.Lax, E.Burstein [Phys. Rev. (USA) vol.97 (1955) p.39]
[6]   R.C.Newman, R.S.Smith [J. Phys. & Chem. Solids (GB) vol.30 (1969)
        p.1493]
[7]   P.G.Dawber, R.J.Elliott [Proc. Royal Soc. London Ser. A (GB)
        vol.273 (1963) p.222]
[8]   P.G.Dawber, R.J.Elliott [Proc. Phys. Soc. London (GB) vol.81
        (1963) p.453]
[9]   R.C.Newman [in 'Infrared Studies of Crystalline Defects' (Taylor
        & Francis, 1973)]
[10]   C.Changsheng, Z.Zhehua, Y.Yiying [Chin. J. Semicond. (China)
        vol.4 (1983) p.20]
[11]   C.S.Chen, D.K.Schroder [Phys. Rev. (USA) vol.35 no.2 (15 Jan
        1987) p.713-17]
[12]   J.A.Baker, T.N.Tucker, N.E.Moyer, R.C.Buschert [J. Appl. Phys.
        (USA) vol.39 (1968) p.4365]
[13]   Y.Endo, Y.Yatsurugi, N.Akayama [Anal. Chem. (USA) vol.44 (1972)
        p.2258]
[14]   C.Gross, G.Gaetano, T.N.Tucker, J.A.Baker [J. Electrochem. Soc.
        (USA) vol.119 (1972) p.926]
[15]   B.O.Kolbesen, T.Kladenovic [Krist. & Tech. (Germany) vol.15
        (1980) p.K1-3]
[16]   D.W.Vidrine [Anal. Chem. (USA) vol.52 (1980) p.92]
[17]   D.G.Mead, S.R.Lowry [Appl. Spectrosc. (USA) vol.34 (1980) p.167]
[18]   D.G.Mead [Appl. Spectrosc. (USA) vol.34 (1980) p.171]
[19]   M.Briska, A.Schmitt, H.Wagner [IBM Tech. Disclosure Bull. (USA)
        vol.23 (1980) p.225]
[20]   Y.V.Dankovskii, M.A.Ilin, V.Y.Kovarskii [Meas. Tech. (USA) vol.25
        (1982) p.443]
[21]   P.Stallhofer, D.Huber [Solid State Technol. (USA) vol.26 (1983)
        p.233]
[22]   J.L.Regolini, J.P.Stoquert, C.Ganter, P.Siffert [J. Electrochem.
        Soc. (USA) vol.133 (1986) p.2165]
[23]   R.E.Whan [Appl. Phys. Lett. (USA) vol.8 (1966) p.131]
[24]   R.E.Whan [J. Appl. Phys. (USA) vol.37 (1966) p.2435]
[25]   F.L.Vook, H.J.Stein [Appl. Phys. Lett. (USA) vol.13 (1968)
        p.343]
[26]   A.R.Bean, R.C.Newman [Solid State Commun. (USA) vol.8 (1970)
        p.189]
[27]   R.C.Newman, A.R.Bean [Radiat. Eff. (GB) vol.8 (1971) p.189]
[28]   A.K.Ramdas, M.G.Rao [Phys. Rev. (USA) vol.142 (1966) p.451]
[29]   H.J.Stein, F.L.Vook [Radiat. Eff. (GB) vol.1 (1969) p.41]
[30]   A.R.Bean, R.C.Newman, R.S.Smith [J. Phys. & Chem. Solids (GB)
        vol.31 (1970) p.739]
[31]   F.A.Abou-el-Foutah, R.C.Newman [Solid State Commun. (USA) vol.15
        (1974) p.1409]
[32]   G.Davies, E.C.Lightowlers, M.C.do Carmo [J. Phys. C (GB) vol.16
        (1983) p.5503]

[33]   G.Davies, E.C.Lightowlers, M.C.do Carmo, J.G.Wilkes,
         G.R.Wolstenholme [Solid State Commun. (USA) vol.50 no.12 (Jun
         1984) p.1057-61]
[34]   G.Davies, E.C.Lightowlers, R.Woolley, R.C.Newman, A.S.Oates [J.
         Phys. C (GB) vol.17 (1984) p.L499]
[35]   A.S.Oates, M.J.Binns, R.C.Newman, J.H.Tucker, J.G.Wilkes,
         A.Wilkinson [J. Phys. C (GB) vol.17 (1984) p.569]
[36]   B.Pajot, J. von Bardeleben [Proc. 13th ICDS, Coronado, USA
         (Metall. Soc., AIME, 1985) p.685]
[37]   G.Davies [J. Phys. C (GB) vol.19 no.6 (28 Feb 1986) p.841-56]
[38]   R.Woolley, E.C.Lightowlers, A.K.Tipping, M.Claybourn, R.C.Newman
         [Mater. Sci. Forum (Switzerland) vol.10-12 1986) p.929]
[39]   G.Davies, E.C.Lightowlers, M.Stavola, K.Bergman, B.Svensson
         [Phys. Rev. B (USA) vol.35 no.6 (15 Feb 1987) p.2755-65]
[40]   M.R.Brozel, R.C.Newman, D.H.J.Totterdell [J. Phys. C (GB) vol.7
         (1974) p.243]
[41]   J.A.Borders, S.T.Picraux, W.Beezhold [Appl. Phys. Lett. (USA)
         vol.18 (1971) p.509]
[42]   H.J.Stein [Proc. 2nd Int. Conf. Ion Implantation in
         Semiconductors (Springer Verlag 1971) p.2]
[43]   A.G.Cullis, R.Series, H.C.Webber, N.G.Chew [Semiconductor
         Silicon 1981 p.518]
[44]   R.C.Newman [in 'Neutron Transmutation Doped Si' Ed. J.Guldberg
         (Plenum, 1981) p.83]
[45]   A.R.Bean, R.C.Newman [J. Phys. & Chem. Solids (GB) vol.32 (1971)
         p.1211]
[46]   N.Akiyama, Y.Yatsurugi, Y.Endo, Z.Imayoshi, T.Nozaki [Appl. Phys.
         Lett. (USA) vol.22 (1973) p.630]
[47]   R.C.Newman [Mater. Res. Soc. Symp. Proc. (USA) vol.59 (1986)
         p.403-18]
[48]   A.R.Bean, R.C.Newman [J. Phys. & Chem. Solids (GB) vol.33 (1972)
         p.255]
[49]   G.S.Oehrlein, J.L.Lindstrom, J.W.Corbett [Appl. Phys. Lett. (USA)
         vol.40 (1982) p.241]
[50]   R.C.Newman [Rep. Prog. Phys. (GB) vol.45 (1982) p.1163]
[51]   K.P.O'Donnell, K.M.Lee, G.D.Waktins [Physica B&C (Netherlands)
         vol.116 (1983) p.258]
[52]   S.P.Chappell, R.C.Newman [submitted to Semicond. Sci. & Technol.
         (GB), 1987]

12.12 IR ABSORPTION DUE TO N IN Si

by P.J.Ashby

July 1987
EMIS Datareview RN=15789

A. MAJOR INFRA-RED ABSORPTIONS DUE TO NITROGEN IN SILICON

Infra-red absorption by nitrogen in silicon was reported by Abe et al
studying float zone crystals doped by adding nitrogen gas to the argon
crystal growth atmosphere or adding Si3N4 to the polycrystalline
silicon [1]. Absorption peaks of similar intensities were observed at
963 cm*-1 in samples having nitrogen concentrations near the solubility
limit [2] of 4.5 X 10*15 atoms/cm*3.

Itoh et al have derived a calibration relating the absorption
coefficient at 963 cm*-1 to nitrogen concentrations as determined by
charged particle activation analysis [3]. It is expressed as:-

  [nitrogen concentration (atoms/cm*3)]   = (1.83 +- 0.24) X 10*17
                                              (absorption coefficient)

This relationship is only valid for melt doped oxygen free silicon in
the as-grown condition.

The 963 cm*-1 and 766 cm*-1 band peak heights are reduced in float-zone
silicon samples when heat treated in the temperature range 800-1150
deg C. A maximum reduction rate is achieved at 900 deg C [4]. The
kinetics of the reduction are consistent with a nucleation and
precipitation process. The transformed nitrogen centre has no reported
IR absorptions associated with it.

The 963 cm*-1 and 766 cm*-1 bands occur on (N-14)+ ion implanted
float zone silicon. Laser annealing increases the absorption per
implanted ion [5]. For (N-15)+ ion-implanted silicon the band frequencies
shift to 937 cm*-1 and 748 cm*-1. The magnitude of this frequency shift
is near to that calculated for localised mode vibrations of diatomic and
triatomic molecules with nitrogen bonded to silicon [6]. Abe et al [7]
studying melt doped silicon conclude that this Si-N bond is weak since
the low temperature spectrum shows no sub-peak on the 963 cm*-1 band due
to naturally occurring silicon isotopes. This is in contrast to the Si-O
9 micron band which has sub-peaks due to (Si-28)-O-(Si-28), (Si-28)-O-
(Si-29), (Si-28)-O-(Si-30) [8].

Silicon containing both N-14 and N-15 has been produced by ion
implantation and melt doping [6,7]. In addition to the absorption bands
described above for each isotope an additional peak at an intermediate
frequency is found for both the upper and lower bands. This is
indicative of strong pairing between nitrogen atoms or the existence
of molecular nitrogen isolated from the lattice. The anomalous low
electrical activity [9], low segregation coefficient and the strong
interaction with dislocations [7] of nitrogen have been modelled by
postulating the existence of molecular nitrogen in the lattice.

B. ABSORPTION DUE TO SUBSTITUTIONAL NITROGEN

Nitrogen ion implanted samples that are rapidly crystallised, (as occurs in pulsed laser annealing) have shown evidence of a substitutional nitrogen defect. Only 10% or less of the nitrogen atoms are incorporated in this way. Brower [10] in an electron paramagnetic resonance study labelled this site SL5 and deduced it to have an axial symmetry about a <111> direction as a result of distortion of substitutional sites. SL5 centres transform to unknown structures, SL6 and SL7, on annealing in the temperature range 300-500 deg C. No evidence of these centres has been reported in melt doped material.

Stein [11] has observed an IR absorption band at 653 cm*-1 which was correlated with the SL5 site by studying the formation and annealing kinetics. Annealing at 350 deg C reduced the intensity of the 653 cm*-1 band and led to an increase in intensity of a band at 687 cm*-1, this may be related to the transformation to the SL6 and SL7 electron paramagnetic resonance centres. An isotopic shift in the 653 cm*-1 band indicates that the IR absorption band is due to localised vibrational modes of the substitutional N centre.

C. NITROGEN/OXYGEN RELATED ABSORPTION

Recent work on silicon doped with both oxygen and nitrogen has revealed an interaction between these impurities, [12,13]. CZ grown silicon doped with nitrogen showed that a smaller fraction of the nitrogen was incorporated into the 963 cm*-1 active centre than in oxygen free silicon [4]. On samples implanted with nitrogen and oxygen and annealed at 500 deg C new infra-red absorption peaks were observed just above the 766 cm*-1 and 964 cm*-1 bands with a consequent reduction in the independent nitrogen and oxygen peaks [6]. The interactions that cause absorptions are still a matter of speculation.

REFERENCES
----

[1]   T.Abe, K.Kikuchi, S.Shirai, S.Naraoka [Proc. 4th Int. Symp. Silicon Mater. Sci. & Technol. Minneapolis, MI, USA, 10-15 May, 1981, Eds. H.R.Huff, R.J.Kriegler, Y.Takeishi vol.81-5 (Electrochem. Soc., 1981)]

[2]   Y.Yatsurugi, N.Akigama, Y.Endo, T.Nozaki [J. Electrochem. Soc. (USA) vol.120 (1973) p.985]

[3]   Y.Itoh, T.Nozaki, T.Masui, T.Abe [Appl. Phys. Lett. (USA) vol.47 no.5 (1 Sep 1985) p.488-9]

[4]   T.Abe, T.Masui, H.Haradu, J.Chikawa [Proc. 3rd Int. Symp. VLSI Sci. & Technol., Toronto, Canada 13-16 May, 1985, (Electrochem. Soc., Pennington, NJ, USA, 1985) vol.85-7]

[5]   H.J.Stein [Appl. Phys. Lett. (USA) vol.43 no.3 (1 Aug 1983)]

[6]   H.J.Stein [Mater. Res. Soc. Symp. Proc. (USA) vol.59 (1986) p.15]

[7]   T.Abe, H.Harada, N.Ozawa, K.Adomi [Mater. Res. Soc. Symp. Proc. (USA) vol.59 (1986) p.537-44]

[8]   D.R.Bosomworth, W.Hayes, A.R.L.Spray, G.D.Watkins [Proc. R. Soc. London Ser. A (GB) vol.317 (1973) p.975]

[9]   P.V.Pavlov, E.I.Zorin, D.I.Tetelbaum, A.F.Khokhlov [Phys. Status Solidi a (Germany) vol.35 (1976) p.11]

[10]   K.L.Brower [Phys. Rev. B (USA) vol.26 (1982) p.6040]

[11]   H.J.Stein [Appl. Phys. Lett. (USA) vol.47 no.12 (15 Dec 1985)
       p.1339-41]

[12]   M.Suezawa, K.Sumino, H.Harada, T.Abe [Jpn. J. Appl. Phys. Part 2
       (Japan) vol.25 no.10 (Oct 1986) p.L859-61]

[13]   H.D.Chiou, J.Moody, R.Sandfoot, F.Shimara [Proc. 2nd Int. Symp.
       VLSI Sci. & Technol., Cincinnati, OH, USA, 6-11 May 1984, Eds.
       K.E.Bean, G.A.Rozganyi (Electrochem. Soc., Pennington, NJ, USA,
       1984)]

CHAPTER 13

DIFFUSION, SOLID SOLUBILITY AND IMPLANTATION OF
GROUP III AND GROUP V IMPURITIES

13.1 DIFFUSION OF B IN Si

by A.A.Brown, P.B.Moynagh and P.J.Rosser

November 1987
EMIS Datareview RN=17809

A. INTRINSIC BORON DIFFUSION, N2 AMBIENT

The mathematical basis for the diffusion process was laid by Fick whose
first law sets forward the hypothesis that the rate of transfer of a
diffusing particle through unit area is proportional to the magnitude
of the gradient normal to that area, i.e. in the x-direction

$$J = -D \frac{dc}{dx}$$

where the proportionality constant, D (metres*2/second), is known as
the diffusion constant or the diffusivity. The diffusion process occurs
because of the random walk of the impurity atoms and is therefore a
thermally activated process. The temperature dependence of the diffusion
coefficient is Arrhenius in nature:

$$D = Do \exp \frac{-E}{kT}$$

where E is the activation energy (which, for the common dopants in
silicon, is about four electron-volts).

Due to the low concentrations of stable point defects in silicon, it has
not been possible to determine directly whether vacancies or
self-interstitials are dominant in diffusion processes. Numerous kinds
of indirect observations have been made to build cases for vacancies and
for self-interstitials. The current belief is that both types of point
defect exist somewhat independently [19], and that both influence
diffusion. Under diffusive conditions where the silicon is intrinsically
doped and where no surface reaction is taking place, the vacancy and
self-interstitial concentrations are temperature dependent only and are
described approximately by the equilibrium concentrations:

$$Cv[eq] = Cov \exp \frac{-Ev}{kT}$$

$$Ci[eq] = Coi \exp \frac{-Ei}{kT}$$

Dopant diffusion takes place in the silicon by interacting with these
point defects so that the diffusing species are the boron/vacancy pair,
BV, and the boron/Si-interstitial pair, BI. The intrinsic diffusivity

due to the vacancy mechanism is referred to as Dv[eq], and that due to the interstitialcy mechanism as Di[eq]. The intrinsic, inert ambient diffusivity is therefore given by the sum of the equilibrium diffusivities due to the vacancy and the self-interstitial components:

$$D[eq] = Dv[eq] + Di[eq]$$

The wide energy available to the Fermi level in silicon leads to a given lattice defect (i.e. a vacancy or Si-interstitial) appearing in a variety of ionized states [20]. As such dopant/point-defect pairs exist as, for example, B(-)V(+) which will be denoted by BV(+). The intrinsic, inert ambient diffusivity can therefore be written:

$$D[eq] = Dv(+)[eq] + Dv(0)[eq] + Dv(-)[eq] + Dv(--)[eq]$$

$$+ Di(+)[eq] + Di(0)[eq] + Di(-)[eq] + Di(--)[eq]$$

where the intrinsic diffusivity of the boron diffusing as Bv(+) is given by Dv(+)[eq], etc.

The diffusion coefficient in crystals with a cubic lattice should be isotropic. For boron doping levels of less than about 10*20/cm*3 with no surface reaction taking place this is indeed the case [14]. The diffusion coefficient is therefore given by:

$$D = Dn \, exp \frac{-En}{kT}$$

Literature diffusivities for inert diffusion of boron in single crystal silicon are listed below.

(100),(110),(111) Si

temperature (deg C)	Dn(cm*2/s)	En(eV)	reference	
840-1250	24	3.87	Hill	[14]
1100-1250	2.46	3.59	Ghoshtagore	[3]
850-950	11.5	3.77	Schnabel	[2]
900-1200	2.64	3.6	Mathiot	[10]
1100-1250	5.1	3.7	Okamura	[15]

Mathiot et al [10] calculated the intrinsic diffusion coefficient given above from the data in refs [8,14,15 and 21] using a least-squares fitting technique.

B. INTRINSIC BORON DIFFUSION, O2 AMBIENT

Oxidation of silicon under most processing conditions enhances the diffusion of boron by a process of interstitial injection into the silicon substrate [19]. The enhanced diffusion occurs because the number of silicon self-interstitials available to interact with the boron atoms has increased and the diffusivity is described by a general diffusion equation:

$$D = Dv + Di$$

i.e.

$$D = Dv[eq] \frac{Cv}{Cv[eq]} + Di[eq] \frac{Ci}{Ci[eq]}$$

or

$$D = D[eq] \left((1-fi) \frac{Cv}{Cv[eq]} + fi \frac{Ci}{Ci[eq]}\right)$$

where

$$fi = \frac{Di[eq]}{D[eq]} = \frac{Di[eq]}{Dv[eq] + Di[eq]}$$

and fi is known as the fractional interstitialcy component of diffusion. Some injected Si-interstitials will be lost by Frenkel recombination with vacancies so that the thermal oxidation process results in a simultaneous supersaturation of Si-interstitials and undersaturation of vacancies. The effect on total diffusion rate of the increase in BI diffusion species and decrease in BV diffusion species is described by the fractional interstitialcy component of diffusion, fi, literature values of which are given below.

Experimental evidence suggests that the Si-interstitial injection rate during thermal oxidation depends on the oxidation rate. More specifically, it has been found that:

$$Ci-Ci[eq] = k \left(\frac{dXox}{dt}\right)*P \text{ with } P < 1$$

where (dXox/dt) is the oxidation rate.

Since the oxidation rate decreases as oxidation time increases, it is expected that the diffusion enhancement will also decrease with oxidation time. The oxidation rate (and therefore the rate of Si-interstitial injection) is not dependent on the orientation beyond the linear segment. However, the rate of decay of these excess point defects will be dependent on the silicon surface or the silicon-oxide interface recombination rates, and these might be expected to be proportional to the rectilinear density. As such the rate of loss of excess point defects would increase as (100), (110), (111), and the magnitude of oxidation enhanced diffusion would increase as (111), (110), (100) [14].

As is described below, oxidation enhanced diffusion in the intrinsic regime is a function of temperature, oxidation rate, depth into the silicon and substrate orientation.

Literature models of oxidation enhanced boron diffusion are described below.

Taniguchi et al [7] studied the diffusion of boron in (100) Si under dry
O2, wet Ar and wet O2 ambients. In the intrinsic regime, the diffusion
coefficient could, in each case, be expressed as:

$$Do = D(inert) + (1.67 \times 10^{-5})\left[\left(\frac{dXox}{dt}\right)^{0.3}\right]\left(\exp\frac{-X}{25}\right)\left(\exp\frac{-2.07}{kT}\right) \ cm^2/s$$

where:

X = depth into the silicon in microns
(dXox/dt) = oxide growth rate

Hill [14], using marker layers, has conducted an extensive study of
intrinsic diffusion and its dependence on dopant type, crystallographic
orientation, surface reaction, temperature and time. He found under
oxidizing conditions that the diffusion coefficient, Do, can be
expressed in the form

$$Do = D(inert) + dD(oxidizing)$$

$$= Dn \exp\frac{-En}{kT} + dDo \exp\frac{-dEo}{kT}$$

For boron

(100)Si

$$Do = 24 \exp\frac{-3.87 \ eV}{kT} + (4.1 \times 10^{-5}) \exp\frac{-2.34 \ eV}{kT} \ cm^2/s$$

(110)Si

$$Do = 24 \exp\frac{-3.87 \ eV}{kT} + (1.5 \times 10^{-5}) \exp\frac{-2.30 \ eV}{kT} \ cm^2/s$$

(111)Si

$$Do = 24 \exp\frac{-3.87 \ eV}{kT} + (3.9 \times 10^{-6}) \exp\frac{-2.22 \ eV}{kT} \ cm^2/s$$

He also showed that the diffusion coefficient under inert conditions is
constant, while under oxidizing conditions the diffusion coefficient
increases with the unsatisfied surface bond density of the adjacent
oxidizing surface independently of diffusion direction, i.e. under inert
and oxidizing conditions diffusion is isotropic, and under oxidizing
conditions

Do(100) > Do(110) > Do(111)

He also found that the oxidation enhancement decreased with temperature
such that above 1150 deg C for (111) Si retarded diffusion was observed.

Masetti et al [5] have studied boron in intrinsically doped silicon in the temperature range 950-1200 deg C. In agreement with Hill [14], they found that the oxidation enhancement depends on the surface being oxidised and decreases with (100), (110), (111). They expressed the diffusion coefficient as:

(111)Si
$$D = 32.5 \exp \frac{-3.34 \text{ eV}}{kT} \text{ cm*2/s}$$

(110)Si
$$D = 41.7 \exp \frac{-3.33 \text{ eV}}{kT} \text{ cm*2/s}$$

(111)Si
$$D = 6.06 \exp \frac{-3.05 \text{ eV}}{kT} \text{ cm*2/s}$$

Antoniadis et al [19] have made an extensive study of oxidation enhanced diffusion in dry O2 at 1000 deg C in intrinsically doped (100) CZ-silicon for times from 5 to 60 minutes. They have found D/D[eq] for intrinsic boron to decrease steadily from about 5.2 after 5 minutes at 1000 deg C to a nearly constant value of 3.2 after 60 minutes and have calculated the fractional interstitialcy component of diffusion as

    fi = 0.3

They also determined that there exists an energy barrier, of the order of 1.4 eV, to vacancy-interstitial recombination.

Antoniadis et al [6] studied diffusion in inert and dry oxidising ambients in the temperature range 850-1200 deg C. They found that the oxidation enhancement decreased with increasing temperature and was greater for a (100) oxidising surface than for (111). At 850 deg C, for a (100) substrate, the enhancement (Dox/Di) was a factor of 10.

Lin et al [8] found that for (100) Si, Dox/Di increased with the oxidation rate (dXox/dt)*P with 0.4 < P < 0.6 in the temperature range 900-1200 deg C.

Matsumoto [17] has calculated that the fractional interstitialcy component, fi, is given by an Arrhenius function:

    fi = 860*exp(-0.829/kT)

This value is larger than that for As or P, and at high temperatures the interstitialcy component of diffusion dominates.

Fair [18] calculated the fractional interstitialcy component in the temperature range 1000 to 1100 deg C to be 0.17.

Servidori et al [1] studied the diffusion of boron in regions of excess interstitials or vacancies caused by ion implantation and found that it was enhanced in the vicinity of interstitials and retarded in the vicinity of vacancies. They therefore postulated that boron diffuses by a predominantly interstitialcy mechanism.

The temperature, oxidation rate and substrate orientation dependence of oxidation enhanced diffusion is similar to that of oxidation induced stacking faults [8,22].

C. INTRINSIC BORON DIFFUSION, O2/HCl AMBIENT

The addition of a few percent of HCl gas to the oxidizing atmosphere is a silicon processing step that results in a reduction of the oxidation enhanced diffusion of boron and in the shrinkage of near-surface extrinsic stacking faults [12,14]. It is believed that this effect is a result of vacancy injection [12] or reduced interstitial injection [23] at the Si/SiO2 interface caused by the reaction of Cl with Si atoms on lattice sites.

Subrahmanyan [16] has found that the addition of a few percent of HCl to the O2 ambient leads to a reduction in the oxidation enhanced diffusion at temperatures in the range 1000 to 1150 deg C. The oxidation enhancement decreased as the percentage of HCl increased from 0 to 4%.

Hill [14] also found that the addition of HCl decreased the amount of oxidation enhancement. At 1200 deg C, less diffusion was observed for an O2 + 6%HCl ambient than for an inert ambient for (111)Si.

D. INTRINSIC BORON DIFFUSION, O2/H2O; Ar/H2O AMBIENT

Hill [14], using marker layers, has conducted an extensive study of intrinsic diffusion. He found under oxidizing conditions that the diffusion coefficient, Do, can be expressed in the form

$$Do = D(inert) + dD(oxidizing)]$$

$$= Dn \exp \frac{-En}{kT} + dDo \exp \frac{-dEo}{kT}$$

For boron in wet O2

(100) Si, 840-1200 deg C

$$Do = 24 \exp \frac{-3.87 \text{ eV}}{kT} + (7.5 \times 10^{-3}) \exp \frac{-2.85 \text{ eV}}{kT} \text{ cm}^2/s$$

(111) Si, 840-1100 deg C

$$Do = 24 \exp \frac{-3.87 \text{ eV}}{kT} + (1.6 \times 10^{-5}) \exp \frac{-2.28 \text{ eV}}{kT} \text{ cm}^2/s$$

Taniguchi [7] studied the diffusion of boron in dry O2, wet O2, and wet Ar. He found that there was no concentration dependence for total concentrations less than 3 X 10*15/cm*3. In each case, the diffusivity could be expressed as:

$$Do = D(inert) + (1.67 \text{ X } 10*{-}5)\left[\left(\frac{dXox}{dt}\right)*0.3\right]\left(\exp\frac{-X}{25}\right)\left(\exp\frac{-2.07\ eV}{kT}\right) \text{ cm*2/s}$$

where:

  X   = depth into the silicon in microns

  (dXox/dt) = oxide growth rate

Prince [4] expressed the diffusion constant in a wet O2 ambient as:

$$Do = 0.0322 \exp\frac{-3.02\ eV}{kT} \text{ cm*2/s}$$

E. EXTRINSIC BORON DIFFUSION, N2 AMBIENT

When the free carrier concentration exceeds the intrinsic carrier concentration at the diffusion temperature, i.e. when p > ni, the diffusivity is altered due to electric-field effects, h, and mass action effects, p/ni, giving a diffusion coefficient described by:

$$D = h \text{ X } \left[\ Dv(+)[eq]\frac{p}{ni} + Dv(0)[eq] + Dv(-)[eq]\frac{ni}{p} + Dv(--)[eq]\left(\frac{ni}{p}\right)*2 \right.$$

$$\left. + Di(+)[eq]\frac{p}{ni} + Di(0)[eq] + Di(-)[eq]\frac{ni}{p} + Di(--)[eq]\left(\frac{ni}{p}\right)*2\ \right]$$

where

$$h = 1 + \left(\frac{p}{2ni}\right)\left[\left(\frac{p}{2ni}\right)*2 + 1\right]*(-0.5)$$

and

Dv(+)[eq] = intrinsic diffusivity of the boron diffusing as the B(-)V(+) dopant/point-defect pair, etc.

Diffusivity is therefore increased by two mechanisms in the extrinsic regime:

(a) the electric field effect, h, where the high mobility hole increases the diffusion coefficient of the slower moving dopant atom, asymptotically reaching a maximum of 2 at high active doping levels, and

(b) the mass action effect where the numbers of charged vacancies and Si-interstitials, V(+), I(+) increase above their equilibrium values as (p/ni), resulting in more vehicles for dopant diffusion.

As such, when the local hole concentration exceeds ni (about 2 X 10*19/cm*3 at processing temperatures) the diffusivity increases. It has been established that interaction with a singly positively charged point defect results in a (p/ni) concentration dependent diffusion coefficient for extrinsic diffusion of boron in silicon [13]. This extrinsic boron diffusivity is 'locally determined' in the sense that the diffusivity depends only on the local hole concentration.

Literature numerical models of diffusion of high concentration boron profiles are outlined below.

Fair [24] describes the inert-ambient extrinsic diffusivity in a manner which physically accounts for the fact that boron diffuses by interaction with neutral point defects, B(-)P(0), and with singly positively charged point defects, B(-)P(+)

$$D = h \; X \; [\; Dv(0)[eq] + Dv(-)[eq] \; \frac{p}{ni} \;]$$

$$= h \; X \; [(2.22 \; X \; 10*8) \; exp \; \frac{-3.46 \; eV}{kT} \; + \; (4.32 \; X \; 10*9)(exp \; \frac{-3.46 \; eV}{kT}) \; \frac{p}{ni} \;]$$

$$micron*2/min$$

Colclaser [25] describes the inert extrinsic diffusivity of boron as

$$D = h \; X \; [\; Dv(0)[eq] + Dv(-)[eq] \; \frac{p}{ni} \;]$$

$$= h \; X \; [\; 0.091 \; exp \; \frac{-3.36 \; eV}{kT} \; + \; 166.3 \; (exp \; \frac{-4.08 \; eV}{kT})(\frac{p}{ni}) \;] \quad cm*2/s$$

Kim et al [9] studied boron diffusion at concentrations between 4 X 10*19 and 2.4 X 10*20/cm*3 for a temperature range of 990 to 1200 deg C. In the extrinsic regime, where p > ni, the diffusivity was given by the expression:

$$d = 2D[eq] \; \frac{p}{ni}$$

Matsumoto [17] found that the concentration dependence of boron diffusion could be expressed as:

$$D = D_i \left[(1-f_i) \frac{p}{n_i} + f_i \frac{n_i}{p} \right] X h$$

with

$$f_i = 860 \exp \frac{-0.829}{kT}$$

This value of f_i is large for boron at high temperature hence the diffusion coefficient is dominated by the interstitialcy component and has only a weak dependence on the concentration at high temperatures.

Bakowski [11] found only a slight dependence of the diffusivity on concentration in the range $1 X 10*17$ to $5 X 10*19/cm*3$. It should be noted that these concentrations are comparable with n_i at normal processing temperatures. He expressed the concentration dependence as:

$$D = D_{[eq]} h \left(1 + A \frac{p}{n_i} \right) cm*2/s$$

with $A = 0.1$

hence since $p = $ approx. n_i : $D = h D_i$

F. EXTRINSIC BORON DIFFUSION, OXIDIZING AMBIENT

The growth or deposition of thin film on top of silicon at various stages during the manufacture of integrated circuits results in several types of phenomena in the underlying substrate. Of interest here is the effect of oxidation on the diffusion of high concentration boron profiles. This effect has not been well characterised. Some observations on this process are outlined below:

Taniguchi [7] studied the diffusion of boron in dry O2, wet O2, and wet Ar. He found that there was no concentration dependence for total concentrations less than $3 X 10*15/cm*3$. At higher concentrations, the amount of oxidation enhancement decreased with increasing concentration.

REFERENCES
----

[1]   M.Servidori et al [J. Appl. Phys. (USA) vol.61 no.5 (1 Mar 1987)
      p.1834-40]
[2]   H.J.Schnabel et al [Phys. Status Solidi a (Germany) vol.8 no.1
      (1971) p.71-8]
[3]   R.N.Ghoshtagore [Solid-State Electron. (GB) vol.15 (1972)
      p.1113-20]
[4]   J.L.Prince, F.N.Schwettmann [J. Electrochem. Soc. (USA) vol.121
      no.5 (May 1974) p.705-10]
[5]   G.Masetti, S.Solmi, G.Soncini [Solid-State Electron (GB) vol.19
      no.6 (Jun 1976) p.545-6]

[6]   D.A.Antoniadis, A.G.Gonzalez, R.W.Dutton [J. Electrochem. Soc.
        (USA) vol.125 no.5 (May 1978) p.813-19]
[7]   K.Taniguchi, K.Kurosawa, M.Kashiwagi [J. ELectrochem. Soc. (USA)
        vol.127 no.10 (Oct 1980) p.2243-8]
[8]   A.Miin-Ron Lin, D.A.Antoniadis, R.W.Dutton [J. Electrochem. Soc.
        (USA) vol.128 no.5 (May 1981) p.1131-7]
[9]   C.Kim, Z.-Y.Zhu, R.I.Kang, K.Shono [J. Electrochem. Soc. (USA)
        vol.131 no.12 (Dec 1984) p.2962-4]
[10]   D.Mathiot, J.C.Pfister [J. Appl. Phys. (USA) vol.55 no.10 (15 May
        1984) p.3518-30]
[11]   A.Bakowski [J. Electrochem. Soc. (USA) vol.127 no.7 (Jul 1980)
        p.1644-9]
[12]   Y.Nabeta, T.Uno, S.Kubo, H.Tsukamoto [J. Electrochem. Soc. (USA)
        vol.123 no.9 (Sep 1976) p.1416-17]
[13]   F.Gaiseanu [J. Electrochem. Soc. (USA) vol.132 no.9 (Sep 1985)
        p.2287-9]
[14]   C.Hill [Semiconductor Silicon 1981 Eds H.R.Huff et al (Princeton,
        NJ, USA, 1981) p.988-1006]
[15]   M.Okamura [Jpn. J. Appl. Phys. (Japan) vol.8 (1969) p.1440]
[16]   R.Subrahmanyan, H.Z.Massoud, R.B.Fair [J. Appl. Phys. (US) vol.61
        no.10 (15 May 1987) p.4804-7]
[17]   S.Matsumoto, Y.Ishikawa, T.Niimi [J. Appl. Phys. (USA) vol.54 no.9
        no.9 (Sep 1983) p.5049-54]
[18]   R.B.Fair [J. Appl. Phys. (USA) vol.51 no.11 (Nov 1980) p.5828-32]
[19]   D.A.Antoniadis, I.Moskowitz [J. Appl. Phys. (USA) vol.53 no.10
        (Oct 1982) p.6788-96]
[20]   D.Anderson, K.O.Jeppson [J Electrochem. Soc. (USA) vol.131 no.11
        (Nov 1984) p.2675-9]
[21]   A.D.Kurtz [J. Appl. Phys. (USA) vol.31 (1960) p.303]
[22]   S.M.Hu [Appl. Phys. Lett. (USA) vol.27 no.4 (15 Aug 1975)
        p.165-7]
[23]   S.Oh, W.A.Tiller, Soo-Kap Hahn [Appl. Phys. Lett. (USA) vol.48
        no.17 (28 Apr 1986) p.1125-7]
[24]   R.B.Fair [Impurity Doping Processes in Silicon Ed. F.F.Y.Wang
        (North Holland, 1981)]
[25]   R.A.Colclaser [Microelectronics: Processing and Device Design
        (John Wiley and Sons, 1980) p.123]

13.2 DIFFUSION OF Al IN Si

by D. de Cogan

July 1987
EMIS Datareview RN=16101

The interest in aluminium as an impurity arises from the fact that it
diffuses faster than other acceptors. The solubility limit is in most
cases low, which reduces dislocation formation and there is no evidence
of intermetallic reactions with the host element.

The diffusion properties can be characterised by either monitoring
junction depth (using a bevel and stain technique) or by means of a
complete profile determination. The standard method which was used by
Fuller and Ditzenberger [1] consists of measuring sheet resistance,
lapping a known thickness and measuring sheet resistance again. Other
techniques such as Hall effect and spreading resistance are sometimes
used.

The early studies of aluminium diffusion were the subject of much
controversy. Values of diffusion constant differed by up to two orders
of magnitude. The TABLE 1 provides data on the pre-exponential and
activation terms in $D = Do \exp (-E/kT)$ quoted by different authors.
The diffusion constant at 1200 deg C is given for the purposes of
comparison.

TABLE 1

Do cm*2/sec	E eV	D(1200 deg C) cm*2/sec	References
8	3.47	9.24 X 10*-12	Fuller, Ditzenberger [1]
2800	3.8	1.28 X 10*-10	Goldstein [2]
4.8	3.36	1.58 X 10*-11	Miller, Savage [3]
0.5	3.0	2.74 X 10*-11	Kao [4]
1.38	3.41	3.00 X 10*-12	Ghoshtagore [5]
1.8	3.2	2.04 X 10*-11	Rosnowski [6]

These differences were largely due to the nature of the impurity sources
and diffusion conditions. Fuller and Ditzenberger [1] used aluminium
metal in an evacuated sealed tube. Goldstein [2] placed an aluminium
silicon alloy button on top of the silicon substrate within an evacuated
sealed tube. Miller and Savage [3] attempted to avoid any reaction
between the quartz tube and the samples. Accordingly the substrate and
source were placed inside a silicon boat held within a tantalum tube
which acted as a getter. The assembly was heated in an open arrangement
using helium or argon at the rate of 2.5 litres/min. Navon and
Chernyshov [7] have re-examined Goldstein's [2] work and obtained results

which are more in line with those of Fuller and Ditzenberger [1]. Kelin
[8] has evaporated and alloyed aluminium into silicon. Diffusion was
then performed from the alloy front. The results were reported to
confirm Goldstein's values. Ghoshtagore [5] has deposited an aluminium
rich epitaxial layer as the diffusing source. Trimethyl aluminium was
used in the epilayer. There have been several other suggestions for
diffusion sources. Chang and Roesch [9] in a German patent propose
using aluminium coated wafers in an open tube as a vapour source. In a
European patent application Chang et al [10] suggest that aluminium
oxide or sapphire could be used as a source in an open tube with argon
or hydrogen as the ambient. Nisnevich [11] has used aluminium doped
silica in an open tube and has obtained a value of D = 3.01 X 10*-11
cm*2/sec at 1250 deg C.

Bullough et al [12] have shown that the free energies for the formation
of silica and alumina are such that any aluminium coming into contact
with silica will bring about a redox reaction: silica is reduced and
aluminium is oxidised. They suggest therefore that the silica tube
makes a significant contribution. Rai-Choudhury et al [13] presented
thermochemical data to show the effect of silica and water vapour
during aluminium diffusion. The aluminium partial pressure is reduced
and the formation of a stable alumina skin seals the source. They have
shown that if residual water vapour can be reduced to less than 0.01 ppm
then liquid aluminium and solid alumina can co-exist and a lower
diffusion temperature is favoured. This minimises source oxidation. If,
however, a quartz diffusion system has a high vapour content then a high
diffusion temperature is preferred, although the source will oxidise
and lead to reproducibility problems. They suggest the use of alumina
or zirconia boats and point to the fact the Kao [4] obtained an
aluminium surface concentration close to the solubility limit by
allowing his quartzware to be passivated by the formation of a layer of
silicon and alumina prior to diffusion. The use of passivation is
recommended. Rosnowski [6] has used passivated quartz in an open tube,
high vacuum diffusion system. The absence of argon as a carrier or
backfill gas raises the aluminium partial pressure. He found that a
metallic source tended to cause sample degradation due to alloying on
the substrate surface. For this reason the use of a silicon-aluminium
alloy source was recommended. There is the added advantage that in an
alloyed source it is not possible for an impervious oxide skin to form.
This might explain why Goldstein [2] obtained a larger value for
diffusion coefficient than many other workers at that time.

The effects of subsequent oxidation on the diffusion of aluminium in
silicon have been studied by Mizuo and Higuchi [14]. In the range
950-1150 deg C diffusion in (100) silicon is enhanced. The diffusion
in (111) silicon is enhanced at low temperatures and retarded at high
temperatures. These effects are totally inhibited by the addition of
HCl to the oxidising ambient.

The reaction between aluminium and oxide normally inhibits the use of
silica as a mask. However Rosnowski [15] in a US patent describes how a
4 micron layer silica can be used for selective masking. The reaction
to form silicon and alumina on the silica surface effectively masks
against aluminium diffusion. The method proposed by Jayant-Baliga [16]
involves the use of silicon nitride. As interfacial stresses can lead
to cracking in nitride films a multilayer mask is recommended. 100-

2000 A of silica are grown on the silicon surface. A layer of nitride is deposited on top of this (500-3000 A) and that is followed by 1500 A of CVD silicon oxide. Diffusion depths of up to 35 microns were achieved at 1250 deg C. Although the cover oxide (necessary for gallium diffusions) was attacked, there was no evidence of reaction between the dopant and the nitride layer. The ratio of lateral diffusion to junction depth was unity.

In their work on the effects of oxidation on aluminium diffusion Mizuo and Higuchi [14] used a pre-deposit of 10*15/cm*2 aluminium implanted at 150 keV. Leroy et al [17] have used a 10*15/cm*2 100 keV implant as a preliminary to their study of diffusion of aluminium in silicon by semicontinuous laser annealing. The diffusion constant was similar to that reported by others and an activation energy of 3.4 eV was observed. Baranova et al [18] describe the technique of transmissionion (recoil atom) bombardment of thin films of aluminium. 2.3 X 10*15/cm*2 50 keV argon ions were used and a diffusion constant of 4.5 X 10*14 cm*2/sec was measured following a one hour anneal at 800 deg C.

REFERENCES
----

[1]   C.S.Fuller, J.A.Ditzenberger [J. Appl. Phys. (USA) vol.27 (1956) p.544]
[2]   B.Goldstein [Bull. Am. Phys. Soc. Ser.II (USA) vol.1 (1956) p.145]
[3]   R.C.Miller, A.J.Savage [J. Appl. Phys. (USA) vol.27 (1956) p.1430]
[4]   Y.C.Kao [Electrochem. Technol. vol.5 (1967) p.90]
[5]   R.N.Ghoshtagore [Phys. Rev. B (USA) vol.3 (1971) p.2507]
[6]   W.Rosnowski [J. Electrochem. Soc. (USA) vol.125 no.6 (1978) p.957-62]
[7]   D.Navon, V.Chernyshov [J. Appl. Phys. (USA) vol.28 (1957) p.823]
[8]   T.Klein [Mullard Research Labs. (GB) Rp8-28 CVD Annual Report December 1962 (W) AD 295065]
[9]   M.F.Chong, A.Roesch [German Patent 2,932,432 (20 March 1980)]
[10]   M.F.Chong, R.W.Kennedy, D.K.Hartmann, A.Roesch, H.B.Assalit, [Eur. Pat. App. 19,272 (26 November 1980) p.272]
[11]   Ya.D.Nisnevich [Izv. Akad. Nauk. SSSR Neorg. Mater. vol.19 no.6 (1983) p.853-6]
[12]   R.Bullough, R.C.Newman, J.Wakefield [Proc. Inst. Electr. Eng. (GB) no.106B (1959) p.277]
[13]   P.Rai-Choudhury, R.A.Selim, W.J.Takei [J. Electrochem. Soc. (USA) vol.124 no.5 (1977) p.762-6]
[14]   S.Mizuo, H.Higuchi [Jpn. J. Appl. Phys. Part 1 (Japan) vol.21 (1982) p.56-60]
[15]   W.Rosnowski [US Patent 4,029,528 (12 June 1977)]
[16]   B.Jayant-Baliga [J. Electrochem. Soc. (USA) vol.126 no.2 (1979) p.292-6]
[17]   C.Leroy, J.E.Bouree, M.Rodot [J. Phys. Colloq. (France) vol.44 no.C5 p.235-40]
[18]   A.S.Baranova, D.I.Tetel'baum, E.I.Zorin, P.V.Pavlov, E.V.Dobrokhotov, A.K.Kuritsyna [Sov. Phys.-Semicond. (USA) vol.7 no.9 (1974) p.1151-3]

13.3 DIFFUSION OF Ga IN Si

by D. de Cogan

July 1987
EMIS Datareview RN=16102

The earliest studies of diffusion of gallium in silicon were carried
out by Fuller and Ditzenberger [1] using a sealed tube arrangement.
Characterisation was undertaken using successive lapping and sheet
resistance measurements. It was found that the diffusion constant could
be described by

$$D = 3.6 \exp (-3.5 \text{ eV}/kT) \text{ cm}*2/\text{sec}$$

in the range 1105-1360 deg C. Gallium (III) oxide was used as the
source. Frosch and Derick [2] used an open tube system with a similar
source and wet hydrogen as the carrier gas. Kurtz and Gravel [3]
employed metallic gallium in an open tube system with argon. It was
found that the diffusion constant was dependent on the dopant
concentration. Diffusion from a radioactive metallic source was first
studied by Kren, Masters and Wajda [4]. Makris and Masters [5] have
used gallium doped silicon powder as a source in sealed tube experiments
at 900-1050 deg C. It was found that the diffusion coefficient of
gallium in boron doped silicon increased linearly with boron
concentration at boron levels below about 10*19/cm*3. They found that
the diffusion coefficient fitted the expression

$$D = 60 \exp (-3.89 \text{ eV}/kT) \text{ cm}*2/\text{sec}$$

Boltaks and Dzhafarov [6] have heat treated antimony doped wafers in an
atmosphere of gallium vapour in the range 1180-1340 deg C, and found
that the presence of antimony accelerated the diffusion of gallium
compared with control samples where D = 2.1 exp (-3.51 eV/kT).
According to Nakajima and Ohkawa [7] the simultaneous diffusion of
gallium and phosphorus into silicon is retarded by the presence of
arsenic vapour in the range 1000-1150 deg C. They attribute this to the
interactions between the internal electric fields of the impurities. No
interaction between phosphorus and gallium was observed in the absence
of arsenic. The interactions between sequential gallium-arsenic
diffusions in silicon have been studied by Jones and Willoughby [8],
who have also compared sequential gallium-arsenic diffusions with
sequential gallium-phosphorus diffusions [9]. They found that at high
concentrations phosphorus enhanced the gallium diffusion. The effects
of arsenic (enhancement or retardation) depended on the experimental
conditions. This is largely in accord with the work of Okamura [10] who
has, however, noted [11] that profiles determined by neutron activation
analysis give higher gallium concentrations than that suggested by
standard lapping/resistance techniques. This effect is attributed to
gallium precipitation during cooling.

The effect of the diffusion source has received considerable attention.
Fuller and Ditzenberger [1] used the tri-valent oxide. The wet-hydrogen
technique of Frosch and Derick [2] is thought to involve the reduction
of gallium oxide to the more volatile gallium (I) oxide. This has been
the subject of a German patent by Popp and Held [12]. Ghoshtagore [13]
has a German patent for a two zone heat treatment of gallium (II)
oxide in an open tube system involving a stream of argon and carbon
monoxide. He has shown [14] that wet carbon monoxide leads to a
reduction of the oxide so that metallic gallium is the volatile
species. He has also shown [15] that the presence of traces of oxygen
can have profound effects on the diffusion of gallium. For this reason
he has grown a gallium-rich epitaxial layer on clean silicon. He quotes
an intrinsic bulk diffusion coefficient $D = 0.374 \exp (-3.39 \text{ eV}/kT)$.
It is suggested that the bulk diffusion occurs by acceptor-vacancy
pair formation and migration. The correlation between activation energy
and covalent radius for gallium is in agreement with the results for
other acceptors. Gallium can also be diffused from a composite source
consisting of gallium and 60% gallium (III) oxide formed by the thermal
decomposition of gallium acetylacetonate at 400 deg C in 10 mtorr of
oxygen [16]. Haridos et al [17] have perfomed sealed tube measurements
using pure gallium metal in the range 700-1100 deg C. They have found
that the diffusion coefficient is given by $D = 0.05 \exp (-2.7/kT)$ which
is significantly less than the previously quoted values. The difference
is attributed to the absence of intrinsic effects in their experiments,
which is consistent with the observations of Ghoshtagore [15].

Martin and Reuschel [18] have outlined a technique for obtaining a
uniform distribution of gallium impurity by neutron irradiation of a
mixed crystal of silicon containing less than 10% germanium.

It has been found that silica is ineffective for masking silicon during
gallium diffusion [2,19]. Gallium diffuses faster in silica than in
silicon. Diffusion studies of gallium in silica have been undertaken by
Grove et al [20]. Nakajima and Ohkawa [21] have studied the effect of
silicon crystal orientation on gallium diffusion through an oxide
layer. Junction depths were independent of oxide thickness in the range
500-5000 A. Jain et al [22] have compared diffusions into bare and
oxide covered silicon. At 1035 deg C the junction depth under oxide
exceeds that in bare silicon. They have shown that enhancement and
retardation effects depend on the relative temperatures of oxide
formation and gallium diffusion. They propose a model based on strain
gradients at the interface. The experimental observations have been
confirmed by Mizuo and Higuchi [23] who note that these effects are
reduced by the addition of HCl to the oxidising ambient.

Silicon nitride has been suggested as a diffusion mask for gallium [24].
However Jayant Baliga [25] mentions that this is effective only for
shallow diffusion. The nitride layer can either crack or react with the
gallium so that it cannot be subsequently removed. To overcome this he
proposes a sandwich structure. A layer of oxide between the silicon and
nitride inhibits cracking and a layer of oxide over the nitride prevents
attack by gallium. Using this arrangement it has been possible to
achieve junction depths of 275 microns. Dumas [26] describes a
modification of this approach in a US patent.

The effects of 45 keV gallium implantation (dose 10*15/cm*2) have been studied by Bamo et al [27]. An enhanced diffusion was observed at implant temperatures of 150-700 deg C. This was independent of temperature and rate over this range. Gyulai et al [28] have implanted Ga at 80 keV. They used both CVD oxide and nitride as encapsulants. Almost all gallium diffused into the oxide and no snowplough effect was observed. With nitride there was very fast gallium diffusion which was probably due to strain damage at the interface.

It has been observed that gold and copper are gettered during gallium diffusion [29]. Gold is gettered regardless of the cooling conditions. Most copper near the surface layer is removed during diffusion. Bulk copper is gettered during slow cooling.

REFERENCES
----

[1]   C.S.Fuller, J.A.Ditzenberger [J. Appl. Phys. (USA) vol.27 (1956) p.544]

[2]   C.J.Frosch, L.Derick [J. Electrochem. Soc. (USA) vol.104 (1957) p.547]

[3]   A.D.Kurtz, C.L.Gravel [J. Appl. Phys. (USA) vol.29 (1958) p.1456]

[4]   J.G.Kren, B.J.Masters, E.S.Wajda [Appl. Phys. Lett. (USA) vol.5 (1964) p.49]

[5]   J.S.Makris, B.J.Masters [J. Appl. Phys. (USA) vol.42 (1971) p.3750]

[6]   B.I.Boltaks, T.D.Dzhafarov [Sov. Phys.-Solid State (USA) vol.5 (1964) p.2649]

[7]   Y.Nakajima, S.Ohkawa [Jpn. J. Appl. Phys. (Japan) vol.10 (1971) p.1745]

[8]   C.L.Jones, A.F.W.Willoughby [Proc. Electrochem. Soc. vol.77 (1977) p.684]

[9]   C.L.Jones, A.F.W.Willoughby [Inst. Phys. Conf. Ser. (GB) no.31 (1977) (Radiation Effects in Semiconductors) p.194]

[10]   M.Okamura [Jpn. J. Appl. Phys. (Japan) vol.7 (1968) p.1067]

[11]   M.Okamura [Jpn. J. Appl. Phys. (Japan) vol.10 (1979) p.434]

[12]   G.Popp, G.Held [German Patent 2,751,163 (17 May 1979)]

[13]   R.N.Ghoshtagore [German Patent 2,644,879 (21 April 1977)]

[14]   R.M.Ghoshtagore [Solid-State Electron. (GB) vol.22 (1979) p.877]

[15]   R.N.Ghoshtagore [Phys. Rev. B (USA) vol.3 no.8 (1971) p.2507-14]

[16]   G.Mermant [3rd Colloq. Int. Appl. Tech. Vide. Ind. Semicond. Composants Electron. Microelectron (AVISEM 71) (1971) p.72 (Fr.)]

[17]   S.Haridos, F.Beniere, M.Gauneau, A.Rupert [J. Appl. Phys. (USA) vol.51 (1980) p.5833]

[18]   J.Martin, K.Reuschel [German Patent 2,407,697 (Sept 1975)]

[19]   F.K.Heumann, D.M.Brown, E.Mets [J. Electrochem. Soc. (USA) vol.115 (1968) p.99]

[20]   A.S.Grove, O.Leistiko, C.T.Sah [J. Phys. & Chem. Solids (GB) vol.25 (1964) p.985]

[21]   Y.Nakajima, S.Ohkawa [Jpn. J. Appl. Phys. (Japan) vol.11 (1972) p.1742]

[22]   G.C.Jain, A.Prasad, B.C.Chakravarty [Phys. Status Solidi a (Germany) vol.46 (1978) p.151]

[23]   S.Mizuo, H.Higuchi [Denki Kagaku (Japan) vol.50 no.4 (1982) p.338-43]

[24]   V.Y.Doo [IEEE Trans. Electron. Devices (USA) vol.ED-13 (1966)
        p.561]
[25]   B.Jayant-Baliga [J. Electrochem. Soc. (USA) vol.126 no.2 (1979)
        p.292-6]
[26]   G.H.Dumas [US Patent 3,895,976 (22 July 1975)]
[27]   K.Bamo et al [Jpn. J. Appl. Phys. (Japan) vol.12 (1973) p.735]
[28]   J.Gyulai, L.Csepregi, T.Nagy, J.W.Mayer, H.Muller [Vide (France)
        no.174 (1974) p.146]
[29]   N.Momma, H.Taniguchi, M.Ura, T.Ogawa [J. Electrochem. Soc. (USA)
        vol.125 (1975) p.963]

13.4 DIFFUSION OF In IN Si

by D. de Cogan

July 1987
EMIS Datareview RN=16104

Fuller and Ditzenberger [1] using a lap and sheet resistance technique
have measured the diffusion coefficient of indium in silicon and find
that it can be described by D = 16.5 exp (-3.9 eV/kT) cm*2/sec.
Ghoshtagore [2] has obtained D = 0.785 exp (-3.63/kT) cm*2/sec using
indium rich implanted layers and spreading resistance measurement. He
has shown that most acceptors have an activation energy which is
proportional to the impurity covalent radius. Indium fits this
relationship very well and an indium vacancy pair migration mechanism
is proposed.

It had been thought that indium might constitute an alternative
acceptor impurity for VLSI technology. For this reason much work has
concentrated on the properties of implanted layers. To a first
approximation the diffusion constant is the same as that of an
ordinary diffusion in an inert atmosphere [3]. However at higher dose
levels there are differences due to field aided diffusion [4].

   D (10*13/cm*2) = 169 exp (-4.19 eV/kT)

   D (10*14/cm*2) = 204 exp (-4.22 eV/kT)

Gamo et al [5] have observed that there is an enhanced diffusion of
indium when the implantation temperature is in the range 250-700 deg C.
The effect is independent of temperature and dose rate over the range
and an interstitial diffusion mechanism is suggested.

Antoniadis and Moskowitz [6] have compared the drive-in properties of
implanted indium in dry nitrogen and dry oxygen. An oxygen enhanced
diffusion is observed which at 1000 deg C is almost identical to that
for boron. Cerofolini et al [3] report that the segregation between
silicon and oxide favours the former. In the initial stages indium can
be incorporated in the oxide because of the high growth rate compared
with the low diffusivity. After that a snowplough effect gives rise to
indium accumulation at the silicon surface.

According to Schroder et al [7] the determination of indium doping
concentration by Hall measurements appears to give erroneously high
results. Capacitance-voltage and junction breakdown techniques are more
reliable.

REFERENCES
----

[1] C.S.Fuller, J.A.Ditzenberger [J. Appl. Phys. (USA) vol.27 (1956)
     p.544]
[2] R.N.Ghoshtagore [Phys. Rev. B (USA) vol.3 no.8 (1971) p.2507-14]
[3] G.F.Cerofolini, G.U.Pignatel, F.Riva, G.Ferla, G.Ottaviani [Thin
     Solid Films (Switzerland) vol.101 (1983) p.137-44]
[4] G.F.Cerofolini, G.Ferla, G.U.Pignatel, F.Riva [Thin Solid Films
     (Switzerland) vol.101 (1983) p.275-83]
[5] K.Gamo et al [Jpn. J. Appl. Phys. (Japan) vol.12 (1973) p.735]
[6] D.A.Antoniadis, I.Moskowitz [J. Appl. Phys. (USA) vol.53 (1982)
     p.9214]
[7] D.K.Schroder, T.T.Braggins, H.M.Hobgood [J. Appl. Phys. (USA)
     vol.49 no.10 (1978) p.5256-9]

13.5 DIFFUSION OF Tl IN Si

by D. de Cogan

July 1987
EMIS Datareview RN=16103

According to Shulman [1] thallium is a deep acceptor in silicon. The
energy level lies at 0.26 eV above the valence band. It has a very
high vapour pressure at the melting point of silicon (it boils at
1490 deg C). Shulman encountered considerable problems in incorporating
it into growing crystals.

There is only limited information available on the diffusion of
thallium into silicon. Fuller and Ditzenberger [2] have used the oxide
as the source in evacuated sealed tube experiments. Profiles were
obtained using successive lap and sheet resistance measurements. TABLE 1
summarises their results which can be described by the equation
D = 16.5 exp (-3.9 eV/kT) cm*2/sec. Within the limits of their
experimental error this is identical to the equation for indium.

TABLE 1

T (deg C)	D(cm*2/sec)
1360	2.03 X 10*-11
1320	1.2 X 10*-11
1255	8.6 X 10*-12
1170	5.0 X 10*-13
1105	8.0 X 10*-14

Data from Fuller and Ditzenberger [2]

Ghoshtagore [3] has used a doped silicon epitaxial layer as a source.
Unlike his work on gallium and aluminium, it was not possible to use
an organo-thallium compound during epi-growth on account of its vapour
pressure and stability. Accordingly, he used a 99.9999% pure metal
source on the susceptor during growth. He was only able to determine
the diffusion properties at a single value of surface concentration.
Profiles were obtained using spreading resistance techniques. He
estimated that the diffusion constant could be described by the
equation D = 1.37 exp (-3.7 eV/kT) cm*2/sec in the range 1244-1338 deg C,
which fits in well with his correlation between activation energy and
impurity tetrahedral radius. Differences between his results and those
of Fuller and Ditzenberger were attributed to oxidation-reduction
reactions on the silicon surface.

Thallium has several oxides which may contribute in varying degrees to
the dopant partial pressure over the diffusion temperature range
800-1350 deg C. According to the CRC Handbook of Chemistry and Physics
[4] thallium (+3) oxide decomposes to thallium (+1) oxide at 875 deg C
(at atmospheric pressure). At 1080 deg C thallium (+1) oxide further
decomposes to metallic thallium.

A recent study by Dedekaev et al [5] estimates a very much higher activation energy, namely 4.62 eV. The reported pre-exponential factor is 9.37 X 10*10 cm*2/sec. These results appear to be quite different to anything quoted previously and the authors do not offer any explanation.

REFERENCES
----

[1]   R.G.Shulman [J. Phys. & Chem. Solids (GB) vol.2 (1957) p.115]
[2]   C.S.Fuller, J.A.Ditzenberger [J. Appl. Phys. (USA) vol.27 (1956) p.27]
[3]   R.N.Ghoshtagore [Phys. Rev. B (USA) vol.3 no.8 (1971) p.2507-14]
[4]   [Handbook of Chemistry & Physics, 64th Edition (CRC press 1983) p.B-147]
[5]   T.T.Dedegkaev, V.V.Eliseev, D.Kh.Lagkuev, V.A.Fidarov [Sov. Phys. Phys.-Solid State (USA) vol.27 no.9 (Sep 1985) p.1699-1700]

13.6 DIFFUSION OF P IN Si

by P.B.Moynagh and P.J.Rosser

December 1987
EMIS Datareview RN=17874

A. INTRINSIC PHOSPHORUS DIFFUSION, N2 AMBIENT

The mathematical basis for the diffusion process was laid by Fick whose
first law sets forward the hypothesis that the rate of transfer of a
diffusing particle through unit area is proportional to the magnitude
of the gradient normal to that area, i.e. in the x-direction

$$J = -D \frac{dC}{dx}$$

where the proportionality constant, D (metres*2/sec), is known as the
diffusion constant or the diffusivity. The diffusion process occurs
because of the random walk of the impurity atoms and is therefore a
thermally activated process. The temperature dependence of the diffusion
coefficient is Arrhenius in nature:

$$D = Do \exp \frac{-E}{kT}$$

where E is the activation energy (which, for the common dopants in
silicon, is about 4 eV).

Due to the low concentrations of stable point defects in silicon, it has
not been possible to determine directly whether vacancies or self-
interstitials are dominant in diffusion processes. Numerous kinds of
indirect observations have been made to build cases for vacancies and
for self-interstitials. The current belief is that both types of point
defect exist somewhat independently [29], and that both influence
diffusion. Under diffusive conditions where the silicon is intrinsically
doped and where no surface reaction is taking place, the vacancy and
self-interstitial concentrations are temperature dependent only and are
described approximately by the equilibrium concentrations:

$$Cv[eq] = Cov \exp \frac{-Ev}{kT}$$

$$Ci[eq] = Coi \exp \frac{-Ei}{kT}$$

Dopant diffusion takes place in the silicon by interacting with these point defects so that the diffusing species are the phosphorus/vacancy pair, PV, and the phosphorus/Si-interstitial pair, PI. The intrinsic diffusivity due to the vacancy mechanism is referred to as Dv[eq], and that due to the interstitialcy mechanism as Di[eq]. The intrinsic, inert ambient diffusivity is therefore given by the sum of the equilibrium diffusivities due to the vacancy and the self-interstitial components:

D[eq] = Dv[eq] + Di[eq]

The wide energy range available to the Fermi level in silicon leads to a given lattice defect (i.e. a vacancy or Si-interstitial) appearing in a variety of ionized states [12]. As such the dopant/point-defect pairs exist as, for example, P(+)V(-) which we will denote by PV(-). The intrinsic, inert ambient diffusivity can therefore be written:

D[eq] = Dv(+)[eq] + Dv(0)[eq] + Dv(-)[eq] + Dv(--)[eq]

+ Di(+)[eq] + Di(0)[eq] + Di(-)[eq] + Di(--)[eq]

where the intrinsic diffusivity of the phosphorus diffusing as PV(-) is given by Dv(-)[eq], etc.

The diffusion coefficient in crystals with a cubic lattice should be isotropic. For phosphorus doping levels of less than about 1 X 10*20/cm*3 with no surface reaction taking place this is indeed the case [10,28]. The diffusion coefficient is therefore given by:

$$D = Dn \exp \frac{-En}{kT}$$

TABLE 1: diffusivities for inert diffusion of phosphorus in single crystal silicon

(100),(110),(111) Si

Temperature (deg C)	Dn(cm*2/s)	En(eV)	Reference	
800-1300	3.85	3.66	Fair	[16]
950-1150	3.62	3.61	Ishikawa	[2]
1100-1300	3.7	3.69	Ghoshtagore	[13]
1100-1200	1.3	3.5	Masetti	[7]
950-1250	5.3	3.69	Makris	[8]
1000-1200		3.7	Barry	[9]
840-1150	0.6	3.51	Hill	[28]

B. INTRINSIC PHOSPHORUS DIFFUSION, O2 AMBIENT

Oxidation of silicon under most processing conditions enhances the
diffusion of phosphorus by a process of interstitial injection to the
silicon substrate [11]. The enhanced diffusion (which is about 3.5 for
intrinsic phosphorus diffusion at 1000 deg C and is less pronounced at
higher temperatures and for higher dopant concentrations), occurs
because the number of silicon self-interstitials available to interact
with the phosphorus atoms has increased and the diffusivity is described
by a general diffusion equation:

$$D = Dv + Di$$

i.e.

$$D = Dv[eq] \frac{Cv}{Cv[eq]} + Di[eq] \frac{Ci}{Ci[eq]}$$

or

$$D = D[eq] \left((1-fi) \frac{Cv}{Cv[eq]} + fi \frac{Ci}{Ci[eq]} \right)$$

where

$$fi = \frac{Di[eq]}{D[eq]} = \frac{Di[eq]}{Dv[eq] + Di[eq]}$$

and fi is known as the fractional interstitialcy component of diffusion.
Some injected Si-interstitials will be lost by Frenkel recombination
with vacancies so that the thermal oxidation process results in a
simultaneous supersaturation of Si-interstitials and undersaturation
of vacancies. The effect on total diffusion rate of the increase in PI
diffusion species and decrease in PV diffusion species is described by
the fractional interstitialcy component of diffusion, fi, literature
values of which are given below.

Experimental evidence suggests that the Si-interstitial injection rate
during thermal oxidation depends on the oxidation rate. More
specifically, it has been found that:

$$Ci - Ci[eq] = k \left(\frac{dXox}{dt} \right) *P \qquad \text{with P < 1}$$

Since the oxidation rate decreases as oxidation time increases, it is expected that the diffusion enhancement will also decrease with oxidation time. The oxidation rate (and therefore the rate of Si-interstitial injection) is not dependent on the orientation beyond the linear segment. The rate of decay of these excess point-defects will be dependent on the silicon surface or the silicon-oxide interface recombination rates, and these might be expected to be proportional to the rectilinear density. As such the rate of loss of excess point-defects would increase as (100), (110), (111), and the magnitude of oxidation enhanced diffusion would increase as (111), (110), (100) [28].

As is described below, oxidation enhanced diffusion in the intrinsic regime is a function of temperature, oxidation rate, depth into the silicon and substrate orientation.

Literature models of oxidation enhanced phosphorus diffusion are described below.

Taniguchi et al [1] investigated oxidation enhanced diffusion (OED) of (100) silicon in the temperature range 950-1150 deg C. In the intrinsic regime they found OED to be dependent on temperature, on oxide growth rate and on depth into the silicon substrate. It increased with oxidation rate and decreased with depth into the silicon. They found no ambient type dependence of OED.

Diffusivity under oxidizing conditions, Do, is described as the diffusivity under inert conditions, D(inert), plus the increase caused by the oxidizing ambient dD(oxidizing),

$$Do = D(inert) + dD(oxidizing)$$

$$Do = Dn \exp \frac{-En}{kT} + F(\frac{dXox}{dt})*P \exp \frac{-X}{L} \exp \frac{-dEo}{kT} \quad microns*2/hr$$

(100) Si

$$Do = Dn \exp \frac{-En}{kT} + 0.052 (\frac{dXox}{dt})*0.3 \exp \frac{-X}{25} \exp \frac{-2.07}{kT} \quad (microns*2/hr)$$

$$Do = Dn \exp \frac{-En}{kT} + 8.67 \times 10*-4 (\frac{dXox}{dt})*0.3 \exp \frac{-X}{25} \exp \frac{-2.07}{kT} \quad (cm*2/s)$$

with

dXox/dt = oxide growth rate

X = depth into silicon

Eo is a converted value at 1 micron/hr oxidation rate [1]

Criticisms of the Taniguchi model centre around their interpretation of
the depth dependence of OED, with a characteristic length of 25 microns,
which they ascribed to vacancy/self-interstitial annihilation. Hill's OED
experiments [28] indicate an interstitial diffusion length of the order
of 100 microns. The 25 micron characteristic length observed by
Taniguchi is therefore a result of interstitial loss on crystallographic
defects.

Ishikawa et al [2] have found oxidation enhanced diffusion in the
intrinsic regime to be dependent on temperature, time and the crystal
orientation of the substrate. It decreased with time (which corresponds
qualitatively to Taniguchi's increase with oxidation rate [1], and it
was higher for a (100) than a (111) orientation (which is consistent
with the passivating effects of the interface).

$$Do = D(inert) + dD(oxidizing)$$

$$Do = Dn \exp \frac{-En}{kT} + (t*-P) \, dDo \, X \exp \frac{-dEo}{kT} \quad (cm*2/s)$$

(100)Si

$$Do = 3.62 \exp \frac{-3.61 \ eV}{kT} + (t*-0.38) \, (3.19 \ X \ 10*-7) \exp \frac{1.39 \ eV}{kT} \quad (cm*2/s)$$

(111) Si

$$Do = 3.62 \exp \frac{-3.61 \ eV}{kT} + (t*-0.43) \, (9.55 \ X \ 10*-12) \exp \frac{-0.30 \ eV}{kT} \quad (cm*2/s)$$

Hill [28], using marker layers, has conducted an extensive study of
intrinsic diffusion and its dependence on dopant type, crystallographic
orientation, surface reaction, temperature and time. He found under
oxidizing conditions that the diffusion coefficient, Do, can be expressed
in the form

$$Do = D(inert) + dD(oxidizing)$$

$$Do = Dn \exp \frac{-En}{kT} + dDo \exp \frac{-dEo}{kT}$$

For phosphorus

(100)Si

$$Do = 0.6 \exp \frac{-3.51}{kT} + (3.7 \times 10^{-5}) \frac{-2.39}{kT} \qquad (cm^2/s)$$

He also showed that the intrinsic diffusion coefficient under inert conditions is constant, while under oxidizing conditions the diffusion coefficient increases with the unsatisfied surface bond density of the adjacent oxidizing surface independently of diffusion direction, i.e. under inert and oxidizing conditions diffusion is isotropic, and under oxidizing conditions

$$Do(100) > Do(110) > Do(111)$$

Lin et al [3] have shown that oxidation enhanced diffusion of (100) silicon increases with increasing oxidation rate with a sub-linear power dependence of 0.4, i.e.

$$dD(oxidizing) = k \left(\frac{dXox}{dt}\right)^{0.4}$$

This compares with the 0.32 arrived at by Taniguchi et al [1].

Hartmann [6] has investigated phosphorus diffusivity in a dry O2 ambient over the temperature range 1100-1300 deg C and has arrived at an Arrhenius relationship:

$$Do = 4.4 \exp \frac{-3.67}{kT}$$

Antoniadis et al [30] have observed for intrinsic phosphorus diffusion in (100) CZ-silicon that OED decreases as temperature increases.

Matsumoto et al [31] have used the general expression for the diffusion coefficient

$$D = D[eq] \left((1-fi) \left(\frac{Cv}{Cv[eq]}\right) + fi \left(\frac{Ci}{Ci[eq]}\right) \right)$$

and arrived at the temperature dependence of the fractional interstitialcy component of diffusion, fi.

with

$$fi = \frac{Di[eq]}{D[eq]} = \frac{Di[eq]}{Dv[eq] + Di[eq]}$$

$$= 156 \exp (-0.666/kT)$$

This value was calculated from oxidation enhanced diffusion and oxidation induced stacking fault data and it was assumed that

$$dD(oxidizing) = (dXox/dt)*0.5$$

This value of fi is for intrinsic diffusion only.

Mathiot et al [32] have calculated the fractional interstitial component to be

$$fi = \frac{Di[eq]}{D[eq]}$$

$$= 17.1 \exp (-0.511/kT)$$

This value was arrived at using a full calculation accounting for the various couplings between P and the point defects.

Antoniadis et al [29] have made an extensive study of OED in dry O2 at 1000 deg C in intrinsically doped (100) CZ-silicon for times from 5 to 60 minutes. They have found D/D[eq] for intrinsic phosphorus to decrease steadily from about 7 after 5 minutes at 1000 deg C to a nearly constant value of 3.5 after 60 minutes and have calculated the fractional interstitialcy component of diffusion as

$$fi = 0.38$$

They also determined that there exists an energy barrier, of the order of 1.4 eV, to vacancy-interstitial recombination. Mathiot [44] has suggested that the transient in enhanced diffusion is due to a transient in interstitial injection and not to a limited recombination with vacancies.

Fahey [33] has used thermal nitridation processes to assess and quantify diffusion mechanisms. For phosphorus diffusion at 1100 deg C he has calculated the fractional interstitialcy component of diffusion as

$$fi = 0.95$$

Fair [34] has calculated the fractional interstitialcy component of diffusion as

$$fi = 0.12$$

Antoniadis et al [35] have calculated the fractional interstitialcy
component of diffusion as

    fi = 0.40

Gosele et al [36] have calculated the fractional interstitialcy component
of diffusion as

    fi = 0.5 to 1.0

The temperature, oxidation rate and substrate orientation dependence of
oxidation enhanced diffusion is similar to that of oxidation induced
stacking faults [14,15].

C. INTRINSIC PHOSPHORUS DIFFUSION, O2/HCl AMBIENT

The addition of a few percent of HCl gas to the oxidizing atmosphere is
a silicon processing step that results in a reduction of the oxidation
enhanced diffusion of phosphorus and in the shrinkage of near-surface
extrinsic stacking faults [4,28]. It is believed that this effect is a
result of vacancy injection [4] or reduced interstitial injection [37]
at the SiO2/Si interface caused by the reaction of Cl with Si atoms on
lattice sites [4].

Nabeta et al [4] have found that for temperatures below 1000 deg C an
O2/HCl ambient gives the same degree of oxidation enhanced diffusion of
phosphorus as an O2 ambient. This is in good qualitative agreement
with the observation that an O2/HCl ambient has no improved effect on
the SiO2/Si interface until the temperature exceeds 1050 deg C [40].
Diffusivity is therefore described as

    D [T<1000 deg C & HCl = any%] = D [O2]

For temperature greater than 1000 deg C and volume ratio of HCl greater
than 9% oxidation enhanced diffusion of phosphorus was found to drop to
zero giving:

    D [T>1000 deg C & HCl>9%] = D [N2]

D. INTRINSIC PHOSPHORUS DIFFUSION, H2O/Ar AMBIENT

From Taniguchi et al [1]

$$Do = Dn \exp \frac{-En}{kT} + 0.052 \left(\frac{dXox}{dt}\right)*0.3 \exp \frac{-x}{25} \exp \frac{-2.07}{kT} \quad microns*2/hr$$

with

    dXox/dt = oxide growth rate in microns/hr

        x = oxide thickness in micron

E. INTRINSIC PHOSPHORUS DIFFUSION, H2O/O2 AMBIENT

From Taniguchi et al [1]

$$Do = Dn \exp \frac{-En}{kT} + 0.052 \left(\frac{dXox}{dt}\right)*0.3 \exp \frac{-x}{25} \exp \frac{-2.07}{kT} \quad microns*2/hr$$

with

dXox/dt = oxide growth rate in microns/hr

X = depth into the silicon

F. EXTRINSIC PHOSPHORUS DIFFUSION, N2 AMBIENT

When the free carrier concentration exceeds the intrinsic carrier concentration at the diffusion temperature, i.e. when n > ni, the diffusivity is altered due to electric-field effects, h, and mass action effects, n/ni, giving a diffusion coefficient described by:

$$D = h \left[Dv(+)[eq] \frac{ni}{n} + Dv(0)[eq] + Dv(-)[eq]\left(\frac{n}{ni}\right) + Dv(--)[eq]\left(\frac{n}{ni}\right)*2 \right.$$

$$\left. + Di(+)[eq]\left(\frac{n}{ni}\right) + Di(0)(eq) + Di(-)[eq]\left(\frac{n}{ni}\right) + Di(--)[eq]\left(\frac{n}{ni}\right)*2 \right]$$

where

$$h = 1 + \frac{n}{ni} \left[\left(\frac{n}{2ni}\right)*2 + 1 \right]*(-0.5)$$

and

Dv(+)[eq] = intrinsic diffusivity of the phosphorus diffusing as the P(-)V(+) dopant/point-defect pair, etc.

Diffusivity is therefore increased by two mechanisms in the extrinsic regime:

(i)   the electric field effect, h, where the high mobility electron increases the diffusion coefficient of the slower moving dopant atom, asymptotically reaching a maximum of 2 at high active doping levels, and

(ii)   the mass action effect where the numbers of charged vacancies and Si-interstitials, V(-), I(-), and V(--), I(--), increase above their equilibrium values as (n/ni) and (n/ni)*2 respectively, resulting in more vehicles for dopant diffusion.

As such, when the local electron concentration exceeds ni (about 2 X
10*19/cm*3 at processing temperatures) the diffusivity increases. It
has been well established that interaction with a singly negatively
charged point defect, probably the vacancy, results in an (n/ni)
concentration dependent diffusion coefficient for extrinsic diffusion
of arsenic in silicon [16]. For phosphorus however the only
concentration dependent diffusivity observed as n increases above ni
is the self-electric-field enhancement, h, which indicates that
phosphorus interaction with a singly negatively charged point defect
is negligible or nonexistent [17]. (It is argued that this difference
between arsenic and phosphorus is due to the lower phosphorus atom
vibrational frequency and lower entropy change which correlates with
the relative atomic radii [17]). As n increases further approaching
about 1 X 10*20/cm*3 (Ef at about Ec-0.11 eV), phosphorus diffusivity
is seen to increase as (n/ni)*2 (if analysed using the Boltzmann-Matano
diffusion coefficient extraction technique).

The extrinsic phosphorus diffusivity thus far described is locally
determined in the sense that the diffusivity depends only on the local
electron concentration. However it is well documented that local
diffusivity can be significantly affected by the non-local diffusion of
high concentrations of phosphorus. This phenomenon, which occurs for
peak phosphorus concentrations greater than about 1 X 10*20 and which
is effective to depths of about 25 microns in CZ-silicon results in
(i) enhancement of phosphorus diffusion giving a 'tail' region [17]
(but different interpretation in ref. [24]), (ii) enhancement of boron
diffusion giving 'emitter push' [23], and (iii) the enhancement and
retardation of boron and antimony respectively, giving buried layer
movement. The enhancement of the phosphorus 'tail' region and shallow
boron base regions, D/D[eq], has been observed to, (i) increase rapidly
with increasing phosphorus peak concentration above about 7 X 10*19/cm*3
and up to about 4 X 10*20/cm*3, (ii) increase rapidly with decreasing
temperature and (iii) decrease with increasing depth into the silicon. The
enhancement mechanisms are as yet not fully understood but considerable
effort continues to be directed towards arriving at a physically
realistic model for high concentration phosphorus diffusion.

Fair and Tsai [17] argue that the phosphorus 'kink' and 'tail' exist
because of dissociation of phosphorus-vacancy pairs when the Fermi-level
drops to the energy level of the silicon-vacancy second-acceptor (whose
existence has been postulated by Kimerling [17]).

P(+)V(--) <---> P(+)V(-) + e(-) <---> P(+) + V(-) + e(-)

This results in an excess of singly-negatively charged vacancies, V(-),
and a corresponding tail diffusivity increase. The temperature
dependence of the enhancement factor and kink concentration is a
consequence of the fact that the probability for excess vacancy
generation decreases as temperature increases.

Mathiot and Pfister [25] and Mathiot [38] also argue that the vacancy
is the most important point defect for phosphorus diffusion both in the
high concentration and tail regions. It is well established that the
diffusivity of phosphorus is reduced in the plateau and tail region for
peak concentrations in excess of 4 X 10*20/cm*3. Since this is the
concentration regime where SiP precipitation occurs, and since this
precipitation process has been shown to be a very efficient source of
self-interstitials then, they argue, the point defect causing tail
region diffusion must be opposite to those injected by SiP precipitation.
Mathiot and Pfister [39] have also introduced the percolation concept
to explain the high concentration plateau on phosphorus profiles,
because it has also been shown [40,27] that a simple coupling between
P and a point defect cannot explain this plateau.

Nishi, Sakamoto and Ueda [41] have observed enhanced antimony
diffusivity in the high concentration region of a phosphorus profile
indicating an increased vacancy concentration due to the high phosphorus
levels.

Marioton, Gosele and Tan [26] have, using a model based on the
Frank-Turbull and vacancy diffusion mechanisms, suggested that oxidation
enhanced and retarded diffusion can be explained without the need to
assume an interstitialcy diffusion component. They have however accepted
that the high growth rate of oxidation induced stacking faults can only
be explained by the presence of self-interstitials.

Hu, Fahey and Dutton [24] argue that phosphorus diffusion by interaction
with both vacancies and silicon self-interstitials in conjunction with
high concentration dopant precipitation can account for all observed
phosphorus diffusion phenomena. The existence of the precipitates has
been illustrated by Nobili et al [42] and Hu et al quote considerable
buried layer experimental evidence which suggests that a high
concentration phosphorus region generates excess Si-interstitials. They
counter the argument of Mathiot [25] with nitridation experiments which
show Si-interstitial enhancement of phosphorus in the extrinsic regime.

Servidori et al [43] have, using triple-crystal X-ray diffraction,
observed point-defect-excess position and type for various silicon-
implants into silicon, and have monitored their effects on diffusion of
the common dopants. Their conclusions as to the nature of phosphorus
diffusion enhancement mechanisms are in good agreement with those of
Fahey et al [24].

Literature numerical models of diffusion of high concentration
phosphorus profiles are outlined below.

There is much debate over the physical accuracy, and therefore future
usefulness, of the vacancy-only high-concentration phosphorus diffusion
model arrived at by Fair and Tsai [17]. Even their use of the Boltzmann-
Matanno diffusion coefficient extraction method, the use of which they
justified, has been contested [24]. However the ability of the Fair-Tsai
model to quantitatively describe the evolution of phosphorus profiles is
well established, and it remains the most implemented of phosphorus
diffusion models in current process modelling packages. The vacancy-only
model is therefore described in some detail.

For high local electron concentration, n > ne, the diffusivity, D(high) is proportional to (n/ni)*2. When the electron concentration drops below ne, the phosphorus diffusivity changes abruptly and increases as (1/n*2) until the kink concentration, C=nk, is reached. Below this kink concentration the diffusivity assumes a constant value, D(tail). Fair and Tsai observed that the point at which the diffusivity changes from an n*2 to a (1/n*2) dependence, ne, also corresponds to, (i) the point at which some of the dopant concentration becomes inactive, and (ii) the electron concentration where the Fermi level is 0.11 eV below the conduction band (the observations of Kimerling (see ref.17) suggest the second acceptor level of the Si-vacancy exists at Ec-0.11 eV). For diffusion temperatures in excess of 1050 deg C the transition region where D is proportional to (1/n*2) disappears. The constant enhanced diffusivity tail region however exists at all diffusion temperatures with the diffusivity depending on, (i) the peak concentration cubed, ns*3 when ns is greater than about 8 X 10*19/cm*3 and, (ii) the diffusion temperature, with [D(tail)/D[eq]] decreasing as temperature increases. The kink and tail, it is argued, exist as a result of dissociation processes at ne. The magnitude of the enhancement, D(tail)/D[eq], and the concentration of the kink, Ck = nk, are consequences of the fact that the probability for excess vacancy generation decreases as temperature increases. The transition region between ne and nk (which varies in width from about 20 nm at 1200 deg C to 200 nm at 700 deg C) is thought to be a result of additional pair dissociations that occur below ne. Fair and Tsai used both (100) and (111) oriented silicon and a POCl3 diffusion source diluted to various extents in an inert carrier gas. No orientation dependence of diffusivity was reported. The mathematics of the Fair and Tsai phosphorus diffusion model is described below.

(100) and (111) Si

To calculate the diffusivity at any given time, one needs the free electron concentration. The first step in a diffusion simulation is therefore to determine the active dopant concentration, which is less than the total dopant concentration for phosphorus levels in excess of about 8 X 10*19/cm*3.

C = n + (2.04 X 10*-41)/n*3

or

$$n = \frac{C}{2} - [1 + (1 + \frac{2ni}{C})*2]*(-0.5) \qquad (/cm*3) \qquad [20]$$

with

C = the total phosphorus concentration

and

ni = the intrinsic carrier concentration

The diffusion calculation is then undertaken by splitting the profile into three parts, the high concentration region (n > ne), the tail region (n < nk), and the transition region (ne > n > nk).

F1 High concentration region, n > ne
----

ne corresponds to Ef = Ec-0.11 eV

$$ne = 4.65 \ X \ 10*21 \ exp(-0.39 \ eV)/kT \ (/cm*3) \qquad [19]$$

In this high concentration region the diffusivity is determined by the local electron concentration.

$$D(high) = h \ [Dv(O)[eq] + Dv(--)[eq](\frac{n}{ni})*2]$$

with

$$h = 1 + \frac{n}{ni} \ [\ (\frac{n}{2ni})*2 + 1 \ ]*(-0.5)$$

$$Dv[eq](O) \ = 3.85 \ exp \ (-3.66 \ eV/kT) \ cm*2/sec$$

$$Dv[eq](--) = 44.2 \ exp \ (-4.37 \ eV/kT) \ cm*2/sec$$

F2 Tail region, n < nk
----

nk is determined by calculating D(tail) and adhering to the empirical approximation outlined in the following section on diffusivity in the transition region, ne > n > nk. In the tail region the diffusivity is determined by local electron concentration (which gives D(normal)) and by the point defect flux from the high concentration region (which gives D(defect flux)) if that high concentration region exceeds ne.

$$D(tail) = D(normal) + D(defect \ flux)$$

$$= h \ [\ Dv[eq](0) + Dv[eq](-) \ \frac{ns*3}{ne*2 \ ni} \ (\ 1 + exp \ \frac{0.3eV}{kT} \ ) \ ]$$

with
$$Dv[eq](-) = 4.44 \ exp(-4.0 \ eV)/kT \ cm*2/sec$$

ns = the peak electron concentration of the phosphorus profile

The Dv[eq](-) term in D(tail) is the diffusive component due to singly negatively charged vacancies. The concentration of the singly negatively charged vacancies is determined by P(+)V(--) and P(+)V(-) pair dissociations with the latter being more probable by a factor of exp(0.3 eV/kT). This factor varies from 36 at 700 deg C to 9 at 1300 deg C.

F3 Transition region, ne > n > nk

In the transition region the diffusivity is estimated from its
empirically observed (1/n*2) or (ni/n)*2 dependence. The diffusivity
calculated at the end of the high concentration region, i.e. D(n=ne),
is multiplied by (ni/n)*2 as depth into the silicon increases, until
D(transition) = D(tail) is reached [19]. The concentration at this point
is called the kink concentration and at this concentration the tail
region is assumed to start.

Fair [21] has shown that when the peak concentration, ns, is greater
than 4-5 X 10*20/cm*3, both the high and tail diffusivities are
retarded. This he explained by a misfit-induced lattice strain which
changes slowly in depth compared to the phosphorus diffusion length.
Strain induced by tensile stress causes a bandgap narrowing which
reduces the high concentration region diffusivity according to

    D(high)strained = D(high) exp(delEg/kT)

and the tail region diffusivity according to

    D(tail)strained = D(tail) exp(3delEg/kT)

where delEg is the bandgap narrowing at the profile peak with a
concentration dependence given by:

    delEg = -1.5 X 10*22 [C(peak)-(3 X 10*20)] eV

with

C(peak) = peak dopant concentration /cm*3 [21]

and a temperature dependence given by:

    delEg = -7.1 X 10*-10[ni/T]*0.5 eV [22]

For high-dose phosphorus implants, additional bandgap narrowing occurs,
probably because of permanent lattice disorder produced by the
implantation process:

    delEg = -2.3 X 10*-6 X Q*0.25 eV

Bandgap narrowing included, the Fair-Tsai diffusion model is applicable
at very high surface concentrations close to the solid solubility limit.

Having finished the diffusion simulation, one reconverts electron
concentration (which, above ni, correspond to active doping levels)
to total doping concentration, C.

    C = n + (2.04 X 10*-41 X n*3)

The form of this relationship between total and activated dopant levels
i.e.:

    C = n + (K X n*3)

does not produce a flat zone on the phosphorus electron concentration
profile, in contradiction with that observed experimentally [37]. Kroger
[38] has suggested that the relationship is described by:

$$C = n + \frac{2K(n*3)}{1 - K(n*2)}$$

which indicates the existence of an asymptotic limit of the electron
concentration at n = (1/K)*0.5. This constitutes a simple modification
of the Fair-Tsai model.

Jeppson et al [20] have modelled extrinsic phosphorus diffusion in a
similar manner to Fair and Tsai [17] but disregarding the transition
between the high concentration region and the 'tail' region (i.e.
disregarding the region between ne and nk on the Fair-Tsai model).
Their diffusion constant therefore makes a discontinuous 'jump' at the
kink concentration. Jeppson et al justify this approximation by noting
that the transition region is very narrow, varying from about 50 nm at
1000 deg C to 150 nm at 875 deg C [17].

$$n = \frac{C}{2} - \left[1 + \left(\frac{2ni}{C}\right)*2 \right]*(-0.5) \ /cm*3 \quad [20]$$

with

    C = the total phosphorus concentration

    ni = the intrinsic carrier concentration

F4 High concentration region, x < xk and n > nk
----

xk = 1.035(Cs/ni)[Dv(--)[eq]t]*(-0.5) for Cs = constant

nk = 1.0 X 10*23 exp(-0.79 eV)/kT)/cm*3 for Cs = constant

xk = [(Q*2 Dv(--)[eq]t)/ni]*0.25 for Q = constant

where Cs is the surface concentration and Q is the quantity of dopant incorporated into the silicon. (Cs = constant therefore corresponds to a pre-deposition and Q = constant to a drive-up or implant diffusion step). In this high concentration region the diffusivity is determined by local electron concentration.

D(high) = h Dv(--)[eq](n/ni)*2

with

h = 2

Dv(--)[eq] = 44.2 exp(-4.37 eV/kT) cm*2/sec

$$n = \frac{C}{2} \ [\ 1 + (\ 1 + (\frac{2ni}{C})*2 \ ) \ ]*(-0.5) \ /cm*3$$

F5 Tail region, x > xk and n < nk

Here the diffusivity is determined by local electron concentration and by the point defect flux from the high concentration region if that high concentration region exceeds ne.

$$D(tail) = Dv(-)[eq] \ \frac{ns*3}{(ne*2)ni} \ (\ 1 + exp(\frac{0.3 \ eV}{kT}) \ )$$

with

$$Dv(-)[eq] = 4.44 \ exp(\frac{-4.0 \ eV}{kT}) \ cm*2/sec$$

ns = the peak electron concentration of the phosphorus profile

$$ne = (4.65 \ X \ 10*21) \ exp(\frac{-0.39}{kT}) \ /cm*3 \qquad [19]$$

As for the Fair-Tsai model, this diffusion model is applicable at very high surface concentrations close to the solid solubility limit if reduced diffusivity due to bandgap narrowing under lattice strain is included. (see the Fair-Tsai model above). Reconverting electron concentration (which, above ni, corresponds to active dopant levels), one gets the total phosphorus dopant level:

C = n + (2.04 X 10*-41) n*3

Erana et al [18] have described the phosphorus profile resulting from a
high surface concentration in a simplified empirical manner. The high
concentration region and 'tail' region are described as two independent
Gaussians with surface concentrations Cs(high) and Cs(tail) and
diffusivities D(high) & D(tail). They used (111) silicon and a P2O5
diffusion source and had a peak phosphorus concentration in excess of
1 X 10*20/cm*3.

$$D = D(high\ or\ tail)\ exp\ (-En/kT)$$

(111)Si

F6 High concentration region

$$Cs(high) = solid\ solubility\ level\ \ /cm*3$$

$$D(high) = (1.365\ X\ 10*-6)\ exp\ (\frac{-1.829\ eV}{kT})\ \ cm*2/sec$$

F7 Tail region

$$Cs(tail) = (1.732\ X\ 10*21)\ exp\ \frac{-0.79\ eV}{kT}$$

$$D(tail) = (1.907\ X\ 10*-7)\ exp\ \frac{-1.431\ eV}{kT}$$

Massetti et al [7] describe extrinsic phosphorus diffusivity simply as:

$$D = Dn\ exp\ \frac{-En}{kT}$$

(111) CZ-Si

$$Cs = 10*20/cm*3\ then\ Dn = 1.3\ exp\ (-3.5\ eV/kT)\ cm*2/s$$

G. EXTRINSIC PHOSPHORUS DIFFUSION, O2 AMBIENT

The growth and/or deposition of thin films on top of silicon wafers,
at various steps of the integrated circuit fabrication process,
results in several types of phenomena in the underlying silicon substrate.
Of interest here is the effect of oxidation of the silicon surface on
diffusion of high concentration phosphorus profiles in the substrate.
The effect has not been well characterised. Some observations on the
effect of oxidation on high concentration diffusion are outlined below.

Massetti et al [7] have investigated OED in the extrinsic regime in
(111) oriented, CZ silicon. The pre-deposition from a phosphorus
oxychloride source resulted in a profile of depth 0.35 microns and peak
concentrations of about 1 X 10*20/cm*3. Drive-in anneals were carried
out in inert (nitrogen) and oxidizing atmospheres (N2+10%O2, dry O2, and
O2 + steam) at temperatures between 1000 and 1200 deg C, and a single
diffusion coefficient was arrived at for each case using the Kato and
Nishi model. They found an activation energy of 3.5 eV for inert anneals
and 2.5 eV for oxidizing anneals, with the pre-exponential constants for
the oxidizing ambient being ambient type dependent.

$$D = Do \exp \frac{-Eo}{kT}$$

(111) CZ-Si

  D[dry O2] = 4.6 X 10*-4 exp(-2.5 eV/kT) cm*2/sec

Taniguchi et al [1] found that oxidation enhanced diffusion decreases
as the concentration of the diffusing species increases beyond the point
where concentration-dependent diffusion begins, i.e. beyond the intrinsic
carrier concentration at the temperature (see his figure 3). This
effect was explained in terms of the reduction of oxidation produced
self-interstitials by recombination with the increasing supply of
vacancies.

Aleksandrov et al [5], using 1.6 X 10*21/cm*3 phosphorus surface
concentrations, found the diffusion coefficients to vary with time and
with orientation during a 1000 deg C dry O2 anneal. The halving of the
diffusivity over the first 5 hours of the anneal and the subsequent
levelling off were put down to a reduction in the growth rate of
individual dislocations and a corresponding loss of the ability to
absorb point defects.

H. EXTRINSIC PHOSPHORUS DIFFUSION, O2/N2 AMBIENT

From Massetti et al [7] extrinsic phosphorus diffusivity in a 10% O2/N2
ambient is given by

  D = Do exp (-Eo/kT)

(111) CZ-Si

  D[10%O2/N2] = 3.7 X 10*-4 exp (-2.5 eV/kT) cm*2/sec

I. EXTRINSIC PHOSPHORUS DIFFUSION, H2O/O2 AMBIENT

From Massetti et al [7] extrinsic phosphorus diffusivity in a steam/O2
ambient is given by

$$D = Do \exp \frac{-Eo}{kT}$$

(111) CZ-Si

  D[steam/O2] = 5.9 X 10*-4 exp(-2.5 eV/kT) cm*2/sec

REFERENCES
----

[1]   K.Taniguchi, K.Kurosawa, M.Kashiwagi [J. Electrochem. Soc.
        (USA) vol.127 no.10 (Oct 1980) p.2243-8]
[2]   Y.Ishikawa, Y.Sakina, H.Tanaka, S.Matsumoto, T.Niimi [J.
        Electrochem. Soc. (USA) vol.129 no.3 (Mar 1982) p.644-8]
[3]   A.Miin-Ron Lin, D.A.Antoniadis, R.W.Dutton [J. Electrochem.
        Soc. (USA) vol.128 no.5 (May 1981) p.1131-7]
[4]   Y.Nabeta et al [J. Electrochem. Soc. (USA) vol.123 no.9
        (Jun 1976) p.1416-7]
[5]   O.V.Aleksandrov et al [Izv. Akad. Nauk SSSR Neorg. Mater.
        (1983) p.517-20]
[6]   U.Hartman [Wiss. Z. Tech. Hochsch. Ilmenau (Germany) vol.20
        no.12 (Dec 1974) p.75-92]
[7]   G.Massetti, S.Solmi, G.Soncini [Solid-State Electron (GB)
        vol.16 no.12 (Dec 1973) p.1491-21]
[8]   J.S.Makris, B.J.Masters [J. Electrochem. Soc. (USA) vol.120
        no.9 (Sep 1973) p.1252-3]
[9]   M.L.Barry [J. Electrochem. Soc. (USA) vol.117 no.11 (Nov 1970)
        p.1405-10]
[10]   T.C.Chan et al [Proc. IEEE (USA) vol.58 no.4 (1970) p.588-9]
[11]   S.M.Hu [J. Appl. Phys. (US)A vol.38 (1967) p.3066]
[12]   R.B.Fair [Diffus. & Defect Data (Switzerland) vol.37 (1984)
        p.1-24]
[13]   R.N.Ghoshtagore [Solid-State Electron. (GB) vol.15 (1972) p.1113]
[14]   S.M.Hu [Appl. Phys. Lett. (USA) vol.27 no.4 (15 Aug 1975) p.165-7]
[15]   A.Miin-Ron Lin, R.W.Dutton, D.A.Antoniadis, W.A.Tiller [J.
        Electrochem. Soc. (USA) vol.128 no.5 (May 1981) p.1121-30]
[16]   R.B.Fair, J.C.C.Tsai [J. Electrochem. Soc. (USA) vol.122 no.12
        (Dec 1975) p.1689-96]
[17]   R.B.Fair, J.C.C.Tsai [J. Electrochem. Soc. (USA) vol.124 no.7
        (Jul 1977) p.1107-18]
[18]   G.Eranna, D,Kakati [J. Appl. Phys. (USA) vol.54 no.11 (Nov 1983)
        p.6754-6]
[19]   D.A.Antoniadis et al [IEEE J. Solid-State Circuits (USA)
        vol.SC-14 (1979) p.412]
[20]   K.O.Jeppson, D.Anderson [J. Electrochem. Soc. (USA) vol.133 no.2
        (Feb 1986) p.397-400]
[21]   R.B.Fair [J. Appl. Phys. (USA) vol.50 no.2 (Feb 1979) p.860-8]
[22]   S.M.Sze [in 'VLSI Technology (McGraw-Hill, 1983) p.180]
[23]   D.B.Lee [Philips Res. Rept. (Netherlands) suppl. no.5 (1974)]
[24]   S.M.Hu, P.Fahey, R.W.Dutton [J. Appl. Phys. (USA) vol.54 no.12
        (Dec 1983) p.6912-22]
[25]   D.Mathiot, J.C.Pfister [Appl. Phys. Lett. (USA) vol.47 no.9 (1
        Nov 1985) p.962-4]
[26]   B.Marioton et al [Chemtronics (GB) vol.1 no.4 (1986) p.156-60]
[27]   D.Mathiot, J.C.Pfister [J. Appl. Phys. (USA) vol.53 (1982) p.3053]
[28]   C.Hill [in 'Semiconductor Silicon 1981' Eds H.R.Huff, J.Kriegler,
        Y. Takeishi (Electrochem. Soc., Princeton, 1981)]
[29]   D.A.Antoniadis et al [J. Appl. Phys. (USA) vol.53 no.10 (1982)
        p.6788-96]
[30]   D.A.Antoniadis, A.M.Lin, R.W.Dutton [Appl. Phys. Lett. (USA)
        vol.33 no.12 (15 Dec 1978) p.1030-3]
[31]   S.Matsumoto, Y.Ishikawa, T.Niimi [J. Appl. Phys. (USA) vol.54
        no.9 (Sep 1983) p.5049-54]
[32]   D.Mathiot, J.C.Pfister [J. Appl. Phys. (USA) vol.55 no.10 (15
        May 1984) p.3518-30]

[33] P.Fahey, G.Barbuscia, M.Moslehi, R.W.Dutton [Appl. Phys. Lett. (USA) vol.46 no.8 (15 Apr 1985) p.784-6]

[34] R.B.Fair [J. Appl. Phys. (USA) vol.51 no.11 (Nov 1980) p.5828-32]

[35] D.Antoniadis, I.Moskowitz [J. Appl. Phys. (USA) vol.53 no.12 (Dec 1982) p.9214-16]

[36] U.Gosele et al [Proc. Symp. Defects in Semiconductors II Boston, MA, USA Nov 1982 Eds Mahajan Corbett (North-Holland, Amsterdam, 1983)]

[37] S.Oh, W.A.Tiller, Soo-Kap Hahn [Appl. Phys. Lett. (USA) vol.48 no.17 (28 Apr 1986) p.1125-7]

[38] D.Mathiot [Proc. Conf. Semiconductor Silicon 1986 Eds. H.R.Huff, et al (Electrochem. Soc., Pennington, USA 1986) p.556]

[39] D.Mathiot, J.C.Pfister [J. Phys. Lett. (France) vol.43 (1982) p.L453]

[40] M.Yoshida [Jpn. J. Appl. Phys. (Japan) vol.18 no.3 (1979) p.479] Also D.Mathiot, J.C.Pfister [J. Appl. Phys. (USA) vol.53 no.4 (1982) p.3053-8]

[41] N.Nishi, K.Sakamoto, J.Ueda [J. Appl. Phys. (USA) vol.59 no.12 (15 Jun 1986) p.4177-9]

[42] D.Nobili, A.Armigliato, M.Finetti, S.Solmi [J. Appl. Phys. (USA) vol.53 (1 Mar 1982) p.1484-91]

[43] M.Servidori et al [J. Appl. Phys. (USA) vol.61 no.5 (1 Mar 1987) p.1834-40]

[44] D.Mathiot [Appl. Phys. Lett. (USA) vol.48 (1986) p.627]

13.7 DIFFUSION OF As IN Si

by A.A.Brown, P.B.Moynagh and P.J.Rosser

November 1987
EMIS Datareview RN=16193

A. INTRINSIC ARSENIC DIFFUSION, N2 AMBIENT

The mathematical basis for the diffusion process was laid by Fick whose
first law sets forward the hypothesis that the rate of transfer of a
diffusing particle through unit area is proportional to the magnitude
of the gradient normal to that area, i.e in the x-direction

$$J = -D \frac{dC}{dx}$$

where the proportionality constant, D (metres*2/second), is known as
the diffusion constant or the diffusivity. The diffusion process occurs
because of the random walk of the impurity atoms and is therefore a
thermally activated process. The temperature dependence of the diffusion
coefficient is Arrhenius in nature:

$$D = Do \exp \frac{-E}{kT}$$

where E is the activation energy (which, for the common dopants in
silicon, is about four electron volts).

Due to the low concentrations of stable point defects in silicon, it has
not been possible to determine directly whether vacancies or
self-interstitials are dominant in diffusion processes. Numerous kinds
of indirect observations have been made to build cases for vacancies
and for self-interstitials. The current belief is that both types of
point defect exist somewhat independently [11], and that both influence
diffusion. Under diffusive conditions where the silicon is intrinsically
doped and where no surface reaction is taking place, the vacancy and
self-interstitial concentrations are temperature dependent only and are
described approximately by the equilibrium concentrations:

$$Cv[eq] = Cov \exp \frac{-Ev}{kT}$$

$$Ci[eq] = Coi \exp \frac{-Ei}{kT}$$

Dopant diffusion takes place in the silicon by interacting with these
point defects so that the diffusing species are the arsenic-vacancy
pair, AV, and the arsenic-Si-interstitial pair, AI. The intrinsic

diffusivity due to the vacancy mechanism is referred to as Dv[eq], and
that due to the interstitialcy mechanism as Di[eq]. The intrinsic, inert
ambient diffusivity is therefore given by the sum of the equilibrium
diffusivities due to the vacancy and the self-interstitial components:

$$D[eq] = Dv[eq] + Di[eq]$$

The wide energy range available to the Fermi level in silicon leads to
a given lattice defect (i.e. a vacancy or Si-interstitial) appearing
in a variety of ionized states [12]. As such the dopant/point-defect
pairs exist as, for example, A(+)V(-) which we will denote by AV(-).
The intrinsic, inert ambient diffusivity can therefore be written:

$$D[eq] = Dv(+)[eq] + Dv(0)[eq] + Dv(-)[eq] + Dv(--)[eq]$$

$$+ Di(+)[eq] + Di(0)[eq] + Di(-)[eq] + Di(--)[eq]$$

where the intrinsic diffusivity of the arsenic diffusing as AV(-) is
given by Dv(-)[eq], etc.

The diffusion coefficient in crystals with a cubic lattice should be
isotropic. For arsenic doping levels of less than about 10*20/cm*3 with
no surface reaction taking place this is indeed the case [13]. The
diffusion coefficient is therefore given by:

$$D = Dn \exp \frac{-En}{kT}$$

Literature diffusivities for inert diffusion of arsenic in single
crystal silicon are listed below

(100),(110),(111) Si

temperature (deg C)	Dn(cm*2/s)	En(eV)	reference
900-1250	22.9	4.1	Fair [14]
1050-1200	59	4.2	Cerofolini [3]
1000-1300	24	4.08	Matsumoto [4]
840-1250	13	4.05	Hill [13]
950-1150	35.3	4.11	Ishikawa [6]
1050-1400	102.8	4.27	Masters [7]
1167-1394	0.0655	3.44	Ghoshtagore [1]
950-1150	2870	4.58	Kennedy [2]
1100-1400	1.218	3.75	Kendall [5]

B. INTRINSIC ARSENIC DIFFUSION, O2 AMBIENT

Oxidation of silicon under most processing conditions enhances the diffusion of arsenic by a process of interstitial injection into the silicon substrate [11]. The enhanced diffusion occurs because the number of silicon self-interstitials available to interact with the arsenic atoms has increased and the diffusivity is described by a general diffusion equation:

$$D = Dv + Di$$

i.e.

$$D = Dv[eq] \frac{Cv}{Cv[eq]} + Di[eq] \frac{Ci}{Ci[eq]}$$

or

$$D = D[eq] \left((1-fi) \frac{Cv}{Cv[eq]} + fi \frac{Ci}{Ci[eq]}\right)$$

where

$$fi = \frac{Di[eq]}{D[eq]} = \frac{Di[eq]}{Dv[eq] + Di[eq]}$$

and fi is known as the fractional interstitialcy component of diffusion. Some injected Si-interstitials will be lost by Frenkel recombination with vacancies so that the thermal oxidation process results in a simultaneous supersaturation of Si-interstitials and undersaturation of vacancies. The effect on total diffusion rate of the increase of AI diffusion species and decrease in AV diffusion species is described by the fractional interstitialcy component of diffusion, fi, literature values of which are given below.

Experimental evidence suggests that the Si-interstitial injection rate during thermal oxidation depends on the oxidation rate. More specifically, it has been found that:

$$Ci-Ci[eq] = k \left(\frac{dXox}{dt}\right)*P \text{ with } P < 1$$

where (dXox/dt) is the oxidation rate.

Since the oxidation rate decreases as oxidation time increases, it is expected that the diffusion enhancement will also decrease with oxidation time. The oxidation rate (and therefore the rate of Si-interstitial injection) is not dependent on the orientation beyond the linear segment. However the rate of decay of these excess point-defects will be dependent on the silicon surface or the silicon-oxide interface recombination rates, and these might be expected to be proportional to the rectilinear density. As such the rate of loss of excess point-defects would increase as (100), (110), (111) and the magnitude of oxidation enhanced diffusion would increase as (111), (110), (100) [13].

As is described below, oxidation enhanced diffusion in the intrinsic regime is a function of temperature, oxidation rate, depth into the silicon and substrate orientation.

Literature models of oxidation enhanced arsenic diffusion are described below.

Ishikawa et al [6] have found oxidation enhanced diffusion in the intrinsic regime to be dependent on temperature, time and the crystal orientation of the substrate. It decreased with increasing time and temperature for a (100) orientation whereas for a (111) orientation, there is no significant difference between oxidizing or inert ambients.

Above about 1100 deg C, no significant oxidation enhancement was observed.

$$Do = D \text{ (inert)} + dD \text{ (oxidizing)}$$

$$Do = Dn \exp\frac{En}{kT} + (t*-P)dDo \exp\frac{-dEo}{kT}$$

(100)Si

$$Do = 35.3 \exp\frac{-4.11 \text{ eV}}{kT} + (t*-0.48)(9.35 \text{ X } 10*-10) \exp\frac{-0.93 \text{ eV}}{kT} \text{ cm*2/s}$$

(111)Si

$$Do = D(\text{inert}) = 35.3 \exp\frac{-4.11 \text{ eV}}{kT} \text{ cm*2/s}$$

Hill [13], using marker layers, has conducted an extensive study of intrinsic diffusion and its dependence on dopant type, crystallographic orientation, surface reaction, temperature and time. He found under oxidizing conditions that the diffusion coefficient, Do, can be expressed in the form

$$Do = D(\text{inert}) + dD(\text{oxidizing})$$

$$= Dn \exp\frac{-En}{kT} + dDo \exp\frac{-dEo}{kT}$$

For arsenic

(100)Si

$$Do = 13 \exp\frac{-4.05 \text{ eV}}{kT} + (1.9 \text{ X } 10*-6) \exp\frac{-2.34 \text{ eV}}{kT} \text{ cm*2/s}$$

He also showed that the diffusion coefficient under inert conditions
is constant, while under oxidizing conditions the diffusion coefficient
increases with the unsatisfied surface bond density of the adjacent
oxidizing surface independently of diffusion direction, i.e. under
inert and oxidizing conditions diffusion is isotropic, and under
oxidizing conditions

$$Do(100) > Do(110) > Do(111)$$

Antoniadis et al [15] have observed for intrinsic arsenic diffusion
in (100) CZ-silicon that oxidation enhanced diffusion decreases as
temperature increases. Indeed at temperatures above 1100 deg C,
oxidation retarded diffusion was observed, suggesting a fractional
interstitialcy component less than 0.5.

Antoniadis et al [11] have made an extensive study of oxidation
enhanced diffusion in dry O2 at 1090 deg C in intrinsically doped (100)
CZ-silicon for times from 5 to 60 minutes. They have found D/D[eq] for
intrinsic arsenic to decrease steadily from about 2 after 5 minutes at
1000 deg C to a nearly constant value of 1.5 after 60 minutes (that is
Do = 1.5D[eq]) and have calculated the fractional interstitialcy
component of diffusion as

$$fi = 0.35$$

They also determined that there exists an energy barrier, of the order
of 1.4 eV, to vacancy-interstitial recombination.

Matsumoto [16], using Ishikawa's data [6] has calculated that the
fractional interstitialcy component, fi, is given by an Arrhenius
function:

$$fi = 42 \exp \frac{-0.542 \text{ eV}}{kT}$$

Ho et al [22] gave the fractional interstitialcy component as

$$fi = 6.85 \exp \frac{-0.63 \text{ eV}}{kT}$$

The temperature, oxidation rate and substrate orientation dependence
of oxidation enhanced diffusion is similar to that of oxidation induced
stacking faults [17,18].

C. INTRINSIC ARSENIC DIFFUSION, O2/HCl AMBIENT

The addition of a few percent of HCl gas to the oxidizing atmosphere is
a silicon processing step that results in a reduction of the oxidation
enhanced diffusion of arsenic and in the shrinkage of near-surface
extrinsic stacking faults [19,13]. It is believed that this effect is a
result of vacancy injection [19] or reduced interstitial injection [21]
at the Si/SiO2 interface caused by the reaction of Cl with Si atoms on
lattice sites.

Subrahmanyan [20] has found that the addition of a few percent of HCl
to the O2 ambient leads to a reduction in the oxidation enhanced
diffusion at temperatures above 1000 deg C.

D. EXTRINSIC ARSENIC DIFFUSION, N2 AMBIENT

When the free carrier concentration exceeds the intrinsic carrier
concentration at the diffusion temperature, i.e. when n > ni, the
diffusivity is altered due to electric-field effects, h, and mass
action effects, n/ni, giving a diffusion coefficient described by:

$$
D = h \times \left[Dv(+)[eq] \frac{ni}{n} + Dv(0)[eq] + Dv(-)[eq] \frac{n}{ni} + Dv(--)[eq](\frac{n}{ni})*2 \right.
$$

$$
\left. + Di(+)[eq] \frac{ni}{n} + Di(0)[eq] + Di(-)[eq] \frac{n}{ni} + Di(--)[eq](\frac{n}{ni})*2 \right]
$$

where

$$
h = 1 + (\frac{n}{2ni}) \left[(\frac{n}{2ni})*2 + 1 \right]*(-0.5)
$$

and

Dv(+)[eq] = intrinsic diffusivity of the arsenic diffusing as
the A(-)V(+) dopant/point-defect pair, etc.

Diffusivity is therefore increased by two mechanisms in the extrinsic
regime:

(a) the electric field effect, h, where the high mobility electron
     increases the diffusion coefficient of the slower moving dopant
     atom, asymptotically reaching a maximum of 2 at high active
     doping levels, and

(b) the mass action effect where the numbers of charged vacancies and
     Si-interstitials, V(-), I(-), and V(--), I(--), increase above
     their equilibrium values as (n/ni) and (n/ni)*2 respectively,
     resulting in more vehicles for dopant diffusion.

As such, when the local electron concentration exceeds ni (about 2 X
10*19/cm*3 at processing temperatures) the diffusivity increases. It
has been well established that interaction with a singly negatively
charged point defect, probably the vacancy, results in an (n/ni)
concentration dependent diffusion coefficient for extrinsic diffusion
of arsenic in silicon [7,14]. This extrinsic arsenic diffusivity is
locally determined in the sense that the diffusivity depends only on
the local electron concentration.

Literature numerical models of diffusion of high concentration arsenic
profiles are outlined below.

Ho et al [22] describe the extrinsic, inert diffusivity as a
combination of interaction of arsenic with neutral vacancies As(+)V(0)
and singly negatively charged vacancies As(+)V(-)such that:

$$
D = 0.0114 \exp \frac{-3.44 \ eV}{kT} + (\frac{n}{ni}) 31.0 \exp \frac{-4.15 \ eV}{kT} \ cm*2/s
$$

Colclaser [23] also expressed the diffusivity as a combination of two
terms, As(+)V(0) and As(+)V(-):

$$D = 0.38 \exp \frac{-3.58 \text{ eV}}{kT} + \left(\frac{n}{ni}\right) 22.9 \exp \frac{-4.1 \text{ eV}}{kT} \text{ cm*2/s}$$

Wong et al [8] calculated diffusion coefficients for arsenic
concentrations from 4 X 10*18/cm*3 to 4 X 10*19/cm*3 in the temperature
range 1000 to 1200 deg C. They found the activation energy to be 4.1
eV and the pre-exponential factor to be proportional to the arsenic
concentration, that is to (n/ni). The values given were:

$$4 \text{ X } 10*18/cm*3: \quad D = 44 \exp \frac{-4.2 \text{ eV}}{kT} \text{ cm*2/s}$$

$$1 \text{ X } 10*19/cm*3: \quad D = 25 \exp \frac{-4.0 \text{ eV}}{kT} \text{ cm*2/s}$$

$$4 \text{ X } 10*19/cm*3: \quad D = 110 \exp \frac{-4.1 \text{ eV}}{kT} \text{ cm*2/s}$$

Ohkawa et al [9] studied the diffusion of arsenic in the temperature
range 700 to 1100 deg C with surface concentration in excess of
10*20/cm*3. They found that the diffusion coefficient was given by an
Arrhenius relationship:

$$D = 969 \exp \frac{-4.45 \text{ eV}}{kT} \text{ cm*2/s}$$

E. EXTRINSIC ARSENIC DIFFUSION, OXIDIZING AMBIENT

The growth or deposition of thin films on top of silicon at various
stages during the manufacture of integrated circuits results in several
types of phenomena in the underlying substrate. Of interest here is the
effect of oxidation on the diffusion of high concentration arsenic
profiles. This effect has not been well characterized. Some observations
on this process are outlined below.

Fair et al [14] found that in the extrinsic regime, the diffusion of
arsenic was independent of any reaction taking place at the surface.
That is, the same amount of diffusion was observed for annealing
ambients of N2, O2 or H2O/O2. The only difference which was seen was
that the segregation of the arsenic into the silicon at the
oxide/silicon interface resulted in arsenic pile-up.

Ishikawa et al [10] observed oxidation retarded diffusion in the
temperature range 1000 to 1100 deg C. At 950 deg C, no enhancement or
retardation was seen. The retardation was found to decrease with time.
Ishikawa suggested that the injection of excess interstitials during
the oxidation process resulted in a greatly reduced vacancy
concentration by recombination. At high temperatures, this recombination
leads to a very low concentration of vacancies and hence less diffusion.

As the oxidation progresses, the oxidation rate decreases and so the interstitial injection rate decreases, allowing the vacancy concentration to increase towards the equilibrium value. Thus the amount of retardation decreases with time.

REFERENCES
----

[1]   R.N.Ghoshtagore [Phys. Rev. B (USA) vol.3 no.2 (1971) p.397-401]
[2]   D.P.Kennedy et al [Proc. IEEE (USA) vol.59 no.2 (1971)
       p.335-6]
[3]   G.F.Cerofolini et al [Thin Solid Films (Switzerland) vol.135
       no.1 (1986) p.59-72]
[4]   S.Matsumoto, J.F.Gibbons, V.Deline, C.A.Evans,Jr [Appl. Phys.
       Lett. (USA) vol.37 no.9 (1 Nov 1980) p.821-4]
[5]   D.L.Kendall et al [Semiconductor Silicon 1969,
       Eds R.R.Haberecht, E.L.Kern (Electrochem. Soc., NY, USA, 1969)]
[6]   Y.Ishikawa et al [J. Electrochem. Soc. (USA) vol.129 no.3
       (1982) p.644-8]
[7]   B.J.Masters et al [J. Appl. Phys. (USA) vol.40 no.6 (1969)
       p.2390-4]
[8]   J.Wong et al [J. Electrochem. Soc. (USA) vol.119 no.10 (1972)
       p.1413-20]
[9]   S.Ohkawa, Y.Nakajima, Y.Fukukawa [Jpn. J. Appl. Phys. (USA)
       vol.14 no.4 (Apr 1975) p.458-65]
[10]   Y.Ishikawa et al [J. Electrochem. Soc. (USA) vol.130 no.10
       (1983) p.2109-11]
[11]   D.A.Antoniadis, I.Moskowitz [J. Appl. Phys. (USA) vol.53 no.10
       (Oct 1982) p.6788-96]
[12]   D.Anderson, K.O.Jeppson [J. Electrochem. Soc. (USA) vol.131
       no.11 (Nov 1984) p.2675-9]
[13]   C.Hill [Semiconductor Silicon 1981, Eds H.R.Huff et al
       (Electrochem. Soc., 1981) p.988-1006]
[14]   R.B.Fair, J.C.C.Tsai [J. Electrochem. Soc. (USA) vol.122 no.12
       (1975) p.1689-96]
[15]   D.A.Antoniadis, A.M.Lin, R.W.Dutton [Appl. Phys. Lett. (USA)
       vol.33 no.12 (15 Dec 1978) p.1030-3]
[16]   S.Matsumoto, Y.Ishikawa, T.Niimi [J. Appl. Phys. (USA) vol.54
       no.9 (Sep 1983) p.5049-54]
[17]   S.M.Hu [Appl. Phys. Lett. (USA) vol.27 no.4 (15 Aug 1975)
       p.165-7]
[18]   A.Miin-Ron Lin, R.W.Dutton, D.A.Antoniadis, W.A.Tiller [J.
       Electrochem. Soc. (USA) vol.128 no.5 (May 1981) p.1121-30]
[19]   Y.Nabeta et al [J. Electrochem. Soc. (USA) vol.123 no.9 (1976)
       p.1416-7]
[20]   R.Subrahmanyan, H.Z.Massoud, R.B.Fair [J. Appl. Phys. (USA)
       vol.61 no.10 (15 May 1987) p.4804-7]
[21]   S.Oh, W.A.Tiller, S.Hahn [Appl. Phys. Lett. (USA) vol.48 no.10
       (28 Apr 1986) p.1125-6]
[22]   C.P.Ho et al [Tech. Report no.SEL 84-001, Stanford Electronics
       Labs, Stanford Univ., Stanford, California, USA (July 1984)]
[23]   R.A.Colclaser [Microelectronics: Processing and Device Design
       (John Wiley and Sons, 1980)]

13.8 DIFFUSION OF Sb IN Si

by P.B.Moynagh and P.J.Rosser

December 1987
EMIS Datareview RN=17811

A. INTRINSIC ANTIMONY DIFFUSION, N2 AMBIENT

The mathematical basis for the diffusion process was laid by Fick whose
first law sets forward the hypothesis that the rate of transfer of a
diffusing particle through unit area is proportional to the magnitude of
the gradient normal to that area, i.e. in the x-direction

$$J = -D \frac{dC}{dx}$$

where the proportionality constant, D (metres*2/second), is known as
the diffusion constant or the diffusivity. The diffusion process occurs
because of the random walk of the impurity atoms and is therefore a
thermally activated process. The temperature dependence of the diffusion
coefficient is Arrhenius in nature:

$$D = Do \exp \frac{-E}{kT}$$

where E is the activation energy (which, for the common dopants in
silicon, is about four electron-volts).

Due to the low concentrations of stable point defects in silicon, it has
not been possible to determine directly whether vacancies or
self-interstitials are dominant in diffusion processes. Numerous kinds
of indirect observations have been made to build cases for vacancies
and for self-interstitials. The current belief is that both types of
point defect exist somewhat independently [6], and that both influence
diffusion. Under diffusive conditions where the silicon is intrinsically
doped and where no surface reaction is taking place, the vacancy and
self-interstitial concentration are temperature dependent only and are
described approximately by the equilibrium concentrations:

$$Cv[eq] = Cov \exp \frac{-Ev}{kT}$$

$$Ci[eq] = Coi \exp \frac{-Ei}{kT}$$

Dopant diffusion takes place in the silicon by interacting with these point defects so that the diffusing species are the antimony/vacancy pair, SbV, and the antimony/Si-interstitial pair, SbI. The intrinsic diffusivity due to the vacancy mechanism is referred to as Dv[eq], and that due to the interstitialcy mechanism as Di[eq]. The intrinsic, inert ambient diffusivity is therefore given by the sum of the equilibrium diffusivities due to the vacancy and the self-interstitial components:

$$D[eq] = Dv[eq] + Di[eq]$$

The wide energy range available to the Fermi level in silicon leads to a given lattice defect (i.e. a vacancy or Si-interstitial) appearing in a variety of ionized states [13]. As such the dopant/point-defect pairs exist as, for example, Sb(+)V(-) which we will denote by SbV(-). The intrinsic, inert ambient diffusivity can therefore be written:

$$D[eq] = Dv(+)[eq] + Dv(0)[eq] + Dv(-)[eq] + Dv(--)[eq]$$

$$+ Di(+)[eq] + Di(0)[eq] + Di(-)[eq] + Di(--)[eq]$$

where the intrinsic diffusivity of the antimony diffusing as SbV(-) is given by Dv(-)[eq], etc.

The diffusion coefficient in crystals with a cubic lattice should be isotropic [5]. For doping levels of less than about 1 X 10*20/cm*3 with no surface reacton taking place this is indeed the case [1,5]. The diffusion coefficient is therefore given by:

$$D = Dn \exp \frac{-En}{kT}$$

Under intrinsic diffusive conditions, antimony dopant profiles show excellent Fick-type behaviour [7].

Literature diffusivities for inert intrinsic diffusion of antimony in single crystal silicon are listed below.

Fair [8] describes the intrinsic diffusivity of antimony in a manner which physically accounts for the fact that antimony diffuses by interaction with neutral vacancies, Sb(+)V(0), and with singly negatively charged vacancies, Sb(+)V(-).

$$D[eq] = Dv(0)[eq] + Dv(-)[eq]$$

$$= (1.284 \text{ X } 10*9) \exp \frac{-3.65 \text{ eV}}{kT} + (1.0 \text{ X } 10*10) \exp \frac{-4.08 \text{ eV}}{kT} \text{ microns*2/min}$$

Colclaser [9] describes the intrinsic diffusivity of antimony as

$$D[eq] = Dv(0)[eq] + Dv(-)[eq]$$

$$= 0.214 \exp \frac{-3.65 \text{ eV}}{kT} + 13 \exp \frac{-4.0 \text{ eV}}{kT} \text{ cm*2/s}$$

Ghoshtagore [7] describes the intrinsic diffusivity of antimony as

$$D[eq] = Dn \; exp \; \frac{-En}{kT}$$

$$= 0.214 \; exp \; \frac{-3.65 \; eV}{kT} \; cm*2/s$$

B. INTRINSIC ANTIMONY DIFFUSION, O2 AMBIENT

In the process modelling field it is more or less universally accepted that antimony diffuses predominantly by a vacancy assisted mechanism [10]. Oxidation of silicon under most processing conditions retards the diffusion of antimony [11,6,12]. This is because oxidation results in the injection of silicon-interstitials to the silicon substrate [2] (but also see [13]), which, by a Frenkel-type process, reduces the vacancy concentration, thereby reducing the antimony diffusivity. (A small and transient interstitial-excess-related enhancement does occur immediately the oxidation process begins whilst the I-V annihilation reaction is moving to equilibrium [14,15]). The diffusivity is described by the general diffusion equation:

$$D = Dv + Di$$

i.e.

$$D = Dv[eq] \; \frac{Cv}{Cv[eq]} \; + \; Di[eq] \; \frac{Ci}{Ci[eq]}$$

or

$$D = D[eq] \; ((1-fi) \; \frac{Cv}{Cv[eq]} \; + \; fi \; \frac{Ci}{Ci[eq]})$$

where

$$fi = \frac{Di[eq]}{D[eq]} = \frac{Di[eq]}{Dv[eq] + Di[eq]}$$

and fi is known as the fractional interstitialcy component of diffusion. The thermal oxidation process results in a simultaneous supersaturation of Si-interstitials and undersaturation of vacancies. The effect on total diffusion rate of the increase in Sb-I diffusion species and decrease in Sb-V diffusion species is described by the fractional interstitialcy component of diffusion, fi, literature values of which are given below.

Experimental evidence suggests that the Si-interstitial injection rate during thermal oxidation depends on the oxidation rate. More specifically, it has been found that:

$$Ci-Ci[eq] = k \; (\frac{dXox}{dt})*P \quad with \; P < 1$$

where (dXox/dt) is the oxidation rate.

Since the oxidation rate decreases as oxidation time increases, it is expected that the diffusion retardation will also decrease with oxidation time. The oxidation rate (and therefore the rate of Si-interstitial injection) is not dependent on the orientation beyond the linear segment. The rate of decay of these excess point defects will be dependent on the silicon surface or the silicon-oxide interface recombination rates, and these might be expected to be proportional to the rectilinear density. As such the rate of loss of excess point defects would increase as (100), (110), (111), and the magnitude of oxidation retarded diffusion would increase as (111), (110), (100) [5], i.e. diffusion of antimony is at its highest when no oxidation takes place, is lower when a (111) surface is being oxidized, and is lowest when a (100) surface is being oxidized.

Oxidation retarded diffusion of antimony in the intrinsic regime is expected to be a function of temperature, oxidation rate, depth into the silicon and substrate orientation.

Literature models of oxidation retarded antimony diffusion are described below.

Ho et al [16] describe the intrinsic diffusivity of antimony in an oxidizing ambient as

$$D = (Dv(0)[eq] + Dv(-)[eq]) \frac{1}{1 + OED}$$

$$= [(1.284 \times 10*9)\exp \frac{-3.65 \ eV}{kT} + (9.0 \times 10*10)\exp \frac{-4.08 \ eV}{kT}] \times \frac{1}{1 + OED}$$

with

$$OED = [(2.86 \times 10*-16)(\exp \frac{5.64 \ eV}{kT}) \frac{dXox}{kT}]*0.5 \times (ornt \times fi \times dfctlgt)$$

and

    dt = time interval

   dXox = oxide thickness grown in time interval

   ornt = orientation factor
        = 1.0 for (100), 0.833 for (110), 0.667 for (111)

    fi = fractional interstitialcy component

        = 4.09 exp $\frac{-0.48 \ eV}{kT}$

   dfctlgt = Si-interstitial diffusion length
           = 25 microns for CZ-Si, 100 microns for FZ-Si

Antoniadis et al [6] have found that to describe completely the behaviour of antimony in the presence of interstitial injection one must use the full continuity equation for vacancies including non-steady-state recombination (bimolecular annihilation) with excess interstitials. In that way the time dependent vacancy undersaturation can be described reasonably accurately. The form of the diffusivity equation indicates the importance of this complete approach for antimony:

$$D = D[eq] \left((1-fi) \frac{Cv}{Cv[eq]} + fi \frac{Ci}{Ci[eq]} \right)$$

since fi is small then $Cv/Cv[eq]$ is critical and this must be calculated given a description for oxidation induced interstitial excess, delta(Ci), and for the annihilation process, $I + V = 0$.

For long oxidation times the diffusion of antimony can be predicted by assuming steady state bimolecular annihilation of interstitials with vacancies,

$$\frac{Cv}{Cv[eq]} = \frac{Ci[eq]}{Ci}$$

which gives

$$D = D[eq] \left((1-fi) \frac{Ci[eq]}{Ci} + fi \frac{Ci}{Ci[eq]} \right)$$

with

$$fi = 0.015 \text{ at } 1100 \text{ deg C}$$

C. IMPLANT DAMAGE ENHANCED ANTIMONY DIFFUSION

The diffusivity of antimony in the initial anneal stage of 700 deg C after ion implantation is some five orders of magnitude greater than the undamaged lattice diffusivity [17]. At higher temperatures however the degree of enhancement induced by implantation-damage decreases significantly and for temperatures greater than about 1050 deg C the diffusivity enhancement is presently technologically negligible [18].

D. EXTRINSIC ANTIMONY DIFFUSION, N2 AMBIENT

When the free carrier concentration exceeds the intrinsic carrier concentration at the diffusion temperature, i.e. when $n > ni$, the diffusivity is altered due to electric-field effects, h, and mass action effects, n/ni, giving a diffusion coefficient described by:

$$D = h \times \left[Dv(+)[eq] \frac{ni}{n} + Dv(0)[eq] + Dv(-)[eq] \frac{n}{ni} + Dv(--)[eq]\left(\frac{n}{ni}\right)*2 \right.$$

$$\left. + Di(+)[eq] \frac{ni}{n} + Di(0)[eq] + Di(-)[eq] \frac{n}{ni} + Di(--)[eq] \left(\frac{n}{ni}\right)*2 \right]$$

where

$$h = 1 + \left(\frac{n}{ni}\right) \left[\left(\frac{n}{2ni}\right)*2 + 1\right]*(-0.5)$$

and

Dv(+)[eq] = intrinsic diffusivity of the antimony diffusing as the
            Sb(+)V(+) dopant/point-defect pair, etc.

Diffusivity is therefore increased by two mechanisms in the extrinsic
regime:

(a) the electric field effect, h, where the high mobility electron
    increases the diffusion coefficient of the slower moving dopant
    atom, asymptotically reaching a maximum of 2 at high active doping
    levels, and

(b) the mass action effect where the numbers of charged vacancies and
    Si-interstitials, V(-), I(-), and V(--), I(--), increase above
    their equilibrium values as (n/ni) and (n/ni)*2 respectively,
    resulting in more vehicles for dopant diffusion.

As such, when the local electron concentration exceeds ni (about 2 X
10*19/cm*3 at processing temperatures) the diffusivity increases. It
has been well established that interaction with a singly negatively
charged point defect, probably the vacancy, results in an (n/ni)
concentration dependent diffusion coefficient for extrinsic diffusion
of arsenic in silicon [4]. For antimony the same behaviour holds, and
the general diffusion coefficient above is simplfied for inert extrinsic
diffusion:

$$D = h \text{ X } \left[\text{ Dv(0)[eq]} + \text{Dv(-)[eq]} \frac{n}{ni} \right]$$

Antimony diffusivity in an inert ambient is 'locally determined' in the
sense that the local diffusivity depends only on the local electron
concentration. (This is also the case for arsenic and under most
circumstances boron, but is not the case for phosphorus as evidenced by
its 'tailing' characteristics). This diffusion equation therefore
adequately describes inert extrinsic antimony profile evolution. In
heavily doped materials bandgap-narrowing causes the effective intrinsic
carrier concentration to increase, thereby modifying the dopant
diffusivity.

Literature numerical models of diffusion of high concentration antimony
profiles are outlined below.

Fair [8] describes the inert-ambient extrinsic-diffusivity in a manner
which physically accounts for the fact that antimony diffuses by
interaction with neutral vacancies, Sb(+)V(0), and with singly negatively
charged vacancies, Sb(+)V(-).

$$D = h \; X \; [\; Dv(0)[eq] + Dv(-)[eq] \; \frac{n}{ni} \;]$$

$$= h \; X \; [(1.284 \; X \; 10*9)\exp \frac{-3.65 \; eV}{kT} + (9.00 \; X \; 10*10)(\exp \frac{-4.08 \; eV}{kT})(\frac{n}{ni})]$$

$$micron*2/min$$

Colclaser [9] describes the inert extrinsic diffusivity of antimony as

$$D = h \; X \; [\; Dv(0)[eq] + Dv(-)[eq] \; \frac{n}{ni} \;]$$

$$= h \; X \; [\; 0.214 \; \exp \frac{-3.65 \; eV}{kT} + 13 \; (\exp- \frac{-4.0 \; eV}{kT}) \; \frac{n}{ni} \;] \; cm*2/sec$$

Wolf [19] describes antimony diffusivity as

$$D = h \; X \; 12.9 \; \exp \frac{-3.98 \; eV}{kT} \; cm*2/s$$

E. EXTRINSIC ANTIMONY DIFFUSION, O2 AND H2O AMBIENT

Ho et al [16] describe extrinsic antimony diffusivity in an oxidizing
ambient as

$$D = h \; X \; (\; Dv(0)[eq] + Dv(-)[eq] \;) \; (\frac{1}{1 + OED})$$

$$= h \; X \; [(1.284 \; X \; 10*9)\exp \frac{-3.65 \; eV}{kT} + (9.00 \; X \; 10*10)(\exp \frac{-4.08 \; eV}{kT}) \; \frac{n}{ni}]$$

$$X \; \frac{1}{1 + OED}$$

with

$$OED = [(2.86 \; X \; 10*-16)(\exp \frac{5.64 \; eV}{kT}) \; \frac{dXox}{dt}]*0.5 \; X \; (ornt \; X \; fi \; X \; dfctlgt)$$

and

    dt = time interval

  dXox = oxide thickness grown in time interval

  ornt = orientation factor
       = 1.0 for (100), 0.833 for (110), 0.667 for (111)

    fi = fractional interstitialcy component

dfctlgt = Si-interstitial diffusion length
        = 25 microns for CZ-Si, 100 microns for FZ-Si

REFERENCES
----

[1]   T.C.Chan et al [Proc. IEEE (USA) vol.58 no.4 (1970) p.588-9]
[2]   S.M.Hu [J. Appl. Phys. (USA) vol.45 no.4 (1974) p.1567-73]
[3]   R.B.Fair [Diffus. & Defect Data (Switzerland) vol.37 (1984)
      p.1-24]
[4]   R.B.Fair, J.C.C.Tsai [J. Electrochem. Soc. (USA) vol.122 no.12
      (Dec 1975) p.1689-96]
[5]   C.Hill [Semiconductor Silicon 1981, Eds H.R.Huff et al
      (Electrochem. Soc., Princeton, NJ, 1981) p.988-1006]
[6]   D.A.Antoniadis, I.Moskowitz [J. Appl. Phys. (USA) vol.53 no.10
      (Oct 1982) p.6788-96]
[7]   R.N.Ghoshtagore [Phys. Rev. B (USA) vol.3 no.2 (1971) p.397-401]
[8]   R.B.Fair [in 'Impurity Doping Processes in Silicon'
      Ed. F.F.Y.Wang (North-Holland 1981)]
[9]   R.A.Colclaser [Microelectronics: Processing and Device Design
      (John Wiley & Sons, 1980) p.123]
[10] S.M.Hu [Appl. Phys. Lett. (USA) vol.51 (1987) p.308]
[11] S.Mizuo, H.Higuchi [Jpn. J. Appl. Phys. (Japan) vol20 (1981)
      p.739]
[12] T.Y.Tan, B.J.Ginsberg [Appl. Phys. Lett. (USA) vol.42 no.5 (1983)
      p.448-50]
[13] R.Francis, P.S.Dobson [J. Appl. Phys. (USA) vol.50 no.1 (Jan 1979)
      p.280-4]
[14] P.Fahey, R.W.Dutton, M.Moslehi [Appl. Phys. Lett. (USA) vol.43
      (1983) p.683-5]
[15] D.A.Antoniadis, I.Moskowitz [J. Appl. Phys. (USA) vol.53 no.12
      (Dec 1982) p.9214-16]
[16] C.Ho et al [SUPREM-III - A Program for Intergrated Circuit
      Process Modelling and Simulation, Tech. Rpt. no.SEL84-001, Stanford
      Electronics Labs, Stanford Univ. (1984)]
[17] K.Gamo et al [Jpn. J. Appl. Phys. (Japan) vol.9 no.3 (1970)
      p.333]
[18] R.Angelucci, P.Negrini, S.Solmi [Appl. Phys. Lett. (USA) vol.49
      no.21 (1986) p.1468-70]
[19] H.F.Wolf [Semiconductors (Wiley-Interscience, 1971) p.153]

13.9 SOLUBILITY OF B IN Si

by D.Nobili

August 1987
EMIS Datareview RN=15771

Boron dissolves substitutionally in silicon, giving rise to lattice
contraction.

In the lower temperature range, the phase in equilibrium with the
solid solution of boron into silicon is the rhombohedral SiB3. It is
believed to transform peritectoidally [2SiB3 <---> SiB6 + Si] around
1270 deg C, hence the orthorombic SiB6 phase is conjugate up to the
eutectic temperature (1385 +- 15 deg C) [1].

Early solubility data are conflicting; in contrast the results reported
in the last decade based on experiments which comply better with the
equilibrium requirements are in substantial agreement, although they were
obtained with different methods and techniques: electrical measurements
and chemical analysis.

Armigliato et al [2] obtained solubility values by isochronal annealing
of bulk doped single and polycrystalline silicon slices of known boron
content. These values were determined by electrical resistivity
measurements. The isochronal curves showed first the occurrence of
precipitation followed by dissolution taking place at increasing
temperature. Boron solubility Ce, which was determined in the dissolution
stage, showed an exponential dependence on temperature:

$$Ce = A \exp(-E/kT)$$

with A = 9.25 X 10*22 atoms/cm*3 and E = 0.73 +- 0.04 eV, which is
verified in the range 900-1325 deg C. The presence of the rhombohedral
SiB3 precipitated phase was verified by TEM imaging and diffraction
techniques on isothermally annealed specimens. The occurrence of SiB6,
the conjugate phase above approx. 1270 deg C, was not checked and no
effect of phase transition was noticed on the Ce-T curve above that
temperature. Therefore the above solubility data in the very high
temperature range have to be considered as provisional.

Ryssel et al [3] determined solubility values from the shape of boron
profiles after annealing heavily implanted single crystals. The
occurrence of precipitation above a given concentration was shown
by the strong reduction of dopant diffusivity, giving rise to a
shoulder in the concentration profile. The depth distribution of the
dopant was determined by measurements of the energy loss of the alpha
particles resulting from the 10-B (n,alpha) 7-Li reaction. Solubilities
of 0.15, 0.6 and 1.1 X 10*20/cm*3 were found at 800, 900 and 1000 deg C
respectively.

Garben et al [4] determined boron concentration profiles in single
crystal silicon after thermal diffusion, from doped polysilicon films.

Profiles were measured by SIMS (Secondary Ion Mass Spectroscopy) with
an absolute accuracy better than +- 30%. For sufficiently high average
dopant concentrations in the film, the boron concentration on the
single crystal silicon side of the interface depends only on
temperature, thus providing a clear indication of two phase equilibrium.
The reported vlaues are 0.2, 0.45, 0.7 and 1 X 10*20/cm*3 at 800, 900,
950 and 1000 deg C respectively.

Comparison of these data shows that the solubility values at 1000 deg C
practically coincide, and good agreement is still presented by the
ones found at 900 deg C. The values determined in [3] and [4] at 800
deg C are very similar and a solubility of 0.2 X 10*20/cm*3 can be
assessed at this temperature. This is lower than the one (0.34 X
10*20/cm*3) obtained by extrapolation of the equation reported by
Armigliato et al [2], thus indicating that this relation is not
accurate enough below 900 deg C.

Nucleation of precipitates strongly affects the thermal stability of
boron supersaturated solutions. This phenomenon is markedly speeded up
by structural defects, e.g. in implanted layers, while the same
supersaturation results in much lower nucleation rates in bulk doped
specimens or in thermally diffused layers. As an example single crystal
specimens bulk doped with 2 X 10*20 atoms/cm*3 were found to be
practically stable at 900 deg C. The strongly hindered nucleation in
crystals of good perfection explains why well known experimental
investigations by Samsonov et al [5] and by Vick et al [6] provided
solubility data largely exceeding the true ones.

REFERENCES
----

[1] R.W.Olesinski, G.J.Abbaschian [Bull. of Alloy Phase Diagrams
     vol.5 (1984) p.478-483]
[2] A.Armigliato, D.Nobili, P.Ostoja, M.Servidori, S.Solmi
     [Semiconductor Silicon (The Electrochem. Soc., USA, 1977)
     Proc. Series P.V.77-2, p.638-647]
[3] H.Ryssel, K.Muller, K.Harberger, R.Henkelmann, F.Jahnd [Appl.
     Phys. (USA) vol.22 (1980) p.35]
[4] B.Garben, W.A.Orr-Arienzo, R.F.Lever [J. Electrochem. Soc. (USA)
     vol.133 (1986) p.2152]
[5] G.V.Samsonov, V.M.Sleptsov [Zh. Neorg. Khim. vol.8 (1963) p.2209]
[6] G.I.Vick, K.M.Whittle [J. Electrochem. Soc. (USA) vol.116 (1969)
     p.1142]

13.10 SOLUBILITY OF Al IN Si

by D. de Cogan

July 1987
EMIS Datareview RN=16106

There has been considerable variation in the results obtained for the
solubility of aluminium in silicon. The early high temperature work of
Spengler [1] and Goldstein [2] represent the extreme values and has
been discussed critically by Miller and Savage [3]. Subsequent
experiments by Navon and Chernyshov [4] as well as Gudmundsen and
Maserjian [5] give results which are of the same order as the scattered
data of Miller and Savage [3]. At lower temperatures Trumbore's [6]
data is an order of magnitude larger than that of Navon and Chernyshov
[4] and two to three orders larger than that of Gudmundsen and
Maserjian [5]. Some best fit data to Trumbore's curve for aluminium in
silicon is given below.

TABLE 1

T deg C	C(max) atoms/cm*3
1375	6 X 10*18
1300	1.6 X 10*19
1200	2 X 10*19
700	1 X 10*19

The differences are largely due to the fact that aluminium reacts with
quartz containers and with residual oxygen in sealed tube experiments.
This lowers the aluminium partial pressure.

The distribution coefficient k of an impurity at the melting point of a
solvent is an important measure of solubility. Burton [7] has estimated
a value of k = 4 X 10*-3, while the value determined by Hall [8] is
2 X 10*-3. Brice [9] has shown that for aluminium, gallium and indium
there is a correlation between the distribution coefficient and A/Ds
where A is the pseudo diffusion coefficient in the liquid and Ds is the
substitutional diffusion coefficient in the solid, both at the melting
point of silicon. The equation which describes the pseudo diffusion
coefficient contains the heat of evaporation of silicon and the
impurity ionic radius.

A thermodynamic basis for the distribution of impurities between liquid
and solid is provided by Lehovec [10] who discusses the implications of
departure from ideal solution theory. Trumbore's [6] correlation
between k and the tetrahedral radius for aluminium does not fit into
the curve defined by boron, gallium and indium. The introduction of an
activity coefficient to take account of interaction between aluminium
and silicon improves the level of agreement in this case. It does not,
however, improve the correlation between the distribution coefficient
and heat of sublimation for aluminium.

This lack of agreement may well be due to errors in the determination of the heat of sublimation or indeed the tetrahedral radius. Fischler [11] has shown that there is a relationship between the maximum solubility of an impurity and the distribution coefficient. This can be described by the equation

$$C(max) = (5.2 \times 10*21)k \text{ atoms/cm*3}$$

Trumbore's [6] results for aluminium fit this equation very closely. Statz [12], using the approach of Lehovec [10] has provided a theoretical basis for Fischler's equation. He has suggested that the maximum mole fraction for aluminium should be 0.21k. The value derived by Fischler from experimental data was 0.19k.

A more recent estimate of maximum solubility of aluminium in silicon has been made by Bailey and Mills [13]. They believe that the maximum solubility of aluminium in silicon is an order of magnitude higher than the accepted value of 2 X 10*19 atoms/cm*3 and suggest that it is less electrically active at high concentrations.

REFERENCES
----

[1]   H.Spengler [Metall. (Germany) vol.9 (1955) p.181]
[2]   B.Goldstein [Bull. Am. Phys. Soc. (USA) vol.1 (1956) p.145]
[3]   R.C.Miller, A.J.Savage [J. Appl. Phys. (USA) vol.27 (1956)
       p.1430]
[4]   D.Navon, V.Chernyshov [J. Appl. Phys. (USA) vol.28 (1957)
       p.823]
[5]   R.A.Gudmundsen, J.Maserjian,Jr. [J. Appl. Phys. (USA) vol.28
       (1957) p.1308]
[6]   F.A.Trumbore [Bell Syst. Tech. J. (USA) vol.39 (1960) p.205]
[7]   J.A.Burton [Physica (Netherlands) vol.20 (1954) p.845]
[8]   R.M.Hall [General Electric Res. Lab. Report No.58-RL-1874]
[9]   J.C.Brice [Solid-State Electron. (GB) vol.6 (1963) p.673]
[10]   K.Lehovec [J. Phys. & Chem. Solids (GB) vol.23 (1962) p.695]
[11]   S.Fischler [J. Appl. Phys. (USA) vol.33 (1962) p.1615]
[12]   H.Statz [J. Phys. & Chem. Solids (GB) vol.24 (1963) p.699]
[13]   R.F.Bailey, T.G.Mills [Electrochem. Soc. Extended Abstr. (USA)
       (9-13 May 1971) p.159-60]

13.11 SOLUBILITY OF Ga IN Si

by D. de Cogan

July 1987
EMIS Datareview RN=16107

The work of Trumbore [1] quotes three data points for the solubility of
gallium in silicon. These are quoted in terms of solidus mole fraction,
X, which may be converted to concentration by multiplying by the atom
density of silicon.

TABLE 1

T (deg C)	X	C /cm*3
805	3 X 10*-4	1.5 X 10*19
982	5.2 X 10*-4	2.6 X 10*19
1066	6.4 X 10*-4	3.1 X 10*19

The uncertainty in temperature was about 10 deg C.
Atom density = density (Avogadro number/atomic weight)

The distribution coefficient k of an impurity at the melting point of a
solvent is an important measure of solubility. Hall [2] cites a value
of k = 8 X 10*-3 for gallium in silicon. Brice [3] has shown that for
gallium, aluminium and indium there is a correlation between the
distribution coefficient and A/Ds where A is the pseudo diffusion
coefficient in the liquid at the melting point and Ds is the
substitutional diffusion coefficient in the solid at the melting point.
The pseudo diffusion coefficient contains the heat of evaporation of
silicon and the ionic radius of the solute (gallium in this case).

A thermodynamic basis for the distribution of impurities between liquid
and solid is provided by Lehovec [4], who discusses the implications of
departure from ideal solution theory. Trumbore's [1] correlation
between impurity distribution coefficient and impurity tetrahedral
radius shows gallium to be an almost ideal (in a thermodynamic sense)
solute in silicon.

Fischler [5] has shown that there is a relationship between the
maximum solid solubility and the distribution coefficient at solvent
melting point for impurities in silicon and germanium. The maximum
solubility cam be described by the equation.

$$C(max) = (5.2 \times 10*21)k \text{ atoms/cm*3}$$

Gallium fits this equation very closely. Statz [6] using the approach
of Lehovec [4] has provided a theoretical basis for this relationship.
According to him the maximum mole fraction of gallium in solid silicon
should be 0.16k. The value derived from Trumbore's [1] work on gallium
is 0.1k.

Later work by Haridos et al [7] has shown that Trumbore's [1] values are a slight over-estimate at the low temperature range.

A series of experiments using a metal source in a sealed tube arrangement has provided the following results:

TABLE 2

T deg C	C(max) atoms/cm*3
850	7.32 X 10*18
900	4.82 X 10*18
950	1.24 X 10*19
1000	1.58 X 10*19
1050	1.7 X 10*19
1100	3.2 X 10*19

The diffusion studies of Okamura [8] have shown that the surface concentration of gallium in silicon, as determined by lapping/sheet resistance measurements, is less than that determined by neutron activation analysis. This is attributed to gallium precipitation during sample cooling.

Fogarassy et al [9] have investigated the effects of laser treatment on deposited layers of impurity metals on silicon. They have compared their results with the maximum solubility estimated from Trumbore's [1] curve for gallium. This should read 4 X 10*19 atoms/cm*3 not 5 X 10*19 as quoted by Fogarassy. The experimental results can be described by the equation

$$C(max) \ (laser) = (8.6 \ X \ 10*21)k*0.51 \ atoms/cm*3$$

For gallium the maximum solubility after laser treatment is 9 X 10*20 atoms/cm*3. This compares with the value of 4.5 X 10*20 atoms/cm*3 obtained by White [10] who laser annealed implanted layers. For antimony, bismuth and indium, White's [10] results are in much closer agreement with those of Fogarassy et al [9].

REFERENCES
----

[1] F.A.Trumbore [Bell Syst. Tech. J. (USA) vol.39 (1960) p.205]
[2] R.N.Hall [General Electric Res. Lab. Report No.58-RL-1874]
[3] J.C.Brice [Solid-State Electron. (GB) vol.6 (1963) p.673]
[4] K.Lehovec [J. Phys. & Chem. Solids (GB) vol.23 (1962) p.695]
[5] S.Fischler [J. Appl. Phys. (USA) vol.33 (1962) p.1615]
[6] H.Statz [J. Phys. & Chem. Solids (GB) vol.24 (1963) p.699]
[7] S.Haridos, F.Beniere, M.Gauneau, A.Rupert [J. Appl. Phys. (USA) vol.51 no.11 (1980) p.5833-7]
[8] C.Okamura [Jpn. J. Appl. Phys. (Japan) vol.10 (1971) p.434]
[9] E.Fogarassy, R.Stuck, J.J.Grob, A.Grob, P.Siffert [J. Phys. Colloq. (France) vol.41 no.C-4 (May 1980) p.4-41]
[10] C.W.White [quoted as 'Private Communication' in ref [8]]

13.12 SOLUBILITY OF In IN Si

by D. de Cogan

July 1987
EMIS Datareview RN=16108

No solidus data for indium was available for inclusion in Trumbore's [1]
review on the solubility of impurities in silicon. However Backenstoss
[2] did obtain a solution of 4 X 10*17 indium atoms/cm*3 in pulled
silicon crystals.

The distribution coefficient, k of an impurity at the melting point of
a solvent is an important measure of solubility. Burton [3] cites a
value of k = 5 X 10*-4, which compares with k = 4 X 10*-4 quoted by
Hall [4]. Brice [5] has shown that for indium, aluminium and gallium
there is a correlation between the distribution coefficient and A/Ds
where A is pseudo diffusion coefficient in the liquid and Ds is the
substitutional diffusion coefficient in the solid, both at the melting
point of silicon. The pseudo diffusion coefficient contains the heat of
evaporation of silicon and the impurity ionic radius.

A thermodynamic basis for the distribution of impurities between liquid
and solid is provided by Lehovec [6], who discusses the implications of
departure from ideal solution theory. Trumbore's [1] correlations
between impurity distribution coefficient and tetrahedral radius and
between distribution coefficient and impurity heat of sublimation show
indium to be an almost ideal (in a thermodynamic sense) solute in
silicon.

Fischler [7] has shown that there is a relationship between the maximum
solubility of an impurity and the distribution coefficient k. This can
be described by the equation

    C(max) = (5.2 X 10*21)k atoms/cm*3

According to the results presented by Fischler indium fits the equation
very closely. Statz [8] using the approach of Lehovec [6] has provided
a theoretical basis for Fischler's equation. He suggests that the
maximum mole fraction for indium in solid silicon should be 0.14k which
compares well with Fischler's [7] value of 0.1k. This implies that the
maximum solubility of indium in silicon should lie in the range 2.28
2.28 X 10*18 and 2.98 X 10*18 atoms/cm*3 which is higher than that
cited by Backenstoss [2].

Implantation studies by Cerofolini et al [9] suggests that the solid
solubility of indium in silicon might be significantly lower than was
previously thought. They quote the following data

TABLE 1

T deg C	C(max) atoms/cm*3
1000	3 X 10*16
1100	8 X 10*16
1200	2.5 X 10*16

By contrast Fogarassy et al [10] have investigated the effects of laser treatment on deposited layers of impurity metals on silicon. The experimental results can be described by the equation

$$C(max) \ (laser) = (8.6 \ X \ 10*21)k*0.51 \ atoms/cm*3$$

For indium the value 1.5 X 10*20 compares closely with the value 1.0 X 10*20 obtained by White [11] following laser annealing of an implant.

REFERENCES
----

[1]   F.A.Trumbore [Bell Syst. Tech. J. (USA) vol.39 (1960) p.205]
[2]   G.Backenstoss [Phys. Rev. (USA) vol.108 (1957) p.416]
[3]   J.A.Burton [Physica (Netherlands) vol.20 (1954) p.845]
[4]   R.N.Hall [General Electric Res. Lab. Report No.58-RL-1874]
[5]   J.C.Brice [Solid-State Electron. (GB) vol.6 (1963) p.673]
[6]   K.Lehovec [J. Phys. & Chem. Solids (GB) vol.3 (1962) p.695-709]
[7]   S.Fischler [J. Appl. Phys. (USA) vol.33 (1962) p.1615]
[8]   H.Statz [J. Phys. & Chem. Solids (GB) vol.24 (1963) p.699]
[9]   G.F.Cerofolini, G.Ferla, G.U.Pignatel, F.Riva [Thin Solid Films vol.101 (1983) p.275]
[10] E.Fogarassy, R.Stuck, J.J.Grob, A.Grob, P.Siffert [J. Phys. Colloq. (France) vol.41 no.C-4 (May 1980) p.4-41]
[11] C.W.White [quoted as 'Private Communication' in ref. [10]]

13.13 SOLUBILITY OF Tl IN Si

by D. de Cogan

July 1987
EMIS Datareview RN=16105

There has been very little reported on the solubility of thallium in
silicon. Thurmond and Kowalchick [1] have estimated liquidus data. No
solidus data was available to Trumbore [2] although he does provide
details of the solubility of thallium in germanium.

Shulman [3] has recorded considerable problems in obtaining thallium
doped single crystal silicon. Thallium (boiling point 1490 deg C) has
a very high vapour pressure at the melting point of silicon.

Fuller and Ditzenberger [4] have estimated the surface concentration as
a function of temperature as part of their work on the diffusion of
impurities into silicon. Their results are summarised below:

TABLE 1

T (deg C)	C (surface) atoms/cm*3
1360	1.3 X 10*17
1320	2.3 X 10*17
1255	1.2 X 10*17
1170	9 X 10*16
1105	3.7 X 10*17

Ghoshtagore [5] has undertaken diffusion studies using doped epitaxial
layers as impurity sources. In the case of thallium it was not possible
to use an organo-thallium compound in the epi-growth on account of its
vapour pressure and stability. Accordingly, he used a 99.9999% pure
metal source on the susceptor. It was only possible to measure one set
of data and from this he estimated a maximum surface concentration of
6 X 10*15 atoms/cm*3.

The partition coefficient, k, of an impurity between liquid and solid
silicon is often used as a measure of solubility. Fischler [6] has
shown that for the other acceptors the maximum solubility can be
described by the equation C(max) = (5.2 X 10*21)k. If this is used
with Ghoshtagore's [5] estimate of C(max) one obtains a value of
k = 1.15 X 10*-6. This would mean that thallium in silicon would not fit
into the k versus tetrahedral radius correlation which as Trumbore [2]
points out holds for most donors as well as acceptors. Trumbore also
provides values of k for the segregation of impurities in germanium.

From his data it can be seen that the ratio of partition coefficients for the two solvents k(germanium) / k(silicon) follows a progression 36.5, 10.8, 2.5 for aluminium, gallium and indium. If one takes 2.5 X 10*17 atoms/cm*3 as an estimate of C(max) from the results of Fuller and Ditzenberger [4] one obtains a value of 4.8 X 10*5 for k (thallium in silicon) which fits into the progression k(germanium) / k(silicon) noted above. For thallium this ratio is 0.83. The value 4.8 X 10*-5 also fits into Trumbore's [2] correlation between k and impurity tetrahedral radius, which suggests that it is a fair estimate of segregation coefficient.

Recent results by Dedekaev et al [7] suggest a value of 6.7 X 10*15/cm*3 at 1200 deg C as the maximum solubility limit. This, like the diffusion data which they also provide for thallium, is at variance with previously published data and no explanation is offered.

REFERENCES
----

[1] C.D.Thurmond, M.Kowalchick [Bell Syst. Tech. J. (USA) vol.39 (1960) p.169]
[2] F.A.Trumbore [Bell Syst. Tech. J. (USA) vol.39 (1960) p.205]
[3] R.G.Shulman [J. Phys. & Chem. Solids (GB) vol.2 (1957) p.115-18]
[4] C.S.Fuller, J.A.Ditzenberger [J. Appl. Phys. (USA) vol.27 (1956) p.27]
[5] R.N.Ghoshtagore [Phys. Rev. B (USA) vol.3 no.8 (1971) p.2507-14]
[6] S.Fischler [J. Appl. Phys. (USA) vol.33 (1962) p.1615]
[7] T.T.Dedegkaev, V.V.Eliseev, D.Kh.Lagkuev, V.A.Fidarov [Sov. Phys. Solid State (USA) vol.27 no.9 (Sep 1985) p.1699-1700]

13.14 SOLUBILITY OF P IN Si

by D.Nobili

August 1987
EMIS Datareview RN=15770

A. INTRODUCTION

Phosphorus dissolves substitutionally in silicon, giving rise to
lattice contraction.

The phase diagram has been recently evaluated by Olesinski et al [1].
In the lower temperature range the phase which is in equilibrium with
the solid solution of phosphorus in silicon is orthorhombic SiP. A
eutectic transformation takes place at 1131 +- 2 deg C and above this
temperature the conjugate phase is a liquid with a phosphorus content
decreasing from 32.8 at.% to zero at the MP of silicon.

It was previously reported in [2] that phosphorus solubility in silicon
was, until recently, a conflicting subject. Early values reported by
Trumbore [3] and by Kooi [4] are in mutual satisfactory agreement and
markedly higher, about a factor of two, than the carrier density after
thermal equilibration of heavily doped silicon specimens.

B. ELECTRICALLY INACTIVE PHOSPHORUS

A well known hypothesis to account for the electrically inactive
phosphorus is the formation of high equilibrium concentrations of
complex point defects (E centres) which compensate or make the excess
dopant inactive. This hypothesis is the basis of models for the
diffusivity of P proposed by Kendall et al [5] and by Fair et al [6].

The high solubility values reported above are not very reliable:
Trumbore [2] referred to unpublished results of MacKintosh obtained by
sheet resistance and junction depth measurements. The approximation
involved to evaluate surface concentrations, a complementary error
function distribution, is surely not justified, as it was also pointed
out by MacKintosh himself in a further work [7]. Kooi [4] used activation
analysis after equilibration with a PSG layer, i.e. a ternary system
containing oxygen. This situation can hardly comply with the basic
requirement of equilibration with SiP, a criticism which is further
strengthened by the observation by Solmi et al [8] of cubic SiO2
P2O5 at the PSG/silicon interface.

Lower solubility values were determined by Abrikosov et al [9] by
microhardness measurements on bulk doped specimens. The procedure is
correct, but the sensitivity of the technique seems inadequate.

On the other hand large solubility values, exceeding the equilibrium carrier density, were supported by additional arguments:

(i) Rutherford backscattering (RBS) experiments by Fogarassy et al [10] showed that a large fraction of the inactive dopant still occupies substitutional sites in the silicon lattice.

(ii) Although TEM observations by Servidori et al [11] and Armigliato et al [12] detected orthorhombic SiP precipitates their amount could not account for the inactive dopant.

More recent works indicate that precipitation is responsible for the inactive phosphorus and leads to the conclusion that solubility corresponds to the carrier concentration after thermal equilibration.

The following results were reported by Nobili et al [13]:

(i) The carrier density after high temperature annealing depends only on the heating temperature and is insensitive to excess dopant. This finding, which strongly supports a two phase equilibrium, confirms previous results by Masetti et al [14] also based on carrier profile measurements.

(ii) The occurrence of reversion: a transient redissolution taking place by a step increase in temperature of a specimen which is still supersaturated. This phenomenon, which can be only due to the thermodynamic instability of precipitates, is typical of a population of very small particles.

They were in fact detected by TEM, in addition to the orthorhombic SiP precipitates, by using the weak beam dark field technique, a method which allows particle visibility when their density is very high.

The coherency of these small precipitates with the silicon matrix, a feature which hinders the observation of precipitation by RBS, was conclusively confirmed by double crystal X-ray diffraction experiments by Servidori et al [15] and more evidently by high resolution transmission electron microscopy by Armigliato et al [16]. In this latter work the formation of roughly spherical precipitates, about 30 A in diameter, perfectly coherent with the silicon matrix and having the structure of cubic SiP was observed after annealing. This phase, which fits the silicon lattice, was first reported by Osugy et al [17].

The initial phosphorus concentration in the experiments by Armigliato et al [16] was chosen to be 5 X 10*20/cm*3, which corresponds to Trumbore's solubility [3] at 850 deg C. These experiments showed in agreement with previous results [2], that heating at the above temperature reduces the carrier density to the equilibrium value (3 X 10*20/cm*3) and results in precipitation of SiP in an amount which is consistent with the concentration of the inactive dopant.

C. NUCLEATION AND SOLUBILITY

It is worth mentioning that the formation of this cubic SiP phase in phosphorus doped silicon was not previously reported in the literature. This is due to its coherency with the silicon lattice, which greatly enhances the nucleation frequency, thus resulting in a very high density and small size of the precipitates.

Still keeping these features, the density and size of the SiP particles depend on the parameters of the process. It was observed that particle size decreases with increasing supersaturation, behaviour which complies with the classical theory of nucleation (Christian [18]). After annealing specimens with initial concentrations of 2 X 10*20 and 5 X 10*20/cm*3, for 1500 h at 450 deg C. Finetti et al [19] observed precipitate densities of 10*10 and 10*12/cm*2 respectively.

It is worth noticing that the carrier density after annealing at low temperatures depends on the initial dopant concentration, behaviour which is in contrast with that observed after annealing at high temperatures, above approx. 750 deg C. For instance after 1000 h annealing at 550 deg C of specimens with initial phosphorus concentrations 50, 6 and 1.2 X 10*20/cm*3 carrier densities of 1.15, 1 and 0.38 X 10*20 respectively were obtained by profile measurements. This effect was accompanied by an increase in the precipitate size [2]. This phenomenon can be interpreted as being due to the dependence of solubility on the size of the precipitates. This is the well known Gibbs-Thomson effect which is motivated by the free enthalpy of the particle/matrix interface and is responsible for coalescence in a system of particles of different sizes. The dependence of solubility on this effect is limited, except for very small particles, and markedly increases with decreasing temperature (Christian [18]).

The above observations explain the difficulty of obtaining accurate values of phosphorus solubility in silicon. Low temperature values tend in fact to be higher than the true ones, as equilibrium can be hardly attained, even for long annealing times. At the other extreme, above approx. 1000 deg C, the high nucleation frequency causes solubility values to be affected by the cooling rate, thus providing values which tend to be lower than the true ones.

In the range 750-1050 deg C phosphorus solubility Ce follows an exponential dependence on temperature:

$$Ce = A \exp(-E/kT)$$

with A = 1.8 +- 0.2 X 10*22 atoms/cm*3 and E = 0.4 +- 0.01 eV. This result is obtained by fitting the data of Masetti et al [14] and Nobili et al [13] based on carrier density profile measurements after equilibration with SiP precipitates. The previous values at 920 deg C and 1000 deg C had been increased of about 15% according to more recent determinations. Moreover in this fitting even new unpublished results obtained on polysilicon films, were included. It is incorrect to extrapolate this exponential relation to lower temperatures as it provides inaccurate results which exceed the value of 0.4 X 10*19/cm*3 found at 550 deg C [2].

Finally it is worth noticing that since the competing SiP phase in these experiments does not have the equilibrium structure the corresponding solubility exceeds, at least in principle, the true value.

REFERENCES
----

[1]   R.W.Olesinski, N.Kanani, G.J.Abbaschain [Bull. of Alloy Phase Diagrams vol.6 (1985) p.130]

[2]   D.Nobili [Conf. Proc. of Satellite Symp. on 'Aggregation Phenomena of Point Defects in Silicon' Eds. E.Sirtl, J.Goorissen (The Electrochemical Soc., Munich, 1982) p.189-208]

[3]   F.A.Trumbore [Bell Syst. Tech. J. (USA) vol.105 (1960) p.205]

[4]   E.Kooi [J. Electrochem. Soc. (USA) vol.111 (1964) p.1383]

[5]   D.L.Kendall, R.Carpio [Conf. IEEE Sectional Meeting (Monterrey NM 1977) p.1]

[6]   R.B.Fair, J.C.C.Tsai [J. Electrochem. Soc. (USA) vol.124 (1977) p.1107; R.B.Fair [J. Appl. Phys. (USA) vol.50 (1979) p.860]

[7]   J.M.MacKintosh [J. Electrochem. Soc. (USA) vol.109 (1961) p.392]

[8]   S.Solmi, G.Celotti, D.Nobili, P.Negrini [J. Electrochem. Soc. (USA) vol.123 (1976) p.654]

[9]   N.K.Abrikosov, V.M.Glazov, Liu Chen-Yuan [Russ. J. Inorg. Chem. vol.7 (1962) p.429]

[10] E.Fogarassy, R.Stuck, J.C.Muller, A.Grab, J.J.Grab, P.Siffert [J. Electron. Mater. (USA) vol.9 (1980) p.197]

[11] M.Servidori, A.Armigliato [J. Mater. Sci. (GB) vol.10 (1975) p.306]

[12] A.Armigliato, D.Nobili, M.Servidori, S.Solmi [J. Appl. Phys. (USA) vol.47 (1977) p.5489]

[13] D.Nobili, A.Armigliato, M.Finetti, S.Solmi [J. Appl. Phys. (USA) vol.53 (1982) p.1484]

[14] G.Masetti, D.Nobili, S.Solmi [in 'Semiconductor Silicon' Eds. H. Huff, R.Sirtl (The Electrochem. Soc. 1977) p.638-647]

[15] M.Servidori, C.del Monte, Q.Zini [Phys. Status Solidi a (Germany) vol.80 (1983) p.277]

[16] A.Armigliato, P.Werner [Ultramicroscopy (Netherlands) vol.15 (1984) p.61]

[17] J.Osugy, R.Namikawa, Y.Tanaka [Rev. Phys. Chem. Jpn. (Japan) vol.36 (1966) p.35]

[18] J.W.Christian [The Theory of Transformations in Metals and Alloys (Pergamon, NY, 1975) Chapters 6 and 10]

[19] M.Finetti, P.Negrini, S.Solmi, D.Nobili [J. Electrochem. Soc. (USA) vol.128 (1981) p.1313]

13.15 SOLUBILITY OF As IN Si

by D.Nobili

August 1987
EMIS Datareview RN=15790

A. INTRODUCTION

It is known that arsenic dissolves substitutionally in silicon. The
phase diagram was recently evaluated by Olesinski et al [1]. In the
lower temperature range the solid solution of arsenic in silicon is
in equilibrium with SiAs. Two crystal structures were reported for
this compound: an orthorhombic one was identified by Beck et al [2]
on crystals obtained by a vapour transport method, while Wadsten [3]
determined a monoclinic structure on crystals obtained by melting
powdered silicon and arsenic. A eutectic transformation takes place
at 1097 deg C, and above this temperature the conjugate phase is a
liquid one.

Trumbore [4] estimated the solid solubility at temperatures above
1070 deg C, based on scattered data from electrical capacitance
measurements. Values, as deduced from the graph reported therein, are:
1.7, 1.8, 1.6 and 1.2 X 10*21 atoms/cm*3 at 1070, 1150, 1250 and 1350
deg C respectively. Sandhu et al [5] reported solubility values from
vapour pressure measurements as a function of the source composition,
which was determined by radiotracer and chemical methods. This source
was prepared by blending a Si powder with a multiphase Si-As alloy
mixture and, subsequently, homogenising in a furnace for 50 h at
1050 deg C. Their solubility values, as deduced from the graph, are:
0.41, 1.12, 1.33 and 1.5 X 10*21 atoms/cm*3 at 800, 900, 950 and
1009 deg C respectively.

It is well known that the above figures are markedly higher than the
carrier density after thermal equilibration; in fact from a fit of the
literature data Hoyt et al [6] reported that the maximum electrically
active concentration N(max) is given by:

    N(max) = A exp(-E/kT)

with A = 2.2 X 10*22 atoms/cm*3 and E = 0.478 eV.

A further confirmation of this discrepancy is provided by the results
of Ohkawa et al [7]: by diffusing As from an elemental source they found
that, at temperatures in the range 800-1050 deg C, the surface carrier
density changes little (1-3 X 10*20/cm*3) over a wide range of As
vapour pressures, while the dopant content increases and can attain
values exceeding 10*21 atoms/cm*3.

B. ELECTRICALLY INACTIVE ARSENIC

The formation of complex point defects in thermal equilibrium
(clusters) is widely accepted as the cause of the discrepancy between
the density of carriers and that of arsenic atoms at high dopant
concentrations. Several models of clusters, involving a different
number of dopant atoms, have been considered. The most well known are
the following ones:

(i) Tsai et al [8] proposed the reversible reaction:
$3As + e- = As(3)++$, with the complex becoming
neutral at RT.

(ii) The reversible formation of an arsenic-vacancy complex:
$2As+ + V + 2e- = As(2) V$ was proposed by Fair
et al [9].

(iii) A general cluster model, with the participation
of one negative charge, electron or negative
vacancy, was proposed by Guerrero et al [10].

The cluster hypothesis is in agreement with RBS and channelling results
which show that a large fraction of the inactive arsenic still keeps
the substitutional lattice site. Accurate determinations performed by
Chu et al [11] showed a limited displacement (0.15 A) from the
substitutional position, attributed to the formation of the complex
As(2)V. Moreover no observation of diffraction patterns which can be
attributed to the formation of an additional crystal structure have
been until now reported in the literature.

On the other hand Lietoila et al [12] and Nobili et al [13], by thermal
equilibration experiments performed on ion implanted laser annealed
slices doped with arsenic in a wide range of compositions, showed that
the electrically active concentration is insensitive to the excess
dopant. This finding strongly supports the alternative hypothesis that
solubility corresponds to the carrier concentration after equilibration
and that the inactive arsenic is in the form of precipitates. The
formation of a second phase was also shown by the occurrence of
reversion [13], a phenomenon which can only be due to the thermodynamic
instability of precipitates. They were in fact detected by Nobili et al
[13] and Angelucci et al [14] by small angle X-ray scattering performed
on specimens with As concentrations ranging from 1 to $4 \times 10*21$
atoms/cm*3 annealed at 900 deg C.

The occurrence of precipitation was confirmed by Armigliato et al [15]
by TEM examinations both in weak beam and high resolution imaging of As
implanted laser annealed slices, subsequently heated at 450 and
900 deg C. Even in specimens doped with $1 \times 10*21$ atoms/cm*3, which
corresponds to Sandhu's solubility [5] at 900 deg C, they detected,
after heating at that temperature, a high density ($8 \times 10*16$/cm*3)
of particles about 15 A in diameter. These precipitates were coherent
with the silicon matrix, a feature which hinders their observation by
RBS. The amount of the observed precipitates, however, accounted for
only a fraction of the inactive arsenic. The hypothesis advanced by the
authors is that precipitates are invisible due to their limited size or
to the local specimen thickness. In fact Armigliato et al [16], by

simulating high resolution images, found no detectable contrast for spherical coherent SiAs particles which are either in the sub-nanometric range, whilst present in thin regions (70 A), or have a 20 A diameter, being embedded in matrix areas thicker than 100 A.

In conclusion, although precipitates were detected by several techniques for dopant concentrations lower than Sandhu's solubility values, the physical nature of the electrically inactive arsenic in silicon is not well understood.

Even the experiments showing that the carrier density after thermal equilibration is insensitive to excess dopant, although in principle crucial, can hardy attain a sufficient accuracy to rule out the cluster models. In fact these models predict only a slight dependence of carrier density on arsenic concentration, so that experimental errors should be very small.

Further experiments, based on equilibrium chemical measurements, are needed to verify the solid solubility of arsenic, and hence to discriminate between the hypothesis of clusters and that of precipitation.

REFERENCES
----

[1]   R.W.Olesinski, G.J.Abbaschian [Bulletin of Alloy Phase Diagrams vol.6 (1985) p.254-8]
[2]   C.G.Beck, R.Stickler [J. Appl. Phys. (USA) vol.37 (1966) p.4683-7]
[3]   T.Wadsten [Acta Chem. Scand. vol.19 (1965) p.1232]
[4]   F.A.Trumbore [Bell Syst. Tech. J. (USA) vol.39 (1960) p.205-33]
[5]   J.S.Sandhu, J.L.Reuter [IBM J. Res. & Dev. (USA) (1971) p.464-71]
[6]   J.L.Hoyt, J.F.Gibbons [Mater. Res. Soc. Symp. Proc. (USA) vol.52 (1986) p.15-22]
[7]   S.Ohkawa, Y.Nakajima, Y.Fukukawa [Jpn. J. Appl. Phys. (Japan) vol.14 no.4 (Apr 1975) p.458-65]
[8]   M.Y.Tsai, F.F.Morehead, J.E.E.Baglin, A.E.Michel [J. Appl. Phys. (USA) vol.51 no.6 (Jun 1980) p.3230-5]
[9]   R.B.Fair, G.R.Weber [J. Appl. Phys. (USA) vol.44 no.1 (Jan 1973) p.273-9]; R.B.Fair [Semiconductor Silicon 1981, Eds H.R.Huff, R.J.Kriegler, Y.Takeishi (Electrochem. Soc., USA, 1981) p.963-78]
[10]   E.Guerrero, H.Potzl, R.Tielert, M.Grosserbauer, G.Stingeder [J. Electrochem. Soc. (USA) vol.129 no.8 (Aug 1982) p.1826-31]
[11]   W.K.Chu, B.J.Masters [AIP Conf. Proc. (USA) no.50 Proc. Symp. Laser-Solid Interactions and Laser Processing, Boston, MA, USA, 27 Nov-1 Dec 1978 Eds. S.D.Ferris, H.J.Leamy, J.M.Poate (AIP, New York, 1979) p.305-9]
[12]   A.Lietoila, J.F.Gibbons, T.J.Magee, J.Peng, J.D.Hong [Appl. Phys. Lett. (USA) vol.35 no.7 (1 Oct 1979) p.532-4]
[13]   D.Nobili, A.Carabelas, G.Celotti, S.Solmi [J. Electrochem. Soc. (USA) vol.130 no.4 (Apr 1983) p.922-8]
[14]   R.Angelucci, G.Celotti, D.Nobili, S.Solmi [J. Electrochem. Soc. (USA) vol.132 no.11 (Nov 1985) p.2726-30]
[15]   A.Armigliato, D.Nobili, S.Solmi, A.Bourret, P.Werner [J. Electrochem. Soc. (USA) vol.133 no.12 (Dec 1986) p.2560-5]
[16]   A.Armigliato, A.Bourret, S.Frabboni, A.Parisini [Inst. Phys. Conf. Ser. (GB) (1987) Microscopy of Semiconducting Materials in press]

13.16 SOLUBILITY OF Sb IN Si

by D.Nobili

September 1987
EMIS Datareview RN=15769

Antimony dissolves substitutionally in silicon, giving rise to expansion
of the lattice.

An assessment of the phase diagram was recently performed by Olesinski
et al [1]. A eutectic transformation takes place at 629.7 deg C, very
close to the melting point of antimony, hence the eutectic liquid is
almost entirely (99.7 at.%) antimony. Above the eutectic temperature
the solid solution of Sb in silicon is in equilbrium with a liquid
phase, which is very rich in antimony even at high temperatures, e.g.
about 91 at.% at 1100 deg C. Below the eutectic temperature two primary
solid solutions are in equilibrium, with negligible solubility of the
respective solutes.

Trumbore [2] estimated the solid solubility of antimony in silicon and
reported that it is retrograde, with a maximum at about 1320 deg C. He
used his own solubility determinations performed by spectrophotometric
analysis of crystals grown in thermal gradient experiments. These data
are scattered due to experimental difficulties:

TABLE 1

Temperature (deg C)	Solubility (atoms/cm*3)
807 +- 10	3.1 X 10*19
980 +- 10	2.8 X 10*19
991 +- 10	4.05 X 10*19
1066 +- 10	3.1 X 10*19
1066 +- 10	4.65 X 10*19

In addition he used the results of Rohan et al [3] who determined the
surface concentration of the dopant after diffusing antimony in silicon
in quartz ampules containing radioactive Sb203. The average values for
surface concentration at each temperature are given in the TABLE 2.

TABLE 2

Temperature (deg C)	Surface Concentration (atoms/cm*3)
1190	2.1 X 10*19
1280	4.7 X 10*19
1330	4.7 X 10*19
1355	3.5 X 10*19
1398	3.1 X 10*19

The assumption of the above figures as solubility data, which is open to question considering the method used by [3], is supported by the fact that the distribution coefficient k calculated by using Rohan's values [3] and the composition of the liquidus curve by Thurmond et al [4] are in very good agreement with experimental k values successively determined by Trumbore et al [5]. These latter determinations were performed, at temperatures in the range 1373 deg C up to the MP of Si by Hall measurements on heavily doped silicon crystals grown by a solvent evaporation technique.

The solubility values estimated by Trumbore [2], as deduced from the graph, reported therein, are shown in TABLE 3:

TABLE 3

Temperature (deg C)	Solubility (atoms/cm*3)
700	1.7 X 10*19
800	2.3 X 10*19
900	3.0 X 10*19
1000	3.8 X 10*19
1100	4.7 X 10*19
1200	5.8 X 10*19
1300	6.6 X 10*19
1400	3.0 X 10*19

A comparison with the experimental figures previously reported shows that in drawing the Sb solubility/temperature curve Trumbore attributed more weight to the highest ones.

Sung Hae Son et al [6] used neutron activation and sheet resistivity measurements to determine the profiles after diffusion from Sb doped silicon dioxide. They noticed that chemical and electrical profiles coincided at the beginning of the process. The difference, which increases with time, was attributed to a separate phase which was detected by electron microprobe X-ray analyzer and SEM. At 1150 deg C they found a value of 4.5 X 10*19 atoms/cm*3, which is slightly lower than the one by Trumbore [2].

Angelucci et al [7] obtained solubility values by carrier density measurements after equilibration at increasing temperatures of heavily doped poly-silicon films. The validity of this procedure and the correspondence of the carrier density with the solubility was supported by carrier profile measurements performed after equilibration at 1100 deg C of single crystal specimens implanted with different doses. In fact the same figure was obtained, at this temperature, by both methods and the equilibrium carrier density resulted to be independent on the amount of Sb.

Augelucci et al [7] found that, in the temperature range 850-1150 deg C, the equilibrium solubility Ce of antimony follows very closely the exponential dependence:

$$Ce = A \ exp-E/kT$$

with A = 3.8 X 10*21 atoms/cm*3 and E = 0.56 eV.

The values at higher temperatures deviate from this trend, as expected considering the liquidus curve [1]: the figures at 1200, 1250 and 1300 deg C being 4.4, 4.9 and 5.3 X 10*19 atoms/cm*3 respectively.

Comparison of the experimental solubility data of [2,3,6,7], which is only possible in the higher temperature range, shows sufficient agreement.

At temperatures lower than about 1100 deg C the values of [7] are increasingly smaller than those of Trumbore in TABLE 3. It is worth noticing that in this range Trumbore's estimates were based on a very limited amount of information.

Note that some variation of measured concentration values is to be expected: the solubility of Sb will be affected by the presence of other solutes (e.g. oxygen).

REFERENCES
----

[1]   R.W.Olesinski, G.J.Abbaschian [Bull. of Alloy Phase Diagrams vol.6
      (1985) p.445]
[2]   F.A.Trumbore [Bell Syst. Tech. J. (USA) vol.39 (1960) p.205]
[3]   J.J.Rohan, N.E.Pickering, J.Kennedy [J. Electrochem. Soc. (USA)
      vol.106 (1959) p.705]
[4]   C.D.Thurmond, M.Kowalchik [Bell Syst. Tech. J. (USA) vol.39 (1960)
      p.169]
[5]   F.A.Trumbore, P.E.Freeland, R.A.Logan [J. Electrochem Soc. (USA)
      vol.108 (1961) p.458]
[6]   Sung Hae Song, Tatsuya Niimi, Kenji Kobayashi, Kiyoshi Kudo [J.
      Electrochem. Soc. (USA) vol.129 (1982) p.841]
[7]   R.Angelucci, A.Armigliato, E.Landi, D.Nobili, S.Solmi [in ESSDERC
      1987 (Proc. Eur. Solid St. Dev. Res. Conf. Bologna, Sept 1987)
      p.461-4]

13.17 ION-IMPLANTATION OF B IN Si

by D.J.Godfrey

October 1987
EMIS Datareview RN=16191

Ion implantation of boron in the energy range 10 keV to 200 keV and
dose range 10*12/cm*2 to 10*16/cm*2 is used extensively within silicon
device technologies. The majority of models used for determining the
atomic profiles after ion implantation assume that the silicon target
is amorphous in order to simplify the calculations. However, in practice
the silicon substrate is initially crystalline where certain axes and
planes within the crystal are aligned and form channels for the incident
ions. The probability of an ion undergoing scattering whilst in such a
channel is reduced and so channelled ions penetrate more deeply into
the silicon than the amorphous approximation would predict. In an
attempt to lessen this effect, most commercial implanters introduce a
wafer tilt of typically 7 deg between the ion beam and the surface
normal of the wafer. This procedure reduces the level of channelling,
but is unable to completely remove it, particularly for low energies
and light ions such as boron.

The equations describing the atomic profile after implantation into an
amorphous target have been derived by Linhard, Scharff and Shiott [1].
These equations can be solved by taking moments and the resultant
profile can be calculated using a Pearson distribution with 4 moments
(range, standard deviation, skewness and kurtosis) or more simply a
Gaussian distribution with 2 moments (range and standard deviation).

e.g for a Gaussian

$$c(x) = \frac{D}{(SQRT\ 2)\ pi\ sigma} \exp \left[\frac{-(x-Rp)*2}{2\ sigma*2} \right]$$

where D is the dose, x is the depth, Rp is the range and sigma is the
standard deviation. Values for these moments for boron have been
tabulated by Smith [2] for the Gaussian and by Gibbons et al [3] and
have been determined experimentally by Hofker [4] using the
calculations of Winterbon [5,6]. The implanted profile for boron is
best represented by using the Pearson IV distribution.

The effect of ion channelling may be accounted for empirically by
adding an exponential tail to the Pearson IV distribution [7] or by
modifying the moments in the Pearson distribution itself [8]. However,
these approaches are restricted to simulations where the result has
previously been determined and it is found that the exact profile
shape under conditions where channelling is important is sensitive to
the particular experimental conditions used for the implantation.

A number of attempts [9-13] have been made to develop analytical models
for ion channelling, but these approaches are restricted to conditions
where implantation is aligned to particular crystalline directions
rather than 7 deg angle used in most commercial implanters.

It is possible to use Monte Carlo techniques to model the ion
implantation process and a number of codes have been successfully
developed (e.g. MARLOWE [14] and TRIM [15,16]). The Monte Carlo
approach is CPU intensive, but may be used for investigating such
effects as crystalline damage resulting from implantation and is able
to model channelling phenomena in a physical manner. However, at
present the approach is not sufficiently advanced to provide good
agreement between theory and experiment for channelling effects.

An alternative approach to the modelling of the ion implantation
distribution is to solve the Boltzmann Transport Equation (BTE) using
numerical integration [17]. This technique has been extended to
incorporate channelling effects and excellent agreement with experiment
has been obtained [18].

The effect of channelling on the boron distribution can be limited by
implanting into an amorphous surface layer created by a prior high dose
implant of either silicon [19] or germanium [20].

Experimental studies over a wide range of tilt angles [21] and at high
implant energies have also been performed [22].

Modelling has been performed of implantation into silicon through masked
layers. The lateral effects are typically included by a convolution of
the one-dimensional profile with a Gaussian profile described by a
lateral range straggling [23] or by the use of Monte Carlo codes such as
TRIM [15,16,24].

For the production of very shallow boron layers in silicon, low energy
implantation of boron difluoride is commonly used. The resultant
distributions may be fitted empirically using the Pearson IV approach
and ref [8] includes values of the four moments for energies between
50 keV and 100 keV with doses between 5 X 10*14/cm*2 and 3 X 10*15/cm*2.

REFERENCES
----

[1]   J.Lindhard, M.Scharff, H.E.Shiott [Mat.-Fys. Medd. Dan. Vidensk.
       Selsk. (Denmark) vol.33 (1963) p.14]
[2]   B.Smith [Ion implantation range data for silicon and germanium
       device technologies (Learned Information (Europe) Ltd., Oxford,
       1977)]
[3]   J.F.Gibbons, W.S.Johnson, S.W.Mylroie [Projected range
       statistics, 2nd ed. (distributed by Halstead Press, 1975)]
[4]   W.K.Hofker [Philips Res. Rep. (Netherlands) Suppl. no.8 (1975)]
[5]   K.B.Winterbon [Ion Implantation Range and Deposition, vol.2
       (Plenum, New York, 1975)]
[6]   K.B.Winterbon [Report AECL-5536, Computing Moments of Implanted-
       Ion Range and Energy Distributions (Atomic Energy of Canada Ltd.,
       1976)]
[7]   D.A.Antoniadis, R.W.Dutton [IEEE Trans. Electron Devices (USA)
       vol.ED-26 (1979)]
[8]   M.Simard-Normandin, C.Slaby [Can. J. Phys. (Canada) vol.63 no.6
       (1985) p.890-3]
[9]   C.Lehmann, G.Leibfried [J. Appl. Phys. (USA) vol.34 (1963)
       p.2821]
[10]   M.T.Robinson [Phys. Rev. (USA) vol.179 (1969) p.327]
[11]   Y.V.Martynenko [Sov. Phys.-Solid State (USA) vol.13 (1972)
       p.2166]

[12] N.Matsunami, L.M.Howe [Radiat. Eff. (GB) vol.51 (1980) p.111]

[13] A.Corciovei, A.Visinescu [Radiat. Eff. (GB) vol.55 (1981) p.141]

[14] M.T.Robinson [MARLOWE Binary Collision Cascade Simulation
      Program, Version 12 (Oak Ridge National Lab., 1 Jul 1984)]

[15] J.P.Biersack, L.G.Haggmark [Nucl. Instrum. & Methods
      (Netherlands) vol.174 (1980) p.257]

[16] J.F.Ziegler, J.P.Biersack, U.Littmark [The Stopping and Range of
      Ions in Solids, vol.1 (Pergamon Press, New York 1985)]

[17] L.A.Christel, J.F.Gibbons, S.Mylroie [J. Appl. Phys. (USA)
      vol.51 (1980) p.6176]

[18] M.D.Giles, J.F.Gibbons [IEEE Trans. Electron Devices (USA)
      vol.ED-32 no.10 (1985) p.1918-24]

[19] D.J.Godfrey, R.A.McMahon, D.G.Hasko, H.Ahmed, M.G.Dowsett [Mater.
      Res. Soc. Symp. Proc. (USA) vol.36 (1985) p.143]

[20] A.C.Ajmera, G.A.Rozgonyi [Appl. Phys. Lett. (USA) vol.49 no.19
      (10 Nov 1986) p.1269-71]

[21] G.Fuse, H.Umimoto, S.Odanaka, M.Wakabayashi, M.Fukumoto, T.Ohzone
      [J. Electrochem. Soc. (USA) vol.133 no.5 (May 1986) p.996-8]

[22] M.Tamura, N.Natsuaki, Y.Wada, E.Mitani [J. Appl. Phys. (USA)
      vol.59 no.10 (15 May 1986) p.3417-20]

[23] R.Tielert [IEEE Trans. Electron. Devices (USA) vol.ED-27 (1980)
      p.1479]

[24] J.F.Marchiando, J.Albers [J. Appl. Phys. (USA) vol.61 no.4 (15
      Feb 1987) p.1380-91]

13.18 ION-IMPLANTATION OF P IN Si

by D.J.Godfrey

November 1987
EMIS Datareview RN=16199

The basic approach adopted for the modelling of phosphorus implanted in
silicon is equivalent to that described in [10] (Ion implantation of
boron in silicon). For conditions where the LSS [1] theory can be
applied, tabulated values of the moments of the distribution can be
found in refs [2] and [3].

Experimental evaluations of the atomic profiles of phosphorus implanted
into amorphous and (100) silicon in the energy range 50 keV to 400 keV
are provided in ref [4] together with empirical values for the Pearson
IV range, standard deviation, skewness and kurtosis parameters.

At higher energies in the MeV range, experimental data for doses between
5 X 10*11/cm*2 and 7.5 X 10*14/cm*2 are presented in ref [5]. These
results have been simulated using the TRIM Monte Carlo model and good
agreement has been obtained [5,6].

Modelling of two dimensional implantation of phosphorus has been
performed using either a Gaussian or Pearson IV profile in the vertical
direction and a Gaussian distribution for the horizontal direction
[7,8]. However, it has been shown using Monte Carlo simulations [9]
that these approaches can be inaccurate. An important additional feature
of the work described in ref [9] is the production of analytical
expressions which incorporate the energy and spatial variations of the
moments of the vertical and horizontal distributions obtained by fitting
to Monte Carlo predictions. In this way, the two dimensional profiles
may be constructed without further recourse to the Monte Carlo codes.
All the major species implanted into silicon including phosphorus are
covered in this work for energies up to 300 keV.

REFERENCES
----

[1]   J.Linhard, M.Scharff, H.E.Schiott [Mat.-Fys. Medd. Dan. Vidensk.
       Selsk. (Denmark) vol.33 (1963) p.14]
[2]   B.Smith [Ion implantation range data for silicon and germanium
       device technologies (Learned Information (Europe) Ltd, Oxford,
       GB, 1977)]
[3]   J.F.Gibbons, W.S.Johnson, S.W.Mylroie [Projected range statistics,
       2nd Edn. (Halstead Press, 1975)]
[4]   R.G.Wilson [J. Appl. Phys. (USA) vol.60 no.8 (15 Oct 1986)
       p.2797-805]
[5]   W.Skorupa et al [Nucl. Instrum. & Methods Phys. Res. Sect. B
       (Netherlands) vol.B19/20 pt.1 (1987) p.335-9]
[6]   M.Posselt, W.Skorupa [Nucl. Instrum. & Methods Phys. Res. Sect.B
       (Netherlands) vol.B21 no.1 (1987) p.8-13]
[7]   H.Runge [Phys. Status Solidi a (Germany) vol.39 no.2 (16 Feb
       1977) p.595-9]
[8]   H.Ryssel, K.Habeuger, K.Hoffmann, G.Prinke, R.Duemcke, A.Sachs
       [IEEE Trans. Electron. Devices (USA) vol.ED-27 (1980) p.1484]
[9]   G.Hobler, E.Langer, S.Selberherr [Solid-State Electron. (GB)
       vol.30 no.4 (1987) p.445-55]
[10]   D.J.Godfrey [EMIS Datareview RN=16191 (Oct 1987) 'Ion
       implantation of B in Si']

CHAPTER 14

DIFFUSION AND SOLUBILITY OF TRANSITION METALS

14.1 DIFFUSION OF Ag IN Si

by E.R.Weber

January 1988
EMIS Datareview RN=17884

Silver is a fast diffuser in silicon [1], but even after fast quenching
only a small fraction of the total Ag solubility can be detected by
observing electrically active deep level defects [2]. This might be
indicative of rapid interstitial diffusion combined with instability of
Ag in the interstitial site during quenching, as has been observed for
Co, Ni and Cu in silicon [3]. This qualitative picture of Ag diffusion
in silicon appears to be well established, but it is difficult to obtain
reliable quantitative data for the diffusion coefficient of Ag in Si.

The effective diffusion coefficient of Ag in silicon has been measured
by radiotracer analysis in the temperature range from 1100 to 1350 deg
C by Boltaks et al [4]. The diffusion coefficient follows an Arrhenius
relation:

D = Do exp (-E/kT)

with the preexponential 2 X 10*-3 cm*2/s and the activation energy of
1.6 eV [4]. A radiotracer analysis by Sterkhov et al [5], using silicon
with a dislocation density of 5 X 10*3/cm*2, determined as preexponential
1.5 and activation energy 1.39 eV, which corresponds to diffusion
coefficients about four orders of magnitude higher than Boltak's results.
Rollert et al [1] assumed for interstitial Ag in silicon the same
diffusion coefficient as for interstitial Au, with a preexponential of
6 X 10*-1 cm*2/s and an activation energy of 1.15 eV. The sectional
analysis of their neutron activated samples can only be explained by a
diffusion coefficient above 10*-7 cm*2/s at 1325 deg C, in agreement with
their own suggestion and the data of Sterkhov et al [5], but clearly at
variance with the results of Boltaks et al [4]. Uskov [6] determined by
radiotracer analysis for a diffusion temperature of 1200 deg C an even
higher value of 6 X 10*-4 cm*2/s. In addition, the comparison with the
diffusion coefficients of interstitial Cu [7] and interstitial Au in Si
[8], the other elements of this column of the periodic table, makes a
diffusion coefficient in the 10*-5 cm*2/s range at temperatures above
1000 deg C quite probable. This high diffusion coefficient does not
describe the behaviour of the electrically active, probably
substitutional Ag species, which are believed to be formed via a
dissociative diffusion mechanism [1].

This review of the presently available diffusion data shows, that more
work is required in order to obtain reliable values for the diffusion
coefficient of Ag in silicon.

REFERENCES
----

[1]   F.Rollert, N.A.Stolwijk, H.Mehrer [J. Phys. D (GB) vol.20 (1987)
      p.1148]
[2]   N.Baber, H.G.Grimmeiss, M.Kleverman, P.Omling, M.Zafar Iqbal
      [J. Appl. Phys. (USA) vol.62 no.7 (1 Oct 1987) p.2853-7]
[3]   E.R.Weber [Appl. Phys. A (Germany) vol.30 (1983) p.1]
[4]   B.I.Boltaks, H.Shih-yin [Sov. Phys. Solid State (USA) vol.2 (1961)
      p.2383]
[5]   V.A.Sterkhov, V.A.Panteleev, P.V.Pavlov [Sov. Phys. Solid State
      (USA) vol.9 (1967) p.533]
[6]   V.A.Uskov [Inorg. Mater. (USA) vol.11 (1975) p.848]
[7]   E.R.Weber [EMIS Datareview RN=17853 (Dec 1987) 'Diffusion of Cu
      in Si']
[8]   E.R.Weber [EMIS Datareview RN=17886 (Dec 1987) 'Diffusion of Au
      in Si']

14.2 DIFFUSION OF Au IN Si

by E.R.Weber

January 1988
EMIS Datareview RN=17886

Au is a fast diffusing element in silicon. This fast diffusion is
generally ascribed to the motion of interstitial gold. Interstitial Au
could not be directly detected by electrical measurements even after
fast quenching. The electrically active Au species is generally
identified with substitutional Au, formed during prolonged heat
treatment by a change of the interstitial Au onto substitutional sites
[1]. This complex diffusion mechanism involves native lattice defects,
whose concentration in thermal equilibrium in silicon is low. Therefore,
deposition of a Au layer on one side of a Si wafer results in a U-shaped
diffusion profile [3], with a flat centre region and large concentration
increases near the surfaces, which act as sources or sinks for native
lattice defects. In early work [2-4] Au diffusion has been described
by the Frank-Turnbull mechanism [5], in which an interstitial metal
atom reacts with a vacancy to form the substitutional species. Goesele
et al [6] suggested for this process the kick-out mechanism:
interstitial Au kicks a silicon atom out of its site, creating a silicon
self-interstitial. Whereas in the Frank-Turnbull mechanism the rate
limiting process is the supply of vacancies, in the kick-out mechanism
it is the annihilation of self-interstitials. Detailed analyses of the
formation kinetics of substitutional Au in a variety of Si crystals
support the kick-out mechanism [7,8].

The complex diffusion profile of Au in Si does not allow fitting by an
erfc curve in order to deduce diffusion coefficients. Rather, a model
for the diffusion process has to be developed which after fitting allows
calculation of the diffusion coefficient. Based on radiotracer
measurements in the 700-1300 deg C temperature range, Wilcox et al [2]
proposed the diffusion coefficient of interstitial Au in silicon to
follow the Arrhenius relation:

D = Do exp (-E/kT)   (1)

with Do = 2.44 X 10*-4 cm*2/s and E = 0.39 eV. These parameters are very
similar to those of Cu [9] and Ni [10] in silicon. They allow
calculation of a diffusion coefficient of 1.1 X 10*-5 cm*2/s at 1200
deg C, which explains reasonably well the observed fast in-diffusion of
Au in Si. The analysis of tracer data by Huntley et al [4] resulted in a
temperature independent interstitial diffusion coefficient around 3 X
10*-7 cm*2/s in the temperature range of 900-1100 deg C.

Most authors agree today, that true substitutional diffusion is
negligible in the build-up of the substitutional component in the
interior of a silicon crystal. As discussed above, the time dependence
of this process is not determined by substitutional diffusion of Au
atoms, but rather by the availability and mobility of native lattice
defects. Therefore it is indeed possible to use the analysis of Au
diffusion profiles to get information on silicon self-diffusion
[2,4,6-8].

The kick-out model predicts [7] that the concentration Cm in the minimum of the U-shaped diffusion profile, i.e. the Au concentration in the centre of a silicon wafer, increases according to:

$$Cm(t) = \frac{2}{d} \ X \ (pi \ Dt)*0.5 \ X \ Csol \qquad (2)$$

with t being the diffusion time, d the wafer thickness, D the effective diffusion constant and Csol the solubility of substitutional Au in Si at the diffusion temperature [11]. The shape of the Au profile is given in this model by the relationship:

$$erfc \ [ln(C/Cm)*0.5] = 1 - 2s/d \qquad (3)$$

with the Au concentration C at a depth s below the wafer surface. These equations allowed a good fit to experimental diffusion profiles, obtained by spreading resistance measurements and neutron activation analysis combined with serial sectioning. With these fits, the following values were determined for silicon with intrinsic conductivity at the diffusion temperature:

TABLE 1: effective diffusion constant D determining the increase of Cm, Eqn (2)

Temp. (deg C)	Diffusion constant (cm*2/s)
1098	1.6 X 10*-9
1050	5.8 X 10*-10
1000	2.2 X 10*-10
900	2.0 X 10*-11
800	1.6 X 10*-12

It should be noted, that the diffusion constant D does not describe substitutional diffusion; it is rather a parameter with the dimension of a diffusion coefficient describing the kinetics of this specific case of the kick-out process. Within this model, substitutional atoms are assumed to be immobile.

Diffusion profiles following Eqn (3) are only found in dislocation-free material. High dislocation densities result in substantial acceleration of build-up of the substitutional Au concentrations and in changes of the diffusion profile. Spreading resistance measurements of samples with very high dislocation density obtained by plastic deformation [8]

showed diffusion profiles which could be fitted by the normal relation
for diffusion from an unlimited source into a semi-infinite solid:

C = Csol X erfc [s/(2(Dt)*0.5)]

with D being the effective diffusion coefficient for diffusion into a
highly dislocated sample. TABLE 2 lists the values for D thus determined

TABLE 2: effective diffusion constant D for Au diffusion into highly
         dislocated Si [8]

Temp. (deg C)	Diffusion constant (cm*2/s)
1154	5.0 X 10*-7
1101	2.5 X 10*-7
1001	1.1 X 10*-7
953	2.5 X 10*-8
907	9.1 X 10*-9

In summary, the diffusion of Au into silicon cannot be described by a
simple interstitial or substitutional diffusion process, but is rather
determined by the concentration and mobility of native defects like
silicon self-interstitials. The detailed investigations of the last few
years allow prediction in a satisfactory way of the shape of the Au
diffusion profile and its time dependence.

REFERENCES
----

[1]   W.C.Dash [J. Appl. Phys. (USA) vol.31 (1960) p.2275]
[2]   W.R.Wilcox, T.J.LaChapelle [J. Appl. Phys. (USA) vol.35 (1964)
      p.240]
[3]   G.J.Sprokel, J.M.Fairfield [J. Electrochem. Soc. (USA) vol.112
      (1965) p.200]
[4]   F.A.Huntley, A.F.W.Willoughby [Philos. Mag. (GB) vol.28 (1973)
      p.1319]
[5]   F.C.Frank, D.Turnbull [Phys. Rev. (USA) vol.104 (1956) p.617]
[6]   U.Goesele, W.Frank, A.Seeger [Appl. Phys. (Germany) vol.23 (1980)
      p.361-8]
[7]   N.A.Stolwijk, B.Schuster, J.Hoelzl [Appl. Phys. A (Germany)
      vol.33 (1984) p.133]
[8]   N.A.Stolwijk, J.Hoelzl, W.Frank, E.R.Weber, H.Mehrer [Appl. Phys.
      A (Germany) vol.39 (1986) p.37]
[9]   E.R.Weber [EMIS Datareview RN=17853 (Dec 1987) 'Diffusion of Cu
      in Si']
[10] E.R.Weber [EMIS Datareview RN=17858 (Dec 1987) 'Diffusion of Ni
      in Si']
[11] E.R.Weber [EMIS Datareview RN=17887 (Jan 1988) 'Solubility of Au
      in Si']

14.3 DIFFUSION OF Co IN Si

by E.R.Weber

December 1987
EMIS Datareview RN=17841

Cobalt is one of the fastest diffusing elements in silicon. It diffuses
interstitially and stays predominantly on interstitial sites at high
temperatures in thermal equilibrium. During cooling down or even fast
quenching, the interstitial cobalt forms pairs or precipitates, so that
no electrically active defect in concentrations comparable to the total
cobalt concentration has been found [1].

Based on radiotracer measurements [2], the diffusion coefficient of
cobalt in Si has been estimated to be of the order of 10*-6 to 10*-4
cm*2/s in the 1000-1300 deg C temperature range. Estimations based on
neutron activation analysis yielded a lower limit of 7 X 10*-6 cm*2/s at
700 deg C [1]. These estimations were confirmed more recently by the
radiotracer measurement of Gilles [3,4] which yielded the following values:

Temp. (deg C)	Diffusion coefficient (cm*2/s)
1000	4.1 X 10*-5
900	2.2 X 10*-5
800	1.2 X 10*-5

From these values, together with the estimation of Weber [1], they fit
the diffusion coefficient by an Arrhenius relation:

$$D = Do \exp(-E/kT)$$

with the pre-exponential 2 X 10*-2 cm*2/s and the activation energy of
0.69+-0.05 eV [4]. This value of the activation energy for the diffusion
of interstitial Co in Si agrees well with typical values for
interstitial diffusion of transition metals in Si [1], the pre-
exponential factor is in the range typically found for similar elements.

Conflicting radiotracer Co diffusion measurements yielded Do = 9.2 X
10*-4 cm*2/s and an activation energy of E = 2.8 eV [5]. Both the pre-
exponential factor and the activation energy are far out of the range
of other measurements of any interstitial transition elements in silicon
[1]. These data have been ascribed to uncontrolled outdiffusion during
cooling down [6] and do not appear to be reliable. Spreading resistance
measurements by Tomokage et al [7,8] of Co-diffused and quenched silicon
detected an electrically active species in very small concentration [9],
whose diffusion coefficient fitted well to the results of Kitagawa [5].

The diffusion of Co in extrinsically doped silicon has been investigated by Gilles et al [3,4]. They found the following values:

Temp. (deg C)	Diffusion coefficient (cm*2/s)	Doping (/cm*3)
800	1.2 X 10*-7	1 X 10*20 P
700	2 X 10*-9	1 X 10*20 P
700	2.3 X 10*-6	8 X 10*19 B

The extrapolated diffusion coefficient of Co in intrinsic Si at 700 deg C [4] is 5.3 X 10*-6 cm*2/s. The small decrease of the diffusion coefficient in p-Si was ascribed to pairing reactions of Co with B. The pronounced decrease in n-Si was explained by the existence of substitutional Co in strongly n-doped Si and the possibility of pairing reactions with P [3,4].

REFERENCES
----

[1] E.R.Weber [Appl. Phys. A (Germany) vol.30 (1983) p.1]
[2] M.K.Bakhadyrkhanov, B.I.Boltaks, G.S.Kulikov [Sov. Phys. Solid
     State (USA) vol.12 (1970) p.144]
[3] D.Gilles, W.Schroeter [Mater. Sci. Forum (Switzerland) vol.10-1
     (1986) p.169]
[4] D.Gilles [PhD Thesis, Goettingen Germany 1987)]
[5] H.Kitagawa, K.Hashimoto [Jpn. J. Appl. Phys. (Japan) vol.16 no.1
     (Jan 1977) p.173-4]
[6] W.Bergholz [J. Phys. D (GB) vol.14 no.6 (14 Jun 1981) p.1099-113]
[7] H.Tomokage, H.Kitagawa, K.Hashimoto [Mem. Fac. Eng. Kyushu Univ.
     (Japan) vol.41 (1981) p.59]
[8] H.Nakashima, H.Tomokage, H.Kitagawa, K.Hashimoto [Jpn. J. Appl.
     Phys. Part.1 (Japan) vol.23 no.6 (Jun 1984) p.776-7]
[9] E.R.Weber [EMIS Datareview RN=17843 (Dec 1987) 'Solubility of Co
     in Si']

14.4 DIFFUSION OF Cr IN Si

by E.R.Weber

December 1987
EMIS Datareview RN=17850

Cr is a fast diffusing element in silicon. It diffuses interstitially
and stays predominantly on interstitial sites in thermal equilibrium
at high temperatures [1]. Interstitial chromium can be easily retained
in the electrically active interstitial site during quenching or not
too slow cooling down [1,2].

Only a few systematic measurements of Cr diffusion in silicon have
been reported [3,4]. Bendik et al [3] investigated the depth of p/n
junctions formed after Cr diffusion in p-Si. Interstitial Cr forms only
a donor level 0.23 eV below the conduction band edge [1,2,5], so that
this method might yield meaningful results. In the temperature range
of 900-1250 deg C these authors fit the diffusion coefficient by an
Arrhenius relation:

$$D = Do \exp(-E/kT)$$

with the pre-exponential 1 X 10*-2 cm*2/s and the activation energy of
1.0 eV [3]. A radiotracer study by Wurker et al [4] confirmed these
results in the temperature range of 1100-1250 deg C, using the p/n
junction method and in addition for 1250 deg C radiochemical neutron
activation analysis of a sectioned sample. The value obtained from the
Arrhenius relation for D(900 deg C) = 5 X 10*-7 cm*2/s is as well in
good agreement with an estimation by Weber [1] (D(900 deg C) = 4 X
10*-7 cm*2/s), giving these early data further credibility for this
high temperature region.

The value of the activation energy for the diffusion of interstitial
Cr in Si is higher than that found for Mn, Fe or Co in Si [1], but
might express a chemical trend in the 3d series of transition elements,
showing slower diffusion for the lighter elements which can be expected
to have larger atomic volume [6,7].

Interstitial Cr in Si shows some instability at room temperature in low-
doped material, and anneals out of the supersaturated solution in the
range of 100-200 deg C [1-3]. The time constants of this process depend
on the dislocation density and very strongly on the acceptor concentration
in p-Si, as Cr forms Coulombic bound pairs with acceptors, similar to Fe
and Mn [2,3]. The activation energy for this process is similar to that
found for the high temperature diffusion [3], suggesting, similarly as
in the case of Fe, motion of the interstitial ion to be the thermally
activated process.

REFERENCES
----

[1] E.R.Weber [Appl. Phys. A (Germany) vol.30 no.1 (Jan 1983) p.1-22]
[2] H.Conzelmann, K.Graff, E.R.Weber [Appl. Phys. A (Germany) vol.30
     no.3 (Mar 1983) p.169-75]
[3] N.T.Bendik, V.S.Garnyk, L.S.Milevskii [Sov. Phys.-Solid State (USA)
     vol.12 (1970) p.150]

[4]   W.Wurker, K.Roy, J.Hesse [Mater. Res. Bull. (USA) vol.9 (1974)
      p.971]
[5]   E.R.Weber [EMIS Datareview RN=17890 (Jan 1988) 'Deep levels due
      to transition metals in Si']
[6]   E.R.Weber [Mater. Res. Soc. Symp. Proc. (USA) vol.35 (1985) p.3]
[7]   K.Graff [Semiconductor Silicon 1986 Ed. H.R.Huff (Electrochem.
      Soc., Pennington, USA, 1986) p.751]

14.5 DIFFUSION OF Cu IN Si

by E.R.Weber

December 1987
EMIS Datareview RN=17853

Copper is the fastest diffusing transition element in silicon, and
is one of the fastest solid state diffusers known, with diffusion
coefficients reaching 10*-4 cm*2/s, a value characteristic of atomic
motion in liquids [1]. It diffuses interstitially and stays
predominantly on interstitial sites at high temperatures in thermal
equilibrium. During cooling down or even fast quenching, the interstitial
copper forms pairs or precipitates, so that no electrically active
defect in concentrations comparable to the total copper concentration
has been found even after the fastest quenching [1].

The radiotracer measurements of Hall and Racette [2] are the most
detailed measurements of Cu diffusion in silicon. Using heavily p-doped
silicon these authors described the diffusion coefficient in the
temperature range 400-700 deg C by an Arrhenius relation:

D = Do exp (-E/kT)

with the pre-exponential 4.7 X 10*-3 cm*2/s and the activation energy
of 0.43 eV. This value of the activation energy for the diffusion of
interstitial Cu in Si agrees well with typical values for interstitial
diffusion of transition metals in Si. The pre-exponential factor is as
well in the range typically found for similar elements [1]. A radiotracer
measurement of Struthers [3] yielded, for Cu in intrinsic Si at 900 deg C
a value of 5 X 10*-5 cm*2/s, in quite good agreement with the relation
above. Based on neutron activation analysis of intrinsic silicon Weber
estimated a lower limit of 10*-5 cm*2/s at 500 deg C [1], which compares
as well favourably with Hall and Racette's analysis.

Whereas the diffusion of Cu in extrinsically doped p-silicon appears to
be similar to that in intrinsic material, Hall and Racette [2] report
considerably slower diffusion in extriniscally doped n-Si without giving
specific values. They ascribe the slower diffusion in heavily doped n Si
to the existence of a donor level of interstitial Cu and to
substitutional Cu triple acceptors in high concentrations in such
extrinsically doped n-Si.

REFERENCES
----

[1] E.R.Weber [Appl. Phys. A (Germany) vol.30 no.1 (Jan 1983) p.1-22]
[2] R.N.Hall, J.H.Racette [J. Appl. Phys. (USA) vol.35 (1964) p.379]
[3] J.D.Struthers [J. Appl. Phys. (USA) vol.27 (1956) p.1560]

14.6 DIFFUSION OF Fe IN Si

by E.R.Weber

December 1987
EMIS Datareview RN=17842

Iron is a fast diffusing element in silicon. It diffuses interstitially
and stays predominantly on interstitial sites at high temperatures in
thermal equilibrium. Interstitial iron can be retained in the
electrically active interstitial site during quenching or not too slow
cooling down [1]. The high diffusivity of this ubiquitous element,
combined with some stability at room temperature, make Fe one of
the most important contaminants of Si during high temperature processing
[2,3].

The diffusion coefficient at room temperature is near 10*-14 cm*2/s,
resulting in room temperature instability of the supersaturated
interstitial iron. This instability is accelerated by the presence of
shallow acceptors, which can attract the Fe atoms and form Fe-acceptor
pairs [4,5].

The diffusion coefficient of Fe in Si has been established for the
temperature range from 30 to 1250 deg C, using the Arrhenius-relation:

$$D = Do \exp(-E/kT)$$

with Do = 1.3 X 10*-3 cm*2/s and E = 0.68 eV [1]. This result was based
mainly on radiotracer results of samples quenched from the 1100-1250 deg
C range [6] and resistivity measurements of the annealing of electrically
active, interstitial Fe in Si in the 30-100 deg C range [4]. A number of
more recent investigations of the low temperature mobility of Fe, e.g.
by DLTS [5], support this description of the diffusion data. Recent high
temperature outdiffusion and tracer measurements yielded D (920 deg C)
= 1.7 X 10*-6 cm*2/s and D (900 deg C) = 1.4 X 10*-6 cm*2/s
respectively [7], in good agreement with the relation given above.

Early electrical and tracer measurements of Fe diffused and quenched
silicon were interpreted in terms of two iron species, interstitial Fe
and a second, slower diffusing species, coupled by a complex diffusion
mechanism [8]. These conclusions were not confirmed in later work. Today
most authors agree on a purely interstitial diffusion mechanism for Fe
in Si.

The diffusion of Fe in extrinsically doped silicon has been investigated
by Gilles [7], using tracer techniques. He found the following values:

Temp. (deg C)	Diffusion coefficient (cm*2/s)	Doping (/cm*3)
700	4 X 10*-11	1 X 10*20 P
700	2.4 X 10*-7	8 X 10*19 B

The interpolated value of the Fe diffusion coefficient in intrinsic Si
is 4 X 10*-7 cm*2/s [1]. The small decrease of the diffusion coefficient
in p-Si was ascribed to pairing reactions of Fe with B, the pronounced
decrease in n-Si was explained by the existence of substitutional Fe in
strongly n-doped Si and the possibility of pairing reactions with P [7].

REFERENCES
----

[1]   E.R.Weber [Appl. Phys. A (Germany) vol.30 (1983) p.1]
[2]   Y.H.Lee, R.L.Kleinhenz, J.W.Corbett [Appl. Phys. Lett. (USA)
      vol.31 (1 Aug 1977) p.142]
[3]   E.R.Weber, H.G.Riotte [Appl. Phys. Lett. (USA) vol.33 no.5 (1 Sep
      1978) p.433-5]
[4]   W.H.Shepherd, J.A.Turner [J. Phys. & Chem. Solids (GB) vol.23
      (1962) p.1697]
[5]   L.C.Kimerling [in 'Defects in Semiconductors' Eds J.Narayan,
      T.Y.Tan (North Holland, 1981) p.85]
[6]   J.D.Struthers [J. Appl. Phys. (USA) vol.27 (1956) p.1560]
[7]   D.Gilles [PhD Thesis (Goettingen, Germany, 1987)]
[8]   C.B.Collins, R.O.Carlson [Phys. Rev. (USA) vol.105 (1987) p.1409]

14.7 DIFFUSION OF Mn IN Si

by E.R.Weber

December 1987
EMIS Datareview RN=17857

Manganese is a fast diffusing element in silicon. It diffuses
interstitially and stays predominantly on interstitial sites at high
temperatures in thermal equilibrium. Interstitial manganese can be
easily retained in the electrically active interstitial site during
quenching or cooling which is not too slow [1].

The diffusion coefficient of Mn in Si has been measured by radiotracer
analysis for the temperature range from 900 to 1200 deg C [2]. It
follows the Arrhenius-relation:

D = Do exp (-E/kT)

with Do = (6.9 +- 2.2) X 10*-4 cm*2/s and E = (0.63 +- 0.03) eV [2].
Both activation energy of diffusion and the pre-exponential factor are
within the range of typical values for interstitial transition elements
in Si. The surface concentrations of Mn obtained from the diffusion
profiles were found to be in excellent agreement with the Mn solubility
[1], which is indicative of the precision of this analysis. The
extrapolation of this Arrhenius relation to room temperature yields a
value of D (300 deg K) = 1.8 X 10*-14 cm*2/s in quite good agreement
with a value of about 10*-15 cm*2/s which can be estimated from the
pairing kinetics of manganese with boron acceptors near room temperature
[3]. Further high temperature estimations of D(1100 deg C) = 2 X 10*-6
cm*2/s based on neutron activation analysis [1], D(1000-1350 deg C) =
10*-6 to 2 X 10*-5 cm*2/s, based on radiotracer measurements [4], and
the tracer measurements of Senk and Borchardt [5], yielding D(900-1200
deg C) = 3 X 10*-7 to 5 X 10*-6 cm*2/s are in quite good agreement with
the data of Gilles et al [2]. However, the values for the activation
energy of diffusion, determined to be 1.3 eV [4] or 1.08 eV [5] fail to
extrapolate to room temperature values consistent with the low-
temperature pairing kinetics. Gilles et al [2] pointed out that the
surface concentrations in these publications differ considerably from
the solubility values, so that these diffusion experiments might have
used time-dependent boundary conditions, making their data less precise.

The diffusion of Mn in extrinsically doped silicon has been investigated
by Gilles [6], using tracer techniques. He found the diffusion
coefficient at 700 deg C in extrinsic n-Si with 10*20 P/cm*3 reduced by
several orders of magnitude, which was ascribed to the existence of
substitutional Mn in strongly n doped Si.

REFERENCES
----

[1] E.R.Weber [Appl. Phys. A (Germany) vol.30 no.1 (Jan 1983) p.1-22]
[2] D.Gilles, W.Bergholz, W.Schroeter [J. Appl. Phys. (USA) vol.59
     no.10 (15 May 1986) p.3590-2]
[3] H.Lemke [Phys. Status Solidi a (Germany) vol.76 (1983) p.223]
[4] M.K.Bakhadyrkhanov, B.I.Boltaks, G.S.Kulikov [Sov. Phys.-Solid
     State (USA) vol.14 (1972) p.1441]
[5] D.Senk, G.Borchardt [Mikrochim. Acta vol.II (1983) p.477]
[6] D.Gilles [PhD Thesis, Goettingen (Germany) 1987]

14.8 DIFFUSION OF Ni IN Si

by E.R.Weber

December 1987
EMIS Datareview RN=17858

Nickel is one of the fastest diffusing elements in silicon. It diffuses interstitially and stays predominantly on interstitial sites at high temperature in thermal equilibrium. During cooling down or even fast quenching, interstitial nickel forms pairs or precipitates, so that no electrically active defect in concentration comparable to the total nickel concentration has been found [1]. A small fraction of electrically active Ni, typically not more than 10*-4 of the total Ni concentration, has been ascribed to substitutional Ni [2,3]. However, direct microscopic proof of this identification is still lacking. In the following, the diffusion of the dominant interstitial Ni will be reviewed.

The radiotracer analysis of Bakhadyrkhanov et al [4] described the diffusion coefficient in the temperature range of 800-1300 deg C by an Arrhenius relation:

D = Do exp (-E/kT)

with the pre-exponential 2 X 10*-3 cm*2/s and the activation energy of 0.47 eV. This value of the activation energy for the diffusion of interstitial Ni in Si agrees well with typical values for interstitial diffusion of transition metals in Si and the pre-exponential factor is as well in the range typically found for similar elements [1]. A number of estimations of the Ni diffusion coefficient are in agreement with these data: Aalberts and Verheijke [5] estimate from neutron activation analysis (NAA) for temperatures between 1100 and 1300 deg C a lower limit of 5 X 10*-5 cm*2/s; radiotracer measurements of Yoshida and Furusho [6] cite for 1200 deg C a diffusion coefficient of 10*-4 to 10*-5 cm*2/s; Weber gives for 700 deg C an estimate for a lower limit of 10*-5 cm*2/s [1].

A recent study of Thompson et al [7] with the Rutherford backscattering technique deduces rather indirectly a pre-exponential value of 6.3 X 10*-4 cm*2/s and an activation energy of Ni diffusion E = 0.76 eV. These values are not in agreement with all the results mentioned before, and the authors ascribe the difference to the impurity content of the Czochralski silicon used in that study. Earlier radiotracer investigations by Bonzel [8] obtained Do = 0.1 cm*2/s and E = 1.91 eV. However, this result appears to be flawed because of the assumption of a dissociative diffusion mechanism. Berning and Levenson [9] obtained by Auger microscopy of samples diffused in the 250-350 deg C range apparent diffusion coefficients of Do = 10*-13 cm*2/s and E = 0.27 eV for diffusion into (100) Si. For diffusion into Si(111) two sets of data were obtained, yielding Do = 10*-1 cm*2/s, E = 1.88 eV and Do = 10*-6 cm*2/s, E = 1.22 eV, respectively. All these values as well as an earlier result of Yoon and Levenson [10] (Do = 0.5 cm*2/s, E = 1.52 eV) are at variance with the diffusion coefficients extrapolated from higher temperatures.

REFERENCES
----

[1] E.R.Weber [Appl. Phys. A (Germany) vol.30 no.1 (Jan 1983) p.1-22]

[2] M.Yoshida, K.Saito [Jpn. J. Appl. Phys. (Japan) vol.6 (1967)
     p.573]

[3] H.Kitagawa, K.Hashimoto, M.Yoshida [Jpn. J. Appl. Phys. Part 1
     (Japan) vol.21 (1982) p.276]

[4] M.K.Bakhadyrkhanov, S.Zainabidinov, A.Khamidov [Sov. Phys.
     Semicond. (USA) vol.14 (1980) p.243]

[5] J.H.Aalberts, M.L.Verheijke [Appl. Phys. Lett. (USA) vol.1 (1962)
     p.19]

[6] M.Yoshida, K.Furusho [Japan J. Appl. Phys. (Japan) vol.3 (1964)
     p.521]

[7] R.D.Thompson, D.Gupta, K.N.Tu [Phys. Rev. B (USA) vol.33 no.4
     (15 Feb 1986) p.2636-41]

[8] H.P.Bonzel [Phys. Status Solidi (Germany) vol.20 (1967) p.493]

[9] G.L.P.Berning, L.L.Levenson [Thin Solid Films (Switzerland) vol.55
     (1978) p.473]

[10] K.H.Yoon, L.L.Levenson [J. Electron. Mater. (USA) vol.4 (1975)
     p.1249]

14.9 DIFFUSION OF Pt IN Si

by E.R.Weber

January 1988
EMIS Datareview RN=17888

The diffusion of Pt in Si has been investigated only in a few papers
[1-3]. All diffusion profiles obtained were complex and could not be
fitted by a simple erfc profile. Therefore it is generally concluded
that Pt diffusion into Si is similar to Au diffusion [4], with mobile
interstitial Pt changing slowly into substitutional sites combined
with a higher substitutional than interstitial solubility. In such a
case the diffusion behaviour cannot be described by a simple diffusion
coefficient. Fitting experimentally determined diffusion profiles
requires the use of a model for the process of metal atoms changing
from interstitial to substitutional sites. This process can be
described by the Frank-Turnbull mechanism [5], assuming that
interstitial metal atoms consume vacancies. If the concentration of
vacancies in thermal equilibrium is too low a kick-out mechanism [6]
might be dominant, in which an interstitial metal atom kicks a silicon
atom out of its site, creating a silicon self-interstitial.

The substitutional atoms are electrically active, so that profiles of
the distribution of substitutional metal atoms may be determined by
electrical measurements. It should be noted that in the case of Pt in
Si two electrically active species have been detected [7]; Pt I is
generally considered to be isolated substitutional Pt [8], the
distinctly different Pt II might be due to a complex, but no clear
model has been developed. The concentration of Pt II is typically at
least an order of magnitude smaller than the Pt I concentration, so that
experiments measuring Pt I concentrations might be representative
for the concentration of substitutional Pt.

Mantovani et al [1] determined diffusion profiles of Pt in Si in the
700-850 deg C range by analysing DLTS scans. For a diffusion
temperature of 800 deg C, even after 96 hrs of diffusion the
concentration in the bulk of a wafer turned out to be an order of
magnitude lower than the substitutional solubility expected from the
near-surface concentration. The resulting concentration profiles $C(s,t)$
of electrically active Pt could be well defined with an equation
derived for the kick-out model of Au diffusion in Si by Seeger [9]:

$$C(s,t) = Csol/[1 + s/(Dt)*0.5] \qquad\qquad (1)$$

with the solubility of Pt in Si at this temperature Csol, the distance s
from the surface and a parameter D which can be regarded as an effective
diffusion coefficient. Mantovani et al [1] did not list the fitting
parameters obtained from their diffusion profiles: however, an analysis
of Figs 4-6 in ref [1] allows one to estimate the following values for D:

TABLE 1: diffusion parameter D for diffusion profiles of Pt in Si
(Eqn (1))

Temp. (deg C)	Diffusion parameter D (cm*2/s)
850	3.6 X 10*-14
800	2.0 X 10*-14
750	9.0 X 10*-15
700	9.8 X 10*-16

The diffusion profiles of Mantovani et al [1] agree very well with the profiles determined with a similar method by Prabhakar et al [2]. These profiles are in shape very similar to those of Au in Si, but the diffusion constants are about two orders of magnitude smaller. It should be noted that D does not represent substitutional diffusion: it rather describes the kinetics of the kick-out process. Within this model, substitutional atoms are assumed to be immobile.

Summarizing, the diffusion of Pt in Si has not yet been thoroughly investigated. The available data are based on electrical measurements which might not detect the total Pt concentration. These measurements indicate that Pt has the same diffusion mechanism as Au in Si, presumably the kick-out mechanism, with diffusion parameters smaller by several orders of magnitude.

REFERENCES
----

[1] S.Mantovani, F.Nava, C.Nobili, G.Ottaviani [Phys. Rev. B (USA) vol.33 (1986) p.5536]
[2] A.Prabhakar, T.C.Gill, M.A.Nicolet [Appl. Phys. Lett. (USA) vol.43 (1983) p.1118]
[3] R.F.Bailey, T.G.Mills [Proc of the Electrochem. Soc. Meet. on Semiconductor Silicon, Eds R.R.Haberecht and E.L.Kern (Electrochem. Soc., New York, 1969) p.481-9]
[4] E.R.Weber [EMIS Datareview RN=17886 (Jan 1988) 'Diffusion of Au in Si']
[5] F.C.Frank, D.Turnbull [Phys. Rev. (USA) vol.104 (1956) p.617]
[6] U.Goesele, W.Frank, A.Seeger [Appl. Phys. (Germany) vol.23 (1980) p.361]
[7] H.H.Woodbury, G.W.Ludwig [Phys. Rev. (USA) vol.126 (1962) p.466]
[8] R.F.Milligan, F.G.Anderson, G.D.Watkins [Phys. Rev. B (USA) vol.29 (1984) p.2819]
[9] A.Seeger [Phys. Status Solidi a (Germany) vol.61 (1980) p.521]

14.10 DIFFUSION OF Ti IN Si

by E.R.Weber

January 1988
EMIS Datareview RN=17867

Titanium is among the slowest diffusing transition metals in silicon
[1]. Recently, an electron paramagnetic resonance signal from
interstitial titanium in the single positive charge state has been
identified [2,3]. This finding makes it likely that titanium diffuses
predominantly as an interstitial and stays in this electrically active
interstitial site during cooling down, like the other 3d metals in
silicon. Because of its small diffusion coefficient, this interstitial
titanium is stable at room temperature and resistant even against higher
temperature processing. It forms three deep level defects [4] and acts
thus as an efficient recombination/generation centre, shortening the
minority carrier lifetime or increasing the leakage currents of silicon
devices, whenever it is present even in small concentrations [5,6].

Information about the diffusion coefficient of titanium in silicon is
very unsatisfactory. Boldyrev et al [7] performed a radiotracer study
in the temperature range 1000-1250 deg C and found diffusion
coefficients in the 10*-11 to 10*-10 cm*2/s range. They fitted their
values to an Arrhenius relation:

D = Do exp (-E/kT)

with the pre-exponential 2 X 10*-5 cm*2/s and the activation energy of
1.50 eV. The value for the pre-exponential factor is very small for
interstitial diffusion in silicon and the value for the activation
energy appears to be very high. Diffusion coefficients calculated from
these parameters are more than three orders of magnitude smaller than
the known diffusion coefficients of other 3d transition metals in
silicon [8].

A recent DLTS study of junction leakage ascribed to Ti contamination
from TiSi2 Schottky contacts allowed determination of a diffusion
profile, with D(800 deg C) = 2.7 X 10*-12 cm*2/s [6] in quite good
agreement with Boldyrev's results. Other observations are at variance
with Boldyrev's results. Rohatgi et al [9] estimated Ti diffusion
coefficients from out-diffusion measurements. Hansen et al [10]
reported a diffusion coefficient based on DLTS measurements. TABLE 1
shows a comparison of all available data together with results
calculated from Boldyrev's parameters:

TABLE 1

Temp. (deg C)	Diff. coeff.[7] (cm*2/s)	Diff. coeff.[6] (cm*2/s)	Diff. coeff.[9] (cm*2/s)	Diff. Coeff. [10] (cm*2/s)
1100	6.4 X 10*-11		1.4 X 10*-9	2.5 X 10*-8
900	7.5 X 10*-12		1.7 X 10*-10	
825	2.7 X 10*-12		4.1 X 10*-11	
800	1.9 X 10*-12	2.7 X 10*-12		

In conclusion, the values presently available for diffusion coefficients of titanium in silicon appear to be questionable and may allow only crude estimations of titanium diffusion lengths.

REFERENCES
----

[1]   E.R.Weber [in 'Impurity Diffusion and Gettering in Semiconductors', Eds R.B.Fair, C.B.Pearce, J.Washburn (Mater. Res. Soc., 1985) p.3]

[2]   D.A.van Wezep, C.A.J.Ammerlaan [J. Electron. Mater. (USA) vol.14a (1985) p.863]

[3]   D.A.van Wezep, R.van Kemp, E.G.Sieverts, C.A.J.Ammerlaan [Phys. Rev. B (USA) vol.32 no.11 (1 Dec 1985) p.7129-38]

[4]   E.R.Weber [EMIS Datareview RN=17890 (Jan 1988) 'Deep level due to transition metals in Si'

[5]   A.Rohatgi, J.R.Davis, R.H.Hopkins, P.Rai-Chaudhury, P.G.McMullin [Solid-State Electron. (GB) vol.23 no.5 (May 1980) p.415-22]

[6]   K.Nauka, J.Amano, M.P.Scott, E.R.Weber, J.E.Turner, R.Tsai [in 'Materials Issues in Silicon Integrated Circuit Processing,' Eds M.Wittmer, J.Stimmell, M.Strathman (Mater., Res. Soc. USA,1986) p.319]

[7]   V.P.Boldyrev, I.I.Pokrovskii, S.G.Romanovskaya, A.V.Tkach, I.E.Shimanovich [Sov. Phys.-Semicond. (USA) vol.11 (1977) p.709]

[8]   E.R.Weber [Appl. Phys. A (Germany) vol.30 no.1 (Jan 1983) p.1-22]

[9]   A.Rohatgi, J.R.Davis, R.H.Hopkins, P.G.McMullin [Solid-State Electron. (GB) vol.26 no.11 (Nov 1983) p.1039-51]

[10] J.Hansen, D.Gilles, W.Schroeter [Verhandig DPG (Germany) vol.18 (1983) p.613; J.Hansen - private communication]

14.11 SOLUBILITY OF Ag IN Si

by E.R.Weber

January 1988
EMIS Datareview RN=17883

Silver is a fast diffuser in silicon [1], but even after fast quenching
only a small fraction of the total Ag solubility can be detected by
observing electrically active deep level defects [2]. This might be
indicative of rapid interstitial diffusion combined with predominantly
interstitial solubility and instability of Ag in the interstitial site
during quenching, as has been observed for Co, Ni and Cu in silicon [3].
The small concentrations of electrically active species detected in
several studies [2,4-6] might be substitutional Ag [1].

The eutectic temperature of Ag in Si is 840 deg C [7]. With such a low
eutectic temperature, Ag is expected to show retrograde solubility in
Si, with a solubility maximum between the eutectic temperature and the
melting point [3].

Radiotracer measurements of the solubility of Ag in silicon were
performed by Boltaks et al [6]. Their data show a solubility maximum at
1350 deg C. The following results were taken from a Fig.4 in [6].

TABLE 1: radiotracer measurements of Ag solubility in Si [6]

Temp. (deg C)	Solubility (/cm*3)
1390	1.6 X 10*16
1380	2.3 X 10*16
1370	5.0 X 10*17
1350	2.0 X 10*17
1300	1.0 X 10*17
1250	1.7 X 10*16
1200	6.0 X 10*15

The study of Rollert et al [1] by neutron activation analysis yielded
quite different results:

TABLE 2: neutron activation analysis of Ag solubility in Si [1]

Temp. (deg C)	Solubility (/cm*3)
1325	9.2 X 10*15
1288	6.5 X 10*15
1250	4.2 X 10*15
1216	2.8 X 10*15
1185	2.1 X 10*15
1142	1.2 X 10*15
1102	1.1 X 10*15
1068	9.0 X 10*14
1024	8.1 X 10*14

These authors did not observe a solubility maximum in the investigated
temperature range. However, their data indicate the existence of a
minimum Ag concentration of about 7 X 10*14/cm*3 ascribed to Ag atoms
trapped in some unknown configuration [1]. The solubility of
interstitial Ag in Si is then defined as the difference of the
experimentally determined total Ag concentration and this temperature
independent trapped concentration. For the temperature range from 1024
to 1325 deg C, this difference can be described by an Arrhenius
relation:

C = Co exp (-E/kT)

with Co = 5.36 X 10*24/cm*3 and E = 2.78 eV [1]. A Neutron Activation
Analysis by Smith et al [8] resulted in Ag concentration of 2 X 10*15
/cm*3 after diffusion at 1200 deg C, in good agreement with the results
of Rollert et al [1].

The maximum concentrations of electrically active Ag in silicon, which
might be representative of the substitutional solubility of Ag in Si,
are reported in the 10*12-10*13/cm*3 range after diffusion near 1200
deg C [2,4-6]. No systematic investigation of the temperature
dependence of the concentration of this species is known.

In summary, the available data on the solubility of Ag in Si indicate
qualitatively a dominant interstitial solubility at high temperatures
and only a small substitutional fraction. However, the quantitative
results known up to now are quite conflicting.

REFERENCES
----

[1]   F.Rollert, N.A.Stolwijk, H.Mehrer [J. Phys. D (GB) vol.20 (1987)
      p.1148]
[2]   N.Baber, H.G.Grimmeiss, M.Kleverman, P.Omling, M.Zafar Iqbal [J.
      Appl. Phys. (USA) vol.62 no.7 (1 Oct 1987) p.2853-7]]
[3]   E.R.Weber [Appl. Phys. A (Germany) vol.30 (1983) p.1]
[4]   H.Lemke [Phys. Status Solidi a (Germany) vol.94 (1986) p.K55]
[5]   S.J.Pearton, A.J.Tavendale [J. Phys. C (GB) vol.17 (1984) p.6701]
[6]   B.I.Boltaks, H.Shih-yin [Sov. Phys. Solid State (USA) vol.2
      (1961) p.2383]
[7]   J.P.Hager [Trans. Metall. Soc. AIME (USA) vol.227 (1963) p.1000]
[8]   P.C.Smith, A.G.Milnes [J. Electrochem. Soc. (USA) vol.117 no.8 (Aug
      1970) p.260C]

14.12 SOLUBILITY OF Au IN Si

by E.R.Weber

January 1988
EMIS Datareview RN=17887

Au is a fast diffusing element in silicon. This fast diffusion is
generally ascribed to the motion of interstitial gold. However, the
solubility of Au in the interstitial site is small. The dominant,
electrically active Au species is generally identified with
substitutional Au, formed during prolonged heat treatment by a change
of the interstitial Au onto substitutional sites [1]. The different
models for this complex diffusion process are discussed elsewhere [2].
The electrically active substitutional Au is a recombination centre
and therefore decreases the carrier lifetime in silicon in a
well-controlled way [3]. Because of this technological application, the
solubility of Au in silicon has been investigated in great detail by
numerous researchers.

The concentration of Au in silicon was measured by Collins et al [4],
using radiotracer techniques. Diffusion times were about 50 hrs. Their
results are shown in TABLE 1.

TABLE 1: Au concentrations in Si determined by Collins et al [4]

Temp. (deg C)	Au concentration (/cm*3)
1400	3.4 X 10*16
1300	1.1 X 10*17
1200	8.4 X 10*16
1000	9.1 X 10*15

These data reflect retrograde solubility with a maximum near 1300 deg C,
as observed for many other transition metals in silicon [5].

Wilcox et al [6] fitted their radiotracer measurements of Au diffusion
profiles in the 700 to 1200 deg C temperature range with solubility
values according to an Arrhenius relation:

C = Co exp (-E/kT)

The solubility parameters were:

for interstitial Au: Co = 6.0 X 10*24/cm*3 and E = 2.5 +- 0.4 eV,
for substitutional Au: Co = 8.2 X 10*22/cm*3 and E = 1.8 eV,
resulting in concentrations of 1.7 X 10*16 interstitial Au/cm*3 and
5.7 X 10*16 substitutional Au/cm*3 at 1200 deg C.

One problem of the determination of solubility values is the question of
whether the slow process of interstitial Au changing onto substitutional
sites has been completed. A later radiotracer study by Brown et al [7]
took great care to ensure saturation conditions, using diffusion times
in the range of 70-138 hrs. Their results for near-intrinsic and heavily
B doped silicon are given in TABLE 2.

TABLE 2: Au solubility in intrinsic and heavily B doped Si [7]

Temp. (deg C)	Au Solubility [/cm*3] ('intrinsic' Si)	Au Solubility [/cm*3] (Si with 9 X 10*19 B/cm*3)
1300	1.6 X 10*17	1.7 X 10*17
1200	1.0 X 10*17	1.3 X 10*17
1100	4.7 X 10*16	7.0 X 10*16
1000	1.6 X 10*16	3.1 X 10*16

The solubility of Au in intrinsic Si was found to be only slightly higher than originally reported by Collins et al [4], TABLE 1, and calculated by Wilcox et al [6]. The solubility enhancement in heavily B doped Si could be described quantitatively by the Fermi level effect, enhancing the solubility due to the presence of charged species. Even stronger solubility enhancement has been reported by Cagnina for P and As doped n-Si [8]. The analysis of Cagnina's results by Brown et al [7] shows that the solubility enhancement up to a factor of 5 found in heavily As-doped Si can be explained by the Fermi level effect, whereas the enhancements by more than an order of magnitude in P-doped silicon might be caused by ion pairing.

Au concentrations very consistent with the results of Brown et al [7] for intrinsic silicon were found in a recent study by Stolwijk et al [9], using neutron activation analysis and also saturation conditions (TABLE 3).

TABLE 3: Au solubility in intrinsic Si [9]

Temp. (deg C)	Au solubility (/cm*3)
1300	1.6 X 10*17
1200	1.0 X 10*17
1098	4.8 X 10*16
1050	2.7 X 10*16
1000	1.5 X 10*16
900	3.4 X 10*15
800	5.8 X 10*14

The eutectic temperature of the Si-Au phase diagram is near 370 deg C [10]. Therefore, all solubility values reported here correspond to an equilibrium with the liquid phase. In this case, only the distribution coefficient, the ratio of solid solubility to the liquidus concentration, but not the solid solubility itself, is expected to follow an Arrhenius law [5]. Wackwitz [11] fitted the Au concentrations reported by Collins et al [4] (TABLE 1) with the Arrhenius relation for the distribution coefficient:

k = ko exp (-E/kT)

with the parameters ko = 60 and E = 2.12 eV. The distribution coefficient for Au in Si at the melting point can be extrapolated with these data to be 2.7 X 10*-5.

REFERENCES
----

[1]   W.C.Dash [J. Appl. Phys. (USA) vol.31 (1960) p.2275]
[2]   E.R.Weber [EMIS Datareview RN=17886 (Jan 1988) 'Diffusion of Au
        in Si']
[3]   W.M.Bullis [Solid State Electron. (GB) vol.9 (1966) p.143]
[4]   C.B.Collins, R.O.Carlson, C.J.Gallagher [Phys. Rev. (USA) vol.105
        (1957) p.1168-73]
[5]   E.R.Weber [Appl. Phys. A (Germany) vol.30 (1983) p.1-22]
[6]   W.R.Wilcox, T.J.LaChapelle [J. Appl. Phys. (USA) vol.35 (1964)
        p.240]
[7]   M.Brown, C.L.Jones, A.F.W.Willoughby [Solid-State Electron. (GB)
        vol.18 (1975) p.763]
[8]   S.F.Cagnina [J. Electrochem. Soc. (USA) vol.116 (1969) p.498]
[9]   N.A.Stolwijk, B.Schuster, J.Hoelzl [Appl. Phys. A (Germany) vol.33
        (1984) p.133]
[10]   F.A.Shunk [Constitution of Binary Alloys, 2nd Suppl. (McGraw-Hill,
        1969)]
[11]   R.C.Wackwitz [1966, unpublished, presented by Bullis [3]]

14.13 SOLUBILITY OF Co IN Si

by E.R.Weber

December 1987
EMIS Datareview RN=17843

Cobalt diffuses interstitially into silicon and stays predominantly on
interstitial sites at high temperatures in thermal equilibrium. During
cooling down or even fast quenching, the interstitial cobalt forms
pairs or precipitates, so that no electrically active defect in
concentrations comparable to the total cobalt concentration has
been found [1]. Solubility data obtained by tracer methods or neutron
activation analysis reflect the solubility of interstitial cobalt. The
following values are obtained by neutron activation analysis of Si with
intrinsic conductivity at the diffusion temperature [1]:

Temp. (deg C)	Solubility (/cm*3)
1300	3.3 X 10*16
1200	2.5 X 10*16
1100	3.9 X 10*15
1000	5.5 X 10*14
900	6.3 X 10*13
800	6.5 X 10*12
700	4 X 10*11

The eutectic temperature of the Si-rich Co-Si phase diagram is 1259 deg
C [2]. Co shows retrograde solubility in Si, with the maximum occurring
near 1300 deg C [1]. Below the eutectic temperature, the solid solution
of Co in Si can reach equilibrium with the silicon-rich silicide CoSi2.
For this temperature region, the solid solubility can be expressed by an
Arrhenius relation

    C = Co exp(-E/kT)

with Co = 1.0 X 10*26/cm*3 and E = 2.83 eV [1].

The radiotracer measurements of Kitagawa et al [3] can be fitted to
Co = 9 X 10*25/cm*3 with E = 2.99 eV. Earlier radiotracer measurements
by Collins and Carlson [4] yielded C(1200 deg C) = 1 X 10*16/cm*3,
tracer measurements by Gilles [5,6] resulted in C(700 deg C) = 4 X
10*11/cm*3. Radiotracer measurements by Bakhadyrkhanov [7] showed a
maximum solid solubility of 2 X 10*16/cm*3 at 1240 deg C. The values
obtained in the 1000-1240 deg C temperature range [7] can be described
with Co = 2 X 10*21/cm*3 and E = 1.45 eV, which is inconsistent with
most other publications. Above the eutectic temperature, the distribution
coefficient of Co in silicon was estimated as:

 Cs/Cl = exp (S/k - E/kT)

with Cs being the Co concentration in the solid, Cl in the liquid state,
S = 1 k the entropy of formation and E = 1.76 eV the enthalpy of
formation. The distribution coefficient at the melting point can be
extrapolated as Cs/Cl (1412 deg C) = 1.5 X 10*-5 [1].

Electrical measurements of Co-diffused and quenched silicon [8] detect electrically active species in very small concentrations compared with the total solubility, which might be substitutional cobalt, a cobalt-related complex or other impurity contaminations [1,9]. The concentration of this defect has been described in the 1000-1250 deg C range by Co = 3 X 10*27/cm*3 and E = 4.0 eV [8].

The solubility of Co in extrinsically doped Si has been investigated by Gilles et al [5,6]. They found the following values:

Temp. (deg C)	Solubility (/cm*3)	Doping (/cm*3)
800	4.5 X 10*14	1 X 10*20 P
700	1.8 X 10*15	1 X 10*20 P
700	1.5 X 10*13	8 X 10*19 B

The enhanced Co concentrations were ascribed to pairing reactions and the formation of a large concentration of substitutional Co in strongly n-doped Si [5,6].

REFERENCES
----

[1] E.R.Weber [Appl. Phys. A (Germany) vol.30 (1983) p.1]
[2] M.Hansen, K.Anderko [Constitution of Binary Alloys (McGraw-Hill, 1958)]
[3] H.Kitagawa, K.Hashimoto [Jpn. J. Appl. Phys. (Japan) vol.16 no.1 (Jun 1977) p.173-4]
[4] C.B.Collins, R.O.Carlson [Phys. Rev. (USA) vol.106 (1957) p.1409]
[5] D.Gilles, W.Schroeter [Mater. Sci. Forum (Switzerland) vol.10-1 (1986) p.169-74]
[6] D.Gilles [PhD Thesis, Goettingen, Germany (1987)]
[7] M.K.Bakhadyrkhanov, B.I.Boltaks, G.S.Kulikov [Sov. Phys. Solid State (USA) vol.12 (1970) p.144]
[8] H.Kitagawa, H.Nakashima, K.Hashimoto [Jpn J. Appl. Phys. Part 1 (Japan) vol.24 (1985) p.373-4]
[9] E.R.Weber [EMIS Datareview RN=17890 (Jan 1988) 'Deep levels due to transition metals in Si']

14.14 SOLUBILITY OF Cr IN Si

by E.R.Weber

December 1987
EMIS Datareview RN=17854

Chromium diffuses interstitiallly into Si and stays predominantly on
interstitial sites at high temperature in thermal equilibrium. Interstitial
Cr can be easily retained in the electrically active interstitial site
by rapidly cooling down after the diffusion [1]. This possibility of
retaining interstitial chromium in solid solution by quenching from the
diffusion temperature allows in principle the performance of meaningful
solubility measurements by electronic characterisation techniques such
as electron paramagnetic resonance (EPR) or DLTS [2,3,4]. However, such
measurements can be flawed by the strong affinity of Cr to oxygen, which
can result in a diffusion barrier at the interface [2], or by the pairing
of Cr with acceptors in p-Si [3].

TABLE 1 shows solid solubility values obtained by neutron activation
analysis of Si with intrinsic conductivity at the diffusion temperature
[1,2]:

TABLE 1

Temp. (deg C)	Solubility (/cm*3)
1200	2.1 X 10*15
1150	9.4 X 10*14
1100	3.1 X 10*14
1040	1.1 X 10*14
1000	6.1 X 10*13
900	7.2 X 10*12

These values agree within a factor of two with the EPR results of Weber
[1] and the DLTS results of Graff [3]. Radiotracer measurements of Wurker
et al [5] show as well agreement within a factor of two for temperatures
above 950 deg C; but below 950 deg C the agreement is only to within
a factor of four. EPR and Hall effect solubility measurements of
Feichtinger [4] appear to be unreliable due to the use of scratched-on
Cr as diffusion source. It is not surprising that the measurements of
the interstitial chromium concentration after quenching yield up to a
factor of two smaller values than neutron activation analysis, as some
of the interstitial species might complex or precipitate during the
quenching [1].

The eutetic temperature of the Si-rich Cr-Si phase diagram is 1335 deg
C [6]. Therefore, Cr does not show retrograde solubility in Si, as a
thermodynamic analysis of the solid solubility predicts a solubility
maximum right at the eutectic temperature [1]. Below the eutectic
temperature, the solid solution of Cr in Si can reach equilibrium with
the silicon-rich silicide CrSi2. For this temperature region, the solid
solubility can be expressed by an Arrhenius relation

C = Co exp (-E/kT)

with Co = 5.5 X 10*24/cm*3 and E=2.79 eV [1].

Above the eutectic temperature the distribution coefficient was estimated as:

Cs/Cl = exp (S/k - E/kT)

with Cs being the Cr concentration in the solid, Cl in the liquid state S = 1 k the entropy of formation and E = 2 eV the enthalpy of formation. The distribution coefficient at the melting point can be extrapolated as Cs/Cl (1412 deg C) = 3 X 10*-6 [1].

REFERENCES
----

[1] E.R.Weber [Appl. Phys. A (Germany) vol.30 no.1 (Jan 1983) p.1-22]
[2] N.Wiehl, U.Herpers, E.Weber [J. Radioanal. Chem. (Switzerland)
     vol.72 no.1-2 (1982) p.69-78]
[3] H.Conzelmann, K.Graff, E.R.Weber [Appl. Phys. A (Germany) vol.30
     no.3 (Mar 1983) p.169-75]
[4] H.Feichtinger, R.Czaputa [Appl. Phys. Lett. (USA) vol.39 no.9 (1981)
     p.706-8]
[5] W.Wurker, K.Roy, J.Hesse [Mater. Res. Bull. (USA) vol.9 (1974)
     p.971]
[6] F.A.Shunk [Constitution of Binary Alloys, 2nd Suppl. (McGraw-Hill
     New York, USA, 1969)]

14.15 SOLUBILITY OF Cu IN Si

by E.R.Weber

December 1987
EMIS Datareview RN=17855

Copper is the fastest diffusing transition element in silicon. It
diffuses interstitially and stays predominantly on interstitial sites
at high temperature in thermal equilibrium. During cooling down or even
fast quenching the interstitial copper forms pairs or precipitates, so
that no electrically active defect in concentrations comparable to the
total copper concentration has been found [1]. Solubility data obtained
by tracer methods or neutron activation analysis reflect the solubility
of interstitial copper. TABLE 1 shows values obtained by neutron
activation analysis (NAA) of Si with intrinsic conductivity at the
diffusion temperature [1]:

TABLE 1

Temp. (deg C)	Solubility (/cm*3)
1250	1.4 X 10*18
1200	1.2 X 10*18
1150	9.5 X 10*17
1100	8.8 X 10*17
1050	6.3 X 10*17
1000	4.6 X 10*17
950	3.0 X 10*17
900	1.9 X 10*17
800	5.8 X 10*16
700	1.1 X 10*16
600	1.7 X 10*15
500	1.2 X 10*14

The eutectic temperature of the Cu-Si phase diagram is 802 deg C [2].
Cu shows retrograde solubility in Si, with the solubility maximum
occurring near 1300 deg C [1]. Below the eutectic temperature, the solid
solution of Cu in Si can reach equilibrium with the silicide Cu3Si. For
this temperature region the solid solubility can be expressed by an
Arrhenius relation

$$C = Co \exp (-E/kT)$$

with Co = 5.5 X 10*23/cm*3 and E = 1.49 eV [1].

The samples used for the neutron activation analysis [1] were Cu-plated
by evaporation and the formation of the Cu3Si boundary phase was not
controlled. An earlier study by Dorward and Kirkaldy [3] measured by NAA
Cu concentrations in Si with a Co3Si diffusion source. Their results
between the eutectic temperature and 650 deg C can be described with Co
= 3.3 X 10*24/cm*3 and E = 1.71 eV, yielding about a factor of two
smaller values than obtained in [1]. Results of the radiotracer study
of Hall and Racette [4] can be fitted in the 400-800 deg C range by Co
= 5.5 X 10*22/cm*3 and E = 1.39 eV, which gives even smaller solubility
concentrations than those obtained by Dorward and Kirkaldy. Struthers
[5] obtained by radiotracer measurements for 600 deg C a solubility
value of 4.3 X 10*15/cm*3, higher than the NAA determination (TABLE 1).

The solubility data for Cu in Si obtained below the eutectic temperature show a great variation. This is probably due to the problem of obtaining a well-defined boundary phase at such low temperatures.

At temperatures above the eutectic temperature the distribution coefficient follows an Arrhenius relation as:

$$Cs/Cl = exp (S/k - E/kT)$$

with Cs being the Cu concentration in the solid and Cl in the liquid state. The results of the NAA study [1] can be fitted with the entropy of formation S = -0.35 k and the enthalpy of formation E = 1.17 eV. A best fit to all available data [1] results in S = 1k and E = 1.35 eV. From this, the distribution coefficient at the melting point can be extrapolated as Cs/Cl (1412 deg C) = 2.5 X 10*-4 [1]. TABLE 2 lists some solubility values calculated with these parameters (concentrations in the liquid are given in atoms/atoms):

TABLE 2

Temp. (deg C)	Cl conc. in liquid	Cs/Cl distr. coeff.	Solubility (/cm*3)
1300	0.197	1.3 X 10*-4	1.3 X 10*18
1200	0.323	6.5 X 10*-5	1.1 X 10*18
1100	0.434	3.0 X 10*-5	6.5 X 10*17
1000	0.532	1.2 X 10*-5	3.3 X 10*17
900	0.620	4.3 X 10*-6	1.3 X 10*17
830	0.680	1.8 X 10*-6	6.3 X 10*16

These values compare very well with the results of the NAA analysis of Dorward and Kirkaldy [3] and the radiotracer data of Struthers [5].

Electrical measurements of Cu-diffused and quenched silicon detect electrically active species in the 10*13/cm*3 range after diffusion at temperatures between 950 and 1200 deg C [6,7], which might be substitutional copper, copper-related complexes or other impurity contaminations. No reliable solubility values are available for this species.

The solubility of Cu in extriniscally doped Si has been investigated in the radiotracer analysis of Hall and Racette [4]. They found comparable solubility values in intrinsic and extrinsic p-type silicon. In extriniscally doped n-Si at 600 and 700 deg C a strong solubility enhancement proportional to the cube of the extrinsic electron concentration was observed. It was explained by the formation of substitutional Cu acting as a triple acceptor.

REFERENCES
----

[1] E.R.Weber [Appl. Phys. A (Germany) vol.30 no.1 (Jan 1983) p.1-22]
[2] M.Hansen, K.Anderko [Constitution of Binary Alloys (McGraw-Hill,
     1958)]
[3] R.C.Dorward, J.S.Kirkaldy [Trans. Metall. Soc. AIME (USA) vol.242
     (1968) p.2055]
[4] R.N.Hall, J.H.Racette [J. Appl. Phys. (USA) vol.35 (1964) p.379]
[5] J.D.Struthers [J. Appl. Phys. (USA) vol.27 (1956) p.1560]
[6] S.J.Pearton, A.J.Tavendale [J. Appl. Phys. (USA) vol.54 no.3 (Mar
     1983) p.1375-9]
[7] H.Lemke [Phys. Status Solidi a (Germany) vol.95 no.2 (1986)
     p.665-78]

14.16 SOLUBILITY OF Fe IN Si

by E.R.Weber

December 1987
EMIS Datareview RN=17851

Iron is a fast diffusing element in silicon. It diffuses interstitially
and stays predominantly on interstitial sites at high temperatures in
thermal equilibrium. Maximum concentration values are obtained after
very short diffusion times and do not further increase, even by
prolonged diffusion treatments. Interstitial iron can be retained in
the electrically active interstitial site by rapidly cooling down or
quenching after the diffusion [1]. This possibility to retain
interstitial iron in solid solution by quenching from the diffusion
temperature allows one to perform meaningful solubility measurements by
electronic characterization techniques, such as deep level transient
spectroscopy (DLTS) [2], electron paramagnetic resonance (EPR) [3,4] or
Hall effect [5].

The following values are obtained by neutron activation analysis of Si
with intrinsic conductivity at the diffusion temperature [4]:

Temp. (deg C)	Solubility (/cm*3)
1200	1.5 X 10*16
1100	3.1 X 10*15
1050	1.1 X 10*15
1000	4.6 X 10*14
950	1.4 X 10*14

These values agree within a factor of two with DLTS results of Graff [2],
EPR results of Lee et al [3], and Weber et al [4], and the EPR and Hall
effect measurements of Feichtinger [5]. It is not surprising that these
measurements of the electrically active interstitial iron concentration
after quenching yield up to a factor of two smaller values than neutron
activation analysis, as some of the interstitial species might complex or
precipitate during the quenching [1]. Struthers [6] obtained up to a
factor of 5 smaller concentrations by radiotracer measurements, probably
due to insufficient active iron on the sample surface.

The eutectic temperature of the Si-rich Fe-Si phase diagram is 1206 deg C
[9]. Fe shows retrograde solubility in Si, with the maximum occurring
near 1300 deg C [1,5]. Below the eutectic temperature, the solid solution
of Fe in Si can reach equilibrium with the silicon-rich silicide FeSi2.
For this temperature region, the solid solubility can be expressed by
an Arrhenius relation

$$C = C_0 \exp(-E/kT)$$

with $C_0 = 1.8 \times 10*26$ /cm*3 and E = 2.94 eV [1].

Above the eutectic temperature, the distribution coefficient was estimated as:

$$Cs/Cl = \exp (S/k - E/kT)$$

with Cs being the Fe concentration in the solid, Cl in the liquid state, S = 1 k the entropy of formation and E = 1.86 eV the enthalpy of formation. The distribution coefficient at the melting point can be extrapolated as Cs/Cl (1412 deg C) = 7 X 10*-6 [1].

An EPR study by Colas et al [7] of silicon containing oxygen precipitates introduced at 700 deg C showed strongly reduced Fe solubilities at temperatures above 900 deg C. This effect was ascribed to the strain-assisted formation of an oxygen-metal-silicon compound replacing the silicide as thermodynamic boundary phase.

The solubility of Fe in extrinsically doped Si has been investigated by Gilles [8]. He reported the following values, obtained by tracer measurements:

Temp. (deg C)	Fe Solubility (/cm*3)	Doping (/cm*3)
700	5 X 10*14	1 X 10*20 P
700	3 X 10*13	8 X 10*19 B

The extrapolated solubility of Fe in intrinsic Si is 1 X 10*11/cm*3 at 700 deg C [1]. The enhanced Fe concentrations were ascribed to pairing reactions and the formation of a large concentration of substitutional Fe in strongly n-doped Si [8].

REFERENCES
----

[1] E.R.Weber [Appl. Phys. A (Germany) vol.30 no.1 (Jan 1983) p.1-22]
[2] K.Graff, H.Pieper [J. Electrochem. Soc. (USA) vol.128 (1981) p.669]
[3] Y.H.Lee, R.L.Kleinhenz, J.W.Corbett [Appl. Phys. Lett. (USA) vol.31 no.3 (1 Aug 1977) p.142-4]
[4] E.R.Weber, H.G.Riotte [J. Appl. Phys. (USA) vol.51 no.3 (Mar 1980) p.1484-8]
[5] H.Feichtinger [Acta Phys. Austriaca (Austria) vol.51 (1979) p.161]
[6] J.D.Struthers [J. Appl. Phys. (USA) vol.27 (1956) p.1560]
[7] E.G.Colas, E.R.Weber [Appl. Phys. Lett. (USA) vol.48 (1987) p.1371]
[8] D.Gilles [PhD Thesis, Goettingen (Germany) 1987]
[9] M.Hansen, K.Anderko [Constitution of Binary Alloys (McGraw-Hill, 1958)]

14.17 SOLUBILITY OF Mn IN Si

by E.R.Weber

January 1988
EMIS Datareview RN=17869

Manganese is a fast diffusing element in silicon. It diffuses
interstitially and stays predominantly on interstitial sites at high
temperatures in thermal equilibrium. Maximum concentration values are
obtained after quite short diffusion times and do not further increase
even by prolonged diffusion treatments. Interstitial manganese can be
retained in the electrically active interstitial site by rapidly cooling
down or quenching after the diffusion [1]. This possibility of retaining
interstitial manganese in solid solution by quenching from the diffusion
temperature allows the performance of meaningful solubility measurements
by electronic characterization techniques, such as deep level transient
spectroscopy (DLTS) [2] or electron paramagnetic resonance (EPR) [1].

TABLE 1: values are obtained by neutron activation analysis (NAA) of Si
         with intrinsic conductivity at the diffusion temperature [1]

Temp. (deg C)	Solubility (/cm*3)
1250	1.7 X 10*16
1200	1.3 X 10*16
1150	8.2 X 10*15
1100	3.9 X 10*15
1050	1.5 X 10*15
1000	5.4 X 10*14
900	6.8 X 10*13

Recent radiotracer measurements of the diffusion and solubility of
manganese in silicon [2] found manganese concentrations in excellent
agreement with these values. The results of tracer measurements of
Bakhadhyrkhanov et al [3] and Senk et al [4] are up to an order of
magnitude at variance with these data which might be due to time
dependent boundary conditions during the diffusion, as pointed out
by Gilles et al [2]. EPR measurements of Weber [1] found concentrations
of interstitial manganese typically 40-50% below the total
concentrations as determined by NAA. DLTS measurements of Gilles et al
[2] found concentration values within only 20% deviation from the
tracer results. It is not surprising, that the measurements of
electrically active interstitial manganese concentrations after
quenching yield up to a factor of two smaller values than neutron
activation of radiotracer analysis, as some of the interstitial
species might complex or precipitate during the quenching [1,2].

The eutectic temperature of the Si-rich Mn-Si phase diagram is 1142
deg C [5]. Manganese shows retrograde solubility in Si, with the
maximum occurring near 1300 deg C [1]. Below the eutectic temperature,
the solid solution of Mn in Si can reach equilibrium with the

silicon-rich silicide MnSi2. For this temperature region, the solid solubility can be expressed by an Arrhenius relation:

$C = C_o \exp(-E/kT)$

with $C_o = 7.4 \times 10^{25}/cm^3$ and $E = 2.81$ eV [1].

Above the eutectic temperature, the distribution coefficient was fitted to:

$Cs/Cl = \exp(S/k - E/kT)$

with Cs being the Mn concentration in the solid, Cl in the liquid state, $S = 1.2$ k the entropy of formation and $E = 1.92$ eV the enthalpy of formation. The distribution coefficient at the melting point can be extrapolated as Cs/Cl (1412 deg C) = 6×10^{-6} [1].

The solubility of Mn in extrinsically doped Si has been investigated by Gilles [6]. He reported the following values, obtained by tracer measurements:

TABLE 2

Temp. (deg C)	Solubility (/cm*3)	Doping (/cm*3)
1038	1.2×10^{15}	2×10^{19} B
1038	8.9×10^{14}	2×10^{19} As
1038	8.6×10^{14}	8×10^{13} P
854	3.1×10^{13}	2×10^{19} B
854	2.0×10^{13}	2×10^{19} As
854	8.9×10^{12}	8×10^{13} P
700	1.6×10^{13}	2×10^{19} B
700	1.1×10^{13}	5×10^{18} P
700	3.0×10^{12}	8×10^{13} P

The extrapolated solubility of interstitial Mn in intrinsic Si is $2.3 \times 10^{11}/cm^3$ at 700 deg C [1]. The enhanced Mn concentrations at temperatures below 900 deg C were ascribed to the formation of substitutional Mn, which might be present at higher temperatures in a fraction smaller than 0.1 of the concentration of interstitial Mn, but is dominant at lower temperatures and further enhanced in concentration by extrinsic doping [6].

REFERENCES
----

[1] E.R.Weber [Appl. Phys. A (Germany) vol.30 no.1 (Jan 1983) p.1-22]
[2] D.Gilles, W.Bergholz, W.Schroeter [J. Appl. Phys. (USA) vol.59 no.10 (15 May 1986) p.3590-2]
[3] M.K.Bakhadyrkhanov, B.I.Boltaks, G.S.Kulikov [Sov. Phys.-Solid State (USA) vol.14 (1972) p.1441]
[4] D.Senk, G.Borchardt [Mikrochim. Acta vol.1983 II (1983) p.477]
[5] F.A.Shunk [Constitution of Binary Alloys, 2nd Suppl. (McGraw-Hill, 1969)]
[6] D.Gilles [PhD thesis, Goettingen (Germany, 1987)]

14.18 SOLUBILITY OF Ni IN Si

by E.R.Weber

December 1987
EMIS Datareview RN=17859

Nickel is one of the fastest diffusing elements in silicon. It diffuses
interstitially and stays predominantly on interstitial sites at high
temperatures in thermal equilibrium. During cooling down or even fast
quenching, interstitial nickel forms pairs or precipitates, so that no
electrically active defect in concentrations comparable to the total
nickel concentration has been found [1]. A small fraction of electrically
active Ni, typically not more than 10*-4 of the total Ni concentration,
has been ascribed to substitutional Ni [5]. However, direct microscopic
proof of this identification is still lacking. Solubility data obtained
by tracer methods or neutron activation analysis reflect the solubility
of interstitial nickel. TABLE 1 shows values obtained by neutron
activation analysis (NAA) of Si with intrinsic conductivity at the
diffusion temperature [1]:

TABLE 1

Temp. (deg C)	Solubility (/cm*3)
1300	6.3 X 10*17
1250	6.8 X 10*17
1200	6.9 X 10*17
1100	5.1 X 10*17
1000	2.4 X 10*17
900	9.0 X 10*16
800	1.9 X 10*16
700	2.9 X 10*15
600	2.2 X 10*14
550	7.7 X 10*13
500	1.4 X 10*13

The eutectic temperature of the Ni-Si phase diagram is 993 deg C [2]. Ni
shows retrograde solubility in Si, with the solubility maximum occurring
near 1250 deg C [1]. Below the eutectic temperature, the solid solution
of Ni in Si can reach equilibrium with the silicide NiSi2. For this
temperature region, the solid solubility can be expressed by an Arrhenius
relation:

C = Co exp (-E/kT)

with Co = 1.5 X 10*24/cm*3 and E = 1.68 eV [1].

The samples used for the neutron activation analysis [1] were Ni-plated
by evaporation. The formation of the NiSi2 boundary phase was not
controlled and might not have been reached especially in the case of
the lower temperatures. Radiotracer measurements of Yoshida and Furusho
[3] found about a factor of 2 smaller solubility values in the
temperature range between 800 deg C and the eutectic temperature.

At temperatures above the eutectic temperature, the distribution coefficient follows an Arrhenius relation as:

Cs/Cl = exp (S/k - E/kT)

with Cs being the Ni concentration in the solid and Cl in the liquid state. The results of the NAA study [1] can be fitted with the entropy of formation S=-1.4k and the enthalpy of formation E = 1.08 eV. A best fit to all available data [1] results in S = -0.17 k and E = 1.25 eV. From this, the distribution coefficient at the melting point can be extrapolated as Cs/Cl (1412 deg C) = 1.5 X 10*-4 [1]. Some solid solubility values calculated with these parameters are shown in TABLE 2 (concentrations in the liquid are given in atoms/atoms):

TABLE 2

Temp. (deg C)	Cl conc. in liquid	Cs/Cl distr. coeff.	Solubility (/cm*3)
1300	0.181	8.3 X 10*-5	7.5 X 10*17
1200	0.280	4.5 X 10*-5	6.2 X 10*17
1100	0.348	2.2 X 10*-5	3.8 X 10*17
1000	0.398	9.5 X 10*-6	1.9 X 10*17

These values compare very well with the results of the NAA analysis of Aalberts and Verheijke [4] and the radiotracer data of Yoshida and Furusho [3].

Electrical measurements of Ni-diffused and quenched silicon detect electrically active species in concentrations up to 10*14/cm*3 [3,5,6], which might be substitutional nickel, nickel-related complexes or other impurity contaminations. Yoshida and Saito [5] determined the concentration of an electrically active species in the temperature range of 850 to 1100 deg C by resistivity and Hall effect measurements. Their results follow in this whole temperature range, i.e. below as well as above eutectic temperature, an Arrhenius relation with the parameters Co = 10*26/cm*3, E = 3.1 eV.

REFERENCES
----

[1] E.R.Weber [Appl. Phys. A (Germany) vol.30 no.1 (Jan 1983) p.1-22]
[2] M.Hansen, K.Anderko [Constitution of Binary Alloys (McGraw-Hill, 1958)]
[3] M.Yoshida, K.Furusho [Jpn. J. Appl. Phys (Japan) vol.3 (1964) p.521]
[4] J.H.Aalberts, M.L.Verheijke [Appl. Phys. Lett. (USA) vol.1 (1962) p.19]
[5] M.Yoshida, K.Saito [Jpn. J. Appl. Phys. (Japan) vol.6 (1967) p.573]
[6] S.J.Pearton, A.J.Tavendale [J. Appl. Phys. (USA) vol.54 no.3 (Mar 1983) p.1375-9]

14.19 SOLUBILITY OF Pt IN Si

by E.R.Weber

January 1988
EMIS Datareview RN=17889

Only a few papers with solubility studies of Pt in Si are available
[1-3]. Pt shows a complex diffusion mechanism, similar to Au in Si:
relatively fast interstitial diffusivity and negligible substitutional
mobility combined with high substitutional solubility, compared to the
lower concentrations of interstitials. Because of the long times
necessary to reach equilibrium substitutional concentrations [4],
solubility values extrapolated from concentrations near the sample
surface might be more reliable than bulk values determined after too
short diffusion times.

Substitutional Pt atoms are electrically active in silicon [5]. It
should be noted, that in the case of Pt in Si two electrically active
species have been detected [6]; Pt I is generally considered to be
isolated substitutional Pt [6,7], the distinctly different Pt II might
be due to a complex, but no clear model has been developed. The
concentration of Pt II is typically at least an order of magnitude
smaller than the Pt I concentration, so that measurements dealing with
Pt I might be representative for the concentration of substitutional
Pt.

Mantovani et al [1] determined diffusion profiles of Pt in Si in the
700-850 deg C range by analysing DLTS scans. The resulting
concentration profiles C(s,t) of electrically active Pt could be well
fitted with an equation derived for the kick-out model of Au diffusion
in Si by Seeger [8]:

$$C(s,t) = Csol/[1 + s/(Dt)*0.5] \qquad\qquad (1)$$

with the solubility of Pt in Si at this temperature Csol, s the distance
from the surface and a parameter D which can be regarded as an
effective diffusion coefficient. Mantovani et al [1] did not quote
the fitting parameters used for their diffusion profiles; however, an
analysis of Fig. 2 in ref [1] allows estimation of the following values
for the solubility of electrically active Pt in Si.

TABLE 1: Pt solubility in Si derived from ref [1]:

Temp. (deg C)	Solubility (/cm*3)
850	1.6 X 10*16
800	2.5 X 10*15
750	3.5 X 10*14
700	1.9 X 10*14

These values are more than an order of magnitude higher than the corresponding data for Au in Si [9]. Earlier work of Lisiak et al [2] found with Neutron Activation Analysis and Hall effect measurements were about an order of magnitude smaller than those given in TABLE 1. The two techniques yielded comparable results, justifying the use of electrical measurements for the determination of Pt concentrations. However, these authors used diffusion times between one and four days for the 800 – 1250 deg C range and did not measure concentration profiles, but rather bulk concentrations. Therefore those low results might not reflect saturation concentrations. Neutron activation analysis by Bailey et al [3] in the diffusion temperature range from 800 to 1000 deg C, yielded Pt solubilities roughly an order of magnitude higher than the data of Mantovani et al [1]; however, Lisiak et al [2] pointed out, that Bailey et al might have measured too high Pt concentrations by counting Au-199 activities. A very efficient interfering reaction from small Au contaminations could have increased the calculated Pt concentrations.

Summarizing, Pt solubility concentrations reported in the literature vary up to an order of magnitude from the data presented in TABLE 1. At present it is not possible to distinguish effects of insufficient diffusion time from systematic measurement errors. It is to be hoped that future work will help to clarify the situation.

REFERENCES
----

[1] S.Mantovani, F.Nava, C.Nobili, G.Ottaviani [Phys. Rev. B (USA) vol.33 (1986) p.5536]
[2] K.P.Lisiak, A.G.Milnes [Solid State Electron (GB) vol.18 (1975) p.533]
[3] R.F.Bailey, T.G.Mills [Proc of the Electrochem. Soc. Meet. on Semiconductor Silicon, Eds R.R.Haberecht and F.L.Kern (Electrochem. Soc., New York, 1969) p.481-9]
[4] E.R.Weber [EMIS Datareview RN=17888 (Jan 1988) 'Diffusion of Pt in Si']
[5] E.R.Weber [EMIS Datareview RN=17890 (Jan 1988) 'Deep levels due to transition metals in Si']
[6] H.H.Woodbury, G.W.Ludwig [Phys. Rev. (USA) vol.126 (1962) p.466]
[7] R.F.Milligan, F.G.Anderson, G.D.Watkins [Phys. Rev. B (USA) vol.29 (1984) p.2819]
[8] A.Seeger [Phys. Status Solidi a (Germany) vol.61 (1980) p.521]
[9] E.R.Weber [EMIS Datareview RN=17887 (Jan 1988) 'Solubility of Au in Si']

14.20 SOLUBILITY OF Ti IN Si

by E.R.Weber

January 1988
EMIS Datareview RN=17868

Titanium is among the slowest diffusing transition metals in silicon
[1]. Recently, an electron paramagnetic resonance signal from
interstitial titanium in the single positive charge state has been
identified [2,3]. This finding makes it likely that titanium diffuses
predominantly as an interstitial and stays in this electrically active
interstitial site at high temperature, like the other 3d metals in
silicon. Because of its small diffusion coefficient most of the
titanium in silicon is found in the electrically active state after
cooling down. This allows its solid solubility to be obtained from
electrical measurements.

The solubility of Ti in Si was determined by deep level transient
spectroscopy (DLTS) in the diffusion temperature range from 1000 to 1250
deg C [4]. It follows an Arrhenius relation

$C = Co \exp (-E/kT)$

with the parameters Co = 2.5 X 10*24/cm*3 and E = 3.0 eV [4] calculated
from Fig.3 in this paper.

The pre-exponential factor and the activation energy representing the
enthalpy of solution of Ti in Si with respect to a silicide boundary
phase are similar to those found for other interstitial 3d metals in
silicon [5] which gives this determination some additional support.

The parameters of Chen et al [4] allow calculation for 1100 deg C of a
solid solubility of Ti in Si of 2.4 X 10*13/cm*3. DLTS results of Graff
show for this temperature a value of 6 X 10*13/cm*3 [6]. Hansen et al
determined by DLTS for this temperature a Ti concentration of 4 X
10*13/cm*3 [7].

Ti has a eutectic temperature of 1330 deg C which can be expected
to coincide with the temperature of maximum solid solubility [5].
Extrapolating the data of Chen et al [4] the maximum solubility of
electrically active Ti in Si is C(1330) = 9 X 10*14/cm*3.

All these determinations of Ti concentrations in Si have still to be
treated with reservation, as they are not yet backed up by radiotracer
or neutron activation analyses of total Ti concentrations. However, it
is noteworthy that the values found by various authors are quite
consistent.

REFERENCES
----

[1] E.R.Weber [in 'Impurity Diffusion and Gettering in Semiconductors',
     Eds R.B.Fair, C.B.Pearce, J.Washburn [Mater. Res. Soc. (1985)
     vol.35 p.3-11]
[2] D.A.van Wezep, C.A.J.Ammerlaan [J. Electron. Mater. (USA) vol.14a
     (1985) p.863]
[3] D.A.van Wezep, R.van Kemp, E.G.Sieverts, C.A.J.Ammerlaan [Phys.
     Rev. B (USA) vol.32 no.11 (1 Dec 1985) p.7129-38]
[4] J.W.Chen, A.G.Milnes, A.Rohatgi [Solid-State Electron. (GB) vol.22
     no.9 (Sep 1979) p.801-8]
[5] E.R.Weber [Appl. Phys. A (Germany) vol.30 no.1 (Jan 1983) p.1-22]
[6] K.Graff [Semiconductor Silicon 1986 Eds H.R.Huff, T.Abe,
     B.Kolbesen (Electroche. Soc. USA, 1986) p.751]
[7] J.Hansen, D.Gilles, W.Schroeter [Verhandig DPG (Germany) vol.18
     (1983) p.613; J.Hansen - private communication]

CHAPTER 15

DIFFUSION AND SOLUBILITY OF ALKALI METALS

15.1 DIFFUSION OF Li IN Si

by L.T.Canham

August 1987
EMIS Datareview RN=16124

A wealth of information, both experimental [1-10] and theoretical
[11-15], is available on the diffusivity of Li in Si. Lithium diffuses
extremely rapidly through the Si lattice, jumping from one interstitial
equilibrium site to another with very little lattice distortion, due to
its small ionic size. It moves along the (111) channels in a zig-zag
fashion, alternately passing through the tetrahedral and hexagonal
interstitial sites of the lattice, which are respectively its
equilibrium and saddle points for diffusion.

The diffusion of Li was first measured by Fuller and Ditzenberger [1]
in 1953. They studied Li diffusing into p-type Si at temperatures
between 450 and 1000 deg C by the p-n junction technique. The rapid
formation of hemispherical n-type regions showed that Li donors
exhibited a high diffusion coefficient. Fuller and Severiens [3] later
studied these p-n junction positions under applied electric fields and
extracted diffusion coefficients from the measured mobilities using the
Einstein relationship. Although this method is in principle capable of
high accuracy, widely used techniques based on in-diffusion can have
certain shortcomings for the case of Li diffusion in Si. For
temperatures below roughly 900 deg C the Li solubility exceeds the
intrinsic carrier concentration, resulting in built-in fields which can
drastically influence diffusion rates. Also, for diffusions above
600 deg C, precipitation processes near the surface become increasingly
difficult to eliminate by rapid quenching techniques. To avoid these
problem areas Pell [6] conducted a careful re-examination of the
diffusion rate at high temperatures using an out-diffusion technique,
and later developed an ion-drift method [7] to both extend the data to
lower temperatures and test some of Maita's findings [4] from
ion-pairing experiments.

It is likely that Pell's data [6-7] are the most accurate to date,
despite the more recent work of Pratt and Friedman [8] and Yuskeselieva
and Antonov [9], using a thin layer lapping and an electrophotographic
method respectively, which yielded slightly higher diffusion rates. In
any case, Li diffusion in Si has been fitted to a relationship of the
type

$$D = Do \exp(-E/kT)$$

over a very wide temperature range, and as TABLE 1 demonstrates,
the agreement between 9 of the 10 independently measured values of
Arrhenius parameters is remarkably good.

TABLE 1

Temp. range studied (deg C)	Do (cm*2/s)	E (eV)	Ref
450-1000	0.0094	0.78	[1]
150-851	0.0019	0.64	[2]
360-877	0.0023	0.66	[3]
0-877	0.0023	0.72	[4]
420-800	0.0022	0.70	[5]
25-1350	0.0025	0.655	[6,7]
300-500	0.00265	0.63	[8]
180-560	0.0021	0.57	[9]
300-550	0.0038	0.66	[10]

Whereas some of the available theories of interstitial diffusion [11-15] can predict the pre-exponential factor Do for Li diffusion (0.0025 +- 0.0002 cm*2/s) quite well, they are not so successful in obtaining the experimental activation energy (0.655 +- 0.01 eV).

Just as the solubility of Li is affected by the presence of other impurities and defects at low temperature, so is its diffusivity. Appreciable Li-acceptor, Li-oxygen and Li-vacancy interactions lower the diffusion rate and lead to activation energies different from those of free Li. In CZ Si, for example, the presence of 10*18 O atoms/cm*3 will reduce the room temperature diffusion rate of Li to 0.1% of its value in pure FZ Si [16]. Such interactions are absent at high temperatures, however, due to the relatively small binding energies involved.

REFERENCES
----

[1]   C.S.Fuller, J.A.Ditzenberger [Phys. Rev. (USA) vol.91 (1953) p.193]
[2]   J.C.Severiens, C.S.Fuller [Phys. Rev. (USA) vol.92 (1953) p.1322]
[3]   C.S.Fuller, J.C.Severiens [Phys. Rev. (USA) vol.96 (1954) p.21]
[4]   J.P.Maita [J. Phys. & Chem. Solids (GB) vol.4 (1958) p.68]
[5]   J.M.Shashkov, I.P.Akimchenko [Sov. Phys.-Dokl. (USA) vol.4 (1959) p.1155]
[6]   E.M.Pell [Phys. Rev. (USA) vol.119 (1960) p.1014]
[7]   E.M.Pell [Phys. Rev. (USA) vol.119 (1960) p.1222]
[8]   B.Pratt, F.Friedman [J. Appl. Phys. (USA) vol.37 (1966) p.1893]
[9]   L.G.Yuskeselieva, A.S.Antonov [Sov. Phys.-Solid State (USA) vol.8 (1966) p.2025]

[10]   J.C.Larue [Phys. Status Solidi a (Germany) vol.6 (1971) p.143]
[11]   K.Weiser [Phys. Rev. (USA) vol.126 (1962) p.1427]
[12]   R.A.Swalin [J. Phys. & Chem. Solids (GB) vol.23 (1962) p.154]
[13]   M.F.Millea [J. Phys. & Chem. Solids (GB) vol.27 (1966) p.315]
[14]   P.Gosar [Nuovo Cimento (Italy) vol.31 (1964) p.781]
[15]   S.M.Hu [in 'Atomc Diffusion in Semiconductors' Eds P.Shaw (Plenum
       Press, London, 1973) p.255]
[16]   E.M.Pell [J. Appl. Phys. (USA) vol.32 (1961) p.1048]

15.2 DIFFUSION OF Na IN Si

by L.T.Canham

August 1987
EMIS Datareview RN=16125

Published data on Na diffusivity in Si concern mainly ion implanted
Na, reflecting the difficulties encountered when attempting to
incorporate Na in the bulk by purely thermal means. The higher
volatility and reactivity of Na compared with Li promotes higher
out-diffusion, surface precipitation, ionic adsorption, and evaporation
when in-diffusing from the vapour phase.

Electrically inactive Na resulting from thermal injection diffuses very
slowly in Si. McCaldin et al [1] estimated:

$$D = 6 \times 10^{-12} \text{ cm}^2/\text{s at 800 deg C}$$

In contrast, electrically active Na resulting from ion injection can
diffuse relatively rapidly. McCaldin [2] implanted 2 keV Na+ ions into
p-type Si at 460 deg C, which created a 10 microns thick n-type surface
layer. From subsequent 1 hr anneals which increased the junction depth,
he estimated the following diffusivities:

$$D = 8 \times 10^{-11} \text{ cm}^2/\text{s at 600 deg C}$$

$$D = 1.5 \times 10^{-11} \text{ cm}^2/\text{s at 500 deg C}$$

He extrapolated his data to 800 deg C and found a diffusivity for ion
injected Na that is about 100 times higher than that of thermally
injected Na.

More recent work has yielded even higher diffusion coefficients for ion
implanted Na. Korol et al [3] used the same technique to study the
diffusion of Na ions implanted at 50 keV into 1-10 ohm cm p-type Si at
room temperature. Their diffusion coefficients were fitted to the
following Arrhenius expression:

$$D = 0.0147 \exp(-1.27/kT(eV)) \text{ cm}^2/\text{s}$$

over the temperature range 650-900 deg C. This yields a diffusivity at
800 deg C, for example, of about 10^{-8} cm^2/s, which is 200 times slower
than that of Li at the same temperature. Their activation energy of
1.27 eV is higher than that normally observed (0.4-0.8 eV) for
impurities diffusing in Si by totally interstitial mechanisms [4]. The
ionic radius of Na (0.95 A), whilst sufficiently small to be
incorporated into the interstitial cavities of the Si lattice, is
comparable with the size of the hexagonal (111) tunnels available for
interstitial diffusants. The maximum 'hard sphere radius' that such
tunnels can accommodate without appreciable lattice dilation is 1.05 A
[5]. From solely geometrical considerations, one would expect
interstitial Na to be significantly less mobile than interstitial Li,
since the latter has an ionic radius of only 0.60 A.

Belikova et al [6] found that the temperature of the Si during
implantation had a significant influence on Na diffusion during
subsequent anneals. After room temperature implantation, Na diffusion
at temperatures below 650 deg C is lowered by interactions with
residual damage. Following hot (> 500 deg C) implantation however,
they reported that subsequent diffusion at temperatures as low as
500 deg C obeyed the above Arrhenius expression of Korol et al [3].

Also using the p-n junction technique, Svob [7] has reported that a
very small fraction of thermally injected Na is electrically active and
diffuses according to:

$$D = 0.00165 \exp(-0.77/kT(eV)) \; cm*2/s$$

over the temperature range 800-1000 deg C. This expression yields
diffusivities much higher than those reported for ion injected Na, and
at 800 deg C for example, a value that is less than 4 times lower than
that of Li. In fact there is a strong possibility that Svob's data
reflect trace Li contamination, as suggested by Parry et al [8].
Further work on the diffusion of Na thermally introduced under
equilibrium conditions is clearly required. Stojic et al [9] have
recently reported preliminary findings using a radioactive Na-24 tracer
method.

REFERENCES
----

[1]   J.O.McCaldin, M.J.Little, A.E.Widmer [J. Phys. & Chem. Solids
      (GB) vol.26 (1965) p.1119]
[2]   J.O.McCaldin [Nucl. Instrum. & Methods (Netherlands) vol.38
      (USA) vol.9 (1975) p.815]
[4]   S.M.Hu [in 'Atomic Diffusion in Semiconductors' Ed. D.Shaw
      (Plenum Press, London, 1973) p.255]
[5]   R.A.Swalin [J. Phys. & Chem. Solids (GB) vol.23 (1962) p.154]
[6]   M.N.Belikova, A.V.Zastavnyi, V.M.Korol [Sov. Phys.-Semicond.
      (USA) vol.10 no.3 (Mar 1976) p.319]
[7]   L.Svob [Solid-State Electron. (GB) vol.10 (1967) p.991]
[8]   E.P.Parry, M.S.Porter, J.O.McCaldin [Solid-State Electron. (GB)
      vol.12 (1969) p.500]
[9]   M.Stojic, D.Kostic, B.Stosic [Physica B&C (Netherlands) vol.138
      nos 1&2 (Mar 1986) p.125-8]

15.3 DIFFUSION OF K IN Si

by L.T.Canham

August 1987
EMIS Datareview RN=16126

The ionic radius of K (1.33 A) exceeds the tetrahedral covalent radius
(1.17 A) of Si and consequently interstitial K is expected to have a
diffusivity far lower than that of Na and Li.

Zorin et al [1] implanted 50 keV K ions into both FZ and CZ Si at room
temperature. K diffusion in the FZ material during 500-800 deg C
anneals was fitted by the Arrhenius expression:

$$D = 1.1 \ X \ 10*-8 \ exp(-0.80/kT(eV)) \ cm*2/s$$

Thus at 800 deg C, for example, the diffusivity of ion injected K was
measured to be about 2 X 10*-12 cm*2/s, about 5000 times lower than that
of ion injected Na, and about a million times lower than that of Li.

Svob [2] has reported much higher diffusion coefficients for K that were
fitted by

$$D = 1.1 \ X \ 10*-3 \ exp(-0.76/kT(eV)) \ cm*2/s$$

In agreement with Parry et al [3], Zorin et al [1] have ascribed these
values to the effects of trace Li contamination of the K dopant.

REFERENCES
----

[1] E.I.Zorin, P.V.Pavlov, D.I.Tetel'baum, A.F.Khokhlov [Sov.
     Phys.-Semicond. (USA) vol.6 (1972) p.21]
[2] L.Svob [Solid-State Electron. (GB) vol.10 (1967) p.991]
[3] E.P.Parry, M.S.Porter, J.O.McCaldin [Solid-State Electron. (GB)
     vol.12 (1969) p.500]

15.4 SOLUBILITY OF Li IN Si

by L.T.Canham

October 1987
EMIS Datareview RN=16188

A series of detailed experimental studies [1-7] on Li solubility in Si
carried out in the 1950's form the basis of well-established data used
today. The low binding energy of the Li 2s valence electron and very
small Li ionic radius (0.60 A) promote exclusive incorporation as a
singly-ionized interstitial donor of relatively high solubility [8].

The earliest measurements [1-2] of Li solubility were based on
resistivity data following rapid quenches from thermal equilibria. At
elevated temperatures bulk Si was readily saturated with Li by thermal
in-diffusion from molten Li-Si alloys on its surface. Subsequent work
by Pell [4-5] showed that such electrical measurements, whilst capable
of providing accurate solubility data below about 650 deg C, grossly
underestimate Li solubilities at higher temperatures. The problem is
that even using ultrafast quenching techniques, some Li is unavoidably
precipitated when the solid solution is cooled from temperatures above
650 deg C, due to increased vacancy concentrations at such temperatures.

Pell [4-5] therefore employed flame photometry to measure total Li
content i.e. a chemical analysis technique sensitive to both
electrically inactive Li precipitates and Li donors. His data for Li
solubilities from 592 to 1382 deg C in nominally undoped Si are widely
used today and specific values are listed below:

    1400 deg C : 3 X 10*19/cm*3

    1200 deg C : 7 X 10*19/cm*3

    1000 deg C : 5 X 10*19/cm*3

     800 deg C : 2 X 10*19/cm*3

     600 deg C : 5 X 10*18/cm*3

It is also generally believed that Li solubilities in pure Si below
635 deg C can be predicted from the expression:

    [Li](Equil) = 1.3 X 10*22 exp(-0.56/kT(eV))/cm*3

which is derived from a data compilation given by Pell [7]. The maximum
solubility of 7 X 10*19/cm*3 at 1200 deg C drops steadily with falling
temperature, the above expression yielding a room temperature
equilibrium solubility of about 10*13/cm*3. It is important to emphasize
that the above equation refers to Si containing Li as the sole major
impurity. Since Li is electrically active, its concentration is
controlled by the existing state of the electron-hole equilibrium.
Through this effect and those of related ion-pairing processes, any
other residual impurities at comparable concentrations will markedly
influence Li solubilities at low temperatures. Thus, for example, Li
solubilities in strongly p-type Si are raised above those in n-type
material [6], and are higher in CZ Si than in FZ Si [7].

REFERENCES
----

[1] C.S.Fuller, J.A.Ditzenberger [Phys. Rev. (USA) vol.91 (1953) p.193]

[2] H.Reiss, C.S.Fuller, A.J.Pietruszkiewicz [J. Chem. Phys. (USA) vol.25 (1956) p.650]

[3] C.S.Fuller, H.Reiss [J. Chem. Phys. (USA) vol.27 (1957) p.318]

[4] E.M.Pell [Bull. Am. Phys. Soc. Ser. II (USA) vol.2 (1957) p.135]

[5] E.M.Pell [J. Phys. & Chem. Solids (GB) vol.3 (1957) p.77]

[6] H.Reiss, C.S.Fuller [J. Met. (USA) (Feb 1956) p.276]

[7] E.M.Pell [Solid State Physics in Electronics and Telecommunications, Proc. IUPAP Conf., Brussels 1958, vol.I Semics, Part 1 (Acad. Press, New York, 1960) p.261]

[8] K.Weiser [J. Phys. & Chem. Solids (GB) vol.17 (1960) p.149]

15.5 SOLUBILITY OF Na IN Si

by L.T.Canham

October 1987
EMIS Datareview RN=16189

From the alkali metals of group I, only the properties of Li in Si have
been thoroughly investigated. Despite the well-documented effect of Na
contamination of the SiO2/Si interface, its state of incorporation and
solubility in bulk Si are far from being established.

In analogy with Li, one might expect the small binding energy of the Na
3s valence electron to promote donor activity, despite the larger Na
ionic radius (0.95 A). Indeed, the early theoretical study of Kaus
[1] predicted that interstitial Na, like interstitial Li, would exhibit
shallow donor behaviour in the bulk, and the alkali ion implantation
studies of McCaldin and coworkers [2-4] in the 1960's confirmed this
simple picture; thin Na implanted layers exhibited similar donor
activity to those that had been Li implanted. Channeling of 5-10 keV Na
ions into Si at 300-500 deg C gave rise to donor concentrations as high
as 10*19-10*20/cm*3. Sodium ion injection was employed in fabricating
shallow n+/p junction of high quality [5-6].

However, when McCaldin et al [7] tried thermal in-diffusion to determine
the solubility of Na under equilibrium conditions, markedly different
behaviour from that of Li was observed. Following exposure to Na vapour
at 800 deg C for about 1 week neutron activation analysis revealed that
Na penetration was limited to thin (50 micron) surface layers and in
the interior, the Na concentration was at or below their detection limit
of 10*15/cm*3. More surprisingly, no donor activity could be found by
thermoelectric type testing; in the surface layers containing Na
concentrations of 3 X 10*17/cm*3, the Si had remained intrinsic or even
become slightly p-type. They concluded that under equilibrium
conditions, Na is only sparingly soluble in Si (3 X 10*17/cm*3 at 800
deg C) and that only a minute fraction can be present as interstitial
donors, the vast majority of diffused Na being electrically inactive.
No evidence of precipitation was found.

In direct contradiction Svob [8-9] reported much higher solubilities
for thermally introduced Na and observed donor activity. He employed
two techniques; diffusion from a very thin layer of metallic Na in
intimate contact with the Si surface, and diffusion from the
electrolytical deposition of NaI. His reported Na solubilities,
determined by flame photometry, rise from 10*18/cm*3 at 600 deg C to a
maximum of 9 X 10*18/cm*3 at 1100-1200 deg C. His value for the
solubility of Na at the single temperature (800 deg C) investigated
by McCaldin et al [7] is about 14 times higher. Electrical measurements
revealed that the active donor concentration was up to 3 orders of
magnitude lower than the total Na content. This was ascribed to
precipitation.

In an attempt to explain these discrepancies, Parry et al [10]
remeasured Na concentrations in Si preserved from their previous work
[7], but used flame photometry, the same analytical technique used by
Svob [9]. They found that none of their samples contained the uniform
high Na concentration (approximately 4 X 10*18/cm*3) predicted by the
diffusivity and solubility data of Svob [9] at 800 deg C. It was

demonstrated that insufficient cleaning prior to flame photometric analysis can yield anomalously high Na levels. They also suggested that the donor activity observed by Svob [8,9] was the result of trace Li contamination of the dopant sources. His quoted detection limit (10*-4%) for Li in the Na foil is not low enough to exclude such a possibility. No details were given as to the purity of the NaI used in electrolysis, but it was claimed that this technique yielded similar results.

The question as to whether or not interstitial Na can be thermally injected into the bulk remains controversial. Doubrava [11-12] repeated the electrolytical technique of Svob [9] and observed donor activity. Thermal in-diffusion from the vapour phase may be ineffective at creating Na interstitials due to surface reactivity and precipitation, rather than their instability in the bulk. More recent work on implanted Na [13-15] has supported the early conjectures of McCaldin [3] that an appreciable fraction remain in interstitial sites and having overcome the Si surface barrier, can diffuse through large distances without precipitating. Further work on thermally-diffused Si is clearly needed to resolve these issues.

The high volatility and reactivity of Na suggests that in-diffusion from an ultrapure NaI electrolyte is the most promising technique. Sensitive spectroscopic analysis with low temperature photoluminescence can identify any trace Li contamination.

REFERENCES
----

[1]   P.E.Kaus [Phys. Rev. (USA) vol.109 (1958) p.1944]
[2]   J.O.McCaldin, A.E.Widmer [Proc. IEEE (USA) vol.52 (1964) p.301]
[3]   J.O.McCaldin [Nucl. Instrum. & Methods (Netherlands) vol.38 (1965) p.153]
[4]   J.O.McCaldin [Prog. Solid State Chem. (GB) vol.2 ch.2 Ed. H.Reiss (Pergamon, Oxford, England, 1965) p.9]
[5]   M.Waldner, P.E.McQuaid [Solid-State Electron. (GB) vol.7 (1964) p.925]
[6]   R.P.Ruth [Proc. 2nd Int. Conf. Electron & Ion Beam Sci. & Technol., New York, USA, 1966, vol.2 Ed. R.Bakish (AIMMPE, 1969) p.1117]
[7]   J.O.McCaldin, M.J.Little, A.E.Widmer [J. Phys. & Chem. Solids (GB) vol.26 (1965) p.1119]
[8]   L.Svob [Phys. Status Solidi (Germany) vol.7 (1964) p.K1]
[9]   L.Svob [Solid-State Electron. (GB) vol.10 (1967) p.991]
[10] E.P.Parry, M.S.Porter, J.O.McCaldin [Solid-State Electron (GB) vol.12 (1969) p.500]
[11] P.Doubrava [Phys. Status Solidi (Germany) vol.34 (1969) p.K9]
[12] P.Doubrava [Phys. Status Solidi (Germany) vol.40 (1970) p.483]
[13] E.I.Zorin, P.V.Pavlov, D.I.Tetel'baum, A.F.Khokhlov [Sov. Phys.-Semicond. (USA) vol.6 (1972) p.344]
[14] N.A.Skakin, N.P.Dikii, P.P.Matyash, V.M.Korol [Sov. Phys.-Solid State (USA) vol.15 (1973) p.123]
[15] V.M.Korol, A.V.Zastavnyi [Sov. Phys.-Semicond. (USA) vol.11 (1977) p.974]

15.6 SOLUBILITY OF K IN Si

by L.T.Canham

October 1987
EMIS Datareview RN=16190

Very little experimental data exist on the behaviour of K in Si. Svob [1] reported that as with Na, he had successfully thermally introduced K donors into bulk Si. Both diffusion from metallic K and from the electrolysis of molten KI were employed. The total K content, as determined from flame photometry, rose from 5 X 10*17/cm*3 at 550 deg C to a maximum of 7 X 10*18/cm*3 at 1100 deg C. His work was criticized by Parry et al [2], however Ho [3] has also recently reported that interstitial K donors can be thermally introduced under electrolysis. Ho claims that whereas in-diffusion from K vapour or liquid K gave rise to no donor activity, electrolytically deposited K+ ions were able to diffuse appreciable distances into the bulk. Hall data indicated that previously 3 kohm cm Si contained a donor concentration of 10*15/cm*3 after treatment. Thirty minutes electrolysis at an unspecified temperature created a p-n junction 0.9 mm below the surface. Spectroscopic analysis of such material is clearly required.

Like Li and Na, ion implanted K can introduce shallow donor levels in the Si band gap. Medved et al [4] channeled 5-20 keV K ions into high resistivity material at 350 deg C. Junction depths were typically 0.2-0.4 micron and donor concentrations of 10*16 - 10*17/cm*3 were achieved. As expected from ionic size considerations, K (ionic radius of 1.33 A) is more effectively accommodated in the Si lattice than Cs (1.69 A), but significantly less than Na (0.95 A) and Li (0.60 A). Zorin et al [5] and Korol and Zastavnyi [6,7] have studied the incorporation of K ions implanted at 50 keV and observed donor concentrations of similar magnitude under various annealing conditions.

REFERENCES

[1] L.Svob [Solid-State Electron. (GB) vol.10 (1967) p.991]
[2] E.P.Parry, M.S.Porter, J.O.McCaldin [Solid-State Electron. (GB) vol.12 (1969) p.500]
[3] L.T.Ho [Mater. Sci. Forum (Switzerland) vol.10-12 pt.1 (1986) p.175]
[4] D.B.Medved, J.Perel, H.L.Daley, G.P.Rolik [Nucl. Instrum & Methods (Netherlands) vol.38 (1965) p.175]
[5] E.I.Zorin, P.V.Pavlov, D.I.Tetel'baum, A.F.Khoklov [Sov. Phys.-Semicond. (USA) vol.6 (1972) p.21]
[6] V.M.Korol, A.V.Zastavnyi [Sov. Phys.-Semicond. (USA) vol.11 (1977) p.926]
[7] V.M.Korol, A.V.Zastavnyi [Sov. Phys.-Semicond. (USA) vol.14 (1980) p.845]

CHAPTER 16

OXIDATION

16.1 THERMAL OXIDATION OF Si

by P.B.Moynagh and P.J.Rosser

December 1987
EMIS Datareview RN=17828

A. INTRODUCTION

The stable, controllable and highly reproducible characteristics of
thermally grown silicon dioxide layers and their interfaces are
largely responsible for the dominance of silicon as the workhorse of
the semiconductor industry. During processing these thermally grown
oxide layers are used for selective masking during diffusion and
implantation, for surface protection, and for impurity gettering. In
device structures they are used as surface passivators, device
isolators, interconnect separators, and as the gate dielectric in MOS
transistors. The thermal oxidation of silicon is a process in which
silicon is introduced into an ambient containing oxidant species at
high temperature. The oxidant species include oxygen and water vapour
at various pressures, at various dilutions in nitrogen or argon, with
or without HCl. The oxidation temperatures of interest to integrated
circuit technologies range from 600 deg C to 1300 deg C and oxidation
at room temperature. Each of the thermal oxidation variables are
qualitatively and quantitatively analysed below.

B. THE OXIDATION PROCESS

Thermal oxidation of silicon results in the formation of a layer of SiO2
on a silicon surface in the presence of an oxidizing ambient. Since the
oxide formed separates the two reacting species, diffusion of one or
both reactants through the oxide layer must occur prior to the chemical
reaction to form the SiO2. The best experimental evidence to date
indicates that for dry oxygen ambients, in the post-30 nm thickness
regime at least, the diffusing species is neutral O2 which does not
interact with the oxide. For oxidation in wet ambients the diffusing
species is molecular H2O which rapidly exchanges with the SiO2 network.

C. LINEAR-PARABOLIC OXIDATION

The most important single contribution to the modelling of oxide growth
was that of Deal and Grove in 1965 [6]. They arrived at a phenomenological
model which describes the thickness evolution of thermal oxides by a
linear-parabolic relationship which has, since its publication, served
as the basis for the thermal oxidation model of all commonly used process
model packages. This model is reproduced and discussed below.

Oxidation occurs through the transport of the oxidizing species from
the ambient gas to the surface of the existing oxide layer and across
the oxide layer as non-ionized molecules to react with silicon at the
SiO2/Si interface to form oxide. As such there exist three fluxes in
series, that due to transport of the oxidizing species across the
ambient-SiO2 interface, that due to the diffusion of the oxidizing

species across the existing oxide, and that due to the consumption of
the oxidizing species at the SiO2/Si interface where the oxide actually
grows. These fluxes are described as

        F1 = h(Ceq-Co)
        F2 = D(Co-Ci)/Xox
        F3 = k Ci

where

        Ceq = equilibrium concentration of oxidant in the SiO2
            = 5 X 10*16/cm*3 O2 in oxide; 3 X 10*19/cm*3 H2O in oxide
        Co = concentration of oxidant at the outer SiO2 surface
        Ci = concentration of oxidant at the SiO2/Si interface
        h   = gas phase mass transfer coefficient
        D   = diffusion constant for the oxidant in the oxide
        k   = chemical surface reaction rate constant
        Xox = the present oxide thickness

The growth rate will be governed by the slowest of the three fluxes and
under steady state conditions these three fluxes will be equal

        F1 = F2 = F3

After some straightforward manipulation one finds [16]

    (Xox)*2 - (Xi)*2 + A(Xox - Xi) = Bt

or   (Xox)*2 + A Xox = B(t + tau)

where

    A = 2D [(1/k) + (1/h)] = 2D/k

    B = 2D Ceq/N

    N = number of oxidant molecules incorporated per unit volume

      = 2.2 X 10*22/cm*3 O2 in oxide; 4.4 X 10*22/cm*3 H2O in oxide

    tau = (Xi)*2 + A Xi/B

and Xi = Xox(initial)

tau is a time coordinate shift to correct for a non-zero initial oxide
thickness due to the presence of an initial oxide at the start of the
oxidation process and to correct for any initial rapid oxide growth not
well modelled by this phenomenological model.

Solving for Xox as a function of time gives

        (Xox)*2   t   (Xi)*2 + A Xi
        ------- = [1 + --------- + -------------]*0.5 - 1
           A   (A*2)/4B   (A*2)/4

or   (Xox)*2   t + tau
        ------- = [1 + ---------]*0.5 - 1
           A   (A*2)/4B

One limiting case is for long oxidation times (thick oxides) where
t >> tau and t >> (A*2)/4B. Here the oxidation process is rate limited
by the diffusion of the oxidant through the existing oxide, and the
oxide growth exhibits parabolic characteristics.

        dXox/dt = B/2Xox

or   (Xox)*2 = Bt

with B being the parabolic rate constant described by

        B = 2D Ceq/N

One would expect the parabolic rate constant to depend a little or not
at all on substrate conditions.

The other limiting case is for short oxidation times (thin oxides)
where (t + tau) << (A*2)/4B. Here the oxidation process is rate
limited at the oxide/silicon interface by the reaction of the oxidant
with the silicon, and the oxide growth exhibits linear characteristics

        Xox = (B/A)(t + tau)

with B/A being the linear rate constant described by

            k h Ceq
    B/A = --------
            (k + h)N

One would expect the linear rate constant to be strongly dependent on
substrate conditions, e.g. orientation [17] and doping concentration
[18].

D. COMPARING THE LINEAR-PARABOLIC MODEL WITH EXPERIMENTAL RESULTS

The linear-parabolic law of Deal and Grove explains satisfactorily the
results obtained in the following cases:

(a) in the case of wet oxidation for all thicknesses and for substrate
    doping concentration less than about 10*20/cm*3.

(b) in the case of dry oxidation for all thicknesses above 20-30 nm and
    for substrate doping concentration less than about 10*20/cm*3.

Values of B, B/A, and tau are obtained by fitting oxidation data to the
linear-parabolic model. The activation energy of the linear rate
constant, B/A, is found to be about 2 eV for the oxidation of silicon
in dry and wet oxygen. This value compares well with the Si-Si bond
breaking energy of 1.83 eV, which supports the Deal-Grove assumption
that the rate limiting step in the linear regime is the reaction of the
oxidizing species with silicon at the SiO2/Si interface. The activation
energy of the dry oxidation parabolic rate constant is found to be
about 1.2 eV which is close to the activation energy for the diffusion
of O2 through fused silica of 1.18 eV [7]. The activation energy of the
wet oxidation parabolic rate constant is found to be about 0.79 eV which

is close to the activation energy for the diffusion of water through fused silica of 0.8 eV. These values support the Deal-Grove assumption that the rate limiting step in the parabolic regime is diffusion through the oxide.

For high doping concentrations the oxidation rate is enhanced and this fact is accounted for by appropriate adjustment of the oxidation rate constants [19].

For dry oxidation in the thin regime, oxidation rate enhancement is also observed and can be accounted for in two ways depending on the thickness regime. For dry oxidation of oxides thicker than about 30 nm one need only account for the thin regime enhancement as an increased oxidation time, tau [6]. For oxidation within this 30 nm enhanced regime oxide growth rate can be modelled by the addition of two terms which decay exponentially with thickness [3,24].

E. DRY OXIDATION

Dry oxidation refers to oxidation in an O2 ambient.

E1 Dry Oxidation of Lightly Doped Silicon in the Thick Regime at
----
     Atmospheric Pressure

Lightly doped refers to substrate doping levels of less than 5 X 10*19/cm*3. Thick regime refers to oxide thicknesses greater than 30 nm. This oxidation is described well by the linear parabolic model which is described below along with coefficient values from the literature.

$$(Xox)*2 + A\ Xox = B\ (t + tau)$$

$$or \quad \frac{2Xox}{A} = [\ 1 + \frac{4B(t + tau)}{A*2}\ ]*0.5 - 1$$

and in discretized form, after each oxidation time step, the total oxide thickness is given by [4]

$$Xox = Xox(old) + \frac{B\ dt}{2Xox(old) + A}$$

with

     dt   = current time step for the discrete solution of the
                 oxidation equation
     Xox(old) = oxide thickness prior to current time step
     B/A   = linear rate constant
     B   = parabolic rate constant

The oxidation parameters are quantified below. [111], [110] and [100] refer to substrate orientation.

The linear rate constants are described by

B/A [T<950 deg C] [111] = (1.479 X 10*6)exp(-1.74 eV/kT) micron/hr [1]
B/A [T<950 deg C] [110] = (3.571 X 10*7)exp(-2.10 eV/kT)   [1]
B/A [T<950 deg C] [100] = (4.666 X 10*5)exp(-1.76 eV/kT)   [1]

   B/A [T>950 deg C]   = (5.997 X 10*9)exp(-2.75 eV/kT)   [1]

        B/A [111] = (6.240 X 10*6)exp(-2.00 eV/kT)   [4]
        B/A [110] = (5.190 X 10*6)exp(-2.00 eV/kT)   [4]
        B/A [100] = (3.708 X 10*6)exp(-2.00 eV/kT)   [4]

        B/A [111] = (6.230 X 10*6)exp(-2.00 eV/kT) micron/hr [9]
        B/A [100] = (3.708 X 10*6)exp(-2.00 eV/kT)   [9]

        B/A [111] = (9.056 X 10*6)exp(-2.00 eV/kT)   [5]
        B/A [100] = (5.390 X 10*6)exp(-2.00 eV/kT)   [5]

The parabolic rate constants are described by

B [T<950 deg C] [111] = (9.000 X 10*4)exp(-1.71 eV/kT) micron*2/hr [1]
B [T<950 deg C] [110] = (6.371 X 10*4)exp(-1.70 eV/kT)   [1]
B [T<950 deg C] [100] = (1.373 X 10*7)exp(-2.22 eV/kT)   [1]

    [T>950 deg C] = (2.484 X 10*2)exp(-1.10 eV/kT)   [1]

        B = (7.720 X 10*2)exp(-1.23 eV/kT) micron*2/hr [9]

        B = (7.722 X 10*2)exp(-1.23 eV/kT)   [4]

        B = (1.016 X 10*3)exp(-1.23 eV/kT)   [5]

    B [T<1000 deg C] = (8.04 X 10*3)exp(-1.71 eV/kT)   [3]

The time scale adjust is described by

        tau = (Xi)*2 + A Xi/B   hr

given

        Xi = Xi(start of oxidation) + 0.023 microns   [6]
                                    + 0.016   [7]
                                    + 0.004   [8]

- 473 -

E2 Dry Oxidation of Lightly Doped Silicon Across the Entire Thickness
----

Regime at Atmospheric Pressure

After each oxidation time step, the total oxide thickness is given
by [4]

$$X_{ox} = X_{ox}(old) + \left[\frac{B}{2X_{ox}(old) + A} + thin.rate \right] dt$$

with

dt   = current time step for the discrete solution of the
              oxidation equation
Xox(old) = oxide thickness prior to current time step
B/A   = linear rate constant (see section E1)
B   = parabolic rate constant (see section E1)
thin.rate = thin regime constant

and with the thin regime oxidation rate constant given by [1,3,24]

$$thin.rate = C0 \exp \frac{-CE}{kT} \exp \frac{-X_{ox}(old)}{L} \quad micron/hr$$

$$(111) = (3.948 \times 10^8) \exp \frac{-2.33\ eV}{kT} \exp \frac{-X_{ox}(old)}{7.8 \times 10^{-3}\ microns}$$

$$(110) = (3.192 \times 10^6) \exp \frac{-1.800\ eV}{kT} \exp \frac{-X_{ox}(old)}{6.0 \times 10^{-3}\ microns}$$

$$(100) = (4.88 \times 10^8) \exp \frac{-2.38\ eV}{kT} \exp \frac{-X_{ox}(old)}{6.9 \times 10^{-3}\ microns}$$

where (111), (110) and (100) refer to the substrate orientation.

E3 Pressure Dependence of Dry Oxidation

If Henry's law were valid, then the concentration of oxidizing species
in the oxide would be linearly proportional to the pressure of the
oxidizing species in the ambient, C = HP. As such one would expect the
rate constants to be linearly dependent on the pressure of the oxidizing
ambient since B/A = h Ceq/N and B = D Ceq/N. In practice Henry's law would
seem to hold for wet oxidation and for the parabolic rate constant of dry
oxidation.

The linear rate constant for dry oxidation is dependent on pressure, P,
in a sub-linear manner with

B/A [at pressure P] = (B/A)P*0.7 [10]
B/A [at pressure P] = (B/A)P*0.5 [11]
B/A [at pressure P] = (B/A)P*0.6 [12]
B/A [at pressure P] = (B/A)P*1.0 [13]
B/A [at pressure P] = (B/A)P*0.75 [4]

The parabolic rate constant for dry oxidation is linearly dependent on pressure, P, i.e.

$$B \text{ [at pressure P]} = BP \quad [4,10,11,13]$$

E4 Dry Oxidation in a Chlorine Containing Ambient

The addition of chlorine-containing compounds to the oxidizing ambient stabilizes the oxide against sodium drift [1,20], reduces the growth of oxidation induced stacking faults [21], increases the oxide dielectric strength [22], and removes some metallic impurities thereby improving MOS minority carrier lifetime [23]. Both the linear and parabolic oxidation rate constants are affected by chlorine in the ambient in general causing enhanced growth rates [1].

The effect of HCl addition to a dry oxygen ambient can be described approximately as follows [7]

$$(B/A)(HCl) = (B/A) \text{ X } (HCl\% + 1) \quad \text{for HCl\% up to 2\%}$$

$$= (B/A) \text{ X } 2 \qquad \text{for HCl\% > 2\%}$$

and

$$B(HCl) = B \text{ X } [(HCl\% + 1)/11] \quad \text{for HCl\% up to 10\%}$$

E5 Dry Oxidation of Heavily Doped Silicon

Heavily doped silicon refers to substrate doping concentrations of in excess of $1 \text{ X } 10^{19}/cm^3$. It is believed [19] that high doping levels significantly increase the vacancy concentration resulting in increased sites for the oxidation reaction at the SiO2/Si interface. It would therefore be expected that the linear rate constant be enhanced and that the enhancement be substrate concentration dependent with a dependence on substrate orientation and no dependence on oxidizing ambient. It is also expected that substrate concentration could affect the parabolic rate constant through segregation and structural effects on the oxide. Heavy n-type doping increases B/A throughout the processing temperature range, and heavy p-type doping increases B/A only at higher temperatures [2].

The dependence on substrate surface doping concentration, Cs, of the linear oxidation rate constant is given by [4]

$$B/A \text{ (Cs)} = (B/A) \text{ X } (1 + gamma \text{ X } (vacancy - 1))$$

with

$$gamma = (2.63 \text{ X } 10^3)exp(-1.1 \text{ eV}/kT)$$

$$vacancy = \frac{1 + [C(V+) \text{ X } (ni/n)] + [C(V-) \text{ X } (n/ni)] + [C(V--) \text{ X } (n/ni)^2]}{1 + C(V+) + C(V-) + C(V--)}$$

```
C(V+)   = exp[ (E(V+) - Ei)/kT ]
C(V-)   = exp[ (Ei -E(V-))/kT ]
C(V--)  = exp[ (2Ei - E(V-) - E(V--))/kT ]
Ei      = Eg/2
Eg      = 1.17 - (4.73 X 10*-4)(T*2)/(T + 636) eV
E(V+)   = 0.35 eV
E(V-)   = Eg - 0.57 eV
E(V--)  = Eg - 0.12 eV
```

The dependence on substrate surface doping concentration, Cs, of the parabolic oxidation rate constant is given by [4]

$$
B(Cs) = B \text{ X } [1 + delta \text{ X } (Cs*1.28 \exp \frac{-0.176 \text{ eV}}{kT})]
$$

with

```
    delta = 0                                    Cs < 1 X 10*19/cm*3

          = (9.63 X 10*-18/cm*3)exp(2.83 eV/kT)   Cs > 1 X 10*19/cm*3
```

F. WET OXIDATION

Wet oxidation refers to oxidation in a water vapour ambient. Although water molecules have a smaller diffusivity in oxide than oxygen molecules [10], their solid solubility concentration in the oxide is some three orders of magnitude higher (3 X 10*19/cm*3 compared to 5 X 10*16/cm*3) because of the difference in the size of their molecules [1]. Both the linear and parabolic rate constants are proportional to this concentration and, as a result, they are both larger for oxides grown in wet than in dry oxygen. This highlights the importance of the control of the water vapour content of a dry oxygen ambient [14].

F1 Wet Oxidation of Lightly Doped Silicon

Lightly doped refers to substrate doping levels of less than 5 X 10*19/cm*3.

$$
(Xox)*2 + A \text{ } Xox = B(t + tau)
$$

$$
\text{or} \quad \frac{2(Xox)}{A} = [1 + \frac{4B(t + tau)}{A*2}]*0.5 - 1
$$

and in discretized form, after each oxidation time step, the total oxide thickness is given by [4]

$$
Xox = Xox(old) + \frac{B \text{ } dt}{2Xox(old) + A}
$$

with

```
    dt         = current time step for the discrete solution of the
                 oxidation equation
    Xox(old) = oxide thickness prior to current time step
    B/A        = linear rate constant
    B          = parabolic rate constant
```

The oxidation parameters are quantified below. [111], [110] and [100] refer to substrate orientation.

The linear rate constants are described by

B/A [T<900 deg C] [111] = (2.070 X 10*6)exp(-1.60 eV/kT) micron/hr [4]
B/A [T<900 deg C] [110] = (1.722 X 10*6)exp(-1.60 eV/kT)   [4]
B/A [T<900 deg C] [100] = (1.236 X 10*6)exp(-1.60 eV/kT)   [4]

B/A [T>900 deg C] [111] = (1.770 X 10*8)exp(-2.05 eV/kT)   [4]
B/A [T>900 deg C] [110] = (1.476 X 10*8)exp(-2.05 eV/kT)   [4]
B/A [T>900 deg C] [100] = (1.056 X 10*8)exp(-2.05 eV/kT)   [4]

$\qquad$ B/A [111] = (1.262 X 10*8)exp(-2.05 eV/kT)   [5]
$\qquad$ B/A [100] = (7.512 X 10*7)exp(-2.05 eV/kT)   [5]

640 torr $\qquad$ B/A = (1.630 X 10*8)exp(-2.05 eV/kT)   [9]

The parabolic rate constants are described by

$\qquad$ B [T<950 deg C] = (1.698 X 10*4)exp(-1.17 eV/kT) micron*2/hr [4]

$\qquad$ B [T>950 deg C] = (4.200 X 10*2)exp(-0.78 eV/kT)   [4]

640 torr $\qquad$ B = (3.860 X 10*2)exp(-0.78 eV/kT)   [9]

The time scale adjust is

$\qquad$ tau = (Xi)*2 + A Xi/B hr

with $\quad$ Xi = Xi(start of oxidation) microns

F2 Thin Regime Enhancement During Wet Oxidation

During wet oxidation, unlike during dry oxidation, there is no thin regime oxidation rate enhancement.

F3 Pressure Dependence of Wet Oxidation

The linear rate constant for wet oxidation is linearly dependent on pressure, P, i.e.

$\qquad$ B/A [at pressure P] = (B/A)P [4,15]

The parabolic rate constant for wet oxidation is linearly dependent on pressure, P, i.e.

$\qquad$ B [at pressure P] = BP [4,15]

F4 Wet Oxidation in a Chlorine Containing Ambient

The addition of chlorine containing compounds to the oxidizing ambient stabilizes the oxide against sodium drift [1,20], reduces the growth of oxidation induced stacking faults [21], increases the oxide dielectric strength [22], and removes some metallic impurities thereby improving MOS minority carrier lifetime [23]. These effects are of most concern during the growth of the gate dielectric of MIS devices. Both the linear and parabolic oxidation rate constants are affected by chlorine in the ambient in general causing enhanced growth rates [1].

F5 Wet Oxidation of Heavily Doped Silicon

Heavily doped silicon refers to substrate doping concentrations of in excess of $1 \times 10^{19}/cm^3$. It is believed [19] that high doping levels significantly increase the vacancy concentration resulting in increased sites for the oxidation reaction at the SiO2/Si interface. It would therefore be expected that the linear rate constant be enhanced and that the enhancement be substrate concentration dependent with a dependence on substrate orientation and no dependence on oxidizing ambient. It is also expected that substrate concentration could affect the parabolic rate constant through segregation and structural effects on the oxide. Heavy n-type doping increases B/A throughout the processing temperature range, and heavy p-type doping increases B/A only at higher temperatures [2].

The dependence on substrate surface doping concentration, Cs, of the linear oxidation rate constant is given by [4]

$$B/A (Cs) = (B/A) \times (1 + gamma \times (vacancy - 1))$$

with

$$gamma = (2.63 \times 10^3)exp(-1.1 \ eV/kT)$$

$$vacancy = \frac{1 + [C(V+) \times (ni/n)] + [C(V-) \times (n/ni)] + [C(V--) \times (n/ni)^2]}{1 + C(V+) + C(V-) + C(V--)}$$

```
C(V+)  = exp[ (E(V+) - Ei)/kT ]
C(V-)  = exp[ (Ei - E(V-))/kT ]
C(V--) = exp[ (Ei - E(V-) - E(V--) ]/kT
Ei     = Eg/2
Eg     = 1.17 - (4.73 X 10*-4)(T*2)/(T + 636)eV
E(V+)  = 0.35 eV
E(V-)  = Eg - 0.57 eV
E(V--) = Eg - 0.12 eV
```

The dependence on substrate surface doping concentration, Cs, of the parabolic oxidation rate constant is given by [4]

$$B(Cs) = B \times [\ 1 + delta \ (Cs^*1.28exp \frac{-0.176 \ eV}{kT}) \ ]$$

with

$$delta = 0 \qquad\qquad Cs < 1 \times 10^{19}/cm^3$$

$$= (9.63 \times 10^{-18}) \ exp \ (2.83 \ eV/kT) \quad Cs > 1 \times 10^{19}/cm^3$$

G. OXIDE GROWTH ON SILICON IN AIR AT ROOM TEMPERATURE

Wafers from a manufacturer's container have an oxide thickness of approximately 1.5 nm as measured using a laser ellipsometer. A 10 sec dip in 10:1 hydrofluoric acid, followed by a 5 minute rinse in high purity water and a spin dry, results in an oxide thickness one minute after removal from the spin dry of about 0.6 nm. Subsequent exposure to the a class 100 clean room atmosphere at room temperature results in a growth rate of approximately 0.1 nm per decade [5].

REFERENCES

[1] H.Z.Massoud [Tech. Rprt.No.G502-1 (Stanford Electronics Labs., Stanford University, USA 1983)]

[2] E.A.Irene, D.W.Wong [J. Electrochem. Soc. (USA) vol.125 no.7 (1978) p. 1146-51]

[3] H.Z.Massoud, J.D.Plummer, E.A.Irene [J. Electrochem. Soc. (USA) vol.132 no.7 (Jul 1985) p.1745-53]

[4] C.P.Ho et al [Tech. Rprt. No.SEL-84-001 (Stanford Electronics Labs, Stanford University, USA 1984)]

[5] P.J.Rosser [private communication]

[6] B.E.Deal, A.S.Grove [J. Appl. Phys. (USA) vol.36 (1965) p.3770]

[7] D.W.Hess, B.E.Deal [J. Electrochem. Soc. (USA) vol.124 no.5 (1977) p.735-9]

[8] L.N.Lie, R.R.Razouk, B.E.Deal [J. Electrochem. Soc. (USA) vol.129 (1982) p.2828]

[9] B.E.Deal [J. Electrochem. Soc. (USA) vol.125 no.4 (Apr 1978) p.576-9]

[10] B.E.Deal et al [Electrochem. Soc. Fall Meeting, Colorado, 1981 (Electrochem. Soc., Princeton, NJ, USA 1981), Ext. Abst. vol.81-2 no.372]

[11] Y.Kamigaki [J. Appl. Phys. (USA) vol.48 (1977) p.2891]

[12] M.A.Hopper, R.A.Clarke, L.Young [J. Electrochem. Soc. (USA) vol.122 no.9 (1975) p.1216-22]

[13] D.Hess, B.E.Deal [J. Electrochem. Soc. (USA) vol.122 no.4 (1975) p.579-81]

[14] E.A.Irene, R.Ghez [J. Electrochem. Soc. (USA) vol.124 no.11 (1977) p.1757-61]

[15] R.Razouk, L.N.Lie, B.E.Deal [J. Electrochem. Soc. (USA) vol.128 no.10 (Dec 1981) p.2214-20]

[16] Grove, Sze [MOS Physics & Technology, Eds E.Nicollian, J.Brews (J.Wiley & Sons, 1982)]

[17] E.A.Irene [J. Electrochem. Soc. (USA) vol.121 no.12 (1974) p.1613-16]

[18] C.P.Ho, J.D.Plummer [J. Electrochem. Soc. USA) vol.126 (1979) p.1516 & p.1523]

[19] C.P.Ho, J.D.Plummer, J.D.Meindl, B.E.Deal [J. Electrochem. Soc. (USA) vol.125 no.4 (1978) p.665-71]

[20] A.Rohatgi, S.R.Butler, F.J.Feigl [J. Electrochem. Soc. (USA) vol.126 no.1 (1979) p.149-54]

[21] S.P.Murarka, H.J.Levinstein, R.B.Marcus, R.S.Wagner [J. Appl. Phys. (USA) vol.48 no.9 (1977) p.4001-3]

[22] C.M.Osburn [J. Electrochem. Soc. (USA) vol.121 no.6 (1974) p.809-15]

[23] J.W.Swart et al [J. Electrochem. Soc. (USA) vol.128 (1981) p.1383]

[24] P.J.Rosser et al [Mater. Res. Soc. Symp. Proc. (USA) vol.54 (1986) p.611]

16.2 OXIDATION RATES OF Si OXIDISED BY LASER-BASED METHODS

by I.W.Boyd

Feb 1988
EMIS Datareview RN=15131

Various methods involving the use of lasers as the primary energy
source have been followed in recent years to prepare different oxide
layers on Si (TABLE 1). These films have been made by for example
implanting O+ atoms into the Si lattice, and then pulse laser annealing
the oxygen-enriched amorphous layer [1], or simply by annealing with
either pulsed [2] or CW laser [3] any previously damaged or amorphous
layer in an oxygen-rich environment. Even by deliberately melting [4,5],
amorphizing [6] or heating [7,10] a crystalline silicon sample in air
or O2, such layers can readily be formed. Blum et al [11] have recently
prepared SiO2 films by photo-oxidation of SiO, while deposition from
the gas phase by photolytic [12,13] or pyrolytic [14] methods has also
been reported.

The use of argon [15,16] and krypton [17] lasers as secondary sources
of energy for the reaction has been shown to enhance the usual thermal
oxidation rate obtained in conventional furnaces. This enhancement
effect is presently subject to further study and interpretation [15-18].

TABLE 1

Material/oxygen source	Method	Wavelength (microns)	Growth Rate	Ref
O+ Implanted Si	Pulse Laser Anneal	1.06	–	[1]
a-Si/Air or O2	Pulse Laser Anneal	0.694	–	[2]
a-Si/O2	CW Laser Anneal	10.6	–	[3]
c-Si/Air or O2	Pulse Laser Melting	1.06 0.530	–	[4]
c-Si/O2	Pulse Laser Melting	0.308	more than 1000 A/s	[5]
c-Si/O2	Pulse Laser Amorphization	0.266	300 A with 15 ns pulse	[6]
c-Si/O2	CW Laser Heating	0.500	c. A/s	[7]
c-Si/O2 or Air	CW Laser Heating	10.6 0.500	c. A/s	[8,9,19]
c-Si/O2	Pulse Laser Heating	0.694	6-5 A in less than 10 microsec	[10]
SiO/Air	Photo-Oxidation	0.193	–	[11]
c-Si/SiH4&N2O	Pulse Laser Photodeposition	0.193	50 A/s	[12,13]
c-Si/SiH4&N2O	CW Laser Pyrolysis	0.531	2 X 10*4 A/s	[14]
c-Si/O2	Furnace + Laser Heating	0.500 0.647,0.413 c.0.350	c. A/s	[15,17]
c-Si/O2	Furnace + Laser	c.0.500	c. A/s	[16,20]

REFERENCES
----

[1]   S.W.Chiang, Y.S.Liu, R.F.Feihl [Appl. Phys. Lett. (USA) vol.39
        (1981) p.752]
[2]   A.Garulli, M.Servidori, I.Vecchi [J. Phys. D (GB) vol.13 no.10
        (14 Oct 1980) p.L199-202]
[3]   I.W.Boyd [Appl. Phys. A (Germany) vol.31 (1983) p.71]
[4]   K.Hoh, H.Koyama, K.Uda, Y.Miura [Jpn. J. Appl. Phys. vol.19
        (1980) p.L375]
[5]   T.E.Orlowski, H.Richter [in 'Laser-Controlled Chemical Processing
        of Surfaces' Eds. A.W.Johnson, D.J.Ehrlich (North-Holland, New
        York, 1984)]
[6]   Y.S.Liu, S.W.Chiang, F.Bacon [Appl. Phys. Lett. (USA) vol.38
        (1981) p.1005]
[7]   J.F.Gibbons [Jpn. J. Appl. Phys. Suppl. (Japan) vol.19 (1981)
        p.121]
[8]   I.W.Boyd, J.I.B.Wilson, J.L.West [Thin Solid Films (Switzerland)
        vol.83 (1981) p.L173]
[9]   I.W.Boyd, J.I.B.Wilson [Appl. Phys. Lett. (USA) vol.41 (1982)
        p.162]
[10]   A.Cros, F.Salvan, J.Derrien [Appl. Phys A (Germany) vol.28 (1982)
        p.241]
[11]   S.E.Blum, K.Brown, R.Srinivasan [Appl. Phys. Lett. (USA) vol.43
        (1983) p.1026]
[12]   P.K.Boyer, G.A.Roche, W.H.Ritchie, G.J.Collins [Appl. Phys. Lett.
        (USA) vol.40 (1982) p.715]
[13]   P.K.Boyer, W.H.Ritchie, G.J.Collins [J. Electrochem. Soc. (USA)
        vol.129 no.9 (Sep 1982) p.2155-6]
[14]   S.Szikora, W.Krauter, D.Bauerle [Mater. Lett. (Netherlands)
        vol.2 (1984) p.263]
[15]   S.A.Schafer, S.A.Lyon [J. Vac. Sci. & Technol. (USA) vol.19
        (1981) p.494]
[16]   S.A.Schafer, S.A.Lyon [J Vac. Sci. & Technol. (USA) vol.21
        (1982) p.422]
[17]   E.M.Young, W.A.Tiller [Appl. Phys. Lett. (USA) vol.42 (1983)
        p.63]
[18]   I.W.Boyd [Appl. Phys. Lett. (USA) vol.42 (1983) p.728]
[19]   F.Micheli, I.W.Boyd [Appl. Phys. Lett. (USA) vol.51 (1987)
        p.1149
[20]   E.M.Young, W.A.Tiller [J. Appl. Phys. (USA) vol.62 (1987) p.2086

16.3 OXIDATION RATES OF Si AT HIGH PRESSURE

by S.P.Tay, J.P.Ellul and M.I.H.King

November 1987
EMIS Datareview RN=17832

A. INTRODUCTION

Increases in packing density for integrated circuits are routinely
achieved by scaling down device feature sizes. As pattern sizes shrink,
scaling theory dictates that vertical device parameters such as
junction depth, lateral impurity diffusion and film thicknesses have to
be scaled accordingly. These are most practically realized by
reductions in process temperature rather than process time. High-pressure
equipment is being increasingly utilized to achieve practical lower
temperature process cycles. Early use of high-pressure processing was
mainly for bipolar production and shows slow growth until 1982.
Utilization of high pressure in MOS processes is demonstrated by a sharp
increase in the number of tubes in production since 1982. In addition
to this rapid growth for IC production applications the reported rates
of development and modifications of high-pressure equipment further
reflect a steady growth in the number of potential applications of
high-pressure technology in R&D for IC fabrication in the VLSI and ULSI
regimes. These applications, namely, field isolation, FIPOS, glass
reflow, gate and interpoly oxides, thermal nitrides and nitrided oxides
have been reviewed in the light of their merits over conventional
processing techniques [1].

Razouk, Lie and Deal [2,3] have investigated the kinetics of
high-pressure oxidation in pyrogenic steam and dry oxygen for ambient
pressure up to 20 atm and for temperatures between 800 to 1000 deg C.
Application of the Grove-Deal model [4] shows that, for steam oxidation,
both the parabolic and linear rate constants (B and B/A respectively)
are directly proportional to the steam pressure. The results indicate
the presence of different activation energies at 900 deg C and below,
particularly in the case of the parabolic rate constant. For B, the
activation energy increases from 75 kJ/mol (0.78 eV) above 900 deg C
to approximately 113 kJ/mol (1.17 eV) below this temperature, while
for B/A the activation energy change is less pronounced and varies from
198 kJ/mol (2.05 eV) above 900 deg C to about 159 kJ/mol (1.65 eV)
below this temperature.

For dry oxidation, the parabolic rate constant is still directly
proportional to pressure while the linear rate constant is proportional
to oxygen pressure to the power of 0.7. A change in activation energy
for B occurs around 900 deg C. It is 134 kJ/mol (1.39 eV) in the range
900 to 1000 deg C and 172 kJ/mol (1.78 eV) in the range 800 to 900
deg C. These values can be related to the diffusivity of oxygen through
fused silica [5] with an activation energy of 113 kJ/mol (1.17 eV) in
the range 950 to 1078 deg C. Little variation can be observed in the
activation energy for T in the range 800 to 1000 deg C, which is 181 to 183
kJ/mol (1.87-1.90 eV). This value can be compared to the energy of 177
kJ/mol (1.83 eV) required to break a Si-Si bond [6].

The changes in activation energies at about 900 deg C could be an indication of a change in the reaction mechanism or structural changes in the oxide at these growth temperatures. The latter possibility could be supported by the work of EerNisse [7] who observed the onset of viscous flow of thermal SiO2 grown in steam or dry oxygen to be around 965 deg C. Lie, Razouk and Deal [2,3] have attempted to correlate pressure dependence of oxidation rate constants with pressure dependence of oxidant diffusivity and solubility in the oxide as well as, in the case of dry O2 oxidation, with the pressure dependence of chemical surface reaction constant. While the analysis of Lie et al is based entirely on the linear-parabolic Deal-Grove oxidation model, various other oxidation mechanisms have been proposed in the literature. They deal with a variety of diffusing species and electrical or electrochemical fields present during oxidation. Applications of these models to oxidation data may lead to varying modes of pressure dependence of oxidation rate constants. Meanwhile, more oxidation data at pressures as high as 1000 atm have emerged in the past few years, thus enabling a more accurate picture for oxidation modelling.

This article will focus on data from 1982 to 1987, and review models that various authors have proposed in the literature. Data published before 1982 have been reviewed and summarized by Zeto et al [8], Katz et al [9], Miyoshi et al [10], and Bussmann [11], and will be mentioned here only briefly.

B. STEAM OXIDATION OF LIGHTLY DOPED SILICON

In the early 1960s Ligenza et al [12-15] found that oxidation kinetics at 650 deg C with steam pressures up to 90 atm in heated, gold-lined, sealed Inconel X ampoules exhibited linear thickness increase with time. This is indicative of a surface-controlled reaction, and growth rate was found to be proportional to steam pressure [15]. Oxidation rates were dependent on silicon crystal orientations as follows: (110) > (311) > (111) > (100), while activation energies decreased in the following silicon orientation order: (110) < (311) < (111) < (100) [14]. The rate controlling step in the oxidation process was identified as the reaction of interstitial water molecules in silica films with silicon bonds at the Si/SiO2 interface. In 1974, Powell et al [16] reported that, at 675 deg C and 60 atm, (111) silicon oxidized at a constant rate of 5.4 nm/min while, at 90 atm, the rate increased to 8 nm/min. Silicon (100) was found to oxidise at about 2.7 nm/min at 675 deg C and 60 atm. Thus the (111)/(100) oxidation rate ratio is about 2:1. It was also found that constant growth rates were obtained up to 6 microns, after which growth became diffusion limited. From 1973 to 1979, various authors [17-21] reported on the design and use of multiwafer systems for high pressure oxidation. A typical configuration for these systems was an inner fused silica reactor surrounded by a furnace, the entire assembly being contained in a pressurized metal chamber. Steam was formed within the reactor either by pyrogenic reaction of H2 and O2 [18-20], or via the use of a water-boiler [17] or a water pump [21].

TABLE 1 summarizes the normalized rate constants for steam oxidation of silicon. Due to practical limits of process equipment, data are generally obtained in the steam pressure range of 1 to 25 atm. Since Razouk et al [2] reported their work, the task of modelling high pressure steam oxidation for IC process engineering purposes may be considered largely completed. Few kinetic data have been reported since 1981. Nevertheless, it is apparent that the Deal-Grove linear-parabolic relationship adequately describes the kinetics of steam oxidation at temperatures above 800 deg C. At temperatures lower than 750 deg C, Bussmann [11] has reported a higher value of activation energy, 2.52 eV, for the parabolic rate process, and a lower value of activation energy, 1.31 eV, for the linear rate process. These authors [11] also noted the close resemblance between these activation energies and those for dry O2 oxidation below 850 deg C, [E(B) = 2.24 eV, E(B/A) = 1.39 eV], but did not offer any physical explanation for the similarity.

TABLE 1: rate constants and activation energies for steam oxidation of silicon

T deg C	P atm	Orient.	Rate Constant B/P micron*2 /h. atm	B/AP micron /h.atm	Activation Energy E(B) eV	E(B/A) eV	Ref
675	60	(111)	0.03	0.003	-	2	[16]
600-1000	1-20	(100), (111)			0.78	1.83	[22]
800	6.5	(100)	0.074	0.015			[18]
800	6.5	(111)	0.093	0.021			[18]
900	6.5	(100)	0.15	0.085(?)			[18]
900	6.5	(111)	0.16	0.14			[18]
1000	6.5	(100)	0.23	0.5			[18]
1000	6.5	(111)	0.27	0.67			[18]
800-1000	6.5	(100), (111)			0.7	2.03	[18]
800	6.5	(100)	0.1	0.032			[10]
900	6.5	(100)	0.15	0.23 (?)			[10]
950	6.5	(100)	0.21	0.42			[10]
800-950	6.5	(100)			0.6	2.0	[10]
550-750	9.9	?			2.52	1.31	[11]
750-990	9.9	?			0.79	1.77	[11]
725	20	(100)	0.095	0.01			[21]
725	20	(111)	0.16	0.011			[21]
750	10,25	(100)	-	0.008			[9]
750	10,25	(111)	0.025	0.011			[9]
920	10,25	(100)	-	0.19			[9]
920	10,25	(111)	0.21	0.27			[9]

T deg C	P atm	Orient	Rate Constant		Activation Energy		Ref
			B/P micron*2 /h.atm	B/AP micron /h.atm	E(B) eV	E(B/A) eV	
800- 900	1-20	(100), (111)			1.17	1.65	[2]
900- 1000	1-20	(100), (111)			0.78	2.05	[2]

C. DRY O2 OXIDATION OF LIGHTLY-DOPED SILICON

Work on high pressure oxidation of Si using dry oxygen was pioneered by
Zeto et al [23]. Oxidation kinetic studies at 140 atm and 800 deg C [24]
revealed that the linear rate constant is not directly proportional
to pressure. Subsequently, Lie et al [3] showed that the linear rate
constant is proportional to oxygen pressure to the power of 0.7 .
This was recently substantiated by Camelin et al [25,26] who investigated
oxidation kinetics over the range of 14 to 1000 atm at 600 deg C to
780 deg C. Zeto and co-workers [27,28] have also reported their work
in the range of 2 to 500 atm at 635 to 1100 deg C. Although they have
presented the electrical characteristics of their dry oxides, dry
oxidation kinetic data have not been published in detail.

TABLE 2 summarizes normalized rate constants for dry O2 oxidation of
silicon. Values of B and B/A in ref [26] were published in terms of their
rate ratios to B(1 atm, T) and B/A (1 atm, T) respectively. The authors
[26] obtained rate constants B(1 atm, T) and B/A(1 atm, T) at 650 deg C
and 700 deg C by means of extrapolating from Massoud's data [29] which
contain rate constants for 800 to 1000 deg C only (TABLE 2). The
present reviewers have attempted to derive the rate constants of Camelin
et al [26] by using the following extrapolated values from ref [29], B/A
(1 atm, 650 deg C) = 0.000114 micron/h, B/A(1 atm, 700 deg C) = 0.000335
micron/h, and B(1 atm, 700 deg C) = 0.0000331 micron*2/h. The recovered
B/A values were normalized to 140 atm, whenever necessary, and verified
on the ln B/A (140 atm) versus 1/T plot in ref [26] with good agreement.

Although the high pressure oxidation data in TABLE 2 have been modelled
reasonably well with the Deal-Grove relationship no attempt has been
reported by these authors [3,11,25,26] to examine growth kinetics
at the initial stage of dry O2 oxidation. Growth rate enhancement
in excess of that predicted by the Deal-Grove model has been observed
for atmospheric dry oxidation under various O2 partial pressures
[30] as well as for high pressure dry O2 oxidation [31]. Massoud
et al [32] modified the Deal-Grove linear-parabolic general oxidation
relationship [4] by introducing two rate-enhancement terms which
decay exponentially with time, thus describing oxide growth in dry
oxygen from the onset of oxidation and beyond the native oxide
with better than 1-2% fit to their oxidation data. On the other hand,
Reisman et al [31] found a simple power law which fits the oxidation of
silicon in dry O2 at oxygen partial pressures ranging from 1.0 X 10*-5
atm up to 20 atm, and in all thickness ranges reported in the literature
since 1965. The power law equation can be expressed as X = a(tg + to)*b,
where X is the final measured thickness, a and b are constants, tg is
actual growth time, and to is the time required to grow an oxide of
thickness Xo, already present at the silicon surface, and/or found as a
consequence of furnace ramp-up and ramp-down sequences. Values for to,
and Xo are determined readily from X-t data. Employing this power-law
model, Reisman et al [31] concluded that there is no evidence of an

'anomalous' oxidation region at small thickness, or of a linear
oxidation region in the earlier stages of oxidation. The critical
parameter 'b' was found to range from less than 0.25 to about 0.8, the
value decreasing with decreasing partial or total pressure of oxygen.
The constant 'a' represents the value of oxide thickness at unit time
and, for constant values of b and to, represents a shift of the X-t
curves along the thickness axis. Reisman et al [31] have not yet offered
a physical interpretation of their model. They have, however, considered
that a diffusion-limited oxidation may be dominant over the entire
thickness regime and suggested that the physical explanation of the
model lies in the variation of the oxidant diffusion coefficient with
time, and therefore with oxide thickness. It is clear that accurate
thickness measurements of very thin oxides is a pre-requisite for a
correct model to account for the 'enhanced growth' at the initial stage
of oxidation.

TABLE 2: rate constants and activation energies for dry O2 oxidation
         of silicon

| T deg C | P atm | Orient. | Rate Constants | | Activation Energy | | Ref |
			B/P micron*2 /h.atm	B/(A.P*0.7) micron /h.atm*0.7	E(B) eV	E(B/A) eV	
800	140	(111)	0.000714	0.00305			[24]
600	500	(100)	-	0.0000336			[25]
780	14	(100)	-	0.00158			[25]
780	140	(100)	0.000257	0.00189			[25]
650	1000	(100)	-	0.000128			[26]
700	140	(100)	0.0000334	0.000411			[26]
700	500	(100)	0.0000321	0.000422			[26]
600- 780	140- 500	(100)			-	1.7	[25], [26]
<850	9.9	?			2.24	1.39	[11]
850- 900	9.9	?			1.36	2.36	[11]
800- 900	1-20	(100), (111)			1.78	1.87	[3]
900- 1000	1-20	(100), (111)			1.39	1.90	[3]
800	1	(100)	0.000396	0.00258			[29]
850	1	(100)	0.00113	0.00558			[29]
900	1	(100)	0.00335	0.0125			[29]
950	1	(100)	0.00726	0.0241			[29]
1000	1	(100)	0.0172	0.0519			[29]
800- 950	1	(100)			2.22	1.76	[29]
>950	1	(100)			1.0 +-0.2	2.75 +-0.22	[29]

D. OXIDATION OF HEAVILY DOPED SILICON

Fuoss and Topich [33] studied the effects of heavy doping on oxide
growth rates in steam ambient at 10 atm and 850 deg C. The enhancement
of growth rates in steam for silicon heavily doped with arsenic compared
to lighter doping was smaller at the high pressure than at 1-atm. On the
other hand, for heavily boron-doped Si(100) high pressure steam appeared
to enhance slightly the oxidation rate compared to that for 1-atm steam.
The overall difference in growth rate is not as dramatic as in the case
of arsenic doping. Fuoss and Topich [33] also examined the segregation
and redistribution of these impurities at high pressure and found less
pile-up of arsenic occurs at the Si/SiO2 interface and less depletion of
boron from silicon compared to 1-atm oxidation. The varying surface
concentrations of arsenic and boron in turn dictate the different
effects of doping on oxide growth at high pressure relative to 1-atm
oxidation.

Shimoda et al [34] investigated HCl-added steam oxidation of silicon at
5-10 atm and 700-800 deg C. They found that the addition of 5% (volume)
HCl to steam did not change the oxidation rate of lightly-doped silicon,
but reduced the oxidation rate of heavily arsenic doped silicon. This
retarded oxidation was explained as a consequence of oxidation-enhanced
diffusion of arsenic being reduced due to suppression of oxidation-
induced stacking faults by chlorine. On the contrary, the presence of
HCl in a dry oxygen ambient was shown to enhance oxide growth by
increasing both the reaction-controlled linear rate constant and the
diffusion-controlled parabolic rate constant [35]. The enhanced reaction
at the Si/SiO2 interface results from the catalytic action of HCl. The
enhanced diffusion is due to a larger number of nonbridging bonds in the
O2-HCl grown oxide relative to a dry O2 grown oxide [36].

Redistribution of arsenic in silicon during high pressure dry O2
oxidation was found by Choi et al [37] to be dependent on the ratio of
oxidation rate to the diffusivity of As+ in silicon, B/SQRT D, as well
as on the thermodynamic equilibrium segregation coefficient. Si(100)
implanted with 3.0 X 10*15 As+/cm*2 (25 keV) was oxidized at 600 and
800 deg C using oxygen pressure of 38 atm up to 143 atm. The conditions
were chosen to maintain constant arsenic diffusivity while varying
oxidation rates. Above a critical B/SQRT D value, the oxidation rate
dominates over other factors such as the thermodynamic equilibrium
segregation coefficient and diffusion rate. Consequently most of the
arsenic impurity becomes trapped in SiO2. Below the critical B/SQRT D
value, snowplowing of arsenic results, which means a dominance of
thermodynamic segregation coefficient over oxidation rate.

The physical model given above is in agreement with the subsequent
observation by Choi et al [38] that 1-atm oxidation of very heavily As+
(> 5.0 X 10*16/cm*2) doped silicon at 850 deg C was faster than at
950 deg C. The anomalous increase in oxidation rate at lower temperature
was attributed to arsenic snowplowing, which was more pronounced at
lower temperature, i.e. B/SQRT D diminished and crosses over the
critical value with temperature decreasing from 950 to 850 deg C. Due to
supesaturation of arsenic (approx. 10*22/cm*3 or approx. 28 at. %) in
silicon a Si - O - As residual layer is formed which enhances a surface
reaction-controlled oxidation rate at the initial stage of oxide growth.

E. OXIDATION OF POLYCRYSTALLINE SILICON

One advantage of high pressure oxidation technology is the capability
to achieve a large thickness ratio between oxide films simultaneously
grown on a heaviy doped polysilicon layer and a lightly doped silicon
substrate. Such a capability has been demonstrated as a key step in the
fabrication of a MOS DRAM device with a double-polysilicon structure
[10,39,40]. Hirayama et al [39] found that the thickness ratio depended
on the phosphorus concentration in polysilicon, and the thickness ratio
rapidly increases with decreasing oxidation temperatures. For steam
oxidation in the temperature range of 700 to 850 deg C and at 3.6 atm,
the thickness ratio of the oxide film on the polysilicon film with a
phosphorus concentration of 7.0 X 10*20/cm*3 to the oxide film on a
boron-doped (1 X 10*15/cm*3) Si(100) substrate changed from 7.9 (at
700 deg C) to 3.3 (at 850 deg C). Sakamoto et al [40] observed that
the oxidant pressure had no remarkable effect on the thickness ratio.

Zhang and Wang [41] reported the characteristics of undoped, heavily
P-doped and heavily B-doped polysilicon films oxidized in high pressure
(5 atm) steam at temperatures of 700 to 940 deg C, and compared
them with those of Si(111) and Si(100). A characteristic polysilicon
oxide thickness has been defined as one that is present at the end of an
initial characteristic period of oxidation during which the polysilicon
oxidation rate is faster than those of Si(111) as well as Si(100). With
increasing time the oxidation rate falls between that of Si(111) and
Si(100). For undoped polysilicon, the oxidation characteristics are
similar to those in normal wet oxidation but the oxidation time
corresponding to the characteristic oxide thickness is shorter than that
in normal wet oxidation. For heavily phosphorus-doped polysilicon,
accelerated oxidation was observed and it was possible to obtain a large
thickness ratio of the oxide film grown on the polysilicon layer to the
oxide film grown on the lightly doped Si(111) substrate, being 2.7:1 for
steam oxidation at 5 atm and 800 deg C. For heavily boron-doped
polysilicon the oxidation rate is faster than that of heavily boron-doped
Si(111). This is opposite to the results for the case of heavy
phosphorus doping.

Tay et al [1,42,43] have reported the growth of high pressure dry
oxide on Si(100) and on polysilicon under conditions where the oxygen
partial pressure increases with time. At 10 atm and 700 deg C, oxide
growth on Si(100) substrates is practically limited to 12 nm while the
thickness ratio of polysilicon oxides (26 to 32 nm) to oxides grown
on Si(100) (10 to 12 nm) is about 3:1. At 10 atm and 800 deg C, the
oxide growth rate on Si(100) increases significantly, being approximately
0.3 nm/min in the thickness range of 5 to 32 nm whereas polysilicon has
oxide growth rate of 0.7 nm/min. It should be noted that with the oxygen
partial pressure increasing with time, the ratio of oxide thickness on
polysilicon to that on Si(100) approaches the ratio of these oxidation
rates, i.e., 2.3:1, with increasing oxidation time.

REFERENCES
----

[1]   S.P.Tay, J.P.Ellul, M.I.H.King [Mater. Res. Soc. Symp. Proc.
        (USA) vol.71 (1986) p.467-78]
[2]   R.R.Razouk, L.N.Lie, B.E.Deal [J. Electrochem. Soc. (USA) vol.128
        no.10 (1981) p.2214-20]
[3]   L.N.Lie, R.R.Razouk, B.E.Deal [J. Electrochem. Soc. (USA) vol.129
        no.12 (1982) p.2828]
[4]   B.E.Deal, A.S.Grove [J. Appl. Phys. (USA) vol.36 (1965) p.3770]
[5]   F.J.Norton [Nature (GB) vol.171 (1961) p.701]
[6]   L.Pauling [The Nature of the Chemical Bond, 3rd edition (Cornell
        Univ. Press, Ithaca, NY, 1960)]
[7]   E.P.EerNisse [Appl. Phys. Lett. (USA) vol.35 no.1 (1979) p.8-10]
[8]   R.J.Zeto, N.D.Korolkoff, S.Marshall [Solid State Technol. (USA)
        vol.22 no.7 (1979) p.62]
[9]   L.Katz, B.F.Howells, L.P.Adda, T.Thompson, D.Carlson [Solid State
        Technol. (USA) vol.24 no.12 (1981) p.87]
[10]   H.Miyoshi, M.Hirayama, N.Tsubouchi, H.Abe [Semiconductor
        Technologies, Japan Annu. Rev. Electron. Comput. & Telecommun.
        (Netherlands) (1982) p.82-99]
[11]   E.Bussmann [Siemens Forsch. & Entwicklungsber. (Germany) vol.10
        no.6 (1981) p.357]
[12]   J.R.Ligenza, W.G.Spitzer [J. Phys. & Chem. Solids (GB) vol.14
        (1960) p.131]
[13]   W.G.Spitzer, J.R.Ligenza [J. Phys. & Chem. Solids (GB) vol.17
        (1961) p.196]
[14]   J.R.Ligenza [J. Phys. Chem. (USA) vol.65 (1961) p.2011]
[15]   J.R.Ligenza [J. Electrochem. Soc. (USA) vol.109 (1962) p.73]
[16]   R.J.Powell, J.R.Ligenza, M.S.Schneider [IEEE Trans. Electron.
        Devices vol.ED-21 (1974) p.636]
[17]   P.T.Panousis, M.Schneider [Electrochem. Soc. Meeting, Chicago,
        IL., May 1973 Ext. Abstr. no.53]
[18]   N.Tsubouchi, H.Miyoshi, A.Nishimoto, H.Abe [Jpn. J. Appl. Phys.
        vol.16 (1977) p.855]
[19]   N.Tsubouchi, H.Miyoshi, A.Nishimoto, H.Abe, R.Satoh [Jpn. J.
        Appl. Phys. (Japan) vol.16 (1977) p.1055]
[20]   R.Champagne, M.Toole [Solid State Technol. (USA) vol.20 no.12
        (1977) p.61]
[21]   L.E.Katz, B.F.Howells, Jr [J. Electrochem. Soc. (USA) vol.126
        (1979) p.1822]
[22]   M.Maeda, H.Kamioka, M.Takagi [Electrochem. Soc. Spring Meeting,
        Seatle, Washington, May 1978, Ext. Abstr. no.178]
[23]   R.J.Zeto, E.Hryckowian, C.D.Bosco, G.J.Iafrate, R.W.Brower,
        C.G.Thornton [Electrochem Soc. Fall Meeting, New York, NY, Oct
        1974, Ext. Abstr. no.206]
[24]   R.J.Zeto, C.G.Thornton, E.Hryckowian, C.D.Bosco [J. Electrochem.
        Soc (USA) vol.122 no.10 (1975) p.1409]
[25]   C.Camelin, G.Demazeau, A.Straboni, J.L.Buevoz [Chemtronics (GB)
        vol.1 no.1 (Mar 1986) p.27-31]
[26]   C.Camelin, G.Demazeau, A.Straboni, J.L.Buevoz [Appl. Phys. Lett.
        (USA) vol.48 no.18 (15 May 1986) p.1211-13]
[27]   J.A.Costello, R.J.Zeto [Electrochem. Soc. Meeting, Toronto,
        Ontario, Canada, May 1985, Ext. Abstr. no.237]
[28]   L.P.Trombetta, R.J.Zeto, F.J.Feigl, M.E.Zvanut [J. Electrochem.
        Soc. (USA) vol.132 no.11 (1985) p.2706-12]
[29]   H.Z.Massoud [PhD dissertation, Stanford Electronics Labs, Tech.
        Report G502-1 (Stanford University, Stanford, CA 1983) p.132]

[30] H.Z.Massoud, J.D.Plummer, E.A.Irene [J. Electrochem. Soc. (USA) vol.132 no.11 (1985) p.2685-92]

[31] A.Reisman, E.H.Nicollian, C.K.Williams, C.J.Merz [J. Electron Mater. (USA) vol.16 no.4 (15 Oct 1987) p.45-56]

[32] H.Z.Massoud, J.D.Plummer [J. Appl. Phys. (USA) vol.62 no.8 (15 Oct 1987) p.3416-23]

[33] D.Fuoss, J.A.Topich [Appl. Phys. Lett. (USA) vol.36 no.4 (15 Feb 1980) p.275-7]

[34] H.Shimoda, M.Maeda, M.Takagi [Electrochem. Soc. Meeting, Montreal, Canada, May 1982 RNP no.743; also in J. Electrochem. Soc. (USA) vol.129 no.6 (1982) p.213C]

[35] K.Hirabayashi, J.Iwamura [J. Electrochem. Soc. (USA) vol.120 (1973) p.1595]

[36] R.J.Kriegler [Thin Solid Films (Switzerland) vol.13 (1972) p.11]

[37] S.S.Choi, M.Z.Numan, W.K.Chu, J.K.Srivastava, E.A.Irene [Appl. Phys. Lett. (USA) vol.50 no.11 (16 Mar 1987) p.688-90]

[38] S.S.Choi, M.Z.Numan, W.K.Chu, E.A.Irene [Appl. Phys. Lett. (USA) vol.51 no.13 (28 Sep 1987) p.1001-3]

[39] M.Hirayama, H.Miyoshi, N.Tsubouchi, H.Abe [IEEE Trans. Electron Devices (USA) vol.ED-29 no.4 (1982) p.503]

[40] M.Sakamoto, K.Hamano [IEDM (IEEE, New York, USA, 1980) Abstr. no. 6.2 Washington, DC 1980 p.136]

[41] Zhang Ai-Zhan, Wang Yang-Yuan [Proc. 3rd Int. Symp. VLSI Sci. & Technol., Toronto, Canada, May 1985 Eds. W.M.Bullis, S.Broydo (Electrochem. Soc., Pennington, NJ, 1985) p.391]

[42] S.P.Tay, J.P.Ellul, M.I.H.King, J.J.White [Electrochem. Soc. Fall Meeting, Las Vegas, NV, Oct, 1985 (Electrochem. Soc. Pennington, NJ, 1985) Ext. Abtrs. no.257]

[43] S.P.Tay, J.P.Ellul, J.J.White, M.I.H.King [J. Electrochem. Soc. (USA) vol.134 no.6 (1987) p.1484-7]

16.4 REACTION MECHANISMS IN Si

by I.W.Boyd

December 1986
EMIS Datareview RN=15704

A. INTRODUCTION

The electronic states, normally arising as a result of free radicals at
the surface of clean crystalline silicon (c-Si), can be reduced by more
than five orders of magnitude under ideal conditions by the growth of
a thin oxide layer on the surface. This layer has contributed
significantly to the development of the planar process and the MOS
transistor as well as the high level of performance of many present day
advanced microelectronic devices. The natural oxide of silicon, SiO2,
is an excellent insulator, can withstand very large electrical fields,
and is chemically very stable.

Oxidation of c-Si has been intensively studied for several decades, and
yet the simple thermal reaction of solid silicon and dry oxygen gas

    Si + O2 ---> SiO2   (1)

is not yet fully understood. Whilst many models can successfully
predict oxidation rates of the silicon under certain conditions, very
few have come close to satisfactorily explaining the observed behaviour
of the reaction under a wider set of experimental environments.

Thermal oxidation of c-Si is known to occur by the diffusion of the
oxidising species through any pre-existing oxide layer to the SiO2/Si
interface, where it reacts with the available silicon atoms. Several
marker experiments with O-18 gas have traced the long range diffusion of
this species through thick oxide layers without exchange with the
already formed network, and found a 93% incorporation at the SiO2/Si
interface, with a small amount present at the outer SiO2/O2 surface,
which is probably due to network oxygen diffusivity from a surface
layer [1]. The fact that there is no interchange of diffusing oxygen
with the network when dry oxygen is used, is completely the opposite
of what happens when H2O is used in the process known as 'wet'
oxidation. As oxidation proceeds, more silicon is consumed as the
SiO2/Si interface retreats further into the bulk. In fact for every
unit thickness of oxide grown, 0.45 units of c-Si have been reacted.

The nature and charge state of the oxidising species has been the
source of much controversy over the years. From the recent work of
Modlin and Tiller [2], it now appears that neutral species are
responsible for oxidation when the oxide layer already formed is
thicker than 300 A. This seems to eliminate previous suggestions that
charged (O2)-- or (O2)- may play a dominant role. It is not yet clear,
however, whether charged species play an active role during the
initial stages of the reaction, when the intervening oxide layer is
thinner than 300 A.

After many independent and unrelated oxidation studies had been
reported, and many different phenomenological models had been
suggested, Deal and Grove (D-G) [3] presented in 1965 not only results
of a comprehensive study of silicon oxidation, but also a most
successful model that continues to be applied today in many oxidation

- 491 -

regimes. The model considers three consecutive physical and chemical processes relevant to the reaction, whose fluxes are equal under steady state conditions. These are, (1) the transfer of the oxidising species from the gas phase into the solid surface, (2) Fickian diffusion of the species through any oxide present to the reacting interface, and (3) reaction of the species with the silicon at the SiO2/Si interface. The general oxidation relationship is thus derived to be

$$x*2 + Ax = B(t + to) \qquad\qquad (2)$$

where x is oxide thickness, t is oxidation time, B/A is the linear rate constant, B is the parabolic rate constant, and

$$A = 2 \ Deff \ (1/k + 1/h) \qquad\qquad (3)$$

$$B = 2 \ Deff \ (C*/N) \qquad\qquad (4)$$

$$to = (xi*2 + Axi)/B \qquad\qquad (5)$$

and Deff is the effective oxidant diffusion constant in the oxide, k and h represent rate constants at the SiO2/Si and O2/SiO2 interfaces respectively, C* is the equilibrium concentration of the oxidant in the oxide, N is the number of oxidant molecules in the oxide unit volume, and xi is the initial oxide thickness at the onset of the reaction.

The two well known limiting forms of Eqn (2) are the PARABOLIC oxidation regime, when t >> A*2/4B, and t >> to,

$$x*2 = B \ (t + to) \qquad\qquad (6)$$

and the LINEAR oxidation regime where t << A*2/4B

$$x = B/A \ (t + to) \qquad\qquad (7)$$

Therefore at high temperatures, or for the growth of thick layers, parabolic kinetics dominate, and the reaction is strongly affected by diffusion and oxidant solubility in the oxide which is known to be proportional to the ambient pressure. At lower temperatures, or for the growth of thin layers, the linear rate constant B/A is important. The rate is also controlled by solubility, but is also now strongly affected by the reaction rate constants at the interfaces. The activation energy for these controlling rate constants has been found to be approximately 1.23 eV and 2.0 eV respectively, for temperatures above about 1000 deg C, suggesting that in this regime, a single elementary process controls the reaction. These energies correspond closely to the known diffusion of oxygen through fused silica (1.17 eV), and the Si-Si bond energy of 1.83 eV, although it is not clear whether the latter association in particular is merely coincidental.

B. LIMITATIONS OF THE DEAL-GROVE (D-G) MODEL

Although a significant contribution to the knowledge of silicon oxidation, the D-G model is somewhat limited for dry oxidation of c-Si. Firstly, it requires an initial oxide thickness of xi = 200 A in the formulae above in order to successfully fit the measured reaction rates, thereby denying the application of the model in its present form to ultrathin layers. The successful use of the correction term, however, led to misleading references to 'abnormally rapid oxidation' for the early stages of growth. Since the oxides used commercially in semiconductor devices at that time were typically 1000 A thick, it was convenient to assume this essentially instantaneous oxide growth, and the effect was not extensively studied. Today however, it is clear that oxides of 200 A and less will be required to enable the drive towards ever-diminishing geometries to continue and it is becoming more crucial to understand the optimum criteria for producing high quality thin oxide layers.

Many models have been proposed to attempt to explain the oxidation kinetics of this regime. Some of these are significantly different from the original D-G model, as shown in TABLE 1. Initially, Deal and Grove proposed that space charge effects may be important during the oxidation of films up to about 150 A [3]. Other ideas based on simple modifications of the original D-G model include the incorporating of changes in diffusion resulting from stress [4-6] or different diffusion mechanisms, due for example to micropores or microchannels [6,7] or the presence of additional oxidising species whose influence is affected during growth [8-13,22-23].

The influence of charged species has been continually suggested over the years. Grove [8] suggested field effects arising from the creation of (O2)- and a hole could assist the initial reaction, since the hole could diffuse faster and then pull the slower O ion through the layer. Other charged species, such as (O2)-- [21,22] and O- [22,23] have been considered. Alternatively, Ghez and van der Meulen [12] and Blanc [13] suggest that an equilibrium dissociation of O2 and 2-O may exist at the reacting interface and introduce the required growth characteristics. There is evidence that the activation energy of the reaction may not be described by a single Arrhenius process after all, but may be a combination of several parallel or serial mechanisms [14-16]. Recent low temperature studies indicate that although the previous activation energies appear to be valid for temperatures above 1000 deg C, below this B appears to increase, while B/A decreases. Furthermore the pressure dependence of B/A is not linear [16-18] and varies with temperature and crystal orientation [19], again indicating the possibility of a more complicated multistep reaction.

It appears that at least two important factors must be included in any further analysis. The first involves the contribution of compressive stress built into the growing oxide layer at the Si interface during the transformation of the regular Si structure into the amorphous oxide layer. It must be remembered that each unit of SiO2 occupies around twice the volume of the reacting silicon. Viscous flow above 970 deg C may relax this stress [20]. Upon cooling, further strain could develop because of the mismatch of the thermal coefficients of the Si and its oxide, but this should not affect any previous single stage oxidation. In a recent work shop on the oxidation of silicon [25] it is argued that the observed strains are orders of magnitude smaller than one might expect from the differences in volume, and Mott [26] has suggested that while these may not be important, the bonding misfit between the Si and the SiO2 can play an important role. Any structural changes, as well as the inbuilt strain, could influence not only the transport of the oxidant, but also, indirectly, the reaction mechanism.

Murali and Murarka [27] claim to explain 'all phenomena observed in silicon oxidation', by invoking an oxygen-rich phase in the Si, arising from a build-up of unreacted oxygen during the initial stages of the reaction, which enhances the proceeding reaction. However, once an oxide layer of sufficient thickness has grown, it will more strongly resist diffusion, so that any oxygen reaching the interface will immediately react, rather than accumulate in the bulk. Two reaction

controlled linear regimes and a diffusion controlled parabolic regime
have been proposed, but the precise kinetics of the linear regimes have
yet to be formulated. The interface oxygen concentration, the presence
of positive surface charge states or changes in diffusion species may
be responsible.

It is also known that different types of surface pre-cleaning
treatments can dramatically alter the subsequent reaction rates as well
as the physical properties of the grown layers. The residual chemical
species can clearly accelerate or retard the interface reaction and
therefore play a dominant role during the early stages of the reaction,
as well as severely affect, less directly, the later stages of the
reaction. In particular, the presence of Cl or F, or a heavily doped
substrate can have a very significant effect on the reaction.

It is clear that the presence of many oxidation models is a consequence
of a severe lack of reliable experimental data. The required
experiments are by no means straightforward to perform, and the field
must wait until the measurements are performed before the true
oxidation model is unanimously acclaimed.

TABLE 1: summary of silicon oxidation models

Description of silicon oxidation model	Ref
Diffusion of O2 & reaction with Si + space charge effect & tunnelling/thermionic emission	Deal & Grove [3,8]
Enhanced diffusion of O2 through micropores	Revesz & Evans [7] Irene [6]
Stress-induced diffusion effects	Doremus [5], Fargiex & Ghibaudo [4]
Creation of (O2)-/hole pair, leading to enhanced diffusion of (O2)-	Grove [8], Tiller [21]
Influence of (O2)-- and O-	Lora-Tamayo et al [22]
Influence of O-	Hu [23]
Fixed charge in oxide assists diffusion of charged species (O2)-, and perhaps O-	Lu & Cheng [10]
Fixed charge in oxide reduces available Si bonds by reducing holes with band-bending	Schafer & Lyon [11]
Both O2 and 2-O react with silicon	Ghez & Van der Meulen [12]
Equilibrium between O2 and 2-O, only O reacts with Si for thin oxides	Blanc [13]
Two-step oxidation: Si -> alpha-SiO2 + Si(I) & Si(I) -> SiO2	Tiller [24]
Creation of an extensive oxygen-rich zone in the Si near the Si-SiO2 interface during thin oxide growth, due to rapid diffusion	Murali & Murarka [27]

REFERENCES
----

[1]   J.C.Mikkelsen,Jr [Appl. Phys. Lett. (USA) vol.45 no.11
        (1 Dec 1984) p.1187-9]; and references therein, E.Rosencher,
        A.Straboni, S.Rigo, G.Amsel [Appl. Phys. Lett. (USA) vol.34 (1979)
        p.254]; F.Rochet, B.Agius, S.Rigo [J. Electrochem. Soc. (USA)
        vol.131 (1984) p.914]; N.Mott [Philos. Mag. A (GB) vol.45 (1982)
        p.323]
[2]   D.N.Modlin, W.A.Tiller [J. Electrochem. Soc. (USA) vol.132 no.7
        (Jul 1985) p.1659-63]
[3]   B.E.Deal, A.S.Grove [J. Appl. Phys. (USA) vol.36 (1965) p.3770]
[4]   A.Fargeix, G.Ghibaudo [J. Appl. Phys. (USA) vol.54 no.12 (Dec
        1983) p.7153-8]
[5]   R.H.Doremus [Thin Solid Films (Switzerland) vol.122 (1984)
        p.191]
[6]   E.A.Irene [J. Appl. Phys. (USA) vol.54 (1983) p.5416]
[7]   A.G.Revesz, R.J.Evans [J. Phys. & Chem. Solids (GB) vol.30 (1969)
        p.551]
[8]   A.S.Grove [Physics and Technology of Semiconductor Devices
        (Wiley, New York, USA, 1967)]
[9]   M.Hamasaki [Solid State Electron. (GB) vol.25 (1982) p.479]
[10]   Y.Z.Lu, Y.C.Cheng [J. Appl. Phys. (USA) vol.56 no.6 (15 Sep 1984)
        p.1608-12]
[11]   S.A.Schafer, S.A.Lyon [Appl. Phys. Lett. (USA) vol.47 no.2
        (15 July 1985) p.154-6]
[12]   R.Ghez, Y.J.van der Meulen [J. Electrochem. Soc. (USA) vol.119
        (1972) p.1100]
[13]   J.Blanc [Appl. Phys. Lett. (USA) vol.33 (1978) p.424]
[14]   E.A.Irene, D.W.Dong [J. Electrochem. Soc. (USA) vol.125 no.7
        (Jul 1978) p.1146-51]
[15]   R.Razouk, L.N.Lie, B.E.Deal [J. Electrochem. Soc. (USA) vol.128
        no.10 (Oct 1981) p.2214-20]
[16]   H.Z.Massoud, J.D.Plummer, E.A.Irene [J. Electrochem. Soc. (USA)
        vol.132 no.7 (Jul 1985) p.1745-53]
[17]   T.Smith, A.J.Carlan [J. Appl. Phys. (USA) vol.43 (1972) p.2455]
[18]   M.A.Hopper, R.A.Clarke, L.Young [J. Electrochem. Soc. (USA)
        vol.122 no.9 (Sep 1975) p.1216-22]
[19]   Y.J.van der Meulen [J. Electrochem. Soc. (USA) vol.119 (1972) p.580]
[20]   E.P.Eernisse [Appl. Phys. Lett. (USA) vol.35 no.1 (1979) p.8]
[21]   W.A.Tiller [J. Electrochem. Soc. (USA) vol.119 (1980) p.591]
[22]   A.Lora-Tamayo, E.Dominguez, E.Lora-Tamayo, J. Llabres [Appl.
        Phys. (Germany) vol.17 (1978) p.79]
[23]   S.M.Hu [Appl. Phys. Lett. (USA) vol.42 (1983) p.672]
[24]   W.A.Tiller [J. Electrochem. Soc. (USA) vol.128 no.3 (Mar 1981)
        p.689-91]
[25]   Proc. Paris Workshop on Oxidation of Silicon, May 1986,
        1986 Philos. Mag. B (GB) to be published]
[26]   N.Mott [in ref 25]
[27]   V.Murali, S.P.Murarka [J. Appl. Phys. (USA) vol.60 no.6 (15 Sept
        1986) p.2106-14]

CHAPTER 17

INSULATING LAYERS ON SILICON SUBSTRATES

(continued

17.1 INTERFACE TRAPS ON Si SURFACES

by C.T.Sah

November 1987
EMIS Datareview RN=16170

A. INTRODUCTION

Due to the tremendous amount of literature and data, the review on the
electronic properties of interface traps on silicon surfaces and
interfaces is split up into several Datareviews [1-3]. This Datareview
provides historical and physical backgrounds and a discussion of the
terminology that will be followed in [1-3]. Early work on interface
traps on 'clean' silicon surfaces covered with little or no oxide film
will also be reviewed.

B. BACKGROUND AND CLASSIFICATION

This introductory section attempts to unify the conflicting terminology
that has been used by scientists and engineers.

The spatial distributions (or wave functions) and the energies (or
the states) of the electrons (and holes) near and on a clean surface
of a chemically pure and otherwise physically perfect crystalline solid
in vacuum differ from those in the bulk due to the interruption of the
periodic potential at the surface. Near and on the surface, some of the
wave functions are not localized while others are localized or bound
with discrete energies. In contrast, wave functions in the bulk are all
spread out over the entire crystal (except near the surface) and the
electron energies represented by these wave functions will group into
bands (allowed energy bands) separated by gaps (forbidden energy
gaps). The surface electronic (electron and hole) states are further
modified by the presence of foreign atoms on the solid surface, from
its contact with a gas, a liquid or another solid. The electronic
states on a clean surface in vacuum are known as surface states while
those on a surface in contact with another material are known as
interface states. They may be intrinsic surface or interface states
(or traps) owing to the displacement of the host atoms of the two
materials making the contact or forming the interface, or they may be
extrinsic due to presence of foreign atoms or molecules.

Wave functions of bulk states of a perfect crystalline solid are not
bound or localized and are given by the Bloch functions
$u(x,y,z) \exp[i(k(x)x+k(y)y+k(z)z]$ where $u(x,y,z)$ is cell periodic
(or repeats from one unit cell to the next). However, the localized

surface and interface states can be bound in one, two or three dimensions
(1-d, 2-d and 3-d) having wave functions of the form:

$$F(x) \ u(x,y,z) \ \exp[i(k(x)x+k(y)y+k(z)z)]$$

$$F(x,y) \ u(x,y,z) \ \exp[i(k(x)x+k(y)y+k(z)z))]$$

$$F(x,y,z) \ u(x,y,z) \ \exp[i(k(x)x+k(y)y+k(z)z)]$$

respectively. The envelope functions, F's, are bound in 1-d, 2-d or 3-d.
For example

$$F(x) \quad = \exp \ [-K(x) \ \text{mod} \ x]$$

$$F(x,y) \quad = \exp \ [-K(x) \ \text{mod} \ x - K(y) \ \text{mod} \ y]$$

$$F(x,y,z) = \exp \ [-K(x) \ \text{mod} \ x - K(y) \ \text{mod} \ y - K(z) \ \text{mod} \ z]$$

There has been frequent confusion in terminology and meaning when
authors use the general term, surface state or interface state, without
specifying the dimensionality. This can be avoided by using the terms
such as 1-d bound states in a thin surface conduction channel or thin
film, 2-d bound states in a small diameter wire or thin-and-narrow or
thin-and-short sheet, 3-d bound states at an imperfection centre whose
geometry is a point or one atomic core or a finite-size (in 3-d)
cluster. The term trap will be used throughout for the 3-d bound states
which emphasizes its ability to capture an electron from the conduction
band or hole from the valence band and to emit or release an electron
or a hole which is bound or trapped at an imperfection centre. It is
evident that only 3-d bound states or traps are truly localized in
space which is the subject of this section while 1-d and 2-d bound
states are respectively 2-d and 1-d conduction states and hence are not
localized. The nonlocalized 2-d and 1-d bound states are fundamentally
and practically important, for example, the 1-d bound states or 2-d
conduction states account for the conduction of electrons or holes in
the thin surface inversion or accumulation channels of metal-oxide-
silicon (MOS) field-effect-transistor (FET) which is reviewed in [4,5].

The traps or 3-d bound states in the oxide layer, known as oxide
traps, are reviewed in [6]. The distinction between the oxide traps
and interface traps is not sharp since the interfacial or transition
layer from a crystalline lattice to a non-crystalline lattice may not
occur abruptly or in one atomic layer: for example, the interfacial
layer between the crystalline silicon (c-Si) and the polycrystalline-
granular (pc-SiO2) or amorphous-continuous-random (a-SiO2) silicon
oxide. Instead the interfacial layer may extend over several atomic
layers or lattice constants with a spatially varying average chemical
composition, SiO(x) where x = 0 on the c-Si side and x = 2 on the SiO2
side of the interfacial layer. Such a variable composition is necessary
to release the strain due to lattice mismatch. It is obvious that the
the spatial variation of the chemical bonds or atomic arrangement
of the host atoms in this transition layer will be highly dependent on
the orientation of the substrate (c-Si) and the oxide growth methods
and conditions such as the surface cleaning procedure (chemical
etches, atomic-ionic bombardments, etc), the pre-oxidation heat-up
cycle and ambient; the oxidation ambient, temperature and time; and the

post-oxidation cool-down cycle. A detailed review of the morphology of the interface is given in [7]. To separate the interface and oxide traps into two groups to be discussed in this review and [6], an operational definition will be used. The interface trap will be defined as one which can rapidly capture an electron or a hole from the silicon conduction or valence bands (CB or VB) and can also rapidly emit the trapped electron or hole to the Si CB or VB. An oxide trap will be defined as a centre whose trapped electron or hole density does not change within a finite measurement time in a non-injection thermal experiment, (for example in a few hours during transient or 1 Hz small-signal admittance measurements of the interface traps). This operational definition is thus dependent on the persistence and longevity of the experimenter as well as the stability of the MOSC system controlled by the other traps and causes.

This subsection will briefly review the data of the 3-d bound states or traps on clean silicon surfaces; but an intensive review will be made of silicon surfaces covered by a layer of thermally grown oxide, namely, the interface traps at the oxide/silicon interface of oxidized crystalline silicon surfaces in [1-3]. The emphasis on the thermally grown oxide/silicon interface is made in this review since it is one of the most important solid/solid interfaces in electronic applications (diodes, transistors, integrated circuits and integrated optics). The interface trap properties in other forms of oxides on c-Si are not included since they have not been investigated and characterized in sufficient detail to allow fundamental description with confidence and to give reliable electrical characterization data. However, it is well known that the oxide/Si interfaces of these non-thermally grown oxides (such as anodic, chemical vapour deposited, evaporated and other non-thermal processes) have very high densities of interface traps, approaching one dangling bond per silicon surface atom (approx. 10*14 states/cm*2). It is also well known that their densities and atomic configurations or electronic transition rates are highly unstable on shelves without the presence of thermal or electric field stresses.

Traps on the clean (uncovered) silicon surface (known as surface states in the literature) will be briefly reviewed only as an introduction to traps on oxidized silicon since the atomically clean surface is the limiting form of an oxidized surface aside from atomic rearrangments (reconstruction) on the clean surface.

The interface traps at the SiO2/Si interface can be categorized into two groups, the intrinsic traps and the extrinsic traps. The intrinsic traps are those due to extra, missing or displaced host atoms (Si or O). The extrinsic traps are those due to foreign impurities, including hydrogen. Extrinsic interface traps due to foreign impurities have not been investigated except in two cases, Na and Au, the latter masked by the bulk Au trapping levels in the silicon surface space charge layer next to the interface.

The areal density and the charging and generation-annealing kinetics of the interface traps at the SiO2/Si interface will be covered in this review. The density notations to be used here are those defined by the Deal Committee [8], which is modified to take into account the limited character set-available, for example, a(A)(x,t) = A(A)(x)+a(a)(x,t). Here a(A)(x,t) is the total value at a position x at time t, A(A)(x) is the time-averaged value when it exists and a(a)(x,t) is the time dependent transient with zero time-average. For a sinusoidal steady-state then a(a)(x,t) = A(a)(x)exp(jwt) where A(a)(x) is the peak or rms value.

Repeated upper case double subscript from the IEEE convention will be used to denote the total concentration. For example, the total concentration may consist of charged and uncharged centres of one trap or impurity species. For a bulk trap, N(TT) is used to denote the volume density or concentration of one species of electrically active trap and it is given by N(TT)(x) = N(T)(x)+P(T)(x) = n(T)(x,t)+p(T)(x,t). For an electron trap, N(T) is the density of the traps each occupied by a trapped electron and P(T) is that of the traps not occupied by a trapped electron. A similar notation can be used for a hole trap. The symbol, a(A)(x,t), and its components are macroscopic variables (such as particle density or electric potential) which are physically meaningful only in region from x to x+dx (or a volume element enclosed by x,y,z to x+dx,y+dy,z+dz) that contains a large number of particles or transition events (scattering, capture, emission, generation, recombination events or transitions) so that the 'number fluctuation' is small, say 1%, which would require a minimum volume containing more than 10*4 particles or quantum mechanical transition events.

The term 'oxide fixed charges' introduced by the Deal Committee will be discarded since it has been known that oxide charges are not fixed and they can be changed by five fundamental mechanisms: (i) diffusion; (ii) drift; (iii) bond breaking by energetic electrons or holes or by thermal hole capture; (iv) charging-discharging via electron or hole capture or emission; and (v) electrical deactivation or activation (absence or presence of electronic bound states) via reaction with hydrogen. These instabilities are commonly observed not only on oxide traps but also interface traps even in the most stable silicon integrated-circuit transistors. Notations were introduced by Sah [9] for the density of the oxide traps and various macroscopic transition probability rates at the oxide traps which followed the universally adopted notations introduced by Shockley and Read for bulk traps. These transition or 'reaction' rates were not considered by the Deal Committee.

Thus, D(IT)[E(T)] (/cm*2 eV) is the areal and energy density of interface trap (subscripts I and T) in the energy range E(T) to E(T)+dE(T) in the silicon energy gap. N(IT)(/cm*2) is the areal density of the interface traps in the energy range E(T1) to E(T2). The density of the trapped charges at the interface traps will be denoted by Q(IT) = qN(IT)(C/cm*2). The four charging rate coefficients are denoted by c(n) and c(p) (in units of cm*3/s) for the capture of an electron or a hole by the trap and by e(n) and e(p) (in /cm*3) for the emission of a trapped electron or trapped hole. The chemical reaction rates of the traps with other atomic species are denoted by c(T) and e(T) where T denoted the other species, for example, hydrogen. The capture rate, c(T), is the rate at which the other atomic or chemical species (for example hydrogen) is captured by the trap while e(T) is the release of the chemical species from the trap. This notation and the concept of trapping an impurity or foreign chemical species follows that of trapping electrons and holes. They are particularly powerful in notation simplification while keeping the physics in focus. For example, e(H)*n would be the emission rate of the hydrogen from the trap by energetic electron impact. The recombination rates of two identical traps or impurities are denoted by g(T) and r(T) where T denotes a trap or related species (such as hydrogen, then T is replaced by H or H2). This notation was first introduced by Sah for analyzing the complex

hydrogenation kinetics of shallow dopant acceptors in silicon in which
the concept of protonic traps (analogous to electronic traps) was
introduced (see [10]). The annealing or annihilation rate of the
interface (as well as oxide) traps is generally a combination of the
capture of an impurity species such as hydrogen and the recombination
of two traps or two parts of a complex trap so that a unique symbol is
not needed. Characterization of these kinetic rate coefficients is still
in its infancy and there is a scarcity of consensus data. In discussing
the generation and annealing kinetics, the concentration of the
participating elements will be denoted by their chemical symbol without
the square bracket (for examples, H and H2, instead of [H] and [H2])
to simplify notation.

The density of states of surface and interface traps, D(IT) has been
observed universally to have two types of dependence on the energy in
the gap. One has a structureless U-shape D(IT) and the other, a peak
with a width of the order of 0.1 eV. For the oxide/silicon interface
traps, one or more peaks have been observed to superimpose onto a
U-shaped component that increases towards the two band edges. The
U-shaped component indicates the presence of some imperfections
(intrinsic or extrinsic) whose atomic potential covers a wide range of
magnitude so that the conduction and valence band states are perturbed
by various amounts and shifted into the energy gap at various depths.
Both Si dangling bonds with different atomic environment, and bonded Si
and O with random variations of the Si-Si and Si-O bond lengths and
angles, could give sufficiently large perturbations to localize the
conduction and valence band states. The peaked density of states is
usually associated with a high density of a dangling bond species whose
atomic environment surrounding each isolated centre is quite similar. The
oxygen dangling bond may be a suitable species although silicon dangling
bond has not been ruled out. The peaked density of states can also be
associated with an impurity, however, due to the large potential barrier
between the oxide and silicon (about 3.1 eV from the silicon conduction
band edge), s-like envelope functions are not allowed. Thus, the most
tightly bound state would be p-like with 1/4 or smaller binding energy
compared with the s-state in the Si bulk. This would result in gap
energy levels near the silicon conduction or valence band edges at the
SiO2/Si interface. The band-edge gap levels may have escaped detection
since the conventional MOS capacitor technique and most other techniques
can accurately probe only the middle half of the energy gap and cannot
give reliable results for traps with energy levels closer than about
0.2 eV from the band edge. Traps due to missing host atoms could also
have highly directed dangling bond orbitals pointing into the cavity
left by the missing host atom.

C. HISTORICAL SUMMARY

The study of electron states on solid surfaces began with Tamm [11]. Tamm
interrupted the ideal one-dimensional Kronig-Penny (periodic square
well) potential by a finite potential step at the edge of the leftmost
square well and found the existence of localized or bound states when the
perturbation or potential step is large. Shockley [12] extended this

model in 1939 by moving the potential step to the centre of the peak of the Kronig-Penny potential well and showed that surface bound states will appear if the potential step is small provided two adjacent energy levels of an isolated well are crossed as the lattice spacing is reduced. These are known as the Tamm and Shockley states. The presence and absence of a high density of surface states on compound and elemental semiconductors had been frequently interpreted using the Shockley surface state model. Tamm's theory was improved and extended by Maue and Goodman before Shockley's 1939 work. A major push of research on semiconductor surface states began when Bardeen [13] successfully interpreted the Schottky barrier height's independence of the metal type, experimentally observed on point contact microwave diode detectors by Mayerhoff [14] and previous researchers. Shockley and Pearson [15] similarly interpreted the smaller-than-expected surface conductance modulation in a silicon field-effect experiment based on the presence of a high density of surface states. Application of the thermally grown oxide to stabilize and passivate silicon transistors was first demonstrated by Atalla in 1959 [16], who with Kahn demonstrated the first modern silicon metal-oxide-semiconductor field-effect transistor (MOSFET) operation in 1960 [17]. Hoerni [18] then made use of the diffusion masking as well as passivation properties of the thermally grown oxide on silicon to manufacture small geometry (down to 5 micron line width in 1960) silicon diodes and transistors and this oxide masking technique was later used by Noyce [19] to manufacture integrated circuits. These started an intense effort to investigate the properties of the oxidized silicon and its interfaces which has continued for the last twenty-five years.

The early quantum theory and experiments on surface states, up to 1970, on clean surfaces were reviewed in depth by Davison and Levine [20]. Surface chemistry up to 1977 was discussed by Morrison [21]. Status reports of the current understanding of the properties of real silicon surfaces covered by silicon oxide were given in two summaries by Deal in 1974 [22] and 1977 [23]. Some of the understanding and concepts have been updated and revised by later work reported at the two silicon dioxide-silicon interface physics conferences in 1978 and 1979 [24,25]. A summary up to 1982 concerning the mechanisms of oxidation of silicon which determines the interface and oxide trap densities was given by Katz in [26]. A most critical review of the methods of measurements of the interface and oxide traps and the research literature of these on oxidized silicon was given by Nicollain and Brews in a book in 1982 [27] which has formed a basis for the re-interpretation of some of the earlier data on the interface traps. Research on the microscopic nature of the interface traps based on electron spin resonance experiments was reviewed by Poindexter and Caplan in 1983 [28] who made the major breakthroughs in experiments that formed the basis of our current understanding of the microscopic configurations of the interface traps on oxidized silicon. Electrical data on the interface and oxide traps of oxidized silicon, pertinent to silicon integrated circuit design and manufacturing (such as the generation and annealing rates of the traps and their process and crystallographic orientation dependences and others) were tabulated and graphed by Goodwin and Meyer in a handbook in 1985 [29].

D. SURFACE STATES (TRAPS) ON CLEAN SURFACES

A fraction of the intrinsic interface states (traps) on oxidized
electronic (VLSI) grade crystalline silicon may come from the remnant
surface states on the clean surface when some of the silicon dangling
bonds on the clean surface are not tied up by oxygen during thermal
oxidation. The experiments on clean silicon surfaces prepared in
ultra-high vacuum (UHV) and the quantum theory (mainly one
dimensional) up to 1970 were reviewed in detailed by Davison and
Levine [20]. Handler [30] showed that the surface state density of a
clean germanium surface approached the density of the surface
dangling bonds. Using Handler's model, the dangling bond densities on
the (100), (110) and (111) silicon surfaces are given in TABLE 1 where
the edge of the cubic cell, a, has a value of 5.430 A.

TABLE 1: atomic and cut-bond densities on an un-reconstruced Si surface
         (300 deg K)

Face	2-D cell shape	One Cell has area	atom	cut-bond	Areal Density atom	cut-bond	Comment
		a*2			10*14/cm*2	10*14/cm*2	
(100)	square	1	2	4	6.780	13.56	Peroxide bond Si-O-O-Si
(100)	square	1	2	2	6.780	6.780	Monoxide bond Si-O-Si
(110)	rectangle	SQRT 2	4	4	9.589	9.589	Exclude bonds in plane
(110)	rectangle	SQRT 2	4	8	9.589	19.178	Include bonds in plane
(111)	diamond	SQRT(3/4)	1	1	7.832	7.832	Face A
(111)	diamond	SQRT(3/4)	1	3	7.832	23.498	Face B
(111)	diamond	SQRT(3/4)	1	2	7.832	15.665	Average (A+B)/2

Allen and Gobeli [31] measured the intrinsic surface trap density as a
function of energy in the silicon energy gap due to silicon dangling
bonds on clean (111) surfaces of silicon. They changed the bulk Fermi
level by varying the phosphorus donor or boron acceptor density in order
to probe the energy gap. The experimental trapped charge density was given
by $Q(IS)[V(S)] = -2.44 \times 10*12 \sinh(U(S)+9.7)$ states/cm*2 where the normalized
surface potential, $U(S)=(Ef-Ei)/kT$, is the position of the Femic level at
the surface measured from its intrinsic position at the surface. This
result also gives the neutral Fermi level position at $Efn=9kT$ or 220 mV
below the Si midgap, Emg. When the Fermi level at the surface is at the
neutral position, the charges residing on all the surfaces states add up
to zero. The neutral Fermi level, denoted by Efn, first used by Bardeen

to explain the independence of Schottky barrier height on the metal type, was also used by Bardeen and Brattain on germanium surface experiments [20]. The energy density of the interface states that gives such a sinh distribution can be expressed either by a U-shaped distribution or by two energy levels one in the upper and one in the lower half of the Si energy gap. The data of Allen and Gobeli gave a U-shaped distribution of about 2 X 6kT = 0.3 eV wide or it could also be fitted to two square-shape energy level distributions separated by less than 0.3 eV, each with a density less than or equal to the (111)Si bond density (about 7.8 X 10*14/cm*2 from TABLE 1). The energy density of states would be 7.8 X 10*15/cm*2 eV if each of the two distributions is 0.1 eV wide. Extensive experimental data on thermally oxidized silicon using the MOSC method have shown the existence of the following four cases all with a U-shaped background: (i) no peak, (ii) one peak at Emg-0.3 eV; (iii) one peak at Emg+0.3 eV, and (iv) two peaks roughly 0.5-0.7 eV apart symmetrically located about the silicon midgap Emg. The recently determined U-shaped energy density of states extends closer to the band edges than the older clean surface data. Thus, a U-shaped energy density of state is probably representative of remnant and highly random Si dangling bonds from the clean Si surface. The data of Allen and Gobeli [31] can be fitted to N(IT) = 9.5 X 10*13 cosh(U(S)+9.7) states/(cm*2 eV) to be evaluated at room temperature where kT/q = 25.85 mV. This could be used as the limiting case of a highly damaged or mismatched SiO2/Si created by severe oxidation conditions or exposure of the oxidized silicon to an intense and high dosage of ionizing radiation which breaks up a large number of the Si-Si and Si-O bonds in the SiO2/Si interfacial layer.

REFERENCES
----

[1]   C.T.Sah [EMIS Datareview RN=16180 (Nov 1987) 'Interface traps on oxidized Si from electron spin resonance']

[2]   C.T.Sah [EMIS Datareview RN=16181 (Nov 1987) 'Interface traps on oxidized Si from X-ray photoemission spectroscopy, MOS diode admittance, MOS transistor and photo-generation measurements']

[3]   C.T.Sah, M.S.C.Luo, C.C.H.Hsu, T.Nishida, A.J.Chen, C.Y.P.Chao [EMIS Datareview RN=16182 (Nov 1987) 'Interface traps on oxidized Si from two-terminal dark capacitance-voltage measurements on MOS capacitors']

[4]   C.T.Sah, T.Nishida [EMIS Datareview RN=17831 (Nov 1987) 'Electron mobility in oxidized Si surface layers']

[5]   C.T.Sah, T.Nishida [EMIS Datareview RN=17813 (Nov 1987) 'Hole mobility oxidized Si surface layers']

[6]   C.T.Sah, C.C.H.Hsu [EMIS Datareview RN=16185 (Nov 1987) 'Oxide traps on oxidized silicon']

[7]   S.M.Goodnick [EMIS Datareview RN=15725 (Oct 1987) 'Structure of the interface in oxide-on-Si structures']

[8]   B.E.Deal [J. Electrochem. Soc. (USA) vol.127 no.4 (Apr 1980) p.979-81]; [IEEE Trans. Electron Devices vol.ED-27 no.3 (USA) (Mar 1980) p.606-8]

[9]   C.T.Sah [EMIS Datareview RN=17815 (Nov 1987) 'Generation-recombination-trapping rates and lifetimes of electrons and holes in SiO2 films on Si - a general classification of the fundamental mechanisms']

[10]   C.T.Sah [EMIS Datareview RN=17845 (Nov 1987) 'Hydrogenation dehydrogenation rates of shallow acceptors and donors in Si: fundamental phenomena and survey of the literature']

[11] Ig. Tamm ['Uber eine mogliche art der elektronenbrindung an
      kristalloberflachen' Physik. Zeits. Sowjet Union (USSR) vol.1
      (1932) p.733; Z. Phys. (Germany) vol.76 (1932) p.849-50]
[12] W.Shockley [Phys. Rev. (USA) vol.56 (15 Aug 1939) p.317]
[13] J.Bardeen [Phys. Rev. (USA) vol.71 (15 May 1947) p.717]
[14] W.E.Mayerhoff [Phys. Rev. (USA) vol.71 (15 May 1947) p.727]
[15] W.Shockley, G.L.Pearson [Phys. Rev. (USA) vol.74 (15 July 1948)
      p.232-3]
[16] M.M.Atalla, E.Tannenbaum, E.J.Scheibner [Bell Syst. Tech. J.
      (USA) vol.38 (May 1959) p.749]; M.M.Atalla [in
      'Properties of Elemental and Compound Semiconductors', vol.5
      H.Gatos Ed. (Interscience, New York, USA, 1960) p.163-81]
[17] D.Kahng, M.M.Atalla ['Silicon-silicon dioxide field-induced
      surface devices', IRE-AIEE Solid-State Device Res. Conf., Carnegie
      Inst. Technol., Pittsburg, PA USA, June, 1960]
[18] J.A.Hoerni [Planar silicon transistors and diodes IEEE Electron
      Devices Meeting, Washington, USA DC 27-29 October 1960];
      [Technical Article and Paper Series, No.TP-14 (Fairchild
      Semiconductor Corp. 645 Whisman Rd., Mountain View, CA, USA, 1961)
      p.9]
[19] R.N.Noyce [Semiconductor device-and-lead structure' US Patent
      no.2 2,981,877]; J.S.Kilby [IEEE Trans. Electron Devices
      (USA) vol.ED-23 no.7 (July 1976) p.648-54]
[20] S.G.Davison, J.D.Levine [Solid-State Phys. (USA) vol.25 (1970)
      p.1]
[21] S.R.Morrison ['The Chemical Physics of Surfaces' (Plenum Press,
      New York, USA, 1977)]
[22] B.E.Deal [J. Electrochem. Soc. (USA) vol.121 no.6 (June 1974)
      p.198C-205C]
[23] B.E.Deal [Semiconductor Silicon-1977, Eds H.R.Huff, E.Sirtl
      (Electrochem., Soc., Princeton, NJ, USA) p.276-96]
[24] S.T.Pantelides, Ed ['Phys. of SiO2 and Its Interfaces' (Pergamon
      Press, New York, USA 1978)]
[25] G.Lucovski, S.T.Pantelides, F.L.Galeener ['Phys. of MOS
      Insulators', (Pergamon Press, New York, USA, 1980)]
[26] L.E.Katz [in 'VLSI Technology' ch.4 Ed. S.M.Sze (McGraw-Hill,
      New York, USA, 1983) p.131-69]
[27] E.H.Nicollain, J.R.Brews [MOS Phys. & Technol. (John Wiley,
      New York, USA, 1982)]
[28] E.H.Poindexter, P.J.Caplan [Prog. Surf. Sci. (GB) vol.14 (1983)
      p.201]:
[29] C.A.Goodwin, W.G.Meyer [in 'Quick Ref. Manual for Silicon
      Integrated Circuit Technol. Eds W.E.Beadle, J.C.C.Tsai,
      R.D.Plumber (John Wiley & Sons, New York, USA, 1985) ch.14]
[30] P.Handler [in 'Semicond. Surf. Phys.' Ed. R.H.Kingston (Univ.
      of Pennsylvania Press, Philadelphia, USA, 1957) p.23-51];
      S.G.Davison, J.D.Levine [ref.11 p.92-102] and Katz [ref 11
      Table 5 p.144 which has typographical errors]
[31] F.G.Allen, G.W.Gobeli [Phys. Rev. (USA) vol.127 no.1 (1962)
      p.150-8] For a comparison with III-V semiconductors, see
      G.W.Gobeli, F.G.Allen, [Phys. Rev. (USA) vol.137 (1965)
      p.A245]

17.2 INTERFACE TRAPS ON OXIDIZED Si FROM ELECTRON SPIN RESONANCE

by C.T.Sah

November 1987
EMIS Datareview RN=16180

Electron paramagnetic or spin resonance (ESR) of trapped electrons is the only viable experimental method than can provide sufficiently detailed information to determine the atomic composition of the trap. It has two requirements which limit its utility to some extent, an unpaired spin and a high density of the spin. Digital averaging of repeated scans has made it possible to detect spin densities down to about 10*12/cm*2 on samples with surface area about 1 cm*2.

ESR measurements of the oxide/silicon interface states up to 1982 were reviewed in detail and physical depth by Poindexter and Caplan [1]. The first successful ESR measurement on oxidized silicon was reported by Nishi [2] in 1971. Our current understanding of the microscopic properties of the intrinsic interface trap detected by ESR and its correlation with the D(IT) measured by the capacitance-voltage methods came mainly from the extensive experiments of Poindexter and associates at Fort Monmount [3-7], Lenahan and Dressendorf at Sandia [8-11], and Warren and Lenahan at Penn State [12]. The principal interfacial defect identified is the P(b) centre which was recently designated as P(b0) when the P(b1) centre was isolated by Poindexter [6,7]. The P(b) label was originally designated by Nishi (see table 1 on p.221 of Poindexter-Caplan's review [1]) and the other paramagnetic centres, P(a), P(c) and P(d) centres were shown to come from Si-bulk traps. From detailed analysis of the ESR spectra, Poindexter et al showed that the P(b0) interfacial trap is the trivalent silicon dangling bond, Si3:::Si. , pointing into the oxide at the SiO2/Si interface on a (111)Si surface or at an angle on (110) and (111)Si surfaces. This SiO2-pointing Si-orbital direction on a (111) interface may at first glance be surprising in view of the 3 eV oxide potential barrier or the large electronegativity and negative charge on the oxygen which should direct the bound electron wave function or orbital towards the Si lattice instead of the oxide film. However, pointing into the 'cavity' on the oxide side may well be the preference due to the space or cavity left by the missing oxygen atom next to the silicon dangling bond. Recent measurements by Stesmans [13] on specially grown 46 A oxide to give high density of interface traps (about 4 X 10*12/cm*2) have shown the existence of P(b0) dangling bonds along the [1 bar1 1] [bar1 1 1] and [1 1 bar1] directions making an angle of 19.5 deg with the (111) SiO2/Si interfacial plane with four-times smaller density (1 X 10*12/cm*2). The less prominent or lower density P(b1) centre has been tentatively attributed to the monovalent silicon dangling bond Si2O:::Si. by Poindexter [7]. The two centres were separated [7] by resolving the experimental twin-peaked D(IT) into two sets of twin-peaked D(IT), one for P(b0) and the other for P(b1). However, experimental twin-peaks are near the band edges and were obtained by the high-frequency/low-frequency capacitance-voltage method and could arise from analysis error instead of real peaks [14]. Further concern on this uncertainty comes from the D(IT) data reported by all other researchers using the capacitance-voltage method which data have always showed a featureless U-shaped D(IT) without peaks. Occasionally, one peak above or below the Si midgap, or two peaks of different areas or amplitudes, were observed when the MOS capacitor underwent special processing or stressing cycles. Notwithstanding this uncertainty, there is little doubt that the U-shaped component of the D(IT) observed by Poindexter and other researchers is associated with or originates from the trivalent silicon dangling bonds, Si3:::Si. .

A more detailed model containing both random and constant dangling bonds was considered previously [15-19] which could account for the U-shaped D(IT) as well as the peaks. This model consists of two components, a completely random distribution and a non-random or nearly constant distribution of the neighbouring environment of one or more of the four centres Si(x)O(3-x):::Si. (where x = 0, 1, 2 or 3) at the SiO2/Si interface. The centres with the random environment (such as the random variations of the nearest and second nearest bond angles and lengths) would give rise to a U-shaped distribution owing to the random perturbation potential associated with a random distribution of the environment. A distinct experimental peak would then suggest that one of the four centres of Si(x)O(3-x):::Si. has a more regular or constant environment at all of its sites.

A great deal of data on the effects of oxidation condition, post-oxidation treatment, orientation, substrate conductivity type, dopant density, and electric stresses has been obtained on the midgap D(IT) of P(b) centres by Poindexter and Lenahan and the reader is referred to the original references cited [3-13,20] and the comprehensive review by Poindexter and Caplan [1]. For additional discussion on the ESR results emphasizing the D(IT) peaks observed by Johnson from transient capacitance MOS measurements see ref [34], but note the uncertainties on the existence of the D(IT) peaks just discussed. See also [35] in a discussion on the reliability of these peaks.

The composition, orientation and environment of the P(b) orbital at the SiO2/Si interface has been investigated by Brower of Sandia [21-23] using hyperfine structure and Zeeman resonance in ESR. From the Si-29 hyperfine interaction, he showed that 80% of the unpaired electron density is localized at a substitutional Si on the Si side of the interface and that this orbital is 12% s-like and 88% p(111)-like [21]. Zeeman linewidth analyses by Brower on the oxidized (111)Si/SiO2 interface [22] suggested that the excess line broadening is associated with strain (or random bond angle, 0.5 deg, and length 0.02 A) neighbouring the dangling bond and that [23] the oxidized (111)Si/SiO2 interface is abrupt and atomically flat over distances from 15 to 300 A, which is interrupted by steps of one atomic plane height.

ESR measurements of the interface traps at the SiO2/Si interface have also been reported by Makino and Takahashi [24] and Carlos [25] on traps at the Si/SiO2/Si (substrate) interfaces of a buried layer created by oxygen implantation. They showed that the dominant ESR centre is the P(b0) centre at the interface between the SiO2 precipitates and the regrown single crystal Si film. ESR measurements were also made on oxide films exposed to X-rays by Triplett, Takahashi and Sugano [26]. As expected, the interface trap they found was the P(b) centre. A particularly informative ESR measurement was reported by Holzenkampfer, Stuka and Voget-Grote [27] on SiO(x) (0<x<2) in which they analyzed the line shape in terms of the four Si-dangling bond trap components, Si(4-y)O(y):::Si.

One would intuitively assume that some of the traps appearing in the bulk or film of SiO2 would also appear as interface traps at the SiO2/Si interface. This has not been observed thus far. None of the three major bulk centres in the SiO2 have been positively identified to appear

at the SiO2/Si interface although they are present in the ESR signal
taken on device grade SiO2 film thermally grown on c-Si by Triplett et
al [26] which also contained the signal from the interfacial P(b)' centre.
The three major bulk intrinsic defect centres observed from ESR signals
of natural and synthetic fused silica and quartz and silica fibre as well
as oxides grown thermally on crystalline silicon are the E' centre, the
peroxy radical (PR) and the non-bridging oxygen hole centre (NBOHC).
The E' centre is a bridging oxygen vacancy with the two unbonded silicon
displaced, (SiO)3:::Si.   .Si:::(O-Si)3. The peroxy radical (PR)
consists of one dangling oxygen bond on the second oxygen of the two
oxygens attached to one silicon and the adjacent silicon dangling bond,
Si-O-O.   .Si-O-Si. The non-bridging oxygen hole centre is a dangling
oxygen bond (SiO)3:::Si-O. . These and other traps in the various forms
of SiO2 have been carefully and extensively reviewed by Friebele and
Griscom [28] and by Griscom [29]. More recent theoretical and experimental
developments on the peroxy radical [30] and the bridging oxygen vacancy
(E' centre) have been reported by Fowler, Edwards and Rudra [31]. The
presence of the E' centre [32] and the peroxy radical [33] and their
distributions in the bulk of pure silica optical fibres have been
demonstrated recently by Hanafusa, Hibino and Yamamoto [32,33].

REFERENCES

[1]   E.H.Poindexter, P.J.Caplan [Prog. Surf. Sci. (GB) vol.14 (1983)
        p.201]
[2]   Y.Nishi [Jpn. J. Appl. Phys. (Japan) vol.10 no.1 (1971) p.52-62]
[3]   P.J.Caplan, E.H.Poindexter, B.E.Deal, R.R.Razouk [J. Appl. Phys.
        (USA) vol.50 no.9 (Sep 1979) p.5847-54]
[4]   E.H.Poindexter, P.J.Caplan, B.E.Deal, R.R.Razouk [J. Appl. Phys.
        (USA) vol.52 no.2 (Feb 1981) p.879-84]
[5]   P.J.Caplan, E.H.Poindexter, S.R.Morrison [J. Appl. Phys. (USA)
        vol.53 no.1 (Jan 1982) p.541-5]
[6]   E.H.Poindexter, G.J.Gerardi, M.-E.Rueckel, P.J.Caplan, N.M.Johnson,
        D.K.Biegelsen [J. Appl. Phys. (USA) vol.56 no.10 (15 Nov 1984)
        p.2844-9]
[7]   G.J.Gerardi, E.H.Poindexter, P.J.Caplan, N.M.Johnson [Appl. Phys.
        Lett. (USA) vol.49 no.6 (11 Aug 1986) p.348-50]
[8]   P.M.Lehahan, P.V.Dressendorfer [Appl. Phys. Lett. (USA) vol.41
        no.6 (15 Sep 1982) p.542-4]
[9]   P.M.Lehahan, P.V.Dressendorfer [J. Appl. Phys. (USA) vol.54 no.3
        (Mar 1983) p.1457-60]
[10]   P.M.Lehahan, P.V.Dressendorfer [J. Appl. Phys. (USA) vol.55 no.10
        (15 May 1984) p.3495-9]
[11]   R.E.Mikawa, P.M.Lenahan [Appl. Phys. Lett. (USA) vol.46 no.6 (15
        Mar 1985) p.550-2]
[12]   W.L.Warren, P.M.Lehahan [Appl. Phys. Lett. (USA) vol.49 no.19
        (10 Nov 1986) p.1296-8]
[13]   A.Stesmans [Appl. Phys. Lett. (USA) vol.48 no.15 (14 Apr 1986)
        p.972-4]
[14]   E.H.Nicollain, J.R.Brews [MOS Physics & Technology (John Wiley,
        New York, 1982)]
[15]   C.T.Sah [IEEE Trans. Nucl. Sci. (USA) vol.NS-23 no.6 (Dec 1976)
        p.1563-8]
[16]   C.T.Sah [Tech. Report No.TRS-100 (Semicond. Res. Corp., Research
        Triangle Park, NC USA, (31 Jul 1984) p.78]
[17]   C.T.Sah, J.Y.-C.Sun, J.J.-T.Tzou [J. Appl. Phys. (USA) vol.54
        no.5 (May 1983) p.2547-55]

[18]   C.T.Sah, J.Y.-C.Sun, J.J.-T.Tzou [J. Appl. Phys. (USA) vol.54
        no.10 (Oct 1983) p.5864-79]
[19]   C.T.Sah, J.Y.-C.Sun, J.J.-T.Tzou [J. Appl. Phys. (USA) vol.55
        no.6 (15 Mar 1984) p.1525-45]
[20]   C.Jorgensen, C.Svensson, K.-H.Ryden [J. Appl. Phys. (USA) vol.56
        no.4 (15 Aug 1984) p.1093-6]
[21]   K.L.Brower [Appl. Phys. Lett. (USA) vol.43 no.12 (15 Dec 1983)
        p.1111-13]
[22]   K.L.Brower [Phys. Rev. B (USA) vol.33 no.7 (1 Apr 1986) p.4471-8]
[23]   K.L.Brower, T.J.Headley [Phys. Rev. B (USA) vol.34 no.6 (15 Sep
        1986) p.3610-19]
[24]   W.E.Carlos [Appl. Phys. Lett. (USA) vol.50 no.20 (18 May 1987)
        p.1450-2]
[25]   T.Makino, J.Takahashi [Appl. Phys. Lett. (USA) vol.50 no.5 (2
        Feb 1987) p.267-9]
[26]   B.B.Triplett, T.Takahashi, T.Sugano [Appl. Phys. Lett. (USA)
        vol.50 no.23 (8 Jun 1987) p.1663-5]
[27]   E.Holzenkampher, F.-W.Richter, J.Stuke, U.Voget-Grove [J.
        Non-Cryst. Solids (Netherlands) vol.32 (1979) p.327-38]
[28]   E.J.Friebele, D.L.Griscom [in 'Treatise Material Science &
        Technology Eds M.Tomozawa, R.H.Doremus (Academic Press, New York,
        USA, 1979 p.257-351]
[29]   D.L.Griscom [J. Non-Cryst. Solids (Netherlands) vol.73 (1985)
        p.51]
[30]   A.H.Edwards, W.B.Fowler [Phys. Rev. B (USA) vol.26 no.12 (15 Dec
        1982) p.6649-60]
[31]   J.K.Rudra, W.B.Fowler [Phys. Rev. B (USA) vol.35 no.15 (15 May
        1987) p.8223-30]
[32]   H.Hanafusa, Y.Hibino, F.Yamamoto [J. Appl. Phys. (USA) vol.58
        no.3 (1 Aug 1985) p.1356-61]
[33]   Y.Hibino, H.Hanafusa [J. Appl. Phys. (USA) vol.62 no.4 (15 Aug
        1987) p.1433-6]
[34]   N.M.Johnson, E.H.Poindexter [EMIS Datareview RN=16169 (Nov 1987)
        'Microscopic identity of characteristic gap states at the
        interface in oxide-on-Si structures']
[35]   C.T.Sah [EMIS Datareview RN=16181 (Nov 1987) 'Interface traps
        on oxidized Si from X-ray photoemission spectroscopy, MOS diode
        admittance, MOS transistor and photogeneration measurements]
[36]   C.T.Sah, C.C.H.Hsu [EMIS Datareview RN=16185 (Nov 1987) 'Oxide
        traps on oxidized silicon']

17.3 INTERFACE TRAPS ON OXIDIZED Si FROM X-RAY PHOTOEMISSION SPECTROSCOPY,
MOS DIODE ADMITTANCE, MOS TRANSISTOR AND PHOTOGENERATION MEASUREMENTS

by C.T.Sah

November 1987
EMIS Datareview RN=16181

A. INTRODUCTION

Major reported efforts on the experimental characterization of the
interface traps on oxidized silicon have employed two experimental
techniques: (i) electron spin resonance (ESR) of unpaired electrons
at traps on oxidized silicon samples with very high interface trap
density (N(IT) approximately 10*13/cm*2); (ii) trap related dark
capacitance measurements on MOS capacitors spanning the density range
from the lowest experimentally obtainable (< 10*9/cm*2) to the highest
(> 10*13/cm*2). Trap density is varied by controlling oxidation
processes and by electrical injection and photo or keV electron-impact
generation of hot electrons and holes in the oxide. (See [22] for a
discussion of the various non-fabrication methods of trap generation).
Due to the substantial differences in the oxide film fabrication
processes to give the very high N(IT) for ESR experiments and the very
low N(IT) for VLSI circuit chip production, there is no assurance that
the interface traps are of the same species.

Due to the tremendous amount of literature and data, the review on
the electronic properties of interface traps on silicon surfaces and
interfaces is split up into several Datareviews [23-25]. The first
Datareview [23] provides historical and physical background and a
discussion of the terminology followed in the other Datareviews. Early
work on the interface traps on 'clean' silicon surfaces covered with
little or no oxide film is reviewed in [23]. The microscopic structures
of the interface traps are probed by electron spin resonance which is
summarized in [24]. This review will describe the data on the interface
density of states, D(IT) on thermally oxidized silicon measured by
methods other than the dark capacitance-voltage (C-V) method on MOS
capacitors for example, X-ray photoemission spectroscopy, the
admittance-voltage frequency method using MOS diodes, and the
capacitance voltage method on MOS transistors. The data obtained by the
dark MOS C-V method cover very many material and processing parameter
variations and are summarized in [25].

B. INTERFACE TRAPS ON OXIDIZED SILICON FROM X-RAY PHOTOEMISSION

    SPECTROSCOPY (XPS)

It is intuitively obvious that the interface traps on oxidized silicon
should be strongly affected by the oxidation process which determines
the atomic structure and hence the dangling bond distribution in the
interfacial layer. Using X-ray photoemission spectroscopy (XPS),
Grunthaner, Hetch and Grunthaner [1] at the Jet Propulsion Laboratory
have made extensive and careful measurements of the interface morphology
and chemical structure of the (100) and (111) SiO2/Si interface. They

observed all the suboxide states in the transition layer, with the mono-
and di-valent silicon, Si+ and Si++, localized in a 6-10 A interfacial
layer and the tri-valent silicon, Si+++, distributed 30 A into the SiO2
film. They investigated the oxidation process dependences which are
given in TABLE 1 where the theoretical cut (dangling) bond densities
on the ideal, unconstructed clean silicon surface, listed in a previous
table [23] are also given for comparison.

TABLE 1: process dependence of the suboxide density at the SiO2/Si
         interface

Face	Oxidation condition	Density 10*14/cm*2						Monolayer covered %
		atom	cut-bond	Si+	Si++	Si+++	Total	
(100)	theory (peroxide)	6.78	13.56	0	13.56	0	13.56	100
(100)	theory	6.78		0	6.78	0	6.78	100
(100)	900C, dry			0.2	2.7	3.5	6.4	94
(100)	900C, wet			0.1	3.0	2.7	5.8	85
(100)	900C, wet POA			0.5	2.1	2.5	5.1	75
(100)	1000C, dry			0	3.2	2.7	5.9	87
(100)	1000C, dry POA			0.5	2.3	2.6	5.4	79
(111)	theory, (face-A)	7.83	7.83	7.83	0	0	7.83	100
(111)	theory, (face-B)	7.83	24.498	0	0	7.83	7.83	100
(111)	theory, (A+B)/2	7.83	15.66	7.83	0	7.83	15.66	100
(111)	900C, dry			3.6	0.2	2.7	6.5	83

The notations used here for the fully bonded silicon are as follows.
Si+ is the monovalent silicon bonded to one oxygen and three silicons,
Si3:::Si-O where the three Si's are bonded to silicon in the bulk Si
and the oxygen is bonded to Si-O in the SiO2. Si++ is the divalent Si
bonded to two oxygen and two silicon, Si2=Si=(O-Si)2. Si+++ is the
trivalent silicon bonded to three oxygen and one silicon, Si-Si:::(O-Si)3.
In the silicon bulk, the covalent silicon is Si(O) or Si2=Si=Si2 and in
the SiO2 film, the 'fully' ionized silicon is Si++++ or (Si-O)2=Si=(O-Si)2.

Both theory and experiments have indicated that interface traps are
associated with silicon and oxygen dangling bonds. Thus, one would expect
a complementary density relationship between the fully suboxide species,
Si(4-x)O(x)==Si, and the partially bonded species whose dangling bonds
give the interface traps. The partially bonded species may have eight
single-dangling-bond configurations given by Si3O(y):::Si. and Si3O(y):::
SiO. where y = 0,1,2 or 3; eighteen double-dangling-bond configurations;
and eight triple-dangling-bond configurations. The four-dangling-bond
configuration would be an unbonded interstitial and may or may not be an
electronic trap. The single Si dangling bonds may be dominant with the
trivalent silicon dangling bond Si3:::Si. , having the largest
concentration which was also identified as the P(bO) centre by electron
spin resonance [24]. The percentage monolayer coverage (last column of
TABLE 1) suggests that the (100) interface (94% bonded) would have
fewer dangling Si bonds than the (111) interface (83% bonded). This
is consistent with the well known results of interface trap density

measurements of oxidized silicon. The interface trap density calculated
from the percentage coverage, for example (1 - 0.94 X 6.8 X 10*14 =
4 X 10*13 trap/cm*2, is substantially higher than that in oxides grown
in a silicon integrated circuit manufacturing line. Such a low density
is due to two reasons: (i) a slow cool at the end of high temperature
oxidation that heals most of the dangling bonds; and (ii) the presence
of trace amounts of water vapour in the oxidation furnace whose hydrogen
could tie up most of the remaining not-healed dangling bonds.

The morphology of the SiO2/Si interface determines not only the
densities and the electron and hole trapping rates of the interface
traps but also these transition rates at oxide traps as well as the
transport properties (electron and hole mobilities in the oxide). It
has been a subject of renewed vigorous investigation by several veteran
researchers during the last five years (1982-1987) due to recent
advances in molecular beam epitaxy that could give oxide films
epitaxially grown on crystalline substrates and a well-known long-held
intuition (at least as far back as the late 1950s when this reviewer
began silicon MOS research and was asked the question 'Is the thermally
grown SiO2 film on crystalline silicon amorphous?'). Early intuition
from silicon-film on crystalline-silicon (c-Si) vapour-phase epitaxial
growth experiences in 1960, which just went into production for the
first generation silicon integrated circuits (known by the trade name,
Micrologic, of Fairchild Semiconductor Corporation), suggested that the
thermally grown SiO2 film on crystalline silicon substrate could also
be crystalline but probably with large strain and stress due to the
difference in atomic spacing between SiO2 and c-Si. For a detailed review
of the recent progresses on the morphology of the SiO2 film, see
[26].

Based on recent evidences of the infrared vibrational spectra of SiO2,
Phillips [2,3] has shown that vitreous silica and thermally grown SiO2
film may not be amorphous as anticipated by the continuous random
network (CRN) model but are crystallites of about 60 A size with the
beta cristoballite (cubic) structure. Phillip's initial (1982) model
[2] invoked oxygen double bonds and dangling bonds in the grain boundary
surfaces joining adjacent crystallites. In his 1987 model [3], the grain
boundaries are completely bonded. The lack of any traps in production
oxide films in Si VLSI circuit chip favours the completely bonded grain
boundary which turned out to be the very model assumed by Herman[4] in
an early SiO2 energy band calculation to alleviate computational
complexity of the hexagonal alpha-quartz model. However, it is possible
that the dangling bonds at the grain boundaries are electrically
deactivated by hydrogen bonding since in silicon oxidation furnaces
there is one part per million residual water vapour and water vapour
is also present in synthetic quartz or silica optical fibre manufacturing
apparatus. Microcrystallites were finally observed in 1987 via grazing
incidence X-ray scattering experiments by Fuoss, Norton, Brennan and
Fischer-Colbrie in thermally grown oxides of 10 to 250 A thickness [5].
A recent detailed analysis of the infrared absorption peak around
1075 cm*-1 in oxide films of 28-450 A thickness by Boyd [6] gave
further support to Phillips' non-CRN or crystallite model of SiO2. From
high resolution transmission electron microscopy measurements, Ourmazd
and Bevk [7] reported that their thermally grown oxide film on
crystalline silicon substrate appears to be an atomically ordered
crystalline layer which they called strained layer epitaxy. The presence
of a highly strained interfacial layer was further indicated by Boyd and
Wilson [8] in their infrared spectra near 1075 cm*-1 as a function of
oxide film thickness (10-600 A). The existence of strain and stress

at the SiO2/Si interface has been well known intuitively due to the
large difference in the interatomic spacings between the crystalline
silicon and the SiO2. There has been considerable interest concerning
the effects of the strain and stress on the growth kinetics of the
first few atomic layers of oxide [9,10,11] on crystalline silicon
substrate. It is evident that the dangling bonds can release stress and
strain. Thus, generation of interface traps via breaking of the strained
bonds by energetic electrons or ionizing radiation may be more prevalent
in highly stressed oxide. A review is given by Ma [27].

C. INTERFACE TRAPS FROM TWO-TERMINAL MOS DIODE ADMITTANCE MEASUREMENTS

One of the less popular and more tedious interface trap measurement
techniques, is the two-terminal admittance-frequency (conductance-
and-capacitance) method. This method is particularly useful when the
interface density is high so that the conductance component, due to
trapping losses, is high and the capacitance component has a large
frequency dispersion in order that contributions to the admittance
from the bulk traps in the silicon surface space charge layer are
negligible. Interpretation of the experimental data (conductance and
capacitance as a function of frequency) is much more complex due to the
distribution in energy of the interface traps for both the U-shaped and
peaked D(IT) as illustrated by Eaton [12] and Katto [13] in contrast to
the general belief. However, admittance-frequency measurements do give
an evaluation of the trapping rate constants, namely, the capture and
and emission rate coefficients of electrons and holes at the interface
traps as a function of trap energy level $c(n)[E(T)]$, $c(p)[E(T)]$,
$e(n)[E(T)]$ and $e(p)[E(T)]$ (See [23] for definition of terms). The
trapping rates cannot be obtained from the dark C-V method or the
other non-electrical methods (such as the ESR and XPS methods).

Measurements were made in 1972 by Eaton [12] on 83 A oxide giving a
D(IT) peak of 2.2 X 10*13/cm*2/eV at Ev+0.25 eV and total peak density
(area under the D(IT) peak curve) of 6 X 10*12/cm*2, and by Katto
[13] on 35 A oxide giving a D(IT) peak of 1.4 X 10*13/cm*2/eV at
Ev+0.28 eV and a total peak density of 1 X 10*12/cm*2. They both
also observed a high density U-shaped D(IT) centred around midgap.
A simple relationship for the trapping rates versus trap energy level
cannot be obtained e.g the empirical exponential energy dependence,
$c(n) = exp(-E(T)/kT)$, frequently assumed by other researchers. This
is expected since the traps are distributed into the oxide film to
give rise to a wide distribution of time constants from tunnelling as
assumed by Eaton and Katto in their data analyses.

D. INTERFACE TRAPS FROM THREE-TERMINAL MOS TRANSISTOR MEASUREMENTS

Another unpopular interface states measurement method relies on the
gate capacitance versus voltage variation on three-terminal MOS
transistors. This allows probing states in the half energy gap towards
the minority carrier band edge in contrast with the two-terminal MOS
C-V method which is reliable on the majority carrier half of the energy
gap.

Measurements on first generation enhancement and depletion mode as well
as epitaxial channel silicon MOSFETs were reported by Sah and Pao [14].
On 6200 A thermal gate oxide, they reported a D(IT) peak of 7 X
10*11/cm*2/eV at Emg-0.3 eV and total peak density of 2.11 X 10*11/cm*2.
Similar results were also obtained in a 1700 A anodic oxide subsequently
annealed at 750 deg C.

Additional interface state measurements on MOSFETs were made by Fu and
Sah in a series of experiments on a correlation of interface density
with the 1/f noise power spectra [15]. In both cases, a U-shaped
interface trap distribution was also observed.

E. INTERFACE TRAPS MEASURED WITH LIGHT (GENERATION VIA CAPTURE OF

     PHOTOGENERATED HOLES)

A third and not frequently used method for interface measurements is
the use of light. Optical responses could provide data as a function
of the trap energy but again it is more difficult to perform and few
attempts have been reported. Another optical method, reported by Sah et
al [16], involved the generation of interface traps via hole capture.
The holes were generated at the aluminium-gate/oxide interface by
exposing thin-Al-gate (about 50 A) MOS capacitors to 10.2 eV photons
from a hydrogen discharge source in vacuum at 300 deg K. The vacuum
ultra-violet (VUV) light generated holes are then injected into the
interior of the oxide by a positive voltage applied to the gate and
these holes migrate through the oxide under the applied electric field
to reach the oxide/silicon interface where they are captured and break
the strain intrinsic bonds and hydrogenated bonds. The principal
advantage of this method is that there are few hot or high energy holes
(and even fewer hot electrons) since the hole mobility in SiO2 is
extremely low so that the interface trap generation mechanism is
predominately limited to only one process, namely, hole capture to
rupture the weak-strained host bonds and impurity (hydrogen) bonds in
the interfacial layer. The density of the interface traps was monitored
by the dark high-frequency C-V (HFCV) characteristics.

Five components were analyzed from the D(IT) versus Si gap energy,
consisting of three donor-like (positively charged) peaks, P1, P2 and
P3 about 0.3 to 0.4 eV above the Si midgap; a low-density U-shaped
component, UO (amphoteric or can be in the positive donor-like, neutral
or negative acceptor-like charge states); and a midgap component, M
(amphoteric), with relatively constant D(IT) over as much as 3/4 of
the midgap range which should increase towards the two band edges. Since
the U-shaped D(IT) of the UO interface trap was observed to be nearly
independent of injected hole fluence (hole/cm*2), it was tentatively
attributed to the band tail states from splitting off the Si conduction
and valence states into the Si gap by weak perturbations to the periodic
potential from random bond angle and length fluctuations. These weakly
strained random bonds would not be broken by hole capture permanently
because they would heal immediately after hole capture. The generation
kinetics (N(IT) versus injected hole fluence) of the remaining four
components were linear; thus, their generation mechanism was attributed
to the creation of dangling bonds by the release of bonded hydrogen during
hole capture. The M trap (U-shaped) was thought to come from randomly
distributed trivalent silicon dangling bonds [16] first proposed by Sah

- 516 -

[17] and now associated with the P(b) centre observed in ESR (see [24], Si3:::Si.). This trap is thought to be generated by the hole capture reaction at the interface [16], Si3:::Si-H + h+ ---> Si3:::Si. + H, in which the released hydrogen ion rapidly (compared with the rate of hole capture at the hydrogen saturated trivalent silicon, Si3:::Si-H) captures a valence or conduction electron from Si (e- + H+ ---> H.) and then recombines to form hydrogen molecules (H. + ---> H-H). The three peaks, P1, P2 and P3 were tentatively attributed to the oxygen dangling bond [16] with three different second nearest neighbours since they were observed in oxides annealed in oxygen after oxidation, a process known as the post oxidation annealing (POA). The hole capture reactions were thought to be $Si(x)O(y):::SiO-H + h+ ---> Si(x)O(y):::SiO. + H+$ where $x + y = 3$. The H+ ion also forms a hydrogen molecule following the two-stage reaction just indicated. From electro-negativity consideration, the deepest level, P1, was thought to be the oxygen dangling bond with $x = 3$ and $y = 0$ or Si3:::SiO. and similarly, P2 to be Si2O:::SiO. , while P3 to be SiO3:::SiO. . The generation efficiency, generation cross section and steady-state density were measured [16] and are given in TABLE 2.

TABLE 2: generation parameters of five interface traps on oxidized silicon via capture of thermal holes at 300 deg K [16]

Trap Label	Efficiency (10*3 trap/hole)	Generation Cross-section (10*-16 cm*2)	Steady-State Density (10*11/cm*2)
U0	0	0	0.45
M	4.3	80	5.2
P1	6.1	140	4.2
P2	1.4	23	4.0
P3	0.3	-	-

The traps in the chemical model suggested by Sah, Sun and Tzou just described [16] could come from the very suboxides detected by Grunthaner [1], $Si(x)O(3-x):::Si-O(3-x)Si(x)$ where $x = 0,1,2,3$ if one Si or oxygen bond is ruptured resulting in the Si dangling bonds, $Si(x)O(3-x):::Si.$, or the oxygen dangling bonds, $Si(x)O(3-x):::SiO.$.

An alternative microscopic model was also considered by Sah [18] attributing the peaks to the multiple charge states of a single (one of the oxygen or silicon dangling bonds) centre. Estimates were made of the binding energies of the electrons at silicon, oxgyen and nitrogen dangling bonds at the oxide/silicon interface using dielectrically screened atomic levels and they are given in TABLE 3. Using an arithmetical average for the dielectric constant, $(4+12)/2 = 8$, for the SiO2/Si interface layer, the first donor level from oxygen dangling bond is then $16.618/8*2 = 0.253$ eV below the Si conduction band edge according to the effective mass theory. This is not far from the observed P1 peak energy of Ec-0.25 eV.

TABLE 3: selected list of electron affinity, ionization potential and binding energy of electrons in atoms and solids (all in eV) (data from Handbook of Chemistry and Physics) and estimated binding energies of electrons at the oxide/silicon interface [18]

	Binding energies in solids and atoms					Binding energies at SiO2/Si			
	SiO2	c-Si	Si	O	N	State	Si.	O.	N.
Electron-Affinity	0.9	4.018	1.427	1.466	0.0007	1st A	0.022	0.023	0.000
First IP		5.188	8.151	13.618	14.534	1st D	0.127	0.213	0.227
Second IP			16.345	35.116	29.601	2nd D	0.255	0.548	0.463
Third IP			33.492	54.943	47.448	3rd D	0.523	0.858	0.742
Fourth IP			45.141	77.412	77.742	4th D	0.705	1.210	1.216

where A = acceptor
D = donor

The participation of hydrogen was further supported by the second order annealing kinetics of the peaked D(IT). A rapid anneal of the P1 and P3 peaks was observed by Sah, Sun and Tzou [16] which lasted about 24 hours at 300 deg K immediately after their generation via hole trapping. The trap density decreased as reciprocal time suggesting a second order kinetics which is consistent with the reaction H-H ---> 2H. as the rate limiting process. This is plausible since the other reaction of the annealing process, given by Si(x)O(y):::SiO. + H. ---> Si(x)O(y):::-SiO-H :::SiO-H should proceed much faster.

Annealing of the M trap (probably U-shaped with relatively constant D(IT) over the entire Si gap) occurs at higher temperatures when additional hydrogen is available to compensate for lateral diffusion and evaporation losses. The silicon dangling bond model (Si3:::Si or P(b) in ESR) for the M centre is again consistent with the oxygen dangling bonds of the peaked centres, Si(x)O(3-x):::Si. , since oxygen dangling bonds of the peaked centres are more electronegative and hence have higher reaction probabilities with hydrogen so that its hydrogenation occurs at lower temperature than the silicon dangling bonds. A detailed analysis was presented [18] on the bases of assigning the four traps to silicon and oxygen dangling bonds and the native UO trap to the random fluctuation of the bond length and angle.

Other peaked interface traps have also been observed by Sah, Sun and Tzou [19-21] and summarised [18]. These peaks are present singularly and occasionally in pairs in the D(IT) spectra in MOS capacitor samples under a variety of electrical and keV electron beam stresses in addition to fabrication or oxidation conditions prior to stress. The peaks are located in the lower as well as the upper half of the Si gap and have annealing temperatures higher than 300 deg K. These are discussed with those observed by other researchers on thermally grown SiO2 on c-Si substrate in [25].

REFERENCES
----

[1]   P.J.Grunthaner, M.H.Hecht, F.J.Grunthaner, N.M.Johnson [J. Appl.
      Phys. (USA) vol.61 no.2 (15 Jan 1987) p.629-38]
[2]   J.C.Phillips [Solid State Phys. (USA) vol.37 (1982) p.93-171]
[3]   J.C.Phillips [Phys. Rev. B (USA) vol.35 no.12 (15 Apr 1987)
      p.6409-13]
[4]   F.Herman, I.P.Batra, V.Kasowski [IEEE Trans. Electron Devices
      (USA) vol.ED-19 no.2 (Feb 1972) p.333-9]
[5]   P.H.Fuoss, L.J.Norton, A.Fischer-Colbrie [Bull. Am. Phys. Soc.
      (USA) vol.32 no.3 (March 1987) p.453]
[6]   I.W.Boyd [Appl. Phys. Lett. (USA) vol.51 no.6 (10 Aug 1987)
      p.418-20]
[7]   A.Ourmazd, J.Bevk [Bull. Am. Phys. Soc. (USA) vol.32 (Mar
      1987) p.774]
[8]   I.W.Boyd, J.I.B.Wilson [Appl. Phys. Lett. (USA) vol.50 no.2 (9
      Feb 1987) p.320-2]
[9]   A.M.Carim, R.Sinclair [J. Electrochem. Soc. (USA) vol.134 no.3
      (Mar 1987) p.741-6]
[10] R.H.Doremus [J. Electrochem. Soc. (USA) vol.134 no.8 (Aug 1987)
      p.2001-3]
[11] B.J.Mrstik, A.G.Revesz, M.Ancona, H.L.Hughes [J. Electrochem. Soc.
      (USA) vol.134 no.8 (Aug 1987) p.2020-7]
[12] D.H.Eaton, C.T.Sah [Phys. Status Solidi a (Germany) vol.12 (1972)
      p.95-109 (See Fig.9 which shows a D(IT) peak about 0.25 eV above
      the valence band edge or 0.3 eV below midgap in 83 A thick oxide]
[13] H.Katto, C.T.Sah [Phys. Status Solidi a (Germany) vol.13 (1972)
      p.417-26; (See Fig.5 which shows two D(IT) peaks about 0.1 and
      0.3 eV above the silicon valence band edge in 35 A thick oxide)]
[14] C.T.Sah, H.C.Pao [IEEE Trans. Electron Devices (USA) vol.ED-13
      no.4 (Apr 1966) p.393-409; (See Figs. 9c and 16b on D(IT) vs
      energy giving a peak at about 0.3 eV below the midgap)]
[15] Horng-sen Fu, C.T.Sah [IEEE Trans. Electron Devices (USA)
      vol.ED-19 no.2 (Feb 1972) p.273-85; (See Figs. 7,8 and 15c which
      again showed D(IT) peak at about 0.3 eV below midgap)]
[16] C.T.Sah, J.S-C.Sun, H.J-T.Tzou [J. Appl. Phys. (USA) vol.53
      no.12 (Dec 1982) p.8886-93]
[17] C.T.Sah [IEEE Trans. Nucl. Sci. (USA) vol.NS-23 no.6 (Dec 1976)
      p.1563-8]
[18] C.T.Sah [Studies of Reliability Physics of Silicon VLSI
      Transistors, Tech. Report No.TRS-100 (Semicond. Res. Corp.,
      Research Triangle Park, NC, USA, 31 Jul 1984) p.78]
[19] C.T.Sah, J.Y-C.Sun, J.J-T.Tzou [J. Appl. Phys. (USA) vol.54 no.5
      (May 1983) p.2547-55]
[20] C.T.Sah, J.Y-C.Sun, J.J-T.Tzou [J. Appl. Phys. (USA) vol.54 no.10
      (Oct 1983) p.5864-79]
[21] C.T.Sah, J.Y-C.Sun, J.J-T.Tzou [J. Appl. Phys. (USA) vol.55 no.6
      (15 Mar 1984) p.1525-45]
[22] C.T.Sah [EMIS Datareviews on Generation-recombination-trapping
      rates and lifetimes of carriers in silicon oxide films on Si
      (Nov 1987) RN=17815-22,17839]
[23] C.T.Sah [EMIS Datareview RN=16170 (Nov 1987) 'Interface traps on
      silicon surfaces']

[24] C.T.Sah [EMIS Datareview RN=16180 (Nov 1987) 'Interface traps on
      oxidized silicon from electron spin resonance']
[25] C.T.Sah, M.S.C.Luo, C.C.H.Hsu, T.Nishida, A.J.Chen, C.Y.P.Chao
      [EMIS Datareview RN=16182 (Nov 1987) 'Interface traps on
      oxidized silicon from two-terminal, dark capacitance-voltage
      measurements on MOS capacitors']
[26] S.M.Goodnick [EMIS Datareview RN=15725 (Oct 1987) 'Structure of
      the interface in oxide-on-Si']
[27] T.P.Ma [EMIS Datareview RN=15733 (Oct 1987) 'Stresses in SiO2-on-Si
      structures']

17.4 INTERFACE TRAPS ON OXIDIZED Si FROM TWO-TERMINAL, DARK
CAPACITANCE-VOLTAGE MEASUREMENTS ON MOS CAPACITORS

by C.T.Sah, M.S.C.Luo, C.C.H.Hsu, T.Nishida, A.J.Chen, and C.Y.P.Chao

November 1987
EMIS Datareview RN=16182

A. INTRODUCTION

Major reported efforts on the experimental characterization of the
interface traps on oxidized silicon have employed two experimental
techniques: (i) electron spin resonance (ESR) of unpaired electrons at
traps on oxidized silicon samples with very high interface trap density
(N(IT) approx. 10*13/cm*2); and (ii) trap related dark capacitance
measurements on MOS capacitors spanning the density range from the
lowest experimentally obtainable (< 10*9/cm*2) to the highest (> 10*13
/cm*2). Trap density is varied by controlling oxidation processes and
by electrical injection and photo or keV electron-impact generation of
hot electrons and holes in the oxide. (See [51] for a discussion of
the various non-fabrication methods of trap generation.) Due to the
substantial differences in the oxide film fabrication processes to give
the very high N(IT) for ESR experiments and the very low N(IT) for VLSI
circuit chip production, there is no assurance that the interface traps
are of the same species.

Due to the tremendous amount of literature and data, the review on the
electronic properties of interface traps on silicon surfaces and
interfaces is split up into several Datareviews, [52-54]. The Datareview
[52] provides historical and physical background and a discussion of
the terminology followed in all the reviews. Early work on interface
traps on 'clean' silicon surfaces covered with little or no oxide
film is reviewed in the first review [52]. The microscopic structures
of the interface traps are probed by electron spin resonance which
is summarized in ref [53]. The data on the interface density of states,
D(IT), on thermally oxidized silicon measured by methods other than
the dark capacitance-voltage (C-V) method on MOS capacitors (for
example, the X-ray photoemission spectroscopy, the admittance-voltage-
frequency method using MOS diodes, and the capacitance voltage method
on MOS transistors) are reviewed in ref [54]. This Datareview reviews
the data obtained by the dark MOS C-V method which covers very many
material and processing parameter variations.

MOS C-V measurement methods, principally a combination of two-terminal
high-frequency, low-frequency, and quasi-static capacitance versus DC
voltage, are the most convenient to peform since they use simple test
device geometry which is easily fabricated and whose fabrication
conditions can be readily varied under controlled conditions, and since
the measurements themselves can be carried out easily using computer-
controlled instruments with sensitivities approaching 10*8 q/cm*2 eV
or one millionth of a monolayer. Their only limitation is the
relatively long measurement time (a few readings per second at most
and usually a few seconds to ten seconds per reading) which is the
result of high sensitivity and accuracy requirements. This hinders the
characterization of the fast phenomena such as the initial generation
or anneal of the traps at times shorter than about one second.

Characterization of the fast interface traps near the band edges is
also hindered by the high frequency limit (about one MHz) owing to
the high accuracies required. In spite of these two limitations,
computer automation and resultant high accuracy of the MOS capacitance
method have facilitated the acquisition of a tremendous amount of data
on interface as well as oxide traps at many academic and industrial
laboratories and routinely used in manufacturing lines for silicon VLSI
chips world-wide. This section will summarize the measurement results
of the density of the U-shaped and peaked interface traps, including
the dependence of the trap density on the silicon surface orientation,
the substrate dopant impurity type and density, the oxidation processes,
and the electrical stresses.

B. ORIENTATION DEPENDENCES

With the fabrication process of the MOS capacitor kept constant [see
section D on its effects], a strong dependence of D(IT) and N(IT)
on the silicon surface orientation has been observed [1-3] which could
be anticipated from the dangling bond model as indicated by the cut-bond
densities in TABLE 1 in [52]. The fabrication process would strongly
influence the initial D(IT) since the dangling bonds can be deactivated
by annealing and hydrogenation processes whose effectiveness could depend
on the surface orientation or the distance between the dangling bonds. A
consistent trend showing the same orientation dependence of D(IT) in
TABLE 1 and the CUT-BOND-DENSITY (table 1 in [52]) is observed
experimentally.

TABLE 1: dependence of D(IT) at midgap on silicon surface orientation

Variables	General Features of D(IT)	D(IT) (E=Emg) (/cm*2/eV)	References
Substrate orientation	(100) < (110) < (111)	approx. 1:2:3	[1-3]

C. DEPENDENCE ON SUBSTRATE DOPANT CONCENTRATION AND TYPE

A dependence on the substrate dopant type (n-type or p-type; P, As,
Sb or B, Al, Ga, In) would be expected if the interface traps are
extrinsic, i.e. consisting of an impurity-defect complex involving the
dopant impurity. One would also expect a peaked D(IT) to indicate the
impurity signature and higher density at higher dopant density.
Furthermore, D(IT) should be higher on n-Si than on p-Si since the
n-type dopants pile up at the oxide/silicon interface during themal
oxidation while the p-type dopants deplete. There have not been many
experiments on these dependences. TABLE 2 lists two experimental
measurements at the midgap, indicating D(IT)(n-Si) > D(IT)(p-Si), which
is not inconsistent with the pile-up/depletion idea but it would imply
that the measured D(IT)s were not intrinsic dangling bonds but were
impurity-defect complexes. More detailed experiments are needed to
ascertain any substrate dopant impurity dependences and the atomic
nature (intrinsic or extrinsic) of any impurity-related interface
traps.

TABLE 2: dependence of D(IT) at the midgap on silicon substrate dopant
density-type

Variables	General Features D(IT)	D(IT) (E=Emg) (/cm*2/eV)	Ref
dopant type	n-Si > p-Si	1 10*11 - 3 10*12	[2,4]
dopant density	D(IT) increases when N(AA) > 1 10*16	1 10*11 - 3 10*12	"
	D(IT) increases when N(DD) > 1 10*17	1 10*11 - 3 10*12	"

D. FABRICATION PROCESS DEPENDENCES

The density of the interface traps, D(IT) and N(IT), is strongly
dependent on the oxide fabrication processes. These include the
following list in the normal fabrication sequence:

(i)   the silicon surface cleaning procedure prior to oxidation
        (chemical, ion bombardment or even heating in oxygen or hydrogen
        containing ambient);

(ii)   the heating rate and ambient to reach the oxidation temperature;

(iii)   the oxidation temperature, time, and ambient;

(iv)   the post-oxidation annealing (POA) temperature, time and ambient;

(v)   post-oxidation cooling (POC) rate and ambient to room temperature;

(vi)   any chemical or non-chemical treatment of the oxide surface (OST
        for Oxide Surface Treatment) prior to gate metallization;

(vii)   the type of conductor used as the gate electrode on the oxide
        (metal such as Al, Au or others, refractory metal, refractory
        metal silicides, or polycrystalline silicon doped n-type or
        p-type);

(viii) the method of putting down the gate (resistance, RF or electron-
        gun heated evaporation; sputtering; CVD; etc.);

(ix)   the temperature, time, ambient of impurity diffusion into
        the gate conductor to control its conductivity;

(x)   the method used to define the gate electrode geometry (shadow
        mask during evaporation , and photo, X-ray or keV-electron-beam
        lithography and the associated chemical etches, baking temperatures
        and baking times);

(xi)   post-metallization annealing (POA) temperature, time and ambient
        before or after gate definition, including chip bonding on a
        transistor header.

Few of these fabrication variables have been characterized systematically to determine their effects on the interface traps. For example, the POC condition (item v) is one of the most critical influences on both the initial value of the D(IT) (before any electrical or ionizing radiation stress) and also the total value or the magnitude, N(ITT). N(ITT) consists of electrically active or dangling bonds plus electrically inactive strained bonds and electrically inactive hydrogen saturated bonds. However, most published data do not give the POC ambient and rate. In addition, OST (item (vi) above) is also critical but seldom given except in some specific early Bell Lab experiments, and recent Urbana experiments in which OST was the controlled experimental variable. In view of the large number of experimental variables and incomplete specifications of these fabrication variables, only a short summary of current results with reasonably certain process dependences is given in TABLE 3 which we shall elaborate. The acronyms are defined as follows:

OXD = Oxidation
POA = Post Oxidation Anneal
POC = Post Oxidation Cooling
PMA = Post Metallization Anneal

TABLE 3: dependence of D(IT) at midgap on oxide fabrication variables

Variables	General features of D(IT)	D(IT) (E=Emg) (/cm*2/eV)	Ref
OXD ambient	H2O < HCl(TCE,Cl2)+O2 < O2	1 X 10*11 - 3 X 10*12	[1-2,5-9]
OXD temp	(note 2) dry and wet oxide (900-1200 deg C)	1 X 10*11 - 3 X 10*12	[2,5]
POA ambient	(note 1) in O2 or vacuum (note 2) N2, Ar, He	1 X 10*11 - 4 X 10*12	[1,2,10]
POA temp	Depends on ambient	1 X 10*11 - 4 X 10*12	[1,2,10]
POA time	Depends on ambient	1 X 10*11 - 4 X 10*12	[1,2,10]
POC rate	(note 2) at lower cooling rate	1 X 10*11 - 4 X 10*12	[11,12]
PMA ambient	H2 < N2, air	< 1 X 10*10 - 3 X 10*12	[13-19]
PMA temp	(note 2) 80 - 500C minimum at < 500C	1 X 10*10 - 3 X 10*12	[17-20]
	(note 1) above 500C	1 X 10*10 - 3 X 10*12	
Gate material	Al < Si < Au	1 X 10*10 - 3 X 10*12	[15-17]

note 1: D(IT) increases
note 2: D(IT) decreases

The D(IT) depends strongly on the water content of the oxidation ambient [1,2,5,6] owing to hydrogenation and hence electrical deactivation of the silicon and oxygen dangling bonds first proposed by Balk in 1962 [15]. Even using the driest oxygen gas from liquid oxygen source, there is still about 1 ppm water in the oxidation ambient in the fused silica furnace tube due to permeation through the tube wall at the oxidation temperature. The initial D(IT) (prior to stress generation of additional interface traps) in 'dry oxide' is in the range of 10*10 - 3 X 10*12/cm*2 and one order of magnitude smaller in steam oxide, the latter being due to hydrogenation. Wet oxide has a much higher density of hydrogenated bonds which could be ruptured by hot electrons or ionizing radiation during stress to give higher densities of new interface traps. Oxides grown in chlorine containing oxygen ambient (3% HCl in dry oxygen or an equivalent amount of TCA, TCE or Cl2) has lower D(IT) than dry oxide [2,7-9] due to the presence of hydrogen in the oxidation ambient which ties up the Si and O dangling bonds. The effect of chlorine itself has not been extensively determined but it can conceivably tie up some of the Si and O dangling bonds forming Si-Cl and SiO-Cl and resulting in lower D(IT) [21].

A new layer of oxide is formed at the SiO2/Si interface instead of the outer surface of the oxide when oxygen diffuses through the existing oxide film and reacts with the silicon from the silicon substrate. Thus, D(IT) due to silicon dangling bonds depends on the relative magnitude of the oxygen diffusion rate and oxygen-silicon reaction rate. Higher D(IT) or more Si dangling bonds will result when the oxide growth is limited by oxygen diffusion through the oxide since then the silicon-oxygen reaction rate is so much higher than the arrival rate of oxygen at the interface that the interfacial layer is oxygen deficient or having an excess silicon or silicon dangling bonds. In contrast, if the growth is limited by the O-Si reaction rate at the interface, then there are fewer Si dangling bonds and lower D(IT). The excess oxygen at the SiO2/Si interface in a reaction limited growth could be in molecular form and electrically inactive. Experimental evidence in all the cited references in TABLES 1-4 are consistent with this model, but most of the published data probably contain combined effects of oxygen and silicon dangling bonds in both the oxide (oxide traps) and at the SiO2/Si interface (interface traps). These oxide trap contributions to the MOS C-V characteristics have not been separated out except in some data taken recently by Sah, Sun and Tzou [22-25]. A new experimental technique to separate the oxide traps from the interface traps was introduced recently [26,27] but only preliminary data have been reported [28,29].

'Annealing' or post oxidation annealing and cooling (POA and POC) is frequently performed in the range of 500-1200 deg C to reduce the oxide trap density [1,2]. This post oxidation anneal (POA) can either increase or decrease the D(IT) depending on the annealing ambient, temperature and time [1,2,6,10,30]. Annealing in vacuum inevitably increases D(IT) because the interface is chemically reduced resulting in higher silicon dangling bond density. POA in oxygen at temperatures below the oxidation temperature should decrease the D(IT) as has been observed by Sah, Ning and Tschopp [31] but other published data have shown the opposite trend, possibly due to a mixing of the oxide trap contributions into D(IT). Annealing in He, or N2 at temperatures higher than about 700 deg C can reduce D(IT), however, tremendous D(IT) (greater than 10*13/cm*2) has always been observed when POA is performed above 1100 deg C in an inert gas with a fast POC (post oxidation-annealing cooling) in the inert gas ambient.

Post metallization anneal (PMA) with metal gates significantly reduces D(IT). This was first observed by Cheroff, Fang, Hochberg [13] and Leman [14] but it was Balk who provided the model of hydrogenation of the silicon dangling bonds [15] in 1964 which is now the consensus model on the effect of hydrogen on the electrical activity of the interface traps. Meanwhile, Kooi performed extensive experiments [16] further supporting Balk's Si dangling bond hydrogenation model. Al gate serves as a source of hydrogen while a polcrystalline gate does not if the poly-Si gate is heated to about 900 deg C as shown by Sah, Sun and Tzou [24]. The D(IT) can vary greatly with PMA temperature, ambient and time [2,5,6,12,14-19,30]. Appropriate annealing temperature (400-450 deg C), time (about 30 s to a few minutes) and ambient (forming gas of 10%H2-90%N2) can reduce D(IT) to < 1 X 10*10/cm*2 [17]. Controlled chip bond experiments in 80-400 deg C [20,30,32,33] and in 380-460 deg C [34-36] on aluminium gate MOS capacitors provided further supporting evidence that hydrogen from the aluminium gate is tieing up the silicon dangling bonds at the SiO2/Si interface. The recent experiments of Reed and Plummer [34] futher confirmed that the simple hydrogen-silicon dangling bond reaction model, first proposed and applied by Sah, Sun and Tzou [22-25,34], is the rate limiting mechanism instead of the complex, multistaged radiolytic molecular hydrogen model proposed by Griscom based on Waite's diffusion limited model [37-39].

E. INTERFACE TRAP GENERATION DURING ELECTRICAL STRESS

It is well known and established that interface and oxide traps are generated during electrical stress of the silicon MOS capacitors when the electric field across the oxide is high. DiMaria showed recently that the generation threshold is about 1.5 MV/cm [40]. In addition to trap generation, existing neutral oxide traps can also be charged. The traps are generated by breaking the hydrogen bonds (Si-H and SiO-H) and the strained intrinsic bonds (Si..Si and Si..O) without healing when the stress is removed. Bond breaking can occur by hot electrons impacting the four bonds or by capture of thermal holes by the four bonds. Impact by hot hole is less likely due to the lack of hot holes owing to the very low hole mobility in the SiO2. Hot electrons in SiO2 can be created by injection of electrons into the oxide and acceleration of the injected electrons by the high oxide electric field.

Electrons and holes can be injected into the oxide from either the c-Si substrate or the gate conductor. They can be injected either over or through the potential barrier between the gate-conductor/oxide or the oxide/c-Si potential barrier. Injection through the barrier is known as the Fowler-Nordheim tunnelling electron (or hole) injection (FN-TEI or FN-THI). Injection over the barrier would require the presence of hot or energetic carriers in the gate or in c-Si in order to have sufficient kinetic energy to surmount the barriers. The Al/SiO2 barrier is 3.3 eV for electrons and 5.7 eV for holes. The SiO2/Si barrier is 3.1 eV for electrons and 4.7 eV for holes. There are several methods to generate the hot carriers at the interfaces so that they can surmount the interfacial potential barriers. These are the interband optical generation in c-Si and poly-Si gates and intraband optical generation in metal-gates, known as internal photoemission; minority carrier injection by interband impact generation in c-Si and possibly in poly-Si gates, known as avalanche electron or hole injection (AEI or AHI); and minority carrier injection in c-Si by optical injection (photo-transistor or photodiode action) and by forward biased p-n junction. In applications of the Si VLSI circuit chips, the AEI, AHI, FN-TEI, and FN-THI are the most dominant processes which cause degradation and limit reliability. In addition, hot electrons (or holes) travelling in

parallel to the SiO2/Si interface may also break bonds to create
interface states. These glancing or tangential hot electrons (or holes)
will be present in the 'depletion' region of the drain junction of a
MOS transistor when the drain voltage, V(DS), is greater than the gate
voltage, V(GS)-V(T). A detailed discussion of these generation and
charge injection processes are given in Datareviews [51] and in two
invited conference papers in 1987 [41,42].

In the trap characterization experiments, these four methods are also
the ones most commonly employed. Due to the strong dependences of the
interface trap (also oxide trap) generation rates on the oxide
fabrication conditions, only selected publications are tabulated in
TABLE 4 in which the authors gave sufficient oxidation details. The list
is further reduced since in many articles, the oxide trap contributions
were not separated out from the raw data in order to give just the
interface trap contributions. For example, the oxide trap contributions
were contained in the threshold and midgap voltage-shift data from the
capacitance-voltage curves. Thus, in TABLE 4 this is indicated as NA
which stands for Not Available or Not Analyzed. A brief discussion of
TABLE 4 will be given.

TABLE 4: dependences of interface trap generation density and rate
during electric stress

Variable	Fabrication condition gate material	Stress method	D(IT)(Emg,U) D(IT)(peak) (/cm*2/eV)	Cross section (cm*2)	Ref
OXD	Dry oxide	AEI	No change	NA	[43]
	Steam oxide	AEI	Larger	NA	
OXD	Dry	AEI	NA	1.1 X 10*-18	[44]
	H2O 30 mmHg	AEI	NA	2.1 X 10*-18	
	H2O 50 mmHg	AEI	NA	2.2 X 10*-18	
OXD	Dry	AEI	0.25 eV fewer	NA	[23-25]
	H2O 25 atm	AEI	0.25 eV more	2 X 10*-17 HIPOX	
POA	N2 30 min	AEI	NA	Present	[45]
	N2 16hr	AEI	NA	Absent	
POA	Ar 20 min	AEI	0.25 eV slower	Fewer	[23-25]
	None	AEI	0.25 eV faster	Present	
PMA	Al gate no PMA	AEI	NA	1.1 X 10*-18	[44]
	PMA	AEI	NA	1.0 X 10*-18	
PMA	Al Gate no PMA	AEI	0.25 eV	Present	[23-25]
	PMA	AEI	0.25 eV	Present	
	Al/Si gate no PMA	AEI	Absent	Absent	
	PMA	AEI	0.25 eV	Present	

Variable	Fabrication condition gate material	Stress method	D(IT)(Emg,U) D(IT)(peak) (/cm*2/eV)	Cross section (cm*2)	Ref
Gate	Al	AEI	NA	Present	[45]
	Si(poly)	AEI	NA	Absent	
Gate	Al	AEI	0.25 eV	Present	[23-25]
	Au	AEI	0.25 eV	Present	
	Al/Si/no PMA	AEI	Absent	Absent	
	Al/Si/PMA	AEI	0.25 eV	Present	
	Au/Si/PMA	AEI	Absent	Absent	
Gate	Dry Oxide and POA				
	Au	AEI	NA	Present	[46]
	Al/Mg	AEI	NA	Present	
	Al/Mg	AEI	NA	Absent	
Gate	Dry Oxide and POA				
	Al	AEI	-0.2 eV	Present	[47]
	Au	AEI	NA	Present	
	Au/Ti	AEI	NA	Slowest	
	Si	AEI	Absent	Absent	
Temp	Dry Oxide, Al-gate, PMA(N2)				
	295 deg K	AEI	NA	Present	[45]
	> 373 deg K	AEI	NA	Absent	
	77 deg K	AEI	NA	Absent	
Temp	Dry Oxide, Al-gate, PMA				
	295 deg K	AEI	Present	Present	[48]
	96 deg K	AEI	Smaller	Present	
	393 deg K	AEI	Smaller	Present	
Temp	Dry Oxide 3%HCl, POA, Al-gate, PMA				
	90 deg K	FNTEI	Absent	NA	[49]
	After Warmup	FNTEI	Increased	NA	
Temp	Wet Oxide, POA, Al-gate, PMA				
	77 deg K	FNTEI	Absent	Present	[50]
	295 deg K	FNTEI	Present	Present	
	Warmup after 77 deg K stress	FNTEI	Present	Present	

Interface traps with U-shaped D(IT) are generated during the electrical stress due to bond breaking by energetic electrons at the interface or by capture of holes injected from the gate electrode or c-Si. The presence of the U-shaped interface traps and their increasing D(IT) with increasing stress are universally observed in all the references cited in TABLE 4. However, interface traps with peaked D(IT) are not always observed to increase during stress. In fact, Sah, Sun and Tzou [24,25] observed that some of the D(IT) peaks increased first and then decreased as the high field AEI stress was continued. The peaks eventually disappeared at very high AEI fluences. This 'reverse generation' of interface traps was attributed to the annealing of the interface traps by the atomic hydrogen released from the Al-gate by hot electron impact during the stress. They also observed a peaked D(IT) trap at E_{mg} + 0.2 eV in boron-diffused poly-Si gate dry oxide which is nearly independent of stress to fluences greater than 25 C/cm*2. (See Fig.20 of reference [25]).

In summary, most of the interface traps are dangling silicon and oxygen bonds with the exception that a residual interface trap with U-shaped density of states may arise from a random distribution of bond lengths and angles. The dangling bond interface traps are generated during electrical stress at an oxide electric field greater than about 1.5 MV/cm. Interface trap generation is caused by breaking of four types of weak bonds: Si-H, SiO-H, Si..Si and Si..O (the last two are strained or stretched intrinsic bonds). Broken bonds, if not healed after the stress is removed, give dangling bonds or interface states. The bonds can be broken by hot electron impact or by capture of a thermal hole (or electron) at the bond site. The bond breaking energy is supplied by the energetic impacting hot electron or by the recombination energy from a hole capture or an electron capture. The annealing of the interface states proceeds mainly by hydrogenation of the dangling bonds or by thermal healing of two adjacent dangling bonds.

REFERENCES
----

[1]   F.Motillo, P.Balk [J. Electrochem. Soc. (USA) vol.118 (1971) p.1463]
[2]   R.R.Razouk, B.E.Deal [J. Electrochem. Soc. (USA) vol.126 (1979) p.1573]
[3]   E.Arnold, J.Ladell, G.Abowitz [Appl. Phys. Lett. (USA) vol.13 (1968) p.413]
[4]   J.Snel [Int. Conf. Insulating Films on Semiconductors 1979, Inst. Phys. Conf. Ser. (GB) no.50 (1979) p.719]
[5]   D.J.Breed, R.P.Kramer [Solid-State Electron. (GB) vol.19 (1976) p.897]
[6]   P.O.Hahn, M.Henzler [J. Vac. Sci. & Technol. A (USA) vol.2 (1984) p.574]
[7]   G.Baccarani, M.Severi, G.Soncini [J. Electrochem. Soc. (USA) vol.120 (1973) p.1436]
[8]   R.J.Kriegler, Y.C.Cheng, P.R.Colton [J. Electrochem. Soc. (USA) vol.119 (1972) p.388]
[9]   D.K.Rao, J.Majhi [J. Phys. D (GB) vol.18 no.8 (14 Aug 1985) p.1627-36]
[10]   D.J.Mountain, K.F.Galloway, T.J.Russell [J. Electrochem. Soc. (USA) vol.134 no.3 (Mar 1987) p.747-9]
[11]   S.Kar, D.Shanker, K.S.Chari [Appl. Phys. Lett. (USA) vol.47 no.11 (1 Dec 1985) p.1203-5]
[12]   A.G.Revesz, R.H.Zaininger [RCA Rev. (USA) vol.29 (1968) p.22]

[13] G.Cheroff, F.Fang, F.Hochberg [IBM J. Res. & Dev. (USA) vol.8 (1964) p.416]

[14] H.S.Lehman [IBM J. Res. & Dev. (USA) vol.8 (1964) p.402]

[15] P.Balk [Extended Abstr. Electrochem. Soc. (All Div) (USA) vol.14 no.1 Abst. 109 p.237-40 (Electrochemical Society Spring Meeting, San Francisco, 9-13 May 1965); Abstr. 111 Electrochemical Society Fall Meeting, Buffalo, 1965]

[16] E.Kooi [Philips Res. Rep. (Netherlands) vol.20 no.5 (1965) p.578-94]

[17] P.L.Castro, B.E.Deal [J. Electrochem. Soc. (USA) vol.118 (1971) p.280]

[18] T.W.Hickmott [J. Appl. Phys. (USA) vol.48 (1977) p.723]

[19] B.J.Fishbein, J.T.Watt, J.D.Plummer [J. Electrochem. Soc. (USA) vol.134 no.3 (Mar 1987) p.674-81]

[20] M.S.-C.Luo, C.T.Sah [Proc. Int. Symp. VLSI Technol. Syst. & Applications Chu Tuhng, Hsinchu, Taiwan (31015) 13-15 May 1987 p.111-14 (TSPD(C60),ERSO,ITRI;No.195-4(C60)] To be published in [J. Appl. Phys. (USA) (1987)]

[21] A.J.Chen, S.Dadgar, C.C.H.Hsu, S.Pan, C.T.Sah [J. Appl. Phys. (USA) vol.60 no.4 (15 Aug 1986) p.1391-8]

[22] C.T.Sah, J.Y.C.Sun, J.J-T.Tzou [J. Appl. Phys. (USA) vol.53 no.12 (Dec 1982) p.8886-93]

[23] C.T.Sah, J.Y.C.Sun, J.J.T.Tzou [J. Appl. Phys. (USA) vol.54 no.5 (May 1983) p.2547-55]

[24] C.T.Sah, J.Y.C.Sun, J.J.T.Tzou [J. Appl. Phys. (USA) vol.54 no.10 (Oct 1983) p.5864-79]

[25] C.-T.Sah, J.Y.C.Sun, J.J.-T.Tzou [J. Appl. Phys. (USA) vol.55 no.6 (15 Mar 1984) p.1525-45]

[26] C.T.Sah, W.W.L.Lin, S.Pan, C.C-H.Hsu [Appl. Phys. Lett. (USA) vol.48 no.12 (Mar 1986) p.782-4]

[27] C.T.Sah, W.W.L.Lin, S.Pan, C.C-H.Hsu [Appl. Phys. Lett. (USA) vol.48 no.25 (23 Jun 1986) p.1736-8]

[28] W.W.L.Lin, C.T.Sah [Tech. Report C87257 (Semicond. Res. Corp. (Research Triangle Park, NC, USA, 6 June 1987)]

[29] W.W.L.Lin, C.T.Sah [Effects of hydrogen chloride on the annealing kinetics of interface and oxide traps in oxidized silicon stressed by avalanche electron injection, Submitted for publication as Technical Report (Semicond. Res. Corp., Research Triangle Park, NC, USA)]

[30] D.H.Eaton [PhD dissertation Univ. Illinois, Urbana, USA (Univ. Microfilm, Ann Arbor, Michigan USA)]

[31] C.T.Sah, T.H.Ning, L.L.Tschopp [Surf. Sci. (Netherlands) vol.32 no.9 (Sep 1972) p.561-75]

[32] D.H.Eaton, C.T.Sah [Phys. Status Solidi a (Germany) vol.12 (1972) p.95-109 (See Fig.9 which shows a D(IT) peak about 0.25 eV above the valence band edge or 0.3 eV below midgap in 83 A thick oxide)]

[33] H.Katto, C.T.Sah [Phys. Status Solidi a (Germany) vol.13 (1972) p.417-26 (See Fig.5 which shows two D(IT) peaks about 0.1 and 0.3 eV above the silicon valence band edge in 35 A thick oxide)]

[34] M.L.Reed, J.D.Plummer [IEEE Trans. Nucl. Sci. (USA) vol.NS-33 (1986) p.1198-201]

[35] M.L.Reed, J.D.Plummer [Mater. Res. Soc. Symp. Proc. (USA) vol.52 (1986) p.333-40]

[36] M.L.Reed, J.D.Plummer, [Appl. Phys. Lett. (USA) vol.51 no.7 (17 Aug 1987) p.514-16]

[37] D.L.Griscom [J. Appl. Phys. (USA) vol.58 no.7 (1 Oct 1985) p.2524-33]

[38] D.L.Griscom [J. Non-Cryst. Solids (Netherlands) vol.68 (1984)
      p.301-25]
[39] R.A.B.Devine [Phys. Rev. B (USA) vol.35 no.18 (15 June 1987-II)
      p.9783-9 (see in particular p.9788)]
[40] D.J.DiMaria [Appl. Phys. Lett. (USA) vol.51 no.9 (31 Aug 1987)
      p.655-7]
[41] C.T.Sah [Proc. Int. Symp. VLSI Technol. Syst. & Applications,
      Chu Tuhng, Hsinchu, Taiwan (31015), 13-15 May 1987 (TSPD(C60),
      ERSO,ITRI; no.195-4(C60) p.153-62]
[42] C.T.Sah [Models and experiments on degradation of oxidized
      silicon, Proc. INFOS-87, April 1987, To be published in Appl. Phys.
      Surf. Sci. (Netherlands) (Oct 1987)]; [Publication No.C87209
      (Semicond. Res. Corp. Research Triangle Park, NC, USA, 28 April
      1987) p.71]
[43] E.H.Nicollian, C.N.Berglund, P.F.Schmidt, J.M.Andrews [J. Appl.
      Phys. (USA) vol.42 (1971) p.5654]
[44] F.J.Feigl, D.R.Young, D.J.DiMaria, S.Lai, J.Calise [J. Appl.
      Phys. (USA) vol.52 (1981) p.5655]
[45] D.R.Young, E.A.Irene, R.F.DeKeersmaecker [J. Appl. Phys. (USA)
      vol.50 (1979) p.6336]
[46] M.V.Fischetti, Z.A.Weinberg, J.A.Calise [J. Appl. Phys. (USA)
      vol.57 no.2 (15 Jan 1985) p.418-25]
[47] L.DoThanh, P.Balk [in 'Insulating Films on Semiconductors',
      Eds J.J.Simone, J.Buxo (North-Holland, 1986) p.177]
[48] M.V.Fischetti, R.Gastaldi, F.Maggiono, A.Modelli [J. Appl. Phys.
      (USA) vol.53 (1982) p.3136]
[49] G.Hu, W.C.Johnson [Appl. Phys. Lett. (USA) vol.36 (1980) p.590]
[50] M.V.Fischetti, B.Ricco [J. Appl. Phys. (USA) vol.57 no.8 Pt 1
      (15 Apr 1985) p.2854]
[51] C.T.Sah [EMIS Datareview RN=17815 (Nov 1987) 'Generation-
      recombination-trapping rates and lifetimes of electrons and holes
      in SiO2 films on Si: a general classification of the fundamental
      mechanisms']
[52] C.T.Sah [EMIS Datareview RN=16170 (Nov 1987) 'Interface traps on
      silicon surfaces']
[53] C.T.Sah [EMIS Datareview RN=16180 (Nov 1987) 'Interface traps on
      oxidized silicon from electron spin resonance']
[54] C.T.Sah [EMIS Datareview RN=16181 (Nov 1987) 'Interface
      traps on oxidized silicon from X-ray photoemission spectroscopy,
      MOS diode admittance, MOS transistor and photogeneration
      measurements']

17.5 OXIDE TRAPS ON OXIDIZED Si

by C.T.Sah and C.C.H.Hsu

November 1987
EMIS Datareview RN=16185

A. GENERAL

General background and definitions of terminology and notation were
discussed in detail in Datareview [53] concerning interface traps on
the silicon surface. A brief summary is given here. Traps refer to
electronic traps unless explicitly stated as protonic traps, while
describing their generation and annealing kinetics due to interaction
or reaction with proton, hydrogen atom and hydrogen molecule. Traps are
three dimensional bound states. Traps on oxidized silicon are
categorized according to their origin (intrinsic or extrinsic) and
location (bulk SiO2 or SiO2/Si and gate-conductor/SiO2 interfaces).
Intrinsic traps are missing and interstitial host atoms (Si or O)
and are thought to be due to the dangling bonds. Extrinsic traps are
those from interstitial or substitutional impurities (such as H, B,
P, Na and others). Datareviews [53-56] reviewed interface traps. This
section reviews the oxide (bulk) traps. The distinction between
interface and oxide traps is based on an operational or measurement
definition. The trap is an oxide trap if its location in the oxide is
more than a few lattice constants from the interface and its charge is
a constant in one period of the probing sinusoidal small-signal voltage.
The term 'oxide fixed charges' introduced by the Deal Committee [1] will
be discarded since it has been known for several decades that oxide
charges are not fixed and they can change owing to five fundamental
mechanisms: (i) diffusion, (ii) drift, (iii) bond breaking by energetic
electron or hole impact or by thermal hole capture, (iv) charging-
discharging via electron or hole capture or emission, and (v) electrical
deactivation or activation (absence or presence of electronic bound
states) via reaction with hydrogen (protonic trap action). The steady-
state volume density or concentration of oxide traps will be denoted by
$N(OT)(x)$ while the instantaneous density is $n(OT)(x,t)$. The electron and
hole capture and emission rates at an oxide trap will be denoted by $c(n)$,
$c(p)$, $e(n)$ and $e(p)$ respectively which give the charging and discharging
rates of the trap. The generation (creation or production) rate by an
external stimulant or stress (via bond breaking without healing from
energetic electron or hole impact, thermal hole or even thermal electron
capture, and keV electron or X-ray exposure) will be denoted by an
operational or effective cross section sigma (cm*2). It is defined by
$sigma = [dn(OT)(x,t)/dt]/F$ (partial differential) where F is the fluence
of the stimulant or stress in particle-number/cm*2. Thus, the cross
section is a unique constant if $n(OT)$ is exponentially dependent on the
time of stimulation or $n(OT)$ consists of several single time constant
exponentials which are separated by least-squares-fit (LSF) of the
$n(OT)(x,t)$ versus stress time curve at a constant stimulant or stress
intensity. For example, a constant stimulant or stress is probably
achieved if the gate oxide current is a constant during the stress while
the voltage or the electric field across the oxide is held constant.
Then, $F = J(G)t/q$ where $J(G)$ is the areal density of current passing
through the gate oxide and t is the cumulative duration of the oxide
current flow, excluding the interruptions during which the trap
densities are measured. If $J(G)$ is a function of time, for example, the

- 532 -

gate voltage during the stress is kept constant only, then
F = integral j(G)(t) dt/q is integrated or summed over the times when
the current j(G)(t) is flowing through the gate oxide.

The oxide trap data to be reviewed are all from older and recent
two-terminal capacitance measurements of MOS capacitors by monitoring
the flat-band and/or midgap shift with cumulated stress fluence. Thus,
there is no assurance that the data do not contain a contribution from
the interface traps but these workers have generally assumed that the
interface trap contributions are probably not significant. In contrast,
the data reviewed for the interface traps in reviews [53-56] are all from
interface traps alone whose oxide trap contributions were subtracted out
experimentally. Furthermore, the voltage shift data do not distinguish
a trap generation-annealing process from a trap charging process. Thus,
in most of the reported data, one cannot be certain if the increase or
decrease in the densities of the charged oxide traps is due to the
generation or annealing of the traps or due to the charging or
discharging of existing traps. Unambiguous oxide trap data with
interface trap subtracted out and with a distinction made between
generation-annealing and charging-discharging processes, using the new
measurement technique demonstrated by Sah and associates [2-6], have
just begun to be accumulated and must be left for a future update of
this Datareview.

The atomic origins and chemical bond structures of the oxide traps have
not been directly correlated to the charging-generation-annealing
measurements and one can only speculate. This is due to the very low
density of the traps that are presented in VLSI grade crystalline
silicon (VLSI c-Si). Qualitative speculations of chemical bond
structure of the traps on oxidized silicon bonds were first made by
Hall in 1966 and extended by Sah in 1976 [7] and referenced by
Deal in 1977 [8]. Other reviews on the phenomenological properties of
oxide traps were given by W.C.Johnson [9] and DiMaria [10]. However,
the detailed atomic structures of the electronic traps in oxide film on
silicon have only been indirectly inferred by comparing their electron
spin (paramagnetic) resonance (ESR) spectra with those accurately
obtained on the various forms of bulk SiO2 over the past two decades,
such as glass or vitreous silica, silica optical fibre, synthetic and
natural quartz, and polycrystalline as well as amorphous silicon. Even
this indirect inference must be taken with caution, again due to the
very low density of the traps in VLSI grade oxides. In such an indirect
identification ESR experiment, the oxide on c-Si must be thermally grown
under special conditions, such as high water content, high temperature
heating in reducing or inert (Ar or N2) ambient and very fast quench
to room temperature, in order to give a very high density of interface
traps (approx. 10*13 traps/cm*2 in 1000 A of oxide of > 1 cm*2 area),
due to the low sensitivity of the ESR method. Thus, there is no
assurance that the atomic structures of the high density of oxide traps
thus deduced from ESR are the same as the residual traps in the
low-trap-density (< 10*9 to 10*10 traps/cm*2) VLSI oxides. But, the
only reliable information available to date on the atomic structure of
oxide traps in VLSI oxides is obtained from such an indirect deduction
from the accurate ESR data taken on fused or vitreous silicon and
quartz. Comprehensive and authoritative reviews of the atomic structure
of the intrinsic and extrinsic or impurity related oxide traps in silica
and quartz have been given by the following researchers: Friebele and
Griscom up to and including 1979 [11], Griscom in 1984 [12], and Edwards
and Fowler in 1985 [13]. A summary follows.

There are three major oxide traps detected by ESR experiments in fused silica and quartz. The El' centre is a bridging oxygen vacancy in which a dangling bond is on one Si while the second Si is relaxed backwards to a nearly planar $O3:::Si$. configuration with a hole trapped at the second silicon. The chemical formula can be written as $(Si-O)3:::Si$. The large space between the two dangling bonds emphasizes the lattice distortion or the Si-O bond angle change in one of the two one-dangling bond Si's. The El' centre has apparently a number of charge states [14], -1, 0, +1, +2; each might have a different atomic configuration due to lattice relaxation when a hole or an electron is trapped. It is an intrinsic trap. It is evident that this centre should be present at high concentrations in oxides exposed to an oxygen deficient ambient at high temperatures, for example, post oxidation annealing (POA) in dry Ar or N2 ambient followed by rapid cooling to room temperature in the inert ambient.

The second abundant oxide trap is known as the peroxy radical (PR) consisting of two oxygens attached to one silicon next to an adjacent silicon dangling bond. It is thought that a hole is trapped at the two oxygens. This is an intrinsic trap. Its chemical formula is Si-O-O. Si-O-Si. It is evident that this centre should be abundant in oxides not heated in inert ambient but heated and cooled in oxygen ambient, for example, an oxide heated between 500 deg C and 800 deg C in oxygen after oxidation, a technique used by Sah, Ning and Tschopp to study mobility [58].

The third prominent centre in oxides is the Non-Bridging Oxygen Hole Centre (NBOHC) which is an oxygen dangling bond adjacent to a second oxygen dangling bond which is hydrogenated. Thus, it is a complex extrinsic or impurity centre in which the impurity is a hydrogen. Its chemical formula can be written as $(Si-O)3:::SiO$. $H-OSi:::(O-Si)3$. It is evident that the density of this centre would be high in oxides grown in hydrogen containing ambient such as the wet or steam oxides and HCl oxide (3-9% of HCl is mixed into the dry oxygen during oxidation and HCl has been recently replaced by TCA to prevent corrosion of the stainless steel line feeding the HCl).

In addition to the above references just cited, more recent theoretical and experimental ESR works on the peroxy radical [13] and E' centres [14] have been described by Edwards, Rudra and Fowler. The presence of the E' centre [15] and the peroxy radical [16] and their spatial distributions in the bulk of pure silica optical fibers have been reported recently by Hanafusa, Hibino and Yamamoto of NTT. The presence of the E' and a hydrogen related 74G (ESR spectra with a doublet of 74-Gauss separation) has been recently observed in thermally oxidized VLSI c-Si by Triplett, Takahashi and Sugano [17]. The 74G and a 10.4G doublet were analyzed in detail in fused silica containing high water content by Tsai and Griscom [18] who proposed two new atomic structures: the 74G is a silicon of two dangling bonds one of which is bonded to a hydrogen, $[(SiO)2H]:::Si$. and the 10.4G is a silicon of two dangling bonds one of which is bonded to a OH, $[(SiO)2OH]:::Si$. The annealing kinetics via hydrogen bonding of these oxide traps in vitreous or fused silica detected by ESR have been studied recently in detail by Devine and co-workers [19] and by Griscom [20] who applied a three-dimensional local diffusion-limited kinetic model due to Waite to interpret his one-dimensional isochronal annealing experiments. (See also Datareview [54] for ESR measurements of interface traps.)

There is no doubt that extrinsic oxide traps due to impurities should exist in thermally grown oxide. It is well known that the silicon dopant impurities (B, Al, Ga, P, As and Sb) diffuse into the oxide during oxidation. Thus, an extrapolation of the ESR results from these impurity traps observed at high densities in quartz and fused silica (see the excellent review by Friebele and Griscom [11]) would suggest that these donor and acceptor dopant impurities could give rise to the impurity E' centres (boron E' centre, phosphorus E' centre, etc.); the nonbridging oxygen-boron and oxygen-phosphorus hole centres (also known as the boron-oxygen and phosphorus-oxygen hole centres); and the bridging boron-oxygen hole centre. One would also expect a bridging phosphorus-oxygen hole centre to exist in which the phosphorus substitutes for an oxygen. Electron traps from the donor impurities (P, As, Sb) and hole traps from the acceptor impurities (B, Al, Ga) should also be present when these impurities substitute for silicon in the SiO2. However, they have not been identified during electrical stress of VLSI grade oxides on c-Si. A possible explanation is that their density is too low in normal VLSI-grade thermal oxide on c-Si and their densities in ion implanted (P, As, Sb, B ions) oxides are very high but their presence as electron or hole traps are masked by the still higher densities of the intrinsic traps from displacement damage of the host atoms (Si and O) by the energetic (keV) implanting ions. Still, for ion implanted oxides, the intrinsic traps should be largely annealed out at an elevated temperature between 800-1100 deg C so that the remaining traps due to the donor and acceptor impurities should be observable. No designed experiments have been performed to search for them. In addition, it is possible that some of the energy levels of these impurity traps are outside the range that can be conveniently charged or discharged by thermal electrons or holes injected into or generated in the oxide.

After Frosch and Derrick of Bell Telephone Laboratories introduced oxide as diffusion mask in 1957 [21], Atalla, Tannenbaum and Scheibner of Bell Telephone Laboratories demonstrated successful stabilization of the electric properties of silicon surfaces by a thin layer of oxide [22]. The stability of the electrical characteristics of silicon diodes and transistors protected by the oxide has been under intensive investigation ever since. This interest was further fueled when Atalla and Kahng [23] reported the first succesful operation of the modern version of the metal-oxide-silicon field-effect transistor (MOSFET) using a metal electrode over a thermally grown oxide on c-Si as the gate. Further interest followed when Hoerni [24] disclosed the planar transistor structure using the oxide passivation technique and Noyce [25] invented the interconnection technique for the monolithic integration of diodes, transistors, resistors and capacitors on one piece of silicon.

The principal degradation mechanisms of the transistor and diode characteristics from the presence of oxide traps are due to the charges on the trap. The charges trapped in the oxide will (i) scatter the electrons and holes in the MOSFET surface channel which reduces their mobility and hence the transconductance or gain of the MOSFET [58], and (ii) change the density of the electrons and holes in the MOSFET surface channel which changes the output current of the MOSFET. Effect (ii) is measured by an equivalent shift of the applied gate voltage, known as the gate threshold voltage shift. Effect (ii) will also increase the current through the surface or perimeter region of the space charge layer of a p-n junction, thereby, affecting the leakage current of a diode and degrading the current gain (alpha or beta) of a bipolar transistor. The charged oxide trap can also increase the fringe electric field intensity at the perimeter of the p-n junction causing a serious reduction of the breakdown voltage of the junction.

Fundamental characterizations of the generation-annealing-charging-discharging kinetics of the oxide traps have used several methods to inject or generate the electrons and holes in the oxide so they can be trapped or they can create traps via impact or capture. Most of the detailed but exploratory studies of the kinetics have taken place in the last ten years and have employed methods which simulate the actual operating conditions of a MOSFET. Williams and Goodman [26] reported the first use of the internal photoemission to inject electrons or holes from silicon into the SiO_2 film by raising the electron energy in Si by light to overcome the SiO_2/Si potential barrier (3.1 eV for electrons and about 4.7 eV for holes). The more practical and real-life method was the avalanche electron (or hole) injection (AEI or AHI) method introduced in 1969 by Nicollian, Goetzberger, and Berglund [27] in which the surface space charge layer is biased into current breakdown so that a high density of energetic (or hot) electrons (or holes) are generated by interband impact generation of electron-hole pairs whose electron kinetic energy is greater than 3.1 eV to enable the electron injection into the oxide from Si (or hole kinetic energy greater than 4.7 eV). A nonavalanche electron (or hole) injection method was first demonstrated in 1973 by Verwey [28] who used a forward biased p-n junction in the silicon substrate below the oxide/silicon interface to inject the electrons (or holes) into the high-field and thin Si surface space charge layer where the hot electrons (or holes) can be injected into the oxide by surmounting the SiO_2/Si potential barrier. This method will be known as the BIMOSEI and BIMOSHI (bipolar MOS electron injection and bipolar MOS hole injection) and it is the most promising to give extensive and unambiguous data from which fundamental or quantum-statistical-mechanical generation-annealing-charging-discharging rates of the electronic oxide traps can be obtained. Ning and Yu were the first [29] to use light (or photo-bipolar transistor) instead of forward biased p-n junction to inject the minority carriers. Injection of electrons and holes into the oxide by tunnelling through the potential barrier at the SiO_2/Si interface is also a real-life method which is known as Fowler-Nordheim Tunnelling Electron (or Hole) Injection (FN-TEI and FN-THI). Finally, exposure to keV electrons or X-rays is also a practical method to introduce electrons and holes into the oxide. The effect of these ionizing radiations is especially important since they are the lithography technology for volume production of future generations of integrated circuits whose dimensions are decreasing to below about 0.3 micron, rendering optical lithography useless. A comprehensive review with illustrations of these oxide trap generation-annealing-charging methods and mechanisms was given by Sah [30] and a condensed review is given in [57]. This Datareview will be limited to the currently most popular methods with the main focus on data obtained from AEI since data from AIH, FN-TEI, FN-THI, BIMOSEI and BIMOSHI are still sparse and intensive efforts have just begun to provide a large data base.

In view of the difficulty or impossibility of direct identification of the atomic structures of the oxide traps on oxidized VLSI silicon owing to the very low trap density, this review will focus on the data of trap density and the effective trap generation-annealing-charging-discharging cross-section measured by the high-frequency capacitance voltage (HFCV) method, using the flat band or/and midgap voltage shift as the monitor of the trap density. We will first review the dependences of the density and cross section of electron traps on the oxidation conditions, post oxidation processes, gate materials, stress temperature, impurities present in the oxide and ionizing radiation. For hole traps, their dependences on processing conditions will be reviewed at the end of this section.

B. ELECTRON TRAPS IN SiO2

Early results are summarized in TABLE 1. Some of the process parameters may not have been controlled and were not specified by the authors. For example, the POC (post oxidation cooling) rates and ambient were not always given although presumably they are kept constant in a group of oxidized silicon wafers when experiments were designed to isolate the effect of one process variable. The acronyms used in the tables are:

```
OXD     = Oxidation
HIPOX   = High Pressure Steam Oxidation
HIPDOX  = High Pressure Dry Oxidation
NA      = Not Available
ND      = Not Detected
POA     = Post Oxidation Annealing
PMA-FG  = Post Metallization Annealing in Forming Gas (10%H2, 90%N2)
AEI     = Avalanche Electron Injection
AHI     = Avalanche Hole Injection
FN-TEI  = Fowler-Nordheim Tunnelling Electron Injection
PTEI    = Photo Transistor Electron Injection
VUVEI   = Vacuum Ultra-Violet Light Electron Injection
```

TABLE 1: early measurements of oxide electron traps

Variable	Fabrication condition & gate material	Stress method	Density N(OT) (/cm*2)	Effective cross section (cm*2)	Refs
OXD	p-Si, Dry	AEI	NA	1.5 X 10*-17	[31]
PMA	p-Si, Dry, FG	PTEI	1.6 X 10*11	3.3 X 10*-13	[29]
			1.0 X 10*12	2.4 X 10*-19	
POA	p-Si, Dry, FG	PTEI	-	0.2 X 10*-13	[32]

Extensive measurements of the density and effective generation-charging cross section of electron traps during avalanche electron injection have been made in the last ten years as a function of the following conditions or parameters: oxidation, post oxidation annealing, post metallization annealing, gate material, oxide impurity, and ionizing irradiation. These results are summarized in the following tables. The trap signature is its effective generation-charging cross section listed in the fifth column.

TABLE 2: effects of oxide growth condition on oxide electron traps

Variable	Fabrication condition & gate material	Stress method	Density N(OT) (/cm*2)	Effective cross section (cm*2)	Refs
OXD	p-Si, no POA, <100>	AEI			[33]
	Dry		1.7 X 10*12	NA	
	Dry, 3% HCl		1.8 X 10*12	NA	
	Dry, 3% H20		2.3 X 10*12	NA	
	Steam		3.9 X 10*12	NA	
	CVD		5.6 X 10*12	NA	
	CVD, B-doped		5.7 X 10*12	NA	
OXD	p-Si, POA	AEI			[3]
	HIPOX, <111>		9.8 X 10*11	2.5 X 10*-18	
			2.6 X 10*11	2.8 X 10*-19	
			7.8 X 10*11	1.5 X 10*-20	
			NA	NA	
	Dry, <100>		< 10*10	NA	
			1.4 X 10*11	2.8 X 10*-19	
			9.9 X 10*11	3.4 X 10*-20	
			1.5 X 10*11	4.0 X 10*-21	
OXD 685C	Dry-HIPDOX no POA	AEI	3.0 X 10*12 1.0 X 10*12	4.0 X 10*-18 4.0 X 10*-19	[34]
	Dry-HIPDOX POA	AEI	3.0 X 10*12 9.0 X 10*11	3.0 X 10*-18 2.0 X 10*-19	
	Wet-HIPDOX no POA	AEI	4.0 X 10*12 3.0 X 10*12	1.0 X 10*-17 1.0 X 10*-18	
	Wet-HIPDOX POA	AEI	2.0 X 10*12 2.0 X 10*12	1.0 X 10*-17 1.0 X 10*-18	
1000C	760 mm H20 no POA	AEI	8.0 X 10*11 NA	5.0 X 10*-18 10*-19	
	Dry, POA	AEI	1.1 X 10*12 NA	4.0 X 10*-18 10*-19	

TABLE 3: effects of post oxidation processes on oxide electron traps

Variable	Fabrication condition and gate material	Stress method	Density N(OT) (/cm*2)	Effective cross section (cm*2)	Refs
POA	n-Si, Dry	VUVEI			[35]
	no POA		7.2 X 10*12	2.6 X 10*-13	
	POA-02, 600C, 1hr		5.5 X 10*12	3.0 X 10*-13	
	POA-02, 600C, 10hr		5.0 X 10*12	3.3 X 10*-13	
	POA-02, 800C, 1hr		2.5 X 10*12	4.8 X 10*-13	
POA	p-Si, Dry	AEI			[36]
	POA-FG, 30 min, PMA		1.5 X 10*12	2.9 X 10*-18	
			4.1 X 10*11	5.0 X 10*-19	
			1.1 X 10*12	4.7 X 10*-20	
	POA-N2, 30 min, PMA		6.7 X 10*11	2.9 X 10*-18	
			2.2 X 10*11	4.4 X 10*-19	
			6.5 X 10*11	1.0 X 10*-19	
	POA-N2, 60 min, PMA		5.0 X 10*11	1.9 X 10*-18	
			1.2 X 10*11	4.9 X 10*-19	
			6.3 X 10*11	9.0 X 10*-20	
	POA-N2, 30 min, no PMA		1.1 X 10*12	7.0 X 10*-18	
			1.8 X 10*11	5.3 X 10*-19	
			5.8 X 10*11	9.0 X 10*-20	
POA	p-Si, Dry	AEI			[37]
	None		1.5 X 10*11	1.0 X 10*-17	
			5.0 X 10*11	2.0 X 10*-18	
	H20 diffused		1.0 X 10*13	1.0 X 10*-17	
			1.0 X 10*12	2.0 X 10*-18	
	FG annealed		3.0 X 10*10	1.0 X 10*-17	
			1.5 X 10*12	2.0 X 10*-18	
H20 Diffused	p-Si, Dry, POA None		1-10 X 10*10	1-3 X 10*-17	[38]
			5-15 X 10*11	1-4 X 10*-18	
			1 X 10*12	2-3 X 10*-19	
	H20 diffused		2-3 X 10*12	1.0 X 10*-17	
AEI-temp POA-time POA-temp	p-Si, dry, <100>	AEI			[39]

Variable	Fabrication condition and gate material	Stress method	Density N(OT) (/cm*2)	Effective cross section	Refs
	POA 0h	400K	1.0 X 10*10	1.4 X 10*-17	
			7.0 X 10*11	1.6 X 10*-18	
			6.3 X 10*11	3.5 X 10*-19	
	5h	400K	1.0 X 10*10	1.4 X 10*-17	
			2.5 X 10*10	2.4 X 10*-18	
			1.8 X 10*11	4.8 X 10*-19	
	10h	400K	1.0 X 10*10	1.4 X 10*-17	
			2.3 X 10*10	2.3 X 10*-18	
			1.5 X 10*11	5.0 X 10*-19	
	POA 0h	100K	< 1 X 10*10	10*-15	
			< 1 X 10*10	10*-15	
			7.0 X 10*10	1.0 X 10*-16	
	5h	100K	5.4 X 10*10	3.6 X 10*-15	
			1.5 X 10*11	9.0 X 10*-16	
			3.5 X 10*11	1.5 X 10*-16	
	10h	100K	1.2 X 10*11	3.6 X 10*-15	
			2.3 X 10*11	9.0 X 10*-16	
			4.7 X 10*11	1.0 X 10*-16	
	POA 1h 900C	100K	< 1 X 10*10	10*-15	
			2.7 X 10*10	9.0 X 10*-15	
			1.2 X 10*10	1.0 X 10*-16	
	1000C		1.3 X 10*10	6.0 X 10*-15	
			9.8 X 10*11	3.0 X 10*-15	
			2.7 X 10*11	1.6 X 10*-16	
	1050C		4.9 X 10*10	1.4 X 10*-14	
			1.8 X 10*11	1.8 X 10*-15	
			4.7 X 10*11	2.4 X 10*-16	

TABLE 4: effects of gate materials on oxide electron traps

Variable	Fabrication condition & gate material	Stress method	Density N(OT) (/cm*2)	Effective cross section (cm*2)	Refs
GATE MAT	p-Si, X-ray/PMA-FG/AEI-295K				[40]
	Al gate	+VG AEI	8.0 X 10*11	2.2 X 10*-13	
			1.0 X 10*12	5.0 X 10*-14	
	Poly-Si gate	+VG AEI	5.4 X 10*11	2.6 X 10*-13	
			1.2 X 10*12	4.8 X 10*-14	
	Al gate	-VG AEI	NA	10*-13	
			1.9 X 10*11	7.0 X 10*-14	
	Poly-Si gate	-VG AEI	4.4 X 10*11	2.7 X 10*-13	
			6.0 X 10*11	3.7 X 10*-14	
GATE MAT	p-Si <100> POA AEI				[41]
	Al		4.9 X 10*11	1.9 X 10*-17	
			3.4 X 10*11	3.7 X 10*-18	
			2.8 X 10*11	2.1 X 10*-19	
	Au		1.0 X 10*12	1.9 X 10*-17	
			NA	10*-18	
			3.3 X 10*11	5.0 X 10*-19	
	Mg-Al		3.6 X 10*11	3.6 X 10*-17	
			9.6 X 10*11	1.3 X 10*-18	
			NA	10*-19	
	Mg-Au		7.4 X 10*11	4.4 X 10*-17	
			6.4 X 10*11	1.9 X 10*-18	
			NA	10*-19	
GATE MAT	p-Si, Dry AEI				[42]
	no POA, Al-gate		6.1 X 10*11	2.8 X 10*-18	
			2.1 X 10*11	3.2 X 10*-17	
	POA, Al-gate		3.6 X 10*11	2.5 X 10*-18	
			3.0 X 10*11	1.7 X 10*-17	
	no POA, n+ poly-Si		3.9 X 10*11	5.8 X 10*-19	
			4.9 X 10*10	5.0 X 10*-18	

TABLE 5: effects of impurity in SiO2 on oxide electron traps

Variable	Fabrication condition and gate material	Stress method	Density N(OT) (/cm*2)	Effective cross section (cm*2)	Refs
NaCl	p-Si, Dry, PMA-FG	AEI	[NaCl] (note 1)	2.0 X 10*-15	[43]
			[NaCl] (note 2)	1.0 X 10*-17	
			[NaCl] (note 1)	2.0 X 10*-19	
			[NaCl] (note 1)	5.0 X 10*-20	
			[NaCl] (note 2)	6.0 X 10*-21	
As+ Ion Implant	p-Si <100>, Dry, POA-N2	AEI			[44]
			6-74 X 10*11	4-16 X 10*-15	
			2-6 X 10*12	1-2 X 10*-15	
			3-6 X 10*12	2-6 X 10*-16	
			5-7 X 10*12	1-4 X 10*-17	

note 1: N(OT) is proportional to [NaCl]
note 2: N(OT) is not proportional to [NaCl]

TABLE 6: effects of ionizing radiation on oxide electron traps

Variable	Fabrication condition & gate material	Stress Method	Density N(OT) (/cm*2)	Effective cross section (cm*2)	Refs
X-ray	p-Si, PMA-FG	AEI 295K	< 1 X 10*10	NA	[45]
			7.0 X 10*10	1.0 X 10*-17	
			2.5 X 10*11	3.8 X 10*-18	
			3.6 X 10*11	4.7 X 10*-19	
	p-Si, no PMA	X-ray +	2.5 X 10*12	> 4 X 10*-14	
	p-Si, PMA-FG	AEI 295K	4.9 X 10*11	0.8 X 10*-15	
			5.9 X 10*11	3.9 X 10*-18	
			1.2 X 10*11	5.5 X 10*-19	

Variable	Fabrication condition & gate material	Stress method	Density N(OT) (/cm*2)	Effective cross section	Refs
E-beam	p-Si, PMA-FG control-no-E-beam	AEI 77K	1.0 X 10*10 4.7 X 10*10 5.2 X 10*11 7.9 X 10*11 7.8 X 10*11	1.6 X 10*-14 1.6 X 10*-15 1.0 X 10*-16 1.7 X 10*-17 1.0 X 10*-18	[46]
	p-Si p-Si, PMA-FG	E-beam AEI 77K	3.0 X 10*10 9.3 X 10*10 6.1 X 10*11 8.9 X 10*11 1.1 X 10*12	1.0 X 10*-14 1.7 X 10*-15 1.4 X 10*-16 1.9 X 10*-17 1.0 X 10*-18	
	p-Si, PMA-FG	AEI 295K	5.0 X 10*9 NA NA 1.9 X 10*11 2.6 X 10*11	1.0 X 10*-14 10*-15 10*-16 3.0 X 10*-17 1.4 X 10*-18	
	p-Si, no PMA p-Si, PMA-FG	E-Beam AEI 295K	2.0 X 10*10 4.4 X 10*10 8.4 X 10*10 2.6 X 10*11 4.7 X 10*11	1.2 X 10*-14 1.6 X 10*-15 2.0 X 10*-16 7.0 X 10*-17 1.0 X 10*-18	
E-beam	p-Si,Dry, HCl,	Wet 77K	NA	1.0 X 10*-15	[47]
E-beam	p-Si, <100>, Dry no POA, PMA-FG	AEI 0 keV 2 keV 5 keV 8 keV	2.1 X 10*12 2.9 X 10*12 2.7 X 10*12 7.4 X 10*12	2.7 X 10*-18 3.5 X 10*-18 3.1 X 10*-18 4.5 X 10*-17	[48]

C. HOLE TRAPS IN SiO2

TABLE 7 summarizes the effects of oxide fabrication processing on the density and effective generation-charging cross-section of oxide hole traps. Since hole capture at the SiO2/Si interface also generates a high density of interface traps, some of which are positively charged, a careful separation of the oxide hole traps from the interface traps in the flat-band and midgap voltage shift data is mandatory to characterize oxide hole traps. Unfortunately, this separation has not been made since an experimental technique was not available until recently [5,6]. Thus, the data in TABLE 7 reflect the total negative shift of the flat-band or/and midgap voltages or the build-up of the net positive charge in the oxide. It may contain a significant contribution from the donor-like positively charged interface traps.

TABLE 7: effects of processing on oxide hole traps

Variable	Fabrication condition and gate material	Stress method	Density N(OT) (/cm*2)	Effective cross section (cm*2)	Refs
POA	n-Si, Steam,HCl POA-He-1000C POA-H2-500C	Corona discharge	4.0 X 10*12	NA	[49]
None	n-Si <100>, dry, POA-N2, PMA-FG	IPE	1.4 X 10*13	3.1 X 10*-13	[50]
AHI-temp	n-Si, dry, POA-N2,PMA-FG	AHI 295K	1.7 X 10*12	2.7 X 10*-14	[51]
		AHI 77K	7.0 X 10*11	1.0 X 10*-13	
		AHI 77K	2.0 X 10*12	1.9 X 10*-14	
POA-temp	n-Si, dry	VUV			[35]
	POA-02 600C 0h		7.7 X 10*12	8.0 X 10*-14	
	POA-02 600C 1h		6.4 X 10*12	6.0 X 10*-13	
	POA-02 800C 1h		1.6 X 10*12	6.0 X 10*-14	
			3.4 X 10*12	1.1 X 10*-15	
POA	n-Si <100>, dry	AHI			[52]
	no POA		6.0 X 10*11	4.0 X 10*-14	
			4.0 X 10*11	2.0 X 10*-13	
	POA-N2 30 min		2.7 X 10*12	4.0 X 10*-14	
			1.2 X 10*12	2.0 X 10*-13	
OXD-temp	n-Si <100>	AHI			[34]
	HIPDOX 635C		2.2 X 10*12	0.9 X 10*-15	
			3.8 X 10*12	1.6 X 10*-14	
	685C		1.4 X 10*12	2.2 X 10*-15	
			3.8 X 10*12	1.6 X 10*-14	
	800C		1.2 X 10*12	1.4 X 10*-15	
			2.6 X 10*12	1.3 X 10*-14	
	900C		1.2 X 10*12	1.6 X 10*-15	
			2.6 X 10*12	1.0 X 10*-14	
	1000C		1.3 X 10*12	2.6 X 10*-15	
			2.6 X 10*12	1.7 X 10*-14	
	1100C		7.0 X 10*11	2.0 X 10*-15	
			6.8 X 10*12	2.8 X 10*-14	

REFERENCES
----

[1]   B.E.Deal [J. Electrochem. Soc. (USA) vol.127 no.4 (1980)
       p.979-81]
[2]   C.T.Sah, J.Y.-C.Sun, J.J.-T.Tzou [J. Appl. Phys. (USA) vol.54
       no.5 (May 1983) p.2547-55]
[3]   C.T.Sah, J.Y-C.Sun, J.J.-T.Tzou, [J. Appl. Phys. (USA) vol.54
       no.10 (Oct 1983) p.5864-79]
[4]   C.T.Sah, J.Y.-C.Sun, J.J.-T.Tzou [J. Appl. Phys. (USA) vol.55
       no.6 (15 Mar 1984) p.1525-45]
[5]   C.T.Sah, W.W-L.Lin, S.Pan, C.Hsu [Appl. Phys. Lett. (USA) vol.48
       no.12 (24 Mar 1986) p.782-4]
[6]   C.T.Sah, W.W-L.Lin, S.Pan, C.Hsu [Appl. Phys. Lett. (USA) vol.48
       no.25 (23 Jun 1986) p.1736-8]
[7]   C.T.Sah [IEEE Trans. Nucl. Sci. (USA) vol.23 no.6 (Dec 1976)
       p.1563-8]
[8]   B.E.Deal [J. Electrochem. Soc. (USA) vol.121 no.6 (Jun 1974)
       p.198C-205C]; [Semiconductor Silicon-1977, Eds H.R.Huff, E.Sirtl
       (The Electrochem. Soc., Princeton, New Jersey, 1977) p.276-96]
[9]   W.C.Johnson [IEEE Trans. Nucl. Sci. (USA) vol.NS-22 no.6 (Dec
       1975) p.2144-50]
[10]   D.J.DiMaria [The Physics of SiO2 and Its Interfaces (Pergamon
       Press, New York, 1978) p.160-77]
[11]   E.J.Friebele, D.L.Griscom [in 'Treatise Materials Science and
       Technology' vol.17 Eds M.Tomozawa, R.H.Doremus (Academic Press,
       New York, 1979) p.257-351]
[12]   D.L.Griscom [J. Non-Cryst. Solids (Netherlands) vol.73 (1985)
       p.51]
[13]   A.H.Edwards, W.B.Fowler [Phys. Rev. B (USA) vol.26 no.12 (15 Dec
       1982) p.6649-60]; [in 'Structure and Bonding in Non-crystalline
       Solids' Eds G.E.Walrafen, A.G.Revesz (Plenum Press, New York,
       1986) p.157-83]; A.H.Edwards, W.B.Fowler [ibid p.139-55]
[14]   J.K.Rudra, W.B.Fowler [Phys. Rev. B (USA) vol.35 no.15
       (15 May 1987) p.8223-30]
[15]   H.Hanafusa, Y.Hibino, F.Yamamoto [J. Appl. Phys. (USA) vol.58
       no.3 (1 Aug 1985) p.1356-61]
[16]   Y.Hibino, H.Hanafusa [J. Appl. Phys. (USA) vol.62 no.4 (15 Aug
       1987) p.1433-6]
[17]   B.B.Triplett, T.Takahashi, T.Sugano [Appl. Phys. Lett. (USA)
       vol.50 no.23 (8 Jun 1987) p.1663-5]
[18]   T-E.Tsai, D.L.Griscom [J. Non-Cryst. Solids (Netherlands) vol.91
       (1987) p.170]
[19]   R.A.B.Devine [Phys. Rev. (USA) vol.35 no.18 (15 June 1987)
       p.9783-9] and cited references of the author's past publications
[20]   D.L.Griscom, M.Stapelbroek, E.J.Friebele [J. Chem. Phys. (USA)
       vol.78 no.4 (15 Feb 1983) p.1638-51]; D.L.Griscom [J. Non-Cryst.
       Solids (Netherlands) vol.68 (1984) p.301-25]
[21]   C.J.Frosch, L.Derrick [J. Electrochem. Soc. (USA) vol.104 no.5
       (May 1957) p.547-52]
[22]   M.M.Atalla, E.Tannenbaum, E.J.Scheibner [Bell Syst. Tech. J.
       (USA) vol.38 (May 1959) p.749-83] See also M.M.Atalla [in
       'Properties of Elemental and Compound Semiconductors' Ed. H.Gatos,
       vol.5 (Interscience, New York, 1960) p.163-81]
[23]   D.Kahng, M.M.Atalla [IRE-AIEE Solid-State Device Research Conf.
       (Carnegie Institute of Technology, Pittsburg, PA, June 1960)]

[24] J.A.Hoerni [IEEE Electron Devices Meeting (Washington, DC, October
27-29 1960) Technical Article and Paper Series, No.TP-14 1961
Fairchild Semiconductor Corporation, (645 Whisman Road, Mountain
View, California, USA)]

[25] R.N.Noyce [IEEE Trans. Electron Devices (USA) vol.ED-23 no.7
(July 1976) p.648-54]

[26] R.Williams [Phys. Rev. (USA) vol.140 no.2A (18 Oct 1965)
p.A569-75]; A.M.Goodman, [Phys. Rev. (USA) vol.152 no.2
(9 Dec 1966) p.780-4]

[27] E.H.Nicollian, A.Goetzberger, C.N.Berglund [Appl. Phys. Lett.
(USA) vol.15 no.6 (15 Sep 1969) p.174-7]

[28] J.F.Verwey [J. Appl. Phys. (USA) vol.44 no.6 (1 Jun 1973)
p.2681-7]

[29] T.H.Ning, H.N.Yu [J. Appl. Phys. (USA) vol.45 no.12 (12 Dec 1974)
p.5373-8]

[30] C.T.Sah [Proc. INFOS-87 (April 1987), Special Issue, Applied
Surface Science, Oct 1987. Also distributed as publication
No.C87209, by Semiconductor Research Corporation (Research Triangle
Park, North Carolina, USA, 28 April 1987) p.71]

[31] E.H.Nicollian, C.N.Berglund, P.F.Schmidt, J.M.Andrews [J. Appl.
Phys. (USA) vol.42 no.13 (Dec 1971) p.5654-64]

[32] T.N.Ning, C.M.Osburn, H.N.Yu [Appl. Phys. Lett. (USA) vol.26 no.5
(1 Mar 1975) p.248-50]

[33] R.A.Gdula [J. Electrochem. Soc. (USA) vol.123 no.1 (Jan 1976)
p.42-7]

[34] L.P.Trombetta, R.J.Zeto, F.J.Feigl, M.E.Zvanut [J. Electrochem.
Soc. (USA) vol.132 (Nov 1985) p.2706-13]

[35] A.R.Stivers, C.T.Sah [J. Appl. Phys. (USA) vol.51 no.12 (Dec 1980)
p.6292-304]

[36] D.R.Young [J. Appl. Phys. (USA) vol.52 no.6 (Jun 1981) p.4090-4]

[37] A.Harstein, D.R.Young [Appl. Phys. Lett. (USA) vol.38 no.8
(15 Apr1981) p.631-3]

[38] F.J.Feigl, D.R.Young, D.J.DiMaria, S.Lai, J.Calise [J. Appl.
Phys. (USA) vol.52 no.9 (Sep 1981) p.5665-82]

[39] P.Balk, M.Aslam, D.R.Young [Solid-State Electron. (GB) vol.27
no.8/9 (1984) p.709-19]

[40] J.M.Aitken, D.J.DiMaria, D.R.Young [IEEE Trans. Nucl. Sci.
(USA) vol.23 no.6 (Dec 1976) p.1526-33]

[41] M.V.Fischetti, Z.A.Weinberg, J.A.Calise [J. Appl. Phys. (USA)
vol.57 no.2 (15 Jan 1985) p.418-25]

[42] L.Dori, M.Arienzo, T.N.Nguyen, M.V.Fischetti, K.J.Stein [J. Appl.
Phys. (USA) vol.61 no.5 (1 Mar 1987) p.1910-15]

[43] D.J.DiMaria, J.M.Aitken, D.R.Young [J. Appl. Phys. (USA) vol.47
no.6 (Jun 1976) p.2740-3]

[44] R.F.DeKeersmaecker, D.J.DiMaria [Inst. Phys. Conf. Ser. (GB)
no.50 (1980) p.40-7]

[45] J.M.Aitken, D.R.Young, K.Pan [J. Appl. Phys. (USA) vol.47 no.3
(Mar 1976) p.1196-8]

[46] J.M.Aitken, D.R.Young, K.Pan [J. Appl. Phys. (USA) vol.49 no.6
(Jun 1978) p.3386-91]

[47] T.H.Ning [J. Appl. Phys. (USA) vol.49 no.12 (Dec 1978)
p.5997-6003]

[48] C.T.Sah, J.Y-C.Sun, J.J-T.Tzou [J. Appl. Phys. (USA) vol.54 no.8
(Aug 1983) p.4378-81]

[49] M.H.Wood, R.Williams [J. Appl. Phys. (USA) vol.47 no.3 (Mar 1976)
p.1082-9]

[50] T.H.Ning [J. Appl. Phys. (USA) vol.47 no.3 (Mar 1976) p.1079-81]

[51] J.M.Aitken, D.R.Young [IEEE Trans. Nucl. Sci. (USA) vol.NS-24 no.6 (Dec 1977) p.2128-34]

[52] M.Aslam, P.Balk [Insulating Films on Semiconductors (North Holland Publishing Co., Amsterdam, 1983) p.103-6]

[53] C.T.Sah [EMIS Datareview RN=16170 (Nov 1987) 'Interface traps on Si surfaces']

[54] C.T.Sah [EMIS Datareview RN=16180 (Nov 1987) 'Interface traps on oxidized Si from electron resonance')]

[55] C.T.Sah [EMIS Datareview RN=16181 (Nov 1987) 'Interface traps on oxidized Si from X-Ray photoemission spectroscopy, MOS diode admittance, MOS transistor and photogeneration measurements']

[56] C.T.Sah, M.S.C.Luo, C.C.H.Hsu, T.Nishida, A.J.Chen, C.Y.P.Chao [EMIS Datareview RN=16182 (Nov 1987) 'Interface traps on oxidized Si from two-terminal, dark capacitance-voltage measurements on MOS capacitors']

[57] C.T.Sah [EMIS Datareview RN=17815 (Nov 1987) 'Generation-recombination-trapping rates and lifetimes of electrons and holes in SiO2 films on Si : a general classification of the fundamental mechanism']

[58] C.T.Sah, T.Nishida [EMIS Datareview RN=17831 (Nov 1987) 'Electron mobility in oxidized Si surface layers']

17.6 MICROSCOPIC IDENTITY OF CHARACTERISTIC GAP STATES AT THE INTERFACE IN
OXIDE-ON-Si STRUCTURES

by N.M.Johnson and E.H.Poindexter

November 1987
EMIS Datareview RN=16169

Electrical measurements on metal-oxide-silicon (MOS) devices reveal
both a continuous distribution of interface states that extends
throughout the silicon forbidden energy band and a random spatial
distribution of fixed charge situated in the oxide near the interface.
In MOS devices fabricated for integrated-circuit applications, the
interface-state distribution is generally found to be U-shaped with the
density increasing monotonically with energy from midgap toward each
band edge [1]. Such distributions are obtained by the combination of a
high-temperature, post-oxidation anneal and a low-temperature, post-
metallization anneal, with process optimization typically yielding
midgap densities of less than 1 X $10*10/eV/cm*2$ on (100)-oriented
silicon. Without these anneals the SiO2/Si interface possesses residual
electronic defects that may be considered a consequence of or
characteristic of the thermal oxidation of silicon. These defects
contribute bands of interfacial electronic states and are paramagnetic.
They can be ranked among the better understood deep-level defects in
semiconductors and provide the only example where interfacial electronic
gap states have been microscopically identified. This Datareview
summarizes the current understanding of these interfacial defects, with
references to the primary literature for further study.

Characteristic interface states were first observed [2,3] in MOS
capacitors fabricated on (100)-oriented silicon wafers but have been
most extensively studied on (111)-oriented silicon where the highest
areal density of interfacial defects has been obtained. Thermal
oxidation in dry O2 and rapid cooling to room temperature in the same
ambient contribute to the maximization of the density of these defects.
Test-device fabrication must be completed without further perturbation
of the interface. In particular, evaporation of the gate metal must be
performed without exposing the wafer to ionizing radiation (e.g. soft
X-rays in an electron-beam evaporator) since, in order to study the
oxidation-related defects, a post-metallization anneal cannot be
utilized to remove the radiation damage.

Both capacitance-voltage (CV) measurements and deep-level transient
spectroscopy (DLTS) reveal that on as-oxidized (111) silicon the
interface state distribution is dominated by two broad peaks that have
equal peak densities within experimental uncertainty: one is centred
approximately 0.3 eV above the silicon valence-band maximum (i.e.,
Ev + 0.3 eV) and the other at approximately 0.25 eV below the
conduction-band minimum (i.e. Ec-0.25 eV) [4,5]. Each of these
characteristic bands of interface states may be approximated by a
Gaussian distribution centred at the peak energy with a standard
deviation of approx. 0.1 eV. Peak densities between 1 and 2 X $10*13/eV$
$/cm*2$ have been reported, and both peaks are removed by a post-
metallization anneal at 450 deg C.

The above oxidation conditions also maximize the density of an
interfacial paramagnetic defect as detected by electron spin resonance
(ESR). The defect is historically termed the 'P'sub'b' centre, hereafter
designated P(b), and displays a distinct resonance spectrum that is

anisotropic with respect to the direction of the magnetic field [6-9]. As-oxidized (111)Si typically possesses approximately 2 X 10*12 defects per cm*2. The principal parameters for the Zeeman interaction, when compared to many different well-characterized paramagnetic defects in bulk Si and SiO2, strongly indicate that the P(b) defect is a trivalent silicon atom with a nonbonding or 'dangling' orbital aligned along the single [111] direction, perpendicular to the interface. The principal components of the g-dyadic for the Zeeman interaction are g(parallel) = 2.0016 +- 0.0003 and g(perpendicular) = 2.0090 +- 0.0003 [10]. Weaker resonances indicative of dangling orbitals associated with the other <111> bonds canted with respect to the (111) interface have been obtained from low-pressure oxidation [11]. Placement of the P(b) defect at the interface is indicated by controlled chemical etch back of the SiO2 layer, by the intimate coupling of the P(b) centre to Si inversion-layer electrons (i.e. spin relaxation), and by the absence of a plausible explanation for the unique orientation of the defect if it were situated in either the amorphous SiO2 or the Si bulk.

Observation by ESR of the hyperfine spectrum of the P(b) centre was essential to confirm and extend, in combination with model calculations, the chemical identity and electronic structure of these interfacial defects. The principal parameters of the A-dyadic for the hyperfine interaction between the unpaired electron and a single Si-29 atom are A(parallel) = 146 (+-5) X 10*-4/cm and A(perpendicular) = 85 (+-8) X 10*-4/cm [10]. The hyperfine spectrum for oxidized (111)Si confirmed that the chemical species on which 80% of the unpaired electron density is localized in the P(b) centre is a substitutional Si atom on the silicon side of the Si-SiO2 interface and that the hybrid orbital at this Si site is 12% s-like and 88% p-like with the p-orbital pointed along the [111] direction [10,12,13]. Similar results have been reported for P(b) centres on the surface of SiO2 precipitates in Si which were introduced by high-dose oxygen implantation with a subsequent furnace anneal to form a buried oxide layer [14]. Theoretical calculations of the hyperfine interactions for the trivalent silicon defect model of the P(b) centre are in excellent agreement with experiment [15,16]. However, an alternative model has been proposed in which the P(b) centre consists of a five fold-coordinated Si atom, with the unpaired electron in a floating-bond state [17].

The characteristic interface states are an electrical manifestation of an oxidation-induced interfacial defect. A correlation between these interface states and P(b) centres was demonstrated with isochronal anneals of oxidized Si (coated with Al) which showed that the spin density and the densities of the characteristic peaks have the same annealing kinetics [4]. In addition, the areal density of interface states obtained from integration of the energetic distribution for either band of characteristic gap states, agrees with the P(b) spin density to within experimental uncertainty. However, if P(b) centres contribute localized states in the Si band gap, it must be possible to change the electron occupancy of the dangling orbital by electrical or optical techniques. This was indeed accomplished by use of the electric-field effect in the MOS structure to shift the intercept of the Fermi level at the SiO2/Si interface while monitoring the spin density by ESR [18-21,4]. The existence of gap states associated with P(b) centres was further demonstrated by spin-dependent recombination [22] and by optically-detected magnetic resonance [23], although these techniques do not establish the distribution of the deep levels. In addition, theoretical calculations in general predict the existence of gap states for the interfacial trivalent silicon defect [24-30,16].

Systematic studies have been reported for large-area (e.g. 0.6 cm*2) MOS capacitors in which the interface-state occupancy was controlled with a gate voltage applied during the ESR measurement: sweeping the Si band gap at the interface through the Fermi level Ef revealed that the spin density is approximately constant as Ef shifts through the central region of the band gap but decreases as Ef approaches the band edges [4,5]. This identifies an amphoteric defect with both electronic transitions in the band gap: when Ef intersects the interface near the Si valence band, the centre is diamagnetic; for Ef near midgap, the centre contains an unpaired electron and yields the P(b) spin signal; and as Ef approaches the conduction band, the centre acquires a second electron which pairs with the first to again produce a diamagnetic centre. Conventional CV analysis was used to relate Ef at the interface to the gate voltage (via the Si surface potential; see ref [1]), which revealed that the electronic transition energies for the P(b) centre coincide with the characteristic peaks in the interface-state distribution [4,5]. Thus, both the annealing kinetics and the voltage dependence of the P(b) centre support the assignment of the 0-1 electron (or 2-1 hole) transition of the amphoteric centre to the Ev + 0.3 eV peak and the 1-2 electron transition to the Ec-0.25 eV peak; the singly occupied centre is charge neutral. The characteristic peaks are separated by an effective correlation energy Ueff of approximately 0.6 eV, which is the difference between the energies required to add an electron to a singly occupied orbital and to an unoccupied centre; it is the sum of two terms, a Coulomb repulsion and a compensating lattice relaxation energy. The measured Ueff is in good agreement with theoretical calculations for the trivalent silicon defect at the SiO2/Si interface [27,30].

Further support for the association of the characteristic peak at Ev + 0.3 eV with an electron transition of the P(b) centre was provided by spin-dependent DLTS [31]. However, this study reported a Ueff of approx. 0.1 eV; this discrepancy with the above cited work has not been resolved, although a theoretical interpretation has been suggested [16].

Two different P(b) centres are observed on (100)Si:P(b0) which is essentially identical to the sole P(b) centre observed on (111)Si and P(b1) which is as yet unidentified but clearly different in nature from P(b0). The following are the principal values of the g-dyadics as reported in ref [32] (in parentheses are specified the Si crystal axes along which the values were observed): for P(b0), g1 = 2.0015 (parallel to [111]), g2 = 2.0080 (between [001] and [bar1 bar1 1] axes), and g3 = 2.0087 (parallel to [1 bar1 0]); and for P(b1), g1 = 2.0012 (between [001] and [bar1 bar1 1] axes), g2 = 2.0081 (parallel to [111]), and g3 = 2.0052 (parallel to [1 bar1 0]). Somewhat different values are reported in ref [9].

The bandgap energy distribution of P(b) centres on (100)Si has been measured and compared with the interface-state density [33]. From electric-field controlled ESR and C-V measurements, it was found that the electron transitions of P(b0) are essentially the same as on (111)Si while P(b1) had the 0-1 electron transition at Ev + 0.45 eV and the 1-2 transition at Ev + 0.8 eV. The P(b) density was found to quantitatively correlate with the interface-state density from CV analysis; the nonbonding P(b) orbitals accounted for approx 50% of the characteristic gap states on the (100)Si wafers. The relative densities of P(b0) and P(b1) centres vary with wafer processing and device-operation stresses. Since their superimposed densities can be major contributions to the interface-state distribution, the shape of this distribution on (100)Si can vary widely, resembling the nearly symmetrical distribution of (111)Si for small P(b1) densities, or becoming asymmetrical for large P(b1) densities with a peak in the upper half of the gap.

Because the ESR signals are weaker on (100) as compared to (111)Si, the hyperfine structure of P(b0) and P(b1) centres has been only partially analyzed at this time [34]. The P(b0) hyperfine parameters strongly confirm the model of the P(b) centre derived on (111)Si; slight differences reflect the second-order influence of the canted dangling-orbital orientation on the (100) face. The limited information available for P(b1) indicates that the unpaired electron is highly localized on the host Si atom, is more tightly bound than the electron in P(b0), and does not show evidence of an adjacent bonded or nonbonded oxygen atom.

Optical absorption by P(b) centres on (111)Si has been measured by electro-absorption spectroscopy [35,36]. On MOS capacitors, electric-field modulation of the occupancy of the characteristic gap states was directly detected as a synchronous modulation of the absorption of subband-gap light. Both photon energy and surface-potential controlled trap occupancy were used to probe the interface-state distribution and indicated the broad peak at Ev + 0.3 eV. In addition, the results provided an estimate of 10*-17 cm*2 (at 1 eV) for the effective optical cross section of the 0->1 electron transition of the interfacial defect; that is, the absorption is weak and dominated by hole emission. This is consistent with an estimate from photothermal deflection spectroscopy that the cross section for electron emission is < < 10*-17 cm*2, which was inferred from the absence of detectable interfacial-defect absorption [37]. An unresolved issue pertains to a theoretical prediction [38] that optical transitions from the Si valence band to unoccupied P(b) orbitals (i.e. 2 -> 1 hole transition) should be the weakest of the several possible transitions, which disagrees with experiment.

In addition to the issues raised above regarding the properties and theory of P(b) centres, other general directions of research can be identified. In particular, there is strong technological motivation to understand the role of P(b) centres in process-induced or operation-induced degradation of MOS devices [21,39-42]. Finally, the P(b) centre should find use as a controllable, characterized interfacial defect to study other interface-related phenomena (e.g. low-dimensional transport in Si MOS devices or the detection of interface states by scanning tunnelling microscopy).

REFERENCES
----

[1]   E.H.Nicollian, J.R.Brews [MOS (Metal Oxide Semiconductor)
        Physics and Technology (Wiley, New York 1982)]
[2]   N.M.Johnson, D.J.Bartelink, M.Schulz [Proc. Conf. Physics of
        SiO2 and Its Interfaces, Yorktown Heights, New York, USA, 22-24
        June 1978, Ed. S.T.Pantelides (Pergamon, New York, 1978) p.421-7]
[3]   N.M.Johnson, D.K.Biegelsen, M.D.Moyer [Proc. Conf. Physics of
        MOS Insulators (Raleigh, North Carolina, USA, 18-20 June 1980) Eds
        G.Lucovsky, S.T.Pantelides, F.L.Galeener (Pergamon, New York,
        1980) p.311-15]
[4]   N.M.Johnson, D.K.Biegelsen, M.D.Moyer, S.T.Chang, E.H.Poindexter,
        P.J.Caplan [Appl. Phys. Lett. (USA) vol.43 no.6 (15 Sep 1983)
        p.563-5]
[5]   E.H.Poindexter, G.J.Gerardi, M.-E.Rueckel, P.J.Caplan,
        N.M.Johnson, D.K.Biegelsen [J. Appl. Phys. (USA) vol.56 no.10
        (15 Nov 1984) p.2844-9]

[6]   Y.Nishi [Jpn. J. Appl. Phys. (Japan) vol.10 no.1 (1971) p.52-62]

[7]   E.H.Poindexter, E.R.Ahlstrom, P.J.Caplan [Proc. Conf. Physics
        of SiO2 and Its Interfaces, Yorktown Heights, New York, USA, 22-24
        June 1978, Ed. S.T.Pantelides (Pergamon, New York, 1978) p.227-31]

[8]   P.J.Caplan, E.H.Poindexter, B.E.Deal, R.R.Razouk [J. Appl. Phys.
        (USA) vol.50 no.9 (Sep 1979) p.5847-54]

[9]   A.Stesmans, J.Braet, J.Witters, R.F.Dekeersmaecker [Surf. Sci.
        (Netherlands) vol.141 no.1 (1984) p.255-84]

[10]   K.L.Brower [Appl. Phys. Lett. (USA) vol.43 no.12 (1983)
        p.1111-13]

[11]   A.Stesmans [Appl. Phys. Lett. (USA) vol.48 no.15 (14 Apr 1986)
        p.972-4]

[12]   K.L.Brower [Phys. Rev. B (USA) vol.33 no.7 (1 Apr 1986)
        p.4471-8]

[13]   K.L.Brower, T.J.Headley [Phys. Rev. B (USA) vol.34 no.6 (15 Sep
        1986) p.3610-19]

[14]   W.E.Carlos [Appl. Phys. Lett. (USA) vol.50 no.20 (18 May 1987)
        p.1450-2]

[15]   M.Cook, C.T.White [Phys. Rev. Lett. (USA) vol.59 no.15 (2 Oct
        1987) p.1741-4]

[16]   A.H.Edwards [Phys. Rev. B in press]

[17]   S.T.Pantelides [Phys. Rev. Lett. (USA) vol.57 no.23 (8 Dec 1986)
        p.2979-82]

[18]   E.H.Poindexter, P.J.Caplan, J.J.Finnegan, N.M.Johnson,
        D.K.Biegelsen, M.D.Moyer [Proc. Conf. Physics of MOS Insulators,
        Raleigh, North Carolina, USA, 18-20 June 1980, Eds G.Lucovsky,
        S.T.Pantelides, F.L.Galeener (Pergamon, New York, 1980) p.326-30]

[19]   C.Brunstrom, C.Svensson [Solid State Commun. (USA) vol.37 no.5
        (Feb 1981) p.399-404]

[20]   P.M.Lenahan, P.V.Dressendorfer [Appl. Phys. Lett. (USA) vol.41
        no.6 (15 Sep 1982) p.542-4]

[21]   P.M.Lenahan, P.V.Dressendorfer [J. Appl. Phys. (USA) vol.55
        no.10 (15 May 1984) p.3495-9]

[22]   B.Henderson [Appl. Phys. Lett. (USA) vol.44 no.2 (15 Jan 1984)
        p.228-30]

[23]   K.M.Lee, L.C.Kimerling, B.G.Bagley, W.E.Quinn [Solid State
        Commun. (USA) vol.57 no.8 (Feb 1986) p.615-17]

[24]   R.B.Laughlin, J.D.Joannopoulos, D.J.Chadi [Phys. Rev. B (USA)
        vol.21 no.12 (1980) p.5733-44]

[25]   K.L.Ngai, C.T.White [J. Appl. Phys. (USA) vol.52 no.1 (Jan 1981)
        p.320-37]

[26]   T.Sakurai, T.Sugano [J. Appl. Phys. (USA) vol.52 no.4 (Apr 1984)
        p.2889-96]

[27]   A.H.Edwards [Proc. 13th Int. Conf. Defects in Semiconductors,
        Coronado, CA, USA, 12-17 Aug, 1984 vol.14a Eds L.C.Kimerling,
        J.M.Parsey,Jr. (Metallurg. Soc., AIME, Warrendale, PA, USA, 1985)
        p.491-7]

[28]   A.S.Carrico, R.J.Elliott, R.A.Barrio [Phys. Rev. B (USA) vol.34
        no.2 (15 Jul 1986) p.872-8]

[29]   R.A.Barrio, R.J.Elliott, A.S.Carrico [Phys. Rev. B (USA) vol.34
        no.2 (15 Jul 1986) p.879-85]

[30]   W.B.Fowler, R.J.Elliott [Phys. Rev. B (USA) vol.34 no.8 pt.II
        (15 Oct 1986) p.5525-9]

[31]   M.C.Chen, D.V.Lang [Phys. Rev. Lett. (USA) vol.51 no.5 (1 Aug
        1983) p.427-9]

[32]   E.H.Poindexter, P.J.Caplan, B.E.Deal, R.R.Razouk [J. Appl. Phys.
        (USA) vol.52 no.2 (Feb 1981) p.879-84]

[33]   G.J.Gerardi, E.H.Poindexter, P.J.Caplan, N.M.Johnson [Appl. Phys.
        Lett. (USA) vol.49 no.6 (11 Aug 1986) p.348-50]

[34] K.L.Brower [Z. Phys. Chem. Neue Folge (Germany) vol.151 pt 1-2
      (1987) p.177-89]
[35] N.M.Johnson, W.B.Jackson, M.D.Moyer [Phys. Rev. B (USA) vol.31
      no.2 (15 Jan 1985) p.1194-7]
[36] W.B.Jackson, N.M.Johnson [Mater. Res. Soc. Symp. Proc. (USA)
      vol.46 (1985) p.545-51]
[37] C.H.Seager, P.M.Lenahan, K.L.Brower, R.E.Mikawa [Mat. Res. Soc.
      Symp. Proc. (USA) vol.46 (1985) p.539-44]
[38] A.H.Edwards, W.B.Fowler [Mat. Res. Soc. Symp. Proc. (USA) vol.46
      (1985) p.533-8]
[39] R.E.Mikawa, P.M.Lenahan [IEEE Trans. Nucl. Sci. (USA) vol.NS-31
      no.6 (1984) p.1573-5]
[40] R.E.Mikawa, P.M.Lenahan [Appl. Phys. Lett. (USA) vol.46 no.6
      (15 Mar 1985) p.550-2]
[41] J.M.Sung, S.A.Lyon [Appl. Phys. Lett. (USA) vol.50 no.17 (27 Apr
      1987) p.1152-4]
[42] M.L.Reed, J.D.Plummer [Appl. Phys. Lett. (USA) vol.51 no.7
      (17 August 1987) p.514-6]

17.7 GENERATION-RECOMBINATION-TRAPPING RATES AND LIFETIMES OF ELECTRONS AND
HOLES IN SiO2 FILMS ON Si - A GENERAL CLASSIFICATION OF THE FUNDAMENTAL
MECHANISMS

by C.T.Sah

November 1987
EMIS Datareview RN=17815

The generation-recombination-trapping (g-r-t) rates of electrons and
holes are fundamental macroscopic material parameters which can be
derived from first principles (quantum and statistical mechanics)
while the electron and hole lifetimes are phenomenological parameters
which are defined to characterize the response of a multi-terminal
solid-state device (diode, transistor, resistor, capacitor, etc.)
to an external excitation. The lifetimes are defined and measured
from the DC sinusoidal steady-state or transient characteristics (DC
current-voltage, impedance-frequency and current-time or voltage-time)
of a device. For DC steady state, the lifetime and the g-r-t rate are
simply related to each other by the phenomenological definition:
(lifetime) = (g-r-t rate)/(excess carrier density). For time dependent
excitations, the relationship is more complex and less explicit. For
high-frequency small-signal sinusoidal steady-state measurement, the
lifetime is given by omega tau = angular frequency X lifetime = 1 from
a set of data or curves consisting of rms (root-mean-square) voltage,
rms current, impedance or admittance versus signal frequency, omega. In
the infrared-visible optical and higher frequency ranges (THz or
10*12 Hz), tau is the reciprocal scattering rate instead of the g-r-t
rate. For large-signal transients, the lifetime is defined by the
magnitude of the carrier density decay rate, partial differential
d ln(carrier density)/dt and its dependence on the reciprocal g-r-t rate
can be highly nonlinear and complex. This complexity is determined by
the magnitude and temporal variations of the excitation and the
recombination mechanisms.

In crystalline silicon, the g-r-t rates and lifetimes are the material
parameters which determine the initial performance of silicon
transistors and integrated circuits. In contrast, in silicon dioxide
films on crystalline silicon substrates, the g-r-t rates and lifetimes
determine the reliability or operating life of silicon integrated
circuits, or the final, terminus performance as well as their
manufacturing yields since they are a measure of the rates of
generation, annealing and charging of the traps in the oxide film and
at the oxide/silicon interface. It is well known that these traps, when
charged, will drastically change and degrade the electrical
characteristics of the transistors and hence the operating life or
endurance of integrated circuit chips.

The magnitude of the g-r-t rates and lifetimes of electrons and holes in SiO2 films are controlled by a variety of widely different fundamental physical phemonena [1,2]. These phemonena are similar to those occuring in other insulators (Si3N4, NaCl and other ionic crystals), and in semiconductors such as the bulk of crystalline silicon and at the interfacial layer of the oxide/silicon interface. A systematic classification of these phenomena in semiconductors and at the insulator/semiconductor interfaces was given by Sah in 1971 [3]. This classification was based on two fundamental factors which control the quantum mechanical transition rates and hence the g-r-t rates or lifetimes of the electrons and holes. These are the energy exchange mechanisms and the initial-final states of the electrons and holes. The mechanisms were listed in two matrix tables, a 3 X 3 for non-tunnelling and a 3 X 4 for tunnelling transitions [3]. This classification scheme was extended by Sah in 1987 [1,2] to insulator films on semiconductor substrates such as the oxide film grown on crystalline silicon which is the subject of this Datareview. These matrix tables, TABLES 1 and 2, are used here to present the data on SiO2 films and to provide their physical interpretations. The transition mechanisms will be labelled by the column and row numbers, cr, of the matrix table. The universal terms and symbols for the transition processes and macroscopic rates, initially employed by Sah [1-3], are also given in the tables. The macroscopic rate is defined as the averaged microscopic or quantum mechanical transition rate. The average is taken over the electron or hole energy distribution function. When the singular term, electron or hole, is used in the following descriptions, we mean the macroscopic average of the parameter of the electrons or holes.

The macroscopic rates or kinetic coefficients enter into the transport or semiconductor equations, also known as the Shockley equations, for the analysis of the transport properties of semiconductor materials and semiconductor devices. For example g(n)t [= g(p)t] and g(n)o are the thermal and optical interband electron-hole pair generation rates respectively while g(n)n [= g(p)n] and g(p)p [= g(n)p] are the interband electron-hole pair generation rates due to electron and hole impact respectively. The electron-hole generation rates by bulk and surface plasmon decay are then g(n)bp [= g(p)bp] and g(n)sp [= g(n)sp], while the bulk and surface plasmon generation rates by fast electrons and X-rays are g(bp)n, g(bp)x, g(sp)n and g(sp)x. The rate coefficients denoted by r(x)y are the interband electron-hole recombination rates of the interband transitions. c(n)t and c(n)o are the thermal and optical or radiative capture rates of an electron by a trap. c(n)n and c(n)p are the Auger capture rates of an electron by a trap with an electron or a hole carrying away the energy respectively. Their inverses, e(x)y, are the thermal, optical and impact emission rates of trapped electrons or holes which are frequently known as the thermal, photo and impact ionization rates of the traps. A standard notation has not been established for the trap-to-trap or intertrap transitions since their presence has been established in only a few cases, i.e. donor-acceptor pair spectra in ZnS, GaP, GaAs and In-Li doped Si.

TABLE 1: non-tunnelling generation-recombination-trapping transitions

energy exchange mechanisms	initial and final state of the electron and hole		
	band-to-band	band-to-trap	trap-to-trap
thermal	11	21	31
optical	12	22	32
Auger-impact	13	23	33
plasma	14	24	34
terminolgy	e-h pair recombination or generation	electron or hole capture or emission	particle transition
rate symbols x=n,p; y=t,o,n,p,m	g(x)y r(x)y	c(x)y e(x)y	t12 t21

TABLE 2: tunnelling generation-recombination-trapping transitions

energy exchange mechanisms	initial and final state of the electron and hole		
	band-to-band	band-to-trap	trap-to-trap
thermal	11	21	31
optical	12	22	32
Auger-impact	13	23	33
plasma	14	24	34
elastic	15	25	35
terminology	e-h pair recombination or generation	electron or hole capture or emission	particle transition
rate symbols	g(x)y	c(x)y	t12
x=n,p; y=t,o,n,p,m	r(x)y	e(x)y	t21

As indicated in these two tables, two fundamental factors which determine the g-r-t rates and lifetimes are: (i) the energy conservation or exchange mechanisms; and (ii) initial and final states of the electron or hole. These transitions may proceed via four elementary energy exchange mechanisms, the thermal (phonon), optical (photon), Auger impact (a third electron or hole), and plasma (plasmon formation and decay), to conserve the total energy. The plasma mechanism involves the generation of plasmons by high energy electrons and decay of the plasmons into electron-hole pairs. A plasmon is a collective oscillation wave composed of all the valence electrons moving cohesively [4] in a volume element of the material under consideration.

These tables are further extensions of those given in [1,2] in order to include the plasma mechanism (bp denotes bulk plasmon and sp denotes surface or interface plasmon) which is the dominant energy exchange mechanism for electrons and holes of high kinetic energies. These high energy electrons and holes are present when the SiO2 film is exposed to X-ray and keV electron radiations (encountered in X-ray and electron beam lithography and in space and other radiation environments) or stressed by very high electric fields or applied voltages. Surface and interface plasmon energies are about 10 eV and bulk plasmon energies are about 17 eV in materials used in silicon integrated circuits, such as silicon, silicon dioxide, aluminium, gold and refractory metals. Thus, the plasmon mechanism can become important when the applied voltage exceeds about 20 V.

The second factor, the initial and final states, determines the necessary energy and momentum conservation between the electron (and/or hole) and the participating particles (phonon, photon, another electron or hole, and plasmon) during the transition. Band-to-band transitions are known as interband transitions in which the initial and final states of the electron are in two bands separated by an energy gap (such as the gap between the conduction and highest valence band or between valence bands. The latter, as we shall see, is important in the determination of the interband impact generation rate of electron-hole pairs and the intrinsic dielectric breakdown of the oxide film based on a new idea proposed by Sah [8]) The word 'interband' is redundant when used together with the words 'electron-hole pairs'. For example, Auger recombination or impact generation of electron-hole pairs can only mean an interband process since it involves both an electron and a hole simultaneously during transition. But, we shall include the word 'interband' at the beginning of this discussion. We include this redundancy since there is so much confusion in the literature in which these interband transitions are not explicitly distinguished from the impact and Auger transitions at the localized states of imperfection centres. The latter are designated in this review as the band-bound transitions: for example, the Auger recombination of an electron with a trapped hole at an imperfection centre and the impact release of a trapped electron at an imperfection centre by an energetic electron or hole. Band-trap or band-bound transitions have the initial (or final) electron state in either the conduction or valence band and the final (or initial) state localized at a trap. Bound-bound or trap-trap transition involves the transition of a trapped electron (or hole) at one trap to an adjacent trap either directly by tunnelling or in several steps involving a combination of phonon, photon and another electron or hole.

Since the energy gap of insulators is much larger than those of semiconductors, some of the fundamental transition processes are important in semiconductors but not in insulators. For example, the interband thermal generation rate of electron-hole pairs is negligible in insulators since their large energy gaps (9 eV in SiO2) are much larger than the lattice oscillation or phonon energies (tens of meV). For example, to create an electron-hole pair in SiO2 thermally by the highest energy polar optical phonons (0.153 eV) would require a most improbable process of simultaneous absorption of 9/0.153 = 60 phonons. The mechanisms and the data of these g-r-t mechanisms are discussed in the Datareviews [5-12].

REFERENCES
----

[1]   C.T.Sah ['VLSI Device Reliability Modeling', Invited paper 5-1
       presented at the 1987 International Symposium on VLSI Technology
       Systems and Applications, 14 May 1987. Proceedings of Technical
       Digest p.153-162 Publishers: TSPD (C60), ERSO, ITRI, No.195(C60)
       Chung Hsing Rd., Sec.4, Chu Tung, Hsinchu, Taiwan (31015) Republic
       of China]

[2]   C.T.Sah [Appl. Surf. Sci. (Netherlands) vol.30 nos 1-4 (Oct 1987)
       p.311-12]

[3]   C.T.Sah [Phys. Status Solidi a (Germany) vol.7 (1971) p.541]

[4]   D.Pines [Elementary Excitations in Solids Chapter 3 (W.A.Benjamin
       Inc., New York, 1963)] See also C.Kittel [Introduction to Solid
       State Physics, 6th Ed. Ch.10 (John Wiley & Sons, New York, 1986)]
       and P.M.Platzman, P.A.Wolff [Waves and Interactions in Solid
       State Physics Supplement 13 (Academic Press, New York, 1973)]

[5]   C.T.Sah [EMIS Datareview RN=17816 (Nov 1987) 'Non-tunnelling
       interband thermal generation and recombination of electron-hole
       pairs in SiO2 films on Si']

[6]   C.T.Sah [EMIS Datareview RN=17817 (Nov 1987) 'Non-tunnelling
       interband optical generation and radiative recombination of
       electron-hole pairs in SiO2 films on Si'

[7]   C.T.Sah [EMIS Datareview RN=17818 (Nov 1987) 'Non-tunnelling
       interband Auger recombination and impact generation of electron-
       hole pairs in SiO2 films on Si']

[8]   C.T.Sah [EMIS Datareview RN=17819 (Nov 1987) 'Intrinsic
       dielectric breakdown of SiO2 films on Si due to interband impact
       generation of electron-hole pairs']

[9]   C.T.Sah [EMIS Datareview RN=17839 (Nov 1987) 'Interband electron-
       hole generation in SiO2 films on Si by keV electrons']

[10]   C.T.Sah [EMIS Datareview RN=17820 (Nov 1987) 'Interband electron-
       hole pair generation in SiO2 by plasmon decay']

[11]   C.T.Sah, C.Y.-P.Chao, A.J.Chen, C.C.-H.Hsu, T.Nishida [EMIS
       Datareview RN=17821 (Nov 1987) 'Generation-recombination-trapping
       of electrons and holes localised at imperfection centres in SiO2']

[12]   C.T.Sah [EMIS Datareview RN=17822 (Nov 1987) 'Tunnelling
       transitions of electrons and holes in SiO2 films on Si']

17.8 NON-TUNNELLING INTERBAND THERMAL GENERATION AND RECOMBINATION OF ELECTRON-HOLE PAIRS IN SiO2 FILMS ON Si

by C.T.Sah

November 1987
EMIS Datareview RN=17816

Interband thermal generation of electron-hole pairs across the SiO2
energy gap (non-tunnelling) is negligible due to its large energy gap
(9 eV). This is process (11) in Table 1 of [31].

Interband thermal recombination of electron hole pairs (non-
tunnelling) has been identified as the main cause of low yield (yield
is equal to the number of electron-hole pairs generated per incident
particle) and it has accounted for the electric field dependence of
yield. Three sets of yield measurements were reported. Photoconductivity
yield in 5000 A thermally grown SiO2 film induced by vacuum ultraviolet
(VUV) light (8-11 eV photons) was reported by DiStefano and Eastman
[1]. Yield of collected charge from current transients in 200-microns
thick fused quartz (Suprasil II of Amersil, Inc.) exposed to pulsed
X-rays (about 300 keV energy range generated by 600 keV electrons)
was reported by Hughes [2]. Photoconductivity yield in SiO2 exposed to
10.2 eV VUV photons was reported by Powell [3]. The observed low yield
and its linear increase with applied electric field were attributed to
geminate or initial recombination of electrons with holes before they
are separated either by phonon scattering or by an electric field. A
summary of this work was given by Hughes [4].

The theory of geminate or initial recombination was developed by Onsager
first for the ionization of electrolytes in 1934 [5] and then extended
to gas by all types of ionizing radiation (alpha-ray, beta-ray and
photons) in 1938 [6]. Onsager's theory has been extensively used to
explain the magnitude as well as the electric field and temperature
dependences of the photoconductive yield of photoconductors such as
amorphous silicon and germanium, selenium, chalcogenides [7,8], organic
semiconductors [9] and polymers [10] such as anthracene [11,12],
poly-N-vinylcarbazole (PVK) [13,14] and trinitrofluorenone
poly-N-vinylcarbazole (PVK:TNF) [15,16].

Before giving a description of Onsager's theory, the valence band
structure of SiO2 is briefly reviewed, since the microscopic or atomic
location of the hole in SiO2 is important for a basic understanding and
a zeroth order analysis of the magnitude of the geminate recombination
rate. There is now a general consensus that the valence electrons (or
holes when the valence electrons are missing) near the top of the
valence band are from the two non-bonding (lone pair) 2p electrons
of the four oxygen 2p electrons. This is based on recent calculations
of the energy band of the various polymorphs of SiO2, cubic beta-
cristoballite by Pantelides and Harrison [17], and Ciraci and Batra
[18]; alpha-quartz by Laughlin, Joannopoulos and Chadi [20] and
Chelikowsku and Schluter [19]; and all polymorphs by Li and Ching [21].
The computed results are also consistent with experiments, such as the
X-ray and UV photoemission data of Fischer et al [22] from which they
concluded that the valence band width is about 3.3 eV and largely due
to the oxygen 2p wavefunction overlap. The computed bands are also
consistent with the optical absorption and photoconductivity data
taken on a thin SiO2 film by Powell and Morad [23]. The recent analyses
of the Raman spectra by Phillips led him to suggest a beta-cristoballite

granular model for the SiO2 [24]. Thus, calculated bands of the cubic beta-cristoballite are employed as the model in this review.

The Onsager theory applied to SiO2 describes the following events concerning the valence electrons on the non-bonding oxygen 2p orbitals. When the SiO2 is exposed to ionizing radiation or enegetic electrons, one of these two valence electrons near the top of the valence band is excited to the edge of the conduction band 9 eV above, leaving behind a hole or unoccupied non-bonding 2p orbital on the oxygen. This hole can be considered fixed or immobile during the initial recombination event since for the hole to move it must tunnel to the adjacent and occupied non-bonding oxygen 2p orbital site which is separated from the hole located on the first oxygen by a silicon atom. This large separation and hence small wavefunction or orbital overlap is also reflected by the rather narrow valence band width of this highest filled valence band in SiO2. The small hole tunnelling probability or rate is akin to the small atomic diffusion rate and is the reason for very low hole mobility in the SiO2 [29].

The excited electron can encounter two fates which are described by the Onsager theory. In both, it is first scattered by lattice vibrations or phonons during its initial diffusive motion to escape from the coulomb potential well due to the hole. In the first fate, if it loses so much energy while diffusing to a distance rd that its final kinetic energy is less than kT(L), then it will fall back into the hole by either radiative or nonradiative recombination. In SiO2, this capture is via the nonradiative or thermal recombination mechanism of transition (11). It is an experimental fact that interband radiative recombination, transition (21) is negligible in SiO2 and this is not inconsistent with the band calculations which showed that both alpha-quartz and beta-cristoballite SiO2 have multi-peak valence band structures with indirect energy gap [17-21].

The energy exchange mechanism of this nonradiative interband thermal recombination mechanism has been the subject of many theoretical analyses. It was at once recognized that the multi-phonon model is untenable since it would require the simultaneous emission of many phonons. For SiO2, the number of phonons required would have been (energy gap)/(polar LO phonon energy) = 9 eV/0.153 eV = 58.8. The currently accepted model is that due to Dexter [25] who removed the restriction of purely harmonic oscillation in the multi one-phonon model by taking into account the large anharmonic vibration around the electron-hole pair using the configuration coordinate approach.

In the second fate, the photo or impact excited O(2p) lone-pair electron diffuses to a distance of rd and still has sufficient kinetic energy left to surmount the remaining Coulomb potential barrier and breaks free. An ionization then results or a free electron and a free hole are generated. Onsager set this kinetic or escape energy to kT(L) and defined the escape radius, re, from kT(L) = e*2/(4 pi epsilon re) where epsilon is the static dielectric constant (3.9 X 8.854 X 10*-14 F/cm for SiO2). A simple illuminating derivation was given by Mott [26] where re is the radius of the spherical surface on which the number of electrons diffusing outwards just balances the number of electrons diffusing inwards. For SiO2 at 300 deg K, the Onsager escape radius is re = 142 A, which suggests that many photogenerated electron-hole pairs will undergo recombination instead of separation and hence will not contribute to the photoconductivity yield.

The diffusion radius or thermalization distance of an electron excited
to a kinetic energy Ek has been calculated by Davis and Knight [27] for
amorphous selenium. This may be extended to crystalline or
polycrystalline SiO2 by using the polar LO phonon generation rate
instead of the reciprocal phonon frequency, 1/nu(p), used by Davis and
Knight for amorphous materials. Using D/mu = kT(L)/e and mu = (e/m)tau
then the thermalization time of an electron with kinetic energy
Ek = h nu - Eg due to diffusion or random motion is

$$tau(t) = \frac{(rd)*2}{D} = [(Ek + \frac{e*2}{4 pi epsilon rd})/(h nu(LO))] X tau(mu)$$

where h nu is the photon energy. Assuming free electron mass and that
the h nu(LO) = 63 meV polar LO phonon dominates the scattering and
taking the equilibrium mobility of electrons in SiO2 at 300 deg K from
[30], mu = 20 cm*/V.s, then Ek = (rd/30.28)*2 - (3.43/rd) where Ek is
in eV and rd in A. In order that an electron can escape recombination with
the hole, we must have rd > re = 142 A just computed. This would give a
minimum electron kinetic energy of Ek > 22.2 eV or a photon energy of
h nu = Ek + Eg = 22.2 + 9 = 31.2 eV. At such a high kinetic energy, the
equilibrium diffusion model just described is no longer valid. A better
estimate would be to use Shockley's ballistic model for the hot
electrons (see [30]) which gives a drift distance of

$$rd = [(Ek - \frac{e*2}{4 pi epsilon rd})/h nu(LO)] lambda(Rn)$$

where the h nu(LO) = 153 meV polar LO phonon dominates electron
scattering and energy loss with an experimental hot electron mean free
path of lambda(Rn) = 30 A. To escape, the drift distance must be greater
than the escape distance or the kinetic energy must be greater than

$$Ek = h nu(LO) \frac{rd}{lambda(Rn)} - kT(L) > h nu(LO) \frac{re}{lambda(Rn)} - kT(L)$$

$$= 0.153 X \frac{143}{30} - 0.026 = 0.70 eV,$$

giving a quantum and photoconduction yield threshold of
h nu = Eg + Ek(min) = approx. 9 + 0.7 = 9.7 eV.

The quantum and photoconductivity yield is then given by Y =
tau(r)/[tau(r) + tau(i)] . Here 1/tau(r) is the thermal recombination
rate of the electron-hole pair which can be obtained from the Dexter
anharmonic electron-phonon interaction model. The ionization rate was
given by Knight and Davis [27],

$$\frac{1}{tau(i)} = \frac{1}{tau(t)} exp \frac{-Ed}{kT(L)} = \frac{1}{tau(t)} exp \frac{-e*2}{4 pi epsilon kT(L) rd}$$

$$= \frac{1}{tau(t)} exp \frac{-re}{rd}$$

where re =(e*2)/(4 pi epsilon kT(L)).

The diffusion or drift radius, rd, was left as an unspecified characteristic thermalization length in Onsager's theory. The effect of electric field on the thermalization probability was given by Onsager [10] and reduced to a function of re/rd by Pai and Enck [28].

Excitons may modify the geminate or initial recombination rate and the separation rate. In SiO2 the exciton is thought to be a resonant direct exciton whose binding energy is several eV larger than the SiO2 energy gap and which spontaneously decays into an electron-hole pair. Thus, exciton effects could be important at higher incident radiation or electron energies.

REFERENCES
----

[1]   T.H.DiStefano, D.E.Eastman [Solid State Commun. (USA) vol.9 (1971) p.2259]

[2]   R.C.Hughes [Phys. Rev. Lett. (USA) vol.30 no.26 (25 Jun 1973) p.1333-6]

[3]   R.J.Powell [IEEE Trans. Nucl. Sci. (USA) vol.NS-22 (1975) p.2240] See also reference [27]

[4]   R.C.Hughes [Solid-State Electron. (GB) vol.21 no.1 (Jan 1978) p.251-8]

[5]   L.Onsager [J. Chem. Phys. (USA) vol.2 (1934) p.599-615]

[6]   L.Onsager [Phys. Rev. (USA) vol.54 (1938) p.554-7]

[7]   N.F.Mott [Adv. Phys. (GB) vol.26 no.4 (1977) p.363-91]

[8]   N.F.Mott, E.A.Davis [Electronic Processes in Non-Crystalline Materials 2nd Ed. (Clarendon Press, Oxford, 1979) p.263-9,392, 542-3]

[9]   F.Gutmann, H.Keyzer, L.E.Lyons [Organic Semiconductors, Part A (1967) and Part B (1983) (Robert E. Krieger Publishing Co., Malabar, FL, USA)]

[10] J.Mort, G.Pfister [Electronic Properties of Polymers (John Wiley and Sons, New York 1982)]

[11] R.C.Hughes [J. Chem. Phys. (USA) vol.55 no.12 (15 Dec 1971) p.5442-7 and refs]

[12] R.R.Chance, C.L.Braun [J. Chem. Phys. (USA) vol.59 no.5 (1 Sep 1973) p.2269-72 and refs]

[13] D.M.Pai [J. Chem. Phys. (USA) vol.52 no.5 (1 Mar 1970) p.2285-91]

[14] R.C.Hughes [IEEE Trans. Nucl. Sci. (USA) vol.NS-18 (1971) p.281-7; Chem. Phys. Lett. (Netherlands) vol.8 no.5 (1 Mar 1971) p.403-6]

[15] P.J.Melz [J. Chem. Phys. (USA) vol.57 no.4 (15 Aug 1972) p.1694-9]

[16] R.C.Hughes [Appl. Phys. Lett. (USA) vol.21 no.5 (1 Sep 1972) p.196-8; J. Chem. Phys. (USA) vol.58 no.6 (15 Mar 1973) 1973) p.2212-19]

[17] S.T.Pantelides, W.A.Harrison [Phys. Rev. B (USA) vol.13 no.6 (15 Mar 1976) p.2667-91]

[18] S.Ciraci, I.P.Batra [Phys. Rev. B (USA) vol.15 no.10 (15 May 1977) p.4923-34]

[19] J.R.Chelikowsky, M.Schluter [Phys. Rev. B (USA) vol.15 no.8 (15 Apr 1977) p.4020-9]

[20] R.B.Laughlin, J.D.Joannopoulos, D.J.Chadi [Phys. Rev. B (USA) vol.20 no.12 (15 Dec 1979) p.5228-37]

[21] Y.P.Li, W.Y.Ching [Phys. Rev. B (USA) vol.31 no.4 (15 Feb 1985) p.2172-9]

[22] B.Fischer, R.A.Pollack, T.H.DiStefano, W.D.Grobman [Phys. Rev.
      B (USA) vol.15 no.6 (15 Mar 1977) p.3193-9]

[23] R.J.Powell, M.Morad [J. Appl. Phys. (USA) vol.49 no.4 (Apr
      1978) p.2499-502]

[24] J.C.Phillips [Phys. Rev. B (USA) vol.35 no.12 (15 Apr 1987)
      p.6409-13; Solid-State Phys. (USA) vol.37 (1982) p.93-171]
      See ref [30] on electron mobility where our analysis
      predicted a crystallite size of 72 A calculated from the
      temperature independent electron mobility of 40 cm*2/V.s in
      fused silica which is consistent with Phillip's model which
      gives 66 A computed from Raman spectra of SiO2

[25] D.L.Dexter, C.C.Klick, G.A.Russell [Phys. Rev. (USA) vol.100
      no.2 (15 Oct 1955) p.603-5]; D.L.Dexter [Solid State Phys.
      (USA) vol.6 (1958) p.355-411 (see section 13 on nonradiative
      transitions starting on p.405)]

[26] N.F.Mott [Philos. Mag. (GB) vol.36 no.2 (1977) p.413-20]

[27] E.A.Davis [J. Non-Cryst. Solids (Netherlands) vol.4 (1970)
      p.107-16]; J.C.Knights, E.A.Davis [J. Phys. & Chem. Solids
      (GB) vol.35 (1974) p.543-54]

[28] D.M.Pai, R.C.Enck [Phys. Rev. B (USA) vol.11 no.12 (15 Jul
      1975) p.5163-73]

[29] C.T.Sah, A.J.Chen, C.C.-H.Hsu, T.Nishida [EMIS Datareview
      RN=16192 (Nov 1987) 'Hole mobility in SiO2 films on Si']

[30] C.T.Sah, A.J.Chen, C.C.-H.Hsu, T.Nishida [EMIS Datareview
      RN=17808 (Nov 1987) 'Electron mobility in SiO2 films on Si']

[31] C.T.Sah [EMIS Datareview RN=17815 (Nov 1987) 'Generation-
      recombination-trapping rates and lifetimes of electrons and holes
      in SiO2 films on Si - a general classification of the fundamental
      mechanisms']

17.9 NON-TUNNELLING INTERBAND OPTICAL GENERATION AND RADIATIVE RECOMBINATION
OF ELECTRON-HOLE PAIRS IN SiO2 FILMS ON Si

by C.T.Sah

November 1987
EMIS Datareview RN=17817

Interband optical recombination or radiative recombination of
electron-hole pairs in SiO2, process 12 in Table 1 of [12], is
negligible due to its indirect energy band. Optical generation rate of
electron-hole pairs can be computed from the optical absorption
coefficient measured by Powell and Morad [3] on 560 A thick thermally
grown SiO2 films on crystalline silicon substrate. They obtained the
transmission coefficient through a 560 A of 1 mm X 1 mm area etched
into the oxidized silicon slice using Philipp's data on reflectance and
complex refractive index obtained for fused quartz [4]. The data
implicity includes the geminate or initial recombination discussed in
[10]. The data are read off from an enlargement of the figure given by
Powell and Morad and tabulated in TABLE 1.

TABLE 1: optical absorption coefficient of a thermally grown SiO2 film
as determined from transmission through a 580 A film by Powell
and Morad [3]

Photon energy (eV)	Absorption coefficient (10*5/cm)
8.2	0.056
8.3	0.11
8.4	0.14
8.5	0.17
8.6	0.26
8.7	0.31
8.8	0.40
8.9	0.48
9.0	0.57
9.1	0.73
9.2	0.86
9.3	1.07
9.4	1.3
9.5	1.50
9.6	1.85
9.7	2.25
9.8	2.8
9.9	3.9
10.0	5.2
10.1	7.25
10.2	10.6
10.3	12.2
10.4	13.3
10.5	13.5
10.6	13.0
10.7	11.9
10.8	9.9
10.9	9.3
11.0	9.5
11.1	10
11.2	11.2

As an example to illustrate the computation of the electron-hole pair generation rate, we assume a 1 milliwatt/cm*2 intensity of 10.2 eV (1216 A) photons from the Lyman alpha line of a hydrogen discharge light source which was used by Stivers et al [5,6]. The pair production rate is given by g(np)o = alpha phi where alpha = 1.06 X 10*6/cm from TABLE 1, and phi = P/(h nu) = (10*-3)/((1.602 X 10*-19) X 10.2) = 6.12 X 10*14 photon/cm*2 s. Thus, the pair production rate at the SiO2 surface is g(np)o = 6.5 X 10*20 e-h/cm*3 s and most photons are absorbed in a surface layer of 1/alpha = 95 A thick. In Stivers' capacitor, there is a 150 A aluminium gate which would reduce the photon flux intensity at the Al/SiO2 interface from the assumed 1 mW/cm*2. He observed an injected electron or hole current of 1.6 micro-Amp/cm*2 at a 10.2 eV photon flux of 1.0 X 10*13/cm*2 s.

Electron-hole pairs generated in the SiO2 by photons with energy greater than the energy gap are used to determine the absorption edge of the SiO2 film [2,3]. They also give rise to trap generation, annealing and charging in the SiO2 film [5-9]. In trapping experiments, photons are strongly absorbed near the SiO2 surface due to electron-hole pair generation. The electron or the hole can then be injected into the oxide by applying a voltage of appropriate polarity to the gate of the metal (thin Au or Al)-oxide-silicon (MOS) capacitor structure. When holes are injected into the oxide by a positive gate voltage, they appear to be trapped near the oxide-silicon interface and create a high density of donor-like interface states peaked at 0.3 eV above the silicon midgap. These interface states have been associated with the oxygen dangling bond (2p orbital) from rupturing of the Si-O bond by the energy supplied during hole capture [6]. Electrons are injected into the oxide by a negative gate voltage which are then captured at electron traps or recombine with trapped holes at hole traps [5].

The absorption [3] and photoconductivity spectra [2,3] taken over the photon energy range of 7.5-11.5 eV are also useful in the determination of the valence band structure of SiO2. A detailed analysis of the absorption data of Powell and Morad of TABLE 1 by this reviewer shows the following photon energy dependences. For 7.775 < h nu < 8.6 eV, alpha = (3.63 X 10*4) (h nu - 7.775)*2/cm; for 8.6 < h nu < 9.8 eV, alpha = (1.34 X 10*-3) exp (h nu/0.51 eV)/cm; and for 9.8 < h nu < 10.2 eV, alpha = (1.83 X 10*-9) exp (h nu/0.03 eV)/cm. The parabolic threshold is sharp and consistent with an indirect energy gap with few or no tail states. There are two distinct exponential dependences with a sharp break at 9.80 eV which appear to have not been noted previously. The exponential dependence is indicative of band tail states. I suspect that they are from two lower valence band peaks below the highest indirect peak since the upper valence band edges or peaks of the O(2p) lone pair band are more sensitive to random O-O bond length fluctuations. The reciprocal slopes, 0.51 and 0.3 eV, are reduced to 0.31 and 0.24 eV if the parabolic absorption is included. These energies are indicative of the width of the tail states which are split off from the upper O(2p) valence band edges.

Generation of electron-hole pairs in SiO2 films by high energy photons, i.e. X-rays, is of increasing reliability concern since X-ray lithography appears to be the most promising future submicron silicon integrated circuit technology. The initial energy loss is via plasmon generation [1], discussed in [11], and the plasmons then decay into electron-hole pairs.

REFERENCES
----

[1]   D.Pines [Elementary Excitations in Solids, Chapter 3
        (W.A.Benjamin Inc., New York 1963)] See also C.Kittel
        [Introduction to Solid State Physics, 6th Edtn, Ch.10 (John
        Wiley & Sons, New York, 1986)] and P.M.Platzman, P.A.Wolff [Waves
        and Interactions in Solid State Plasma, Solid State Physics,
        Supplement 13 (Academic Press, New York, 1973)]

[2]   T.H.DiStefano, D.E.Eastman [Solid State Commun. (USA) vol.9
        (1971) p.2259-61]

[3]   R.J.Powell, M.Morad [J. Appl. Phys. (USA) vol.49 no.4 (Apr 1978)
        p.2499-502]

[4]   H.R.Philipp [J. Phys. & Chem. Solids (GB) vol.32 (1971)
        p.1935-45]

[5]   A.R.Stivers, C.T.Sah [J. Appl. Phys. (USA) vol.51 no.12 (Dec
        1980) p.6292-304]

[6]   C.T.Sah, J.Y-C.Sun, J.J-T.Tzou [J. Appl. Phys. (USA) vol.53 no.12
        (Dec 1982) p.8886-93]

[7]   R.J.Powell, G.F.Derbenwick [IEEE Trans. Nucl. Sci. (USA) vol.NS-18
        (1971) p.99]

[8]   D.J.DiMaria, Z.A.Weinberg, J.M.Aitkin [J. Appl. Phys. (USA) vol.48
        (1977) p.898]

[9]   Z.A.Weinberg, G.W.Rubloff, E.Bassous [Phys. Rev. B (USA) vol.19
        no.6 (1979) p.3107-17]

[10]   C.T.Sah [EMIS Datareview RN=17816 (Nov 1987) 'Non-tunnelling
        interband thermal generation and recombination of electron-hole
        pairs in SiO2 films on Si']

[11]   C.T.Sah [EMIS Datareview RN=17839 (Nov 1987) 'Interband electron-
        hole generation in SiO2 films on Si by keV electrons']

[12]   C.T.Sah [EMIS Datareview RN=17815 (Nov 1987) 'Generation-
        recombination-trapping rates and lifetimes of electrons and holes
        in SiO2 films on Si - a general classification of the fundamental
        mechanisms']

17.10 NON-TUNNELLING INTERBAND AUGER RECOMBINATION AND IMPACT GENERATION OF ELECTRON-HOLE PAIRS IN SiO2 FILMS ON Si

by C.T.Sah

November 1987
EMIS Datareview RN=17818

A. INTRODUCTION

Interband Auger recombination of electron-hole pairs is not important in SiO2 under normal low densities of electrons and holes. It should become important at high excitation levels which give high densities of electrons and holes, such as exposure to an intense electron beam of 10-100 eV. We have not found any data which are explained in terms of interband Auger recombination of electron-hole pairs.

The inverse mechanism, interband impact generation of electron-hole pairs, dominates in three situations which are discussed in this and other Datareviews: impact generation rate in the hot electron range, of about 0.2 eV to 9 eV (energy gap or impact threshold) or 10 eV (surface-interface plasmon h nu (sp) = 10 eV) [see below]; dielectric breakdown from 10 eV to a few hundred electron volts range [8], and X-ray/electron-beam lithography ranging from 1 to 30 keV [9]. The interband Auger recombination and impact generation transitions are listed as process 13 in Table 1 of [10].

B. IMPACT GENERATION OF ELECTRON-HOLE PAIRS BY HOT ELECTRONS OR HOT HOLES

Pair production may become important in SiO2 films when a high electric field and a high voltage (a few hundred volts) is applied to accelerate the electrons and holes to their pair production threshold energies, about Eg (9 eV) to 2Eg (18 eV). We expect the pair production rate by electron impact to be substantially larger than that by hole impact due to the very large hole mass and narrow and disjointed valence bands in SiO2. This interband impact generation phenomenon has been studied and characterized extensively in semiconductors (Ge, Si, III-V) owing to the engineering importance of the maximum voltage and electric field that can be applied to a diode or transistor before electronic breakdown or current runaway occurs. The pair production rate data in these semiconductors have been obtained in sufficient detail (as a function of electric field's magnitude and orientation relative to the crystallographic direction, and lattice temperature) that most of the basic physics (scattering and energy loss rates by phonons, impurities and other interactions) are determined. The semiconductor data, however, have been reported under the guise of avalanche breakdown in p-n junctions, charge, current or avalanche multiplication, ionization, impact, and others. Each of these terms describes a specific property or phase of the phenomenon and none focuses on the fundamental properties, namely, interband, electron-hole pairs and energetic electron (or hole) impact.

Data on SiO2 are scarce and unreliable due to two problems. First, true electric field at the defect site, where impact generation is localized and has the highest rate, is not known in a SiO2 sample. Unlike semiconductors, especially Si and lately GaAs, geometry control to make the electric field areally constant and fabrication control to make the

SiO2 defect-free have yet to be attained in state-of-the-art silicon
VLSI (very large scale integration) oxidation processes. Second,
carrier density at the localized generation site can be very high and
would be more appropriately described by the diffusion model of Wolff
[1] instead of the ballistic model of Shockley [2] or by the
intermediate model of Baraff [3], making a fit of the data to a theory
less reliable. Only two sets of pair production rate data from electron
impact are available to-date for SiO2, one obtained by Solomon and Klein
[4] for thermally grown thin films on crystalline silicon and the other
by Bloembergen and associates [5] on fused silica from laser breakdown.
Solomon and Klein obtained a much smaller impact generation rate and
pair production coefficient by electron impact, alpha(n) (/cm), and a
much faster and abrupt rise with electric field, beginning at about
8 MV/cm, than the laser breakdown measurements of Bloembergen
and associates. Solomon and Klein fitted their data to Shockley's
ballistic model and obtained alpha(n) = (6.5 X 10*11)exp(-180/F)/cm
where the electric field, F, is in MV/cm. The threshold field,
180 MV/cm, is theoretically given by In/lambda(Rn) based on Shockley's
model where In is the threshold kinetic energy of electrons for impact
generation of an electron-hole pair and it can be as low as the energy
gap of SiO2 (9 eV) if the hole mass is much larger than the electron
mass. It can be as high as 2Eg (18 eV) if the hole mass is much
smaller than the electron mass, an unlikely situation. It is equal to
E(I) = Eg[1 + mn/(mn + mp)] = (3/2)Eg = 13.5 eV if the electron and
and hole masses, mn and mp, are equal. In either the heavy hole or equal
mass case, the mean free path of the hot electrons, limited by optical
phonon scattering (lambda(Rn) = 5 A for heavy hole or 7.5 A for equal
mass) at high fields, is too small compared with that expected from
extrapolation of the experimental low-field hot electron mobility
since the mean free path is expected to increase from the hot electron
mobility value of 30 A to larger values at higher electric fields owing
to the decreasing energy dissipation rate of electrons with increasing
electric field via polar optical phonon emission. The data of Solomon
and Klein were later fitted by Ferry to Wolff's diffusion model [6].
The fit was poor and the magnitudes of the mean free path parameters
were not physical as noted also by Fischetti. We think that the Solomon-
Klein data may reflect pair production at defective spots since they
show characteristic features of localized breakdown, for example, the
electric fields at the defects are higher than the average they used in
presenting their data. Use of the higher actual field would reduce the
abrupt rise and its curvature when plotted as a function of 1/E, and
the threshold field would be considerably smaller than 180 MV/cm. The
laser breakdown data of Bloembergen and associates [5] are probably
more representative of uniform or intrinsic breakdown, that is, their
electric field values computed from the laser field intensities are
probably closer to the actual field strengths of the electric field
which accelerated the electrons to the impact threshold, In. In fact
their four data points can be fitted to Shockley's ballistic model to
give very reasonable parameter values. Two fits are made by this
reviewer [7], a fit to the original Shockley model (electron drift with
zero-phonon scattered) and a fit to the extended Shockley model
(electron drift with many forward scatterings of the polar LO phonon,
0.153 eV, before pair production by impact). These are respectively,

alpha(n)/F = (10*6)/[76.5 exp(45/F)+13.5] /MV (zero phonon) and
alpha(n)/F = 1200/([(1.86 X 10*-4) + (0.51/F)]exp(45/F)+1] /MV (many
                                                                phonons)
where the electric field is in MV/cm. The threshold field, 45 MV/cm,
from the fits now gives very reasonable values of the hot electron mean
free path due to the 153 meV polar LO phonon scattering. For electron

impact threshold energies of 13.5 eV and 9.0 eV, the mean free paths are 30 A and 45 A respectively. These fits also give the electron mean free path ratio due to pair production or ionization and phonon scattering, rn = lambda(In)/lambda(Rn) = 76.5/0.153 = 500 (zero phonon) and 5400 (many phonons). These ratios are phenomenological and seem high but they are not inconsistent with the inherent low probability of pair production by electron impact expected in large energy gap insulators such as SiO2. The two models give essentially the same values of alpha(n) below 20 MV/cm. Above 20 MV/cm, the many phonon-scattering model saturates to alpha(n)/F = 1200/MV which agrees with the trend of the laser data while the zero-phonon model increases cotinually to an asymptotic value of 11000/MV.

REFERENCES
----

[1]   P.A.Wolff [Phys. Rev. (USA) vol.95 no.6 (1954) p.1415]
[2]   W.Shockley [Solid-State Electron. (GB) vol.2 no.1 (Jan 1961) p.35-67]
[3]   G.A.Baraff [Phys. Rev. (USA) vol.128 no.6 (1962) p.2507]
[4]   P.Solomon, N.Klein [Solid State Commun. (USA) vol.17 (1975) p.1397]
[5]   W.L.Smith, J.H.Bechtel, N.Bloembergen [Phys. Rev. B (USA) vol.15 no.8 (15 Apr 1977) p.4039-55 and refs]
[6]   D.K.Ferry [J. Appl. Phys. (USA) vol.50 no.3 (Mar 1979) p.1422-7]
[7]   C.T.Sah [unpublished]
[8]   C.T.Sah [EMIS Datareview RN=17819 (Nov 1987) 'Intrinsic dielectric reakdown of SiO2 films on Si due to interband impact generation of electron-hole pairs']
[9]   C.T.Sah [EMIS Datareview RN=17839 (Nov 1987) 'Interband electron-hole generation in SiO2 films on Si by keV electrons']
[10] C.T.Sah [EMIS Datareview RN=17815 (Nov 1987) 'Generation-recombination-trapping rates and lifetimes of electrons and holes in SiO2 films on Si - a general classification of the fundamental mechanisms']

17.11 INTRINSIC DIELECTRIC BREAKDOWN OF SiO2 FILMS ON Si DUE TO INTERBAND
IMPACT GENERATION OF ELECTRON-HOLE PAIRS

by C.T.Sah

November 1987
EMIS Datareview RN=17819

The current state-of-the-art performance of silicon MOS integrated
circuits appears to be limited by the highest electric field that can
be applied across the increasingly thinner SiO2 film before dielectric
breakdown of the SiO2 films occurs. The highest electric fields observed
to-date in thermally grown SiO2 film on crystalline silicon, before
the onset of oxide current runaway or breakdown, has reached 25 MV/cm
[12a] to 30 MV/cm [12b], which were still a factor of two smaller than
the highest laser breakdown field of fused silica, > 50 MV/cm [10]. The
thermal oxide film data (25-30 MV/cm) are twice the maximum or
breakdown field, 10-12 MV/cm, reported by most manufacturers of silicon
MOS integrated circuits today in their statistical sampling of
production runs. But at the 25-30 MV/cm field, most researchers believe
that electrical breakdown of the oxide is still localized at defects
and not caused by areally uniform intrinsic interband generation of
electron-hole pairs. Thus, a most interesting and still unanswered
fundamental and applied question is 'what is the intrinsic breakdown
field of the SiO2 film or does the SiO2 film break down at all via the
interband impact ionization mechanism?'. Two related questions are:
what is the oxide thickness dependence of the interband impact intrinsic
breakdown field and can breakdown or runaway occur when the total
potential drop is less than the impact ionization threshold or is there
a minimum voltage drop below which breakdown via interband impact
cannot occur? Consider the following scenario. If all the defects and
thin edges are eliminated so that the SiO2 film is completely uniform
microscopically, current runaway will occur at some given applied
voltage across the thin SiO2 film if it is theoretically possible. If
breakdown is not possible, then the current will increase only
exponentially with applied electric field or voltage due to Fowler-
Nordheim tunnelling until the field is so high that the electrons in
the Si-O bonds are ripped off and the lattice explodes.

A concentrated effort has been undertaken recently at IBM to characterize
the energy distribution function of the hot electrons in SiO2 both
experimentally [13-16] and theoretically [17]. Their main motivation
has been to determine the theoretical maximum sustainable electric field
in SiO2 since it poses a fundamental limit on the speed and density of
silicon MOS transistors. The search for the origin of dielectric
breakdown in SiO2 during the last fifty years has been influenced by
Frohlich's 1937 theory [18] which suggested that current runaway or
breakdown can be expected at a few MV/cm in ionic crystals such as NaCl
and by the notion that Frohlich's theory should also be applicable to
SiO2 since SiO2 is partially ionic. Frohlich's theory was based on
electron scattering by one polar optical phonon species to dissipate the
electron energy and it predicted a current runaway, designated also as
breakdown by Frohlich, at a few MV/cm electric field. This runaway was
defined as the electric field at which the electron energy distribution
becomes unstable or the electrons will continue to gain kinetic energy
without limit since the electron energy loss by scattering diminishes to
zero as the electron kinetic energy increases. Thus, above this electric
field all valence electrons will be accelerated to the Newtonian
velocity of the applied field or to the energy given by the voltage drop

across the thin SiO2 film. The current-voltage characteristics will be those of a vacuum diode tube modified by the field, tunnelling or thermionic emission current at the cathode. However, the experimental breakdown field of SiO2 in silicon integrated circuits has kept increasing, from 10 MV/cm in the late 1970s to 25 MV/cm in 1983 [12a] and 30 MV/cm in 1979 [12b] and vacuum behaviour has not been observed at fields higher than the Frohlich runaway field. A discussion of the fundamental difficulties will follow which shows that Frohlich's theory is incomplete since it includes the emission of only one type of polar longitudinal optical (LO) phonons to account for electron scattering and energy loss. Inclusion of the zone boundary and long wavelength polar longitudinal acoustic or piezoelectric phonons will eliminate the Frohlich instability.

Another problem in using Frohlich's polar LO phonon instability model to explain current runaway or breakdown is the neglect of a mandatory positive feedback mechanism required for current runaway at some constant applied voltage. This will be elaborated and followed by an estimate of the intrinsic dielectric breakdown field of SiO2 not previously available.

The basic or mandatory requirement to have an intrinsic electronic breakdown or current runaway, defined as unlimited or infinite current passing through a two-terminal device at some fixed applied voltage, is the presence of two species of conduction or charge carrying mobile particles (electron and hole in semiconductors or insulators and electron and ion in gases). By intrinsic electronic breakdown or runaway, we imply that there is no localized heating, diffusion, alloying and melting and the current-voltage characteristics are not permanently or irreversibly changed as long as the current power dissipation is limited by an external resistor and prevented from becoming very high. The presence of two species of carriers is a necessary condition to produce positive feedback in the electron-hole pair generation process. The two species of carriers will both gain kinetic energy in the applied electric field until they reach their respective threshold energies for pair production. Then, secondary pairs with lower-than-threshold kinetic energies are generated which can then also be accelerated to energies above the threshold to cause additional pairs to be generated. This positive feedback will give an infinite current when the probability of generating an electron-hole pair by a secondary electron (or hole) reaches unity as the applied voltage is increased to the breakdown value. If there is only one species of carrier or if the second species has an infinite mass or zero pair production probability, then the current will increase only exponentially with a applied voltage or electric field, and the true infinite current or breakdown-runaway condition cannot be attained. This two-carrier species requirement has not been appreciated by all insulator breakdown researchers. For example, during the fifty years since Frohlich first proposed a dielectric breakdown theory for ionic solids in 1937 [18] based on a one-carrier (electron) current runaway and one-phonon [polar longitudinal optical (LO) phonon] energy loss model and since Seitz reviewed and extended the theory in 1949 [19-20], there have been persistent and repeated searches for a dielectric breakdown mechanism based on the one-carrier (electron) and one-phonon model. The futile search seems to have been motivated by a unique feature of Frohlich's theory, namely, the decreasing rate of energy loss of the electron to the polar LO phonon with increasing electric field or energy. That intrinsic breakdown in solids is untenable using this one-phonon and one-particle model was clearly indicated by the rate of electron energy loss via acoustical phonon emission given in Shockley's

1951 paper on hot electrons [21]. Shockley showed that the electron energy loss to acoustical phonons will dominate at high electric fields or high electron kinetic energy which would prevent runaway. He later suggested that experimentally observed breakdowns or current runaway in silicon p-n junction diodes are likely to be associated with localized high electric fields at defects [9].

The rate of electron energy loss has the following dependences on the kinetic energy of the electron (hence the electric field or the applied voltage) at high energies. It varies as (Ek)*(3/2) for acoustical phonon emission and (Ek)*(1/2) for nonpolar optical or intervalley phonon emission as shown by Shockley [21], and as log(Ek)/(Ek)*(1/2) for polar optical phonon emission from Frohlich's original theory [19] and the improved theory by Ehrenreich [22,23]. Thus, the suggested runaway due to the polar optical phonon alone, whose energy loss rate decreases with increasing kinetic energy, as (Ek)*(-1/2), is prevented by the increasing energy loss rate to the acoustic phonon, as (Ek)*(3/2). This simple physical consideration on the impossibility of current runaway or dielectric breakdown in a one-carrier system has also been demonstrated and rediscovered recently by Fischetti [17] in a Monte Carlo computation of the scattering rate. His results show a rapidly increasing scattering rate above an electron energy of 3 eV due to acoustical phonon emission while the polar LO phonon emission rate is decreasing. This gave an electron distribution peak between 2 and 4 eV in the electric field range of 4 to 10 MV/cm which is consistent with the experiments by DiMaria and coworkers [13-16]. The latest experimental results at electric fields below 2 MV/cm [15,16] showed the dominance of polar optical phonon scattering as expected by theory in the low field range and the nearly ballistic or scatterless transport from Shockley's ballistic hot electron model [9], below an applied voltage of 1 V across a 55 A SiO2 film or below an electric field of 1.8 MV/cm [16].

Experiments thus far have probably not observed the areally uniform interband generation of electron-hole pairs by impact of hot electrons accelerated in a high electric field applied across the SiO2 film, except perhaps in the laser breakdown data for fused silica reported by Bloembergen and associates [10]. In view that the plasmon energy in SiO2 is about 25 eV [1] while the impact ionization threshold is about 13 eV if electron and hole masses are equal (3Eg/2 = 3 X 9/2 = 13.5 eV) or about 9 eV if the hole is much heavier than the electron, one would expect pair production by energetic electrons as well as holes. There is no fundamental reason to prohibit interband electron-hole pair production and it must be significant when the SiO2 is exposed to ionizing radiation (such as X-rays [24] and keV electron beam [26]). However, in SiO2, purely electronic current breakdown or runaway is either not possible or has a very high threshold field. The fundamental reason is that the pair production rate by hot hole impact is expected to be extremely small in SiO2 since the maximum hole energy in the highest filled SiO2 valence band is limited to about 4 eV by its band width. Theoretical estimates [2-6] gave: (i) 2-4 eV bandwidth for the upper filled valence band constructed from the oxygen 2p [O(2p)] nonbonding orbital near the top and bonding O(2p)/Si(3p-3s-bonding) orbital near the bottom, (ii) 4-6 eV bandwidth for the second or next lower filled valence bands, constructed from bonding O(2p)/Si(3p-3s-antibonding) orbitials and (iii) an energy gap between the two of 2-4 eV. The calculated locations of the two valence bands, the energy gaps, and the deeper O(2s) band (about 20-25 eV below the highest peak of the filled valence band) are all consistent with experiments [7]. Thus, in order to have holes with kinetic energies greater than the threshold energy (assume Ip = Eg = 9 eV), electrons near the bottom of the second lower valence band, about 8-14 eV below the top edge of the top filled valence band, must be excited. Thus, holes must tunnel across the energy gap of about 2 eV between the two uppermost filled valence bands in order to be accelerated to 9 eV kinetic energy and reach the bottom or middle of the second lower valence band. An estimate of the intrinsic breakdown field has been made for the first time [11] using the experimental pair production coefficient of hot electrons, alpha(n) given earlier [25] and theoretical beta-cristoballite energy bands calculated for SiO2 [2,3,6] based on Phillip's beta-cristoballite crystallite model for SiO2 [8]. This calculation gives an intrinsic current breakdown or runaway field of 100 MW/cm for a 1000 A thick oxide or 1000 V breakdown voltage. This numerical result is insensitive to the phonon scattering mean free path of holes, the pair-production threshold energy by hole impact, and the pair-production mean free path of holes, none of which have been determined experimentally. The principal dependence is on the size of the energy gap between the two upper valence bands through which the hole must tunnel across in order to reach the pair-production threshold energy. The free electron mass is employed for holes in Fowler-Nordheim's tunnelling formulae through the triangular potential barrier of the 2 eV energy gap. The narrow widths of the two filled SiO2 valence bands would suggest the use of a larger hole mass which would increase the breakdown field further. However, the tunnelling rate is a delta function that peaks along the low mass path in k-space, thus, the use of the free electron mass for holes could be a better approximation.

The estimated breakdown field of 100 MV/cm suggests that electrical breakdowns of SiO2 films observed to-date (10-30 MV/cm) in factories are likely to be due to defects and that intrinsic breakdown, if it exists, may still be more than four times larger than the highest breakdown field of silicon dioxide films observed to-date, 25-30 MV/cm. This estimate of 100 MV/cm is approaching the atomic electric field due to the silicon and oxygen ions. It may be possible that the SiO2 would disintegrate or explode when its bond electrons are ripped off by such a high electric field before current runaway occurs.

REFERENCES
----

[1]   D.Pines [Elementary Excitations in Solids Chapter 3 (W.A.Benjamin Inc., New York 1963)] See also C.Kittel [Introduction to Solid State Physics 6th edtn, ch.10 (John Wiley & Sons, New York, 1986)] and P.M.Platzman, P.A.Wolff [Waves and Interactions in Solid State Plasma, Solid State Physics, Supplement 13 (Academic Press, New York, 1973)]
[2]   S.T.Pantelides, W.A.Harrison [Phys. Rev. B (USA) vol.13 no.6 (15 Mar 1976) p.2667-91]
[3]   S.Ciraci, I.P.Batra [Phys. Rev. B (USA) vol.15 no.10 (15 May 1987) p.4923-34]
[4]   J.R.Chelikowsky, M.Schluter [Phys. Rev. B (USA) vol.15 no.8 (15 Apr 1977) p.4020-9]
[5]   R.B.Laughlin, J.D.Joannopoulos, D.J.Chadi [Phys. Rev. B (USA) vol.20 no.12 (15 Dec 1979) p.5228-37]
[6]   Y.P.Li, W.Y.Ching [Phys. Rev. B (USA) vol.31 no.4 (15 Feb 1985) p.2172-9]
[7]   B.Fischer, R.A.Pollack, T.H.DiStefano, W.D.Grobman [Phys. Rev. B (USA) vol.15 no.6 (15 Mar 1977) p.3193-9]

[8]   J.C.Phillips [Phys. Rev. B (USA) vol.35 no.12 (15 April 1987)
        p.6409-13; Solid State Phys. (USA) vol.37 (1982) p.93]
        See ref [27] on electron mobility where our analysis predicted a
        crystallite size of 72 A calculated from the temperature
        independent electron mobility of 40 cm*2/Vs in fused silica which
        is consistent with Phillip's model which gives 66 A computed from
        Raman spectra of SiO2]

[9]   W.Shockley [Solid-State Electron. (GB) vol.2 no.1 (Jan 1961)
        p.35-67]

[10]   W.L.Smith, J.H.Bechtel, N.Bloembergen [Phys. Rev. B (USA) vol.15
        no.8 (15 Apr 1977) p.4039-55 and refs]

[11]   C.T.Sah [unpublished]

[12a]   S.S.Cohen [J. Electrochem. Soc. (USA) vol.130 no.4 (Apr 1983)
        p.929-32]

[12b]   E.Harari [J. Appl. Phys. (USA) vol.49 no.4 (Apr 1978) p.2478-89]

[13]   D.J.DiMaria, T.N.Theis, J.R.Kirtley, F.L.Pesavento, D.W.Wong,
        S.D.Brown [J. Appl. Phys. (USA) vol.57 no.4 (15 Feb 1985)
        p.1214-38]

[14]   S.D.Brorson, D.J.DiMaria, M.V.Fischetti, F.L.Pesavento,
        P.M.Solomon, D.W.Wong [J. Appl. Phys. (USA) vol.58 no.3 (1 Aug
        1985) p.1302-13]

[15]   D.J.DiMaria, M.V.Fischetti, M.Arienzo, E.Tierney [J. Appl. Phys.
        (USA) vol.60 no.5 (1 Sep 1986) p.1719-26]

[16]   D.J.DiMaria, M.V.Fischetti, J.Batey, L.Dori, E.Tierney,
        J.Stasiak [Phys. Rev. Lett. (USA) vol.57 no.25 (22 Dec 1986)
        p.3213-16]

[17]   M.V.Fischetti, D.J.DiMaria, S.D.Brorson, T.N.Theis, R.R.Kirtley
        [Phys. Rev. B (USA) vol.31 no.12 (15 Jun 1985) p.8124-42]

[18]   H.Frohlich [Proc. R. Soc. London Ser.A (GB) vol.160 (1937)
        p.230 and vol.188 (1947) p.521-32]

[19]   F.Seitz [Phys. Rev. (USA) vol.76 no.9 (1949) p.1376-93]

[20]   H.Frohlich, F.Seitz [Phys. Rev. (USA) vol.79 no.3 (1950) p.526-7]

[21]   W.Shockley [Bell Syst. Tech. J. (USA) vol.30 (1951) p.990]

[22]   H.Ehrenreich [J. Phys. & Chem. Solids (GB) vol.2 (1957) p.131];
        see also p.153-9 of ref [23]

[23]   E.M.Conwell [High Field Transport in Semiconductors (Academic
        Press, New York, 1967)]

[24]   C.T.Sah [EMIS Datareview RN=17817 (Nov 1987) 'Non-tunnelling
        interband optical generation and radiative recombination of
        electron-hole pairs in SiO2 films on Si']

[25]   C.T.Sah [EMIS Datareview RN=17818 (Nov 1987) 'Non-tunnelling
        interband Auger recombination and impact generation of electron-
        hole pairs in SiO2 films on Si']

[26]   C.T.Sah [EMIS Datareview RN=17839 (Nov 1987) 'Interband electron-
        hole generation in SiO2 films on Si by keV electrons']

[27]   C.T.Sah, A.J.Chen, C.C.-H.Hsu, T.Nishida [EMIS Datareview
        RN=17808 (Nov 1987) 'Electron mobility in SiO2
        films on Si']

17.12 INTERBAND ELECTRON-HOLE GENERATION IN SiO2 FILMS ON Si BY KEV ELECTRONS

by C.T.Sah

November 1987
EMIS Datareview RN=17839

This subject is of considerable practical interest in the manufacturing of silicon integrated circuits. Either keV electron-beam or X-ray (see [10]) lithography may be required to reduce the transistor structures to below 300 nm in order to increase the bit density beyond 16 Mbit/chip. Traps can be generated by keV electron impact which ruptures the bonds or by X-rays via the secondary electron-hole pairs generated by the X-rays. The pair generation rates in oxidized silicon during the exposure to keV electrons [1-5] had a magnitude which scaled roughly with the keV electron beam energy or the kinetic/threshold energy ratio. This is indicative of secondary pair production as the intermediate step which may proceed via plasmon decay as discussed in [11]. Operation of silicon integrated circuits in ionizing environments in space is another application area where interband electron-hole generation by X-ray and keV-electrons is important. The energy loss mechanisms during X-ray or keV-electron exposures via plasmon decay are discussed in [11].

An empirical formula has been developed by Everhart and Hoff [6] which gives a universal curve for the volume generation rate of electron-hole pairs in any material for incident electron energy in the range of 5 keV to 25 keV. In terms of this universal curve, the volume generation rate of electron-hole pairs by an incident electron beam of incident energy E(B)(keV) and current density of I(B) is

```
                 f I(B)   E(B)    1             x
g(n)n (x) = g(p)n = (------)(------)(----) lambda(----) pair/cm*3
                 e     E(AVE)  R(G)          R(G)
```

where (1-f) is the fraction of electrons backscattered on the surface (f = 0.9, but may be energy dependent at lower energies), E(AVE) is the average energy loss of the electron beam to create one electron-hole pair (see [11] on the physics of this term) and R(G) is the modified Gruen range given by R(G) = (3.98[E(B)]*1.75)/rho where rho is the mass density of the material in microgram/cm*3: rho(Al) = 2.70, rho(SiO2) = 2.20 and rho(Si) = 2.33 gram/cm*3. The electron energy loss or deposition per unit absorber thickness, dE(x)/dx, is given by the universal curve, lambda(y) = 0.6 + 6.21 y - 12.40y*2 + 5.69y*3 where y =x/R(G), i.e.

```
dE(x)     E(B)  d[(E(x)/E(b)]   E(B)         x
----- = (----) ------------- = ---- lambda(----)
dx        R(G)  d(x/R(G))       R(G)        R(G)
```

In VLSI applications subjecting to radiations, the energy deposited or dose in units of rad (for example, rad-Si) is frequently used. This is given by

```
                 Q(B)  [E(x1) - E(x2]
D(rad-material) = (----)(--------------)
                 rho        DELTA x
```

Here E(x1) is the energy of the electron beam entering the surface at x1. E(x2) is the exit electron beam energy at x2. DELTA x is the layer

thickness, x2 - x1, if the layer is thinner than the range and
DELTA x = R(G) if it is thicker. Q(B) is the total charge during the
electron beam exposure given by Q(B) = eI(B)t(B) for an exposure time of
t(B) if the beam current is constant. The fractional energy absorbed by
a film of thickness y1 can be computed from 1 - phi(y1) where phi(y1) =
integral lambda(y) dy = 0.6 y1 + 3.10 (y1)*2 - 4.13 (y1)*3 + 1.42 (y1)*4
is integrated through the layer from y = 0 to y = y1. The Everhart-Hoff
curve has been used in many studies of the generation kinetics of oxide
and interface traps in silicon MOS capacitors by Everhart and McDonald
during 1968-1971 [6,7], Curtis, Srour and Chu in 1974 [8] and Ausman
and McLean in 1975 [9]. Ausman and McLean reanalysed the data of Curtis,
Srour and Chu and showed for E(B) = 4 keV, E(A) is 18.4 eV and columnar
recombination of the electrons can account for the current-electric
field curve with a columnar radius of 81.3 A. The initial or geminate
recombination which dominates the photogeneration rate discussed in
another Datareview [12] was not important in the keV electron range.
This was understandable since the initial energy loss by the keV
electrons proceeds via plasmon generation to be discussed in [11].

REFERENCES
----

[1]   C.T.Sah, J.Y-C.Sun, J.J-T.Tzou [J. Appl. Phys. (USA) vol.54
        (1983) p.944, fig.1]
[2]   C.T.Sah, J.Y-C.Sun, J.J-T.Tzou [J. Appl. Phys. (USA) vol.54 no.8
        (1983) p.4378-81]
[3]   C.T.Sah, J.Y-C.Sun, J.J-T.Tzou [Appl. Phys. Lett. (USA) vol.43
        no.10 (1983) p.962-4]
[4]   C.T.Sah, W.W-L.Lin, S.C-S.Pan,, C.C.-H.Hsu [Appl. Phys. Lett.
        (USA) vol.48 no.12 (24 Mar 1986) p.782-4]
[5]   S.C-S.Pan, C.T.Sah [J. Appl. Phys. (USA) vol.60 no.1 (1986)
        p.156-62]
[6]   T.E.Everhart, P.H.Hoff [J. Appl. Phys. (USA) vol.42 no.13 (Dec
        1971) p.5837-46]
[7]   N.C.MacDonald, T.E.Everhart [J. Appl. Phys. (USA) vol.39 (1968)
        p.2433]
[8]   O.L.Curtis,Jr, J.R.Srour, K.Y.Chu [J. Appl. Phys. (USA) vol.45
        no.10 (Oct 1974) p.4506-13]
[9]   G.A.Ausman,Jr [F.B.McLean [Appl. Phys. Lett. (USA) vol.26 no.4
        (15 Feb 1975) p.173-5]
[10]   C.T.Sah [EMIS Datareview RN=17817 (Nov 1987) 'Non-tunnelling
        interband optical generation and radiative recombination of
        electron-hole pairs in SiO2 films on Si']
[11]   C.T.Sah [EMIS Datareview RN=17820 (Nov 1987) 'Interband electron-
        hole pair generation in SiO2 by plasmon decay']
[12]   C.T.Sah [EMIS Datareview RN=17816 (Nov 1987) 'Non-tunnelling
        interband thermal generation and recombination of electron-hole
        pairs in SiO2 films on Si']

17.13 INTERBAND ELECTRON-HOLE PAIR GENERATION IN SiO2 BY PLASMON DECAY

by C.T.Sah

November 1987
EMIS Datareview RN=17820

The energy exchange mechanism between plasmons and electron-hole pairs
is gaining importance in present and future transistor and integrated
circuit design owing to the presence of electrons with kinetic energies
greater than the plasmon energies, about 10 eV. The bulk and interface
plasmon energies are [1-3]: Al (15.3 eV), SiO2 (25 eV), Si (16.9 eV),
Al/SiO2 (8.55 eV) and SiO2/Si (9.45 eV). High energy electrons may enter
the SiO2 film from exposures of the silicon integrated circuit to keV
electrons or X-rays during fabrication using electron-beam or X-ray
lithography and during operation in an ionizing radiation environment.
The fast electrons lose their energy mainly by generation of plasmons
which then decay into electron-hole pairs of tens of electron-volt
energy. Rothwarf was the first to present an illuminating elementary
analysis in 1973 [4] which showed that one electron-hole pair is
generated in a direct energy gap material by the decay of one plasmon
if the bulk plasmon energy, Epb, is between Eg and 4Eg. Three pairs
are generated if 4Eg < Epb < 12Eg. For SiO2 Epb/Eg = 25/9 = 2.8 eV
and the threshold energy is 16-18 eV. Using the details of the SiO2
valence bands just described, Ausman and McLean [9] estimated that on
the average, three plasmons (they used a plasmon energy of 22.4 eV
in SiO2 instead of the experimental value of 25 eV cited by Pines [2])
will decay into four electron-hole pairs. Of the three plasmons, two
will decay by generating two electron-hole pairs from the lower filled
oxygen-2p-bonding valence band while the third will decay by generating
one electron-hole pair from the upper-filled oxygen-2p-nonbonding or
lone pair valence band. The mean free path for plasmon production by
the fast incident electrons is given by [5]

$$\lambda(bp) = \frac{(a(B)/2)(4E/Epb)}{\ln(4E/Epb)}$$

where a(B) = 0.53 A is the Bohr radius. For a 8 keV incident electron,
lambda(bp) = 52.1 A, and at 1.5 keV, lambda(bp) = 12.7 A. Since four
electron-hole pairs are generated by the decay of three plasmons, the
mean free path of electron-hole pair generation by a 1.5 keV electron
is 12.7 X 3/4 = 9.52 A and by a 8 keV electron, 39 A.

The energetic electrons that create plasmons may also arise from the
high voltage applied to the gate electrode of MOS structures. This
voltage would accelerate the electrons injected into the oxide by
avalanche or Fowler-Nordheim tunnelling injection from either the
silicon substrate or the metal or poly-Si gate.

In addition to bulk plasmons just described, interface and surface
plasmons can also be generated by the fast keV electrons or by the
avalanche or tunnelling injected electrons since their energies are
lower than the bulk plasmons [6,7,1-3]. The surface or interface plasmon
can then decay into electron-hole pairs which can then be trapped at the
oxide traps or cause trap generation by bond breaking. Fischetti [3] has
recently showed that interfacial plasmons may be generated at the
gate-metal/SiO2 interface as well as the SiO2/Si interface by hot

electrons injected from the cathode at the opposite interface. These surface or interfacial plasmons can then decay into electron-hole pairs and the holes can drift to and be trapped at the SiO2/Si interface to generate both positive oxide traps and interface states. This creates a serious reliability concern. Interfacial plasmon generation efficiency by hot electrons was estimated to be as high as 0.3 at an Al/SiO2 interface [3]. As much as 1% of the generated hot holes can be injected into the SiO2 from the interface [3]. This is now thought to be the dominant mechanism of positive charge generation or trapping in the oxide during avalanche or tunnelling injection of electrons into the oxide.

REFERENCES
----

[1]   D.Pines [Elementary Excitations in Solids, Chapter 3 (W.A.Benjamin
      Inc., New York, 1963)] See also, C.Kittel [Introduction to Solid
      State Physics, 6th edtn, ch.10 (John Wiley & Sons, New York,
      1986) and P.M.Platzman, P.A.Wolff [Waves and Interactions in Solid
      State Plasma, Solid State Physics, Supplement 13 (Academic Press,
      New York, 1973)]
[2]   D.Pines [Rev. Mod. Phys. (USA) vol.28 (1956) p.184]
[3]   M.V.Fischetti [Phys. Rev. B (USA) vol.31 no.4 (15 Feb 1985)
      p.2099-113]
[4]   A.Rothwarf [J. Appl. Phys. (USA) vol.44 no.2 (Feb 1973) p.752-6]
[5]   D.Pine [Phys. Rev. (USA) vol.92 (1953) p.626]
[6]   R.H.Ritchie [Phys. Rev. (USA) vol.106 no.5 (1 Jun 1957) p.874-81]
[7]   E.A.Stern, R.A.Ferrell [Phys. Rev. (USA) vol.120 no.1 (Oct 1960)
      p.130-6]
[8]   C.T.Sah [EMIS Datareview RN=17819 (Nov 1987) 'Intrinsic dielectric
      breakdown of SiO2 films on Si due to interband impact generation
      of electron-hole pairs']
[9]   G.A.Ausman,Jr, F.B.McLean [Appl. Phys. Lett. (USA) vol.26 no.4
      (15 Feb 1975) p.173-5]

17.14 GENERATION-RECOMBINATION-TRAPPING OF ELECTRONS AND HOLES LOCALIZED AT IMPERFECTION CENTRES IN SiO2

by C.T.Sah, C.Y.-P.Chao, A.J.Chen, C.C.-H.Hsu, M.S.-C.Luo and T.Nishida

November 1987
EMIS Datareview RN=17821

A. THERMAL GENERATION-RECOMBINATION-TRAPPING OF ELECTRONS AND HOLES AT

IMPERFECTION CENTRES IN SiO2 (TRANSITION 21 IN [6])

The lifetimes of electrons and holes in a SiO2 film are controlled mainly by the presence of oxide traps via the thermal or phonon energy exchange mechanism [1-3]. In semiconductors, this is known as the Shockley-Read-Hall mechanism [4]. From the capture cross section and the trap density data reviewed in [5] for SiO2 the lifetime can be estimated. For example, consider a 1000 A thick oxide with a total or areal trap density of $10*12/cm*2$ uniformly distributed so that the volume trap density is $(10*12)/(1000 \times 10*-8) = 10*17/cm*3$. For a large electron trap with an electron capture cross section of $10*-14$ $cm*2$, the electron lifetime is then $10*-10$ s if we assume an electron thermal velocity of $10*7$ cm/s. For a small electron trap with an electron capture cross section of $10*-19$ $cm*2$, the electron lifetime would be $10*-5$ s. Similar estimates may be made for the hole lifetimes controlled either by hole capture at the oxide hole traps or hole recombination with trapped electrons at the oxide electron traps.

The electron-phonon energy exchange during a transition localized at an imperfection centre follows the Dexter anharmonic mechanism [3] since the energy change is very large and equal to the energy of tens of phonons (about 30 to 60). However, a variety of optical transitions have also been observed in SiO2 but the luminescence efficiency is low.

B. OPTICAL GENERATION-RECOMBINATION-TRAPPING OF ELECTRONS AND HOLES AT

IMPERFECTION CENTRES IN SiO2 (TRANSITION 22 IN [6])

There is a vast literature on this localized optical mechanism in bulk SiO2 such as quartz, but little on SiO2 films on Si. However the data for the bulk SiO2 should apply to the SiO2 films if the same imperfection centres exist in the SiO2 films. The bulk data will not be reviewed here due to space-time limitation.

C. AUGER CAPTURE AND IMPACT EMISSION OF ELECTRONS AND HOLES AT

IMPERFECTION CENTRES IN SiO2 (TRANSITION 23 IN [6])

Impact emission of trapped electrons and holes by hot electrons or holes is thought to be an important device degradation mechanism [1,2], however, there have not been definitive experiments which identify these as the dominant processes nor indicate their existence.

D. TRAP-TRAP TRANSITIONS (TRANSITIONS 31,32,33 IN [6])

The following three transitions have not been identified in experiments
on SiO2 films, although optical transitions between traps, known as
donor-acceptor pair spectra, have been extensively studied and
characterized in compound semiconductors (GaP, GaAs, ZnS and others)
and Si (In-O pair):

Thermal transitions between traps in SiO2 (transition 31 in [6])
Optical transitions between traps in SiO2 (transition 32 in [6])
Auger-impact transitions between traps in SiO2 (transition 33 in [6])

REFERENCES
----

[1]   C.T.Sah [Proc. INFOS-87, Leuven, Belgium (15 April 1987),
      Appl. Surf. Sci. (Netherlands) vol.30 nos.1-4 (Oct 1987) p.311-12]
[2]   C.T.Sah ['VLSI Device Reliability Modeling', Invited paper 5-1
      presented at the 1987 International Symposium on VLSI Technology,
      Systems and Applications, 14 May 1987. Proceedings of Technical
      Digest p.153-162, Publisher: TSPD (C60), ERSO, ITRI, No.195-4(C60),
      Chung Hsing Rd., Sec. 4, Chu Tung, Hsinchu Taiwan (31015), Republic
      of China]
[3]   D.L.Dexter, C.C.Klick, G.A.Russell [Phys. Rev. (USA) vol.100 no.2
      (15 Oct 1955) p.603-5; Solid-State Phys. (USA) vol.6 (1958) p.355]
      (See section 13 on nonradiative transition starting on p.405)
[4]   W.Shockley, W.T.Read,Jr [Phys. Rev. (USA) vol.87 (1952) p.835]
[5]   C.T.Sah, C.C.-H.Hsu [EMIS Datareview RN=16185 (Nov 1987) 'Oxide
      traps on oxidised Si']
[6]   C.T.Sah [EMIS Datareview RN=17815 (Nov 1987) 'Generation-
      recombination-trapping rates and lifetimes of electrons and holes
      in SiO2 films on Si - a general classification of the fundamental
      mechanisms']

17.15 TUNNELLING TRANSITIONS OF ELECTRONS AND HOLES IN SiO2 FILMS ON Si

by C.T.Sah

November 1987
EMIS Datareview RN=17822

A. INTRODUCTION

Tunnelling transitions are listed in [10] and are technologically
important generation-recombination-trapping mechanisms in which
tunnelling is the primary rate controlling step. We shall discuss only
two elastic cases listed in [10], transitions (15) and (25), since they
have the highest transition rates in the interband and band-trap
categories respectively. Reliable experimental data are available only
for the interband tunnelling, transition (15), however, theoretical
rates are available for both cases.

B. ELASTIC INTERBAND TUNNELLING TRANSITION BETWEEN SiO2 AND Si

(TRANSITION 15 IN [10])

The elastic interband tunnelling transitions between the conduction
bands of SiO2 and Si, known as the Fowler-Nordheim tunnelling electron
(or hole) injection (FN-TEI or FN-THI) and listed as (15) in [10]
have been identified as two of the ultimate causes of silicon integrated
circuit failure as the transistor size shrinks and the oxide thickness
decreases to below 10 nm [1,2]. The instability leading to failure
arises from trapping the electron, which has tunnelled into the SiO2
conduction band from the Si conduction band, by an oxide trap [11].
However, this is also the very mechanism employed to store a bit of
information in the Electrical Erasable Programmable Read Only Memory
(EEPROM) in which FN-TEI is used to charge or discharge a floating
gate buried in the oxide. The FN-TEI current between Si and SiO2 has
been determined accurately by Weinberg [3] and verified world-wide. As
expected from the anisotropy of the two-dimensional silicon electron
effective mass, there is a strong orientation dependence in the
tunnelling rate. However, the tunnelling current due to holes via the
FN-THI mechanisms at the anode (positively biased electrode) has not
been measured in the SiO2/Si barrier system owing to the larger hole
potential barrier (4.7 eV) than the electron potential barrier
(3.1 eV) so that the latter dominates at the cathode (negatively
biased electrode). Thus, scaled theoretical formulae for FN-THI current
were proposed [1,2]. These are listed below.

$J(N)TI = (2.6 \times 10^6)(Eo^2)\exp(-238.5/Eo)$ A/cm^2 <100> (Weinberg)

$J(N)TI = (1.8 \times 10^6)(Eo^2)\exp(-285.0/Eo)$ A/cm^2 <111> (Weinberg)

$J(P)TI = (2 \times 10^6)(Eo^2)\exp(-440/Eo)$ A/cm^2 (scaled)

where (N) indicates FN-TEI and (P) indicates FN-THI

Eo is the constant oxide electric field (in MV/cm) at the
SiO2/Si interface which determines the thickness of the triangular
tunnelling barrier.

C. ELASTIC TUNNELLING BAND-TRAP TRANSITIONS BETWEEN Si AND SiO2

    (TRANSITION 25 IN [10])

Electron tunnelling between a band state and an impurity trap state,
denoted by transition (25) in [10] was first identified under controlled
experimental condition in gold-doped silicon tunnelling diodes [4]. It
was known as excess current in semiconductor (Si, Ge, GaAs, etc.)
tunnelling diodes. Surveys of the earlier [5] and recent [6] experimental
work on semiconductor tunnel junctions were given. Structures in the
conductance-voltage curve and peaks in its derivative for thin MIM
(Metal-Insulator-Metal) and MIS (Metal-Insulator-Semiconductor)
diodes were attributed to band-trap transitions with the assistance of
phonon (transition 21), photon (transition 22), and plasmon
(transition 24) [7]. The practical importance of an electron
tunnelling transition between an oxide trap and the conduction (or
valence) band of Si and SiO2 on the reliability of silicon MOS field
effect transistors and integrated circuits was recently emphasized
[1,2]. Unlike the abundant although earlier (prior to 1969)
experiments on semiconductor junctions, experimental data on the trap-
band transition in the metal/SiO2/Si junctions are non-existent. A
theoretical tunnel-to-trap transition rate was obtained by Sah in 1961
[4] based on Price's formulation [8]. This formulae was rederived
recently by Anderson and Hoffman in 1982 [6] and extended to more
general trap potential well shapes by Chaudhuri, Coon and Karunasiri
in 1983 [9]. The original tunnelling transition rate [4] was recently
applied to a triangular tunnelling barrier of a neutral (or delta-
function) potential well of an oxide trap by Sah [1,2] who obtained
the following formula

    Tx = (pi*2/h*3)[(my mz)*(1/2)](Ft/Et)(W*2)exp(-2 THETA)

       = (7.93264 X 10*-10)(Ft/Et)exp[-68.28995[(Et)*(3/2)]/Ft] /s

where the tunnelling integral taken over the forbidden path is

    THETA = mod(integral kc dx) = (4 pi/3)[(2 mx)*(1/2)][(Et)*(3/2)]/hF,

Ft is the constant electric field of the triangular tunnelling barrier
(in MV/cm), and Et is the trap depth (in eV). In the numerical
formula, it was assumed that mx = my = mz (free electron mass) and
W2 = 10*-24 V*2 cm*3 for the square of the tunnelling transition matrix.
Formulae for the tunnelling current of electrons into or out of an
oxide trap, against (triangular barrier) or following (trapezoidal
barrier) the force from the electric field and into or out of the
silicon or SiO2 conduction band, were obtained by Sah [2] from scaling
the results of the tunnelling currents observed in gold-doped silicon
tunnel diodes [4]. The current due to a silicon electron, tunnelling
into an oxide trap located at phi(T) below the oxide conduction band
edge, can be estimated from [2]

J(NTB) = [(10*-14) P(T) xT] exp[-(240/Eo)[1 - (1-(xT/xB))*(3/2)] A/cm*2

and the tunnelling current of an electron trapped at an oxide trap,
making an elastic tunnelling transition into the conduction band of
the oxide, can be estimated from [2]

J(NTB) = [(10*-14) NT(xT)] exp[-(240/Eo)(phi(T)/phi(B))*(3/2)] A/cm*2

In the first formula, xB = phi(B)/Eo is the barrier width, P(T) is the
density of the empty traps (not occupied by electrons) at a distance
xT from the oxide/silicon interface which is also the tunnelling
distance or the forbidden path. In the second formula, N(T) is the
occupied (by electron) trap density at xT. Sah also gave a scaled
formula for hole tunnelling by replacing the factor 240 by 440 to take
into account the hole/electron barrier height ratio at the SiO2/Si
interface, (4.7/3.1)*(3/2) = 1.85. Hole/electron effective mass
differences will also change the 440 number and an orientation
dependence could also be expected due to anisotropic electron mass in
the two-dimensional silicon conduction band at the SiO2/Si interface.
These scaled formulae provide a vehicle for interpreting experiments
which are yet to be performed in the SiO2/Si system but whose
importance to oxide trap charging has been universally recognized.

REFERENCES
----

[1]   C.T.Sah [Proc. INFOS-87 Leuven Belgium, 15 April 1987, Appl.
       Surf. Sci. (Netherlands) vol.30 nos.1-4 (Oct 1987) p.311-12]
[2]   C.T.Sah ['VLSI Device Reliability Modeling' Invited paper 5-1
       presented at the 1987 International Symposium on VLSI Technology,
       Systems and Applications, 14 May 1987, Proceedings of Technical
       Digest p.153-62, Publisher: TSPD (C60), ERSO, ITRI, No.195-4(C60)
       Chung Hsing Rd., Sec.4, Chu Tung, Hsinchu, Taiwan (31015),
       Republic of China   ]
[3]   Z.A.Weinberg [Solid-State Electron. (GB) vol.20 (1977) p.11;
       J. Appl. Phys. (USA) vol.53 no.7 (Jul 1982) p.5052-6]
[4]   C.T.Sah [Phys. Rev. (USA) vol.123 no.5 (1 Sep 1961) p.1594-612]
[5]   C.T.Sah [in 'Tunneling Phenomena in Solids' ch.14
       Eds E.Burstein, S.Lundqvist (Plenum Press, New York, 1969) p.193-
       205]
[6]   W.W.Anderson, H.J.Hoffman [J. Appl. Phys. (USA) vol.53 no.12
       (Dec 1982) p.9130-45]
[7]   See a survey in C.B.Duke [Tunneling in Solids, Solid State
       Physics, Suppl. 10 (Academic Press, New York, 1969)]
[8]   P.J.Price, J.M.Radcliff [IBM J. Res. & Dev. (USA) vol.3 (1959)
       p.364]
[9]   S.Chaudhuri, D.D.Coon, R.P.G.Karunasiri [J. Appl. Phys. (USA)
       vol.54 no.9 (Sep 1983) p.5476-8]
[10]   C.T.Sah [EMIS Datareview RN=17815 (Nov 1987) 'Generation-
       recombination-trapping rates and lifetimes of electrons and holes
       in SiO2 films on Si - a general classification of the fundamental
       mechanisms', table 2]
[11]   C.T.Sah, C.C.-H.Hsu [EMIS Datareview RN=16185 (Nov 1987) 'Oxide
       traps on oxidized Si']

17.16 HYDROGENATION AND DEHYDROGENATION OF SHALLOW ACCEPTORS AND DONORS
      IN Si: FUNDAMENTAL PHENOMENA AND SURVEY OF THE LITERATURE

by C.T.Sah

December 1987
EMIS Datareview RN=17845

A. HISTORICAL BACKGROUND

Electrical deactivation or disappearance of the hole bound states at
the shallow acceptor impurities is a new phenomenon. Boron acceptor
deactivation in silicon was first observed by Sah and Fu in 1972 when
p-Si MOS capacitors were exposed to a 10 keV electron beam [1].
Phosphorus donor deactivation was not observed however. The mechanism
of boron acceptor deactivation was first identified by Sah, Sun and
Tzou in 1982 as a hydrogenation phenomena in which the boron-hydrogen
pair is formed [2]. They then showed that the deactivation also occurs
with other group-III acceptors (Al, Ga and In) [3]. Its practical
importance on the reliability and endurance or operating life of
silicon integrated circuits has also been discussed by these
researchers [4,5].

Twenty four experiments were designed by Sah, Sun and Tzou to test
the hydrogen-acceptor bonding model during the first year using MOSC
(metal-oxide-silicon capacitors) with atomic hydrogen released from
the gate conductor (Al or polycrystalline Si) and oxide (rinsed
in deionized water or wet oxidation) layers by hot electron impact,
thermal hole capture or keV electron-beam exposure [1-8]. These were
followed by a series of non-MOSC experiments by Pankove using plain
p-type silicon wafers exposed to atomic hydrogen from a plasma [9].
Theoretical modeling of the atomic configurations of the hydrogen bond
were then considered by Sah and Pankove [10] and additional experiments
were undertaken by them and others. A chronology of these discovery
experiments by Sah's and Pankove's groups [1-18] up to December 1984 is
given in TABLE 1. Later experiments, theories and reviews not tabulated
in TABLE 1 will be discussed in subsequent paragraphs.

TABLE 1: a chronology of studies on acceptor hydrogenation in silicon
        [18-20]

Legend: JAP=J.Appl.Phys.(USA); APL=Appl.Phys.Lett.(USA); PRL=Phys.Rev.
        Lett.(USA); SRC=Semiconductor Research Corp.; e=experiment;
        m=model; [n]=reference; AEI=Avalanche Electron Injection; AHI=
        Avalanche Hole Injection; PMA=Post Metallization Anneal; POA=
        Post Oxidation Anneal. ':' is an electron pair bond.

| Item No | Authors | Dates | | Descriptions |
		Submit	Publish	
0	Sah, Fu	Mar 72	Feb 83 JAP [1]	10 keV e-beam, Cinv drops in p-Si MOSC and does not change in n-Si MOSC.
1	Sah, Sun Tzou	Aug 82	Feb 83 JAP [1]	(1e) AEI, Cinv drops in p-Si MOSC only.
				(2e) AHI, Cinv stays constant in n-Si MOSC.
				(3e) AEI, hole depth profile, P(x) in p-Si MOSC
				(4e) Second order annealing kinetics of boron dehydrogenation, 1.07 eV.
				(1m) Bond breaking model postulated after 1976 paper by Sah in IEEE Trans. NS-23(6), 1563 for keV electron and ionizing radiation. Si:OH + e* --> Si. + OH. + e
				(2m) Hydrogen bond model first proposed, p.946. Consistent with constant & low P(x) p.854.
				(3m) Oxygen dangling bond donor model p.953 SiO:H + e*--> SiO. (donor)+H. (defunct)
				(4m) Hydrogen molecule formation/dissoc. p.954 2(Si:O) + H:H--> 2(Si:OH) implies H. + H. <--> H:H (2nd order)
2	Sah, Sun Tzou	Sep 82	Dec 82 IEDM-82 [4]	(5m) Hydrogen source demonstrated from hydroxyl in oxide and at oxide/silicon interface.
				(5e) Boron hydrogenation rate: NOPOA > POA
				(6e) Boron hydrogenation rate: (111) > (100)

Item No	Authors	Dates		Descriptions
		Submit	Publish	
3	Sah, Sun Tzou	Oct 82	Mar 83 VLSITSA [5]	(7e) Absent in 900 deg C heated Si gate p-Si MOSC. (8e) Present in boron-diffused n-Si MOSC. (9e) Absent in not-boron diffused n-Si MOSC.
4	Sah, Sun Tzou	Feb 83	Jul 83 APL [2]	(6m) First paper on boron acceptor deactivated by hydrogen. (8e) Present in boron-diffused n-Si MOSC. (9e) Absent in not-boron-diffused n-Si MOSC. (10e) Observed in Al/p-Si Schottky barrier with 5 keV e-beam for the first time. (7m) Hydrogen bond breaking model by thermal hole capture.
5	Sah, Sun Tzou	Feb 83	Aug 83 JAP [7]	(11e) Enhanced boron hydrogenation rate by 5 keV and 8 keV e-beam pre-AEI irradiation.
6	Sah, Sun Tzou	Feb 83	Oct 83 JAP [16]	(12e) Oxygen eliminated. FZ= CZ (13e) Hydroxyl in oxide. No POA >> POA (N2 or Ar) (14e) Hydrogen ambient. PMA > no PMA (400 deg C FG) (15e) H from Al-gate. PMA = no PMA (16e) Wet oxide surface. DI >> NODI (22 deg C) (17e) Wet oxide HIPOX, HIPOX >> DRYOX (8m) 2BH <--> 2B + H. + H. <--> 2B + H:H; Bond Breaking model for hydrogen release. (9m) Al:H + e*-->Al. + H. (Al-gate) (10m) H.(gate) --> H. (Si) H migrate in SiO2 (11m) H. + B. --> B:H Hydrogenation of B (12m) B:H --> H. + B. Dehydrogenation of B-H. (13m) H. + H. --> H:H Molecular Hydrogen.

Item No	Authors	Dates		Descriptions
		Submit	Published	
6 (continued)				(14m) H+B- symbol and bond model suggested. See Dickerson-Gray-Haight 1978 or Pauling 1960.
7	Sah,Sun Tzou	Jun 83	Mar 84 JAP [8]	(18e) No hydrogenation in 950 deg C heated Si-gate (19e) Hydrogenation in not-heated Si-gate p-MOSC. (20e) No hydrogenation in 950 deg C heated Si-gate is independent of gate-dopant (B,P). (21e) No hydrogenation in 950 deg C heated Si-gate is independent of phosphorus gate-dopant type. (22e) Poly-Si gate is a barrier (H-sink) to hydrogen penetration from overlay Al-gate.
8	Sah, Sun Tzou	Jul 83	Nov 83 APL[3]	(23e) Hydrogenation of B, Al,Ga,In during 8-keV (24e) Chemical trend.
9P1	Pankove Carlson Berkeyheiser, Wance	Sep 83	Dec 83 PRL[9]	Neutralization of boron from RF plasma H source.
9P2	Pankove	Jan 84	to Sah [10a]	Pankove described his Si3:::B H:Si model to Sah and interpreted Sah's model (letters).
9P3	Sah	Feb 84	to Pankove [10b,c,d,e]	Sah described his hydrogen bridging bo model Si.H.B to Pankove with figures (letters).
9P4	Pankove Wance	Jul 84	Nov 84 APL[11]	Neutralization of B,Al,Ga,In acceptors by atomic hydrogen from plasma source.
9H1	Hall	Aug 84 ICDS	Aug 84 [12]	Emphasize H. + H.--> H:H on the kinetics of hydrogen-defect complex formation in Ge and Si from earlier work (letter from Hall to Sah).

Item No	Authors	Dates Submit	Publish	Descriptions
10	Sah	Jul 84	Oct 84 SRC[13]	(15m) Atomic configurations of Si.H.B:::Si3 given for five symmetry locations and four multi-atom centers (2c2e,3c2e,4c4e,5c6e). (16m) Lattice distortion and 'reconstruction'.
11P5	Thewalt, Steiner, Pankove	Jul 84	Jan 85 APL [14]	Bound exciton spectra of In and Tl in p-Si with atomic hydrogen from plasma.
11P6	Pankove, Zansucchi, Magee, Lucovsky	Sep 84	Feb 85 JAP [15]	IR peak at 1875 cm*-1 an and Si3:::B H:Si.
11P7	Thewalt, Ligthowlers, Pankove	Dec 84	Jan 85 APL [16]	Bound exciton spectra show neutralization of B,In and Tl by atomic hydrogen from plasma.
12	Sah, Pan, Hsu	Dec 84	Jun 85 JAP[17]	(17m) Protonic bound state model proposed. (18m) Proton capture and emission rates at the group-III impurity proton traps analyzed. (25e) Two-time-phase hydrogenation and dehydrogenation kinetics of all four group-III acceptors observed (AEI stress). (19m) Analytical solution of two-phase kinetics developed and correlated with experiments.
13	Sah	Aug 84	Jul 85 SRC [18]	(15m) Multi-atomic-core and multi-electron centre model of hydrogen-acceptor pair. (16m) Lattice relaxation around H. (17m) Protonic bound state and trapping kinetics

The fundamental simplicity of the hydrogenation phenomenon at the shallow level acceptors compared with that at the deep levels and its practical importance in silicon integrated circuits have led to intensive world-wide research interests [4-60]. Due to its recent discovery and identification,

the fundamental phenomena and definition of terms are reviewed first so
that a consistent set of terms and the basic physics and chemistry can be
presented to the Datareview users. Historical backgrounds are then given,
followed by tabulations of available hydrogenation-dehydrogenation kinetic
rate data in the complementary Datareview [61].

B. FUNDAMENTAL PHENOMENA AND DEFINITION OF TERMS

Hydrogenation of a shallow acceptor in silicon refers to the electrical
deactivation of a group-III substitutional acceptor impurity centre or
hole trap (B,Al,Ga,In or Tl) when the centre captures a hydrogen atom
(or a proton, see remark in the following paragraphs) to form an
acceptor-hydrogen pair. During the capture of a hydrogen, the thermally
released hole from the acceptor is annihilated by the electron from the
hydrogen atom or both the electron and hole are swept away by an electric
field. After the proton or hydrogen is captured by the acceptor, the hole
bound state at the acceptor centre disappears because the acceptor-
hydrogen pair apparently does not produce a sufficiently large
perturbation potential to maintain the hole bound state nor to create a
new hole bound state in the sense of a bound solution of the one-electron
effective-mass Schroedinger equation.

The concept of proton or protonic trap was introduced by Sah [13,17,18]
to visualize, conceptualize and analyze this acceptor hydrogenation
phenomena. The protonic trap concept was derived from a one-to-one analog
with the electronic trap (consisting of electron trap and hole trap)
used in the well-known generation-recombination-trapping (grt) kinetics
of electrons and holes at three-dimensional or truly localized bound
states by Shockley, Read and Hall. The protonic trapping kinetics is
somewhat simpler than the electronic trapping kinetics since it lacks
an electronic hole counter part in solids, namely the non-existence of
the anti-proton, the negative proton or proton-hole at the thermal energy
(0.026 eV) of interest here. Thus, there are only two proton trapping
transition: the proton capture by a protonic trap and the emission of
a trapped proton from the protonic trap; while there are four grt
transition at the electronic trap. Take an electron trap as an example.
There are two electron trapping transitions, the capture and emission of
an electron at the trap; and there are the two recombination-generation
transitions involving the recombination of a hole with a trapped
electron (or the emission of the trapped electron into a hole in an
unoccupied electron state in the valence band) and the generation of a
hole and a trapped electron (or the capture of a valence band electron
by the electron trap).

There is also a one-to-one analogy of the direct generation and
recombination processes between two protons and between an electron and a
hole, with some differences. A proton can capture an electron to form a
hydrogen atom and two hydrogen atoms can recombine to form a hydrogen
molecule. The reverse sequence would give two hydrogen atoms and two
protons and two electrons from the dissociation of a hydrogen molecule.
In the electronic case, an electron and hole can recombine and no
identifiable point-like or localized end product is designated, except to
say a perfect lattice results. The reverse transition involves the
excitation, emission or release of a valence band or bond electron
resulting in the generation of an electron and a hole.

These grt reactions are central to an understanding of the acceptor
deactivation phenomenon and in the application of the rate data to
reliability analysis, prediction and design of silicon integrated

circuits. The importance of molecular hydrogen formation and dissocation
on the hydrogenation kinetic was recognized in the boron dehydrogenation
experiments [1-3] and in earlier hydrogen experiments in germanium by
R.N.Hall [12].

The chemical formula of the hydrogenated acceptor is Si(3):::A.H.Si.
The four Si's are the four nearest silicon atomic cores. The colon
':' is a Lewis electron pair bond or an electron orbital fully occupied
by two electrons and the dot '.' is an electron in the bond or orbital.
A is the atomic core of the acceptor impurity and H is the hydrogen core
or proton. An atomic core is an atom without its valence electrons. The
positive charges on these atomic cores are not labeled explicity at
times in this Datareview and are understood. For example, Si would be
Si++++ and H would be H+. A neutral atom is denoted by a core with its
valence electrons in 'dots'. For example, H. (or H.o or Ho) is the
neutral hydrogen atom, :B. (or Bo) is the neutral boron atom and
:Si: (or Sio) is the neutral silicon atom. The atomic core symbols are
also used to denote the concentration of the atom. The bracket
convention, [], to denote concentration is deleted in this review to
simplify the notations in the differential equations and their solutions.
For example, H and A instead of [H] and [A] are used to denote the
volume density of the unbounded atomic hydrogen and the substitutional
acceptor. When confusion can exist between the ion and the atom, such as
[H]+ and [H:]-, then their densities will be explicity expressed by H+
and Ho. Otherwise, for simplicity of symbol, H means the density of
hydrogen atom and A means the density of :A. or [:A.]. In all the
experiments to-date, the ionic form of hydrogen or proton has not been
detected unambiguously although there were experimental indications that
the positively charged hydrogen atom, or proton, is the diffusion species
in the high-field space-charge layer of a reverse-biased p-n junction or
Schottky barrier. Isolated B+++ and Si++++ could not exist in crystalline
silicon. More explicit formulae will be used in some of the following
discussion.

C. ATOMIC CONFIGURATIONS OF HYDROGENATED ACCEPTORS IN SILICON

The chemical composition formulae, (.Si)3:::A.H.Si::: or Si3:::A.H.Si,
where the continuations of the two ends are understood give neither the
location of the hydrogen nor the atomic configurations of the
hydrogen-acceptor centres. Several atomic configurations for the
locations of the hydrogen were proposed by Sah [2] and Pankove [15].
The original hydrogen bridge model of Sah is described accurately
by the chemical symbol with the hydrogen situated on or next to
the line joining the acceptor and a nearest neighbour silicon [2]
forming the well-known bridging hydrogen bond given by Fig.2(c)
of reference [15]. Pankove's interpretation of Sah's original
model consisted of the hydrogen terminating in a covalent bond with
the boron and the dangling bond on the two adjacent silicons form a
second covalent bond by reconstruction or lattice relaxation. This model
was given by Pankove in Fig.2(b) of reference [15] and attribute to Sah
but is at variance with the original hydrogen bridge model proposed by
Sah and described above. However, current experiments have not
demonstrated its complete absence and a more detailed 3-d figure
be given to show the reconstructed locations of the atomic cores and
hydrogen involved. Pankove also provided another configuration,
frequently referred to as the Pankove model by subsequent theorists
who have been doing numerical computations using a finite cluster of Si
cores containing one boron core and one proton. This Pankove model has a

large lattice relaxation and was given Fig.2(c) of reference [15]. It can be represented by the formula [Si3:::B H:Si] in which the 3-fold boron is coplanar with the three silicons and the large space between B and H reflects the displacement to form the coplanar Si3:::B. It also shows that the proton is attached entirely to the Si by a covalent or electron-pair bond, H:Si, and there is no B.H bond. Pankove arrived at this model from his IR absorption data of a hydrogenated boron centre [15] by noting the similarity between the infrared frequency of the hydrogenated boron acceptor (1875 cm*-1) and a similar frequency observed in molecules containing the Si-H bonds (2270 cm*-1). This Pankove model has a 2-electron/2-centre configuration [13,18] and its atomic displacements to give the coplanar Si-3(triple bond)-B bond are similar to the E1 centre or the oxygen vacancy centre in SiO2 and in alpha quartz. In the oxygen vacancy or E1 centre, one of the Si (instead of the substitutional boron in the boron-hydrogen pair) is pulled back to form a planar SiO3 while a hydrogen is bonded to the dangling silicon bond. Furthermore, there is an electron or hole bound state at the silicon dangling bond in the oxygen or E1 centre while in Pankove's model of a trivalent boron coplanar bonded to three Si cores, there is no dangling or unbonded electron. There is no assurance that this triavlent boron has no electronic bound state. Another deficiency in the Pankove model is its inability to provide any chemical trend of the dehydrogenation energy observed experimentally as B<Al<Ga<In. This trend suggests that the hydrogen is bonded to the acceptors as well as the silicon rather than covalent bonded to the silicon only in Pankove's model.

A detailed elaboration was given by Sah on the possible atomic configurations of the hydrogen on all four group-III acceptor sites [13,18,21]. This analysis included the five B.H.Si 3-centre/2-electron configurations with various displacements of the B and surrounding Si cores shown in Figures 18(a1), 18(a2) on page 60 and Figures 19(a), 19(b) and 19(c) on page 61 of [13] or Figure 1 on page 8 of [18]; the 2-centre/2-electron configuration of Pankove, shown in Figure 19(d) on page 61 of [13] or Figure 1 on page 8 of [18]; the 4-centre-4-electron configuration where the hydrogen is at the C-site shown in Figure 18(b) on page 60 of [13] or Figure 1 on page 8 of [18]; and the 5-centre/6-electron configuration where the hydrogen is along the backbond direction either at the Td site or displaced from it, shown in Figure 18(c) on page 60 of [13] or Figure 1 on page 8 of [18]. These configurations were also presented and distributed on one page by Pan in two invited papers given at the Gordon Research Conferences [19,20].

D. THEORETICAL CALCULATIONS OF THE ATOMIC CONFIGURATION

Recently, there have been three theoretical cluster calculations to determine which of these hydrogenated boron configurations postulated by Sah and Pankove have the lowest minimum (not saddle point) in total energy. DeLeo and Fowler [21,22] found that lowest energy-minimum configuration is the 3-centre/2-electron configuration with the H situated on the Si-B line and the B and Si cores are displaced away from H. These displacements or lattice relaxations were anticipated in five 3-centre/2-electron configurations of Sah, shown in Figures 19(a), 19(b) and 19(c) on page 61 of [13] based on simple electrostatics. However, Assali and Leite [23] computed a minimum energy at the Td antibonding site, shown in Figure 18(c) on page 60 of [13] and insisted that both Pankove's and Sah's configurations were incorrect. In addition, Assali also stated that there was a different Pantelides configurations which was also incorrect. These assertions appear to have come from a misunderstanding of what actual atomic configurations were postulated by Sah and a lack of appreciation of the generality of Pantelides' considerations of the proton trapping kinetics which are not limited to a specific atomic configuration. DeLeo and Assali could not resolve their differences [24,25]. It is now resolved by the latest calculations of Amore Bonapasta, Lapiccirella, Tomassini and Capizzi [26] which showed that in the rigid lattice model the antibonding Boron-Td line has a saddle point instead of a minimum energy as deduced by Assali and Leite. Such a saddle point in Assali's result was also suggested by DeLeo [24]. Bonapasta et al also located an absolute minimum energy position for the hydrogen at 1.22 A from the boron and 34 deg C from the Si-B line in the rigid lattice which is identical to one of the 3-centre/2-electron configuration drawn by Sah based on simple physical (electrostatic charge) and chemical bond length considerations given by Figure 18(a2) on page 60 of [13] or Figure 1 on page 8 of [18]. Lattice relaxation or outward displacements of the B and adjacent Si cores, anticipated by Sah [13,18], will suck the H into the B-Si line and push the B and Si cores away from the H. Indeed the final results of the Bonapasta et al cluster calculations [26] concluded that H is at 1.46 A from Si and 1.59 A from B on the B-Si line. The sum of these separations is 0.5 A [from (1.46+1.59)-2.54 = 0.50 A] which is > the Si-Si distance and requires the displacement of both the boron and silicon atoms away from the the hydrogen. This equilibrium configuration, Si3:::B.H.Si:::Si3, agrees with the calculation of DeLeo and Fowler and confirms confirms the original hydrogen bridge configuration proposed by Sah based on simple physical and qualitative electrostatic considerations as shown by Figures 19(a), (b) and (c) on page 61 of [13] or page 8 of [18]. These three configurations have the bridging hydrogen located on the B-Si bond and three possible atomic displacements of the B and Si cores:

(i)   Si is moved out of its lattice site away from the hydrogen and the boron is at the lattice site shown in Fig.19(a) of [13]

(ii) both Si and B are moved from their lattice sites away from the hydrogen in Fig.19(b) of [13]

(iii) B is moved out of its lattice site away from the hydrogen while Si is at the lattice site as shown in Figure 19(c) of [13].

The three Si cores adjacent to the B are displaced towards B due to the smaller B core than Si in all configurations while the three Si cores adjacent to the Si of the Si.H. are displayed away from the Si of the Si.H. at least in configurations (i) and (ii). The absolute displacements of the Si and B core next to the H and displacements of these second neighbour Si cores have not been precisely determined by the cluster calculations but the second and general configuration, (ii), is probably the most likely case. There are four equivalent bond-centre positions due to the four Si-B bonds. The bonded hydrogen would tunnel between the four positions which further favours the four bridging hydrogen bond-central sites [27]. The Pankove model with a coplanar B:::Si3 would have no equivalent positions for hydrogen tunnelling. The Bonapasta et al [26] calculations also gave the charge distribution of the two electrons in the three centre configuration. It is 0.07 au for the H-B bond and 0.26 au for the H-Si bond while Pankove's model implies 0.0 au for the H-B bond, i.e. the two electrons are completely concentrated on the H-Si bond.

For the larger group-III acceptors (Al, Ga and In) similar m-centre/n-electron configurations were also proposed by Sah, some of which are shown in Fig.20 on page 63 of [13]. Due to the larger atomic cores of Al, Ga and In than B, somewhat larger displacements of the Al, Ga and In and surrounding Si cores are expected but the hydrogen will still be on the Acceptor-Si line. Larger binding energies of the hydrogen or proton to these larger proton traps (Al, Ga and In) are also expected due to the larger perturbation. Indeed the thermal activation energy measured by the thermal dehydrogenation experiments performed by Sah, Sun, Tzou and Pan showed such a chemical trend, B(1.2 eV) < Al(1.6 eV) < Ga(2.2 eV) < In(>2.2 eV) [1,3,17]. Cluster calculations for Al have been made by DeLeo and Fowler [22]. However,those for Ga, In and Tl are yet to be reported. Estimated configurations were reported by Baranowski and Tatarkiewicz [28] using a qualitative model which predicted an 'outside' or antibonding position for the hydrogen at the boron acceptor, contradicting the bond-central bridging site, but an 'inside' or bridging position for the hydrogen at the other heavier and larger acceptors centres, such as Al and Ga.

E. EXPERIMENTS ON ATOMIC CONFIGURATIONS

Three types of non-transport experiments have been carried out to determine indirectly the atomic configuration of the acceptor-hydrogen pair and the location of the hydrogen. The most widely reported and also first used configuration determination method is the IR absorption spectra of the localized vibration mode of the A-H pair. Only one set of experiments has been reported using the other two types of experiment, which were the disappearance of the localized or bound exciton spectra and the ion channelling and nuclear-reaction analysis. The results of these experiments will be summarized.

The disappearance of the bond exciton lines from the shallow acceptors in the photoluminescence spectra of p-type silicon was first reported by Thewalt, Steiner, Lightowlers and Pankove [14,15] in 1984 when the samples were exposed to atomic hydrogen from a plasma. These are probably the most decisive direct experimental proof which showed that the reduction of hole concentration during exposure to atomic hydrogen was due to the removal of the acceptor bound state as proposed by Sah, Sun and Tzou [2,3] and not to charge compensation by the presence of a hydrogen related or assure that there is absolutely no hydrogen related or deactivable donor formation but if it exists, its concentration is much less than 1% of the hydrogen-acceptor pair. It would have been detected readily at 1% level by the capacitance transient spectroscopy method developed by Sah and students during 1965-1972. Unfortunately, the hydrogenated acceptor or A.H. does not have an electronic bound state to allow a detailed determination of its atomic configuration and orientation from the dependence of the exciton spectra on an magnetic field, a strain or other anisotropic or uniaxial forces. The lack of an electron bound state on the A.H. pair also eliminates the powerful electron spin resonance as a technique to determine its precise atomic configuration.

The ion channelling and nuclear-reaction analysis experiment is the latest experiment on the atomic configuration of the boron-hydrogen complex [29]. This method has somewhat limited resolution but is probably the only direct method to determine the positions of the atoms in a crystal lattice. The results on boron implanted silicon at high dose, (1-5) X

10*15 B/cm*2, showed that after hydrogenation, the boron core is displayed
0.22 +- 0.04 A from the lattice site and the hydrogen preferentially
occupies the bond centre position between the Si and the B cores. This
is precisely the three-center bridging hydrogen configuration considered
by Sah (Fig.19(a)-(c) on page 61 of [13]) instead of the two-centre
covalent hydrogen configuration of Pankove (Figures 1(a) or 4 of [15]).
The high concentration of boron could cause lattice contraction, making
the measured displacement different from that in a low boron
concentration sample. Similar experiments on the remaining four acceptors
(Al, Ga, In and Tl), have not been reported.

The infrared absorption experiment is the most widely reported. It
measures the absorption frequency peak of the H-B complex in Si and from
its difference from the H-B and H-Si vibration frequencies in isolated
bond or molecule in vacuum, an inference can be made of the atomic
configuration of the Si.H.B complex in crystalline silicon. The first IR
experiment was reported by Pankove [15] at 1875 cm*-1 for the stretching
mode. Comparison with the 2280 cm*-1 for the isolated H-Si led to
Pankove's two-centre model with a planar trivalent boron, Si3B::: H:Si
with no B-H bonding. It is evident that the IR spectra or IR absorption
peaks cannot determine a precise configuration as this case illustrates
since is was later proved (just reviewed) by theoretical computations
that there is finite B-H bonding and hence it is a three-center complex
of Si3:::B.H.Si:::Si3 instead of a two-centre complex (Si-H).

Other IR experiments included the following. Du, Zhang
and Qin [30] reconfirmed the IR peak due to BH at 1873/cm*3 and reported
also the increase of hole mobility when the implanted boron acceptors are
hydrogenated. The isotope shift of the IR peak from 1870 cm*-1 of Si.H.B
to 1360 cm*-1 of Si.D.B in boron-doped p-Si exposed to deuterium plasma by
Johnson further confirmed the hydrogen-boron bond model. (The
effect of holes given in this Johnson paper will be discussed in the
later paragraph on transport experiments). The shift gives (1870/1360)
= 1.375 which is close to the mass root ratio, SQRT 2=1.414. The
difference is another indication of the importance of the H-B bond and
the electron charge distribution on the three-centre/two-electron
configuration postulated by Sah [10,13,18] and demonstrated in the
numerical calculations of DeLeo and Fowler [22] which is completely
missing in Pankove's [15] two-centre/two-electron model of Si:H B:::Si3
since it has no electron charge between H and B. Liquid He temperature IR
measurements were made on B, Al, Ga and In implanted silicon wafers by
Stavola, Pearton, Lopata and Dautremont-Smith [32]. The ion doses were
in the range of (3-10) X 10*15/cm*2 which were thermally activated at
1200 deg C for 60s and the wafers were then exposed to H2 or D2 plasma.
The IR absorption peaks in cm*-1 taken at liquid helium temperature
were: Si.H.B (1907); Si.D.B (1392); Si.H.Al (2201); Si.D.Al (1596);
Si.H.Ga (2171); Si.D.Ga (1577). The indium concentration was too low for
the IR to detect. The Si.H.Al (2201) peak is in excellent agreement with
the theory of DeLeo and Fowler who predicted a value of 2220 cm*-1 [22].
An additional peak was observed at 2160 cm*-1 at 77 deg K which
disappears at 77 deg K in the Al-implanted and H2-plasma exposed samples.
This was attributed to the thermal occupation of a ladder of closely
spaced levels which forms the side-band of the vibrational ground state.
Again the (Si.D.A)/(Si.H.A) frequency ratios are B(1.370), Al(1.379) and

Ga(1.377) in agreement with the D/H mass root, SQRT 2=1.414. The more
exact estimate from the linear chain formulae is

$$\frac{\text{H frequency}}{\text{D frequency}} = \text{SQRT} \left[\frac{(mD/mH)(2mSi-mD)}{2mSi-mH} \right]$$

where m denotes mass, which gives 1.401. The difference and the ratio's
near independence of chemical species is again a clear indication that
the isotope effect is influenced by the displacement of the surrounding
Si atoms owing to the acceptor core size differences. The chemical trend
of the vibration itself, B(1907), Al(2201) and Ga (2171), further
supports Sah's 3-centre model and the importance of the electron
charge distribution on the A-H bond. The reversal is indicative of the
complex effect of the core charge distribution, reminescent of the
reversal of the chemical trend of the electron ground state energies of
the shallow donor electron traps in Si (P,As,Sb,Bi) [33]. The frequency
of the H-Al (2201 cm*-1) is closer to that of the isolated H-Si
(2270 to 2285 cm*-1) or the interstitial hydrogen at the Td or
antibond position of Si (2210 cm*-1) than H-B (1907) and H-Ga (2171).
This is an indication that the core charge distribution is
important since the Al core is identical to the Si core while the cores
of B and Ga are quite different from the Si core which is analogous to
the effect of core charge distribution on the electronic bound states
at the group-III acceptors and group-V donors in semiconductors [33].
Stutzmann [34,35] recently reported the Raman or IR absorption spectra
of both the hydrogen-boron pair and the effect of hydrogen on the boron
local mode in very heavily boron-doped p-type Si (6 X 10*19 B/cm*3)
as a function of temperatures from 5 to 290 deg K. The heavy doping is
necessary owing to the detection sensitivity of the Raman spectra, but
it no doubt would shift the vibration spectra to some extent; but the H
and B local mode frequencies are probably not affected too much at this
0.1% atomic percent level (one B in 125 eight-Si cubic unit cells)
since the local strain due to the small B core compared with the Si core
is localized within a few shells of Si. The results were: B-H at 1880
cm*-1 confirming earlier results of Pankove and later researchers just
reviewed, and the B local mode of the isotopes B-11 (80% abundance) at
620 cm*-1 and B-10 (20%) at 643 cm*-1 whose ratio 643/620 = 1.037 is
consistent with the mass root ratio, SQRT (11/10) = 1.049. (The linear
chain formulae would give SQRT [(11/10)(56-22)/(56-20)] = 1.019). The
absorption peaks of these local modes decreased by five when the above
is hydrogenated, further confirming the H-B model instead of the charge
compensation model, the latter assumes the presence of a hydrogen donor
in the silicon energy gap postulated by Sah, Sun and Tzou and recalled
later by Pantelides but rejected earlier by Sah, Sun and Tzou [2]. A new
vibration band at around 650 cm*-1 was also reported by Stutzmann [34]
which was attributed to a boron mode associated with the B-H pair.
Temperature dependences of the shift and broadening of the Raman lines
in boron-doped silicon exposed to H and D were also reported by Stutzmann
and Herrero [35]. Over the whole temperature range of 5-290 deg K, the
H/D frequency ratio was nearly constant at 1.371 compared with the mass
root ratio of SQRT (mD/mH) = SQRT 2 = 1.414. The peak position and line
width changed with temperature linearly above about 200 deg K and saturated
at low temperatures. However, their changes with temperature are
proportional to each other. For hydrogenated samples, the slope is unity
while for the deuterated samples it is approximately given by

peak shift = SQRT 2 X width shift

The temperature shift of the peak could be related at least in part to lattice expansion-contraction while the width increase with temperature could reflect the thermal hopping of the proton among the four equivalent Si-B bond-central sites as well as thermal hopping to the secondary and higher energy minima whose existence was suggested by the calculations of DeLeo and Fowler [22] and Bonapasta et al [26].

The infrared spectra is the only microscopic technique for locating the impurities in a crystalline lattice if there are no electronic bound states. Detailed atomic configurations and locations can be deduced if the dependence of the IR peaks on the direction and magnitude of a uniaxial stress or strain is measured. This has not been reported so far for the hydrogen-acceptor bonds although isolated impurity local modes, such as boron in Si, have been investigated.

F. TRANSPORT EXPERIMENTS

Transport experiments on acceptor hydrogenation provide data which can be directly used to analyze and predict the reliability and operational life of silicon transistors as affected by acceptor hydrogenation. The acceptor hydrogenation phenomena in silicon was in fact discovered in transport experiments, first in MOS capacitor measurements [1,2,3] in which the hydrogen was demonstrated indirectly to come from the wet oxide, the aluminium/silicon dioxide interface, the distilled-water (DI) rinsed silicon surface prior to forming an Al, and the silane-deposited polycrystalline gate. It was then confirmed by resistivity profile measurements on silicon wafers exposed to atomic hydrogen from a hydrogen plasma [9,11]. These transport experiments are tabulated in TABLE 1. The IR Raman lattice vibration and bound exciton electronic spectra, just discussed, came afterwards to give even more direct proof of the formation of the acceptor-hydrogen pair and to eliminate as a contender the model of charge compensation (also a transport phenomena) by a hydrogen or hydrogen-related donor. An excellent review of the literature up to about December 1985 was given by Pearton [36] which discussed many unresolved questions at that time concerning the hydrogen diffusion and reaction species which are still being investigated. In addition, Pearton has reviewed the importance of acceptor hydrogenation in silicon integrated circuit manufacturing since the silicon wafer is exposed to hydrogen gas or hydrogen containing gas or liquid in many if not all processing technologies or steps used in the manufacturing of submicron silicon integrated circuits. Detailed although still preliminary hydrogenation and dehydrogenation rate data have been obtained on MOS capacitors and these are reported and tabulated in another Datareview [61]. Here, some fundamental transport questions are reviewed in chronologcial sequence.

One of the questions was whether the presence of holes is necessary to produce acceptor hydrogenation, as raised by Johnson [31] using two boron-implanted phosphorus-doped Si samples, one with boron density more than phosphorus and one less, and by Pankove [37] using an As-diffused boron-doped n+/p junction. The question of the necessity of holes was previously considered in detail by Sah, Sun and Tzou in forming their models to account for the hydrogenation and dehydrogenation data [6,7,8] which showed no effects from holes. It was also demonstrated that holes are absent in the decisive first experiment leading to the hydrogenation model by them using an n-MOSC on boron-diffused n-Si [2]. In this experiment, hydrogenation of the diffused boron-acceptor occurred in the depleted surface space charge layer of the n-type silicon surface where holes were absent. The appearance of holes in the hydrogenation and dehydrogenation kinetics was first incorporated in the chemical equations by Sah [13,18] and by Sah, Pan and Hsu [17], and they stated that the reaction involving holes was not rate limiting. The unimportance of holes was further demonstrated by Chao, Luo, Pan and Sah [38] using Fowler-Nordheim tunnelling electron injection from silicon into the oxide in which holes are completely absent in the depleted and inverted (to n-type) silicon surface space charge layer where hydrogenation occurred. In contrast, Johnson's SIMS profile of deuterium in boron-implanted phosphorus-doped n-Si indicated a deuterium concentration tracked to the implanted boron acceptor concentration when the boron concentrations exceeds the uniform phosphorus donor concentration [31]. However, no boron hydrogenation was evident when the phosphorus donor concentration (2 X 10*19/cm*3) is higher than the peak boron acceptor (3 X 10*18/cm*3). This was attributed to the necessity of holes in the hydrogenation reaction A- + h+ + Ho <---> (AH)o by Johnson [31] which is inconsistent with the results of Sah's group just described. The following two chemical reaction equations will reconcile these two sets of data and all other data to be discussed.

Si3:::B:Si- + H. <---> Si3:::B.H.Si + e-

　　　　e- + h+ <---> o

In Johnson's [31] boron-undercompensated and hence still n-type Si (10*19 phosphorus/cm*3), there are many electrons to prevent the hydrogenation of boron occurring as indicated by the first equation. In Johnson's boron overcompensated and hence p-type Si, boron hydrogenation was observed which is enhanced since the electrons released from the hydrogenation reaction were taken up by the holes present in the sample. In the experiment of Sah, Sun and Tzou [1-8] using avalanche electron injection from a p-Si substrate (positive gate voltage), the large electric field in the silicon surface space charge layer causes the electron released from the hydrogenation reaction to be swept to the SiO2/Si interface and this electron depletion enhances the hydrogenation reaction given by the above equation. In their experiments using avalanche hole injection from boron-undercompensated n-silicon (negative gate voltage) [2], the electric field in the silicon surface space charge layer again depletes the electron released from the hydrogenation reaction by sweeping them out into the quasi-neutral n-type Si bulk. In fact, their acceptor and donor density profile measurements [1-8] confirmed that hydrogenation occurred only in the junction space charge layer. In Chao's [38] Fowler-Nordheim tunnelling electron injection experiment on p-Si, the space charge layer is thin but there are many holes in the p-type region ready to combine with the electrons released by the boron hydrogenation reaction, hence boron hydrogenation is also enhanced.

This model also accounts for the boiling water experiments of Tavendale, Williams and Pearton [39]. They observed an inverse square root dependence of the hydrogenation depth in excellent agreement with our theory based on neutral atomic H diffusion, SQRT [2DtHo/N(AA)]. The theory also predicts the observed total density of boron hydrogenated and the temperature dependences from the data on unboiled (22 deg C) and boiled (100 deg C) devices. The temperature dependence comes from the proton concentration or PH in the water, SQRT [H+(100 deg C)/H+(22 deg C)] = 3.4 (computed from the ionization energy of water which is about 0.58 eV) which compares well with their experimental ratio of about 3.

In the experiments performed by Sah and associates using a keV electron beam [3,7,40], high densities of electrons and holes are present so there will always be an enhanced hydrogenation reaction.

The explanation just given cannot account for the lack of hydrogenation observed in the n+/p junction by Pankove (As diffused n+ layer) [37], Tavendale (P diffused n+ layer) [41,42] and Johnson (P ion implanted n+ layers) [31]. These samples showed that the hydrogen or deuterium cannot penetrate through the n+ layer. A satisfactory explanation was given by Tavendale [41] who indicated that the high-density As or P layers were loaded with proton traps, mainly silicon dangling bonds from dislocations and defects rather than from the As and P donors themselves, so that the hydrogen is trapped in the n+ layer and cannot migrate across the layer.

The simple hydrogenation model based on electrons and holes and neutral atomic hydrogen diffusion can also account for the detailed concentration profiles of the H-B pair during thermal dehydrogenation in MOS capacitors reported by Sah, Sun and Tzou [1-8]. The accuracy of the H-B concentration obtained from the MOS capacitance-voltage curve was recently verified [43]. However, the model cannot simply account for the thermal dehydrogenation profiles of the B-H pair in plasma exposed [41,42,44] or hydrogen implanted [45] Al/p-Si Schottky barrier and n+/p-Si junction diodes in which a reverse bias voltage is applied to the diode during the thermal dehydrogenation. These results showed an increasing B-H density with annealing time and a position peak which grows and moves deeper into the quasi-neutral p-Si [41,42,45]. The detailed profile of B-H also has a strong bias voltage dependence. These authors suggested that the complex B-H profile is evidence of proton or hydrogen ion diffusion and drift in the presence of the electric field in the space charge layer of the diode, preventing the proton from reaching the Al layer which is a source and sink of hydrogen. However, the H-B profiles near the Al/p-Si interface and the n+/p junction seem to be consistent with the earlier results that Al is a source of protons while the n+ layer is a sink or diffusion barrier of protons. In addition, the complex profile and its complicated changes with annealing time and bias voltage may also be influenced by the large surface damage from the plasma exposure or proton implant and these have not been considered in the model. In MOS capacitors, the profile behaviour during thermal annealing seems normal and expected with monotonic position variations of the B-H density. MOS structures use AEI, AHI or keV electron beams to inject the proton into the silicon and do not have the large surface damage of the plasma and proton implantation experiments.

A donor level in the Si energy gap from a hydrogen ion was postulated by Capizzi and Mittiga [46] and Pantelides [47] but rejected by earlier workers [Sah, Sun, Tzou, Tavendale and Pearton] since a donor level was never observed by the most sensitive capacitance transient spectrometers and since the hydrogenated acceptor profiles can be accounted for by the neutral hydrogen diffusion model just described.The surface spike of the deuterium profile attributed by Capizzi and Mittiga to a fast diffusion species, D+ or (H-2)+ can be readily accounted for instead by the presence of D2 molecules which are detected by SIMS profiling. At present, all evidence point to the participation of a neutral atomic hydrogen species during hydrogenation. The complex B-H profile data from dehydrogenation under reverse bias could be accounted for by the drift and diffusion of the hydrogen ion or proton via a chain of bonding sites in the junction electric field, without a donor bound state in the Si band gap from the proton, in analogy to the migration of a proton in water.

Other hydrogenation experiments and results will now be summarized. Experiments with SIMS profiling by Mikkelson [48] and Johnson and Moyer [49] have shown that oxygen diffusion is absent during the acceptor hydrogenation and dehydrogenation reactions. This was postulated by Hansen, Pearton and Haller [50]. Chen showed that hydrogenation of the boron acceptor is retarded in oxides grown in hydrogen chloride due to the concentration of chlorine at the SiO2/Si interface which acts as a sink for hydrogen [51]. Boron acceptor hydrogenation was also observed in Pd-gate Si MOSCs by Fare, Lundstrom, Zemel and Feygenson [52] which releases the hydrogen from the Pd gate layer. Hydrogenation of the boron acceptor has also been observed during 500V proton bombardment of crystalline silicon by Horn, Heddleson and Fonash [53] and 1500V proton implantation of polycrystalline silicon by Martinuzzi, Sebbar and Gervais [54] and by Rodder, Antoniadis, Scholz and Kalnitsky [55]. The increase of the contact resistance in transistor wire bonds to highly doped p+ silicon has also been attributed to the hydrogenation of the boron acceptor in the p+ region by Cohen [56].

G. HYDROGENATION OF GROUP-V SHALLOW DONORS IN SILICON

Johnson has claimed that hydrogenation of shallow donors such as phosphorus has been observed [57,58]. However, it has not been observed by other researchers under conditions and in samples similar to Johnson's. Hydrogenation of deep-level donors such as S, Se and Te have been positively identified, however [59]. Preliminary data of Hsu on specially designed MOS capacitors has indicated detectable hydrogenation of phosphorus donors in silicon with rather weak binding energy [60]. A positive identification of the hydrogenation of shallow group-V donors awaits more detailed and controlled experiments. There is no funamental reason for its absence. However, the hydrogen - phosphorus bond configuration is not expected to be the hydrogen bridging bond of the group-V acceptors but instead an interstitial-substitutional pair with a binding energy or proton affinity not unlike the electron affinity.

REFERENCES

[1]   C.T.Sah , J.Y.-C.Sun, J.J.-T.Tzou [J. Appl. Phys. (USA) vol.54 no.2 (Feb 1983) p.944-56]

[2]   C.T.Sah, J.Y.-C.Sun, J.J.-T.Tzou [Appl. Phys. Lett. (USA) vol.43 no.2 (15 Jul 1983) p.204-6]

[3]   C.T.Sah, J.Y.-C.Sun, J.J.-T.Tzou, S.C.-S.Pan [Appl. Phys. Lett. (USA) vol.43 no.10 (15 Nov 1983) p.962-4]

[4]   C.T.Sah, J.Y.-C.Sun, J.J.-T.Tzou [Proc. Int. Electron Device Meeting, IEDM82 San Francisco, CA, USA, 13-15 Dec 1982 (IEEE, New York, NY, USA, 1982) p.753-5]

[5]   C.T.Sah, J.Y.-C.Sun, J.J.-T.Tzou [Proc. 1983 Int. Symp. on VLSI Technol. Syst. & Appl. Mar 30 - Apr 1, 1983. (ERSO, ITRI, Hsinchu, Taiwan 311) p.165-9]

[6]   C.T.Sah, J.Y.-C.Sun, J.J.-T.Tzou [J. Appl. Phys. (USA) vol.50 no.10 (Oct 1983) p.5864-79]

[7]   Chih-Tang Sah, J.Y.-C.Sun, J.J.-T.Tzou [J. Appl. Phys. (USA) vol.54 no.8 (Aug 1983) p.4378-81]

[8]   Chih-Tang Sah, J.Y.-C.Sun, J.J.-T.Tzou [J. Appl. Phys. (USA) vol.55 no.6 (15 Mar 1984) p.1525-45]

[9]   J.I.Pankove, D.E.Carlson, J.E.Berkeyheiser, R.O.Wance [Phys. Rev.
        Lett. (USA) vol.54 no.24 (12 Dec 1983) p.2224-5. See also Phys.
        Rev. Lett. (USA) vol.53 no.8 (20 Aug 1984) p.855 and J.I.Pankove,
        D.E.Carlson, J.E.Berkeyheiser, R.O.Wance, ibid. p.856]

[10a]   J.I.Pankove [private communication to C.T.Sah dated January 4,
        1984 in which the three atomic configurations (see Figures 1(a),
        (b) and (d) of [15]) were sketched. Configuration of Figure 1(c)
        in which the hydrogen serves as the bridge between Si and B was
        not recognized]

[10b]   C.T.Sah [private communication (with a copy of Pankove letter)
        to his students D.Jackson, A.Wang, L.Lu, S.Pan, C.Hsu, W.Lin,
        J.Sun, J.Tzou and visiting associate M.K.Lee, dated January 10,
        1984, pointing out Pankove's models and the difference of his
        interpretation of our model and our hydrogen bridge model]

[10c]   C.T.Sah [private communication to J.I.Pankove dated February 4,
        1984 in which the original Sah model is illustrated with a
        diagram (see Figure 1(c) of reference [15]) to show its
        difference from Pankove's interpretation of Sah's model (see
        Figure 1(b) of reference [15]). The importance of core size was
        also indicated. The hydrogen banana bond, well-known to chemists
        as pointed out by M.K.Lee, was also mentioned. This bridging
        hydrogen model was also presented by Sah in an invited paper
        presented to the closed IEEE Solid State Circuit Committee
        meeting on February 1, 1983 to an audience of more than 100]

[10d]   C.T.Sah [private communication to W.L.Hansen dated February 5,
        1984 with a copy of Pankove's letter and Sah's response given in
        [10a] and [10c] attached, discussing the Hanson O-H model and
        pointing out that the Si-OH bond is nearly 9 eV while the Si-H
        bond is only 4 eV]

[10e]   J.I.Pankove [private communication to C.T.Sah dated July 9, 1984
        responding to Sah's 5-Feb letter and showing all four
        configurations as finally appeared in Figures1(a) to 1(d) of [15].
        The letter states 'We have neutralized other acceptors: Al, Ga,
        In and Tl. The OH model of Hansen et al does not apply to our
        case (the O concentration is too low). The IR vibrational
        spectrum shows a softened mode of Si-H but no trace of B-H. Jerry
        Lucovsky does not think the bridging mode to be responsible.']

[11]   J.I.Pankove, R.O.Wance, J.E.Berkeyheiser [Appl. Phys. Lett.
        (USA) vol.45 no.10 (15 Nov 1984) p.1100-2]

[12]   R.N.Hall [private communication dated August 30, 1984 on
        hydrogenation and especially the H + H <---> H2 reaction as a
        possible rate limiting. Proceeding of the 1984 International
        Conference on Defects in Semiconductors, held in Coronado, CA.
        See also R.N.Hall, 'HP Ge: Purification, crystal growth and
        annealfing properties', IEEE Trans. NS-31(1), 320-325, 01 Feb
        1984 and 'Kinetics of hydrogen-defect complex formation in Ge and
        Si', J. Electron. Mat. (USA) vol.14a (1985) p.759-65]

[13]   C.T.Sah ['Reliability Physics of Silicon VLSI Transistors',
        Annual Report No.1 May 1983 to 31 Jul 1984, TRS-100 (Semicondcutor
        Res. Corp. P.O.Box 12053, Research Triangle Park, NC 27709 USA]

[14]   M.L.W.Thewalt, T.Steiner, J.I.Pankove [J. Appl. Phys. (USA)
        vol.57 no.2 (15 Jan 1985) p.498-502]

[15]   J.I.Pankove, P.J.Zanzucchi, C.W.Magee, G.Lucovsky [Appl. Phys.
        Lett. (USA) vol.46 no.4 (15 Feb 1985) p.421-3]

[16]   M.L.W.Thewalt, E.C.Lightowlers, J.I.Pankove [Appl. Phys. Lett.
        (USA) vol.46 no.7 (1 Apr 1985) p.689-91]

[17]   C.T.Sah, S.C.-S.Pan, C.C.-H.Hsu [J. Appl. Phys. (USA) vol.57
        no.12 (15 Jun 1985) p.5148-61]

[18] C.T.Sah ['Reliability Physics of Silicon VLSI Transistors',
Annual Report No.2 1 Aug 1983 to 31 Jul 1984, T85076
(Semiconductor Res. Corp. PO Box 12053, Research Triangle Park,
NC 27709 USA)]

[19] C.T.Sah, S.Pan ['Summary of Chronology of Boron and Group-III
Acceptor Hydrogenation in Crystalline Silicon', in Gordon Res.
Conf. on Line Defects and Interfaces in Semiconductors, Plymouth
State College, NH, USA, July 8-12, 1985]

[20] C.T.Sah, S.Pan ['Experiments and Model on Deactivation of Boron
and Other Group III Acceptors at the Si-SiO2 Interface', Gordon
Res. Conf. Metal Insulator Semiconductor Systems, Tilton School,
Tilton, New Hampshire, USA, July 14-18 1986]

[21] G.G.DeLeo, W.B.Fowler [Proc. 13th Int. Conf. Defects in
Semiconductors Coronado, California, USA, 12-17 Aug 1984, Eds.
L.C.Kimerling, J.M.Parsey (Metall. Soc. AIME, Warrendale, PA,
USA, 1985); J. Electron. Mater. (USA) vol.14a (1985) p.745]

[22] G.G.DeLeo, W.B.Fowler [Phys. Rev. (USA) vol.31 no.10 (15 May
1985) p.6861-4]

[23] L.V.C.Assali, J.R.Leite [Phys. Rev. Lett. (USA) vol.55 no.9
(26 Aug 1985) p.980-2]

[24] G.G.DeLeo, W.B.Fowler [Phys. Rev. Lett. (USA) vol.56 no.4 (27
Jan 1986) p.402]

[25] L.V.C.Assali, J.R.Leite [Phys. Rev. Lett. (USA) vol.56 no.4
(27 Jan 1986) p.403]

[26] A.A.Bonapasta, A.Lapiccirella, N.Tomassini, M.Capizzi [Phys.
Rev. B (USA) vol.36 no.11 (15 Oct 1987) p.6228-30]

[27] E.E.Haller, B.Joos, L.M.Falicov [Phys. Rev. B (USA) vol.21 no.10
(15 May 1980) p.4729-39. Also B.Joos, E.E.Hallerand, L.M.Falicov,
Phys. Rev. B (USA) vol.22 no.2 (15 Jul 1980) p.832-40]

[28] J.M.Baranowski, J.Tatarkiewicz [Phys. Rev. B (USA) vol.35 no.14
(15 May 1987) p.7450-3]

[29] A.D.Marwick, G.S.Oehrlein, N.M.Johnson [Phys. Rev. B (USA)
vol.36 no.8 (15 Sep 1987) p.4539-42]

[30] Yong-Chang Du, Yu-Feng Zhang, Guo-gang Qin, Shi-Fu Weng [Solid
State Commun. (USA) vol.55 no.6 (Aug 1985) p.501-3]

[31] N.M.Johnson [Phys. Rev. B (USA) vol.31 no.8 (15 Apr 1985)
p.5525-8]

[32] M.Stavola, S.J.Pearton, J.Lopata, W.C.Dautremount-Smith [Appl.
Phys. Lett. (USA) vol.50 no.16 (20 Apr 1987) p.1086-8]

[33] T.H.Ning, C.T.Sah [Phys. Rev. B (USA) vol.4 no.10 (15 Nov 1971)
p.3469-81]; S.T.Pantelides, C.T.Sah [Phys. Rev. B (USA) vol.10
no.2 (15 Jul 1974) p.621-37]; [ibid. p.638-58]]

[34] M.Stutzmann [Phys. Rev. B (USA) vol.35 no.11 (15 Apr 1987)
p.5921-4] This paper contained several erroneous statements and
concepts. It asserted erroneously that there are three atomic models.
It erroneously called Pankove's model a 'bridging B-H-Si site'
which is really not the non-bridging 2-centre configuration proposed
by Pankove, Si3:::B HS:Si, but is the bridging 3-centre configuration
postulated by Sah, given by Si3:::B.H.Si. Stutzmann's chemical
symbol of B-H-Si using bars '-' or two electron bonds between
B-H and H-Si is also wrong since it would imply that there are four
electrons shared by three centres instead of two electrons.
Stutzmann then went on to assert that a third model was
proposed by Sah, Sun and Tzou and Pantelides denoted by the
Coulomb-bound Hi+Bs- pair. This was never implied by the Sah-
Sun-Tzou papers which used to notation H+B- in the initial papers

to denote the pair formation as distinct from charge compensation
and the more explicit notation of Si.H.B was used in their
later papers. Instead of the detailed atomic configuration
of the hydrogen-boron pair, Pantelides was more concerned
with the diffusion species (neutral hydrogen or charged hydrogen
or proton) and the intermediate transition steps of the
boron-hydrogen formation kinetics as well as whether a donor
state of the proton exists in silicon which Sah-Sun-Tzou had
felt to be very unlikely or to have a very short liftime based
on simple physical considerations.]

[35]   M.Stutzmann, C.P.Herrero [Appl. Phys. Lett. (USA) vol.51 no.18
        (20 Apr 1987) p.1413-5]. As commented in [34], the IR data in this
        paper good but the descriptions of the literature results are
        erroneous and misleading. The authors now call the Pankove model the
        3-centre model hydrogen bridge still with the wrong hydrogen
        bridge symbol, B-H-Si, while Pankove emphatically rejected the
        3-centre model [15]. The authors then attribute the substitutional
        boron/interstitial hydrogen model, B-/H+ model to Pantelides
        whose paper was concerned with diffusion species and proton-donor
        bound state and energy level position. These two Stutzmann papers
        reflect the confusion and misunderstanding that exist among some
        recent researchers on this subject.]
[36]   S.J.Pearton [Mater. Res. Soc. Symp. Proc. (USA) vol.59 (1986)
        p.457-68] and ibid p.11-13 which are relevant to the details
        of the hydrogenation and dehydrogenation of acceptors in silicon.]
[37]   J.I.Pankove, C.W.Magee, R.O.Wance [Appl. Phys. Lett. (USA) vol.47
        no.7 (1 Oct 1985) p.748-50]
[38]   C.Y.-P.Chao, M.S.-C.Luo, S.C.-S.Pan, C.T.Sah [Appl. Phys.
        Lett. (USA) vol.50 no.4 (26 Jan 1987) p.180-1]
[39]   A.J.Tavendale, A.A.Williams, S.J.Pearton [Appl. Phys. Lett. (USA)
        vol.48 no.9 (3 Mar 1986) p.590-2]
[40]   S.C.-S.Pan, C.-T.Sah [J. Appl. Phys. (USA) vol.60 no.1 (1 Jul
        1986) p.156-62]
[41]   A.J.Tavendale, D.Alexiev, A.A.Williams [Appl. Phys. Lett. (USA)
        vol.47 no.3 (1 Aug 1985) p.316-18]
[42]   A.J.Tavendale, A.A.Williams, D.Alexiev, S.J.Pearton [Mater. Res.
        Soc. Symp. Proc. (USA) vol.59 (1986) p.469-74]
[43]   S.C-S.Pan, C.T.Sah [Appl. Phys. Lett. (USA) vol.51 no.3
        (Aug 1987) p.334-6]
[44]   N.M.Johnson [Appl. Phys. Lett. (USA) vol.47 no.8 (15 Oct 1985)
        p.874-6]
[45]   T.Zundel, E.Courcelle, A.Mesli, J.C.Muller, P.Siffert [Appl.
        Phys. A (Germany) vol.40 no.2 (Jun 1986) p.67-70]
[46]   M.Capizz, A.Mittiga [Appl. Phys. Lett. (USA) vol.50 no.4 (6 Apr
        1987) p.918-20]
[47]   S.T.Pantelides [Appl. Phys. Lett. (USA) vol.50 no.15 (13 Apr
        1987) p.995-7]
[48]   J.C.Mikkelsen, Jr [Appl. Phys. Lett. USA) vol.46 no.9 (1 May
        1985) p.882-4]
[49]   N.M.Johnson, M.D.Moyer [Appl. Phys. Lett. (USA) vol.46 no.8
        (1 Apr 1985) p.787-9]
[50]   W.L.Hansen, S.J.Pearton, E.E.Haller [Appl. Phys. Lett. (USA)
        vol.44 no.6 (15 Mar 1984) p.606-8]
[51]   A.J.Chen et al [J. Appl. Phys. (USA) vol.60 no.4 (15 Aug 1986)
        p.1391-7]
[52]   T.L.Fare, I.Lundstrom, J.N.Zemel, A.Feygenson [Appl. Phys.
        Lett. (USA) vol.48 no.10 (10 Mar 1986) p.632-4]

[53]   M.W.Horn, J.M.Heddleson, S.J.Fonash [Appl. Phys. Lett. (USA)
        vol.51 no.7 (17 Aug 1987) p.490-2]
[54]   S.Martinuzzi, M.A.Sebbar, J.Gervais [Appl. Phys. Lett. (USA)
        vol.47 no.4 (15 Aug 1985) p.376-8]
[55]   M.Rodder, D.A.Antoniaadis, F.Scholz, A.Kalnitsky [IEEE Electron
        Device Lett. (USA) vol.EDL-8 no.1 (Jan 1987) p.27-9]
[56]   S.S.Cohen [J. Appl. Phys. (USA) vol.59 no.6 (15 Mar 1986)
        p.2072-5]
[57]   N.M.Johnson, C.Herring, D.J.Chadi [Phys. Rev. Lett. (USA)
        vol.56 no.7 (17 Feb 1986) p.769-72]
[58]   N.M.Johnson, S.K.Hahn [Appl. Phys. Lett. (USA) vo.48 no.11
        (17 Mar 1986) p.709-11]
[59]   G.Pensl, G.Roos, C.Holm, E.Sirtl, N.M.Johnson [Appl. Phys. Lett.
        (USA) vol.51 no.6 (10 Aug 1987) p.451-3]
[60]   C.C.-H.Hsu [unpublished]
[61]   C.T.Sah, C.Y-C.Chao, A J Chen, C.C-H.Hsu, M.S-C.Luo, T.Nishida
        [EMIS Datareview RN=17848 (Dec 1987) 'Hydrogenation and
        dehydrogenation of shallow acceptors and donors in Si: kinetics
        rate data']

17.17 HYDROGENATION AND DEHYDROGENATION OF SHALLOW ACCEPTORS AND DONORS
      IN Si: KINETICS RATE DATA

by C.T.Sah, C.Y-C.Chao, A.J.Chen, C.C-H.Hsu, M.S-C.Luo, T.Nishida

December 1987
EMIS Datareview RN=17848

A. GENERAL

The only hydrogenation-dehydrogenation rate data available are those
taken by Sah and his graduate students on MOS capacitors. Other papers
on transport experiments (cited in [1]) did not use MOS capacitors.
They were exploratory in nature and the authors did not provide an
analysis to extract the fundamental rates from their data. Part of the
reason was the complexity of the boundary conditions of the diode
structure used (Al/n+/p, Al/p+/n, and Al/n-Si or Al/p-Si
Schottky barrier diodes) and the initial conditions from the
hydrogenaton method or sequence (hydrogen plasma, proton implant,
boiling water, and others). For example, Al is a source of atomic
hydrogen, implanted Si surface is a sink of hydrogen, and n+ is a sink
and diffusion barrier of hydrogen. These properties complicate the
boundary conditions and solutions of the reaction-transport equations
significantly. The MOS capacitor structure has fewer complications. Its
only drawback is that a deep profile of the hydrogenated acceptor
concentration can be obtained from the C-V curve only at a sufficiently
low temperatures so that the silicon surface at the SiO2/Si interface
does not invert. In fact, the only data to-date (although still very
incomplete) were obtained in the first experiments on hydrogenation
performed by Sah, Sun and Tzou [2]. A substantially broader data
base on the reaction rates covering wider range of operating
conditions is needed for VLSI transistor design and reliability
prediction

The chemical reactions, transport equations and definition of the rate
constants will be given first. The data measured using MOS capacitors
are then given in tables. The notation of the macroscopic rate
constants is that introduced by Sah using the protonic trap concept [3].

The complete set of reaction-transport equations for the MOSC structure
is given as follows. This is an extension of the original set given by
Sah [3]. Some of the detailed transitions or reactions are explicity
written out in the silicon surface layer to illustrate the recent
controversy on the importance of hole and electric field. In the past,
these intermediate reactions and transitions were understood and shown
to be not rate limiting.

In a general MOSC structure (glass/gate-conductor/poly-Si-gate/oxide/Si)
which frequently appears in silicon integrated circuits, the reaction
and transport of hydrogen occur in ten thin-film and interfacial layers.

The ten layers are as follows:

(i)   ambient-gas/glass interface
(ii)   glass layer
(iii)   glass/gate-conductor interface
(iv)   gate-conductor layer
(v)   gate-conductor/poly-Si-gate interface
(vi)   poly-Si-gate layer
(vii)   poly-Si-gate/SiO2 interface
(viii) SiO2 layer
(ix)   SiO2/Si interface
(x)   Si surface layer

The last can also be divided into a space charge layer next to the
SiO2/Si interface and a quasi-neutral layer extending deep into the
Si bulk. Although the Schottky barrier diode has fewer layers, the
general MOSC diode can be considerably simplified in controlled
experiments designed to measure the fundamental rates. The example to
be given is the aluminium-gate (Al/SiO2/Si) in which the Al and the
Al/SiO2 interface are lumped into one layer since there is no electric
field penetration into the Al layer and Al can be thought of as an
infinite source of hydrogen.

Thus, the layers are:

(1) Al and Al/SiO2 interface
(2) SiO2
(3) SiO2/Si interface
(4) Si surface

One dimensional geometry is assumed. Two-dimensional areal
inhomogeneities due to edge effects would prevent accurate data analysis
to give the rate constants. Experimental test device structures can be
designed to minimize the two- and three-dimensional effects.

The chemical reaction and transport equations in the four layers are now
given and their chemical physics will be briefly discussed. The rate
equations from these chemical reaction and transport equations will then
be described.

B. GATE AND GATE-BOUNDARY-CONDUCTOR/SiO2 INTERFACIAL LAYERS

$$e** + H-X <---> e* + X. + H. \qquad (1.1)$$
$$e* + H. + H. <---> e** + H:H \qquad (1.2)$$

Here X denotes the Al. It also denotes the Si and O dangling bonds at
the SiO2 surface which are hydrogenated during oxdidation or subsequent
exposure to hydrogen from gas plasma used in dry etching or liquid
during wet processing before putting on the aluminium contact electrode.
e** and e* are the hot electrons injected from the Si substrate or
generated by a keV electron beam or X-rays. There are other possible
reactions such as the interfacial plasmon generation by the hot
electron, the decay of the plasmon into hot electrons and hot holes,
and injection of the hot holes into the oxide. These are discussed in
[8]. Note that the hydrogen recombination and generation rates of Eqn
(1.2) could be significantly increased in the presence of hot electrons
of 10 eV kinetic energy since the H:H bond energy is only 4.52 eV.

C. SiO2 LAYER

$$(\text{at gate}) \quad e** + H. \quad <---> \quad H. + e* \quad (\text{at SiO2/Si interface}) \quad (2.1)$$

$$e* + H. + H. \quad <---> \quad H:H + e** \quad (2.2)$$

$$e* + (SiO)3:::Si. \quad + H. \quad <---> \quad (SiO)3 :::Si:H + e** \quad (2.3A)$$

$$e* + (SiO)3:::SiO. \quad + H. \quad <---> \quad (SiO)3 :::SiO:H + e** \quad (2.3B)$$

$$e* + (SiO)3:::X. \quad + H. \quad <---> \quad (SiO)3:::X:H + e** \quad (2.4)$$

$$e* + (SiO)3:::Y-OSi + H. \quad <---> \quad (SiO)3:::Y.H.OSi+ + e- + e** \quad (2.5)$$

X. denotes the dangling bond of an impurity Z at the Si site, Z. , or the O site, SiZ. . Y is a substitutional impurity. ':' is an electron-pair bond. '+' and '-' are signs of charge. Eqn (2.5) gives only one of four alternatives which are: Y substitutes for Si or O and H bridges Y-O or YO:Si.

D. SiO2/Si INTERFACIAL LAYER

The reactions at the oxide/silicon interfacial layer are identical to those given above for the oxide layer. Some will dominate over others depending on the structure of the interface determined by the oxidation conditions and the impurity types and densities. In addition, strained intrinsic bonds, Si..O..Si and Si..Si, exist at the interfacial layer due to the transition from a crystalline silicon lattice to a SiO2 lattice. The symbol of two horizontal dots (.. or . . with a gap) means a strained (stretched or weak) electron-pair bond while two vertical dots (:) means a strain-free average electron-pair bond. These strained intrinsic bonds can be readily broken by the hot electrons to give rise to dangling bonds which can then trap hydrogen. In addition to the equations given in the oxide layer, ionization and deionization of an electronic-protonic trap due to electronic transitions (capture or emission of an electron or a hole from or to the conduction or valence band of silicon) at the interfacial layer could also be important in affecting the hydrogenation and dehydrogenation rates or proton (or hydrogen) capture and emission rates at the trap. These are given by additional equations in the silicon surface layer presented next. The relevance of these electronic transitions is the very question raised by Johnson, Pankove, Pantelides and others discussed earlier [1] concerning the importance of holes and the existence and importance of an electronic bound state in the silicon energy gap due to the ionized hydrogen or proton.

E. SILICON SURFACE LAYER

$$e* + Si3:::A.Si \quad <---> \quad h+ + Si3:::A:Si- + e* \quad (4.1)$$

$$e* + H. \quad <---> \quad e- + H+ + e* \quad (4.2A)$$

$$e* + H. + h+ \quad <---> \quad H+ + e* \quad (4.2B)$$

$$e* + Si3:::A:Si- + H+ \quad <---> \quad Si3:::A.H.Si + e* \quad (4.3)$$

$$e* + H. + H. \quad <---> \quad H:H + e* \quad (4.4)$$

$$e- + h+ \quad <---> \quad 0 \quad (4.5)$$

These equations in the silicon surface layer explicity include the
ionization of hydrogen, (4.2A) and (4.2B), and the hydrogenation
reaction by proton capture (4.3). However, all experimental evidences
on the existence or detectability of a charged hydrogen or proton are
inconclusive or negative thus far. This null result seems physically
reasonable to Sah, Sun, Tzou, Pan and Hsu [4-7] because a bare hydrogen
nucleus or proton is not likely to exist for a very long time in a solid
since there are so many electrons in a solid. Thus, even if the
ionization reactions Eqns (4.2A) and (4.2B) can exist in a solid (say
enhanced by hot electron impact), the lifetime of the proton, H+, would
be very short compared with the observed acceptor hydrogenation and
dehydrogenation rates (time constants of seconds, minutes to hundreds
of hours). Thus, the hydrogenation-dehydrogenation reaction involving a
proton Eqn (4.3), can be combined with either the atomic hydrogen
ionization reaction (4.2A) or the hole capture reaction at the atomic
hydrogen (4.2B) to give the following two alternative dominant reactions:

$$e* + Si3:::A:Si- + H. \ <---> \ Si3:::A.H.Si + e- + e* \qquad (4.3A)$$

$$e* + h+ + Si3:::A:Si + H. \ <---> \ Si3:::A.H.Si + e* \qquad (4.3B)$$

These two reactions together with the hydrogen and electron-hole
recombination-generation reactions, (4.4 and (4.5), can then explain
the results concerning the effects of hole, electron (minority carrier)
and electric field on the hydrogenation-dehydrogenation rates of boron
acceptor described in [1] without invoking a proton bound state in
the silicon energy gap nor an electric-field enhanced ionic diffusivity
of hydrogen. In some cases, the presence of the high electric field will
sweep out the electron in the hydrogenation reaction (4.3A) which greatly
enhances the hydrogenation rate. In other cases, the depletion of
electrons depends on recombination with holes, either via interband
recombination (4.5) or a localized electronic bound state (Shockley-
Read-Hall process). The boron dehydrogenation rate has a thermal
activation energy near the silicon energy gap, 1.2 eV, suggesting that
the electron-hole thermal recombination-generation reaction could be
rate controlling or an accidental similarity of the gap and
boron-hydrogen bond energy

F. THERMAL DEHYDROGENATION

All of the above equations include the hot electrons, e** or e*, which
are present during acceptor hydrogenation. The hot electrons are absent
during thermal dehydrogenation so these equations can also apply to
dehydrogenation if the hot electrons on both sides of the equations are
removed.

G. MEASUREMENT DETAILS

The fundamental macroscopic rate constants of these equations are
measured using MOS capacitors. Dehydrogenation measurements give the most
detailed data on the thermal rates. Hydrogenation rates are also
measured but they are highly dependent on the hot electron density and
flux. The symbols of the rate constants follow the notation we adopted
from the electron-hole grt kinetics (see Datareview [13]). For example,
the hydrogen capture rate by an boron acceptor (or proton trap) is
$c_{Ht} X H X A$, where $H(r,t)$ is the hydrogen concentration, $A(r,t)$ is the
acceptor concentration and c_{Ht} is the proton thermal capture coefficient

(in units of cm*6/s) by the proton trap (or the boron acceptor). For another example, the thermal emission rate of trapped proton or hydrogen at a boron acceptor is given eHt X (N(AA)-A), where N(AA) is the concentration of all the substitutional borons. Some of the substitutional boron atoms are active electronic acceptors and the rest are hydrogenated and electronically inactive. The macroscopic thermal emission rate coefficient of a trapped proton, eHt (in units of s*-1), is the rate at which trapped protons are thermally released from the proton traps (the hydrogenated borons). The thermal recombination rate of two hydrogens to form a hydrogen molecule is (rHt)x(H2) and thermal dissociation rate of a hydrogen molecule (or thermal generation rate of atomic hydrogen by dissociation) is then (gHt)x(H2) and 2(gHt)x(H2) respectively. When hot electrons are present, the dehydrogenation rate is increased owing to continued hydrogen bond rupture by hot electron impact. Thus, the total rate is now given by (eHt + eHn)(N(AA)-A) where eHn is the impact release rate of protons or hydrogens trapped at the protonic trap or boron by bond rupture. Generally eHn is proportional to the hot electron flux above the bond breaking threshold energy and an impact release cross section so it can be controlled in an experiment and measured from integrating the oxide current over time. The hot electron impact release rate of a trapped proton eHn, will increase with increasing electric field, not because of the impact cross section but the higher concentration of hot electrons above bond breaking threshold at higher fields. Similarly, the dissociation rate of molecular hydrogen, given by (gHt + Hn)H2 will increase due to hot electron impact causing gHn to increase with elecric field. Both of these two parameters, eHn and gHn, are expected to increase with the electric field as exp(-Eb/kTe). Eb is the bond-breaking threshold energy (about 4.52 eV for H:H and 3.42 eV for B:H bonds in vacuum). Te is the hot electron temperature and is approximately proportional to the electric field above the critical field (see Shockley's theory on hot electrons [9]). The thermal capture and recombination rates are not affected by the electric field until the field intensity is comparable to the periodic field from Si++++ atomic cores (10*8-10*9 V/cm). The inverse or Auger processes of the hot electron impact release and dissocation transitions are insignificant. A similar set of processes and rate coefficients can also be operative with hot holes in SiO2. However, there are few hot holes in SiO2 due to the low hole mobility and the disjointed and narrow SiO2 valence bands (see discussion in [9-11]). Capture of thermal holes by a hydrogenated trap can release the trapped proton or hydrogen. The rate would be cp(tp)(N(AA)-A) where p is the hole concentration and cp(t) is the thermal capture rate of a hole by the hydrogenated trap.

The hydrogenation rates are tabulated in TABLES 1, 2, 3 and 4. These are obtained in p-Si MOS capacitors which contain one of the four acceptors B, Al, Ga and In and have dry oxide and aluminium gate. TABLE 1 gives the acceptor hydrogenation rate parameters during Avalanche Electron Injection (AEI) [3] in which the hydrogen was released from the aluminium gate by hot electron impact collision. TABLE 2 gives the same hydrogenation rate data taken during 8 keV Electron Beam Irradiation (EBI) [12]. The dependence of the AEI hydrogenation cross section on pre-electron beam irradiation is given in TABLE 3 [5]. TABLE 4 gives the temperature dependence of the hydrogenation rate during AEI from 120 deg K to 300 deg K [5].

Thermal dehydrogenation of the acceptor-hydrogen pair has been measured and was found to follow two phases [2,3]. The initial phase is a first order reaction corresponding to the breakup of the hydrogen-acceptor bond given by Eqn (4.3A) or (4.3B), while the second and long-time phase

is second order and limited by hydrogen recombination to form a hydrogen
molecule described by Eqn (4.4). The initial rate from the A(t) vs time
data is the thermal emission rate of trapped hydrogens at the acceptor,
eHt. The long-term rate for a complete second order process (when
hydrogen recombination-generation reaction dominates the hydrogen-acceptor
formation-breakup reaction) is the combined second order rate
(eHt/cHt)rHt, where cHt is the hydrogen capture rate by the acceptor and
rHt is the recombination rate of atomic hydrogen to form a molecule. The
incomplete second order long-time phase has a somewhat more complex time
dependence and was analyzed in detail by Sah [3].

Extensive and accurate data for both phases are available only for the
boron acceptor [2] while extensive and moderately accurate second order
rate data of the aluminium acceptor have also been obtained. These data
are tabulated in TABLE 5 which follows the chemical bond trend, B(1.5
eV) < Al(1.55 eV) < Ga(2.2 eV). Data for indium at this time are meager
and not accurate. The data shows that the first order dehydrogenation
rate for the boron-hydrogen pair, eHt, has essentially the same thermal
activation energy as the second order rate, (eHt/cHt)rht. This is
expected since the hydrogen capture rate, cHt, and the hydrogen
recombination rate, rHt, should both be relatively insensitive to
temperature or not thermally activated, while the thermal emission rate
of a trapped hydrogen, eHt, is thermally activated. The dehydrogenation
kinetics observed thus far for aluminum and gallium seemed more complex
and requires further and more detailed experiments.

H. TABLES OF HYDROGENATION DATA

TABLE 1: kinetic parameters of hydrogenation of group-III acceptors
during avalanche electron injection at room temperature
in silicon MOS capacitors [3].

	Unit	B	Al	Ga	In
AEI voltage	V	60	65	65	65
AEI frequency	kHz	50	300	300	300
Oxide Thickness	A	1000	1000	900	800
N(AA)	10*16/cm*3	4.08	0.633	0.652	0.799
A(o)	10*16/cm*3	4.07	0.605	0.644	0.799
A(inf)	10*16/cm*3	0.25	0.069	0.165	0.465
Initial Jump	1-A(o)/N(AA)	0.005	0.044	0.012	0.005
Initial Delay, td	s	310	460	1000	2500
Initial Delay, Fd	10*17 e/cm*2	0.32	1.5	25	--
Cross section	10*-20 cm*2	6.72	4.47	1.23	3.80
Rate constant, kH	10*-6 s*-1	8.9	8.7	5.3	2.4
H emission rate, eH	10*-6 s*-1	0.54	0.95	1.34	1.4
H capture rate, CH x HG	10*-6 s*-1	8.3	7.8	4.0	1.0
H release rate, eX	10*-6 s*-1	3200	2200	1000 +-500	400 +-100

TABLE 2: kinetic parameters of hydrogenation of group-III acceptors during 8 keV electron irradiation at room temperature in silicon MOS capacitors [12].

	Unit	B	Al	Ga	In
EBI voltage	keV	8	8	8	8
EBI Current Density	micro-A/cm*2	1.4	1.4	1.4	1.4
Oxide Thickness	A	1000	1000	900	800
N(AA)	10*16 cm*3	3.66	0.72	0.65	1.05
Initial Jump	1-A(o)/N(AA)	0	0.02	...	0.02
Initial Delay tD	s	0	0.4	...	0.4
Initial Delay, FD	10*12 e/cm*2	0	...	3.5	3.5
Final Density	A(inf)/N(AA)	0.14	0.19	0.78	0.24
Cross Section	10*-14 cm*2	1.94	1.25	...	...
Rate, eH + CH x HG	s*-1	0.17	0.11	...	...
eH/(eH + CH x HG)		0.25	0.30	...	...
H emission rate, eH	s*-1	0.04	0.03	...	...
H capture rate, cH(HG)	s*-1	0.13	0.08	...	...

TABLE 3: dependence of the AEI hydrogenation cross section on pre-electron beam irradiation energy [11].

		Hydrogenation Cross Section (10*-19 cm*2)			
		B	Al	Ga	In
EBI Voltage (keV)	0	1.11+-0.02			
	2	1.09+-0.01			
	5	1.54+-0.01			
	8	5.6 +-0.1			

TABLE 4: temperature dependence of the hydrogenation rate of the B acceptor in silicon from avalanche injection at 60 volts in MOS capacitors with 1000A dry oxide and aluminium gate [2].

Injection Temperature (deg K)	Steady-State H-B pairs A(o)-A(inf) (10*17/cm*3)	Hydrogenation cross section (10*-20 cm*2)
Initial	1.14	
120	0.135	
180	0.84 +- 0.06	0.685
200	0.84 +- 0.06	1.12
220	0.91 +- 0.02	2.03 +- 0.10
240	0.85 +- 0.02	4.1 +- 0.2
260	1.0 +- 0.1	5.6 +- 0.3
298	1.112 +- 0.007	11.1 +- 0.2

$$\text{Least squares fit: cross section} = 8.6 \times 10^{*}{-18} \exp \frac{-0.113 \text{ eV}}{kT} \text{ cm*2}$$

TABLE 5: thermal dehydrogenation rates of hydrogenated acceptors in Si.

| Temperature (deg C) | Second Order Dehydrogenation Rate (eHt/cHt)rHt (10*-4 s) | | | |
	Boron	Aluminium	Gallium	Indium
47.7	0.007			
83	0.256			
97.8	1.68			
108.4	2.88			
118.0	8.05			
128.4	16.0			
138.2	40.0			
140		0.075 +- 0.0078		
145		0.140 +- 0.027		
150		0.126	0.78	
155		0.417 +- 0.044		
160		0.833		
162			4.6	
170		2.30 +- 0.36		
175		3.33	20.	
185		4.38 +- 0.25		
185		5.29 +- 0.43		
190		9.91 +- 0.73		
195		11.1 +- 1.5		
195		13.1 +- 0.7		
205		31.9 +- 3.4		

| Hydrogenation Method | Thermal Activation Energy of Dehydrogenation (eV) | | | |
	Boron	Aluminium	Gallium	Indium
AEI	1.15	1.56	2.2	...
8 keV EBI	2.1	1.5	...	...

REFERENCES
----

[1]   C.T.Sah [EMIS Datareview RN=17845 (Nov 1987) 'Hydrogenation and
        dehydrogenation of shallow acceptors and donors in Si:
        fundamental phemonena and survey of the literature']
[2]   C.T.Sah, J.Y.C.Sun, J.J.T.Tzou [J. Appl. Phys. (USA) vol.54
        no.2 (Feb 1983) p.944-56]
[3]   C.T.Sah, C.S.Pan, C.H.Hsu [J. Appl. Phys. (USA) vol.57 no.12
        (15 Jun 1985) p.5148-61]
[4]   C.T.Sah, Y.C.Sun, J.J.T.Tzou [J. Appl. Phys. (USA) vol.50 no.10
        (Oct 1983) p.5864-79]
[5]   C.T.Sah, J.Y.C.Sun, J.J.T.Tzou [J. Appl. Phys. (USA) vol.54 no.8
        (Aug 1983) p.4378-81]
[6]   C.T.Sah, J.Y.C.Sun, J.J.T.Tzou [J. Appl. Phys. (USA) vol.55 no.6
        (15 Mar 1984) p.1525-45]
[7]   C.T.Sah, S.C-S Pan, C.C-H Hsu [J. Appl. Phys. (USA) vol.57 no.12
        (Jun 1985) p.5148-61]
[8]   C.T.Sah [EMIS Datareview RN=17820 (Dec 1987) 'Interband
        electron-hole pair generation in SiO2 by plasmon decay']
[9]   C.T.Sah [EMIS Datareview RN=17819 (Dec 1987) 'Intrinsic dielectric
        breakdown of SiO2 films due to interband impact generation of
        electron-hole pairs']
[10]   C.T.Sah [EMIS Datareview RN=17817 (Nov 1987) 'Non-tunnelling
        interband optical generation and radiative recombination of
        electron-hole pairs in SiO2 films on Si']

[11]   C.T.Sah [EMIS Datareview RN=17818 (Dec 1987) 'Non-tunnelling
        interband Auger recombination and impact generation of
        electron-hole pairs in SiO2 films on Si']
[12]   S.C.S.Pan, C.T.Sah [J. Appl. Phys. (USA) vol.60 no.1 (1 July
        1986) p.156-62]
[13]   C.T.Sah [EMIS Datareview RN=17815 (Nov 1987) 'Generation-
        recombination-trapping rates and lifetimes of electrons and
        holes in SiO2 films on Si: a general classification of the
        fundamental mechanisms']

17.18 ELECTRON MOBILITY IN SiO2 FILMS ON Si

by C.T.Sah, A.J.Chen, C.C-H.Hsu and T.Nishida

November 1987
EMIS Datareview RN=17808

A. INTRODUCTION

Electron and hole mobilities in thermally grown SiO2 film on crystalline
silicon are important transport parameters which affect the reliability
of silicon integrated circuits. They influence the charging and
generation rates of traps in the SiO2 film which cause the transistor
characteristics to drift and deteriorate. Hughes reported the first and
only measurements of electron drift mobility in SiO2 using a transit-
time technique during 1971-1978 [1-3]. The sample was a 200 microns thick
fused quartz or vitreous silica (Suprasil II from Amersil, Inc.) with
aluminium electrodes on both surfaces. A 200 micron film is too thick
to grow on Si. The electron transit-time would be too short to measure
in a 1 micron oxide film that can be readily grown on Si. There are
fundamental reasons and several experimental results that suggest a
significant difference in the electron mobility in these two forms of
SiO2. However, due to lack of data, Hughes' results on vitreous silica
are used to estimate the electron mobility in thermally grown oxide
films.

In Hughes' experiments, electron-hole pairs were generated in fused
quartz by a short high-energy X-ray pulse (3 ns half-width) from a
600 keV electron-pulse Febetron. The electron mobility, mu, was extracted
by fitting the photo-current decay curve to:

$$i = e \; n(o) \; mu \; E \; [\; 1 - (\; mu \; E/d)t \;] \; e*(-t/tau)$$

where n(o) is about 4.5 X 10*12 carriers/cm*3.rad [1] at a dose of
about 0.1 rad [3], e = 1.6 X 10*-19 C, mu is the mobility (cm*2/V.s), E
is the electric field (V/cm), d = 200 micron, t is the observation time
and tau is the electron lifetime (about 8 ns [3]).

B. EXPERIMENTAL DATA

The electron mobility and drift velocity were measured by Hughes both as
a function of temperature (113-379 deg K) [1] and electric field (1 X
10*4 - 1 X 10*6 V/cm at 298 deg K) [2]. The electron drift velocity
was found to increase linearly with electric field up to 2 X 10*5 V/cm
giving a low-field mobility of 21 +- 2 cm*2/V.s [2,3] at 298 deg K. The
low-field electron mobility increases with decreasing temperature and
saturates to 40 cm*2/V.s below 200 deg K [1]. The experimental data are
read from enlargements of Hughes' figures and are given in TABLES 1 and
2.

The temperature dependence of the low-field electron mobility can be fitted to

```
 1        1          1
--- = ------- + ------
 mu     mu(GL)     mu(LO)
```

according to Matthiessen's rule giving mu(GL) = 40 and mu(LO) = (3.88 +- 0.30)[exp(63 meV/kT(L)-1] cm*2/V.s. Here k is the Boltzmann constant and T(L) is the lattice temperature. The electric field dependence can be represented by mu = 20.6 cm*2/V.s for 0 < E < 6.0 X 10*5 V/cm and vd = mu E = vs = 1.24 X 10*7 cm/s for E > 6.0 X 10*5 V/cm. It can also be fitted to an empirical equation of the form

```
                 mu(o)
mu = -------------------------------------
      [ 1 + (mu(o) E/vs)*gamma ]*(1/gamma)
```

where gamma is larger than 2 (about 5 to 10).

The significant figures listed above are from numerical fitting. They are given only to reproduce the data in the tables and are not meant to represent the true accuracy. The data have random errors of about 15% and an unknown amount of draughtsman's error. The reading error of the enlarged figures is about 2%.

C. INTERPRETATION OF EXPERIMENTAL DATA

The decrease of the low-field (up to 6 X 10*5 V/cm) electron mobility with increasing temperature above 200 deg K is entirely accountable by a one-phonon model based on scattering of the electrons by the long wavelength (q -> 0) polar longitudinal optical (LO) phonon of 63 meV (508 cm*-1, 731 deg K). The polar LO phonon model was first suggested by Frohlich concerning dielectric breakdown of insulators [4]. The mobility formula due to LO phonon scattering cited by Hughes [1] may be in error [5-8]. It has an extra temperature dependence of T(L)*n where n = 1 and if included would give a phonon energy of 100 meV (807 cm*-1 or 24.2 THz). Although there is a LO phonon at this energy, its coupling to the electron is much weaker than the 63 and 153 meV phonons [9-12] and amorphousness would not increase this coupling significantly [10]. No contribution from the 153 meV (1235 cm*-1, 1777 deg K) LO phonon is observable in Hughes' data when fitted to our mobility formulae above or to Hughes' formulae with the extra T(L) in the pre-exponential factor. Long wavelength longitudinal acoustic (LA) phonons, which are important in nonpolar semiconductors and were first treated by Seitz [13], Bardeen and Shockley [14], are not as effective as the polar LO phonons. The LA phonon limited mobility, mu(LA) (about 500 cm*2/V.s), is one order higher than mu(LO). Since the conduction band minimum of SiO2 is at the zone centre (k = 0) [15], intervalley phonon scattering is ruled out. Only intravalley scattering of thermal electrons is allowed. The change of the electron momentum is small and phonons with small wave numbers dominate the scattering events.

The temperature independent low-field mobility below 200 deg K (mu(GL) about 40 cm*2/V.s) observed by Hughes is suggestive of a temperature and energy independent scattering mechanism. This is characteristic of a disorder scattering in granular materials. When the scatterer size is comparable to the electron wavelength, the scattering cross section can be estimated by sigma(GL) = pi a(s) lambda(e) where a(s) is the size of the scatterer and lambda(e) is the electron wavelength given by 2 pi/k. Then, the scattering rate is

1/tau = pi h a(s) N(s)/m

and the mobility is

mu(GL) = e / [pi h N(s) a(s)] = 7.7 X 10*13 / [N(s) a(s)] cm*2/V.s

For 40 cm*2/V.s, we then have N(s) a(s) = 1.92 X 10*12/cm*2. This gives an electron mean free path of:

lambda(mfp)   = v tau = 1 / [pi N(AM) a(s) lambda(e)] or

lambda(mfp) lambda(e) = 1 / [pi N(AM) a(s)] = (40.6 A)*2 (fused quartz)

Using E   = 3kT(L)/2 = 26 meV at 200 deg K, then

lambda(e)   = 2pi/k = h/SQRT (2mE) = 76 A   and

lambda(mfp)   = 68 A

The density of the scatterer can be estimated by using the beta-cristoballite as a model SiO2 glass. Its cubic unit cell contains 8 SiO2 molecules and has an edge of 7.16 A. Consider the isolated discrete scatter model and assume that its size, a(s), is 7.16 A (i.e. the cubic unit cell is a coherent cube or all atoms in the cube will scatter as a whole) then the density of the scatterer is N(s) = 2.7 X 10*19/cm*3 and there would be one imperfect cube that scatters among 100 perfect cubes which do not scatter or a 1% fluctuation. The separation of the scatterer is then N(s)*-1/3 = 33 A. This model is evidently inconsistent with the original assumption that the size of the scatterer (7.16 A) and the electron wavelength are comparable (76 A). Thus, consider the opposite extreme of a distributed scatterer then all adjacent cubes are effective scatterers since each has a different scattering strength. Then the coherent length, a(s) = 72 A, is the average size of the cubic crystallite and N(s) = 2.7 X 10*18/cm*3 is the crystallite density. There are then 1038 beta-cristoballite unit cells (8-SiO2 per unit cell) in each crystallite or coherent cube on the average. It is evident that the distributed scatterer model is consistent with the initial assumption that the size of the scatterer (72 A) is comparable to the electron wavelength (76 A). In addition, they are also similar to the electron mean free path (68 A). The coherent length, a(s), should be sensitive to the structure or the degree of amorphousness of the fused quartz which can be varied by the cooling rate. No mobility experiments testing this idea have been reported. However, the magnitude of the scattering potential (which did not enter our simple mu(AM) model) and the size of the scatterer and its density should also affect other properties, such as the line shape of the infrared spectra and other properties which depend on the band edge electron states. Indeed the size of the distributed scatterers or crystallites we just estimated from electron mobility data, 72 A, is nearly equal to the size of the grain, 66 A, estimated by J.C.Phillips who arrived at a new polycrystalline or granular beta-cristoballite model of the vitreous silica or SiO2 glass based on a detailed analysis of its infrared spectra in terms of the bulk and localized surface or interfacial-boundary modes of snugly fitted clusters or grains [16]. The result suggests that the low temperature electron mobility in thermally grown SiO2 film on crystalline silicon could be substantially higher than that in fused silica we just analyzed. This may be anticipated when the film is thinner than the grain size but also when it is thicker since the SiO2 film is grown on a crystalline silicon seed and hence may be more crystalline than glass. A test of this prediction could also be made by electron mobility data taken on crystalline quartz which is not available at present.

An analysis of the mechanisms controlling the electric field dependence of the electron mobility given in TABLE 2 can be divided into two electric field ranges separated at about 1 MV/cm. Below this field, the field dependences can be readily predicted by energy loss to the 63 meV polar LO phonons based on Shockley's simple hot electron theory [17]. Using the free electron mass, mo, this theory predicts a saturation drift velocity of v(sat) = SQRT(h nu(LO)/m) = 1.05 X 10*7 cm/s for the 63 meV LO phonon but higher experimental drift velocities were observed by Hughes (see TABLE 2). However, an electron effective mass of 0.5mo has been deduced from other experiments, such as Fowler-Nordheim tunnelling current-voltage characteristics of thermally grown thin oxides. This gives v(sat) = 1.49 X 10*7 cm/s which is satisfactorily above Hughes' highest experimental value of 1.16 X 10*7 cm/s at the highest attained field, 785 kV/cm. At this field, electron heating barely begins and the electron temperature is less than about twice the lattice temperature (2kT = 52 meV) so that the one phonon (63 meV) model is adequate. At still higher fields when the electron temperature is many times the lattice temperature, the second polar LO phonon (153 meV) will contribute to and begin to dominate the decrease of the mobility.

There have been considerable theoretical efforts on delineating the mechanisms that control the electric field dependence of the electron drift velocity at fields above 1 MV/cm in ionic solids and SiO2 insulator. These were motivated by their applications which may be limited by the maximum dielectric breakdown strength. For nearly fifty years, it has been assumed that the observed breakdown or current runaway in ionic insulators may be accounted for by electron scattering and energy loss via the emission of one polar LO phonon. This model was first suggested by Frohlich in 1937 [18-20]. The one polar LO-phonon theory will provide the required positive feedback mechanism to give the current or velocity runaway. The positive feedback arises since the electrons are scattered less by the polar LO phonons as the electron kinetic energy or velocity increases. Hence, the rate of electron energy loss due to polar LO-phonon emission will decrease with increasing electron energy. Several recent numerical calculations of the breakdown field from the analysis of the velocity-field characteristics were based on this theory [21-23]. A calculation made by Lynch in 1972 [22] showed that current runaway occurs at about 10 MV/cm if the 153 meV polar LO phonon is assumed for electron energy loss. This was in good agreement with the highest experimental breakdown field observed in thermally grown oxide on crystalline silicon at that time. However, breakdown fields as high as 25 and 30 MV/cm have been reported by Simmon and Harari recently (see Datareview [30]). Ferry indicated both the 63 and 153 meV LO phonons are important in his 1979 calculations of the velocity-field characteristics [23]. His computed velocity-field curve followed only part of Hughes' data in the high field range. The still higher field range where current runaway would occur was not given.

The one or two polar LO phonon models could not be the correct theory for current runaway or electronic breakdown of the insulator since it does not include acoustic phonons. Without the acoustic phonon, a breakdown or runaway will always occur whenever the applied field is higher than the critical field, regardless of dielectric thickness. The basic reason is that the rate of energy loss to acoustic phonons will increase with increasing electron energy and will overwhelm the energy loss to the LO phonons at high fields. Thus, the acoustic phonon prevents current or velocity runaway. This was clearly evident in Shockley's 1950 simple hot electron theory on mobility [17] which showed that power dissipation due to acoustic phonon emission increases as (energy)*3/2. But, acoustic phonons have not been taken into account in theoretical calculations until Fischetti's work in 1985 [24]. His results demonstrated no velocity runaway as expected. A detailed discussion of the dielectric breakdown data and theory in SiO2 is given in [30].

Since 1982, several experimental techniques (carrier separation, electroluminescence and vacuum emission) have been employed by DiMaria and collaborators at IBM [24-29] to observe the instability of the electron-energy distribution and the velocity runaway in SiO2. Their data demonstrated that the electron energy distribution is stable and there is no velocity runaway. They showed that their data are consistent with electron energy stabilization by LO phonon scattering below 1.5 MV/cm and by acoustic phonon scattering at high electric fields.

TABLE 1: temperature dependence of the low-field electron drift mobility in Suprasil II fused quartz from Hughes [1]. Hughes indicated a LO phonon temperature of 670 deg K (58 meV) instead of 731 deg K (63 meV)

Temperature (deg K)	1000/T (/deg K)	Mobility (cm*2/V.s)
113	8.80	39.2
124	8.05	39.6
150	6.67	38.6
176	5.69	34.5
208	4.81	30.2
266	3.76	24.9
300	3.33	20.2
333	3.00	16.8
379	2.64	14.6

TABLE 2: electric field dependence of the electron drift velocity at
298 deg K in Suprasil II fused quartz from Hughes [2]

Electric field (10*3 Volt/cm)	Drift velocity (10*6 cm/sec)	Drift mobility (cm*2/V.s)	Thickness (cm)
12.2	0.259	21.2	0.02
26.3	0.531	20.2	0.02
38.3	0.797	20.8	0.02
49.9	1.03	20.6	0.02
54.0	1.07	19.8	0.04
82.0	1.81	22.0	0.04
90.0	1.54 +- 0.25	17.1	0.02
106	2.36	22.3	0.04
130	2.71	20.9	0.04
160	3.37	21.1	0.04
211	4.26	20.2	0.04
11-point average		20.6+-1.4	
8-point average		20.6+-0.5	
249	6.22	25.0	0.10
297	6.97	23.5	0.10
400	7.98	20.0	0.10
500	9.56	19.3	0.10
592	10.5	17.7	0.10
648	10.9	16.8	0.10
716	11.4	16.0	0.10
785	11.6	14.7	0.10

REFERENCES
----

[1]   R.C.Hughes [Phys. Rev. Lett. (USA) vol.30 no.26 (1973) p.1333-6]
[2]   R.C.Hughes [Phys. Rev. Lett. (USA) vol.35 no.7 (1975) p.449-52]
[3]   R.C.Hughes [Solid-State Electron. (GB) vol.21 (1978) p.251]
[4]   H.Frohlich, N.F.Mott [Proc. R. Soc. London Ser. A (GB) vol.171
       (1939) p.496; See also a review by Frohlich [5] and a discussion by
       Seitz [6] and Frohlich and Seitz [7] on polar LO and acoustical
       phonon scattering]
[5]   H.Frohlich [Adv. Phys. (GB) vol.3 no.11 (July 1954) p.325-61] See
       Eqns (7.19) and (7.22) on p.360; See also Conwell [8], Eqn
       (3.6.23) on p.157, for the case of electron kinetic energy smaller
       than the LO phonon energy (63 meV, 731 deg K) which is satisfied
       below 400 deg K]
[6]   F.Seitz [Phys. Rev. (USA) vol.76 no.9 (1949) p.1376]
[7]   H.Frohlich, F.Seitz [Phys Rev. (USA) vol.79 no.3 (1950) p.526-7
[8]   E.M.Conwell [High Field Transport in Semiconductors (Academic
       Press, New York, 1967)]
[9]   F.L.Galeener, G.Lucovsky [Phys. Rev. Lett. (USA) vol.37 no.22
       (1976) p.1474-8]
[10]   P.H.Gaskell, D.W.Johnson [J. Non-Cryst. Solids (Netherlands)
       vol.20 (1976) p.153-69, 171-91]
[11]   J.F.Scott, S.P.S.Porto [Phys. Rev. (USA) vol.161 no.3 (1967)
       p.903-10]
[12]   W.G.Spitzer, D.A.Kleinman [Phys. Rev. (USA) vol.121 no.5 (1961)
       p.1324-35]
[13]   F.Seitz [Phys. Rev. (USA) vol.73 (1948) p.550]
[14]   J.Bardeen, W.Shockley [Phys. Rev. (USA) vol.80 (1950) p.69]

[15] J.R.Chelikowsky, M.Schluter [Phys. Rev. B (USA) vol.15 no.8
      (1977) p.4020]
[16] J.C.Phillips [Solid State Phys. Adv. Res. & Appl. (USA) vol.37
      (1982) p.93-171; Phys. Rev. (USA) vol.35 no.12 (1987) p.6409]
[17] W.Shockley [Bell Syst. Tech. J. (USA) vol.30 (1951) p.990-1034]
[18] H.Frohlich [Proc. R. Soc. London Ser.A (GB) vol.160 no.900 (15
      June 1937) p.230-41]
[19] H.Frohlich [Proc. R. Soc. London Ser.A (GB) vol.172 (1939)
      p.94]
[20] H.Frohlich [Proc. R. Soc. London Ser.A (GB) vol.188 (1947)
      p.521-32]
[21] K.K.Thornber, R.P.Feynman [Phys. Rev. B (USA) vol.1 no.10 (1970)
      p.4099-114]
[22] W.T.Lynch [J. Appl. Phys. (USA) vol.43 no.8 (1972) p.3274-8]
[23] D.K.Ferry [J. Appl. Phys. (USA) vol.50 no.3 (1979) p.1422-7]
[24] W.V.Fischetti, D.J.DiMaria, S.D.Brorson, T.N.Theis, J.R.Kirtley
      [Phys. Rev. B (USA) vol.31 no.12 (1985) p.8124-42]
[25] D.J.DiMaria, T.N.Theis, J.R.Kirtley, F.L.Pesavento, D.W.Wong,
      S.D.Brorson [J. Appl. Phys. (USA) vol.57 no.4 (15 Feb 1985)
      p.1214-38]
[26] S.D.Brorson, D.J.DiMaria, M.V.Fischetti, F.L.Pesavento,
      P.M.Solomon, D.W.Wong [J. Appl. Phys. (USA) vol.58 no.3 (1 Aug
      1985) p.1302-13]
[27] D.J.DiMaria, M.V.Fischetti, M.Arienzo, E.Tierney [J. Appl. Phys.
      (USA) vol.60 no.5 (1 Sep 1986) p.1719-26]
[28] D.J.DiMaria, M.V.Fischetti, J.Batey, L.Dori, E.Tierney, J.Stasiak
      [Phys. Rev. Letts. (USA) vol.57 (1986) p.3213-5]
[29] D.J.DiMaria [Appl. Phys. Letts (USA) vol.51 no.9 (31 Aug 1987)
      p.655-8]
[30] C.T.Sah [EMIS Datareview RN=17819 (Nov 1987) 'Intrinsic dielectric
      breakdown of SiO2 films due to interband impact generation of
      of electron-hole pairs']

17.19 HOLE MOBILITY IN SiO2 FILMS ON Si

by C.T.Sah, A.J.Chen, C.C-H.Hsu and T.Nishida

November 1987
EMIS Datareview RN=16192

Hole and electron mobilities in thermally grown SiO2 film on crystalline silicon are important transport parameters which affect the reliability of silicon integrated circuits. They influence the charging and generation rates of traps in the SiO2 film and the charged traps cause the transistor characteristics to drift and deteriorate. It has been experimentally established that hole trapping gives positive oxide charge, generates interface traps at the SiO2/Si interface, and releases bonded hydrogen at the gate-conductor/oxide interface and in the oxide. These were reviewed in Datareviews [17-23]. In crystalline silicon (c-Si), the properties of holes are very similar to those of electrons but they are vastly different in SiO2 whether it is fused or vitreous silica or thermally grown oxide. Electron mobility in SiO2 is rather similar to that in c-Si which was reviewed in [24]. Except for a smaller magnitude due to strong polar phonon scattering, the electron-mobility-limiting scattering mechanisms and their physical and mathematical descriptions in SiO2 are very similar to those in the c-Si. Even the electric field dependences of the electron mobility in SiO2 are completely accounted for using the same theories as that used for the hot electrons in c-Si.

The complexity of the properties of holes in SiO2, including their mobility, arises from the complex valence band structure of SiO2 due to the presence of the bridging oxygen and the ionicity of SiO2. To provide an illustrative description of the movements of holes and the hole mobility, the bond and band models of the SiO2 valence band are reviewed first. Hole mobility data as a function of the time of observation after the creation of a hole, the sample temperature and the electric field will then be tabulated.

The bond picture is as follows. A hole near the top of the SiO2 valence band or a thermal hole in SiO2 is a half occupied oxygen 2p orbital which has lost one of its two electrons. There are four 2p electrons on the bridging oxygen. Two of these form the two electron-pair bonds with the 3s-3p electrons from the two adjacent silicon atoms and their energies are in the lower part of the valence band. The remaining two p electrons are in the remaining four nonbonding (known as lone pair) p-orbital lobes. These lobes are perpendicular to the line joining the Si-O-Si atoms in the case of cubic beta-cristoballite. A hole is present when one of these non-bonding lone pair 2p electrons is released from the non-bonding oxygen 2p orbitals.

It is immediately obvious that the hole cannot move along the valence band or the non-bonding oxygen 2p orbitals in SiO2 as readily as a hole in c-Si, since in SiO2, the hole must overcome a significant potential barrier which separates one oxygen site from the next oxygen site. The distance between two nearest neighbour oxygens in the beta cristoballite is 2.53 A (almost equal to the Si-Si distance in c-Si) and about 2.57 A to 2.67 A in the other forms of SiO2. The radius of the O(2p) radial charge density is about 0.413 A. Thus, the probability of hole motion along the SiO2 valence band or along the non-bonding oxygen 2p orbital sites is reduced by at least $\exp(2.53/0.413)/\cos(54.7)*2 = 1370$ because of the very small overlap of the non-bonding oxygen 2p orbitals on the

adjacent oxygens. The hole movement between the adjacent oxygen sites
is mainly by tunnelling at low temperatures and by thermally activated
hopping at higher temperatures. Thus, one would expect the holes in SiO2
to have a very low intrinsic mobility. One would also expect it to be
relatively insensitive to the electric field since the local atomic
field from the atomic potential is of the order 500 MV/cm or 5 V/A.

A second complexity of the transport of holes in SiO2 comes from its
ionicity or the strong coulomb interaction of the hole with the
surrounding Si and O atoms which polarizes the surrounding lattice.
When the hole is present or a nonbonding oxygen 2p electron is missing,
the oxygen is then positively charged. This positive charge would cause
a significant asymmetric local distortion or deformation of the
surrounding lattice by pushing the two adjacent positively charged
silicon atomic cores away and pulling the six adjacent negatively
charged oxygen atomic cores closer. This lattice deformation or
polarization is known as a polaron. In SiO2 it is a small polaron sinces
the hole is tightly attached to the oxygen and the lattice distortion is
concentrated in about one SiO4 tetrahedron. When the hole moves to an
adjacent non-bonding oxygen orbital, it would drag the lattice
deformation with it giving rise to what is known as polaron transport
or polaron mobility. If the lattice deformation is sufficiently large,
the polaron could be localized or spatially trapped, forming a bound
state and the hole is known to be self-trapped.

A third complexity arises from the slight randomness of the Si-O bond
angle and the distance between the adjacent oxygen atoms. Such a
randomness is expected to exist in noncrystalline SiO2 of various forms.
This would produce a random variation of the hole transition rate from
one oxygen site to the next. This gives rise to a random variation of
the mobility which is empirically known as the mobility edge. The
mobility edge can be estimated from the random variation of the valence
band width due to the random spacing between adjacent oxygens since the
valence band width is determined by the overlap of the non-bonding 2p
orbital on the adjacent oxygen. Thus, a 0.05 A variation of the
oxygen-to-oxygen distance would result in a exp(0.05/0.413) or 10%
variation of the non-bonding oxygen 2p valence band width or 0.1 eV for
a valence band width of 1 eV. This is unlikely to give a static hole
bound state but the random potential would cause further hole
scattering and mobility reduction.

The bond model for holes in SiO2 proposed by Sah and just described to
provide an understanding of the small hole mobility and its temperature
dependences (thermally activated at high temperatures and constant at
lower temperatures) is consistent with the energy band calculations of
the valence bands of the various polymorphs of SiO2 (cubic-beta-
cristoballite by Pantelides and Harrison [1], and Ciraci and Batra
[2]; alpha-quartz by Laughlin, Joannopoulos and Chadi [3], and
Chelikowski and Schluter [4]; and all the polymorphs by Li and Ching
[5]). The computed bands are also consistent with experiments, such as
the X-ray and VUV photoemission data of Fisher, Pollack, DiStefano and
Grobman [6] from which they concluded that the upper valence band width
is about 3.3 eV and largely due to the overlap of the oxygen non-bonding
2p wavefunction on the two adjacent oxygens. The computed bands are
also consistent with the optical absorption and photoconductivity data
taken on thin SiO2 film by Powell and Morad [7] and earlier optical
absorption data of H.R.Philipp on fused quartz [8]. For an introductory
background on polarons, see Kittel's 1986 edition [9] and the book by
Mott and Davis [10]. The bond model of hole described above also applies
to the low hole mobilities in compound semiconductors.

Experimental data on hole mobility were obtained by Hughes [11] in thin
thermally grown oxides (about 4000 A thickness) on c-Si in high
electric field. Holes were generated by a 3-ns wide pulsed X-ray source.
The high electric field sweeps out the electrons in about 4000 X
10*-8/10*7 = 4 ps after the X-ray pulse. The total number of electrons
collected, known as the prompt current, also gives the total number of
holes generated since electrons and holes are generated in pairs by
X-rays. After the initial electron current, which ceases in about 10 ns
(determined by the load capacitance and resistance) the current is
entirely carried by hole motion. There was little recombination loss of
holes at oxide and interface traps in the oxide used by Hughes who noted
that the total integrated hole current is nearly equal to the total
integrated electron current.

The hole motion can be divided into three time ranges [12]. During the
initial interval about a lattice oscillation frequency or 10*-13 sec
after the hole is created, there is not enough time for the lattice to
relax or a small polaron to form. Thus, the hole moves with high
mobility, about 1 cm*2/V.s. This initial motion occurs in too short a
time to be observed. Hughes called this the 'dry hole'. The polaron
formation time was deduced from the mobility-lifetime product
measurement which gave (mu tau) = 7 X 10*-13 cm*2/V. For a mobility
of about 1 cm*2/V.s, tau = 7 X 10*-13 s which is the right order of
magnitude for polaron formation in a lattice or phonon oscillation
frequency. In the second time range, the polaron is formed and the hole
moves at a much smaller velocity and mobility, about 2 X 10*-5 cm*2/V.s
at 298 deg K. This is designated as the intrinsic hole-polaron mobility.
It is relatively independent of electric field as we anticipated earlier.
In the third and longer time range, the hole motion is further impeded
due to trapping. In this long time range, the time, temperature and
electric field dependences of the hole current are all described by the
continuous time random walk (CTRW) model developed by Scher and Montroll
deg K, about 40 cm*2/V.s, can be accounted for by a neutral scatterer
of a size of about 72 A with a density of about 2.7 X 10*18/cm*3. Using
the random-walk model to describe the hole transport kinetics, the trapping
rate is given by 1/tau = 4 pi DN(TT) [R(T)] where D is the diffusivity
computed from measured mobility and given by D = (kT(L)/q)mu = 0.026 X 2
X 10*-5. The experimental lifetime is 7 X 10*-8 s, so that N(TT)R(T) =
= 2.2 X 10*12 cm*2. This is in excellent agreement with the electron
scatterer density just cited, 72 X 10*-8 X 2.7 X 10*18 = 1.9 X 10*12/cm*2.
Thus, both the hole and the electron mobilities in SiO2 are accounted for
by the granular or paracrystalline model of SiO2 recently proposed by
J.C.Phillips [14] from analzying its Raman spectra.

The experimental data obtained by Hughes [11] are as follows. The
instrinsic hole polaron mobility in the second time range was measured
from 77 deg K to 298 deg K, showing a thermally activated mobility at
high temperatures given by mu = 0.012 exp(-0.16 eV/kT(L)) cm*2/V.s which
has a value of 2 X 10*-5 cm*2/V.s at 298 deg K. At temperatures below
about 120 deg K, the mobility drops to 10*-7 cm*2/V.s and becomes almost
independent of temperature. There was no electric field dependence
from 0.7 to 5.3 MV/cm in the entire temperature range. The mobility data
of Hughes is given in TABLE 1.

TABLE 1: temperature dependences of the intrinsic mobility of the
         hole-polaron in SiO2 [11]

Temperature (deg K)	Hole mobility (cm*2/V.s)
298	2 X 10*-5
273	1 X 10*-5
230	3 X 10*-6
200	7 X 10*-7
175	4 X 10*-7
140	2 X 10*-7
100	1 X 10*-7
75	1 X 10*-7

Hole transport during the long time range was analyzed by Curtis and
Srour [15] using a multiple-trapping model with a distribution of hole
trapping energy levels near the SiO2 valence band. Good fits can be
obtained. However, the reality of the parameters of the traps was yet
to be tested. The hole transport mechanisms in the long time range were
also investigated by McLean, Boesch and McGarrity using flat band
voltage shift of the MOS CV curves as a function of time after exposure
to an electron beam [16]. They interpreted their results based on the
CTRW (Continuous Time Random Walk) model of polaron hopping between
randomly distributed hole traps. Their data fits a trap separation of
about 9 +- 2 A which is an order of magnitude smaller than Phillip's
paracrystallite or grain size, 70 A, and implies an unlikely trap
density greater than 10*21/cm*3.

REFERENCES
----

[1]   S.T.Pantelides, W.A.Harrison [Phys. Rev. B (USA) vol.13 no.6
      (15 Mar 1976) p.2667-91]
[2]   S.Ciraci, I.P.Batra [Phys. Rev. B (USA) vol.15 no.10 (15 May
      1977) p.4923-34]
[3]   J.R.Chelikowsky, M.Schluter [Phys. Rev. B (USA) vol.15 no.8
      (15 April 1977) p.4020-9]
[4]   R.B.Laughlin, J.D.Joannopoulos, D.J.Chadi [Phys. Rev. B (USA)
      vol.20 no.12 (15 Dec 1979) p.5228-37]
[5]   Y.P.Li, W.Y.Ching [Phys. Rev. B (USA) vol.31 no.4 (15 Feb 1985)
      p.2172-9]
[6]   B.Fischer, R.A.Pollack, T.H.DiStefano, W.D.Grobman [Phys. Rev.
      B (USA) vol.15 no.6 (15 Mar 1977) p.3193-9]
[7]   R.J.Powell, M.Morad [J. Appl. Phys. (USA) vol.49 no.4 (Apr
      1978) p.2499-502]
[8]   H.R.Philipp [J. Phys. & Chem. Solids (GB) vol.32 (1971)
      p.1935-45]
[9]   C.Kittel [Introduction to Solid State Physics 6th ed (John &
      Wiley & Sons, NY, 1986) p.281-3]
[10] N.F.Mott, E.A.Davies [Electronic Processes in Non-Crystalline
      Materials (Clarendon Press, Oxford, 1979) chapter 3];
      N.F.Mott [Adv. Phys. (GB) vol.26 no.4 (1977) p.363-91]
[11] R.C.Hughes [Phys. Rev. (USA) vol.15 no.4 (15 Feb 1977)
      p.2012-20]
[12] R.C.Hughes, D.Emin [Proc. Conf. Small polaron formation and motion
      of holes in a-SiO2 Ed. S.T.Pantelides (Peragmon Press, 1978)
      p.14-18]

[13]   H.Scher, E.W.Montroll [Phys. Rev. B (USA) vol.12 no.6 (15 Sep
       1975) p.2455-77]

[14]   J.C.Phillips [Phys. Rev. B (USA) vol.35 no.12 (15 April 1987)
       p.6409-13] See Datareview on electron mobility where our
       analysis predicted a crystallite size of 72 A calculated from
       the temperature independent electron mobility of 40 cm*2/V.s in
       fused silica which in consistent with Phillips' model which gives
       66 A computed from Raman spectra of SiO2]

[15]   O.L.Curtis,Jr, J.R.Srour [J. Appl. Phys. (USA) vol.48 no.9
       (Sep 1977) p.3819-28]

[16]   F.B.McLean, H.E.Boesch,Jr, J.M.McGarrity [Prog. Conf.,Field-
       dependent hole transport in amorphous SiO2 Ed. S.T.Pantelides
       (Pergamaon Press, 1978) p.19-23]

[17]   C.T.Sah [EMIS Datareview RN=16170 (Nov 1987) 'Interface traps on
       Si surfaces']

[18]   C.T.Sah [EMIS Datareview RN=16180 (Nov 1987) 'Interface traps on
       oxidized Si from electron spin resonance']

[19]   C.T.Sah [EMIS Datareview RN=16181 (Nov 1987) 'Interface traps on
       oxidized Si from X-ray photoemission spectroscopy, MOS diode
       admittance, MOS transistor and photogeneration measurements']

[20]   C.T.Sah [EMIS Datareview RN=16182 (Nov 1987) 'Interface traps on
       oxidized Si from two terminal dark capacitance-voltage
       measurements on MOS capacitors']

[21]   C.T.Sah, C.C.C.Hsu [EMIS Datareview RN=16185 (Nov 1987) 'Oxide
       traps on oxidized silicon']

[22]   C.T.Sah [EMIS Datareview RN=17815 (Nov 1987) 'Generation-
       recombination-trapping rates and lifetimes of electrons and holes
       in SiO2 films on Si: a general classification of the fundamental
       mechanisms']

[23]   C.T.Sah [EMIS Datareview RN=17845 (Nov 1987) 'Hydrogenation and
       dehydrogenation rates of shallow acceptors and donors in Si:
       fundamental phenomena and survey of the literature']

[24]   C.T.Sah, C.C.H.Hsu, T.Nishida [EMIS Datareview RN=17808 (Nov 1987)
       'Electron mobility in SiO2 films on Si']

17.20 ELECTRON MOBILITY IN OXIDIZED Si SURFACE LAYERS

by C.T.Sah and T.Nishida

November 1987
EMIS Datareview RN=17831

A. INTRODUCTION

Both majority and minority carrier mobilities on silicon surfaces have
been measured on samples with different physical geometries (shape and
crystallographic orientations) and material compositions (oxide
thicknesses and substrate impurity concentration). They have also been
measured as a function of operating conditions such as the transverse
and longitudinal DC electric fields and lattice temperature. Transverse
DC electric field creates two electrical geometries, the inversion and
accumulation surface channels, in addition to the physical geometry of
no channel under the electrical flat band condition. A longitudinal
electric field creates the hot carrier effect which reduces the
mobility as the electric field is increased. Surface mobility is
anisotropic, i.e. dependent on the crystallographic directions of the
longitudinal and transverse electric fields. This orientation dependence
arises from the anisotropic constant energy surface of the conduction
and valence band edges and the removal of cubic symmetry by one-
dimensional quantization from the confinement of the electrons or holes
in the narrow surface channel potential well. Orientation dependences
of experimental mobility data have been observed even at room
temperatures. Furthermore, different surface mobilities have been
deduced from different measurement techniques, such as the channel
conductance, field-effect and Hall effect methods. These are known
as conductivity or effective mobility, field-effect mobility and Hall
mobility respectively. The field-effect mobility was further complicated
by two different measurement methods; one employs the transconductance
and the other the derivative of the channel conductance with respect to
DC gate voltage. Mobility measurements under such a large variety of
situations and by many researchers have been motivated by silicon
transistor and integrated circuit applications. However, two completely
opposite approaches have been employed to represent or fit the mobility
data. One is based on fundamental scattering mechanisms and the
confinement of the electrons or holes in a thin surface channel layer
which quantizes one direction of the electron or hole motion resulting
in a two-dimensional (2-d) conduction system. The other, used often by
integrated circuit design engineers, is based on highly empirical multi-
parameter least-squares fits. This approach is valid only for a limited
range of geometrical, material and operating parameters.

This Datareview will focus on only one type of mobility data, namely the
conductivity mobility of minority carriers in a surface inversion
channel covered by a thermally grown silicon dioxide layer. This
mobility and its measurement conditions are most relevant to integrated
circuit applications. The fundamental approach, fitting the data to the
physics-based formulae, will be employed so that the data can be
accurately extended over a wider range of operating conditions than the
data-taking conditions, with confidence and known limitations. This
choice is motivated by the need of an accurate and extendable mobility
characterization for use in the design of silicon integrated circuits.

For a comprehensive treatment of the basic physics underlying mobility and scattering mechanisms, see the review by Ando, Fowler and Stern [1]. Extensive earlier data were also presented by Fang and Fowler [2]. Electron mobility is covered in this Datareview. Hole mobility is covered in [20].

The conductivity mobility of electrons in the silicon surface inversion channel has been measured on carefully designed metal(aluminium)-oxide-silicon field-effect transistors (MOSFET) which have large geometries (long channel and wide gate) so that edge effects are unimportant [3]. The electron surface mobility is computed from mu(n) = G(n)/Q(N). G(n) is the DC conductance of the channel measured between the source and the drain contact at zero DC drain-to-source voltage and it can also be measured as a small-signal AC conductance at low or zero signal frequency. Q(N) is the total DC areal charge density of the minority carriers (electrons) in the inversion channel. For strong inversion, Q(N) is accurately computed from Cox(Vg-Vt) where Cox is the capacitance of the gate oxide (pF/cm*2), Vg is the DC voltage applied to the gate and Vt is the gate threshold voltage at which the surface channel just begins to conduct. Vt is obtained from the intercept of the tangent to the G(n) versus Vg curve [3]. The samples have very low interface state density so that the dependences of Vt on Vg via charging or discharging the interface states is negligible and Vt is independent of Vg. At low inversion, the G(n)-Vg data is curved with a knee or tail and Vt is no longer definable by the tangent extrapolation method just stated. This region covers the threshold and sub-threshold ranges. In these ranges, Q(N) is obtained directly without extrapolation by measuring the input admittance of the MOSFET channel treated as a transmission line [4]. The low-frequency open-circuit input capacitance, Cin, of the transmission line is defined as the imaginary part of the admittance divided by the angular frequency measured between the drain and substrate with the gate AC shorted to the substrate and the source open-circuited. It gives the channel charge which is computed from Q(n) = C(n)(kT(L)/q) where C(n) = Cin-Cox-Csub, Cin is the measured input capacitance and Csub is the sum of the overlap and silicon surface depletion layer capacitances. The real part of the short-circuit input admittance gives the channel conductance, G(n).

The experimental data were taken over a wide range of temperatures (30-300 deg K) and at low longitudinal and transverse electric fields. The experimental mobility is decomposed into three components from three fundamental scattering mechanisms: the acoustical phonon, the intervalley phonon and the ionized oxide charge. To isolate each component, the measurement temperature and oxide charge density were varied, the latter by varying the post oxidation heating temperature from 600 deg C to 1000 deg C in either oxygen or argon. The interface trap density is minimized in all samples. The data were fitted to the theoretical formula of phonon and ionized impurity scattering of a classical two-dimensional (2-d) electron gas. The classical formula resulted in excellent fits to the experimentally observed temperature dependences. This was expected since the channel is thick (from low applied gate voltage) and the temperatures (30-300 deg K) are not too low, so that the two-dimensional energy bands due to quantization are merged into a two-dimensional classical continuum. The thick channel makes the 2-d energy gaps small and the not-too-low temperature thermally broadens the band widths. In the very strong inversion range when the channel is thin, the data were fitted to a simple surface roughness scatterer model [5,6] which again provided excellent agreement between the data and the theoretically expected transverse electric field dependence.

The data presented in this review are in the form of fundamental
equations in which the numerical coefficients are obtained by fits to
experimental data. Mathiessen's rule is used to separate each fundamental
scattering mechanism from the data. For example,

$$\frac{1}{mu(n)} = \frac{1}{mu(nLA)} + \frac{1}{mu(nLO)} + \frac{1}{mu(nOX)} + \frac{1}{mu(nDP)} + \frac{1}{mu(nSR)} + \frac{1}{mu(nCI)}$$

Here, mu(n) is the measured electron mobility in the surface inversion
channel. The mobilities due to each of the surface scattering mechanisms
are: mu(nLA)[T(L)] and mu(nLO)[T(L)] for the intravalley and intervalley
longitudinal acoustic phonon scattering where we use LO to show the
analogy to the longitudinal optical phonon scattering of holes; mu(nOX)
for oxide charge scattering; mu(nDP) for dipole scattering; mu(nSR) for
surface roughness scattering; and mu(nCI) for scattering by channel ions
in the bulk part of the surface channel. T(L) is the lattice temperature
in degrees Kelvin.

Experimental variations of the geometrical, material and operational
parameters are by no means complete, leaving some data gaps which can
be extrapolated at present only by theoretical considerations. The
limitations and possible extrapolation procedures for use in silicon
transistor simulation and design will be discussed. The mobility formulae
have been tested satisfactorily in MOSFET transistor simulations using
2-d MINIMOS program [6]. These formulae are listed as follows by
individual scattering mechanisms for thick channels. The thin channel
or transverse electric field effect and the hot electron or longitudinal
electric field effect are discussed in later individual subsections.

B. LONGITUDINAL-ACOUSTIC PHONON-SCATTERING MOBILITY [3]
----

mu(nLA) = 7.4 X 10*5 / T(L) (cm*2/V.s)   (1)

The 1/T(L) dependence comes from the two-dimensional classical electron
gas theory and the numerical coefficient was obtained by first subtracting
out the intervalley phonon and oxide charge scattering components from
the data using their temperature and oxide charge density dependences
respectively and Mathiessen's rule. Then the remaining mobility is
fitted to the 1/T(L) dependence.

C. INTERVALLEY-PHONON-SCATTERING MOBILITY [3]
----

mu(nLO) = 1.0 X 10*8 / T(L)*1.9 (cm*2/V.s)   (2)

The T(L)*-1.9 dependence comes from an empirical fit to the 2-d theory
of electron scattering by surface intervalley longitudinal acoustic
phonons whose analytical formula was given in [3]. The simpler
empirical expression is given here since there may be two important
intervalley phonons, one for 135 deg and one for 90 deg valleys, and
each has a complex formula with two terms, a phonon emission and a
phonon absorption term. The exact formula given in [3] can be used for a
more exact simulation.

D. OXIDE-CHARGE-SCATTERING MOBILITY [3]

$$mu(nOX) = \frac{1000 \; X \; (3 \; X \; 10*11)}{N(OX)} \; X \; \frac{T(L)}{80} \quad (cm*2/V.s) \qquad (3)$$

The T(L) temperature dependence comes from the 2-d Coulomb scattering
theory. N(OX) is the areal density (trap/cm*2) of the charged oxide
traps located at the oxide/silicon interface. The assumption of the
oxide charge location as a sheet at the oxide/silicon interface is a
good model for VLSI oxides initially. After high electric field stress,
charges may be injected and trapped in the interior of the oxide away
from the oxide/silicon interface. Then, the mobility will be higher than
that given by (3) if the stress value of the effective N(OX) referred to
the interface from threshold voltage shift measurements is used. For a
distributed oxide charge in the oxide, see the original formula given
in [3].

E. DIPOLE SCATTERING MOBILITY [6]

$$mu(nDP) = \frac{mu(dpO) \; (T(L)/100)*(3/2)}{N(DP) \; L(DP)*2} \quad (cm*2/V.s) \qquad (4)$$

N(DP) is the areal density of the dipoles (dipole/cm*2) and L(DP) is the
average dipole length (cm). The coefficient mu(dpO) has not been
measured and a theoretical estimate gives a value of about 100 cm*2/V.s.
The possible importance of dipole scattering was noted [6] since it has
been well known that interface traps are electrically deactivated by
hydrogen bonding or hydrogenation of the silicon and oxygen dangling
bonds at the oxide-silicon interface [7]. These hydrogen bonds are
electrical dipoles which are randomly directed and located. Taking the
high end as an example, we assume there are 10*13 interface traps per
cm*2 and an H-O or H-Si bond length of 2 A then

$$mu(nDP) = 100 \; (\frac{T(L)}{100})*3/2 \; \frac{1}{10*13 \; X \; 4 \; X \; 10*-16} = 2.5 \; X \; 10*4 \; (\frac{T(L)}{100})*3/2$$

$$= 25,000 \; cm*2/V.s$$

at 100 deg K. Thus, molecular dipole scattering is not very important
unless the temperature is very low and the dipole density approaches a
monolayer, about 4 X 10*14/cm*2 interface traps. However, if the dipole
is extended instead of molecular in size, significant scattering and
mobility reduction would occur. Scattering by large dipoles would
approach that of isolated ions whose mobility was given for oxide charge
scattering in the previous paragraph. The temperature dependence then
drops to T(L) from T(L)*3/2. In conventional VLSI oxides, the density of
the interface traps, whether hydrogenated or not, is made as low as
possible, below 10*10/cm*2, so that dipole scattering from molecular-size
hydrogenated silicon and oxygen dangling bonds would not be important.

F. SURFACE ROUGHNESS SCATTERING MOBILITY [5,6]

$$mu(nSR) = \frac{1.5 \times 10*29}{[\ N(DEP) + N(INV) \ ]*2} \qquad (cm*2/V.s) \qquad (5)$$

N(DEP) is the depletion layer ion density which may be approximated by
N(DEP) = N(I) x(d) (/cm*2) where N(I) (/cm*3) is the net concentration
of dopant impurity ions in the silicon surface layer and the surface
depletion layer thickness is given by

$$x(d) = SQRT \ [\ \frac{4 \ epsilon(Si)}{qN(I)} \ \frac{kT(L)}{q} \ [\ U(F)-(3/2) + (1/2) \ ln(U(F)-3) \ ] \ ]$$

Here, epsilon(Si)=11.7 X 8.85 X 10*-14 F/cm for Si, q=1.602 X 10*-19 C,
T(L) is the lattice temperature and kT(L)/q = 0.02585 V at 300 deg K,
U(F)=qV(F)/kT(L)=ln(N(I)/ni) is the normalized Fermi level in the bulk
and ni is the intrinsic carrier concentration of silicon. N(INV) is the
inversion layer electron density (/cm*2) and it can be computed from
Cox(Vg-Vgt)/q where q is the electron charge. It is also related to the
transverse electric field at the interface, N(INV)=epsilon(Si) Es/q
where Es is the magnitude of the transverse (to current flow in the
channel) electric field normal to and at the oxide/silicon interface.
This mobility expression is based on a simple theory of a scatterer
with short-range square-well scattering potential at the oxide-silicon
interface [5,6]. The numerical coefficient, 1.5 X 10*29, was obtained
by fitting to low temperature mobility data (7-77 deg K) in strong
inversion. This gives an average size (edge of a cube) of 5.4 A and
density of 9.3 X 10*13/cm*2 for the randomly distributed scatterers at
the oxide/silicon interface [5]. In the 77-300 deg K range,
experimental data at smaller inversions showed significantly smaller
mobility than that predicted by (5) which is partially attributable to
the channel thickness dependences of the other scattering mechanisms
(phonon and channel ion). Both theories and experiments on surface
roughness scattering have been investigated by a number of researchers
[8-11] but these were all at 4.2 deg K. The 4.2 deg K data of Hartstein
[10,11] can be used as a guide to extrapolate to higher temperatures of
practical interest.

G. CHANNEL ION SCATTERING MOBILITY [6]

$$mu(nCI) = mu(nCIO) \ [\ 1 + \frac{N(IO)}{N(I)} \ (\ \frac{T}{300} \ ) \ ] \ *alpha \qquad (cm*2/V.s) \qquad (6)$$

where mu(nCIO) = 90 cm*2/V.s, N(IO) = 2.0 X 10*18/cm*3 and alpha = 3/2
or 1 have been employed. N(I) is the ion density (/cm*3). This formula
is identical to that of the ion scattering in the silicon bulk following
the Conwell-Weisskopp and Brook-Herring formulae. The mobility due to
scattering by ions in the surface space charge layer is higher than that
given by (6) due to the physical separation of the electron from the
scatterers. This increase can be estimated using the theory developed
for the distributed oxide charge [3]. The increase in the ion scattering
mobility in the surface channel can also be seen from a new analysis

made by us of the experimental mobility data of Sun and Plummer [12]
described as follows [12]. The channel ion scattering mobility is now
extracted from their mobility data on the oxide charge and substrate
doping dependence by extrapolating their mobility data to zero oxide
charge density. The mobility values were read from their figure and
also verified using the empirical fit formula for zero oxide charge
[12]. Now the new analysis step is to subtract the contributions of
longitudinal-acoustic phonon, Eqn (1), and intervalley-phonon, Eqn (2),
using Mathiessen's rule. Then the mobility due to scattering by ions in
the surface space charge layer is obtained. It shows a N(I)*(-1/3)
dependence for the impurity concentration range of 10*15 to 10*18/cm*3.
A fit gives

mu(nCI) = (0.85 X 10*9) [N(I)]*(-1/3) (cm*2/V.s)   (6A)

This weaker dependence on N(I) compared to the 1/N(I) dependence in
the bulk given by Eqn (6) is expected since the ion scatterers are
distributed and is predicted by the 2-d theory [3]. In contrast, Sun and
Plummer empirically fitted to a logarithm dependence and gave an
unphysical cause.

H. TRANSVERSE ELECTRIC FIELD EFFECTS

Increasing the transverse electric field normal to the oxide/silicon
interface will make the inversion channel thinner. The increased
confinement of the electron to the oxide/silicon interface will cause a
change of the phonon scattering from bulk to the surface phonons. To
facilitate numerical estimates, the bulk phonon scattering mobilities
are listed below [6].

mu(nLA)(bulk) = 2.18 X 10*7/T*(3/2)   (cm*2/V.s)   (7)

mu(nLO)(bulk) = 1.22 X 10*11/T*(3.13) (cm*2/V.s)   (8)

For example, at 300 deg K, the values are mu(nLA)(surface) = 2466,
mu(nLO)(surface) = 1965, mu(nLA)(bulk) = 4195 and mu(nLO)(bulk) = 2153
cm*2/V.s. These show that the mobilities decrease as the transverse
field is increased causing the channel to thin and the electrons to be
further confined.

The ionized impurity scattering and oxide charge scattering have opposite
dependences as the transverse field is increased or the channel is
thinned. The oxide charge scattering will increase with decreasing
channel thickness width and hence lower the mobility since the sheet of
electrons is closer to the oxide charges. Scattering by the ions in the
silicon surface space charge layer will decrease due to increased
separation as the inversion channel is thinned, hence, the electron
mobility in the channel will increase.

The interfacial roughness scattering will increase as the transverse
field is increased or the electrons are confined nearer to the interface.
This is already included in the surface roughness scattering mobility
formula.

Dipole scattering will have a similar dependence on the transverse
electric field as the oxide charge for the same geometrical or
increasing proximity reason.

Experimental separation of transverse electric field effects due to
increasing confinement in the channel and closer proximity to the
oxide/silicon interface from each mobility or scattering component has
not been made in the practical temperature range. The separation made
by Hartstein at 4.2 deg K [10,11] can serve as a guide for future
efforts at higher temperatures. Cooper and Nelson have given some
electron drift velocity data at 300 deg K on <100> silicon with
specified current direction [13,14]. The transverse field dependence
of their electron mobility at low longitudinal field mobility can be
approximated by

$$
mu(nL) = \frac{mu(nLO)}{[1 + E(T)/E(TOmu)]*C}
$$

where mu(nLO) = 1350 cm*2/V.s is the bulk mobility, C=1.0, E(TOmu) = 54
kV/cm and E(T) is the average transverse or surface field which is half
of the peak surface field and was obtained from Cox(Vg-Vgt)/2 epsilon(Si).
The saturation velocity extrapolated to zero transverse field is 0.95 X
10*7 cm/s instead of the bulk value of (1.05 to 1.1) X 10*7cm/s. Cooper
and Nelson [14] gave mu(nLO) = 1105 cm*2/V.s instead of the bulk value
of 1350, resulting in a smaller exponent of C = 0.657.

I. LONGITUDINAL ELECTRIC FIELD EFFECTS

As the electric field increases, the electron mobility decreases due to
the increasing rate of loss of the electron kinetic energy or drift
velocity by phonon emission during the scattering. At low fields, the
loss is mainly from the emission of the longitudinal acoustic intravalley
phonon. At high fields, the loss is mainly from the emission of the two
longitudinal acoustic intervalley phonons. At high fields, the drift
velocity saturates and the mobility decreases as the reciprocal of the
electric field. The electric field dependences due to phonon emission
were fully predicted by Shockley's simple theory on hot electrons in the
bulk of semiconductors [15]. Shockley's simple formula has been used as
the basis to fit experimental bulk mobility data. This fit can also be
used to extrapolate the electric field dependences of the surface
mobilities since no reliable experimental data are available and a large
deviation from the bulk theory is not expected. The theory is given by

$$
mu(n) = \frac{mu(nO)}{[1 + (mu(nO)E(L)/v(ns))*gamma]*1/gamma} \tag{8}
$$

where mu(nO) is the low field phonon scattering mobility, E(L) is the
magnitude of the longitudinal electric field, v(ns) is the saturation
velocity theoretically given by SQRT[h nu(LO)/m(e)] where h nu(LO) is
the LO phonon energy and m(e) is the electron effective mass, and f
is a numerical factor between 1 and 2. An adequate approximation for
transistor design is to take the bulk values [6], v(ns) = 1 X 10*7 cm/s
and gamma = 2.0. These values are consistent with the data of Cooper
and Nelson at low transverse field (30 kV/cm) [13] who gave [14]
v(ns)=0.923 X 10*7 cm/s and gamma=1.92 from empirical fit. They reported

experimental electron drift velocity data taken at 300 deg K on <100>
silicon surface alone the <110> in a longitudinal field range of 2.6-42
kV/cm and a transverse field range of 30-180 kV/cm. A smaller drift
velocity and hence mobility at increasingly higher transverse field is
evident and expected. When the transverse field is increased from 80 to
180 kV/cm, the drift velocity at 300 deg K dropped from 8.56 X 10*6 to
7.75 X 10*6 cm/s. A linear fit gives v(ns) = v(nsO)/(1 + E(T)/E(Tns))
where v(nsO) = 0.95 X 10*7 cm/s and E(Tns) = 1340 kV/cm. Cooper and Nelson
[14] assumed no change of the saturation velocity with the transverse
field based on an assertion by Thornber. However, they did not consider
the transition from bulk to surface phonon scattering and energy loss
which produces transverse field dependence in the saturation velocity.

J. THRESHOLD AND SUBTHRESHOLD MOBILITY [4]

Near threshold when the surface channel conduction just begins, the
volume density of the electrons is very low. Both the carrier density
and the electrical thickness of the channel are highly nonuniform. These
random spatial variations arise from the microscopic and macroscopic
variations of (i) the chemical bonds at the oxide/silicon interface,
(ii) the oxide charge and (iii) the dopant impurity ion density in the
silicon surface layer. These inhomogeneities give rise to a spatially
random interfacial potential seen by the electron. The conducting
channel layer can be viewed as a piece of Swiss cheese whose holes are
high potential hills where electron movement is forbidden. The effective
conductivity mobility defined by mu = G/Q will decrease due to three
factors: the smaller effective channel width, the smaller carrier density,
and the additional boundary scattering of the electrons by the walls of
the individual protruded and immersed islands of forbidden potential
wells or walls of the holes or voids in the Swiss cheese-like channel
layer. Mobility reduction near and below the conduction threshold was
first described by Stern and Howard when they considered the two-
dimensional quantization effect in weakly inverted surface layers [16].
Chen and Muller presented data taken on both p-channel and n-channel
Si MOS transistors and attempted to correlate the data with a surface
potential fluctuation model [17]. Their data showed mobility drop off
near the threshold; however, they used a computed carrier density for
Q(N) to obtain the conductivity mobility from the measured channel
conductance, G(N) = mu(n) Q(N) since the carrier density is too low to
be measured by the conventional capacitance method. Brews improved on
the theory [18] which was further extended by Guzev, Gurtov, Rzhanov
and Frantsuzov [19] to include also Hall mobility. Since the carrier
density in the channel was measured directly from the Hall voltage,
their Hall mobility data were free of the uncertainty in the data of
Chen & Muller. Their data also showed a mobility drop off, reaching a
constant plateau when the carrier density in the channel drops below
10*10/cm*2. In addition, they demonstrated that the surface potential
fluctuation can be screened by the presence of the metal gate electrode.
This screening increases when the oxide thickness decreases or the gate
conductor is closer to the channel.

In the surface potential fluctuation model employed by these authors,
neither the smaller effective channel width nor the additional boundary
scattering were taken into account. Furthermore, the location of the
random charges which give rise to the potential fluctuation does not
enter, and the theory gives a divergent result when the sheet of
electrons (or holes) are concentrated in an infinitesimal layer at

the oxide/silicon interface. Thus, it does not take into account the random charges distributed into the oxide away from the interface plane. The smaller effective channel width and the distributed random charge sheets were considered by Shiue and Sah [4] using a percolation model which also neglected boundary scattering. However, the additional boundary scattering probably has a second order effect compared with the narrower electrical width unless the density of the inhomogeneity scatterers is very high.

These works just cited do not provide very reliable conductivity mobility data. Chen & Muller computed rather than measured the carrier density which could contain large error and Guzev et al measured the Hall mobility. To alleviate the difficulty of measuring very low carrier densities, Shiue and Sah developed a new technique [4]. In this method, the admittance of the MOSFET transmission line was measured. The carrier density is then computed from the imaginary part of the measured admittance. Electron densities as low as 10*8 /cm*2 can be determined.

The subthreshold mobility is highly dependent on the fabrication process variables as well as the subsequent amount of electrical stress which could produce highly nonuniform microscopic and macroscopic oxide charge distributions. Extensive experimental data have yet to appear. Sample results [4] showed the conductivity mobility reduction from 1000 cm*2/V.s between N(INV) = 10*10 to 10*11/cm*2 to 100 cm*2/V.s at N(INV) = 10*8/cm*2 at 300 deg K and from 4000 to 600 cm*2/V.s at 77 deg K in a low oxide charge transistor (N(OX) = 10*11/cm*2). The reduction is much larger in a high oxide charge transistor (N(OX) = 6.5 X 10*11/cm*2), from 600 to less than 1 cm*2/V.s at 300 deg K and from 300 to less than 0.1 cm*2/V.s at 77 deg K. This large decrease reflects the randomness of the oxide charges which creates large nonuniformity of channel width in the subthreshold-threshold conduction ranges. These observed conductivity mobilities decrease continuously with decreasing channel electron density without reaching a plateau as suggested by the computed conductivity mobilities of Chen and Muller [17] and the Hall mobilities of Guzev et al [19].

K. ORIENTATION DEPENDENCES

Considerable orientation dependences have been observed at low temperatures in strongly inverted deep surface channel wells. This anisotropy is expected from the effective mass differences of the multi-valley ellipsoidal energy surface of the silicon conduction band edge when the two-dimensional quantization effects become important [1]. However, orientation dependences in the range of 30-300 deg K have not been observed on the mobility components due to oxide charge as well as acoustic and intervalley phonon scattering [3]. Current and future applications to the design of submicron three-dimensional transistor structures, such as the proposed high density dynamic memory cell using the capacitor and transistor in a trench below the silicon surface, may require a careful re-measurement of the electron mobility in the 77-400 deg K range on several surface orientations and along several current flow directions.

L. APPLICATION OF MOBILITY TO SUB-MICRON TRANSISTOR DESIGN

The mobility data cited were obtained on large devices to avoid errors
due to two- and three-dimensional boundary effects. Thus, they are
fundamental data and should be valid for the design of small transistors.
Boundary scattering from the edges of small transistors can be computed
by incorporating these mobility formulae into the two- or three-
dimensional transistor simulators. The applicability limit is reached
when the transistor dimension is so small that the macroscopic
quantities, such as the carrier density and mobility lose physical
significance. This limit is reached when the differential volume
elements used in the device simulator contain less than about 1000
electrons, particles or scatterers which would result in more than 3%
number fluctuation in a real device.

REFERENCES

[1]   T.Ando, A.B.Fowler, F.Stern [Rev. Mod. Phys. (USA) vol.54 no.2
        (Apr 1982) p.437-672]
[2]   F.F.Fang, A.B.Fowler [Phys. Rev. (USA) vol.169 no.3 (15 May
        1968) p.619-31]
[3]   C.T.Sah, T.H.Ning, L.L.Tschopp [Surf. Sci. (Netherlands) vol.32
        no.3 (Sep 1972) p.561-75]
[4]   C.C.Shiue, C.T.Sah [Phys. Rev. B (USA) vol.19 no.4 (15 Feb 1979)
        p.2149-62]
[5]   C.T.Sah [Bull. Amer. Phys. Soc., (USA) vol.18 (1973) p.344
        Illinois State Electronics Tech. Rep. no.25 (25 March 1973)]
[6]   T.Nishida, C.-T.Sah [IEEE Trans. Electron Devices (USA) vol.ED-34
        no.2 pt.1 (Feb 1987) p.310-20]
[7]   P.Balk [Spring Meeting of the Electrochemical Society, 9-13 May
        1965 Extended Abstracts of the Electronics Division vol.14 Abstr.
        109 p.237-240 (Electrochem. Soc. Princeton, NJ)]
[8]   Y.Matsumoto [US Japan Seminar Surface Quantization & Transport
        Honolulu, Hawaii, August 14-16 1972 paper no.T-4 Sponsor US,
        National Science Foundation, Washington, DC] See ref. [1] for a
        review
[9]   Y.C.Cheng, E.A.Sullivan [Surf. Sci. (Netherlands) vol.34 no.3
        (1973) p.717-31]
[10]   A.Harstein, T.H.Ning, A.B.Fowler [Surf. Sci. (Netherlands)
        vol.58 (1976) p.178-81]
[11]   A.Harstein, T.B.Fowler, M.Albert [Surf. Sci. (Netherlands) vol.98
        (1980) p.181-90]
[12]   S.C.Sun, J.D.Plummer [IEEE Trans. Electron Devices (USA)
        vol.ED-27 no.8 (Aug 1980) p.1497-508]
[13]   J.A.Cooper, D.F.Nelson [IEEE Electron Device Lett. (USA)
        vol.EDL-2 no.7 (Jul 1981) p.171-3]
[14]   J.A.Cooper,Jr. D.F.Nelson [J. Appl. Phys. (USA) vol.53 no.3
        (March 1983) p.1445-56]
[15]   W.Shockley [Bell Syst. Tech. J. (USA) vol.30 no.10 (Oct 1951)
        p.990-1034]
[16]   F.Stern, W.E.Howard [Phys. Rev. (USA) vol.163 no.3 (15 Nov 1967)
        p.816-35]

[17]   J.T.C.Chen, R.S.Muller [J. Appl. Phys. (USA) vol.45 no.2 (Feb
        1974) p.828-34]
[18]   J.R.Brews [J. Appl. Phys. (USA) vol.46 no.5 (May 1975) p.2193-203]
[19]   A.A.Guzev, V.A.Gurtov, A.V.Rzhanov, A.A.Frantsuzov [Phys. Status
        Solidi a (Germany) vol.56 (1979) p.61-73]
[20]   C.T.Sah, T.Nishida [EMIS Datareview RN=17813 (Nov 1987) 'Hole
        mobility in oxidized Si surface layers']

17.21 HOLE MOBILITY IN OXIDIZED Si SURFACE LAYERS

by C.T.Sah and T.Nishida

November 1987
EMIS Datareview RN=17813

Both majority and minority carrier mobilities on silicon surfaces have
been measured on samples with different physical geometries (shape and
crystallographic orientations) and material compositions (oxide
thicknesses and substrate impurity concentration). They have also been
measured as a function of operating conditions such as the transverse
and longitudinal DC electric fields and lattice temperature. A transverse
DC electric field creates two electrical geometries, the inversion and
accumulation surface channels, in addition to the physical geometry of
no channel under the electrical flat band condition. A longitudinal
electric field creates the hot carrier effect which reduces the mobility
as the electric field is increased. Surface mobility is anisotropic,
i.e. dependent on the crystallographic directions of the longitudinal
and transverse electric fields. This orientation dependence arises from
the anisotropic constant energy surface of the conduction and valence
band edges and the removal of the cubic symmetry by the one-dimensional
quantization from the confinement of the electrons or holes in the
narrow surface channel potential well. Orientation dependences of
experimental mobility data have been observed even at room temperatures.
Furthermore, different surface mobilities have been deduced from
different measurement techniques, such as the channel conductance,
field-effect and the Hall effect methods. These are known as the
conductivity or effective mobility, field-effect mobility and Hall
mobility respectively. The field-effect mobility is further complicated
by two different measurement methods, one employs the transconductance
and the other, the derivative of the channel conductance with respect
to DC gate voltage. Mobility measurements under such a large variety of
situations and by many researchers have been motivated by silicon
transistor and integrated circuit applications. However, two completely
opposite approaches have been employed to represent or fit the mobility
data. One is based on fundamental scattering mechanisms and the
confinement of the electrons and holes in a thin surface channel layer
which quantizes one direction of the electron or hole motion resulting
in a two-dimensional (2-d) conduction system. The other, used often by
integrated circuit design engineers, is based on highly empirical multi-
parameter least-squares fits which can cover only a limited range of
geometrical, material and operating parameters within which the data
were obtained.

This Datareview will focus on only one type of mobility data, namely the
conductivity mobility of minority carriers in a surface inversion
channel covered by a thermally grown silicon dioxide layer. This
mobility and its measurement conditions are most relevant to integrated
circuit applications. The fundamental approach, fitting the data to the
physics-based formulae, will be employed so that the data can be
accurately extended over a wider range of operating conditions than the
data-taking conditions, with confidence and known limitations. This
choice is motivated by the need of an accurate and extendable mobility
characterization for use in the design of silicon integrated circuits.

For a comprehensive treatment of the basic physics underlying mobility and scattering mechanisms, see the review by Ando, Fowler and Stern [1]. Hole mobility data is discussed in this section. Electron mobility is presented in [23].

Research on the mobility of holes in oxidized silicon surface layers began before electron mobility. This lead in the early 1960's was a consequence of the fact that the enhancement mode p-channel MOSFET was easier to manufacture with high yields than the enhancement mode n-channel MOSFET due to the positive sign of the residual charges trapped in the oxide. A mass-produced p-channel enhancement mode MOSFET was marketed by Fairchild Semiconductor Corporation of Palo Alto, California in late 1962. However, detailed mobility measurements, such as those made for electrons by Fang and Fowler [2] and by Sah, Ning and Tschopp [3], have not yet been carried out for holes in surface inversion channels. These measurements are necessary to separate the various scattering contributions. A limited amount of hole mobility data can be found in articles by Murphy (Hall-effect, 1964) [4]; Leistiko, Grove and Sah (conductivity, 1965) [5]; Colman, Bate and Mize (Hall, 1968) [6]; Pierret and Sah (conductivity, 1968) [7]; Sato, Takeishi and Hara (field-effect via transconductance, 1969) [8]; Murphy, Berz and Flinn (field-effect via differential-conductance, 1969) [9]; Chen and Muller (conductivity with computed instead of measured carrier or hole density Q(P), 1974) [10]; Guzev, Gurtov, Rzhanov and Frantsuzov (Hall, 1979) [11]; Coen and Muller (conductivity, 1980) [12] and Su, Wei and Ma (method not stated, 1985)[13]. The limited number of fundamental measurements is a reflection of the complex valence band structure which discourages simple theoretical-experimental correlations.

The available measurements were made on carefully designed metal(aluminium)-oxide-silicon field-effect transistors (MOSFET) with large geometries (long channel and wide gate) so that the edge effects are unimportant. The conductivity mobility of holes in the surface inversion channel is computed from mu(P) = G(P)/Q(p) G(p) is the DC conductance of the channel measured between the source and the drain contact at zero DC drain-to-source voltage and it can also be measured as a small-signal AC conductance at low or zero signal frequency. Q(P) is the total DC areal charge density of the minority carriers (holes) in the inversion channel. For strong inversion Q(P) can be accurately computed from Cox(Vg -Vt) where Cox is the capacitance of the gate oxide (pF/cm*2), Vg is the DC voltage applied to the gate and Vt is the gate threshold voltage at which the surface channel just begins to conduct. Vt is taken as the intercept of the tangent to the G(p) versus Vg curve [3] in the range where the G(p)-Vg curve is nearly linear. The interface state density of the samples used was not reported but presumably was not high so that the dependence of Vt on Vg via charging or discharging of the interface states can be neglected and Vt is independent of Vg. At low inversion, the G(p)-Vg data is curved with a knee or tail and Vt is no longer definable by the tangent extrapolation method just stated. This region covers the threshold and subthreshold ranges. In these ranges, the gate voltage is near or below the extrapolated threshold voltage and Q(P) cannot be obtained from Cox(Vg-Vt). Two alternative methods were used, one from the Hall-effect voltage and the other by theoretical calculation (see Chen and Muller [10]). Neither method gives the true conductivity mobility which enters the MOSFET design equations. However, Q(P) can be accurately obtained by a new transmission line input admittance method even at very low carrier densities in the subthreshold region. This was demonstrated by Shiue and Sah for electrons in n-channels (see [18]) but it has not been applied to measure the hole mobilities in p-channels.

The experimental data of Hall, field-effect and conductivity mobilities have covered a wide range of temperatures from 77 deg K [6,8] to 350 deg K [10-12], transverse electric fields or inversion hole densities from 5 X 10*7 to 9 X 10*12/cm*2 [11], and longitudinal fields from 0.5 to 80 kV/cm [12]. In addition, a broad range of surface and current directions were investigated by Bate et al [6], Sato et al [8], and Coen and Muller [12]. Their data have shown a 5:1 anisotropy at 77 deg K [6,8] and 2:1 at 300 deg K. The oxide thickness (31 to 240A) and transverse electric field were investigated by Ma and his students [13] at 300 deg K.

Due to the uncertainty in the accuracy of the limited number of published conductivity mobility data of holes in silicon inversion channel, a detailed analysis of the data to separate out the phonon, oxide charge, channel ion, surface roughness and surface potential fluctuation contributions has not been performed as those for electrons presented in [23].

However, some specific data on the longitudinal electric field dependence of the hole mobility in the surface inversion channel or the hot carrier effect were reported by Coen and Muller [12] in the form of an empirical equation suggested by Shockley's simple hot carrier theory [14]. This was given by mu = mu(o)/[1 + (mu(o)E(L)/vs)*gamma]*(1/gamma) where mu(o) is the mobility at low longitudinal electric field, E(L) is the magnitude of the longitudinal electric field and v(s) is the saturated drift velocity at high fields. Coen and Muller reported values of v(s) = 2 X 10*6 cm/s and gamma = 2.8 for holes and 5 X 10*6 cm/s and 2.9 for electrons. Their saturated velocity for electrons is a factor of two smaller than the more recent data of Cooper and Nelson (see reference [23]) which was extrapolated to 9.5 X 10*6 cm/s at zero transverse electric field by us. This nearly equaled the bulk value of 1.1 X 10*7 cm/s. Thus, the data of Coen and Muller for holes are probably also low and a better approximation would be to use the bulk values for holes in inverted surface channels also, v(s) = 8.2 X 10*6 cm/s and gamma=1. The discrepancy is probably due to the effect of the transverse field which was clearly demonstrated by Cooper and Nelson and also the data of Ma [13]. A severe degradation or reduction of the hole mobility has been reported by Ma [13] in large MOSFETs as their gate oxide decreases from 240 A to 31 A. The extrapolated zero transverse oxide field mobility is 90 cm*2/V.s at 31 A and saturated to 170 cm*2/V.s at 240 A while the bulk value is 410-450 cm*2/V.s. Thus, there was some unknown interface hole scattering centers in the MOSFETs used by Ma [13] or there was significant measurement or data analysis error. However, hole mobility reduction in thin oxide MOSFETs has also been observed by submicron CMOS designers at NTT Electrical Communications Laboratories [15] and NEC Microelectronics Research Laboratories [16]. Extrapolated values are 140 cm*2/V.s at 35 A and 0.2 micron channel [15] and 80-100 cm*2/V.s at 120 A and 0.25 micron channel [16]. A microscopic theory of random surface potential scattering was developed by Li and Ma [17,18] assuming a random oxide thickness due to random penetration of the aluminium or polycrystalline silicon gate material into the oxide. Although the reduction of mobility with thinner oxide was predicted the parameters values were unphysical and macroscopic instead of microscopic.

In transistor design and simulation, the hole-to-electron bulk mobility ratio can be used as a scale factor [19] to approximate the surface mobility of holes from the surface mobility of electrons given in [23]. This constant-scale-factor approach has been employed for the

simulation of p-channel Si MOSFET using MINIMOS 2.2 [20]. The computed
drain current using the scaled hole mobility formula is nearly the
same as that computed using the empirical hole mobility formula given
by the MINIMOS with the following deviations: 10% in the subthreshold
range, less than 20% near and above threshold, and about 5% in strong
inversion. A comparison of the computed drain current using the scaled
hole mobility with other empirical hole mobility formulae, such as those
employed by MINIMOS 3.0 [21] and PISCES [22] has not been made. However,
the MINIMOS-2 comparison just cited suggests that the basic formula
presented in this Datareview should be more reliable and accurate over
a wider range of operating parameters than the empirical formulae.

REFERENCES
----

[1]   T.Ando, A.B.Fowler, F.Stern [Rev. Mod. Phys. (USA) vol.54 no.2
        (April 1982) p.437-672]
[2]   F.F.Fang, A.B.Fowler [Phys. Rev. (USA) vol.169 no.3 (15 May 1968)
        p.619-31]
[3]   C.T.Sah, T.H.Ning, L.L.Tschopp [Surf. Sci. (Netherlands) vol.32
        no.3 (Sep 1972) p.561-75]
[4]   N.St.J.Murphy [Surf. Sci. (Netherlands) vol.2 (1964) p.89-92]
[5]   O.Leistiko, A.S.Grove, C.T.Sah [IEEE Trans. Electron Devices
        (USA) vol.ED-12 no.5 (May 1965) p.248-54]
[6]   D.Colman, R.T.Bate, J.P.Mize [J. Appl. Phys. (USA) vol.39 no.4
        (1968) p.1923-31]
[7]   R.F.Pierret, C.T.Sah [Solid-State Electron. (USA) vol.11 (1968)
        p.279]
[8]   T.Sato, Y.Takeishi, H.Hara [Jpn. J. Appl. Phys. (Japan) vol.8
        no.5 (May 1969) p.588-98]
[9]   N.St.J.Murphy, F.Berz, I.Flinn [Solid-State Electron. (GB) vol.12
        (1969) p.775]
[10]   J.T.C.Chen, R.S.Muller [J. Appl. Phys. (USA) vol.45 no.2 (Feb
        1974) p.828-34]
[11]   A.Guzev, V.A.Gurtov, A.V.Rzhanov, A.A.Frantsuzov [Phys. Status
        Solidi a (Germany) vol.56 (1979) p.61]
[12]   R.W.Coen, R.S.Muller [Solid-State Electron (GB) vol.23 (1980)
        p.35]
[13]   H.-Q.Su, C.-C.Wei, T.-P.Ma [IEEE Trans. Electron Devices (USA)
        vol.ED-32 no.3 (March 1985) p.559-61]
[14]   W.Shockley [Bell Syst. Tech. J. (USA) vol.30 no.10 (1951)
        p.990-1034]
[15]   M.Miyake, T.Kobayashi, K.Deguchi, M.Kimizuka, S.Horiguchi,
        K.Kiuchi [IEEE Electron Device Letts. (USA) vol.EDL-8 no.6 (June
        1987) p.266-8]
[16]   N.Kasai, N.Endo, H.Kitajima [1987 Int. Elect. Device Conf. Proc.
        (USA) (Dec 1987) p.367-70]
[17]   Jia Li, T.-P.Ma [J. Appl. Phys. (USA) vol.61 no.4 (15 Feb 1987)
        p.1664-6]
[18]   Jia Li, T.-P.Ma [J. Appl. Phys. (USA) vol.62 no.10 (15 Nov 1987)
        p.4212-5]
[19]   T.Nishida, C.T.Sah [IEEE Trans. Electron Devices (USA) vol.34
        no.2 (Feb 1987) p.310-20]
[20]   T.Nishida, C.-T.Sah [unpublished and computed for this review]
[21]   W.Hansch, S.Selberherr [IEEE Trans. Electron Devices (USA)
        vol.ED-34 no.5 (May 1987) p.1074-8]
[22]   M.R.Pinto, C.S.Rafferty, R.W.Dutton [PISCES-II User's Manual
        (1984) Stanford Electronics Labs, Stanford, CA, USA]
[23]   C.T.Sah, T.Nishida [EMIS Datareview RN=17831 (Nov 1987) 'Electron
        mobility oxidized Si surface layers']

17.22 BREAKDOWN BEHAVIOUR OF SiO2 FILMS

by P.Pan

November 1987
EMIS Datareview RN=17834

The capacitor-voltage (C-V) stability and breakdown behaviour are two
key issues in determining the potential of using a certain insulator
in integrated circuits (ICs). Thermally grown SiO2 films on Si
substrates have been widely used in the IC industry for 30 years due
to their superior C-V stability and good breakdown behaviour. Thin SiO2
films of < 15 nm in thickness are required for the future very large
scale integrated (VLSI) circuits and the breakdown behaviour of thin
SiO2 films is known to be difficult to control. In this review, the
effects of processing on breakdown behaviour of SiO2 films and the SiO2
breakdown mechanism will be discussed.

The breakdown behaviour of SiO2 films depends on the quality of Si
substrate. Crystal defects, oxygen precipitates, metal and mobile ion
contaminations in Si crystals or on Si surfaces were reported to degrade
the SiO2 integrity [1-6]. A correlation between the formation of defects
in SiO2 films and the metal contaminations (or defects) on Si
substrates has been observed physically [7,8]. The surface roughness,
caused by wafer polishing, reactive ion etching (RIE), etc., also
degraded the SiO2 breakdown [9,10]. By having sacrificial oxide films
grown and then stripped before gate SiO2 growth, it was possible to
reduce the degradation caused by surface roughness. The wafer cleaning
before oxidation was very critical on controlling SiO2 quality.
Different cleaning processes [11-14] have different effects on different
contaminations and a multi-cleaning process is normally applied before
SiO2 growth. Kooi's effect [15] which was generated during the growth of
local oxide on Si (LOCOS), was found to cause poor gate dielectric
breakdown [16]. The use of contamination-free Si substrate, Si backside
gettering, high purity cleaning solutions, and ultraclean tools and
environments will improve the quality of thin SiO2 dielectrics.

The breakdown behaviour of SiO2 films also strongly depends on the
oxidation processes. Pan and Schaefer [17] reported different breakdown
behaviours for SiO2 films grown in wet O2 ambients with different H2O
flow rates. The quality of SiO2 films grown in wet O2 ambient was found
to be better than that grown in dry O2 ambient [18,19]. Different
densities of defects, e.g., micropores [20] or Si-dangling bonds for
SiO2 films grown in different ambients may cause different breakdown
behaviours. The quality of SiO2 films grown in an ambient containing
'Cl' was found to be superior to that grown without having 'Cl' [21-23].
The presence of 'Cl' in the oxidation ambient was effective to eliminate
stacking fault generation, to reduce metal contaminations, and to
passivate mobile ion diffusion [24,25]. Contaminations induced by the
oxidation furnace also effect the SiO2 quality. The breakdown behaviour
of SiO2 films grown in a double-layer furnace was found to be better
than that grown in a single-layer furnace [26-27]. After pre-cleaning
the oxidation furnace with HCl, the quality of SiO2 films grown in O2
ambients was found to be as good as that grown in O2/HCl ambients [11].

High quality SiO2 films can also be grown in a rapid thermal oxidation
system [28], which have a cold wall reactor. The key reason for having
these high quality SiO2 films was the reduction of metal contaminations
which were diffused through the wall of the oxidation furnace during high
temperature oxidation processes. The oxide quality was also found to be
affected by the inert gas annealing, either between oxidations [29] or
after SiO2 growth [17,30]. The growth of high quality SiO2 films in a
low pressure ambient was also demonstrated [31].

Post-oxidation processing was also found to affect the SiO2 breakdown.
In order to obtain a fine line without having a large etching bias,
the RIE process was commonly used to define the gate electrodes. The
SiO2 breakdown degraded after the gate electrode was defined by RIE
[17]. Charging effects during RIE process, which were similar to voltage
stressing, enhanced the existing defects under gate electrodes and
resulted in a poor breakdown behaviour. Because the ratio of etching
rates between gate electrode materials and SiO2 was finite, a thinning
of SiO2 films and a generation of defects on SiO2 along edges of
electrodes occurred after RIE; this resulted in a poor breakdown
behaviour. The degradation in breakdown caused by the charging effects
may be reduced by adjusting the RIE process conditions and the defects
generated on SiO2 along edges of electrodes can be reduced by applying
an oxidation after RIE process to regrow damaged SiO2 layers.

SiO2 films are often cleaned before electrode deposition or after
through-oxide ion implantation. Solutions containing NH4OH are commonly
used for film cleaning. After thin SiO2 films were dipped into these
solutions, many low field defects appeared and the breakdown distribution
was very broad [17]. No significant change in breakdown behaviour was
observed after thick (> 40 nm) SiO2 films were dipped into the same
solution. The etching of SiO2 films in these solutions, which generated
or enlarged defects, was believed to cause the degradation. Recently, Wu
et al [19] reported that the total charge to breakdown (Q(bd)) of
thin SiO2 films degraded after these films being cleaned with RCA
cleaning solutions (solutions containing NH4OH).

Ion implantation technique has been widely used to dope Si substrates
or poly Si films. In order to adjust the field-effect-transistor (FET)
channel doping concentration, ions (As or B) were implanted through
gate dielectrics into channel regions. Defects were generated in SiO2
films after ion implantation [32]. A degradation in breakdown was
observed after implanting either As ions [33] or BF2 ions [33] through
thin SiO2 gate dielectrics. The degree of degradation was a function of
ion mass, ion dose, and ion energy. The degradation was caused by
charging effects or by collisions between ions and atoms in SiO2 matrix.
Charging effects can be reduced by applying electrons flooding during
ion implantation. In order to form self-aligned source/drain junctions,
poly Si electrodes were used as masks during junction implantations.
Osburn et al [35] observed a degradation in gate dielectric breakdown
when poly Si electrodes without having spacers were used as masks
during ion implantation. This was caused by the generation of damage
on SiO2 dielectrics along edges of poly Si electrodes. The best way to
reduce the edge effects is to form spacers on the poly Si sidewall
before carrying out ion implantation.

In order to have highly conductive electrodes, polycide, salicide, or
metal was used for gate electrodes on SiO2 films. The breakdown behaviour
of thin SiO2 gate dielectrics may be affected by the new electrode
structures or materials. Ting et al [36] reported a degradation on SiO2

breakdown after TiSi2/poly Si/SiO2 structures annealed at a temperature > 850 deg C. Ito et al [37] observed a severe SiO2 breakdown degradation when Mo/poly Si electrodes were applied and no degradation when Mo/Si-N/poly Si electrodes were used. The breakdown behaviour was found to be normal for W electrodes directly deposited on thin SiO2 gate dielectrics [38]. The reaction between electrode materials or impurities in electrodes with thin SiO2 films, the stress from electrodes and/or the electrode deposition process may be the cause of of degrading breakdown of thin SiO2 gate dielectrics.

The effects of processing on the quality of SiO2 films were found to be more severe for thin SiO2 than for thick SiO2. As reported by Wu et al [19], the SiO2 quality degraded after the deposition and definition of the first interconnection metal, and degraded more after applying the second interconnection metal. This suggested that the SiO2 integrity was affected not only by processes as decribed above, but also affected by other processes, which included passivation, contact hole open, and the deposition and defining of interconnection metal.

The first step to qualify SiO2 breakdown behaviour is to measure the breakdown distribution. The voltage ramping technique is normally used. Defect density, defect mode and maximum breakdown field can be obtained by the breakdown distribution. The maximum breakdown field was found to decrease with increasing SiO2 film thickness [11,39] and this thickness dependence was found to be more pronounced for the lightly doped substrate [11]. The breakdown distribution became broader and the maximum breakdown field became lower when a lower voltage-ramping rate was applied [40]. The maximum breakdown field was found to increase slightly as the measurement temperature increased [11]. The second step to qualify SiO2 breakdown is to measure the time dependent dielectric breakdown (TDDB) [41,42]; this measurement will provide the information on reliability. The ramping techniques [40,43] were also reported to forecast the dielectric life-time accurately. The Q(bd) was found to increase with decreasing SiO2 film thickness [44-47] during stressing at low and medium current density. This indicates that thin SiO2 films are more reliable than thick SiO2 films during low and medium current operations. When SiO2 films were stressed at high current density, different results were observed; Liang and Choi [46] reported that thinner SiO2 films became less reliable, but Kusaka et al [45] and Chen et al [46] found that thinner SiO2 films were more reliable. These conflicting observations may be caused by the difference in their SiO2 quality. The dependency and values of electric field and temperature acceleration factors for thin SiO2 films can be obtained in the literature [41-50]. By applying DC and AS stressing, Fazan et al [50] reported that charge to breakdown for thin SiO2 films under AC stressing is significantly higher than that under DC stressing.

The mechanism of SiO2 breakdown is a complicated issue. The SiO2 breakdown at low and medium fields is a defect-related behaviour. Defects include pinholes, micropores and impurities. Defects can be found in the as-grown SiO2 films, which were caused by contaminations on the Si surface, defects in Si crystals and the use of improper growth processes. Defects can also be generated by post-oxdiation processing. Pinholes and 'large' weak spots are the main causes of low-field breakdown. During low-field measurements, pinholes and large weak spots became electrical shorts or super hot spots and the catastrophic breakdown occurred. The 'small' weak spots (impurities, micropores) are the key causes of breakdown at medium-field. When a voltage is applied, the field across these localized weak spots is higher than that in other areas. The catastrophic breakdown occurred as the localized field at weak spots become high enough to cause a destructive thermal runaway. Electron trapping may improve SiO2 breakdown during low and medium field stressing. When traps with a small cross section were distributed uniformly in SiO2 films, the trapping of electrons induced a buildup of retarding field which reduced current injection or electron energy near the anode. This resulted in a higher breakdown field and a tighter breakdown distribution.

Breakdown at high-field is somewhat confusing. Is it an intrinsic or a defect-related behaviour? Because the maximum breakdown field, Q(bd) and t(bd) depended on process technologies, it is difficult to define an 'intrinsic' breakdown. Therefore, breakdown at high field is also believed to be a defect-related behaviour. The super fine defects (e.g. Si dangling bonds, impurities, and very small micropores which act as effective thinner spots or as 'seeds' for the generation of 'large' defects during high current injections are believed to play an important role in causing dielectric breakdown at high field.

Localized electron trapping and Si-O bond breaking was proposed by Harari [51] to explain the high-field SiO2 breakdown behaviour. Deep electron traps near the injecting electrodes were generated locally during high current injection. When localized trap sites with large cross sections were filled with electrons, the internal field at that spot became very high. A catastrophic breakdown occurred when these internal fields were large enough to cause a Si-O bond breaking. The observations of traps generating and filling were also reported by others [52,53]. Measurements on electron heating [54-56] showed a steady increase (rather than a saturation or a decay) in electron energy even at fields near the breakdown strength. Because the trap generation is a localized event and the probability of trap generation is low, it is reasonable to expect that the electron energy is not affected noticeably by the process of trap generation and filling. As reported by Pan and Schaefer [17], only small percentages of tested MOS capacitors across a wafer showed trap generation behaviour. This indicated that the trap generation and filling was a defect-related behaviour, rather than an intrinsic property. The Si-O bond breaking is not a necessary criterion for SiO2 breakdown. Breakdown in SiO2 may simply be caused by forcing a high current injection at localized effective thinner areas.

Localized hole trapping was another mechanism being proposed [39,57-59] to explain high-field SiO2 breakdown. During the electron injection, holes were generated by the impact ionization and drifted toward the cathode. The accumulation of holes near the cathode enhances current injections and the enhancement of current injection induces more hole generation and more hole trapping near the cathode. This feedback effect continued and a catastrophic breakdown occurred when the field and injected current at the local spots were high enough to cause a destructive runaway. Recently, Chen et al [60] showed that Q(bd) decreased with increasing gate current, but the substrate hole fluence at breakdown was independent of the gate current. This strongly supported the idea that the breakdown of SiO2 gate dielectrics was caused by the hole trapping mechanism. However, several reports in the literature conflict with this model. The electron-phonon scattering length, determined by either electron heating measurements or by photo-injection technique, was found to be 3-4 nm [54-56,61]. This is about 20 times larger than the critical value (0.173 nm) required for the impact ionization [39]. Measurements on electron heating [54-56] showed no evidence of having current runaway either. Recently, Weinberg and Nguyen

[62] reported that a large increase in current was observed initially for oxides with large hole trapping efficiencies, but this current increase is followed by a fast current decay. This indicates that it is impossible to have the feedback effect to cause a current runaway.

In conclusion, the breakdown behaviour of thin SiO2 films was found to be sensitive to the processing and the SiO2 breakdown is a defect-related behaviour. Thin SiO2 films were found to be more reliable than thick films during low and medium current operations. However, it is difficult to obtain high yields for thin SiO2 films. An ultraclean manufacturing environment and new processing technologies are required to grow high quality thin SiO2 films. Alternative insulators [63-65] have been intensively studied; these insulators generally showed excellent breakdown behaviour, insensitivity to processing, and a slight C-V instability. The use of alternative insulators on areas (e.g. storage node and interpoly insulator) which does not require having a superior C-V stability, will help to improve the overall yields. The use of dual insulators as gate dielectrics will also play an important role for future VLSI.

REFERENCES
----

[1]   C.M.Osburn, D.W.Ormond [J. Electrochem. Soc. (USA) vol.121 no.9 (1974) p.1229-33]

[2]   P.S.D.Lin, R.B.Marcus, T.T.Sheng [J. Electrochem. Soc. (USA) vol.130 no.9 (1983) p.1878-83]

[3]   T.H.DiStefano [J. Appl. Phys. (USA) vol.44 no.1 (1973) p.527-8]

[4]   C.M.Osburn, D.W.Ormond [J. Electrochem. Soc. (USA) vol.121 no.9 (1974) p.1195-8]

[5]   H.Abe, F.Kiyosumi, K.Yoshioka, M.Ino [Tech. Digest Int. Electr. Dev. Meeting, Washington, DC, USA, 1-4 Dec 1985 (IEEE, New York, USA, 1985) p372-5]

[6]   A.Ohsawa, K.Honda, N.Toyokura [J. Electrochem. Soc. (USA) vol.131 no.12 (1984) p.2964-9]

[7]   M.Kobayashi, T.Ogawa, K.Wada [Electrochem. Soc. Spring Meeting, Toronto, Ontario, Canada (Electrochem. Soc., 1985) Ext. Abstr. no.66]

[8]   S.I.Raider, K.Hofmann, G.W.Rubloff [Electrochem. Soc. Fall Meeting, San Diego, CA, USA (Electrochem. Soc., 1986) Ext. Abst. no.395]

[9]   N.Lifshitz [J. Electrochem. Soc. (USA) vol.130 no.7 (1983) p.1549-50]

[10]   J.Lee, C.D.Wong, C.Y.Tung, W.L.Smith, S.Hahn, M.Arst [Appl. Phys. Lett. (USA) vol.51 no.1 (1987) p.54-6]

[11]   C.M.Osburn, D.W.Ormond [J. Electrochem. Soc. (USA) vol.119 (1972) p.597]

[12]   R.L.Meek, T.M.Buck, C.F.Gibbon [J. Electrochem. Soc. (USA) vol.120 no.9 (1973) p.1241-6]

[13]   S.Iwamatsu [J. Electrochem. Soc. (USA) vol.129 no.1 (1982) p.224-5]

[14]   T.K.Ito, R.Sugino, T.Yamazaki, S.Watanabe, Y.Nara [Electrochem. Soc. Fall Meeting, Honolulu, Hawaii (Electrochem. Soc., 1987) Ext. Abstr. no.751]

[15]   E.Kooi, J.G. van Lierop, J.A.Appels [J. Electrochem. Soc. (USA) vol.123 no.7 (1976) p.1117-20]

[16]   M.Itsumi, F.Kiyosumi [J. Electrochem. Soc. (USA) vol.129 (1982) p.800]

[17] P.Pan, C.Schaefer [J. Electrochem. Soc. (USA) vol.133 no.6 (1986)
      p.1171-5]
[18] E.A.Irene [J. Electrochem. Soc. (USA) vol.125 no.10 (1978)
      p.1708-14]
[19] I.W.Wu, J.C.Mikkelsen,Jr., M.Koyanagi [Electrochem. Soc. Fall
      Meeting, Honolulu, Hawaii (Electrochem. Soc., 1987) Ext. Abstr.
      no.616]
[20] J.M.Gibson, D.W.Dong [J. Electrochem. Soc. (USA) vol.127 no.12
      (1980) p.2722-8]
[21] C.M.Osburn [J. Electrochem. Soc. (USA) vol.121 no.6 (1974)
      p.809-15]
[22] C.Hashimoto, S.Muramoto, N.Shiono, O.Nakajima [J. Electrochem.
      Soc. (USA) vol.127 (1980) p.129]
[23] D.N.Chen, Y.C.Cheng [J. Electrochem. Soc. (USA) vol.132 no.10
      (1985) p.2510-12]
[24] B.R.Singh, P.Balk [J. Electrochem. Soc. (USA) vol.125 no.3 (1978)
      p.453-61]
[25] A.Rohatg, S.R.Butler, F.J.Feigl [J. Electrochem. Soc. (USA)
      vol.126 no.1 (1979) p.149-54]
[26] P.F.Schmit [J. Electrochem. Soc. (USA) vol.130 (1983) p.196]
[27] R.F.De Keersmaecker, M.W.Hillen, M.M.Heyns, S.K.Haywood,
      I.S.Darakchiev [Electrochem. Soc. Spring Meeting, Cincinnati,
      Ohio, USA (Electrochem. Soc., 1984) Ext. Abstr. no.64]
[28] J.Nulman, J.P.Krusius, A.Gat [IEEE Electron Device Lett. (USA)
      vol.EDL-6 no.5 (1985) p.205-7]
[29] A.Bhattacharya, C.Vorst, A.H.Carim [J. Electrochem. Soc. (USA)
      vol.132 no.8 (1985) p.1900-3]
[30] L.Dori, M.Arienzo, Y.C.Sun, T.N.Nguyen, J.Wetzel [Mater. Res.
      Soc. Sym. Proc. (USA) vol.76 (1987) p.259-63]
[31] A.C.Adams, T.E.Smith, C.C.Chang [J. Electrochem. Soc. (USA)
      vol.127 no.8 (1980) p.1787-94]
[32] R.A.B.Devine, A.Golanski [J. Appl. Phys. (USA) vol.55 no.7 (1984)
      p.2738-40]
[33] S.Sugiura, S.Shinozaki [J. Electrochem. Soc. (USA) vol.134 no.3
      (1987) p.681-4]
[34] C.Y.Wong, T.Nguyen, Y.Taur, D.Quinlan, F.S.Lai [Electrochem. Soc.
      Fall Meeting, San Diego, CA, USA (Electrochem. Soc., 1986) Ext.
      Abstr. no.403]
[35] C.M.Osburn, A.Cramer, A.M.Schweigart, M.R.Wordeman [Electrochem.
      Soc. Fall Meeting, Detriot, Michigan, USA (Electrochem. Soc., 1982)
      Ext. Abstr. no.177]
[36] C.Y.Ting, F.M.d'Heurle, S.S.Iyer, P.M.Fryer [J. Electrochem. Soc.
      (USA) vol.133 no.12 (1986) p.2621-5]
[37] T.Ito, H.Horie, T.Fukano, H.Ishikawa [IEEE Trans. Electron.
      Devices (USA) vol.ED-33 no.4 (1986) p.464-8]
[38] S.Iwata, N.Yamamoto, N.Kobayashi, T.Terada, T.Mizutani [IEEE
      Trans. Electron Devices (USA) vol.ED-31 (1984) p.1174]
[39] T.H.DiStefano, M.Shatzkes [J. Vac. Sci. & Technol. (USA) vol.13
      no.1 (1976) p.50-4]
[40] M.Shatzkes, M. Av-Ron [Thin Solid Films (Switzerland) vol.91 no.3
      (1982) p.217-30]
[41] D.L.Crook [Proc. Int. Reliab. Phys. Symp., San Francisco, CA,
      vol.P-1 (1979)]
[42] C.Hu [Tech. Digest Int. Electron Device Meeting, Washington, DC,
      USA, 1-4 Dec 1985 (IEEE, New York, USA, 1985) p.368-71]
[43] A.Berman [Proc. Reliab. Phys. Sym, Orlando, FA (1981) p.204]

[44]   Y.Hokari, T.Baba, N.Kawamura [IEEE Trans. Electron Devices (USA)
       vol.ED-32 no.11 (1985) p.2485-91]
[45]   T.Kusaka, Y.Ohji, K.Mukai [IEEE Electron Device Lett. (USA)
       vol.EDL-8 no.2 (1987) p.61-3]
[46]   I.C.Chen, S.Holland, C.Hu [Tech. Digest Int. Electron Device
       Meeting, Los Angeles, CA, USA 1986 (IEEE, New York, USA, 1986)
       p.660]
[47]   M.-S.Liang, J.Y.Choi [Appl. Phys. Lett. (USA) vol.50 no.2 (1987)
       p.104-6]
[48]   J.W.McPherson, D.A.Baglee [J. Electrochem. Soc. (USA) vol.132
       no.8 (1985) p.1903-8]
[49]   K.Yamabe, K.Taniguchi [IEEE Trans. Electron Devices (USA)
       vol.ED-32 no.2 (1985) p.423-8]
[50]   P.Fazan, M.Dutoit, J.Manthey, M.Ilegems, J.M.Moret [Electrochem.
       Soc. Fall Meeting, San Diego, CA, USA, 1986 (Electrochem. Soc.,
       USA) Ext. Abstr. no.399]
[51]   E.Harari [Appl. Phys. Lett. (USA) vol.30 no.11 (1977) p.601-3]
[52]   P.Solomon [J. Appl. Phys. (USA) vol.48 no.9 (1977) p.3843-9]
[53]   M.S.Liang, C.Hu [Tech. Digest Int. Electron Device Meeting,
       Washington, DC, USA, 1981 (IEEE, New York, USA, 1981) p.396]
[54]   D.J.DiMaria, T.N.Theis, J.R.Kirtley, F.L.Pesavento, D.W.Wong [J.
       Appl. Phys. (USA) vol.57 no.4 (1985) p.1214-38]
[55]   M.V.Fischetti, D.J.DiMaria, S.D.Brorson, T.N.Theis, J.R.Kirtley,
       [Phys. Rev. B (USA) vol.31 no.12 (1985) p.8124-42]
[56]   S.D.Brorson, D.J.DiMaria, M.V.Fischetti, F.L.Pesavento,
       P.M.Solomon, D.W.Wong [J. Appl. Phys. (USA) vol.58 no.3 (1985)
       p.1302-13]
[57]   M.Shatzkes, M.Av-Ron [J. Appl. Phys. (USA) vol.47 no.7 (1976)
       p.3192-202]
[58]   N.Klein [Thin Solid Films (Switzerland) vol.50 (1978) p.223-32]
[59]   I.-C.Chen, S.E.Holland, C.Hu [IEEE Trans. Electron Devices (USA)
       ED-32 no.2 (1985) p.413-22]
[60]   I.-C.Chen, S.Holland, K.K.Young, C.Chang, C.Hu [Appl. Phys. Lett.
       (USA) vol.49 no.11 (1986) p.669-71]
[61]   C.N.Berglund, R.J.Powell [J. Appl. Phys. (USA) vol.42 (1971)
       p.573]
[62]   Z.A.Weinberg, T.N.Nguyen [J. Appl. Phys. (USA) vol.61 no.5 (1987)
       p.1947-56]
[63]   T.Ito, T.Nozaki, H.Ishikawa [J. Electrochem. Soc. (USA) vol.127
       no.9 (1980) p.2053-7]
[64]   P.Pan, J.Abernathey, C.Schaefer [J. Electron. Mater. (USA) vol.14
       no.5 (1985) p.617-32]
[65]   T.Watanable, A.Menjoh, M.Ishikawa, J.Kumagai [Tech. Digest Int.
       Electron Device Meeting, San Francisco, CA, USA, 1984 (IEEE, New
       York, USA) p.173]

17.23 STRUCTURE OF THE INTERFACE IN OXIDE-ON-Si STRUCTURES

by S.M.Goodnick

October 1987
EMIS Datareview RN=15725

This Datareview summarises the current knowledge of the interface
between Si and SiO2. The focus is on the interface formed during
thermal oxidation, and thus deposited and anodic oxide interfaces are
not considered.

The SiO2/Si interface is one of the most widely studied semiconductor
interfaces due to its tremendous technological importance in
semiconductor devices. Perhaps the most characteristic feature of this
interface is the abruptness with which the transition from crystalline
Si to stoichiometric SiO2 occurs. Experimental evidence in the late
1970's [1] suggested that the interface width is only one or two
monolayers, and work since that time has substantiated this fact. An
extensive review of the subject has been given by Grunthaner et al [2].

A variety of different techniques have been employed in the study of
the SiO2/Si interface, none of which provides a complete picture of the
chemical and morphological nature of this region. Care must be taken in
the interpretation of experimental data as many techniques such as ion
milling, chemical etching, etc. introduce artifacts which change the
structure of the interface being observed.

Most studies have concentrated on oxides grown on either the (100) or
(111) Si surface. For the diamond Si lattice, the former surface is
terminated with two dangling bonds per surface atom while the (111)
surface contains only one which influences the stoichiometry of the
transition region. The structure of bulk SiO2 appears to be well
described by a continuous-random network (CNR) model [3] in which
tetrahedrally coordinated Si is linked to other Si atoms via a bridging
oxygen atom, the bond angle of which is distributed about an average
angle of 144 deg. The Si tetrahedra then form a ring structure in
bulk SiO2 which is predominantly 6 member.

Due to differences in the thermal expansion coefficients between Si and
SiO2, the interface is strained. On the Si side of the interface,
Haight et al [4] used Rutherford Backscattering (RBS) on (111) surfaces
to infer that the first 1-2 monolayers of Si atoms were distorted,
although less so than on relaxed unoxidized Si surfaces. Other than
this distortion, evidence suggests that the crystalline Si surface
terminates abruptly over a distance of one or two atomic layers. This
has been ascertained from TEM [5,6], photoemission spectroscopy [7-10],
ellipsometry [11], X-ray diffraction [12] and pulsed laser atom probe
[13,14]. Pantelides and Long [15] first demonstrated using a ball and
stick model that a continuous-random network model for the Si/SiO2
interface could be constructed in which the interface was abrupt and
free of dangling bonds. Recently, more sophisticated models have been
employed in which the oxide atoms are allowed to relax to their
equilibrium positions according to a Keating type potential for the
elastic energy [16-18]. Such models show that the nonstoichiometric

region should actually extend over 1-3 monolayers. To effect the change from bulk SiO2 to crystalline Si, the ring size of the SiO2 is thought to decrease from primarily six-fold to four-fold near the interface with a corresponding decrease in the average bridge bond angle [7].

Some disagreement still exists on the exact stoichiometry of SiOx within the transition region. The last crystalline layer of Si must be bonded differently than either bulk Si or bulk SiO2 giving rise to suboxide species. Raider et al [19] found a 10 A thick nonstoichiometric region between Si and SiO2 employing XPS on ultra-thin and chemically thinned oxides. More recently, numerical deconvolution and curve fitting of the Si 2p core level XPS spectra has been used to infer the distribution of suboxide states at the interface [2,7,8,20]. These studies indicate the expected orientation dependence of the suboxide species, that is Si+ on Si(111) surfaces and Si++ near Si(100) interfaces corresponding to the number of Si dangling bonds on each surface respectively. However, in all cases, excess Si+++ was detected which implies Si dangling bonds, Si-Si bonds or hydrogen bonded Si in the oxide. Grunthaner et al [2,7] find that this excess Si+++ is located within 30 A of the interface, while Hattori et al [8,20] show that this bonding state extends into the bulk SiO2. Differences may arise from the numerical procedures used as the suboxide peaks in the Si 2p line are not resolved in the raw data. High resolution UPS data using tunable synchrotron radiation clearly shows the suboxide shifts in binding energy of the Si 2p peak for ultrathin oxides grown in UHV [9,10]. However, Hollinger et al [9] find no orientation dependence in the suboxide concentrations, measuring almost equal ratios of Si+ and Si++ uniformly distributed in the thin oxide on both (111) and (100) surfaces. Discrepancies could be due to the nature of the ultra-thin vacuum grown oxides used. For (100) surfaces, Braun et al [10] have found that the concentration of lower oxidation state species is higher at the interface in contrast to Hollinger et al [9].

The issue of interface stoichiometry is further complicated by the fact that the interface is not atomically flat, but instead contains atomic steps or 'roughness'. A step or terrace in the interface results in a suboxide species different from the ideally terminated surface and may be associated with interface states in the band gap. Hahn et al [21] used model fits to the broadened LEED pattern from Si(111) surfaces (in which the oxide was chemically removed) to show that the interface roughness occurs over one monolayer. Changes in the measured roughness from LEED with surface preparation (before oxidation) were correlated with the measured Hall mobility and surface state densities. High resolution TEM (HRTEM) of Si interfaces also appear to be quite flat with rms deviations of the interface of the order of 1-2 A [5,6, 22]. Goodnick et al [22] have estimated the statistical fluctuation of the interface from HRTEM micrographs and compared to model fits to the Hall mobility limited by roughness scattering. There it was shown that the actual interface roughness is larger than suggested from HRTEM due to projection effects in the measurement, an effect also shown using ball and stick models for the interface [23]. The measurement of suboxide species in photoemission studies (compared to what is expected for an atomically flat surface) has been suggested as a measure of interface roughness also [24]. Recent HRTEM studies on atomically flat Si grown by molecular beam epitaxy have suggested the presence of an ordered SiO2 layer which was assigned to a tridymite phase of the bulk oxide [25].

REFERENCES
----

[1] S.T.Pantelides (Ed) [in 'The Physics of SiO2 and Its Interfaces'
     (Pergamon, New York, USA 1978)]
[2] F.J.Grunthaner, P.J.Grunthaner [Mater. Sci. Rep. (Netherlands)
     vol.1 (1986) p.65]
[3] R.J.Bell, P.Dean [Philos. Mag. (GB) vol.25 (1972) p.1381-98]
[4] R.Haight, L.C.Feldman [J. Appl. Phys. (USA) vol.53 no.7 (1982)
     p.4884-7]
[5] O.L.Krivanek, D.C.Tsui, T.T.Sheng, A.Kamgar [in 'The Physics of
     SiO2 and Its Interfaces' Ed. S.T.Pantelides (Pergamon, New York,
     1978) p.351-5]
[6] J.H.Mazur, J.Washburn [AIP Conf. Proc. (USA) no.122, Physics of
     VLSI (Xerox, Palo Alto, USA, 1984) p.52-5]
[7] F.J.Grunthaner, P.J.Grunthaner, R.P.Vasquez, B.F.Lewis,
     J.Maserjian, A.Madhukar [Phys. Rev. Lett. (USA) vol.43 no.22
     (1979) p.1683-6]
[8] T.Hattori, T.Suzuki [Appl. Phys. Lett. (USA) vol.43 no.5
     (1 Sep 1983) p.470-2]
[9] G.Hollinger, F.J.Himpsel [Appl. Phys. Lett. (USA) vol.44 no.1
     (1984) p.93-5]
[10] W.Braun, H.Kuhlenbeck [Surf. Sci. (Netherlands) vol.180 (1987)
     p.279]
[11] D.E.Aspnes, J.B.Theeten [J. Electrochem. Soc. (USA) vol.127 no.6
     (Jun 1980) p.1359-65]
[12] I.K.Robinson, W.K.Waskiewicz, R.T.Tung, J.Bohr [Phys. Rev. Lett.
     (USA) vol.57 no.21 (24 Nov 1986) p.2714-17]
[13] C.R.M.Grovenor, A.Cerezo, G.D.W.Smith [Mat. Res. Soc. Symp.
     Proc. (USA) vol.37 (1985) p.199-204]
[14] T.Adachi, M.Tomita, T.Kuroda, S.Nakamura [J. Phys. Colloq.
     (France) vol.47 no.C-7 (1986) p.315-19]
[15] S.T.Pantelides, M.Long [in 'The Physics of SiO2 and Its
     Interfaces' Ed. S.T.Pantelides (Pergamon, New York, USA 1978)
     p.339-43]
[16] K.Hubner, E.D.Klinkenberg, A.Stern [Phys. Status Solidi b
     (Germany) vol.135 no.2 (1986) p.475-85]
[17] V.Drchal, J.Malek [Philos. Mag. B (GB) vol.54 no.1 (Jul 1986) p.61
     -70]
[18] I.Ohdomari, H.Akatsu, Y.Yamakoshi, K.Kishimoto [J. Non-Cryst.
     Solids (Netherlands) vol.89 no.1 (Jan 1987) p.239-48]
[19] S.I.Raider, R.Flitsch [in 'The Physics of SiO2 and Its
     Interfaces' (Pergamon, New York, 1978) p.384-8]
[20] T.Suzuki, M.Muto, M.Hara, K.Yamabe, T.Hattori [Jpn. J. Appl.
     Phys. Pt 1 (Japan) vol.25 no.4 (Apr 1984) p.544-51]
[21] P.O.Hahn, M.Henzler [J. Vac. Sci. & Technol. A (USA) vol.2 no.2
     (1984) p.574-83]
[22] S.M.Goodnick, D.K.Ferry, C.W.Wilmsen, Z.Liliental, D.Fathy,
     O.L.Krivanek [Phys. Rev. B (USA) vol.32 no.12 (15 Oct 1985)
     p.8171-86]
[23] I.Ohdomari, T.Mihara, K.Kai [J. Appl. Phys. (USA) vol.60 no.11
     (1 Dec 1986) p.3900-4]
[24] P.J.Grunthaner, M.H.Hecht, F.J.Grunthaner, N.M.Johnson [J. Appl.
     Phys. (USA) vol.61 no.2 (1987) p.629-38]
[25] A.Ourmazd, D.W.Taylor, J.A.Rentschler, J.Bevk [Phys. Rev. Lett.
     (USA) vol.59, no.2 (15 Jan 1987) p.213-16]

17.24 STRESSES IN SiO2-ON-Si STRUCTURES

by T.P.Ma

June 1987
EMIS Datareview RN=15733

SYMBOLS

Due the limited character set available, the following symbols are used
in this Datareview:

a1 = thermal expansion coefficient of Si
a2 = thermal expansion coefficient of oxide
E1 = Young's modulus for Si
E2 = Young's modulus for oxide
p1 = Poisson's ratio for Si
p2 = Poisson's ration for oxide
R = radius of curvature
S1 = maximum stress in Si
S2 = average stress in oxide
t1 = thickness of Si substrate
t2 = thickness of oxide film

A. INTRODUCTION

The mechanical stresses developed in SiO2/Si structures during
semiconductor processing are known to affect the yield and quality of
the devices and circuits being produced. Problems such as cracking of
the deposited SiO2 films [1-3] and stress-induced point defects and
dislocations in Si [4,5] have been reported. Some high-temperature
processing parameters such as the oxidation rate of Si [6,33] and the
impurity diffusion behaviour in Si [7,8] are also known to be affected
by such stresses. In addition, some transistor characteristics have
been reported to be sensistive to the stress distribution near the
SiO2/Si interface [9], and the susceptibility of MOS devices to
ionising radiation and hot carrier damage [10-12] has been shown to be
a function of the build-in stresses.

B. MEASUREMENT TECHNIQUES

Most of the stress measurements reported in the literature are based on
the determination of (1) sample bending or warpage, (2) lattice strain
or (3) deflection of a suspended SiO2 membrane, bridge, or cantilever
beam. More specifically, in item (1) above the bending of the SiO2/Si
structure can be measured mechanically [13], by optical reflection
[10,14], or by optical interference [15]. The lattice strain is usually
measured by an X-ray diffraction technique [16,17] which detects the
changes in the lattice spacing. The deflection of a SiO2 membrane was
first measured by a balloon method reported in ref [13]. Subsequently,
measurements on suspended SiO2 bridges and beams based on optical
techniques [18,19] and scanning electron microscopy [20] were reported.

In general, the bending-plate methods are relatively easy to use
without complicated sample preparation procedures. The key parameter
obtained from such measurements is the curvature of the sample surface.
Knowing the curvature, the stresses can be calculated, based on
applicable stress theories. The most widely used equation that relates
the curvature to the average stress in SiO2 is based on the theory
developed by Stoney [21] with slight modifications by Jaccodine and
Schlegel [13]:

$$S2 = \frac{E1}{6(1-p1)} \times \frac{(t1)*2}{t2} \times \frac{1}{R} \qquad (1)$$

For t1 >> t2 the corresponding maximum stress in the Si crystal, S1, is
related to the oxide stress by

$$S1 = -4 \frac{S2 \times t2}{t1} \qquad (2)$$

The bending-plate methods, however, typically yield average values over
a relatively large area, and therefore do not offer good spatial
resolution. Since in an integrated circuit there are many edges and
other topographical variations, the local stress values of interest may
be quite different from the average. Techniques that allow measurements
of small areas typically involve creating micro-bridges or small
membranes of the samples by photolithography and etching, followed by
optical or electron microscopy to determine the magnitude of the
deflection [18-20].

C. RELEVANT ELASTIC PARAMETERS

The elastic parameters for SiO2 and Si used by the various investigators
vary over a considerable range, depending on the material and the
measurement method. TABLES 1 and 2 list the more commonly used values
for the Young's modulus, Poisson's ratio and thermal expansion
coefficient. Note that the thermal expansion is a function of
temperature, and the values cited in TABLES 1 and 2 represent those
selected by the authors for their particular experiments.

D. DATA FOR THERMAL SiO2/Si

The stress distribution in the thermal SiO2/Si has been found to depend
on many parameters, including the thicknesses of the SiO2 film and the
Si substrate, the oxide growth conditions, post-oxidation heat
treatment, the crystalline orientation of Si, and the temperature at
which the stress is measured.

Two major components contributing to the overall stress in the thermal
SiO2 film are the intrinsic stress and the thermal stress. The relative
magnitudes of these two components also depend on the parameters cited
above.

TABLE 3 shows the data reported in the literature.

It should be noted that the compressive stresses in the oxide reported for films thinner than 700 A are typically much higher than those for thicker films. Several papers reported the thickness dependence of the oxide stress in the thinner range [19,20,24,26,31], and the results indicate a systematic increase of the oxide stress as the oxide gets thinner. Since the gate oxide thickness is reducing as the integrated circuit technology advances, these findings may be of technological importance. It has been suggested that the stress distribution in thermal SiO2 is a nonlinear function of the distance from the SiO2/Si interface: the stress is at its highest value at the interface, which decreases rapidly with the distance away from the interface within 1000 A [26]. This model provides a reasonable explantation for the observed thickness dependence mentioned early. The reason for the stress gradient present in the thermally grown SiO2 was attributed to the relaxation effect [26] due to the viscous flow of the SiO2 at high temperatures [6,10,17].

The viscous flow model has been used to explain the fact that relatively small intrinsic stress exists in thick (> 1000 A) thermal SiO2 films grown at high temperatures (> 1000 deg C). Without some kind of stress relaxation process such as viscous flow, one would expect a very large intrinsic stress due to the substantial volume change resulting from the oxidation reaction.

A detailed study of the intrinsic stress in thermal SiO2 has been published [6], and TABLE 4 lists some of the results. These data were obtained by subtracting out the thermal stress component from the measured overall oxide stress. The thermal stress was calculated using the corresponding thermal expansion coefficients for SiO2 and Si listed in TABLES 1 and 2.

From the data in TABLE 4 it is apparent that the intrinsic compressive stress in thermal SiO2 increases with decreasing oxidation temperature. This is consistent with the viscous flow model which predicts that the stress relaxation time is a strong function of temperature, and for oxidation at low temperatures (< 1000 deg C), the relaxation times are longer than the normal oxidation times, thereby resulting in significant intrinsic stresses. This also explains the substantial reduction of the intrinsic stress after annealing at 1000 deg C in oxides grown at lower temperatures (700-800 deg C). Another fact worth noting is that the wet oxides grown at 700-800 deg C exhibit lower intrinsic stresses than dry oxides grown at the same temperatures. This is thought to be due to the decreased viscosity of oxides containing Si-OH groups [6,32]. The observed orientation dependence was attributed to the combined effects due to the orientation dependent E1/(1-p1) values and the role that the surface steps play in the oxidation process [6,33].

E. DATA FOR CVD SiO2/Si

The chemical vapour deposition of SiO2 can be done at atmospheric
pressure (APCVD), low pressures (LPCVD), or in a plasma environment
(PECVD). For a given CVD reactor, stress in the CVD SiO2 has been
found to depend on many parameters [1,25,34-37], including the
deposition temperature, deposition rate, post-deposition heat treatment,
film composition, and water content in the film. The incorporation of
phosphorus in SiO2 also has a significant effect on the film stress.

In contrast to the thermal SiO2, most of the CVD SiO2 films exhibit a
tensile stress at the deposition temperature, and films deposited by
the APCVD method tend to retain a tensile stress upon cooling down to
room temperature. Also significantly different from thermal SiO2 is
the large hysteresis in the stress-temperature curves.

TABLE 5 lists some representative results published in the literature
for CVD SiO2 formed by oxidation of SiH4, and TABLE 6 lists the
published data for CVD SiO2 containing phosphorus (PSG).

It should be noted that, because of the large hysteresis in the
stress-temperature curves, the stress data listed in TABLES 5 and 6
only apply to the specific heat cycles cited in the related references.
It should also be noted that the film stress measured at room
temperature depends on the storage ambient prior to the measurement,
due to the absorption of water molecules in the film [1,34-36].

F. TABLES

TABLE 1: elastic parameters of Si

Si orientation	E1 10*12 dyne/cm*2	p1	E1/(1-p1) 10*12 dyne/cm*2	a1 10*-6/deg C	Ref
(111)	-	0.34	-	-	[15]
(111)	1.87	0.279	-	4.2	[9]
(111)	1.69	0.262	2.29	-	[22]
(111)	1.88	0.181	-	-	[24]
(111)	-	-	2.3	2.52	[25]
(111)	1.72	-	-	4.5	[26]
(100)	1.3	0.279	1.805	-	[22]
(100)	1.3	0.279	-	4.2	[9]
(100)	-	-	1.81	2.52	[25]
(110)	-	-	2.187	-	[6]
(110)	1.69	0.279	-	4.2	[9]

TABLE 2: elastic parameters of SiO2

E2 10*12 dyne/cm*2	p2	E2/(1-p2) 10*12 dyne/cm*2	a2 10*-6/deg C	Ref
0.66	0.18	0.805	0.28	[13]
0.7	-	-	0.52	[6]
0.73	-	-	0.48	[9]
-	-	1.64	0.6	[25]
0.72	0.17	0.85	-	[27]
0.72	0.17	-	0.35	[26]

TABLE 3: overall stress in thermal SiO2/Si structure

Oxidation Temp. deg C	Oxide Thickness kA	Si Orientation	S1 10*7 dyne/cm*2 (tensile)	S2 10*9 dyne/cm*2 (compressive)	Ref
wet oxide:					
875	3-6	(100),(111)	-	2.5	[13]
900	2-6	(100),(111)	-	2.1	[13]
1000	2-6	(100),(111)	-	2.72	[13]
1050	8	(100),(111)	-	2.4-2.8	[13]
1100	8	(100),(111)	-	2.8-3.2	[13]
1200	3-20	(100),(111)	-	3.1-3.4	[13]
1000	2.5	(111)	1.2	-	[15]
1000	4.5-6.5	(111)	2-2.2	-	[15]
1000	12	(111)	4.4	-	[15]
1200	8.4	(111)	8.2-8.5	4.5-4.7	[28]
1200	8.4	(110)	8.5-8.6	3.9	[28]
1200	8.4	(100)	9	3.8-4	[28]
1000	0.5	(111)	1.5	-	[24]
1000	1	(111)	2.6	-	[24]
1000	1.5	(111)	4.2	-	[24]
1000	5	(111)	4.7	-	[24]
1000	10	(111)	3.7	-	[24]
1050	0.5	(111)	0.7	-	[24]
1050	1.5	(111)	2.7	-	[24]
1050	> 5	(111)	2.2	-	[24]
1100	0.5	(111)	0.6	-	[24]
1100	1.7	(111)	1.2	-	[24]
1100	> 5	(111)	0.5-1	-	[24]
850-900	0.1-2	(100),(111)	-	7	[30]
975-1000	0.1-2	(100),(111)	-	near 0	[30]

TABLE 3 (continued)

Oxidation Temp. deg C	Oxide Thickness kA	Si Orientation	S1 10*7 dyne/cm*2 (tensile)	S2 10*9 dyne/cm*2 (compressive)	Ref
dry oxide:					
1200	8.6-9.2	(110)	7.2-8.1	2.7-3.7	[28]
1150	1.5	(111)	-	3.02	[19]
1150	1	(111)	-	4.55	[19]
1100	0.27	(100)	-	17.94	[19]
1000	0.5-0.7	(100)	-	15-51	[20]
900	0.5	(100)	-	6.5	[38]
900	1.0	(100)	-	3.2	[38]
900	1.8	(100)	-	3.6	[38]
999	1.5	(100)	-	7.5	[38]
999	2.4-3.75	(100)	-	4.4-4.9	[38]
1095	1-7	(100)	-	3-3.6	[38]
1100	6.5-10	(100)	-	3.75-4.25	[38]

NB Data are for room temperature measurements, except for the data from ref [30] where the measurements were at the oxidation temperature

TABLE 4: intrinsic stress in thermal SiO2 [6]

(SiO2 thickness range: 1000-10000 A)

Si Orientation	Wet or dry Oxide	Oxidation Temp. (deg C)	Post Oxidation Annealing in N2	intrinsic stress 10*9 dyne/cm*2 (compressive)
(100)	Dry	700	None	3.7
(100)	Dry	800	None	3.5
(100)	Dry	900	None	2.5
(100)	Dry	1000	None	1.7
(100)	Dry	1100	None	1.2
(100)	Dry	700	1 hr, 1000 deg C	0.5
(100)	Dry	800	1 hr, 1000 deg C	1.7
(100)	Wet	700	None	1.6
(100)	Wet	800	None	3
(111)	Dry	700	None	2.8
(111)	Dry	800	None	2.5
(111)	Dry	900	None	2.0
(111)	Dry	1000	None	1.5
(111)	Dry	1100	None	1.2
(111)	Dry	700	1 hr, 1000 deg C	1.2
(111)	Dry	800	1 hr, 1000 deg C	1.3
(111)	Wet	700	None	1.6
(111)	Wet	800	None	2.1
(110)	Dry	700	None	3.8
(110)	Dry	800	None	4.2
(110)	Dry	900	None	3
(110)	Dry	1000	None	2
(110)	Dry	1100	None	1.5

TABLE 5: stress in CVD SiO2/Si

T(D) deg C	r(D) A/min	T(PDM) deg C	stress/(10*9 dyne/cm*2) S(RT1)	S(RT2)	S(DT)	ref
APCVD SiO2/Si:						
450	490	-	+1.7	-	+2.4	[1]
450	1900	-	+2.2	-	+3.7	[1]
340	950-1400	-	+2.7-3.4	-	-	[34]
400	1240-1760	-	+2.9-3.3	-	-	[34]
450	1070-1260	-	+2.6-2.8	-	-	[34]
450	5000	-	+3.4	-	-	[34]
415	600	450	+0.4	+2.4	+3.2	[35]
415	600	900	+0.8	-1	+2.0	[35]
415	60	450	-2.7	+0.1	+0.8	[35]
480	-	500	+0.1	+0.8	+2.8	[25]
500	500	1000	+1.3	-0.8	+4.5	[36]
LPCVD SiO2/Si:						
430	50	1000	-3.5	-2.8	+0.5	[36]
400	-	450	-2.7	-2.1	-1.6	[37]
PECVD SiO2/Si:						
200	-	450	-0.5	+1.3	-0.2	[37]

T(D) = deposition temperature
r(D) = deposition rate
T(PDM) = post-deposition maximum temperature
S(RT1) = stress at room temperature before heat treatment
S(RT2) = stress at room temperature after heat treatment
S(DT) = stress at deposition temperature
+ sign indicates tensile stress
- sign indicates compressive stress

TABLE 6: stress in CVD PSG/Si

% of P205	T(D) deg C	r(D) A/min	T(PDM) deg C	stress/(10*9 dyne/cm*2)			ref
				S(RT1)	S(RT2)	S(DT)	
APCVD PSG/Si:							
4	450	–	–	+2	–	+3	[1]
0	450	10,000	–	+3.1	–	–	[34]
3	450	10,000	–	+2.7	–	–	[34]
5	450	10,000	–	+2.2	–	–	[34]
6.8	450	10,000	–	+1.4	–	–	[34]
8.5	450	10,000	1000	near 0	-1.3	–	[34]
0	415	600	450	+0.4	+2.4	+3.2	[35]
5	415	600	450	+0.2	+1.3	+2	[35]
7	415	600	450	+0.7	+1.3	+1.7	[35]
2	380	500	1000	-0.2	-1.3	+2.3	[36]
4	380	500	1000	-0.4	-1.3	+1.7	[36]
2	500	500	1000	-1.1	-1.2	+2.4	[36]
6	500	500	1000	-0.9	–	-0.3	[36]
LPCVD PGS/Si:							
5	415	600	900	near 0	-1.4	+0.65	[35]
2.6	430	50	100	-1.8	-2	+2	[36]
4.9	430	50	1000	-0.8	-1	+1.8	[36]
0	400	–	450	-2.85	-2.14	+1.59	[37]
6	400	–	450	+0.03	+1.02	+1.38	[37]
8	400	–	–	+0.84	–	+1.54	[37]
8	449	–	–	+0.71	–	+1.29	[37]

[symbols are explained in the notes at the end of TABLE 5]

REFERENCES
----

[1]   H.Sunami, Y.Itoh, K.Sato [J. Appl. Phys. (USA) vol.41 (1970)
        p.5115]
[2]   A.K.Sinha, H.J.Levinstein, T.E.Smith [J. Appl. Phys (USA) vol.49
        no.4 (1978) p.2423-6]
[3]   H.Miyoshi, N.Tsubouchi, A.Nishimoto [J. Electrochem. Soc. (USA)
        vol.125 (1978) p.1824]
[4]   K.Bulthuis [Philips Res. Rep. (Netherlands) vol.20 (1965) p.415]
[5]   A.Bohg, A.K.Gaind [Appl. Phys Lett. (USA) vol.33 (1978) p.895]
[6]   E.Kobeda, E.A.Irene [J. Vac. Sci. & Technol. B (USA) vol.5 no.1
        (1987) p.15-19]
[7]   S.M.Hu [in 'Atomic Diffusion in Semiconductors' Ed. D.Shaw
        (Plenum Press, London, 1973) p.217]
[8]   V.I.Sokolov, A.S.Tregubova, N.A.Fedorovich, V.A.Shelenskevich,
        I.L.Shulpina [Sov. Phys. Solid State (USA) vol.21 no.5 (1979)
        p.814-17]
[9]   T.Sugano, K.Katemoto [Proc. 2nd Microelectronic Meeting of INEA,
        Munich, Germany, Oct 1966]
[10]   E.P.EerNisse, G.P.Derbenwick [IEEE Trans. Nucl. Sci. (USA)
        vol.NS-23 (1976) p.1534-9]
[11]   V.Zekeriya, T.P.Ma [IEEE Trans. Nucl. Sci. (USA) vol.NS-31 (1984)
        p.1261-6]
[12]   T.B.Hook, T.P.Ma [Appl. Phys. Lett. (USA) vol.48 no.18 (1986)
        p.1208-10]
[13]   R.J.Jaccodine, W.A.Schlegel [J. Appl. Phys. (USA) vol.37 (1966)
        p.2429-34]
[14]   E.Kobeda, E.A.Irene [J. Vac. Sci. & Technol. (USA) vol.4 no.3
        (1986) p.720-2]
[15]   E.P.Jacobs, G.Dorda [Surf. Sci. (Netherlands) vol.73 (1978)
        p.357-64]
[16]   E.S.Meieran, I.A.Blech [J. Appl. Phys. (USA) vol.36 (1965)
        p.3162-7]
[17]   N.Kato, J.R.Patel [J. Appl. Phys. (USA) vol.44 (1973) p.965-77]
[18]   C.H.Lane [IEEE Trans. Electron. Dev. (USA) vol.ED-15 (1968)
        p.998-1003]
[19]   S.C.H.Lin, I.Pugaez-Muraszkiewicz [J. Appl. Phys. (USA) vol.43
        (1972) p.119-25]
[20]   H.W.Conru [J. Appl. Phys. (USA) vol.47 (1976) p.2079-81]
[21]   G.Stoney [Proc. R. Soc. London Ser.A (GB) vol.A82 (1909) p.172]
[22]   W.A.Brantley [J. Appl. Phys. (USA) vol.44 (1973) p.534-5]
[23]   D.F.Gibbons [Phys. Rev. (USA) vol.112 (1951) p.136-40]
[24]   H.Iechi, S.Satoh [Jpn. J. Appl. Phys. (Japan) vol.23 (1984)
        p.743-5]
[25]   A.K.Sinha, H.J.Levinstein, T.E.Smith [J. Appl. Phys. (USA) vol.49
        no.4 (1978) p.2423-6]
[26]   V.I.Sokolov, N.A.Fedorovich [Phys. Status Solidi a (Germany)
        vol.99 (1987) p.151-8]
[27]   R.J.Charles [J. Appl. Phys. (USA) vol.31 (1960) p.741]
[28]   M.V.Whelan, A.H.Goemans, L.M.C.Goossens [Appl. Phys. Lett. (USA)
        vol.10 (1967) p.262-4]
[29]   P.G.Borden [Appl. Phys. Lett. (USA) vol.36 no.10 (1980)
        p.829-31]
[30]   E.P.EerNisse [Appl. Phys. Lett. (USA) vol.35 no.1 (1979) p.8-10]
[31]   G.E.Davis, M.E.Taylor [J. Vac. Sci. & Technol. (USA) vol.19 no.4
        (1981) p.1024-9]
[32]   R.Bruckner [J. Non-Cryst. Solids (Netherlands) vol.5 (1970)
        p.123-75]

[33] B.Leroy [Philos. Mag. B (GB) vol.55 (1987) p.159-99]

[34] W.Kern, G.L.Schnable, A.W.Fisher [RCA Rev. (USA) vol.37 (1976)
      p.3-53]

[35] A.Shintari, S.Sugaki, H.Nakashima [J. Appl. Phys. (USA) vol.51
      no.8 (1980) p.4197-205]

[36] M.Shimbo, T.Matsuo [J. Electrochem. Soc. (USA) vol.130 (1983)
      p.135-8]

[37] G.Smolinsky, T.P.H.F.Wendlig [J. Electrochem. Soc. (USA) vol.132
      (1985) p.950-4]

[38] K.Kobeda, E.A.Irene [J. Vac. Sci. & Technol. B (USA) vol.4 no.3
      (May-June 1986) p.720-2]

17.25 CHARACTERISTICS OF SILICON OXYNITRIDE FILMS GROWN BY NITRIDATION OF
SiO2 FILMS

by P. Pan

October 1987
EMIS Datareview RN=16184

Silicon oxynitride films SiN(x)O(y), which were grown by annealing
silicon oxide films in NH3 gas at a temperature > 1000 deg C was first
reported by Ito et al [1]. These nitrided oxide films have attracted
a lot of attention in integrated circuits (IC) industry, because these
films showed a better ability to reduce impurity diffusion, a better
breakdown behaviour, a higher dielectric constant, and a lower
interface trap generation rate. In this paper, characteristics of
nitrided oxide films, such as the incorporation of nitrogen and
hydrogen, the flatband voltage shift, the bulk and interface traps,
and the dielectric breakdown behaviour, will be discussed.

The nitrogen concentration in nitrided oxide films affects both physical
and electrical properties of these films. The amount of nitrogen
incorporated into nitrided oxide films increased with increasing the
nitridation temperature, the nitridation time, and the NH3 partial
pressure, and it decreased with increasing the initial SiO2 film
thickness and the degree of oxidant contamination in the nitridation
ambient [1-8]. During the first stage of nitridation, nitrogen was
observed on the top surface and its concentration increased with
increasing the nitridation temperature. The first stage of nitridation
occurred during the first 10-20 seconds as nitridation temperature >
1100 deg C. During the second stage of nitridation, a pile-up of
nitrogen at the SiO2/Si interface was found and the amount of
nitrogen inside the SiO2 films was very low. Nitridation models
proposed in the literature [2,9,10] explain these observations
qualitatively well. When the nitridation process proceeds, the nitrogen
nitrogen concentration inside the SiO2 films begins to increase and
the increase of nitrogen on the top surface and at the interface is
found to be slower than that inside the oxide film. This inhomogeneous
incorporation of nitrogen is always observed, regardless of which
nitridation technique is used [2-10]. Because of the inhomogeneous
distribution in nitrogen concentration, it is difficult to use the
refractive index to indicate the total amount of nitrogen in the
nitrided oxide films; it is also difficult to measure the film
thickness by using an ellipsometer.

In order to get a homogeneous nitrogen concentration throughout the
whole nitrided film, a high temperature and long-time annealing are
required. By nitriding SiO2 films in an NH3 plasma ambient or in an
NH3/CHF3 plasma ambient, an enhancement of nitrogen incorporation into
nitrided oxide films was observed [5,7]; this is probably due to the
increase of neutral and charged radicals (NH(x) or H) in the plasma
ambient. The incorporated nitrogen was found to increase with
increasing the nitridation pressure [10], ranging from 0.001 to 5 atm.

However, the amount of nitrogen in the nitrided films was found to
be nearly the same for films nitrided at a pressure ranging from 1 atm
to 25 atm [11].

The presence of nitrogen at the top surface fo SiO2 films will help to
reduce the diffusion of impurity (Na, B, ... etc.) through the thin
SiO2 films [3,7]. The higher the nitrogen concentration, the better the
ability to act as a diffusion barrier. However, the incorporation of
the nitride inside the SiO2 will increase the traps. For the purpose of
using nitrided oxide films as a gate dielectric, a structure with high
nitrogen concentration at the top surface and very low nitrogen in
SiO2 is desired. Rapid thermal nitridation at temperature > 1100 deg C
for 10-20 sec was reported to produce high nitrogen at the top surface
and very low nitrogen at the interface and inside the SiO2. For the
application of using nitrided oxide films as a storage capacitor, an
interpoly dielectric or an oxidation barrier, a nitrided film with a
uniform and high nitrogen concentration is desired. A low temperature
(< 900 deg C) or short time nitridation technique to grow a
homogeneous $SiN(x)O(y)$ film will be attractive for these applications.

Studies on the amount of hydrogen and the bonding of hydrogen in
nitrided oxide films help to understand the nitridation mechanism. By
using the nuclear reaction analysis, the amount of hydrogen incorporated
into nitrided oxide films was found to be a function of the nitridation
conditions [8,12] and the initial SiO2 film thickness [12]. The hydrogen
concentration in the top layers of a nitrided film was higher than that
inside the film. No hydrogen pile-up at the $SiN(x)O(y)$/Si interface
was observed for films nitrided at a temperature > 800 deg C. By using
infrared (IR) analysis, Habreken et al [2] did not observe the presence
of N-H or O-H bonds and Lai et al [13] reported the presence of O-H
bonds. These were different from the results found by Pan [8] and by
Roggles et al [14], which showed that most of the hydrogen in $SiN(x)O(y)$
films was in the N-H bonds. The reason for having these different
observations may be due to different degrees of oxidant (O-H, O2 ..
etc) contamination in different nitridation ambients. When a
nitridation was done in ambient with high oxidant contamination,
especially O-H, a growth of SiO2 films will occur and the incorporated
nitrogen concentration will be low. Because of the increase in SiO2
thickness and the decrease in nitrogen concentration, an increase in
O-H bonds after nitridation will be observed in IR spectra. The use of
different analysis techniques may be another reason for these
different observations. By using Auger analysis and nuclear reaction
analysis, both nitrogen and hydrogen concentrations were found to
decrease after nitrided films were reoxided [15]. This can be
explained by the observation that hydrogen in $SiN(x)O(y)$ films is
mainly in the $NH(x)$ bonds; the loss of nitrogen concentration, after
being reoxidized, simultaneously induces the loss of hydrogen
concentration. The amount of hydrogen in the $SiN(x)O(y)$ films nitrided
at 25 atm was found to be about 1 order higher than that in films
nitrided at 1 atm [11].

A Generation of positive charges in nitrided oxide films has been
reported by many authors [8,16,19]. The amount of charge generated after
nitridation was a function of nitridation temperature, nitridation time,
and initial SiO2 film thickness. At a given time and SiO2 film thickness,
the amount of flatband voltage shift increased with increasing the
annealing temperature from 700 deg C to 950 deg C and then decreased
with increasing the annealing temperature from 1000 deg C to 1200 deg C.
At a given temperature and SiO2 film thickness, the amount of flatband
voltage shift was found to increase with increasing the annealing time.

After reaching a maximum shift, the amount of shift decreased with
increasing the annealing time [8,18,19]. The annealing time to reach the
maximum flatband shift decreased with increasing the annealing
temperature. In order to obtain a nearly zero flatband shift, a high
temperature (> 1000 deg C) and long-time annealing (> 90 min) is
required. The amount of flatband voltage shift was found to be nearly
independent of the process to grow initial SiO2 films [19] and also
independent of the nitridation pressure [11]. The post nitridation
annealing may effect the flatband voltage of the nitrided oxide films.
The positive charge in the nitrided films was found to be stable even
after being annealed at 1000 deg C in N2 for 60 min [8,17]. However, an
increase in the flatband voltage shift was observed after nitrided
films were reoxidized [15,19], especially in a wet O2 ambient. The
distribution of the generated positive charges was also found to depend
on the nitridation conditions and initial SiO2 film thickness [8,18].

Models proposed in references [16,17], which include the generation of
Si-dangling bonds by dissociating Si-O bonds with hydrogen, could
explain the flatband shift for films nitrided at different temperatures.
A model proposed by Pan [8], which includes not only the dissociation
of Si-O bonds by radicals, but also the reaction between NH(x) radicals
and Si dangling bonds, can be used to explain qualitatively the flatband
shift as a function of the annealing temperature, annealing time and
initial SiO2 film thickness. The generation of positive charge is
believed to be due to the formation of defects (Si-dangling bonds)
which was caused by the presence of radicals (H and/or NH(x)) to
dissociate Si-O bonds. The reduction of positive charge was observed
after the generated Si-dangling bonds were reduced by the formation of
Si-N bonds. Both the generation of Si-dangling bonds and the formation
of Si-N bonds occurred and competed with each other during the nitridation
process. For a low temperature (< 1000 deg C) and/or short-time
nitridation, the rate of defect generation was larger than that of Si-N
formation; this resulted in larger flatband shift. For a high temperature
and/or long-time nitridation, the rate of Si-N formation is faster than
that of defect generation; this resulted in a small flatband shift. Using
electron spin resonance to study properties of nitrided films during
nitridation will help to provide important information on the causes of
positive charge generation.

The density of interface traps at SiN(x)O(y)/Si interfaces was higher
than that at SiO2/Si interfaces [8,16,17,19,20]. The amount of increase
in midgap surface state density decreased with increasing the
nitridation temperature [16,17]. After high temperature (> 1000 deg C)
and long-time (> 60 min) annealing, the surface state density was
found to be nearly the same as that for initial SiO2/Si interfaces.
These suggested a generation of defects (Si- dangling bonds) at the
interface during nitridation, by breaking either Si-H or Si-O bonds. A
high temperature and prolonged annealing helps to heal these defects,
by forming either Si-N or Si-O bonds. For SiO2/Si interfaces with high
interface trap density initially, the nitridation process seems to help
to reduce the density of interface traps. In contrast to the
observation of higher surface state density, the generation rate of the
interface trap at SiN(x)O(y)/Si interfaces was much lower than that at
SiO2/Si interfaces after being stressed by high field [5,20,21] or by
high dose radiation [22]. A decrease in the number of Si-H bonds and an
increase in the number of Si-N bonds at the interface is expected after
high temperature nitridation. This and the fact that a Si-H bond is

weaker than a Si-N bond indicates that under the same stress conditions, the generation rate of Si- dangling bonds at SiO2/Si interfaces, by breaking Si-H bonds, will be higher than that at SiN(x)O(y)/Si interfaces, by breaking Si-N bonds. This may be the cause for having different interface trap generation rates before and after nitridation.

The electron trap density for nitrided oxide films was higher than that for initial SiO2 films [5,6,20,21,23-25]; a trap density of 2 X 10*12 [23], 2 X 10*18 [21] or 1 X 10*19/cm*2 [24] was reported. The capture cross section, ranged from 3 X 10*-19 to 1 X 10*-14 cm*2 [21,23-25], and a trap level at > 2 eV below SiN(x)O(y) conduction band [24,25] was reported. The reason for having these disagreements is probably due to the use of different nitridation conditions for preparing samples and/or due to different degrees of oxidant contamination in different nitridation ambients. A decrease in electron traps was observed after nitrided oxide films were reoxidized [21,26]. The origin of traps in the nitrided oxide films was not clearly understood. The O-H traps were reported by Lai et al [13], but were not observed by others [5,6,20,22,25]. It is believed that the origin of traps in the nitrided oxide films should be similar to that found in Si3N4 films [27]. The fact of increasing electron traps after nitridation should be considered carefully prior to using nitrided oxide films as a gate dielectric.

Ito et al [28] reported an improvement in breakdown behaviour of SiO2 films after being annealed in NH3 gas. The breakdown current and the breakdown field increased with increasing nitridation temperature and nitridation time. Lai et al [21] reported that nitrided oxide films had a better endurance during a low-field or low-current stressing and a poor endurance during a high-field or high-current stressing. They also observed that the breakdown behaviour of nitrided oxide films after being reoxidized was in between SiO2 and nitrided oxide films. Pan [8] reported that the breakdown distribution of nitrided oxide films depended on the quality of initial SiO2 films. When different quality SiO2 films were nitrided simultaneously, an improvement in breakdown distribution was observed for poor quality SiO2 films, and a degradation or no improvement in breakdown was observed for good quality SiO2 films. Pan [8] also reported that the breakdown behaviour varied during different periods of nitridation. The breakdown behaviour for films nitrided at 25 atm was found to be worse than that nitrided at 1 atm [11]. Moslehi and Saraswat [6] found that there was no significant improvement in breakdown charge density before and after nitridation.

As reported by Harari [29], local bond breaking due to the trapping of electrons was the breakdown mechanism of SiO2 films. Electron traps in nitrided oxide films may play dual roles. The localized large traps will degrade the breakdown of SiN(x)O(y) films and the uniform small traps will help to improve the breakdown behaviour. In other words, the trap density, the trap cross section, and the uniformity of the distribution of traps play roles in breakdown behaviour of SiN(x)O(y) films. The reduction of localized defects (pin hole, weak, spot, local trap site) after nitridation was the reason for having an improvement in breakdown as reported by Ito et al [28] and Pan [8]. Ito et al [28] also showed that the ability of nitrided oxide films to reduce mobile ion diffusion helped to prevent the breakdown behaviour caused by mobile ions. As reported by DiMaria and Abernathey [30], the electron temperature for the low pressure chemcial vapour deposited silicon oxynitride films was cooler than that for SiO2 films. The presence of uniform shallow traps in nitrided oxide films, which resulted in a reduction of

electron heating, may be the cause for having an improvement on
endurance during low-field and medium-field stressing. As reported by
Solomon [31], electron traps were generated in SiO2 films during
high-field stress. Pan and Schaefer [32] reported that the trap
generation in SiO2 at high-field was a localized defect-related
behaviour. The generation of localized traps and/or the localized
enhancement of the existing trap sites may be the cause for having a
degradation of endurance during high-field stressing.

In conclusion, the advantages of using nitrided oxide films are high
yield, low impurity diffusion, and radiation hardness; the disadvantages
are a slight decrease in channel mobility and a high trap density. In
order to use nitrided oxide films in future VLSI circuits, the
development of a low temperature (< 900 deg C) and/or a short-time
nitridation technique is most attractive. Many important film properties
have been reported in the literature. However, more systematic studies
on film compositions and electrical properties are necessary to clarify
some inconsistent results.

REFERENCES
----

[1]   T.Ito, T.Nozaki, H.Ishikawa [J. Electrochem. Soc. (USA) vol.127
        (1980) p.2053]
[2]   F.H.P.M.Habraken, A.E.T.Kuiper, Y.Tamminga [J. Appl. Phys. (USA)
        vol.53 (1982) p.6996]
[3]   S.S.Wong, C.G.Sodini, T.W.Ekstedt, H.R.Grinolds, K.H.Jackson,
        S.H.Hwan [J. Electrochem. Soc. (USA) vol.130 (1983) p.1139]
[4]   J.Nulman, J.P.Krusius, L.Rathbun [Tech. Digest IEEE IEDM
        (San Francisco, CA, 1984) p.169]
[5]   S.S.Wong, W.G.Oldham [IEEE Trans. Electron Devices (USA) vol.ED-32
        (1985) p.978]
[6]   M.M.Moslehi, K.C.Saraswat [IEEE Trans. Electron. Devices (USA)
        vol.ED-32 (1985) p.106]
[7]   T.Ito, H.Ishikawa [Proc. of Fifth Inter. Sym. on Silicon
        Materials Sci. and Tech., Eds H.R.Huff, T.Abe (Electrochem.
        Soc., Pennington, NJ, 1986) p.487]
[8]   P.Pan [J. Appl. Phys. (USA) vol.61 (1987) p.284]:
[9]   R.P.Vasquez, A.Madhukar [J. Appl. Phys. (USA) vol.60 (1986)
        p.234]
[10]   Y.Hayafuji, K.Kajiwara [J. Electrochem. Soc. (USA) vol.129 (1982)
        p.2102]
[11]   P.Pan, C.Paquette [Proc. of Fifth Intern. Sym. on Silicon
        Materials Sci. and Tech., Eds H.R.Huff, T.Abe (Electrochem.
        Soc., Pennington, NJ, 1986) p.508]
[12]   F.H.P.M.Habraken, E.J.Evers, A.E.T.Kuiper [Appl. Phys. Lett.
        (USA) vol.44 (1984) p.62]
[13]   S.K.Lai, D.W.Dong, A.Hartstein [J. Electrochem. Soc. (USA) vol.129
        (1982) p.2042]
[14]   G.A.Ruggles, R.Koba, R.E.Tressler [J. Electrochem. Soc. (USA)
        vol.133 (1986) p.2549]
[15]   P.Pan [Extend Abs., (The Electrochem. Soc. Fall Meeting, San
        Diego, CA, 1986) p.859]
[16]   T.Ito, T.Nakamura, H.Ishikawa [J. Electrochem. Soc. (USA) vol.129
        (1982) p.184]
[17]   C.T.Chen, F.C.Tseng, C.Y.Chang [J. Electrochem. Soc. (USA)
        vol.131 (1984) p.875]
[18]   P.Pan, C.Paquette [Appl. Phys. Lett. (USA) vol.47 (1985) p.473]

[19] G.A.Ruggles, J.R.Monkowski [J. Electrochem. Soc. (USA) vol.133 (1986) p.787]

[20] M.M.Moslehi, C.Y.Fu, K.C.Saraswat [Tech. Digest 1985 Sym. on VLSI Tech., (Kobe, Japan, 1985) p.14]

[21] S.K.Lai, J.Lee, V.H.Dham [Tech. Digest IEEE IEDM (Washington DC, 1983) p.190]

[22] F.L.Terry Jr, R.J.Aucoin, M.L.Naiman, S.D.Senturia [IEEE Electron Device Lett. (USA) vol.EDL-4 (1983) p.191]

[23] S.S.Wong, S.H.Hwan, H.R.Grinolds, W.G.Oldham [Proc. of Sym. on Silicon Nitride Thin Insulating Films, Eds V.J.Kapoor, H.J.Stein (Electrochem. Soc., Pennington, NJ, 1983) p.346]

[24] S.T.Chang, N.M.Johnson [Appl. Phys. Lett. (USA) vol.44 (1984) p.316]

[25] F.L.Terry,Jr, P.W.Wyatt, M.L.Naiman, B.P.Mathur, C.T.Kirk [42nd Dev. Research Conf., (Santa Barbara, CA, 1984) paper IIIB-5]

[26] R.Jayaraman, W.Yang, C.G.Sodini [Tech. Digest IEDM (Los Angeles, CA 1986) p.668]

[27] S.Fujita, A.Sasaki [J. Electrochem. Soc. (USA) vol.132 (1985) p.398]

[28] T.Ito, H.Arakawa, T.Nozaki, H.Ishikawa [J. Electrochem. Soc. (USA) vol.127 (1980) p.2248]

[29] E.Harari [J. Appl. Phys. (USA) vol.49 (1978) p.2478]

[30] D.J.DiMaria, J.R.Abernathey [J. Appl. Phys. (USA) vol.60 (1986) p.1727]

[31] P.Solomon [J. Appl. Phys. (USA) vol.48 (1977) p.3843]

[32] P.Pan, C.Schaefer [J. Electrochem. Soc. (USA) vol.133 (1986) p.1171]

17.26 CHARACTERISTICS OF THIN LPCVD SILICON OXYNITRIDE FILMS

by P.Pan

October 1987
EMIS Datareview RN=16183

Silicon oxynitride SiN(x)O(y) films can be deposited at a relatively
low temperature by using chemical vapour deposition (CVD) technique. The
deposition and properties of CVD SiN(x)O(y) have been studied for years
[1,2]. Recently, Pan et al [3,4] reported the properties of high quality
ultrathin low pressure CVD (LPCVD) SiN(x)O(y) films. Pan et al [5] also
reported that thin LPCVD oxynitride films have a high potential of being
used as a gate dielectric, an interpoly insulator, or a storage
capacitor for future very large scale integrated (VLSI) circuits. The
advantages of these films are flexible film compositions, good impurity
diffusion barrier, high dielectric constant, and excellent breakdown
behaviour. In this paper, the physical and electrical properties of thin
LPCVD SiN(x)O(y) films will be discussed.

LPCVD oxynitride films can be deposited at a temperature < 800 deg C in
a conventional low pressure chemical vapour deposition system. The most
common reactant gases are NH3, SiH2Cl2 and N2O [2-6]. By varying the
N2O/NH3 gases ratio, oxynitride films with different compositions,
ranging from SiO2 to Si3N4, can be easily deposited. The film composition
as a function of depth was found to be uniform even for a film of about
6 nm in thickness. No carbon contamination in the deposited films was
detected by Auger analysis (< 1 at%). The refractive index of the
deposited films decreased with increasing N2O/NH3 gas ratio, or with
increasing the oxygen concentration in these films. For LPCVD SiN(x)O(y)
films deposited at a high (N2O + NH3)/SiH2Cl2 flow ratio (> 3), no
Si clusters were detected by transmission electron microscopy (TEM)
even after these films were annealed at 1000 deg C for 60 min [4]. No
detectable change in refractive index was found before and after
annealing at 1000 deg C for 30 min. These observations suggested that
the density and the structure of the deposited SiN(x)O(y) films were
stable.

By using X-ray photoelectron spectroscopy (XPS), the binding energy of
the Si 2p peak for SiN(x)O(y) films was found to be in between the
binding energy of SiO2 and Si3N4 films [4]. The width of the Si 2p peak
was found to be nearly the same for SiO2, SiN(x)O(y) [4]. Infrared (IR)
spectra, ranging from 600 cm*-1 to 1100 cm*-1, showed the presence of a
single broad peak and the wavenumber of this peak increased with
increasing the oxygen concentration in these films [1,4,6]. This is
different from the observation of having two distinct peaks (Si-O
and Si-N) for dual dielectrics (Si3N4 (300 nm)/SiO2 (300 nm))
[4]. From XPS and IR results, it is clear that the structure of LPCVD
oxynitride films contained mainly the mixed matrix of Si, N and O,
rather than the mixture of clusters of SiO2 and Si3N4.

Hydrogen was detected in LPCVD oxynitride films. By using IR analysis,
the N-H bonds (3380 cm*-1) were found and no Si-OH, H-OH or Si-H bonds
were detected [1,4]. After H+ ion implantation, Si-H bonds at 2200 cm*-1
appeared. This behaviour was similar to that observed for LPCVD Si3N4
films [7,8] and was believed to be caused by the transfer of N-H bonds
to Si-H bonds or by the reaction of H+ ions with Si- dangling bonds

during ion implantation. By using nuclear reaction analysis [9], the amount of hydrogen was found to be of the order of 5 X 10*20/cm*3 and the distribution of hydrogen as a function of depth was uniform [4].

The film stress of SiN(x)O(y) films was found to be in between the film stress of SiO2 and Si3N4 [1,10]. The stress is tensile and is of the order of 5 X 10*9 dynes/cm*2. The stress of annealed films was slightly higher than that of as-deposited films [10]; this was believed to be due to the loss of hydrogen after annealing. The ability of oxynitride films to act as an oxidation barrier increased with increasing the nitrogen concentration in these films [1,10]. Similar to the observation for Si3N4 films [8], the oxidant cannot diffuse through the nitrogen rich SiN(x)O(y) layer. After the top SiN(x)O(y) layer was converted into an oxygen rich layer of a SiO2 film during oxidation, the oxidant then had a chance to diffuse through this layer and react with the nearest SiN(x)O(y) layer. The ability of SiN(x)O(y) films to act as an impurity (Na or B) diffusion barrier was also found to increase with increasing the nitrogen concentration in these films [1,5].

The etching characteristics of oxynitride films, either in a wet-etching solution or in a dry-etching ambient, were found to be in between the etching characteristics of SiO2 and Si3N4 films. The etching rate of oxynitride films in buffered HF increased sharply with decreasing refractive index [1,4,10]. On the other hand, the etching rate of SiN(x)O(y) films in hot H3PO4 increased with increasing refractive index [1,10]. The etching rate of oxynitride films in either a CF4/O2 plasma [5] or a SF6/He plasma [6] decreased with increasing the refractive index.

The flatband voltage of poly Si/SiN(x)O(y)/Si structures was found to be between -0.9 V and -1.2 V, which indicated the presence of positive charge in oxynitride films [1,3-6]. The amount of positive charge was reduced by N2 or O2 annealing [4]. Negative charges were observed [6] in oxygen rich films. After electrical stressing, a positive shift in capacitor-voltage (C-V) curve and a negative shift in current-voltage (I-V) curve were observed [4-6], which indicated the trapping of electrons. The trap density was found to increase with increasing the nitrogen concentration in SiN(x)O(y) films. By depositing oxynitride films directly on the Si substrate, a high surface state density D(IT) (8 X 10*10 to 5 X 10*11/cm*2/eV) was reported [1,3-6]. Structures having CVD films (SiO2, SiN(x)O(y), or Si3N4) on Si substrates generally showed high surface state density. This is believed to be due to the presence of a native oxide film between the CVD film and the Si substrate. The surface state density can be reduced by applying an in situ clean to take away the native oxide films right before the deposition or by annealing the deposited CVD films in an O2 ambient to have a thin SiO2 layer grown at the CVD film/Si interface. Because oxynitride films are a good oxidant diffusion barrier, an in situ cleaning process is a better way to reduce D(IT) at the SiN(x)O(y)/Si interface. The dielectric constant of oxynitride films was found to increase to 3.9 to 7 as the film composition was changed from SiO2 to to Si3N4 [1,5,6].

The dual dielectrics, which include a thin SiN(x)O(y) film (6 nm) on a thin SiO2 film (10 nm), were found to have high potential of being used as a gate dielectric [5]. The flatbound voltage for poly Si/SiN(x)O(y)/SiO2/Si structures was the same as that for poly Si/SiO2/Si structures. No detectable C-V shift was observed after the dual dielectrics were stressed at 6 MV/cm for 60 min. The observation of stable C-V for dual dielectrics was due to: (i) the SiN(x)O(y) film was thin and the number of electrons trapped in this film was small, and (ii) the trapped electrons were far away from the SiO2/Si interface. Low surface state density and high yield were also observed for poly Si/SiN(x)O(y)/SiO2/Si structures.

The conduction behaviour of LPCVD SiN(x)O(y) films depended on film composition [1,3-5]. For films with high oxygen concentration, the conduction mechanism is close to the Fowler-Nordheim mechanism [11]. For films with high nitrogen concentration, the conduction behaviour is dominated by the Poole-Frenkel mechanism [12]. The conduction behaviour for SiN)x)O(y) films with refractive index between 1.55 and 1.7 seems to be controlled by both mechanisms. Different amount of traps in films with different film compositions is the cause for different conduction behaviour. Under the same applied electric field, the film with higher oxygen concentration showed a lower leakage current.

Excellent breakdown behaviour for thin oxynitride films has been reported [3-5]. The average breakdown field is > 12 MV/cm and the deviation of the breakdown field is less than 2 MV/cm. Only a small percentage (< 5%) of tested devices showed low field breakdown. These results indicated that the film quality and thickness uniformity are excellent for thin LPCVD SiN(x)O(y) films. The breakdown behaviour was found to degrade slightly for films with refractive index < 1.6 or > 1.9. More than 90% of the tested samples showed only current conduction, rather than catastrophic breakdown, even measured at 12 MV/cm; this indicated a good endurance of these thin oxynitride films. The effect of using a thin SiN(x)O(y) film as an interpoly insulator or as a storage capacitor in deep trenches (> 9 micron in depth) was also found to be excellent [5]. By using the multi-ramping technique, more than 95% of the tested Al/SiN(x)O(y) (10 nm)/Si structures were found to be able to survive for 100 years at 3.5 V at 60 deg C [5].

Breakdown at low field (< 2 MV/cm) for oxynitride films is caused by the physical defects (pin hole, weak spot). Harari [13] reported that the trapping of electrons at a localized site induced a high field at that spot, and the breakdown of SiO2 occurred when this localized field was high enough to break Si-O bonds. Traps in oxynitride films play dual roles. The localized 'large' traps will degrade the breakdown of SiN(x)O(y) films and the uniform 'small' traps will help to improve the breakdown behaviour. In other words, the trap density, trap cross section and the uniformity of the distribution of traps play roles in causing breakdown of oxynitride films. As reported by DiMaria and Abernathey [14], the electron temperature in LPCVD SiN(x)O(y) films was cooler than that in SiO2 films. The presence of uniform shallow traps in oxynitride films which reduces the electron temperature in these films may be the cause for having a good endurance during electrical stressing. As reported by Solomon [15], the generation of traps in SiO2 films was observed during high-field stressing. Pan and Schaefer [16] reported that the trap generation at high-field in SiO2 is localized defect-related behaviour, rather than intrinsic behaviour of SiO2 films. Localized trap generation or localized enhancement of existing trap sites at high field, which induced bonding breaking as suggested by Harari [13], is believed to the cause of final breakdown of SiN(x)O(y) films.

In conclusion, high quality thin SiN(x)O(y) films can be deposited in a conventional LPCVD system at a relatively low temperature. These films can be easily integrated into the present IC process and fabrication. Good yield and reliability of these films have been demonstrated. Except using these films for passivation, thin SiN(x)O(y) films have the potential of being used as an interpoly insulator, a storage capacitor in planar or 3-dimensional structures. Dual dielectric (SiN(x)O(y)/SiO2) were found to be very promising for gate dielectrics.

REFERENCES
----

[1]   M.J.Rand, J.F.Roberts [J. Electrochem. Soc. (USA) vol.12 (1973) p.446]

[2]   A.E.T.Kuiper, S.W.Kooi, F.H.P.M.Habraken, Y.Tamminga [J. Vac. Sci. & Technol. (USA) vol.1 (1983) p.62]

[3]   P.Pan, J.R.Abernathey, C.Schaefer [Electronic Materials Conf. (Santa Barbara, CA, 1984) Paper E-6]

[4]   P.Pan, J.R.Abernathey, C.Schaefer [J. Electron. Mater. (USA) vol.14 (1985) p.617]

[5]   P.Pan, J.R.Abernathey, P.Geiss [Recent Newspaper 1852, The Electrochem. Soc. Fall Meeting, Honolulu, Hawaii, 1987)]

[6]   J.Remmerie, H.E.Maes [in 'Insulating Films on Semiconductors' Eds. J.J.Simonne, J.Buxo (North-Holland, 1986)]

[7]   H.J.Stein [J. Electrochem. Soc. (USA) vol.129 (1982) p.1786]

[8]   P.Pan, W.Berry [J. Electrochem. Soc. (USA) vol.132 (1985) p.3001]

[9]   W.A.Lanford, H.P.Trautvetter, J.F.Ziegler, J.Keller [Appl. Phys. Lett. (USA) vol.28 (1976) p.566]

[10]   A.K.Gaind, E.W.Hearn [J. Electrochem. Soc. (USA) vol.125 (1978) p.139]

[11]   M.Lenzinger, E.H.Snow [J. Appl. Phys. (USA) vol.40 (1969) p.278]

[12]   S.M.Sze [J. Appl. Phys. (USA) vol.38 (1967) p.295]

[13]   E.Harari [J. Appl. Phys. (USA) vol.49 (1978) p.2478]

[14]   D.J.DiMaria, J.R.Abernathey [J. Appl. Phys. (USA) vol.60 (1986) p.1727]

[15]   P.Solomon [J. Appl. Phys. (USA) vol.48 (1977) p.3843]

[16]   P.Pan, C.Schaefer [J. Electrochem. Soc. (USA) vol.133 (1986) p.1171]

CHAPTER 18

SILICON ON INSULATING SUBSTRATES

18.1 ELECTRONIC DEFECTS IN Si FILMS RECRYSTALLISED ON OXIDISED Si SUBSTRATES

by A.Chantre

July 1987
EMIS Datareview RN=15776

A. INTRODUCTION

Zone-melting recrystallisation of polycrystalline Si films on oxidised
Si substrates is one of the most promising techniques to produce
silicon-on-insulator (SOI) material for electronic device fabrication.
Recrystallised films usually consist of elongated grains containing
sub-boundaries. Grain sizes and sub-boundary spacing depend on the heat
source used to produce the molten zone, i.e. CW lasers [1], or strip
heaters [2] and lamps [3]. Typical values are indicated in TABLE 1
below. Techniques have been developed to avoid the random distribution
of the defects (grain or subgrain boundaries, (S)GB's) [4,5]. The
localisation of (S)GB's is controlled by a photolithographic step, and
leaves single crystal strips of material in which devices are to be
placed. Again, the maximum width of the defect-free Si strips is
determined by the crystallisation system:

TABLE 1

Heat source	CW Ar laser	strip heater/lamps
Grain size	5 microns X 10 microns	3 mm X 1 cm
SGB spacing	-	25 microns
Width of defect-free strips	15 microns	40 microns

B. ELECTRICAL CHARACTERISATION

The evaluation of residual electronic defects in crystallised Si thin
films has necessitated the implementation of specially designed test
structures. Inverted MOS capacitors [6,7], lateral p-n junction diodes
[8], and MOS capacitors [9,10] and Schottky barriers [11] with a self-
aligned gate and surrounding ohmic contacts have been fabricated to
characterise large area (S)GB-containing material. Depletion-mode
transistors have been used for the study of defects in small defect-
free regions of SOI [12,13].

For recent comprehensive descriptions of electrical characterisation
techniques for SOI, the reader should consult the papers by Johnson
[14] and Vu [15].

C. DEFECTS

Crystallised Si films on SiO2 present primarily three distinct sources of electronic defects (i) defects at grain and subgrain boundaries, (ii) defects at the SiO2/Si interface, and (iii) point defects within grains.

C1 (Sub) Grain Boundaries

(S)GB's are dominant imperfections in this SOI material. Three major areas in which their role has been clearly demonstrated are listed below.

C1.1 Minority carrier lifetime

TABLE 2 below summarises the influence of (S)GB's on minority carrier generation lifetimes, as deduced from measurements which have been performed on the different SOI materials discussed above.

TABLE 2

Material	GB's + SGB's	SGB's	Defect-free
minority carrier lifetime (microseconds)	< 0.1 [8]	0.1 - 1 [9-11,16]	10 - 100 [11,12,16]

These studies have revealed that (S)GB's are the lifetime-controlling imperfections in crystallised SOI films [16]. Remarkably, the values quoted for (S)GB-free material are comparable to typical lifetimes for processed bulk Si wafers.

C1.2 Mobility

Electron and hole mobilities reported for defect-free SOI are also typical of similarly doped crystalline bulk Si [17,18]. However, GB's oriented perpendicular to current flow cause significant (approx. 15%) bulk conductivity degradation [18,19]. In contrast, it has been found that SGB's have negligible influence on carrier mobility [10,19,20].

C1.3 Enhanced dopant diffusion

Enhanced dopant diffusion along (S)GB's may alter the performance of a thin film transistor, ultimately causing an electrical short circuit between drain and source contacts. The significance of the phenomenon has been well demonstrated for arsenic [8,14]. Specifically, a four-hour diffusion at 1100 deg C resulted in protrusions of arsenic measuring approximately 3-5 microns along GB's and only 1-2 microns along SGB's [21].

C2 SiO2/Si Interfaces

The interface between the recrystallised Si layer and the underlying
insulator is an integral part of this SOI material. The dependence of
its electrical characteristics on encapsulants for crystallisation and
underlying oxides has been reported [22,23]. Fixed charge densities
of 10*10/cm*2, and interface state densities of 10*10/eV cm*2 are
typical of strip-heater [24] and lamp [13] crystallisation. The defect
densities are an order of magnitude larger for CW laser crystallised
material [8,22]. Despite its good electrical quality, this buried
SiO2/Si interface remains the dominant source of trapping centres in
the films [14].

The interface between crystallised Si and an overgrown gate oxide has
also been characterised [10,16]. Its properties have been found
comparable to the state of the art in thermally oxidised bulk
single-crystal Si.

C3 Defects Within Grains

Little is known about defects and impurities within grains. In
particular, the source of the natural doping of the films has not
been much discussed. Halogen lamp-recrystallised layers usually
display n-type conductivity [10], which has been tentatively ascribed
to a shallow donor impurity (phosphorus or arsenic) [11].

Whereas oxygen, carbon and nitrogen have been identified by SIMS [9,25],
no electrically active impurities have been detected with DLTS. The data
put an upper limit of 10*14/cm*3, 10*13/cm*3 and 10*12/cm*3 to the
concentration of residual deep level defects in CW laser - [8], strip
heater - [24] and lamp - [16] crystallised SOI films, respectively. In
the latter case, MeV electron irradiations have been performed to
activate oxygen and carbon impurities. It has been concluded that
oxygen is segregated in the vicinity of (S)GB's and SiO2/Si interfaces
[11].

REFERENCES

[1]   K.F.Lee, J.F.Gibbons, K.C.Saraswat, T.I.Kamins [Appl. Phys.
       Lett. (USA) vol.35 (1979) p.173]
[2]   E.W.Maby, M.W.Geis, Y.L. Le Coz, D.J.Silversmith, R.W.Mountain,
       D.A.Antoniadis [IEEE Electron Device Lett. (USA) vol.ED-L2
       (1981) p.241]
[3]   D.P.Vu, M.Haond, D.Bensahel, M.Dupuy [J. Appl. Phys. (USA)
       vol.54 (1983) p.437]
[4]   J.P.Colinge, E.Demoulin, D.Bensahel, G.Auvert [Appl. Phys. Lett.
       (USA) vol.41 (1982) p.346]
[5]   M.Haond, D.Dutartre, D.Bensahel [Proc. Mater. Res. Soc. Europe
       Meeting, Strasbourg, France, 13-15 May 1985, Eds. V.T.Nguyen,
       A.G.Cullis (Editions de Physique, Les Ulis, France, 1985)
       p.417-26]
[6]   T.I.Kamins, K.F.Lee, J.F.Gibbons [IEEE Electron Device Lett.
       (USA) vol.EDL-1 (1980) p.5]
[7]   N.M.Johnson, M.D.Moyer, L.E.Fennell [Appl. Phys. Lett. (USA)
       vol.41 (1982) p.560-3]

[8]   N.M.Johnson, D.K.Biegelsen, M.D.Moyer [Appl. Phys. Lett. (USA)
        vol.38 (1981) p.900-2]
[9]   B.Y.Tsaur, J.C.C.Fan, M.W.Geis [Appl. Phys. Lett. (USA) vol.41
        (1982) p.83-5]
[10]   D.P.Vu, A.Chantre, H.Mingam, G.Vincent [J. Appl. Phys. (USA)
        vol.56 (1984) p.1682-6]
[11]   D.Ronzani, D.P.Vu, M.Haond, A.Chantre [ibid ref. 5, p.513-8]
[12]   D.P.Vu, J.C.Pfister [Appl. Phys. Lett. (USA) vol.47 no.9 (1985)
        p.950-2]
[13]   D.P.Vu, J.C.Pfister [Appl. Phys. Lett. (USA) vol.48 no.1 (1986)
        p.50-2]
[14]   N.M.Johnson [Proc. Mater. Res. Soc. Meeting, Boston, USA, 2-7
        Dec 1985, Eds A.Chiang, M.W.Geis, L.Pfeiffer (North Holland, New
        York, USA, 1986) p.337-47]
[15]   D.P.Vu [Proc. Mater. Res. Soc. Europe Meeting, Strasbourg,
        France, 17-20 June 1986, Eds G.G.Bentini, E.Fogarassi, A.Golanski
        (Editions de Physique, Les Ulis, France, 1986) p.369-78]
[16]   A.Chantre, D.Ronzani, D.P.Vu [ibid ref. 14, p.349-55]
[17]   D.P.Vu, A.Chantre, D.Ronzani, J.C.Pfister [ibid ref. 14,
        p.357-61]
[18]   J.P.Colinge, E.Demoulin, D.Bensahel, G.Auvert [Proc. Mater. Res.
        Soc. Europe Meeting, Strasbourg, France, 25-27 May 1983), J. Phys.
        Colloq. (France) vol.44 no.C5 (Oct 1983) p.409-13]
[19]   E.W.Maby, D.A.Antoniadis [Appl. Phys. Lett. (USA) vol.40 (1982)
        p.691]
[20]   B.Y.Tsaur, J.C.C.Fan, M.W.Geis, D.J.Silversmith, R.W.Mountain
        [IEEE Electron. Device Lett. (USA) vol.ED-L3 (1982) p.79]
[21]   E.W.Maby, H.A.Atwater, A.L.Keigler, N.M.Johnson [Appl. Phys.
        Lett. (USA) vol.43 (1983) p.482]
[22]   H.P.Le, H.W.Lam [IEEE Electron Device Lett. (USA) vol.ED-L3
        (1982) p.161]
[23]   J.C.Sturm, J.D.Plummer, J.F.Gibbons [Appl. Phys. Lett. (USA)
        vol.46 no.12 (1985) p.1171-3]
[24]   N.M.Johnson, M.D.Moyer, L.E.Fennell, E.W.Maby, H.Atwater [Proc.
        Mater. Res. Soc. Meeting, Boston, USA, 1-4 Nov 1982, Eds. J.
        Narayan, W.L.Brown, R.A.Lemons (North Holland, New York, USA,
        1983) p.491-7]
[25]   C.I.Drowley, T.I.Kamins [ibid ref. 24, p.511-16]

18.2 ELECTRON MOBILITY IN Si FILMS RECRYSTALLISED ON AMORPHOUS OXIDES

by G.F.Hopper

September 1987
EMIS Datareview RN=16177

A. INTRODUCTION

Silicon films recrystallised on amorphous oxides (usually silicon
dioxide) are of interest principally for the fabrication of silicon-
on-insulator (SOI) metal-oxide-semiconductor field effect transistors
(MOSFETs). Accordingly, almost all of the carrier mobility data reported
in the literature refer to field-effect mobility, derived from the
current-voltage characteristics of enhancement or depletion mode MOSFETs
formed in the recrystallised films. A collection of such electron
mobility data is presented in TABLE 1.

A convention which is widely, though not universally adopted is to
obtain mobility from the following MOS transistor equation:

$$dId/dVg \; (Vgm) = mu \; C(Vg) \; (w/L) \; Vd$$

where Id - source to drain current

      Vg - gate voltage

      mu - mobility

   C(Vg) - gate capacitance

      w - channel width

      L - channel length

      Vd - drain voltage

Vgm is that value of gate bias at which the transconductance
dId/dIg is a maximum.

The measurement is carried out at low drain bias, typically 0.1 V. In
cases where carrier conduction is confined to the surface, as in an
enhancement-mode device, Eqn (1) simplifies to :-

$$dId/dVg \; (Vgm) = mu \; Cox \; (w/L) \; Vd$$

where Cox is gate oxide capacitance.

B. THE RECRYSTALLISATION PROCESS

Before considering carrier mobility in recrystallised silicon films it
is appropriate to summarise the formation and properties of these films.

In general the silicon layer is initially polycrystalline, having been deposited over an oxidised silicon wafer. Recrystallisation is achieved by melting the layer, which is usually capped with a deposited dielectric film, and allowing it to resolidify. In order to do this a heat source which may be a laser beam [2,3], a graphite strip heater [4,5], focussed lamp [6,7], or electron beam [8,9], is swept across the surface of the sample, creating a molten zone in the film which moves along with the heat source. Reference [10] provides a general review of such liquid phase recrystallisation techniques. The moving heat source usually takes the form of a line, although in the laser case a spot is used.

Unless precautions are taken the recrystallised film is not single-crystal, but contains a number of grains of different orientation. Several techniques have been devised to suppress the growth of multiple grains. In one method (seeding) holes are etched in the oxide layer, allowing the polycrystalline silicon (polysilicon) to communicate with the single-crystal silicon wafer beneath; with this method it is possible to seed the recrystallisation and hence control the orientation of the film. An alternative strategy (grain boundary entrainment) is to define a pattern in the capping dielectric layer which spatially modulates the absorption of heat, providing a means of imposing temperature gradients and hence entraining or controlling the location of grain boundaries. Various other techniques have been adopted, resulting in a variety of recrystallised film types. For simplicity, the films are classified in this review according to four types, as follows.

(1) Multiply-grained layers of typical grain size 5-10 microns, where the grains have widely differing texture (crystallographic orientation perpendicular to the film surface) and in-plane orientation; unseeded laser and electron-beam produced material falls into this category.

(2) Multiply-grained material in which grains have a common texture (usually <100>) but varying in-plane orientation, e.g. unseeded lamp or strip-heater films.

(3) Single-crystal films with randomly-located low-angle grain boundaries, typical of certain lamp and strip-heater films.

(4) Single-crystal films containing low-angle boundaries only at pre-determined locations; such material may be produced by a laser or electron beam using a seeding method, or by a lamp or strip heater using the entrainment technique.

An indication is given in TABLE 1 of the category which most closely describes the material in question.

C. FACTORS AFFECTING CARRIER MOBILITY

Carrier mobility as defined above is an experimental quantity which is influenced by three factors (i) the proximity of the carriers to the silicon surface and the resultant mobility degradation, (ii) any mechanical stress present in the film, and (iii) crystalline defects within the film. Brief remarks will be made about each of these in turn.

C1 Mobility Degradation due to Proximity of the Surface

The value of carrier mobility obtained from MOS transistor measurements using Eqn (1) depends upon whether the carriers flow within the bulk of the silicon layer (depletion mode transistor), or at a surface inversion layer (enhancement mode). In the latter case the mobility is reduced as a result of carrier scattering at the surface [11]. Furthermore the mobility in the enhancement device is dependent upon the nature of the surface, and the magnitude of the surface electric field. Thus one would expect the measured carrier mobility to depend upon properties of the transistor (e.g. silicon doping level, gate oxide thickness, threshold voltage) as well as on more general properties of the silicon film. For this reason values of gate oxide thickness and threshold voltage are reproduced in TABLE 1, where this information is available. As a general rule, a transistor of positive threshold voltage is expected to be enhancement mode (surface conduction) while one of negative threshold is depletion mode (bulk conduction). This is an over-simplification, however, particularly for devices which have threshold voltages close to zero. The polarities given here refer to electron surface and bulk conduction: the reverse polarities apply to hole conduction.

C2 Mechanical Stress

The majority of cases reported involve polysilicon recrystallised over a thin layer (approx. 1 micron) of amorphous silicon dioxide grown on a bulk silicon wafer. Under these conditions the stress in the silicon layer is likely to be low [12] and to have a negligible effect on carrier mobility. Tsaur et al [12] have recrystallised silicon over oxide films which were supported by sapphire or quartz wafers instead of silicon. Relatively high values of stress were measured in the recrystallised silicon, which were found to affect carrier mobility values significantly.

C3 Crystalline Defects

Values of carrier mobility measured in MOS transistors fabricated in a multiply-grained silicon film might be expected to differ from the values measured in equivalent single crystal bulk silicon devices because of (i) the presence of different crystal orientations, and (ii) the effects of grain boundaries. Long-channel devices (with lengths in the range 10-100 microns) in certain films show reduced mobility in comparison to equivalent devices fabricated in bulk silicon wafers [13,14]; this mobility reduction is associated with boundaries perpendicular to the direction of current flow. Shorter-channel length transistors in the same material exhibit higher mobility; at channel lengths of 3-5 microns where individual transistors may frequently escape grain boundaries altogether, mobility is comparable to bulk silicon devices values.

Under certain circumstances - a transistor channel region containing a small number of low-angle grain boundaries perpendicular to current flow - it is possible to derive anomalously high values of carrier mobility from the usual measurement of device transconductance [15,16]. In such cases the effect of grain boundaries is to suppress the drain current at low gate voltage, allowing it to rise rapidly at higher voltage. Consequently the transistor has a higher transconductance, and hence derived mobility, than an equivalent bulk silicon device, although it actually passes less drain current.

TABLE 1: electron mobility

I   - Technique

II   - Nature of Material (refers to the classification described in the text)

III - Silicon thickness (micron) (Values are those of the recrystallised film at the device fabrication stage)

IV   - Gate Oxide thickness (nm)

V   - Threshold Voltage (V)

VI   - Mobility (cm*2/V.s)

VII - Comments

VIII- References

I	II	III	IV	V	VI	VII	VIII
Heat Source - Electron Beam (EB)							
seeded	3	0.4-0.5	38	0.2-0.4	300-600		[17]
unseeded	1				280-450		
unseeded	1	0.5	?	?	140+-10		[20]
seeded	3/4				550+-100		
seeded EB-line	4	1.0	56	0.1+-0.2	644+-45	409+-55 (note 1)	[22]
unseeded EB-line	1		70	0.2+-0.2	433+-108	163+-120 (note 1)	
unseeded EB-spot	1	?	68	1.0	300		[26]
unseeded EB-line	1	0.6	70	-3.5 +2.3	700 120	effect of grain boundary orientation described	[33]

- 682 -

I	II	III	IV	V	VI	VII	VIII
Heat Source - Laser							
entrainment	4	0.5	70	1.5+-0.2	400+-60		[18]
no entrainment	1				350+-80		
seeded	3	0.35	?	?	600		[19]
patterned Si film	4	0.2	25	0.4	650+-50		[2]
no seeding or entrainment	1	1.0 (thinned by oxidation to 0.35 for device fab.)	27	?	650 at short L	mu decreases as L increases	[13]
patterned Si film		0.5	50	-2.0	300		[25]
entrainment	4	0.5	?	?	650		[28]
entrainment	4	0.5	100	0.3	500	mu measured as a function of channel doping	[29]
no seeding or entrainment	1	0.5	50	0.95+-0.2	300-600	mu given as a function of laser power	[30]
heat sink entrainment	4	0.5	?	-1.0+-0.5	500		[32]
no seeding or entrainment	1	0.5	35-40	0.35-0.45 -0.5to-0.7	170 215		[34]
no seeding or entrainment	1	0.5	100	?	500-800 at short L	mu decreases as L increases	[14]
entrainment	4	0.4	40	-0.6	490		[36]

I	II	III	IV	V	VI	VII	VIII

Heat Source - Lamp

I	II	III	IV	V	VI	VII	VIII
entrainment	4	0.4	80	depletion mode	< 1000 (note 2)	mu vs Si depth obtained	[21]
no seeding or entrainment	3	0.15 (thinned by etching for device fab)	90	-4.0	700	mu vs Si depth given	[24]
no seeding or entrainment	3	0.5	45	1.4	480		[27]
entrainment	4	0.5	-	-	690	Hall mobility measured in film of 1.2 ohm cm resistivity	[37]

Heat Source - Graphite Strip heater

I	II	III	IV	V	VI	VII	VIII
unseeded	2/3	0.5	94	-2.2+-0.2	520+-40		[23]
seeded	4	0.5	100	-2.0+-0.3	600-700		[31]
no seeding or entrainment	2	?	?	?	580-600		[35]

note 1: these figures measured at the back channel
note 2: reaching a maximum in the centre of the film

REFERENCES
----

[1]   S.M.Sze [Physics of Semiconductor Devices 2nd Ed. (Wiley, 1981)]
[2]   A.J.Auberton Herve, J.P.Joly, P.Jeuch, J.Gautier, J.M.Hode [IEDM
       Tech. Digest, Sect.34.5 (IEEE, 1984) p.808]
[3]   C.I.Drowley, P.Zorabedian, T.I.Kamins [Mater. Res. Soc. Symp.
       Proc. (USA) vol.23 (1984) p.465]
[4]   M.W.Geis, H.I.Smith, D.L.Silversmith, R.W.Mountain, C.V.Thompson
       [J. Electrochem. Soc. (USA) vol.130 no.5 (1983) p.1178-83]
[5]   L.Pfeiffer, K.W.West, D.C.Joy, J.M.Gibson, A.E.Gelman [Mater.
       Res. Soc. Symp. Proc. (USA) vol.53 (1986) p.29]
[6]   D.P.Vu, M.Haond, D.Bensahel, M.Dupuy [J. Appl. Phys. (USA) vol.54
       no.1 (Jan 1983) p.437-9]
[7]   T.J.Stultz, J.F.Gibbons [Appl. Phys. Lett. (USA) vol.41 no.9
       (1 Nov 1982) p.824-6]

[8]   J.R.Davis, R.A.McMahon, H.Ahmed [J. Electrochem. Soc. (USA)
       vol.132 no.8 (1985) p.1919]
[9]   J.A.Knapp [J. Appl. Phys. (USA) vol.58 no.7 (1 Oct 1985)
       p.2584-92]
[10]   [Single-crystal silicon on non-single-crystal insulators J.
       Cryst. Growth (Netherlands) vol.63 no.3 (1983)]
[11]   S.C.Sun, J.D.Plummer [IEEE Trans. Electron Devices (USA)
       vol.ED-27 no.8 (Aug 1980) p.1497-1508]
[12]   B.-Y.Tsaur, J.C.C.Fan, M.W.Geis [Appl. Phys. Lett. (USA) vol.40
       vol.40 no.4 (15 Feb 1982) p.322-4]
[13]   K.K.Ng, G.K.Celler, E.I.Povilonis, R.C.Frye, H.J.Leamy, S.M.Sze
       [IEEE Electron. Device Lett. (USA) vol.EDL-2 no.12 (1981)
       p.316-18]
[14]   M.A.Bosch et al [IEE Proc. I (GB) vol.131 no.4 (1984) p.121-4]
[15]   J.G.Fossum, A.Ortiz-Conde [IEEE Trans. Electron Devices (USA)
       vol.ED-30 no.8 (1983) p.933-40]
[16]   J.P.Colinge [Microelectron. J. (GB) vol.14 no.6 (1983) p.58-65]
[17]   D.B.Rensch, J.Y.Chen [Microelectron. J. (GB) vol.14 no.6 (1983)
       p.66-73]
[18]   H.E.Lu, J.T.Boyd, H.E.Jackson, J.L.Janning [J. Appl. Phys. (USA)
       vol.60 no.12 (15 Dec 1986) p.4273-6]
[19]   M.Miyao, M.Ohkura, T.Tokuyama [in 'Silicon-on-Insulator: its
       technology and applications' Ed. S.Furukawa (KTK, Tokyo, 1985)
       p.269-81]
[20]   Y.Hayafuji et al [Mater. Res. Soc. Symp. Proc. (USA) vol.45
       (1985) p.311-16]
[21]   D.P.Vu, A.Chantre, D.Ronzani, J.C.Pfister [Mater. Res. Soc.
       Symp. Proc. (USA) vol.53 (1986) p.357-61]
[22]   G.F.Hopper, J.R.Davis, R.A.McMahon, H.Ahmed [Electron. Lett. (GB)
       vol.20 (1984) p.500]
[23]   E.W.Maby, M.W.Geis, Y.L.LeCoz, D.J.Silversmith, R.W.Mountain,
       D.A.Antoniadis [IEEE Electron Device Lett. (USA) vol.EDL-2 (1981)
       p.241]
[24]   D.P.Vu, A.Chantre, H.Mingam, G.Vincent [J. Appl. Phys. (USA)
       vol.56 no.6 (15 Sep 1984) p.1682-6]
[25]   H.W.Lam, A.F.Tasch, T.C.Holloway, K.F.Lee, J.F.Gibbons [IEEE
       Electron Device Lett. (USA) vol.EDL-1 no.6 (1980) p.99-100]
[26]   Y.Ohumura, K.Shibata, T.Inoue, T.Yoshi, T.Horike [IEEE Electron
       Device Lett. (USA) vol.EDL-4 no.3 (1983) p.57-9]
[27]   D.P.Vu, C.Leguet, M.Haond, D.Bensahel, J.P.Colinge [Electron.
       Lett. (GB) vol.20 no.7 (1984) p.298-9]
[28]   J.P.Colinge, E.Demoulin, D.Bensahel, G.Auvert [J. Phys. Colloq.
       (France) vol.44 no.C-5 (1983) p.409-13]
[29]   J.P.Colinge, E.Demoulin, D.Bensahel, G.Auvert, H.Morel [IEEE
       Electron Device Lett. (USA) vol.EDL-4 no.4 (1983) p.75-7]
[30]   H.W.Lam, A.F.Tasch, T.C.Holloway [IEEE Electron Device Lett.
       (USA) vol.EDL-1 no.10 (1980) p.206-8]
[31]   B.-Y.Tsaur, J.C.C.Fan, M.W.Geis, D.J.Silversmith, R.W.Mountain
       [Appl. Phys. Lett. (USA) vol.39 no.7 (1 Oct 1981) p.561-3]
[32]   S.Kawamura, N.Sasaki, M.Nakano, M.Takagi [J. Appl. Phys. (USA)
       vol.55 no.6 (15 Mar 1984) p.1607-9]
[33]   K.Shibata, Y.Ohmura, T.Inoue, K.Kato, Y.Horiike, M.Kashiwagi
       [Jpn. J. Appl. Phys. Suppl. (Japan) vol.22 no.22-1 (1983)
       p.213-16]

[34] A.F.Tasch, T.C.Holloway, K.F.Lee, J.F.Gibbons [Electron. Lett. (GB) vol.15 (1979) p.435-7]

[35] H.W.Lam, R.F.Pinizzotto, S.D.S.Mahli, B.L.Vaandrager [Appl. Phys. Lett. (USA) vol.41 no.11 (1 Dec 1982) p.1083-5]

[36] R.Mukai, N.Sasaki, M.Nakano [J. Electron Mater. (USA) vol.15 no.6 (Nov 1986) p.339-43]

[37] W.Scharff, C.Weissmantel [J. Vac. Sci. & Technol. A (USA) vol.4 no.6 (Nov/Dec 1986) p.3160-4]

18.3 HOLE MOBILITY IN Si FILMS RECRYSTALLISED ON AMORPHOUS OXIDES

by G.F.Hopper

October 1987
EMIS Datareview RN=16172

Values of hole mobility derived from the characteristics of metal-
semiconductor field effect transistors (MOSFETs) made in recrystallised
silicon films are presented in TABLE 1. For discussion see reference [1].

TABLE 1: hole mobility

I   - Technique

II   - Nature of Material (refers to the classification described in the
        text of ref. [1]).

III   - Silicon thickness (micron) (values are those of the recrystallised
        film at the device fabrication stage).

IV   - Gate Oxide thickness (nm)

V   - Threshold Voltage (V)

VI   - Mobility (cm*2/V.s)

VII   - Comments

VIII - References

I	II	III	IV	V	VI	VII	VIII
Heat Source - Electron Beam (EB)							
seeded	4	1.0	56	-1.5+-0.4	187+-43	81+-17 (note 1)	[4]
unseeded	1		70	-2.1+-0.4	190+-39	84+-34 (note 1)	
EB-line							
no seeding	1	0.55	75	-3.0	120		[8]
EB-spot							
Heat Source - Laser							
entrainment	4	0.5	70	?	200+-40		[2]
no	1				180+-50		
entrainment							
patterned	4	0.2	25	-0.4	280+-50		[5]
silicon							
film							
no seeding	1	0.55	75	-1.3 to -2.0	120+-40		[6]
or							
entrainment							

I	II	III	IV	V	VI	VII	VIII
entrainment	4	0.4	80	depletion mode	< 340 (note 2)	mobility vs silicon depth given	[3]
no seeding or entrainment	3	0.5	45	-3.0	180		[7]

Heat Source - Graphite Strip heater

| no seeding or entrainment | 2 | ? | ? | ? | 220-300 | | [9] |

note 1: figures measured at the back channel.
note 2: reaching maximum value at the back interface.

REFERENCES
----

[1] G.F.Hopper [EMIS Datareview RN=16171 (Oct 1987) 'Electron mobility in silicon films recrystallised on amorphous oxides']
[2] H.E.Lu, J.T.Boyd, H.E.Jackson [J. Appl. Phys. (USA) vol.60 no.12 (1986) p.4273-6]
[3] D.P.Vu, A.Chantre, D.Ronzani, J.C.Pfister [Mater. Res. Soc. Symp. Proc. (USA) vol.53 (1986) p.357-61]
[4] G.F.Hopper, J.R.Davis, R.A.McMahon, H.Ahmed [Electron. Lett. (GB) vol.20 (1984) p.500]
[5] A.J.Auberton Herve, J.P.Joly, P.Jench, J.Gautier, J.M.Hode [IEEE 1984 IEDM Tech. Digest, Sect 34.5 p.808]
[6] T.I.Kamins [IEEE Electron. Device Lett. vol.EDL-3 no.11 (1982) p.341-3]
[7] D.P.Vu, C.Leguet, M.Haond, D.Bensahel, J.P.Colinge [Electron Lett. (GB) vol.20 no.7 (1984) p.298-9]
[8] T.I.Kamins [IEEE Electron Device Lett. (USA) vol.EDL-2 no.12 (1981) p.313-5]
[9] H.W.Lam, R.F.Pinizzotto, S.D.S.Mahli, B.L.Vaandrager [Appl. Phys. Lett. (USA) vol.41 no.11 (1982) p.1083-5]

18.4 RECRYSTALLISATION OF POLY-Si TO FORM SOI SUBSTRATES BY ELECTRON BEAM HEATING

by H.Ahmed and R.A.McMahon

September 1987
EMIS Datareview RN=16166

Devices made in an electrically isolated silicon film on a substrate offer significant advantages over devices made in bulk silicon. A number of e-beam methods of forming silicon-on-insulator by using liquid phase recrystallization of deposited polysilicon films have been explored. Virtually all structures used for electron-beam recrystallization to form silicon-on-insulator substrates have been of the following kind. The isolating dielectric layer of $SiO2$ is formed on a bulk silicon substrate, usually by thermal oxidation. Next polycrystalline silicon is deposited by chemical vapour deposition. The deposited polysilicon is capped to prevent agglomeration; both oxide and nitride films have been tried but oxide is preferred. Typically the thickness of each of these layers is 1 micron.

Some work has been done, with e-beam heating, on unseeded regrowth of the polysilicon which results in a large grain structure, typically with a grain size of several tens of microns. However, most work has focused on seeded regrowth, in which the deposited polysilicon film makes contact with the bulk through windows in the isolating dielectric. Seeding gives lateral epitaxy of single crystal regions and the precise localization of defects through heat flow control. Recrystallization from the liquid phase of polysilicon layers with a spot beam, was reported by Shibata et al [1]. They used a 1 micron thick thermally grown $SiO2$ layer coated with a 0.35 micron polycrystalline silicon layer. Recrystallization was performed at 10 to 160 cm/s with a maximum beam current of 3 mA, energy between 5 and 30 keV and a spot size of 70-80 microns. Both single and repetitive scans were tried. After recrystallization, grain sizes up to 20 micron were reported. Capping layers between 0.1 and 1.0 micron thick were tried, with no significant increase in size. A similar approach was used by Kamins and von Herzen [2] to prepare substrates for MOSFET fabrication. They recrystallized 0.55 micron thick films on a 1.4 micron thick $SiO2$ isolating layer, capped with a 6 nm layer of $Si3N4$. As the silicon layer did not make contact with the bulk, again unseeded regrowth occurred, giving a large grain structure. P-channel MOSFET's were made using a modified LOCOS process and compared with standard devices. The transistors had well-behaved electrical properties but mobilities were about 70% of values in bulk Si. Ohmura et al [3] fabricated n-channel devices in similar large grain material, and again noted reduced mobilities compared with transistors in bulk Si. Further studies on substrate preparation have been reported using scan speeds of 35 to 180 cm/s to obtain <111> textured films with about 20 micron grain size [4].

Performance of devices manufactured in unseeded substrates with grain sizes in the tens of micron range was restricted as the obtainable mobilities were low, less than silicon-on-sapphire, and the risk of excessive diffusion along the unpredictably sited grain boundaries was considered unacceptable. Consequently, interest moved to seeded regrowth of polysilicon, with precisely oriented films and predictable positions of grain boundaries. Seeded recrystallization, using electron beams, was first reported by Davis et al [5]. Recrystallization of a structure comprising a 1 micron deposited polycrystalline Si layer capped with 1.0 micron of deposited oxide was performed with a dual electron beam system. The deposited Si layer made contact with the bulk Si wafer through periodically patterned slots cut in the isolating oxide, thus acting as seed windows. Processing of this material by shaping a round spot into a line beam about 2 mm long and 100 micron wide, resulted in lateral epitaxy for up to 200 micron from a seed edge. Scanning was at about 10 cm/s. A notable feature of this work was the high background temperature of around 1000 deg C, produced by a uniform electron beam, incident on the back of the substrate. Seeded recrystallization was also explored by Sedgwick et al [6] who obtained similar growth distances, using a round spot and a background temperature of 600 deg C.

For high throughput of wafers, and because they give more controlled crystal growth, line sources have been preferred; they also offer a more uniform heat source. Davis et al [7] adopted a line beam synthesised by rapid scanning, combining it with uniform background heating in a dual electron beam apparatus. Detailed characterization of the material prepared by this method has been described [8]. Seeded single crystal regions, 18 micron wide extending over several mm were obtained, with no defects other than a central subgrain boundary where melting fronts meet. The characteristics of CMOS devices made in this material show that mobilities are close to those of devices in bulk silicon. Other parametric measurements also gave results close to those obtained in standard CMOS processing [9].

The use of a synthesised line was also reported by Ishiwara et al [10]. In developments of this work, Horita et al [11] showed that greater distances of lateral epitaxy were possible with <100> rather than <110> oriented seed windows, and that scanning obliquely to the seed window direction also increased the defect free zone. Discontinuous seeding was also explored and found to eliminate voids and improve material quality [12]. Suguro et al [13] also used a synthesised or pseudoline approach, with improved structures using tapered seed windows and a cap with an overlayer of tungsten. Areas up to 300 micron by 1.3 mm were obtained. Problems of cap stability have also been studied [14].

In parallel with the synthesised line approach, the preparation of material by a line focus beam, with seeded regrowth, has been extensively studied by Knapp et al [15,16]. High scanning speeds, 100 to 600 cm/s which approach the limit of single crystal growth, with a 1 X 20 mm beam size and a background temperature of about 450 deg C, were used. With seeding, regions up to 50 micron by 350 micron and 1 micron thick were obtained but device results were not reported. Hayafuji [17] also used a line focus beam, and studied the effect of seed orientation on the film quality.

Seeded recrystallization using line electron beams can reliably produce SOI material with approximately 40-50 micron between seed windows and tens of mm length. Device processing trials are underway. A development from the work on a single level of SOI formed by e-beams has been towards stacked SOI layers to form 3-D circuits, following the work of Gibbons [18] using lasers. Sequential seeding of two silicon-on-insulator layers for three-dimensional integration has been demonstrated by Hamasaki et al [19], using the pseudoline e-beam system. Employing a seeded structure, with thickened polycrystalline silicon around the seed window, successful epitaxy was obtained for two 0.6 micron thick

layers. Sequential regrowth of two layers has also been reported by
Williams et al [20] who have also recrystallized two stacked SOI layers
simultaneously with a 0.5 micron lower silicon film and a 1.0 micron
upper silicon film. Studies on device structures formed in mezzanine
layers of SOI or fully stacked SOI layers formed by e-beams have not
yet been reported.

REFERENCES
----

[1]   K.Shibata, T.Inoue, T.Takigawa, S.Yoshii [Appl. Phys. Lett. (USA)
      vol.39 (1981) p.645]
[2]   T.I.Kamins, B.P.von Herzen [IEEE Electron Device Lett. (USA)
      vol.EDL-2 no.12 (1981) p.313-5]
[3]   Y.Ohmura, K.Shibata, T.Inoue, T.Yoshii, Y.Horiike [IEEE Electron
      Device Lett. (USA) vol.EDL-4 (1983) p.57]
[4]   H.Hada, S.Saitoh, H.Okabayashi [Extended Abstracts 17th Conf.
      Solid State Devices & Mater., Tokyo, Japan, 1985 p.139]
[5]   J.R.Davis, R.A.McMahon, H.Ahmed [Electron. Lett. (GB) vol.18
      (1982) p.163]
[6]   T.O.Sedgwick et al [J. Electrochem. Soc. (USA) vol.129 (1982)
      p.2802]
[7]   J.R.Davis, R.A.McMahon, H.Ahmed [Mater. Res. Soc. Symp. Proc.
      (USA) vol.13 (1983) p.563]
[8]   J.R.Davis, R.A.McMahon, H.Ahmed [J. Electrochem. Soc. (USA)
      vol.132 no.8 (Aug 1985) p.1919-24]
[9]   G.F.Hopper, J.R.Davis, R.A.McMahon, H.Ahmed [Electron Lett. (GB)
      vol.20 (1984) p.500]
[10] H.Ishiwara, M.Nakano, H.Yamamoto, S.Furukawa [Jpn. J. Appl.
      Phys. Suppl. (Japan) vol.22 no.22-1 (1983) p.607]
[11] S.Horita, H.Ishiwara [J. Appl. Phys. (USA) vol.61 no.3 (1 Feb
      1987) p.1006-14]
[12] S.Horita, H.Ishiwara [Appl. Phys. Lett. (USA) vol.50 (1987)
      p.748]
[13] K.Suguro et al [Appl. Phys. Lett. (USA) vol.47 no.7 (1 Oct 1985)
      p.696-9]
[14] R.Angelucci, G.Lulli, P.G.Merli [Mater. Lett. (Netherlands)
      vol.4 no.4 (Jun 1986) p.185-8]
[15] J.A.Knapp, S.T.Picraux [Mater. Res. Soc. Symp. Proc. (USA) vol.13
      (1983) p.557]
[16] J.A.Knapp [J. Appl. Phys. (USA) vol.58 no.7 (1 Oct 1985)
      p.2584-92]
[17] Y.Hayafuji et al [Mater. Res. Soc. Symp. Proc. (USA) vol.23
      (1984) p.491]
[18] J.F.Gibbons, K.F.Lee [IEEE Electron Device Lett. (USA) vol.EDL-1
      (1980) p.117]
[19] T.Hamasaki, T.Inoue, M.Yoshimi, T.Yoshii, H.Tango [J. Appl. Phys.
      (USA) vol.62 no.1 (1 Jul 1987) p.126-30]
[20] D.A.Williams et al [Mater. Res. Soc. European Symposium,
      Strasbourg, France, June 1987; to be published in J. Phys.
      Colloq. (France)]

18.5 ELECTRON MOBILITY IN Si FILMS ON SAPPHIRE

by G.A.Garcia and R.E.Reedy

May 1987
EMIS Datareview RN=15712

Early measurements by Dumin and Robinson [1] of carrier mobilities in
relatively thick SOS films showed electron Hall mobilities near bulk
silicon levels for films greater than 6 micrometres thick. As the films
were made thinner by successive oxidation and oxide removal the electron
mobility was found to decrease by about 50% at a thickness of 1.5
micrometre. Hall measurements of the type described in this early work
were hampered by the strong dependence of the properties of SOS on the
distance from the Si/sapphire interface and by depletion of the
carriers from the surface of lightly doped films by surface charge.
These effects were particularly troublesome in thin films (< 1
micrometre) as described by Ham [2] who used these mechanisms to
explain the observed decrease in electron Hall mobility with decreasing
carrier concentration [3]. The use of MOS gated Hall structures allowed
the Si surface potential to be controlled thereby permitting measurement
of average electron Hall mobility and mobility as a function of distance
from the Si/sapphire interface. Elliot and Anderson [4] used this
technique to observe a rapid reduction in electron mobility near the
Si/sapphire interface in approximately 1 micrometre thick SOS films.
Similarly, Ipri [5] found average electron Hall mobility in undepleted
1 micrometre films of 575 cm*2/V.s with a factor of 2 reduction at
0.3 micrometre from the SiO2/Si interface. In an effort to understand
the dependence of the electron mobility on distance from the Si/sapphire
interface Ipri and Zemel [6] used the MOS Hall bar method to measure
the temperature dependence of the mobility at various distances from the
interface. An increasing density of impurity scattering sites ((SiO4)+
complexes) with depth was invoked to explain the depth dependence of
electron mobility. Hsu and Scott [7] have used a deep depletion MOS
transistor to probe the absolute electron field-effect mobility as a
function of depth into the film. They obtained near bulk Si values in
the range of 500-700 cm*2/V.s at the surface of 0.6 micrometre SOS
down to very low values in the range of 10-20 cm*2/V.s near the
Si/sapphire interface in 0.2 micrometre films at room temperature. The
variation of the electron mobility with temperature led these authors
to conclude that an increasing concentration of crystal defects was
primarily responsible for the decline in electron mobility as the
Si/sapphire interface was approached. Cross-sectional transmission
electron microscopic studies of SOS by Abrahams and Buiocchi [8]
corroborated this conclusion by revealing the presence of several
defect types including dislocations, microtwins and stacking faults
whose density decreased with distance from the Si/sapphire interface.
The early work described above has been reviewed extensively by Ipri
[9] and by Cullen [10] who include discussions of the effects of Si
deposition rates and temperature, Si film thickness, carrier
concentration and mechanical stress and stress anisotropy on electron
mobility in SOS.

More recent studies of the effect of Si film stress on electron
mobility in SOS have generally corroborated the findings of earlier
workers cited in [9] and [10], who attribute a decrease in electron
mobility to the large lateral compressive stress present in SOS. Onga
et al [11] fabricated enhancement and depletion MOS devices oriented
along different crystal axes in 1 micrometre thick SOS and found

approximately 40% higher electron surface channel mobility in the [010] Si direction compared to the [100] direction at 4.2 deg K which they attribute to the effect of compressive stress on the electron effective mass. Tsaur et al [12] were able to correlate electron mobility with tensile and compressive lateral stress using zone melting recrystallized 0.5 micrometre Si films on various substrates. Tensile stress and increased electron mobility were observed in Si films on SiO2 on fused quartz while Si on SiO2 on sapphire displayed compressive stress and reduced electron mobility; the baseline was stress-free Si on SiO2 on Si. Using laser annealing of 0.5 micrometre SOS Kobayashi et al [13] improved the electron Hall mobility by about a factor of 2 but further heat treatment at temperatures greater than 600 deg C resulted in a reduction in mobility to as-grown levels. This behaviour was related to the presence of tensile stress, as monitored by Raman spectroscopy, following laser melting of the surface of the Si layer. Subsequent furnace anneals at 600 deg C resulted in a return to as-grown compressive stress levels and reduced electron mobility.

Since crystalline imperfections are a major cause of the carrier scattering that degrades electron mobility in SOS, a number of workers have focused on defect reduction as a means of improving carrier transport. Yoshii et al [14] used solid phase epitaxy (SPE) after Si ion implantation induced amorphization of the highly defective near-interfacial zone to improve the crystallinity of this region [15] followed by a second amorphization and regrowth tailored to improve the Si surface region. Using this double regrowth process (DSPE) in 0.3 micrometre SOS these authors obtained 30% higher electron field-effect mobility than measured in as-grown samples. Refinements of this technique to deal primarily with the problem of aluminum autodoping resulting from implantation damage of the sapphire substrate have included the use of oxygen ion and oxygen/silicon ion implantation to suppress Al outdiffusion. Yamamoto et al [16] report 80% and 40% higher electron Hall mobility at 77 deg K and RT respectively using this approach. Reedy et al [17] have shown that careful selection of Si implant energy and dose can lead to a low defect near-interfacial region without Al contamination of the Si film. Using the implant prescription described in [17] Garcia and Reedy [18] reported electron field-effect mobilities in the range of 500 cm*2/V.s at distances as small as 80 nm from the Si/sapphire interface in DSPE regrown SOS. These results suggest that very thin (< 100 nm) films improved by DSPE may be usable for sub-micrometre CMOS/SOS VLSI circuitry. A three-step process for improvement of the crystallinity of SOS has been developed by Amano and Carey [19] who follow solid phase regrowth of implantation amorphized 0.2 micrometre SOS with an additional epitaxial growth of Si to increase the film thickness to 0.6 micrometre. An adaptation of this technique used by Mayer et al [20] to obtain improved 0.3 micrometre SOS yielded increases in low-field electron mobility of 44% over as-grown SOS of the same film thickness.

Another avenue to improvement in carrier transport in SOS has been to depart from conventional chemical vapor deposition to achieve better epitaxial Si film growth. Such an approach has been taken by Itoh and Takai [21] who used a combination of plasma dissociation of silane coupled with implantation of various ionic species into the sapphire substrate prior to epitaxial deposition of Si. Electron Hall mobility in these films was found to be highest, about 75% of bulk Si, in the films with the lowest Al concentration. These films were grown on sapphire substrates implanted with oxygen ions. Substrates implanted with Si ions yielded the lowest defect concentration, as determined by ion channelling measurements, but electron mobility was only 65%

of bulk Si. Ishida et al [22,23] have grown high quality SOS films by predepositing very thin buffer layers of amorphous Si prior to normal chemical vapor deposition of Si. SOS grown by this method does not display the strong dependence of electron mobility on CVD growth rate observed with films grown without the predeposited amorphous layer. Best results were obtained with predeposited layers 20 A thick which yielded field-effect electron mobility as much as 50% higher than was measured in films grown on bare sapphire.

REFERENCES
----

[1]   D.J.Dumin, P.H.Robinson [J. Cryst. Growth (Netherlands) vol.3 (1968) p.214]
[2]   W.E.Ham [Appl. Phys. Lett. (USA) vol.21 no.9 (1 Nov 1972) p.440-3]
[3]   D.J.Dumin, P.H.Robinson [J. Appl. Phys. (USA) vol.39 (1968) p.2759
[4]   A.B.M.Elliot, J.C.Anderson [Solid-State Electron. (GB) vol.15 (1972) p.531]
[5]   A.C.Ipri [Appl. Phys. Lett. (USA) vol.20 (1972) p.1]
[6]   A.C.Ipri, J.N.Zemel [J. Appl. Phys. (USA) vol.44 no.2 (Feb 1973) p.744-51]
[7]   S.T.Hsu, J.H.Scott [RCA Rev. (USA) vol.36 (1975) p.240]
[8]   M.S.Abrahams, C.J.Buiocchi [Appl. Phys. Lett. (USA) vol.27 no.6 (15 Sept 1975) p.325-7]
[9]   A.C.Ipri [in 'Applied Solid State Science, Advances in Materials and Devices Research, Supplement 2, Silicon Integrated Circuits, Part A' Ed. D.Kahng (Academic Press, 1981) p.253]
[10] G.W.Cullen [in 'Heteroepitaxial Semiconductors for Electronic Devices', Eds. G.W.Cullen, C.C.Wang (Springer-Verlag, New York, USA, 1978) p.6]
[11] S.Onga, K.Hatanaka, S.Kawaji, Y.Yasuda [Jpn. J. Appl. Phys. (Japan) vol.17 no.2 (Feb 1978) p.413-22]
[12] B-Y.Tsaur, J.C.C.Fan, M.W.Geis [Appl. Phys. Lett. (USA) vol.40 no.4 (15 Feb 1982) p.322-4]
[13] Y.Kobayashi, M.Nakamura, T.Suzuki [Appl. Phys. Lett. (USA) vol.40 no.12 (15 June 1982) p.1040-2]
[14] T.Yoshii, S.Taguchi, T.Inoue, H.Tango [Jpn. J. Appl. Phys. Suppl. (Japan) vol.21 no.21-1 (1982) p.175]
[15] S.S.Lau et al [Appl. Phys. Lett. (USA) vol.34 (1979) p.76-8]
[16] Y.Yamamoto, H.Kobayashi, T.Takahashi, T.Inada [Nucl. Instrum. & Methods Phys. Res. Sect. B (Netherlands) vol.7 (1985) p.273-7]
[17] R.Reedy, T.W.Sigmon, L.A.Christel [Appl. Phys. Lett. (USA) vol.42 no.8 (15 Apr 1983) p.707-9]
[18] G.A.Garcia, R.E.Reedy [Electron. Lett. (GB) vol.22 no.10 (8 May 1986) p.537-8]
[19] J.Amano, K.Carey [Appl. Phys. Lett. (USA) vol.39 no.2 (15 Jul 1981) p.163-5]
[20] D.C.Mayer, P.K.Vasudev, J.Y.Lee, Y.K.Allen, R.C.Henderson [IEEE Electron Device Lett. (USA) vol.EDL-5 (1984) p.156]
[21] T.Itoh, H.Takai [Jpn. J. Appl. Phys. Part 1 (Japan) vol.22 (1983) p.597]
[22] M.Ishida, Y.Yasuda, H.Wakamatsu, H.Abe, T.Nishinaga, T.Nakamura [Jpn. J. Appl. Phys. Part 2 (Japan) vol.22 no.7 (Jul 1983) p.L438-40]
[23] M.Ishida, Y.Yasuda, H.Ohyama, H.Wakamatsu, H.Abe, T.Nakamura [J. Appl. Phys. (USA) vol.59 no.12 (15 Jun 1986) p.4073-8]

18.6 HOLE MOBILITY IN Si FILMS ON SAPPHIRE

by G.A.Garcia

July 1987
EMIS Datareview RN=15753

Measurements of film-average hole mobility and hole mobility as a
function of depth in SOS have been made using the same Hall and gated
MOS structure described in [1] for use in measurements of electron
mobility. Similarly, the effect of film deposition rate, substrate
temperature, film thickness and carrier concentration on hole mobility
generally follow the same trends observed in measurements of electron
mobility [2,3].

However, a notable difference has been observed in the effect of Si
film stress on the mobility of the two carrier types. Ross and Warfield
[4] calculated the energy splitting of the valence bands of SOS under
the influence of the large (10*9-10*10 dyne/cm*2) compressive stresses
present in these films [5]. An increase in the population of the
light-hole band was used to explain the observed increase in hole
mobility with increasing electric field. The piezoresistive effect in
strained Si films was studied by Schlotterer [6] who concluded that
hole mobility in compressively strained (111) Si films should
significantly exceed bulk Si hole mobility. Experimental confirmation of
this mobility increase was hampered by high grown-in defect densities.
In an attempt to understand reduced carrier mobility at elevated
temperature in SOS Ipri and Zemel [7] also applied the phenomenological
piezoresistance formalism to (100) SOS films. They concluded that
compressive stress should not affect hole mobility and instead
suggested that a shear stress may be responsible for reduced Hall hole
mobility. Hughes [8] examined the piezoresistive effect on carrier
mobility for a number of crystallographic orientations of SOS including
several not previously considered. While electron mobility was predicted
to be enhanced or reduced depending on the Si/sapphire epitaxial growth
mode and film orientation, hole mobility was found to be enhanced for
all cases. For some orientations and growth modes the stressed hole
mobility was calculated to be as much as 2.4 times unstressed levels.

Application of solid-phase epitaxial (SPE) regrowth techniques,
described in [1], to the reduction of defect concentrations in SOS has
yielded a somewhat smaller improvement in p-channel mobility as compared
to the n-channel improvement. Yoshii et al [9] reported a 10% increase
in p-channel mobility following double-solid-phase epitaxial (DSPE)
regrowth of 0.3 micron SOS. This modest increase in hole mobility may
be explained as the composite result of two competing effects: reduced
defect scattering combined with reduced compressive stress in DSPE
regrown SOS as reported by Ohmura et al [10].

Departing from conventional chemical vapour deposition of Si on sapphire,
Bean [11] has found that SOS films grown by molecular beam epitaxy have
approximately 20% higher Hall hole mobility for film thicknesses of 0.5
and 1.0 micrometre than has been reported for films of the same
thickness and doping levels prepared by CVD.

REFERENCES
----

[1] G.A.Garcia, R.E.Reedy [EMIS Datareview RN=15712 (May 1987)
     'Electron Mobility in Silicon-on-Sapphire']
[2] A.C.Ipri [in 'Applied Solid State Science, Advances in Materials
     and Device Research, Suppl 2, Silicon Integrated Circuits,
     Part A' Ed D.Kahng (Academic Press, 1981) p.253]
[3] G.W.Cullen [in 'Heteroepitaxial Semiconductors for Electronic
     Devices', Eds G.W.Cullen, C.C.Wang (Springer-Verlag, New York,
     1978) p.6]
[4] E.C.Ross, G.Warfield [J. Appl. Phys. (USA) vol.41 (1970) p.2657]
[5] C.Y.Ang, H.M.Manasevit [Solid-State Electron. (GB) vol.8 (1965)
     p.994]
[6] H.Schlotterer [Solid-State Electron. (GB) vol.11 (1968) p.947]
[7] A.C.Ipri, J.N.Zemel [J. Appl. Phys. (USA) vol.44 (1973) p.744]
[8] A.J.Hughes [J. Appl. Phys. (USA) vol.46 (1975) p.2849]
[9] T.Yoshii, S.Taguchi, T.Inoue, H.Tango [Jpn. J. Appl. Phys. Suppl.
     (Japan) vol.21 no.21-1 (1982) p.175]
[10] Y.Ohmura, T.Inoue, T.Yoshii [Solid State Commun. (USA) vol.37
     (1981) p.583]
[11] J.C.Bean [Appl. Phys. Lett. (USA) vol.36 (1980) p.741]

18.7 CHARACTERISATION OF Si FILMS ON SAPPHIRE

by A.M.Hodge and C.Pickering

July 1987
EMIS Datareview RN=15766

A. INTRODUCTION

The material which forms the subject of this review is single
crystalline silicon grown epitaxially on single crystal sapphire [1].
It is used as the starting material for the fabrication of CMOS SOS
devices. At present CMOS SOS is the most advanced of all the
technologies being used for the fabrication of silicon-on-insulator
(SOI) devices [2]. CMOS SOS components including memories (16k by 1
bit SRAM), 16 bit microprocessors, controllers, multipliers, and
10,000 gate arrays incorporating 1.25 micron design rules are
available from certain commercial manufacturers [3]. The CMOS SOS
devices are produced in the silicon films by the same processes as
conventional MOS devices in bulk silicon wafers.

CMOS SOS devices potentially offer several performance advantages over
bulk silicon devices, for example higher speed, greater packing
density, improved radiation tolerance, lower power consumption and
freedom from latch-up [2]. In practice these are not fully realised,
principally because defects in the silicon degrade operation of the
devices. The defects arise mainly because of lattice and thermal
expansion mismatch between the two materials. The mismatch between
(100) silicon and (1 bar1 0 2) sapphire is about 12.5% in the
[1 1 bar2 0] direction and 4.2% in the [1 bar1 0 1] direction, i.e.
the strain is anisotropic [4]. The defect density is minimised by
optimisation of the sapphire preparation and silicon growth, or, less
ideally, by appropriate treatment after manufacture [5-7]. Current
designs dictate that silicon films between 0.3 and 0.5 microns thick
are required.

A wide range of complementary techniques have been used to characterise
SOS and hence numerous physical and electrical parameters derived
[8-10]. Many of these are used for bulk silicon but most methods
require optimisation for SOS. It is essential that the material
properties are correlated with the electrical performance of the
devices after fabrication. Measured parameters quoted in this review
should be considered as representative as material quality is still
variable, being dependent on precise conditions during manufacture and
processing. Within about the last 5 years, considerable advances have
been made in the specification, manufacture and quality control of SOS
material. It is by continued characterisation of materials and
performance that improvements are being made.

B. DEFECT STRUCTURE

Cross-section transmission electron microscopy (XTEM), including more
recently the lattice imaging capability of the highest resolution
instruments, is an essential technique for the study of the
microstructure of materials. In SOS films, dislocations, microtwins and
stacking faults were all identified but with density decreasing with
distance from the silicon-sapphire interface [11,12,18]. For films

approximately 1 micron thick, the defect density was about 10*4/cm. The dominant defect structure comprised (221) family oriented material in the (100) matrix [13]. Microtwins on the four possible planes of the (111) family of [001] silicon were non-uniformly distributed, the majority subset having the twinning planes on (111) and (bar1 1 1), the minority on (1 bar1 1) and (bar1 bar1 1) [67,68]. XTEM was also used to study the defect reduction in SOS films after solid phase epitaxial regrowth (SPEG) [20,21], one of the methods by which SOS films have been improved after deposition of the silicon film but before device fabrication. Scanning electron microscopy was also used to observe microtwins in SOS films [22]. An approximately constant density of 10*9/cm*2 was recorded in films thicker than 600 nm but for thinner films the defect density increased. For films less than 400 nm thick the density was so high that counting the defects was difficult, i.e. the resolution limit of the method had been reached. TEM of surface replicas [19,23], particularly after selective etching, was used to estimate surface defect densities; values of 10*9/cm*2 were thus recorded [23]. Phonon boundary scattering at the silicon-sapphire interface was also consistent with a dislocation density at this interface of > 10*9/cm*2 [24]. It was suggested that this technique, which uses thermal conductivity measurement [25], could be developed for non-destructive monitoring of average dislocation density at the interface.

C. SILICON-SAPPHIRE INTERFACE STRUCTURE

The silicon-sapphire interface has been studied, by XTEM in particular, because of its influence on the formation of the defects. The interface was seen to be abrupt on an atomic scale [14]. Additionally, the earliest stages of silicon film growth were investigated [15]. Misfit edge dislocations were recorded in silicon islands prior to complete coverage of the sapphire substrate, lying close to the silicon-sapphire interface, in the (100) plane and in the [011] and [0 1 bar1] directions [16]. However, the absence of dislocations locally at the immediate silicon-sapphire interface [17] was used as evidence of the minimal role of lattice mismatch on the defect structure of the silicon film. Interface phenomena and the influence of various growth conditions including the use of pre-deposited amorphous silicon layers have also been studied by a wider variety of techniques [19]. RHEED (reflection high-energy electron diffraction) patterns from SOS films grown with a thin amorphous silicon film deposited and annealed before epitaxial growth, showed no spots due to twins whereas those from films without the amorphous inter-layer showed twin spots. Chemical bonding at the silicon-sapphire interface was studied using Auger spectroscopy and electron spectroscopy for chemical analysis (ESCA) and reported in the same publication [19]; silicon atoms of the heat treated, pre-deposited amorphous silicon layers were shown to bind chemically with oxygen atoms in the sapphire surface.

D. X-RAY DIFFRACTION

X-ray diffraction also provides data on the crystallography of the films. It is non-destructive, so wafers can be studied then processed and the device performance assessed and compared with the X-ray data from the same wafer [26]. A direct correlation between reduced carrier mobility and X-ray diffraction peak broadening was recorded. Complementary information was also obtained from X-ray pole-figure analysis [27] as the volume fraction of each twin component correlated directly with carrier mobility. The volume percentage of microtwinned

material was so measured to be 2.7% in as-grown SOS films but this was
reduced to less than 1% by a rapid thermal heat treatment [7]. X-ray
double crystal diffractometry was also used to assess the dislocation
density in SOS films; for these 0.45 micron films the dislocation
density increased from 4 X 10*10/cm*3 at the silicon surface to 20 X
10*10/cm*3 at the silicon-sapphire interface [28].

E. DEFECT PROFILES

Ion channelling is also used to compare the depth profile of defects in
SOS films with relatively perfect bulk silicon wafers [31]. The
technique has been exploited by many workers to study the relative
improvement in SOS film quality after SPEG. By way of example, the
back-scattering minimum yields, chi(min), in aligned spectra
(indicating crystallinity), for 0.2 micron as-grown films were 0.2-0.3
for the surface and 0.4-0.5 at the interface; after improvement,
chi(min) values were 0.14 at the interface and 0.03 at the surface [32].
It was proposed that the majority defects after SOS film improvement
were dislocations, rather than twins [33,21], and that their density was
10*10/cm*3 [34].

Quantification of densities of atomic species in SOS films is achieved
by secondary ion mass spectrometry (SIMS). It was shown that diffusion
of, for example, boron is enhanced by a factor of 1.1-2.4 times in SOS
in comparison with bulk silicon because of the crystallographic defects
[35]. Obtaining accurate profiles in the thin films on insulating
substrates is, however, not trivial [36]. It is important to be able
to characterise not only the dopant species in SOS samples but also
potential contaminants in the films. For example, oxygen and aluminium
from the sapphire substrate migrate into the silicon films during
manufacture of the SOS wafers [37]. The measurements are again prone to
erroneous results [38] but aluminium concentrations were shown to be
close to the detection limits of the SIMS equipment, i.e. about 10*15
atoms/cm*3. During wafer processing involving ion implantation and heat
treatments, aluminium may diffuse into the silicon films leaving even
higher concentrations (3 X 10*16 atoms/cm*3) particularly at the
silicon-sapphire interface [21,39].

F. OPTICAL METHODS OF CHARACTERISATION

Optical methods are especially valuable as characterisation tools since
they are non-destructive. Studies by the RCA group in particular, have
led to one technique, ultra-violet reflectometry (UVR) [63-66], being
developed and used extensively for the routine quality control of SOS
wafers prior to device fabrication. The visible haze and hence the
specular reflectance from the silicon films depends on both surface
texture and crystalline quality. The reflectance, relative to bulk
silicon, was measured at 280 nm (near a crystallinity peak) and
400 nm, the latter being used to allow for surface scattering. The
derived parameter, relative reflectance, was correlated with twin
concentration from X-ray pole-figure analysis, X-ray diffraction
line-width broadening, surface photovoltage measurements and circuit
yield and device performance. Higher haze was correlated with higher
defect levels and poorer device performance. Although there has been
some debate recently over the fundamental physical phenomena (surface
scattering versus degraded crystallinity) responsible for the
variation in reflectance of SOS films [40-43], the UVR technique

remains valuable for characterisation of SOS films. UV wavelengths only probe the silicon surface because of the dielectric properties of the film; longer wavelengths (visible and near IR) are not absorbed so strongly in the silicon and have therefore been used to probe deeper in the films [44]. From the reflectivity parameters so obtained, it was deduced that the silicon was partially amorphous at the silicon-sapphire interface. This is apparently in conflict with the more direct observations using XTEM which, in most recent reports [13,14], have indicated that no intermediate layer was present between the silicon and sapphire. Optical absorption in the silicon film as a function of wavenumber has also been shown to be dependent on the SOS film quality and indicative in turn of the performance of devices fabricated subsequently [69].

G. STRESS

Early work [45] estimated stress in SOS films, from the deformation of the wafers, to be $10*8 - 10*9$ N/m*2, and a similar figure was calculated from the different thermal expansion coefficients of silicon and sapphire [46]. Experimentally, stress effects have been studied by Raman scattering. The compressive stress in SOS films at room temperature was found to be $6-13$ X $10*8$ N/m*2 at the silicon-sapphire interface and $5.5-7.5$ X $10*8$ N/m*2 at the free silicon surface [47]. The change in the stress in the layer and the change in the crystallinity on pulsed laser irradiation were followed by Raman probing; the improved electron Hall mobility (best value 550 cm*2/V.s) was correlated with a change to tensile stress in the film [48]. X-ray double crystal diffractometry was also used to show that the principal stresses in the plane of the silicon film were $(-0.92 +- 0.16)$ X $10*9$ N/m*2 in the [100] direction and $(-0.98 +- 0.17)$ X $10*9$ N/m*2 in the [010] direction [29], and that the in-plane strain was $(-4.2 +- 0.2)$ X $10*-3$ [30].

H. ELECTRICAL MEASUREMENTS

The data above were obtained essentially by non-electrical methods. The remainder of this review will summarise results of electrical measurements. Standard techniques have been exploited and reported [1,9]. They include four point probing, spreading resistance (see also work by Pawlik [49]), Hall measurements, capacitance-voltage techniques etc. For most of these, appropriate test structures have to be fabricated in the silicon films by conventional processing techniques. Such structures may be included as test inserts on chips or wafers and hence parameters such as conductivity, mobility, and transistor characteristics were obtained. Representative data for SOS devices with effective channel length of 1.4 microns, produced in 0.3 micron films are as follows [50]: electron mobility = 300 cm*2/Vs and hole mobility = 185 cm*2/V.s in as-grown material; these have been increased to 435 and 250 cm*2/V.s respectively by suitable improvement techniques. n- and p-channel subthreshold slopes are close to 100 and 145 mV/decade respectively. Leakage currents can be reduced to 0.3 pA/micron. Drive currents are of order 50 mA/mm. Inverter delays, for devices with L(eff) = 0.5 micron in improved material, can be less than 50 ps. Power dissipation per stage at V(dd) = 3V averaged 105 microwatts L(eff) = 0.5 micron. By thinning the sapphire and constructing a metal insulator

semiconductor structure, it was shown that a native polarisable
interfacial layer, equivalent to a reduced charge of 10*11
electrons/cm*2 occurred at the silicon-sapphire interface and that the
interface state density at midgap was 1 X 10*12/cm*2/eV [52].

SOS films are stressed because of the mismatches with the sapphire
substrate. This stress, which is anisotropic, deforms the electronic
sub-band structure and degrades device performance. The effects have
been examined theoretically [53]. The stresses in the SOS films are
dependent on the orientation, dimensions and aspect ratios of the
silicon island mesas etched during device structure fabrication.
Anisotropy in hole Hall mobility in particular samples was measured to
vary from 40 to 44 cm*2/V.s at 300 deg K depending on orientation [51].
Elastoresistance was measured using Hall structures [54] and
conductivity measurements made in a magnetic field [55]. Piezoresistance
coefficients in SOS and bulk also differ markedly; this has been
explained by the residual strain in the SOS films and the phenomenon
has been discussed in detail [56,70]. The piezoresistance coefficients
of SOS inversion layers were measured and the modulus of pi11/pi12
was shown to be about unity for the [100]Si channel direction instead of
about 2 in bulk silicon [56]. Magnetoresistance coefficients for SOS
films were reported [57,71]. The longitudinal effect is higher than the
transverse effect, for example with a 10 T magnetic field, the
longitudinal magnetoresistance coefficient is about 1.3 but the
transverse value is less than 1.2 [57].

I. ENERGY LEVELS

Photoconductivity and optical absorption measurements have shown that a
donor level at 0.3 eV from the valence band edge and an acceptor level
at 0.25 eV from the conduction band edge exist with approximately equal
concentrations of 10*17 - 10*18/cm*3 [58]. Deep level transient
spectroscopy has also been applied to SOS in order to characterise the
defect levels in the forbidden gap of the silicon [59,60]. Surface
photovoltage spectroscopy permits the determination of band structure,
trapping centres and deep levels in SOS films but without the need to
form ohmic contacts to the sample [61]. Deep levels close to the centre
of the enery gap (0.6 eV) were recorded but their density was
dependent on treatment (equivalent to device processing) of the films.
The photoelectric response was dominated by trapping phenomena similar
to those seen in bulk silicon having a high density of dislocations.
The refractive index of the films as a function of photon energy was
very similar to bulk silicon. All films showed a slight degree of
amorphisation as they all exhibited an energy gap at about 1.45 eV in
addition to one at 1.1 eV.

J. CARRIER LIFETIME

From time-resolved reflectivity and transmission measurements obtained
following excitation with femtosecond optical pulses, the carrier
lifetime in SOS films was measured to be > 300 ps for unimplanted
wafers but this was reduced by ion implantation to a limiting value of
about 600 fs [62]. This very short carrier lifetime, due to
recombination at defects in the material, is exploited particularly for
picosecond optoelectronic devices but is the reason for the general
unsuitability of SOS for bipolar operation. Conversely, the material
can be improved to increase the lifetime [50].

REFERENCES
----

[1]   A.C.Ipri [Appl. Solid State Sci. (USA), Advances in Materials
       and Device Research suppl.2 pt.A Ed. D.Kahng (Academic Press,
       1981)]
[2]   S.L.Partridge [IEE Proc. E (GB) vol.133 no.3 (1986) p.106-16]
[3]   T.Naegele [Electronics (USA) (8 Jan 1987) p.38]
[4]   G.W.Cullen [J. Cryst. Growth (Netherlands) vol.9 (1971)
       p.107-25]
[5]   S.S.Lau et al [Appl. Phys. Lett. (USA) vol.34 (1976) p.76]
[6]   I.Golecki, M.-A.Nicolet [Solid-State Electron. (GB) vol.23 (1980)
       p.803]
[7]   L.Pfeiffer et al [Appl. Phys. Lett. (USA) vol.50 no.8 (1987)
       p.466]
[8]   P.J.Zanzucchi [in 'Heteroepitaxial Semiconductors for Electronic
       Devices' Eds G.W.Cullen, C.C.Wang (Springer-Verlag, 1978) ch.5]
[9]   W.E.Ham [in 'Heteroepitaxial Semiconductors for Electronic
       Devices' Eds G.W.Cullen, C.C.Wang (Springer-Verlag, 1978) ch.6]
[10]   S.Cristoloveaunu, G.Ghibaudo, G.Kamarinos [Rev. Phys. Appl.
       (France) vol.19 (1984) p.161-85]
[11]   M.S.Abrahams, C.J.Buiocchi [Appl. Phys. Lett. (USA) vol.27 no.6
       (1975) p.325-7]
[12]   W.E.Ham, M.S.Abrahams, C.J.Buiocchi, J.Blanc [J. Electrochem.
       Soc. (USA) vol.124 no.4 (1977) p.634-6]
[13]   G.W.Cullen et al [J. Cryst. Growth (Netherlands) vol.56 (1982)
       p.281-95]
[14]   F.A.Ponce, J.Aranovich [Appl. Phys. Lett. (USA) vol.38 no.6
       (1981) p.439-41]
[15]   K.C.Paus, J.C.Barry, G.R.Booker, T.B.Peters, M.G.Pitt [Inst.
       Phys. Conf. Ser. (GB) no.76 sect.1 (1985) p.35-40]
[16]   M.S.Abrahams, C.J.Buiocchi, J.F.Corboy, G.W.Cullen [Appl. Phys.
       Lett. (USA) vol.28 (1976) p.275]
[17]   F.A.Ponce [Appl. Phys. Lett. (USA) vol.41 no.4 (1982) p.371-3]
[18]   J.Blanc [in 'Heteroepitaxial Semiconductors for Electronic
       Devices' Eds G.W.Cullen, C.C.Wang (Springer-Verlag, 1978) ch.8]
[19]   M.Ishida, Y.Yasuda, H.Ohyama, H.Wakamatsu, H.Abe, T.Nakamura [J.
       Appl. Phys. (USA) vol.59 no.12 (1986) p.4073-8]
[20]   M.A.Parker, R.Sinclair, T.W.Sigmon [Appl. Phys. Lett. (USA)
       vol.47 no.6 (1985) p.626-8]
[21]   A.M.Hodge, A.G.Cullis, N.G.Chew [Mater. Res. Soc. Symp. Proc.
       (USA) vol.35 (1985) p.393-9]
[22]   M.L.Zorrilla Carfagnini, J.Trilhe [J. Electrochem. Soc. (USA)
       vol.128 no.2 (1981) p.385-8]
[23]   T.Yoshii, S.Taguchi, T.Inoue, H.Tango [Jpn. J. Appl. Phys. Suppl.
       (Japan) vol.21 no.21-1 (1982) p.175-9]
[24]   M.N.Wybourne, C.G.Eddison, M.J.Kelly [J. Phys. C (GB) vol.17
       (1984) p.L607-12]
[25]   M.N.Wybourne [GEC J. Res. Incorp. Marconi Rev. (GB) vol.3 no.4
       (1986) p.251-5]
[26]   C.E.Weitzel, R.T.Smith [J. Electrochem. Soc. (USA) vol.125 no.5
       (1978) p.792-8]
[27]   R.T.Smith, C.E.Weitzel [J. Cryst. Growth (Netherlands) vol.58
       (1982) p.61-72]
[28]   A.Rey, J.Trilhe, J.Borel [J. Cryst. Growth (Netherlands) vol.60
       (1982) p.264-74]
[29]   T.Vreeland [J. Mater. Res. (USA) vol.1 no.5 (1986) p.712-6]
[30]   I.Golecki, H.L.Glass, G.Kinoshita [Appl. Phys. Lett. (USA) vol.40
       no.8 (1982) p.670-2]

[31] J.H.Chang [Nucl. Instrum. & Methods (Netherlands) vol.173 (1980) p.565-70]

[32] J.Amano, K.Carey [Appl. Phys. Lett. (USA) vol.39 no.2 (1981) p.163-5]

[33] Y.Yamamoto, T.Sugiyama, A.Hara, T.Inada [J. Appl. Phys. (USA) vol.53 no.1 (1982) p.793-6]

[34] T.Inoue, T.Yoshi [Nucl. Instrum. & Methods (Netherlands) vol.182/183 (1981) p.683-90]

[35] F.Rudolf, C.Jaccard, M.E.Roulet [Thin Solid Films (Switzerland) vol.59 (1979) p.385-91]

[36] D.S.McPhail, M.G.Dowsett, E.H.C.Parker [J. Appl. Phys. (USA) vol.60 no.7 (1986) p.2573-9]

[37] Q.Chen, X.Cai, R.Shi, Q.Wang [Mater. Lett. (Netherlands) vol.3 no.9,10 (1985) p.372-4]

[38] G.D.Robertson, P.K.Vasudev, R.G.Wilson, V.R.Deline [Appl. Surf. Sci. (Netherlands) vol.14 (1982-1983) p.128-33]

[39] M.G.Dowsett, E.H.C.Parker, D.S.McPhail [Springer Ser. Chem. Phys. (Germany) vol.5 (1986) p.340-2]

[40] C.Pickering, S.Dixon, D.B.Gasson, D.J.Robbins, A.M.Hodge [J. Cryst. Growth (Netherlands) (1987)]

[41] G.D.Robertson, R.Baron, P.K.Vasudev, O.J.Marsh [J. Cryst. Growth (Netherlands) vol.68 (1984) p.691-7]

[42] C.Pickering, A.M.Hodge, A.C.Daw, D.J.Robbins, P.J.Pearson, R.Greef [Mater. Res. Soc. Symp. Proc. (USA) vol.53 (1986) p.317-22]

[43] M.T.Duffy, G.W.Cullen, R.A.Soltis, G.Harbeke, J.R.Sandercock [RCA Rev. (USA) vol.46 (1985) p.19-33]

[44] J.Lagowski, L.Jastrzebski, G.W.Cullen [J. Electrochem. Soc. (USA) vol.130 no.8 (1983) p.1744-8]

[45] D.J.Dumin [J. Appl. Phys. (USA) vol.36 (1965) p.2700]

[46] A.J.Hughes, A.C.Thorsen [J. Appl. Phys. (USA) vol.44 (1973) p.2304-10]

[47] K.Yamazaki, M.Yamada, K.Yamamoto, K.Abe [Jpn. J. Appl. Phys. (Japan) vol.23 (1984) p.681-6]

[48] M.Nakamura, Y.Kobayashi, K.Usami [Jpn. J. Appl. Phys. (Japan) vol.23 no.6 (1984) p.687-94]

[49] M.Pawlik, R.D.Groves [Appl. Phys. Lett. (USA) vol.44 no.5 (1984) p.542-4]

[50] P.K.Vasudev [Mater. Res. Soc. Symp. Proc. (USA) vol.53 (1986) p.121-7]

[51] A.V.Beloglazov, V.M.Stuchebnikov, V.Vkhasikov, V.N.Chernitsyn [Sov. Phys.-Semicond. (USA) vol.16 no.8 (1982) p.946-7]

[52] P.Krusius, C.Dube, J.Frey [Appl. Phys. Lett. (USA) vol.38 no.7 (1981) p.547-9]

[53] A.J.Hughes [J. Appl. Phys. (USA) vol.46 no.7 (1975) p.2849-63]

[54] J.Hynecek [J. Appl. Phys. (USA) vol.45 no.6 (1974) p.2631-5]

[55] K.von Klitzing, T.Englert, G.Landwehr, G.Dorda [Solid State Commun. (USA) vol.24 (1977) p.703-6]

[56] S.Zaima, Y.Yasuda, S.Kawaguchi, M.Tsuneyoshi, T.Nakamura, A.Yoshida [J. Appl. Phys. (USA) vol.60 no.11 (1986) p.3959-66]

[57] J.Lee, S.Cristoloveneau, A.Chovet [Solid-State Electron. (GB) vol.25 no.9 (1982) p.947-53]

[58] D.J.Dumin [Solid-State Electron. (GB) vol.13 (1970) p.415-24]

[59] J.Chen, R.J.Ko, D.W.Brzezinski, L.Forbes, C.J.Dell'oca [IEEE Trans. Electron Devices (USA) vol.ED-28 no.3 (1981) p.299-304]

[60] G.Holmen, S.Peterstrom, G.Alestig [Appl. Phys. Lett. (USA) vol.45 no.7 (1984) p.741-3]

[61] J.Lagowski, L.Jastrzebski, G.W.Cullen [J. Electrochem. Soc. (USA) vol.128 no.12 (1981) p.2665-70]

[62]   F.E.Doany, D.Grischkowsky, C.-C.Chi [Appl. Phys. Lett. (USA)
       vol.50 no.8 (1987) p.460-2]
[63]   M.T.Duffy et al [J. Cryst. Growth (Netherlands) vol.58 (1982)
       p.10-18]
[64]   M.T.Duffy, P.J.Zanzucci, W.E.Ham, J.F.Corboy, G.W.Cullen,
       R.T.Smith [J. Cryst. Growth (Netherlands) vol.58 (1982)
       p.19-36]
[65]   L.Jastrzebski, M.T.Duffy, J.F.Corboy, G.W.Cullen, J.Lagowski [J.
       Cryst. Growth (Netherlands) vol.58 (1982) p.37-43]
[66]   D.E.Passoja, D.McLeod, L.G.Dowell, H.F.Hillery, J.E.A.Maurits,
       L.R.Rothrock [J. Cryst. Growth (Netherlands) vol.58 (1982)
       p.44-52]
[67]   R.Lihl, H.Oppolzer, P.Pongratz, P.Skalicky, W.Svanda [J. Microsc.
       (GB) vol.118 pt.1 (1980) p.89-95]
[68]   C.E.Weitzel, R.T.Smith [J. Electrochem. Soc. (USA) vol.124
       (1977) p.1080]
[69]   M.Druminski, C.Kuhl, E.Preuss, F.Schwidefsky, H.Splittgerber,
       D.Takacs [Jpn. J. Appl. Phys. Suppl. (Japan) vol.15 no.15-1
       (1976) p.217-20]
[70]   S.Onga, T.Yoshii, K.Hatanaka, Y.Yasuda [Jpn. Appl. Phys. Suppl.
       (Japan) vol.15 no.15-1 (1976) p.225-31]
[71]   Y.Ohmura [Jpn. J. Appl. Phys. Suppl. (Japan) vol.15 no.15-1
       (1976) p.233-8]

18.8 STRUCTURE OF THE INTERFACE IN SOS STRUCTURES

by M.Ishida

July 1987
EMIS Datareview RN=15754

In early work [1-3], Schlotterer [1] indicates that an intermediate
layer (a transition layer) exists between sapphire substrates and
silicon films from the results of reflectance measurements and
secondary ion mass spectroscopy (SIMS) measurements. The thickness of
this layer varied between 20 and 40 nm for different samples. The
composition of the intermediate layer was assumed to contain a large
amount of oxygen, similar to aluminium silicates. The first direct
observation of the cross sectional structure of SOS films was achieved
by Abrahams et al [2]. They reported the presence of large densities
of stacking faults and microtwins. However, misfit dislocations were
not present at the interface. They also indicated that there was no
evidence for the presence of an Al-bearing phase (transition layer)
in the proximity of the interface. Abrahams et al [3] showed that
misfit dislocations were observed in silicon islands prior to complete
substrate coverage.

From 1981, the detailed structure of the silicon-sapphire interface was
observed by high-resolution transmission electron microscopy [4,5].
Ponce et al [4] presented lattice images of the silicon-sapphire
interface region of heteroepitaxial (001) silicon on (1 bar1 0 2)
sapphire. The silicon layer was deposited at 960 deg C from the
pyrolysis of silane in H2 at a rate of 1.4 micron/min to a total
thickness of 0.65 micron. The specimen was prepared in cross-section
with the foil perpendicular to [110]Si and to [bar2 0 2 1]sapphire.
From the picture, they pointed out that the interface was planar and
abrupt to the limit of resolution (less than 3 A); no interface
compound of phases was observed. Ponce [5] directly demonstrated
that most of the silicon immediately next to the silicon-sapphire
interface was in fact locally free of faults and, in particular, free
of misfit dislocations. These fault-free regions exist over interface
areas as large as 2000 A in diameter. It was shown that the silicon-
sapphire interface was incoherent, and that the structure of the
silicon layer was unperturbed right up to the interface plane. An
analysis of the nature of the chemical bond at the interface showed
that for bonding to occur, the last aluminium layer in the sapphire
must rearrange resulting in Al-O-Si chains [aluminosilicate structure]
where the O atom has a higher degree of freedom than in the bulk of
Al2O3.

Ishida et al [6] showed the existence of Si-O bonds at the interface
of SOS from the results of Auger electron spectroscopy and electron
spectroscopy for chemical analysis. Recent results [7,8], obtained by
high-resolution electron microscopy lattice images of cross-sectional
view of as grown SOS, also show that the interface is abrupt to within
a few atomic layers.

REFERENCES
----

[1] H.Schlotterer [J. Vac. Sci. & Technol. (USA) vol.13 no.1 (Jan/Feb
     1976) p.29-36]
[2] M.S.Abrahams, C.J.Buiocchi [Appl. Phys. Lett. (USA) vol.27 no.6
     (15 Sept 1975) p.325-7]
[3] M.S.Abrahams, C.J.Buiocchi, J.F.Corboy,Jr, G.W.Cullen [Appl.
     Phys. Lett. (USA) vol.28 no.5 (1 Mar 1976) p.275-7]
[4] F.A.Ponce, J.Aranovich [Appl. Phys. Lett. (USA) vol.38 no.6 (15
     Mar 1981) p.439-41]
[5] F.A.Ponce [Appl. Phys. Lett. (USA) vol.41 no.4 (15 Aug 1982)
     p.371-3]
[6] M.Ishida, Y.Yasuda, H.Ohyama, H.Wakamatsu, H.Abe, T.Nakamura [J.
     Appl. Phys. (USA) vol.59 no.12 (15 June 1986) p.4073-8]
[7] D.J.Smith, L.A.Freeman, R.A.McMahon, H.Ahmed, M.G.Pitt, T.B.Peters
     [J. Appl. Phys. (USA) vol.56 no.8 (15 Oct 1984) p.2207-12]
[8] M.A.Parker, R.Sinclair, T.W.Sigmon [Appl. Phys. Lett. (USA)
     vol.47 no.6 (15 Sept 1985) p.626-8]

18.9 STRESS AT THE INTERFACE IN SOI STRUCTURES

by A.Ogura

June 1987
EMIS Datareview RN=15746

Due to residual stress, carrier mobilities in SOI are usually different
from those in bulk Si. Pioneering studies extensively examined electron
mobility in SOS (silicon on sapphire) film and showed that it is
lower than that in bulk Si because of compressive stress. Hughes [1]
calculated a change of mobility based on the piezoresistance effect and
demonstrated a mobility anisotropy which is due to thermal expansion
anisotropy of sapphire.

Recently, a large number of substrates which have various thermal
expansion coefficients were used in place of sapphire for the SOI
substrate. Tsaur et al [2] showed that electron mobility is enhanced and
hole mobility is reduced under tensile stress in a Si/SiO2/fused-quartz
structure, while, on the other hand, electron mobility is reduced and
hole mobility is enhanced under compressive stress in a Si/SiO2/sapphire
structure. Thermal stress in SOI also acts as a cause of crystal
imperfections, such as small angle grain boundaries [3] or film cracks
[4].

Some techniques for relaxation of residual stress after SOI fabrication
have been investigated, including high temperature annealing [5],
pulsed-laser annealing [6-9], CW-laser annealing [10], patterning [11],
furnace annealing after ion implantation [11,12], etc.

Residual stress is mainly caused by the thermal expansion difference
between silicon and the substrates, even in the epitaxial case, such as
with SOS or silicon on spinel. Thus, for reduction of built-in stress,
when selecting substrates, it is necessary to take their thermal
expansion into consideration. With this in mind, residual stress in SOI
was measured with Raman spectroscopy. This SOI had been fabricated by
laser annealing on various insulating substrates kept at 300 deg C.
These results are shown in TABLE 1 [14].

TABLE 1: residual stress versus thermal expansion coefficients of
         substrates

Substrate	Thermal Expansion Coefficient $10*-6$/deg C	Residual Stress $10*9$ dyne/cm*2
quartz	0.55	-15.7
Corning 7740 glass	3.25	-8.5
silicon	3.5	-7.5
alumina	6.7	3.5

In the experiment, an alumina substrate was coated with Si3N4 to prevent
impurity diffusion, and then 7740 glass, silicon, and an Si3N4/alumina
substrate were lightly coated with SiO2. Residual stress values seen in

TABLE 1 were obtained using the relationship between stress and Raman frequency shift demonstrated by Englert et al [15]. Minus signs indicate that residual stress is tensile. From TABLE 1, it can be seen that even when a Si substrate is used, stress still remains, because of the temperature difference between the Si film and Si-substrate during laser annealing. If a stress-free SOI is desired, a substrate which has a slightly larger thermal expansion coefficient (5.4 X 10*-6/deg C) than Si would be required.

Taking into account the temperature difference between the Si film and substrate during SOI fabrication, the residual stress in Si film(s) which is caused by thermal construction can be estimated as follows

$$s = k (a[Si](Tm-Tr) - a[sub](Ts-Tr)) \qquad (1)$$

where

    k   : constant
    Tm   : melting point of Si
    Tr   : room temperature
    Ts   : substrate temperature
    a[Si]   : thermal expansion coefficient of Si
    a[sub] : thermal expansion coefficient of substrate.

In the usual laser annealing case, a[Si] < a[sub] is necessary to obtain stress-free Si film, since Tm is much higher than Ts.

If the substrate consists of multi-layers such as a Si/insulator/Si-substrate structure, which is very common especially for 3-dimensional integrated circuits, a[sub] in Eqn (1) should be thought of as the mean value of the coefficients of an insulator (a[i]) and Si (a[Si]).

$$a[sub] = xa[i] + ya[Si] \quad (x+y=1) \qquad (2)$$

where x and y are weight factors, which are functions of both insulator thickness and its thermal conductivity. In the Si/1 micron thick SiO2/Si-substrate structure, which has been fabricated by laser annealing at 400 deg C substrate temperature, residual stress in the Si film was 5.0 X 10*9 dyne/cm*2. As SiO2 thickness increases, residual stress also increases. In the 3 micron-thick SiO2 case, residual stress is 8.0 X 10*9 dyne/cm*2, moreover, film cracks caused by stress are observed. For stress-free SOI, it is desired that a[sub] = 5.4 X 10*-6/deg C. Thus, it can be seen from Eqn (2) that the most suitable insulating material is that for which thermal expansion coefficient (a[i]) is larger than that of Si. With this in mind, AlN, of which the thermal expansion coefficient is 4.0 X 10*-6/deg C, was examined as an insulating layer. Its residual stress was minimised to 5.2 X 10*9 dyne/cm*2, even on 3 micron thick AlN [14]. Operating with similar ideas, Shiozaki et al [16] presented an application of SiON as a SOI insulating material.

An SOI structure essentially consists of multi-layers of different materials, each of which has different thermal characteristics, i.e. expansion, conductivity, etc. New material design, then, needs to consider those thermal properties. Moreover, a low temperature SOI fabrication process would be even better as a fundamental solution to these thermal problems.

REFERENCES
----

[1]   A.J.Hughes [J. Appl. Phys. (USA) vol.46 (1975) p.2849]
[2]   B.-Y.Tsuar, J.C.C.Fan, M.W.Geis [Appl. Phys. Lett. (USA) vol.40
      (1982) p.322]
[3]   J.M.Gibson, L.N.Pfeiffer, K.W.West, D.C.Joy [Mater. Res. Soc.
      Symp. Proc. (USA) vol.53 (1986) p.289]
[4]   R.A.Lemons, M.A.Bosch, A.H.Dayem, J.K.Grogan, P.M.Mankiewich
      [Appl. Phys. Lett. (USA) vol.40 (1982) p.469]
[5]   M.Rivier, F.Reidinger, G.Goetz, J.McKitterick [Mater. Res. Soc.
      Symp. Proc. (USA) vol.35 (1985) p.681]
[6]   K.Yamazaki, M.Yamada, K.Yamamoto, K.Abe [Jpn. J. Appl. Phys.
      (Japan) vol.20 (1981) p.L299]
[7]   K.Yamazaki, M.Yamada, K.Yamamoto, K.Abe [Jpn. J. Appl. Phys.
      (Japan) vol.20 (1981) p.L371]
[8]   Y.Kobayashi, M.Nakamura, T.Suzuki [Appl. Phys. Lett. (USA)
      vol.40 (1982) p.1040]
[9]   M.Nakamura, Y.Kobayashi, K.Usami [Jpn. J. Appl. Phys. Part 1
      (Japan) vol.23 (1984) p.687]
[10] I.Golecki, H.L.Glass, G.Kinoshita [Appl. Surf. Sci. (Netherlands)
      vol.9 (1981) p.299]
[11] K.Yamazaki, R.K.Uotani, K.Nambu, M.Yamada, K.Abe [Jpn. J. Appl.
      Phys. Part 2 (Japan) vol.23 (1984) p.L403]
[12] I.Goleki, H.L.Glass, G.Kinoshita [Appl. Phys. Lett. (USA) vol.40
      (1982) p.670]
[13] F.Moser, R.Besermen [J. Appl. Phys. (USA) vol.54 (1982) p.1033]
[14] A.Ogura, K.Egami, M.Kimura [Jpn. J. Appl. Phys. Part 2 (Japan)
      vol.24 (1985) p.L699]
[15] T.Englert, G.Abstreiter, J.Pontcharra [Solid-State Electron.
      (GB) vol.23 (1980) p.31]
[16] K.Shiozaki, H.Otake, M.Koba, K.Awane [Preprints 33rd Spring
      Meeting Japan Society of Applied Physics and of the Related
      Societies (Japan) (1986) p.490]

18.10 INTERNAL OXIDATION OF Si BY ION BEAM SYNTHESIS

by P.L.F.Hemment

November 1987
EMIS Datareview RN=17837

Oxidation of the near surface layer of bulk silicon may be achieved by
the non-equilibrium process of ion implantation provided that a
sufficiently high dose (areal density) of O+ ions is implanted to achieve
a volume concentration commensurate with SiO2. Synthesis of SiO2 in this
way was first reported in 1966 by Wantanabe and Tooi [1]. The depth of
the oxide layer beneath the surface will vary monotonically with ion
energy [2], whilst the thickness will depend upon the implanted dose.

TABLE 1

Ion Energy (keV)	Projected range (A)	Straggle (A)
15	313	116
200	4515	858

The instantaneous implanted depth profile [2] may be approximated to a
rectangular distribution of constant volume concentration which is
related to the implanted dose by

$$\text{constant volume concentration (/cm*3)} = \frac{0.4 \text{ X implanted dose (/cm*2)}}{\text{straggle (cm)}}$$

Stoichiometric SiO2 contains 4.8 X 10*22 O/cm*3 and from the above
equation a minimum dose of 0.14 X 10*18 O+/cm*2 at 15 keV will be
required to achieve SiO2. The increased straggle at 200 keV necessitates
a higher dose of 1.0 X 10*18 O+/cm*2. For a Gaussian distribution Hemment
[6] defines a critical dose as that dose which is required to form a
continuous oxide at the profile maxima. An experimental value of 1.4 X
10*18 O/cm*2 or 200 keV O+ ions is reported.

Using values from thermal oxides for the density (2.3 gm/cm*3) and
composition (SiO2), the dose of O+ ions required to form an ideal planar
structure with abrupt interfaces will be 4.8 X 10*22 X 10*-8 X t, where t
is the thickness of the oxide in Angstrom units (e.g a layer of thickness
4000 A contains 1.9 X 10*18 O atoms/cm*3). The dose dependence of the
oxide thickness before annealing has been reported by Namavar et al [3]
for 150 keV O+ ions. They determine by Rutherford Backscattering (RBS)
an oxide thickness of 4000 A for a dose of 1.9 X 10*18 O+/cm*2 and
thus conclude the composition to be SiO2. A linear dependence on dose
has been confirmed by Dubus et al [4] and Mao [5] who measured annealed
oxides (1150 deg C, 1275 deg C). They conclude that the oxide is
stoichiometric (see below).

Kilner et al [7] have used SIMS to determine the depth profiles of 200
keV O+ ions and find that abrupt interfaces (SiO2/Si) are not formed
during implantation as the instantaneous depth profile is a broad skew
Gaussian, as discussed by Ryssel [8] in the context of impurity doping.
Previously, Gill and Wilson [9], using RBS, found that the volume

concentration saturates at a value commensurate with SiO2. Namavar et al [3] and Tuppen et al [10] observe that the interface (gradient of the oxygen concentration profile) becomes more abrupt with increasing dose. This effect has been modelled by Maydell-Ondrusz [11] (see below).

In recent work Kilner et al [7,12,13] studied oxide layers formed by 15 keV, 150 keV, 200 keV and 300 keV O+ ion implantation. They found that the evolution of the oxygen distribution from a buried skew Gaussian [8] (low dose) to a flat topped distribution [7] (onset of oxide formation), which for very high doses may extend up to the silicon surface, is similar over this energy range. The ion dose (/cm*2) needed to form a surface oxide, rather than a buried layer, will be a function of energy, as this parameter controls both the ion range and straggle. Thus Dylewski and Joshi [14] and Hensel et al [15] used doses of < 10*18 O+/cm*2 to form surface oxides with 30 keV ions, whilst Izumi et al [16] formed buried layers by implanting 150 keV ions to a dose of 2 X 10*18 O+/cm*2. Formation of deep buried insulating layers by 2 MeV O+ ions was reported by Schwuttke et al [17] in 1969.

Compounds cannot be synthesised if they are unstable under ion bombardment [28,29]. Fortunately, SiO2 is chemically stable and, as discussed by Wilson [30], the electro-negativity and small size of the oxygen atom will readily lead to Si-O bonding. The reduction in bond length, compared to Si-Si, will limit the lattice strain. The enthalpies of formation of bonds [30] in the O/Si system are O-O (119 kJ/mole), O-Si (192 kJ/mole) and SiO2 (200 kJ/mole), so that existence of the second phase, SiO2, is energetically favourable providing

(i)   the volume concentration of oxygen is high enough

(ii) the energy barrier to the stable state can be surmounted by
      thermal activation or by defect barrier lowering [30].

The high bond strengths of C-O (257 kJ/mole) and N-O (150 kJ/mole) implies that the presence of high concentrations of carbon or nitrogen impurities will inhibit the formation of Si-O bonds during the early stages of the implantation.

The synthesis of SiO2 has been inferred by many workers using IR spectroscopy (Das et al [30], Dylewski and Joshi [14], Harbeke et al [31]). Stein [19] has examined silicon samples implanted at 500 deg C with 200 keV O+ at doses of 10*17/cm*2 and 2 X 10*18/cm*2. He observes strong absorption between 850 and 1100 wavenumbers cm*-1 in the low dose sample and associates the broad signal with two defects:

(i)   oxygen around Vox centres

(ii) Int(ox) in crystalline silicon.

The high dose sample (which has a flat topped oxygen concentration profile) has a narrow peak consistent with absorption by Si-O-Si bonds in the stretching mode. This absorption (9 microns line) is characteristic of amorphous SiO2 [32]. Kim et al [32] find that the oxide is formed during implantation at -181 deg C, 68 deg C and 535 deg C with the absorption peak moving to higher wave numbers with increasing temperature. This has subsequently been confirmed by Harbeke et al [31] who find that, for a given dose and energy, the position and width of the peak may be used to determine retrospectively the substrate temperature during implantation. This peak and other lines around

800 cm*-1 and 450 cm*1 are found to sharpen with increasing anneal temperature [30,19], which is an indication of structural changes in the buried oxide. Kim et al [32] note that the shift of the main peak from 1020 cm*-1 to 1100 cm*-1 is similar to the effect observed for evaporated SiO(x) when x increases from 1.5 to 2.0. However, such a large change in composition is not confirmed by SIMS [7], RBS [6,3], Auger [10] analyses. Harbeke [31] offers the alternative explanation that the shift is mainly due to relaxation of intrinsic compressive stress coupled with redistribution of oxygen bonds, in the wings of the oxygen depth profile. Stress relief [33] during annealing is expected as viscous flow of the oxide will occur above 950 to 1000 deg C. Harbeke notes that this process may be responsible for tensile stress relief in the silicon overlay (above a buried oxide), which is detected by Raman scattering. Vreeland and Jayader [34] find no residual strain normal to the surface after annealing an SOI structure at 1100 deg C in argon.

The physical evolution of the buried oxide has been studied by Hemment et al [35]. This group quantified the movement of the upper and lower interfaces of a buried oxide with increasing dose, using as a reference a buried Si3N4 marker layer formed by ion beam synthesis. Doses of upto 2.4 X 10*18 O+/cm*3 of 200 keV O+ ions were implanted at 500 deg C into the single crystal silicon overlayer above the marker. The structures were analysed by RBS. For implantations above the critical dose (about 1.4 X 10*18 O+/cm*2), growth of the oxide occurred predominantly at the upper interface. Subsequent SIMS analysis by Chater et al [36] confirmed that no growth occurred at the lower SiO2/Si interface. A similar phenomena has been reported by Todorov (1986) [37] for low energy implantations. Thus the growth of the oxide occurs through the rapid redistribution of each incremental dose of oxygen to the upper interface, where internal oxidation of unsaturated silicon atoms occurs. This is consistent with Kilner's [7] earlier O-18 tracer experiments, which showed that the oxygen exchanges with matrix oxygen in the preformed SiO2. Recently van Ommen (1986) [38] has proposed that a high concentration of self-interstitials (Int Si) builds up below the oxide as soon as it becomes a continuous layer and this inhibits further oxidation in this region (diffusion data can be found in [20]). The computer modelling by Jager et al [39] predicts this behaviour. Van Ommen's [38] postulation is supported by Hemment et al [35] who find that the Si3N4 marker layer which is a diffusional barrier to point defects, does not distort the oxygen profile from that formed in bulk silicon.

Precise determination of the physical properties of synthesised SiO2 has proved difficult as the material exists as a thin film (thickness typically 1000-3000 A) which, before annealing, has rather ill defined interfaces [6]. Estimates of the atomic density have been made by Dubus et al [4] using nuclear reaction analysis to determine the oxygen profile. They conclude that the density is similar to a thermal oxide (2.3 X 10*22 mole/cm*3). Bunker et al [44] determined the refractive index to be 1.48 for oxides formed at 500 deg C by implanting 1.9 X 10*18 O+/cm*2 at 160 keV. Levy et al [45] found that the refractive index is essentially constant with depth and is close to the ideal value of 1.45. Higher values near the interfaces may be associated with silicon inclusions [36]. Izumi (1986) [16] used C-V methods to determine the properties of a surface oxide formed by implanting 1 X 10*18 O+/cm*2 at 32 keV. Even before annealing the dielectric strength was as high as 60% of the value for a thermal oxide and was greatly improved by a modest anneal (900 deg C) whilst the interface charge density was reduced by annealing at 1150 deg C.

TABLE 2

Property	none	Anneal	
		900 deg C	1150 deg C
Boundary charge density (/cm*2)		4.7 X 10*11	2.1 X 10*11
Dielectric strength (V/cm)	5.1 X 10*6	8.5 X 10*6	8.5 X 10*6
Dielectric constant	1.40	3.82	3.83

The identity of defects in the synthesised SiO2 has recently been
studied by Barklie et al [46] using electron paramagnetic resonance.
These workers have identified the E1' centre, which is characteristic of
irradiated thermal SiO2 and is assumed to be an oxygen vacancy bonded to
an adjacent silicon atom. In samples implanted with 1.8 X 10*18 O+/cm*2
at 520 deg C a volume concentration of 2.5 X 10*18/cm*3 is found, which
is significantly smaller than the saturated value of 1.5 X 10*19/cm*3
seen in room temperature implanted oxides [47]. The concentration of
these centres is found to depend upon the oxygen dose and implantation
temperature over the range 250 to 600 deg C [48]. The centres are
annihilated at temperatures above 450 deg C.

The composition of the synthesised oxides is generally assumed to be
stoichiometric SiO2 as both RBS [3,53] and SIMS [6,7] data show signal
saturation at the values appropriate for SiO2 for doses greater than the
critical dose. However, experimental errors are typically +- 3% and
often as high as +- 5% to 10%, which introduces uncertainties in the
estimation of the composition. Hensel et al [15] report the formation of
overstoichiometric surface oxides, although the composition is not
conclusively established. The problem of interpretation of IR absorption
spectra has been discussed by Harbiki [31]. Namavar et al [3]
have suggested that the high mobility of free oxygen and silicon in SiO2
[6] inhibits the build up of excess concentrations during implantation
even at 77 deg K [3]. However, recently Scanlon et al [49] have
implanted O-18 (+) into SiO2 at 6 deg K and using RBS have found a build
up of oxygen above the stoichiometric value. During warm up to 300 deg K
the excess oxygen redistributes.

Transmission electron microscopy (TEM) has been used to study
the microstructure of SIMOX substrates [27,50-52]. The synthesised
oxide is found to be amorphous homogeneous with no voids
or bubbles. There is no evidence for recrystallisation, even during
an anneal at 1405 deg C. However, several groups [27,36,50] report that
silicon islands can be included in oxides formed with near critical doses.
The interfaces are ill defined before annealing but at high temperature,
above 1250 deg C, segregation of oxygen occurs, causing the thickness of
the oxide layer to increase, and so covering silicon islands may be
entrapped [54], especially at the lower SiO2/Si interface.

Nondestructive characterisation in terms of thickness, depth beneath the
surface, and interfacial abruptness of the synthesised oxide is highly
desirable. Narayan et al [55] have applied spectroscopic ellipsometry to
annealed samples implanted with 150 keV O+ ions to doses 1.6 X 10*18/cm*2
and 1.8 X 10*18/cm*2. Excellent agreement with TEM data, in terms of
the thicknesses of the multilayers (including a native oxide), was
achieved. From a study of the interference pattern it was shown that the
composition and thickness of the interfacial layers may be determined.
Narayan et al conclude that in the spectral range 1.36 eV to 4.3 eV it
is possible to measure layer thickness to a depth of 1.7 microns below

the surface. Levy et al [45] have highlighted the problems of correctly interpreting optical data (ellipsometric) and the need to use at least a four layer model to describe buried oxide structures. Bunker et al (1987) [44] have evaluated similar samples using optical reflectrometry and report oxide thicknesses which agree with SIMS and cross sectional TEM data to better than +- 1%. The optical data was analysed using a graded index, in the manner reported by Habler et al [56] for MeV N+ ion implantation into silicon.

Computer models describing the evolution of the oxygen depth profile from a skew Gaussian to a flat topped distribution have been reported by Hayashi et al [40] Ohwada et al [41], Dobson et al [42], Maydell-Ondrusz et al [11] and Jager et al [39,43]. The most detailed study is by Jager et al who consider diffusional transport of oxygen in the SiO2 and include the processes of sputtering, swelling and changes in the stopping power. Diffusion of excess oxygen to the wings of the distribution (interfaces), where internal oxidation occurs at a constant rate to form stoichiometric SiO2, is permitted. Jager has generalized the model to follow the redistribution of tracer implants of O-18. This permits critical tests of the model to be made by comparing the predictions with experimental SIMS depth profiles. Much better fits were achieved when skew-implantation profiles are used rather than symmetric Gaussian profiles.

The small oxide precipitates [26] in the wings of the oxygen distribution which form during implantation, may be dissolved during high temperature annealing. This was first reported by Jaussaud et al [52] and subsequently Stoemenos and Margail [26] and others [51,52] showed that the oxygen redistributes by a mechanism of diffusion controlled precipitate growth [57] and segregates on the buried oxide, causing the layer to grow in thickness. The resulting structure is planar and has abrupt SiO2/Si interfaces [35,54].

Izumi [18] pioneered the use of synthesised buried layers as a technology for the production of silicon-on-insulator (SOI) substrates for small geometry devices. This group proposed the acronym 'SIMOX' (separation by implanted oxygen). Currently much effort is directed to the study of these materials. Recent reviews are by Hemment (1986) [6], Stein (1986) [19] and Wilson (1987) [20]. High performance CMOS devices being realised by Davis [21], Chen et al [22] and Colinge [23]. Ruffell has described a dedicated high current (100 mA) O+ implanter which is suitable for commercial production of SIMOX substrates [24].

REFERENCES
----

[1]   M.Wantanabe, A.Tooi [Jpn. J. Appl. Phys. (Japan) vol.5 (1966)
       p.737]
[2]   J.F.Ziegler, J.Biersack, V.Littmark [The Stopping and Range
       of Ions in Solids (Pergamon, 1985)]
[3]   F.Namavar, J.I.Budnick, F.H.Sanchez, H.C.Hayden [Mater. Res.
       Soc. Symp. Proc. (USA) vol.45 (1985) p.317-21]
[4]   M.Dubus, J.Margail, P.Martin [Nucl. Instrum. & Methods Phys.
       Res. Sect B (Netherlands) vol.B15 (1986) p.59-62]
[5]   B-Y.Mao, P-H.Chang, C.E.Chen, H.W.Lain [J. Appl. Phys.
       (in press)]
[6]   P.L.F.Hemment [Mater. Res. Soc. Symp. Proc. (USA) vol.53 (1986)
       p.207]

[7]   J.A.Kilner et al [Nucl. Instrum. & Methods Phys. Sect.B
      (Netherlands) vol.B7/8 (1985) p.293-8]
[8]   H.Ryssel [Ion Implantation Techniques Range Distribution, Eds.
      H.Ryssel, H.Glawischnig (Springer-Verlag, Berlin 1982)]
[9]   S.S.Gill, I.H.Wilson [Mater. Res. Soc. Symp. Proc. (USA) vol.27
      (1984 p.275]
[10] C.G.Tuppen, G.J.Davis [J. Electochem. Soc. (USA) vol.131 (1984)
      p.1423]
[11] E.A.Maydell-Ondrusz, I.H.Wilson [Thin Solid Films (Switzerland)
      vol.114 no.4 (27 Apr 1984) p.357-66]
[12] P.L.F.Hemment, E.A.Maydell-Ondrusz, K.G.Stephens, J.A.Kilner,
      J.B.Butcher [Vacuum (GB) vol.34 (1984) p.203]
[13] S.D.Littlewood, J.A.Kilnes, J.P.Gold [in press]
[14] J.Dylewski, M.C.Joshi [Thin Solid Films (Switzerland) vol.35
      (1976) p.227 and vol.42 (1977) p.227]
[15] E.Hensel, K.Wollschlager, D.Schulge, U.Kressig, W.Skorupa,
      J.Finster [Surf. & Interface Anal. (GB) vol.7 no.5 (Oct 1985)
      p.207-10]
[16] K.Izumi [Nucl. Instrum. & Methods Phys. Res. Sect.B (Netherlands)
      vol.B21 no.2 (1987) p.124-8]
[17] G.H.Schwuttke, K.Brack [Acta Crystallogr. (Denmark) vol.A25 pt.3
      (1969) p.S43]
[18] K.Izumi, M.Dohen, H.Ariyoshi [Electron. Lett. (GB) vol.14 (1978)
      p.593]
[19] H.J.Stein [Fall Meeting Electrochem. Soc., San Diego, USA 1986]
[20] I.H.Wilson [in 'Ion Beam Modification of Insulators' Ch.7 Eds.
      P.Mazzoldi, G.Arnold (Elsevier Science Publication, Netherlands,
      1987)]
[21] J.R.Davis, A.E.Glaccum, K.Reeson, P.L.F.Hemment [IEEE Electron
      Device Lett. (USA) vol.EDL-7 no.10 (Oct 1986) p570-2]
[22] C.E.Chen, T.G.W.Blake, L.R.Hite, S.D.S.Malk, B.-Y.Mao, H.W.Lain
      [Int. Electron. Device Meeting Tech. Digest, (1984) p.702]
[23] J.P.Colinge,, T.I.Kamins [IEEE SOS/SOI Technology Workshop,
      Durango, Oct 1987]
[24] J.P.Ruffell, D.H.Douglas-Hamilton, R.E.Kaim, K.Izumi [Nucl.
      Instrum. & Methods Phys. Res. Sect.B (Netherlands) vol.B21 no.2
      (1979) p.229-34]
[25] M.Guerra [Ibis Technology Corp., 32A Cherry Hill Drive, Danvers
      MA 01923, USA]
[26] J.Stoemenos, J.Margail [Thin Solid Films (Switzerland) vol.135
      (2 Jan 1986) p.115-28]
[27] C.D.Marsh, J.L.Hutchinson, G.R.Booker, K.J.Reeson, P.L.F.Hemment,
      G.K.Celler [Microscopy of Semiconductor Materials, Oxford, 1987
      to be published in Inst. Phys. Conf. Ser. (GB)]
[28] H.M.Nagub, R.Kelly [Radiat. Eff. (GB) vol.25 (1975) p.1]
[29] R.Kelly [Radiat. Eff. (GB) vol.64 (1982) p.205]
[30] K.Das, J.B.Butcher, K.V.Anand [J. Electron Materials (USA) vol.13
      no.4 (1984) P.635]
[31] G.Harbeke, E.F.Steigmeier, P.L.F.Hemment, K.J.Reeson, L.Jastrzebski
      [Semicond. Sci. & Technol. (GB) vol.2 no.10 (10 Oct 1987) p.687-90]
[32] M.J.Kim, D.M.Brown, M.Garfunkel [J. Appl. Phys. (USA) vol.54 no.4
      (Apr 1983) p.1991-9]
[33] E.F.Steigmeier, R.Loudon, G.Harbeke, H.Auderset, G.Scheiber [Solid
      State Commun. (USA) vol.17 (1975) p.1447]
[34] T.Vreeland, T.S.Jayader [Mater. Res. Soc. Symp. Proc. (USA)
      vol.53 (1986) p.263]
[35] P.L.F.Hemment et al [Nucl. Instrum. & Methods Phys. Res. Sect.B
      (Netherlands) vol.B21, no.2 (1987) p.129-33]

[36] R.J.Chater, J.A.Kilner, P.L.F.Hemment, K.J.Reeson, R.F.Pearl
      [Electrochem. Soc., Spring Meeting, Boston, May 1986 ; Appl.
      Surf. Sci.(Netherlands) vol.30 (1987) p.390]

[37] S.S.Todorov, C.F.Yu, E.R.Fossum [Vacuum (GB) vol.36 nos.11/12
      (9186) p.129-32]

[38] A.H.van Ommen, B.H.Koek, M.P.A.Viegers [Appl. Phys. Lett. (USA)
      vol.49 no.11 (15 Sep 1986) p.628-30]

[39] H.U.Jager [Nucl. Instrum. & Methods Phys. Sect.B (Netherlands)
      vol.B15 (1986) p.748-51]

[40] T.Hayashi, H.Okamoto, Y.Homma [Inst. Phys. Conf. Ser. (GB) vol.59
      (1981) p.559]

[41] K.Ohwada, K.Izumi, T.Hayashi [Jpn. Ann. Rev. Electron Telecom.
      Semicon. Technol. vol.21 (1982) p.25]

[42] R.Dobson, R.P.Arrowsmith, A.E.Glaccum, P.L.F.Hemment [Int. Symp.
      Electron Ion Photon Beams, Los Angeles, USA, 1983]

[43] H.V.Jager, J.A.Kilner, R.J.Chater, P.L.F.Hemment, R.F.Peart,
      K.J.Reeson [Int. Conf. Energy Pulse and Particle Beam
      Modification of Materials, Dresden, GRD, Sept 1987]

[44] S.N.Bunker, P.Siashanse, M.M.Sanfacon, S.P.Tobin [Appl. Phys.
      Lett. (USA) vol.50 no.26 (29 Jun 1987) p.1900-2]

[45] M.Levy, E.Scheid, S.Cristoloveanu, P.L.F.Hemment, [Thin Solid
      Films (Switzerland) vol.148 no.2 (13 Apr 1987) p.127-34]

[46] R.C.Barklie, A.Hobbs, P.L.F.Hemment, K.Reeson [J. Phys. C (GB)
      vol.19 no.32 (20 Nov 1986) p.6417-32]

[47] R.A.B.Devine, A.Golanski [J. Appl. Phys. (USA) vol.54 no.7 (Jul
      1983) p.3833-8]

[48] R.c.Barklie, T.J.Ennis, P.L.F.Hemment, K.J.Reeson [4th Int. Conf
      Radiation Effects in Insulators Lyon, France July 1987]

[49] P.J.Scanlon, P.L.F.Hemment [Fall Meeting Mater. Res. Soc. Boston,
      USA, Nov 1987]

[50] B.-Y.Mao, P.-H.Chang, H.W.Lam, B.W.Shen, J.A.Keenan [Appl. Phys.
      Lett. (USA) vol.48 no.12 (24 Mar 1986) p.794-6]

[51] S.J.Krause, C.O.Jung, S.R.Wilson, R.P.Lorgan, M.E.Burnham [Mater.
      Res. Soc. Symp. Proc. (USA) vol.53 (1986) p.257]v

[52] C.Jaussaud, J.Stoemenos, J.Margail, M.Dupuy, B.Blanchard, M.Bruel
      [Appl. Phys. Lett. (USA) vol.46 no.11 (1 Jan 1985) p.1064-6]

[53] F.Namavar, J.I.Budnick, F.H.Sanchez, H.C.Hayden [Mater. Res. Soc.
      Symp. Proc. (USA) vol.45 (1985) p.317-21]

[54] P.L.F.Hemment et al [Vacuum (GB) vol.36 nos.11/12 (1986)
      p.877-82]

[55] J.Narayan, S.Y.Kim, K.Vedam, R.Manukonda [Appl. Phys. Lett. (USA)
      vol.51 no.5 (3 Aug 1987) p.343-5]

[56] G.K.Hubler, P.R.Malmberg, T.P.Smith, III [J. Appl. Phys. (USA)
      vol.50 no.11 (Nov 1979) p.7147-55]

[57] R.C.Newman et al [Physica B&C (Netherlands) vol.116 no.1-3 (Feb
      1983) p.264-30]

18.11 INTERNAL NITRIDATION OF Si BY ION BEAM SYNTHESIS

by P.L.F.Hemment

November 1987
EMIS Datareview RN=17838

Formation of Si3N4 by the implantation of N+ ions into silicon was
reported by Pavlov et al 1966 [1], with more detailed work subsequently
reported by these and other authors during the following decade [2-5].
More recent experiments have spanned the energy range from eV to MeV
and may be grouped:

(i)   high energy implantation (100 keV to 2 MeV) to synthesise
        buried nitride layers for optical elements [12] and silicon
        on insulator (SOI) structures [13-18, 48]

(ii)   medium energy (10-40 keV) to achieve local control of
        oxidation rates during device processing [9-11] and surface
        nitrides [40]

(iii)   low energy implantation (200 keV to 5 keV) to form a thin
        surface dielectric layer as a high integrity gate dielectric
        [6,7] or for surface barrier modification [8].

The synthesis of dielectric films by ion implantation has been reviewed
by Wilson (1987) [17] who concludes, from a consideration of the chemical
and structural stability [21] and bond energies (N-N, 227 kJ/mol: N-Si,
105 kJ/mol: Si3N4, 180 kJ/mol), that amorphous Si3N4 should be formed
during ion implantation of high doses (typically 10^{17}-10^{18} N+/cm*2)
of nitrogen. Stein [19] has discussed the partition of the energy of the
ions between ionisation and atomic collision processes. Depth
distributions of these implanted ions are generally skew Gaussians [18]
depending upon the incident-energy, and have a peak which is about 20%
deeper than the damage profile [18]. The displacement damage causes
cascades of vacancies and interstitials which at low temperature
(ambient) and for high doses (> 10^{15} N+/cm*2) will result in
amorphisation of the silicon.

Synthesis of Si3N4 has been inferred from IR absorption spectra
[2,5,13,22]. Stein [19] reports spectra from Si samples implanted at
550 deg C with 100 keV N+ ions over the dose range 2 X 10^{15}/cm*2 to 8
X 10^{17}/cm*2 for which the peak volume concentration lies between
10^{20}/cm*3 and 4 X 10^{22}/cm*3. The low dose samples show absorption
peaks at 760 cm*-1 and 960 cm*1, which Stein associates with N-N pairs
in crystalline silicon [20]. With increasing dose a broad absorption
peak occurs at about 850 cm*-1, indicating Si-N absorption in amorphous
regions, due to the formation of a buried layer of SiN(x) (centred near
the peak of the Gaussian depth profile). Skorupa et al [13] report
similar data for 330 keV N+ implants at about 500 deg C over the dose
range 0.9 X 10^{18}/cm*3 to 1.5 X 10^{18}/cm*2. Upon annealing at 1200 deg C
these workers and others [22,5], all observe sharp phonon bands at
about 845 cm*-1, 840 cm*-1 and 930 cm*-1, which are indicative of
crystalline nitride. Electron microscopy shows that the annealed nitride
exists as a well defined buried layer of predominantly poly-crystalline
alpha Si3N4 [13,15,16,23].

In two critical reviews [24,25] of the literature on silicon nitride Jennings et al [25] describe the synthesis of Si3N4 under thermal (equilibrium) conditions. They note that under these conditions silicon and nitrogen only react at temperatures > 1000 deg C and that in the binary system only Si3N4 exists, although the presence of impurities will result in the formation of tertiary compounds (e.g. Si3C3N [25], Si2N2O [26]). Si3N4 can exist in two forms (alpha-, beta-) both being stable upto at least 1500 deg C, and may be spontaneously formed provided the activation barriers to the reaction is overcome. Both have hexagonal symmetry [25].

beta   Si3N4 a = 7.608 A, c = 2.911 A, unit cell Si6N8

alpha Si3N4 a = 7.766 A, c = 5.6 A, unit cell Si12N16.

The presence of oxygen as a volume impurity may be required to stabilize the alpha phase [25].

Si3N4 can exist in various morphologies [25,27]; however the most common is a needle like structure (alpha-phase) with other forms being a fine grained structure (alpha-phase) and large grains and spikes (beta-phase). Both morphologies have been reported for Si3N4 layers synthesised by ion implantations [15,28].

The formation of buried nitride layers by implanting 100 keV to 300 keV N+ ions has been reviewed by Reeson [29] and Hemment [30]. A critical dose phi(c) must be exceeded to achieve a continuous buried layer (phi(c) approx. 1 X 10*18 N+/cm*2 for 200 keV ions). In order to maintain the crystallinity of the silicon overlay the substrates are implanted at an elevated temperature, typically 500 deg C, to achieve dynamic annealing. Meekison et al [15] have used TEM techniques and find the nitride layer to be amorphous. Other workers using SIMS [29,31] and Rutheford backscattering (RBS) [13,29] report that the volume concentration does not saturate at a value commensurate with stoichiometric Si3N4 but exceeds that value for doses > phi(c). The low diffusivity of nitrogen in Si3N4, even in the presence of excess point defects created by the ion bombardment (typical beam flux > 10 micro-amp/cm*2), is offered as the explanation for excess nitrogen. Stein [19] quotes equilibrium values of the diffusivity, where D = 0.87 exp -3.29/kT cm*2/sec [32] for nitrogen in silicon. For nitrogen in alpha- Si3N4 be quotes SQRT(D) = 10*-7 micron (hr)*(1/2) at 900 deg K, that is to say the nitrogen is essentially immobile. This model is supported by Namavar et al [33] who report that during implantation at a low flux (< (micro-amp/cm*2), when they postulate the supply of nitrogen does not exceed the out-diffusion rate, a uniform (homogeneous) nitride layer may be formed.

For high doses (> phi(c)) Meekison [15] and Petruzzello [23] report that the excess nitrogen is accommodated as bubbles which form a layer of porous nitride with amorphous nitride regions on either side in the wings of the nitrogen depth profile. This structure does not change significantly during a 7 day anneal at 600 deg C [34]. However, the nitride layer shows the onset of crystallisation at 1000 deg C and this process is complete after 1200 deg C for 2 hrs [34,23,15,16]. All of these groups report the formation of alpha-Si3N4 in samples implanted at temperatures of < 600 deg C, energies in the range 100 keV to 300 keV, and doses in the range 0.1 X 10*18 N/cm*2 to 1.5 X 10*18 N/cm*2.

The morphology of Si3N4 has been discussed by Jennings et al [24], see above. Recently Bussman et al [28] have implanted 150 keV N+ at higher temperatures and report the formation of a buried layer of beta- Si3N4. At an implantation temperature of 650 deg C precipitates of 1000 A-5000 A are observed by electron diffraction. At 800 deg C the precipitates are 2000 A - 8000 A in diameter. The minimum dose at which beta grains are observed is 0.5 X 10*18 N/cm*2 [28,35]. During high temperature annealing (1200 deg C) the beta phase mainly reverts to alpha-Si34N4, although some beta-grains remain imbedded in the matrix of alpha-Si3N4 [28].

Redistribution of the implanted nitrogen occurs during high temperature annealing [16,36,29], resulting in the formation of well defined buried layers of Si3N4 [15,16,48]. For low doses (< phi(c) this layer consists of polycrystalline Si3N4 and, because of this microstructure, the film is resistive in nature with current flow occurring along grain boundaries [37]. Higher doses (> phi(c)) enable single crystal regions to be formed [29] and these structures provide good vertical isolation and have been shown by Zimmer and Vogt [14] to be suitable as SOI substrates for small geometry CMOS devices. Unfortunately, if the dose is too high the stress associated with the nitrogen bubbles (see above) will cause mechanical failure by delamination [29,30]. For these reasons synthesised buried nitride layers are attracting little attention as SOI substrates.

The evolution and formation kinetics of Si3N4 by ion implantation has been discussed by Lezheiko et al [38]. These authors have prepared mechanisms for the phase build up. They propose that the nitride initially nucleates at the peak of the atomic depth profile and that nitrogen in the wings of the distribution diffuses to the phase boundary. Recently, Jager et al [39] have extended their phenomenological model of the build up of oxygen during synthesis of SiO2 to include the analogous N/Si system [40]. This detailed model includes diffusional transport of nitrogen along with sputtering, swelling and changes in the stopping power. These workers have determined profiles for 20 keV N+ ions implanted into silicon upto a dose of 0.45 X 10*18 N+/cm*2. The mean range and straggle of 0.25 keV to 2.5 keV N+ ions in silicon have been determined from Auger depth profiles by Malherbe [41].

Hobbs et al [47] have made EPR measurements on SOI substrates formed by the implantation of 200 keV N+ ions into silicon at a substrate temperature of 520 deg C. They observe only a single line (g = 2.0039+-0.0002 and line width 0.80+-0.04 mT) which they attribute to silicon dangling bonds within a nitrogen rich buried amorphous layer.

The formation of near surface nitride layers to inhibit the oxidation rate of silicon has been reported by Berruyer and Bruel [11]. These workers implanted 110 keV to 40 keV N2+ ions to doses between 10*16 to 10*17/cm*2. The resulting nitride layers were found to be efficient diffusional barriers to oxygen. Nitridation of silicon by ion implantation has also been reported by Theimer et al [10] and Lieb [42], who report a reduced oxidation rate in masked device structures implanted with, typically, doses of 10*15 to 10*16 N2+/cm*2 at 40 keV. The formation of silicon nitride during co-evaporation of silicon and ion implantation of 5 keV to 20 keV N+ ions has been confirmed by Auger and IR absorption [43]; again these layers find application for oxidation control. The synthesis of thin gate dielectric layers by low energy ion implantation has been discussed by Chiu et al [6].

The study of ultra thin surface nitrides formed by direct implantation or by energetic ions in a reactive ion etcher has been discussed by Chiu et al [6], Ringel and Ashok [8] and other workers [44,45].

Synthesis of buried oxynitride layers by the implantation of O+ and N+ or NO+ ions at 200 keV over the dose range 0.1 X 10*18/cm*2 to 1.6 X 10*18/cm*2 has been reported by Reeson et al [46].

REFERENCES
----

[1]   P.V.Pavlov, E.I.Zorin, D.I.Telel'baum, Yu.S.Popov [Sov. Phys.-Dokl. (USA) vol.10 (1966) p.786]

[2]   P.V.Pavlov, T.A.Kruze, D.I.Telel'baum, E.I.Zorin, E.W.Shitova, N.V.Godkova [Phys. Status. Solidi a (Germany) vol.A36 (1976) p.81]

[3]   F.L.Edelman, O.N.Kuymetsov, L.V.Leyhata, E.V.Lubopytova [Radiat. Eff. (GB) vol.29 (1976) p.13]

[4]   Y.Wada, M.Ahikawa [Jpn. J. Appl. Phys. (Japan) vol.15 (1976) p.1725]

[5]   R.J.Dexter, S.B.Watelski, S.T.Picraux [Appl. Phys. Lett. (USA) vol.23 no.8 (1973) p.455]

[6]   T.Y.Chiu, W.G.Oldham, C.Hovland, H.Ryssel, H.Glawischnig [Proc. IV Int. Conf. on Ion Implantation Equipment & Techniques, Berchtesgaden, 1982 (Springer-Verlag) p.465-72]

[7]   R.Hezel, N.Lieske [J. Electrochem. Soc. (USA) vol.29 no.2 (1982) p.379]

[8]   S.A.Ringel, S.Ashok [J. Electochem. Soc. (USA) vol.134 no.6 (1987) p.1494]

[9]   W.J.M.J.Josquin [Nucl. Instrum. & Methods (Netherlands) vol.209/210 pt.2 (1982) p.581]

[10]   J.Theimer, E.W.Maby, S.Lieb, R.K.MacCrone [J. Appl. Phys. Rev. (USA) vol.61 no.2 (1987) p.795]

[11]   P.Berruyer, M.Bruel [Appl. Phys. Lett. (USA) vol.50 no.2 (1978) p.89]

[12]   G.K.Hubler et al [Radiat. Eff. (GB) vol.48 (1980) p.81]

[13]   W.Skorupa, K.Wollschlaeger, U.Kreissig, R.Grolyschel [Nucl. Instrum. & Methods (Netherlands) vol.B19/20 (1987) p.285]

[14]   G.Zimmer, H.Vogt [IEEE ED-30 (1983) p.1515]

[15]   C.D.Meekison et al [Vacuum (GB) vol.36 nos.11/12 (1986) p.425]

[16]   J.Belz, E.H.te Kaat, G.Zimmer, H.Vogt [Nucl. Instrum. & Methods (Netherlands) vol.B19/20 (1987) p.279]

[17]   I.H.Wilson [in 'Ion Beam Modifications of Insulator' Eds P.Maggoldi, G.Arnold (Elsevier Sci. Publication BV, 1987) ch.7 p.245]

[18]   J.F.Ziegler, J.Biersack, V.L.Littmark [The Stopping and Range of Ions in Solids (Permagon, 1985)]

[19]   H.J.Stein [Fall Meeting Electrochem. Soc., San Diego, 1986]

[20]   H.J.Stein [Mater. Res. Soc. Symp. Proc. (USA) vol.59 (1986) p.523]

[21]   R.Kelly [Radiat. Effect. (GB) vol.64 (1982) p.205]

[22]   T.Tsujide, M.Nojiri, H.Kitagawa [J. Appl. Phys. (USA) vol.51 no.3 (1980) p.1605]

[23]   J.Petruzzello, T.F.McGee, M.H.Frommer [J. Appl. Phys. (USA) vol.58 no.12 (1985) p.4605]

[24]   H.M.Jennings [J. Mater. Sci. (GB) vol.18 no.4 (1984) p.451]

[25]   H.M.Jennings, J.O.Edwards, M.H.Richman [Inorganic Chimica Acta vol.20 (1976) p.167]

[26] K.J.Reeson [Radiation Effects in Insulators-4, Lyon, July 1987)]

[27] P.V.Pavlov, N.V.Belov [Sov. Phys.-Dokl. (USA) vol.23 no.8 (1978)
      p.534]

[28] U.Bussman, F.H.J.Meerback, E.H.te Kaat [Nucl. Instrum. & Methods
      (Netherlands) vol.B (1987)]

[29] K.J.Reeson [Nucl. Instrum. & Methods (Netherlands) vol.B19/20
      pt.1 (1987) p.269]

[30] P.L.F.Hemment [MRS European Meeting, Strasbourg May 1985]

[31] J.A.Kilner et al [7th Int. Conf. Ion Beam Analysis, Berlin, 1985]

[32A] A.H.Clark, J.D.MacDougall, K.E.Manchester, P.E.Roughan,
      F.W.Anderson [Bull. Am. Phys. Soc. vol.13 (1968) p.376]

[32B] K.Kijima, Sein-ichi Shirasaki [J. Chem. Phys. (USA) vol.65 (1976)
      p.2668]

[33] F.Namavar, J.I.Budmick, A.Fasihuddin, F.H.Sanchez, H.C.Hayden
      [Mater. Res. Soc. Symp. Proc. (USA) vol.53 (1986) p.281]

[34] P.H.Chang, C.Slawinski, B.-Y.Mao, H.W.Lam [J. Appl. Phys. Phys.
      (USA) vol.61 no.1 (1987) p.166]

[35] S.P.Wong, M.C.Poon [Nucl. Instrum. & Methods (Netherlands) vol.B22
      (1987) p.512]

[36] L.Nesbit, S.Stiffler, G.Slusser, H.Vinton [J. Electrochem. Soc.
      (USA) vol.132 (1985) p.2713]

[37] M.B.Kerger, R.Kwor, M.Zeller, P.L.F.Hemment, K.J.Reeson [Mater.
      Res. Soc. Fall Meeting, Boston, Dec 1987]

[38] L.V.Lezheiko, E.V.Lyubopytova, L.S.Smirmov [Z Tekh. Fiz (USSR)
      vol.51 no.4 (1981) p.818]

[39] H.U.Jager [Nucl. Instrum. & Methods (Netherlands) vol.B15 nos.1-6
      (1986) p.748]

[40] E.Sobeslavsky, H.U.Jager, W.S.Korupa, R.Wollschlager [Int. Conf.
      Energy Pulse & Particle Beam Modification of Materials, Dresden,
      GDR, Sep. 1987]

[41] J.B.Malherbe [Radiat. Eff. (GB) vol.70 nos.1-4 (1983) p.261]

[42] S.Lieb, R.K.MacCrone, J.Theimer, E.W.Maby [J. Mater. Res. vol.6
      (1986) p.792]

[43] F.Sugawara, T.Ajioka, F.Ichikawa, S.Ushio [Mater. Res. Soc. Symp.
      Proc. (USA) vol.54 (1986) p.535]

[44] H.Nakamura, M.Kaneko, S.Matsumoto, S.Fujita, A.Sarsaki [Appl.
      Phys. Lett. (USA) vol.43 no.7 (1983) p.691]

[45] R.Hezel, N.Lieske [J. Electrochem. Soc. (USA) vol.129 no.2 (1982)
      p.379]

[46] K.J.Reeson, et al [Radiaton Effect in Insulators, Lyon, July
      1987]

[47] A.Hobbs, R.C.Barklie, P.L.F.Hemment, K.Reeson [J. Phys. C (GB)
      vol.19 no.32 (20 Nov 1986) p.6433-9]

[48] E.H.te Kaat, J.Belz [Mat. Res. Soc. Symp. Proc. vol.45 (1985)
      p.329]

CHAPTER 19

STRUCTURE OF METAL/Si INTERFACES

19.1 STRUCTURE OF THE Ag/Si INTERFACE

by J.F.McGilp

July 1987
EMIS Datareview RN=13928

A. INTRODUCTION

It is now generally agreed that Ag is one of the few metals which forms
abrupt, unreactive junctions with Si (for recent reviews see Le Lay
[1], Calandra et al [2], and Hanbucken and Le Lay [3]). The extensive
intermixing found with other noble and near-noble metal/Si systems is
absent. There is no good evidence for silicide formation except,
possibly, for the top double layer of Si atoms. The Si(111)-Ag and
Si(100)-Ag systems have been extensively studied, but little work is
reported from the less technologically important Si(110) surface. The
Ag/Si interaction appears to be stronger on the (100) face than the
(111) face [3].

B. HIGH TEMPERATURE GROWTH

Above about 500 deg K both faces favour the Stranski-Krastanov (SK)
growth mode of initial 2-dimensional layer formation followed by
3-dimensional island growth [1-4]. For the (111) surface, the 2-D layer
is complete at about 2/3 monolayer (ML), while the 2-D layer on the
(100) surface is complete at about 1/4 ML.

C. ROOM TEMPERATURE GROWTH

At room temperature the mode appears to be closer to Frank-van der
Merwe (FVW), layer by layer, growth, but this is due to SK growth with
a high density of small, low aspect-ratio islands [4,5]. Small surface
area probes such as UHV-SEM [6], micro-AES [6], micro-RHEED [7] and
STM [8], have a major role to play in investigating the complicated
structure and growth of metal/Si interfaces. The recent results of
Tokutaka et al [9] show how the lateral averaging of conventional large
area surface probes can be misleading. It is also becoming clear that
the degree of surface perfection can have a substantial effect on
growth modes for these systems [4,10].

D. CRYSTALLOGRAPHY OF Si(111)-Ag

The complex nature of the crystallography of Si(111) surfaces is well
known, and this applies also to Si(111)-metal systems. Vacuum-cleaved
Si(111) surfaces show a 2 X 1 reconstructed LEED pattern. Depositing
less than 1 ML of Ag at room temperature superimposes a SQRT 7 X SQRT
7 - R 19.1 deg pattern. At higher coverages a 1 X 1 pattern appears,
corresponding to epitaxial Ag(111)//Si(111), with azimuthal orientation
[1 bar1 0]Ag//[1 bar1 0]Si [1] (or, presumably, anti-parallel to
this [11]). In contrast, initial deposition onto Si(111) 7 X 7 at room
temperature gives no additional LEED spots, but gives the increased
background intensity associated with disordered adsorption. Above about
1 ML the Ag(111) islands associated with the SK growth mode are detected
with predominantly parallel ([1 bar1 0]Ag//[1 bar1 0]Si), but some
anti-parallel, epitaxy [11]. Annealing sub-monolayer Ag films to above

- 725 -

450 deg K, or depositing above this temperature, gives a SQRT 3 X SQRT 3 - R 30 deg pattern, and annealing this to above 800 deg K, which is in the Ag desorption region, gives a 3 X 1 pattern. On cooling, a 6 X 1 pattern can be observed, which is thought to be a systematically depleted 3 X 1 phase [1]. The SQRT 3 structure is the most extensively studied, but no consensus has been reached. Takayanagi [12] has recently discussed various likely models. LEED IV analysis indicates that simple models of the longer range order are unlikely to be correct [13]. As regards the local bonding, SEXAFS [14], synchrotron photoemission [15, 16], and ISS [17,18] favour locating the Ag atoms in the 3-fold hollow sites of the Si surface, but there is disagreement over whether the Ag atoms are imbedded. The majority of work favours slight embedding [14-17,19], with the result that the interface could be considered as a double layer of silicide [15], but the recent ISS work of Aono et al [18] using Li ions, favours the Ag atoms located just above the topmost Si layer. New STM measurements [20,21] support the general features of the Si honeycomb model of Horio and Ichimaya [22], but there is still debate over whether the trimer of Ag atoms is embedded [21] or not [20]. Photoemission cross-section measurements indicate that the Si-Ag bonding is predominantly of s- and p-character [23].

E. CRYSTALLOGRAPHY OF Si(100)-Ag

The Si(100) surface is less complex than the (111) surface, and there is a consensus emerging that the 2 X 1 reconstruction revealed by LEED arises (mainly) from the asymmetric pairing of Si atoms at the surface [24]. Deposition of Ag does not lift the 2 X 1 reconstruction of the Si(100) surface [3]. The 3-D Ag islands formed above 1/4 ML have been reported to be Ag(111)//Si(100) [25], Ag(100)//Si(100) [26], and both [3], and it appears that surface perfection and evaporation rates are important in determining the epitaxy. The current model of the 1/4 ML 2-D film favours the Ag atoms sitting in the troughs which are present in the asymmetric dimer model of the Si(100) 2 X 1 surface [3,4]. This is consistent with the observation of a 2 X 1 reconstruction with modified diffraction intensities.

REFERENCES
----

[1]   G.Le Lay [Surf. Sci. (Netherlands) vol.132 (1983) p.169]
[2]   C.Calandra, O.Bisi, G.Ottaviani [Surf. Sci. Rep. (Netherlands)
       vol.4 nos 5/6 (1985) p.271-364]
[3]   M.Hanbucken, G.Le Lay [Surf. Sci. (Netherlands) vol.168 (1986)
       p.122]
[4]   M.Hanbucken, M.Futamoto, J.A.Venables [Surf. Sci. (Netherlands)
       vol.147 nos 2/3 (1984) p.433-50]
[5]   J.A.Venables, J.Derrien, A.P.Janssen [Surf. Sci. (Netherlands)
       vol.95 (1980) p.411]
[6]   M.Futamoto, M.Hanbucken, C.J.Harland, G.W.Jones, J.A.Venables
       [Surf. Sci. (Netherlands) vol.150 no.2 (1985) p.430-50]
[7]   M.Ichikawa, T.Doi, K.Hayakawa [Surf. Sci. (Netherlands) vol.159
       no.1 (1985) p.133-48]
[8]   F.Salvan, H.Fuchs, A.Baratoff, G.Binnig [Surf. Sci. (Netherlands)
       vol.162 (1985) p.634]
[9]   H.Tokutaka, K.Nishimori, N.Ishihara, T.Seki [Jpn. J. Appl.
       Phys. (Japan) vol.25 no.10 (1986) p.1584-88]
[10]   Q.-G.Zhu, A.-D.Zhang, E.D.Williams, R.L.Park [Surf. Sci.
       (Netherlands) vol.172 no.2 (1986) p.433-41]
[11]   E.J.van Loenen, M.Iwami, R.M.Tromp, J.F.van der Veen [Surf. Sci.
       (Netherlands) vol.137 (1984) p.1]

[12] K.Takayanagi [Mat. Res. Soc. Symp. Proc. (USA) vol.56 (1986)
      p.129-38]

[13] W.S.Yang, S.C.Wu, F.Jona [Surf. Sci. (Netherlands) vol.169
      nos 2/3 (1986) p.383-93]

[14] J.Stohr, R.Jaeger, G.Rossi, T.Kendelewicz, I.Lindau [Surf. Sci.
      (Netherlands) vol.134 (1983) p.813]

[15] G.Rossi, I.Abbati, I.Lindau, W.E.Spicer [Appl. Surf. Sci.
      (Netherlands) vol.11/12 (1982) p.348]

[16] F.Houzay, G.M.Guichar, A.Cros, F.Salvan, R.Pinchaux, J.Derrien,
      [Surf. Sci. (Netherlands) vol.124 (1983) p.L1]

[17] M.Saitoh, F.Shoji, K.Oura, T.Hanawa [Surf. Sci. (Netherlands)
      vol.112 (1981) p.306]

[18] M.Aono, R.Souda, C.Oshima, Y.Ishizawa [Surf. Sci. (Netherlands)
      vol.168 nos 1-3 (1986) p.713-23]

[19] S.Nishigaki, K.Takao, T.Yamada, M.Arimoto, T.Komatsu [Surf. Sci.
      (Netherlands) vol.158 (1985) p.473-81]

[20] R.J.Wilson, S.Chiang [Phys. Rev. Lett. (USA) vol.58 no.4 (1987)
      p.369-72]

[21] E.J.van Loenen, J.E.Demuth, R.M.Tromp, R.J.Hamers [Phys. Rev.
      Lett. (USA) vol.58 no.4 (1987) p.373-6]

[22] Y.Horio, A.Ichimaya [Surf. Sci. (Netherlands) vol.164 nos 2/3
      (1985) p.589-601]

[23] J.-J.Yeh, K.A.Bertness, R.Cao, J.Hwang, I.Lindau [Phys. Rev. B
      (USA) vol.35 no.6 (1987) p.3024-7]

[24] R.M.Tromp, R.J.Hamers, J.E.Demuth [Phys. Rev. Lett. (USA) vol.55
      (1985) p.1303]

[25] M.Hanbucken, H.Neddermayer [Surf. Sci. (Netherlands) vol.114
      (1982) p.563]

[26] T.Hanawa, K.Oura [Jpn. J. Appl. Phys. (Japan) vol.16 (1977)
      p.519]

19.2 STRUCTURE OF THE Al/Si INTERFACE

by C.A.Sebenne

June 1987
EMIS Datareview RN=15749

A. GENERAL FEATURES OF THE Si-Al INTERFACE

On clean single crystal silicon, aluminium forms an abrupt interface at
the atomic scale. The Al atoms of the first layer are covalently bonded
to the Si surface atoms. All the other Al atoms form a metallic layer
which may be an oriented single crystal with respect to the Si substrate
without any true epitaxial relation between the two. The details of the
atomic arrangement at the interface are not known but it is clear that
it depends upon substrate orientation and surface roughness on the one
hand and upon the thermal treatment applied to the system on the other
hand, as explained later. Besides the interface structure, Al also has
a p-type doping effect on silicon and a much higher diffusion constant
along both dislocation lines and grain boundaries than in the perfect
Si crystal; this will not be discussed any further here. In the
following, the structural properties during and after interface
formation are reported as a function of the preparation conditions,
limited to the cases where aluminium is vacuum deposited onto clean
crystalline Si surfaces, within the general structural frame given
above.

B. ROOM TEMPERATURE DEPOSITION OF Al ON Si(111) 2 X 1

Some time ago, Lander and Morrison [1] showed that the deposition of
Al on clean cleaved 2 X 1 - reconstructed Si(111) surfaces induces a
new surface reconstruction with a SQRT 3 X SQRT 3 R 30 deg unit mesh. It
seems at its best at 1/3 of a monolayer coverage (1 ML = 7.8 X 10*14
atoms/cm*2, the density of Si atoms in the last atomic plane of the
unreconstructed (111) face) and leads to a model where each Al atom
sits in a ternary site, either the hollow site H3 or the one above the
Si atom of the second layer T4. It will be discussed in the next
section. As shown by P.Chen et al [2], at 1/3 ML of Al, the work
function is decreased by 0.3 eV from its after-cleavage value of
4.87 eV, and the ionisation energy by 0.42 eV from 5.32 eV giving an
increase of the surface barrier (energy difference between the
conduction band edge and Fermi level at the surface) from 0.67 eV to
0.79 eV. It is reasonable to consider that the dangling bond surface
states are essentially removed at 1/3 ML of Al and replaced by a double
structure as deduced from results of Kobayashi et al [3,4]. However the
LEED diagrams always display a background indicating many misplaced Al
atoms and the surface state structure remains difficult to
understand [2].

When the Al coverage increases toward 1 ML, the surface structure
becomes 1 X 1. Kobayashi et al [3,4] favour Al in top position with
respect to Si surface atom. At this coverage, the surface does not show
any metallic character. Chen et al [2] give a further decrease of both
the work function and the ionisation energy by the same amount of
0.2 eV, keeping the same value of 0.79 eV for the surface barrier. The

double structure of the surface state band (0.4 eV and about 1.5 eV below the valence band edge from refs [2] and [4]) characterises this submonolayer Al coverage.

Upon further Al deposition, the metallic character of the layer starts to show up, the growth is island-like since the 1 X 1 LEED diagram of Si remains visible after a few ML. The islands are oriented: the 1 X 1 diagram of the Al(111) plane parallel to Si(111) becomes observable after 4 ML, with the Al[110] direction parallel to Si[110]. A thick Al layer gives a Schottky barrier height of 0.69 eV on n-type Si, as measured by Van Otterloo [5].

C. DEPOSITION OF Al ON Si(111) 7 X 7

The 7 X 7 reconstructed Si(111) clean surface is more stable than the cleaved 2 X 1, then upon Al deposition at room temperature the adsorption energy is not sufficient to change the surface unit mesh: in the half-monolayer coverage range, Lander and Morrison [6] showed that the 7 X 7 symmetry was kept while spot intensity changes indicated that Al atoms occupy definite positions and do not move at random along a fixed surface. It remains so upon heating up to 300 deg C. The electronic properties of this system have been studied by Margaritondo et al [7]; as on the cleaved surface, the ionisation energy and the work function decrease (by about 0.3 eV for less than 1 ML); up to 1 ML the layer does not display any metallic character and the dangling bond double structure (at the valence band edge and 0.6 eV below) of the clean surface transforms into a new double structure at about 0.3 and 1 eV below the valence band edge.

Upon annealing of a sample with less than 1/2 ML of Al at temperature between 400 and 700 deg C or upon deposition of the same amount on a similarly heated substrate, a well ordered SQRT 3 X SQRT 3 R 30 deg surface is observed [6]. It seems to correspond to 1/3 ML of Al and the atomic geometry is very likely the same as on the cleaved surface, as shown by Kelly et al [8].

The electronic properties of this system have been studied both experimentally and theoretically. Angular resolved photoemission by Uhrberg et al [9] and Kinoshita et al [10] gives slightly different results which lead to strongly different interpretations. The former find two dispersive surface states which agree with calculation by Northrup [11] where the Al atom is adsorbed in the ternary site T4 (on top of the second layer Si atom). Kinoshita et al [10] find three structures, two being surface related and prefer that with the Al atom adsorbed in the hollow ternary site H3. High resolution electron energy loss spectroscopy by Kelly et al [12] does not show any local vibration which could remove the uncertainty. The only remark is that if Al is on top of the second layer Si, the distance between Al and this Si underneath cannot be shorter than the sum of the covalent radii, that is 2.43 A since R(Al) = 1.26 A and R(Si) = 1.17 A; this is hardly compatible with a 2.43 A distance between the Al atom and the three Si surface atoms without large atomic displacements. Recent X-ray diffraction measurements on similar systems like Ge SQRT 3 X SQRT 3 - In by Feidenhans'l et al [19] show that the T4 adsorption site is the one to consider in this family of SQRT 3 X SQRT 3 reconstruction at 1/3 ML coverage.

Upon deposition close to the one ML range on heated substrates or after annealing, another structure shows up: a SQRT 7 X SQRT 7 surface unit first observed by Hansson et al [13] for which they propose a model where the Al coverage is 3/7 ML. The corresponding photoemission spectra measured by Kinoshita et al [10] do not indicate any drastic change of the electronic structure as compared to the SQRT 3 X SQRT 3 one.

Large deposition leads to a well oriented growth of crystalline Al as studied in detail by Legoues et al [14] using high resolution transmission electron microscopy. Room temperature deposition results in a multigrained film a large portion of which has the (111) orientation with Al[110]//Si[110] and Al(111)//Si(111). After annealing, it becomes a true pseudoepitaxy with the same orientation (perfectly oriented growth without matching of the two crystal parameters). Al deposition by ionised cluster beam, as done by Yamada et al [15] leads to the same pseudoepitaxial growth.

D. Al ON OTHER Si SURFACES AND Si ON Al

On Si(100), the growth process is comparable to that on the (111) face. Here again, after the first layer which removes the Si dangling bonds, a pseudoepitaxial growth tends to occur, the crystalline quality of the Al layer being improved by annealing beyond 400 deg C. Yamada et al [15] have found two orthogonal orientations, (i) Al(110)//Si(100) with Al[001]//Si[011] and (ii) Al(110)//Si(100) with Al[110]//Si[011].

On etched Si surfaces, various situations depending on the local composition may occur as demonstrated by the variety of Schottky barrier heights which have been reported by Rhoderick [16].

Reciprocally, Si can be deposited onto clean Al. Munoz et al [17] reported on Al(111) - first a 3 X 3 surface below half monolayer coverage continued by a layer by layer growth up to 2 ML, with the formation of a covalent bonding at the interface, as expected. This is in agreement with photoemission measurements of Si adsorbed on polycrystalline Al by Niles et al [18].

REFERENCES
----

[1]   J.J.Lander, J.Morrison [J. Appl. Phys. (USA) vol.36 (1965)
      p.1706]
[2]   P.Chen, D.Bolmont, C.A.Sebenne [J. Phys. C (GB) vol.17 (1984)
      p.4897]
[3]   K.L.I.Kobayashi, F.Gerken, J.Barth, C.Kunz [Solid State Commun.
      (USA) vol.39 (1981) p.851]
[4]   K.L.I.Kobayashi, Y.Shiraki, F.Gerken, J.Barth [Phys. Rev. B
      (USA) vol.24 (1981) p.3575]
[5]   J.D.Van Otterloo [Surf. Sci. (Netherlands) vol.104 (1981)
      p.L205]
[6]   J.J.Lander, J.Morrison [Surf. Sci. (Netherlands) vol.2 (1964)
      p.553]
[7]   G.Margaritondo, J.E.Rowe, S.B.Christman [Phys. Rev. B (USA)
      vol.14 (1976) p.5396]
[8]   M.K.Kelly, G.Margaritondo, J.Anderson, D.J.Frankel, G.J.Lapeyre
      [J. Vac. Sci. & Technol. A (USA) vol.4 (1986) p.1396]

[9]   R.I.G.Uhrberg, G.V.Hansson, J.M.Nicholls, P.E.S.Persson,
       S.A.Flodstrom [Phys. Rev. B (USA) vol.31 (1985) p.3805]
[10]   T.Kinoshita, S.Kono, T.Sagawa [Phys. Rev. B (USA) vol.32 no.4
       (15 Aug 1985) p.2714-16]
[11]   J.E.Northrup [Phys. Rev. Lett. (USA) vol.53 (1984) p.683]
[12]   M.K.Kelly et al [Phys. Rev. B (USA) vol.32 (1985) p.2693]
[13]   G.V.Hansson, R.Z.Bachrach, R.S.Bauer, P.Chiaradia [Phys. Rev.
       Lett. (USA) vol.46 (1981) p.1033]
[14]   F.K.Legoues, W.Krakow, P.S.Ho [Philos. Mag. A (GB) vol.53 (1986)
       p.833; and Mater. Res. Soc. Symp. Proc. (USA) vol.37 (1985)
       p.395]
[15]   I.Yamada, H.Inokawa, T.Takagi [J. Appl. Phys. (USA) vol.56
       (1984) p.2746] and [Thin Solid Films (Switzerland) vol.124
       (1985) p.179]
[16]   E.H.Rhoderick [Metal Semiconductor Contacts (Clarendon Press,
       Oxford, 1980) p.53]
[17]   M.C.Munoz, J.L.Sacedon, F.Soria, V.Martinez [Surf. Sci.
       (Netherlands) vol.172 (1986) p.422]
[18]   D.W.Niles, N.Tache, D.G.Kilday, M.K.Kelly, G.Margaritondo [Phys.
       Rev. B (USA) vol.34 (1986) p.967]
[19]   R.Feidenhans'l, J.S.Pedersen, M.Nielsen, F.Grey, R.L.Johnson
       [Surf. Sci. (Netherlands) vol.178 (1986) p.927]

19.3 STRUCTURE OF THE Au/Si INTERFACE

by J.F.McGilp

July 1987
EMIS Datareview RN=13934

A. GENERAL

The Au/Si interface is reactive, with substantial intermixing, even at
liquid nitrogen temperatures. An alloy rather than a stoichiometric
silicide tends to form (for recent reviews see Le Lay [1], Hiraki [2],
Calandra et al [3], and Hanbucken and Le Lay [4]). Si(111)-Au is one
of the most widely studied Si-metal systems, due to its technological
importance, and a significant amount of work has also been done on
Si(100)-Au. Little work has been reported for Si(110)-Au. Deposition
of thick Au films (200 nm) onto clean Si wafers, and subsequent heating
to 400 deg K in air, produces a 100 nm thick SiO2 film on the surface
of the Au within 10 minutes [2]. Rapid and extensive diffusion of Si
through the Au film to the surface is occurring, with subsequent
oxidation. This diffusion is significantly faster for the (100) face
than the (111) face [5]. A native oxide layer on the Si surface, if
defect-free, will inhibit this diffusion process [6]. Au films on these
oxidised Si surfaces do not adhere well, unless they are ion- or
electron-bombarded [7,8], in contrast to Au films on clean Si surfaces.
Recently Hung et al [9] have used a-CoTa films, successfully, as
diffusion barriers for Si/Au contacts on Si(100). In contrast, Ito and
Gibson [10] have shown that sequential deposition of Ag and Au films on
Si(111) does not inhibit diffusion, even though the Ag/Si interface is
unreactive. Au has been found to diffuse even more easily through
a-Si than c-Si [11], and to enhance crystallisation of both a-Si [12]
and polysilicon [13]. The specific growth mode of Au on c-Si differs
with crystal face and with temperature, and is strongly inhomogeneous,
necessitating further studies using small area surface probes. Much
remains to be done, also, in establishing crystallographic structure
and bonding.

B. HIGH TEMPERATURE GROWTH

Above about 700 deg K Si(111)-Au favours Stranski-Krastanov (SK) growth
of 3-dimensional islands on top of a 2-D intermediate layer. The
micro-AES studies of Calliari et al [14] show that 1-2 micron islands
occupy about 10% of the surface. The Au islands contain not more than
0.1% Si, and are covered by a Au-rich intermixed Au/Si film (not a
simple accumulation of Si atoms at the surface), estimated to be 3-4
ML thick. The intermediate layer between the islands, although also a
Au-rich alloy, is not the same as the island film, as determined from
Auger line-shape and oxygen-adsorption characteristics [14]. Cros et al
[15,16] had shown previously that intermixed, disordered Au/Si films
show enhanced oxidation tendencies, while ordered reconstruction of the
Au/Si layer passifies the surface. Calliari et al's results are
consistent with an ordered 2-D intermediate layer, but with a disordered
intermixed film on top of the islands. Clearly, the lateral averaging
which occurs with conventional large area surface probes will tend to
obscure this type of inhomogeneity [17], and this can be a serious
limitation with Si/metal systems. Si(100)-Au also favours SK growth
above 700 deg K [1,4] and, in general, appears to behave similarly to
Si(111)-Au.

C. ROOM TEMPERATURE GROWTH

Room temperature deposition onto both cleaved and annealed Si(111)
surfaces [1,3,18], and onto Si(100) [4], produces quasi-Frank-
van-der-Merwe (FVW), or layer-by-layer, growth. Demuth and Persson's
HREELS results for Si(111)-Au [19] provide good evidence for a
continuous film above 1 ML coverage. For this system it is generally
agreed that above 4 ML strong reaction and intermixing occurs, and
that a 1-2 ML thick surface layer comprising about 20% Si (close to
the Au/Si eutectic composition) is always formed. This alloy layer has
an enhanced reactivity to oxygen, relative to the clean surface [15]. A
similar alloy is formed when Si is deposited on Au [20], which supports
the view that the intermixed layer is not interface-specific. It has
been proposed that the room temperature Si(111)-Au system comprises a
Au film sandwiched between two Au/Si intermixed layers on the Si
substrate [1,3], but Taleb-Ibrahimi et al [18] have recently suggested,
on the basis of Auger intensity analysis and photoemission yield
spectroscopy, that the Si(111)-Au interface under the Au film is
abrupt and not intermixed. Support for the abrupt interface model
comes from surface second harmonic generation, a new technique for
probing the structure of buried interfaces [21,22]. It is not clear
whether the model also applies to the Si(100)-Au system.

At less than 4 ML there is controversy over whether a delayed onset of
intermixing occurs: the evidence is summarised by Hiraki [2] and
Calandra et al [3]. While it seems clear that strong bonding has
occurred by 1 ML, possibly involving embedding of the Au atoms, it is
less clear that a more complete intermixing, which moves Si atoms a
substantial distance from their lattice sites, has also occurred.
Recent ion channelling studies indicate Si atom movement at 1.5-2 ML for
Si(111) [13], and at 4 ML for Si(100) [23]. An interesting speculation,
within the context of the above interface model [18], is that the Si
atom movement seen in the channelling studies corresponds to the
segregation of the alloy layer away from the (111) substrate, as the Au
coverage exceeds 2 ML.

D. CRYSTALLOGRAPHY OF Si(111)-Au

Si(111)-metal systems have a complex crystallography. Vacuum-cleaved
Si(111) surfaces show a 2 X 1 reconstructed LEED pattern. Au deposition
at room temperature removes the superstructure spots, leaving the
substrate 1 X 1 pattern by 0.6 ML, and the LEED pattern fades completely
by 2 ML: disordered adsorption is occurring [18]. Deposition onto the
annealed Si(111) 7 X 7 structure gives similar results. At higher
coverages, recent probing of the crystallographic structure of the
buried interface using second harmonic generation indicates that,
although abrupt, the interface is largely disordered in the substrate
plane [21,22]. Parallel epitaxial growth at room temperature may occur
at high coverages with fast evaporation rates [24]. Deposition above
about 700 deg K or annealing to this temperature, produces a variety of
ordered structures, depending on coverage: up to 0.4 ML a 5 X 1
structure is seen; by 1 ML a SQRT 3 X SQRT 3 - R 30 deg pattern
dominates; above 1.5 ML a 6 X 6 structure is found [1-3].

The 5 X 1 structure has been studied extensively. There is general
agreement that an early suggestion by Lipson and Singer [25] is
correct. The basic crystallographic unit is 5 X 2, arising from strings

of Au atoms running along [110]. An ordered phase relationship between the strings gives the observed 5 X 1 pattern. Recent elegant micro-RHEED work by Ichikawa et al [26], and UHV-REM and TEM work by Takayanagi [27], shows that these structures nucleate at steps. The local bonding has been probed by ISS [28], AREELS [29], and X-ray standing waves [30]. The Au atoms are embedded in hollows just below the top layer of Si atoms. There is less agreement over the local bonding of the SQRT 3 structure. Recent ISS results suggest a triplet cluster of Au atoms slightly above top Si layer [31]. ARUPS [32] and inverse photoemission [33] show well-developed interface electronic states. STM work indicates local regions of 6 X 6 order in the overall SQRT 3 structure, which are attributed to lateral variations in the Au concentration [34]. The 6 X 6 structure of the 2-D intermediate layer of the SK growth mode, seen above 1.5 ML, has been described as a compressed Au(111) film [35] but, on the basis of the micro-AES results of Calliari et al [14], this is more likely to be an alloy film of this structure. The 3-D islands at higher coverages grow epitaxially, Au(111)//Si(111) and Au[1 bar1 0]//Si[1 bar1 0] [1-3].

E. CRYSTALLOGRAPHY OF Si(100)-Au

The Si(100) surface is less complex than the (111) surface. The 2 X 1 reconstruction arises mainly from the asymmetric pairing of Si atoms at the surface [36]. Hanbucken et al [37] have shown that Au deposition at room temperature results in the superstructure spots disappearing between 0.4 ML and 1 ML, leaving the substrate 1 X 1 pattern, and the LEED pattern disappears totally by 5 ML: disordered adsorption is again occurring. At higher coverages ex-situ TEM reveals the Au atoms in a channel structure, and parallel epitaxial growth occurs, Au(100)// Si(100) with Au[100]//Si[100]. Earlier work reported Au(111)//Si(100) [38], and it appears this orientation may occur with higher evaporation rates [37]. Above 700 deg K a number of ordered structures are found depending on coverage and annealing. Recent work reports 4 X 2, 8 X 2 and, above 1.5 ML, SQRT 26 X 1 patterns, the latter being the intermediate layer on which the 3-D islands grow [3]. A SQRT 26 X 3 structure has also been reported [1]. Further work is needed on this system to determine local structure and bonding.

REFERENCES
----

[1]   G.Le Lay [Surf. Sci. (Netherlands) vol.132 (1983) p.169]
[2]   A.Hiraki [Surf. Sci. Rep. (Netherlands) vol.3 no.7 (1984) p.357-412]
[3]   C.Calandra, O.Bisi, G.Ottaviani [Surf. Sci. Rep. (Netherlands) vol.4 nos 5/6 (1985) p.271-364]
[4]   M.Hanbucken, G.Le Lay [Surf. Sci. (Netherlands) vol.168 (1986) p.122]
[5]   C.Chang, G.Ottaviani [Appl. Phys. Lett. (USA) vol.44 (1984) p.901]
[6]   R.Anton [Thin Solid Films (Switzerland) vol.120 no.4 (1984) p.293-311]
[7]   A.E.Berkowitz, R.E.Benenson, R.L.Fleischer, L.Wielunski, W.A.Lanford [Nucl. Instrum. & Methods Phys. Res. Sect. B (Netherlands) vol.B7/8 (1985) p.877-80]
[8]   H.Dellaporta, A.Cros [Appl. Phys. Lett. (USA) vol.48 no.20 (1986) p.1357-9]

[9]   L.S.Hung, E.G.Colgan, J.W.Mayer [J. Appl. Phys. (USA) vol.60
       no.12 (1986) p.4177-81]
[10]   T.Ito, W.M.Gibson [J. Vac. Sci. & Technol. A (USA) vol.2 (1984)
       p.561]
[11]   K.Takita et al [Jpn. J. Appl. Phys. Part 2 (Japan) vol.24 no.12
       (1985) p.L932-4]
[12]   L.G.Hultman, P.A.Psaras, H.T.G.Hentzell [Mat. Res. Soc. Symp.
       Proc. (USA) vol.54 (1986) p.109-14]
[13]   C.R.M.Grovenor, D.A.Smith [Mat. Res. Soc. Symp. Proc. (USA)
       vol.25 (1984) p.305-8]
[14]   L.Calliari, M.Sancrotti, L.Braicovich [Phys. Rev. B (USA) vol.30
       (1984) p.4885]
[15]   A.Cros, J.Derrien, F.Salvan [Surf. Sci. (Netherlands) vol.110
       (1981) p.471]
[16]   A.Cros, F.Houzay, G.M.Guichar, R.Pinchaux [Surf. Sci.
       (Netherlands) vol.116 (1982) p.L232]
[17]   A.Taleb-Ibrahimi, C.A.Sebenne, F.Proix [J. Vac. Sci. & Technol.
       A (USA) vol.4 no.5 (1986) p.2331-5]
[18]   A.Taleb-Ibrahimi, C.A.Sebenne, D.Bolmont, P.Chen [Surf. Sci.
       (Netherlands) vol.146 (1984) p.229]
[19]   J.E.Demuth, B.N.J.Persson [Appl. Surf. Sci. (Netherlands)
       vol.22/23 (1985) p.415]
[20]   A.Franciosi, D.W.Niles, G.Margaritondo, C.Quaresima, M.Capozi,
       P.Perfetti [Phys. Rev. B (USA) vol.32 (1985) p.6917]
[21]   J.F.McGilp, Y.Yeh [Solid State Commun. (USA) vol.59 no.2 (1986)
       p.91-4]
[22]   J.F.McGilp [Semicond. Sci. & Technol. (GB) vol.2 (1987) p.102]
[23]   H.S.Jin, T.Ito, W.M.Gibson [J. Vac. Sci. & Technol. A (USA)
       vol.3 no.3 pt I (1985) p.942-5]
[24]   A.K.Green, E.Bauer [J. Appl. Phys. (USA) vol.47 (1976)
       p.1284]
[25]   H.Lipson, K.E.Singer [J. Phys. C. (GB) vol.7 (1974) p.12]
[26]   M.Ichikawa, T.Doi, K.Hayakawa [Surf. Sci. (Netherlands) vol.159
       no.1 (1985) p.133-48]
[27]   K.Takayanagi [Mat. Res. Soc. Symp. Proc. (USA) vol.56 (1986)
       p.129-38]
[28]   Y.Yabuuchi, F.Shoji, K.Oura, T.Hanawa [Surf. Sci. (Netherlands)
       vol.131 (1983) p.L412]
[29]   S.T.Li, S.Hasegawa, H.Iwasaki, Y.Mizokawa, S.Nakamura [Appl.
       Surf. Sci. (Netherlands) vol.22/23 (1985) p.275-9]
[30]   S.M.Durbin, L.E.Berman, B.W.Batterman, J.M.Blakely [Phys. Rev. B
       (USA) vol.33 no.6 (1986) p.4402-5]
[31]   K.Oura, M.Katayama, F.Shoji, T.Hanawa [Phys. Rev. Lett. (USA)
       vol.55 (1985) p.1486]
[32]   F.Houzay, G.M.Guichar, A.Cros, F.Salvan, R.Pinchaux, J.Derrien
       [J. Phys. C (GB) vol.15 (1982) p.7065]
[33]   J.M.Nicholls, F.Salvan, B.Reihl [Surf. Sci. (Netherlands) vol.178
       (1986) p.10]
[34]   F.Salvan, H.Fuchs, A.Baratoff, G.Binnig [Surf. Sci. (Netherlands)
       vol.162 (1985) p.634]
[35]   A.K.Green, E.Bauer [Surf. Sci. (Netherlands) vol.103 (1981)
       p.L127]
[36]   R.M.Tromp, R.J.Hamers, J.E.Demuth [Phys. Rev. Lett. (USA) vol.55
       (1985) p.1303]
[37]   M.Hanbucken, Z.Iman, J.J.Metois, G.Le Lay [Surf. Sci.
       (Netherlands) vol.162 (1985) p.628]
[38]   K.Oura, T.Hanawa [Surf. Sci. (Netherlands) vol.82 (1979)
       p.202]

19.4 STRUCTURE OF THE Ce/Si INTERFACE

by G.Rossi

August 1987
EMIS Datareview RN=15755

A. INTRODUCTION

Ce overlayers on n-type Si(111) 2 X 1 and Si(100) have been studied
by synchrotron radiation photoemission, LEED and angle resolved Auger
spectroscopies.

B. INTERFACE FORMATION ON Si(111) AT ROOM TEMPERATURE

A complex interface growth mechanism was derived from valence band and
medium-high resolution core level photoelectron spectroscopy. At the
submonolayer coverages two valence photoemission peaks associated with
Ce 4f and 5d states overlap the Si sp emission features; at 2-3 ML the
VB spectrum represents a metallic Ce-Si mixture; from 4 ML upwards sharp
Ce-related features grow towards a metallic Ce density of states which
is reached at 8 ML. The Si 2p core photoemission shows three components
after the rigid shift of eV due to the Schottky barrier, a chemically
shifted Si 2p doublet appears at 0.67 eV lower binding energy starting
at 0.6 ML, and another doublet shifted by 1.2 eV (lower BE with respect
to the unreacted doublet) appears for coverages larger than 3 ML [1].
From the selective coverage dependent attenuation of the three Si 2p
doublets it is proposed that: 1) Ce clusters in weak interaction with
the substrate form at the lowest coverages; 2) at 0.6 ML reactive
intermixing begins forming reacted patches of interface on the Si(111)
substrate; 3) above 3 ML the reaction, which procedes via lateral
growth of the reacted islands, saturates and pure Ce overgrows with
some Si segregated into it. LEED shows the disappearance of the 2 X 1
pattern at submonolayer coverages, which leaves a Si 1 X 1 pattern
in between the reacted islands. Polar profiles of the Auger emission
intensities confirm the heterogeneity of the interface [2]. The
clustering has been tentatively explained by considering that the heat
of formation of Ce-Ce clusters might be higher than the heat of
formation of Ce-Si bonds [3]. The clusters should ripen up to a critical
size before stimulating the reactive intermixing [1-6]. The first Ce-Si
reacted phase has a Si concentration of 0.55 +- 0.05 and exists in the
Ce coverage range 0.6-9 A. The diluted Si in metallic Ce has a Si
concentration of 0.08 +- 0.05 and exists in the Ce coverage range
2-40 A [7].

C. ANNEALED LAYERS ON Si(111) 7 X 7

SQRT 3 X SQRT 3 and 2 X 2 superstructures are obtained by annealing of
up to 1 ML of Ce on Si(111) at about 350 deg C (effective ordered
coverage = 0.35 ML) and about 450 deg C (= 0.2 ML) respectively [8].
Valence band emission features similar to CeSi2 are found for the SQRT 3
phase, but with less intense bonding peaks, interpreted as lower Ce-Si
coordination than in the silicide (Ce-6 Si). A stronger bonding
structure is suggested on the basis of the photoemission for the
2 X 2 phase.

D. INTERFACE FORMATION AT RT ON Si(100)

A sequence of Ce-silicides is found to form on the Si(100) substrate
as a function of the Ce coverage exceeding 1 ML [9]. The maximum Si 2p
core level chemical shift is about 1.3 eV (lower BE).

Magnetic behaviour is expected for Ce-Si reacted interfaces of average
stoichiometry lower than the disilicide [10,11].

E. THE Ce/Ge(111) INTERFACE

The Ce/Ge(111) interface shows the formation of three reacted phases
coexisting in the Ce coverage range 32-8 A; above 8 A the interface
becomes a two phase system [5,7]. The first reacted phase (90% Ge),
deduced from energy shifted Ge 3d photoemission, appears from the
lowest Ge coverage. A second reacted component (40% Ge) starts to
develop at above 2 A and this two phase system grows until the onset
of a diluted third phase (11% Ge) above 3 A. The first phase is
saturated at 8 A, the second phase at 19 A, and the final phase at
over 30 A. The growth of the Ce/Ge(111) interface appears to be a
sequence of diluted Ce in Ge phase, Ce-Ge compound formation, and
dilution of Ge in Ce phase [5,7].

Photoemission data exist on the related Ce/GaAs(110) interface system
[12].

For a review of rare earth/semiconductor interfaces see [13].

REFERENCES
----

[1]   M.Grioni, J.Joyce, S.A.Chambers, D.G.O'Neill, M.del Giudice,
        J.Weaver [Phys. Rev. Lett. (USA) vol.53 (1984) p.2331]
[2]   M.Grioni, J.Joyce, M.del Giudice, D.G.O'Neill, J.Weaver [Phys.
        Rev. B (USA) vol.30 (1984) p.7370]
[3]   A.Fujimori, M.Grioni, J.Weaver [Phys. Rev. B (USA) vol.33
        (1986) p.726]
[4]   A.Fujimori, M.Grioni, J.J.Joyce, J.Weaver [Phys. Rev. B (USA)
        vol.31 (1985) p.8291]
[5]   M.Grioni, M.del Giudice, J.J.Joyce, J.Weaver [J. Vac. Sci. &
        Technol. A (USA) vol.3 (1985) p.907]
[6]   M.Grioni, J.J.Joyce, J.Weaver [Phys. Rev. B (USA) vol.32 (1985)
        p.962]
[7]   R.A.Butera, M.del Giudice, J.Weaver [Phys. Rev. B (USA) vol.33
        (1986) p.5435]
[8]   A.Fujimori, M.Grioni, J.J.Joyce, J.Weaver [Phys. Rev. B (USA)
        vol.36 (1987) p.1075]
[9]   F.U.Hillebrecht, F.J.Himpsel [Proc. 8th Int. Conf. Vacuum
        Ultraviolet Radiation Phys., Lund, Sweden, 4-8 Aug 1986, Abstract
        vol.II Ed. P.O.Nillson]
[10]   H.Yashima, T.Satoh [Solid State Commun. (USA) vol.41 (1982)
        p.723]
[11]   H.Yashima, N.Sato, H.Mori, T.Satoh [Solid State Commun. (USA)
        vol.43 (1982) p.595]
[12]   J.Weaver, M.Grioni, J.J.Joyce, M.del Giudice [Phys. Rev. B (USA)
        vol.31 (1985) p.5290]
[13]   G.Rossi [Surf. Sci. Rep. (Netherlands) vol.7 (1987) p.128]

19.5 STRUCTURE OF THE Co/Si INTERFACE

by L.J.Chen

July 1987
EMIS Datareview RN=15761

A. SILICIDE FORMATION, FORMATION MECHANISM

Cobalt forms three silicides - Co2Si (formation temperature 350 -
500 deg C), CoSi (425 - 500 deg C) and CoSi2 (550 deg C) on
single-crystal silicon substrate [1-2].

B. KINETICS, DIFFUSING SPECIES

Co2Si and CoSi were found to grow according to diffusion-controlled
kinetics [2-5]. However, on single crystal silicon, CoSi2 grows with
complex kinetics, which results from a superposition of nucleation
and diffusion-controlled mechanisms [6,7]. TEM revealed that nuclei of
CoSi2 form at triple junctions of CoSi grain boundary and spread both in
normal and lateral directions [8]. The dominant diffusing species were
identified to be Co and Si for the growth of Co2Si and CoSi,
respectively [9-12]. The formation of CoSi2 was found to occur mostly
through the motion of metal atoms [6]. The activation energies of Co2Si,
CoSi and CoSi2 were measured to fall into the ranges of 1.5 +- 0.1-1.75,
1.75-1.9 +- 0.1, 2.3 +- 0.1-2.8 eV, respectively [2-7,13].

C. EPITAXY, INTERFACE STRUCTURE

Under non-ultrahigh vacuum deposition and annealing conditions, epitaxial
CoSi2 was found to be mixtures of type A and type B orientations, which
refer to orientations identical to that of the (111)Si substrate and
rotated 180 deg about the normal of the (111)Si surface, respectively
[14-16]. Single-crystalline CoSi2 was grown either by UHV deposition or
by a preannealing-laser annealing-postannealing scheme in a liquid phase
epitaxy regime [17,18]. Single-crystalline CoSi2 was grown on (111)Si
by non-UHV electron gun deposition of Co thin films followed by rapid
thermal annealing [19]. Buried single-crystal CoSi2 layers in silicon
have been formed by high dose implantation of cobalt followed by
annealing [20].

D. ORIENTATION EFFECT

The growth of Co2Si and CoSi as well as the nucleation and growth of
CoSi2 were found to be uninfluenced by the (001) or (111) orientation
of the silicon substrates [2-6]. However, single-crystal CoSi2 was
successfully grown on (111)Si but not on (001)Si by conventional
thermal annealing [17].

E. STABILITY

When a CoSi2 layer is placed between a single-crystal Si substrate and
an amorphous Si overlayer, annealed at about one-half the melting
temperature of the CoSi2, the silicide is seen to move through the

amorphous Si and rise to the surface, leaving below it an epitaxially
grown layer of Si [21].

F. SILICIDE LAYER IN OXIDISING AMBIENTS

When a film of CoSi2 on a SiO2 substrate is thermally annealed in an
oxidising ambient in the temperature range 750-1000 deg C, a SiO2
layer grows on the surface by depleting the film of its Si. The
oxidation of a CoSi2 film on (111)Si substrate establishes that the
displacement of Si atoms against Co films in the silicide does not
create a morphological instability [22]. In the temperature range of
650-1100 deg C, the growth rate was found to be independent of the
substrate orientation and the thickness of the CoSi2 layer. Activation
energies for the dry and wet oxidation are 1.49 +- 0.05 and 1.05 +-
0.05 eV, respectively [23]. Radioactive tracer experiments showed that
during wet oxidation of CoSi2 at 1000 deg C, silicon from the substrate
completely mixes with the silicon in the silicide layer before
oxidation at silicide-SiO2 interface, owing to the high self-diffusion
coefficient of silicon in CoSi2 at the oxidation temperature. In CoSi2,
silicon diffusion includes a substitutional mechanism [24].

G. INTERFACE BONDING, ELECTRONIC STRUCTURE, SURFACE STRUCTURE

The TEM images of the interfaces of type B CoSi2 with (111)Si appear to
agree better with the fivefold structure model, which refers to the five
coordination numbers of the atoms contained in the first CoSi2 layer at
the interface. However, some uncertainties surround the TEM data on
CoSi2 due to the possibility of structure relaxation in thin samples
[25].

A combination of in situ LEED, ARUPS and XPS measurements showed that at
room temperature the Co/Si(111) interface structure was a very thin
silicide layer covered by an unreacted Co metal film with a few Si atoms
dispersed in the metallic matrix and/or at the top surface. At high
temperature (about 600 deg C) nearly perfect epitaxial CoSi2 was
obtained on the (111)Si surface [26]. During the course of the CoSi2
formation, the presence of an intermediate silicide phase displaying
two-dimensional structure characteristics was detected [27]. With further
study, the two-dimensional structure was interpreted in terms of CoSi2
formation with a Co-rich (111) surface [28]. However, a LEED and AES
study of the reactions of a clean (111)Si surface with Co revealed the
presence of three 1 X 1 phases with increasing Co coverage. The
intermediate phase was found to be CoSi2, whereas the other two phases
were not identified [29].

Surface electron energy-loss fine-structure investigation of the
Co-(111)Si interface suggested the formation at room temperature of an
initial silicide phase followed by a nearly pure Co film. Upon
annealing at progressively higher temperature, sequential silicide
formation (Co2Si, CoSi, CoSi2) occurred [30].

Experimental and theoretical angular distribution of the Co LMM Auger
intensity revealed that a type B CoSi2 phase forms for one to thirty-
monolayer coverages of Co on (111)Si (annealed at 500 deg C). At
low coverages, the CoSi2 phase is in the form of clusters which appear
to be two to three CoSi2 layers in thickness and are terminated by a
(111)Si layer [31].

H. ION BEAM MIXING

Irradiations of Ar, Kr, and Xe ions on Co/Si at room temperature produced a layer of Co2Si near the original Co/Si interface [9]. The thickness of the Co2Si layer was found to be proportional to the square root of the dose [32]. Ar ion beam mixing of Co thin films thermally evaporated in poor vacuum on (111)Si was found to promote the growth of a uniform CoSi2 layer [33]. Implanted nitrogen atoms exhibited an interesting fast diffusion in Co films evaporated onto a crystalline silicon substrate. A nitride barrier layer forms at the Co-Si interface with a saturation areal density and prevents reaction between Co and Si for annealing temperatures up to 550 deg C [34].

I. ELECTRICAL PROPERTIES

The lowest values of resistivity listed for Co2Si and CoSi are 66.2 and 86 +- 16 micro-ohm cm, respectively [35]. The resistivity of CoSi2 is one of the lowest among all transition metal silicides. A value of 15 micro-ohm cm was reported for thin epitaxial film on (111)Si [36]. Schottky barrier heights for CoSi and CoSi2 on n-type silicon were measured to be 0.68 and 0.64 eV, respectively [37-39].

REFERENCES

[1]   K.N.Tu, J.W.Mayer [in 'Thin Film - Interdiffusion and Reactions'
      Eds. J.M.Poate, K.N.Tu, J.W.Mayer (Wiley, New York, USA, 1978)
      p.359-405]
[2]   G.J.van Gurp, C.Langereis [J. Appl. Phys. (USA) vol.46 no.10
      (Oct 1975) p.4301-7]
[3]   S.S.Lau, J.W.Mayer, K.N.Tu [J. Appl. Phys. (USA) vol.49 no.7
      (Jul 1978) p.4005-10]
[4]   K.N.Tu, G.Ottaviani, R.D.Thompson, J.W.Mayer [J. Appl. Phys.
      (USA) vol.53 no.6 (Jun 1982) p.4406-10]
[5]   C.D.Lien, M.A.Nicolet, C.S.Pai, S.S.Lau [Appl. Phys. A (Germany)
      vol.36 no.3 (Mar 1985) p.153-7]
[6]   F.M.d'Heurle, C.S.Petersson [Thin Solid Films (Switzerland)
      vol.128 no.3/4 (28 Jun 1985) p.283-97]
[7]   F.M.d'Heurle, P.Gas [J. Mater. Res. (USA) vol.1 no.1 (1986)
      p.205-21]
[8]   A.Appelbaum, R.V.Knoell, S.P.Murarka [J. Appl. Phys. (USA) vol.57
      no.6 (15 Mar 1985) p.1880-6]
[9]   M.A.Nicolet, S.S.Lau [in 'Materials and Process Characterisation'
      Eds. N.G.Einspruch, G.B.Larrabee (Academic, New York, USA, 1983)
      p.329-464]
[10] G.J.van Gurp, D.Sigurd, W.F.van der Weg [Appl. Phys. Lett. (USA)
      vol.29 no.3 (1 Aug 1976) p.159-61]
[11] G.J.van Gurp, W.F.van der Weg, D.Sigurd [J. Appl. Phys. (USA)
      vol.49 no.7 (Jul 1978) p.4011-20]
[12] A.P.Botha, R.Pretorius [Thin Solid Films (Switzerland) vol.93
      no.1/2 (9 Jul 1982) p.127-33]
[13] C.-D.Lien, M.A.Nicolet, S.S.Lau [Appl. Phys. A (Germany) vol.34
      no.4 (Aug 1985) p.249-51]
[14] R.T.Tung, J.M.Poate J.C.Bean, J.M.Gibson, D.C.Jacobson [Thin
      Solid Films (Switzerland) vol.93 no.1/2 (9 Jul 1982) p.77-90]
[15] L.J.Chen, J.W.Mayer, K.N.Tu, T.T.Sheng [Thin Solid Films
      (Switzerland) vol.93 no.1/2 (9 Jul 1982) p.91-7]

[16] L.J.Chen, J.W.Mayer, K.N.Tu [Thin Solid Films (Switzerland) vol.93 no.1/2 (9 Jul 1982) p.135-41]

[17] R.T.Tung, J.M.Gibson, J.M.Poate [Phys. Rev. Lett. (USA) vol.50 no.6 (15 May 1983) p.429-32]

[18] R.T.Tung, J.M.Gibson, J.M.Poate [Appl. Phys. Lett. (USA) vol.42 no.10 (26 Jan 1983) p.888-90]

[19] H.C.Cheng, I.C.Wu, L.J.Chen [Appl. Phys. Lett. (USA) vol.50 no.4 (26 Jan 1987) p.174-6]

[20] A.E.White, K.T.Short, R.C.Dynes, J.P.Garno, J.M.Gibson [Appl. Phys. Lett. (USA) vol.50 no.2 (12 Jan 1987) p.95-7]

[21] S.S.Lau, Z.L.Liau, M.-A.Nicolet [Thin Solid Films (Switzerland) vol.47 no.3 (1977) p.313-22]

[22] S.-J.Kim, T.C.Banwell, R.Shima, M.A.Nicolet [Proc. SPIE Int. Soc. Opt. Eng. (USA) vol.530 (1985) p.152-8]

[23] M.Bartur, M.A.Nicolet [Appl. Phys. A (Germany) vol.29 no.2 (1982) p.69-70]

[24] R.Pretorius, W.Strydom, J.W.Mayer, C.Comrie [Phys. Rev. B (USA) vol.22 no.4 (1980) p.1885-91]

[25] J.M.Gibson, J.C.Bean, J.M.Poate, R.T.Tung [Appl. Phys. Lett. (USA) vol.41 no.9 (1 Nov 1982) p.818-20]

[26] C.Pirri, J.C.Peruchetti, G.Gewinner, J.Derrien [Phys. Rev. B (USA) vol.29 no.6 (15 Mar 1984) p.3391-7]

[27] C.Pirri, J.C.Peruchetti, G.Gewinner, J.Derrien [Phys. Rev. B (USA) vol.30 no.10 (15 Nov 1984) p.6227-9]

[28] C.Pirri, J.C.Peruchetti, D.Bolmont, G.Gewinner [Phys. Rev. B (USA) vol.33 no.6 (15 Mar 1986) p.4108-13]

[29] S.C.Wu, Z.Q.Wang, Y.S.Li, F.Jona, P.M.Marcus [Phys. Rev. B (USA) vol.33 no.4 (15 Feb 1986) p.2900-2]

[30] E.Chainet, M.de Crescenzi, J.Derrien, T.T.A.Nguygen, R.C.Cinti [Surf. Sci. (Netherlands) vol.168 no.1-3 (Mar 1986) p.801-9]

[31] S.A.Chambers, S.B.Anderson, H.W.Chen, J.H.Weaver [Phys. Rev. B vol.34 no.2 (15 Jul 1986) p.913-20]

[32] B.Y.Tsaur [Ph.D. Thesis, California Institute of Technology (1980)]

[33] S.C.Edwards, R.A.Collins, G.Dearnaley [Vacuum (GB) vol.34 no.10/11 (1984) p.1017-9]

[34] K.T.Ho, M.-A.Nicolet [Thin Solid Films (Switzerland) vol.127 no.3/4 (24 May 1985) p.313-22]

[35] G.V.Samsonov, I.M.Vinitskii [Handbook of Refractory Compounds (IFI/Plenum Press, New York, USA, 1980)]

[36] S.Saitoh, H.Ishiwara, T.Asano, S.Furukawa [Jpn. J. Appl. Phys. (Japan) vol.20 no.9 (Sep 1981) p.1649-56]

[37] G.J.van Gurp [J. Appl. Phys. (USA) vol.46 no.10 (Oct 1975) p.4308-11]

[38] R.T.Tung, J.M.Gibson [J. Vac. Sci. & Technol. A (USA) vol.3 no.3 Part 1 (1985) p.987-91]

[39] S.M.Sze [Physics of Semiconductor Devices (Wiley-Interscience, New York, USA, 1981)]

19.6 STRUCTURE OF THE Cr/Si INTERFACE

by L.J.Chen

July 1987
EMIS Datareview RN=15765

A. SILICIDE FORMATION, FORMATION MECHANISM

CrSi2 is the only bulk silicide known to form on heat treatment of the
Cr/Si interface. The formation temperature of CrSi2 is 450 deg C, lower
than that of most refractory metal silicides (usually about 600 deg C)
[1-3]. Reference [4] and [5] reported a formation temperature of less
than 415 deg C.

B. KINETICS, DIFFUSING SPECIES

The thickness of CrSi2 was usually presented to be linearly dependent
on the annealing time in the temperature range of 415-525 deg C [4,6,7].
However, based on the results of a cross-sectional TEM study of the
growth kinetics and those shown in references [4,6] and [7], Natan and
Duncan argued that the data could also fit parabolic curves well [5].
The activation energies for linear kinetics were measured to be
(1.5-2.1) +- 0.1 eV, whereas a value of 2.3 +- 0.1 eV was obtained for
parabolic kinetics [4-6]. Experiments on the annealed Cr/Si interface
have shown that Si is the diffusing species during CrSi2 formation
[4,8-11]. The diffusion mechanism was likely to be Si substitutional
(vacancy) diffusion with a high self-diffusion coefficient [8].

C. EPITAXY, INTERFACE STRUCTURE

LEED data showed oriented growth of CrSi2 on (111)Si [12,13]. Epitaxial
CrSi2 was found to grow locally on (111), (001), and (011)Si [14,15].
In (111)Si the orientation relationships were found to be [0001]CrSi2//
[111]Si and (2 0 bar2 0)CrSi2//(2 0 bar2)Si. Interfacial dislocations
were identified to be of edge or 60 deg type with 1/6<112> Burgers
vectors [14]. The quality of the epitaxial CrSi2 layers, in terms of
size, extent of the silicon surface coverage, and the regularity of
interfacial dislocation of the epitaxial regions, was found to be the
best in (111)Si and the poorest in (011)Si. The quality of CrSi2 epitaxy
was found to correlate directly with the lattice match at the interfaces
[15].

D. ORIENTATION EFFECT

The growth kinetics of CrSi2 were found to be independent of the
substrate orientation [4,6].

E. STABILITY

When a CrSi2 layer is placed between a single-crystal Si substrate and
an amorphous Si overlayer, annealed at about one-half the melting
temperature of the CrSi2, the silicide is seen to move through the
amorphous Si and rise to the surface, leaving below it an epitaxially
grown layer of Si [16].

F. SILICIDE LAYER IN OXIDISING AMBIENTS

Synchrotron radiation photoemission showed that for Cr coverages on
(111)Si below and above a critical threshold coverage the overlayer
is not affected substantially and sharply enhances the oxygen
adsorption rate of silicon, respectively [17].

G. INTERFACE BONDING, ELECTRONIC STRUCTURES, SURFACE STRUCTURE

Formation of an intermixed phase showing definite metallic character at
the Cr/Si interface has been observed at room temperature for Cr
coverages up to 10-12 monolayers [18]. At higher coverage a Cr film is
formed on top of the intermixed phase and elemental Si segregation is
observed at the surface. Further studies of the metallic intermixed
phase have shown it to have a Si-rich average composition and to have
fundamentally different properties from CrSi2 [19].

LEED and AES studies showed oriented growth of CrSi2 on (111)Si by a
complicated procedure [12]. Cr films with 0.5-28 monolayer thickness
were deposited in a vacuum of 5 X 10*-10 torr and then annealed at
500 deg C for 1-7 min [13]. For 0.5-8 monolayer depositions, the Si
surface underwent reconstruction and Cr phases with two distinct
structures were found. For 10-28 monolayer Cr film deposition,
epitaxial CrSi2 film was formed.

H. REDISTRIBUTION OF DOPANTS, DOPANT AND CONTAMINANT EFFECTS

The redistributions of As and Kr atoms were studied for the cases where
As atoms were introduced in the Cr films or in the silicon substrate
[20,21]. When As or Kr was initially in Cr, it was incorporated in
CrSi2 during silicide formation and significantly reduced the reaction
rate; when As or Kr was initially in silicon, it accumulated at the
silicide/silicon interface with a less pronounced retarding effect than
that if As or Kr was present in Cr [20,21].

I. ION BEAM MIXING

Irradiations of Ar, Kr and Xe ions on Cr/Si at 250 deg C produced a
layer with composition CrSi2 near the original Cr/Si interface [9,22].
The thickness of the CrSi2 layer was found to be proportional to the
irradiation dose [22]. Si was found to be the faster diffusing species
in CrSi2 formation by Ar and Xe ion beam mixing [11,23]. Ion beam
induced CrSi2 formation by low energy light ions such as H, He and Ar
was found to be proportional to the number of knock ons [24]. The depths
of ion beam mixing after Si implantation of Cr thin films on silicon
showed linear dose dependence [25].

J. ELECTRICAL PROPERTIES

The electrical resistivities of CrSi2 thin films were measured to be about 600 micro-ohm cm [26,27]. Schottky barrier height of CrSi2 was measured to be 0.57 eV on n-type silicon [28]. CrSi2 was determined to be a p-type semiconductor with an energy gap of 0.27 eV [29].

REFERENCES

[1]   K.N.Tu, J.W.Mayer [in 'Thin Films - Interdiffusion and Reactions'
        Eds J.M.Poate, K.N.Tu, J.W.Mayer (Wiley, New York, USA, 1978)
        p.359-405]

[2]   K.E.Sandstrom, S.Petersson, P.A.Tove [Phys. Status Solidi a
        (Germany) vol.20 no.2 (1973) p.653-68]

[3]   R.W.Bower, J.W.Mayer [Appl. Phys. Lett. (USA) vol.20 no.9
        (1 May 1972) p.359-61]

[4]   A.Martinez, D.Esteve, A.Guivarc'h, P.Auvray, P.Henoc, G.Pelous
        [Solid-State Electron. (GB) vol.23 no.1 (Jan 1980) p.55-64]

[5]   M.Natan, S.W.Duncan [Thin Solid Films (Switzerland) vol.123 no.1
        (4 Jan 1985) p.69-85]

[6]   J.O.Olowolafe, M.-A.Nicolet, J.W.Mayer [J. Appl. Phys. (USA)
        vol.47 no.12 (Dec 1976) p.5182-6]

[7]   C.-D.Lien L.S.Wielunski, M.-A.Nicolet, K.M.Stika [Thin Solid Films
        (Switzerland) vol.104 no.1/2 (17 Jan 1983) p.235-42]

[8]   A.P.Botha, R.Pretorius, S.Kritzinger [Appl. Phys. Lett. (USA)
        vol. 40 no.5 (1 Mar 1982) p.412-4]

[9]   M.-A.Nicolet, S.S.Lau [in 'Materials and Process Characterisation'
        Eds. N.G.Einspruch, G.B.Larrabee (Academic, New York, USA, 1983)
        p.329-464]

[10]   F.M.d'Heurle, P.Gas [J. Mater. Res. (USA) vol.1 no.1 (1986)
        p.205-21]

[11]   L.S.Hung, J.W.Mayer, C.S.Pai, S.S.Lau [J. Appl. Phys. (USA)
        vol.58 no.4 (15 Aug 1985) p.1527-36]

[12]   K.Oura, S.Okada, T.Hanawa [Proc. 8th Int. Vacuum Congress,
        Cannes, France, 1981, Eds. F.Abeles, J.Croset p.181-7]

[13]   V.G.Lifshits, V.G.Zavodinskii, N.I.Plyusnin [Phys. Chem. & Mech.
        Surf. (GB) vol.2 no.3 (1984) p.784-801]

[14]   F.Y.Shiau, H.C.Cheng, L.J.Chen [Appl. Phys. Lett. (USA) vol.45
        no.5 (1 Sep 1984) p.524-6]

[15]   F.Y.Shiau, H.C.Cheng, L.J.Chen [J. Appl. Phys. (USA) vol.59 no.8
        (15 Apr 1986) p.2784-7]

[16]   S.S.Lau, Z.L.Liau, M.A.Nicolet [Thin Solid Films (Switzerland)
        vol.47 no.3 (15 Dec 1977) p.313-22]

[17]   A.Franciosi, S.Chang, P.Philips, C.Caprile, J.Joyce [J. Vac.
        Sci. & Technol. A (USA) vol.3 no.3 Part 1 (May/Jun 1985) p.933-7]

[18]   A.Franciosi, D.J.Peterman, J.H.Weaver [J. Vac. Sci. & Technol.
        (USA) vol.19 no.3 (Sep/Oct 1981) p.657-60]

[19]   A.Franciosi, D.J.Peterman, J.H.Weaver, V.L.Moruzzi [Phys. Rev.
        B (USA) vol.25 no.8 (15 Mar 1982) p.4981-93]

[20]   L.R.Zheng, J.W.Mayer [Appl. Phys. Lett. (USA) vol.45 no.6
        (15 Sep 1984) p.636-8]

[21]   L.R.Zheng, L.S.Hung, S.H.Chen, J.W.Mayer [J. Appl. Phys. (USA)
        vol.59 no.6 (15 Mar 1986) p.1998-2001]

[22]   B.Y.Tsaur [Ph.D. Thesis, California Institute of Technology
        (1980)]

[23]   K.Affolter, X.-A.Zhao, M.-A.Nicolet [J. Appl. Phys. (USA) vol.58
        no.8 (15 Oct 1985) p.3087-93]

[24] Y.Horino, N.Matsunami, N.Itoh, Y.Yamamura [Appl. Phys. Lett.
      (USA) vol.47 no.9 (1 Nov 1985) p.967-9]
[25] D.B.Poker, B.R.Appleton [J. Appl. Phys. (USA) vol.57 no.4
      (15 Feb 1985) p.1414-6]
[26] S.P.Murarka [J. Vac. Sci. & Technol. (USA) vol.17 no.4 (1980)
      p.775-92]
[27] S.P.Murarka, M.H.Read, C.J.Doherty, D.B.Fraser [J. Electrochem.
      Soc. (USA) vol.129 no.2 (Feb 1982) p.293-301]
[28] S.M.Sze [Physics of Semiconductor Devices (Wiley-interscience,
      New York, USA, 1981)]
[29] F.Nava, T.Tien, K.N.Tu [J. Appl. Phys. (USA) vol.57 no.6
      (15 Mar 1985) p.2018-25]

19.7 STRUCTURE OF THE Cu/Si INTERFACE

by J.F.McGilp

July 1987
EMIS Datareview RN=13929

A. GENERAL

The Cu/Si interface is reactive, with substantial silicide formation,
even at liquid nitrogen temperatures (for recent reviews see Calandra
et al [1] and Hanbucken and Le Lay [2]). The Si(111)-Cu system has
been quite widely studied, the Si(100)-Cu system less so, and there
has been no work reported on the Si(110)-Cu system. The Cu/Si
interaction appears to be stronger on the (100) face than the (111)
face [2]. Growth modes differ with crystal face, and with temperature.
In marked contrast to Au, there is no evidence of significant
penetration of Cu into the bulk (except for the (100) face at high
temperatures [2]), or surface segregation of Si after Cu deposition.
Further work is needed to establish both local bonding and
crystallographic structure.

B. HIGH TEMPERATURE GROWTH

Above about 550 deg K Si(111)-Cu favours Stranski-Krastanov (SK)
growth of 3-dimensional islands on top of a 2-dimensional intermediate
layer. At coverages above 20 monolayers (ML) there is good evidence,
from the spatially-resolved AES work of Calliari et al [3], that the
islands comprise copper silicides of undetermined composition, but with
an outer region close to the Cu/Si eutectic composition of Cu3Si. At
the lower coverages of 2 to 8 ML, Chambers and Weaver [4] have reported,
on the basis of SEM and micro-AES studies, that annealing to above 900
deg K gives islands of recrystallised Si, also covered by a Cu3Si skin.
At these higher annealing temperatures significant amounts of Cu are
lost from the surface, and bulk thermodynamics favours segregation into
Si and Cu3Si at low Cu concentrations [4]. A different growth mode,
involving significant penetration of Cu into the bulk, has been found
for Si(100)-Cu between 500 and 900 deg K by Hanbucken et al [5], and
this may result from the stronger Cu/Si interaction on the (100) face.
At 100 ML coverage, SEM and TEM show islands of undetermined
composition, but principally Cu, implanted in the silicon surface, and
covered with a 5 nm outer skin of Si. This, with the shallow Auger
escape depth, explains the remarkable absence of Cu AES signal up to
this coverage. The radically different results from the different
faces, and the dependence on coverage and annealing conditions, merits
further investigation using small area probes.

C. ROOM TEMPERATURE GROWTH

Room temperature deposition onto both cleaved and annealed Si(111) surfaces [6-9], and onto Si(100) [5], produces quasi-Frank-van-der-Merwe (FVM), or layer-by-layer, growth. Island formation at 3 ML and above has not been excluded and, indeed, a discontinuous film has been observed on the (100) surface by ex-situ TEM [5]. Above 1 ML the Si(111)-Cu system shows strong intermixing, and at higher coverages either a pure Cu phase, or an alpha-Cu(1-x)Si(x) alloy phase containing less than 10% Si, is formed. Taleb-Ibrahimi et al [9] have suggested a 2-step growth process involving 2-D alloy formation followed by Cu growth. Chambers et al [10] have used HREELS, combined with atomic hydrogen as a chemical probe, to show that, below 1 ML coverage, the Cu atoms cluster on the Si(111) surface, and that it is only above 1 ML that the surface is disrupted. This agrees with their previous work using angle-resolved AES [11]. The Si(100) surface appears to behave similarly, in general, to the Si(111) surface, except that no evidence has been found for this delayed onset of reaction [2,5].

D. CRYSTALLOGRAPHY OF Si(111)-Cu

Si(111) surfaces have a complex crystallography, and this also applies to Si(111)-metal systems. Vacuum-cleaved Si(111) surfaces show a 2 X 1 reconstructed LEED pattern. Deposition of Cu at room temperature has been reported to produce 4 X 1, 4 X 2, SQRT 3 X SQRT 3 - R 30 deg and 1 X 1 patterns as the coverage increases to 10 ML, although the spots are not sharp [9]. Above 10 ML a sharper 1 X 1 - R 30 deg overlayer pattern develops. In contrast, deposition onto Si(111) 7 X 7 at room temperature suppresses the superstructure spots, leaving the 1 X 1 substrate pattern at 1 ML [8]. This pattern disappears progressively, to be replaced by the same rotated 1 X 1 overlayer pattern as is observed with the cleaved surface. Epitaxial growth of either Cu(111) or alpha-Cu(1-x)Si(x)(111) is thought to be occurring. Daugy et al [8] have suggested that the epitaxy can be understood if the Cu atoms occupy 3-fold hollow sites on the unreconstructed substrate surface. Annealing at 900 deg K produces a quasi- 5 X 5 pattern, which may be incommensurate [8], for the intermediate layer of the high temperature SK growth mode. Chambers et al [12] have used Auger electron diffraction to probe the short range order of this structure and conclude that the Si(111) surface reconstructs to a planar geometry with the Cu atoms slightly below the surface in 6-fold hollow sites. Evidence of strong bonding between the Si and Cu at these annealed surfaces comes from their relative passivity to oxygen [7] and atomic hydrogen [10]. Inverse photoemission results indicate that the local bonding is similar in ordered Cu, Ag and Au overlayers [13].

E. CRYSTALLOGRAPHY OF Si(100)-Cu

Little work has been done here [2]. Hanbucken et al [5] have shown that deposition of Cu at room temperature removes the superstructure spots of the Si(100) 2 X 1 surface by 3 ML coverage. The substrate 1 X 1 pattern progressively weakens and has gone by 8 ML. Random adsorption appears to be occurring. At 800 deg K a 2 X 1 pattern can still be seen at 100 ML coverage. At higher coverages a 4 X 4 pattern appears [5].

REFERENCES
----

[1]   C.Calandra, O.Bisi, G.Ottaviani [Surf. Sci. Rep. (Netherlands)
      vol.4 nos 5/6 (1985) p.271-364]
[2]   M.Hanbucken, G.Le Lay [Surf. Sci. (Netherlands) vol.168 (1986)
      p.122]
[3]   L.Calliari, F.Marchetti, M.Sancrotti [Phys. Rev. B (USA) vol.34
      no.2 (1986) p.521-5]
[4]   S.A.Chambers, J.H.Weaver [J. Vac. Sci. & Technol. A (USA) vol.3
      no.5 (1985) p.1929-34]
[5]   M.Hanbucken, J.J.Metois, P.Mathiez, F.Salvan [Surf. Sci.
      (Netherlands) vol.162 (1985) p.622]
[6]   F.Ringeisen, J.Derrien, E.Daugy, J.M.Layet, P.Mathiez, F.Salvan
      [J. Vac. Sci. & Technol. B (USA) vol.1 (1983) p.546]
[7]   E.Daugy, P.Mathiez, F.Salvan, J.M.Layet, J.Derrien [Surf. Sci.
      (Netherlands) vol.152/153 (1985) p.1239]
[8]   E.Daugy, P.Mathiez, F.Salvan, J.M.Layet [Surf. Sci. (Netherlands)
      vol.154 no.1 (1985) p.267-83]
[9]   A.Taleb-Ibrahimi, V.Mercier, C.A.Sebenne, D.Bolmont, P.Chen
      [Surf. Sci. (Netherlands) vol.152/153 (1985) p.1228]
[10] S.A.Chambers, M.del Giudice, M.W.Ruckman, S.B.Anderson,
      J.H.Weaver, G.J.Lapeyre [J. Vac. Sci. & Technol. A (USA) vol.4
      no.3 pt II (1986) p.1595-8]
[11] S.A.Chambers, G.A.Howell, T.R.Greelee, J.H.Weaver [Phys. Rev. B
      (USA) vol.31 (1985) p.6402]
[12] S.A.Chambers, S.B.Anderson, J.H.Weaver [Phys. Rev. B (USA)
      vol.32 (1985) p.581]
[13] J.M.Nicholls, F.Salvan, B.Reihl [Surf. Sci. (Netherlands)
      vol.178 (1986) p.10]

19.8 STRUCTURE OF THE Eu/Si INTERFACE

by G.Rossi

August 1987
EMIS Datareview RN=15756

A. INTRODUCTION

Eu is a divalent rare earth metal (RE), with a 4f7 subshell
configuration; it forms a pseudo-disilicide EuSi(2-x) with the
alpha-ThSi2 crystal structure [1].

Photoemission and Auger spectroscopy studies on the growth of the
Eu/Si(111) 2 X 1, 7 X 7, and annealed interfaces are reported [2,4].

B. INTERFACE GROWTH AT ROOM TEMPERATURE (RT)

The interface formation on both Si(111) reconstructions proceeds via
the formation of one weakly interacting monolayer (ML) of chemisorbed
Eu, and the subsequent onset of the reactive intermixing at 2 ML. The
Si 2p core photoemission lineshape analysis for the interface on
cleaved Si [3,5] shows that a reacted Si signal chemically shifted by
1.6 eV towards higher binding energies (BE) appears at 2 ML. For higher
Eu coverages the whole Si emission is attenuated by the overgrowth of
an unreacted Eu layer; at 15 ML the interface is completely buried [2].
The Eu 4f final state photoemission multiplets are better seen with
soft X-rays [4]: broadening of the Eu 4f distribution is observed for
coverages exceeding 2 ML, and a gradual convergence to bulk Eu is
observed up to 15 ML. The interface is characterised by the formation
of a narrow mixed region rich in Eu which is not directly related to
precursor stages of the formation of the stable pseudo-disilicide.
Different thicknesses for the intermixed interface region are reported
in the case of cleaved Si or annealed 7 X 7 substrates, 5 ML and 10-15
ML respectively. Such differences could in fact be due to differences
in the deposition conditions and coverage determination accuracy.

C. ANNEALING OF Eu/Si INTERFACES

Annealing at 250-300 deg C of 30 ML Eu on Si(111) produces a divalent
non-metallic Eu/Si mixture of average composition EuSi(x) (x = approx.
4) which is stable versus annealing time [2]. The divalent character
of this diluted phase is consistent with the Mossbauer data for
implanted Eu in Si at concentrations varying between 1% and 10% [6].

For a review of rare earth/semiconductor interfaces see [7].

REFERENCES
----

[1]   J.Evers, G.Oehlinger, A.Weiss, F.Hullinger [J. Less Common Met.
       (Switzerland) vol.90 (1983) p.L19]
[2]   G.Rossi, J.Nogami, I.Lindau, J.J.Yeh [Surf. Sci. (Netherlands)
       vol.152/153 pt.2 (1985) p.1247]

[3]   J.Nogami, C.Carbone, J.J.Yeh, I.Lindau, S.Nannarone [Proc. 17th
       Int. Conf. Physics of Semiconductors, San Francisco, CA, USA
       6-10 Aug, 1984 Eds. J.Chadi, W.Harrison (Springer, New York, 1985)
       p.201]
[4]   J.Nogami [PhD Thesis, Stanford 1986]
[5]   G.Rossi, J.J.Yeh, I.Lindau, J.Nogami [Surf. Sci. (Netherlands)
       vol.152/153 pt.2 (1985) p.743]
[6]   J.A.Sawicki, T.Tyliszczak, B.D.Sawica, J.Kowalski [Phys. Lett. A
       (Netherlands) vol.91 (1982) p.414]
[7]   G.Rossi [Surf. Sci. Rep. (Netherlands) vol.7 (1987) p.128]

19.9 STRUCTURE OF THE Fe/Si INTERFACE

by R.B.Jackman

July 1987
EMIS Datareview RN=15792

A. INTRODUCTION

This Datareview considers the physico-chemical properties of the Fe/Si
interface, the onset of silicide formation and the influence of the
mode of overlayer generation. Ion-beam induced silicide formation is
not discussed. Many metal-silicon systems have been studied, with
particular reference to electronic device fabrication needs (contacts,
gate materials, etc.), and data on this system may be compared to
refractory metal junctions discussed in other Datareviews [1].

B. IRON OVERLAYER FORMATION

Despite early reports [2] of the structural influence of evaporating
Fe onto an Si(111) surface, data upon the initial growth of iron
monolayers upon a range of silicon crystals is still incomplete. Grant
and Haas [2] found evidence for a new surface structure, from LEED
measurements, following formation of a few monolayers of deposit on an
Si(111) 7 X 7 reconstructed surface. Chemical analysis of such layers,
and a study of the onset of silicide formation, has been carried out by
Zhu et al [3]. Following evaporation of 30 A Fe layers onto Si(111) AES
and ELS measurements suggested the iron remained inert to reaction with
the silicon. Interestingly, both Auger intensity data and plasmon loss
features indicated that a range of silicide formation reactions took
place on heating this surface. These are summarised in TABLE 1. No
evidence was found for the formation of Fe_3Si which has been studied by
ELS previously [4].

An alternative means of interface formation is Chemical Vapour
Deposition (CVD) from the metal carbonyl. Such experiments have been
widely demonstrated [5,6] to deposit Fe on a range of silicon
crystals. Decomposition of $Fe(CO)_5$ may be thermal, photolytic or
induced by electron impact. Only the influence upon the resultant
interface is considered here, detailed reviews being available
elsewhere [7]. Most studies have revealed significant incorporation of
carbon impurities at the interface and within the growing iron layer.
This results in electrical resistivities some 20-30 times the bulk
value of iron [6,8]. Such contamination was, however, observed to
prevent significant penetration of the Fe, leading to a more abrupt
interfacial region.

Reaction pathways following such Fe(CO)5 decomposition on Si(100) by
thermal, UV-photolytic and electronic means have been studied by AES,
TDS and LEED [9]. Interestingly, during UV irradiation at temperatures
of 300 deg K, the silicon was found to remain unperturbed. Most of the
carbonyl species desorbed from the surface; resulting in a largely
contaminant free (< 4% C) Fe/Si interface. Thus, the incorporation of
large quantities of interfacial carbon observed in some experiments
appears to be due to handling/vacuum considerations. Significant
interfacial desorption was reported from electronic or thermal
dissociation. In the former significant quantities of carbon arose
from unavoidable CO degradation by the impinging beam, whilst in the
latter the required temperatures were observed to lead to significant
Fe migration into the silicon bulk. Diffusion of iron within Si may
occur via interstitial sites accompanied by a second slower,
substitutional diffusion process [10]. However, when considering
diffusion at an Fe/Si interface it has been found [9] that the
relatively low bulk solubility value observed for iron in silicon
(about $10*16$ atoms/cm*3 [11]), and the the relatively high solubility
of Si in Fe (about 25% [12]), may inhibit the speed of effective
removal of surface Fe into the Si bulk.

C. SILICIDE FORMATION

Silicide deposits have been formed by evaporation of up to 1000 A Fe
layers upon a range of silicon crystals, and evidence for FeSi, Fe3Si
and FeSi2 observed [12,13]. Both orthorhombic (beta) and tetragonal
(alpha) FeSi2 are formed. TEM analysis by Cheng et al [14], employed
an N2 or low vacuum environment to anneal, in a single step, 300 A
layers or at temperatures up to 1100 deg C. At temperatures of around
400 deg C, FeSi was observed, in broad agreement with the thin layer
results of Zhu et al [3], but by 450-500 deg C small quantities of
Fe3Si were found. Si atoms have been found to be the moving species
during growth of an FeSi layer [15]. It seems likely that impurities
which were not present in thin-film studies carried out at UHV, may
inhibit such Si diffusion and encourage the formation of the metal-rich
silicide form Fe3Si. Beta-FeSi2 grains were dominant with little FeSi
remaining after annealing at 600 deg C. For samples heated above
800 deg C some alpha-FeSi2 was also detected. With regard to crystal
orientation, Si(111) substrates induced epitaxial FeSi2 growth above
900 deg C which accounted for 40% of the surface area. Both epi-beta-
FeSi2 and epi-alpha-FeSi2 were found, the former being dominant. The
observed epitaxial relationships are given in TABLE 2. In samples with
an Si(100) substrate only polycrystalline FeSi2 grains were found.
Silicide formation rates have been found to be slightly faster upon
Si(111) than Si(100) [12].

A two step annealing process was found [14] to enhance epitaxial FeSi2
growth on Si. This effect was attributed to the removal of impurities
from the interface during the first, lower temperature, step.
Variations in the exact ratios of beta and alpha FeSi2 in different
studies may be due to the differing amounts of oxygen present at the
interface during the various experimental approaches. It has been
suggested [14] that high O2 levels stabilise beta-FeSi2 whilst making
it more difficult to form alpha-FeSi2.

A study of the bonding states and bulk crystal structures of the range
of observed iron silicides has been presented by Egert and Panzner [4].

TABLE 1: the effect of annealing upon 30 A Fe layers on Si(111)

Temperature (deg C)	Stable Phase Observed by AES and ELS
280 or below	Fe
440-530	FeSi
590-670	FeSi2
670 or above	Some Si segregation to the top of the FeSi2 phase.

From Zhu et al [3]

TABLE 2: epitaxial relationships for FeSi2

Silicide Form	Expitaxial Relationship Observed
beta-FeSi2	(220)FeSi2//(111)Si (044)FeSi2//(004)Si
alpha-FeSi2	(100)FeSi2//(100)Si (010)FeSi2//(010)Si

From Cheng et al [14].

REFERENCES
----

[1]   See EMIS Datareviews on metal/Si interface structure
[2]   J.T.Grant, T.W.Haas [Surf. Sci. (Netherlands) vol.23 (1970)
        p.347]
[3]   Q.-G.Zhu, H.Iwasaki, E.D.Williams, R.L.Park [J. Appl. Phys. (USA)
        vol.60 no.7 (1986) p.2629-30]
[4]   B.Egert, G.Panzner [Phys. Rev. B (USA) vol.29 (1984) p.2091]
[5]   D.J.Ehrlich, R.M.Osgood,Jr, T.F.Deutsch [J. Electrochem. Soc.
        (USA) vol.128 (1981) p.2039]
[6]   S.D.Allen, A.B.Tringubo [J. Appl. Phys. (USA) vol.54 (1983)
        p.1641]
[7]   R.M.Osgood,Jr [Ann. Rev. Phys. Chem. vol.34 (1983) p.77]
[8]   P.J.Love, R.T.Loda, R.A.Rosenberg, A.K.Green, V.Rehn [Proc.
        SPIE Int. Soc. Opt. Eng. (USA) vol.459 (1984) p.25]
[9]   J.S.Foord, R.B.Jackman [Surf. Sci. (Netherlands) vol.171 no.1
        (1986) p.197-207]; R.B.Jackman, J.S.Foord [In prep. for Surf.
        Sci. (Netherlands)]
[10]   A.Seeger, W.Frank, V.Gosele [Diffusion in Elemental
        Semiconductors (Inst. Phys. Conf. Ser., 1979) ch.1]
[11]   C.B.Collins, R.O.Carlson [Phys. Rev. (USA) vol.108 (1957)
        p.1409]
[12]   S.S.Lau, J.S.Y.Feng, J.O.Olowolafe, M.A.Nicolet [Thin Solid
        Films (Switzerland) vol.25 (1975) p.415]
[13]   H.C.Cheng, L.J.Chen, T.R.Your [Mater. Res. Soc. Symp. Proc.
        (USA) vol.25 (1984) p.441]
[14]   H.C.Cheng, T.R.Yew, L.J.Chen [J. Appl. Phys. (USA) vol.57 no.12
        (1985) p.5246-50]
[15]   W.K.Chu, S.S.Lau, J.W.Mayer, H.Muller, K.N.Tu [Thin Solid Films
        (Switzerland) vol.25 (1975) p.393]

19.10 STRUCTURE OF THE Ga/Si INTERFACE

by C.A.Sebenne

June 1987
EMIS Datareview RN=15751

A. GENERAL FEATURES OF THE Si-Ga INTERFACE

Gallium forms an abrupt interface, at the atomic scale on clean single
crystal Si surfaces. At the interface the first atomic layer of Ga is
covalently bonded to the first atomic layer of Si and is essentially
non-metallic. The actual atomic geometry at the interface is usually
unknown, except in a few cases, and depends on gallium coverage, Si
surface orientation, initial reconstruction and initial roughness and
thermal history. Beyond the first layer, at room temperature and above,
Ga forms liquid metal islands covering a portion of the surface which
depends on the Ga average thickness and thermal treatment.

Ga is known to be a p-type doping element in Si; its diffusion into the
bulk either in the perfect crystal or along dislocations and grain
boundaries will not be discussed here.

B. ROOM TEMPERATURE DEPOSITION OF Ga ON Si(111) 2 X 1

Like Al (see [11] section B) room temperature deposition of Ga on a
cleaved 2 X 1 - reconstructed Si(111) surface transforms the initial
structure into a SQRT 3 X SQRT 3 R 30 deg which is at its best
between 0.3 and 0.4 ML (monolayer 1 ML = 7.8 X 10*14 atoms/cm*2, the
density of Si atoms in the last atomic plane of the unreconstructed
(111) face). It is convincingly modelled by a Ga atom covalently
bonded to three Si surface atoms, removing all dangling bonds at a 1/3
ML coverage as discussed by Bolmont et al [1]. The discussion about
the actual ternary adsorption site reported in the case of Al (see
[11]) is also valid in the case of Ga.

Freeouf et al [2] reported a decrease of the work function upon
submonolayer deposition of Ga. This is extended to the measurement of
the ionisation energy decrease as a function of calibrated Ga coverage
by Bolmont et al [1]. At 1/3 ML coverage the work function is decreased
by 0.27 eV from 4.87 down to 4.60 eV while the ionisation energy is
decreased by 0.38 eV from 5.33 down to 4.95 eV. At the same time the Si
surface dangling bond band is removed and replaced by a double
structure, one slightly inside the valence band at 0.15 eV below the
edge, the other at roughly 1.2 eV below the valence band edge.

On increasing Ga coverage close to 1 ML, the SQRT 3 X SQRT 3
structure is replaced by a Si 1 X 1 with a small complementary decrease
of both the work function and ionisation energy by about 0.1 eV leading
to a barrier height (distance between the Fermi level and conduction
band edge at the surface) of 0.78 eV at 1 ML coverage. This value is
close to the 0.72 eV Schottky barrier found by Van Otterloo [3] using
C-V measurement. The electronic structure remains essentially the same,
with no metallic character at 1 ML coverage.

Beyond 1 ML, liquid Ga metallic islands start to grow and coalescence is not yet observed at 100 ML since the 1 X 1 surface structure of Si remains observable.

C. Ga DEPOSITION ON Si(111) 7 X 7 AND ANNEALING

At coverage below 1 ML, room temperature deposition of Ga keeps the trace of the initial 7 X 7 reconstruction as mentioned by Margaritondo et al [4] who found a decrease of the ionisation energy similar to the one on the cleaved surface and a smaller decrease of the work function simply because the work function of the clean 7 X 7 surface is 4.70 eV as compared to 4.87 eV on the 2 X 1 surface. The surface state band is also replaced by a double structure, the position of which is very close to the one found on the cleaved surface at low coverages.

On annealing between 400 and 700 deg C or deposition on a heated substrate, the surface switches to the SQRT 3 X SQRT 3 R 30 deg reconstruction as shown by Kawazu et al [5]. It seems to be the same atomic geometry as on the cleaved surface, as well as in the case of Al deposition (see [11]), well modelled by 1/3 ML coverage with Ga in a ternary T4 site. Angular resolved UV photoemission spectroscopy by Kinoshita et al [6] displays a more detailed structure of the Ga induced surface state band; besides the non-dispersive surface state peak close to the valence band edge, as on the cleaved surface, there is a double peak at 1.0 and 1.25 eV below the valence band edge, the situation being the same as with Al deposition and, therefore, these authors favour again the hollow site H3 as the position of the Ga adsorbed atom. Angular XPS and Auger electron diffraction by Higashiyama et al [7] bring results compatible with either of the two adsorption sites.

Higher coverages do not bring any interesting new features.

D. OTHER Si-Ga INTERFACES

Some work has been done on clean Si(100) substrates. Sakamoto and Kawanami [8] found by RHEED measurements at increasing Ga coverages on a substrate heated in the 400-700 deg C range a series of successive surface unit meshes: first the initial 2 X 1 structure is replaced by a 3 X 2 from 0.15 to 0.33 ML, then a 5 X 2 is seen around 0.4 ML, it changes into a 2 X 2 at 0.5 ML which transforms into a 8 X 1 after 0.65 ML.

Room temperature deposition of Ga on clean Si(100) has been studied by Andriamanantenasoa [9]: slight indication of the 2 X 2 reconstruction is observable in the 0.5 ML coverage range; a smaller decrease of both the work function and ionisation energy than on the (111) surface is observed, completed at 1/2 ML coverage: its value is 0.4 eV and the dangling bond state band transforms smoothly into a double peak band at 1 ML coverage (0.25 and 0.65 eV below the valence band edge) with no metallic character yet. The Ga-Ga dimer model where each Ga is bonded to two Si substrate atoms and one Ga neighbour at 1/2 ML coverage as proposed by Knall et al [10] seems appropriate in the present case.

REFERENCES
----

[1]   D.Bolmont, P.Chen, C.A.Sebenne, F.Proix [Surf. Sci.
       (Netherlands) vol.137 (1984) p.280]
[2]   J.L.Freeouf, M.Aono, F.J.Himpsel, D.E.Eastman [J. Vac. Sci. &
       Technol. (USA) vol.19 (1981) p.681]
[3]   J.D.Van Otterloo [Surf. Sci. (Netherlands) vol.104 (1981)
       p.L205]
[4]   G.Margaritondo, J.E.Rowe, S.B.Christman [Phys. Rev. B (USA)
       vol.14 (1976) p.5396]
[5]   A.Kawazu, K.Akimoto, T.Oyama, G.Tominaga [Proc. 4th Int. Conf.
       Solid Surfaces 1980, Ed. Degras-Costa, Suppl. to Vide les Couches
       Minces (France) vol.201 p.1015]
[6]   T.Kinoshita, S.Kono, T.Sagawa [Solid State Commun. (USA) vol.56
       (1985) p.681]
[7]   K.Higashiyama, S.Kono, T.Sagawa [Surf. Sci. (Netherlands)
       vol.175 (1986) p.L794]
[8]   T.Saramoto, H.Kawanami [Surf. Sci. (Netherlands) vol.111 (1981)
       p.177]
[9]   I.Andriamanantenasoa [Thesis, Univ. Paris VII, Paris, France
       (1987) unpublished]
[10]   J.Knall, J.Sundgren, G.Hansson, J.Greene [Surf. Sci.
       (Netherlands) vol.166 (1986) p.512]
[11]   C.A.Sebenne [EMIS Datareview RN=15749 (June 1987) 'Structure of
       the Al/Si interface']

19.11 STRUCTURE OF THE Gd/Si INTERFACE

by G.Rossi

August 1987
EMIS Datareview RN=15757

A. INTRODUCTION

Photoemission data on the Gd/n-type, cleaved Si(111) interface at room
temperature have been reported [1].

B. ROOM TEMPERATURE GROWTH

The intermixing between Gd and Si takes place for coverages of the
order of one monolayer (ML), and is saturated for 7 ML [2]. The Si core
level intensity profile versus Gd coverages suggests heterogeneous
growth of the interface similar to the Ce/Si(111) interface [3].
Surface segregation of Si is observed for coverages larger than 3 ML
[2]. The Si 2p chemical shift for the intermixed phase is about 1.2 eV
and it is explained by charge transfer from Gd to Si. The Gd 4f7
electrons are found at higher binding energy in the mixed region with
respect to pure trivalent Gd. The top of the valence band, dominated by
the Gd 5d contribution, is directly measured in Gd intermetallics since
the 4f subshell lies at relatively high binding energy (> 6 eV BE).
A peak of the density of states is found at 2.5 eV BE which is
interpreted as being due to bonding Gd 5d - Si 3p states [1,2]. Ionicity
is considered to play a dominant role in the formation of Gd-Si bonds;
it is suggested that the charge transfer is compatible with a molecular
orbital bonding configuration having a small gap between the bonding
states and the empty antibonding states [1,2].

C. ANNEALING

Chemical reaction and silicide formation is completed, for Gd/Si bilayer
interfaces, for annealing at 350 deg C [4]. The Schottky barrier height
for such reacted interfaces is reported as 0.4 eV on n-type Si and
0.7 eV on p-type Si [3].

The disilicide GdSi2 is an orthorhombic deformation of the tetragonal
ThSi2-type, with lattice parameters a = 3.877 A, c = 4.172 A [5] which
give a 0.83% mismatch to Si(111) allowing pseudomorphic growth in
bombarded interfaces [6]. A defect structure for the pseudo-disilicide
has been suggested having a composition GdSi(1.4) [7].

A discussion of the rare earth-silicon bonding structure is found in
ref [8].

REFERENCES
----

[1]   C.Carbone, J.Nogami, I.Lindau [J. Vac. Sci. & Technol. A (USA)
      vol.3 (1985) p.972]
[2]   J.Nogami [PhD Thesis, Stanford 1986]
[3]   G.Rossi [EMIS Datareview RN=15755 (Aug 1987) 'Structure of the
      Ce/Si interface']
[4]   I.C.Cheng, S.S.Lau, R.D.Thompson, K.N.Tu [Mater. Res. Soc. Symp.
      Proc. (USA) vol.25 (1984) p.39]
[5]   A.Iandelli, A.Palenzona, G.Olcese [J. Less-Common Met.
      (Switzerland) vol.64 (1979) p.213]
[6]   J.A.Knapp, S.T.Picraux [Mater. Res. Soc. Symp. Proc. (USA) vol.54
      (1986) p.261]
[7]   J.A.Perri, E.Banks, B.Post [J. Phys. Chem. (USA) vol.63 (1959)
      p.2073]
[8]   G.Rossi [Surf. Sci. Rep. (Netherlands) vol.7 (1987) p.128]

19.12 STRUCTURE OF THE In/Si INTERFACE

by C.A.Sebenne

June 1987
EMIS Datareview RN=15752

A. GENERAL FEATURES OF THE Si-In INTERFACE

On clean single crystal silicon, indium forms an abrupt interface at
the atomic scale. The indium atoms of the first layer are covalently
bonded to the Si surface atoms. All the other In atoms form a more
or less uniform metallic layer, usually polycrystalline, which, in
some cases, shows a preferentially oriented growth. The details of the
various atomic arrangements at the interface are not known: it is clear
that they must depend on (i) substrate orientation, (ii) accommodation
of two strongly different covalent radii (1.44 A for In against 1.17 A
for Si), (iii) preparation conditions, including annealing, (iv)
surface roughness. Let us recall that In is a p-type doping
substitutional impurity in bulk Si with a low solubility and a small
diffusion coefficient (see [15]): the start of the diffusion process
at the interface will not be discussed here. In the following, the
structural properties, together with the corresponding electronic
properties, during and after interface formation are reported for
various cases, as they are presently known. In all cases, In is
thermally evaporated in vacuum on a clean Si surface.

B. ROOM TEMPERATURE DEPOSITION OF In ON Si(111) 2 X 1

As for the case of Al, Lander and Morrison [1] showed some time ago
that deposition of In on the clean cleaved, 2 X 1 reconstructed Si
surface induced a SQRT 3 X SQRT 3 R 30 deg surface unit mesh. It seems
at its best around 1/3 ML of In (where one monolayer, 1 ML = 7.8 X
10*14 atoms/cm*2, the density of Si atoms in the last atomic plane of
the unreconstructed (111) face) and leads to a model where In atoms
sit in ternary sites, either H3 or T4 (see the discussion in the case
of Al, [16]). The SQRT 3 X SQRT 3 reconstruction does not stay, it is
replaced by a 2 X 2 structure beyond 1/2 ML coverage while new spots
in low energy electron diffraction start to be observable beyond
1.2 ML indium coverage, which can be attributed to an oriented growth of
metallic indium, as claimed by Bolmont et al [2]. These authors, using
Auger electron spectroscopy, show that the first monolayer is
essentially uniform while islands start to form beyond that coverage.
The same group studied the electronic properties: at 1/3 ML, the work
function is decreased by 0.4 eV from its after-cleavage value of
4.87 eV, and the ionisation energy by 0.6 eV from 5.32 eV after
cleavage, leading to an increase of the distance between the Fermi level
and the conduction band edge at the surface from 0.67 eV to 0.87 eV: it
is interpreted as the removal of the surface dangling bond of three Si
atoms by each In atom, bringing a change in the dipole moment normal to
the surface which reduces the ionisation energy, and an empty surface
state deeper in the band gap of Si which pushes the Fermi level towards
the valence band edge. The electronic properties do not change much

upon increasing coverage up to 1 ML. However some dipole moment and/or empty state distribution changes must occur at higher coverages since Schottky barrier measurements by Van Otterloo [8] give a lower value at 0.73 eV compared to 0.87 eV below 1 ML. The shape of the surface state band upon increasing In coverage is difficult to understand because at any coverage between 0 and 1 ML, there is always a mixture of various structures, even at 1/3 ML.

The metallic character of the In layer appears only beyond 1 ML coverage. The pseudoepitaxial relation between oriented In metal crystallites and the Si substrate is not yet established.

C. DEPOSITION OF In ON Si(111) 7 X 7

The 7 X 7 reconstruction of clean Si(111) is more stable than the 2 X 1 and room temperature deposition of In, up to 1/2 ML, does not change it. However, Lander and Morrison [4] showed that spot intensities were changed, meaning preferential adsorption sites of In atoms in the 7 X 7 unit mesh (this is understandable knowing the atomic arrangement of the 7 X 7 reconstructed surface (see [17]). From the work of Margaritondo et al [5] it seems that no other reconstruction is observed at higher In coverages. As on the cleaved surface, these authors find a decrease of the ionisation energy and a lesser one of the work function, the values of which are comparable to the ones on the cleaved surface. However it seems to need one monolayer of In to be completed. Their UPS measurements indicate a removal of the clean silicon surface states close to Fermi level Ef and the growth of a peak about 1.2 eV below Ef which saturates above 1 ML coverage.

When a submonolayer of deposited In is annealed at least at 300 deg C the 7 X 7 superstructure transforms into a 4 X 1 unit mesh. Further annealing at increased temperature, in the 500 deg C range leads to SQRT 31 X SQRT 31 and finally to the SQRT 3 X SQRT 3 R 30 deg reconstructed surface, as reported in part by Lander and Morrison [4] and fully by Baba, Kawaji and Kinbara [6,7]. The accepted model for the SQRT 3 X SQRT 3 reconstruction is the same as for Al and Ga and for the deposition on the cleaved surface: 1/3 ML of In, each T4 site of the ideal Si(111) surface being occupied (see the discussion in [16]). This is confirmed by recent high resolution electron energy loss spectra measured by Kelly et al [8]. Surface state studies, using angular resolved UV photoemission spectroscopy compared to calculations were reported by Nicholls et al [9]: a main double peak mostly between 1 and 2 eV below Ef and small peak about 0.3 eV below Ef are observed, in good agreement with the observations of Bolmont et al [2] on the cleaved surface. The dispersion in k-space fits with the T4 adsorption site.

The other reconstructions, either the 4 X 1 or the SQRT 31 X SQRT 31, are not as clear. The absolute In coverage is not determined since it is somewhat below 1 ML and In migration occurs upon transformation into the SQRT 3 x SQRT 3, as discussed by Zhou [10] from reflection high energy electron diffraction: one part of In coverage (probably at the origin of the SQRT 3 X SQRT 3) is strongly fixed, the other part is much more mobile along the surface.

D. DEPOSITION OF In ON Si(100) 2 X 1

The surface structure and the growth process in the early stages of
In deposition on the clean, 2 X 1 reconstructed (100) face of Si have
been studied by Knall et al [11] as a function of In coverage from less
than 10*-1 ML to more than 10 ML and as a function of temperature
between 300 and 920 deg K. From reflection high energy electron
diffraction, low energy electron diffraction, Auger electron
spectroscopy and scanning electron microscopy, these authors propose the
phase diagram of the system which complements the results of a much less
extensive work reported earlier by Kuwata et al [12]. In brief, a
temperature threshold is found around 420 deg K.

Below it, upon increasing In coverage, the Si 2 X 1 is progressively
replaced by a 2 X 2 which is fully established at 1/2 ML coverage (on
Si(100), 1 ML = 6.8 X 10*14 atoms/cm*2). Then a 2 X 1 structure
recovers and remains the only observable one from 1 ML to about 3 ML.
At higher coverages, [011] oriented polyhedral islands of In metal
start to be observed. Above 420 deg K, the 2 X 1 initial structure is
progressively replaced by a 3 X 4 surface unit mesh which remains the
only one observable at 1/2 ML coverage and above. Hemispherical islands
show up above 3 ML. At coverage above 1/2 ML, and temperatures above
700 deg K a disordered In layer is added. Some Si(310) facets are
observed when the temperature is raised above 820 deg K.

Some work has been done on the electronic properties of the Si 2p core
level by Rich et al [13] and of the surface state band in the valence
band edge region by Andriamanantenasoa [14]. Both results focus on the
1/2 monolayer coverage as the breaking point where the clean Si(100)
dangling bonds are saturated and strongly support the formation of In
dimers bonded to four Si surface atoms, as proposed by Knall et al [11]
to model the 2 X 2 reconstruction. This model is also in agreement with
the smaller decrease of the ionisation energy observed in the present
case (less than 0.5 eV in ref [14]) as compared to more than 0.6 eV
when the same amount (or even less) is deposited on Si(111) 2 X 1 to
form the SQRT 3 X SQRT 3 (see section B).

REFERENCES
----

[1]   J.J.Lander, J.Morrison [J. Appl. Phys. (USA) vol.36 (1965)
        p.1706]
[2]   D.Bolmont, P.Chen, C.A.Sebenne, F.Proix [Surf. Sci.
        (Netherlands) vol.137 (1984) p.280]
[3]   J.D.Van Otterloo [Surf. Sci. (Netherlands) vol.104 (1981)
        p.L205]
[4]   J.J.Lander, J.Morrison [Surf. Sci. (Netherlands) vol.2 (1964)
        p.533]
[5]   G.Margaritondo, J.E.Rowe, S.B.Christman [Phys. Rev. B (USA)
        vol.14 (1976) p.5396]
[6]   M.Kawaji, S.Baba, A.Kinbara [Appl. Phys. Lett. (USA) vol.34
        (1979) p.748]
[7]   S.Baba, M.Kawaji, A.Kinbara [Surf. Sci. (Netherlands) vol.85
        (1979) p.29]
[8]   M.K.Kelly, G.Margaritondo, J.Anderson, D.J.Frankel, G.J.Lapeyre
        [J. Vac. Sci. & Technol. A (USA) vol.4 (1986) p.1396]
[9]   J.M.Nicholls, P.Martenson, G.V.Hansson, J.E.Northrup [Phys.
        Rev. B (USA) vol.32 (1985) p.1333]

[10] Zhou Jun-ming [Chin. Phys. (USA) vol.4 (1984) p.251]

[11] J.Knall, J.E.Sundgren, G.V.Hansson, J.E.Greene [Surf. Sci.
      (Netherlands) vol.166 (1986) p.612]

[12] N.Kuwata, T.Asai, K.Kimura, M.Mannami [Surf. Sci. (Netherlands)
      vol.143 (1984) p.L393]

[13] D.H.Rich et al [Phys. Rev. Lett. (USA) vol.58 (1987) p.579]

[14] I.Andriamanantenasoa [Thesis, Univ. Paris VII, Paris, France
      (1987) unpublished]

[15] D.De Cogan [EMIS Datareviews RN=16108 (July 1987) 'Solubility of
      In in Si; and RN=16104 (July 1987) 'Diffusion of In in Si']

[16] C.A.Sebenne [EMIS Datareview RN=15749 (June 1987) 'Structure
      of the Al/Si interface'

[17] P.J.Dobson [EMIS Datareview RN=11850 (Oct 1987) 'Surface
      structure of clean Si']

19.13 STRUCTURE OF THE Mo/Si INTERFACE

by L.J.Chen

July 1987
EMIS Datareview RN=15760

A. SILICIDE FORMATION, FORMATION MECHANISM

Hexagonal MoSi2 (h-MoSi2) and tetragonal MoSi2 (t-MoSi2) were found to
be the low temperature and high temperature phase formed in the
interfacial reactions of Mo thin films on silicon, respectively [1-5].
The transition temperature from h-MoSi2 to t-MoSi2 was found to vary
from 525 to 1000 deg C. Factors such as film thickness, deposition
method, doping impurity, and annealing ambient were found to exert
strong influence on the phase formation as well as the transition
temperature [5-10]. However, the transition temperature was generally
reported to be below about 800 deg C.

B. KINETICS, DIFFUSING SPECIES

Square root, linear and square time dependences for the thickness of
MoSi2 were reported. The substrate cleanness, substrate temperature,
deposition method, film thickness and annealing ambient were all found
to exert strong influence on the formation of silicides and the growth
of the phases. The activation energies were measured to vary from 2.2 -
4.1 eV [2-5,7,11-15]. The sample preparation methods were rather
diversified in these measurements. A more recent study by the cross-
sectional transmission electron microscopy showed that the thickness of
the growth phase, h-MoSi2, at 560-580 deg C is proportional to the
square root of annealing time. The activation energy was measured to be
2.3 eV [13]. A comprehensive list is given in reference [13]. Si was
found to be the dominant diffusing species for MoSi2 growth [11,16-18].

C. EPITAXY, INTERFACE STRUCTURE

Oriented growth of MoSi2 on silicon wafers heated to high temperatures
(about 1100 deg C) during Mo chemical vapour deposition was reported
[19]. Growth of t-MoSi2 with preferential orientation on (001)Si was
found to occur on heated substrates during Mo deposition under ultrahigh
vacuum conditions [20]. Epitaxial tetragonal and hexagonal MoSi2 were
grown locally on (001), (111) and (011)Si, by electron gun deposition of
Mo films on silicon followed by annealing at high temperatures in vacuum
[21,22]. Five different epitaxial modes, referring to sets of definite
orientation relationships between silicides and the substrate Si, were
identified for t-MoSi2, whereas three distinct modes were found for
h-MoSi2. Variants of epitaxy, required by the symmetry consideration,
were also observed. The roles of lattice match in the growth of
epitaxial MoSi2 were explored [23].

D. SILICIDE LAYER IN OXIDISING AMBIENTS

Heating of MoSi2 layers in dry and wet oxygen was investigated [24,25]. The oxidation in dry oxygen leads to an initial loss of metal in the form of volatile MoO3, which in turn leads to the formation of a SiO2 cap on the rest of the silicide. With continued oxidation, the thickness of SiO2 layer on top increases with time. The Mo/Si ratio in the underlying silicide was found to decrease with time. The wet oxidation of MoSi2 causes the silicide layer to become metal rich when an SiO2 layer grows. The formation of the volatile MoO3 was found to disrupt the silicide layer [25].

E. REDISTRIBUTION OF DOPANTS, DOPANT AND CONTAMINANT EFFECTS

Arsenic atoms implanted in the Si substrate were found to redistribute toward the MoSi2/Si interface, but not into the underlying Si [26].

F. INTERFACE BONDING, ELECTRONIC STRUCTURE, SURFACE STRUCTURE

Mo/(111)Si interface grown at approx. 85 deg K was studied by UPS. In spite of the sharpness of the interface the electron states in the transition region show the same qualitative features as refractory metal silicides [27]. Intermixing was found to take place at room temperature by synchrotron radiation photoemission measurement [28]. However, a combined UPS, LEED and AES study of the first stage of Mo/(111)Si interface formation indicated a layer by layer growth where no intermixing occurs at room temperature. At 300 deg C, interface reaction starts to take place. MoSi2 forms at about 600 deg C [29]. XPS and XAES have been used to follow the evolution of the electronic structure of the interface formed when submonolayer amounts of Mo atoms are evaporated onto a clean (111)Si surface. In comparison with the more extensively investigated Ni/Si interfaces, the main differences are a weaker chemical shift of the Mo core-levels, an opposite charge transfer in the metal atoms and an indication of a much more abrupt interface [30].

G. ION BEAM MIXING

Ar ion bombardment on Mo thin films evaporated on silicon was found to induce the migration of Mo atoms into the underlying substrate [31]. Hexagonal MoSi2 was produced by P- and As-ion implantation through Mo films deposited on Si substrates [32,33]. Tetragonal and hexagonal MoSi2 as well as hexagonal Mo5Si3 were found to form mainly along Mo grain boundaries in CVD deposited Mo thin films on (001)Si as a result of Ar ion beam mixing [34]. The backscattering results indicated that As atoms are snowplowed into Si during the formation of the silicide [6]. Uniform, stoichiometric, and low-resistivity MoSi2 layers were formed by implanting As ions through Mo films deposited on silicon followed by rapid thermal annealing [35]. B ion, a light ion, was found to be ineffective for smooth film formation. Heavy ions such as As or P or molecular ions such as BF2 were found to be effective [36]. The amount of Si atoms intermixed into the silicide by Ar ion beam mixing was found to be proportional to the square root of the dose [37].

H. ELECTRICAL PROPERTIES

The electrical resistivities of MoSi2 thin films on silicon were measured
to be in the range of 60-200 micro-ohm cm [32,38-40]. Schottky barrier
height of t-MoSi2 was measured to be 0.55 eV on n-type silicon [40].

REFERENCES

[1]   K.N.Tu, J.W.Mayer [in 'Thin Films - Interdiffusion and Reactions'
       Eds. J.M.Poate, K.N.Tu, J.W.Mayer (Wiley, New York, USA, 1978)
       p.359-405]
[2]   R.W.Bower, J.W.Mayer [Appl. Phys. Lett. (USA) vol.20 no.9
       (1 May 1972) p.359-61]
[3]   B.Oertel, R.Sperling [Thin Solid Films (Switzerland) vol.37 no.2
       (1 Sep 1976) p.185-94]
[4]   B.I.Fomin, A.E.Gershinskii, E.I.Cherepov [Talanta (GB) vol.24
       no.3 (1977) p.192-4]
[5]   A.Guivarc'h et al [J. Appl. Phys. (USA) vol.49 no.1 (Jan 1978)
       p.233-7]
[6]   F.M.d'Heurle, C.S.Petersson, M.Y.Tsai [J. Appl. Phys. (USA)
       vol.51 no.11 (Nov 1980) p.5976-80]
[7]   S.Yanagisawa, T.Fukuyama [J. Electrochem. Soc. (USA) vol.127
       no.5 (May 1980) p.1150-6]
[8]   K.Shibata, S.Shima, M.Kashiwagi [J. Electrochem. Soc. (USA)
       vol.129 no.7 (Jul 1982) p.1527-31]
[9]   T.P.Chow, C.S.Grant, W.Katz, G.Gildenblat, R.F.Reihl [J.
       Electrochem. Soc. (USA) vo.130 no.4 (Apr 1983) p.933-8]
[10]   H.C.Cheng, T.R.Yew, L.J.Chen [J. Appl. Phys. (USA) vol.57 no.12
       (15 Jun 1985) p.5246-50]
[11]   P.R.Gage, R.W.Bartlett [Trans. Metall. Soc. AIME (USA) vol.233
       (Apr 1965) p.8324]
[12]   M.J.Rice,Jr, K.R.Sarma [J. Electrochem. Soc. (USA) vol.128
       no.6 (Jun 1981) p.1368-73]
[13]   J.Y.Cheng, H.C.Cheng, L.J.Chen [J. Appl. Phys. (USA) vol.61 no.6
       (15 Mar 1987) p.2218-24]
[14]   M.A.Nicolet, S.S.Lau [in 'Materials and Process Characterisation'
       Eds. N.G.Einspruch, G.B.Larrabee (Academic, New York, USA, 1983)
       p.329-464]
[15]   F.M.d'Heurle, P.Gas [J. Mater. Res. (USA) vol.1 no.1 (1986)
       p.205-21]
[16]   J.Baglin, F.d'Heurle, S.Petersson [Appl. Phys. Lett. (USA)
       vol.33 no.4 (15 Aug 1978) p.289-90]
[17]   J.Baglin, J.Dempsey, W.Hammer, F.d'Heurle, S.Petersson, C.Serrano
       [J. Electron. Mater. (USA) vol.8 no.5 (Sep 1979) p.641-61]
[18]   J.Baglin, F.d'Heurle, W.N.Hammer, S.Petersson [Nucl. Instrum. &
       Methods (Netherlands) vol.168 no.1-3 (1980) p.491-7]
[19]   J.J.Casey, R.R.Verderber, R.R.Garnache [J. Electrochem. Soc.
       (USA) vol.114 (1967) p.201]
[20]   A.Perio, J.Torres, G.Bomchil, F.A.d'Avitaya, R.Pantel [Appl.
       Phys. Lett. (USA) vol.45 no.8 (15 Oct 1984) p.857-9]
[21]   W.T.Lin, L.J.Chen [Appl. Phys. Lett. (USA) vol.46 no.11
       (1 Jun 1985) p.1061-3]
[22]   W.T.Lin, L.J.Chen [J. Appl. Phys. (USA) vol.59 no.5 (1 Mar 1986)
       p.1518-24]
[23]   W.T.Lin, L.J.Chen [J. Appl. Phys. (USA) vol.59 no.10 (15 May 1986)
       p.3481-7]

[24] T.Inoue, K.Koike [Appl. Phys. Lett. (USA) vol.33 no.9 (1 Nov 1978)
      p.826-7]

[25] T.Mochizuki, M.Kashiwagi [J. Electrochem. Soc. (USA) vol.127 no.5
      (May 1980) p.1128-35]

[26] I.Ohdomari et al [J. Appl. Phys. (USA) vol.59 no.9 (1 May 1986)
      p.3073-6]

[27] I.Abbati, L.Braicovich, B.DeMicheles, A.Pasana, E.Puppin,
      A.Rizzi [Solid State Commun. (USA) vol.52 no.8 (Aug 1984) p.731-4]

[28] G.Rossi et al [Physica B&C (Netherlands) vol.117/8 (1983)
      p.795-7]

[29] H.Balaska, R.C.Cinti, T.T.A.Nguyen, J.Derrien [Surf. Sci.
      (Netherlands) vol.168 no.1-3 (Mar 1986) p.225-33]

[30] T.T.A.Nguyen, R.C.Cinti [J. Phys. Colloq. (France) vol.45 no.C-5
      (1984) p.435-9]

[31] H.Nishi, T.Sakurai, T.Akamatsu, T.Furuya [Appl. Phys. Lett. (USA)
      vol.25 no.6 (15 Sep 1974) p.337-9]

[32] M.Y.Tsai, C.S.Petersson, F.M.d'Heurle, V.Maniscalco [Appl. Phys.
      Lett. (USA) vol.37 no.3 (1 Aug 1980) p.295-8]

[33] S.W.Chiang, T.P.Chow, R.F.Reihl, K.L.Wang [J. Appl. Phys. (USA)
      vol.52 no.6 (Jun 1981) p.4027-32]

[34] L.J.Chen, L.S.Hung, J.W.Mayer [Appl. Surf. Sci. (Netherlands)
      vol.11/12 (1982) p.202-8]

[35] D.L.Kwong, D.C.Meyers, N.S.Alvi, L.W.Li, E.Norbeck [Appl. Phys.
      Lett. (USA) vol.47 no.7 (1 Oct 1985) p.688-91]

[36] M.I.Beale, V.G.Deshmukh, N.G.Chew, A.G.Cullis [Physica B&C
      (Netherlands) vol.129 no.1-3 (Mar 1985) p.210-4]

[37] T.Kanayama, H.Tanoue, T.Tsurushima [Jpn. J. Appl. Phys. Part 1
      (Japan) vol.23 no.3 (Mar 1984) p.277-82]

[38] S.P.Murarka [J. Vac. Sci. & Technol. (USA) vol.17 no.4 (1980)
      p.775-92]

[39] S.P.Murarka, M.H.Read, C.J.Doherty, D.B.Fraser [J. Electrochem.
      Soc. (USA) vol.129 no.2 (Feb 1982) p.293-301]

[40] S.M.Sze [Physics of Semiconductor Devices (Wiley-Intersicience,
      New York, USA, 1981)]

19.14 STRUCTURE OF THE Ni/Si INTERFACE

by R.B.Jackman

July 1987
EMIS Datareview RN=16109

A. INTRODUCTION

Ni/Si interfaces and the silicides which form at such a junction are
of considerable importance within the field of microelectronic device
fabrication, and have received considerable recent attention. This
Datareview will outline the initial chemical reactions which occur,
the role of silicide formation upon interfacial structure and the
formation of Schottky barriers. Specific surface modification by, for
instance, ion bombardment, is not considered.

B. Ni/Si INTERFACES AT ROOM TEMPERATURE

Despite intensive study, some debate exists over the exact nature of
the processes involved during the deposition of Ni onto silicon. The
following discussion considers the evaporation of nickel at room
temperature onto atomically clean Si. It is clear that spontaneous
mixing occurs at the interface under these conditions [1]. However,
the composition of this layer has remained controversial, and the
chemical nature of the species present must be considered in the light
of the apparent low temperature disruption of the surface Si-Si
bonding. Studies performed by Field Ion Microscopy (FIM) [2], Ion
Channelling [3] and several photoemission experiments [4-6] have
suggested that the initial interface corresponds to the metal rich
silicide form, Ni_2Si. Other photoemission work [7-8] has shown evidence
for a graded composition varying from NiSi to Ni. A five monolayer
thick $NiSi_2$ like film was proposed from SEXAFS measurements by Comin
et al [9]. Observation of an apparent $NiSi_4$ stoichiometry, accounted
for in terms of interstitial Ni atoms in an Si(100) substrate, has
been made by Chang and Erskine [10]. However, recent studies of the
Ni/Si(111) system by a combination of high resolution ion scattering,
in-situ Auger measurements, RHEED and TEM [11-13] propose a new model
for the mixing process which appears to explain many of the previous
observations. The composition of monolayer thickness films following
evaporation of Ni onto Si(111) by ion scattering correlated to Ni_2Si,
in agreement with some of the previous data collected upon the system
[2-6]; moreover, Auger chemical shifts only revealed the presence of
the metal rich silicide form. This silicide was, however, formed in
disordered islands upon the surface, with further Ni coverage then
leading to coalescence of such silicide clusters. This behaviour was
proposed to result from the following model. The initial arrival of
Ni atoms upon the clean Si surface may lead to clustering, the energy
from such a process (bonding of Ni approx. 4.3 eV/atom) then being
adequate to exceed the activation barrier for silicide formation (1.5
eV for Ni_2Si) and silicide islands result. During this (exothermic)
reaction, energy is available to promote Si diffusion over the islands
which may then further react until they coalesce. The reaction must now
slow, since Si or Ni diffusion through the disordered mixed film is
required, which is slow at room temperature. This model is discussed in
more detail within references [11] and [12].

An alternative model to explain the apparent low temperature silicide
formation has been discussed by Erskine and Chang [14]. Here, nickel
atoms are assumed to diffuse, via substitution, into interstitial sites
of the silicon lattice. This defect then increases the number of
nearest neighbours of its host lattice, and hence changes the nature of
the chemical bonding. This may be manifested in the reduced Si-Si bond
strength.

C. SURFACE ANNEALING

Annealing at temperatures greater than 300 deg C leads to the
transformation of Ni_2Si to the monosilicide, NiSi [15,16]. However, if
direct evaporation is carried out at this temperature a delay occurs,
whereby the formation of Ni_2Si preceeds the NiSi phase. The growth rate
of this monosilicide depends upon the orientation of the substrate,
(100) showing faster kinetics than the (111) surface. The measured
activation energies are 1.23 and 1.83 eV respectively in the 300-370
deg C temperature range. Majni et al [15] have discussed these
differences in terms of substitutional diffusion being the major
control of growth for Si(111), whilst the kinetics of NiSi formation
upon Si(100) are mainly controlled by grain boundary mechanisms.
Observations made at higher temperatures (> 400 deg C) show that
substitutional diffusion becomes the dominant mechanism for both
substrates [15,17].

Further silicide types are formed upon annealing the deposited nickel
layer at still higher temperatures. These have been summarised by
Ottaviani [18] and are shown in TABLE 1. At 750 deg C the onset of
$NiSi_2$ formation is apparent, which may grow epitaxially. This is
discussed further below.

D. NiSi2 GROWTH

High quality epitaxial films of $NiSi_2$ (CaF_2 structure) can be grown
by vacuum evaporation of a thick (> 200 A) Ni film on either (100)
or (111)Si followed by annealing at 750-800 deg C [20]. However, if the
'template' method is used (discussed further below) for thickening very
thin Ni overlayers the epitaxial silicide can also be grown as a single
crystal.

Overlayer growth on Si(111) can either have the same orientation as
the substrate (A-type) or be rotated by 180 deg about the surface
normal (B-type). For thin layers, this orientation has been shown to
be a function of the original Ni overlayer thickness [20], and Tung et
al [21] have tabulated the available data, some of which are reproduced
in TABLE 2. Thin layers formed with either 1-5 A or 9-11 A Ni are type B
orientated whilst those with 16-20 A of Ni are type A. TEM observations
[22] indicate uniform layer growth with few misfit dislocations, whilst
LEED studies [21] indicate that the $NiSi_2$ may be thickened by annealing
Ni deposited on top of the thin silicide layer. The original
orientation is then maintained leading to a high quality single crystal
$NiSi_2$ layer. This has been termed the 'template' method for single
crystal epitaxy. The role of lattice mismatch criteria has been
discussed by Hamm et al [23].

The Si dangling bonds of the Si(111) interface can either be attached
to the Si or Ni atoms of the silicide, resulting in seven or five fold
co-ordinated Ni atoms respectively. Van Loenen et al [24] investigated
B-type overlayers with ion blocking techniques and found the Ni atoms
at the interface to be seven-fold co-ordinated. This is in agreement
with the TEM results of Cherns et al [25] and the X-ray standing wave
(XSW) analysis of Vlieg et al [26] who found seven-fold co-ordination
for both A and B type orientations. Akimoto et al [27], however,
concluded from XSW on A-type overlayers that the Ni atoms were
five-fold co-ordinated.

TEM/LEED analysis of annealed thin Ni layers upon Si(100) [21]
suggests a surface of NiSi2(100) plus some areas of bare Si. At
thicknesses greater than 20 A initial Ni, inclined (111) facets are
observed and the interface becomes non-uniform with large exposed
silicon regions.

Thin epitaxial layers of NiSi2 may be grown upon Si(110) and employed
to fabricate a thick single crystal disilicide phase as with Si(111)
[28]. RBS channeling and TEM analysis have revealed near perfect
crystallinity in the films.

E. SCHOTTKY BARRIER HEIGHTS

Schottky barriers (SB's) resulting from nickel silicide - silicon
interfaces exhibit particularly interesting features which are not yet
fully understood. Despite the use of the Schottky-Mott model, with the
addition of interface states and a dipole layer, the detailed orgin
of the observed barriers is still unclear.

Erskine and Chang [14] and others [29,30] have suggested the Schottky
barrier height (SBH) is independent of the silicide phase which forms
the metal contact or its structure. However, data collected by a number
of groups [21,31,32] have revealed the silicide phases to display a SBH
dependent upon the orientation of the interface. For example type A
NiSi2 differs by 0.14 eV from that of type B NiSi2 upon Si(111).
Figures produced by Tung et al [21] are reproduced in TABLE 3. No
dependence upon substrate doping was observed.

Whilst recent work by Liehr et al [29] has observed decreased barrier
heights for certain orientations it was suggested that the cause was
defect states localised at the interface or oxygen impurity
incorporation. They concluded that a single SBH of 0.78 eV exists.
However, further work by Tung et al [33] has reproducibly demonstrated
the dependence upon orientation, and whilst no mechanism for such SB
formation was proposed, a possible explanation for the apparent
discrepancies was presented. The experimental procedures employed by
Liehr et al [30] involve annealing the Si substrate to a high
temperature in vacuum and this may induce a surface p-n junction [34].
In such a case, this may then dominate the electrical transport
behaviour perpendicular to the interface and affect the observed barrier
height.

TABLE 1: formation temperatures of nickel silicides following the
deposition of nickel onto silicon at room temperature [18]

Silicide Phase	Formation Temperature (deg C)
Ni3Si	450
Ni2Si	< 300
Ni5Si2	400
NiSi	350-750
NiSi2	750

TABLE 2: orientation of thin NiSi2 layers upon Si(111) from initial Ni
deposition at room temperature followed by an anneal at
500 deg C

Ni deposit thickness (A)	Deposition rate (A/s)	Orientation of NiSi2
1 - 5	0.5 - 2.0	B (islands)
6 - 8	0.5 - 2.0	A+B
9 - 11	0.5 - 2.0	B
12 - 14	0.5 - 2.0	A+B
16 - 20	0.5 - 2.0	A
> 24	0.5 - 2.0	A+B
8 - 11	0.1 - 2.0	B
13 - 30	0.1 - 2.0	A+B
14 - 50	1.5	A

TABLE 3: Schottky barrier heights of single crystal silicide layers on
n-type silicon [21]

Silicide	Orientation	Substrate	SBH (eV)
NiSi2	A	(111)	0.65
NiSi2	B	(111)	0.79
NiSi2	-	(100)	0.48

REFERENCES
----

[1]   E.J.van Loenen [J. Vac. Sci. & Technol. A (USA) vol.4 no.3
      (May/Jun 1986) p.939-43]
[2]   O.Nishikawa, M.Shibata, T.Yoshimura, E.Nomura [J. Vac. Sci. &
      Technol. B (USA) vol.2 (1984) p.21]
[3]   N.W.Cheung, R.J.Culbertson, L.C.Feldman, P.J.Silverman, K.W.West,
      J.W.Mayer [Phys. Rev. Lett. (USA) vol.45 (1980) p.120]
[4]   Y.Shiraki, K.L.I.Kobayashi, H.Daimon, A.Ishizuka, S.Sugaki,
      Y.Murata [Physica B&C (Netherlands) vol.117/118 (1983) p.843]
[5]   K.L.I.Kobayashi, S.Sugaki, A.Ishizaka, Y.Shiraki, H.Diamon,
      Y.Murata [Phys. Rev. B (USA) vol.25 (1982) p.1377]
[6]   I.Abbati, L.Braicovich, B.de Michelis, U.del Pennino, S.Valeri
      [Solid State Commun. (USA) vol.43 (1982) p.199]

[7]   P.J.Grunthaner, F.J.Grunthaner, J.W.Mayer [J. Vac. Sci. &
      Technol. (USA) vol.17 (1980) p.924]
[8]   A.Franciosi et al [J. Vac. Sci. & Technol. (USA) vol.21 (1982)
      p.624]
[9]   F.Comin, J.E.Rowe, P.H.Citrin [Phys. Rev. Lett. (USA) vol.51
      (1983) p.2402]
[10] Y.Chang, J.L.Erskine [Phys. Rev. B (USA) vol.28 (1983) p.5766]
[11] E.J.van Loenen, J.W.M.Frenken, J.F.van der Veen [Appl. Phys.
      Lett. (USA) vol.45 (1 Jul 1984) p.41]
[12] E.J.van Loenen, J.F. van der Veen, F.K.LeGoues [Surf. Sci.
      (Netherlands) vol.157 no.1 (1985) p.1-16]
[13] A.E.M.J.Fischer, P.M.J.Maree, J.F.van der Veen [Appl. Surf. Sci.
      (Netherlands) vol.27 no.2 (1986) p.143-50]
[14] J.L. Erskine, Y.J.Chang [Mater. Res. Soc. Symp. Proc. USA)
      vol.25 (1984) p.353]
[15] G.Majni, M.Costato, F.D.Valle [Nuovo Cimento D (Italy) vol.4
      no.1 (Jul 1984) p.27-38]
[16] E.C.Cahoon, C.M.Comrie, R.Pretorius [Mater. Res. Soc. Symp.
      Proc. (USA) vol.25 (1984) p.57]
[17] J.E.E.Baglin, H.A.Atwater, D.Gupta, F.M.d'Heurle [Thin Solid Films
      (Switzerland) vol.93 (1982) p.255]
[18] G.Ottaviani [J. Vac. Sci. & Technol. (USA) vol.16 (1979) p.1112]
[19] H.Ishiwara, K.Hikosaka, M.Nagatono, S.Furukawa [Surf. Sci.
      (Netherlands) vol.86 (1979) p.711]
[20] R.T.Tung, J.M.Gibson, J.M.Poate [Phys. Rev. Lett. (USA) vol.50
      no.6 (1983) p.429-32]
[21] R.T.Tung, A.F.J.Levi, J.M.Gibson [J. Vac. Sci. & Technol. B
      (USA) vol.4 no.6 (Nov/Dec 1986) p.1435-43]
[22] J.M.Gibson, R.T.Tung, J.M.Poate [Mater. Res. Soc. Symp. Proc.
      (USA) vol.25 (1984) p.405]
[23] R.A.Hamm, J.M.Vandenberg, J.M.Gibson, R.T.Tung [Mater. Res. Soc.
      Symp. Proc. (USA) vol.37 (1985) p.367-74]
[24] E.J.van Loenen, J.W.M.Frenken, J.F.van der Veen, S.Valeri
      [Phys. Rev. Lett. (USA) vol.54 no.8 (25 Feb 1985) p.827-30]
[25] D.Cherns, G.R.Anistis, J.L.Hutchison, J.C.H.Spence [Philos.
      Mag. A (GB) vol.46 (1982) p.849]
[26] E.Vlieg, A.E.M.J.Fischer, J.F.van der Veen, B.N.Dev, G.Materlik
      [Surf. Sci. (Netherlands) vol.178 (1986) p.36-46]
[27] K.Akimoto, T.Ishikawa, T.Takahashi, S.Kikuta [Jpn. J. Appl.
      Phys. Part 2 (Japan) vol.22 no.12 (Dec 1983) p.L798-800] and
      [Jpn. J. Appl. Phys. Part 1 (Japan) vol.24 no.11 (Nov 1985)
      p.1425-31]
[28] R.T.Tung, S.Nakahara, T.Boone [Appl. Phys. Lett. (USA) vol.46
      no.9 (1 May 1985) p.895-7]
[29] M.Liehr, P.E.Schmid, F.K.LeGoues, P.S.Ho [J. Vac. Sci. &
      Technol. A (USA) vol.4 no.3 pt.1 (May/June 1986) p.855-9]
[30] M.Liehr, P.E.Schmid, F.K.LeGoues, P.S.Ho [Phys. Rev. Lett.
      (USA) vol.54 no.19 (13 May 1985) p.2139-42]
[31] R.T.Tung, J.M.Gibson [J. Vac. Sci. & Technol. A (USA) vol.3 no.3
      pt.1 (May/June 1985) p.987-91]
[32] B.D.Hunt et al [Mater. Res. Soc. Symp. Proc. (USA) vol.54 (1986)
      p.479-84]
[33] R.T.Tung, K.K.Ng, J.M.Gibson, A.F.J.Levi [Phys. Rev. B (USA)
      vol.33 no.10 (15 May 1986) p.7077-80]
[34] L.N.Aleksandrov, R.N.Lovyagin, P.A.Simonov, I.S.Bzinkovskaya
      [Phys. Status. Solidi a (Germany) vol.45 (1978) p.521]

19.15 STRUCTURE OF THE Pd/Si INTERFACE

by L.J.Chen

July 1987
EMIS Datareview RN=15764

A. SILICIDE FORMATION, FORMATION MECHANISM

The Pd/Si interface is one of the most widely studied metal/
semiconductor interfaces. The interface is reactive at room temperature,
the silicide phase Pd2Si being formed, and at temperatures above
700 deg C PdSi is formed [1-8]. Pd2Si formation has been observed to
occur spontaneously at temperatures as low as 180 deg K [9].

B. KINETICS, DIFFUSING SPECIES

Pd2Si was found to grow according to diffusion controlled kinetics
[4,6,8,10-15]. The values of activation energy obtained for Pd2Si
growth ranged from 0.9-1.5 eV [4,6,8,10,12,13,15]. PdSi was shown to
grow via nucleation-controlled reaction [8]. For the growth of Pd2Si
and PdSi, both Pd and Si atoms were generally reported to be diffusing
species [16-18]. However, there were reports that a re-interpretation
of the experimental data from references [16] and [17] is possible and
an agreement can be reached with new results that Si is the dominant
diffusing species in polycrystalline Pd2Si [18-20]. Pd, on the other
hand, was found to be the moving species in epitaxial Pd2Si [21].

C. EPITAXY, INTERFACE STRUCTURE

High resolution TEM studies of the Pd2Si/Si interface formed by
annealing Pd/Si in He showed an apparently abrupt interface with
interfacial steps associated with misfit dislocations. The interface
region contained a mixture of Si-rich and Si-deficient Pd2Si planes
adjacent to the Si side of the interface [22-24]. Epitaxial Pd2Si was
formed by implantation of ions through Pd thin films on (111)Si [25].

D. ORIENTATION EFFECT

Epitaxial Pd2Si was formed on annealing of Pd-covered UHV-sputter-
cleaned (111)Si at 200 deg C for 1 hr and on Pd-covered chemically-
cleaned (111)Si at 215 deg C for 2 hr. Polycrystalline Pd2Si was formed
when Pd deposited on chemically cleaned (100)Si was heat treated at
215 deg C for 2 hr [26,27].

E. STABILITY

Thin Pd2Si layers (thickness 20 A) on (111)Si have been shown to be
unstable to annealing at 400 deg C, coalescing to form islands at the
Si surface [28]. Other workers have shown that Pd deposited onto (111)Si
at room temperature and at 100 deg C forms 3-dimensional Pd2Si
crystallites. On deposition at 300-400 deg C island formation was found
to precede the growth of 3-D crystallites [29]. When a Pd2Si layer is
placed between a single-crystal Si substrate and an amorphous Si
overlayer, annealed at about one-half the melting temperature of the
Pd2Si, the silicide is seen to move through the amorphous Si and rise
to the surface, leaving below it an epitaxially grown layer of Si [30].

F. REDISTRIBUTION OF DOPANTS, DOPANT AND CONTAMINANT EFFECTS

Redistribution of dopant atoms during silicide formation (the
snowplough effect) is important because of its effect on interface
electrical properties. Implanted P atoms pile up at the Pd2Si/Si
interface during Pd2Si formation by annealing at 250 deg C. For 100 keV
P implantation doses 5 X 10*13/cm*2 and 1 X 10*13/cm*2, P concentrations
at the Pd2Si/Si interface were 2.5 X 10*19/cm*3 and 4 X 10*18/cm*3,
respectively. These concentrations were one order of magnitude higher
than those at the Si surface before Pd2Si formation [31]. Uniformly
doped As atoms (initial concentration 4 X 10*19/cm*3) pile up near
the Pd2Si/Si interface during Pd2Si formation by annealing at 250 deg C.
The maximum As concentration at the interface in the annealed samples is
about 2 X 10*20/cm*3 [32]. Rapid thermal annealing has been used to form
Pd2Si by reacting thin layers of Pd metal on As-implanted Si. For
implant concentration > 1 X 10*15/cm*2 As is found to accumulate at the
surface of the Pd2Si and at the Pd2Si/Si interface [33].

G. INTERFACE BONDING, ELECTRONIC STRUCTURE, SURFACE STRUCTURE

The existence of a critical Pd thickness for intermixing at the Pd/Si
interface at room temperature has been suggested. It has been reported
that EELS and AES data are consistent with Pd atoms being chemisorbed
to satisfy the Si dangling bonds for coverages up to 1 monolayer and
intermixing taking place for coverages above about 3 monolayers [34].
Other workers found that for Pd coverages exceeding one monolayer a
mixing reaction was observed on both (111)Si and (001)Si with some
differences in behaviour for the two surfaces [35]. On (001)Si the
mixing reaction appeared to occur even at coverages below 1 monolayer
and the interface, unlike Pd/(111)Si, was not sharp, but probably faceted.

Evidence has been obtained for the existence of a thin (2-3 atomic
layers) Si-rich interfacial layer separating Pd2Si from the Si
substrate at room temperature [36]. X-ray absorption resonance
spectroscopy provided evidence of the existence of empty antibonding
d-like states in the electronic structure of Pd/Si and Pd2Si/Si
interfaces [37].

A LEED study revealed strong evidence for the dependence of the
nucleation and growth of Pd2Si on the initial surface reconstruction
on cleaved (111)Si - 2 X 1 and 7 X 7 surfaces [38].

H. ION BEAM MIXING

Irradiations of Ar, Kr and Xe ions on Pd/Si at room temperature produced a layer of Pd2Si near the original Pd/Si interface [10,39]. The thickness of the Pd2Si layer was found to be proportional to the square root of the dose [39]. A more detailed study showed that Pd2Si formation is reaction limited when the ion energy-deposition density at the Pd/Si interface is low and is diffusion limited when it is high [40]. Si was found to be the faster diffusing species in Pd2Si formation by Ar ion beam mixing [41]. PdSi was also obtained by implanting Xe ions through a thin Pd (or Pd2Si) film on a Si substrate [42]. The depths of ion beam mixing after Si implantation of Pd thin films on silicon showed linear dose dependence [43].

I. ELECTRICAL PROPERTIES

The resistivities of Pd2Si thin films on silicon were measured to be 9-35 micro-ohm cm [44-46]. Epitaxial Pd2Si thin film on (001)Si was found to possess a resistivity of 25 micro-ohm cm [47]. The Schottky barrier height of Pd2Si was measured to be 0.74 eV [48].

REFERENCES
----

[1]   K.N.Tu, J.W.Mayer [in 'Thin Films - Interdiffusion and Reactions'
      Eds. J.M.Poate, K.N.Tu, J.W.Mayer (Wiley, New York, USA, 1978)
      p.359-405]
[2]   C.J.Kircher [Solid-State Electron. (GB) vol.14 no.6 (1971)
      p.507-13]
[3]   W.D.Buckley, S.C.Moss [Solid-State Electron. (GB) vol.15 no.12
      (1972) p.1331-7]
[4]   D.J.Fertig, G.Y.Robinson [Solid-State Electron. (GB) vol.19 no.5
      (Apr 1976) p.407-13]
[5]   S.S.Lau, D.Sigurd [J. Electrochem. Soc. (USA) vol.121 no.11
      (Nov 1974) p.1538-40]
[6]   R.W.Rower, R.E.Scott, D.Sigurd [Solid-State Electron. (GB) vol.16
      no.12 (1973) p.1461-71]
[7]   D.Sigurd, R.W.Bower, W.F.van der Weg, J.W.Mayer [Thin Solid Film
      (Switzerland) vol.19 no.2 (17 Dec 1973) p.319-28]
[8]   G.A.Hutchins, A.Shepela [Thin Solid Films (Switzerland) vol.18
      no.2 (Nov 1973) p.343-63]
[9]   G.W.Rubloff, P.S.Ho, J.F.Freeouf, J.E.Lewis [Phys. Rev. B (USA)
      vol.23 no.8 (1981) p.4183-96]
[10] M.A.Nicolet, S.S.Lau [in 'Materials and Process Characterisation'
      Eds. N.G.Einspruch, G.B.Larrabee (Academic, New York, USA, 1983)
      p.329-464]
[11] F.M.d'Heurle, P.Gas [J. Mater. Res. (USA) vol.1 no.1 (1986)
      p.205-21]
[12] R.W.Bower, J.W.Mayer [Appl. Phys. Lett. (USA) vol.20 no.9
      (1 May 1972) p.359-61]
[13] N.W.Cheung, M.-A.Nicolet, M.Wittmer, C.A.Evans,Jr, T.T.Sheng
      [Thin Solid Films (Switzerland) vol.79 no.1 (8 May 1981) p.51-68]
[14] J.F.Ziegler, J.W.Mayer, C.J.Kircher, K.N.Tu [J. Appl. Phys.
      (USA) vol.44 no.9 (Sep 1973) p.3851-7]

[15] B.Coulman, H.Chen [J. Appl. Phys. (USA) vol.59 no.10 (15 May 1986)
p.3467-74]

[16] W.K.Chu, S.S.Lau, J.W.Mayer, H.Muller, K.N.Tu [Thin Solid
Films (Switzerland) vol.25 no.2 (Feb 1975) p.393-402]

[17] R.Pretorius, C.L.Ramiller, M.A.Nicolet [Nucl. Instrum. & Methods
(Netherlands) vol.149 no.1-3 (1978) p.629-33]

[18] H.Foll, P.S.Ho [J. Appl. Phys. (USA) vol.52 no.9 (Sep 1981)
p.5510-6]

[19] D.M.Scott [Ph.D. Thesis, California Institute of Technology
(1982)]

[20] K.T.Ho, C.D.Lien, U.Shreter, M.-A.Nicolet [J. Appl. Phys. (USA)
vol.57 no.2 (15 Jan 1985) p.227-31]

[21] C.D.Lien, M.A.Nicolet, C.S.Pai [J. Appl. Phys. (USA) vol.57
no.2 (15 Jan 1985) p.224-6]

[22] D.Cherns, D.A.Smith, W.Krakow, P.E.Batson [Philos. Mag. A (GB)
vol.45 no.1 (1982) p.107-25]

[23] D.Cherns, W.Krakow, D.A.Smith [Inst. Phys. Conf. Ser. (GB) no.60
(1981) p.409-14]

[24] W.Krakow [Thin Solid Films (Switzerland) vol.93 no.1/2
(9 Jul 1982) p.109-25]

[25] H.Ishiwara, K.Kuzuta [Appl. Phys. Lett. (USA) vol.37 no.7
(1 Oct 1980) p.641-3]

[26] G.W.Rubloff [Festkorperprobleme (Germany) vol.XXIII (Friedr.
Viewig, Wiesbaden, 1983) p.179-206]

[27] P.S.Ho [J. Vac. Sci. & Technol. A (USA) vol.1 no.2 (Apr/Jun 1983)
p.745-57]

[28] R.Tromp, E.J.van Loenen, M.Iwami, R.Smeenk, F.W.Saris [Thin Solid
Films (Switzerland) vol.93 no.1/2 (9 Jul 1982) p.151-9]

[29] A.Oustry, J.Berty, M.Caumont, M.J.David, A.Escaut [Thin Solid
Films (Switzerland) vol.97 no.4 (26 Nov 1982) p.295-300]

[30] S.S.Lau, Z.L.Liau, M.-A.Nicolet [Thin Solid Films (Switzerland)
vol.47 no.3 (15 Dec 1977) p.313-22]

[31] A.Kikuchi [J. Appl. Phys. (USA) vol.54 no.7 (Jul 1983) p.3998-4000]

[32] I.Ohdomari et al [J. Appl. Phys. (USA) vol.54 no.8 (Aug 1983)
p.4679-82]

[33] N.S.Alvi, D.L.Kwong, C.G.Hopkins, S.G.Bauman [Appl. Phys. Lett.
(USA) vol.48 no.21 (26 May 1986) p.1433-5]

[34] K.Okuno, T.Ito, M.Iwami, A.Hiraki [Solid State Commun. (USA)
vol.44 no.2 (Oct 1982) p.209-12]

[35] R.M.Tromp et al [Surf. Sci. (Netherlands) vol.124 no.1 (1983)
p.1-25]

[36] G.W.Rubloff [Surf. Sci. (Netherlands) vol.132 no.1-3 (1983)
p.268-314]

[37] G.Rossi, P.Roubin, D.Chanderis, J.Lecante [Surf. Sci.
(Netherlands) vol.168 no.1-3 (Mar 1986) p.787-94]

[38] J.G.Clabes [Surf. Sci. (Netherlands) vol.145 no.1 (1984)
p.87-100]

[39] B.Y.Tsaur [Ph.D. Thesis, California Institute of Technology
(1980)]

[40] Y.Horino, N.Matsunami, N.Itoh [Nucl. Instrum & Methods Phys.
Res. Sect. B (Netherlands) vol.16 no.1 (May 1986) p.50-55]

[41] L.S.Hung, J.W.Mayer, C.S.Pai, S.S.Lau [J. Appl. Phys. (USA)
vol.58 no.4 (15 Aug 1985) p.1527-36]

[42] B.Y.Tsaur, S.S.Lau, J.W.Mayer [Appl. Phys. Lett. (USA) vol.35
no.3 (1 Aug 1979) p.225-7]

[43] D.B.Poker, B.R.Appleton [J. Appl. Phys. (USA) vol.57 no.4
      (15 Feb 1985) p.1414-6]
[44] S.P.Murarka [J. Vac. Sci. & Technol. (USA) vol.17 no.4 (1980)
      p.775-92]
[45] M.Wittmer, D.L.Smith, P.W.Lew, M.-A.Nicolet [Solid-State Electron.
      (GB) vol.21 no.3 (Mar 1978) p.573-80]
[46] S.P.Murarka, M.H.Read, C.J.Doherty, D.B.Fraser [J. Electrochem.
      Soc. (USA) vol.129 no.2 (Feb 1982) p.293-301]
[47] S.Saitoh, H.Ishiwara, T.Asano, S.Furukawa [Jpn. J. Appl. Phys.
      (Japan) vol.20 no.9 (Sep 1981) p.1649-56]
[48] S.P.Muraraka [Silicides for VLSI Applications (Academic,
      Orlando, USA, 1983)]

19.16 STRUCTURE OF THE Pt/Si INTERFACE

by L.J.Chen

July 1987
EMIS Datareview RN=15762

A. SILICIDE FORMATION, FORMATION MECHANISM

Annealing of a Pt film on a Si substrate leads to formation of the
silicides Pt2Si (formation temperature 200-500 deg C) and PtSi
(formation temperature 300 deg C) [1-5].

B. KINETICS, DIFFUSING SPECIES

It has been shown that the growth of both Pt2Si and PtSi is diffusion
limited [3,4,6-12]. Using radioactive Si tracer, it was found that Pt
is the dominant diffusing species for both Pt2Si and PtSi formation
[13,14]. However, there were also reports that both Pt and Si atoms
are diffusing species [15,16]. Activation energies of Pt2Si and PtSi
formation were found to be 1.04-1.6 +- 0.1 and 1.47-1.7 eV, respectively
[3,4,8,9,11,12,17].

C. EPITAXY, INTERFACE STRUCTURE

Pt2Si was found to grow epitaxially on (001)Si [18]. PtSi was grown
epitaxially on (001), (011), and (111)Si [5,19-22]. Small PtSi domains
(20-20 nm) which have three equivalent positions on (111)Si grow
epitaxially all over the substrate with an atomically uneven interface.
However, the transition from a PtSi lattice to a Si lattice was found
to be abrupt [23].

D. ORIENTATION EFFECT

The growth kinetics of both Pt2Si and PtSi were found to be independent
of the substrate orientation [3,4,8,9,11,12,17].

E. STABILITY

Pt diffusion into Si has been observed on annealing the Pt/Si interface
at 800 deg C. For annealing up to 700 deg C there was no observable Pt
diffusion from the PtSi layer, showing that this processing temperature
can be safely used if formation of Pt-related traps is to be avoided
[24]. When a PtSi layer is placed between a single-crystal Si substrate
and an amorphous Si overlayer, annealed at about one-half the melting
temperature of the PtSi, the silicide is seen to move through the
amorphous Si and rise to the surface, leaving below it an epitaxially
grown layer of Si [25].

F. SILICIDE LAYER IN OXIDISING AMBIENTS

An oxide layer was found to form well below the Pt surface if excess amounts of oxygen are present in the heat treatment ambient [26]. Radioactive tracer experiments showed that, during wet oxidation of PtSi at 800 deg C, silicon from the substrate completely mixes with the silicon in the silicide layer before oxidation at silicide-SiO2 interface, owing to the high self-diffusion coefficient of silicon in PtSi at the oxidation temperature. In PtSi, silicon diffusion includes a substitutional mechanism [27].

G. REDISTRIBUTION OF DOPANTS, DOPANT AND CONTAMINANT EFFECTS

The presence of impurities, both contaminants and intentional dopants, influence silicide growth. Kinetic studies of PtSi formation on annealing of Pt film evaporated onto Si containing a high concentration of As (5 X 10*20/cm*3) showed the effect of the high As concentration on the PtSi growth kinetics. The activation energy was the same as that for Pt films on Si substrates with low As concentration, but the pre-exponential factor was lowered by a factor of two [12]. The activation energies for the growth of Pt2Si and PtSi are not affected by the substrate crystallinity and doping with phosphorous [28]. A study of low dose (5 X 10*14/cm*2) Kr++ implantation effects on silicide formation at the Pt/Si interface showed the occurrence of random nucleation of both Pt2Si and PtSi and the subsequent formation of a layer containing both phases rather than a Pt2Si layer. Continued annealing led to formation of a fine-grained and randomly oriented PtSi phase instead of the highly textured PtSi found in the absence of Kr++ implantation [29]. It has been shown that redistribution of As occurs during Pt2Si formation, the effect not being as strong as during PtSi formation [30].

H. INTERFACE BONDING, ELECTRONIC STRUCTURE AND SURFACE STRUCTURE

Synchrotron radiation photoemission experiments on the Pt/(111)Si interface showed that very strong chemical interaction occurred at room temperature [31]. XPS study showed that the core level signal from low coverage of Pt on (001)Si coincides exactly with that for stoichiometric PtSi at room temperature [32].

X-ray absorption resonance spectroscopy provided evidence of the existence of empty antibonding d-like states in the electronic structure of Pt/Si and PtSi/Si interfaces [33].

I. ION BEAM MIXING

Irradiations of Ar, Kr and Xe ions on Pt/Si at room temperature produced a layer of Pt2Si near the original Pt/Si interface [6,34].

The thickness of the Pt2Si layer was found to be proportional to the square root of the dose [34]. Si was found to be the faster diffusing species in Pt2Si formation by Xe ion beam mixing [35].

J. ELECTRICAL PROPERTIES

The resistivities of PtSi thin films were measured to be 28-80 micro-
ohm cm [5,36,37]. The Schottky barrier heights for Pt2Si and PtSi were
measured to be 0.78 and 0.87 eV, respectively [38].

REFERENCES
----

[1]   K.N.Tu, J.W.Mayer [in 'Thin Films - Interdiffusion and Reactions'
        Eds. J.M.Poate, K.N.Tu, J.W.Mayer (Wiley, New York, USA, 1978)
        p.359-405]
[2]   A.Hiraki, M.A.Nicolet, J.W.Mayer [Appl. Phys. Lett. (USA) vol.18
        no.5 (1971) p.178-81]
[3]   H.Muta, D.Shinoda [J. Appl. Phys. (USA) vol.43 no.6 (Jun 1972)
        p.2913-7]
[4]   J.M.Poate, T.C.Tisone [Appl. Phys. Lett. (USA) vol.24 no.8 (15 Apr
        1974) p.391]
[5]   A.K.Sinha, R.B.Marcus, T.T.Sheng, S.E.Haszka [J. Appl. Phys.
        (USA) vol.43 no.9 (Sep 1972) p.3637-43]
[6]   M.A.Nicolet, S.S.Lau [in 'Materials and Process Characterisation'
        Eds N.G.Einspruch, G.B.Larrabee (Academic, New York, USA, 1983)
        p.329-464]
[7]   F.M.d'Heurle, P.Gas [J. Mater. Res. (USA) vol.1 no.1 (1986)
        p.205-21]
[8]   C.Canali, C.Catallani, M.Prudenziati, W.H.Wadlin, C.A.Evans,Jr
        [Appl. Phys. Lett. (USA) vol.31 no.1 (1 Jun 1977) p.43-5]
[9]   C.A.Crider, J.M.Poate [Appl. Phys. Lett. (USA) vol.36 no.6 (1980)
        p.417-19]
[10]   D.M.Scott [Ph.D Thesis, California Institute of Technology (1982)]
[11]   G.Ottaviani, G.Majni, C.Canali [Appl. Phys. (Germany) vol.18 no.3
        (1979) p.285-9]
[12]   M.Wittmer [J. Appl. Phys. (USA) vol.54 no.9 (Sep 1983) p.5081-6]
[13]   R.Pretorius, C.L.Ramiller, M.A.Nicolet [Nucl. Instrum & Methods
        (Netherlands) vol.149 no.1-3 (1978) p.629-33]
[14]   R.Pretorius, A.P.Botha, J.C.Lombard [Thin Solid Films
        (Switzerland) vol.79 no.1 (8 May 1981) p.61-8]
[15]   K.N.Tu [Appl. Phys. Lett. (USA) vol.27 no.4 (15 Aug 1975) p.221-4]
[16]   J.E.E.Baglin, F.M.d'Heurle, W.N.Hammer, S.Petersson [Nucl. Instrum.
        Methods (USA) vol.168 nos.1-3 (1980) p.491-7]
[17]   J.T.Pan, I.A.Blech [Thin Solid Films (Switzerland) vol.113 no.2
        (16 Mar 1984) p.129-34]
[18]   H.Ishiwara, K.Hikosaka, S.Furukawa [J. Appl. Phys. (USA) vol.50
        no.8 (Aug 1979) p.5302-6]
[19]   H.B.Ghozlene, P.Beaufrere, [J. Appl. Phys. (USA) vol.49 no.7
        (Jul 1978) p.3998-4004]
[20]   R.M.Anderson, T.M.Reith [J. Electrochem. Soc. (USA) vol.122 no.10
        (Oct 1975) p.1337-47]
[21]   H.Ishiwara, K.Hikosaka, M.Nagatomo, S.Furukawa [Surf. Sci.
        (Netherlands) vol.86 (1979) p.711-7]
[22]   H.Foll, P.S.Ho, K.N.Tu [J. Appl. Phys. (USA) vol.52 no.1
        (Jan 1981) p.250-5]
[23]   H.Kawarada, M.Ishida, J.Nakanishi, I.Ohdomari, S.Horiuchi
        [Philos. Mag. A (GB) vol.54 no.5 (Nov 1986) p.729-41]

[24] A.Prabhakar, T.C.McGill, M.A.Nicolet [Appl. Phys. Lett. (USA)
      vol.43 no.12 (15 Dec 1983) p.1118-20]
[25] S.S.Lau, Z.L.Liau, M.-A.Nicolet [Thin Solid Films (Switzerland)
      vol.47 no.3 (15 Dec 1977) p.313-22]
[26] R.J.Blattner, C.A.Evans,Jr, S.S.Lau, J.W.Mayer, B.M.Ullrich [J.
      Electrochem. Soc. (USA) vol.122 no.12 (Dec 1975) p.1732-6]
[27] R.Pretorius, W.Strydom, J.W.Mayer, C.Comrie [Phys. Rev. B (USA)
      vol.22 no.4 (1980) p.1885-91]
[28] H.Takai, P.A.Psaras, K.N.Tu [J. Appl. Phys. (USA) vol.58 no.11
      (1 Dec 1985) p.4165-71]
[29] T.W.Orent, C.I.Knudson, J.A.Sartell, S.Lee, T.L.Brewer [J.
      Electrochem. Soc. (USA) vol.130 no.3 (Mar 1983) p.687-91]
[30] M.Wittmer, K.N.Tu [Phys. Rev. B (USA) vol.29 no.4 (15 Feb 1984)
      p.2010-20]
[31] G.Rossi, I.Abbati, L.Braicovich, I.Lindau, W.E.Spicer [Phys. Rev.
      B (USA) vol.25 no.6 (15 Mar 1982) p.3627-36]
[32] P.J.Grunthaner, F.J.Grunthaner [Proc. SPIE Int. Soc. Opt. Eng.
      (USA) vol.463 (1984) p.25-32]
[33] G.Rossi, P.Roubin, D.Chanderis, J.Lecante [Surf. Sci.
      (Netherlands) vol.168 no.1-3 (Mar 1986) p.787-94]
[34] B.Y.Tsaur [Ph.D. Thesis, California Institute of Technology
      (1980)]
[35] K.Affolter, X.-A.Zhao, M.-A.Nicolet [J. Appl. Phys. (USA) vol.58
      no.8 (15 Oct 1985) p.3087-93]
[36] S.P.Murarka, [J. Vac. Sci. & Technol. (USA) vol.17 no.4 (1980)
      p.775-92]
[37] S.P.Muraraka, M.H.Read, C.J.Doherty, D.B.Fraser [J. Electrochem.
      Soc. (USA) vol.129 no.2 (Feb 1982) p.293-301]
[38] S.P.Muraraka [Silicides for VLSI Applications (Academic, Orlando,
      USA, 1983)]

19.17 STRUCTURE OF THE Sm/Si INTERFACE

by G.Rossi

August 1987
EMIS Datareview RN=15759

A. INTRODUCTION

The Sm/Si(111) (and Sm/Ge(111)) interface has been the object of core
and valence level synchrotron radiation photoemission studies [1,2].

B. CHEMISORPTION AT ROOM TEMPERATURE

The chemisorption stage is characterised by a weak perturbation of the
substrate, since only minor chemical shifts are observed for the Si 2p
(and Ge 3d) core levels, and by divalency of the Sm. This stage is
completed for 2-3 A Sm coverage, at room temperature, possibly because
of the low heat of formation of the Sm-Si(Ge) bonds [1]. For coverages
larger than 3 A reactive interdiffusion occurs with Sm trivalent
features appearing, and large chemical shifts of the semiconductor core
levels. The maximum valence number is obtained for the intermixed
region produced by condensing 8-10 A Sm onto Si(111) with a Si 2p
chemical shift of 1.2 eV. Charge transfer from Sm to semiconductors may
account for this shift, according to the Pauling's electronegativity
differences. For Sm coverages exceeding 15 A the interface reaction
saturates and the surface becomes richer in pure Sm with photoemission
features of the pure metal. The diffusion of Si into the metallic
overlayer involves roughly 25 A of the semiconductor (40 A in the case
of Ge). In the reacted phase the Sm atoms are mostly in the 4f3
configuration.

Sm/Si(001): From XPS data on Sm 3d core levels [6] divalent Sm is
observed at submonolayer coverages, while an average valence of 2.75 is
observed at one monolayer. The Sm valence is determined by the Sm-Sm
interaction.

C. SCHOTTKY BARRIER

The values of the Schottky barrier measured at the native interfaces
on n-type substrates are scattered between 0.7 eV (Sm/Si(111)),
0.63 eV (Sm/GaAs(110)) [3], and 0.45 eV (Sm/Ge(111)). The value
for Si obtained from photoemission is much higher than the typical
values obtained by transport measurements on low vacuum prepared
technological-like interfaces [4].

D. OXIDATION

Ultrathin Sm/Si(111) interfaces have a coverage and valence state
dependent catalytic effect on the semiconductor oxidation [5].
Monolayer range trivalent Sm-Si mixed phases are found to be much
more effective oxidation promoters than submonolayer divalent
Sm/Si(111) interfaces.

A discussion of the rare earth/semiconductor interfaces is found in ref
[7].

REFERENCES

----

[1] A.Franciosi, J.H.Weaver, P.Perfetti, A.Katnani, G.Margaritondo
     [Solid State Commun. (USA) vol.47 (1983) p.427]
[2] A.Franciosi, J.H.Weaver, P.Perfetti, A.Katnani, G.Margaritondo
     [Phys. Rev. B (USA) vol.29 (1984) p.5611]
[3] M.Grioni, J.J.Joyce, J.H.Weaver [Phys. Rev. B (USA) vol.32 (1985)
     p.962]
[4] R.D.Thompson, K.N.Tu [Thin Solid Films (Switzerland) vol.93
     (1982) p.265]
[5] S.Chang, P.Philip, A.Wall, A.Raisanen, N.Trouillier, A.Franciosi
     [Phys. Rev. B (USA) vol.35 (1987) p.3013]
[6] A.Faldt, H.P.Myers [Phys. Rev. B (USA) vol.33 (1986) p.1424]
[7] G.Rossi [Surf. Sci Rep. (Netherlands) vol.7 (1987) p.128]

19.18 STRUCTURE OF THE Sn/Si INTERFACE

by C.A.Sebenne

June 1987
EMIS Datareview RN=15726

A. GENERAL FEATURES OF THE Sn/Si INTERFACE

On clean single crystal silicon, tin forms an abrupt interface at the
atomic scale. The Sn atoms of the first layer are covalently bonded
to the Si surface atoms. Then the metallic character of the tin layer
starts to appear, at a coverage which is never far beyond 2 ML (one
monolayer is 7.8 X 10*14 atoms/cm*2 on Si(111) and 6.8 X 10*14
atoms/cm*2 on the (100) face). Sn would be an isoelectronic
substitutional impurity in bulk silicon; however, because of its
large covalent radius of 1.40 A as compared to 1.175 A for Si, its
solubility is very small. Such bulk effects will not be considered
here. In the following, some details of the forming interface, either
on the (111) or on the (100) face of silicon, are given.

B. ROOM TEMPERATURE DEPOSITION OF Sn ON Si(111) 2 X 1

One study has been reported on this system, by Taleb-Ibrahimi et al
[1] who performed LEED, Auger electron and photoemission yield
spectroscopies. Summarising the results, the first 2 ML of Sn adsorb as
a uniform layer on cleaved Si, then metallic Sn islands grow. The
adsorption stage displays successively a SQRT 3 X SQRT 3 R 30 deg and
a 1 X 1 surface reconstruction upon increasing Sn coverage, which are
completed at 1/3 and 1 ML respectively. The ionisation energy decreases
by about 0.45 eV, from 5.35 eV to 4.90 eV upon 1 ML deposition, while
the Fermi level gets slightly closer to the valence band edge at the
surface. Up to 2 ML coverage the system remains essentially covalent;
the double-structured dangling bond band of clean cleaved Si(111) is
replaced at 1/3 ML coverage by a single surface state band at 0.3 eV
below the valence band edge and about 0.5 eV wide. If the sample is
heated beyond about 50 deg C, the SQRT 3 reconstruction is not observed.
1 ML coverage appears as a singularity (enhanced photoemission yield,
work function discontinuity, sharp 1 X 1 LEED pattern) which may
indicate either a well ordered adsorbed Sn layer or a homogeneous
two-dimensional liquid.

The discussion of the SQRT 3 reconstruction at 1/3 ML coverage has been
given in the case of the Si-Al system [8].

C. DEPOSITION OF Sn ON Si(111) 7 X 7

Long ago, Estrup and Morrison [2] recognized through LEED measurements
that i) upon room temperature deposition of Sn, in the initial stage, a
new atomic arrangement keeping the 7 X 7 periodicity establishes, then
the surface becomes disordered, ii) upon annealing of the 7 X 7-Sn, it
transforms into a first SQRT 3-Sn at 175 deg C, which becomes a
2 SQRT 3-Sn at 700 deg C which changes into a second SQRT 3-Sn at 800

deg C, the corresponding Sn coverages remaining unknown. More recently, RHEED measurements were performed by Ichikawa [3] who demonstrated the complexity of the system. First, room temperature deposition shows the same results as in [2]; moreover, after a slight temperature rise, there is the evidence for white Sn oriented and crystallised islands, at coverages of several monolayers, with (110)Sn//(111)Si and [001]Sn//[110]Si. Deposition beyond 1 ML of Sn on substrates at 120-150 deg C keeps the 7 X 7 surface periodicity while Sn islands grow. On substrates at 240-320 deg C, the 7 X 7 is replaced by a 1 X 1 at 1.3 ML and, under RT cooling, a (2 SQRT 3 X 2 SQRT 3) R 30 deg is observed. On substrates at 390-480 deg C, a SQRT 3 X SQRT 3 R 30 deg appears at 0.2 ML, otherwise the behaviour is the same as at lower substrate temperatures.

The observations are different when a RT deposited layer is annealed. Deposition in the 2/3 ML range keeps the 7 X 7 periodicity which transforms into a SQRT 3 beyond 330 deg C which converts into a 1 X 1 at 860 deg C. Upon cooling, it goes back to the SQRT 3 then, below 190 deg C, the 2 SQRT 3 is observed. Below 1/3 ML coverage, only the SQRT 3 is observable, meaning that the 2 SQRT 3 implies a higher coverage. Upon initial coverages beyond 1 ML and annealing, superstructures identified as 2 SQRT 91 X 2 SQRT 91 - R(30 +- 10.9 deg), 3 SQRT 7 X 3 SQRT 7 - R(30 +- 3 deg) and SQRT 133 X 4 SQRT 3 were observed. Modelling of such systems is presently not possible. Angle resolved photoemission measurements performed on the SQRT 3 - Sn surface by Kinoshita et al [4] show the presence of two surface states, one about 0.3 eV below Ef, which may correspond to the one observed on the cleaved surface [1], and a seemingly double-structured one, more dispersive, between 1 and 2 eV below Ef. These observations agree with a simple model of SQRT 3 - Sn on Si (111), complete at 1/3 ML coverage, and common to Al, Ga, In, Ge and Sn adsorption (see Si/Al for discussion, [8]).

D. DEPOSITION OF Sn ON Si(100)

RHEED observations by Kuwata et al [5], together with LEED and AES measurements by Ueda et al [6] have shed some light on the growth process of Sn on Si(100). At room temperature, the 2 X 1 reconstruction vanishes in the 1 ML coverage range where only the 1 X 1 remains observable. Highly polycrystalline and less and less uniform metallic Sn grows beyond 3 ML.

Upon annealing several surface structures are observed depending on coverage: 0.3 and 0.4 ML annealed at 570 deg C induce respectively a c(4 X 4) and a (6 X 2) reconstruction: 0.8 and 1.6 ML of Sn display respectively a c(8 X 4) and a (5 X 1) after the same 570 deg C annealing. Annealing at increasing temperature of a 1.2 ML covered sample brings the following sequence: 570 deg C brings the 5 X 1 without significant changes of both Sn and Si room temperature Auger signals; 640 deg C transforms the surface into the c(8 X 4) with a decrease (increase) of the Sn(Si) Auger signal; and 670 deg C brings the 6 X 2 with further decrease (increase) of the Sn(Si) Auger signal. Beyond 700 deg C, the 2 X 1 reconstruction is restored and the Sn Auger signal has almost vanished.

The electronic property changes have been reported by Andriamanantenasoa et al [7] upon room temperature deposition, using photoemission yield spectroscopy. They show that the metallic character of the tin layer does not show up below 2 ML coverage. The ionisation energy and the work function decrease in parallel with an overall variation of about 0.35 eV (less than on the cleaved surface); most of the decrease occurs at 0.25 ML. New features start to grow in the band of filled surface states beyond 0.25 ML.

REFERENCES
----

[1] A.Taleb-Ibrahimi, C.A.Sebenne, F.Proix, P.Maigne [Surf. Sci. (Netherlands) vol.163 no.2/3 (1985) p.478-88]

[2] P.J.Estrup, J.Morrison [Surf. Sci. (Netherlands) vol.2 (1964) p.465]

[3] T.Ichikawa [Surf. Sci. (Netherlands) vol.140 (1984) p.37]

[4] T.Kinoshita, S.Kono, T.Sagawa [Phys. Rev. B (USA) vol.34 no.4 (15 Aug 1986) p.3011-14]

[5] N.Kuwata, T.Asai, K.Kimura, M.Mannami [Surf. Sci. (Netherlands) vol.143 (1984) p.L393]

[6] K.Ueda, K.Kinoshita, M.Mannami [Surf. Sci. (Netherlands) vol.145 (1984) p.261]

[7] I.Andriamanantenasoa, J.P.Lacharme, C.A.Sebenne [Surface. Sci. (Netherlands) (1987) in press]

[8] C.A.Sebenne [EMIS Datareview RN=15749 (June 1987) 'Structure of the Al/Si interface']

19.19 STRUCTURE OF THE Ti/Si INTERFACE

by P.J.Rosser

November 1986
EMIS Datareview RN=13931

The Ti/Si interface is unreacted at room temperature within the
resolution of RBS. Thermal annealing results in silicide formation,
with TiSi forming at about 500 deg C and TiSi2 at about 600 deg C. The
C49 (orthorhombic ZrSi2 structure) disilicide phase is the first to
form followed closely by the C54 (orthorhombic, ABCDA stacking of
distorted hexagonal planes [13]) phase. Interest in this interface
results from the wide use of titanium in metallisation systems because
of its excellent adhesion to both silicon and silicon dioxide [1-3].

Using more sensitive medium energy ion scattering in combination with
ion shadowing and blocking techniques it has been demonstrated that
mixing of Ti and Si already occurs at room temperature for coverages
below 2.8 X 10*15 Ti atoms/cm*2. The composition of this uniformly
mixed film is close to TiSi. For higher coverages no further mixing
takes place and pure Ti is present on top of the ultrathin mixed film
[4]. The sample preparation for this study included light sputter
etching of the silicon substrate, the cleanliness of the surface being
monitored by LEED, prior to deposition of titanium in the same UHV
system. A slightly higher base pressure during Ti deposition (1 X 10*-9
torr compared with 5 X 10*-11) results in the onset of silicide
formation being delayed until 250 deg C [5]. In this case the silicide
formed is nearly amorphous and if it is more than 10 nm thick then
significantly higher reaction temperatures are required to reach the
larger grain silicide phase than if no such low temperature phase is
formed.

Using still less stringent sample preparation and film deposition
techniques, and Raman spectroscopy to determine the phases formed after
annealing, the onset of silicon diffusion through the titanium film was
reported to be at 400 deg C along with the probable formation of both
titanium oxides and silicide. At temperatures above 700 deg C the
disilicide is the only phase observed [6].

Annealing of Ti films on silicon (111) at 550 deg C for 30 min leads to
formation of a Ti/Ti5Si3/TiSi/TiSi2/Si structure. On annealing at 600
deg C and above only TiSi2 is formed [7].

The presence of an adsorbed O2 or an oxide layer on the silicon surface
modifies the Ti/Si interface structural and electrical properties. In
one study [8] Ti deposited on clean Si(100) showed a sharp interface.
Silicide formation occurred on annealing at 400-600 deg C. For Ti
deposited with adsorbed O2 or with a thin oxide layer (thickness less
than 2 nm) or a thick thermal oxide a strong reaction between Ti and O
occurred at room temperature. Annealing of samples with a thin oxide
layer resulted in silicide formation, as for a clean surface. For
samples with a thick oxide layer annealing effects were not observed
below 700 deg C. Annealing at this temperature led to the formation of

titanium oxide. Another study showed similar results, but for the clean Si/Ti interface atomic mixing was observed at temperatures as low as 300 deg C, and for thin titanium films deposited on adsorbed oxygen the Ti-O peak present at room temperature is replaced by a Si-O peak as silicide formation takes place [9].

Studies of annealing of Ti covered heavily implanted n-Si have shown that for Sb, As and P implants the thickness of the TiSi2 layer is about half of that for unimplanted samples. In the case of O and Ar implants TiSi2 is not formed, the annealing resulting in Ti atom precipitation at 700 deg C and flaking off of the titanium layer at higher temperatures [10].

Comparison of silicide formation kinetics for Ti/crystalline Si and for sequentially deposited Ti/amorphous Si showed that reproducible kinetics could be obtained for TiSi2 growth by thermal annealing of Ti/a-Si at 475-550 deg C in vacuo. TiSi2 growth rate was proportional to the square root of time. It was shown that surface cleanliness rather than the structural nature of the silicon was the primary factor in determining reaction kinetics and the lower formation temperature of TiSi2 on a-Si [2].

By annealing clean titanium films on silicon and analysing the resulting layers by TEM, epitaxial TiSi2 has been observed on both (100) [11] and (111) [12] silicon with a grain size of up to 5 microns. The orientation relationships on (100) are typically [110]TiSi2//[110]Si; [001]TiSi2//[1 bar1 0]Si; [010]TiSi2//[001]Si. Of the possible projections of the TiSi2 lattice onto the (100) Si lattice the most promising for more extensive epitaxy is the (010).

REFERENCES
----

[1]   K.N.Tu, J.W.Mayer ['Thin Films - Interdiffusion and Reactions' Ed. J.M.Poate et al; Wiley, NY (1978) p.359-405]
[2]   L.S.Hung, J.Gyulai, J.W.Mayer, S.S.Lau, M.-A.Nicolet [J. Appl. Phys. (USA) vol.54 (1983) p.5076-80]
[3]   R.Beyers, R.Sinclair [J. Appl. Phys. (USA) vol.57 (1985) p.5240-5]
[4]   E.J.Van Loenen, A.E.M.J.Fischer, J.F.van der Veen [Surf. Sci. (Netherlands) vol.155 no.1 (1985) p.65-78]
[5]   R.Butz, G.W.Rubloff, T.Y.Yan, P.S.Ho [Phys. Rev. B (USA) vol.30 (1984) p.5421-9]
[6]   R.J.Nemanich, R.T.Fulks, B.L.Stafford, H.A.Vander Plas [Appl. Phys. Lett. (USA) vol.46 (1985) p.670-2]
[7]   Y-C.Liu, C-J.Lin, J-S.Maa [Jpn. J. Appl. Phys. (Japan) vol.18 (1979) p.991-2]
[8]   M.A.Taubenblatt, C.R.Helms [J. Appl. Phys. (USA) vol.53 (1982) p.6308-15]
[9]   R.Butz, G.W.Rubloff, P.S.Ho [J. Vac. Sci. & Technol. A (USA) vol.1 no.2 pt 1 (1983) p.771-5]
[10]   P.Revesz, J.Gyimesi, E.Zsoldos [J. Appl. Phys. (USA) vol.54 (1983) p.1860-4]
[11]   R.Nipoti, A.Armigliato [Jpn. J. Appl. Phys. Part 1 (Japan) vol.24 no.11 (1985) p.1421-4]
[12]   L.J.Chen, H.C.Cheng, W.T.Lin, L.J.Chou, M.S.Fung [Mater. Res. Soc. Symp. Proc. (USA) vol.37 (1985) p.375-80]
[13]   F.M.d'Heurle [Proc. Spring ECS meeting (1982) p.194]

19.20 STRUCTURE OF THE V/Si INTERFACE

by L.J.Chen

July 1987
EMIS Datareview RN=15763

A. SILICIDE FORMATION, FORMATION MECHANISM

Only one stable vanadium silicide, VSi2, forms by chemical interaction
at the clean refractory metal/silicon interface at about 550 deg C
[1-5]. This silicide has a hexagonal C40 structure.

B. KINETICS, DIFFUSING SPECIES

The thickness of VSi2 is often reported to increase linearly as a
function of time on (001)Si [3,6-9]. Square root time dependence was
also reported in an earlier work and during a later stage of the
formation which is attributed to the influence of oxygen contamination
[2,9]. The activation energies for linear kinetics were measured to be
1.7 +- 0.2-2.2 eV [3,8,9]. Si was found to be the dominant diffusing
species [10,11].

C. EPITAXY, INTERFACE STRUCTURE

Epitaxial VSi2 was found to grow locally on (111)Si. The orientation
relationships were found to be [0001]VSi2//[111]Si and (2 0 bar2 0)VSi2
//(2 0 bar2)Si. Interfacial dislocations were identified to be of edge
or 60 deg type with 1/6<112> Burgers vectors [5].

D. STABILITY

Island structure was found in V thin films on (111)Si after annealing at
1100 deg C for 1 hr [5]. When a VSi2 layer is placed between a single-
crystal Si substrate and an amorphous Si overlayer, annealed at about
one-half the melting temperature of the VSi2, the silicide is seen
to move through the amorphous Si and rise to the surface, leaving below
it an epitaxial layer of Si [12].

E. REDISTRIBUTION OF DOPANTS, DOPANT AND CONTAMINANT EFFECTS

Studies on silicide formation in V/Si:As structures [13] showed that
annealing at 550 deg C led to complete silicide formation.
Redistribution of As did not take place during VSi2 formation
(contrasting with As redistribution observed in near-noble-metal/Si:As
systems).

F. INTERFACE BONDING, ELECTRONIC STRUCTURES, SURFACE STRUCTURE

Combined UPS, AES, XPS, LEED and TEM studies have shown that no
intermixing occurs at the V/Si interface at room temperature [14].
An IS, AES and SIMS study also showed that vanadium deposited at
300 deg K did not induce silicide formation [15]. On annealing at about
500 deg C, the silicide formation temperature, single-phase VSi_2 is
produced. Reaction begins at temperatures well below the silicide
formation temperature, an intermixed layer (about 10 nm or more)
being formed at 350 deg C [15].

G. ION BEAM MIXING

VSi_2 was found to form after Ar, Kr and Xe ion beam mixing [6,16]. The
depths of ion beam mixing after Si implantation of V thin films on
silicon showed linear dose dependence [17].

H. ELECTRICAL PROPERTIES

The electrical resistivities of VSi_2 thin films were measured to be
50-55 micro-ohm cm [18,19]. Schottky barrier height of VSi_2 was
measured to be 0.55 eV on n-type silicon [20].

REFERENCES

[1]   K.N.Tu, J.W.Mayer [in 'Thin Films - Interdiffusion and Reactions'
       Eds. J.M.Poate, K.N.Tu, J.W.Mayer (Wiley, New York, USA, 1978)
       p.359-405]
[2]   K.N.Tu, J.F.Ziegler, C.J.Kircher [Appl. Phys. Lett. (USA) vol.23
       no.9 (1 Nov 1973) p.493-5]
[3]   H.Krautle, M.-A.Nicolet, J.W.Mayer [J. Appl. Phys. (USA) vol.45
       no.8 (Aug 1974) p.3304-8]
[4]   G.W.Rubloff [Surf. Sci. (Netherlands) vol.132 nos.1-3 (1983)
       p.268-314]
[5]   C.J.Chien, H.C.Cheng, C.W.Nieh, L.J.Chen [J. Appl. Phys. (USA)
       vol.57 no.6 (15 Mar 1985) p.1887-9]
[6]   M.A.Nicolet, S.S.Lau [in 'Materials and Process Characterisation'
       Eds. N.G.Einspruch, G.B.Larrabee (Academic, New York, USA, 1983)
       p.329-464]
[7]   F.M.d'Heurle, P.Gas [J. Mater. Res. (USA) vol.1 no.1 (1986)
       p.205-21]
[8]   B.I.Fomin, A.E.Gershinskii, E.I.Cherepan, F.L.Edelman [Phys.
       Status Solidi a (Germany) vol.36 no.1 (1976) p.K89-91]
[9]   R.J.Wagner, S.S.Lau, J.W.Mayer [Thin Solid Films (Switzerland)
       vol.45 no.1 (15 Aug 1977) p.123-4]
[10]   W.K.Chu, H.Krautle, J.W.Mayer, H.Muller, M.A.Nicolet, K.N.Tu
       [Appl. Phys. Lett. (USA) vol.25 no.8 (15 Oct 1974) p.454-7]
[11]   J.E.Baglin, F.M.d'Heurle, W.N.Hammer, S.Petersson [Nucl. Instrum.
       & Methods (Netherlands) vol.168 nos.1-3 (1980) p.491-7]
[12]   S.S.Lau, Z.L.Liau, M.A.Nicolet [Thin Solid Films (Switzerland)
       vol.47 no.3 (15 Dec 1977) p.313-22]
[13]   M.Wittmer, K.N.Tu [Phys. Rev. B (USA) vol.29 no.4 (15 Feb 1984)
       p.2010-20]

[14]   J.G.Clabes, G.W.Rubloff, T.Y.Tan [Phys. Rev. B (USA) vol.29 no.4
        (15 Feb 1984) p.1540-50]
[15]   C.Achete, H.Niehus, W.Losch [J. Vac. Sci. & Technol. B (USA)
        vol.3 no.5 (Sep/Oct 1985) p.1327-31]
[16]   B.Y.Tsaur [Ph.D. Thesis, California Institute of Technology
        (1980)]
[17]   D.B.Poker, B.R.Appleton [J. Appl. Phys. (USA) vol.57 no.4
        (15 Feb 1985) p.1414-6]
[18]   S.P.Murarka [J. Vac. Sci. & Technol. (USA) vol.17 no.4 (1980)
        p.775-92]
[19]   S.P.Murarka, M.H.Read, C.J.Doherty, D.B.Fraser [J. Electrochem.
        Soc. (USA) vol.129 no.2 (Feb 1982) p.293-301]
[20]   J.L.Freeouf [Solid State Commun. (USA) vol.33 no.10 (Mar 1980)
        p.1059-61]

19.21 STRUCTURE OF THE W/Si INTERFACE

by R.B.Jackman

July 1987
EMIS Datareview RN=15780

A. INTRODUCTION

Tungsten and tungsten silicide layers upon silicon have become
increasingly important within VLSI technology as diffusion barriers,
gate materials and electrical contacts. This Datareview will
concentrate upon the physicochemical properties of the silicon (100)
and (111)/tungsten interface, the onset of silicide formation and the
influence of the method of interface formation. The specific formation
of silicides by such techniques as ion-beam mixing or rapid-thermal-
annealing (RTA) will not be considered.

B. INTERFACES FORMED BY EVAPORATION OR SPUTTERING OF TUNGSTEN

The initial evaporation of tungsten onto a silicon surface at room
temperature has been reported to lead to an inert and abrupt interface
followed by a layer by layer growth mechanism. The work, by Azizan and
co-workers [1], employed surface spectroscopic probes in ultra high-
vacuum (UHV), to study the properties of this interface and the
effect of subsequent annealing. The silicon studied was (111) (7 X 7)
in orientation. The abrupt interface, suggested by XPS and Auger signal
intensity and energy analysis, was confirmed by work function
measurements. These show a rapid variation in the monolayer region and
reach a constant value, slightly lower than that of pure tungsten, for
coverages > 0.5 nm. The silicon band bending variation was complete by
approx. 0.1 monolayer, suggesting the Schottky barrier has already
formed at this point. Annealing thin deposits (approx. 1 monolayer)
at temperatures of around 500 deg C was found, by LEED, to remove the
(7 X 7) Si superstructure and some evidence for a diffuse (1 X 1)
pattern was suggested. At this point W-Si intermixing and bonding
became evident. At temperatures > 750 deg C thicker deposits (5 - 500
monolayers) developed into silicide of epitaxial (0001) WSi2 form.

Torres et al [2] have considered structural effects upon the interface
and subsequent silicide formation, for two modes of evaporated layer
formation. Analysis was performed by XRD and TEM, subsequent to use
of an UHV reaction environment. Room temperature evaporation of 15-60 nm
tungsten layers onto Si(100) (2 X 1) wafers was found to lead to a
randomly orientated polycrystalline metal film which remained disordered
following high temperature annealing. However, direct deposition onto
the silicon at 600-800 deg C gave rise to the formation of epitaxially
aligned tetragonal WSi2. Extensive investigations into such epitaxial
growth have been carried out from TEM observations by Lin and Chen [3].
Since a less rigorous vacuum environment was employed deposition of a
30 nm amorphous silicon cap onto the Si(111) or Si(100) wafers followed
evaporation of 30 nm tungsten films to prevent metal oxidation during
further heat treatments. Upon annealing it was suggested that this
amorphous layer led to WSi2 grain which coalesced with the underlying
WSi2 as it formed. However, little investigation into the diffusion
behaviour of this layer prior to the onset of silicide formation was

presented and consequently its possible structural influence upon the
work was not fully established. Both tetragonal and hexagonal silicide
forms were found after annealing W/Si(100) in two steps to 1100 deg C
for prolonged periods (approx. one hour), with t-WSi2 being the most
dominant species. Epitaxial growth of these forms was found to occur in
a range of modes as summarised in TABLE 1.

All modes except mode C of t-WSi2 plus B and 1 of h-WSi2 epitaxy were
found after annealing a W/Si(111) interface. It was suggested that the
absence of these may be associated with experimental difficulties in
the observation of weak signals. Interestingly, it was found that
epitaxial growth was achieved with lattice mismatches in excess of 4%,
and the frequency of occurrence of a mode could not be predicted from
lattice match criteria alone.

C. OTHER INTERFACE FORMATION METHODS

The formation of a W/Si interface may be achieved by means of Chemical
Vapour Deposition (CVD) of tungsten from tungsten hexafluoride [4-8].
In addition such a reaction may be enhanced by photon or particle
irradiation [8-12]. This approach has achieved significant
technological importance and the mechanistic details pertaining to
reaction pathways have recently begun to be evaluated [12]. However,
only information upon the effect of this method upon the resultant
interfacial properties will be considered here. Tungsten layers of
between 50 and 200 nm, formed by either WF6/H2 reaction with silicon
or a direct WF6/Si reaction pathway have been characterised by XRD.
Such studies indicate [13] a slight (100) preferred orientation within
the metal film. Layers of this form have been found to display
resistivities of 12-17 micro-ohm cm, around 2-3 times the bulk tungsten
value [4]. Following annealing at temperatures above 650 deg C
tetragonal WSi2 can be detected [4,6,7] with no evidence for the
hexagonal species seen following promotion of an Si/W interface by
evaporation [3,14,15]. In addition this annealing temperature is
relatively low compared with the conversion of tungsten films produced
by such evaporation. It has been suggested that oxygen atoms, which
may be incorporated during evaporation, play a role in the silicide
formation. Such species may diffuse towards the W/Si interface
preventing the formation of a disilicide until higher temperatures. The
interface formed by WF6 CVD is free from such impurities [4,12]. It has
been found that the surface resistivity initially rises with annealing
temperatures of 600-700 deg C, then falls to a level of around 70-85
micro-ohm cm by 900 deg C. Whilst Murarka and co-workers [14] have
discussed this change in terms of h-WSi2, the absence of such a species
during similar resistivity measurements by Kamins et al [7] lead them
to suggest the change is related to grain size and the presence of
defects.

The mechanisms for diffusion within the interfacial region prior to any
silicide formation are still not yet fully characterised, and are
discussed in references [16] and [17]. Reports have been made of both
mobility of Si into the W layer and W penetration into the silicon bulk
[12,16,18,19] and, if present, oxygen impurities appear to have a
significant role in the process. An effective activation energy for the
total dissolution of 2.6 eV has been proposed by Chang and Quintana
[18].

TABLE 1: modes of epitaxial growth observed by Lin and Chen [3]
following high temperature annealing of a 30nm tungsten layer
on Si(100)

Silicide type	Mode	
t-WSi2	A	(110)WSi2//(001)Si and (0 0 bar4)WSi2//(2 bar2 0)Si
	B	(111)WSi2//(001)Si and (1 1 bar2)WSi2//(2 bar2 0)Si
	C	(100)WSi2//(001)Si and (004)WSi2//(220)Si
	D	(110)WSi2//(111)Si and (004WSi2)//(2 0 bar2)Si
	E	(111)WSi2//(111)Si (1 1 bar2)WSi2//(2 0 bar2)Si
h-WSi2	A	(0001)WSi2//(001)Si and (2 0 bar2 0)WSi2//(2 bar2 0)Si
	B	(bar2 4 bar2 3)WSi2//(001)Si and (2 bar1 bar1 2)WSi2//(bar2 bar2 0)Si
	C	(0001)WSi2//(111)Si and (2 0 bar2 0)WSi2//(2 0 bar2)Si

REFERENCES
----

[1]   M.Azizan, T.A.Nguyen Tan, R.Cinti, R.Baptiest, G.Chauvet [Surf.
       Sci. (Netherlands) vol.178 (1986) p.17-26]
[2]   J.Torres, A.Perio, R.Pantel, Y.Campidelli, F.A.D'Avitaya [Thin
       Solid Films (Switzerland) vol.126 no.3/4 (26 Apr 1985) p.233-9]
[3]   W.T.Lin, L.J.Chen [J. Appl. Phys. (USA) vol.59 no.10 (15 May
       1986) p.3481-8]. Also [J. Appl. Phys. (USA) vol.58 no.4 (15 Aug
       1985) p.1515-8]; L.J.Chen, H.C.Cheng, W.T.Lin, L.J.Chou,
       M.S.Fung [Mater. Res. Soc. Symp. Proc.(USA) vol.37 (1985) p.375]
[4]   Y.Pauleau, Ph.Lami, A.Tissier, R.Pantel, J.C.Oberlin [Thin
       Solid Films (Switzerland) vol.143 no.3 (15 Oct 1986) p.250-68]
[5]   M.L.Green, Y.S.Ali, B.A.Davidson, L.C.Feldman, S.Nakahara
       [Mater. Res. Soc. Symp. Proc. (USA) vol.54 (1986) p.723]
[6]   P.J.Codella, F.Adar, Y.S.Liu [Appl. Phys. Lett. (USA) vol.46
       no.11 (1 Jun 1985) p.1076-8]
[7]   T.I.Kamins, S.S.Laderman, D.J.Coulman, J.E.Turner [J.
       Electrochem. Soc. (USA) vol.133 no.7 (Jul 1986) p.1438-42]
[8]   E.K.Broadbent, C.L.Ramiller [J. Electrochem. Soc. (USA) vol.131
       (1984) p.1427]
[9]   S.D.Allen, A.B.Tringubo [J. Appl. Phys. (USA) vol.54 (1983)
       p.1641]
[10]   T.F.Deutsch, D.D.Rathman [Appl. Phys. Lett. (USA) vol.45 (1984)
       p.623]

[11]   K.Gamo, K.Takehara, Y.Hamamura, M.Tomita, S.Namba [Microelectron.
       Eng. (Netherlands) vol.5 (1986) p.163]
[12]   R.B.Jackman, J.S.Foord [Eur. Mater. Res. Soc. Symp. Proc.
       vol.Xl 'Laser Processing and Diagnostics (II)' Eds D.Baverle,
       K.L.Kompa, L.Lavde (1986) p.37]; R.B.Jackman, J.S.Foord [to be
       published in Surf. Sci.]
[13]   N.D.McMurray, R.M.Singleton, K.E.Muszar,Jr., D.R.Zimmerman [J.
       Met. (USA) vol.17 (1965) p.600]
[14]   S.P.Murarka, M.H.Read, C.C.Chang, [J. Appl. Phys. (USA) vol.52
       no.12 (1981) p.7450]
[15]   F.M.D'Heurle, C.S.Peterson, M.Y.Tsai [J. Appl. Phys. (USA)
       vol.51 (1980) p.7450]
[16]   K.N.Tu, J.W.Mayer [in 'Thin Films - Interdiffusion and
       Reactions.' Eds. J.M.Poate, K.N.Tu, J.W.Mayer (J. Wiley,
       New York, 1978) chapter 10]
[17]   F.M.D'Heurle, R.T.Hodgson, C.Y.Ting [Mater. Res. Soc. Symp.
       Proc. (USA) vol.52 (1986) p.261]
[18]   C.C.Chang, G.Quintana [J. Electron. Spectrosc. & Relat. Phenom.
       vol.2 (1973) p.363]
[19]   V.I.Zonii, G.N.Kartmazov, N.S.Poltavtser, N.A.Semenov, [Izv.
       Akad. Nauk. SSSR, Neorg. Mater. vol.17 (1981) p.916]

19.22 STRUCTURE OF THE Yb/Si INTERFACE

by G.Rossi

August 1987
EMIS Datareview RN=15758

A. INTRODUCTION

The Yb/Si system is by far the most studied by spectroscopic
techniques: UPS, XPS, photon-energy dependent (resonant) soft X-ray
photoemission, X-ray absorption, Auger lineshape, ion scattering, LEED;
and is the only case, so far, where bulk single phase standards have
been available for photoemission studies. See ref [1] for an extensive
description of the Yb/Si interface system and reference list.

B. CHEMISORPTION AT ROOM TEMPERATURE

The chemisorption of Yb on Si(111) at submonolayer and monolayer
coverage proceeds via the occupation of Si surface sites that leave the
Yb adsorbed atoms in the divalent ionic configuration [2,3]. The Si
surface reconstructions are removed without formation of ordered Yb
induced superstructures at room temperature (RT), and with a small
perturbation of the Si-Si bonds. The divalence of 1 ML Yb/Si(111) has
been tentatively interpreted as the occupation by Yb of the Si(111)
threefold sites with Yb-Si bond length similar to the Yb silicide bond
lengths (2.95 +- 0.05 A) but with much lower than silicide-like
coordination, i.e. only 3 Si-Yb neighbours in the hollow sites [3]. ISS
established threefold hollow sites for annealed 1 ML Yb/Si(111) at
700-800 deg K and displaying a 2 X 1 LEED pattern (3 X 1 and 5 X 1
phases are observed for higher temperature annealings) [4]. The ISS
establishes the position of the Yb at 1.9 +- 0.3 A height above the Si
surface plane, which is compatible with a bond length of 2.9 +- 0.4 A.
The different surface patterns are correlated with different
occupation of the same threefold sites, with the excess Yb being
thermally diluted into the substrate. The occupied DOS for < 1 ML
Yb/Si(111) is dominated by a 4f13 final states doublet that shifts
towards Ef with coverage, and broadens [2,5]. XARS show at 0.2 ML a
narrow white line indicating an atomic-like 5d band [3]. The white line
spectrum for 1 ML Yb-Si(111) is a much broader 5d band attributed to
antibonding 5d hybrid states, in a molecular orbital like Yb 5d - Si 3p
bonding scheme. Results for 1 ML Yb/Si(100) and 1 ML Yb/a-Si show a
much higher reactivity due to the availability of higher co-ordination
sites, and a mixed valent configuration for the interdiffused Yb-Si
interface layers [3].

At two monolayers the three-dimensional character of the interface
region is developed also for the least reactive of the three Yb/Si
interfaces reported: the Yb/Si(111). The Si 2p core levels show a
chemically shifted component at -2 eV (maximum shift) due to the Si
atoms that penetrate the Yb overlayer [2,5,7]. XARS shows mixed
valency averaging at 2.11 for 2 ML Yb/Si(111), the bonding structure
being attributed to orbital mixing between Yb 5d and Si 3p with the
unoccupied part of the 5d states pushed and spread well above the Fermi
level. Through the mixed orbitals a net charge transfer from Yb to Si
would occur due to the electronegativity difference and accounting for
the core level shifts [2,5-7].

C. INTERFACE SATURATION AND ANNEALING

The reaction reaches a saturation above 2 ML on the Si(111) surface at
room temperature, whilst annealings at 450 deg C promote further
intermixing and the formation of a silicide-like phase with 2.37
average valence [5,14]. Thicker layers of Yb on Si(111) annealed at 300
and 500 deg C show mixed valent silicide-like extended interfaces with
the top layer Yb atoms in divalent configuration due to the reduced
Yb-Si co-ordination at the surface of the silicide [8,7]. At RT for 4 ML
Yb/Si(111) the interface is saturated and covered by unreacted metallic
Yb [2,3,9,10]. The Yb interfaces at RT on the Si(100) and a-Si
substrates at 2 and 4 ML coverages evolve spontaneously towards
silicide-like mixed valent phases. The reported values for the Schottky
barrier on n-type Si(111) are of the order of 0.4 eV [11], and on
p-type Si(100) are 0.64-0.67 eV [12].

The lattice mismatch between YbSi(1.8) and Si(111) is -1.59%.

Photoemission data have been reported for bulk Yb2Si3 which has a bulk
mixed valence of 2.52 +- 0.2 and a surface and sub-surface divalent Yb
configuration [13].

D. THE Yb/Ge(111) INTERFACE

The growth of the Yb/Ge(111) interface proceeds from the chemisorption
regime with weak interaction between Yb and the substrate (divalent
phase) to a complex intermixed phase which is established for coverages
between 1 and 3 ML as derived from a lineshape analysis of both Ge 3d
core levels and Yb 4f13 (divalent) final state multiplets [7,15]. The
intermixed phase is stable at RT and appears to prevent further
outdiffusion of Ge in the Yb overlayer. A diluted phase is then
established in the Yb rich overlayer which becomes more and more
diluted until, for 20 ML Yb/Ge(111) at RT a pure Yb film ends the
interface region. The interface width is larger than in the case of
Yb/Si(111).

REFERENCES

[1]   G.Rossi [Surf. Sci. Rep. (Netherlands) vol.7 (1987) p.128]
[2]   G.Rossi, J.Nogami, I.Lindau, L.Braicovich, U.del Pennino,
        S.Nannarone [J. Vac. Sci. & Technol. A (USA) vol.1 (1983) p.781]
[3]   G.Rossi, D.Chandesris, P.Roubin, J.Lecante [Phys. Rev. B (USA)
        vol.33 (1986) p.2926]
[4]   J.Kofoed, I.Chorkendorff, J.Onsgaard [Solid State Commun. (USA)
        vol.52 (1984) p.283]
[5]   L.Braicovich, I.Abbati, C.Carbone, J.Nogami, I.Lindau [Surf. Sci.
        (Netherlands) vol.168 (1986) p.193]
[6]   J.Nogami [PhD Thesis, Stanford 1986]
[7]   G.Rossi [These d'Etat, Paris 1985]
[8]   G.Rossi, J.Nogami, J.J.Yeh, I.Lindau [J. Vac. Sci. & Technol. B
        (USA) vol.1 (1983) p.530]
[9]   G.Rossi [Proc. 17th Int. Conf. Physics of Semiconductors, San
        Francisco, CA, USA, 6-10 Aug. 1984, Eds J.Chadi, W.A.Harrison
        (Springer, New York, 1985) p.149]

[10] I.Chorkendorff, J.Kofoed, J.Onsgaard [Surf. Sci. (Netherlands)
      vol.152/153 pt.2 (Apr 1985) p.749-56]
[11] R.D.Thompson, K.N.Tu [Thin Solid Films (Switzerland) vol.93
      (1982) p.265]
[12] H.Norde, J.de Sousa Pires, F.d'Heurle, F.Pesavento, S.Peterson,
      P.A.Tove [Appl. Phys. Lett. (USA) vol.38 (1981) p.865]
[13] I.Abbati et al [Phys. Rev. B (USA) vol.34 (1986) p.4150]
[14] I.Abbati et al [Solid State Commun. (USA) vol.60 (1986) p.595]
[15] J.Nogami, C.Carbone, D.J.Friedman, I.Lindau [Phys. Rev. B (USA)
      vol.33 (1986) p.864]

CHAPTER 20

BARRIER HEIGHTS AND CONTACT RESISTANCES: METAL/Si

20.1 BARRIER HEIGHT AT THE Ag/Si INTERFACE

by K.K.Ng

September 1987
EMIS Datareview RN=16143

The available data for Ag/Si barrier heights (BH) are summarized in
TABLE 1 below.

TABLE 1

BH (eV)	Measurement	Substrate	Reference
0.79	I-V, C-V	n(111)	[1]
0.67 +- 0.02	I-V, C-V	n(111)(100)	[4,6]
0.78-0.81	I-V, C-V	n(111)	[2]
0.65	C-V	n(111)	[9]
0.78	C-V	n(111)	[8]
0.62		n	[7]
0.47	I-V, C-V	p(111)(100)	[4]
0.54 +- 0.01	I-V, C-V	p(111)	[3]
0.37	C-V	p(111)	[8]

Turner and Rhoderick obtained their BH (above) on cleaved surfaces,
and they found that the BH on chemically cleaned surfaces changed with
time to reach an equilibrium value of 0.56 eV [1]. Hirose et al obtained
BH on both n-type and p-type substrates [4], and the sum of the two
gave the value of the Si band gap. Smith and Rhoderick obtained good
characteristics on p-Si by heating the substrate during metal deposition
[3]. Van Otterloo obtained Schottky barriers on cleaved surfaces, and
he showed that the BH could vary in the presence of cleavage steps [8].
Kobayashi showed that when mechanical stress was applied to the
Ag/n-Si Schottky, the I-V characteristics were different but the C-V
characteristics remained unchanged [9]. The effects of surface
preparation on barrier formation have been examined by Varma et al [5].

REFERENCES
----

[1]   M.J.Turner, E.H.Rhoderick [Solid-State Electron (GB) vol.11
      (1968) p.291]
[2]   J.D. Van Otterloo, J.G.De Groot [Surf. Sci. (Netherlands)
      vol.57 (1976) p.93]
[3]   B.L.Smith, E.H.Rhoderick [Solid-State Electron. (GB) vol.14
      (1971) p.71]

[4]   M.Hirose, N.Altaf, T.Arizumi [Jpn. J. Appl. Phys. (Japan)
       vol.9 (1970) p.260]

[5]   R.R.Varma, A.McKinley, R.H.Williams, I.G.Higginbotham [J. Phys.
       D (GB) vol.10 (1977) p.L171]

[6]   T.Arizumi, M.Hirose [Jpn. J. Appl. Phys. (Japan) vol.8 (1969)
       p.749]

[7]   N.T.Tam, Tran.Chot [Phys. Status Solidi a (Germany) vol.93 no.1
       (1986) p.K91-5]

[8]   J.D.van Otterloo [Surf. Sci. (Netherlands) vol.104 (1981)
       p.L205]

[9]   Y.Kobayashi [IEEE Trans. Electron Devices vol.ED-26 (1979)
       p.993]

20.2 BARRIER HEIGHT AT THE Al/Si INTERFACE

by K.K.Ng

September 1987
EMIS Datareview RN=16141

Al is the most popular metal for VLSI. It is commonly used as the final
interconnect metallization material. The reported barrier heights (BH)
are summarized below. For Al, it is generally believed that the Si
surface preparation is very critical.

TABLE 1

BH (eV)	Measurement	Substrate	Reference
0.76	I-V, C-V	n(111)	[1]
0.68 +- 0.02	I-V, C-V	n	[2]
0.69 +- 0.01	I-V, photo	n(111)	[5]
0.72	I-V, C-V	n(111)	[3]
0.81	I-V, C-V	n(100)	[3]
0.71-0.72	I-V, C-V	n(111)	[11]
0.71	I-V	n	[16]
0.71-0.76	I-V, C-V	n(111)	[4]
0.74	I-V	n	[7]
0.72 +- 0.02	I-V	n	[15]
0.70-0.72	C-V, photo	n	[8]
0.69	C-V	n(111)	[17]
0.72-0.73		n	[21]
0.48	I-V, C-V	p	[2]
0.57-0.58	I-V, C-V	p(111)	[6]
0.47	C-V	p(111)	[17]
0.39-0.40	photo	p(100)	[18]

Turner and Rhoderick obtained their BH (above) on cleaved surfaces,
and they found that the BH on chemically cleaned surfaces changed with
time to reach an equilibrium value of 0.50 eV [1]. Hirose et al obtained
BH's on both n-type and p-type substrates [2], and the sum of the two
gave the value of the Si band gap. Gutknecht and Strutt compared the

BH obtained on different Si orientations [3]. They found that a higher
BH was formed on (100) surfaces compared to (111) surfaces. Wilkinson
et al [15] prepared Schottky diodes on substrates with different
concentrations, and they found them to be similar. Smith and Rhoderick
obtained good characteristics on p-Si by heating the substrate during
metal deposition [6]. Van Otterloo obtained his BH value on cleaved
surfaces, and he showed that it could vary in the presence of cleavage
steps [17].

The change of BH for temperature above 500 deg C was studied by different
researchers [9-13,19,20]. Generally an increase of BH could be observed.
This has been attributed to the Si surface doping by Al.

Al alloy metallization for VLSI was studied by Ghate [14].

REFERENCES
----

[1]   M.J.Turner, E.H.Rhoderick [Solid-State Electron. (GB) vol.11
      (1968) p.291]
[2]   M.Hirose, N.Altaf, T.Arizumi [Jpn. J. Appl. Phys. (Japan)
      vol.9 (1970) p.260]
[3]   P.Gutknecht, M.J.O.Strutt [Appl. Phys. Lett. (USA) vol.21 (1972)
      p.405]
[4]   M.R.Namordi, H.W.Thompson,Jr [Solid-State Electron. (GB) vol.18
      (1975) p.499]
[5]   A.Y.C.Yu, C.A.Mead [Solid-State Electron. (GB) vol.13 (1970)
      p.97]
[6]   B.L.Smith, E.H.Rhoderick [Solid-State Electron. (GB) vol.14 (1971)
      p.71]
[7]   A.Rusu, C.Bulucea, C.Postolache [Solid-State Electron. (GB) vol.20
      (1977) p.499]
[8]   J.D. van Otterloo, L.J.Gerritsen [J. Appl. Phys. (USA) vol.49
      no.2 (Feb 1978) p.723-9]
[9]   H.C.Card [IEEE Trans. Electron Devices (USA) vol.ED-23 (1976)
      p.538]
[10] C.M.Wu, E.S.Yang [J. Appl. Phys. (USA) vol.52 no.7 (Jul 1981)
      p.4700-3]
[11] K.Chino [Solid-State Electron. (GB) vol.16 (1973) p.119]
[12] J.Basterfield, J.M.Shannon, A.Gill [Solid-State Electron. (GB)
      vol.18 (1975) p.290]
[13] T.M.Reith, J.D.Schick [Appl. Phys. Lett. (USA) vol.25 (1974) p.524]
[14] P.B.Ghate [Thin Solid Films (Switzerland) vol.83 (1981) p.195]
[15] J.M.Wilkinson, J.D.Wilcock, M.E.Brinson [Solid-State Electron. (GB)
      vol.20 (1977) p.45]
[16] E.Demoulin, F. van de Wiele [Solid-State Electron. (GB) vol.17 (1974)
      p.825]
[17] J.D. van Otterloo [Surf. Sci. (Netherlands) vol.104 (1981) p.L205]
[18] O.Engstrom, H.Pettersson, B.Sernelius [Phys. Status Solidi a
      (Germany) vol.95 (1986) p.691]
[19] G.C.McGonigal, H.C.Card [IEEE Electron Device Lett. (USA) vol.EDL-2
      (1981) p.149]
[20] C.M.Wu, E.S.Yang [J. Appl. Phys. (USA) vol.51 (1980) p.5889]
[21] R.N.Sreenath, M.M.Chandra, G.Suryan [IEE Proc. I (GB) vol.131 no.2
      (Apr 1984) p.63-5]

20.3 BARRIER HEIGHT AT THE Au/Si INTERFACE

by K.K.Ng

September 1987
EMIS Datareview RN=16142

Au on n-Si forms a high potential barrier. It can also be used as a
good ohmic contact material on p-Si. However, due to the fast diffusion
of Au in Si, and also its detrimental effect on the recombination
lifetime of Si, it is not commonly used in VLSI (other than being
used as the wafer back contact). Available data on barrier height (BH)
are summarized below.

TABLE 1

BH (eV)	Measurement	Substrate	Reference
0.79 +- 0.02	I-V, C-V, photo	n(111)	[3]
0.81-0.82	I-V, C-V	n(111)	[1]
0.80 +- 0.02	I-V, C-V	n(111)(100)	[2,12]
0.78-0.83		n	[5]
0.75-0.77	I-V, C-V	n(111)	[13]
0.81	C-V	n(111)	[11]
0.81		n	[9]
0.25	I-V, C-V	p(111)(100)	[2]
0.34	I-V, C-V	p(111)	[4]
0.34	C-V	p(111)	[11]
0.32-0.35	photo	p(100)	[10]

Turner and Rhoderick compared the Schottky barriers prepared on cleaved
surfaces and chemically prepared surfaces [1]. They showed that the
BH obtained on chemically cleaned surfaces changed with time but
saturated to the same BH as the cleaved surfaces. Hirose et al compared
the barriers on (111) and (100) surfaces, and showed they were similar
[2]. They also obtained BH's on both n-type and p-type substrates [2],
and the sum of the two gave the value of the Si band gap. Smith and
Rhoderick obtained good characteristics on p-Si by heating the
substrate during metal deposition [4]. Panayotatos and Card obtained
their BH by a novel open-circuit technique [5]. The effects of surface
preparation on the BH were examined by Varma et al [6] and Williams et
al [7,14]. Van Otterloo obtained his BH value on cleaved surfaces, and
he showed that it could vary in the presence of cleavage steps [11].
The aging of a Au/n-Si Schottky diode was examined by Ponpon and Siffert
[8].

REFERENCES

[1]   M.J.Turner, E.H.Rhoderick [Solid-State Electron. (GB) vol.11
       (1968) p.291]
[2]   M.Hirose, N.Altaf, T.Arizumi [Jpn. J. Appl. Phys. (Japan) vol.9
       (1970) p.260]
[3]   D.Kahng [Solid-State Electron. (GB) vol.6 (1963) p.281]
[4]   B.L.Smith, E.H.Rhoderick [Solid-State Electron. (GB) vol.14
       (1971) p.71]
[5]   P.Panayotatos, H.C.Card [13th IEEE Photovoltaic Specialists'
       Conf. Record (IEEE, New York, USA, 1978) p.634]
[6]   R.R.Varma, A.McKinley, R.H.Williams, I.G.Higginbotham [J. Phys.
       D (GB) vol.10 (1977) p.L171]
[7]   R.H.Williams, V.Montgomery, R.R.Varma, A.McKinley [J. Phys. D
       (GB) vol.10 (1977) p.L253]
[8]   J.P.Ponpon, P.Siffert [J. Appl. Phys. (USA) vol.49 no.12
       (Dec 1978) p.6004-11]
[9]   N.T.Tam, Tran.Chot [Phys. Status Solidi a (Germany) vol.93 no.1
       (1986) p.K91-5]
[10]   O.Engstrom, H.Petterson, B.Sernelius [Phys. Status Solidi a
       (Germany) vol.95 (1986) p.691]
[11]   J.D. van Otterloo [Surf. Sci. (Netherlands) vol.104 (1981)
       p.L205]
[12]   T.Arizumi, M.Hirose [Jpn. J. Appl. Phys. (Japan) vol.8 (1969)
       p.749]
[13]   M.J.Howes, D.V.Morgan, K.D.Al-Baidhawi [IEEE Trans. Electron
       Devices vol.ED-26 (1979) p.1262]
[14]   R.H.Williams, R.R.Varma, V.Montgomery [J. Vac. Sci. & Technol.
       (USA) vol.16 (1979) p.1418]

20.4 BARRIER HEIGHT AT THE Cr/Si INTERFACE

by K.K.Ng

September 1987
EMIS Datareview RN=16144

The available data for Cr Schottky barriers on Si are summarized in
TABLE 1. The data are not very consistent with one another and show
a maximum difference of approximately 0.1 eV.

TABLE 1

BH (eV)	Measurement	Substrate	Reference
0.60 +- 0.02	I-V, C-V	n	[1]
0.56 +- 0.02	I-V, C-V	n	[3]
0.53 - 0.57	I-V	n	[2]
0.52	I-V	n(111)	[4]

BH - Barrier Height

Saltich has demonstrated the independence of BH on the substrate doping
concentration [3]. Martinez et al studied the Cr/n-Si Schottky at
different annealing temperatures [4]. They showed that up to 400 deg C,
the BH remained constant but the ideality factor could be improved.

REFERENCES
----

[1] A.M.Cowley, R.A.Zettler [IEEE Trans. Electron Device (USA)
     vol.ED-15 (1968) p.761]
[2] C.Rhee, J.Saltich, R.Zwernemann [Solid-State Electron. (GB) vol.15
     (1972) p.1181]
[3] J.Saltich [in 'Ohmic Contacts to Semiconductors' Ed. B.Schwartz
     (Electrochem. Soc. 1969) p.187]
[4] A.Martinez, D.Esteve, A.Guivarc'h P.Auvray, P.Henoc, G.Pelous
     [Solid-State Electron. (GB) vol.23 no.1 (Jan 1980) p.55-64]

20.5 BARRIER HEIGHT AT THE Cu/Si INTERFACE

by K.K.Ng

September 1987
EMIS Datareview RN=16146

The reported barrier heights (BH) of Cu on n-Si are widely different,
as shown in TABLE 1. The two values reported on p-Si are much closer.

TABLE 1

BH (eV)	Measurement	Substrate	Reference
0.79	I-V, C-V	n(111)	[1]
0.61 +- 0.02	I-V, C-V	n(111)(100)	[2,5]
0.62 +- 0.01	I-V, photo	n(111)	[4]
0.75 +- 0.01	C-V	n(111)	[4]
0.64		n	[6]
0.50	I-V, C-V	p(111)(100)	[2]
0.46	I-V, C-V	p(111)	[3]

Turner and Rhoderick obtained their BH (above) on cleaved surfaces,
and they found that the BH on chemically cleaned surfaces changed with
time to reach an equilibrium value of 0.69 eV [1]. The BH's obtained
by Hirose et al on both n-Si and p-Si were consistent in that the sum
of the barriers was the Si band gap [2]. Hirose et al also compared
BH's on both (111) and (100) surfaces, and showed they were similar
[2]. Smith and Rhoderick obtained good characteristics on p-Si by
heating the substrate during metal deposition [3]. Thanailakis and
Rasul discussed the different barrier heights obtained by the I-V, C-V
and photoelectric techniques [4].

REFERENCES
----

[1] M.J.Turner, E.H.Rhoderick [Solid-State Electron. (GB) vol.11
     (1968) p.291]
[2] M.Hirose, N.Altaf, T.Arizumi [Jpn J. Appl. Phys. (Japan) vol.9
     (1970) p.260]
[3] B.L.Smith, E.H.Rhoderick [Solid-State Electron. (GB) vol.14
     (1971) p.71]
[4] A.Thanailakis, A.Rasul [J. Phys. C (GB) vol.9 (1976) p.337]
[5] T.Arizumi, M.Hirose [Jpn. J. Appl. Phys. (Japan) vol.8 (1969)
     p.749]
[6] N.T.Tam, Tran Chot [Phys. Status Solidi a (Germany) vol.93 no.1
     (1986) p.K91-5]

20.6 BARRIER HEIGHT AT THE Mg/Si INTERFACE

by K.K.Ng

September 1987
EMIS Datareview RN=16145

The reported barrier heights (BH) of Mg on Si are summarized in TABLE 1. The BH on n-Si is around half of the Si band gap and so it is expected that the BH on p-Si is similar.

TABLE 1

BH (eV)	Measurement	Substrate	Reference
0.52 +- 0.03	I-V	n(111)	[1]
0.55 +- 0.02	I-V	n	[2]
0.55 - 0.58	C-V, photo	n	[3]
0.62	C-V	n(111)	[5]
0.56	C-V	p(111)	[5]

Van Otterloo and Gerritsen compared and commented on the barrier heights obtained by I-V, C-V and photoelectric measurements [3]. Van Otterloo fabricated Schottky diodes on cleaved surfaces, and he showed that the BH could vary in the presence of cleavage steps [5]. Akiya and Nakamura observed Mg2Si formation as low as 230 deg C at which point the BH dropped to 0.46 eV on n-Si [4].

REFERENCES
----

[1] C.R.Crowell, H.B.Shore, E.E.LaBate [J. Appl. Phys. (USA) vol.36
     (1965) p.3843]
[2] J.M.Wilkinson, J.D.Wilcock, M.E.Brinson [Solid-State Electron
     (GB) vol.20 (1977) p.45]
[3] J.D. van Otterloo, L.J.Gerritsen [J. Appl. Phys. (USA) vol.49 no.2
     (Feb 1978) p.723-9]
[4] M.Akiya, H.Nakamura [J. Appl. Phys. (USA) vol.59 no.5 (1 Mar 1986)
     p.1596-8]
[5] J.D.van Otterloo [Surf. Sci. (Netherlands) vol.104 (1981) p.L205]

20.7 BARRIER HEIGHT AT THE Mo/Si INTERFACE

by K.K.Ng

September 1987
EMIS Datareview RN=16138

Molybdenum gives a relatively low barrier height (BH) on n-Si. Available
data are shown in TABLE 1. The data seem to cluster around two values,
namely approx. 0.55 eV [2,3,7] and approx. 0.68 eV [1,4,8,10].

TABLE 1

BH (eV)	Measurement	Substrate	Reference
0.57 +- 0.02	I-V, C-V, photo	n(111)	[7]
0.68	I-V	n(111)	[1]
0.55 +- 0.02	I-V	n	[2]
0.53-0.59	C-V	n	[2]
0.64-0.67	I-V, C-V	n(111)	[8]
0.68 +- 0.01	C-V	n	[4]
0.55	I-V	n(100)	[3]
0.69 +-0.02	I-V	n	[10]
0.70 +- 0.03	C-V	n	[10]

Zettler and Cowley, by using p-n junction guard ring, produced Mo/n-Si
Schottky diodes with nearly ideal I-V characteristics (n = 1.01),
even in the reverse bias [1]. Saltich studied the influence of substrate
doping on barrier height [2]. He showed that the BH only varied within
0.03 eV for a doping of 4 X 10*14/cm*3 to 6.5 X 10*16/cm*3. The effects
of a surface layer due to sputtering of Mo was studied by Miyamoto [8]
and Auret et al [9]. Kano et al [7] and Simeonov et al [10] were able to
obtain near ideal diode characteristics by chemical deposition of Mo.
Calleja et al studied the thermal stability of Mo/n-Si diodes [4]. They
found that the idealics factor improved with temperature until 350 deg C
and the BH from I-V characteristics increased at the same time. At 450
deg C, the diode characteristics started to degrade.

REFERENCES
----

[1]   R.A.Zettler, A.M.Cowley [IEEE Trans Electron Devices (USA)
       vol.ED-16 (1969) p.58]
[2]   J.Saltich [in 'Ohmic Contacts to Semiconductors' Ed. B.Schwartz
       (Electrochem. Soc., 1969) p.187-99]
[3]   K.T.-Y.Kung, I.Suni, M.-A.Nicolet [J. Appl. Phys. (USA) vol.55
       no.10 (15 May 1984) p.3882-5]

[4]   E.Calleja, J.Garrido, J.Piqueras, A.Martinez [Solid-State
        Electron. (GB) vol.23 (1980) p.591]
[5]   F.H.Mullins, A.Brunnschweiler [Solid-State Electron. (GB) vol.19
        (1976) p.470]
[6]   S.J.Fonash, S.Ashok, R.Singh [Appl. Phys. Lett. (USA) vol.39
        no.5 (1 Sep 1981) p.423-5]
[7]   G.Kano, M.Inoue, J.Matsuno, S.Takayanagi [J. Appl. Phys. vol.37
        (1966) p.2985]
[8]   S.Miyamoto [Jpn. J. Appl. Phys. (Japan) vol.16 (1977) p.101]
[9]   F.D.Auret, O.Paz, N.A.Bojarczuk [J. Appl. Phys. (USA) vol.55
        (1984) p.1581]
[10]   S.S.Simeonov, E.I.Kafedjiiska, A.L.Guerassimov [Thin Solid Films
        (Switzerland) vol.115 (1984) p.291]

20.8 BARRIER HEIGHT AT THE Ni/Si INTERFACE

by K.K.Ng

September 1987
EMIS Datareview RN=16140

The reported barrier heights (BH) of Ni on n-Si are widely scattered
between 0.55 eV and 0.74 eV, as shown in TABLE 1. Half of the data are
around 0.60 eV.

TABLE 1

BH (eV)	Measurement	Substrate	Reference
0.70	I-V, C-V	n(111)	[1]
0.59 +- 0.01	I-V, photo	n(111)	[3]
0.74 +- 0.02	C-V	n(111)	[3]
0.70 +- 0.02	I-V, C-V	n(111)	[4]
0.66	I-V	n(100)	[5]
0.61-0.62	I-V	n	[10]
0.55-0.57	photo	n(100)(111)	[8]
0.61	I-V	n(111)	[6]
0.59 +- 0.01	I-V	n(100)	[9]
0.60		n	[7]
0.50-0.51	I-V, C-V	p(111)	[2]

Turner and Rhoderick obtained their BH (above) on cleaved surfaces,
and they found that the BH on chemically cleaned surfaces changed with
time to reach an equilibrium value of 0.67 eV [1]. Thanailakis and
Rasul found that the BH obtained by the C-V method was always higher
than that by the I-V and photoelectric methods [3]. This difference
was also observed by Coe and Rhoderick [4]. Smith and Rhoderick [2]
obtained good characteristics on p-Si by heating the substrate during
metal deposition.

The thermal stability of Al contact to Ni Schottky was studied by
Bartur and Nicolet with and without Cr as barrier layer [6], while
Finetti et al used TiN as the barrier layer [9].

REFERENCES
----

[1]   M.J.Turner, E.H.Rhoderick [Solid-State Electron (GB) vol.11
       (1968) p.291]
[2]   B.L.Smith, E.H.Rhoderick [Solid-State Electron (GB) vol.14
       (1971) p.71]
[3]   A.Thanailakis, A.Rasul [J. Phys. C (GB) vol.9 (1976) p.337]
[4]   D.J.Coe, E.H.Rhoderick [J. Phys. D (GB) vol.9 (1976) p.965]
[5]   G.Ottaviani, K.N.Tu, J.W.Mayer [Phys. Rev. B (USA) vol.24 no.6
       (15 Sep 1981) p.3354-9]
[6]   M.Bartur, M.-A.Nicolet [J. Electrochem. Soc., (USA) vol.131
       no.5 (May 1984) p.1118-22]
[7]   N.T.Tam, Tran Chot [Phys. Status Solidi a (Germany) vol.93 no.1
       (1986) p.K91-5]
[8]   P.E.Schmid, P.S.Ho, H.Foll, T.Y.Tan [Phys. Rev. B (USA) vol.28
       no.8 (15 Oct 1983) p.4593-601]
[9]   M.Finetti, I.Suni, M.-A.Nicolet [J. Electron. Mater. vol.13
       no.2 (Mar 1984) p.327-40]
[10]   T.Chot [Phys. Status Solidi a (Germany) vol.70 (1982) p.311]

20.9 BARRIER HEIGHT AT THE Pd/Si INTERFACE

by K.K.Ng

September 1987
EMIS Datareview RN=16147

Due to the fact that Pd reacts with Si to form Pd2Si at very low
temperature (< 200 deg C), it is very difficult to confirm that a
metal Pd Schottky does not contain Pd2Si, formed during metal
deposition . The available data without intentional heat treatment
are summarized in TABLE 1.

TABLE

BH (eV)	Measurement	Substrate	Reference
0.74	I-V, C-V	n(100)	[1]
0.76	I-V	n(111)	[4]
0.71 +- 0.02	I-V, photo	n(100)	[2]
0.75	I-V	n(111)	[8]
0.73-0.75	photo	n(100)(111)	[5]
0.75-0.77	I-V, photo	n(100)(111)	[7]
0.73-0.76	C-V	n(100)	[9]
0.74	I-V	n(111)	[6]
0.49 +- 0.06	I-V	p(100)	[2]
0.31-0.32	photo	p(100)	[3]

BH - Barrier Height

Tongson et al indeed showed traces of Pd2Si in the as-deposited Pd
Schottky [1]. The thermal stability of Pd/Si Schottky was studied by
Ottaviani et al [8]. They showed that the BH started to decrease at a
temperature of 350 deg C. The effects of using Cr as a barrier between
Al and Pd was studied by Bartur and Nicolet [6].

REFERENCES
----

[1] L.L.Tongson, B.E.Knox, T.E.Sullivan, S.J.Fonash [J. Appl. Phys.
     (USA) vol.50 (1979) p.1535]
[2] E.Hokelek, G.Y.Robinson [Solid-State Electron. vol.24 no.2
     (Feb 1981) p.99-103]
[3] O.Engstrom, H.Pettersson, B.Sernelius [Phys. Status Solidi a
     (Germany) vol.95 (1986) p.691]

[4]   M.Eizenberg, G.Ottaviani, K.N.Tu [Appl. Phys. Lett. (USA) vol.37
      no.1 (Jul 1980) p.87-9]
[5]   R.J.Purtell, P.S.Ho, G.W.Rubloff, P.E.Schmid [Physica B & C
      (Netherlands) vol.117 (1983) p.834]
[6]   M.Bartur, M.-A.Nicolet [J. Electrochem. Soc. (USA) vol.131 no.5
      (May 1984) p.1118-22]
[7]   P.E.Schmid, P.S.Ho, H.Foll, T.Y.Fan [Phys. Rev. B (USA) vol.28
      no.8 (Oct 1983) p.4593-601]
[8]   G.Ottaviani, K.N.Tu, J.W.Mayer [Phys. Rev. (USA) vol.24 no.6
      (15 Sep 1981) p.3354-9]
[9]   A.Diligenti, M.Stagi, V.Ciuti [Solid State Commun. (USA) vol.45
      (1983) p.347]

20.10 BARRIER HEIGHT AT THE Pt/Si INTERFACE

by K.K.Ng

September 1987
EMIS Datareview RN=16137

Pt on n-Si yields one of the highest metal barrier heights on silicon.
Consequently the barrier height (BH) on p-Si is very low such that
deduction from data is difficult. Most available results are on (111)
orientation surfaces and they are summarized below.

TABLE 1

BH (eV)	Measurement	Substrate	Reference
0.83-0.84	I-V, C-V, photo	n(111)	[1]
0.71 +- 0.01	I-V, photo	n(111)	[2]
0.82 +- 0.02	C-V	n(111)	[2]
0.85	I-V	n	[5]
0.80	I-V	n(111)	[3]
0.80 +- 0.02	I-V	n	[4]
0.85 +- 0.02	C-V	n	[4]
0.84	I-V	n(111)	[8]
0.80	I-V	n(111)	[7]
0.84-0.85	photo	n(111)	[6]

Roberts and Crowell obtained a consistent barrier height using I-V, C-V
and photoelectric measurement techniques [1]. Thanailakis and Rasul on
cleaved Si surfaces obtained different barrier heights when different
measurement techniques were used [2]. Toyama et al combined photoelectric
and I-V techniques to conclude that the effective Richardson constant was
a function of heat treatment [6]. Eizenberg et al [3] and Calleja et al
[4] had studied the stability of Pt/n-Si Schottky barriers as a function
of temperature. They both found that the BH increased with temperature
up to approx. 400 deg C, and then decreased. This particular temperature
corresponds to the beginning of silicide formation. Bartur and Nicolet
studied Cr as a diffusion barrier between Pt and Al at elevated
temperatures [7].

REFERENCES
----

[1] G.I.Roberts, C.R.Crowell [Solid-State Electron. (GB) vol.16
     1973) p.29]
[2] A.Thanailakis, A.Rasul [J. Phys. C (GB) vol.9 (1976) p.337]

[3] M.Eizenberg, G.Ottaviani, K.N.Tu [Appl. Phys. Lett. (USA) vol.37
     no.1 (1 Jul 1980) p.87-9]

[4] E.Calleja, J.Garrido, J.Piqueras, A.Martinez [Solid-State
     Electron (GB) vol.23 (1980) p.591]

[5] Y.Anand [IEEE Trans. Electron Devices (USA) vol.ED-24 (1977)
     p.1330]

[6] N.Toyama, T.Takahashi, H.Murakami, H.Koriyama [Appl. Phys. Lett
     (USA) vol.46 no.6 (15 Mar 1985) p.557-9]

[7] M.Bartur, M.-A.Nicolet [J. Electrochem. Soc. (USA) vol.131 no.5
     (May 1984) p.1118-22]

[8] G.Ottaviani, K.N .Tu, J.W.Mayer [Phys. Rev. B (USA) vol.24 no.6
     (15 Sep 1981) p.3354-9]

20.11 BARRIER HEIGHT AT THE Ti/Si INTERFACE

by K.K.Ng

September 1987
EMIS Datareview RN=16139

Ti is one of the few metals that gives higher barrier height (BH) on p-Si
than on n-Si. It gives very consistent BH from different researchers as
indicated in TABLE 1. Partly because it is a reducing agent for SiO2, the
Si substrate surface preparation is not as critical as for other metals.
Recently there has been great interest in Ti because of the application
of TiSi2 for MOSFET source/drain contacts. The stability of TiN is also
of great potential for VLSI applications.

TABLE 1

Metal	BH (eV)	Measurement	Substrate	Reference
Ti	0.48 +- 0.02	I-V	n	[2]
Ti	0.51 +- 0.05	C-V	n	[2]
Ti	0.50	I-V	n	[1]
Ti	0.51-0.52	I-V, C-V	n	[6]
Ti	0.48	I-V	n(111)	[4]
Ti	0.48-0.50	I-V	n	[10]
Ti	0.51 +- 0.01	photo	n(100)	[5]
Ti	0.515	I-V	n(100)	[3]
Ti	0.61 +- 0.01	I-V, C-V	p	[1]
Ti	0.60 +- 0.01	photo	p(100)	[5]
Ti	0.596	I-V	p(100)	[3]
TiN	0.49 +- 0.01	I-V	n(100)	[8,9]
TiN	0.55	I-V	n(100)	[7]
TiN	0.57	I-V	p(111)	[7]

Kato studied the thermal stability of Ti/n-Si Schottky diodes [4]. He
showed that up to 500 deg C the BH remained unchanged. Aboelfotoh and
Tu showed that the diode characteristics were improved by an anneal at
370 deg C [3]. Wilkinson et al prepared Schottky diodes on substrates
with different concentrations [10], and they found them to be similar.
Taubenblatt et al studied the effects of Si surface preparation before
metal evaporation [5].

On TiN Schottky formation, Wittmer et al found that the as-deposited TiN gave non-rectifying characteristics [8]. They found that the BH increased with annealing temperature, and it saturated at approx. 500 deg C. Finetti studied the damage due to sputtering of TiN, and the thermal annealing of the damage [7]. Wittmer and Melchior studied the thermal stability of TiN/n-Si [9]. They showed that TiN is a very efficient barrier for Si diffusion.

REFERENCES
----

[1]   A.M.Cowley [Solid-State Electron (GB) vol.13 no.4 (Apr 1970)
       p.403-14]
[2]   J.Saltich [in 'Ohmic Contacts to Semiconductors' Ed. B.Schwartz
       (Electrochem. Soc., 1969) p.187-99]
[3]   M.O.Aboelfotoh, K.N.Tu [Phys. Rev. B (USA) vol.34 no.4
       (15 Aug 1986) p.2311-8]
[4]   H.Kato, Y.Nakamura [Thin Solid Films (Switzerland) vol.34 no.1
       (4 May 1976) p.135-8]
[5]   M.A.Taubenblatt, D.Thomson, C.R.Helms [Appl. Phys. Lett. (USA)
       vol.44 no.9 (1 May 1984) p.895-7]
[6]   J.M.Wilkinson [Solid-State Electron. (GB) vol.17 (1974) p.583]
[7]   M.Finetti, I.Suni, M.Bartur, T.Banwell, M.-A.Nicolet [Solid-State
       Electron. (GB) vol.27 (1984) p.617]
[8]   M.Wittmer, B.Studer, H.Melchior [J. Appl. Phys. (USA) vol.52 no.9
       (Sep 1981) p.5722-6]
[9]   M.Wittmer, H.Melchior [Thin Solid Films (Switzerland) vol.93
       (1982) p.397]
[10]   J.M.Wilkinson, J.D.Wilcock, M.E.Brinson [Solid-State Electron.
       vol.20 (1977) p.45]

20.12 CONTACT RESISTANCE OF THE Al/Si INTERFACE

by K.K.Ng

September 1987
EMIS Datareview RN=16155

The contact resistance rho(c) of Al to Si is of great importance due to
the common use of Al as the final metallization material. It is generally
observed that when the Al/Si contact is heated to above approx. 400 deg C,
metallurgical reaction begins such that if a junction is underneath,
shorting of the diode occurs. Addition of Si (approx. 1%) in Al helps
to a certain extent, but the problem remains. Available data on the
rho(c) are summarised below. Most data are based on Al doped with a few %
of Si which hinders reaction between Al and the Si substrate at elevated
temperatures.

TABLE 1

Metal	rho(c) (ohm cm*2)	Substrate	Surface Concn (/cm*3)	Ref
Al	5 X 10*-3	n+(100)		[1]
Al	c. 6 X 10*-7	n+(100)	c. 1.5 X 10*20	[16]
Al	c. 4 X 10*-6	n+(100)	1.5 X 10*20	[15]
Al	1.5 X 10*-7	p+(100)	2 X 10*19	[8]
Al	c. 4 X 10*-7	p+(100)	6 X 10*19	[15]
Al(Si)	c. 1 X 10*-2	n+(100)		[1]
Al(Si)	c. 1 X 10*-6	n+(100)	2 X 10*20	[3]
Al(Si)	3.2 X 10*-6	n+	8.5 X 10*19	[18]
Al(Si)	c. 1 X 10*-7	n+(100)	3 X 10*20	[2]
Al(Si)	1-3 X 10*-6	n+	1.6 X 10*19	[19]
Al(Si)	1-2 X 10*-6	n+(100)	2 X 10*20	[21]
Al(Si)	8 X 10*-7	n+(100)	3.7 X 10*20	[17]
Al(Si)	4 X 10*-7	n+(100)	5 X 10*20	[17]
Al(Si)	2.8 X 10*-7	n+(100)		[4]
Al(Si)	1.5 X 10*-7	n+(100)		[6]

Metal	rho(c) (ohm cm*2)	Substrate	Surface Concn (/cm*3)	Ref
Al(Si)	c. 1 X 10*-7	n+(100)	3 X 10*20	[5]
Al(Si)	c. 1 X 10*-7	p+(100)	4 X 10*19	[3]
Al(Si)	9-1.3 X 10*-5	p+(100)	2 X 10*19	[21]
Al(Si)	2.8 X 10*-7	p+(100)		[4]
Al(Si)	c. 7 X 10*-8	p+(100)	7 X 10*19	[5]

Finetti et al reported very high rho(c) [1], probably because of low surface concentrations. Cohen et al reported Mo/Al/Si structures with improved thermal stability [5]. The reduction of rho(c) upon thermal annealing was studied by Dascalu et al [7] and Naguib and Hobbs [14]. Naguib and Hobbs showed that the optimized temperature for minimum rho(c) was 450 deg C above which the rho(c) on p+-Si was constant and the rho(c) on p+-Si was increased. Non-conventional annealing techniques were studied by different authors. Hara et al [8], Alvi and Kwong [9] and Pai et al [16] studied lamp annealing. Armigliato et al [10] and Finetti et al [11] used laser and electron beam annealing while Fukano et al explored microwave annealing [12].

The dependence of contact hole size on rho(c) was also examined by Mori [13] and Proctor et al [18,19]. The rho(c) of Al with varying Si doping percentage was studied by Faith et al [20].

REFERENCES
----

[1]   M.Finetti, P.Ostoja, S.Solmi, G.Soncini [Solid-State Electron.
      (GB) vol.23 no.3 (Mar 1980) p.255-62]
[2]   J.A.Mazer, L.W.Linholm, D.Pramanik, S.Tsai, A.N.Saxena [IEEE
      1983 Custom Integrated Circuits Conf., Rochester, NY, USA, 23-25
      May 1983 p.291-4]
[3]   S.S.Cohen, G.Gildenblat, M.Ghezzo, D.M.Brown [J. Electrochem.
      Soc. (USA) vol.129 no.6 (Jun 1982) p.1335-8]
[4]   J.M.Ford [IEEE Trans. Electron. Devices (USA) vol.ED-32 no.4
      (Apr 1985) p.840-2]
[5]   S.S.Cohen, G.S.Gildenblat [IEEE Electron. Devices (USA)
      vol.ED-34 no.4 (Apr 1987) p.746-52]
[6]   J.A.Mazer, L.W.Linholm, A.N.Saxena [J. Electrochem. Soc. (USA)
      vol.132 no.2 (Feb 1985) p.440-3]
[7]   D.Dascalu, G.Brezeanu [Appl. Phys. Lett. (USA) vol.37 no.2
      (15 Jul 1980) p.215-17]
[8]   T.Hara, N.Ohtsuka, S.Enomoto, T.Hirayama, K.Amemiya, M.Furukawa
      [Jpn. J. Appl. Phys. Part 2 (Japan) vol.22 (1983) p.L683]
[9]   N.S.Alvi, D.L.Kwong [IEEE Electron. Device Lett. (USA)
      vol.EDL-7 no.2 (1986) p.137-9]
[10] A.Armigliato, P.De Luca, M.Finetti, S.Solmi [in 'Laser and
      Electron-Beam Interactions and Materials Processing' Eds Gibbons,
      Hess, Sigmon (North-Holland) p.329-336]
[11] M.Finetti, S.Solmi, G.Soncini [Solid-State Electron. (GB) vol.24
      (1981) p.539]
[12] T.Fukano, T.Ito, H.Ishikawa [IEEE IEDM Tech. Digest (1985)
      p.224]

[13] M.Mori [IEEE Trans. Electron Devices (USA) vol.ED-30 (1983) p.81]

[14] H.M.Naguib, L.H.Hobbs [J. Electrochem. Soc. (USA) vol.124 (1977)
      p.573]

[15] L.Van den Hove, R.Wolters, K.Maex, R.F.De Keersmaecker,
      G.J.Declerck [IEEE Trans. Electron Devices (USA) vol.ED-34 no.3
      (Mar 1987) p.554-61]

[16] C.S.Pai, E.Cabreros, S.S.Lau [Appl. Phys. Lett. (USA) vol.46 no.7
      (1 Apr 1985) p.652-3]

[17] J.G.J.Chern, W.G.Oldham [IEEE Electron Device Lett. (USA)
      vol.EDL-5 (1984) p.178]

[18] S.J.Proctor, L.W.Linholm [IEEE Electron Device Lett. (USA)
      vol.EDL-3 (1982) p.294]

[19] S.J.Proctor, L.W.Linholm, J.A.Mazer [IEEE Trans. Electron Devices
      (USA) vol.ED-30 (1983) p.1535]

[20] T.J.Faith, R.S.Irven, S.K.Plante, J.J.O'Neill,Jr [J. Vac. Sci.
      & Technol. A (USA) vol.1 no.8 (Apr-Jun 1983) p.443-8]

[21] A.Scorzoni, M.Finetti, G.Soncini [Alta Freq. (Italy) vol.53 (1984)
      p.282]

20.13 CONTACT RESISTANCE OF THE Ti/Si INTERFACE

by K.K.Ng

September 1987
EMIS Datareview RN=16156

Because of the potential application of Ti on Si VLSI, the contact
resistance (rho(c)) of Ti and TiN to Si is naturally very important. The
available data are summarised below. The main advantage of TiN is its
thermal stability with Si and metals but generally the data indicate that
the rho(c) of TiN on Si is relatively high. The high rho(c) of TiN on
Si is probably not fundamental, but due to imperfect interface caused
by the deposition procedure.

TABLE 1

Metal	rho(c) (ohm cm*2)	Substrate	Surface Concn (/cm*3)	Ref
Ti	2 X 10*-6	n+	1 X 10*20	[1]
Ti	3.6 X 10*-7	n+	2 X 10*20	[1]
Ti	c. 5 X 10*-7	n+(100)		[6]
Ti	2 X 10*-7	n+(100)	2 X 10*20	[2]
TiN	c. 1 X 10*-4	n+(100)	2 X 10*19	[4]
TiN	7 X 10*-5	n+(100)		[9]
TiN	c. 1.5 X 10*-5	n+(100)	2 X 10*20	[3]
TiN	3 X 10*-5	n+	2 X 10*20	[5]
TiN	6 X 10*-5	p+	1 X 10*20	[5]
TiN	c. 3 X 10*-4	p+(100)	1 X 10*20	[3]

Ting and Crowder showed that the Al/Ti/n+-Si contact system is stable
up to approx. 450 deg C [1]. The thermal stability of both Al/n+-Si and
Al/Ti/n+-Si was studied by Suni et al [6]. The contact resistance of a
TiW alloy was studied by Kim et al [7,8], Cohen et al [10] and Hara et
al [11].

Finetti et al showed that the rho(c) of TiN on Si changed with annealing
temperature, in the opposite direction for n+-Si compared to rho(c)-Si
[5]. The rho(c) on p+-Si decreased with anneal while it increased on n+
-Si with anneal.

REFERENCES
----

[1]   C.Y.Ting, B.L.Crowder [J. Electrochem. Soc. (USA) vol.129 (1982)
       p.2590]
[2]   I.Suni, M.Finetti, K.Grahn [Mater. Res. Soc. Symp. Proc. (USA)
       vol.71 (1986) p.363-8]
[3]   M.Finetti, I.Suni, M.-A.Nicolet [Sol. Cells (Switzerland) vol.9
       (1983) p.179]
[4]   M.Wittmer, B.Studer, H.Melchior [J. Appl. Phys. (USA) vol.52
       no.9 (Sep 1981) p.5722-6]
[5]   M.Finetti, I.Suni, M.-A.Nicolet [Solid-State Electron. (GB)
       vol.26 no.11 (Nov1983) p.1065-7]
[6]   I.Suni, M.Blomberg, J.Saarilahti [J. Vac. Sci. & Technol. A
       (USA) vol.3 no.6 (Nov/Dec 1985) p.2233-6]
[7]   M.J.Kim, D.M.Brown, S.S.Cohen, P.Piacente, B.Gorowitz [IEEE
       Trans. Electron. Devices (USA) vol.ED-32 no.7 (July 1985)
       p.1328-33]
[8]   M.J.Kim, S.S.Cohen, D.M.Brown, P.A.Piacente, B.Gorowitz [IEEE
       IEDM Tech. Digest (1984) p.134-7]
[9]   M.Maenpaa, I.Suni, M-A.Nicolet, F.Ho, P.Iles [15th IEEE
       Photovoltaic Specialists Conf., Kissimmee, FL, USA, 12-15 May
       1981, p.518-21]
[10]   S.S.Cohen, M.J.Kim, B.Gorowitz, R.Saia, T.F.McNelly [Appl. Phys.
       Lett. (USA) vol.45 (1984) p.414]
[11]   T.Hara, N.Ohtsuka, K.Sakiyama, S.Saito [IEEE Trans. Electron
       Devices (USA) vol.ED-34 no.3 (Mar 1987) p.593-8]

by K.K.Ng

September 1987
EMIS Datareview RN=16157

W is an important metal for contacts in Si VLSI, mostly because of its
thermal stability with Si and other metals. The additional feature of
low pressure chemical vapour deposition (LPCVD) enables conformal films
to be deposited in the presence of contact holes. Selective CVD
deposition is also potentially useful when W is desired only on Si or
metal but not on oxide. The available data on contact resistance
(rho(c)) for W on Si are summarised in TABLE 1.

TABLE 1

rho(c) (ohm cm*2)	Substrate	Surface Concn (/cm*3)	Ref
1-4 X 10*-6	n+(100)	7.5 X 10*19	[1]
1 X 10*-7 (Fe-W)	n+	c. 2 X 10*20	[10]
2-5 X 10*-8	n+	2 X 10*20	[2]
1-2 X 10*-7	n+	8 X 10*19	[2]
0.7-2.8 X 10*-7 (10% Ti)	n+(100)	3 X 10*20	[8]
1-2 X 10*-7	n+	c. 1 X 10*20	[4]
< 3.5 X 10*-7 (30% Ti)	n+(100)		[9]
1 X 10*-7	n+	2 X 10*20	[3]
c. 1 X 10*-7	n+		[5]
0.6-3 X 10*-6	p+(100)	1 X 10*20	[1]
2.8 X 10*-6 (Fe-W)	p+	c. 1 X 10*20	[10]
2-8 X 10*-7 (10% Ti)	p+(100)	7 X 10*19	[8]
2-3 X 10*-7	p+	c. 1 X 10*20	[4]
3-8 X 10*-7 (30% Ti)	p+(100)		[9]
3 X 10*-7 (30% Ti)	p+(100)	2 X 10*20	[3]

The rho(c) reported by Kumar included a large contribution from the
spreading resistance, and thus it represents the upper limit of rho(c)
[1]. Swirhun and Swanson measured the rho(c) as a function of temperature
[3], and showed that it was quite weakly temperature dependent. Shioya
et al [5] and Hara et al [6] studied the thermal stability when Al was

used as the top contact, while Kim et al used Mo as the final metal contact [9,11]. Different authors had studied the rho(c) of W alloys such as 10% Ti [8], 30% Ti [9], and Fe-W [10]. The effects of fabrication parameters on rho(c) using selective CVD of W were examined by Levy et al [7].

REFERENCES
----

[1]   V.Kumar [J. Electrochem. Soc. (USA) vol.123 (1976) p.262]
[2]   S.E.Swirhun, K.C.Saraswat, R.M.Swanson [IEEE Electron Device
      Lett. (USA) vol.EDL-5 (1984) p.209]
[3]   S.E.Swirhun, R.M.Swanson [IEEE Electron Device Lett (USA)
      vol.EDL-7 no.3 (Mar 1986) p.155-7]
[4]   D.Brown, B.Gorowitz, R.Wilson, B.Stoll, R.Saia [IEEE Electron
      Device Lett. (USA) vol.EDL-6 no.8 (Aug 1985) p.408-9]
[5]   Y.Shioya, M.Maeda, K.Yanagida [J. Vac. Sci. & Technol. B (USA)
      vol.4 no.5 (Sep/Oct 1986) p.1175-9]
[6]   T.Hara, N.Ohtsuka, K.Sakiyama, S.Saito [IEEE Trans. Electron
      Devices (USA) vol.ED-34 no.3 (Mar 1987) p.593-8]
[7]   R.A.Levy, M.L.Green, P.K.Gallagher, Y.S.Ali [J. Electrochem.
      Soc. (USA) vol.133 no.9 (Sep 1986) p.1905-12]
[8]   S.S.Cohen, M.J.Kim, B.Gorowitz, R.Saia, T.F.McNelly [Appl. Phys.
      Lett. (USA) vol.45 (1984) p.414]
[9]   M.J.Kim, D.M.Brown, S.S.Cohen, P.Piacente, B.Gorowitz [IEEE
      Trans. Electron Devices (USA) vol.ED-32 no.7 (Jul 1985)
      p.1328-33]
[10] M.Finetti, E.T.S.Pan, I.Suni, M.-A.Nicolet [Appl. Phys. Lett.
      (USA) vol.42 (1983) p.987]
[11] M.J.Kim, S.S.Cohen, D.M.Brown, P.A.Piacente, B.Gorowitz [IEEE
      IEDM Tech. Digest (IEEE, New York, USA, 1984) p.134]

CHAPTER 21

BARRIER HEIGHTS AND CONTACT RESISTANCES: SILICIDE/Si

21.1 BARRIER HEIGHT AT THE CoSi2/Si INTERFACE

by K.K.Ng

September 1987
EMIS Datareview RN=16152

CoSi2 can be formed by deposition of Co on Si followed by heat
treatment in the range of 550-800 deg C. Unreacted Co on SiO2 can be
preferentially removed in HCl/H2O2 solution. The available data are
very consistent and are given in TABLE 1 below.

TABLE 1

BH (eV)	Measurement	Substrate	Reference
0.64	I-V	n(100)(111)	[1,2]
0.64 - 0.65	I-V	n(100)	[6]
0.64	I-V, C-V	n(111)	[3]
0.65 +- 0.02	I-V, C-V	n(111)	[4,5]
0.405	I-V	p(100)(111)	[1]

BH - Barrier Height

Van Gurp studied the BH as a function of annealing temperature [1]. He
showed that the barrier heights of Co and CoSi on n-Si were the same
(approx. 0.68 eV). When heated above 550 deg C, the phase becomes CoSi2
and its BH dropped to 0.64 eV. Van Gurp and Reukers studied the thermal
stability of Al contact on CoSi2 Schottky diodes [2]. They showed that
improvement could be made with AlSi alloy, or with a W or Ti/W layer in
between the silicide and Al. The BH results reported by Tung and Gibson
[3], Rosencher et al [4] and D'Avitaya et al [5] were on epitaxial single-
crystalline CoSi2.

REFERENCES
----

[1] G.J. van Gurp [J. Appl. Phys. (USA) vol.46 no.10 (Oct 1975)
    p.4308-11]
[2] G.J.van Gurp, W.M.Reukers [J. Appl. Phys. (USA) vol.50 no.11
    (Nov 1976) p.6923-6]
[3] R.T.Tung, J.M.Gibson [J. Vac. Sci. & Technol. A (USA) vol.3 no.3
    (May/June 1985) p.987-91]
[4] E.Rosencher, S.Delage, F.A.D'Avitaya [J. Vac. Sci. & Technol.
    B (USA) vol.3 no.2 (Mar/Apr 1985) p.762-5]
[5] F.A.D'Avitaya, S.Delage, E.Rosencher, J.Derrien [J. Vac. Sci.
    & Technol. B (USA) vol.3 no.2 (Mar/Apr 1985) p.770-3]
[6] C.-D.Lien, M.Finetti, M-A.Nicolet [Appl. Phys. A (Germany) vol.35
    (1984) p.47]

21.2 BARRIER HEIGHT AT THE MoSi2/Si INTERFACE

by K.K.Ng

September 1987
EMIS Datareview RN=16149

MoSi2 Schottky diodes can be formed by Mo deposition followed by heat
treatment which is usually between 500-650 deg C. Unreacted Mo on SiO2
can be preferentially removed in a NH40H/H2O2 solution. The barrier
heights (BH) reported below are much more consistent than those of Mo/n-Si.
These values for MoSi2 are similar to the high range value reported for
Mo metal Schottky barriers.

TABLE 1

BH (eV)	Measurement	Substrate	Reference
0.69	I-V	n(100)	[2]
0.68	I-V	n(111)	[3]
0.63	I-V	n(100)	[1]
0.66	C-V	n(100)	[1]

Van Gurp and Reukers [2] studied the thermal stability of Al/MoSi2/Si
and Al/W/MoSi2/Si diodes. They showed that for the Al/MoSi2/Si structure,
the barrier height decreased with temperature treatment, and then
increased above approx. 400 deg C. In the presence of the W layer, they
showed that the BH is stable up to approx. 500 deg C. Yamamoto et al [3]
found that when a 2-percent-Si:Al electrode contact was used on a MoSi2/Si
diode, the characteristics were stable up to 500 deg C.

REFERENCES
----

[1] A.K.Kapoor, M.E.Thomas, M.B.Vora [IEEE Trans. Electron Devices
     (USA) vol.ED-33 no.6 (June 1986) p.772-8]
[2] G.J.van Gurp, W.M.Reukers [J. Appl. Phys. (USA) vol.50 no.11
     (Nov 1979) p.6923-6]
[3] Y.Yamamoto, H.Miyanaga, T.Amazawa, T.Sakai [IEEE Trans. Electron
     Devices (USA) vol.ED-32 no.7 (July 1985) p.1231-9]

21.3 BARRIER HEIGHT AT THE NiSi2/Si INTERFACE

by K.K.Ng

September 1987
EMIS Datareview RN=16151

NiSi2 can be formed by deposition of Ni on Si followed by heat
treatment. The silicidation temperature is typically between 400-600
deg C. There has been great interest in epitaxial NiSi2 Schottky diodes
since Tung's original work leading to the observation of different
barriers on type A and type B oriented NiSi2 [5]. (Type A had identical
orientation to the substrate while type B is rotated 180 deg.) The
available data on poly-crystalline and single-crystalline NiSi2 are
summarised below.

TABLE 1

BH (eV)		Measurement	Substrate	Ref
0.66		I-V	n(111)	[1]
0.68 +- 0.02		C-V, I-V	n(111)	[4]
0.62		I-V, C-V	n(111)	[2]
0.66		I-V	n(111)	[3,17]
0.66		I-V	n(100)	[18]
0.67 +- 0.01		I-V	n(100)	[16]
0.45		I-V	p(111)	[1]
Epitaxial NiSi2				
0.64 +- 0.01		I-V	n(111)	[8]
0.65	type A	I-V, C-V	n(111)	[5,13]
0.79	type B	I-V, C-V	n(111)	[5,13]
0.78	type A,B	I-V, photo	n(111)	[9,14]
0.48 +- 0.02		I-V, C-V	n(100)	[12,19]
0.62 +- 0.01	type A	I-V, photo	n(111)	[7,10,15]
0.69 +- 0.01	type B	I-V	n(111)	[7,10,15]
0.77 +- 0.05	type B	photo	n(111)	[7,10,15]
0.47 +- 0.02	type A	I-V, C-V	p(111)	[6,13]
0.34 +- 0.03	type B	I-V	p(111)	[13]

BH - Barrier Height

Andrews and Koch [1] and Coe and Rhoderick [4] found that below the annealing temperature of 430-450 deg C, a non-equilibrium mixture of Ni2Si and NiSi could be detected. Coe and Rhoderick also showed that the BH remained quite constant between the annealing temperatures of 230-430 deg C [4]. The thermal stability of NiSi2 on Si has been studied. Hokelek and Robinson found that when Al was used to contact the NiSi2 layer, the BH increased with annealing time (at temperatures up to 500 deg C) and then saturated [2]. Different barrier layers between Al and NiSi2 such as W [3], Cr [17] and TiN [16] have been studied.

On epitaxial single-crystalline NiSi2, Tung showed that the BH on type A (111) surface differed from that on type B by 0.14 eV [5,13]. This was consistent for both n-type substrate [5,13] and p-type substrate [6,13]. This finding was also supported by Hunt et al [7], and Hauenstein et al [10,15]. Liehr et al, on the other hand, found only one value for the BH between type A and type B materials [9,14]. They also found that the NiSi phase on n-Si(111) yielded the same BH. Tung et al showed that the BH grown on (100) surfaces was significantly different from that on (111) surfaces [12,19]. The impact of epitaxial NiSi2 growth conditions on BH was addressed by Shiraki et al [11].

REFERENCES
----

[1]   J.M.Andrews, F.B.Koch [Solid-State Electron. (GB) vol.14 (1971) p.901-8]

[2]   E.Hokelek, G.Y.Robinson [Thin Solid Films (Switzerland) vol.53 no.2 (1 Sep 1978) p.135-40]

[3]   M.Bartur, M-A.Nicolet [Appl. Phys. Lett. (USA) vol.39 no.10 (15 Nov 1981) p.822-4]

[4]   D.C.Coe, E.H.Rhoderick [J. Phys. D (GB) vol.9 (1976) p.965]

[5]   R.T.Tung [Phys. Rev. Lett. (USA) vol.52 no.6 (Feb 1984) p.461-4]

[6]   R.T.Tung [J. Vac. Sci. & Technol. B (USA) vol.2 (1984) p.465]

[7]   B.D.Hunt et al [Mater. Res. Soc. Symp. Proc. (USA) vol.54 (1986) p.479-84]

[8]   T.R.Harrison, A.M.Johnson, P.K.Tien, A.H.Dayem [Appl. Phys. Lett. (USA) vol.41 no.8 (15 Oct 1982) p.734-6]

[9]   M.Liehr, P.E.Schmid, F.K.LeGoues, P.S.Ho [Phys. Rev. Lett. (USA) vol.54 no.19 (13 May 1985) p.2139-42]

[10]   R.J.Hauenstein, T.E.Schlesinger, T.C.McGill [Appl. Phys. Lett. (USA) vol.47 no.8 (15 Oct 1985) p.853-5]

[11]   Y.Shiraki, T.Ohshima, A.Ishizaka, K.Nakagawa [J. Cryst. Growth (Netherlands) vol.81 (1987) p.476]

[12]   R.T.Tung, J.M.Gibson [J. Vac. Sci. & Technol. A (USA) vol.3 no.3 pt.1 (May/Jun) p.987-91]

[13]   R.T.Tung, K.K.Ng, J.M.Gibson, A.F.J.Levi [Phys. Rev. B (USA) vol.33 no.10 (15 May 1986) p.7077-90]

[14]   M.Liehr, P.E.Schmid, F.K.LeGoues, P.S.Ho [J. Vac. Sci. & Technol. A (USA) vol.4 no.3 pt.1 (May/Jun 1986) p.855-9]

[15]   R.J.Hauenstein, T.E.Schlesinger, T.C.McGill [J. Vac. Sci. & Technol. A (USA) vol.4 no.3 pt.1 (May/Jun 1986) p.860-4]

[16]   M.Finetti, I.Suni, M-A.Nicolet [J. Electron. Mater. (USA) vol.13 no.2 (Mar 1984) p.327-40]

[17]   M.Bartur, M-A.Nicolet [J. Electrochem. Soc. (USA) vol.131 no.5 (May 1984) p.1118-22]

[18]   G.Ottaviani, K.N.Tu, J.W.Mayer [Phys. Rev. B (USA) vol.24 no.6 (15 Sep 1981) p.3354-9]

[19]   R.T.Tung, A.F.J.Levi, J.M.Gibson [J. Vac. Sci. & Technol. B (USA) vol.4 no.6 (Nov/Dec 1986) p.1435-43]

21.4 BARRIER HEIGHT AT THE Pd2Si/Si INTERFACE

by K.K.Ng

September 1987
EMIS Datareview RN=16153

Pd2Si can be formed by deposition of Pd on Si followed by heat treatment
in the range 200-600 deg C. The low temperature for silicidation is quite
unique. For self-aligned structures, unreacted Pd on oxide can be removed
by an aqueous solution of KI + I2. Available data on the barrier height
are summarized in TABLE 1 below. The reported data are quite consistent
and the barrier height (BH) is similar to that of Pd on Si.

TABLE 1

BH (eV)	Measurement	Substrate	Reference
0.73-0.76	I-V, C-V	n(111)	[1]
0.75 +- 0.01	I-V, C-V	n	[3]
0.72 +- 0.02	I-V, C-V	n(111)	[2]
0.73 +- 0.02	I-V, C-V	n(111)	[5]
0.72	I-V, photo	n(100)	[11]
0.73-0.74	I-V	n(100)(111)	[21,22]
0.73-0.74	I-V	n	[4]
0.735 +- 0.01	I-V	n(100)	[10]
0.73	I-V	n(111)	[16]
0.75	I-V	n(100)	[18]
0.74		n(111)	[14]
0.71-0.73	photo	n(100)(111)	[15]
0.71-0.73	I-V, photo	n(100)(111)	[17]
0.74-0.75	I-V	n(100)	[8,9]
0.74	I-V	n(111)	[19]
0.337	photo	p	[20]
0.378	I-V	p	[20]
0.35 +- 0.01	I-V, photo	p(111)(100)	[6]
0.34	photo	p(100)	[23]

Eizenberg et al [4] and Kritzinger and Tu [21,22] studied the formation of Pd2Si by co-deposition of Pd-Si alloys while Wittmer et al studied silicidation by laser anneal [10]. Canali et al studied thin film Pd2Si Schottky barriers [5]. They showed that the BH was independent of silicide thickness for thickness greater than 100 A. The BH from Huang was obtained by a non-conventional electron-beam induced voltage technique [14]. The thermal stability of Al on Pd2Si was studied by Grinolds and Robinson [12] and Parekh et al [13]. Different barrier metals on Pd2Si were studied. These include TiW alloy [7], Ti [16] and Cr [19]. The redistribution of phosphorus during Pd2Si formation was studied by Kikuchi [8,9] and Studer [11].

REFERENCES
----

[1]   C.J.Kircher [Solid-State Electron. (GB) vol.14 no.6 (1971) p.507-3]
[2]   D.J.Fertig, G.Y.Robinson [Solid-State Electron. (GB) vol.19 no.5 (Apr 1976) p.407-13]
[3]   R.F.Broom [Solid-State Electron. (GB) vol.14 (1971) p.1087]
[4]   M.Eizenberg, H.Foell, K.N.Tu [J. Appl. Phys. (USA) vol.52 no.2 (Feb 1981) p.861-8]
[5]   C.Canali, F.Catellani, S.Mantovani, M.Prudenziati [J. Phys. D (GB) vol.10 (1977) p.2481]
[6]   R.C.McKee [IEEE Trans. Electron Devices (USA) vol.ED-31 (1984) p.968]
[7]   V.F.Drobny [J. Electron. Mater. (USA) vol.14 no.3 (May 1985) p.283-96]
[8]   A.Kikuchi [J. Appl. Phys. (USA) vol.54 no.7 (Jul 1983) p.3998-4000]
[9]   A.Kikuchi [Jpn. J. Appl. Phys. Part 1 (Japan) vol.23 no.10 (Oct 1984) p.1345-7]
[10] M.Wittmer, W.Luthy, B.Studer, H.Melchior [Solid-State Electron. (GB) vol.24 (1981) p.141]
[11] B.Studer [Solid-State Electron. (GB) vol.23 (1980) p.1181-4]
[12] H.Grinolds, G.Y.Robinson [J. Vac. Sci. & Technol (USA) vol.14 (1977) p.75]
[13] P.C.Parekh, R.C.Sirrine, P.Lemieux [Solid-State Electron. (GB) vol.19 (1976) p.493]
[14] H.-C.W.Huang, C.F.Aliotta, P.S.Ho [Appl. Phys. Lett. (USA) vol.41 (1982) p.54]
[15] R.J.Purtell, P.S.Ho, G.W.Rubloff, P.E.Schmid [Physica B & C (Netherlands) vol.117 (1983) p.834]
[16] G.Salomonson, K.E.Holm, T.G.Finstad [Phys. Scr. (Sweden) vol.24 (1981) p.401]
[17] P.E.Schmid, P.S.Ho, H.Foll, T.Y.Tan [Phys. Rev. B (USA) vol.28 no.8 (15 Oct 1983) p.4593-601]
[18] G.Ottaviani, K.N.Tu, J.W.Mayer [Phys. Rev. B (USA) vol.24 no.6 (15 Sep 1981) p.3354-9]
[19] M.Bartur, M.-A.Nicolet [J. Electrochem. Soc. (USA) vol.131 no.5 (May 1984) p.1118-22]
[20] H.Elabd, T.Villani, W.Kosonocky [IEEE Electron. Devices. Lett. (USA) vol.EDL-3 (1982) p.89]
[21] S.Kritzinger, K.N.Tu [J. Appl. Phys. (USA) vol.52 (1981) p.305]
[22] S.Kritzinger, K.T.Tu [Appl. Phys. Lett. (USA) vol.37 (1980) p.205]
[23] O.Engstrom, H.Pettersson, B.Sernelius [Phys. Status Solidi a (Germany) vol.95 (1986) p.691]

21.5 BARRIER HEIGHT AT PtSi/Si INTERFACE

by K.K.Ng

September 1987
EMIS Datareview RN=16148

PtSi forms one of the highest silicide potential barriers on Si. Generally reported, its barrier height (BH) is slightly larger than that of Pt on Si before reaction. When Pt is deposited on Si, silicide can be formed in the temperature range 250-625 deg C. After heat treatment, Pt on field oxide is not reacted and it can be removed by aqua regia while the silicide remains intact. The reported barrier heights are fairly consistent and they are summarized below.

TABLE 1

BH (eV)	Measurement	Substrate	Reference
0.81-0.86	I-V, C-V	n	[12]
0.85-0.87	I-V	n	[13]
0.84+-0.01	I-V, C-V	n(111)(100)	[16]
0.85	I-V	n	[25]
0.843	I-V	n(100)(111)	[8]
0.82-0.83	I-V, C-V	n(111)	[7]
0.82 +- 0.02	I-V, C-V	n(111)	[14]
0.87	I-V	n(100)	[4,23]
0.83	I-V	n(111)	[19]
0.87-0.88	I-V	n(111)	[22]
0.85	I-V	n(100)	[5]
0.84		n(111)	[18]
0.82	I-V	n(100)	[6]
0.84	I-V	n	[20]
0.83-0.85	I-V	n(111)	[21]
0.85-0.88	I-V	n(100)	[1-3]
0.85	I-V, C-V	n(100)	[26]
0.25	I-V	p	[13]
0.22	I-V, photo	p(111)	[9]

Gutknecht and Strutt compared directly the barrier heights on both (111) and (100) surfaces, and found that they were the same [16]. Canali et al studied thin film silicide Schottky barriers for solar cell applications [14]. They showed that the BH was independent of silicide thickness for thickness greater than 100 A. The BH from Huang was obtained by a non-conventional electron-beam induced voltage technique [18].

Murarka et al [6] and Eizenberg et al [15] formed silicide Schottky diodes by co-sputtering of Pt and Si followed by heat treatment. Tsuar et al have formed PtSi Schottky from alternating depositions of Pt layers and Si layers, followed by heat treatment [4,23]. These techniques minimise the consumption of Si from the substrate when silicide is formed.

After formation of silicide from deposited metal, Lew and Helms observed that arsenic from the Si substrate redistributed and it segregated at both the PtSi/Si interface and Pt/Si surface [10]. Kikuchi and Sugaki also observed the piling up of phosphorus at the PtSi/Si interface [5].

Murarka et al studied the stability of PtSi/Si structures at high temperatures [6]. They started to observe electrical degradation at 700 deg C. Hosack studied the thermal stability of aluminium on PtSi/Si Schottky diodes [24] while Merchant and Amano used Si doped Al [11]. Other authors used different materials as the diffusion barriers between Al and PtSi such as TiW [17], Ti [19], TiN [20] and Cr [21].

REFERENCES
----

[1]   C.-A.Chang [J. Appl. Phys. (USA) vol.59 no.9 (1 May 1986) p.3116-21]

[2]   C.-A.Chang et al [J. Vac. Sci. & Technol. B (USA) vol.4 no.3 (May/Jun 1986) p.745-54]

[3]   C.-A.Chang et al [J. Electrochem. Soc. (USA) vol.133 no.6 (Jun 1986) p.1256-60]

[4]   B.L-Y.Tsuar, D.J.Silversmith, R.W.Mountain, L.S.Hung, S.S.Lau, T.T.Sheng [J. Appl. Phys. (USA) vol.52 (1981) p.5243]

[5]   A.Kikuchi, S.Sugaki [J. Appl. Phys. (USA) vol.53 (1982) p.3690]

[6]   S.P.Murarka, E.Kinsbron, D.B.Fraser, J.M.Andrews, E.J.Lloyd [J. Appl. Phys. (USA) vol.54 (1983) p.6943]

[7]   M.Severi, E.Gabilli, S.Guerri, G.Celotti [J. Appl. Phys. (USA) vol.48 (1977) p.1998]

[8]   M.J.Rand [J. Electrochem. Soc. (USA) vol.122 (1975) p.811]

[9]   J.Siverman et al [Mater. Res. Soc. Symp. Proc. (USA) vol.54 (1986) p.515]

[10]   P.W.Lew, C.R.Helms [J. Appl. Phys. (USA) vol.56 no.12 (15 Dec 1984) p.3418]

[11]   P.Merchant, J.Amano [J. Vac. Sci. & Technol. A (USA) vol.1 (1983) p.459]

[12]   J.Saltich [in 'Ohmic Contacts to Semiconductors' Ed. B.Schwartz (Electrochem. Soc., 1969) p.187-99]

[13]   J.M.Andrews, M.P.Lepselter [Solid-State Electron. (GB) vol.13 (1970) p.1011]

[14]   C.Canali, F.Catellani, S.Mantovani, M.Prudenziati [J. Phys. D (GB) vol.10 (1977) p.2481]

[15]   M.Eizenberg, H.Foell, K.N.Tu [J. Appl. Phys. (USA) vol.52 (1981) p.861]

[16] P.Gutknecht, M.J.O.Strutt [Appl. Phys. Lett. (USA) vol.21 (1972)
      p.405]

[17] C.Canali, F.Fantini, E.Zanoni [Thin Solid Films (Switzerland)
      (1982) vol.97 p.325]

[18] H.-C.W.Huang, C.F.Aliotta, P.S.Ho [Appl. Phys. Lett. (USA) vol.41
      (1982) p.54]

[19] G.Salomonson, K.E.Holm, T.G.Finstad [Phys. Scr. (Sweden) vol.24
      (1981) p.401]

[20] R.J.Schutz [Thin Solid Films (Switzerland) vol.104 (1983) p.89]

[21] M.Bartur, M.-A.Nicolet [J. Electrochem. Soc. (USA) vol.131 (1984)
      p.1118]

[22] G.Ottaviani, K.N.Tu, J.W.Mayer [Phys. Rev. B (USA) vol.24 (1981)
      p.3354]

[23] B.-Y.Tsuar, D.J.Silversmith, R.W.Mountain, C.H.Anderson,Jr
      [Thin Solid Films (Switzerland) vol.93 (1982) p.331]

[24] H.H.Hosack [J. Appl. Phys. (USA) vol.44 (1973) p.3476]

[25] E.Demoulin, F. van de Wiele [Solid-State Electron. (GB) vol.29
      (1986) p.825]

[26] M.J.Hargrove, R.L.Anderson [Solid-State Electron. (GB) vol.29
      (1986) p.365]

21.6 BARRIER HEIGHT AT THE TaSi2/Si INTERFACE

by K.K.Ng

September 1987
EMIS Datareview RN=16154

TaSi2 films are usually formed by co-sputtering of the silicide
materials followed by a relatively high temperature anneal of approx.
900 deg C. The data on barrier height (BH) for TaSi2 on Si are very
limited, as shown in TABLE 1. Unfortunately the references cited do
not give similar values. However, they do indicate that TaSi2 may give
a higher barrier height on p-Si than on n-Si.

TABLE 1

BH (eV)	Measurement	Substrate	Reference
0.58	I-V	n(100)	[2,3]
0.373	C-V	n	[1]
0.715	C-V	p	[1]

The BH on n-Si reported by Hsu and Maa was obtained with a thin-oxide
MOS diode to avoid excessive leakage [1]. The thin oxide of 10-20 A was
assumed to have negligible effects on C-V characteristics. Their reported
barrier heights on n-Si and p-Si matched the Si band gap value. The
reaction of Al contact with TaSi2 which changed the BH was discussed by
Neppl et al [2,3].

REFERENCES
----

[1] S.T.Hsu, J.S.Maa [RCA Rev. (USA) vol.46 (1985) p.163]
[2] F.Neppl, F.Fischer, U.Schwabe [Thin Solid Films (Switzerland)
     vol.120 no.4 (26 Oct 1984) p.257-66]
[3] F.Neppl, U.Schwabe [Mater. Res. Soc. Symp. Proc. (USA) vol.25
     (1984) p.587-92]

21.7 BARRIER HEIGHT AT THE TiSi2/Si INTERFACE

by K.K.Ng

September 1987
EMIS Datareview RN=16150

TiSi2 is usually formed by reaction of deposited Ti on Si, in the
temperature range 600-800 deg C. In a self-aligned process, the unreacted
Ti on field oxide can be preferentially removed by NH4OH/H2O2 solution.
When Ti is simultaneously deposited on SiO2, reaction above 700 deg C
should be avoided because Ti reacts with the oxide to form titanium oxides
which are difficult to remove. The BH (barrier height) of TiSi2 on n-Si
is slightly larger than that of Ti, by approx. 0.1 eV. Available data are
shown below.

TABLE 1

BH (eV)	Measurement	Substrate	Reference
0.57		n(111)	[7]
0.58	photo	n(100)	[8]
0.60	I-V	n(100)	[1]
0.58 +- 0.01	I-V	n(100)	[2]
0.51	I-V	p(100)	[1]

The BH from Huang et al was obtained by a non-conventional electron-beam
induced voltage technique [7].

The thermal stability of Al/TiSi2/Si was studied by Chen et al [4] and
Yu and Wittmer [3]. They both showed that without a diffusion barrier
layer, Si diffused through TiSi2 and reacted with Al at 400-450 deg C,
a reaction typical of direct contact of Al to Si.

The redistribution of doping during TiSi2 formation was studied by
Amano et al [5,6].

REFERENCES
----

[1] M.O.Aboelfotoh, K.N.Tu [Phys. Rev. B (USA) vol.34 no.4 (15 Aug
     1986) p.2311-8]
[2] A.Kikuchi [Jpn. J. Appl. Phys. Part 2 (Japan) vol.25 no.11
     (Nov 1986) p.L894-5]
[3] C.Yu Ting, M.Wittmer [J. Appl. Phys. (USA) vol.54 (1983)
     p.937]
[4] D.C.Chen, P.Merchant, J.Amano [J. Vac. Sci. & Technol. A (USA)
     vol.3 no.3 (May/June 1985) p.709-13]

[5]   J.Amano, P.Merchant [Appl. Phys. Lett. (USA) vol.44 (1984)
      p.744-6]

[6]   J.Amano, P.Merchant, T.R.Cass, J.N.Miller [J. Appl. Phys. (USA)
      vol.59 no.8 (15 Apr 1986) p.2689-93]

[7]   H.-C.W.Huang, C.F.Aliotta, P.S.Ho [Appl. Phys. Lett. (USA) vol.41
      (1982) p.54]

[8]   M.A.Taubenblatt, D.Thomson, C.R.Helms [Appl. Phys. Lett. (USA)
      vol.44 (1984) p.895]

21.8 CONTACT RESISTANCE OF THE CoSi2/Si INTERFACE

by K.K.Ng

September 1987
EMIS Datareview RN=16160

The data on the contact resistivity of CoSi2 on Si are limited as
shown in TABLE 1.

TABLE 1

rho(c) (ohm cm*2)	Substrate	Surface Concn (/cm*3)	Ref
< 5 X 10*-7	n+(100)		[4]
c. 1 X 10*-7	n+(100)	1.5 X 10*20	[2,5]
c. 1 X 10*-6	n+(100)(111)		[3]
c. 1 X 10*-6	p+(100)	2 X 10*19	[1]
c. 1.5 X 10*-7	p+(100)	6 X 10*19	[2,5]
4-7 X 10*-7	p+(100)(111)	c. 1 X 10*20	[3]

rho(c) - contact resistance

Van Gurp studied the rho(c) as a function of annealing temperature [1].
He observed a minimum value at the temperature of approx. 520 deg C.
Van den Hove et al found that the Si substrate doping profile was
essentially unchanged by silicidation [2]. The dopant redistribution
caused by silicide formation was also studied by Tabasky et al [3].
This is important because the Si surface concentration is a critical
parameter for rho(c).

REFERENCES
----

[1] G.J.van Gurp [J. Appl. Phys. (USA) vol.46 no.10 (1975) p.4308-11]
[2] L.Van de Hove, R.Wolters, K.Maex, R. De Keersmaecker, G.Declerck
     [IEEE Trans. Electron Devices (USA) vol.ED-34 no.3 (Mar 1987)
     p.554-61]
[3] M.Tabasky, E.S.Bulat. B.M.Ditchek, M.A.Sullivan, S.C.Shatas
     [IEEE Trans. Electron Devices (USA) vol.ED-34 (1987) p.548]
[4] S.Vaidyam R.J.Schutz, A.K.Sinha [J.Appl. Phys. (USA) vol.55
     (1984) p.3514]
[5] L.Van den Hove. R.Wolters, K.Maez, R. De Keersmaecker, G.Declerck
     [J. Vac. Sci. & Technol. B (USA) vol.4 (1986) p.1358]

21.9 CONTACT RESISTANCE OF THE Pd2Si/Si INTERFACE

by K.K.Ng

September 1987
EMIS Datareview RN=16161

The available data of contact resistivity of Pd2Si on Si are very
limited as shown in TABLE 1.

TABLE 1

rho(c) (ohm cm*2)	Substrate	Surface Concn (/cm*3)	Ref
4 X 10*-6	n+(111)	2 X 10*20	[5]
2 X 10*-6	n+(111)	c. 1 X 10*20	[1]
7 X 10*-8	n+	4 X 10*20	[2]
c. 2 X 10*-7	n+(100)		[3]
3-6 X 10*-6	p+		[2]
3 X 10*-7	p+(100)		[3]

The redistribution of substrate doping during silicide formation was
observed by Ohdomari et al [1], Finetti et al [2] and Wittmer et al [4].
This is important because the Si surface concentration is a critical
parameter for rho(c). The data on rho(c) from Singh et al [3] showed a
dependence on contact size, indicating some non-uniformity of the
interface. The thermal stability of rho(c) with Al alloy contacting the
Pd2Si layer was studied by Sugerman et al [6].

REFERENCES
----

[1] I.Ohdomari et al [J. Appl. Phys. (USA) vol.54 no.8 (1983)
     p.4679-82]
[2] M.Finetti, S.Guerri, P.Negbrini, A.Scorzoni, I.Suni [Thin Solid
     Films (Switzerland) vol.130 no.1/2 (16 Aug 1985) p.37-45]
[3] R.N.Singh, D.W.Skelly, D.M.Brown [J. Electrochem. Soc. (USA)
     vol.133 no.11 (Nov 1983) p.2390-3]
[4] M.Wittmer, C.-Y.Ting, K.N.Tu [J. Appl. Phys. (USA) vol.54 (1983)
     p.699]
[5] C.J.Kircher [Solid-State Electron. (GB) vol.14 (1971) p.507]
[6] A.Sugerman, H.J.Tsai, J.S.Kristoff, J.Regh [J. Electron. Mater.
     (USA) vol.11 (1982) p.943]

21.10 CONTACT RESISTANCE OF PTSI/SI INTERFACE

by K.K.Ng

September 1987
EMIS Datareview RN=16158

Since the barrier height of Si/PtSi is high on n-type (approximately
0.85 eV), the contact resistance (rho(c)) on n+ is expected to be much
higher that on p+ for the same surface doping. The available data on
rho(c) are summarised in TABLE 1.

TABLE 1

rho(c) (ohm cm*2)	Substrate	Surface Concn (/cm*3)	Ref
9 X 10*-7	n+(100)	1 X 10*20	[4]
6 X 10*-6	n+(100)	1.5 X 10*19	[4]
1-2 X 10*-5	n+(100)	5 X 10*19	[2]
c. 8 X 10*-8	n+(100)	2.1 X 10*20	[1]
c. 4 X 10*-8	n+(100)	3 X 10*20	[1,3]
1 X 10*-6	p+(100)	1.5 X 10*20	[4]
4 X 10*-6	p+(100)	3 X 10*19	[4]
c. 2 X 10*-7	p+(100)	3.2 X 10*19	[1]
c. 7 X 10*-8	p+(100)	7 X 10*19	[1,3]

The results reported by Sinha included contributions from spreading
resistance [4], and thus they represented the upper limit of rho(c).
Sinha also showed that when using W as the contact material to the PtSi,
the system was stable up to 700 deg C. Cohen et al studied the thermal
stability of Si/PtSi contacts [3]. They observed that when molybdenum
was used as the final metallization material, the contacts were stable
up to 700 deg C. The redistribution of As during silicide formation
was examined by Lew and Helms [5]. This is important because the Si
surface concentration is a critical parameter for rho(c).

REFERENCES
----

[1] S.S.Cohen, P.A.Piacente, G.Gildenblat, D.M.Brown [J. Appl. Phys.
     (USA) vol.53 (1982) p.8856]
[2] G.Boberg, L.Stolt, P.A.Tove, H.Norde [Phys. Scr. (Sweden) vol.24
     (1981) p.405]
[3] S.S.Cohen, P.A.Piacente, D.M.Brown [Appl. Phys. Lett. (USA)
     vol.41 (1982) p.976]
[4] A.K.Sinha [J. Electrochem. Soc., (USA) vol.120 (1973) p.1767]
[5] P.W.Lew, C.R.Helms [J. Appl. Phys. (USA) vol.56 (1984) p.3418]

21.11 CONTACT RESISTANCE OF THE TiSi2/Si INTERFACE

by K.K.Ng

September 1987
EMIS Datareview RN=16159

The contact resistance (rho(c)) of TiSi2 on Si is of great interest
because of the potential application in VLSI. Available data are
summarized in TABLE 1.

TABLE 1

rho(c) (ohm cm*2)	Substrate	Surface Concn (/cm*3)	Ref
8 X 10*-8	n+(100)	2 X 10*20	[1,12]
4 X 10*-8	n+	6 X 10*20	[4]
2.5 X 10*-7	n+	1.5 X 10*20	[4]
5 X 10*-6	n+	1.9 X 10*20	[5]
3 X 10*-7	n+	1 X 10*20	[2]
2 X 10*-6	n+	3 X 10*19	[2]
c. 3 X 10*-8	n+		[3]
c. 1 X 10*-7	n+(100)	1.5 X 10*20	[9,13]
7 X 10*-5	p+(100)	1 X 10*19	[1,12]
8 X 10*-7	p+	1 X 10*20	[4]
1 X 10*-5	p+	2 X 10*19	[4]
1 X 10*-6	p+	5 X 10*19	[2]
5 X 10*-7	p+		[3]
c. 3 X 10*-7	p+(100)	6 X 10*19	[9,13]

Taur et al measured the rho(c) at low temperatures and confirmed the weak
temperature dependence [2]. Scott et al [3] showed that the silicidation
temperature was critical for a low rho(c), although in their experiment
the surface concentration was not given. The thermal stability of TiSi2
contact was studied by Ogawa et al using W contact metal [5]. A TiSi2
contact formed by rapid thermal annealing was studied by Kwong et al
[6,7]. Park et al [8] studied the effect of high doping concentration on
the formation of TiSi2. They concluded that silicidation was retarded
for Si substrate doping above a critical value. The redistribution of
dopants during TiSi2 formation was studied by Amano et al [10,11]. This
is important because the Si surface concentration is a critical parameter
for rho(c).

REFERENCES
----

[1]   M.Finetti, S.Guerri, P.Negrini, A.Scorzoni, I.Suni [Thin Solid
      Films (Switzerland) vol.130 no.1/2 (16 Aug 1985) p.37-45]
[2]   Y.Taur et al [IEEE Trans. Electron Devices (USA) vol.ED-34
      no.3 (Mar 1987) p.575-80]
[3]   D.B.Scott, R.A.Chapman, C.-C.Wei, S.S.Mahant-Shetti, R.A.Haken,
      T.C.Holloway [IEEE Trans Electron Devices (USA) vol.ED-34
      (1987) p.562]
[4]   J.Hui, S.Wong, J.Moll [IEEE Electron Device Lett. (USA)
      vol.EDL-6 no.9 (Sep 1985) p.479-81]
[5]   S.Ogawa, K.Yamazaki, S.Akiyama, Y.Terui [IEEE IEDM Tech. Digest
      (1986) p.62-5]
[6]   D.L.Kwong, N.S.Alvi [J. Appl. Phys. (USA) vol.26 no.2 (15 July 1986)
      p.688-91]
[7]   D.L.Kwong, N.S.Alvi, Y.H.Ku, A.W.Cheung [Mater. Res. Soc. Symp.
      Proc. (USA) vol.52 (1986) p.241-9]
[8]   H.K.Park, J.Sachitano, M.McPherson, T.Yamaguchi, G.Lehman [J.
      Vac. Sci. & Technol. A (USA) vol.2 (1984) p.264]
[9]   L. Van den Hove, R.Wolters, K.Maex, R.F. De Keersmaecker,
      G.J.Declerck [IEEE Trans. Electron Devices (USA) vol.ED-34 (1987)
      p.554]
[10] J.Amano, P.Merchant, T.Koch [Appl. Phys. Lett. (USA) vol.44
      (1984) p.744]
[11] J.Amano, P.Merchant, T.R.Cass, J.N.Miller, Y.Koch [J. Appl. Phys.
      (USA) vol.59 (1986) p.2689]
[12] M.Finetti, A.Scorzoni, G.Soncini [IEEE Electron Device Lett. (USA)
      vol.EDL-5 (1984) p.524]
[13] L. Van den Hove, R.Wolters, K.Maex, R. De Keersmaecker,
      G.Declerck [J. Vac. Sci. & Technol. B (USA) vol.4 (1986) p.1358]

CHAPTER 22

SURFACE STRUCTURE

22.1   Surface structure of clean Si

22.1 SURFACE STRUCTURE OF CLEAN Si

by P.J.Dobson

October 1987
EMIS Datareview RN=11850

A. INTRODUCTION

There has been intense world-wide effort to seek an understanding of the
complex reconstructions that are seen on clean surfaces of silicon. Most
of this effort has been directed towards the two reconstructions 2 X 1
and 7 X 7 which occur on the (111) surface, and the 2 X 1 reconstruction
on the (001) surface. The following summary treats each of these in turn
and refers briefly to some observations on the (110) surface. In all
cases we restrict discussion to atomically clean surface structures.

B. THE Si(111) SURFACE

Si(111) displays two important reconstructions, viz: the 2 X 1 which
is produced by cleaving in ultra high vacuum and the 7 X 7 which
results from heating the 2 X 1 surface to above 200-400 deg C. Further
heating to 870 deg C results in a 1 X 1 structure which can be
stabilised at room temperatures by rapid cooling; otherwise a reversible
transformation to the 7 X 7 is seen. Early results were reviewed by
Miller et al [1] and the subject has been recently covered by Haneman
[2] and Olmstead [3].

B1 The 2 X 1 Reconstruction on the Si(111) Surface

Original ideas were based on a buckling of the surface with a raising
and lowering of the atomic layers [4,5], which was apparently confirmed
by LEED [6] and total energy calculations [7,8]. The surface state
dispersion from UPS observations also lent support to this model [9-11].
However Northrup et al [12] questioned the energy calculations, as did
Pandey [13,14], and he proposed a new model which is now favoured. This
model is based on pi-bonded chains along the short axis of the 2 X 1
cell i.e. there is re-bonding of atoms at the surface. This model was
also consistent with the UPS results [9-11] and [15-17]. Chadi [18]
proposed a pi-bonded pair model but this has a higher energy
configuration than the pi-bonded chain [19]. Northrup and Cohen [20,21]
have suggested that the atom pairs in the chain are tilted and this is
finding wide experimental support from ion scattering [22-24] and LEED
[25,26]. Light scattering and optical studies also support a pi-bonded
chain model [27-29,3].

B2 The 7 X 7 Reconstruction on the Si(111) Surface

In the last three years there has been a convergence of opinion, based
on a wide variety of experimental techniques, that the dimer adatom
stacking fault (DAS) model is the most plausible for this surface.
There have been many models suggested in the past. Cardillo [30]

suggested arrays of steps might be responsible for the 7 X 7 cell.
Pandey [13,14] invoked a pi-bonded chain model and there have been many
'cluster' models e.g. the tripedal model [31], pyramid cluster model
[32] and modified milk stool model [33]. There were also many models
based on adatoms and vacancies [34-38,30]. Yamaguchi [39-41] has
suggested relaxed adatom models to minimise the strain energy on the
basis of the adatom model of Binnig et al [38]. Himpsel and Batra [42]
developed a stacking fault model with re-bonding to give 12 adatoms per
7 X 7 unit cell. This model had much in common with that of McRae and
Petroff [43] and that of Bennett et al [44]. The transmission electron
microscopy and diffraction analysis by Takayanagi et al [45,46]
provided the key which has now established the dimer adatom stacking
fault model, which has features from several of the earlier ideas. In
this model, the 7 X 7 cell is divided into two triangular halves: one
has a regular diamond stacking, the other has a stacking fault in the
sub-surface layer. There are 12 Si adatoms; each is bonded to 3 Si
surface atoms and there are six further bonded atoms around the corner
of the unit cell. The number of dangling bonds is relatively low, with
19 per 7 X 7 unit cell. The energy of this model has been examined by
Qian and Chadi [47,48] and it has been shown to be the most stable of
all the suggested models. A similar conclusion was reached by Tersoff
[49].

Evidence from X-ray diffraction at grazing incidence [50] and further
STM observations [51-53] have added more support to the DAS model. In
particular the current imaging tunnelling spectroscopy [52,53] which
produces real space images of the tunnelling current associated with
the surface states adds to the knowledge of the electronic structure
of this surface. Attempts have been made to determine the atomic
displacements normal to the surface using X-ray standing wave techniques
[54]. The situation with regard to the electronic surface states has
been summarized by Northrup [55]. Other points of note are that the
transition between the 7 X 7 and 1 X 1 has been observed directly in a
low energy electron reflection microscope by Telieps and Bauer [56] and
the 7 X 7 structure has been observed to persist, even when 'buried'
under an amorphous silicon layer [57]. In this case the structure
differs from the free, clean surface reconstruction, in that the ordered
array of adatoms is absent.

C. THE Si(001) SURFACE

The clean Si(001) surface usually shows a 2 X 1 reconstruction such
that both 2 X 1 and 1 X 2 domains are in evidence. These two domains
exist because flat terraces which differ in height by a/4 usually occur.
It has been demonstrated that single domain 2 X 1 (001) surfaces can be
prepared [58,59] and this is particularly important for epitaxial
growth, e.g. GaAs on Si(001) [60]. There is evidence for other
reconstructions [61-65] with the c 4 X 2 and various other 2 X n
structures being reported.

The models used to explain the reconstructions have been based on
regular arrays of missing rows or vacancies [66-68], dimers [69-75] and
various conjugated chain models [76,77]. The evidence is now strongly
in favour of models based on dimers [73]. Chadi [74,75] outlined the
expected atomic and electronic structures associated with dimer bond
reconstruction. There is good reason to believe that the dimer is
asymmetric i.e. the bond axis is tilted with respect to the surface.

Calculations [81] show that a symmetric dimer would lead to a metallic
surface state which is at variance with angle resolved UPS measurements
[78-80] which indicate an energy gap in the surface bands i.e.
semiconducting behaviour. Chadi's asymmetric dimer model is supported
by STM results [82], helium diffraction [63], EELS [83-85], ion
scattering [86,87] and LEED [88]. Yang et al [89] had suggested, on the
basis of LEED results, that the dimers may also have been subject to an
in-plane twist displacement but this has been shown to be unnecessary by
Duke et al [90]. Most workers agree that the lattice distortion extends
for several layers below the top dimerized layer. There is still some
doubt regarding the most stable surface unit cell and a second order
phase change from a c 4 X 2 to a 2 X 1 above 200 deg K has been seen by
Tabata et al [91]. Farrell et al [85] also suggest that the unit cell
is larger than 2 X 1 and this is supported by helium diffraction [63]
and STM [82], along with the presence of disorder. The disorder involves
the sense of tilt of the dimers, their registry and (according to STM
[82]) the presence of symmetric dimers.

Theoretical studies of the surface lattice vibrations have been made
[92] which suggest that the reconstruction is the result of local
chemistry rather than Fermi instability and some insight is offered into
the origin of the c 4 X 2   structure.

The structure and occurrence of steps on the Si(001) surface has
received attention with UHV reflection electron microscopy studies [93],
transmission electron microscopy/diffraction [94], and RHEED [58].

Bi-layer steps are commonly observed and this is important for producing
single domain structures for epitaxial growth. The implications of
biatomic steps have been discussed by Aspnes and Ihm [95].

D. THE Si(110) SURFACE

Relatively little work has been performed on this surface and the
situation with regard to stable reconstruction is not clear. Jona [96]
and Olshanetsky and Shklyaev [97] have reported 4 X 5, 2 X 1, 5 X 1,
7 X 1 and 9 X 1 structures and also some tendency to form facets.
Yamamoto et al [98] have used RHEED to demonstrate the presence of some
of these structures and a more complex '16-structure' on sputtered and
annealed surfaces.

REFERENCES
----

[1]   D.J.Miller, D.Haneman, L.W.Walker [Surf. Sci. (Netherlands)
      vol.94 (1980) p.555]
[2]   D.Haneman [Rep. Prog. Phys. (GB) vol.50 (1987) p.1045]
[3]   M.A.Olmstead [Surf. Sci. (Netherlands) vol.6 nos.4/5 (1987)]
[4]   D.Haneman [Phys. Rev. (USA) vol.121 (1961) p.1093]
[5]   D.E.Eastman [J. Vac. Sci. & Technol. (USA) vol.17 (1980)
      p.492]
[6]   R.Feder, W.Moench, P.P.Auer [J. Phys. C (GB) vol.12 (1979) p.L179]
[7]   D.J.Chadi [Phys. Rev. Lett. (USA) vol.41 (1978) p.1062]
[8]   D.J.Chadi [J. Vac. Sci. & Technol. (USA) vol.18 (1981) p.856]
[9]   F.Houzay, G.M.Guichar, R.Pinchaux, Y.Petroff [J. Vac. Sci. &
      Technol. (USA) vol.18 (1981) p.860]
[10] J.E.Rowe, M.M.Traum, N.V.Smith [Phys. Rev. Lett. (USA) vol.33
      (1974) p.1333]

[11]   G.V.Hansson, R.Z.Bachrach, R.S.Bauer, W.Goepel [Surf. Sci.
         (Netherlands) vol.99 (1980) p.13]
[12]   J.E.Northrup, J.Ihm, M.L.Cohen [Phys. Rev. Lett. (USA) vol.47
         (1981) p.1910]
[13]   K.C.Pandey [Phys. Rev. Lett. (USA) vol.47 (1981) p.1913]
[14]   K.C.Pandey [Phys. Rev. Lett. (USA) vol.49 (1982) p.223]
[15]   R.I.G.Uhrberg, G.V.Hansson, J.M.Nicholls, S.A.Flodstrom [Phys.
         Rev. Lett. (USA) vol.48 (1982) p.1032]
[16]   G.V.Hansson, R.I.G.Uhrberg, J.M.Nicholls [Surf. Sci.
         (Netherlands) vol.132 (1983) p.31]]
[17]   F.Houzay et al [Surf. Sci. (Netherlands) vol.132 (1983) p.40]
[18]   D.J.Chadi [Phys. Rev. B (USA) vol.26 (1982) p.4762]
[19]   K.C.Pandey [Physica B&C (Netherlands) vol.117/118 (1983) p.761]
[20]   J.E.Northrup, M.L.Cohen [Phys. Rev. Lett. (USA) vol.49 (1982)
         p.1349]
[21]   J.E.Northrup, M.L.Cohen [J. Vac. Sci. & Technol. (USA) vol.21
         (1982) p.333]
[22]   R.M.Tromp, L.Smit, J.F. van der Veen [Phys. Rev. Lett. (USA)
         vol.51 (1983) p.1672]
[23]   R.M.Tromp, L.Smit, J.F. van der Veen [Phys. Rev. B (USA) vol.30
         (1984) p.6235]
[24]   L.Smit, R.M.Tromp, J.F. van der Veen [Surf. Sci. (Netherlands)
         vol.163 nos.2/3 (1985) p.315-34]
[25]   F.J.Himpsel et al [Phys. Rev. B (USA) vol.30 (1984) p.2257]
[26]   R.Feder, W.Moench [Solid State Commun. (USA) vol.50 (1984) p.311]
[27]   M.A.Olmstead, N.M.Abil [Phys. Rev. Lett. (USA) vol.52 (1984)
         p.1148]
[28]   T.F.Heinz, M.M.T.Loy, W.A.Thompson [Phys. Rev. Lett. (USA) vol.54
         no.1 (1 Jan 1985) p.63-6]
[29]   P.Chiaradia, A.Cricenti, S.Selci, G.Chiarotti [Phys. Rev. Lett.
         (USA) vol.52 (1984) p.1145]
[30]   M.J.Cardillo [Phys. Rev. B (USA) vol.23 (1981) p.4279]
[31]   K.Higashiyama, S.Kono, H.Sakurai, T.Sagawa [Solid State Commun.
         (USA) vol.49 (1984) p.253]
[32]   M.Aono, Y.Hou, C.Oshima, Y.Ishizawa [Phys. Rev. Lett. (USA)
         vol.49 (1982) p.567]
[33]   L.C.Snyder [Surf. Sci. (Netherlands) vol.140 (1984) p.101]
[34]   J.J.Lander, J.Morrison [J. Chem. Phys. (USA) vol.37 (1962)
         p.720]
[35]   W.A.Harrison [Surf. Sci. (Netherlands) vol.55 (1976) p.1]
[36]   W.Moench [Surf. Sci. (Netherlands) vol.63 (1979) p.79]
[37]   S.Ino [Jpn. J. Appl. Phys. (Japan) vol.19 (1980) p.1277]
[38]   G.Binnig, H.Rohrer, Ch.Gerber, E.Weibel [Phys. Rev. Lett. (USA)
         vol.50 (1983) p.120]
[39]   T.Yamaguchi [Phys. Rev. B (USA) vol.30 (1984) p.1992]
[40]   T.Yamaguchi [Phys. Rev. B (USA) vol.31 no.8 (15 Apr 1984)
         p.5297-304]
[41]   T.Yamaguchi [Phys. Rev. B (USA) vol.32 no.4 (15 Aug 1985)
         p.2356-70]
[42]   F.J.Himpsel, I.P.Batra [J. Vac. Sci. & Technol. A (USA) vol.2
         (1984) p.952]
[43]   E.G.McRae, P.M.Petroff [Surf. Sci. (Netherlands) vol.147 nos.2/3
         (1984) p.385-95]
[44]   P.A.Bennett, L.C.Feldman, Y.Kuk, E.G.McRae, J.E.Rowe [Phys. Rev.
         B (USA) vol.28 (1983) p.3696]
[45]   K.Takayanagi, Y.Tanishiro, M.Takahashi, S.Takahashi [J. Vac. Sci.
         & Technol. A (USA) vol.3 no.3 pt.2 (May/June 1985) p.1502-6]
[46]   K.Takayanagi, Y.Tanishiro, S.Takahashi, M.Takahashi [Surf. Sci.
         (Netherlands) vol.164 nos.2/3 (1985) p.367-92]

[47] G-X.Qian, D.J.Chadi [Phys. Rev. B (USA) vol.35 no.3 (15 Jan 1987)
p.1288-93]

[48] G-X.Qian, D.J.Chadi [J. Vac. Sci. & Technol. A (USA) vol.5 (1987)
p.906]

[49] J.Tersoff [Phys. Rev. Lett. (USA) vol.56 no.6 (10 Feb 1986)
p.632-5]

[50] I.K.Robinson, W.K.Waskiewicz, P.H.Fuoss, J.B.Stark, P.A.Bennett
[Phys. Rev. B (USA) vol.33 no.10 (15 May 1986) p.7013-16]

[51] R.S.Becker, J.A.Golovchenko, E.G.McRae, B.S.Swartzentruber [Phys.
Rev. Lett. (USA) vol.55 no.19 (4 Nov 1985) p.2028-31]

[52] R.S.Becker, J.A.Golovchenko, D.R.Hamann, B.S.Swartzentruber [Phys.
Rev. Lett. (USA) vol.55 no.19 (4 Nov 1985) p.2032-4]

[53] R.J.Hamers, R.M.Tromp, J.E.Demuth [Phys. Rev. Lett. (USA) vol.56
no.18 (5 May 1986) p.1972-5]

[54] B.N.Dev, G.Materlik, F.Grey, R.L.Johnson, M.Clausnitzer [Phys.
Rev. Lett. (USA) vol.57 no.24 (15 Dec 1986) p.3058-61]

[55] J.E.Northrup [Phys. Rev. Lett. (USA) vol.57 no.1 (7 Jul 1986)
p.154-7]

[56] W.Telieps, E.Bauer [Surf. Sci. (Netherlands) vol.162 (1985)
p.163-8]

[57] J.M.Gibson, H.-J.Gossmann, J.C.Bean, R.T.Tung, L.C.Feldman [Phys.
Rev. Lett. (USA) vol.56 no.4 (27 Jan 1986) p.355-8]

[58] T.Sakamoto, G.Hashiguchi [Jpn. J. Appl. Phys. Part 2 (Japan)
vol.25 no.1 (Jan 1986) p.L78-80]

[59] R.D.Bringans, R.I.G.Uhrberg, M.A.Olmstead, R.Z.Bachrach [Phys.
Rev. B (USA) vol.34 no.10 (15 Nov 1986) p.7447-50]

[60] W.I.Wang [Appl. Phys. Lett. (USA) vol.44 no.12 (15 Jun 1984)
p.1149-51]

[61] J.J.Lander, J.Morrison [J. Chem. Phys. (USA) vol.37 (1962)
p.229]

[62] M.J.Cardillo, G.E.Becker [Phys. Rev. B (USA) vol.21 (1980)
p.1497]

[63] M.J.Cardillo, W.R.Lambert [Surf. Sci. (Netherlands) vol.168
(1986) p.724-33]

[64] J.Ihm, D.H.Lee, J.D.Joannopoulos, A.N.Berker [J. Vac. Sci. &
Technol. B (USA) vol.1 (1983) p.705]

[65] J.A.Martin, D.E.Savage, W.Moritz, M.G.Lagally [Phys. Rev. Lett.
(USA) vol.56 no.18 (5 May 1986) p.1936-9]

[66] J.C.Phillips [Surf. Sci. (Netherlands) vol.40 (1973) p.459]

[67] W.A.Harrison [Surf. Sci. (Netherlands) vol.55 (1976) p.1]

[68] T.D.Poppendieck, T.C.Ngoc, M.B.Webb [Surf. Sci. (Netherlands)
vol.75 (1978) p.287]

[69] R.E.Schlier, H.F.Farnsworth [J. Chem. Phys. (USA) vol.30 (1959)
p.917]

[70] M.Green, R.Seiwatz [J. Chem. Phys. (USA) vol.37 (1962) p.458]

[71] J.D.Levine [Surf. Sci. (Netherlands) vol.34 (1973) p.90]

[72] J.A.Appelbaum, D.R.Hamann [Surf. Sci. (Netherlands) vol.74
(1978) p.21]

[73] J.A.Appelbaum, G.A.Baraff, D.R.Hamann [Phys. Rev. B (USA) vol.14
(1976) p.588]

[74] D.J.Chadi [Phys. Rev. Lett. (USA) vol.43 (1979) p.43]

[75] D.J.Chadi [J. Vac. Sci. & Technol. (USA) vol.16 (1979) p.1290]

[76] R.Seiwatz [Surf. Sci. (Netherlands) vol.2 (1964) p.493]

[77] F.Jona, H.D.Shih, A.Ignatiev, D.W.Jepsen, P.M.Marcus [J. Phys.
C (GB) vol.10 (1977) p.L67]

[78] J.E.Rowe, H.Ibach [Phys. Rev. Lett. (USA) vol.32 (1974) p.421]

[79] F.J.Himspel, D.E.Eastman [J. Vac. Sci. & Technol. (USA) vol.16 (1979) p.1297]

[80] M.A.Bowen, J.D.Dow, R.E.Allen [Phys. Rev. B (USA) vol.26 (1982) p.7083]

[81] A.Mazur, J.Pollmann [Phys. Rev. B (USA) vol.26 (1982) p.7086]

[82] R.M.Tromp, R.J.Hamers, J.E.Demuth [Phys. Rev. Lett. (USA) vol.55 no.12 (16 Sep 1985) p.1303-6]

[83] H.Iwasaki, S.Nakamura [Surf. Sci. (Netherlands) vol.131 (1983) p.448]

[84] S.Maruno, H.Iwasaki, K.Horioka, S-T.Li, S.Nakamura [Phys. Rev. B (USA) vol.127 (1983) p.4110]

[85] H.H.Farrell, F.Stucki, J.Anderson, D.J.Frankel, G.J.Lapeyre, M.Levinson [Phys. Rev. B (USA) vol.30 (1984) p.721]

[86] R.M.Tromp, R.G.Smeenk, F.W.Saris [Phys. Rev. Lett. (USA) vol.46 (1981) p.939]

[87] M.Aono, Y.Hou, C.Oshima, Y.Ishizawa [Phys. Rev. Lett. (USA) vol.49 (1982) p.567]

[88] A.Ignatiev, F.Jona, M.Debe, D.E.Johnson, S.J.White, D.P.Woodruff [J. Phys. C (GB) vol.10 (1977) p.1109]

[89] W.S.Yang, F.Jona, P.M.Marcus [Phys. Rev. B (USA) vol.28 no.4 (15 Aug 1983) p.2049-59]

[90] C.B.Duke, A.Paton, B.W.Holland [Proc. 17th Int. Conf. on Physics of Semiconductors, Eds. D.J.Chadi, W.A.Harrison (Springer, Berlin 1984) p.59]

[91] T.Tabata, T.Aruga, Y.Murata [Surf. Sci. (Netherlands) vol.179 no.1 (1987) p.L63-70]

[92] O.L.Alerhand, E.J.Mele [Phys. Rev. B (USA) vol.35 no.11 (15 Apr 1987) p.5533-46]

[93] N.Inoue, Y.Tanishiro, K.Yagi [Jpn. J. Appl. Phys. Part 2 (Japan) vol.26 no.4 (Apr 1987) p.L293-5]

[94] T.Nakayama, Y.Tanishiro, K.Takayanagi [Jpn. J. Appl. Phys. Part 2 (Japan) vol.26 no.4 (Apr 1987) p.L280-2]

[95] D.E.Aspnes, J.Ihm [Phys. Rev. Lett. (USA) vol.57 no.24 (15 Dec 1986) p.3054-7]

[96] F.Jona [IBM J. Res. & Dev. (USA) vol.9 (1965) p.375]

[97] B.Z.Olshanetsky, A.A.Shklyaev [Surf. Sci. (Netherlands) vol.67 (1977) p.581]

[98] Y.Yamamoto, S.Ino, T.Ichikawa [Jpn. J. Appl. Phys. (Japan) Part 2 vol.25 no.4 (Apr 1986) p.L331-4]

CHAPTER 23

ETCHING RATES

23.1 ETCHING RATES OF CHEMICALLY ETCHED Si

by Y.H.Lee and M.R.Polcari

September 1987
EMIS Datareview RN=16119

Chemical etching includes wet etching, gas phase vapour etching and hot
molecular beam etching for which the etching mechanisms are basically
chemical in nature. The wet chemical etching is a consequence of
oxidation-reduction reactions followed by dissolution of the oxidation
products. The gas phase etching, both vapour and hot molecular beam
etchings, involves chemical reactions of highly reactive radicals
(normally halogen atoms) with the Si surface followed by desorption
of the etch products in the gas phase. Kern [1,2] has reviewed and
tabulated the chemical etching data until 1978. Tuck [3] also reviewed
the chemical polishing of semiconductors. Our emphasis will be on
recently published data as well as the technologically important wet
etching solutions frequently used in microelectronics and materials
characterisation. For convenience we subdivide the chemical etching
rates into isotropic wet etching (TABLE 1), anisotropic wet etching
(TABLE 2), preferential etching for defect delineations (TABLE 3) and
gas and vapour phase etching (TABLE 4).

The most popular etchant for isotropic wet etching is the $HF-HNO_3$
system. HNO_3 is the oxidation agent and HF is the complexant. The etch
rate is determined by the HNO_3 concentration in a high HF concentration
(the oxidation limited process), while it is proportional to the HF
content in a high HNO_3 etchant (the dissolution limited process).
Silicon etch rates in $HF-HNO_3$ are independent of dopants and doping
concentration, but very sensitive to temperature and agitation. Either
H_2O or CH_3COOH is normally used as the diluent. Excessive addition of
CH_3COOH can cause an erratic etch rate due to a small change in the
diluent. Addition of H_2O becomes critical, especially, in a high HF
solution. Buffered HF can also etch silicon even though etch rates are
very slow, around 0.04 nm/min [10].

Aqueous alkaline solutions exhibit anisotropic etching depending on
crystallographic orientation. Normally, etch rates decrease in the
order of (100), (110) and (111) plane-families, reflecting the density
of free bonds (1, 0.71 and 0.58 for (100), (110) and (111),
respectively). The oxidation rate goes in the reverse order. Thus,
the (111) surface instantly forms a hydrated oxide film, resulting in
a slower etch rate [11]. Also, the atomic packing density decreases in
the order of (111), (100) and (110), so that etch rates are faster on
(110) than on (100) plane [7]. Anisotropic etching eventually stops
when a slow etching plane is encountered. Hence, the etch depth is
normally determined by the intersection of the [111] planes,
particularly when a hole dimension is very small.

Preferential etching is a simple and quick method of investigating
defects such as dislocations, stacking faults, and swirls and
striations in crystalline silicon. The CrO_3-HF-H_2O system is an etchant
frequently used, in which the concentration ratio between CrO_3 and HF
controls the characteristics of dislocation delineation [20]. Etching
is sensitive to crystallographic orientations, temperature and
ultrasonic agitation. Sirtl etch works well on the (111) surface, while
Secco etch, Schimmel etch and Wright etch are for the (100) wafers.

Yang etch has been tested for (100), (110) and (111) surfaces. In preferential etching special care must be taken to avoid artifacts such as mounds formed by CrF3 precipitates [20]. The Wright etch [22] is very stable on room temperature storage for several weeks and always produces a smooth surface due to CH3COOH.

Gas phase vapour etching requires a very high temperature, usually above 1000 deg C. SF6 and HCl have been most extensively studied for in-situ etching prior to epitaxial silicon growth, in order to reduce dislocation densities in the epitaxial film. SF6 yields a poor etch selectivity over SiO2, unsuitable for the selective epitaxial growth. Both HCl and Cl2 gases are commonly used when the selectivity over SiO2 is required. Etch rates are less sensitive to temperature above 1000 deg C, but depend very strongly on partial pressure of the etch gas, flow rate of the carrier gas (usually H2) and reactor configuration. The HCl/H2 system results in anisotropic etching forming (111), (311) and (110) plane-family facets in a hole defined by either SiO2 or Si3N4 window [34,35]. The etched surface often becomes rough and forms etch pits, when the temperature is low, below 1000 deg C [32].

TABLE 1: isotropic etching

Etchant	Etch Rate (micron/min)	Temp. (deg C)	Remarks	Ref
7 HNO3 (69.5%): 3 HF (49.25%)	127	25		[4]
21 HNO3: 4 HF	13.8 8.39 6.17	25	w/large Si area w/SiO2 mask w/AZ mask	[5]
25 HNO3: 1 HF: 25 H2O	0.36 0.54 0.69	25	c-Si n+ poly Si poly Si	[6]
7 HNO3 (69%): 35 HF (49%): 8 H2O	40			[4]
75 HNO3: 8 HF: 17 CH3COOH	5	25	'Planar etch'	[7]
5 HNO3: 3 HF: 3 CH3COOH	50-75	25	'CP-4'	[8]
40 HNO3: 1 HF: 15 CH3COOH	0.15 (111) 0.20 (100)	25	'B-etch'	[7]

Etchant	Etch Rate (micron/min)	Temp. (deg C)	Remarks	Ref
3 HNO3: 1 HF: 10 CH3COOH	3	25	sensitive to type and density of dopants	[7]
27 HNO3: 27 HF: 46 CH3COOH	25	25		[5]
HF (48%)	0.3 A/min 0.7 A/min	25	c-Si a-Si	[9]
1 HF (48%): 4 H2O	0.47 A/min	25		[9]
1 HF (48%): 7NH4F (40%)	0.4 A/min	25	buffered HF	[10]

TABLE 2: anisotropic etching

Etchant	Etch Rate (micron/min)	Temp (deg C)	Remarks		Ref
44 w% KOH: 56 w% H2O	5.8 11.7 0.02	120	(100) (110) (111)		[11]
35 w% KOH: 65 w% H2O	2.2 neg.	85	(100) (111)		[5]
23.4 w% KOH: 63.3 w% H2O: 13.3 w% IPA	1.0 0.06	80	(100) (111)	sensitive to boron concn	[12]
4 m% C6H4(OH)2: 46.4 m% NH2(CH2)2NH2: 49.4 m% H2O	0.83 0.50 0.017	118	(100) (110) (111)	'Pyrocat' neg. etch on p++	[13]
45g C6H4(OH)2: 250cc NH2(CH2)2NH2: 120cc H2O	1.1 neg.	110	(100) (111)		[5]
50% N2H4: 50% H2O	2.0 neg.	100	(100) (111)	sensitive to temp. and dopant concn	[14,15]
1 m/l Cu(NO3)2: 4 m/l NH4F	0.185 0.117 0.012	22	(100) (110) (111)		[16]

TABLE 3: preferential etching for defect delineation

Etchant	Etch Rate (micron/min)	Temp. (deg C)	Remarks	Ref
1 HF: 3 HNO3: 10 CH3COOH	0.13 0.005	25	(100) 'Dash' etch (111) works on (100), (111)	[7,17]
0.3 CrO3: 0.36 HF (100%): 1 H2O	3.0	25	'Sirtl' etch works well on (111)	[18]
1 CrO3 (1 M): 2 HF: 1.5 H2O	1.7-3.2	25	'Shimmel' etch sensitive to doping density and agitation, works on (100)	[19]
1 CrO3 (1.5 M): 1 HF (49%)	1.5	25	'Yang' etch works on (100), (110), (111)	[20]
1 K2Cr2O7: 2 HF (49%)	1.5	25	'Secco' etch works on (100), needs agitation	[21]
60cc HF (49%): 30cc HNO3 (69%): 30cc CrO3 (5 M): 2gm Cu(NO3)2.3H2O: 60cc CH3COOH: 60cc H2O	1.0	25	'Wright' etch good for OSF, swirl, striation on (100)	[22]
36 HF (49%): 20 CH3COOH (glacial): 1-2 HNO3 (70%)	5-20	25	'Sopori' etch good for poly-Si	[23]

TABLE 4: gas and vapour phase etching

Etchant	Etch Rate (micron/min)	Temp (deg C)	Pressure (torr)	Flow (1/min)	Ref
F2	0.5	220	2	-	[24]
SF6	0.06 2	915 3000	1.6 X 10*-4 7.5 X 10*-3	0.0023 -	[25] [26]
C12	0.01	1500	0.01	-	[27]
CF3Br	0.01	1500	0.01	-	[27]
H2/SF6 (0.1%)	1	1060	1 atm	100 (H2)	[28]

Etchant	Etch Rate (micron/min)	Temp (deg C)	Pressure (torr)	Flow (l/min)	Ref
H2/HCl (5%)	1.48 (111)	1200	1 atm	-	[29]
	3.0 (110)				
	3.4 (100)				
	4.8 (111)	1280	1 atm	6 (H2)	[30]
	2.0 (100)	1150	1 atm	-	[29]
H2/HBr (2%)	1.0	1260	1 atm	6 (H2)	[30]
H2/H2S (0.55%)	15.1	1200	1 atm	100	[31]
H2/H2O (0.013%)	0.071	1200	1 atm	100	[31]
He/Cl2 (0.2%)	1.0	1100	1 atm	10	[32]
He(20% H2)/HI(1%) + HF(0.001%)	1.0	900	1 atm	10	[33]

REFERENCES
----

[1]   W.Kern [RCA Rev. (USA) vol.39 (1978) p.278]
[2]   W.Kern, C.A.Deckert [in 'Thin Film Processes', Eds J.L.Vossen,
       W.Kern (Academic Press, New York, 1978) ch.V-1]
[3]   B.Tuck [J. Mater. Sci. (GB) vol.10 (1975) p.321]
[4]   B.Schwartz, H.Robbins [J. Electrochem. Soc. (USA) vol.123 (1976)
       p.1903]
[5]   G.Kaminsky [J. Vac. Sci. & Technol. B (USA) vol.3 no.4 (1985)
       p.1015-24]
[6]   S.B.Felch, J.S.Sonico [Solid State Technol. (USA) vol.29 no.9
       (1986) p.70]
[7]   K.E.Bean [IEEE Trans. Electron Devices (USA) vol.ED-25 (1978)
       p.1185]
[8]   B.Schwartz, H.Robbins [J. Electrochem. Soc. (USA) vol.108 (1961)
       p.365]
[9]   S.M.Hu, D.R.Kerr [J. Electrochem. Soc. (USA) vol.114 (1967)
       p.414]
[10]   W.Hoffmeister [J. Appl. Radiat. & Isotopes vol.2 (1969) p.139]
[11]   D.L.Kendall [Annu. Rev. Mater. Sci. (USA) vol.9 (1979) p.373]
[12]   J.B.Price [Semiconductor Silicon-1973 Eds H.R.Huff,
       R.R.Burgess (Electrochem. Soc., Pennington, 1973) p.339]
[13]   E.Bassous [IEEE Trans. Electron Devices (USA) vol.ED-25 (1978)
       p.1178]
[14]   D.B.Lee [J. Appl. Phys. (USA) vol.40 (1969) p.4569]
[15]   M.J.Declercq, L.Gerzberg, J.D.Meindl [J. Electrochem. Soc. (USA)
       vol.122 (1975) p.545]
[16]   W.K.Zwicker, S.K.Kurtz [Semiconductor Silicon-1973 Eds
       H.R.Huff, R.Burgess (Electrochem. Soc., Pennington 1973) p.315]
[17]   W.C.Dash [J. Appl. Phys. (USA) vol.27 (1956) p.1193]
[18]   E.Sirtl, A.Adler [Z. Metallkd. (Germany) vol.52 (1961) p.529]
[19]   D.G.Schimmel [J. Electrochem. Soc. (USA) vol.126 (1979) p.479]
[20]   K.H.Yang [J. Electrochem. Soc. (USA) vol.131 (1984) p.1140]
[21]   F.Secco d'Aragona [J. Electrochem. Soc. (USA) vol.119 (1972)
       p.948]

[22] M.W.Jenkins [J. Electrochem. Soc. (USA) vol.124 (1977) p.948]
[23] B.L.Sopori [J. Electrochem. Soc. (USA) vol.131 (1984) p.667]
[24] M.Chen, V.J.Minkiewicz, K.Lee [J. Electrochem. Soc. (USA) vol.126 (1979) p.1946]
[25] K.Suzuki, K.Ninomiya, S.Nishimasu, O.Okada [Jpn. J. Appl. Phys. Part 2 (Japan) vol.25 (1986) p.L373]
[26] M.W.Geis, N.N.Efremow, S.W.Pang, A.C.Anderson [J. Vac. Sci. & Technol. B (USA) vol.5 (1987) p.363]
[27] M.W.Geis, N.N.Efremow, G.A.Lincoln [J. Vac. Sci. & Technol. B (USA) vol.4 no.1 (1986) p.315-7]
[28] L.J.Stinson, J.A.Howard, R.C.Neville [J. Electrochem. Soc. (USA) vol.123 (1976) p.551]
[29] K.E.Bean, P.S.Gleim [Proc. IEEE (USA) vol.57 (1969) p.1469]; Thin Solid Films (Switzerland) vol.83 (1981) p.173]
[30] L.V.Gregor, P.Balk, F.J.Campagna [IBM J. Res. Dev. (USA) vol.9 (1965) p.327]
[31] P.Rai-Choudhury, A.J.Noreika [J. Electrochem. Soc. (USA) vol.116 (1969) p.539]
[32] J.P.Dismukes, R.Ulmer [J. Electrochem. Soc. (USA) vol.118 (1971) p.634]
[33] E.R.Levin, J.P.Dismukes, M.D.Coutts [J. Electrochem. Soc. (USA) vol.118 (1971) p.1171]
[34] K.Sugawara [J. Electrochem. Soc. (USA) vol.118 (1971) p.110]
[35] M.Druminski, R.Gessner [J. Cryst. Growth (Netherlands) vol.31 (1975) p.312]

23.2 ETCHING RATES OF PLASMA ETCHED Si

by Y.H.Lee and M.R.Polcari

September 1987
EMIS Datareview RN=16120

Halogen gas plasmas are commonly used for plasma etching, with the etch
mechanism being predominantly chemical in nature. Etching involves four
basic steps: (a) generation and diffusion of radicals (b) chemisorption
of radicals on Si, (c) formation of product molecules and (d) desorption
of etch products and finally pumping away. Silicon etch rates are
limited by any of those four basic steps, depending on radical species
and plasma parameters. Plasma etching is loosely defined for the
situation where Si substrates sit on an electrically grounded electrode
and plasmas are generated at an opposite electrode or transported from
a nearby plasma generation chamber (down-stream etching). Under these
circumstances etch rates are directly proportional to RF power and gas
pressure, that is, a number density of radicals generated in plasmas.
In general the physical sputtering is fairly weak due to a high gas
pressure and low plasma floating potential the substrate sees. A
substantial etch rate can be obtained at a high power density (above
1 W/cm*2) and high pressure (above 0.1 torr). In this section we
also include DC discharge etching with no external bias voltage.

Either SF6 or NF3 etches silicon much faster than fluorocarbon gas or
chlorinated gas plasmas such as CF4, CF2Cl2 and CCl4. Fluorine atoms are
responsible for the spontaneous chemical etching and carbon tends to
passivate the surface consequently slowing down etch rates [31]. Also,
fluorocarbon plasmas generate the precursor of polymer formation (for
example, CF and CF2 [32]) which may lead to deposition of a polymeric
layer rather than etching. CHF3, CClF3, C2F6 and C3F8 belong to this
category. The C/F ratio determines the effectiveness of etching and
deposition [31]. Even in plasma etching which is generally isotropic,
the sidewall passivation by carbon, polymer or other foreign species
can lead to anisotropic etch profiles. Zarowin [33] has shown that a
degree of the anisotropy is determined by the E/p ratio, where E refers
to the electric field strength across the plasma sheath and p is a gas
pressure. This means that plasma etching also involves ion bombardment
effects to some extent. This is presumably due to the plasma floating
potential which could be very large, especially at a high RF power.

Plasma etching is inherently more selective because it relies more
heavily on chemistry than physical effect. Hirata [12] observed that the
higher the Cl/F ratio, the better the selectivity that can be obtained
in the Si/SiO2 system. Chlorine tends to more effectively etch Si than
SiO2. Also, Hirata [12] showed a strong indication of preferential
etching along the major crystallographic orientations in chlorine based
gas plasmas.

Etch rates are sensitive to the electrode materials, which may consume
the etching species (loading effect) consequently reducing etch rates.
Flamm [34] measured a temperature dependence of etch rate in a
downstream F2 plasma and obtained a thermal activation energy of
2.5 kcal/mol (0.108 eV) for the spontaneous chemical etching
mechanisms of Si. Lee [35] obtained the same activation energy for
CF4/O2 plasma etching, independent of the structural phase (amorphous,

polycrystalline and crystalline) or the electrical properties (n-type, p-type and doping density). Etch rates increase with increasing doping density in heavily doped p-type Si [35,36]. This doping effect is substantially larger in chlorine based gas plasmas [3].

In RF powered plasmas the frequency is usually 13.56 MHz and otherwise it is specified in the Remarks column of TABLE 1. Also, substrate temperature, one of the important parameters, is omitted from the table, because the majority of etch data were obtained from room temperature experiments.

TABLE 1: plasma etching rates of Si

Etchant	Etch Rate (nm/min)	Power (W/cm*2)	Press. (torr)	Flow (sccm)	Remarks	Ref
CF4	50	0.38	0.6	30	13.5 MHz	[1,2]
C2F6	3	0.32	0.35	200		[3]
C3F8	<40	<0.5	0.3	-	DC, 1 MHz	[1,4]
C6F14	800	5	1	100	50 kHz	[5]
F2	200	(20 W)	0.5	45	downstream	[6]
NF3	1800	0.3	0.15	-	n+Si, 0.38 MHz	[7]
SF6	1750	0.36	0.2	50		[8,9]
	3000	(2 W/cm*3)	0.007	1.5	helicon, 7MHz	[10]
C12	50	0.6	0.3	35		[11]
	600	0.6	1	35	1 MHz	[11]
CC14	45	0.19	0.10	25		[12]
	95	0.38	0.25	40	0.37 MHz	[13]
CF3C1	10-100	0.16	0.35	200		[3,12,14]
CF2C12	130	0.38	0.18	50		[12,15]
CFC13	43	0.19	0.1	25		[12]
C2C1F5	25	0.36	0.15	50		[9]
H2	25-50	(10 W/cm*3)	0.2	-	30 MHz	[16]
CHF3	100				polymer	[1,5]
CBrF3	25-75	0.30	0.5	25-150		[17]
SO2F2	560	1	0.1	-		[18]
CF4/H2 (>10%)	<30	0.16	0.17	20	polymer	[8]

Etchant	Etch Rate (nm/min)	Power (W/cm*2)	Press. (torr)	Flow (sccm)	Remarks	Ref
CF4/O2 (16%)	450	0.16	0.35	200	100 deg C	[19]
(50%)	600	(680 W)	0.12	–	2.45 GHz	[20]
CF4/He (50%)	6	(50 mA)	0.25	55	DC, hollow	[21]
CF4/N2 (20%)	60	(40 W)	0.01	100	barrel	[22]
CF4/N2O (20%)	140	(40 W)	0.01	100	barrel	[22]
CF4/CF3Cl (6%)/ O2 (1%)	1000	5	3.5	–		[23]
C2F6/Cl2 (20%)	120	0.32	0.35	200		[3,14]
C2F6/CF3Cl (50%)	98	0.44	0.40	200		[3]
C6F14/H2 (20%)	5	5	1.0	100	50 kHz	[5]
C6F14/N2 (50%)	500	5	1	100	50 kHz	[5]
CHF3/N2 (20%)	180	5	1.0	100	50 kHz	[5]
F2/He (80%)	200	(250 V)	0.5	–		[24]
NF3/He (50%)	500	(50 mA)	0.25	55	DC, hollow	[21]
NF3/O2 (25%)	2000	(80 mA)	0.10	–	DC, hollow	[25]
SF6/H2 (20%)	280	0.16	0.20	20		[8]
	>1000	1.8	1	27	27 MHz	[26]
(65%)	800	0.5	0.15	450		[27]
SF6/Ar (99%)	100	0.16	0.20	20		[8]
SiF4/O2 (20%)	30	(150 W)	1.0	--		[28]
CCl4/O2 (20%)	100	0.38	0.25	40	10.4 MHz	[13]
CCl4/Cl2 (40%)	55	(500 W)	0.075			[29]
CF2Cl2/ C2H6 (23%)	<10	0.38	0.18	65		[15]
CF2Cl2/H2 (50%)	0	0.19	0.18	50		[15]
CF3Br/He (30%)	100	0.28	0.3	200		[3]
CF3Br/SF6 (50%)	510	0.36	0.15	50		[9]
C2ClF5/SF6 (50%)	510	0.36	0.15	50		[9]
C3F8/CF4 (6%)/ Ar (80%)	50	3	0.23	--	0.1 MHz	[30]

REFERENCES
----

[1]   A.R.Reinberg [in 'Etching for Pattern Definition' Eds H.G.Hughes,
        M.J.Rand (Electrochem. Soc., Princeton, 1976) p.91]
[2]   H.Mader [in 'Plasma Processing' Eds R.G.Frieser, C.J.Mogab
        (Electrochem. Soc., Pennington, 1981) p.125]
[3]   C.J.Mogab, H.J.Levinstein [J. Vac. Sci. & Technol. (USA) vol.17
        (1980) p.721]
[4]   R.A.H.Heinecke [Solid State Electron. (GB) vol.18 (1975)
        p.1146]
[5]   K.M.Eisele [J. Vac. Sci. & Technol. B (USA) vol.4 (1986)
        p.1227]
[6]   V.M.Donnelly, D.L.Flamm, J.A.Mucha [J. Appl. Phys. (USA) vol.52
        (1981) p.3633]
[7]   D.H.Bower [J. Electrochem. Soc. (USA) vol.129 (1982) p.795]
[8]   K.M.Eisele [J. Electrochem. Soc. (USA) vol.128 (1981) p.123]
[9]   J.P.McYitte, C.Gonzalez [in 'Plasma Processing - 1985' Eds
        G.S.Mathad, G.C.Schwartz, G.Smolinsky (Electrochem. Soc.,
        Pennington, 1985) p.552]
[10]   R.W.Boswell, D.Henry [Appl. Phys. Lett. (USA) vol.47 (1985)
        p.1095]
[11]   V.M.Donnelly, D.Flamm, R.H.Bruce [J. Appl. Phys. (USA) vol.58
        (1985) p.2135]
[12]   K.Hirata, Y.Ozaki, M.Oda, M.Kimizuka [IEEE Trans. Electron
        Devices (USA) vol.ED-28 (1981) p.1323]
[13]   C.S.Korman [J. Vac. Sci. & Technol. (USA) vol.20 (1982) p.476]
[14]   A.C.Adams, C.D.Capio [J. Electrochem. Soc. (USA) vol. 128 (1981)
        p.366]
[15]   M.Kamizuka, K.Hirata [J. Vac. Sci. & Technol. B (USA) vol.3
        (1985) p.16]
[16]   R.P.H.Chang, C.C.Chang, S.Darack [J. Vac. Sci. & Technol. (USA)
        vol.20 (1982) p.45]
[17]   D.L.Flamm, R.L.Cowan, J.A.Golovchenko [J. Vac. Sci. & Technol.
        (USA) vol.17 (1980) p.1341]
[18]   K.M.Eisele [in 'Plasma Processing' Eds J.Dieleman, R.G.Frieser,
        G.Mathad (Electrochem. Soc., Pennington, 1982) p.146]
[19]   C.J.Mogab, A.C.Adams, D.L.Flamm [J. Appl. Phys. (USA) vol.49
        (1978) p.3796]
[20]   Y.Horiike, M.Shibagaki [Semiconductor Silicon Eds H.R.Huff,
        E.Sirtle (Electrochem. Soc., Princeton 1977) p.1071]
[21]   N.J.Ianno, K.E.Greenberg, J.T.Verdeyen [J. Electrochem. Soc.
        (USA) vol.128 (1981) p.2174]
[22]   Y.Tzeng, T.H.Lin [J. Electrochem. Soc. (USA) vol.133 (1986)
        p.1443]
[23]   L.Baldi, D.Beardo, E.Landi [J. Electrochem. Soc. (USA) vol.133
        (1986) p.2202]
[24]   M.J.Vasile [J. Appl. Phys. (USA) vol.51 (1980) p.2510]
[25]   K.E.Greenberg, J.T.Verdeyen [J. Appl. Phys. (USA) vol.57 (1985)
        p.1596]
[26]   R.d'Agostino, D.L.Flamm [J. Appl. Phys. (USA) vol.52 (1981)
        p.162]
[27]   R.W.Light, H.B.Bell [J. Electrochem. Soc. (USA) vol.130 (1983)
        p.1567]
[28]   H.Boyd, M.S.Tang [Solid State Technol. (USA) vol.22 no.4 (1979)
        p.133-8]

[29] H.Kalter, A.Meyer, R.A.M.Walters [in 'Plasma Processing' Eds
J.Dieleman, R.G.Frieser, G.Mathad (Electrochem. Soc., Pennington,
1982) p.154]

[30] A.R.Reinberg, J.Dalle-Ave, G.Steinberg, R.Bruce [in 'Plasma
Processing' Eds J.Dieleman, R.G.Freiser, G.Mathad (Electrochem.
Soc., Pennington, 1982) p.198]

[31] J.W.Coburn [Plasma Chem. & Plasma Process. (USA) vol.2 (1982)
p.1]

[32] M.M.Millard, E.Kay [J. Electrochem. Soc. (USA) vol.129 (1982)
p.160]

[33] C.B.Zarowin [J. Vac. Sci. & Technol. A (USA) vol.2 (1984)
p.1537]

[34] D.L.Flamm, V.W.Donnelly [Plasma Chem. & Plasma Process. (USA)
vol.1 (1981) p.317]

[35] Y.H.Lee, M.M.Chen [J. Vac. Sci. & Technol. B (USA) vol.4 no.2
(1986) p.468]

[36] L.Baldi, D.Beardo [J. Appl. Phys. (USA) vol.57 (1985) p.2225]

23.3 ETCHING RATES OF ION BEAM MILLED AND SPUTTER ETCHED Si

by Y.H.Lee and M.R.Polcari

September 1987
EMIS Datareview RN=16121

Ion beam milling and plasma sputtering with chemically inert gas ions
are based on the physical sputtering, which is caused by a momentum
transfer from an energetic incident particle to a Si lattice atom.
Therefore etch rates are very sensitive to the incident angle of the
incoming ions with respect to the surface normal. Sigmund [1] found
that the angular dependence of the sputtering yield (i.e. the number
of atoms removed per ion) is proportional to 1/(cos theta) for ions
lighter than Si, where theta is the incident angle from the surface
normal. For heavier ions a faster variation with angle is expected.
When the ion beam approaches parallel to the surface (theta = 90 deg),
the ion will be simply reflected by the surface without sputtering.
Thus the yield reaches the maximum at theta = 60-80 deg and then
falls off quickly at a larger theta. Also the yield is proportional to
the mass and energy of incident ions. Andersen [2] reviewed the
sputtering yield of Si. Wehner [10,11] measured the sputtering yield of
Si by DC discharge of He, Ne, Ar, Kr and Xe.

Okajima [6] found a linear dependence of the ion milling rate on the
beam current. Cantagrel [12] observed a slowdown of the sputtering
rate at a high O2 partial pressure of 10*-5 torr due to surface
modification. A similar effect can be seen when H2O content is very
high in the residual background of a vacuum system. For micromachining
applications, ion beam sputtering results in many technical problems
such as faceting, trenching, redeposition, surface roughening and
radiation damage [5]. The etch rate selectivity is generally poor.
Thus, mask erosion and resist cross-linking leads to the difficulty of
choosing a mask material adequate for pattern delineation.

Plasma sputtering is normally performed with the specimen sitting at the
cathode. Thus the etching behaviour is the same as that for ion milling.
Since a medium pressure (10-20 mtorr) is usually required to
maintain a stable plasma, redeposition of etched material creates
serious problems for pattern definitions. However, plasma sputtering has
been widely used for thin film deposition [13]. The tabulated etch rates
were derived from the sputtering yield assuming a linear dependence of
etch rates on ion-current densities. The yield of 1 atom/ion corresponds
to 75.2 nm/min at 1 mA/cm*2.

TABLE 1: ion milling

Ion	Etch Rate (nm/min per mA/cm*2)	Yield (atoms/ion)	Energy (keV)	Angle of incidence (deg)	Ref
H	0.6	0.003	1		[2]
D	1.5	0.02	1		[2]
He	7.5	0.1	1		[2]
N	37.6	0.5	5		[2]
O	120	1.6	8	35	[2]
Al	75	1.0	5		[2]
Ne	34.6	0.46	0.2	0	[3]
	45.9	0.61	0.5		
	62.4	0.83	1		
	59.4	0.79	10		
Ar	24.8	0.33	0.2	0	[3,4]
	51.1	0.68	0.5		
	70	0.93	1		
	119	1.58	10		
	22-50	--	0.5	0	[5]
	36-75	--	1	0	
	75	--	10	0	[6]
	90	1.2	3	0	[7]
	210	2.8	3	45	[7]
	353	4.7	3	60	[7]
Kr	22	0.29	0.2	0	[3]
	53	0.70	0.5		[3]
	68.4	0.91	1		[3]
	158	2.10	10		[3]
Xe	21	0.28	0.2	0	[3,4]
	60	0.80	0.5		[3,4]
	76	1.01	1		[3,4]
	188	2.5	10		[3,4]

TABLE 2: plasma sputtering

Etchant	Etch Rate (nm/min)	Power (W/cm*2)	Pressure (Torr)	Freq. (MHz)	Ref
Ar	13.8	1.3	0.02	13.5	[8]
	1	0.5	0.01	13.5	[9]
	4	0.5	0.03	13.5	[9]

REFERENCES
----

[1]   P.Sigmund [Phys. Rev. (USA) vol.184 (1969) p.383]
[2]   H.H.Andersen, H.L.Bay [in 'Sputtering by Particle Bombardment I'
        Ed. R.Behrisch (Springer-Verlag, Berlin, 1981) ch.4]
[3]   P.C.Zalm [J. Appl. Phys. (USA) vol.54 (1983) p.2660]
[4]   P.Blank, K.Wittmaack [J. Appl. Phys. (USA) vol.50 (1979) p.1519]
[5]   C.M.Mellian-Smith [J. Vac. Sci. & Technol. (USA) vol.13 (1976)
        p.1008]
[6]   Y.Okajima [J. Appl. Phys. (USA) vol.51 (1980) p.715]
[7]   S.T.Kang, R.Shimizu, T.Okutani [Jpn. J. Appl. Phys. (Japan)
        vol.18 (1979) p.1717]
[8]   N.Hosokawa, R.Matsuzaki, T.Asawaki [Jpn. J. Appl. Phys. Suppl.
        (Japan) no.2 pt.1 (1974) p.435]
[9]   H.W.Lehmann, R.Widmer [Appl. Phys. Lett. (USA) vol.32 (1978)
        p.163]
[10]   N.Laegreid, G.K.Wehner [J. Appl. Phys. (USA) vol.32 (1961)
        p.365]
[11]   D.Rosenberg, G.K.Wehner [J. Appl. Phys. (USA) vol.33 (1962)
        p.1842]
[12]   M.Cantagrel, M.Marchal [J. Mater. Sci. (GB) vol.8 (1973) p.1711]
[13]   J.L.Vossen, J.J.Cuomo [in 'Thin Film Processes' Eds J.L.Vossen,
        W.Kern (Academic Press, New York, 1978) ch.II-1 p.12]

23.4 ETCHING RATES OF REACTIVE ION AND REACTIVE ION BEAM ETCHED Si

by Y.H.Lee and M.R.Polcari

November 1987
EMIS Datareview RN=16122

Reactive ion etching (or reactive sputter etching) is defined for a
plasma etching condition where the substrate sits at the RF powered
electrode (cathode). Thus the specimen always receives ion
bombardments primarily due to the self-biased DC potential across the
plasma sheath which can be as high as 1000 V in some cases. Those
energetic ions of either inert or reactive species enhance etching along
the impinging direction, consequently resulting in anisotropic
directional etching regardless of crystallographic orientation of
silicon. This ability of directionality control has attracted a wide
application of reactive ion etching especially in microelectronic
fabrications. Directional etching can also be achieved by reactive ion
beam etching and ion beam assisted chemical etching, both of which
directly utilise the ion gun as a method of accelerating ions. Drawbacks
of those etching techniques are radiation damage by energetic ions and a
relatively poor selectivity of etching between two different materials.

Reactive ion etching rates are normally proportional to RF power (more
often to the self-bias voltage) and gas pressure (number density of
reactive radicals). The substrate temperature greatly influences etch
rates mostly through the isotropic chemical etching component which
partially contributes to total etch rate of reactive ion etching. Also,
etch rates are very sensitive to erosive materials, if any, on the
electrode [2,3]. The silicon loading effect in reactive ion etching
is much higher than that in plasma etching [22]. The relative magnitude
in etch rate between the directional ion assisted etching and the
isotropic chemical etching determines a degree of anisotropy, i.e., an
extent of etch undercuts on the sidewall. In CF_4 plasma, addition of Cl_2
improves the selectivity of Si over SiO_2 [23], while addition of H_2
etches SiO_2 selectively over Si [21]. Further miniaturisation of silicon
devices demands more stringent etching requirements, which in turn has
led to investigation of more Cl or Br based gases and also new plasma
excitation techniques such as magnetron [4,18,25], electron cyclotron
resonance [5], multipole discharge [28] and helicon discharge [43]. The
fundamental question on the physico-chemical origins of the ion assisted
etching mechanism still remains to be unravelled [44,45,46], although
its existence has been experimentally confirmed in a simulated UHV
experiment [47] and also in an actual plasma reactor [22]. Winters [44]
has reviewed the basic mechanisms of reactive ion etching in silicon.
Reactive ion etching produces physical damage within 3-5 nm from the
surface of crystalline silicon as radiation damage by a low energy ion
implantation and also introduces the impurities such as H, C and F deep
into the bulk [48,49]. Reactive ion etching develops extrinsic
dislocation loops in the Si lattice [50] and some defects are found as
deep as 200 nm from the surface [51].

Reactive ion beam etching utilises reactive ions accelerated by an
extraction voltage which usually runs below 1.0 keV. Chemical
sputtering, i.e., chemically enhanced physical sputtering, is
responsible for etching, so that those problems such as faceting,
trenching and redeposition in ion milling (see [67]) still remain in
reactive ion beam etching, limiting its practical applications. Etch

rates are relatively low because the ion beam system has to operate in a low pressure around 10*-4 torr in order to maintain a long mean free path. The etch selectivity of Si over SiO2 is usually below 10 [59]. Chinn [60] has mass-analysed Cl2 and CCl4 ion beams and observed C+, Cl+, CCl+, Cl2+, CCl2+ and CCl3+ in the CCl4 beam. Thus etch rates widely fluctuate from one experiment to another depending on pressure and configuration of the ion source, when the ion beam was a mixture of many ion species. A clear example was observed by Mayer [55] and Harper [56] in the CF4 ion beam with no mass analysis. Etch rates are linearly proportional to the beam current and also increase with the incident ion energy below a few keV.

Tachi [52-54] measured the chemical sputtering yield with a mass analysed beam of reactive gases in UHV system and observed that a carbon beam leads to deposition (a negative yield) instead of etching and threshold energy of deposition varies with the F/C ratio of the feed gas. Because of the complexity of many species involved in reactive ion etching, many fundamental studies of understanding the etching mechanism rely on ion beam assisted chemical etching [62,65,66], where an inert gas ion beam bombards the specimen surface along with exposure to the neutral reactive gases such as Cl2 and XeF2. In ion beam assisted chemical etching, etch rates are not only sensitive to the beam energy and current, but also to the flux of the reactive molecules landing on the silicon surface. In reactive ion beam etchings, the etching yield (in Si atoms per ion) is usually given in the literature. Thus etch rates in TABLES 2 and 3 were derived from the yield for an ion current density of 1 mA/cm*2, i.e., the etch yield of 1 atom/ion corresponds to 75.2 nm/min at 1 mA/cm*2.

TABLE 1: reactive ion etching

Etchant	Etch Rate (nm/min)	Power (W/cm*2)	Freq. (MHz)	Press. (torr)	Flow (sccm)	Remarks	Ref
CF4	90	1.3	13.5	0.02	--		[1]
	40	0.3	13.5	0.03	25	Si-cathode	[2]
	240	1.67	13.5	0.05	43	quartz cathode	[3]
	180	1.0	13.5	0.01	--	magnetron	[4]
	20	(60 W)	2450	0.0005	--	ECR w/0.8 MHz RF bias	[5]
C2F6	14	0.64	13.5	0.03	25	graphite c.	[6]
	10	(60 W)	2450	0.0005	--	ECR,150 V bias	[5]
C3F8	2.5	0.1	13.5	0.02	25		[7]
	10	(60 W)	2450	0.0005	--	ECR, 150 V bias	[5]
C4F8	3	(60 W)	2450	0.0005	--	ECR w/RF bias	[5]
C4F16	9	(60 W)	2450	0.0005	--	ECR w/RF bias	[5]
NF3	300-500	0.2-0.6	13.5	0.05	20		[8]

Etchant	Etch Rate (nm/min)	Power (W/cm*2)	Freq. (MHz)	Press. (torr)	Flow (sccm)	Remarks	Ref
SF6	520	0.16	13.5	0.20	20		[9]
SiF4	10	0.64	13.5	0.03	10		[10]
C12	400	0.67	13.5	0.01	10		[11]
HC1	450-600	2.2	13.5	1.0	600		[12]
BC13	14	0.2	13.5	0.075	90		[13]
CC14	240	0.67	13.5	0.01	10		[11]
SiC14	100	0.30	13.5	0.06	--		[14]
CF3C1	5	(150 W)	13.5	0.20	--		[15]
CF2C12	220	1.3	13.5	0.02	--		[1]
	40	0.27	13.5	0.025	5	Si loaded	[16]
CFC13	167	1.3	13.5	0.02	--		[1]
CHF3	7	0.37	13.5	0.03	30		[17]
	100	1.6	13.5	0.05	--	magnetron	[18]
CHC1F2	143	1.3	13.5	0.02	--		[1]
CHC12F	41	1.3	13.5	0.02	--		[1]
(CC12F)2	128	1.3	13.5	0.02	--		[1]
C2HC13	33	1.3	13.5	0.02	--		[1]
CC13F3	202	1.3	13.5	0.02	--		[1]
(CBrF2)2	185	1.3	13.5	0.02	--		[1]
SO2F2	349	1.0	13.5	0.06	--		[19]
PF5	350	1.0	13.5	0.06	--		[19]
CBrF3	100	0.48	13.5	0.03	15	SiO2 cathode	[20]
CF4/H2 (40%)	<1	0.26	13.5	0.035	28		[21]
CF4/O2 (20%)	310	0.48	13.5	0.03	50		[22]
	300	1.0	13.5	0.01	--	magnetron	[4]
CF4/C12 (60%)	45	0.2	13.5	0.04	30		[23]
CF4/I2 (20%)	53	0.32	13.5	0.02	15		[20]

Etchant	Etch Rate (nm/min)	Power (W/cm*2)	Freq. (MHz)	Press. (torr)	Flow (sccm)	Remarks	Ref
CF4/C2F4 (20%)	2	1.0	13.5	0.02	600		[24]
CF4/C2H4 (10%)	10	0.64	13.5	0.02	28		[6]
NF3/HC1 (25%)	400	0.2	13.5	0.05	20		[8]
NF3/CC14 (25%)	40	0.2	13.5	0.05	20		[8]
NF3/C12 (50%)	500	2.0	13.5	0.01	80	magnetron	[25]
NF3/Ar (60%)	200	0.27	13.5	0.06	10		[26]
NF3/N2 (60%)	180	0.27	13.5	0.06	10		[26]
SF6/Ar (50%)	1600	3.0	13.5	0.01	80	magnetron	[25]
SF6/H2 (20%)	370	0.16	13.5	0.20	20		[9]
SF6/N2 (50%)	100	0.5	13.5	0.015	--		[27]
SF6/O2 (20%)	650	0.16	13.5	0.20	20		[9,10]
	1500	2	13.5	0.01	80	magnetron	[25]
	1100	(2 A)	DC	0.002	--	multipole	[28]
SF6/C12 (10%)	840	0.55	13.5	0.23	28		[29]
SF6/HC1 (25%)	170	0.2	13.5	0.05	20		[8]
SF6/CC14 (20%)	50	0.09	13.5	0.07	--		[30]
SF6/CFC13 (20%)	540	0.55	13.5	0.23	28		[29]
SF6/CF2C12 (50%)	900	3	13.5	0.01	80	magnetron	[25]
SiF4/C12 (50%)	50	0.64	13.5	0.03	10		[10]
SiC14/He (50%)	70	0.34	13.5	0.06	100		[31]

Etchant	Etch Rate (nm/min)	Power (W/cm*2)	Freq. (MHz)	Press. (torr)	Flow (sccm)	Remarks	Ref
SiC14/ He (50%)/ C12 (25%)	100	0.34	13.5	0.06	100		[31]
SiC14/O2 (50%)	110	0.30	13.5	0.02	3		[32, 33]
C12/He (50%)	40	0.34	13.5	0.06	100		[31]
C12/Ar (96%)	140	0.16	13.5	0.09	50		[34]
C12/H2(25%) (43%)	320 800	1.6 1.0	13.5 13.5	0.2 0.05	130 50	magnetron	[35] [36]
C12/H2(2%) /CH4 (4%)	320	1.6	13.5	0.2	100		[35]
C12/CHC13 (45%)	400	1.6	13.5	0.2	105		[35]
BC13/C12 (10%)	33	0.2	13.5	0.075	90		[13]
CC14/O2 (25%)	300	0.69	13.5	---	200		[37]
C2F6/C2H4 (10%)	9	0.64	13.5	0.03	28		[6]
C2F6/He (69%)	250	1.25	13.5	2	58		[38]
CF2C12/ O2 (20%)	500 40	1.3 0.45	13.5 13.5	0.02 0.03	-- 20		[1] [39]
CF2C12/ C12 (50%)	300	2	13.5	0.01	80	magnetron	[25]
CF2C12/ NH3 (30%)	105	0.3	13.5	0.07	63		[40]
CHF3/NH3 (4%)	3	0.5	13.5	0.03	12		[41]
CHF3/O2 (8%)	20	0.25	13.5	0.03	30		[42]
CHF3/CO2 (8%)	5	0.25	13.5	0.03	30		[42]
CHF3/H2 (20%)	100	1.5	13.5	0.05	40	magnetron	[36]

TABLE 2: reactive ion beam etching

Etchant	Etch Rate (nm/min per mA/cm*2)	Yield (atoms/ ion)	Energy (keV)	Pressure (torr)	Remarks	Ref
F+	45	0.6	0.5	10*-8	100 micro-A	[52]
	72	0.96	1	10*-8		[52]
Cl+	83	1.1	0.5	10*-8	100 micro-A	[52]
	98	1.3	1	10*-8	100 micro-A	[52]
Br+	92	1.22	0.5	10*-8	100 micro-A	[52]
	143	1.9	1	10*-8	100 micro-A	[52]
CF+	--	-0.7	0.5	10*-8	deposition	[53,
	23	0.3	1.0	10*-8	I=40 micro-A	54]
CF2+	23	0.3	0.5	10*-8		[53,
	38	0.5	1.0	10*-8		54]
CF3+	45	0.6	0.5	10*-8		[53,
	90	1.2	1.0	10*-8		54]
BF+	--	0	0.5	10*-8	deposition	[53,
	23	0.3	1.0	10*-8	10-100 micro-A	54]
BF2+	23	0.3	0.5	10*-8		[53,
	53	0.7	1	10*-8		54]
PF+	30	0.4	0.5	10*-8		[53,
	68	0.9	1	10*-8		54]
SF+	23	0.3	0.5	10*-8		[53,
	53	0.7	1	10*-8		54]
P+	19	0.25	0.5	10*-8		[53,
	60	0.8	1	10*-8		54]
S+	53	0.7	0.5	10*-8		[53,
	75	1.0	1.0	10*-8		54]
CF4)*	15	0.20	0.5	10*-4	mass-	[55]
	20	0.27	1.0		unresolved	[55]
	60	0.8	1			[56]
	18	--	3	8 X 10*-4		[57]

Etchant	Etch Rate (nm/min per mA/cm*2)	Yield (atoms/ ion)	Energy (keV)	Pressure (torr)	Remarks	Ref
C2F6)*	25-60	0.3-0.8	1	10*-4	mass-	[6,58]
	21	--	3	8 X 10*-4	unresolved	[57]
C3F8)*	1	--	3	8 X 10*-4	mass-unresolved	[57]
SF6)*	39	0.52	3	8 X 10*-4	mass-unresolved	[57]
Cl2)*	90	1.25	0.5	10*-4	mass-	[55]
	130	1.80	1	10*-4	unresolved	[55]
	150	2	1	5 X 10*-4		[59]
CCl4)*	90	1.2	1	5 X 10*-4	mass-unresolved	[59]
XeF2/Ar)*	128	1.7	0.5	10*-4	mass-unresolved	[60]
Cl2/Ar)*	117	1.55	0.45	4 X 10*-4	mass-unresolved	[61]

TABLE 3: ion beam assisted chemical etching

Etchant (ion/gas)	Etch rate (nm/min per mA/m*2)	Yield (atoms/ ion)	Energy (keV)	Mol. flux (molecules/ cm*2/sec)	Remarks	Ref
He+/XeF2	880	11.7	1	3 X 10*16		[62]
He+/Cl2	71	0.95	1	3 X 10*16		[62]
Ne+/XeF2	1572	20.9	1	3 X 10*16		[62]
Ne+/Cl2	202	2.68	1	3 X 10*16		[62]
Ar+/XeF2	128	1.70	0.5	1.5 X 10*16		[60]
	191	2.54	1	1.5 X 10*16		[60]
	2106	28	1	3 X 10*16		[62]
Ar+/Cl2	338	4.5	1	3 X 10*16		[62, 64]
	110	1.46	0.35	4.6 X 10*16		[63]
	406	5.4	1	1.2 X 10*16		[59]
	180	2.4	0.5	1.1 X 10*16		[59]
	150	2.0	1	2.1 X 10*16		[61]
Xe+/Cl2	526	7.0	1	1.0 X 10*16		[59]
Kr+/Cl2	421	5.6	1	10*16		[59]
Cl2)*/Cl2	526	7.0	1	10*16	mass-	[59]
	135	1.8	1	2.3 X 10*16	unresolved	[55]
CCl4)*/Cl2	414	5.5	1	2.3 X 10*16	mass-unresolved	[59]

REFERENCES

----

[1] N.Hosokawa, R.Matsuzaki, T.Asawaki [Jpn. J. Appl. Phys. (Japan) suppl.2 pt.1 (1974) p.435]

[2] J.A.Bondur [J. Vac. Sci. & Technol. (USA) vol.13 (1976) p.1023]

[3] J.L.Mauer, J.S.Logan, L.B.Zielinski, G.C.Schwartz [J. Vac. Sci. & Technol. (USA) vol.15 (1978) p.1734]

[4] I.Lin, D.C.Hinson, W.H.Class, R.L.Sandstrom [Appl. Phys. Lett. (USA) vol.44 (1984) p.185]

[5] K.Suzuki, K.Ninomiya, S.Nishimatsu, S.Okudaira [J. Vac. Sci. & Technol. B (USA) vol.3 (1985) p.1025]

[6] S.Matsuo [J. Vac. Sci. & Technol. (USA) vol.17 (1980) p.587]

[7] M.-M.Chen, Y.H.Lee [in 'Plasma Processing' v.83-10 Eds G.Mathad, G.Schwartz, G.Smolinsky (Electrochem. Soc., Pennington, 1983) p.3]

[8] T.P.Chow, G.M.Fanelli [J. Electrochem. Soc. (USA) vol.132 (1985) p.1969]

[9] K.M.Eisele [J. Electrochem. Soc. (USA) vol.128 (1981) p.123]

[10] M.Zhang, J.Z.Li, I.Adesida, E.D.Wolf [J. Vac. Sci. & Technol. B (USA) vol.1 (1983) p.1037]

[11] G.C.Schwartz, P.M.Shaible [J. Vac. Sci. & Technol. (USA) vol.16 (1979) p.410]

[12] L.Y.Tsou [Jpn. J. Appl. Phys. Part 1 (Japan) vol.25 (1986) p.1594]

[13] S.Schwarzl, W.Beinvogl [in 'Plasma Processing' v.83-10 Eds G.Mathad, G.Schwartz, G.Smolinsky (Electrochem. Soc., Pennington, 1983) p.310]

[14] M.Sato, H.Nakamura [J. Vac. Sci. & Technol. (USA) vol.20 (1982) p.186]

[15] M.F.Leahy [in 'Plasma Processing' v.82-6 Eds J.Dieleman, R.G.Frieser, G.Mathad (Electrochem. Soc., Pennington, 1982) p.176]

[16] L.M.Ephrath, R.S.Bennett [in 'VLSI Science and Technology' Eds C.J.Dell'Oca, W.M.Bullis (Electrochem. Soc., Pennington, 1982) p.108]

[17] J.D.Chinn, I.Adesida, E.D.Wolf, R.C.Tiberio [J. Vac. Sci. & Technol. (USA) vol.19 (1981) p.1418]

[18] Y.Horiike, H.Okano, T.Yamazaki, H.Horie [Jpn. J. Appl. Phys. (Japan) vol.20 (1981) p.L817]

[19] K.M.Eisele [in 'Plasma Processing' v.82-6 Eds J.Dieleman, et al (Electrochem. Soc., Pennington, 1982) p.146]

[20] S.Matsuo [Appl. Phys. Lett. (USA) vol.36 (1980) p.768]

[21] L.M.Ephrath [J. Electrochem. Soc. (USA) vol.126 (1979) p.1419]

[22] Y.H.Lee, M.M.Chen [J. Appl. Phys. (USA) vol.54 (1983) p.5966]

[23] M.Shibakaki, Y.Horiike [Jpn. J. Appl. (Japan) vol.19 (1980) p.1579]

[24] J.W.Coburn, E.Kay [Solid State Technol. (USA) vol.22 no.4 (1979) p.117-24]

[25] G.Brasseur, F.Coopmans [in 'Plasma Processing' v.87-6 Eds G.Mathad, G.Schwartz, R.Gottscho (Electrochem. Soc., Pennington 1987) p.424]

[26] T.P.Chow, S.Ashok, B.J.Baliga, W.Katz [in 'Plasma Processing' v.83-10 Eds G.Mathad, G.Schwartz, G.Smonlinsky (Electrochem. Soc., Pennington 1983) p.101]

[27] H.W.Lehmann, R.Widmer [J. Vac. Sci. & Technol. (USA) vol.17 (1980) p.1177]

[28] T.E.Wicker, T.D.Mantei [J. Appl. Phys. (USA) vol.57 (1985) p1638]

[29] M.Mieth, A.Barker [J. Vac. Sci. & Technol. A (USA) vol.1 (1983) p.629]

[30] Y.Takasu, H.Okada, Y.Todokoro [Dry Process Symposium (IEE Japan, Tokyo, 1985) p.120]

[31] K.O.Park [in 'Plasma Processing' v.83-10 Eds G.Mathad et al (Electrochem. Soc., Pennington, 1983) p.257]

[32] S.M.Cabral, D.Rathman, N.P.Economou, L.A.Stern [in 'Plasma Processing' v.83-10 Eds G.Mathad et al (Electrochem. Soc., Pennington 1983) p.249]

[33] C.M.Horwitz [IEEE Trans. Electron Devices (USA) vol.ED-28 (1981) p.1320]

[34] H.B.Pogge, J.A.Bondur, P.J.Parkhardt [J. Electrochem. Soc. (USA) vol.130 (1983) p.1592]

[35] F.Faili, F.Wong [in 'Plasma Processing' v.87-6 Eds G.Mathad, G.C.Schwartz, R.Gottscho (Electrochem. Soc., Pennington 1987) p.458]

[36] H.Okano, H.Horiike [in 'Plasma Processing' v.82-6 Eds J.Dieleman, R.Frieser, G.Mathad (Electrochem. Soc., Pennington, 1982) p.206]

[37] A.I.T.Pau [in 'Plasma Processing' v.83-10 Eds G.Mathad et al (Electrochem. Soc., Pennington, 1983) p.310]

[38] C.Huynh [in 'Plasma Processing' v.87-6 Eds G.Mathad et al (Electrochem. Soc., Pennington, 1987) p.507]

[39] J.Paraszczak, M.Hatzakis [J. Vac. Sci. & Technol. (USA) vol.19 (1981) p.1412]

[40] S.V.Nguyen, D.Dobuzinsky, F.White, M.Kerbaugh [in 'Plasma Processing' v.87-6 Eds. G.Mathad, et al (Electrochem. Soc., Pennington, 1987) p.552]

[41] G.Smolinsky, D.N.K.Wang, D.Maydan [in 'Plasma Processing' v.82-6 Eds J.Dieleman et al (Electrochem. Soc., Pennington, 1982) p.165]

[42] T.C.Mele, J.Nulman, J.P.Krusius [J. Vac. Sci. & Technol. B (USA) vol.2 (1984) p.684]

[43] R.W.Boswell, D.Henry [Appl. Phys. Lett. (USA) vol.47 (1985) p.1095]

[44] H.F.Winters, J.W.Coburn [J. Vac. Sci. & Technol. B (USA) vol.3 (1985) p.1376]

[45] D.L.Flemm, V.M.Donnelly [Plasma Chem. & Plasma Process. (USA) vol.1 (1981) p.317]

[46] J.Dieleman, F.H.M.Sanders [Solid State Technol. (USA) vol.27 no.4 (1984) p.191-6]

[47] J.W.Coburn, H.F.Winters [J. Appl. Phys. (USA) vol.50 (1979) p.3189]

[48] R.G.Frieser, F.Montillo, N.Zingerman, W.K.Chu, S.Mader [J. Electrochem. Soc. (USA) vol.130 (1983) p.2237]

[49] G.Oehrlein, R.Tromp, Y.H.Lee, E.Petrillo [Appl. Phys. Lett. (USA) vol.45 (1984) p.420]

[50] S.N.Ryabov, S.Kutolin, B.Bondareva [Izv. Akad. Nauk. SSSR Neorg. Mater. vol.21 (1985) p.5]

[51] R.J.Davis, H.Hambermeier, J.Weber [Appl. Phys. Lett. (USA) vol.47 (1985) p.1295]

[52] S.Tachi, S.Okudaira [J. Vac. Sci. & Technol. B (USA) vol.4 no.2 (1986) p.459]

[53] K.Miyake, S.Tachi, K.Yogi, T.Tokuyama [J. Appl. Phys. (USA) vol.53 (1982) p.3214]

[54] S.Tachi, K.Miyake [in 'Semiconductor Technologies' v.13 Ed. J.Nishizawa (North-Holland, Amsterdam, 1984) p.333]

[55] T.M.Mayer, R.A.Barker, L.J.Whitman [J. Vac. Sci. & Technol. (USA) vol.18 (1981) p.349]

[56] J.M.Harper, J.J.Cuomo, P.A.Leary, G.A.Summa [J. Electrochem. Soc. (USA) vol.128 (1981) p.1077]

[57] P.J.Revell, A.C.Evans [Thin Solid Films (Switzerland) vol.86 (1981) p.117]

[58] B.A.Heath [J. Electrochem. Soc. (USA) vol.129 (1982) p.396]

[59] J.D.Chinn, E.D.Wolf [J. Vac. Sci. & Technol. (USA) vol.3 no.4 (1985) p.410]

[60] J.D.Chinn, I.Adesida, E.D.Wolf [J. Vac. Sci. & Technol. B (USA) vol.1 (1983) p.1028]

[61] N.Takasaki, E.Ikawa, Y.Kurogi [J. Vac. Sci. & Technol. B (USA) vol.4 (1986) p.806]

[62] U.Gerlach-Meyer, J.W.Coburn, E.Kay [Surf. Sci. (Netherlands) vol.103 (1981) p.177]

[63] H.Okano, H.Horiike [Jpn. J. Appl. Phys. (Japan) vol.20 (1981) p.2429]

[64] E.E.Krueger, A.L.Ruoff [J. Vac. Sci. & Technol. B (USA) vol.23 (1985) p.1650]

[65] Y.Y.Tu, T.J.Chuang, H.F.Winters [Phys. Rev. B (USA) vol.23 (1981) p.823]

[66] J.Dieleman, F.H.M.Sanders, A.Kolfschoten, P.C.Zalm, A.de Vries, A.Haring [J. Vac. Sci. & Technol. B (USA) vol.3 (1985) p.1384]

[67] Y.H.Lee, M.R.Polcari [EMIS Datareview RN=16121 (Sept 1987) 'Etching rates of ion beam milled and sputter etched Si']

23.5 ETCHING RATES OF Si ETCHED BY LASER-ASSISTED METHODS

by Y.H.Lee and M.R.Polcari

November 1987
EMIS Datareview RN=16123

Silicon etching can be enhanced by photons in a wide range of
wavelengths from infrared to deep UV and also in conjunction with
reactive gas molecules (photon assisted chemical etching) or reactive
gas plasmas (photon enhanced plasma etching). Also silicon can be
etched in a chemically inert gas ambient, if the photon energy is
sufficiently high (photoablation). Photon-induced ablation requires a
threshold energy density of 1 J/cm*2 (70 MW/cm*2) almost equal to the
minimum energy density of 0.75 J/cm*2 necessary to melt silicon [1,17].
Thus photoablation etch rates are usually very high in a range of 20-50
micron/min [1]. Laser enhanced chemical etching involves radical
generation by photolysis and desorption-rate enhancement by
photoexcitation. Sveshnikova [2] observed a very slow etch rate, when
the Ar laser was illuminated parallel to the silicon surface, while
Ehrlich [3] obtained a fast etch rate as high as 6 micron/sec, when the
Ar laser was directed perpendicular to the surface. Chuang [7,18] has
shown photon stimulated desorption processes through the electronic
excitation in UV and the vibrational excitation in IR, which greatly
contributes to etch rates. Houle [4,19] observed doping density
dependence in laser enhanced chemical etching of silicon. Reksten [15]
found that the doping effect varies with laser wavelength, i.e. UV
enhances etch rates more than Visible does due to a difference in the
penetration of the light into the bulk.

Etch rates are normally proportional to the laser power and substrate
temperature. Okano [9] observed that etch rate initially increases with
gas pressure, but saturates at high pressure, and also that etch
rate decreases with increasing the wavelength. Spatial resolution of a
small feature size is inversely proportional to the wavelength
(diffraction-limited) and proportional to gas pressure
(diffusion-limited). The laser etching technique is specially
attractive for maskless etching [9] and direct writing. More
practical applications can be found in review articles by Ehrlich [20]
and Rytz-Froidevaux [21]. Podlesnik [22] showed that unlike ion or
electron beam etching, laser etching is highly anisotropic due to
grazing angle reflection at the sidewall (waveguide effect). Hayasaka
[16] demonstrated a vertical etch profile at 1 micron feature size by
adding a small amount of methylmethacrylate (MMA) to Cl2 plasma for
sidewall passivation.

In pulsed laser beam etchings, etch rates are converted from the etch
yield per pulse if the pulse repetition rate is known; etch rate
(nm/s) = etch yield (nm/pulse) x pulse rate (Hz). Also the radiant
power density (W/cm*2) is estimated from the energy density (J/cm*2)
divided by pulse width (sec). The table does not specify a beam
optics (focused or unfocused) or a beam incidence angle with respect
to the silicon surface. Most etching data were obtained with the
incident angle less than 45 deg from the surface normal except for
the experiment by Sveshnikova [2].

TABLE 1: photoablation etching rates

Laser (wavelength)	Etch rate (micron/sec)	Energy density (J/cm*2)	Gas	Press. (atm)	Remarks	Ref
ArF (193 nm)	0.36 (0.18 micron/ pulse)	2	He	1	14 ns pulse	[1]
KrF (248 nm)	0.8 (0.4 micron/ pulse)	10	He	1	18 ns pulse	[1]

TABLE 2: photon-assisted chemical etching rates

Laser (wavelength)	Etch rate (nm/sec)	Energy or power density	Etchant	Press. (torr)	Remarks	Ref
Ar (448 nm)	0.006	30 W/cm*2	Br2	14	CW	[2]
	0.1	30 W/cm*2	Br2	180	CW	[2]
	6000	(7 W)	C12	200	CW	[3]
	0.6	(< 4 W)	XeF2	10*-5	CW	[4]
	15000	1 MW/cm*2	KOH	liquid	CW	[5]
Lamp (550 nm)	2.6	53 micro-W/cm*2	9 HF (48%) 19 H2O	liquid	0.3 mA plate current	[6]
CO2 (10.6 micron)	0.15 nm/pulse	1 J/cm*2 (20 MW/cm*2)	HF	1	50 ns pulse	[7]
	0.07 nm/pulse	1 J/cm*2	XeF2	1		[7]
N2 (337 nm)	2.7 (a-Si) 1.9 (100) 1.3 (111)	0.12 J/cm*2 (12 MW/cm*2)	C12	0.1	10 ns pulse	[8]
Hg-Xe (313 nm)	0.67	0.275 W/cm*2	C12	10	n+ poly Si	[9]
	0.1				poly Si	
	0.01				p+ poly Si	
ArF (193 nm)	0.13	85 mJ/cm*2	COF2	260	pulsed	[10]
			He	500		

Laser (wavelength)	Etch rate (nm/sec)	Energy or power density	Etchant	Press. (torr)	Remarks	Ref
XeCl (308 nm)	6.7	1.5 W/cm*2	Cl2	17	w/Hg lamp	[11]
			Si(CH3)4 3			
	3 nm/ pulse	0.4 J/cm*2	Cl2	7.6 X 10*-8	10 ns pulse	[12]
KrF (248 nm)	3 nm/ pulse	1 J/cm*2	Cl2	7.6 X 10*-8	10 ns pulse	[12]
D2 lamp (115-350 nm)	0.05	1.61 mW/cm*2	SF6	3.4	depends on flow pattern	[13]

TABLE 3: photon enhanced plasma etching rates

Laser (wavelength)	Etch rate (nm/sec)	radiant power density	Plasma condition	Press. (torr)	Remarks	Ref
Ar (514 nm)	8	120 kW/ cm*2	CF4/20%O2 0.1 W/cm*2 30 kHz	0.12	CW	[14]
	12	10 kW/ cm*2	NF3 0.1 W/cm*2 30 kHz	0.1		[15]
KrF (248 nm)	0.7	0.15 W/ cm*2	Cl2 MMA 2.45 GHz	0.2 0.04	10 ns pulse	[16]

REFERENCES
----

[1] G.B.Shinn, F.Steigerwald, H.Stiegler, R.Sauerbrey, F.K.Tittle, W.L.Wilson,Jr [J. Vac. Sci. & Technol. B (USA) vol.4 (1986) p.1273]
[2] L.L.Sveshnikova, V.I.Donin, S.M.Repinskii [Sov. Phys. Lett. (USA) vol.3 (1977) p.223]
[3] D.J.Ehrlich, R.M.Osgood,Jr, T.F.Deutsch [Appl. Phys. Lett. (USA) vol.3 (1977) p.223]
[4] F.A.Houle [J. Chem. Phys. (USA) vol.79 (1983) p.423 and vol.80 (1984) p.4851]
[5] R.J.von Gutfeld, R.T.Hodgson [Appl. Phys. Lett. (USA) vol.40 (1982) p.352]
[6] H.J.Hoffmann, J.M.Woodall [Appl. Phys. A (Germany) vol.33 (1984) p.243]
[7] T.J.Chuang [J. Vac. Sci. & Technol. (USA) vol.18 (1981) p.638; J. Chem. Phys. (USA) vol.74 (1981) p.146 and vol.74 (1981) p.1453]
[8] W.Sesselmann, T.J.Chuang [J. Vac. Sci. & Technol. B (USA) vol.3 (1985) p.1507]

[9]   H.Okano, Y.Horiike, M.Sekine [Jpn. J. Appl. Phys. Part 1 (Japan)
       vol.24 (1985) p.68]

[10]   G.L.Loper, M.D.Tabat [Appl. Phys. Lett. (USA) vol.46 (1985)
       p.654]

[11]   M.Sekine, H.Okano, Y.Yamabe, N.Hayasaka, H.Horiike [Proc. 6th
       Symp. Dry Process (Inst. Electrical Engineers, Japan, Tokyo, 1984)
       p.74]

[12]   T.Baller, D.J.Oostra, A.E. de Vries, G.N.A. van Veen [J Appl.
       Phys. (USA) vol.60 (1986) p.2321]

[13]   S.Watanabe, S.Ueda, N.Nakazato, M.Takai [Jpn. J. Appl. Phys.
       Part 2 (Japan) vol.25 (1986) p.L881]

[14]   W.Holber, G.Reksten, R.M.Osgood,Jr [Appl. Phys. Lett. (USA)
       vol.46 (1985) p.201]

[15]   G.M.Reksten, W.Holber, R.M.Osgood,Jr [Appl. Phys. Lett. (USA)
       vol.48 (1986) p.551]

[16]   N.Hayasaka, H.Okano, M.Sekine, Y.Horiike [Appl. Phys. Lett. (USA)
       vol.48 (1986) p.1165]

[17]   J.Narayan, C.W.White, M.J.Aziz, B.Stritzker, A.Walthuis [J. Appl.
       Phys. (USA) vol.57 (1985) p.564]

[18]   T.J.Chuang [J. Vac. Sci. & Technol. (USA) vol.21 (1982) p.798]

[19]   F.A.Houle [in 'Laser Chemical Processing of Semiconductor
       Devices' Eds F.A.Houle, T.F.Deutsch, R.M.Osgood,Jr (Materials
       Res. Soc., Pittsburgh, 1984) p.77]

[20]   D.J.Ehrlich, J.Y.Tsao [J. Vac. Sci. & Technol. B (USA) vol.1
       (1983) p.969]

[21]   Y.Rytz-Froidevaux, R.P.Salathe, H.H.Gilgen [Appl. Phys. A
       (Germany) vol.37 (1985) p.121]

[22]   D.V.Podlesnik, H.H.Gilgen, R.M.Osgood,Jr [Appl. Phys. Lett.
       (USA) vol.48 (1986) p.496]

CHAPTER 24

RAPID THERMAL ANNEALING

24.1 ULTRA-SHORT LASER PULSE INTERACTIONS WITH Si

by I.W.Boyd

October 1987
EMIS Datareview RN=15057

A. INTRODUCTION

The discovery of the phenomenon of laser annealing of silicon in the
mid-1970s sparked off a world-wide explosion of interest in the
subject. Many of the studies were involved with determining the
underlying principles behind the interaction of the laser radiation with
both amorphous silicon (a-Si) and single crystal silicon (c-Si). This
review covers the ultra-short pulse regime (i.e. in the picosecond (ps)
range, or less). Q-switched laser annealing of silicon on a nanosecond
(ns) timescale is discussed elsewhere [1].

The majority of studies in the picosecond regime have used the mode-
locked Nd:YAG laser system, operating fundamentally at 1.06 micron
or in the frequency-doubled mode at 0.53 micron. By introducing etalons
into the laser cavities, pulses typically from around 20-250 ps are
available with these systems [2]. Shorter pulses (4-20 ps) can be
obtained from Nd:glass lasers operating at similar wavelengths [2], or
from excimer laser systems (e.g. 15 ps at 248 nm [31]). The shortest
pulses obtainable have been provided using fibre compression and
colliding pulse techniques. Although pulses as short as 6 femtoseconds
(fs) have been produced the shortest thus far used to study interactions
with silicon have been of the order of 55 fs [4].

B. OPTICAL INTERACTIONS, ENERGY REDISTRIBUTION AND PHASE CHANGES

From the wealth of experiments performed in the nanosecond regime with
Q-switched laser irradiation of silicon [1] it has been established that
the energy absorption, carrier relaxation and recombination, and
subsequent transfer of energy to the lattice follows the generally
expected pattern of behaviour. Incident photons are absorbed by the
carrier system, which, on a timescale much less than 1 ns transfers the
energy via phonon scattering to the atomic system, thereby increasing
the sample temperature. Eventually, if sufficient energy is absorbed to
increase the temperature to the melting point of the material and
provide the latent energy for the phase change, the surface region will
melt. Subsequently, the material will resolidify into a phase whose
free energy will be determined by several factors, including the cooling
rate, the presence of any impurities, the crystal type and orientation
of the underlying material. Many studies have since been aimed at
accurately tracking the progress of the energy distribution in optically
excited Si and precisely determining the characteristic timescales
involved.

B1 Interaction with Visible Pulses

In the visible region of the spectrum, where Si absorbs by mainly
indirect optical absorption, several groups have performed many
time-resolved studies involving optical probing (i.e. transmission,
reflectivity and second harmonic generation) [4-17]. Photoemission
studies have provided information on the energy distribution of
electrons at the irradiated Si surface [10]. From the optical
experiments, it was first of all established for pulses of duration
down to around tens of picoseconds, that melting can occur during the
pulse, indicating extremely rapid energy transfer to the lattice [6,14].
On the other hand, with 55 fs pulses [4,11] melting only occurred 100s
of femtoseconds after the initial pulse. For example, Shank et al found
that after irradiation by 55-80 fs pulse, the lattice symmetry of the
silicon crystal disappeared after around 500 fs [12]. Also, by taking
fs-resolved photographs of the Si surface before, during and after
irradiation by these pulses, they found that a highly reflective phase,
a characteristic of liquid-Si, is formed after only a few hundred fs
[4]. Thus, it has become well established that thermal energy is
introduced to the lattice in times of the order of, or less than, one
picosecond.

Evidence has also been presented which indicates that material
can be ejected from the irradiated surface some 5 to 10 ps after
irradiation by the 55 fs pulses [4]. Thus, although melting can occur
as fast as 1 ps, or less, and sufficient energy can be introduced to
cause evaporation, it takes a significant amount of time for a
sufficient density of atoms to be evaporated so as to be optically
detectable.

B2 Interaction with Infrared Pulses

Silicon is most usually a weak absorber of infrared (IR) radiation,
exhibiting a low intensity absorption coefficient of only 10 and 14
cm*-1 at 1.604 and 1.054 micron respectively [23]. However, it absorbs
short pulses of intense radiation non-linearly at these wavelengths,
characterised by an indirect 2-photon coefficient of 1.5 cm/GW and a
free carrier cross-section of 5 X 10*-18 cm*2 [23]. Thus, during
absorption of picosecond IR pulses, the absorption mechanisms change
dramatically, as does the region within the crystal that heats up
rapidly.

The first few studies of the time resolved optical reflectivity around
1 micron found an apparently slow increase in reflectivity after
irradiation, from 300 ps [24] to 5000 ps [25]. This was initially
interpreted as evidence of delayed melting [25,26], in contrast to the
case for visible pulse irradiation. However, it has since been shown
by a pulse-probe imaging technique, that a similar material ejection
phenomenon to that observed by Downer et al [4] with fs visible pulses
operates in the IR for the picosecond pulses [27]. Indeed, experiments
show that melting occurs well within 5 ps of absorbing the appropriate
energy (around 1.6 J/cm*2 for 48 ps pulses, and 0.6 J/cm*2 for 6 ps
pulses [28]), and there it is exceedingly difficult to melt Si with
these pulses and not evaporate material [29]. Thus, evaporation appears
to have obstructed early interpretations of the data in this regime.

A recent study has shown that the energies required for UV and IR radiation to melt c-Si vary by nearly two orders of magnitude, from 24 mJ/cm*2 to 1.7 J/cm*2. However, once the Si has melted, the additional energy needed to initiate evaporation is about 120 mJ/cm*2, which is nearly independent of wavelength [29].

B3 Theory and Calculations

Fundamental calculations and modelling of experimental results in the picosecond regime has led to useful knowledge of the nature of the laser induced electron-hole plasmas [13,18,19] including an accurate determination of the ratio N/m* (where m* is the effective mass of the electron) [9], the relevant carrier diffusion processes [20], recombination processes [21], and second harmonic generation [22].

The mechanism by which silicon, irradiated by ultrashort pulses either in the visible or the infrared (IR), melts, has been modelled by several groups [13,26,30]. Despite the severe demand on the precise knowledge of a large number of optical and thermodynamical properties under the extreme non-equilibrium conditions imposed by such irradiation conditions, some models [26,30] are reportedly successful at accurately predicting melting thresholds and pulsewidth effects etc. [31,32].

C. SURFACE MORPHOLOGY

Picosecond pulsed radiation can induce a variety of surface modifications on crystalline and amorphous silicon in much the same fashion as can the longer nanosecond pulses [24,28,33-45]. There are many reports of laser induced formation of ring patterns on crystal and amorphous substrates [24,28,33-38,40-43]. These have been correlated to specific threshold energy densities for initiating melting followed by either amorphisation or recrystallisation [28,33,34,40-43]. Under the fastest cooling conditions just above the threshold for melting, Em, a-Si is usually formed. Just above Em, this appears as a disc at the centre of the irradiated region. At higher energies, the disc becomes a ring defining the lower limits of melting, and crystalline material is formed at the centre. At even higher energies, further rings may be formed due to thermal shock effects, evaporation, or other surface damaging effects.

Thus, the fact that even IR picosecond pulses can create the necessarily strong thermal cooling following melting that the liquid does not recrystallise, indicates the degree of non-linearity in the absorption during the pulse. Recent studies [39] have examined the cross-sectional patterns induced on c-Si by this irradiation method, and three distinct types of structure are apparent. It appears that not only amorphous, but also large and small grain material regrows from the melted crystal.

By comparison, when amorphous silicon is irradiated, a similar concentric ring pattern is generally produced, but often an additional anomalous recrystallisation ring is formed outside the usually expected amorphous region [41-43]. Within the region, other crystallisation rings are present that are remarkably similar to those obtained in the crystalline material. From observations of small changes in surface height near the edges of the irradiated region, a mechanism induced by shock stress has been invoked to explain the surface transformation.

Laser induced multishot damage has been studied by several groups [28,36,37,40]. Even though the incident laser energy density is well below the usual single shot melting threshold, many such shots on the same area can eventually impart surface damage to the crystal. Ripple formation in the picosecond regime has also attracted some attention [28,35-37,44,45]. There is a general consensus of opinion that some form of surface scattering occurs such that the incident radiation interferes with itself, thereby producing regular patterns which are eventually frozen into the surface. Such patterns are usually strongly related to the wavelength of the radiation incident as well as its degree and type of polarisation. There is evidence, however, that some corrugated patterns may be formed by a variety of different mechanisms [35,44,45].

REFERENCES
----

[1]   C.W.White [EMIS Datareview RN=15734 (June 1987) 'Pulsed laser
        annealing of silicon (2-60 ns pulses)']
[2]   I.W.Boyd et al [Trends in Quantum Electronics, Eds A.M.Prokhorov,
        I.Ursu (Springer-Verlag, 1986) and refs therein]
[3]   P.H.Bucksbaum, J.Bokor [in 'Energy Beam-Solid Interaction and
        Transient Thermal Processing', Eds J.C.C.Fan, N.M.Johnson
        (North-Holland, Amsterdam, 1984) p.93]
[4]   M.C.Downer, R.L.Fork, C.V.Shank [J. Opt. Soc. Am. B (USA) vol.2
        (1985) p.595]
[5]   J.M.Liu, R.Yen, H.Kurz, N.Bloembergen [Appl. Phys. Lett. (USA)
        vol.39 no.9 (1 Nov 1981) p.755-7]
[6]   J.M.Liu, H.Kurz, N.Bloembergen [Appl. Phys. Lett. (USA) vol.41
        no.7 (1 Oct 1982) p.643-6]
[7]   R.Yen, J.M.Liu, H.Kurz, N.Bloembergen [Appl. Phys. A (Germany)
        vol.27 (1982) p.153]
[8]   M.Yu.Aver'yanova et al [Sov. Tech. Phys. Lett. (USA) vol.11 no.7
        (Jul 1985) p.316-8]
[9]   L.-A.Lompre, J.M.Liu, H.Kurz, N.Bloembergen [Appl. Phys. Lett.
        (USA) vol.43 no.2 (15 Jul 1983) p.168-70 and vol.44 no.1 (1 Jan
        1984) p.3-5]
[10]   A.M.Malvezzi, H.Kurz, N.Bloembergen [Appl. Phys. A (Germany)
        vol.36 no.3 (Mar 1985) p.143-6]
[11]   C.V.Shank, R.Yen, C.Hirliman [Phys. Rev. Lett. (USA) vol.50 no.6
        (7 Feb 1983) p.454-7]
[12]   C.V.Shank, R.Yen, C.Hirliman [Phys. Rev. Lett. (USA) vol.51
        no.10 (5 Sep 1983) p.900-2]
[13]   C.V.Shank, M.C.Downer [Mater. Res. Soc. Symp. Proc. (USA) vol.51
        (1986) p.15]
[14]   D.von der Linde, N.Fabricus [Appl. Phys. Lett. (USA) vol.41 no.10
        (15 Nov 1982) p.991-3]
[15]   N.Bloembergen [Mater. Res. Soc. Symp. Proc. (USA) vol.51 (1986)
        p.3]
[16]   D.Hulin, M.Combescot, J.Bok, A.Migus, J.Y.Vinet, A.Antonetti
        [Phys. Rev. Lett. (USA) vol.52 no.22 (28 May 1984) p.1998-2001]
[17]   D.Hulin et al [J. Lumin. (Netherlands) vol.30 (1985) p.262]
[18]   H.M.van Driel, L.-A.Lompre, N.Bloembergen [Appl. Phys. Lett.
        (USA) vol.44 no.3 (1 Feb 1984) p.285-7]
[19]   M.I.Gallant, H.M.van Driel [Phys. Rev. B (USA) vol.26 no.4 (1 Aug
        1982) p.2133-46]
[20]   H.Bergner et al [J. Lumin. (Netherlands) vol.30 (1985) p.114]
[21]   P.M.Fauchet, W.L.Nighan,Jr [Appl. Phys. Lett. (USA) vol.48
        no.11 (17 Mar 1986) p.721-3]

[22] J.A.Litwin, J.E.Sipe, H.M.van Driel [Phys. Rev. B (USA) vol.31 no.8 (15 Apr 1985) p.5543-6]

[23] T.F.Boggess et al [IEEE J. Quantum Electron. (USA) vol.QE-22 no.2 (Feb 1985) p.360-8]

[24] Y.Kanemitsu, H.Kuroda, S.Shionoya [Jpn. J. Appl. Phys. Part 1 (Japan) vol.23 no.5 (15 May 1984) p.618-21]

[25] K.Gamo, K.Murakami, M.Kawabe, S.Namba [in 'Laser and Electron Beam Solid Interactions and Materials Processing', Eds J.F.Gibbons, L.D.Hess, T.W.Sigmon (North-Holland, Amsterdam, 1981) p.97]

[26] A.Lietoila, J.F.Gibbons [in 'Laser and Electron Beam Solid Interactions and Materials Processing', Eds J.F.Gibbons, L.D.Hess, T.W.Sigmon (North-Holland, Amsterdam, 1981) p.23]

[27] I.W.Boyd, S.C.Moss, T.F.Boggess, A.L.Smirl [Appl. Phys. Lett. (USA) vol.46 no.4 (1985) p.366-8]

[28] I.W.Boyd, S.C.Moss, T.F.Boggess, A.L.Smirl [Appl. Phys. Lett. (USA) vol.45 no.1 (1984) p.80-2]

[29] I.W.Boyd et al [Opt. Acta (GB) vol.33 (1986) p.527]

[30] H.M.van Driel [in 'Semiconductors Probed by Ultrafast Laser Spectroscopy', Ed. R.R.Alfano (Academic, New York, 1984) p.57]

[31] P.M.Fauchet, A.E.Siegman [in 'Energy Beam-Solid Interactions and Transient Thermal Processing', Eds J.C.C.Fan, N.M.Johnson (North-Holland, Amsterdam, 1984) p.63]

[32] H.M.van Driel, A.L.Smirl [Appl. Phys. Lett. (USA) vol.49 no.12 (22 Sep 1986) p.743-4]

[33] P.L.Liu, R.Yen, N.Bloembergen, R.T.Hodgson [Appl. Phys. Lett. (USA) vol.34 no.12 (15 Jun 1979) p.864-6]

[34] G.A.Rozgonyi et al [in 'Laser and Electron Beam Interactions with Solids', Eds B.R.Appleton, G.K.Celler (Elsevier, 1982) p.177]

[35] D.Jost, W.Luthy, H.P.Weber, R.P.Salathe [Appl. Phys. Lett. (USA) vol.49 no.11 (15 Sep 1986) p.625-7]

[36] R.M.Walser et al [in 'Laser and Electron Beam Solid Interactions and Materials Processing', Eds J.F.Gibbons, L.D.Hess, T.W.Sigmon (North-Holland, Amsterdam 1981) p.177]

[37] P.M.Fauchet [Phys. Lett. A (Netherlands) vol.93A no.3 (3 Jan 1983) p.155-7]

[38] I.W.Boyd et al [in 'Laser Processing and Diagnostics' Ed D.Bauerle (Springer Verlag, Heidelberg, 1984)]

[39] A.L.Smirl, I.W.Boyd, T.F.Boggess, S.C.Moss, H.M.van Driel [J. Appl. Phys. (USA) vol.60 no.3 (1 Aug 1986) p.1169-82]

[40] I.W.Boyd et al [in 'Energy Beam-Solid Interactions and Transient Thermal Processing', Eds J.C.C.Fan, N.M.Johnson (North-Holland, 1984) p.203]

[41] Y.Kanemitsu, I.Nakada, H.Kuroda [Appl. Phys. Lett. (USA) vol.47 no.9 (1 Nov 1985) p.939]

[42] Y.Kanemitsu, Y.Ishida, I.Nakada, H.Kuroda [Appl. Phys. Lett. (USA) vol.48 no.3 (20 Jan 1986) p.209-11]

[43] Y.Kanemitsu, H.Kuroda, I.Nakada [Jpn. J. Appl. Phys. Part 1 (Japan) vol.25 no.9 (Sep 1986) p.1377-81]

[44] A.A.Bugaev, B.P.Zakharchenya, V.A.Lukoshkin [Sov. Tech. Phys. Lett. (USA) vol.12 no.6 (Jun 1986) p.292-3]

[45] A.A.Bugaev, B.P.Zakharchenya, M.G.Ivanov, I.A.Merkulov [Sov. Phys.-Solid State (USA) vol.28 no.5 (May 1986) p.836-8]

24.2 PULSED LASER ANNEALING OF Si (2-60 ns PULSES)

by C.W.White

November 1987
EMIS Datareview RN=15734

Research sponsored by the Division of Materials Sciences, U.S.
Department of Energy under contract DE-AC05-840R21400 with
Martin Marietta Energy Systems, Inc.

A. INTRODUCTION

The observation in the late 1970s that laser irradiation of ion-
implanted silicon could be used to remove ion implantation damage
gave rise to a field that became known as laser annealing. Much of the
early work in this field was carried out using radiation from pulsed
lasers (Ruby, YAG) operated in the Q-switched mode. The work carried
out using these directed energy sources was referred to as pulsed
laser annealing to distinguish it from annealing that could be achieved
using CW lasers operated in a scanned mode. The annealing mechanisms
are quite different for these different types of laser sources. With
Q-switched (or pulsed) lasers, annealing takes place in the liquid
phase, whereas CW lasers generally cause annealing to take place in the
solid-phase region. In this article we will review some of the basic
concepts associated with pulsed laser annealing. We will restrict our
discussion to lasers with pulse durations in the range of 2 to 60 nsec.
This includes Ruby, YAG, and a variety of pulsed excimer laser sources.
We will also point out several of the related fields that have evolved
from laser annealing. Progress in the field of laser annealing is
beautifully documented in a series of conferences proceedings [1-9].

B. ANNEALING WITH PULSED LASERS

Ion implantation of a semiconductor such as silicon causes the
near-surface region to become damaged or even turned amorphous. The ion
implantation damage can be removed and crystalline order restored to
the near-surface region using a single pulse of radiation (several
nanosecond pulse duration time) from a visible or UV laser. This
radiation is heavily absorbed in the near surface and if the energy
density is sufficient, the crystal can be melted to a depth of several
thousand angstroms. When this occurs, the liquid layer is in contact
with an undamaged substrate and the substrate acts as a seed for
liquid-phase epitaxial regrowth. Velocities of solidification that can
be achieved are 1-20 meters/sec for pulse duration times of 2×10^{-9}
- 60×10^{-9} sec. As a result, the annealing process is over in a time
period of several hundred nanoseconds. Pulsed laser annealing therefore
refers to the rapid deposition of laser energy into the near-surface
region which gives rise to melting to a depth of several thousand
angstroms, followed by liquid-phase epitaxial regrowth from the
underlying substrate [1-9].

In ion implanted silicon, absorption coefficients for visible and UV
radiation are very high, and the laser light is absorbed in the very
near surface. Absorbed energy is deposited initially into the
electronic system but almost instantaneously is transformed into heat
by electron-phonon collisions (on a time scale of 10*-11 - 10*-12 sec)
[10,11]. The absorption of radiation from nanosecond laser pulses
therefore leads to very high temperatures and melting in the
near-surface region of silicon while the total energy used for annealing
(0.5-2.5 Joules/cm*2 for radiation of wavelength less than 0.7 micron)
is not sufficient to raise substantially the temperature of the bulk of
the material. The ability to confine the heat treatment to the
near-surface region (where ion implantation damage is deposited) and
the very short time span required for annealing were two of the primary
factors that made pulsed laser annealing attractive for removing ion
implantation damage in the later 1970's.

Heating and cooling rates that can be achieved during pulsed laser
annealing are far in excess of those which can be achieved by
conventional methods. By irradiating ion-implanted silicon with visible
or UV light (approx. 2-60 nanosecond pulse duration time) and an
energy density of 1-2 Joules/cm*2, we can achieve heating and cooling
rates of 10*8-10*10 deg K/sec, thermal gradients of 10*6-10*8 deg K/cm
and solidification velocities of 1-20 meters/sec [12,13]. Computer
solutions to the one-dimensional heat flow equation can be used to
calculate the response of the material to the laser pulse, to predict
the instantaneous temperature profiles, melt front position versus time,
etc. [14-16]. Elegant experimental techniques have been developed to
provide experimental measurements of the onset and duration of surface
melting [17] (optical reflectivity), the position and velocity of the
liquid/solid interface during and after laser irradiation [18,41]
(transient conductance), and the temperature profile and thermal
gradient during and after laser irradiation [19] (time-resolved X-ray
diffraction).

In pulsed laser annealing of ion-implanted silicon, the annealed region
is free of any extended defects if the laser energy density is
sufficient to melt all the way through the damaged (or amorphous)
region into the undamaged substrate. When this occurs, the substrate
provides the seed for liquid phase epitaxy, and the solidified region
will be free of any extended defects unless the velocity of
solidification becomes so high that the atoms do not have time to
establish their proper bonding configurations at the interface and an
amorphous layer results from solidification. (For silicon, this
velocity is approx. 15 m/sec for the (100) orientation and approx. 12
m/sec for the (111) orientation [20,41,45]). Alternatively, if the melt
depth does not penetrate through the damaged or amorphous region, then
the solidified region will be polycrystalline or contain extended
defects.

C. DOPANT REDISTRIBUTION DURING PULSED LASER ANNEALING

During pulsed laser annealing, implanted impurities undergo substantial
redistribution [21]. This occurs because the near-surface region is
melted during annealing and therefore implanted impurities can diffuse

in liquid silicon during the time that region remains melted. Diffusion coefficients in liquid silicon are orders of magnitude higher than in solid silicon even at the melting point. Therefore, a substantial redistribution can occur even though the near-surface region is molten only for a few hundred nanoseconds. Dopant profiles following solidification can be calculated if the melt depth as a function of time and the liquid phase diffusivity are known. The (nonequilibrium) segregation coefficient can be determined by comparing calculated profiles with those measured by experimental methods [22,42].

D. LATTICE LOCATION OF IMPLANTED IMPURITIES

Group III, IV, and V impurities are found to be substitutional in the silicon lattice following pulsed laser annealing [22,23]. This is true even when the dopant concentration considerably exceeds the equilibrium solid solubility limit. Substitutional fractions that are measured for these impurities by RBS-ion channelling experiments are substantially better than those measured following furnace annealing. The Group III and V substitutional impurities are electrically active.

E. NONEQUILIBRIUM CRYSTAL GROWTH

At the very rapid growth velocities that are achieved during pulsed laser annealing, recrystallisation of the melted region takes place under conditions that are far from equilibrium at the liquid solid interface. The interfacial distribution coefficients (k') which describe the partition of dopant between the solid and liquid phase at the interface are found to be much higher than their equilibrium values [22,24,42]. Values for k' are determined by comparing model calculations for dopant redistributions during pulsed laser annealing to measured dopant profiles. Experiments show that the values of k' for the Group III, IV, and V impurities are all in the range of 0.1 to 1.0 for growth velocities of approx. 3 meters/sec or greater, even though the equilibrium values are as low as 4 X 10*-4 (for In). This provides an indication of the large deviation from equilibrium during crystal growth at these velocities. Values for k' are found to increase toward unity as the velocity increases [37-39,42], and they are also observed to be a function of crystal orientation [37,40,43].

F. FORMATION OF SUPERSATURATED SOLID SOLUTIONS

Substitutional solubilities achieved during pulsed laser annealing can greatly exceed equilibrium solubility limits for Group III, IV, and V impurities in silicon [23,24,42]. These supersaturated alloys are formed because the velocity of the interface during solidification is much greater than the diffusion rate of the impurity away from the interface. As a consequence, even those impurities that would like to remain in the liquid (In and Bi for example) cannot avoid being incorporated into the solid to form a supersaturated solid solution. TABLE I compares equilibrium (Cs[o]) solid solubility limits with those achieved during pulsed laser annealing (Cs[max]) at a solidification velocity of approx. 4.5 meters/sec.

TABLE 1: equilibrium (Cs[o]) and pulsed laser annealing induced (Cs[max])
         solid solubility limits [22]

Dopant	Cs[o](/cm*3)	Cs[max](/cm*3)
As	1.5 X 10*21	6.0 X 10*21
Sb	7.0 X 10*19	2.0 X 10*21
Ga	4.5 X 10*19	4.5 X 10*20
Bi	8.0 X 10*17	4.0 X 10*20
In	8.0 X 10*17	1.5 X 10*20

Solubility limits that can be achieved by pulsed laser annealing can be
increased by increasing the solidification velocity and there is also a
dependence on crystal orientation due to the dependence of the
distribution coefficient on crystal orientation [25].

G. LIMITATIONS TO SUBSTITUTIONAL SOLUBILITY

Substitutional solubilities achieved during pulsed laser annealing are
approaching thermodynamic limits predicted for dopant incorporation
even at infinite growth velocity [25]. The predicted limits [26] are
determined by the locus of points resulting from the intersections of
the solidus and liquidus lines on a plot of the Gibbs Free Energy
versus composition (the To curve). For the impurities listed in TABLE
1, the substitutional solubilities achieved at a velocity of 4.5
meters/sec are all within an order of magnitude of the predicted limits.
In this velocity range substitutional solubility is limited primarily
by interface instability during solidification [25,28]. Interface
instability is caused by constitutional supercooling in front of the
interface. This leads to lateral segregation of the dopant into the
walls of a well defined cell structure in the near-surface region. Both
the concentration where instability develops and the resulting cell size
can be predicted using morphological instability theory modified to
account for the variation of the distribution coefficient with velocity
[27,28,42]. In addition to interface instability, lattice strain and
dopant precipitation in the liquid phase have been found to limit
substitutional solubility for some impurities [25].

H. OTHER APPLICATIONS OF PULSED LASER ANNEALING

In addition to the removal of ion implantation damage, pulsed laser
annealing has been used to promote the indiffusion of dopants deposited
as a thin film on a substrate [29], and as a means to remove diffusion
induced precipitates in the near surface [30]. Extensive research [1-9]
has indicated that pulsed lasers will probably not be widely used to
process ion-implanted semiconductors because melting and solidification
of thin surface layers does not appear to be compatible with
maintaining structural integrity on wafers that have been patterned
with metal or dielectric films. Nevertheless, the work done on pulsed
laser annealing has led to the development of several other areas of

research including (a) Rapid Thermal Annealing, (b) Photochemical Processing, and (c) Silicon on an Insulating Substrate which clearly will have application to device processing.

Pulsed laser annealing has contributed enormously to our understanding of high-speed, nonequilibrium crystal growth phenomena. Techniques developed in this field have been used to measure such fundamental parameters as the melting temperature of amorphous silicon [31,44],and to investigate explosive crystallisation [31], velocity-undercooling relationships during solidification [19,32-34] and to perform direct measurement of melt and solidification dynamics which provides fundamental information on melt nucleation [35,36]. In general, the very high velocities that can be achieved in pulsed laser annealing, and the ability to control the velocity in a predictable way, are providing unique opportunities to study high-speed, nonequilibrium crystal growth phenomena under well defined conditions.

REFERENCES
----

[1]   S.D.Ferris, H.J.Leamy, J.M.Poate (Eds) [AIP Conf. Proc. (USA) no.50, Laser-Solid Interaction and Laser Processing, 1978 (Am Inst. Phys., 1979)]

[2]   C.M.White, P.S.Peercy (Eds) [Laser and Electron Beam Processing of Materials (Academic Press New York, 1980)]

[3]   J.F.Gibbons, L.D.Hess, T.W.Sigmon (Eds) [Mater. Res. Soc. Symp. Proc. (USA) no.1, Laser and Electron-Beam Solid Interactions and Materials Processing (North-Holland, New York, 1981)]

[4]   B.R.Appleton, G.K.Celler (Eds) [Mater. Res. Soc. Symp. Proc. (USA) no.4, Laser and Electron-Beam Interactions with Solids (North-Holland, New York 1982)]

[5]   J.Narayan, W.L.Brown, R.A.Lemons (Eds) [Mater. Res. Soc. Symp. Proc. (USA) no.13, Laser-Solid Interactions and Transient Thermal Processing of Materials (North-Holland, New York 1983)]

[6]   J.C.C.Fan, N.M.Johnson (Eds) [Mater. Res. Soc. Symp. Proc. (USA) no.23, Energy Beam-Solid Interactions and Transient Thermal Processing (North-Holland, New York, 1984)]

[7]   D.K.Biegelsen, G.Rozgonyi, C.Shank [Mater. Res. Soc. Symp. Proc. (USA) no.35, Energy Beam-Solid Interactions and Transient Thermal Processing (Materials Research Society, Pittsburgh, 1985)]

[8]   H.Kurz, G.L.Olson, J.M.Poate (Eds) [Mater. Res., Soc. Symp. Proc. (USA) no.51, Beam-Solid Interactions and Phase Transformations (Materials Research Society, Pittsburgh, 1986)]

[9]   S.T.Picraux, M.O.Thompson, J.S.Williams (Eds) [Mater. Res. Soc. Symp. Proc. (USA) no.74 (in press), Beam-Solid Interactions and Transient Processes]

[10]   N.Bloembergen [AIP Conf. Proc. (USA) no.50 Laser-Solid Interaction and Laser Processing, 1978, Eds S.D.Ferris, H.J.Leamy, J.M.Poate (Am. Inst. Phys., 1979) p.1]

[11]   M.von Allmen [Laser Annealing of Semiconductors Eds J.M.Poate, J.W.Mayer (Academic Press, New York, 1982) ch.3]

[12]   P.Baeri, S.U.Campisano [Laser Annealing of Semiconductors Eds J.M.Poate, J.W.Mayer (Academic Press, New York, 1982) ch.4]

[13]   E.Rimini [Surface Modification and Alloying Eds J.M.Poate, G.Foti, D.C.Jacobson ch.2]

[14]   J.C.Wang, R.F.Wood, P.P.Pronko [Appl. Phys. Lett. (USA) vol.33 (1978) p.455]

[15] P.Baeri, S.U.Campisano, G.Foti, E.Rimini [J. Appl. Phys. (USA)
      vol.50 (1978) p.788]
[16] R.F.Wood, G.E.Jellison [Semicond. & Semimet. (USA) vol.23
      (Academic Press, New York, 1984) ch.4]
[17] D.H.Auston et al [AIP Conf. Proc. (USA) no.50, Laser-Solid
      Interaction and Laser Processing, 1978, Eds S.D.Ferris, H.J.Leamy,
      J.M.Poate (Am. Inst. Phys., 1979) p.11]
[18] G.Galvin, M.O.Thompson, J.W.Mayer, R.B.Hammond, N.Paulter,
      P.S.Peercy [Phys. Rev. Lett. (USA) vol.48 (1982) p.33]
[19] B.C.Larson, J.Z.Tischler, D.M.Mills [J. Mater. Res. vol.1 (1986)
      p.45]
[20] J.M.Poate [Mater. Res. Soc. Symp. Proc. (USA) no.4, Laser and
      Electron-Beam Interactions with Solids, Eds B.R.Appleton,
      G.K.Celler (North-Holland, New York, 1982) p.121]
[21] C.W.White, W.H.Christie, B.R.Appleton, S.R.Wilson, P.P.Pronko,
      C.W.Magee [Appl. Phys. Lett. (USA) vol.33 (1978) p.455]
[22] C.W.White, S.R.Wilson, B.R.Appleton, F.W.Young [J. Appl. Phys.
      (USA) vol.51 (1980) p.738]
[23] C.W.White, P.P.Pronko, S.R.Wilson, B.R.Appleton, J.Narayan,
      R.T.Young [J. Appl. Phys. (USA) vol.50 (1979) p.2967]
[24] C.W.White, D.M.Zehner, S.U.Campisano, A.G.Cullis [Surface
      Modification and Alloying Eds J.M.Poate, G.Foti, D.C.Jacobson
      (Plenum Press, New York, 1983) ch.4]
[25] C.W.White [Semicond. & Semimet. (USA) vol.23 (Academic Press,
      New York, 1984) ch.2]
[26] J.W.Cahn, S.R.Coriell, W.J.Boettinger [Laser and Electron Beam
      Processing of Materials Eds C.M.White, P.S.Peercy (Academic Press,
      New York, 1980) p.89]
[27] J.Narayan [J. Appl. Phys. (USA) vol.52 (1981) p.1289]
[28] A.G.Cullis et al [Appl. Phys. Lett. (USA) vol.38 (1981) p.642]
[29] J.Narayan, R.T.Young, R.F.Wood, W.H.Christie [Appl. Phys. Lett.
      (USA) vol.33 (1978) p.338]
[30] R.T.Young, J.Narayan [Appl. Phys. Lett. (USA) vol.33 (1978)
      p.14]
[31] M.O.Thompson et al [Phys. Rev. Lett. (USA) vol.52 (1984) p.2360]
[32] P.S.Peercy, M.N.Thompson [Mater. Res. Soc. Symp. Proc. (USA)
      no.35, Energy Beam-Solid Interactions and Transient Thermal
      Processing (Materials Research Society, Pittsburgh, 1985) p.53]
[33] P.S.Peercy, M.O.Thompson, J.Y.Tsao [Mater. Res. Soc. Symp. Proc.
      (USA) no.74 (in press), Beam-Solid Interactions and Transient
      Processes, Eds S.T.Picraux, M.O.Thompson, J.S.Williams]
[34] B.C.Larson, J.Z.Tischler, D.M.Mills [Mater. Res. Soc. Symp. Proc.
      (USA) no.51, Beam-Solid Interactions and Phase Transformations,
      Eds H.Kurz, G.L.Olson, J.M.Poate (Materials Research Society,
      Pittsburgh, 1986) p.113]
[35] M.O.Thompson, P.S.Peercy [Mater. Res. Soc. Symp. Proc. (USA)
      no.51, Beam-Solid Interactions and Phase Transformations, Eds
      H.Kurz, G.L.Olson, J.M.Poate (Materials Research Society,
      Pittsburgh 1986) p.99]
[36] P.S.Peercy, M.O.Thompson, J.Y.Tsao, J.M.Poate [Mater. Res. Soc.
      Symp. Proc (USA) no.51, Beam-Solid Interactions and Phase
      Transformations, Eds H.Kurz, G.L.Olson, J.M.Poate (Materials
      Research Society, Pittsburgh, 1986) p.125]
[37] P.Baeri, J.M.Poate, S.U.Campisano, G.Foti, E.Rimini, A.G.Cullis
      [Appl. Phys. Lett. (USA) vol.37 (1980) p.912]
      P.Baeri, G.Foti, J.M.Poate, S.U.Campisano, A.G.Cullis [Appl.
      Phys. Lett. (USA) vol.38 (1981) p.800]

[38] M.J.Aziz, J.Y.Tsao, M.O.Thompson, P.S.Peercy, C.W.White [Phys. Rev. Lett. (USA) vol.56 (1986) p.2489]

[39] A.G.Cullis, H.C.Webber, J.M.Poate, A.L.Simons [Appl. Phys. Lett. (USA) vol.36 (1980) p.320]

[40] M.J.Aziz, C.W.White [Phys. Rev. Lett. (USA) vol.57 (1986) p.2675]

[41] M.O.Thompson et al [Phys. Rev. Lett. (USA) vol.50 (1983) p.896]

[42] S.U.Campisano, J.M.Poate [Appl. Phys. Lett. (USA) vol.47 (1985) p.485]

[43] L.M.Goldman, M.J.Aziz [J. Mater. Res. (USA) vol.2 (1987) p.524]

[44] P.Baeri, G.Foti, J.M.Poate, A.G.Cullis [Phys. Rev. Lett. (USA) vol.45 (1980) p.2036]

[45] A.G.Cullis, H.C.Webber, N.G.Chew, J.M.Poate, P.Baeri [Phys. Rev. Lett. (USA) vol.49 (1982) p.219]

24.3 INCOHERENT LIGHT ANNEALING OF Si

by R.E.Harper

August 1987
EMIS Datareview RN=16128

The field of rapid thermal processing (RTP) has grown enormously over
the last eight years, driven primarily by the requirements on
integrated circuit technology to produce shallow heavily doped p-n
junctions by activating ion-implanted dopants and annealing implant
damage while minimising the dopant diffusion. Of the RTP techniques,
incoherent light annealing using quartz-halogen or arc lamps is the
most promising for most device applications, to such an extent that
there are now at least five commercially available machines capable of
high throughput with large area wafers. An overview of some of these
systems, including a consideration of the problems of (and solutions
to) temperature measurement and control, heating uniformity and wafer
handling has been given by Wilson et al [1]. This Datareview will
consider only the use of incoherent light for annealing of dopants in
crystalline silicon, and cites only a small fraction of the literature.
A comprehensive bibliography which lists 342 references up to 1985
covering a variety of RTP techniques and applications has been compiled
by Marion and Powell [2]; the same volume contains some 50 more papers
on various topics including a review of RTP for silicide formation by
d'Heurle et al [3].

The behaviour during incoherent light annealing of all the common
dopants implanted into silicon has been extensively studied. Arsenic
and boron (including BF2+ implantation and implantation of B+ and BF2+
into pre-amorphised Si) have received most attention [4-6] though
indium [7] and antimony [8] have also been investigated. The studies
have covered many aspects such as the dopant redistribution [4,5,7,8],
the evolution of the implantation related defects [6,8], DLTS studies
of the residual defects [9-11] and measurements of junction leakage
currents [11,13]. Hill [14] has given an excellent review of these
topics from the point of view of device fabrication.

Implantation with a heavy ion such as As at moderate or high doses
produces an amorphous layer and a localised dislocation array just
beyond the amorphous/crystalline interface, which must be removed
during the subsequent thermal treatment. Seidel et al [4] found
activation energies for As diffusion and dislocation removal during
RTP of 4.0 and 5.0 eV respectively. They thus concluded that RTP using
short time, high temperature anneals offered a significant advantage
over conventional lower temperature longer time processing for the
formation of shallow, defect-free As+ junctions. A further advantage
of RTP is that extremely high concentrations of electrically active As
can be obtained with high temperature (> 1100 deg C) anneals;
however, these doping levels are metastable and some de-activation due
to As clustering takes place if the RTP step is followed by a prolonged
low temperature anneal [15]. The detailed mechanism for As diffusion
during RTP has been the subject of much controversy over the existence
of a transient enhanced diffusion during the early stages of the anneal
cycle. This seems to have been resolved by a 'round robin' exercise
conducted by Seidel et al [4]. Most workers now agree that there is no
transient effect and that the observed diffusion can be explained as a
concentration-enhanced effect. Recently however Pennycook et al [8]

reported a large transient enhanced diffusion for Sb, and the model developed from their results also predicts a transient effect for As. However, as pointed out by Hill [14] the Pennycook et al results refer only to microscopic redistribution as detected by TEM or conductivity measurements; the macroscopic redistribution can be explained in terms of concentration-enhanced diffusion only.

Boron implantation into crystalline silicon does not produce an amorphous layer, and the shallowness of the junction is limited by the existence of an ion channelling tail. It is generally accepted that a transient enhanced diffusion occurs in the tail region during RTP. Even during conventional furnace annealing an anomalous diffusion occurs in this tail [16]. Several models have been proposed: Fair et al [17] for example speculated that point defects generated near the maximum concentration region migrate to the channelling tail during RTP, and once there interact with the boron to produce the enhanced diffusion. This is in conflict with Seidel's [5] experimental results which indicate that the enhanced diffusion is a locally controlled phenomenon. A more recent model by Michel [18] treats the anomalous furnace and transient enhanced diffusions as different manifestations of the same phenomenon. He assumed that the boron exists in two forms, an active mobile species and an inactive immobile species which converts to the active form exponentially with time during RTP. The immobile species may be associated with defect clusters which break up at high temperature, generating excess point defects which produce the enhanced diffusions.

The situation for boron implanted into pre-amorphised silicon is very different, since the channelling tail is much reduced or absent altogether. Therefore transient enhanced diffusion is not expected to occur, and indeed at least two groups [4,19] have reported 'normal' boron diffusion. However Armigliato et al [20] have carried out a systematic study of boron diffusion during incoherent light annealing of boron-implanted crystalline, pre-damaged and pre-amorphised Si. They found a weak transient effect even in the pre-amorphised case. They attributed this to interaction with interstitials created during the pre-amorphisation step, and possibly with point defects trapped during the SPE growth. Qualitatively similar results were obtained by Cowern et al [21] using electron beam annealing.

Residual defects remaining after RTP can be detrimental to device performance if they form electronic levels in the gap. These residual defects have been investigated using DLTS. Pensl et al [9] found no traps in the upper half of the bandgap in n-type silicon, but a high trap density at Ev + 300 meV in p-type Si. This trap density could be greatly reduced by a slow ramp down after the high temperature anneal or by a subsequent furnace anneal at 450 deg C. In contrast, Ransom et al [10] observed traps in B+ and BF2+ implanted as well as unimplanted n-type Si after RTP and thus concluded that the incoherent light anneal alone introduced defects. A third study by Brotherton et al [11] examined the effect of both furnace and rapid thermal annealing at temperatures up to 900 deg C on boron-implanted pre-amorphised Si, and correlated particular electronic levels with specific defects. Some of these defects persisted even after 900 deg C annealing. They also found that acceptable junction leakage currents were only obtained when the junction edge was significantly deeper than the original amorphous/ crystalline interface. This was also observed by Kamgar et al [12] for As+ implanted junctions, and by Ho et al [13] for As+, B+ and BF2+ implanted junctions. Furthermore Kamgar et al pointed out that the absence of residual defects as detected by TEM was not a sufficient condition to ensure low leakage current.

In summary, RTP using incoherent light is extremely useful for the fabrication of highly doped junctions with appreciably less dopant diffusion than in furnace annealing. However all the reports to date indicate that some diffusion is essential to obtain low leakage currents. Furthermore as pointed out by Hill [14], the shallowness of the junction will always be limited by the transient and concentration enhanced diffusions described above. The investigation of these mechanisms will no doubt remain a fruitful area of research for some time.

REFERENCES
----

[1]   S.R.Wilson, R.B.Gregory, W.M.Paulson [Mater. Res. Soc. Symp. Proc. (USA) vol.52 (1986) p.181-90]
[2]   R.A.Powell, M.L.Manion [ibid p.441]
[3]   F.M. d'Heurle, R.T.Hodgson, C.Y.Ting [ibid p.261-70]
[4]   T.E.Seidel, D.J.Lischner, C.S.Pai, R.V.Knoell, D.M.Maher, D.C.Jacobson [Nucl. Instrum. & Methods Phys. Res. Sect.B (Netherlands) vol.7/8 (1985) p.251-60]
[5]   T.E.Seidel [Mater. Res. Soc. Symp. Proc. (USA) vol.45 (1985) p.7-20]
[6]   J.Narayan, O.W.Holland [J. Appl. Phys. (USA) vol.56 no.10 (15 Nov 1984) p.2913]
[7]   W.Katz, G.A.Smith, R.F.Reihl, E.F.Koch [Mater. Res. Soc. Symp. Proc. (USA) vol.23 (1984) p.299]
[8]   S.J.Pennycook, J.Narayan, O.W.Holland [J. Electrochem. Soc. (USA) vol.132 no.8 (Aug 1985) p.1962-8]
[9]   G.Pensl, M.Schultz, P.Stolz, N.M.Johnson, J.F.Gibbons, J.Hoyt [Mater. Res. Soc. Symp. Proc. (USA) vol.23 (1984) p.347]
[10]   C.M.Ransom, T.O.Sedgewick, S.Cohen [Mater. Res. Soc. Symp. Proc. (USA) vol.52 (1986) p.153-6]
[11]   S.D.Brotherton, J.P.Gowers, N.D.Young, J.B.Clegg, J.R.Ayres [J. Appl. Phys. (USA) vol.60 no.10 (15 Nov 1986) p.3567-75]
[12]   A.Kamgar, W.Fichtner, T.T.Sheng, D.C.Jacobson [Appl. Phys. Lett. (USA) vol.45 no.7 (1 Oct 1984) p.754]
[13]   C.C.Ho, R.Kwor, C.Araujo, J.Gelpey [Mater. Res. Soc. Symp. Proc. (USA) vol.52 (1986) p.225-32]
[14]   C.Hill [Nucl. Instrum. & Methods Phys. Res. Sect.B (Netherlands) vol.19/20 (1987) p.348-58]
[15]   A.Kamgar, F.A.Baiocchi [Mater. Res. Soc. Symp. Proc. (USA) vol.52 (1986) p.23-30]
[16]   W.Hofker [Philips Res. Rep. Suppl. (Netherlands) vol.69 (1975)]
[17]   R.B.Fair, J.J.Wortman, J.Liu [J. Electrochem. Soc. (USA) vol.131 no.10 (Oct 1984) p.2387-94]
[18]   A.E.Michel [Mater. Res. Soc. Symp. Proc. (USA) vol.52 (1986) p.3-14]
[19]   S.R.Wilson, W.M.Paulson, R.B.Gregory, A.H.Hamdi, F.D.McDaniel [J. Appl. Phys. (USA) vol.55 (1984) p.4162]
[20]   A.Armigliato, S.Guimaraes, S.Solmi, R.Kogler, E.Wieser [Nucl. Instrum. & Methods Phys. Res. Sect.B (Netherlands) vol.19/20 (1987) p.512-15]
[21]   N.E.B.Cowern et al [Mater. Res. Soc. Symp. Proc. (USA) vol.52 (1986) p.65-72]

24.4 ELECTRON BEAM ANNEALING OF Si

by H.Ahmed and R.A.McMahon

September 1987
EMIS Datareview RN=16165

The physical principle underlying the annealing of ion implanted layers
and other thermal processes on semiconductors carried out by means of
electron beams is the deposition of energy into the material. High
energy electrons, usually in the range 5 to 30 keV, are scattered by
electrons in the solid and occasionally by nuclei. The energy is
transferred by phonon emission, raising the lattice temperature of the
substrate. It is a feature of electron beams that the absorbed energy
and its distribution into the thickness of the solid semiconductor such
as silicon is largely independent of surface topography and virtually
the same for crystalline, amorphous or molten states of the material.
Furthermore, the power absorbed is not greatly influenced by metallic
or insulating overlayers on the semiconductor. The depth of penetration
depends on the beam energy and the atomic numbers of the substrate
materials.

Electron beam heating has been most widely used for annealing ion
implanted semiconductors but many other thermal treatments of
semiconductor layers have also been carried out. The thermal processes
that have been developed fall into distinctly different time-temperature
regimes which are also distinguished by the different physical processes
that are caused in the layers. Hill [1] has classified a range of beam
processing systems, including lasers, lamps and electron beams, in terms
of the time for the heating cycle, and Ahmed and McMahon [2] have
classified different types of electron beam equipment in a similar
order.

Electron beam systems may be listed in order of increasing time for the
thermal cycle as follows:

1. Pulsed electron beam heating in which the effective heating cycle
    typically lasts for 100 ns, usually melting the surface layer which
    subsequently freezes. This regime is described as adiabatic since
    the heating is localised to the surface in the timescale of the
    process.

2. Scanned electron beam heating, in which each point on the wafer is
    sequentially exposed to the electron beam, and heated locally for
    times of about 1 ms and cools by conduction of heat into the bulk
    of the wafer. This regime is called a thermal flux regime since the
    heat flow reaches a steady state. The technique is suitable for
    annealing selected areas of the specimen by a maskless method and
    generally the silicon remains in the solid phase during the
    process. Apparatus has been described using round or line shaped
    electron beams.

3. The isothermal annealing regime is one in which the electron
    bombardment cycle ensures that the temperature is essentially the
    same throughout the wafer. Heating times range from about 1 s to
    as long as necessary and cooling is by radiation after the peak
    temperature has been achieved and the electron beam switched off.
    The process is always in the solid state.

4. A further emerging regime involves raising the background
    temperature of a thermally isolated wafer to several hundred
    degrees centigrade, with a second heat source giving an additional
    temperature rise at the surface. Cooling is by heat conduction into
    the bulk of the wafer and by radiation. Additionally, a number of
    intermediate heating regimes have been reported, usually involving
    the formation of special spot shapes by scanning.

Pulsed electron beam annealing uses a large area, planar electron
source to bombard the surface of the wafer. High current pulses with a
wide spread of electron energies up to several tens of kV, giving a
power density of tens of MW/cm*2 for pulse durations around 100 ns are
used. The process and its effects on silicon are closely related to
pulsed laser annealing although the energy absorption mechanisms are
different. The technique was described for solar cell production by
Kirkpatrick et al [3]. The annealing of ion implanted silicon was
achieved by melting the surface layer with an electron beam pulse of
about 1 J/cm*2 fluence lasting for about 50 ns. The annealing mechanisms
were investigated by Greenwald et al [4] and the possibilities for
applying the process in device fabrication were explored by Kamins and
Rose [5]. They made diodes with good characteristics and showed that
e-beam induced states existed in MOS structures. Doping depth profiles
after annealing phosphorus implanted silicon were measured by Inada et
al [6]. These showed full activation for implant concentration which did
not exceed the solid solubility limit and the doping profile obtained
was consistent with regrowth of the layer from a liquid phase. The work
on silicon followed on from experiments on annealing of Se implanted
GaAs where carrier concentrations higher than those obtainable by
furnace processing were reported [7]. Annealing of As implants in
silicon was studied by Barbier et al [8]. The results again were
consistent with liquid phase epitaxy following ion implantation damage.
More recently, supersaturated concentrations of arsenic in silicon have
been reported [9]. The technique has been found useful for annealing
uniformly implanted layers and for fundamental physical studies, but it
is not suitable for processing multiple layers with circuit and device
patterns. The melting and re-solidification sequence causes a large
amount of damage and inter-mixing of dopants in different parts of the
substrate.

Scanning electron beam annealing of silicon can be performed in the
solid phase as in CW laser annealing. The substrate is heated point by
point with a focused electron beam again having an energy of several
tens of kV. Although the beam current is usually as small as a few
milliamps the beam current density is high and together with the dwell
time can be varied over a wide range. Only the region underneath the
spot is annealed and a steady state heat flow away from the beam occurs.
Regolini [10] reported the activation of arsenic implants with a 30 keV
beam with a power of 15 W, focused to 300 micron spot, scanned in a
raster at 2.5 cm/s. With a specimen held at a constant background
temperature, usually somewhere between room temperature and about
450 deg C, the temperature is proportional to the beam power per unit
radius. Krimmel et al [11] did not use a background heat source but
chose beam conditions which did not greatly perturb the background
temperature. More recently Ishiwara et al [12] have investigated the
activation of P implants in silicon and the formation of shallow boron
doped junctions in silicon by this method [13]. The process is
relatively slow and not attractive for annealing full silicon wafers
in production. The high lateral temperature gradients in this process
can cause damage such as slip and a high defect density.

In contrast to these systems where the aim is blanket annealing of large areas, the heating of semiconductors by a finely focused beam to effect localised material changes has been explored as a potential method for maskless fabrication. McMahon and Ahmed [14] annealed selected areas of boron-implanted silicon to form diodes with a resolution of about 50 microns similar to that of the beam diameter used. Ratnakumar et al [15] achieved micron-sized features using a modified SEM, and Sun et al [16] explored pattern formation in detail. The technique has possible applications in the future for the fabrication of special structures, for example to investigate the physics of low dimensionality effects in silicon.

Knapp and Picraux [17] described a line electron beam annealing system with a beam energy of 10-50 keV and a current density up to 12 A/cm*2. Beam dwell times of 150 to 160 microsec were shown to regrow, in the solid phase, amorphous layers formed by As implantation. This time regime is rather faster than for single spot systems, and the heating time is close to the minimum times necessary for useful regrowth in the solid phase. The crystalline quality as measured by RBS is claimed to be as good as bulk silicon. Annealing by a static line beam through which heatsinked wafers were mechanically translated has also been reported [18].

In addition to beams focused into line electron sources beam shapes can also be simulated by scanning. Bentini et al [19] formed a 7 mm diameter spot into a 125 mm long line, so that a point in the line scan was exposed to the spot every 10 ms. Cooled, heatsinked wafers were moved mechanically through this line. Phosphorus implants at 10 keV for solar cells were annealed, and carrier concentrations exceeding the solid solubility of the phosphorus were obtained. A similar scheme was used by others to anneal boron implanted junction diodes [20]. Theoretical analysis of thermal profiles and stresses in the substrate resulting from this type of treatment have been discussed by Correra and Bentini [21]. These methods go some way towards finding practical application in silicon device manufacture but significant developments have not been reported subsequent to the initial publications.

The most notable development in the annealing of ion implanted silicon as part of a device manufacturing process has been to replace furnace cycles lasting several tens of minutes by rapid thermal annealing (RTA) lasting only a few seconds. The most common systems use banks of lamps to give heating times of about 10 s; the temperature rise is sometimes followed by a dwell period and then cooling by radiation. This type of heating cycle was first realized with electron beams by McMahon and Ahmed [22] by rapidly scanning an electron beam in multiple, interlaced scans so as to give isothermal heating of the whole substrate. The heating effect of the beam is equivalent to a uniform power density given by the beam power divided by the scanned area. The spot, beam energy, beam current and scan rates are not important between broad limits and critical adjustment of the system is not necessary to obtain reproducible results. The cycle can be readily controlled with a pyrometer as the infrared radiation from the specimen is easily monitored, McMahon et al [23]. The activation of ion implants of B, P, As, Sb and Ga has been described [24]; sheet resistance values indicated close to full activation could be achieved for doses below solid solubility in most cases, with the exception of high dose B implants.

Annealing B implants to form p+/n junction diodes gave devices which had low leakage, similar to furnace annealed controls, and acceptable forward characteristics [25]. This work was extended to ion-implanted NMOS devices with a range of gate lengths. Electron beam processing reduced the extent of diffusion under the gate, gave low leakage currents and did not cause significant shifts in threshold voltage [26].

Detailed work on annealing Bi and As implants in a similar system has been reported by Bontemps et al [27], and more recently by Zheng and Chen [28]. The shortest reported RTA heating cycles, with heating times of about 0.1 s, were used by McMillan et al to study the annealing of very shallow, Rp < 50nm, implants [29]. However, the shortening of the overall heating time is limited by the fact that heat loss for cooling is by radiation only. The formation of metal silicides [30], contact formation on silicon [31], the improvement of silicon-on-sapphire films by Si-implantation and subsequent annealing [32] have also been investigated by isothermal e-beam heating. E-beam isothermal heating is an efficient process which couples almost the full beam power into the silicon wafer. It gives a reproducible heat cycle which can be precisely controlled using a pyrometrically monitored feedback loop. It can give shorter heat cycles than lamp isothermal annealers but the processing has to be carried out in vacuum so there can be some loss of material from the surface. Capping layer and very rapid heat cycles can however be used to prevent this.

As device dimensions continue to shrink, diffusion during the annealing of ion implants has to be restricted to as little as 0.01 micron which means that the heat-cycle must be reduced to milliseconds. A method which overcomes some important difficulties with short dwell time thermal flux methods, such as point-by-point scanning, is the millisecond annealing approach reported by McMahon et al [33]. This uses a dual electron beam approach. One beam gives uniform background heating to 600-750 deg C, avoiding significant solid state diffusion, and the other beam is formed into a line by rapid scanning. Scanning the line over the wafer gives an additional temperature transient of about 1 ms which activates even fast diffusers, such as boron, with very little dopant movement. The relatively high background temperature and consequently lower thermal stresses is the main advantage over other thermal flux approaches. It is expected that e-beam techniques will continue to be developed as the demand for shorter and more controlled heat-cycles continues to grow as part of advanced processing techniques for sub-micron silicon devices.

REFERENCES
----

[1]   C.Hill [Mater. Res. Soc. Symp. Proc. (USA) vol.1 (1981) p.361]
[2]   H.Ahmed, R.A.McMahon [Mater. Res. Soc. Symp. Proc. (USA) vol.13 (1983) p.653]
[3]   A.R.Kirkpatrick, J.A.Minnucci, A.C.Greenwald [IEEE Trans. Electron Devices (USA) vol.ED-24 (1977) p.429]
[4]   A.C.Greenwald, A.R.Kirkpatrick, R.G.Little, J.A.Minnucci [J. Appl. Phys. (USA) vol.50 (1979) p.783]
[5]   T.I.Kamins, P.H.Rose [J. Appl. Phys. (USA) vol.50 (1979) p.1309]
[6]   T.Inada, T.Sugiyama, N.Okano, Y.Ishikawa [Electron Lett. (GB) vol.16 (1980) p.54]
[7]   T.Inada, K.Tokunaga, S.Taka [Appl. Phys. Lett. (USA) vol.35 (1979) p.546]

[8] D.Barbier, A.Laugier, A.Cachard [J. Phys. Colloq. (France) vol.43 no.C-5 (1982) p.411]

[9] V.P.Popov, A.V.Dvurechenskii, B.P.Kashnikov, A.I.Popov [Phys. Status Solidi a (Germany) vol.94 no.2 (1986) p.569-72]

[10] R.A.Regolini, J.F.Gibbons, T.W.Sigmon, R.F.W.Pease, T.J.Magee, J.Peng [Appl. Phys. Lett. (USA) vol.34 (1979) p.410]

[11] E.F.Krimmel, H.Oppolzer, H.Runge, W.Wondrak [Phys. Status Solidi a (Germany) vol.66 (1981) p.565]

[12] H.Ishiwara, K.Suzuki [Proc. Int. Ion Engineering Congress - ISIAT '83 & IPAT '83 Kyoto, Japan, 1983 p.1753]

[13] H.Ishiwara, S.Horita [Jpn. J. Appl. Phys. Part 1 (Japan) vol.24 no.5 (May 1985) p.568-73]

[14] R.A.McMahon, H.Ahmed [J. Vac. Sci. & Technol. (USA) vol.16 (1979) p.1843]

[15] K.N.Ratnakumar, R.F.W.Pease, D.J.Bartelink, N.M.Johnson, J.D.Meindl [Appl. Phys. Lett. (USA) vol.35 (1979) p.463]

[16] H.T.Sun, R.A.McMahon, H.Ahmed [J. Vac. Sci. & Technol. B (USA) vol.1 (1983) p.827]

[17] J.A.Knapp, S.T.Picraux [Appl. Phys. Lett. (USA) vol.38 (1981) p.873]

[18] T.Tu, K.J.Soda, B.C.Streetman [J. Appl. Phys. (USA) vol.51 (1980) p.4399]

[19] G.G.Bentini et al [J. Appl. Phys. (USA) vol.52 (1981) p.6735]

[20] T.O.Yep, R.T.Fulks, R.A.Powell [Appl. Phys. Lett. (USA) vol.38 (1981) p.162]

[21] L.Correra, G.G.Bentini [J. Appl. Phys. (USA) vol.54 (1983) p.4330]

[22] R.A.McMahon, H.Ahmed [Electron. Lett. (GB) vol.15 (1979) p.45]

[23] R.A.McMahon, D.G.Hasko, H.Ahmed [Rev. Sci. Instrum. (USA) vol.56 (1985) p.1257]

[24] R.A.McMahon, H.Ahmed [IEE Proc. I (GB) vol.129 (1982) p.105]

[25] J.D.Speight, A.E.Glaccum, D.Machin, R.A.McMahon, H.Ahmed [Mater. Res. Soc. Symp. Proc. (USA) vol.1 (1981) p.383]

[26] R.A.McMahon, H.Ahmed, D.J.Godfrey, K.J.Yallup [IEEE Trans. Electron Devices (USA) vol.ED-30 (1983) p.1550]

[27] A.Bontemps, H.J.Smith, R.Danielou [J. Appl. Phys. (USA) vol.53 (1982) p.5258]

[28] L.Zheng, J.Chen [Nucl. Instrum. & Methods Phys. Res. Sect. B (Netherlands) vol.6 no.1/2 (1985) p.321-4]

[29] G.B.McMillan, J.M.Shannon, J.B.Clegg, H.Ahmed [J. Appl. Phys. (USA) vol.59 no.8 (15 Apr 1986) p.2694-703]

[30] E.A.Maydell-Ondrusz, R.E.Harper, I.H.Wilson, K.G.Stephens [Vacuum (GB) vol.34 (1984) p.995]

[31] J.Y.Chen, D.B.Rensch [IEEE Trans. Electron Devices (USA) vol.ED-30 (1983) p.1542]

[32] D.J.Smith, L.A.Freeman, R.A.McMahon, H.Ahmed, M.G.Pitt, T.B.Peters [J. Appl. Phys (USA) vol.56 (1984) p.2207]

[33] R.A.McMahon, D.G.Hasko, H.Ahmed, W.M.Stobbs, D.J.Godfrey [Mater. Res. Soc. Symp. Proc. (USA) vol.35 (1985) p.347]

CHAPTER 25

GETTERING TECHNIQUES

25.1 EXTRINSIC GETTERING IN Si TECHNOLOGY

by K.V.Ravi

June 1987
EMIS Datareview RN=15718

A. INTRODUCTION

The presence of imperfections and unwanted impurities in silicon
crystals has been demonstrated to have adverse effects on the
performance and the yield of devices and circuits fabricated in the
silicon [1]. As a consequence techniques for the elimination and the
control of defects and impurities have been developed. The general
approach to defect and impurity control in silicon can be divided into
three categories:

a) The prevention of defect and impurity introduction into crystals
   and devices during the course of material preparation and device
   fabrication is clearly the preferred method for the creation of
   defect-free structures in silicon. In this context some of the major
   techniques available include the following: The growth of silicon
   crystals in the presence of magnetic fields to minimize thermal
   convection currents has been shown to suppress the formation of point
   defect complexes which have adverse effects on device performance and
   yield [2]. The control over the concentration and the distribution
   of impurities such as the transition metals, oxygen and carbon is
   found to be important in controlling defect formation processes. In
   wafer processing defect-free processes include low temperature
   processing, the use of extremely clean processing ambients, process
   chemicals and gases, and approaches for the selective heating of
   silicon wafers to minimize the operation of unwanted thermally
   activated processes.

b) Notwithstanding the attempts to create materials and process them in
   such a manner as to minimize defect introduction, in practice it is
   found that unwanted contamination of silicon wafers is a frequent
   occurrence in device fabrication. As a result, processes for the
   removal of the unwanted impurities from critical regions of the
   wafers have been developed which are generically termed gettering
   techniques.

c) The third approach for impurity and defect control is the process
   of defect passivation or the electrical inactivation of electrically
   active defect states by the use of appropriate physical and chemical
   techniques.

In this review extrinsic gettering techniques in silicon technology are
discussed. Gettering can be generally defined as the process of moving
unwanted impurities and defects from regions of the wafer which are
critical from the point of view of device operation to regions, either
within the wafer or outside of the wafer, where their presence would
not cause any harm. Gettering techniques can be classified into
extrinsic and intrinsic techniques. Extrinsic gettering techniques
involve moving unwanted impurities from the critical regions of the
wafer to regions outside the wafer as well as to the wafer surfaces.
These processes typically involve the use of external mechanical or
chemical agents. Intrinsic gettering is discussed in another Datareview
[16].

The effectiveness of gettering depends upon the operation of three processes: (i) The defect or the impurity to be gettered has to be dissolved in the silicon in order that it can be subsequently transported to another location in the wafer by solid state diffusion. The impurity or defect nuclei should be in solid solution rather than in the form of a stable, immobile precipitate in the silicon. (ii) The impurity should be transportable through the silicon once it is in solid solution. The temperature of the process should be sufficiently high as to allow solid state diffusion of the relevant impurity and a suitable diffusion gradient has to be available in order that the impurity can be moved away from the critical regions of the wafer. (iii) The impurity should be captured at a position removed from the critical regions of the wafer and not be re-released into the wafer upon subsequent heat treatment.

B. STACKING FAULT ANNIHILATION PROCESSES

One of the critical defects which have a strong impact on the electrical characteristics of microelectronic products is the stacking fault which is generated during the oxidation of silicon [3]. Extrinsic gettering techniques have been effectively utilised to reduce the size and eliminate stacking faults from the critical near-surface regions of silicon wafers. The most effective way of controlling the formation of oxidation induced stacking faults is by the appropriate manipulation of the concentration and distribution of point defects in the crystal. The formation and growth of stacking faults is a result of oxidation enhanced production of a supersaturation of silicon self-interstitials [4]. A reduction in the growth rate of stacking faults requires the creation of an undersaturation of interstitials or a supersaturation of vacancies. Several techniques have been developed for achieving the appropriate vacancy supersaturation in the wafers leading to the annihilation of stacking faults. The most important of these are:

(i)   Oxidation at high temperature for extended periods of time has been demonstrated to eliminate stacking faults from the structure [5]. For a given oxidation time the stacking fault size is found to increase with temperature following the expected Arrhenius relationship, with a peak in the fault size being achieved at some temperature followed by a sharp decrease in fault size with further increase in the temperature. It has been proposed that a change in the diffusion mechanism occurs as the thickness of the oxide increases [6]. At low temperatures and short times SiO2 formation occurs at the oxide/silicon interface resulting in the injection of self-interstitials into the silicon in response to the need of the system to accommodate the volume change associated with the reaction. At higher oxidation temperatures the diffusion mechanism changes from one of anion diffusion through the growing oxide to one of cation diffusion with the silicon from the Si-SiO2 interface counterdiffusing against the flow of oxygen through the oxide, with SiO2 formation occurring somewhere within the oxide film, with the attendant volume expansion being accommodated by viscoelastic relaxation in the oxide without disturbing the interface. The flow of cations, silicon atoms, from the interface into the oxide leads to an interstitial undersaturation and the shrinkage of the stacking faults.

(ii) A reduction in the self-interstitial concentration at the surface
      during oxidation can also be achieved by performing the oxidations
      in a halogen-containing ambient [7]. The addition of small amounts
      of HCl, trichloroethylene, and 1,1,1-trichloroethane has been
      demonstrated to cause the shrinkage and the eventual disappearance
      of stacking faults. Chlorine reacts at the oxide-silicon interface
      leading to a decrease in the injection of silicon self
      interstitials and the shrinkage of stacking faults. The effects
      of chlorine are found to be most pronounced at temperatures in
      excess of 1100 deg C where a dynamical equilibrium between
      vacancies and interstitials is quickly established.

(iii) A self-interstitial undersaturation at the interface can also be
      established by the nitridation of the silicon. The nitridation
      reaction does not take place at the interface but occurs at the
      nitride surface with the silicon cations migrating through the
      nitride and reacting with the nitrogen at the free surface [8].

C. IMPURITY GETTERING

The gettering of fast diffusers and oxygen from critical regions of the
wafer can be achieved by providing an external sink for the impurities
and sufficient energy in the form of temperature to cause the diffusion
of impurities to the sinks at the wafer surfaces. The sinks can be
based on inducing chemical reactions between the impurity and another
element introduced at the surface or the creation of mechanical stress
in the form of mechanical damage at the surface. Several approaches to
extrinsic gettering have been developed and the key ones are shown in
TABLE 1.

Surface abrasion [9], ion implantation [10] and the treatment of the
surface with energetic beams such as the use of lasers or electron
beams [11] have the effect of generating dislocations near the surface.
The dislocations can provide the appropriate sinks for fast-diffusing
impurities when the wafers are heat treated following the back side
damage generation process resulting in the impurities being gettered
from the critical regions of the wafer. Gettering by the diffusion of
phosphorus can involve both physical gettering by the misfit
dislocations generated as a result of the diffusion process, and
chemical gettering whereby a reaction between heavy metals and
phosphorus takes place with the resultant removal of the metals from
the active regions of the wafer [12]. Phosphorus gettering can be
performed on both the front, active surface of the wafer and the back
side of the wafer. Often phosphorus gettering is a natural
accompaniment to device processing operations such as the diffusion
of emitters in transistors, the formation of heavily doped contact
regions, etc.

TABLE 1: extrinsic gettering techniques

Back Surface	Front Surface
Surface abrasion	Controlled misfit dislocation generation (e.g. Si-Ge alloy)
Phosphorus diffusion	Phosphorus diffusion
Silicon nitride deposition	Use of HCl ambients

TABLE 1 (continued)

Ion implantation

Polysilicon deposition

Beam processing (laser,
electron beam)

Extrinsic gettering has also been achieved by the controlled
introduction of misfit dislocations in the front surface of wafers by
the growth of Si-Ge alloys on the silicon substrate [13]. The deposition
of a thin layer of polycrystalline silicon on the back surface has also
been found to be effective in gettering impurities [14]. The grain
boundaries in the polysilicon function as sinks for the impurities and
these grain boundaries are less subject to being annealed out during
subsequent high temperature processing than the damage created by
processes such as back surface abrasion, ion implantation and other
mechanisms of gettering damage in the wafers. Elastic stresses at the
back surface of wafers can also be generated by the deposition of
silicon nitride films by low temperature RF sputtering or by chemical
vapour deposition [15]. Sufficiently large stresses (5 X 10*9 to 5 X
10*10 dynes/cm*2) can be introduced at the nitride-silicon interface
to provide the necessary sink for impurities when the wafers are heat
treated at elevated temperatures.

One of the limitations of extrinsic gettering processes is their lack
of stability. If the gettering process is followed by high temperature
processes it is possible to reverse the gettering action and
re-introduce the gettered impurities into the wafer. Consequently the
point in the fabrication sequence that gettering is introduced is
critical for its optimum operation. In general, temperatures and times
higher than those employed during gettering should be avoided after the
gettering treatment.

REFERENCES
----

[1]   K.V.Ravi [Imperfection and Impurities in Semiconductor Silicon
      (J. Wiley & Sons, New York, 1981)]
[2]   K.Hoshi, N.Isawa, T.Suzuki, Y.Ohkubo [J. Electrochem. Soc. (USA)
      vol.132 (1985) p.693]
[3]   K.V.Ravi, C.J.Varker, C.E.Volk [J. Electrochem. Soc. (USA)
      vol.120 (1973) p.533]
[4]   W.Frank, U.Gosele, H.Hehrer, A.Seeger [in 'Diffusion in
      Crystalline Solids' Eds G.Murch, A.S.Nowick (Academic, New York,
      1984) p.63]
[5]   S.P.Murarka [J. Appl. Phys. (USA) vol.49 (1978) p.2513]
[6]   U.Gosele, T.Y.Tan [Defects in Semiconductors II, Symp. Proc.
      Eds S.Mahajan, J.S.Corbett (North-Holland, New York, 1983) p.45]
[7]   H.Shiraki [Jpn. J. Appl. Phys. (Japan) vol.15 (1976) p.1-83]
[8]   H.Hayafugi, K.Kajiwra, S.Usui [Jpn. J. Appl. Phys. (Japan)
      vol.53 (1982) p.8639]
[9]   W.T.Stacy, M.C.Arst, K.N.Ritz, J.G.de Groot, M.H.Norcott [Defects
      in Silicon Processing, Electrochem. Soc. Symp. vol. 85-9
      Ed. W.M.Bullis, L.C.Kimerling (The Electrochemical Society, 1983)
      p.423]
[10] H.J.Geipel, W.K.Tice [IBM J. Res. & Dev. (USA) vol.24 (1980)
      p.310]

[11] Y.Hayafugi, T.Yanade, A.Aoki [J. Electrochem. Soc. (USA) vol.128
      (1981) p.1975]
[12] A.Ourmazd, W.Schroter [Mater. Res. Soc. Symp. Proc. (USA)
      'Impurity Diffusion and Gettering in Silicon' Eds. R.B.Fair,
      C.W.Pearce, J.Washburn (1985) p.25]
[13] A.S.M.Salih, H.J.Kim, R.F.Davis, G.A.Rozonyi [Semiconductor
      Processing (ASTM STP 850) Ed D.C.Gupta (ASTM, Philadelphia, USA,
      1984)]
[14] G.A.Rozgonyi, C.Pearce [Appl. Phys. Lett. (USA) vol.32 (1978)
      p.747]
[15] P.M.Petroff, G.A.Rozgonyi, T.T.Sheng [J. Electrochem. Soc.
      (USA) vol.123 (1976) p.565]
[16] K.V.Ravi [EMIS Datareview RN=15719 (June 1987) 'Intrinsic
      gettering in Si technology']

25.2 INTRINSIC GETTERING IN Si TECHNOLOGY

by K.V.Ravi

June 1987
EMIS Datareview RN=15719

Intrinsic or internal gettering is a process which is based on the control of oxygen in silicon crystals [1]. The approach utilizes oxide precipitates and associated defects to internally getter unwanted impurities away from regions of the wafer in which the active elements of devices and circuits are fabricated.

The general principles of internal gettering can be described as follows:

1. The initial step is the creation of an oxygen denuded zone at the wafer surface in which the active elements of the circuits are fabricated. The depth of the denuded zone should be such that precipitates that are subsequently formed in the bulk of the crystal do not intrude into the denuded zone since precipitates occurring within this zone will result in electrical aberrations.

2. The second step involves the formation of oxide precipitates by the appropriate heat treatment of the oxygen rich bulk of the crystal. The oxygen, in supersaturation at lower temperatures, will precipitate as silicon dioxide. The misfit stresses generated as a consequence of the precipitation process can result in dislocation generation around the precipitates. Additionally the volume change associated with precipitation results in the injection of excess silicon interstitials into the matrix which can lead to the formation of stacking faults around the precipitates. The resulting precipitate defect colonies (PDC's) function as local gettering agents.

3. The diffusion of impurities to the defect trapping sites completes the gettering process.

Impurities in silicon, with particular reference to transition metals, cause two related problems. They can generate near surface defects and they can introduce deep states in the band gap. Both these processes lead to a variety of electrical aberrations in microelectronic products [2]. Consequently gettering techniques play a very important role in the processing of complex circuits.

The creation of an oxygen denuded zone of a depth of several microns at the surface can be achieved by heating the wafers to elevated temperatures (e.g. 1300 deg C) which has the effect of dissolving any oxide precipitates formed during crystal growth as well as promoting the out diffusion of oxygen from the near surface region. Out diffusion and hence the depth of the denuded zone can be controlled by the choice of the appropriate ambient during heat treatment. Heat treatment in non-oxidising ambients increases the depth of the denuded zone whereas the presence of steam or oxygen in the ambient results in shallow oxygen depleted regions. Dry oxygen has been shown to inhibit the formation of denuded zones since dry oxidation is believed to cause the injection

of silicon interstitials into the surface thus preventing oxygen out-
diffusion [3]. The use of HCl in the oxidising ambient results in the
formation of deep denuded zones since HCl injects vacancies into the
silicon thus promoting oxygen out diffusion. The depth of the denuded
zone must be carefully controlled with the particular device type in
mind. For example dynamic RAM's only need a relatively shallow denuded
zone since much of the electrical activity of such devices takes place
close to the surface. On the other hand certain bipolar structures and
CCD imaging devices require denuded zone depths of 20 to 40 microns [4].

Following the formation of the denuded zone oxide precipitation within
the bulk of the wafer is induced by suitable heat treatment. A two step
heat treatment involving a lower temperature (< 900 deg C) heat
treatment to induce the formation of stable oxygen nuclei followed by a
higher temperature (> 900 deg C) heat treatment to induce
precipitate growth and the formation of secondary defects is found to
be effective in generating the necessary PDC's [5]. The electrical
characteristics of the wafer following the internal gettering process
will be represented by a high lifetime region near the surface(s) and a
low lifetime in the bulk of the wafer.

The effectiveness of intrinsic gettering processes is dependent upon
the concentration and the distribution of oxygen in the silicon.
Homogenising the distribution of oxygen in silicon crystal by the
utilisation of processes such as magnetic Czochralski growth [6] can
result in a more homogeneous distribution of PDC's in the wafer upon
subsequent heat treatment. Turbulent convection in the melt during
crystal growth is found to be effectively suppressed and temperature
fluctuations reduced by the application of transverse magnetic fields
over 2000 G to the melt. A reduction in forced convection has been
shown to have several beneficial effects. The uniformity of the
resistivity along the length and across the radius of the crystal is
found to be improved. Reduction in the thermal convection currents
results in a reduced degree of melt stirring and thus a reduced degree
of reactivity between the melt and the silica crucible, the source of
oxygen in the melt. It has been demonstrated that if the oxygen content
of the crystal is high, oxygen precipitation upon subsequent heat
treatment occurs quickly and the depth of the denuded zone can be
limited. On the other hand if the oxygen concentration is very low
precipitation may be limited thus reducing the effectiveness of the
intrinsic gettering process [7]. Consequently the use of magnetic
Czochralski techniques offers the possibility of controlling the
oxygen concentration to the optimum level for maximising the
effectiveness of the gettering process.

The concentration level of oxygen in the crystal can also be controlled
by the use of continuously replenished melts during crystal growth [8].
In conventional Czochralski growth the melt volume in the crucible
changes continuously as the crystal is grown. Continuously feeding
molten silicon to the crucible can make the process more steady state
in nature and the crystals can be grown from small, constant melt
volumes with low aspect ratios wherein forced convection dominates. The
longitudinal oxygen distribution can be kept constant and by growing
crystals with different aspect ratios the concentration level of oxygen
can be controlled.

An undesirable side effect of intrinsic gettering is the effect of
oxygen precipitation on the mechanical properties of the silicon wafers.
Oxygen in solid solution strengthens silicon and retards the motion of
dislocations. When oxygen is removed from solid solution as a
consequence of the precipitation process the dislocation pinning effect

of oxygen is eliminated with the result that the wafers are more susceptible to warpage at high processing temperatures. Consequently to minimise wafer warpage it is important to avoid excessively high temperature processes following intrinsic gettering.

Although the vast majority of work on intrinsic gettering has been done by the manipulation of the nature of oxygen in silicon it has been suggested that intrinsic gettering is also possible in very low oxygen or oxygen-free crystals such as those grown by the magnetic Czochralski process or the float zone process [9]. It is proposed that the intrinsic gettering centres are composed of excess silicon interstitials which form interstitial dislocation loops which attract and getter metallic impurities from the interstitial denuded region at the wafer surface. The formation of the denuded zone is determined by surface reactions which can be controlled by the appropriate choice of the heat treating ambient.

Intrinsic gettering has been effectively utilised to improve the performance and the yield of a wide class of devices. Refresh loss in dynamic RAM's is markedly reduced by utilising intrinsic gettering processes during device processing [10]. Electrical malfunctions, such as excess reverse currents and emitter collector pipes, have been reduced by intrinsic gettering in bipolar structures [11].

Internal gettering results in a material with differing electrical properties at different regions of the silicon wafer with a high lifetime, defect-free denuded zone near the surface in which the active elements of circuits are resident and a low lifetime region beneath the denuded zone which is the repository of gettering agents and gettered impurities. The low lifetime region is for most devices an unwanted region. However it has been shown that a certain class of devices benefit from the presence of the low lifetime region in the bulk of the wafer. CMOS structures which involve the building of adjacent p- and n-channel devices generally separated from each other by field oxides, are subject to a failure mode referred to as 'latch up'. Due to the close proximity of the n- and the p- channels to each other parasitic lateral pnp and vertical npn transistors are created. These can form parasitic pnpn circuits leading to regenerative switching along this path resulting in the phenomena of latch up which interferes with the normal operation of the circuit. Latch up becomes more prevalent as the spacing between the two complementary devices decreases. An effective way of avoiding latch up is to provide local recombination centres within the bulk of the semiconductor beneath the active regions of the device which restrict or eliminate the flow of parasitic currents by promoting the recombination of carriers that contribute to the flow of parasitic currents. The generation of PDC'S within the bulk during the process of intrinsic gettering has been shown to serve this purpose [12]. A similar process has been shown to cure the phenomena of image cross talk in photo sensing arrays [13]. A reduced lifetime in the bulk of the wafer is also found to be beneficial in reducing soft errors in memory circuits.

REFERENCES
----

[1]   R.A.Craven [Mater. Res. Soc. Symp. Proc. (USA) 'Impurity
        Diffusion and Gettering in Silicon' Eds R.B.Fair, C.E.Pearce,
        J.Washburn (1985) p.159]
[2]   K.V.Ravi [Imperfections and Impurities in Semiconductor Silicon
        (J. Wiley & Sons, New York, 1981)]
[3]   R.B.Fair [J. Electrochem. Soc. (USA) vol.128 (1981) p.1360]
[4]   R.A.Craven [International Electron Devices Meeting 81, Technical
        Digest (IEEE, USA, 1981) p.228]
[5]   W.K.Tice, T.Y.Tan [Appl. Phys. Lett. (USA) vol.28 (1976) p.564]
[6]   M.Ohwa, T.Higuchi, E.Toji, M.Watanabe, K.Homma, S.Takasu
        [Semiconductor Silicon, 1986, Eds H.R.Huff, T.Abe, B.Kolbesen
        (1986) p.117]
[7]   R.A.Craven [Semicond. Int. (USA) vol.8 (1985) p.134]
[8]   G.Fiegl [Solid State Technol. (USA) (1983) p.121]
[9]   K.Nauka, J.Lagowski, H.C.Gatos, O.Ueda [J. Appl. Phys. (USA)
        vol.60 (1986) p.615]
[10]   H.R.Huff, H.F.Schaake, J.T.Robinson, S.C.Barber, D.Wong [J.
        Electrochem. Soc. (USA) vol.130 (1983) p.1551]
[11]   L.Jastrzebski, R.Soydan, B.Goldsmith, J.T.McGinn [J. Electrochem.
        Soc. (USA) vol.131 (1984) p.2944]
[12]   J.O.Borland, R.S.Singh [VLSI Science and Technology/1985 Eds
        W.M.Bullis, S.Bryoydo (The Electrochemical Society,1985) p.77]
[13]   C.N.Anagnostopoulos, E.T.Nelson, J.P.Lavine, K.Y.Wong,
        S.N.Nichols [IEEE Trans. Electron Devices (USA) vol.ED-31
        (1984) p.225]

25.3 HYDROGEN PASSIVATION OF DEFECTS IN Si TECHNOLOGY

by K.V.Ravi

June 1987
EMIS Datareview RN=15720

Metallic impurities and crystallographic defects in silicon represent
the primary causes for the electrical degradation of crystals, films
and devices fabricated in these materials. In particular, the presence
of metallic impurities has been shown to cause a reduction in the
minority carrier lifetime leading to phenomena such as a reduction in
the storage time in MOS devices, a degradation in the energy conversion
efficiency of solar cells and a reduction in the gain of bipolar
transistors. Typically, impurity gettering techniques are utilized to
remove unwanted impurities from the active regions of the structures
to improve device electrical properties. A process which has gained
significant attention in recent years is the electrical passivation or
the neutralisation of impurities and defects in silicon by the use of
ionised hydrogen.

The early use of hydrogen in passivating defects in semiconductors was
in reducing the density of defect states in amorphous silicon films
formed by the plasma aided decomposition of silane. The defect states
which are thought to be predominantly composed of dangling or
unsatisfied silicon bonds are occupied by hydrogen thus electrically
neutralising them. The presence of hydrogen is found to have a
significant effect on a variety of properties of the films including
a reduction in the density of states, an improvement in the
photoconductivity and the ability to fabricate devices such as solar
cells and thin film transistors with very respectable characteristics
[1].

Hydrogen passivation techniques have been extended to cope with the
defect state in crystalline silicon. Passivating defects such as grain
boundaries, dislocation arrays and other extended defects has been
shown to be feasible with hydrogen [2]. In the case of amorphous silicon
films hydrogen introduction is an integral part of the film deposition
process with hydrogen being a by-product of the decomposition of silane
and being incorporated in the film as it is formed. With crystalline
silicon structures hydrogen is added to crystals and devices either by
ion implantation or by the use of plasma techniques. Hydrogen can be
ionised, accelerated and implanted into silicon crystals. The second
approach is the use of inductively or capacitively coupled RF hydrogen
plasmas to generate highly mobile and reactive atomic hydrogen. A
similar effect can be achieved by the use of the abundant hydrogen
contained in plasma deposited silicon nitride films which are generally
used as overcoatings in integrated circuit manufacture.

Using such techniques it has been demonstrated that grain boundaries in
silicon crystals can be passivated leading to improved electrical
characteristics. It has been demonstrated that in polysilicon wafers and
shaped crystals grown by the edge defined film fed growth (EFG) process,
both of which are prone to contain a high population of grain
boundaries and other extended defects, defect passivation is possible
with hydrogen leading to an enhancement in the energy conversion
efficiency of solar cells fabricated in these less than perfect

crystals [3]. Using electron beam induced current display techniques it has been demonstrated that recombination at grain boundaries in polysilicon crystals can be markedly reduced by the presence of hydrogen. Hydrogen passivation in integrated circuit manufacture is of interest in developing the technologies of three dimensional integration. As limits to scaling of the conventional two dimensional integration are approached three dimensional integration offers some advantages. The fabrication of three dimensional circuits involves the building of active elements of circuits on top of each other while electrically isolating them from each other. Electrical isolation requires the presence of layers of insulating films such as SiO2 films between the active layers of semiconductors. This requires the growth of device quality silicon films on top of insulating layers. The growth of silicon films on insulating films results in polycrystalline films which are not suitable for device fabrication because of the presence of grain boundaries in them. Improvement in the characteristics of devices fabricated in the polycrystalline films has been achieved by hydrogen passivation of the grain boundaries [4]. The mechanism proposed for the effectiveness of hydrogen passivation of grain boundaries is based on the proposition that the interface states due to grain boundaries trap carriers and become charged. As a result a depletion region is created surrounding the grain boundary in order that local charge neutrality is maintained. The depletion region functions as a potential barrier reducing mobility. Upon hydrogen passivation atomic hydrogen is postulated to diffuse along grain boundaries and become attached to some of the dangling bonds with the result that the interface state density drops and the potential barrier is lowered. The result is that the polycrystal tends to function more like a single crystal. Grain boundary passivation in MOSFET's built in polycrystalline films has been shown to result in a decrease in the threshold voltage, and improvement in the channel mobility and a reduction of junction leakage currents.

In the absence of extended defects hydrogen has also been shown to be effective in passivating impurities [5]. A reduction in the electrically active concentration of defects associated with Au, Fe, Cu and Ag in silicon has been achieved by exposure to hydrogen plasmas. The effect of hydrogen however has been found to be different for different impurities. The fast diffusing metals such as Cu, Fe, Ag and Au are passivated but slow diffusers such as Ti and V are not. Hydrogen has not been found to be effective in passivating deep states due to the slow diffusers [6]. TABLE 1 shows some of the properties of deep level impurities prior to and following hydrogen passivation. The difference in the passivation behaviour of the fast and the slow diffusers has been explained on the basis that the mechanism is not one of hydrogen passivation but rather one of enhanced gettering of the fast diffusers by a combination of hydrogen enhanced diffusion of the fast diffusers and the creation of ion bombardment damage at the surface due to the implantation of the hydrogen [6]. This mechanism however does not explain the fact that hydrogen implantation induced damage is not likely when hydrogen is introduced into the crystal by the use of plasma techniques such as during the plasma deposition of silicon nitride. In addition the electrical deactivation of slow diffusers such as the group-III acceptors by hydrogen has been demonstrated [7]. It is clear that although the beneficial electrical influence of hydrogen has been amply demonstrated the detailed mechanisms responsible for the effects are not understood.

TABLE 1

Impurity	Concentration of defects/cm*3	
	As grown crystal	Following H passivation
Ti	10*14	10*14
Au	7 X 10*14	1.3 X 10*13
Cr	2.3 X 10*14	1.4 X 10*12
V	10*14	8.6 X 10*13

REFERENCES
----

[1] A.Madan [in 'Silicon Processing for Photovoltaics' Eds
     C.P.Khattak, K.V.Ravi (North-Holland, 1985) p.331]
[2] J.I.Hanoka, C.H.Seager, D.J.Sharp, J.K.G.Panitz [Appl. Phys.
     Lett. (USA) vol.42 (1983) p.618]
[3] K.V.Ravi, R.Gonsiorawski, A.R.Chaudhuri [Proc. 18th IEEE
     Photovoltaic Specialists Conf. Proc. (1985) p.1222]
[4] S.D.S.Malhi et al [IEEE Trans. Electron Devices (USA) vol.ED-32
     (1985) p.258]
[5] A.J.Tavendale, S.J.Pearton [J. Appl. Phys. (USA) vol.54 (1983)
     p.1357]
[6] R.Singh, S.J.Fonash, A.Rohatgi [Appl. Phys. Lett. (USA) vol.49
     (1986) p.29]
[7] S.C.S.Pan, C.T.Sah [J. Appl. Phys. (USA) vol.60 (1986)
     p.156]

CHAPTER 26

POLYCRYSTALLINE SILICON

26.1 CHEMISTRY OF GRAIN BOUNDARIES IN POLY-Si

by C.R.M.Grovenor

September 1987
EMIS Datareview RN=15777

The current understanding of the structure of grain boundaries in
polycrystalline silicon has been described in [24] where it was
demonstrated that the core of some highly symmetric boundaries can be
thought of as a two dimensional array of characteristic structural
units. Because some of these units are highly distorted from the six
membered ring unit characteristic of the perfect silicon lattice they
are expected to offer preferential segregation sites for impurity atoms
present in the grain interiors. Segregation phenomena in semiconductor
grain boundaries have recently been reviewed by Grovenor [1].
Segregation to grain boundaries can alter the electronic properties
of polycrystalline semiconductor materials quite dramatically, and some
authors have suggested that the electrical activity of grain boundaries
is wholly a result of impurity segregation [2]. In this Datareview the
segregation of contaminant species to silicon grain boundaries$ill be
described first, followed by a discussion of the effect of dopant
segregation.

Two classes of experimental techniques have been used to study the
chemistry of grain boundaries in polycrystalline silicon - surface
analysis techniques allied with depth profiling, AES and SIMS [3,4] and
Scanning Transmission Electron Microscope studies [5]. Two dimensional
SIMS and AES maps of grain boundary composition [4] give particularly
clear and detailed information on the chemistry of silicon grain
boundaries.

One of the most widely studied impurity segregation phenomena in
silicon grain boundaries is that of oxygen diffusion to the grain
boundaries during annealing treatments at over 600 deg C [4,6,7]. The
increasing concentration of oxygen at the grain boundaries as the
annealing time or temperature is increased has been linked with a
larger grain boundary potential barrier. This decreases the
conductivity of the polycrystalline aggregate, and also increases the
rate of recombination at the boundaries [2,8]. Fluorine seems to
segregate to silicon grain boundaries as well, and has a similar effect
to that of oxygen [8]. It is thus expected that annealing any large
grained silicon sheet intended for solar cell production will result
in a strong degradation of device performance. Most material of this
kind contains a high oxygen concentration. Whether the oxygen forms
oxide precipitates at the grain boundaries [9], or is present as
individual segregated atoms along the boundary plane, is not clear in
most cases. The more severe the segregation the more likely it will be
that precipitates are formed.

The passivation of grain boundaries in silicon sheet with hydrogen has
also been widely investigated following the identification of dangling
bonds at boundaries with ESR [10]. It has been assumed that the
hydrogen will saturate the dangling bonds and lower the electrical
activity of the boundaries. Clear observations have been made of the
segregation of hydrogen to grain boundaries in silicon [4,7].

Hydrogenation treatments have been shown to decrease the potential
barriers and the recombination rates at grain boundaries [7,11,12], and
increase the efficiency of silicon solar cells [12]. The replacement
of oxygen in silicon grain boundaries by hydrogen has also been
observed [4], although the nature of the exchange reaction between
the two elements is not yet determined. Kazmerski [4] has suggested
that the hydrogen segregating to oxygen rich grain boundaries saturates
the oxygen induced dangling bonds by the formation of hydroxyl groups.
This reaction offers a most valuable method for improving the
performance of polcrystalline silicon solar cells.

Large grained silicon sheet is produced by a number of techniques,
including dendritic web, edge defined film fed, and casting processes
[13]. In most of these sheets the impurity levels will be quite high
since efforts are being made to use less pure, and so cheaper, starting
materials than normally required for VLSI applications. If graphite dies
are used to shape the sheets, carbon will also be present in levels
around 10*17/cm*3. SiC precipitates have been observed to form at grain
boundaries in particularly heavily contaminated material as a result of
a segregation process [14]. These precipitates increase the rate of
recombination around the grain boundaries. Titanium and aluminium have
also been shown to segregate to silicon grain boundaries [3,15], and
the segregated boundaries appear to have a higher potential barrier
than 'clean' boundaries. It is likely that this effect is linked
primarily with the segregated titanium rather than the aluminium.
Sundaresan et al [16] have shown that indiffusion of aluminium along
silicon grain boundaries decreases the local recombination rate by an
order of magnitude. It is clear that many impurities will increase the
potential barrier at silicon grain boundaries, but some reduction in
boundary activity can also be effected by the controlled segregation
of hydrogen, aluminium and possibly some other species as well.
However, in situations where gross segregation and precipitation has
occurred it is unlikely that passivation treatments will be very
effective.

Dopant segregation to silicon grain boundaries has been investigated
primarily by STEM experiments [5,17,18], although Rutherford
Backscattering [19] and AES [4] have also been used. The results from
these experiments are generally in agreement that phosphorus and
arsenic segregate strongly to silicon grain boundaries to a saturation
level between 0.1 and 1 monolayer. This is an equilibrium segregation
process so that the saturation segregation level decreases with
increasing temperature as expected from the McLean model of
interfacial segregation [20]. The activation energies for segregation
calculated from this data are; As - 0.65 eV [5] and P - 0.33 eV [17].
Interestingly, little evidence has yet been found for the segregation
of boron to silicon grain boundaries, although aluminium does seem to
segregate quite freely [3,16].

There is some disagreement over whether the segregated dopant atoms
are ionized in grain boundaries or not. Some authors [21,22] have
assumed that the dopant atoms are inactive once segregated, but then
their Hall mobility data on polycrystalline silicon films heat treated
to promote grain boundary segregation can only be interpreted if 2-5
monolayers of dopant are present at every grain boundary. This is
difficult to reconcile with the direct experimental data which
identifies much lower saturation segregation levels. Wong et al [23]
have suggested that n-type dopants first segregate to grain boundary

sites adjacent to dangling bonds where they will not be ionized . This will result in a decrease in the boundary barrier height. At a later stage in the segregation process the dopant atoms can segregate to other sites in the grain boundaries, where they can be ionized and create a positively charged sheet of material at the boundaries. The full implications of this suggestion on the electronic properties of grain boundaries in silicon have not yet been considered, but it is possible to speculate that heavily segregated grain boundaries in polysilicon may have no significant recombination rate [4] but still present a Coulombic barrier to current flow [23].

We should note that most segregation phenomena only occur after high temperature heat treatments for relatively long times, several hours at 700-900 deg C for example. This is because of the necessity for diffusion of impurity or dopant species to the grain boundaries from the grain interiors. However, as-deposited CVD polysilicon already shows significant arsenic segregation [5], and rapid diffusion along grain boundaries from an external source can occur at relatively low temperatures - from an aluminium contact for instance. We must therefore expect segregation phenomena to have a significant influence on the electrical properties of both large and small grained polysilicon material under most circumstances.

REFERENCES
----

[1]   C.R.M.Grovenor [J. Phys. C (GB) vol.18 (1985) p.4079]
[2]   D.Redfield [Appl. Phys. Lett. (USA) vol.38 (1981) p.174]
[3]   L.L.Kazmerski, P.J.Ireland [J. Vac. Sci. & Technol. (USA) vol.17
       (1980) p.525]
[4]   L.L.Kazmerski [J. Vac. Sci. & Technol. A (USA) vol.4 (1986)
       p.1638]
[5]   C.R.M.Grovenor, P.E.Batson, D.A.Smith, C.Wong [Philos. Mag. A
       (GB) vol.50 (1984) p.409]
[6]   G.J.Russell, M.J.Robertson, B.Vincent, J.Woods [J. Mater. Sci.
       (GB) vol.15 (1980) p.939]
[7]   L.L.Kazmerski, J.R.Dick [J. Vac. Sci. & Technol. A (USA) vol.2
       (1984) p.1120]
[8]   D.S.Ginley [Appl. Phys. Lett. (USA) vol.39 (1981) p.624]
[9]   F.Battistella, A.Rocher, A.George [Mater. Res. Soc. Symp. Proc.
       (USA) vol.59 (1986) p.347]
[10]   N.M.Johnson, D.K.Biegelsen, M.D.Moyer [Appl. Phys. Lett. (USA)
       vol.40 (1982) p.882]
[11]   C.H.Seager, D.S.Ginley [Appl. Phys. Lett. (USA) vol.34 (1979)
       p.337]
[12]   C.H.Seager, D.S.Ginley, J.D.Zook [Appl. Phys. Lett. (USA)
       vol.36 (1980) p.831]
[13]   K.J.Bachmann [Curr. Top. Mater. Sci. (Netherlands) vol.3
       Ed E.Kaldis (North Holland, Amsterdam) p.477]
[14]   R.Sharko, A.Gervais, C.Texier-Hervo [J. Phys. Colloq. (France)
       vol.43 no.C-1 (1982) p.129]
[15]   L.L.Kazmerski [J. Vac. Sci. & Technol. (USA) vol.20 (1982)
       p.423]
[16]   R.Sunderasen, J.G.Fossum, D.E.Burk [J. Appl. Phys. (USA) vol.55
       (1984) p.1162]
[17]   J.M.Rose, R.Gronsky [Appl. Phys. Lett. (USA) vol.41 (1982)
       p.993]

[18]   H.Oppolzer, W.Eckers, H.Schober [Inst. Phys. Conf. Ser. (GB)
       no.76 (1985) p.461]

[19]   B.Swaminathan, E.Demoulin, T.W.Sigmon, R.W.Dutton, R.Reif [J.
       Electrochem. Soc. (USA) vol.127 (1980) p.2227]

[20]   D.McLean [Grain Boundaries in Metals (Oxford Univ. Press,
       Oxford 1957)]

[21]   M.M.Mandurah, K.C.Saraswat, C.R.Helms, T.I.Kamins [J. Appl.
       Phys. (USA) vol.51 (1980) p.5755]

[22]   A.Carebelas, D.Nobili, S.Solmi [J. Phys. Colloq. (France) vol.43
       no.C-1 (1982) p.187]

[23]   C.Y.Wong, C.R.M.Grovenor, P.E.Batson, D.A.Smith [J. Appl. Phys.
       (USA) vol.57 (1985) p.438]

[24]   C.R.M.Grovenor [EMIS Datareview RN=15773 (Sep 1987) 'Grain
       boundary structure in poly-Si']

26.2 GRAIN BOUNDARY STRUCTURE IN POLY-Si

by C.R.M.Grovenor

September 1987
EMIS Datareview RN=15773

A. INTRODUCTION

The study of the structure of grain boundaries in polycrystalline
silicon has been stimulated by the wide use made of this material as
an interconnect in integrated circuit metallization systems, and of
rather larger grained material in silicon solar cells. It is hoped that
a link between the grain boundary structure and electronic properties
of these materials may eventually be determined. The properties which
are of most interest to the device engineer are: the effect of grain
boundaries on the resistivity, the enhanced recombination and diffusion
rates at grain boundaries, and short circuit conduction across p-n
junctions. The observation by Johnson et al [1] that polycrystalline
silicon contained a high density of dangling bonds, about $10*12/cm*2$ of
boundary area, has focussed attention on trying to find the structural
nature of these defects. This density of dangling bonds is quite
sufficient to explain the observed potential barrier heights at silicon
grain boundaries, and models of grain boundary recombination and
resistivity in polycrystalline silicon have been developed which give
reasonable agreement with experimental measurements on the variation of
these parameters with barrier height [2-4]. The influence of grain
boundary chemistry on the electronic properties will be described in
[39].

In the past few years a number of reviews on the structure and
properties of grain boundaries in semiconductors have been published
[2-4], as well as several conference proceedings devoted to the same
topic [5-7]. These references give an excellent general overview of the
state of current knowledge on these defects. One important point which
is not always appreciated is that the kinds of material in which the
grain boundary structure can be determined in the TEM experiments
reviewed below must be either specially grown bicrystal samples or the
large grained material grown by casting or sheet growth techniques. Fine
grained polysilicon films deposited by CVD methods are suitable for the
investigation of the electrical properties of a polycrystalline
aggregate, but not for the detailed analysis of grain boundary
structure by existing experimental techniques. The significance of this
becomes clear when we consider that cast or melt grown silicon material
often contains only highly symmetric grain boundaries which may have
relatively simple structures [8,33]. By far the most common grain boundary
in these sheet materials is the coherent twin which is electrically
active only when decorated by lattice dislocations or impurity precipites
[9,10]. It would be quite wrong to assume that CVD polysilicon contained
grain boundaries only of this kind, and the electrical activities of
individual boundaries in CVD material may be quite different to those
observed in large grained material. The extrapolation of information on
the structure of grain boundaries obtained in large grained material to
describe the electrical properties of CVD polysilicon films must therefore
be treated with caution. In fine grained material it is common to ignore
the problem of identifying the structure of the grain boundaries
altogether and to think of the boundaries as all having the same

potential barrier, Qb. This barrier usually increases with impurity content (see [39]), and also has a maximum value at a doping concentration of about 10*17/cm*3 [37].

Two experimental techniques have proved of particular value in the study of the structure of grain boundaries in silicon: High Resolution Electron Microscopy (HREM), and the alpha-fringe technique [11]. Transmission electron microscopy can be used to reveal the dislocation structure of a grain boundary, HREM imaging gives information on the atomic structure of a grain boundary viewed along a direction which lies in the boundary plane, while the alpha-fringe technique uses characteristic TEM image fringes to determine whether there is any relative translation of the two crystals. HREM imaging can usually reveal the atomic structure of grain boundaries in silicon only when the surface normals of both grains are parallel to either [100] or [110] type directions. This restricts the boundaries which can be studied to perfect tilt boundaries about axes of these indices. X-ray [12] and electron diffraction [13,14,38] techniques have also been used in the study of the structure of silicon grain boundaries, but not yet very widely.

In this review the CSL notation [15] will be used to describe the geometric misorientation between the two crystals on either side of a grain boundary in terms of SIGMA, the ratio of the density of crystal lattice sites to the density of coincidence sites in the CSL unit cell [16], and the angle, theta, and axis, [XYZ], of misorientation. In some cases the grain boundary plane wil also be given, (ABC), and where relevant the observed translation vector at the boundary plane. The model of grain boundary structure which is most valuable in understanding the experimental results is the Structural Unit Model (SUM) [17,18], where all grain boundaries are considered to be made up of suitable combinations of a relatively small number of characteristic structural units, each containing a few atoms.

B. EXPERIMENTAL OBSERVATIONS

Low angle [100] and [110] tilt grain boundaries in both silicon and germanium have been shown to consist of arrays of discrete dislocations [19-22], but these dislocations have been found to have a variety of Burgers vectors and some are dissociated and others not. The character of the dislocation array which accommodates the lattice misorientation may depend on the method used to prepare the samples containing the grain boundaries. The results of D'Anterroches and Bourret [22] on SIGMA = 25, theta = 16.25 deg [100] boundaries show that the boundary core consists of a regular array of a structural unit identified as the core of an a/2 [011] lattice dislocation.

A number of high angle [110] tilt grain boundaries in silicon have been studied by TEM techniques. There is general agreement that the coherent twin, SIGMA = 3 (111), consists of an array of 'boat shaped' stacking fault units replacing the 6-membered ring structural units characteristic of the perfect diamond cubic lattice [22]. Fontaine and Smith [23] and Vlachavas and Pond [24] have suggested a structure for the SIGMA = 3 (211) incoherent twin boundary which consists of a staggered array of 5 and 7 membered rings with the two crystals translated by a/9 [111] + a/18 [211]. However, Bourret et al [25] have

shown that this boundary appears to have a periodicity double that
suggested in these models. The structure of this boundary is still not
completely determined. The SIGMA = 9 (221) symmetric second order twin
boundary has a core structure which consists of a zigzag array of the
5 and 7 membered ring units, similar, but not identical to those found in
the incoherent twin, and arranged in a different manner [27,28,33]. Similar
studies of SIGMA = 27 and SIGMA = 11 tilt boundaries have once again
identified characteristic structural units at the cores of these highly
symmetric boundaries [29,30], and in the case of the SIGMA = 11 (233)
boundary a doubling of the expected period identified as well [28].

The structures of SIGMA = 5 high angle [100] tilt boundaries have been
investigated in both silicon and germanium, and here the structural
unit at the grain boundary seems to be the a/2 [110] edge dislocation.
These dislocation cores are arranged in zig-zag arrays reminiscent of
the SIGMA = 9 (221) boundary [28]. The crystals on either side of the
SIGMA = 5 boundary are translated by 1/8 [100].

It is noticeable that none of these boundary structures contain any
dangling bonds - complete reconstruction has taken place at the
boundary plane to preserve tetrahedral coordination. This may mean that
these highly symmetric grain boundaries will be electrically active only
when they contain dislocations other than those which form the core
structure of the boundary, 'extrinsic dislocations'. It has been shown
that dangling bonds can be found at extrinsic dislocations in some
grain boundaries [22]. Second phase precipitates caused by the
segregation of impurity species are another possible cause of electrical
activity in otherwise inactive boundaries. Conflicting data on the
electrical activity of [110] tilt boundaries in Si has been obtained
[9,31], although the possible effects of contamination have not been well
characterised in these experiments. It is very hard in practice to
prepare a genuinely clean grain boundary in a thin TEM specimen. A
second point which is worth emphasising about the grain boundary
structures described above is that in all cases easily characterised
structural units were observed in the HREM experiments. These units are
sometimes arranged in quite complex arrays, but the basic concept of
grain boundary cores consisting of a small number of structural units
seems to be borne out by the available data. Most of the structural
units can also be considered as the cores of lattice dislocations in the
diamond cubic structure.

C. THEORETICAL MODELLING AND CALCULATIONS

Theoretical modelling of the structure of grain boundaries in silicon
was first attempted by Hornstra [32] who used the core structures of a
few lattice dislocations as the structural units for constructing [110]
tilt boundaries. In the misorientation range 0 < theta < 70 deg some
of Hornstra's models are in surprisingly good agreement with the HREM
structures. The SIGMA = 9 (221) boundary in particular can be
considered to be an array of lattice dislocations. At higher
misorientation angles Hornstra was forced to introduce dangling bonds
into the grain boundary structures. Vaudin et al [33] have extended the
range of boundaries modelled by using the SUM concept that the
structure of any grain boundary can be predicted by combining a suitable
mixture of the structural units characteristic of the two nearest
'favoured boundaries'. A favoured boundary has a core structure made up
of an array of only one structural unit [17,18]. More recently, Papon

and Petit [26] have proposed structures for a number of [110] tilt
boundaries based on a very complete set of possible structural units,
most of which are the cores of dislocations in the diamond cubic
lattice. These units are 5,6,7 and 8 membered rings of a variety of
shapes, and in all of them perfect tetrahedral bonding is preserved.
These results are in agreement with most of the experimentally
determined structures, and it seems clear that we should regard [110]
tilt grain boundaries in silicon in terms of a continuity of structure
as demanded by the SUM model. This means that there is a smooth
transition in grain boundary structure, and so of grain boundary
properties as well, as the angle of misorientation increases – the
mixture of structural units being changed to accommodate the particular
value of theta. This continuity of grain boundary structure has been
further emphasised by the work of Kohyama et al [34] who obtained
relaxed structures of [110] tilt grain boundaries in silicon in a
simple computational routine. These structures are in excellent
agreement with the structures described above, but since the starting
configurations for the relaxation routines were those of Papon and
Petit [26] this is not very surprising. The manner in which the boundary
core structure changes as the value of theta increases to include more
structural units characteristic of boundaries of high misorientation
is particularly well shown in their figures. Whether this kind of
relaxation proceedure can be used to predict the structure of grain
boundaries when no models exist from which to construct the starting
configurations is rather doubtful.

Wetzel et al [40] have used valence force field calculations to model
the structure of SIGMA = 5 [100] boundaries in covalently bonded
materials. Their model of the (310) interface does not agree with the
experimentally determined model of Bacmann et al [28] in that there is
no translation of the crystals parallel to the misorientation axis. The
most recent calculation of grain boundary structure in materials with
diamond cubic structure is on the SIGMA = 5 twist boundary in germanium
using an ab-initio molecular dynamics method [35], and has predicted
a novel structure containing both dangling bonds and a most unusual
4 membered ring unit. There is unfortunately no direct experimental
evidence with which to compare this structure. This kind of rigorous
calculation may be an important way of probing the structure of more
randomly oriented grain boundaries in diamond cubic lattices, although
the amount of computing power needed in any calculation of this kind is
extremely large.

D. DISCUSSION

The experimental and theoretical results on the structure of [110]
tilt grain boundaries in silicon agree in the main very well, although
the data on the [100] boundaries is still rather confused. The evidence
for dangling bonds being present in the cores of [110] tilt boundaries
is slight. We can speculate that the SUM can be used to predict the
core structures of boundaries misoriented about axes other than [100]
and [110], but this would require the identification of the
characteristic structural units of a new set of favoured boundaries.
HREM experiments cannot be used to determine these structural units.
In the search for dangling bonds in grain boundaries we are forced
to assume that both extrinsic dislocations and second phase precipitates
will introduce these, or other electrically active defects, but that
some low symmetry grain boundaries may also contain them as an intrinsic
part of their structure. Dangling bonds are only visible in an HREM

image if they are collected together along the core of a grain boundary dislocation, and so we cannot expect this technique to give very much data on the density of isolated examples of this defect. The modelled structure of the SIGMA = 5 twist boundary [35] contains a high density of dangling bonds - more than would be expected from the ESR results. Whether this will prove to be a characteristic feature of the structure of all twist boundaries remains to be determined.

It seems possible that the electrical activity of highly symmetric grain boundaries in silicon is a wholly extrinsic effect, having nothing to do with the intrinsic structure of the boundaries at all [36]. These boundaries are common in large grained sheet silicon material. On the other hand structural units containing dangling bonds may be common in grain boundaries of a less symmetric nature, or in twist boundaries. Either of these possibilities is hard to investigate in HREM. It is these kinds of boundaries which we expect to be far more common in CVD polysilicon than symmetric tilt boundaries. It is at present not clear whether we expect the electrical properties of grain boundaries in CVD polysilicon to be controlled by extrinsic defects (or contamination) or by structural defects intrinsic to the core structure of the boundaries.

REFERENCES
----

[1]   N.M.Johnson, D.K.Biegelsen, D.K.Moyer [Appl. Phys. Lett. (USA) vol.40 (1982) p.882]
[2]   C.R.M.Grovenor [J. Phys. C. (GB) vol.18 (1985) p.4079]
[3]   C.H.Seager [Annu. Rev. Mater. Sci (USA) vol.15 (1985) p.271]
[4]   H.F.Matare [J. Appl. Phys. (USA) vol.56 (1984) p.2606]
[5]   H.J.Leamy, G.E.Pike, C.H.Seager [Grain Boundaries in Semiconductors (North Holland, New York, 1982)]
[6]   [J. Phys. Colloq. (France) vol.43 no.C-1 (1982)]
[7]   G.Harbeke [Polycrystalline Semiconductors (Springer-Verlag, Berlin, 1985)]
[8]   F.Komninou, T.Karakostas, G.L.Bleris [ibid [6] p.9]
[9]   B.Cunningham, H.P.Strunk, D.G.Ast [ibid [5] p.51]
[10]   C.Dianteill, A.Rocher [ibid [6] p.75]
[11]   R.C.Pond [J. Microsc. (GB) vol.116 (1979) p.105]
[12]   S.L.Sass, P.D.Bristow [Grain Boundaries, Structure and Kinetics (ASM, Pittsburgh 1980) p.71]
[13]   P.Lamarre, S.L.Sass [Scr. Metall. (USA) vol.17 (1983) p.1141]
[14]   F.W.Schapink [Rev. Phys. Appl. (France) vol.21 (1986) p.747]
[15]   P.H.Pumphrey [in 'Grain Boundary Structure and Properties' Eds G.A.Chadwick, D.A.Smith (Academic Press, London 1976) p.139]
[16]   H.Grimmer, W.Bollmann, D.H.Warrington [Acta Crystallogr. Sect A (Denmark) vol.30 (1974) p.197]
[17]   A.P.Sutton, V.Vitek [Philos. Trans. R. Soc. London A (GB) vol.309 (1983) p.1]
[18]   A.P.Sutton [Int. Met. Rev. (GB) vol.29 (1984) p.377]
[19]   A.Bourret, J.Desseaux [Philos. Mag. A (GB) vol.39 (1979) p.405, p.419]
[20]   W.Skrotzki, H.Wendt, C.B.Carter, D.L.Kohlstedt [in Thin Films and Interfaces II Eds J.E.E.Baglin, D.R.Campbell, W.K.Chu (North Holland, 1984) p.299]
[21]   H.Foll, D.G.Ast [Philos. Mag. A (GB) vol.40 (1979) p.589]
[22]   C.d'Anterroches, A.Bourret [Philos. Mag. A (GB) vol.49 (1984) p.783]

[23] C.Fontaine, D.A.Smith [ibid [5] p.39]

[24] D.S.Vlachavas, R.C.Pond [Inst. Phys. Conf. Ser. (GB) no.60 (1981)
      p.159]

[25] A.Bourret, L.Billard, M.Petit [Inst. Phys. Conf. Ser. (GB) no.76
      (1985) p.23]

[26] A.M.Papon, M.Petit [Scr. Metall. (USA) vol.19 (1985) p.391]

[27] O.L.Krivanek, S.Isoda, K.Kobayashi [Philos. Mag. (GB) vol.36
      (1977) p.931]

[28] J-J.Bacmann, A.M.Papon, M.Petit [ibid [6] p.15]

[29] B.Cunningham, H.P.Strunk, D.G.Ast [Scr. Metall. (USA) vol.16
      (1982) p.349]

[30] A.M.Papon, M.Petit, J-J.Bacmann [Philos. Mag. A (GB) vol.49
      (1984) p.573]

[31] P.Ruterana, A.Bary, G.Nouet [ibid [6] p.27]

[32] J.Hornstra [Physica (Netherlands) vol.25 (1959) p.409 and vol.26
      (1960) p.198]

[33] M.D.Vaudin, B.Cunningham, D.G.Ast [Scr. Metall. (USA) vol.17
      (1983) p.191]

[34] M.Kohyama, R.Yamamoto, M.Doyama [Phys. Status Solidi b (Germany)
      vol.137 (1986) p.11 and vol.138 (1986) p.387]

[35] M.C.Payne, P.D.Bristowe, J.D.Joannopoulos [Phys. Rev. Lett.
      (USA) vol.58 (1987) p.1348]

[36] D.Redfield [Appl. Phys. Lett. (USA) vol.38 (1981) p.174]

[37] J.Y.W.Seto [J. Appl. Phys. (USA) vol.46 (1975) p.5247]

[38] M.D.Vaudin, E.Burkel, S.L.Sass [Philos. Mag. A (GB) vol.54 no.1
      (1986) p.1]

[39] G.R.M.Grovenor [EMIS Datareview RN=15774 (July 1987) 'The
      chemistry of grain boundaries in polycrystalline Si']

[40] J.T.Wetzel, A.A.Levi, D.A.Smith [Mat. Res. Soc. Symp. Proc. (USA)
      vol.63 (1985) p.157]

26.3 GROWTH AND MICROCRYSTALLINE STRUCTURE OF POLY-Si

by C.Hill and S.Jones

December 1987
EMIS Datareview RN=17863

A. GROWTH TECHNIQUES

Polycrystalline silicon can be grown by a variety of techniques, each
suited to the needs of particular semiconductor device technologies.
The thinnest layers (10-1000 A) are generally grown by vacuum deposition
(evaporation [1,2], sputtering [3], MBE [4]); the athermal nature of the
deposition processes allows low temperatures (0-600 deg C) to be used
and, therefore, these techniques find application where thermally unstable
substrates are used (e.g. solar cells [5], thick film transistors [6],
hard mask layers [7]). Layers in the thickness range 100 A - 10 microns
are conveniently grown over the temperature range 50-850 deg C by low
pressure chemical vapour deposition (LPCVD) [8,9], and nearly all the
polycrystalline silicon layers incorporated into silicon integrated
circuits are now deposited in this way, because of the uniformity,
reproducibility and conformal nature of the layer [10]. Thicker layers
(1-100 microns) can be grown in atmospheric chemical vapour deposition
systems at temperatures between 900 and 1300 deg C [11,12]; at this
higher pressure, however, uniformity of deposition depends on gas phase
diffusion, and so good uniformities can only be achieved by spacing
substrates wide apart, reducing throughput. The thickest polycrystalline
layers are produced from liquid silicon by either controlled casting or
pulling through a die [13]. These find application as free-standing
starting materials for solar cell applications. In general, the grain
size of the polycrystalline silicon increases with deposition
temperature and with film thickness, increasing from less than 10 A
(amorphous) for thin evaporated films to greater than 1 mm for thick
layers grown from the liquid phase. The microstructure is, however, a
complex function of the deposition conditions, and will be covered in
detail only for chemically vapour deposited polysilicon films. A number
of deposition techniques have been studied in which the thermal
activation energy is supplemented by other energy sources, including
plasmas [14,15,18], and laser irradiation [18,19].

B. CVD DEPOSITION PROCESSES

The transfer of silicon from gaseous phase compounds (e.g. silane SiH_4)
to solid phase incorporation into a growing polycrystalline silicon
layer on a planar substrate has been extensively studied [8,19,20].
Over a wide range of substrate temperatures and gas composition and
pressure, the deposition rate is found not to be limited by the
transport of silicon from gas to substrate surfaces; the rate-limiting
step is the rate of migration of SiH_4 molecules across the polysilicon
surface and their self-nucleation and growth [11,12]. It is this process
which is thought to determine the measured activation energy for
deposition (approx. 1.7 eV/atom) [8,22,23]; the recurrence of values
close to this activation energy in much of the published data for
deposition from silane (see TABLE 1) indicates that this is the rate-

determining step over most of the experimental range (570-900 deg C,
0.004-760 torr pressure). However, TABLE 1 also shows that at high
deposition rates relative to the silane pressure, the gaseous transport
of silicon can become the rate-limiting process. In LPCVD furnace
deposition systems, which typically operate in the range 580-630 deg C,
silane pressures 100-400 millitorr and deposition rates 50-200 A/min.,
gas transport is not the limiting factor, and it is because of this that
the reproducible conformal coatings characteristic of such systems are
obtained.

The migration rate of SiH4 molecules across the surface is reduced by
the presence of hydrogen [20] and phosphine [25], both of which are
strongly adsorbed onto the silicon surface at these temperatures. The
nucleation and growth increases more than linearly with the number of
SiH4 molecules on the surface. This gives [21] a relationship between
growth rate R and the growth parameters P(S) (silane partial pressure)
and P(H) (molecular hydrogen partial pressure) and temperature (T deg K):

$$R = A \ \frac{P(S)*1.4}{P(H)} \ \exp \frac{-Q1}{kT} \qquad (1)$$

where Q1 = 1.73 eV/atom and k = 8.62 X 10*-5 eV/atom/deg K.

Other species are known to affect deposition rate (e.g. carbon [26] and
oxygen [27]) and it is likely that these and other trace contaminants
are responsible for different values of the constant A above which can be
derived from the deposition rates published by different workers (see
TABLE 1). The stability of the nuclei and the reproducibility of their
occurrence are both worse at higher temperatures (above 700 deg C) and
in the presence of HCl [20]. For this reason, LPCVD polycrystalline
silicon deposition systems are generally operated below 650 deg C, and
with pure silane gas. At such temperatures, the nature of the substrate
has little influence on nucleation or growth: the native oxide (5-15 A)
always present on silicon substrates is stable, and so growth effectively
takes place always on an amorphous substrate, whether silicon, silicon
dioxide, or silicon nitride is the nominal underlayer.

At higher temperatures, on silicon substrates, the native oxide is not
stable at low pressures [28] and deposition takes place directly onto
single crystal silicon. Epitaxial growth may then occur, if the density
of crystallographic steps on the surface is sufficient to act as nuclei
for the surface-migrating SiH4 molecules [21]. No activation energy for
nucleation is then required, and the epitaxial regime can be clearly
distinguished from the polycrystalline by the consequently different
deposition rate (R) behaviour:

$$R = B \ \frac{P(S)}{SQRT \ P(H)} \ \exp \frac{-Q2}{kT} \qquad (2)$$

where Q2 = 0.52 eV/atom.

However, at sufficiently high deposition rates, the self-nucleation process dominates, and polycrystalline silicon will be obtained, growth rate being determined as before by Eqn (1). The critical deposition rate for polycrystalline silicon formation is given approximately in [20] as

$$R = approx. \ 3 \times 10^{*}19 \ exp \ \frac{-Q_3}{kT} \ microns/min \qquad (3)$$

where Q_3 = 5 eV/atom.

It is suggested [11] that the value of the critical rate is determined by the rate of arrival of SiH4 molecules at surface steps (J) exceeding the rate of diffusion of vacancies in the underlying bulk silicon (D), so that the continual atomic reconstruction necessary to preserve the surface crystalline perfection is incomplete. The critical rate then occurs when J/D >= 1, and Q3 is determined by the activation energy for self diffusion of silicon, which is close to 5 eV [29]. It must be emphasised that in many growth systems, especially at temperatures below 1000 deg C, epitaxial growth may never occur, even at low deposition rates, because the criterion for an atomically clean surface is not met. This may be because the oxygen species pressure is not low enough to remove surface oxide [28], or because an initially clean surface is contaminated by other gases, e.g. hydrocarbons. At 1000 deg C, 10*19 atoms/cm*3 of incorporated carbon is sufficient to transform epitaxial growth into polysilicon growth [49,50].

At low temperatures and deposition rates high enough that there is a high probability of two migrating SiH4 molecules meeting before they encounter an existing nucleus, each pair becomes a nucleus and an amorphous silicon layer is deposited [11]. The growth process is essentially identical to the polysilicon growth process and follows Eqn (1), so that there is no characteristic change in the deposition rate behaviour by which the amorphous-polycrystalline transition can be detected. It is suggested [11] that the critical deposition rate R, determined by the criterion J/D >= 1 (where D is now the surface diffusion coefficient of the silicon atoms) is given by:

$$R = 4.0 \ exp \ \frac{-Q_4}{kT} \ micron/min \qquad (4)$$

where Q_4 = .5 eV/atom.

This criterion correctly predicts the transition temperature between deposition of amorphous films and deposition of polycrystalline films commonly observed in LPCVD systems (570-590 deg C) where growth rates are 30-70 A/min. At higher temperatures, however, it shold not be used as a guide to the final deposited structure, since solid state transformation of the deposited amorphous film to polycrystalline silicon will be proceeding during deposition [30]. The microstructure of silicon films deposited at temperatures between 550 and 770 deg C in atmospheric pressure system has been convincingly interpreted entirely in terms of fully amorphous deposition and concurrent internal solid state growth of polycrystalline grains [22]. This is discussed further in Section D.

The surface diffusion coefficient D is very much reduced by surface
contaminants [20], including the reaction gases in chemical vapour
deposition systems. Thus, for similar values of J, the amorphous-
crystalline transition (J/D = approx. 1) occurs at about 570-590 deg
C in LPCVD systems, but at 400 deg C in vacuum evaporation systems
[1,2] where the gas pressure is 10*5-10*9 times lower, and the
consequent reduction in contaminant barriers to surface motion offsets
the lower thermal energy available for diffusion. There is evidence
in LPCVD systems [31] that high deposition rates minimise the
incorporation of contaminants (e.g. oxygen) onto and into the growing
film.

C. CVD DEPOSITION RATE

The deposition rates of polycrystalline silicon obtained in real chemical
vapour deposition systems are not usually directly comparable, because
of the effects of contaminants described in the previous section, and
also because the published data is seldom comprehensive enough to enable
comparison for exactly the same values of the important parameters. Key
parameters are the partial pressure and nature of the silicon compound
(e.g. SiH4, SiCl4), the partial pressure of hydrogen, doping gas and
inert gas, and the total pressure in the system, as well as substrate
temperature. The examples of the influence of these parameters that
follow are all chosen from the growth regime controlled by surface
reaction (Section B). In LPCVD systems, it is essential to operate in
this regime to achieve uniform conformal layers that are independent of
substrate numbers or geometry.

The gaseous silicon species chosen has a strong effect on deposition
rate: most of the experimental data for this comes from epitaxial growth
regime, in which a wider variety of gaseous sources are employed.

In one study [11], the growth rates from silane (SiH4), dichlorosilane
(SiH2Cl2), trichlorosilane (SiHCl3) and silicon tetrachloride (SiCl4)
were measured over the temperature range 650-1200 deg C. In the surface
reaction limited regime, all the activation energies controlling the
temperature dependence were the same (1.45 eV) , while the
pre-exponential factors decreased as the chlorine content of the species
increased. The results are summarised in TABLE 2. Comparison of the
growth rates at 850 deg C shows an increasingly large reduction in
growth rate as each chlorine atom is added to the gas molecule. The HCl
interferes with the nucleation process and also reduces the surface
mobility of the adsorbed silicon species [11]. Growth rates can also be
increased by increasing the silicon content of the gas molecule. Growth
rates of polycrystalline silicon from silane (SiH4) and disilane (Si2H6)
were compared under similar conditions (0.13 millitorr) in an LPCVD
reactor [32]. At 850 deg C, the growth rate was increased by a factor of
10; at 482 deg C, by a factor of 12.

At present, silane is almost universally used as the gaseous source in
LPCVD polysilicon deposition systems. Typical values of deposition rates
reported over the temperature range 570-700 deg C, and silane pressure
4-350 millitorr, are given in TABLE 3. The use of hydrogen as a carrier
gas reduces deposition rates (see Section B) and this effect can be
seen in some typical values summarised in TABLE 4. Phosphine (added to

dope the polysilicon with phosphorus) also reduces growth rates [24], as can be seen in the data summarised in TABLE 5. At 620 deg C, 1% of silane replaced by phosphine reduces the growth rate by a factor of 8.

The growth rate of oxygen-doped layers, grown from N2O/SiH4 mixtures is also reduced as the N2O content increases. At 620 deg C and 34 millitorr silane pressure, the growth rate falls by a factor of 2 (from 91 to 45 A/min) as the N2O pressure is increased from 0 to 7 millitorr. The activation energy for growth also changes, from 1.9 to 1.1 eV [27].

The changes in activation energy when dopants are added to the gas stream (as for PH3 and N2O above) indicate that the growth mechanism is also being altered.

At low deposition temperatures, growth rates can be considerably enhanced by supplying additional energy from plasmas [16] or lasers [18]. The degree of enhancement obtainable can be seen in TABLE 1, where deposition rates from LPCVD silane with and without plasma energy are compared. At 620 deg C the deposition rate is increased by a factor of 30.

D. MICROSTRUCTURE OF CVD POLY-Si

The microstructure of polycrystalline silicon films at the end of the deposition process depends both on the nucleation and kinetics of growth at the gas-solid surface, and also on the solid-state transformations taking place inside the film during the whole of the time that the substrate is hot. As outlined in Section B, the deposited film forms an amorphous structure if the arrival rate J of the silicon atoms is greater than the surface diffusion rate D, so that pairs of atoms form individual nuclei rather than migrating to larger nuclei. In CVD systems, this criterion appears to be met over the range 570-770 deg C depending on deposition rate [11]. Only where the structure of the deposited film has been monitored continuously during growth [22] is it possible to determine directly whether the deposition mode is amorphous or polycrystalline, since solid state transformations are occurring during deposition. The final microstructure does, however, reflect the original mode of growth, and four distinct types of microstructure are to be expected:

    (i)   fully amorphous films

    (ii)   partially transformed films

    (iii) fully transformed films

    (iv)   films which have never been amorphous but are deposited directly
           in the polycrystalline form.

The range of microstructures actually obtained after deposition can be seen in TABLE 6, which summarises some of the published data.

Fully amorphous films are obtained up to a temperature (Tc) characteristic of each deposition system and deposition rate. As expected, Tc increases with deposition rate (from 570 at 20 A/min. to 610 deg C at 450 A/min.) as the data from [33,22] in TABLE 6

shows. Comparison with the data also given for the temperature
dependence of the incubation period for nucleation of crystallites at
the amorphous silicon-substrate interface, and the velocity of the
growth of the crystallites up into the amorphous layer, shows that
amorphous layers would be expected under these conditions. The
persistence of Tc up to 650 deg C for plasma-enhanced LPCVD [16], which
would not be expected by the above criteria, indicates that either the
plasma energy or the incorporated hydrogen reduces either or both of the
incubation rate or regrowth velocity. Hydrogenated surfaces are reported
to reduce crystallite nucleation [35]. The presence of other chemical
species during deposition also affects Tc. High phosphine levels in Si2H6
[36] LPCVD systems are reported to increase Tc slightly: from 590 deg C
to 610 deg C for PH = 0 to PH3 = 0.008 Si2H6 partial pressure.
Hydrocarbons also affect Tc. Values of Tc about 800 deg C are reported
for atmospheric CVD deposition from silane-phosphine-acetylene ambients
(PH3 = 10*-4 SiH4, C2H2 = 2 X 10*-2 SiH4, SiH4 = 10*-3 H2) [26].

LPCVD deposited amorphous films are dense and free from voids. The room
temperature optical properties at visible wavelengths are reported as
absorption coefficient X = 8.4 X 10*4 cm*-1 at wavelength = 4416 A, 1.05
X 10*5 cm*-1 at wavelength = 5145 A. Refractive index at both wavelengths
= 3.95. Refractive index at 600 deg C is 3.89 at a wavelength of 11000 A
[35].

Some elastic stress is present in the films as evinced by 'curling-up'
when released by etching from the substrate [9]. Although X-ray [33] and
Raman scattering measurements [23] invariably indicate a structure
indistinguishable from atomically amorphous films, TEM examination
often reveals the presence of small crystallites (approx. 50 A) in
small [23] and large [9] concentrations.

At deposition temperatures above Tc, a series of partially transformed
structures is obtained, as summarised in TABLE 6. X-ray analysis shows
a high density of crystallites with either random orientation or a weak
<311> texture. As the deposition temperature is further increased, an
increasing <100> texture develops which persists up to high deposition
temperatures. The random and <311> texture structures are characterised
by equiaxed grains, which probably result from the simultaneous
deposition and transformation processes described above. In-situ optical
measurements are able to detect an increasing proportion of
polycrystalline silicon as the deposition temperature increases above
Tc ([22], TABLE 6). The remaining amorphous silicon in such partially
transformed structures is often not detected probably because during
the residence time in the furnace after deposition, but before
unloading, transformation continues to completion to polycrystalline
silicon. (For example, at 600 deg C, 10 minutes residence time gives a
regrowth distance of over a 1000 A). The final microstructures reported
vary considerably in detail, with weak <111>, <311>, <331>, <110> textures
being observed in varying relative intensities, but <311> and <110>
usually predominating. In transverse TEM, the structure is made up of
small (50-150 A) heavily faulted grains which increase in size from the
bottom to the top of the layer [9]. The optical absorption coefficient
decreases as deposition temperature increases above Tc. For Tc = approx.
570 deg C [23], it decreases from 1 X 10*5/cm to 4 X 10*4/cm (wavelength
= 5145 A) and from 8.2 to 8.0 X 10*4 cm*-1 (wavelength = 4416 A) as
temperature increases from 580 to 620 deg C. The final class of structures
observed, when a strong <100> texture is evident in the microstructure,

is associated with a marked columnar habit of growth in which grains run from bottom to top of the film, increasing in lateral dimension as they do so [9]. Thicker films thus have a larger average grain size, and also increased <100> texture. The <100> texture and grain shape have been cited [9] as evidence that the film has been deposited directly in the polycrystalline form, since fully amorphous films when transformed to polysilicon at higher temperatures (approx. 800 deg C) develop <111> or <311> texture, but not <100>.

There is evidence that direct deposition of polysilicon in LPCVD systems can occur as low as 600 deg C at low silane pressures (2.5-40 millitorr), from the reported [37,38] increase in crystallite size near the film/substrate interface as pressure decreases in this range. The silane pressure, both through its effect on deposition rate (see TABLE 3) and on nucleation behaviour at the growing surface, is the strongest factor in determining the onset of polycrystalline growth and its subsequent development at a particular temperature. At 630 deg C, this dependence is reported [38] to be:

(i)   at 2.5 millitorr, a perfect columnar structure is grown, with individual grains having a preferred (100) orientation normal to the film and each having a strongly protruding curved outer surface.

(ii) at 10 millitorr a less perfect columnar structure with <111> microtwins (inside less regularly but still strongly protruding grains) occurs.

(iii) at 200 millitorr, a high density of <111> twins occurs randomly oriented in striated grains which terminate in a curved surface of large radius of curvature, which thus protrude only slightly beyond the average surface level.

The electron diffraction data are consistent with the <110> texture normal to the film reported by other authors under these conditions [9,16,23,33]. Thus while the above data is convincing evidence for direct deposition of polysilicon at 630 deg C, it does not support the argument that the occurrence of <110> texture necessarily indicates such a form of growth. In atmospheric polycrystalline silicon deposition, the film texture is reported [39] to change from <110> to <100> to random as deposition temperature increases over the range 650-850 deg C.

E. GRAIN SIZE IN CVD POLY-Si

It will be evident from the microstructures described above that LPCVD deposited films are unlikely to be completely described by a single grain size. However, there is much work describing grain size as a function of deposition conditions, most of it referring to average grain diameters in the plane of the film. Some typical data are summarised in TABLE 7. The continual increase in grain size as film thickness increases has been simplified [40] to the relationship

    grain size L = 0.25 X film thickness Z.

Values of L/Z are also given in TABLE 7, and vary between 0.04 and 0.37. The simple relationship is only a very rough guide, partly because it fails to take into account the effect of ambient pressure [38] and temperature [16] on grain size, but also because the actual grain structure is bimodal, containing equiaxed grains of about 50 A diameter, and columnar grains of about 390 A diameter [45]. The columnar grains studied in more recent publications are fitted better by $L/Z = 0.13$.

Grain size is reported to be affected by the presence of contamination. Oxygen contamination is reduced as deposition rate increases and there is an associated increase of grain size in polysilicon layers 4200 A thick [31]. As deposition rate increased from 120 to 800 A/min, oxygen content fell from 1.3wt% to 0.1wt%, and grain size increased from 160 to 350 A. A similar effect is reported for carbon incorporation in CVD polysilicon growth [47]. Occasional very large grains have been attributed to enhancement of growth by localised contamination [48]. These results suggest that some of the differences in observed grain size may be due to the different levels of purity of the deposition gases used in each investigation, the different surface purities of the starting substrates.

F. MORPHOLOGY OF CVD POLYSILICON

The morphology of deposited silicon films is determined by the mechanism of growth [37]. LPCVD films which are deposited amorphous and do not transform to polycrystalline silicon, are very smooth [23]. Root mean square values of roughness are always typically less than 15 A. Partially transformed films are sometimes as smooth as this, but often of roughness up to $r = 50$ A. Fully columnar films have $r = 50$-60 A, and at higher temperatures (above 880 deg C) values of 100 A are reported. The peak to peak roughness is approximately 2 SQRT 2 r in all these cases. At lower growth pressures when direct growth of polysilicon is occurring, very much larger surface roughness is observed [38], corresponding to the competitive growth rates of individual grains and crystal planes. As silane pressure is reduced from 200 millitorr to 2.5 millitorr at 630 deg C, the surface roughness r increases from about 300 A (3% of film thickness) to 1800 A (15% of film thickness). Atmospheric CVD gives an increasing peak to peak roughness (60-130 A) as deposition temperature increases from 650-850 [39].

Even under conditions where moderately smooth films are produced, occasional surface nodules protrude, which give a characteristic 'hazy' appearance to the polysilicon film. These are clearly a gas-solid growth phenomenon, since they protrude well above the average level of the film (1000 A above a 4000 A film), and are associated with a large underlying grain originating at the bottom interface. Abnormal growth rates catalysed by surface contamination have been given as the explanation for the effect [48].

G. TABLES

TABLE 1: some published data on growth rate behaviour in silane-based
deposition systems including activation energies for the
surface-reaction controlled range of deposition.

SiH4 Press torr	H2 Press torr	Inert Gas Press torr	PH3 Press torr	Temp. Range surface limited deg C	Activation Energy eV	Temp. Range gas transport limited deg C	Ref
0.004	–	–	–	560–680	2.08	680–800	[16]
0.08	–	320	0	610–650	1.39	650–680	[24]
0.08	–	320	8 X 10*-5	610–680	1.56	–	[24]
0.08	–	320	8 X 10*-4	610–680	1.99	–	[24]
0.12	–	–	–	560–650	1.39	–	[23]
0.35	–	–	–	560–620	1.73	–	[23]
0.11	1.89	–	–	680–750	1.73	–	[8]
0.57	9.44	–	–	680–830	1.73	–	[8]
0.76	760	–	–	630–860	1.43	860–1100	[11]
5.65	94	–	–	630–780	1.73	–	[8]
20	760	–	–	560–800	1.76*	–	[22]

* Average value: two activation energies are found: 560–690 2.2 eV
  690–800 1.4 eV

TABLE 2: effect of the chlorine content of the silicon source on
deposition rate for epitaxial growth. All sources are 0.1 mol%
in 1 atmosphere of H2 [11]

Gaseous silicon source	Expt. Temp. Range surface reaction limited deg C	Q eV	Ro micron/ min	Expt. Temp. Range. gas transport limited deg C	Grow. Rate (850 deg C) micron/min
SiH4	630–850	1.45	1.1 X 10*6	850–1100	0.34
SiH2C12	770–920	1.45	3.0 X 10*5	920–1200	0.094
SiHC13	800–1000	1.45	1.3 X 10*5	1000–1150	0.041
SiC14	850–1100	1.45	2.9 X 104	1100–1200	0.0091

TABLE 3: deposition rates from pure silane LPCVD as a function of
temperature and silane pressure. All rates are in A/min.

DEPOSITION TEMPERATURE (DEG C)

SiH4 Pressure (millitorr)	500	570	600	620	650	700	Ref
350	–	41	92	150	310*	–	[23]
120	–	35	65	100	170	–	[23]
4	–	2.5*	6.7	13	33	120	[16]
4+	135	180	200	215	235	270	[16]

+ plasma-enhanced (20 Watts RF power)

* extrapolated value

TABLE 4: deposition rates for silane-hydrogen LPCVD as a function of
gas pressures and temperature. All rates are in A/min.

DEPOSITION TEMPERATURE (DEG C)

H2 Press. torr	SiH4 Press torr	570	600	650	700	800	Ref
2	0.12	–	–	410	1100		[8]
10	0.60	–	–	870	2500		[8]
100	6	–	–	2200	8800		[8]
740	17.5	12.5	340	1450	5090	23,600	[22]

TABLE 5: deposition rates from silane-helium-phosphine LPCVD as a
function of gas pressure and temperature. All rates are in
A/min.

DEPOSITION TEMPERATURE (DEG C)

H2 Press. millitorr	PH3 Press. millitorr	SiH4 Press. millitorr	620	650	675	Ref
320	0	80	86	150	200	[24]
320	0.08	80	20.5	41	70	[24]
320	0.8	80	10.8	27	60	[24]

TABLE 6: microstructures of CVD Si films on amorphous substrates as a function of substrate temperature and total growth time (g.t.) in mins. Published rates of transformation of a-Si to poly-Si and incubation times before transformation start are included for comparison. Structure is measured by X-ray reflection.

DEPOSITION TEMPERATURE (DEG C)

Ref	Final Layer Thick. A	560	570	580	600	620	650	700	750	800	
[16]	2000-5000	–	–	–	300 \<R\>	–	62 \<R\>	40 \<110\>	33 \<110\>	33 \<110\>	g.t. structure
[33]	4000	200 A	– –	140 weak \<311\>	98 mod \<110\>	72 \<110\>	– –	– –	– –	– –	g.t. structure
[23]	5000	170 A	130 A	97 weak \<311\>	55 \<311\>	33 \<110\>	– –	– –	– –	– –	g.t. structure
[22]	16000	170 A	122 A	83 A	48 A	27 10%P	11 40%P	3.2 94%P	1.4 99%P	0.68 100%P	g.t. structure
[16]	5000	29 A	– –	– –	25 A	– –	21 A	18 \<110\>	16 \<110\>	14 \<110\>	g.t. structure
[9]	3200-4500			45 A	68 \<110\>	43 \<110\> W\<311\>	45 \<110\> W\<311\>				g.t. structure
[34]		16	26	41	102	240*				3D	R A/min
[22]							840	4200	1100	crys. grow.	R A/min
[22]		>120	>90	>80	>40	7	2.7	0.6	0.16*		t(o) min

+ = plasma enhanced deposition   * = extrapolated value
A = amorphous   P = polycrystalline

R = linear rate of transformation of a-Si to poly-Si

t(o) = incubation time before transformation starts

Time to completely transform layer thickness x = $t(o) + \dfrac{x}{R}$ for T < 800 deg C

For T > 800 deg C, random homogeneous crystallization dominates.

TABLE 7: grain size in as-deposited undoped LPCVD poly-Si films. Grain
sizes are average values measured in a plane parallel to the
substrate surface.

Deposition Temp deg C	Final Polysilicon Thickness Z A	Grain Size L A	L/Z	Ref
600	3000	390	0.13	[41]
625	10,000	2100	0.21	[42]
632	3000	50-200	0.02-0.07	[43]
632	3000-6000	200	0.02-0.03	[44]
630	3000	Columnar 390 Equiaxed 50	0.2 & 0.13	[45]
625	4600	600	0.13	[46]
600	5000	180-200	0.04	[23]
620		280-320	0.06	[23]
600	2000	300	0.15	[16]
650	5000	700	0.14	[16]
700	5000	400	0.08	[16]
750	5000	350	0.07	[16]
800	5000	400	0.08	[16]
620	2600	500-1000+	0.29	[9]
	4700	1500-2000+	0.37	[9]
	10500	3000-4000+	0.29	[9]

+ measured at the top surface of the film.

REFERENCES
----

[1]   I.Esquivas, J.Sanz-Maudes, J.Sangrador, T.Rodriguez [J. Vac.
       Sci. & Technol. A (USA) vol.3 no.4 (July/Aug 1985) p.1791-6]
[2]   R.Tsu, J.Gonzalez-Hernandez, S.S.Chao, D.Martin [Appl. Phys. Lett.
       (USA) vol.48 no.10 (10 March 1986) p.647-9]
[3]   H.Windischmann, J.M.Cavese, R.W.Collins, R.D.Harris,
       J.Gonzalez-Hernandez [Mater. Res. Soc. Symp. Proc. (USA) vol.47
       (1985) p.187-94]
[4]   J.C.Bean [in 'Impurity Doping Processes in Silicon' Ed. F.F.Y.Wang,
       Ch.4 (North Holland, Amsterdam, 1981) p.175-215]
[5]   Y.Hamakawa [Mater. Res. Soc. Symp. Proc. (USA) vol.49 (1985)
       p.239-50]
[6]   P.G.LeComber [ibid p.341-51]
[7]   A.Marsh [Proc. SPIE Int. Soc. Opt. Eng. (USA) vol.394 (1983)
       p.28-32]
[8]   F.Hottier, R.Cadoret [J. Cryst. Growth (Netherlands) vol.61
       (1983) p.245-58]

[9]   T.I.Kamins M.M.Mandurah, K.C.Saraswat [J. Electrochem. Soc. (USA)
       vol.125 no.6 (June 1978) p.927-32]
[10]   T.Morie, J.Murota [Jpn. J. Appl. Phys. Part.2 (Japan) vol.23 no.7
       (Jul 1984) p.L482-4]
[11]   J.Bloem, L.J.Giling [in Current Topics in Materials Science 1, Ed.
       E.Kaldis, Ch.4 (North Holland, Amsterdam, 1978) p.147-342]
[12]   K.R.Sarma, M.J.Rice, Jr [J. Cryst. Growth (Netherlands) vol.56
       (1982) p.313-32]
[13]   K.V.Ravi, F.V.Wald [Semiconductor Silicon 1977, Eds. H.R.Huff,
       E.Sirtl (Electrochem. Soc. Princeton, USA 1977) p.820-35]
[14]   S.Veprek, S.Iqbal, H.R.Oswald, A.P.Webb [J. Phys. C (GB) vol.14
       no.3 (Jan 1981) p.295-308]
[15]   T.I.Kamins, K.L.Chiang [J. Electrochem. Soc. (USA) vol.129 no.10
       (Oct 1982) p.2331-5]
[16]   J.-J.J.Hajjar, R.Reif, D.Adler [J. Electron. Mater. (USA) vol.15
       no.5 (Sep 1986) p.279-85]
[17]   R.W.Andreatta, C.C.Abele, J.F.Osmundsen, J.G.Eden, D.Lubben,
       J.E.Greene [Appl. Phys. Lett. (USA) vol.40 no.2 (15 Jan 1982)
       p.183-5]
[18]   S.K.Roy, A.S.Vengurlekar, V.T.Karulkar, A.V.Joshi, S.Chandrasekhar
       [J. Electron. Mater. (USA) vol.16 no.4 (Jul 1987) p.211-17]
[19]   J.-O.Carlsson [Thin Solid Films (Switzerland) vol.130 no.3/4
       (1985) p.261-82]
[20]   J.Bloem [J. Cryst. Growth (Netherlands) vol.50 (1980) p.581-604]
[21]   F.Hottier, R.Cadoret [J. Cryst. Growth (Netherlands) vol.52 (1981)
       p.199-206]
[22]   A.M.Beers, H.T.J.M.Hintzen, J.Bloem [J. Electrochem. Soc. (USA)
       vol.130 no.6 (1983) p.1426-33]
[23]   G.Harbeke, L.Krausbauer, E.F.Steigmeir, A.E.Widmer, H.F.Kappert,
       G.Neugebauer [J. Electrochem. Soc. (USA) vol.131 no.3 (Mar 1984)
       p.675-82]
[24]   H.Kurokawa [J. Electrochemn. Soc. (USA) vol.129 no.11 (Nov 1982)
       p.2620-4]
[25]   R.F.C.Farrow [J. Electrochem. Soc. (USA) vol.121 (1974) p.899]
[26]   J.Bloem, W.A.P.Claassen [Appl. Phys. Lett. (USA) vol.40 no.8
       (15 Apr 1982) p.725-6]
[27]   M.L.Hitchman, J.Kane [J. Cryst. Growth (Netherlands) vol.55 no.3
       (Dec 1981) p.485-500]
[28]   J.J.Lander, J.Morrison [J. Appl. Phys. (USA) vol.33 (1962) p.2089]
[29]   R.C.Newman [Rep. Prog. Phys. (GB) vol.45 (1982) p.1163-210]
[30]   G.L.Olson, S.A.Kokorowski, J.A.Roth, L.D.Hess [Mater. Res. Soc.
       Symp. Proc. (USA) vol.1 (1981) p.125-31]
[31]   R.Angelucci [Appl. Phys. Lett. (USA) vol.39 no.4 (1 Aug 1981)
       p.346-8]
[32]   J.M.Blum, D.C.Green, B.S.Meyerson, B.A.Scott [IBM Tech.
       Disclosure Bull. (USA) vol.26 no.3A (Aug 1983) p.921-2]
[33]   H.S.Yoon, C.S.Park, S.-C.Park [J. Vac. Sci. & Technol. A (USA)
       vol.4 no.6 (Nov/Dec 1986) p.3095-3100]
[34]   K.Zellama, P.Germain, S.Squelard, J.C.Bourgoin, P.A.Thomas [J.
       Appl. Phys. (USA) vol.50 (1979) p.6995]
[35]   A.M.Beers, J.Bloem [Appl. Phys. Lett. (USA) vol.41 (Jul 1982)
       p.153]
[36]   W.Ahmed, D.B.Meakin [J. Cryst. Growth (Netherlands) vol.79 nos.1-3
       (Dec 1986) p.394-8]
[37]   P.Joubert, B.Loisel, Y.Chounan, L.Haji [J. Electrochem. Soc. (USA)
       vol.134 no.10 (1987) p.2541-5]
[38]   D.Meakin, J.Stoemenos, P.Migliorato, N.A.Economou [J. Appl. Phys.
       (USA) vol.61 no.11 (June 1987) p.5031-7]
[39]   H.Oppolzer, R.Falckenberg, E.Doering [J. Microsc. (GB) vol.118
       pt.1 (Jan 1980) p.97-103]

[40]   C.P.Ho, J.D.Plummer, S.E.Hansen, R.W.Dutton [IEEE Trans. Electron
       Devices (USA) vol.ED-30 no.11 (1983) p.1438-53]
[41]   S.J.Krause, S.R.Wilson, W.M.Paulson, R.B.Gregory [Appl. Phys.
       (USA) vol.45 no.7 (1 Oct 1984) p.778-80]
[42]   B.Swaminathan, K.C.Saraswat, R.W.Dutton, T.I.Kamins [Appl. Phys.
       Lett. (USA) vol.40 no.9 (May 1982) p.795-8]
[43]   S.J.Krause, et al [Mater. Res. Soc. Symp. Proc. (USA) vol.52 (1986)
       p.145-52]
[44]   S.R.Wilson, et al [Appl. Phys. Lett. (USA) vol.49 no.11 (Sept
       1986) p.660-2]
[45]   S.R.Wilson, et al [J. Electrochem. Soc. (USA) vol.132 no.4 (April
       1985) p.922-9]
[46]   A.H.Reader, F.W.Schapink, S.Radelaar [Proc. Symp.
       Poly-micro-crystalline and Amorphous Semiconductors, Strasbourg,
       (Les Ulis, France, Editions de Phys. June 1984) p.253-8]
[47]   M.Hendriks, R.Delhez, Th.H.de Keigser, S.Radelaar [Solid State
       Chemistry 1982, Eds. R.Meselaar, H.J.M.Heijligr, J.Schoonman,
       Studies in Inorganic Chemistry vol.3 (Elsevier, Amsterdam, 1983)
[48]   R.E.Mallard, D.C.Houghton, G.J.C.Carpenter, F.R.Shepherd [Inst.
       Phys. Conf. Ser. (GB) no.76 (1985) p.145-50]
[49]   K.V.Ravi [Thin Solid Films (Switzerland) vol.31 (1976) p.171]
[50]   F.Mieno, Y.Furumura, T.Nishizawa, M.Maeda [J. Electrochem. Soc.
       (USA) vol.134 no.11 (Nov 1987) p.2862-67]

26.4 DIFFUSION IN POLY-Si

by S.Jones and C.Hill

December 1987
EMIS Datareview RN=17862

A. INTRODUCTION

The diffusion of dopants in polysilicon is important because of the key
role of polysilicon layers in VLSI technology (as an interconnect; as a
gate material; and as a diffusion source to the underlying substrate)
and in solar cell technology. These applications require a knowledge of
the diffusion both vertically and laterally within the polysilicon, and
to the underlying substrate. The diffusivity of all common dopants is
observed to be increased in polycrystalline as compared to single
crystal silicon: this increase is attributed to the presence of grain
boundaries in the former which provide high diffusivity paths through
the material. Measurement of diffusivity of dopants within polysilicon
must therefore take account of two parallel processes: diffusion at and
along the grain boundaries; and diffusion within the bulk interior
grains. The reported measurements of diffusivities in polysilicon can be
divided into two categories: those that apply a single effective
diffusivity to describe the combined effects of grain interior and grain
boundary; and those that analyse the two separate processes, providing
values for the diffusivity at the grain boundary and in the grain
interior. Measurements in the latter category generally require a 2D
analysis of the dopant distribution around a single grain boundary and
so large grain material (cast, or laser enhanced recrystallised CVD)
is used. Effects such as the segregation of dopant between grain and
grain boundary, and the motion of grain boundaries during heat treatment
should also be considered [1]. In general no segregation effects are
considered and the analysis assumes a static network of grain
boundaries. The former category of effective diffusion coefficient are
generally made on thin CVD layers in which the grain size is much
smaller than in cast poly. The diffusivity is deduced from the
redistribution of as-implanted 1-D profiles during heat treatment.

In the following, the data for diffusivity of dopants within polysilicon
from these two distinct categories will be presented separately. There
will then follow sections on the related topics of segregation of
dopants to grain boundaries and diffusion of dopants from a polysilicon
source. For reference a glossary of notation is presented:

D_o	pre-exponent of diffusivity (in Arrhenius form)
E_a	activation energy of diffusivity (in Arrhenius form)
D_{gb}	diffusivity at a grain boundary
D_g	diffusivity in the bulk interior grain
L_g	grain diameter
delta	grain boundary width
k	segregation coefficient between grain and grain boundary
C_g	dopant concentration in the bulk interior grains
C_{gb}	dopant concentration at the grain boundary
N_g	density of bulk interior grain sites
N_{gb}	density of grain boundary sites
X_g	atom fraction of dopant at grain interior sites = C_g/N_g
X_{gb}	atom fraction of dopant at grain boundary sites = C_{gb}/N_{gb}
Q	heat of segregation

B. EFFECTIVE DIFFUSIVITIES IN POLYSILICON

The majority of values for dopant diffusivities quoted in the literature
are single, effective diffusivities averaged over many grains and
generally obtained by fitting to the 1-D redistribution from as-implanted
profiles. Harrison [2] gives the criterion for the validity of such an
approach as being that diffusion lengths in the bulk grain are much
larger than the typical grain diameter, in order that the dopant sees an
effectively homogeneous matrix of grain and grain boundary sites. This
criterion is not satisified for the data presented here: indeed Grovenor
[3] has pointed out that the measured activation energies are more
appropriate to single crystal silicon. This indicates that the grain
boundaries behave as limited source diffusion paths and the rate
limiting step in the process is diffusion from the grain interior to the
boundary. Despite this limitation, the data is useful for describing the
rate at which the implant profile spreads in polysilicon.

B1 Arsenic

Measured data of effective diffusivities of arsenic in polysilcon with
activation energies (where reported) are shown in TABLE 1 with an
indication of material and dopant source used. TABLE 2 gives a comparison
of these diffusivity values over the temperature ranges reported. It can
be seen that there is a wide range in the data (greater than an order
of magnitude) but the variation can be at least partly explained by the
following factors: composition and structure of the deposited poly;
effects of grain growth during heat treatment; and furnace or RTA heat
treatments.

The values reported for lateral diffusivity in polysilicon layers
[10,11] are larger than the majority of vertical diffusivities, although
ref [10] made measurements of both lateral and vertical diffusivity (by
different techniques) and reported no difference. The results of ref
[11] are further complicated in that the implant and diffusion heat
treatments were performed on the material as deposited and so grain
growth occurred simultaneously with diffusion. The authors estimate an
enhancement factor of 10 due to the grain growth during the first hour
of heat treatment at 1100 deg C.

The diffusivity of arsenic in polysilicon deposited on an oxide layer
is higher than that of poly directly deposited on single crystal. The
authors of ref [6] claim this is due to the smaller grain size of
polysilicon deposited on oxide. Ryssel et al [5] made a comparison of
the effect of the silicon/polysilicon interface on diffusivity and
observe a larger diffusivity in samples with a thin oxide (15 A)
interface compared to those with a reduced interface (freshly HF
cleaned before deposition). They also attribute the diffusivity
difference to the structural difference in polysilicon deposited on the
two substrates.

Takai et al [6] made a comparison between furnace and RTA heat treatments
observing an enhanced diffusivity value in the latter. This is explained
by a higher dopant activation occurring during RTA because of reduced
complex formation during the rapid temperature rise.

Swaminathan et al [8], Schubert [9] and Arienzo et al [7] have analysed
their results in terms of diffusion at a grain boundary (rather than an
effective combined diffusivity) even though they have studied 1-D
profiles averaged over many grains. This was done by restricting the
analysis to the diffusion behaviour in the tail region away from the
as-implanted peak. Swaminathan et al [8] observed that the diffusion
length in the tail region of a shallow implanted 1 micron poly layer was
independent of grain size (a range of sizes established by pre-anneal).
The diffusion tail was then interpreted as being solely along grain
boundaries which were supplied by a low diffusivity source (the grain
interior in the as-implanted region). A diffusivity was obtained by
fitting a complementary error function to the profiles. Schubert [9]
performed a similar study but with a more detailed model for the transfer
of dopant from grain interior to grain boundary, and obtained a similar
value for the diffusivity at low temperature (800 deg C). It was
found that the diffusivity decreased for longer anneal times (1 hour or
longer). This could be due to a saturation of grain boundary sites
available for diffusion [9]. Arienzo et al [7] used a bilayer
polysilicon structure with the top layer in-situ doped with arsenic.
Diffusion into the lower layer was assumed to be along grain boundaries
and a gaussian fit to the profile was used to obtain a value for
diffusivity. The values obtained by this method are more than an order
of magnitude lower than refs [8] and [9] above. This was conjectured
[7] as being due to the effects of different deposition techniques on
the impurity composition of the grain boundaries (there is no
quantitative evidence for this, carbon is reported to have a negligible
effect [21]).

The differences between diffusion in an ambient of either N2 or O2 [4,5]
are reported to be negligible. No authors have observed a concentration
dependence for the effective arsenic diffusivity as evidenced by the
success in fitting simple functions to the observed profiles. It should
be emphasised that many of the above results do not take into account
grain growth effects on the redistribution during the early stages of
thermal treatment. For RTA anneals at 900 deg C of arsenic implanted
polysilicon layers, an effective diffusivity of order 10*-13 cm*2/sec
[11] was observed over the first 20 seconds of anneal. This is slightly
larger (< 10) than the majority of values for effective diffusivities
for polysilicon on single crystal silicon. Wilson et al [12] performed
RTA at 1000 deg C on As-implanted polysilicon layers deposited on oxide
and observed a diffusivity of order 8 X 10*-11 cm*2/sec which is about
an order of magnitude larger than the values quoted in TABLE 2. It
should be noted, however, that applying an effective diffusivity in this
time regime is a severe simplification and will not model the detailed
profile features in the polysilicon layer.

B2 Phosphorus

The results for the effective diffusivity of phosphorus in polysilicon
are shown in TABLES 3 and 4. Ref [10] is a measurement of lateral
diffusivity. The data of Kamins et al [13] were obtained for polysilicon
layers deposited at three different temperatures. The temperature
governs the structure of the as-deposited polysilicon and it appears
that the diffusivity is sensitive to this structure, that with maximum
[110] texture having maximum diffusivity. The diffusivity also depends
upon the deposition rate of the poly film and is corrected to the
deposition temperature (lowering the rate gives similar diffusivity
behaviour to lowering the deposition temperature).

B3 Boron

The results for the effective diffusivity of boron in polysilicon are
also shown in TABLES 3 and 4. Kamins [13] reports a dependence of
diffusivity on the deposition temperature: maximum diffusivity occurring
for structures with maximum [110] texture. The data of Coe [15] is for
lateral diffusion, but is in reasonable agreement with the vertical
diffusivity of Michel [14]. Michel observes an interaction effect for
boron in the presence of high concentration arsenic which indicates that
boron at the grain boundaries is negatively charged. The results of
Ghannam et al [41] are for diffusion in as-implanted as-deposited
layers. For a pre-annealed polysilicon (1000 deg C, 45 mins) layer
these authors observe a 30% reduction in the diffusivity due to the
increased grain size.

B4 Other Dopants

The only information is for aluminium at low temperature, displayed in
TABLE 3.

C. DIFFUSION ALONG GRAIN BOUNDARIES IN POLYSILICON

A study of the vertical penetration of dopant into large grained
polysilicon material from an unlimited surface source (doped glass or
chemical) can provide direct information on the diffusivity along
grain boundaries. Many authors use variants of the model presented
by Whipple [17] for diffusion in a semi-infinite polysilicon layer
with a static network of grain boundaries, to interpret the dopant
penetration depth along a grain boundary. In fact only the quantity
P is measured directly, where

$$P = k \text{ delta } D_{gb}$$

(k is the segregation coefficient, delta the grain boundary width).
This quantity, P, is obtained in terms of the diffusivity in the bulk
interior grain, D_g. The latter value is either measured directly or is
assumed to be equal to the single crystal silicon diffusivity. Unless
stated below, it is assumed that a grain boundary width of 5 A and no
segregation (k = 1) are used in deriving a value of D_{gb} from P. It
should be noted that the diffusion activation energy derived in this way
will contain a (negative) contribution from the heat of segregation
(see following section). Values of k are, in general, very much greater
than 1. Published values (in the range 20-1250) are given in the next
section.

TABLES 5 and 6 present the grain boundary diffusion data from common
dopants. The majority of data is for phosphorus. It can be seen that the
measured activation energies are lower than those for effective
diffusivities (see TABLE 3); expected due to the more open structure
of grain boundaries compared to the bulk material. The phosphorus data
can be divided into two groups with significantly different
diffusivities. The higher diffusivities of refs [18], [19], [20] and [21]
were obtained on either cast or recrystallised CVD polysilicon, using
radiotracer measurements [19,21] or EBIC measurements of grain boundary
spikes in junction depths [20,22]. The results of Buonaquisti et al
[18] are a re-analysis of published data (B,P from Kamins [13]; As from

Johnson [24]) using the Whipple model [17] and an extension of this for thin films. The data with a lower diffusivity [22,23] were obtained on LPCVD poly deposited on oxide for which different types of grain boundary may be expected. (Note that the diffusivity values of Losee [23] are enhanced by a factor of 20 if we apply the standard values of segregation and grain boundary width, as defined above).

Data for antimony, boron and hydrogen are also displayed in TABLES 5 and 6. The antimony data [19,21] were obtained from similar samples analysed by radiotracer techniques. The overall diffusivities are in reasonable agreement, but the activation energies are significantly different. The low activation energy of antimony in ref [21] is explained as due to the contribution of the heat of segregation (which counteracts the diffusion activation energy). The boron data [24] are a re-analysis of data presented by Kamins [13] using the Whipple model [17]. The hydrogen data was obtained from cast polysilicon samples and the grain boundary diffusivity is of order 10*3 larger than the bulk diffusivity.

In order to derive an effective diffusion coefficient in terms of a combination of bulk and grain boundary values valid for long anneal times/high temperatures, Harrison [2] suggests the form:

$$D(eff) = (1-f)Dg + f\ Dgb$$

where f is the fraction of time a dopant atom spends at a grain boundary site. For cubic grains a suitable value of f is 3k delta/Lg.

TABLE 1: effective arsenic diffusivities in polysilicon (vertical diffusivity except for data from refs [10] and [11] which refer to lateral diffusivity)

Temp. (deg C)	Do (cm*2/sec)	E (eV)	source+ambient	material* (thickness)	Ref
950-1100	0.63	3.22	implant N2,O2	LPCVD 700 deg C (1 micron)	[4]
800-1000	8.5 X 10*-3	2.74	implant N2,O2	LPCVD 690 deg C HF cleaned interface (1 micron)	[5]
800-1000	1.66	3.22	implant N2,O2	LPCVD 690 deg C 15 A oxide interface (1 micron)	[5]
700-1000	1.81	3.14	implant N2	LPCVD 610 deg C on 500 A oxide (0.54 micron)	[6]
900-1050	0.28	2.84	implant RTA in Ar	LPCVD 610 deg C on 500 A oxide (0.54 micron)	[6]

TABLE 1 (continued)

Temp. (deg C)	Do (cm*2/sec)	E eV	source+ambient	material* (thickness)	Ref
700-850	10	3.36	in-situ doped N2	LPCVD bilayer on HF cleaned Si	[7]
750-950	8.6 X 10*4	3.9	implant N2	LPCVD 625 deg C on SiO2	[8]
800	Dgb=5 X 10*-14 Dg=1.5 X 10*-16		implant oxide cap	LPCVD 640 deg C on 400 A SiO2	[9]
900-1100	(see table 2)		implant N2	LPCVD (0.31 micron) 1000 deg C pre-anneal	[10]
850-1100	(see table 2)		implant N2	LPCVD (0.3 micron)	[11]

* polysilicon layers are deposited either on silicon crystal substrates
  or on thermally grown silicon dioxide layers (SiO2) on such substrates

TABLE 2: effective arsenic diffusivity in polysilicon versus temperature

Diffusivity (cm*2/sec)

700 deg C	800 deg C	900 deg C	1000 deg C	1100 deg C	Ref
		9.2 X 10*-15	1.1 X 10*-13	9.6 X 10*-13	[4]
	1.1 X 10*-15	1.4 X 10*-14	1.2 X 10*-13		[5]
	1.2 X 10*-15	2.4 X 10*-14	3.0 X 10*-13		[5]
1.0 X 10*-16	3.2 X 10*-15	5.9 X 10*-14	6.7 X 10*-13		[6]
		1.8 X 10*-13	1.6 X 10*-12	1.1 X 10*-11	[6]
4.0 X 10*-17	1.7 X 10*-15	3.7 X 10*-14			[7]
5.4 X 10*-16	4.1 X 10*-14	1.5 X 10*-12	3.1 X 10*-11		[8]
		1.6 X 10*-13	2.3 X 10*-12	6.6 X 10*-11	[9]
		1.3 X 10*-12	1.6 X 10*-12	2.4 X 10*-11	[10]

TABLE 3: effective diffusivities of dopants other than arsenic in
        polysilicon (vertical diffusivity unless stated)

Temp. (deg C)	Do (cm*2/sec)	E (eV)	source+ambient	material & thickness in microns	Ref
P dopant:					
1000-1200	(see table 4)		P doped glass N2	CVD 900 deg C 5/15 on (111) Si	[13]
1000-1200	(see table 4)		P doped glass N2	CVD 1040 deg C	[13]
1000-1200	(see table 4)		P doped glass N2	CVD 1100 deg C	[13]
900-1000	(see table 4)		implant N2	LPCVD 0.3	[10]
B dopant:					
1000-1200	4.4	3.48	B doped glass N2	CVD 900 deg C	[10]
1000-1200	1.33 X 10*6	1.47	B doped glass N2	CVD 1040 deg C	[10]
1000-1200	1.2 X 10*-6	1.63	B doped glass N2	CVD 1100 deg C	[10]
800	5.4 X 10*-15		implanted	LPCVD 650 deg C 0.1 on HF cleaned Si	[14]
800-1000	1.183	3.04	implanted Ar	LPCVD 625 deg C 0.86 on 1000A SiO2	[41]
875-1000	2.21	3.25	BN	CVD 770 deg C 0.71 on SiO2	[15]
B dopant (lateral diffusion):					
925-1075	7.06	3.36	BN	CVD 720 deg C	[15]
	3.2	3.31	BN	CVD 770 deg C	[15]
	0.46	3.12	BN	CVD 820 deg C	[15]
Al dopant:					
350-425	1.3 X 10*7	2.64	200 A Al (He)	e beam evaporated on SiO2 800 deg C	[16]

TABLE 4: effective dopant diffusivity in polysilicon versus temperature

		Diffusivity (cm*2/sec)			
800 deg C	900 deg C	1000 deg C	1100 deg C	1200 deg C	Ref

P dopant:

		8.5 X 10*-13	3.4 X 10*-12	2.5 X 10*-11	[13]
		6.3 X 10*-12	1.1 X 10*-11	6.2 X 10*-11	[13]
		7 X 10*-13	1.5 X 10*-12	1.1 X 10*-11	[13]
	1.4 X 10*-12	1.5 X 10*-11	8 X 10*-11		[10]

B dopant:

		7.4 X 10*-14	7.4 X 10*-13	5.5 X 10*-12	[10]
		2.0 X 10*-12	5.3 X 10*-12	1.2 X 10*-11	[10]
		4.2 X 10*-13	1.2 X 10*-12	3.2 X 10*-12	[10]
5.4 X 10*-15					[14]
6.2 X 10*-15	1.0 X 10*-13	1.1 X 10*-12			[41]
1.2 X 10*-15	2.4 X 10*-14	3 X 10*-13			[15]

B dopant (lateral diffusion):

	2.6 X 10*-14	3.5 X 10*-13	3.3 X 10*-12		[15]
	1.9 X 10*-14	2.5 X 10*-13	2.3 X 10*-12		[15]
	1.8 X 10*-14	2.0 X 10*-13	1.6 X 10*-12		[15]

TABLE 5: dopant diffusivities along polysilicon grain boundary (grain boundary width = 5 A; segregation = 1 unless stated)

Temp (deg C)	Do (cm*2/sec)	Q (eV)	source+ambient	material (thickness)	Ref
As dopant:					
1000	10*-12				[18]
P dopant:					
900-1150	120	2.87	H3PO4 (radiotracer)	Wacker CVD recryst. at 1380 deg C	[19]
1040	2.6 X 10*-9		P glass	Wacker SILSO	[20]
900-1100	1.18 X 10*-3	1.4	P2O5 (radiotracer)	CVD recryst. at 1350 deg C	[21]
900-1100	5.6 X 10*-5	1.95	implant N2	LPCVD 0.5 micron, laser recryst.	[22]
*700-900	3.8	3.2	implant N2	LPCVD 580 deg C on SiO2	[23]
1000-1200	4 X 10*-3	1.72		reanalysis of Kamins' data [13]	[18]
B dopant:					
1000-1200	6.6 X 10*-4	1.88		reanalysis of Kamins' data [13]	[18]
Sb dopant:					
930-1150	3.8 X 10*2	2.9	SbCl3	CVD recryst. at 1380 deg C	[19]
900-1100	4.4 X 10*-6	0.83	SbCl3	CVD recryst. at 1350 deg C	[21]
H dopant:					
100-400	8.2 X 10*-5	0.28		cast	[25]

* authors assume delta = 10 A, k = 10

TABLE 6: dopant diffusivities along polysilicon grain boundaries versus
temperature

	800 deg C	900 deg C	1000 deg C	1100 deg C	1200 deg C	Ref
As dopant:						
			10*-11			[18]
P dopant:						
	5.6 X 10*-11	5.2 X 10*-10	3.5 X 10*-9	1.8 X 10*-8		[19]
	1.1 X 10*-9	3.4 X 10*-9	8.6 X 10*-9			[21]
	2.3 X 10*-13	1.1 X 10*-12	3.9 X 10*-12			[22]
3.6 X 10*-15	6.8 X 10*-14					[23]
		6.2 X 10*-10	1.9 X 10*-9	5.2 X 10*-9		[18]
B dopant:						
		2.4 X 10*-11	8.3 X 10*-11	2.4 X 10*-10		[18]
Sb dopant:						
	1.3 X 10-10	1.3 X 10*-9	8.6 X 10*-9	4.6 X 10*-8		[19]
	1.2 X 10*-9	2.3 X 10*-9	4.0 X 10*-9			[21]

note: values from ref [23] should be increased by a factor of 20 to give
standard grain boundary width and segregation

D. DOPANT SEGREGATION TO THE GRAIN BOUNDARY IN POLYSILICON

This section will present data for the segregation of arsenic,
phosphorus and boron to polysilicon grain boundaries. For information
on the segregation behaviour of other impurities in polysilicon the
reader should consult ref [44].

All the data presented here refer to an equilibrium segregation (i.e.
that following heat treatments sufficiently long that diffusion lengths
are greater than the typical grain size). Segregation is reported
quantitatively in three different ways, reflecting the techniques used
to measure the segregation:

(i)   The ratio of atomic fractions of dopant at grain boundary sites to
       that at bulk interior grain sites, i.e. X_{gb}/X_g where $X_{gb} = C_{gb}/N_{gb}$
       is the fraction of grain boundary sites occupied and $X_g = C_g/N_g$ is
       the fraction of bulk interior sites occupied. This is the measured
       quantity in Scanning Transmission Electron Microscope (STEM) X-ray
       microanalysis across a grain boundary (where the enhanced signal
       is interpreted as due to excess dopant at the grain boundary).

(ii)   The ratio of dopant concentration at grain boundary sites to that
       at bulk interior sites, i.e. C_{gb}/C_g, where the concentrations are
       averaged per unit volume. This has the disadvantage of being
       dependent on grain size, but is the natural measurement made by
       Hall Mobility or resistance analysis (where the dopant at the
       grain boundary is assumed to be inactive).

(iii) Since grain boundaries are 2D structures the dopant at a grain
       boundary can be expressed as a fraction of a monolayer (taking
       a monolayer as $6.8 \times 10^{14}/cm^2$: the atomic density of a (100)
       plane).

In order to translate between definitions (i) and (ii), an estimate of
the ratio of available grain boundary sites to bulk sites needs to be
made. For cubic grains in which the grain boundary is a single monolayer
Mandurah et al [26] suggest $N_{gb}/N_g = 16.2/L_g$ (Angstroms).

A heat of segregation, Q, can be defined from the McLean isotherm
(assuming grain boundary saturation at $X_{gb} = 1$):

$$\frac{X_{gb}}{1-X_{gb}} = \left(\frac{X_g}{1-X_g}\right) A e^{Q/kT}$$

where $A = \exp(S/k)$ expresses the difference in vibrational entropy between
grain and grain boundary sites. For dilute solutions this becomes

$$\frac{C_{gb}}{C_g} = A \left(\frac{N_{gb}}{N_g}\right) e^{Q/kT}$$

TABLE 7 presents values for the heat of segregation of As and P
together with values of L_g (the grain size) and A, the pre-exponent
factor, where deduced by the authors. TABLE 8 presents values of the
segregation coefficient (as defined in (i) above) over the temperature
and concentration ranges presented in the literature. There has been no
evidence of any boron segregation to the grain boundaries seen in these
studies [26], although there is some evidence of boron segregation from
1-D profiles of boron in polysilicon (see below).

It can be seen that the activation energies are positive (i.e. lower
enthalpy at the grain boundary), and so segregation effects are greater
at lower temperature. In general, segregation decreases with increased
dopant concentration (with the exception of the As data of Carabelas
et al [27] at high concentrations). The activation energies are in
reasonable agreement despite the very different measurement techniques.
However, the overall segregation values displayed in TABLE 8 are very
different for the two techniques. The resistivity measurements predict
a much larger value of segregation than do the direct STEM measurements:
indeed, for the highest concentrations, the data of Mandurah et al [26]
and Carabelas et al [27] would predict several monolayers of dopant
segregated at the grain boundary. In contrast the STEM results of
Oppolzer [30] give a maximum segregation of approximately one monolayer
of arsenic with a total concentration of $8 \times 10^{20}/cm^3$ at 950 deg C,
whilst Grovenor et al [28] observe saturation of grain boundaries at
only 0.1 to 0.15 of a monolayer. Grovenor [3] has suggested that the
resistivity measurements of Mandurah [26] and Carabelas [27] over-
estimate the amount of segregation because they assume that any grain
boundary potential barriers have no influence on the electrical

conduction process. Such potential barriers reduce the mobility of the charged carriers and so assuming that they are not present leads to an underestimation of the active concentration (and hence overestimation of the inactive dopant which is assumed to be segregated to the grain boundaries). The electrical measurements of segregation are useful, however, for prediction of the electrical resistance of doped polysilicon layers.

Vertical profiles of dopant through polysilicon layers exhibit a concentration peak at the interface of the polysilicon layer with the underlying substrate. This peak has been interpreted as occurring at a grain boundary along the lateral interface and is observed for arsenic, phosphorus and, to a lesser extent, boron. In the case of arsenic, the interface peak has been estimated to contain 3 X 10*14 atoms/cm*2 at 1000 deg C (=14% of the As in the poly layer at average concentration 1.4 X 10*20/cm*3) [31]; 8 X 10*14 atoms/cm*2 at 950 deg C (=4% of the As in the poly layer at average concentration 9 X 10*20/cm*3) [32]. For phosphorus an interface peak of 9.5 X 10*14 atoms/cm*2 at 950 deg C (=8% of the P in the poly layer at concentration 6 X 10*20/cm*2) is observed. These numbers are of order one monolayer and so are consistent with the segregation data presented above. In the case of boron, a smaller peak of 4 X 10*13 atoms/cm*2 at 950 deg C (=1% of B in the poly layer) is reported [32]. It is not clear to what extent this peak is a property of the interfacial oxide as opposed to the grain boundary; however, the peak is relatively larger for high boron doses (above saturation concentration) indicating that the boron in the peak may be in the form of complexes). An overall segregation step at the polysilicon/silicon interface of 0.7 for boron is also reported [32,33]. This is believed to be due to dopant segregation in the form of immobile boron complexes at the polysilicon grain boundaries [32].

TABLE 7: heat of segregation for dopants in polysilicon

Technique	Temp (deg C)	Concentration (/cm*3)	Q (eV)	A	Lg (micron)	Ref
As dopant:						
Hall effect	800-1000	2 X 10*19	0.44	3.02	0.1	[26]
Hall effect	800-1000	6 X 10*19	0.42	2.67	0.12	[26]
Hall effect	800-1000	2 X 10*20	0.41	2.17	0.16	[26]
resistivity + Hall effect	700-1050	1.2 X 10*20	0.54	0.97*	0.38	[27]
resistivity + Hall effect	700-1050	2.8 X 10*20	0.59	1.09*	0.48	[27]
STEM	700-1000	2 X 10*20	0.65**	1	0.4	[28]

Technique	Temp (deg C)	Concentration (/cm*3)	Q (eV)	A	Lg (micron)	Ref
P dopant:						
Hall effect	800-1000	2 X 10*19	0.44	2.46	0.1	[26]
resistivity + Hall effect	700-1050	1 X 10*20	0.54	0.55*	0.4	[27]
resistivity + Hall effect	700-1050	2.7 X 10*20	0.59	0.19*	0.4	[27]
STEM	650-800	3.3 X 10*20	0.33	1	-	[29]

* deduced from fig 3 in ref [27]
** grain boundary saturation at Xgb = 0.1-0.15

TABLE 8: grain boundary segregation coefficient for dopants in polysilicon. (Segregation coefficient defined as the ratio of atom fraction of dopant at grain boundary sites to the atom fraction at grain interior sites).

Concn (/cm*3)	Segregation coefficient (Xgb/Xg)				Ref
	(700 deg C)	(800 deg C)	(900 deg C)	(1000 deg C)	
As dopant:					
2 X 10*19		350	230	170	[26]
6 X 10*19		250	170	120	[26]
2 X 10*20		180	125	90	[26]
1.2 X 10*20	600	330	205	135	[27]
2.8 X 10*20	1250	650	380	240	[27]
2 X 10*20	21-24	19-30	17-25	15-22	[28]
P dopant:					
2 X 10*19		280	190	135	[26]
1 X 10*20	340	190	115	80	[27]
2.7 X 10*20	220	110	65	40	[27]
3.3 X 10*20	38	27			[29]

E. POLYSILICON AS A DIFFUSION SOURCE

Polysilicon layers deposited on a single crystal silicon substrate are
often used as diffusion sources into the substrate. The important
issues in using such sources are: the effects of the interface between
the poly and single crystal silicon both as a barrier to diffusion and
as a barrier to epitaxial regrowth of the polysilicon; and the diffusion
behaviour of dopant in the single crystal in the presence of a
polycrystalline boundary condition.

E1 Interface Effects

The presence of a thin interfacial amorphous oxide region between the
single crystal substrate and the deposited polysilicon layer has been
observed in transverse TEM. The thickness of this layer depends on the
clean and etch applied to the single crystal prior to deposition. For a
standard RCA clean a thickness of 13-14 A is measured by ellipsometry
[34]; 15-20 A in TEM [35]. The oxide layer can be further reduced by
applying a clean in buffered hydrofluoric acid leaving an interfacial
oxide estimated to be between 5 and 10 A [35]. Assuming that the
interface is a uniform layer of silicon dioxide the layer thickness was
estimated as 1.7 A by the authors of ref [1] from the Auger oxygen
signal at the interface of an HF cleaned sample. The composition of this
interfacial layer is estimated to contain oxygen to silicon in the ratio
1.7 to 1.8 [34] (though discontinuities in the interface may distort
this value).

During high temperature heat treatment of the polysilicon layer
epitaxial realignment to the substrate may occur depending on the
thickness of the interfacial layer. For the thinnest interface
(buffered HF treated) epi-realignment occurs for temperatures above
(approximately) 950 deg C and is a function of anneal time and
doping. For the thicker interface oxides (15-20 A) no epitaxial
realignment occurs [36]. If epitaxial realignment occurs partially,
then dopant diffusion is hard to control (though Hoyt et al [37]
obtain well controlled shallow profiles by intentional epi-realignment
using RTA at 1100 deg C).

There is some disagreement in the literature as to whether or not the
thin oxide interface layer presents a barrier to diffusion from the
polysilicon into the substrate. Various authors claim that even the
thinnest interface acts as a barrier [14,34,38] whereas others only see
a significant barrier for thicker, intentionally grown oxides of 25 A
[30,32,33,36]. The disagreement appears to rest upon the interpretation
of a commonly seen peak of dopant at the interface. This peak can be
interpreted as indicating either a diffusion barrier or interfacial
segregation. The area under the peaks observed for arsenic and phosphorus
is consistent with the segregation values presented above if it is
assumed that the interface consists of a continuous lateral grain
boundary. Schaber et al [32] have shown that, if the peak is subtracted,
then the arsenic and phosphorus profiles are continuous across the
interface even for interfaces with approximately 15 A of oxide, although
there is an expected discontinuity in the gradient of the profile at the
interface (due to the enhanced diffusivity in the polysilicon layer).
For the case of boron a discontinuity in the concentration is seen in

the ratio 0.7 of polysilicon to silicon [32]. This has also been studied by Rausch et al [33] and by studying both diffusion of boron from the polysilicon and up-diffusion from the single crystal they have showed that this is indeed a segregation effect, and not a barrier effect. The segregation is assumed to occur in the grain boundaries of the polysilicon where the boron is conjectured to be in the form of immobile complexes [32].

Evidence for an interfacial diffusion barrier may also be seen in a retarded diffusion profile into the single crystal substrate. Schaber et al [32] see no differences between the diffusion profiles for interfaces with and without an HF clean and so deduce that there is no barrier operating. Stork et al [34] found a junction depth dependence for thin chemically grown oxides (approx. 15 A). Sagarat et al [42] observed that an interfacial oxide of 10 A caused a retardation in the diffusion profile of 100 A. This was explained as being due to an increase in the segregation coefficient between the polysilicon and single crystal substrate for samples with a 10 A oxide interface. Jorgensen et al [43] performed TEM studies of the interface for HF cleaned as-deposited samples and observed a high density of pinholes in the oxide layer. The discontinuous nature of such oxide interfaces may explain why they do not act as diffusion barriers.

E2 Diffusion in the Single Crystal Substrate

Some anomalous features of the diffusion profiles within the single crystal substrate from a polysilicon dopant source have been observed. Barbuscia et al [39] have observed enhanced diffusion tails for both arsenic and boron diffusion from polysilicon layers for concentrations below n(i). In the case of arsenic the enhancement was dose dependent. They speculate that the cause may be point defect mechanisms in the polysilicon layer. For high concentration arsenic and phosphorus Schaber et al [32] observe diffusion occurring at concentrations greater than the bulk maximum active concentration, though for boron the profiles never exceed the expected maximum concentration in the single crystal. Lever et al [40] observed an enhanced boron diffusion in the substrate with a cut-off in the concentration dependence at $2.5 \times 10^{19}/cm^3$ at 950 deg C.

REFERENCES
----

[1]   A.O'Neill, C.Hill, J.King, C.Please [Submitted to J. Appl.
       Phys. (USA) (1987)]
[2]   L.G.Harrison [Trans. Faraday Soc. (GB) vol.57 (1961) p.1191]
[3]   C.R.M.Grovenor [J. Phys. C (GB) vol.18 no.21 (30 Jul 1985)
       p.4079-120]
[4]   K.Tsukamoto, Y.Akasaka, K.Horie [J. Appl. Phys. (USA) vol.48
       no.5 (May 1977) p.1815-21]
[5]   H.Ryssel, H.Iberl, M.Bleier, K.Haberger, H.Kranz [Appl. Phys.
       (Germany) vol.24 (1981) p.197]
[6]   M.Takai et al [Nucl. Instrum. & Methods Phys. Res. Sect. B
       (Netherlands) vol.B19/20 (1987) p.603-6]
[7]   M.Arienzo,Y.Komen, A.E.Michel [J. Appl. Phys. (USA) vol.55 no.2
       (1984) p.365-9]

[8]   B.Swaminathan, K.C.Saraswat, R.W.Dutton [Appl. Phys. Lett.
        (USA) vol.40 (1982) p.795]
[9]   W.K.Schubert [J. Mater. Res. (USA) vol.1 (1986) p.311]
[10]   Y.Sato, K.Murase, H.Harada [J. Electrochem. Soc. (USA) vol.129
        (1982) p.1635]
[11]   N.Lewis, G.Gildenblat, M.Ghezzo, G.A.Smith [Appl. Phys. Lett.
        (USA) vol.42 (1983) p.171]
[12]   S.R.Wilson, W.M.Paulson, R.B.Gregory, J.D.Gressett, A.H.Hamdi,
        F.D.McDaniel [Appl. Phys. Lett. (USA) vol.45 (1984) p.464]
[13]   T.I.Kamins, J.Manoliu, R.N.Tucker [J. Appl. Phys. (USA)
        vol.43 (1972) p.83]
[14]   A.E.Michel, R.H.Kastl, S.R.Mader [Nucl. Instrum. & Methods
        Phys. Res. (Netherlands) vol.209/210 (1983) p.719]
[15]   D.J.Coe [Solid-State Electron. (GB) vol.20 (1977) p.985]
[16]   J.C.M.Hwang, P.S.Ho, J.E.Lewis, D.R.Campbell [J. Appl. Phys.
        (USA) vol.51 (1980) p.1576]
[17]   R.T.P.Whipple [Philos. Mag (GB) vol.45 (1954) p.1225];
        A.D.Le Claire [Br. J. Appl. Phys. (GB) vol.14 (1963) p.351];
        T.Suzuoka [J. Phys. Soc. Jpn. (Japan) vol.19 (1964) p.839]
[18]   A.D.Buonaquisti, W.Carter, P.H.Holloway [Thin Solid Films
        (Switzerland) vol.100 (1983) p.235]
[19]   F.H.M.Spit, H.Bakker [Phys. Status Solidi a (Germany) vol.97
        no.1 (1986) p.135-42]
[20]   P.H.Holloway [J. Vac. Sci. & Technol. (USA) vol.21 no.1 (1982)
        p.19-22]
[21]   J.L.Liotard, R.Biberian, J.Cabane [J. Phys. Colloq. (France)
        vol.43 no.C-1 (1982) p.213-18]
[22]   H.Baumgart, H.J.Leamy, G.K.Celler, L.E.Trimble [ibid p.363-8]
[23]   D.L.Losee, J.P.Lavine, E.A.Trabka, S.T.Lee, C.M.Jarman [J. Appl.
        Phys. (USA) vol.55 (1984) p.1218]
[24]   N.M.Johnson, D.K.Bieglesen, M.D.Moyer [Appl. Phys. Lett. (USA)
        vol.38 no.11 (1 Jun 1981) p.900-2]
[25]   L.L.Kazmerski [J. Vac. Sci. & Technol. A (USA) vol.3 no.3 pt.2
        (1985) p.1287-90]
[26]   M.M.Mandurah, K.C.Saraswat, C.R.Helms [J. Appl. Phys. (USA)
        vol.51 no.11 p.5755-63]
[27]   A.Carabelas, D.Nobili, S.Solmi [J. Phys. Colloq. (France) vol.43
        no.C-1 (1982) p.187-92]
[28]   C.R.M.Grovenor, P.E.Batson, D.A.Smith, C.Wong [Philos Mag. A (GB)
        vol.50 (1984) p.409]
[29]   J.H.Rose, R.Gronsky [Appl. Phys. Lett. (USA) vol.41 no.10 (15
        Nov 1982) p.993-5]
[30]   H.Oppolzer, W.Eckers, H.Schaber [J. Phys. Colloq. (France) vol.46
        no.C-4 (1985) p.523-8]
[31]   B.Swaminathan, E.Demoulin, T.W.Sigmon, R.W.Dutton, R.Reif [J.
        Electrochem. Soc. (USA) vol.127 no.10 (1980) p.2227-9]
[32]   H.Schaber, R.V.Criegem, I.Weitzel [J. Appl. Phys. (USA) vol.58
        no.11 (1 Dec 1985) p.4036-42]
[33]   W.A.Rausch, R.F.Lever, R.H.Kastl [J. Appl. Phys. (USA) vol.54
        (1983) p.4405]
[34]   J.M.C.Stork, M.Arienzo, C.Y.Wong [IEEE Trans. Electron Devices
        (USA) vol.ED-32 no.9 (Sep 1985) p.1766-70]

[35] V.Probst, H.J.Bohm, H.Schaber, H.Oppolzer, I.Weitzel
[Semiconductor Silicon 1986, Eds H.R.Huff et al (Electrochem.
Soc., Princeton, NJ, USA, 1986)]

[36] H.J.Bohm, H.Wendt,, H.Oppolzer, K.Masselli, R.Kassing [J.
Appl. Phys. (USA) vol.62 (1986) p.2784-8]

[37] J.L.Hoyt, E.Crabbe, J.F.Gibbons, R.F.W.Peare [Appl. Phys. Lett.
(USA) vol.50 no.12 (23 Mar 1987) p.751-3]

[38] C.Y.Wong, A.E.Michel, R.D.Isaac, R.H.Kastl, S.R.Mader [J. Appl.
Phys. (USA) vol.55 (1984) p.1131]

[39] G.P.Barbuscia, G.Chin, R.W.Dutton, T.Alvarez, L.Arledge [Int.
Electron Devices Meet. Tech. Dig. (USA) (1984) p.757]

[40] R.F.Lever, B.Garben, C.M.Hsieh, W.A.Orr Arienzo [Mater.
Res. Soc. Symp. Proc. (USA) vol.36 (1985) p.95]

[41] M.Y.Ghannam, R.W.Dutton, S.W.Novak [Mater. Res. Symp. Proc. (USA)
vol.76 (1987) p.283-8]

[42] K.Sagara, T.Nakamura, Y.Tomaki, T.Shiba [IEEE Trans. Electron
Devices (USA) vol.ED-34 (1987) p.2286]

[43] N.Jorgensen et al [Inst. Phys. Conf. Ser. no.76 (1985) p.471]

[44] C.R.M.Grovenor [EMIS Datareview RN=15777 (Dec 1987) 'Chemistry
of grain boundaries in poly-Si']

26.5 RECRYSTALLISATION OF POLY-Si

by C.Hill and S.Jones

January 1988
EMIS Datareview RN=17892

A. RECRYSTALLIZATION PROCESSES

A1 Introduction

Heat treatment of polycrystalline silicon films causes changes in
microstructure, in particular, the grain structure. This review
summarises the changes in grain structure that occur in thin (0.01 -
10 micron) LPCVD polycrystalline silicon films by solid state
processes during heat treatments in the range 500-1400 deg C.
Recrystallization can also be effected by localised melting and
regrowth: this important and large field is covered in several reviews
[1-3] and will not be further considered here.

Recrystallization occurs because polycrystalline silicon has a higher
free energy than single crystal silicon, and heating provides the
thermal activation energy to overcome the potential barrier and the
transformation from the higher energy to the lower energy form. Various
features of the microstructure contribute energy terms to the driving
force for transformation (section A2), which transformation proceeds
by nucleation and boundary motion (section A3), and the combination
of these factors in doped and implanted films gives rise to
characteristic modes of recrystallization (Section A4).

A2 Driving Forces for Recrystallization

The microstructure of polycrystalline silicon incorporates regions of
higher stored energy than the same mass of bulk single crystal at the
same temperature. In all cases, the increase in energy results from
distortion or incompletion of the perfect diamond cubic lattice
of bulk single crystal. If all the atoms of a region are in the
higher energy state, then a second phase exists, and the boundary
between the region and the lower energy crystal around is a phase
boundary. The only example of this commonly met with in polysilicon
is regions of amorphous silicon; the stored energy (enthalpy)
released on conversion to crystalline silicon is +0.2 eV per atom [4].
Regions of high dislocation density can approximate to a second phase,
since the elastic energy is shared over all the atoms: the total
stored energy is estimated [5,6] at 2 X 10*-9 Joules/metre of
dislocation line. The driving force F for recrystallization across a
phase boundary remains constant as recrystallisation proceeds, and is
simply given by the change in free energy per unit volume of phase
transformed, DELTA g (approximately equal to the stored energy).

F = A   DELTA g   (1)

Stored energy is also associated with the surfaces of a region: the
important surfaces in polysilicon are the grain boundaries, estimated
to have an energy 1-2 Joules/metre*2 [7] and the bounding surfaces of

the film itself, measured as 1.23, 1.51 and 2.13 Joules/metre*2 for free
crystal surfaces bounded by (111), (110) and (100) crystal planes
respectively [8]. The driving force associated with these stored
energies changes as recrystallization proceeds. In one analysis [9],
the driving force F for grain boundary motion, arising solely from the
grain boundary interfacial energy is given as:

$$F = \frac{a \; \lambda \; b^2}{L} \qquad\qquad (2)$$

where lambda = grain boundary energy per unit area, b is the lattice
constant, L is the average grain size and a is a geometric factor.

Reduction in the energy of the film surfaces cannot occur without grain
boundary motion, so that the driving force also includes terms due to
reduction of total grain boundary area, and to the formation of a new
boundary. In one analysis [10] the driving force for the growth of new
large grains of (111) surface orientation (minimum surface energy)
is given as:

$$F = \frac{2 \; \Delta E}{h} + \frac{\beta \; \lambda}{r} - \frac{2 \; \lambda}{R} \qquad\qquad (3)$$

where r is the radius of the original grains, R is the radius of the
new large grain, DELTA E is the difference in their surface energies
per unit area, lambda is the grain boundary interfacial energy, beta
is a geometrical factor, and h is the film thickness. Note that the
last term is negative, reducing the driving force.

The total stored energies in the microstructural components of a
typical 1000 A thick LPCVD film are estimated and compared with that
of a fully amorphous film of the same thickness in TABLE 1.

A3 Dynamics of Recrystallization

The rate at which recrystallization occurs is determined by the fastest
process that can proceed under the experimental conditions. Even in the
presence of a large driving force, no recrystallization can occur if
suitable stable crystal seeds are not present. The nucleation of these
may introduce an incubation period before recrystallisation begins.
Once begun, the recrystallization velocity (V) is determined by the
product of the driving force (f) and the boundary mobility (m) [11].

$$V = Fm$$

The boundary mobility is approximately of the form:

$$m = \frac{B \; b \; f_o}{kT} \exp \frac{-Q}{kT} \qquad\qquad (5)$$

where B is a constant, b is the lattice constant, fo is the fundamental
lattice vibration frequency, k is Boltzmann's constant, T is the
absolute temperature, and Q is the activation energy required to move an
atom across the boundary. This relation shows that boundary mobility

increases strongly as temperature increases, and as the activation energy to move a silicon atom across the boundary is reduced.

The simplest case of recrystallization front motion is when the driving force and mobility are constant, and the phase boundary is a plane, parallel to the film surfaces, transforming the film structure as it moves at right angles to the surfaces (e.g. solid state epitaxial regrowth of a silicon-implanted amorphised layer in a single crystal silicon surface [12]). The phase boundary then moves at constant velocity (V), and the time (t) taken to completely recrystallize a film of thickness h is given simply by

$$t = zeta(o) + \frac{h}{V} \tag{6}$$

where zeta(o) is the incubation time for crystal nucleii to form, if none already exist. zeta(o) is zero when there is atomic contact between crystalline and amorphous phases, but can be large for amorphous films with boundaries only in contact with gas or amorphous layers (e.g. SiO2, Si3N4). In these cases Eqn (6) still holds as long as the crystal nuclei form at the bounding surfaces. If nucleation occurs homogeneously within the bulk of the layer, then subsequent growth is three-dimensional and analysis must take account of depletion of amorphous material and of crystalline interface area as grains impinge. A recent numerical analysis of nucleation and growth of crystals in amorphous silicon [13] is well approximated by the relation

$$\text{Fraction of layer crystallised, } x = 1 - exp[-(\frac{t}{tau(o)})*3]$$

as long as the nucleation time tau(n) is much less than the characteristic crystal growth time tau(o). Recrystallisation velocities and incubation times derived from literature data on various silicon films, ranging from fully amorphous on single crystal or amorphous substrates to partially or fully amorphised polysilicon on amorphous absence of suitable seeding crystals, nucleation times are always longer than growth times for temperatures up to 900 deg C.

The dynamics of recrystallization are yet more complex when the driving force is the reduction of grain boundary energy: the recrystallisation front is then the three-dimensional surface of each expanding grain. In one analysis for cylindrical grains of diameter L [9], based on Eqns (2),(4) and (5), the rate of increase of grain size (dL/dt) is:

$$\frac{dL}{dt} = mF = \frac{a \ lambda \ b*2}{LkT} \ Dgo \ exp \ \frac{-Qg}{kT} \tag{7}$$

where Dgo and Qg are the pre-exponential and exponential coefficients which determine the diffusion coefficient of silicon (Dg) across the grain boundary. If both lambda and Dg are constant with time, this relationship implies a grain growth given by:

$$L = SQRT [Lo*2 + j*2 \ t] \tag{8}$$

where

$$j = SQRT \left[\frac{2a\ lambda\ b*2\ Dg}{kT} \right]$$

and

Lo = initial grain size

Except at the early stages this gives a grain size which increases
continuously as the square root of annealing time. This simple treatment
does not, however, take into account the change in geometry that occurs
as the grains grow to sufficient size to impinge on each other and on
the film surfaces. The effective driving force is gradually reduced [14].
This effect has been treated empirically in various ways [9,15,16],
either by reducing the grain boundary energy term lambda(o) as the
grains grow) [9]:

$$lambda = lambda(o)\ (1 + \frac{3As}{Ag})*-1 \tag{9}$$

(where As is the free surface area, and Ag the grain boundary area,
per unit area of film)

or by introducing a velocity independent drag term mu [15,16] so that:

$$\frac{1}{L} = \frac{C}{m}\frac{dL}{dt} + \frac{1}{Lf} \tag{10}$$

which has a solution

$$tF = \frac{C}{m}\ (Lf)*2\ ln\ \frac{Lf}{Lf-Lo} \tag{11}$$

where tF is the time at which the grains reach their final diameter Lf,
Lo is the initial grain diameter at t=0, m is the grain boundary
mobility and C is a constant.

Some grains can grow beyond the final average grain size
as defined above, if they have bounding surfaces of
lower minimum energy than the average. These grains grow laterally,
under the driving force given in Eqn (3), consuming the other grains.
The rate of increase in secondary grain diameter, dR/dt, before the
grains impinge on each other is calculated to be [10] approximately:

$$\frac{dR}{dt} = \frac{Dgo\ b*2}{RT}\ (\frac{-2\ DELTA\ E\ -\ lambda}{h})exp\ \frac{-Qgo}{kT} \tag{12}$$

where Dgo, Qgo, b, DELTA E, lambda and k and T are as previously defined,
and R is the gas constant.

This equation has been solved [10] to give the following dependencies of
the fraction of the film consumed by secondary grains (x) on film
thickness (h) and temperature (T):

$$x = 1 - \exp \frac{C1}{h*4} \qquad\qquad (13)$$

$$x = C2 \exp \frac{-2Qgo}{kT} + \ln \frac{C3}{T*2} \qquad\qquad (14)$$

$$tx = C4 \; SQRT(-\ln(1-x)) \; h*2 \qquad\qquad (15)$$

where tx is the time at which a fraction x of the film is transformed and C1 - C4 are terms which include the contributions of the other viables.

These relationships show that the rate of growth of secondary grains increases rapidly as film thickness is reduced. The final size (Rf) of the grains is determined by the area density (p) of seed grains of the minimum surface energy crystal orientation. For an initial microstructure of randomly orientated columnar grains of average diameter h, p is inversely proportional to h and [10]:

$$Rf = C5 \; h$$

where C5 is a constant which depends on the grain shape, and also on the variation of surface energy with crystal orientation.

A4 Characteristic Modes of Recrystallization

In recrystallization of real structures, a number of the foregoing driving forces, nucleation processes and growth mechanisms may be operative simultaneously, and this should always be borne in mind when interpreting experimental data. Often, one mechanism dominates over part of the time-temperature range, and this gives rise to some characteristic recrystallization modes.

As the temperature of a sample is raised (500-800 deg C), the first process to be observed is usually the solid state regrowth of amorphous regions in direct contact with polysilicon grains, because of the high mobility of the a-Si/c-Si boundary and the large driving force. The regrown material reproduces the orientation of the underlying polysilicon seed grains, although competitive solid state growth may give rise to a slighly larger grain size [17]. Tiny grains may be present in deposited films apparently entirely amorphous by X-ray analysis [30]. These can act as nucleii during subsequent annealing, and the grain size and morphology will be determined by the average spacing and location (i.e. surface or bulk) of these nucleii.

At somewhat higher temperatures (850 deg C or above), grain boundary mobility becomes high enough for some grain boundary motion to occur. In polysilicon films with grain size smaller than the film thickness, the main driving force comes from reduction in grain boundary length and curvature, and gives rise to a characteristic growth of most of the grains to an average limiting diameter approximately equal to the film thickness. This process is generally known as primary recrystallization [18].

At still higher temperatures (950 deg C and above), a few grains may grow to a much greater size, eventually consume all the others. This process has been called secondary or even tertiary recrystallization, or secondary grain growth. When the driving force is minimisation of surface energy, the term surface-energy driven secondary grain growth (SEDSGG) has been used to describe this process [19].

There are two modes of recrystallization which are special cases where either no seed crystal or only one dominant seed crystal is present.

Where no seed crystals are present, and the boundaries of the film are amorphous (e.g. a-Si film bounded by SiO2 layers), crystallization is delayed until quite high temperatures, because homogenous nucleation rates at temperatures below 900 deg C are low (see TABLE 2). Once crystallization starts, the nuclei grow very rapidly and dendritically, producing characteristic star-shaped grains with (110) surface normals [20]. A similar type of recrystallization has been extensively studied by inducing nucleation with a transient pulse from mechanical probe, laser or electron beam, and the resulting explosive recrystallization can occur with the substrate at room temperature [21]. It has been shown [22] that in this mode, the propogating front is maintained at a high temperature by the continuous release of heat of crystallization, and that, in fact, a thin liquid silicon phase exists at the moving front. It is not clear whether such a phase is present during dendritic crystallization in uniformly heated films, but it is possible since the final microstructures are remarkably similar [23,24].

If a silicon film is in atomic contact with a single crystal substrate, then the substrate-film interface can act as a single nucleus for crystallization, and solid state epitaxial regrowth (SSER) may occur. The driving force for this is greatest if the film is wholly amorphous [25], but SSER can also occur in polycrystalline films [26]. SSER is often delayed, or prevented altogether, by the presence of atomically thin layers of contaminants (e.g. SiO2) between the substrate and deposited film [27-29].

TABLE 1: estimated stored energies in microstructural components of deposited Si layers 0.1 micron thick (the last two columns are calculated per sq cm of 0.1 micron thick layer)

Micro-structural component	Stored energy relative to single crystal Si bounded by (111) surface planes	Probable maximum density of defects	Equivalent number of defects	Total stored energy (milljoules)
Amorphous silicon	0.2 eV/atom	5 X 10*22/ cm*3	5 X 10*17 at/cm*2	approx. 16
Grain boundaries	1-2 X 10*-4 Joules/cm*2	3 X 10*6/cm*2	30 cm*2	approx. 4
Dislocations	2 X 10*-11 Joules/cm	10*12/cm*2	10*7 cm	approx. 0.2
(100) Surface	9 X 10*-5 Joules/cm*2	2 cm*2/cm*2	2 cm*2	approx. 0.2

TABLE 2: incubation times and recrystallisation velocities for
transformation of fully and partially amorphous silicon
films to polysilicon

Ref	Ro A/sec	Q eV	Velocity of crystallisation front (A/sec) [V]							[temp./deg C]
			(500)	(525)	(600)	(700)	(750)	(800)	(900)	
[12]	8.8 X 10*13	2.26	0.160	0.48	8.0	175*	---	---	---	Si-implanted amorphised [100] single crystal
[58]	5.8 X 10*15	2.62	---	---	4.4*	156	718	2870	32000	Si-implanted amorphised [100]
[58]	5.8 X 10*15	2.62	---	---	4.4*	156	718	2870	32000	Evaporated amorphous silicon
[60]	1.2 X 10*17	2.92	0.011	0.044	1.68	92	---	---	---	Evaporated amorphous silicon
[57]	1.7 X 10*18	3.25	0.001*	0.005	0.30	25.2*	---	---	---	Si-implanted amorphised polysilicon
[35]	---	---	---	0.019	---	---	---	---	---	Si-implanted amorphised polysilicon
[61]	---	---	---	---	0.21 0.32	---	---	---	---	Si-implanted amorphised polysilicon
[20]	---	---	---	---	---	10-20	---	---	---	P-implanted amorphised polysilicon
[47]	---	---	---	---	---	---	---	---	900	As-implanted amorphised polysilicon
	3.7 X 10*12	2.26	0.007	0.020	0.34	7.40	28	91	730	best line through Refs [35,61,20,47]

note: V = Ro exp(-Q/kT)

TABLE 2 (continued)

Ref	tau(o) sec	Q eV	500	525	600	700	750	800	900	[Temp./deg C]
					incubation time (sec)					[tau]
[58]+o			---	---	220000*	1500	175	25	0.7	UHV deposited a-Si in air
[58]o			---	---	2300*	29	4.2	0.7	0.03	UHV deposited a-Si in UHV
[57]*	2.8 X 10*15	2.71	4.8 X 10*6	1.3 X 10*6	45300	1110	229	55	4.5	Si-implanted amorphised polysilicon
[61]			---	---	54000	---	---	---	---	Si-implanted amorphised polysilicon
[47]+			---	---	---	---	---	---	2-5	As-implanted amorphised polysilicon
[59]o			---	---	750	36	9.6	---	---	LPCVD deposited a-Si polysilicon
[35]o			---	<2500	---	---	---	---	---	Si implanted, partially amorphised polysilicon

note: $tau = tau(o) \exp(-Q/kT)$

* extrapolated values

+ incubation time calculated assuming explosive lateral crystallisation of film from nuclei about 1 micron apart.

o derived values

TABLE 3: average grain size as a function of anneal time at various
temperatures for undoped polysilicon films on amorphous
SiO2 layers

Time (sec)	(950)	(1100)	(1250)	(1300) (Temp./deg C)
0	600	500	500	500
20	-	-	780	1750
50	-	-	870	1870
200	-	730	1010	-
500	620	1270	-	-
2000	700	-	-	-
5000	870	-	-	-
20000	900*			
Layer thickness (A)	4600	1500	1500	1500
Ref	[38]	[15]	[15]	[15]

* extrapolated values

TABLE 4: average grain size (GS) as a function of average doping level for
polycrystalline silicon films annealed 60 min at 1050 deg C

notes:
    all dimensions are in Angstroms
    figures in curved brackets refer to average doping in 10*20 atom/cm*3

	GS for annealed film											GS before anneal	poly-Si thickness
	(0.1)	(0.6)	(1.0)	(1.2)	(2)	(4)	(6)	(8)	(10)	(12)	(20)		
Si:B													
	1800	---	---	---	---	---	1800	---	---	---	1800	800	2500 [9]
Si:As													
	---	---	380	---	400	500	650	640	620	580	580	250	4600 [40]
	1800	1800	---	1900	---	---	1500	---	---	---	---	800	2500 [9]
Si:P													
	---	---	380	---	400	500	650	850	1180	---	---	250	440 [40]
	1480	1800	---	1900	---	---	2250	---	---	---	---	800	2500 [9]
	---	1000	---	---	---	---	5000	---	---	---	---	500	500 [17]
	100	---	---	---	---	---	---	---	---	---	3500	100	3000 [41]

TABLE 5: distribution of grain sizes (in %) as a function of
deposition and annealing conditions

Ref	[47]	[47]	[47]	[51]	[51]	[51]	
Deposition temp. (deg C)	620	620	620	620	620	570	
Anneal temp. (deg C)	900	900	900	---	1000	1000	
Anneal time (sec)	2	20	200	---	1800	1800	
Mean grain diam. (A)	100	100	100	320	320	amorph.	(as-deposited)
	460	475	540	---	400	780	(annealed)

Range of grain
size (A)

0-100	0	0	0	up to 11	up to 11	
100-200	9	2	0	21	20	
200-300	28	23	8	19	18	31
300-400	12	25	26	16	15	
400-500	9	15	18	13	12	
500-600	11	12	14	8	10	
600-700	14	4	13	6	8	
700-800	14	8	11	5	8	28
800-900	2	8	3			
900-1000	0	2	3			
1000-1100	0	0	3			
1100-1200	0	0	0			
1200-1300	0	2	1			17
1300-1400	0	0	1			
1400-1500	0	0	0			
1500-1600	0	0	0			
1600-1700						
1700-1800						8
1800-1900						
1900-2000						
2000-2500						5
2500-3000						4
3000-3500						3
3500-4000						2
4000-4500						1.5
4500-5000						1
5000-5500						0.5
5500-6000						0

B. EXPERIMENTAL DATA

B1 Experimental Parameters

Most of the papers reviewed below present their recrystallisation data
in terms of the change in average grain size (from some stated starting
value) as a function of temperature, or doping species and
concentration. While this is a convenient and useful parameter for
making semi-quantitative measurements of the effects of these major
experimental variables on the microstructure of a particular deposited
layer, it does not contain all the information required for comparison
of the results of different workers. The purity of the as-deposited
layer [31,32], the crystallographic texture [33], the size, spacing and
density of small crystalline regions in nominally amorphous layers [33-
35], time dependence of the grain structure [15,36] and the distribution
of grain sizes about the average values are parameters which must be
known for quantitative comparisons. Since only one paper [51] comes close
to reporting all these parameters, the review necessarily can only
indicate trends in recrystallization behaviour.

B2 Recrystallization on Amorphous Substrates

B2.1 Undoped polycrystalline layers

Most recrystallisation data come from technological studies aimed at
increasing the grain size of polysilicon before doping by implantation,
so as to reduce the amount of dopant subsequently lost by segregation
to the grain boundaries, and so increase the film conductivity [37].
Only primary recrystallisation seems to occur, even at high temperature
(1100-1300 deg C) since a limiting average grain size of about the
film thickness is approached as the annneal progresses, as illustrated
in TABLE 3. The rate of approach to the final thickness Lf is inversely
proportional to Lf*2 [15] and to exponential temperature [9], so that
for thick films at lower temperatures, Lf may not be achievable in
experimental times (e.g. in ref [38] h is about 20% of film thickness
after about 6 hours). Qualitatively similar results to those in TABLE 3
are reported [39] for 1 micron thick layers deposited at 625 deg C; for
60 min anneals at 1100 deg C and 1200 deg C, the grain size increased from
an initial 2100 A through 3600 A to 5100 A, while the crystallographic
texture changed from mainly (110) to mainly (311). A similar change in
texture is reported for 0.5 micron thick layers deposited at 600 deg C
and annealed 30 mins at 1000 deg C [51]; although at a higher deposition
temperature, 620 deg C, both the initial and annealed textures were
(110). Films of similar thickness deposited in the same temperature
range are reported to change from initial (110) texture through weak (311)
to weak (111) as anneal temperature is increased from 1100 deg C to
1200 deg C, anneal time being 60 minutes [33].

B2.2 Doped polycrystalline layers

Dopants may be introduced into polycrystalline silicon layers,
intentionally or unintentionally, either by incorporation during
deposition, or by subsequent ion-implantation or diffusion from a chemical

source. The dopants fall into two broad groups (i) substitutional dopants (B, P, As, Sb) which, except at very high doping levels, either increase or do not change recrystallization rates (ii) interstitial dopants (C,O) which reduce crystallization rates.

For As and P the method of doping does not influence recrystallization behaviour, as long as the heat-treatment is long enough at high enough temperature to distribute the dopant uniformly through the film [40]. Boron does not affect recrystallization behaviour [9]. Some reported grain size data for doped films annealed for 60 mins at 1050 deg C are summarised in TABLE 4. Phosphorus has the largest effect on the rate dL/dt of growth of grains [9,40]: qualitatively similar results are obtained for 20 mins 1000 deg C [18], and 60 mins 1050 deg C [42].

The increase in dL/dt is found [18,40] to follow the same superlinear relationship with phosphorus concentration as does the intrinsic diffusion coefficient of silicon in silicon [43], and the same enhancement of the point defect concentrations caused by the shift in the Fermi level due to the electrically active dopant is involved to explain both effects [18]. However, this model predicts arsenic to have the same effects as phosphorus, whereas in fact arsenic causes a similar increase up to about 6 X 10*20/cm*3 [40] or 2 X 10*20/cm*3 [9] and thereafter dL/dt decreases. An alternative model is that the large effect of phosphorus is caused by excess of Si interstitial atoms associated with phosphorus precipitation, and that arsenic precipitated in the grain boundaries reduces dL/dt [40]. At concentrations high enough to cause precipitation, boron is also reported to reduce dL/dt [44].

The data that are available on the time-dependence of dL/dt have been fitted to an L = approx. SQRT(t) form for P-doped [9,18] and As-doped [45,46] annealed layers. In all cases, a reduction in dL/dt at longer times, leading to a limiting value of grain size was also observed. In [45], the increase in grain size of two distinct types of grains in the initial film were measured separately. Both showed a SQRT(t) dependence, and a limiting grain size Lf: both dL/dt and Lf were greater for the initially columnar grains than for the initially smaller and equiaxed grains. Lf for the columnar grains occurred approximately when Lf was about equal to film thickness (3000 A), and for equiaxed grains when they impinged on each other (about 1500 A). A similarly bimodal grain population, with two different growth rates, has been observed in As-doped films at 900 deg C [47]: the time-dependence of grain size over the whole size distribution (TABLE 5) shows most of the distribution increasing in size at an equal rate, but the largest grains grow faster. At lower temperatures (800-950 deg C) the limiting average grain size seems to be smaller (0.3-0.5h) than the film thickness h [38,46,9]: at higher temperatures, slightly greater than h [45]. The data of [18] are exceptional, in that Lf exceeds or equals h even for 800 deg C anneals. The average rate of growth of grains is also very variable from one report to another; at a phosphorus doping level of about 6 X 10*20/cm*3, film thickness 0.25-0.44 micron and temperature 1050 deg C, average values of dL/dt obtained from [9], [40] and [18] are 0.4, 1.7, and 18 A/min; furthermore, the rates are not in order of decreasing film thickness as expected. A similar variation exists in temperature dependence of recrystallization: values from 1.05 [9] to 5.5 [18] eV are reported. Most of these differences can be ascribed to the factors mentioned in Section B1. An additional factor is that

in the absence of comprehensive data over the whole time-temperature
diagram, the mode of crystallization cannot always be ascertained. The
data have been selected (on the basis that Lf does not exceed h)
as being characteristic of primary recrystallization. The bimodal
distributions [45,47] and different rate of growth for L greater than
or equal to Lf [45,18] show, however, that a fraction of the total
grain population is growing as secondary grains, and that this fraction
increases with temperature and growth time.

Clear evidence for secondary grain growth is only reported for heavily-
doped (2 X 10*20 - 10*21 atom/cm*3) n-type films [18,19,45,48] at high
temperatures (1000 deg C or above). The continuing growth of the large
secondary grains (from R = 0.5 Lf to R = approx. 10 Lf) can be clearly
distinguished from the essentially static primary grains (L is approx.
equal to Lf) under these conditions. The preferred orientation of the
secondary grains (111) reported for thin films [19,48] is consistent with
SEDSGG theory predictions [10]. The theory also predicts that secondary
grain growth rate should be inversely proportional to the film thickness
h, and this relationship holds approximately for the data that can be
compared [18,19]. After 20 minute anneals at 1200 deg C, in films doped
with 7-9 X 10*20 phosphorus atom/cm*3, the product hR/Lf varies only 25%
about the mean, as film thickness h changed from 470 to 4000 A and
2R/Lf changed from 45 to 5.6. Kinetic studies of the early stages of
grain growth using rapid transient annealing [48] show that while the
rate of film transformation follows the expected S curve initially,
(maximum dx/dt is about 50%/sec) the rate decreases after about 50% of
the film is transformed. (4.3 seconds at about 1140 deg C, h = 700 A,
phosphorus concentration = 10*21 atoms/cm*3). The activation energy
for grain growth in the earlier stages is 3.8 eV [19]. The mechanism by
which phosphorus doping increases grain growth is clearly shown to be a
Fermi level effect, by co-doping experiments with phosphorus and boron,
in which the temperature T required to transform 50% of the film after
20 minutes anneal was reduced smoothly from 1150 deg C to 1070 deg C as
the net concentration of (donors-acceptors) was increased from 5 X 10*20
to 1 X 10*21/cm*3. Undoped films required T = 1350 deg C or above.

The effect of interstitial dopants on recrystallization has been
reported for oxygen [31,32] and carbon [50]. Both dopants reduced grain
growth rates during anneal at 1000 deg C. For a phosphorus concentration
of 1.5 X 10*19/cm*3, grain size after 50 min decreased with oxygen
content as follows [31]:

TABLE 6: effect of oxygen on grain size after 50 mins at 1000 deg C [31]

grain size (A)	1200	1200	1000	700	400	300
oxygen content (atom/cm*3)	0	5 X 10*14	1 X 10*20	2 X 10*20	5 X 10*20	1 X 10*21

At very high oxygen levels (10-30%), grain size is severely restricted
(e.g. 100-50 A after 10 seconds at 1150 deg C) and becomes independent
of phosphorus doping level [32].

The rate of recrystallization of doped polysilicon can also be changed
by reactions occurring at the polysilicon surface [41]. Polysilicon
films 3000 A thick, homogeneously doped with phosphorus by implantation

and 10 mins 900 deg C anneal, were reacted with 500 A of titanium for
120 min at 700 deg C. This completed the reaction forming 1500 A of
TiSi2 and consuming about 1000 A of polysilicon. The combination of
silicide reaction and high phosphorus doping caused considerable increase
in average grain size, that was not seen in control samples with only
one of these features present (see TABLE 7). A number of possible
mechanisms are proposed, including diffusion-induced grain boundary
migration [52] and oxygen gettering from grain boundaries by titanium.

In VLSI processing, complex influences on recrystallisation can be
produced by oxidation, silicidation and non-planar topography. These
effects are best judged from published transverse TEM micrographs [53,
56].

TABLE 7: increase in grain size during (A) 10 min 900 deg C anneal and
         additional anneals at either (B) 700 deg C 120 min, (C) 1000
         deg C 120 min or (D) 700 deg C 120 min with silicide formation
         reaction occurring [41].

Anneal	Grain size (Angstroms)			
	(0)	(2 X 10*20)	(4 X 10*20)	(1.6 X 10*21) (initial av. P concn/cm*3)
A	< 100	< 100	300-500	3000-4000
A+B	< 100	< 100	300-500	3000-4000
A+C	< 100	–	–	3000-4000
A+D	< 100	<100	1200-1500	4000-8000

B2.3 Amorphous layers

The amorphous layers which most commonly occur in polysilicon in planar
device technology are those resulting from doping the polysilicon with
shallow high-dose implanatations of heavy ions (e.g. P, As).
Characteristically, the ion damage creates an amorphous region with
boundaries parallel to the sample surface, the upper boundary usually
coinciding with that surface, and the lower boundary being somewhere
between the upper and lower surfaces of the polysilicon film [17]. The
phase boundary between the amorphous and polycrystalline silicon is thus
a horizontal plane, and transformation can occur by solid state growth
of the polycrystals into the amorphous layer on an approximately planar
front. The velocity of this front [20] is much lower than that measured
in single crystal material [12], and low-temperature (600-800 deg C)
anneals which completely transform the amorphous layer as measured by
X-ray reflection techniques [17], may leave small untransformed regions
[54]. The differences possibly arise because, although the major
crystallization mechanism is 'epitaxial' solid state regrowth on each
polygrain [17], differences in growth rate of crystal orientations and
nucleation of new grains lead to significant lateral and competitive
growth [20]. In addition, LPCVD films can be expected to have higher
oxygen content than single crystal, which will reduce growth velocities.
The summary of crystallization velocities of amorphous silicon by a
number of mechanisms in TABLE 2 shows that LPCVD films regrow at
velocities 10-40 times slower than films deposited in ultra high vacuum.

The microstructure of implanted and annealed LPCVD polysilicon layers depends on anneal temperature. At temperatures 600-900 deg C, a bimodal grain size distribution exists, with an upper layer of average grain size 1.5-2.0 times greater than the lower layer, the demarcation line being the original amorphous-crystalline boundary [17,20,46]. At higher temperatures, an equiaxed monomodal grain structure (corresponding to completion of primary recrystallization) results [17,47,18,33] and thereafter the recrystallization behaviour is as described for doped polysilicon layers in Section B2.2.

When the implanted amorphous layer thickness exceeds the polysilicon film thickness there is no longer a planar crystalline interface for recrystallisation. There may still, however, be seed crystals in the film, due to imcomplete local amorphisation. The existence of these is inferred from the dose-dependence of crystallisation behaviour for polysilicon films 1000 A thick amorphised with 130 keV Ge implants, and annealed for 30 mins at temperatures 400-900 deg C [34]. A dose of 2 X 10*15 ions/cm*2 gave an amorphous layer (as judged from TEM and X-ray diffraction) which crystallised at 500 deg C to the same morphology as the original polycrystalline material (equiaxed 1100 A grains). A dose of 2 X 10*16 ions/cm*2 gave a layer indistinguishable by TEM and X-ray diffraction, which nevertheless, only annealed at temperatures of 700 deg C and above, with a quite different dendritic and 10 times larger grain structure. These two crystallization behaviours correspond exactly to the growth of existing crystals described earlier, and the nucleation and growth of crystals in fully amorphous films described below. A systematic study of dose-dependence of annealed microstructure in 4400 A thick polysilicon LPCVD layers, implanted with 210 keV Si ions/cm*2, and annealed 48 hours at 600 deg C came to similar conclusions [55]. The critical dose, at 90 deg C implant incidence angle and substrate temperature about -190 deg C was between 0.6 and 1.6 X 10*14 ions/cm*2.

In this transition zone, the recrystallised structures showed large increases in (110) crystallographic texture, interpreted as indicating the selective preservation of the (110) grains in the original polysilicon film, because of ion-channelling of the normal incidence silicon ions in such grains. This effect has been exploited [35] to obtain large grain size (10000 A) in thin films (2400 A) after low temperature annealing (48 hours, 525 deg C). Small seed crystals are often also present in LPCVD films deposited at temperatures just below the amorphous-crystalline transition temperature [33,51]. These give rise to similar recrystallisation from randomly spaced nuclei, the impingement of which determines the grain size. Because the nuclei are often at a wider spacing (about 1000 A) than the grain size (about 100 A) of the polysilicon films deposited just above the critical temperature, and because the velocity of the recrystallisation front is higher in amorphous silicon than in polysilicon (about 1000 A/sec as against 100 A/sec at 900 deg C), grain sizes after 800-1000 deg C anneal are usually much larger (2-3 times) in LPCVD amorphous films as compared with polysilicon films deposited in the same equipment [51]. At higher temperatures (1100-1200 deg C) this difference is not seen [33] because other grain growth mechanisms dominate especially in heavily-doped n-type films.

Fully amorphous layers, produced either by heavy implant amorphisation of polysilicon [20,33,57] or by clean low temperature deposition techniques [58], contain insignificant densities of crystal nuclei,

and a nucleation process most precede the crystallization process.
Systematic studies of the incubation times for nucleation and
recrystallization velocities as a function of temperature have been
made for implanted amorphised films [57] and UHV deposited amorphous
films [58]. These, and data extracted from other sources [57-61,35,47]
are summarized in TABLE 2. These data show that the longest incubation
times (and thus presumably the most homogeneously amorphous films)
are those produced by UHV deposition [58] and implant amorphisation
[47,57,61]. LPCVD deposited amorphous silicon nucleates crystallites
much more quickly [59]. An important observation [58] is that surface
nucleation is much faster (25-50 times) than bulk nucleation and that
the former is almost completely suppressed by annealing in air rather
than UHV.

Once crystallization starts, it propogates dendritically in three
dimensions until the grains impinge on each other [57,20,33]. The
velocity of crystallization can be calculated by an Avrami analysis
of the fraction crystallized [57,58] or estimated from the time tf
for the grains to reach final grain size Df (V = approx. Df/tf)
[57,20,47,61]. These data, summarized in TABLE 2, show that the highest
velocities of crystallization occur in UHV deposited films [58], and
that the velocities for implanted amorphised LPCVD films are all lower
by a factor of 20 (at 600 deg C) to 35 (at 900 deg C). In fact, nearly
all the LPCVD data lie very close to a temperature-dependence (2.26 eV)
identical to that for regowth of amorphous layers in (100) single
crystal silicon, [12], but with a pre-exponential factor 24 times
smaller. The study of [57], over a much smaller temperature range,
obtained an activation energy of 3.25 eV: but their midrange
(600 deg C) values are also very close to the above best fit. These
values are independent of the nature of the amorphising implant, which
include heavy doses of silicon [56,57,61], phosphorus [20] and arsenic
[47]. The latter might be expected to increase crystallization
velocities considerably [12]. Since the LPCVD recrystallization
velocities are so much lower than the UHV deposited layers, and both
oxygen and carbon reduce the recrystallization rate of amorphous
silicon [62,76], it is possible that interstitial contaminants are
responsible for the slow LPCVD recrystallization.

The final average grain size Df in initially fully amphorhous layers
after anneal is shown [57] to be dependent on the crystallization
velocity V, the nucleation rate n and the film thickness h according
to the relationship:

$$Df = C6 \left(\frac{V}{hn}\right)*(1/3) \qquad (17)$$

The temperature dependence of Df for h = 1000 A was determined to
be [57]:

$$Df(\text{Angstroms}) = 6 \exp \frac{0.62 \text{ eV}}{kT} \qquad (18)$$

The weak dependence of Df on both film thickness and temperature
implies that large increases over the grain size already achieved
(about 30000 A at 570 deg C) are not possible by this crystallization
films annealed at 600 deg C achieved a maximum grain size of 50,000 A

after 200 hours anneal [61]. Eqns (17) and (18) indicate an expected
average grain size of 25000 A for the conditions, in reasonable
correspondence with the maximum size observed. Thicker films (2400 A)
fully amorphised by 2 X 10*15 argon ions/cm*2 and annealed for
30 min, gave average dendritic grain sizes of 5500, 6250 and 7500 A
at temperatures of 700, 800 and 900 deg C [63]. Eqns (17) and (18)
would predict 7300, 3700 and 2100 A, suggesting that this theory holds
up to about 700 deg C, but that other processes dominate the temperature
dependence above this temperature.

B3 Layers on Single Crystal Substrates

B3.1 Polycrystalline silicon layers

When polysilicon layers are in direct contact with an underlying silicon
single crystal substrate, the possibility arises of recrystallizatin
from a single planar seed, giving rise to a completely monocrystalline
layer aligned epitaxially with the substrate. Such recrystallization
does occur on (100) silicon substrates [26], and is facilitated by high
concentrations of n-type dopants including phosphorus [64] and arsenic
[27-29,65-69]. In a particular deposited layer, the time for complete
epitaxial regrowth is reduced by increasing the temperature, the time,
and the doping level, although in comparing the results of different
workers, these trends do not always appear to hold (see TABLE 8). The
reason for the discrepancy is probably variations in the thickness and
stability of the very thin interfacial oxide which has been shown to
separate the polysilicon film from the substrate [27,28,64-66,69,70].
Reported values of the thickness of this layer are 1.5 A [29], 1.7 A
[46], 8-14 A [66], 10 A [27] and 30 A [65]. No epitaxial regrowth takes
place until the thin layer becomes unstable and separates into discrete
'balls', which often have a characteristic polygonal morphology [29,66].
Up to this point, the polycrystalline silicon recrystallises by the
mechanisms described above for doped layers (see section B2.2). Beyond
this point, epitaxial regrowth begins at the interface between the oxide
balls, and propogates upwards and laterally until the whole film is
transformed [26]. Initially the film contains many defects [26-28,64]
and the time taken to remove these by further annealing [68] is
usually longer than the time required to disrupt the oxide film. The
times given in TABLE 7 are for essentially defect-free epitaxial layer
formation.

The effect of boron on epitaxial regrowth is not clear. Complete
epitaxial realignment of films doped with 1.7 X 10*20 atom/cm*3 of
boron is reported for 5 sec anneal at 1150 deg C [69] which is much
faster than that reported for undoped films (see TABLE 7). A film doped
with 9.4 X 10*20 boron atoms/cm*3 failed to epitaxially align after
120 secs at 1150 deg C, while similar films, doped to the same level
with As or phosphorus aligned in 30 sec at 1150 deg C [64].

B3.2 Amorphous silicon layers

Amorphous layers on silicon single crystal substrates can also regrow
epitaxially, whether vacuum deposited [71-73], LPCVD deposited [25],
or formed by implant amorphisation of LPCVD polysilicon films [74,75].
The main differences from regrowth of polysilicon films are the nature

and stability of the interfacial contaminant layer, and the additional driving force for regrowth provided by the phase transformation. In UHV depositions and anneals, no effects of contaminants are seen, and the same rate as for implanted amorphous layers in the same substrate material [73]. The temperature dependence of this rate is given in TABLE 2. However, exposure of such UHV deposited layers to air before anneal reduces the regrowth rate considerably, and at high temperatures may inhibit epitaxial regrowth to such an extent that homogeneous nucleation of polycrystalline silicon occurs [73].

Low temperature epitaxial regrowth of implanted amorphised LPCVD films has been reported [25]. Starting with amorphous material enabled full amorphicity thoughout the 1900 A layer to be achieved with only one 190 keV silicon implant. The implant also disrupted the interfacial oxide layer, and in fact, was essential to obtain epitaxial regrowth at 650 deg C. Good quality epitaxial layers were obtained for silicon doses greater than 7 X 10*15 ions/cm*2. Such doses, however, created defects in the underlying substrate. Similar results have been obtained with implant-amorphised polysilicon layers [75]. Good quality epitaxy occurred after 60 hours at 550 deg C.

TABLE 8: time (in seconds) required for complete epitaxial recrystallization of polycrystalline silicon films on cleaned silicon substrates, as a function of temperature and arsenic doping level

Anneal temp. deg C	Epitaxial recrystallisation time (sec)					
	(0)	(3-4)	(5-6)	(7-8)	(9-10)	(Average As concn/ 10*20/cm*3)
900			900 [66]			
950				1800 [28]		
1000		1200 [27]				
1100					<3600 [29]	
1150	7200 [26]	60 [68]		5 [69]	10 [67] 30 [64]	

REFERENCES
----

[1]   H.J.Leamy [Mater. Res. Soc. Symp. Proc. (USA) vol.4 (1982)
       p.459-70]
[2]   J.C.C.Fan, B.-Y. Tsaur, M.W.Geis [J. Cryst. Growth (Netherlands)
       vol.63 no.3 (1983) p.453]
[3]   H.Ahmed, R.A.McMahon [EMIS Datareview RN=16166 (Sept 1987)
       'Recrystallisation of poly-Si to form SOI substrates by electron
       beam heating']
[4]   J.C.C.Fan, H.Anderson [J. Appl. Phys. (USA) vol.52 (1981)
       p.4003]
[5]   A.S.Nandekar, J.Narajan [Philos. Mag. A (GB) vol.56 no.5 (1987)
       p.625-39]
[6]   K.Lodge et al [Philos. Mag. B (GB) vol.49 (1984) p.41]

[7]   H.J.Moller [J. Phys. Colloq. (France) vol.43 no.C-1 (Oct 1982) p.33-43]

[8]   R.J.Jaccodine [J. Electrochem. Soc. (USA) vol.110 (1963) p.524]

[9]   L.Mei, M.River, Y.Kwark, R.W.Dutton [J. Electrochem. Soc. (USA) vol.129 no.8 (1982) p.1791-5]

[10]   C.V.Thompson [J. Appl. Phys. (USA) vol.58 no.2 (1985) p.763-72]

[11]   D.Turnbull [Trans AIME vol.191 (1951) p.661]

[12]   L.Cspregi, E.F.Kennedy, T.J.Gallagher, J.W.Mayer, T.W.Sigmon [J. Appl. Phys. (USA) vol.48 no.10 (1977) p.4234-40]

[13]   H-J.Muller, K-H.Heinig [Nucl. Instrum. & Methods Phys. Res. Sect B (Netherlands) (North Holland, Amsterdam) vol.22 no.4 (1987) p.524-7]

[14]   C.A.Chadwick, D.A.Smith [Grain Boundary Structure and Properties (Academic Press, London, 1976)]

[15]   R.F.Pinizotto, F.Y.Clark, S.D.S.Malhi, R.R.Shah [Mater. Res. Soc. Symp. Proc. (USA) vol.33 (1984) p.169-77]

[16]   E.A.Grey, G.T.Higgins [Acta Metall. (USA) vol.21 (1973) p.309]

[17]   H.Oppolzer, R.Falckenberg, E.Doering [Inst. Phys. Conf. Ser. (GB) no.60 (1981) p.283-7]

[18]   Y.Wada, S.Nishimatsu [J. Electrochem. Soc. (USA) vol.125 no.9 (1978) p.1499-504]

[19]   C.V.Thompson, H.I.Smith [Appl. Phys. Lett. (USA) vol.44 no.6 (1984) p.603-5]

[20]   A.H.Reader, F.W.Schapink, S.Radelaar [Inst. Phys. Conf. Ser. (GB) no.76 (1985) p.151-6]

[21]   A.G.Cullis [Rep. Prog. Phys. (GB) vol.48 (1985) p.1155-233]

[22]   H.J.Leamy, W.L.Brown, G.K.Celler, G.Foti, G.H.Gilmer, J.C.C.Fan [Appl. Phys. Lett. (USA) vol.38 (1981) p.137]

[23]   G.Gotz [Appl. Phys. A (Germany) vol.40 (1986) p.29-36]

[24]   J.Narajan [Mater. Lett. (Netherlands) vol.2 no.3 (1984) p.219-22]

[25]   M.K.Hatalis, D.W.Greve [J. Electrochem. Soc. (USA) vol.134 no.10 (1987) p.2536-5]

[26]   B.Y.Tsaur, L.S.Hung [Appl. Phys. Lett. vol.37 no.7 (1980) p.648-51]

[27]   M.C.Wilson, P.Ashburn, B.Soerwirdjo, G.R.Booker, P.Ward [J. Phys. Colloq. (France) vol.43 no.C-1 (1982) p.253-9]

[28]   C.Y.Wong, A.E.Michel, R.D.Isaac, R.H.Kastl, S.R.Mader [J. Appl. Phys. (USA) vol.55 no.4 (1984) p.1131-4]

[29]   J.C.Bravman, G.L.Patton, R.Sinclair, J.D.Plummer [Mater. Res. Soc. Symp. Proc. vol.37 (1985) p.461-66]

[30]   C.Hill, S.Jones [EMIS Datareview RN=17863 (Dec 1987) 'Growth and microcrystalline structure of poly-Si']

[31]   R.Angelucci, M.Severi, S.Solmi, L.Baldi [Thin Solid Films (Switzerland) vol.103 no.3 (1983) p.275-81]

[32]   T.L.Alford, D.K.Yang, W.Maszara, V.H.Ozguz, J.J.Wortman, G.A.Rozgonyi [J. Electrochem. Soc. (USA) vol.134 no.4 (1987) p.998-1003]

[33]   Y.Komem, I.W.Hall [J. Appl. Phys. (USA) vol.52 no.11 (1981) p.6655-8]

[34]   T.I.Kamins, M.M.Mandurah, K.C.Saraswat [J. Electrochem. Soc. (USA) vol.125 no.6 (1978) p.927-34]

[35]   P.Kwizera, R.Reif [Appl. Phys. Lett. (USA) vol.41 no.4 (1982) p.379-81]

[36]   S.M.Garrison, R.C.Cammarata, C.V.Thompson, H.I.Smith [J. Appl. Phys. (USA) vol.61 no.4 (1987) p.1652-5]

[37] S.J.Krause et al [Mater. Res. Soc. Symp. Proc. (USA) vol.52 (1986) p.145-52]

[38] A.H.Reader, F.W.Schapink, S.Radelaar [Proc. Symp. Polymicrocrystalline and Amorphous Semiconductors, Strasbourg (Mater. Res. Soc., 1984) p.253-8]

[39] B.Swaminathan, K.C.Saraswat, R.W.Dutton, T.I.Kamins [Appl. Phys. Lett. (USA) vol.40 no.9 (1982) p.795-8]

[40] S.Solmi, M.Severi, R.Angelucci, L.Baldi, R.Bilenchi [J. Electrochem. Soc. (USA) vol.129 no.8 (1982) p.1811-18]

[41] T.C.Chou, C.Y.Wong, K.N.Tu [J. Appl. Phys. (USA) vol.62 no.7 (1987) p.2722-6]

[42] M.Taniguchi, M.Hirose, Y.Osaka, S.Hasegawa, T.Shimizu [Jpn. J. Appl. Phys. (Japan) vol.19 no.4 (1980) p.665-73]

[43] J.M.Fairfield, B.J.Masters [J. Appl. Phys. (USA) vol.38 (1967) p.3148]

[44] T.Makino, H.Nakamura [Solid State Electron. (GB) vol.24 (1981) p.49]

[45] S.R.Wilson et al [J. Electrochem. Soc. (USA) vol.132 no.4 (1985) p.922-9]

[46] A.O.Neill, C.Hill, C.Pleese, J.King [J. Appl. Phys. (USA) (1988) in press]

[47] C.Hill, S.Jones [unpublished data]

[48] S.M.Garrison, R.C.Cammarata, C.V.Thompson, H.I.Smith [J. Appl. Phys. (USA) vol.61 no.4 (1987) p.1652-5]

[49] H-J.Kim, C.V.Thompson [Mater. Res. Soc. Symp. Proc. (USA) vol.54 (1986) p.729-32]

[50] J.Bloem, W.A.P.Claassen [Appl. Phys. Lett. (USA) vol.40 no.8 (1982) p.725-6]

[51] G.Harbeke, L.Krausbauer, E.F.Steigmeir, A.E.Widmer, H.F.Kappert, G.Neugebauer [J. Electrochem. Soc. (USA) vol.131 no.3 (1984) p.675-82]

[52] J.C.M.Li, B.B.Rath [Scr. Metall. (USA) vol.19 (1985) p.689]

[53] R.B.Marcus, T.T.Sheng [TEM of Silicon VLSI Circuits and Structures (Wiley, 1983, New York)]

[54] D.L.Black, J.P.Lavine, S-T.Lee, D.L.Losee [Microscopy of Semiconducting Materials, 1985 [Proc. Roy. Microscopical Soc. Conf., Oxford, England, 1985 (Adam Hilger, Bristol., 1985) p.157-62]

[55] K.T-Y.Kung, R.Reif [J. Appl. Phys. (USA) vol.59 no.7 (1986) p.2422-8]

[56] H.Oppolzer, V.Huber [Inst. Phys. Conf. Ser. (GB) no.67 (1983) p.461]

[57] R.B.Iverson, R.Reif [J. Appl. Phys. (USA) vol.62 no.5 (1987) p.1675-81]

[58] J.A.Roth, S.A.Kokorowski, G.L.Olson, L.D.Hess [Mater. Res. Soc. Symp. Proc. (USA) vol.4 (1982) p.169-76]

[59] A.M.Beers, H.T.J.M.Hintzen, J.Bloem [J. Electrochem. Soc. (USA) vol.130 no.6 (1983) p.1426-33]

[60] K.Zellama, P.Germain, S.Squelard, J.C.Bourgoin, P.A.Thomas [J. Appl. Phys. (USA) vol.50 (1979) p.6995]

[61] T.Noguchi, H.Hayashi, T.Ohshima [J. Electrochem. Soc. (USA) vol.134 no.7 (1987) p.1771-7]

[62] E.F.Kennedy, L.Cspregi, J.W.Mayer, T.W.Sigmon [J. Appl. Phys. (USA) vol.48 (1977) p.4241]

[63] A.Bhattacharyya, K.N.Ritz [J. Electrochem. Soc. (USA) vol.131 no.9 (1984) p.2143-5]

[64] M.Tamura, N.Natsuaki, S.Aoki [Jpn. J. Appl. Phys. (Japan) vol.24 no.2 (1985) p.L151-4]

[65] H.Oppolzer [Inst. Phys. Conf. Ser. (GB) vol.76 (1985) p.461-70]

[66] N.Jorgensen et al [ibid 471-476]

[67] J.L.Hoyt, E.Crabbe, J.F.Gibbons, R.F.W.Pease [Appl. Phys. Lett.
      (USA) vol.50 no.12 (1987) p.751-3]

[68] Y.Komem, C.Y.Wong, H.B.Harrison [J. Appl. Phys. (USA) vol.62 no.1
      (1987) p.131-6]

[69] H.J.Bohm, H.Wendt, H.Oppolzer, K.Masseli, R.Kassing [J. Appl.
      Phys. (USA) vol.62 no.7 (1987) p.2784-8]

[70] D.K.Skinner, C.Hill, M.W.Jones [Mater. Res. Soc. Symp. Proc.
      vol.48 (1985) p.179-84]

[71] J.A.Roth, C.L.Anderson [Appl. Phys. Lett. (USA) vol.31 (1977)
      p.689]

[72] M.von Allmen, S.S.Lau, J.W.Mayer, W.F.Tseng [Appl. Phys. Lett.
      (USA) vol.35 (1979) p.280]

[73] J.A.Roth, G.L.Olson, S.A.Kokorowski, L.D.Hess [Mater. Res. Soc.
      Symp. Proc. (USA) vol.1 (1981) p.413-26]

[74] K.L.Wang, G.P.Li, T.W.Sigmon [Appl. Phys. Lett. vol.39 (1981)
      p.709]

[75] N.T.Quach, R.Reif [Appl. Phys. Lett. (USA) vol. 45 (1984) p.910]

[76] P.Ling, J.Washburn [Mater. Res. Soc. Proc. (USA) vol.45 (1985)
      p.47-52]

CHAPTER 27

SILICON-BASED MULTILAYERS

27.1 ELECTRON MOBILITY IN Si/SiGe SUPERLATTICES

by E.Kasper and F.Schaffler

August 1987
EMIS Datareview RN=15778

A. GENERAL

The first indication that electron mobility can be enhanced in Si/SiGe
superlattices was given in 1982 by Manasevit et al [1]. In Si/Si(1-x)Ge(x)
multilayers grown by chemical vapour deposition (CVD) at a temperature of
1000 deg C Hall mobilities of 1350 cm*2/V.s were measured for electron
concentration n = 1 X 10*16/cm*3 and Ge content x = 0.15. The results
show a mobility enhancement of 35% over that of Si layers and of about
100% over that of SiGe alloy layers. The mobility enhancement decreases
with increasing electron concentration approaching Si-mobility when
n = 2 X 10*17/cm*3. The reason for the observed mobility enhancement is
not obvious because (i) the higher growth temperature leads to
interdiffusion, (ii) the layer thicknesses (30 nm - 150 nm) are
above the critical thickness at the growth temperature, which leads
to misfit dislocations, and (iii) the distribution of the n-dopant
phosphorus between the Si and SiGe layers is unknown. For the
following we use only experimental results measured on molecular beam
epitaxy (MBE) material obtained from two sources (AEG Research Centre,
Ulm, FRG [2-5] and Bell Labs, Murray Hill, USA [6,7]). This material
is grown at temperatures low enough to avoid interdiffusion (up to 600
deg C). The layer thicknesses are smaller than the critical thickness
except for some systematic investigations with thickness variation by
Jorke et al [5].

Intentional doping of the AEG samples was done by thin (2 nm - 4 nm)
Sb spikes (delta-doping) creating a nini-doping superlattice, which
was superimposed on the Si/SiGe-alloy superlattice. The phase angle phi
of the doping superlattice was varied from phi = 0 (doping of the Si
layers) to phi = pi (doping of the SiGe layers). The phase angle phi
is given by

     phi = 2 pi DELTA L/L   (1)

with L the period length of the superlattice and DELTA L the distance
of the doping spike from the middle of the Si layer.

Intentional doping of the Bell samples was done by As+ ion implantation
of the Si layers with an undoped region near the interface called
set-back or spacer. The material system Si/SiGe is lattice mismatched,
which results in strained layer superlattices if the layer thickness is
smaller than the critical thickness for misfit dislocation generation.
The strain within the superlattice can be adjusted following
Kasper's concept [8] of a thin, homogeneous buffer layer. The Bell
samples are unsymmetrically strained with Si unstrained and SiGe
compressively strained. The AEG samples are usually symmetrically
strained with Si tensilely strained and SiGe compressively strained.
Some samples are unsymmetrically strained like the Bell samples.
Strained layer heteroepitaxy by Si-MBE is rapidly developing (see
Bean's review [9]) but the early results [2-7] on (100) surfaces are
certainly influenced by background doping, interface roughness and
imperfections like e.g. threading dislocations.

B. UNSYMMETRICALLY STRAINED Si/SiGe

People et al [6] did not find strong 2-D carrier effects for electrons.
They took this as an indication for a negligibly small conduction-band
discontinuity DELTA Ec = 20 meV (flat conduction band [7]). On the
other hand Abstreiter et al [4] found in AEG samples with doped SiGe
layers (note: Bell doped the Si layers) a 2-D electron gas from which
Zeller et al [10] calculated a staggered type II band ordering at the
Si/SiGe interface with DELTA Ec = -100 meV. The Hall mobility in a 10
period Si/Si(0.5)Ge(0.5) superlattice with period length L = 12 nm
(sample VS 94) increases from 300 cm*2/V.s at room temperature to
1200 cm*2/V.s at T = 4 deg K.

TABLE 1: unsymmetrically strained Si/Si(0.5)Ge(0.5) superlattices (ten
periods, period length L = 12 nm) on Si substrate (sample VS
94).

Temperature	300	150	80	20	4	deg K
Hall mobility	550	940	1060	1105	1200	cm*2/V.s

Samples were modulation doped with Sb [4] at phase angle phi = pi.
2-D electron density, ns, is 2 X 10*12/cm*2 per period.

C. SYMMETRICALLY STRAINED Si/SiGe

Symmetrically strained Si/SiGe superlattices were grown by Jorke et al
[2]. Modulation doped superlattices exhibited a strong influence of
doping phase angle on the mobility (TABLE 2).

TABLE 2: room temperature Hall mobility versus doping phase angle for
modulation doped Si/SiGe superlattices.

Doping phase angle	0	pi/4	pi/2	3pi/4	pi
Hall mobility (cm*2/V.s) L = 12 nm	150	210	350	550	615
Hall mobility (cm*2/V.s) L = 20 nm	270	330	470	700	850

Mean doping density amounts to 4 X 10*18/cm*3 [2].
Period length L = 12 nm and 20 nm, respectively.
The doping phase angle was corrected for the penetration depth of the
secondary implanted Sb.

The dependence of electron mobility on doping phase angle is even more pronounced at lower temperatures (TABLE 3).

TABLE 3: Hall mobility versus doping phase angle for different temperatures T compared with bulk Si (n = 4 X 10*18/cm*3).

Temp.	Hall mobility (cm*2/V.s)			
deg K	phi=0	phi=pi/2	phi=pi	bulk Si
300	150	350	615	120
150	110	385	1250	120
80	85	400	1600	80
40	75	415	1900	
20	70	420	2100	

Period length of the Si/SiGe superlattice is L = 12 nm.

The highest mobility of the 2-D electron gas is obtained by doping the smaller gap semiconductor SiGe (phase angle phi = pi). Therefore, the band ordering at the symmetrically strained Si/SiGe interface is of the staggered type II. Abstreiter et al [3] and People [7] agree now that the strain induced splitting of the sixfold degenerate conduction band enlarges the type II band ordering of symmetrically strained (100) Si/SiGe interfaces. The sixfold degenerate conduction band is split by a biaxial strain parallel to the (100) interface in a twofold degenerate valley with wave vector normal to the quantum well and in a fourfold degenerate valley with in-plane wave vector. The twofold valleys are lowered by tensile strain (5.8 meV per kbar) and the fourfold valleys are shifted upwards (2.9 meV per kbar). For symmetrically strained interfaces the band offset DELTA Ec is given by the energy difference between the twofold valley in Si (tensile strain) and the fourfold valley in SiGe (compressive strain). The electronic sub-band structure is treated by Zeller et al [10]. They obtained for the symmetrically strained Si/Si(0.5)Ge(0.5) interface a conduction band offset (type II) of DELTA Ec = 250 meV, which is about 100 meV [8] larger than the offset calculated by People [7].

Dambkes et al confirmed the high mobility in modulation doped Si/SiGe quantum wells by magneto-transconductance measurements in MODFET-device structures [12]. Channel mobilities from 1200 cm*2/V.s to 1500 cm*2/V.s were obtained for gate voltages from 0.4 to -1.5 V, respectively.

D. VERTICAL TRANSPORT

The transport properties of a 2-D electron gas are highly anisotropic
with low conductivity perpendicular to the quantum wells. Jorke et al
[5] investigated the vertical transport through heterostructure barriers
for the thermally activated regime (low contribution of tunnelling).
With symmetrically strained superlattices homogeneously doped to a
doping level of 5 X 10*16/cm*3 the vertical resistance, Rb, of the
heterobarrier was measured with a 7 micron diameter contact under weak
field conditions. For thicker layers (L > 20 nm) the mismatch is
partly accommodated by misfit dislocations reducing the strain: The
conduction band offset DELTA Ec and the vertical mobility mu(v) can be
estimated by the amount of the vertical heterobarrier resistance.

TABLE 4: vertical heterobarrier resistance, strain, period length and
          estimated band offset for Si/Si(0.5)Ge(0.5) superlattices

| Sample | Strain | L | Rb | DELTA Ec |
		nm	ohm	meV
61	high	4	195	190
59	high	10	495	191
58	high	20	490	174
60	low	60	20	72
62	low	100	13	52

Note: Rb = vertical heterobarrier resistance, L = period length,
      DELTA Ec = estimated band offset
      n = 5 X 10*16/cm*3; contact diameter = 7 microns

The value of DELTA Ec = 180 meV for the symmetrically strained
superlattices is within the range of the data given earlier. For the
thicker layers with low strain much lower values of DELTA Ec = 60 meV
are obtained. This is in agreement with Abstreiter's [3,10]
explanation for the influence of strain.

E. COMPARISON WITH THEORY

Gold [12] calculated the mobility (T = 0) of a 2-D electron gas in
modulation doped Si/SiGe quantum wells as a function of background
doping, modulation doping and interface roughness. The experimental
results of TABLE 3 could be explained if one assumes that the
modulation doping was five times higher than the measured carrier
density. An explanation for this discrepancy is lacking.

Krishnamurthy et al [13] investigated the electronic structure of
silicon superlattices with respect to zone folding effects and
effective masses. For impurity scattering limited mobility an
enhancement of about a factor of two is predicted for n = 10*18/cm*3.
An experimental proof is lacking.

REFERENCES
----

[1]   H.M.Manasevit, I.S.Gergis, A.B.Jones [Appl. Phys. Lett. (USA)
        vol.41 no.5 (1 Sept 1982) p.464-6]
[2]   H.Jorke, H.-J.Herzog [J. Electrochem. Soc. (USA) vol.133 no.5
        (May 1986) p.998-1001]
[3]   G.Abstreiter, H.Brugger, T.Wolf, H.Jorke, H.-J.Herzog [Phys.
        Rev. Lett. (USA) vol.54 no.22 (3 June 1985) p.2441-4]
[4]   G.Abstreiter, H.Brugger, T.Wolf, H.Jorke, H.-J.Herzog [Surf.
        Sci. (Netherlands) vol.174 (1986) p.640]
[5]   H.Jorke, H.-J.Herzog, E.Kasper, H.Kibbel [J. Cryst. Growth
        (Netherlands) vol.81 (1987) p.440]
[6]   R.People, J.C.Bean, D.V.Lang [J. Vac. Sci. & Technol. A (USA)
        vol.3 no.3 pt 1 (May/Jun 1985) p.846-50]
[7]   R.People [IEEE J.Quantum Electron. (USA) vol.QE-22 no.9 (Sep
        1986) p.1696-710]
[8]   E.Kasper, H.-J.Herzog, H.Jorke, G.Abstreiter [Superlattices &
        Microstruct. (GB) vol.3 no.2 (1987) p.141-6]
[9]   J.C.Bean ['Silicon Based Heterostructures', in: 'Si-MBE', Eds
        E.Kasper, J.C.Bean (CRC Press, Boca Raton, USA, 1987)]
[10]   Ch.Zeller, G.Abstreiter [Z. Phys. B (Germany) vol.64 no.2 (1986)
        p.137-44]
[11]   H.Daembkes, H.-J.Herzog, H.Jorke, H.Kibbel, E.Kasper [IEEE Trans.
        Electron Devices (USA) vol.ED-33 no.5 (May 1986) p.633-8]
[12]   A.Gold [Phys. Rev. B (USA) vol.35 no.2 (15 Jan 1987) p.723-33]
[13]   S.Krishnamurthy, J.A.Moriarty [Superlattices & Microstruct. (GB)
        vol.1 no.3 (1985) p.209-15]

27.2 HOLE MOBILITY IN Si/SiGe SUPERLATTICES

by E.Kasper and F.Schaffler

August 1987
EMIS Datareview RN=15779

A. GENERAL

An experimental investigation of the hole mobility in Si/SiGe
superlattices was performed by People, Bean et al [1,2]. Selectively
doped Si/Si(0.8)Ge(0.2) strained layer heterostructures have been grown
in a single quantum well configuration on (001)-Si substrates using
molecular beam epitaxy. The wide band gap Si layers were doped with
boron to a level of 10*17/cm*3 to 10*18/cm*3, while the narrow gap
SiGe layers were not intentionally doped. The thickness of the Si
cladding layers was 0.1 micron, the doping was set back from the
interface to the SiGe well by a spacer distance Ls. A systematic study
of the influence of doping set-back, Ls, well width, Lw, and doping
level, Na, was given by People et al [2].

The heterostructure is strained because of the lattice mismatch of 4.2%
between Si and Ge. The only investigated structure consisting of a Si
cladding layer, a SiGe quantum well, and a second Si cladding layer on
top of a Si substrate is strained unsymmetrically. That means
compression of the SiGe layer and no strain in the Si cladding layers.
The strain distribution influences the band ordering [3], the band gap
[4] and a splitting of the degenerate carrier valleys [5,6]. For
unsymmetrically strained Si/Si(0.2)Ge(0.8) [4] calculated the valence
band offset DELTA Ev to be 150 meV and the split-off of the heavy and
light hole band in SiGe to be 23 meV.

In a single quantum well structure holes from ionised boron acceptors
in the Si cladding layer are transferred to the SiGe channel. In the
given structure with 0.1 micron thick cladding layers only a rather
small fraction of the acceptors contribute to this carrier transfer.
The main fraction of the carriers remains in the cladding layer. This
is best demonstrated by a measurement of the sheet carrier density
versus temperature (TABLE 1).

TABLE 1: sheet carrier density, ps, versus temperature, T, for a single
quantum well Si/SiGe/Si heterostructure [2]

T	300	150	75	40	deg K
ps	1.1 X 10*13	5 X 10*12	2 X 10*12	1.4 X 10*12	/cm*2

T	30	15	deg K
ps	1.25 X 10*12	1.25 X 10*12	/cm*2

Cladding layer thickness Lc = 0.1 micron, intentional boron doping of
Na = 10*18/cm*3, doping set-back Ls = 0, and well thickness Lw =
0.2 microns.

At room temperature the sheet carrier density is dominated by 1 X 10*13/cm*2 3-D holes in the cladding layers. With decreasing temperature the expected freeze out of 3-D holes is observed, leaving the 1.25 X 10*12 2-D holes in the quantum well. In the following we concentrate on transport measurements of the 2-D hole gas at T < 50 deg K

B. INFLUENCE OF DOPING DENSITY Na

Carriers are transferred into the quantum well from within an effective width W(eff) on both doped sides. The 2-D carrier sheet density ps is given by

        ps = 2 Na W(eff)   (1)

A simple model [2] of a modulation doped heterojunction, in which band bending within the undoped side is neglected, leads to the following equation for the effective width W(eff)

        W(eff) = (Ls*2 + Wd*2)*(1/2) - Ls   (2)

with the depletion layer width Wd of a one sided abrupt junction given by

        Wd*2 = 2 epsilon X DELTA Ev/e*2 X Na   (3)

TABLE 2 gives measured values of the sheet carrier density ps and the mobiity mu of the 2-D hole gas.

TABLE 2: sheet carrier density, ps, and hole mobility, mu, at 20 deg K
         versus doping density, Na, of the cladding layers.

Na (/cm*3)	1 X 10*17	3 X 10*17	4 X 10*17	7.5 X 10*17	1 X 10*18
ps (/cm*2)	5 X 10*11	6.8 X 10*11	7.5 X 10*11	1.1 X 10*12	1.4 X 10*12
mu (cm*2/V.s)	900	925	650	530	470
Wd (nm)	43.3	25.0	21.7	15.8	13.7
W(eff) (nm)	34.4	16.9	13.9	8.7	7.0

(Lw = 200 nm, Ls = 10 nm). The depletion layer width, Wd, and the effective width W(eff) are calculated with epsilon = 10*-10 As/Vm, DELTA Ev = 150 meV

The simple model of Eqns (1-3) can roughly explain the observed sheet carrier densities, ps. The mobility peaks at a sheet carrier density of 6 X 10*11/cm*2 for Ls = 10 nm.

C. INFLUENCE OF DOPING SET-BACK Ls

The doping set-back reduces the sheet carrier density because the
depletion layer, which extends from the heterojunction into the
cladding layer, is partly undoped (TABLE 3).

TABLE 3: sheet carrier density, ps, and mobility, mu, of the 2-D hole
         gas at 20 deg K versus doping set-back Ls.

Ls (nm)	0	10	25	40	60
ps (/cm*2)	1.4 X 10*12	1.4 X 10*12	1.1 X 10*12	4 X 10*11	2.5 X 10*11
mu (cm*2/V.s)	360	470	780	870	420
W(eff) (nm)	13.7	7.0	3.5	2.3	1.55

(Lw =200 nm, Na = 10*18/cm*3)
W(eff) calculated using Eqns (2) and (3).

The experimental values for the sheet carrier density ps are roughly
explained by the simple model (Eqns (1-3)) for Ls greater than or equal
to 10 nm. The saturation of ps between Ls = 0 and Ls = 10 nm is
unexplained. The hole mobility peaks for Ls = 40 nm with a sheet carrier
density ps = 4 X 10*11/cm*2. The occurrence of an optimum Ls was
interpreted by People et al [2] as resulting from a combination of two
mobility determining mechanisms: increasing Ls improves the mobility due
to a decrease in remote scattering by the modulation doping.
Simultaneously, the sheet carrier density and thereby the average hole
momentum decreases, leading to an increasing effectiveness of existing
scattering centres within the SiGe layer, since Coulomb potentials
scatter predominantly with small wave vectors. But it may be that
reduced screening with lower carrier densities is more effective for
explaining the results. For a qualitative comparison the treatment of a
2-D electron gas by Gold [7] can be used, which roughly shows a square
root dependence of mobility on the sheet carrier density for impurity
scattering.

D. INFLUENCE OF WELL THICKNESS Lw

The well thickness Lw is of minor importance for Lw between 50 nm and
250 nm. The latter value corresponds to the critical thickness for
misfit dislocation generation for these Si/Si(0.8)Ge(0.2) structures
grown at low temperatures (up to 550 deg C). The mobility (TABLE 4)
is reduced for Lw = 10 nm, which is suspected to reflect enhanced
interface scattering.

TABLE 4: hole mobility, mu, of the 2-D hole gas at 20 deg K versus
         well thickness, Lw

Lw (nm)	200	50	10
mu (cm*2/V.s)	1600	1800	500

(Ls = 10 nm, Na = 10*18/cm*3).

The data given in TABLE 4 differ from those given in TABLE 3, presumably due to differences in background doping and interface scattering. Peak mobility values of 6000 cm*2/V.s were reported for T = 4.2 deg K [3]. From this value the background doping is estimated to be about 10*15/cm*3.

E. EFFECTIVE HOLE MASS

The light and heavy hole masses mlh, mhh in Si, Ge and Si(0.8)Ge(0.2) (linearly interpolated) are given in TABLE 5.

TABLE 5: light hole mass mlh and heavy hole mass mhh (in units of electron mass)

	Si	Ge	Si(0.8)Ge(0.2)
mlh	0.16	0.044	0.14
mhh	0.49	0.28	0.45

From the Shubnikov-de Haas measurements People et al [3] derived an effective mass of 0.32 +- 0.03 for a sheet carrier concentration of 6 X 10*11/cm*2.

REFERENCES
----

[1]   R.People et al [Appl. Phys. Lett. (USA) vol.45 no.11 (1 Dec 1984) p.1231-3]
[2]   R.People, J.C.Bean, D.V.Lang [J. Vac. Sci. & Technol. A (USA) vol.3 no.3 pt 1 (May/Jun 1985) p.846-50]
[3]   E.Kasper, H.-J.Herzog, H.Jorke, G.Abstreiter [Superlattices & Microstruct. (GB) vol.3 no.2 (1987) p.141-6]
[4]   R.People [IEEE J. Quantum Electron. (USA) vol.QE-22 no.9 (Sep 1986) p.1696]
[5]   G.Abstreiter, H.Brugger, T.Wolf, H.Jorke, H.-.J.Herzog [Phys. Rev. Lett. (USA) vol.54 no.22 (3 June 1985) p.2441-4]
[6]   C.G. Van de Walle, R.M.Martin [Phys. Rev. B (USA) vol.34 no.8 (15 Oct 1986) p.5621-34]
[7]   A.Gold [Phys. Rev. B (USA) vol.35 no.2 (15 Jan 1987) p.723-33]

27.3 STRUCTURE OF THE Ge/Si INTERFACE

by C.A.Sebenne

June 1987
EMIS Datareview RN=15727

A. GENERAL FEATURES OF THE Ge/Si INTERFACE

Both silicon and germanium are tetravalent elements which crystallise
by forming covalent bonds with four nearest neighbours and tetrahedral
symmetry. They can together form alloyed crystals of Ge(x)Si(1-x) for
any x, from 0 to 1. Their differences can be abruptly summarised by two
numbers; the bond length of Ge (2.450 A) is 4.18 percent larger than
the bond length of Si (2.352 A); the bond strength of Ge (1.94 eV)
is 16 percent weaker than the bond strength of Si (2.32 eV). Then,
whatever the preparation conditions, the structure of the Si-Ge
interface will obey the following rules:

- respect as much as possible of the local tetrahedral
  symmetry

- compensation of the atom size difference by the most
  favourable distribution of defects and compositional
  gradient, within the imposed thermodynamical conditions,
  taking into account the binding energy difference.

Within these rules some variety of interface topologies can be obtained
and the main influences are:

- nature and topology of the initial substrate (Si or Ge,
  crystallinity, cleanliness, orientation, surface
  reconstruction, defects...)

- deposition conditions (process, rate, thickness,
  temperature)

- post-deposition annealing.

Since many groups have dealt with the subject since the late seventies,
the results will be analysed taking the initial substrate as the
leading parameter and focusing on vacuum deposition of the
complementary element.

B. VACUUM DEPOSITION OF Ge ON CLEAVED Si(111) 2 X 1

At room temperature below 300 deg C vacuum deposition of Ge on clean
cleaved 2 X 1 reconstructed Si(111) substrates leads to an abrupt
interface at the atomic scale as first shown by Nannarone et al [1].
Three stages may be distinguished in the Ge layer formation. Coverages
theta being given in monolayer (ML) units (on Si(111), 1 ML = 7.8 X
10*14 atoms/cm*2), these stages are found respectively in the following

ranges; i) 0 < theta < 0.8 ML, ii) 0.8 < theta < 2 ML, iii) theta > 2 ML. Room temperature deposition gives rise to a SQRT 3 X SQRT 3 R 30 deg reconstruction in the first stage, complete in the 1/3 ML coverage range and well observable up to 0.8 ML, and shown by P.Chen et al [2]. This was not observed in layers deposited on substrates heated at 350 deg C, as reported by Perfetti et al [3], meaning that higher Ge surface mobility prevents the formation of an ordered layer at low density (unless heating causes too much deterioration of the cleaved surface). In any case, in the second stage, the LEED pattern becomes the 1 X 1 of Si(111) which remains observable close to 2 ML, the Ge layer being uniform. At higher coverages, a uniform and amorphous layer grows at room temperature while a wavy and maybe highly polycrystalline layer grows at 350 deg C. The structure at 1/3 ML coverage seems well established; from the calculation by Zhang et al [4], the Ge atoms occupy the T4 ternary sites, where the adatom is above the Si atom of the second layer. It agrees with the removal of the Si dangling bond state, essentially complete below 1/2 ML coverage at room temperature [2] as well as at 300 deg C [3].

It is not clear whether Ge atoms stay in the same position upon increasing coverage or not: approaching 2 ML coverage one would expect a tendency towards perfect epitaxy, since the bond length difference can be accommodated through slight displacements normal to the (111) plane. This would imply a reordering at the interface from 1/3 to 2 ML Ge coverage, which has not been demonstrated yet. The electronic properties at the interface can be summarised as follows: i) Ge adsorption does not change the ionisation energy of Si [2]; ii) the valence band edge of Ge shows up about 0.4 eV above the Si valence band edge following high resolution photoemission yield measurements of Chen et al [2]; a probably less accurate band offset value of 0.2 eV being deduced from standard UPS measurements by Perfetti et al [5]; iii) the dangling bond states of Si are replaced by dangling bond states of Ge beyond 1/3 ML coverage at about 0.8 eV below Ef [2] and structures attributable to interface states are seen at 5 and 8 eV below Ef [5].

C. VACUUM DEPOSITION OF Ge ON Si(111) 7 X 7

While the growth of oriented Ge crystallites on heated Si(111) was established long ago through electron microscopy and RHEED observations by Aleksandrov et al [6], the first interface structure study of the Ge-Si(111) 7 X 7 system was presented by Narusawa and Gibson [7] who used a high energy He+ ion channelling technique together with LEED control. Room temperature Ge deposition gives a uniform amorphous layer with an abrupt interface. In the 350 deg C range, the interface is still abrupt, the first 3 ML of Ge form a uniform and strained pseudomorphic layer and further Ge would form strained epitaxial islands; after 30 ML of Ge, more than 10 ML are distorted at the interface. At higher substrate temperature, the general process is the same but signs of alloying during the first 3 ML stage are found and the epitaxy which follows is easier. LEED measurements always show a progressive blurring of the 7 X 7 pattern ultimately replaced by a Ge 1 X 1. In order to get a better insight of the interface structure, two points were needed: i) to improve the knowledge of the actual geometry of the 7 X 7 reconstruction surface, ii) to study in detail the Ge deposition at low coverages. The first point came with the proposal of the DAS model (Dimer-Adatom-Stacking fault model) by Takayanagi et al [8]. The second started with the remark that the 7 X 7 reconstruction is not degraded at low Ge coverage as the 2 X 1 is at room temperature on both

Si(111), as observed by Chen et al [9], and Si(100), as observed by
Gossmann et al [10]; this would remain true on Si(111) 7 X 7 substrates
heated in the 350 deg C range [11,12] in spite of some intermixing.
This has led to recent work using X-ray standing wave measurements by
Dev et al [13,14] and by Patel et al [15]. The conclusions are as
follows: i) the DAS model is respected; ii) Ge atoms occupy the surface
and adatom atop sites of the DAS model with a higher binding energy for
the surface atop site. The corresponding expansion normal to the surface
of about 2 percent agrees with the Si-Ge bond length, intermediate
between the Si-Si and Ge-Ge ones. Let us note, that at 350 deg C,
Gossman et al [12] discard the island growth beyond 3 ML in favour of
the uniform growth of a constant composition alloy. Upon temperature
rising to around 400 deg C and above, alloying effects occur at the
interface; the higher the temperature, the wider the compositional
gradient: the mismatch-induced strains are eased by the alloy formation
and the critical thickness for the pseudomorphic layer increases upon
increasing the Si content of the Ge layer. This has been shown
experimentally by Bean et al [16] and a simple model calculated by
People and Bean [17] brings quantitative agreement. Strain effects are
also revealed by LEED and/or RHEED measurements at Ge coverages from
1.5 to a few ML when the temperature is around 400 deg C and beyond by
the observation of a 5 X 5 reconstruction, as shown by Shoji et al [18],
Ichikawa and Ino [19], and Gossmann et al [20], which can be interpreted
as the sign of a silicon surface strained by compression. (NB: Thicker
Ge(x)-Si(1-x) layers strained by epitaxial growth on Si(111) can also
display 5 X 5 surface reconstruction, as seen by McRae and Malic [21]
and studied by Becker et al [22] who proposed a model derived from the
7 X 7 DAS model.)

The electronic properties of the system have been studied using both
photoemission techniques and electron energy loss spectroscopy. A first
evaluation of the valence band offset by Margaritondo et al [23] gave
0.16 eV for a Ge layer deposited at room temperature while a value of
about 0.4 eV comes from Chen et al [9], which would almost align the
conduction bands. The Si dangling bonds are always removed before 1 ML
Ge coverage and replaced by Ge dangling bonds. At high temperature, the
interface is not easy to characterise since it is non-uniform and/or
alloyed. Recent measurements of the Si 2p core level by Miller et al
[24], of angular resolved photoemission with 10.2 eV photons by
Martensson et al [25] and of angle resolved electron loss spectroscopy
by Hasegawa et al [26] essentially confirm the validity of the atomic
geometries proposed for the 7 X 7-Ge and the 5 X 5 reconstructions
defined above. Electronic features [25] which are identified at 0.2,
1.0 and 1.4 eV below Ef for the 7 X 7-Ge and at 0.12 and 1.2 eV below
Ef for the 5 X 5 are surface structures which may not survive the
interface.

D. VACUUM DEPOSITION OF Ge ON Si(100) 2 X 1

Within the rules given in section A deposition of Ge on the clean 2 X 1
reconstructed (100) face of Si displays some differences as compared to
the 7 X 7 reconstructed (111) substrate. Using LEED, AES and Rutherford
backscattering/channelling techniques Gossmann et al [10,12] reached
the following conclusions. At low (120 deg K) and room temperatures,
Ge forms an abrupt interface and grows as a uniform layer which is
immediately sufficiently disordered to remove the 2 X 1 reconstruction
before 1 ML and to mask the remaining 1 X 1 beyond 2 ML (here, 1 ML =
6.8 X 10*14 atoms/cm*3). On heated substrates (250 deg C), Ge still

forms an abrupt interface and grows as a uniform epitaxial layer, the 2 X 1 reconstruction remaining observable close to 2 ML, a 1 X 1 LEED pattern staying at higher coverages. Asai et al [27] confirm these conclusions and find an epitaxial growth of Ge upon heating in the 350-470 deg C range with a 2 X 1 reconstructed Ge surface. Heating at higher temperatures leads to alloying effects and in the 800 deg C range Ge vanishes into Si. MBE growth, studied by Sheldon et al [28], Fukuda and Kohama [29], shows essentially that significant alloying occurs beyond 600 deg C, that epitaxy occurs above roughly 300 deg C and that increasing the growth temperature decreases the strains and increases the dislocation density.

E. Si DEPOSITION ON Ge

When Si is deposited on clean Ge, a higher tendency towards intermixing is expected since it will be easier for a Si atom to break the weak Ge-Ge bond and to form stronger Ge-Si bonds. Taking the mixing temperature of Ge/Si to be about 600 deg C and its epitaxial growth temperature to be 300 deg C, the corresponding temperatures for Si/Ge will be roughly exchanged; it means that there is no hope of growing an ordered Si layer without intermixing. Mahowald et al [30] used LEED, AES and photoemission spectroscopy to study room temperature and 300 deg C deposited Si layers on cleaved Ge(111). Only room temperature deposition gives a seemingly abrupt interface where the initial 2 X 1 reconstruction is already diffuse at 0.5 ML (here 1 ML = 6.9 X 10*14 atoms/cm*2) and gone at 2 ML. At 300 deg C the initial c-(2 X 8) keeps its brighest spots visible up to 12 ML. Photoemission gives a valence band offset of 0.4 eV, the same as determined by Chen et al [2,9,11] on the reverse system. EXFAS measurements by Woicik et al [31] on a room temperature deposited Si layer on cleaved Ge, subsequently annealed from 370 to 800 deg C, show that the initial disordered layer, with the usual Si-Si bond length, transforms at 370 deg C to a better ordered fifty-fifty alloy with a strained Si-Ge bond length stretched to the Ge-Ge one.

REFERENCES

[1]   S.Nannarone et al [Solid State Commun. (USA) vol.34 (1980) p.409]

[2]   P.Chen, D.Bolmont, C.A.Sebenne [Solid State Commun. (USA) vol.44 (1982) p.1191]

[3]   P.Perfetti et al [J. Vac. Sci. & Technol. (USA) vol.19 (1981) p.319]

[4]   S.B.Zhang, M.L.Cohen, J.E.Northrup [Surface. Sci. (Netherlands) vol.157 (1985) p.L303]

[5]   P.Perfetti et al [Phys. Rev. B (USA) vol.24 (1981) p.6174]

[6]   L.N.Aleksandrov, R.N.Lovyagin, O.P.Pehelyakov, S.T.Stenin [J. Cryst. Growth (Netherlands) vol.24/25 (1974) p.298]

[7]   T.Narusawa, W.M.Gibson [Phys. Rev. Lett. (USA) vol.47 (1981) p.1459]

[8]   K.Takayanagi, Y.Tanishiro, M.Takahashi, S.Takahashi [J. Vac. Sci. & Technol. A (USA) vol.3 (1985) p.1502]

[9]   P.Chen, D.Bolmont, C.A.Sebenne [Solid State Commun. (USA) vol.46 (1983) p.689]

[10]   H.J.Gossman, L.C.Feldman, W.M.Gibson [Phys. Rev. Lett. (USA) vol.53 (1984) p.294]

[11] P.Chen, D.Bolmont, C.A.Sebenne [Thin Solid Films (Switzerland) vol.111 (1984) p.367]

[12] H.J.Gossman, L.C.Feldman, W.M.Gibson [Surf. Sci. (Netherlands) vol.155 (1985) p.413]

[13] B.N.Dev, G.Materlik, R.L.Johnson, W.Kranz, P.Funke [Surf. Sci. (Netherlands) vol.178 (1986) p.1]

[14] B.N.Dev, G.Materlik, F.Grey, R.L.Johnson, M.Clausnitzer [Phys. Rev. Lett. (USA) vol.57 (1986) p.3058]

[15] J.R.Patel, P.E.Freeland, J.A.Golovchenko, A.R.Kortan, D.J.Chadi, Guo-Xin Qian [Phys. Rev. Lett. (USA) vol.57 (1986) p.3077]

[16] J.C.Bean, T.T.Sheng, L.C.Feldman, A.T.Fiory, R.T.Lynch [Appl. Phys. Lett. (USA) vol.44 (1984) p.102]

[17] R.People, J.C.Bean [Appl. Phys. Lett. (USA) vol.47 (1985) p.322]

[18] K.Shoji, M.Hydo, H.Ueba, C.Tatsuyama [Jpn. J. Appl. Phys. Part 2 (Japan) vol.22 (1983) p.L200]

[19] T.Ishikawa, S.Ino [Surf. Sci. (Netherlands) vol.136 (1984) p.267]

[20] H.J.Gossman, J.C.Bean, L.C.Feldman, W.M.Gibson [Surf. Sci. (Netherlands) vol.138 (1984) p.L175]

[21] E.G.McRae, R.A.Malic [Surf. Sci. (Netherlands) vol.163 (1985) p.L702]

[22] R.S.Becker, J.A.Golovchenko, B.S.Swartzentruber [Phys. Rev. B (USA) vol.32 (1985) p.8455]

[23] G.Margaritondo, N.G.Stoffel, A.D.Katnani [Solid State Commun. (USA) vol.36 (1980) p.215]

[24] T.Miller, T.C.Hsieh, T.C.Chiang [Phys. Rev. B (USA) vol.33 (1986) p.6983]

[25] P.Martensson, A.Cricenti, L.S.O.Johansson, G.Hansson [Phys. Rev. B (USA) vol.34 (1986) p.3015]

[26] S.Hasegawa, H.Twaskai, Sung-Te Li, S.Nakamura [Phys. Rev. B (USA) vol.32 (1985) p.6949]; S.Hasegawa, H.Iwasaki, M.Akizuki, Sung-Te Li, S.Nakamura [J. Vac. Sci. & Technol. A (USA) vol.4 (1986) p.2336 and [Solid State Commun. (USA) vol.58 (1986) p.697]

[27] M.Asai, H.Ueba, C.Tatsuyama [J. Appl. Phys. (USA) vol.58 (1985) p.2577]

[28] P.Sheldon, B.G.Yacobi, S.E.Asher, K.M.Jones, M.J.Hafich, G.Y.Robinson [J. Vac. Sci. & Technol. A (USA) vol.4 (1986) p.889]

[29] Y.Fukuda, Y.Kohama [J. Cryst. Growth (Netherlands) vol.81 (1987) p.451]

[30] P.H.Mahowald, R.S.List, W.E.Spicer, J.Woicik, P.Pianetta [J. Vac. Sci. & Technol. B vol.3 (1985) p.1252]

[31] J.C.Woicik. R.S.List, B.B.Pate, P.Pianetta [J. Phys. Colloq. (France) vol.47 no.C-8 pt.1 (1986) p.497]

27.4 GROWTH OF EPITAXIAL CoSi2/Si MULTILAYERS

by B.D.Hunt

October 1987
EMIS Datareview RN=16175

A. INTRODUCTION

CoSi2 has the cubic fluorite structure with a lattice constant only
slightly smaller than that of Si (1.2% at room temperature). As a
consequence, high quality epitaxial CoSi2 films can be grown on the
Si(111) surface using UHV growth techniques [1-4]. CoSi2 is also a good
metal (room temperature resistivity about 15 micro-ohm cm) with
sufficient thermal stability to allow high quality Si overgrowth [4].
Although growth studies and device applications of Si/CoSi2/Si
structures have been investigated by a number of groups [4,6] progress
on the growth of epitaxial CoSi2/Si heterostructures with multiple
silicide and Si layers has only recently been reported [4,7]. Epitaxial
CoSi2/Si multilayers are of interest for fundamental physical studies
of crystalline metal-semiconductor superlattices, as well as for
application to buried electrical interconnects, 3-D integration, and
novel devices [12].

Two particular problems complicate the fabrication of silicide-Si
multilayers. The first of these is the tendency of thin, epitaxial
CoSi2 layers to form with high pinhole densities. Silicide pinhole
formation is undesirable because the holes lead to shorts between
adjacent Si interlayers. The second problem is due to nonuniform Si
growth over CoSi2 layers, which results in discontinuities within a
given Si layer and shorting between the silicide layers. However, it
has recently been found that these problems can be reduced or
eliminated by using growth techniques which utilize the room
temperature deposition of a thin, amorphous Si layer. These techniques
are outlined below.

B. CONTROL OF CoSi2 PINHOLE FORMATION

Studies of CoSi2 grown by standard UHV deposition techniques have shown
that thin, epitaxial CoSi2 layers generally form with pinhole densities
greater than 10*8/cm*2 on Si(111) substrates [4,5,9], although there is
some controversy on this point [6]. However, recent work by Hunt et al
and Tung et al has shown that the pinhole density and pinhole size
distribution in epitaxial CoSi2 films can be controlled by using a new
solid phase epitaxy process [4,5,7,9]. This technique utilizes a low
temperature (20-200 deg C) deposition of a thin Co layer (< 60 A),
followed by the evaporation of an amorphous Si overlayer. The Co/Si
bilayer is then annealed at temperatures of 600-650 deg C to form an
epitaxial CoSi2 film with a thickness 3.5 times the starting Co
thickness. Electron microscopy studies of 70 A and 175 A CoSi2 films
produced in this manner show that the pinhole densities and size
distributions are dependent upon the thickness of the Si overlayer. By
changing the initial Si cap layer thickness, the silicide pinhole
density can be varied over a range from 10*6/cm*2 to > 10*9/cm*2 (see
also Datareview [13]). At an optimum Si/Co thickness ratio of 2-3, the

CoSi2 pinhole density is approximately 2-3 orders of magnitude smaller than in films grown by conventional methods. Such low pinhole density films are well-suited for application to CoSi2/Si multilayers. Recent work has shown that still lower pinhole densities may be obtained using room temperature codepositions of Co and Si [10,11].

C. ELIMINATION OF Si ISLANDING OVER CoSi2

As noted above, a second potential problem associated with the CoSi2/Si system is the nonuniform growth of Si epilayers over CoSi2. It has been demonstrated that Si islanding occurs during high temperature Si epitaxy on CoSi2 [4,7,8]. TEM and SEM observations indicate that the Si islands have either the same orientation as the underlying silicide (type-A), or have the same orientation except for a 180 deg rotation about the (111) surface normal (type B). Hunt et al discovered that islanding in thin Si overlayers could be prevented by depositing a thin (generally 30-50 A) Si layer on the CoSi2 near room temperature (< 200 deg C), before growth of the bulk of the Si overlayer at high temperatures (about 615 deg C) [7]. A similar technique has been reported to control the orientation of Si on CoSi2 [5]. 90 A Si overlayers grown using this Si predeposit technique were found to be uniform and continuous, and had the same orientation as the CoSi2 layer [7]. Recently, Henz and coworkers have found that Si islanding can also be reduced by going to Si growth temperatures below 400 deg C [11].

D. GROWTH OF EPITAXIAL CoSi2/Si MULTILAYERS

The ability to produce reduced-pinhole-density CoSi2 layers and uniform Si epilayers over the silicide films is essential for growing high quality CoSi2/Si multilayer structures. Cross-sectional TEM micrographs of multilayers grown without using the growth techniques described above show that such structures are highly defective with many discontinuities in both the Si and silicide layers [7]. However, when the pinhole-reduction and Si-islanding reduction techniques are utilized, the quality of the multilayer structures is dramatically improved. In that case, the TEM studies indicate that the silicide and Si layers are essentially continuous, with only occasional gaps in individual layers. Hunt and coworkers have succeeded in growing CoSi2/Si multilayers with 70 A CoSi2 layers and 100 A Si interlayers and up to four periods of silicide and Si [4,7]. These are the first reported epitaxial metal/semiconductor multilayers. Henz et al have also recently produced multi-period CoSi2/Si heterostructures [11].

REFERENCES

[1]   S.Saitoh, H.Ishiwara, S.Furukawa [Appl. Phys. Lett. (USA) vol.37 no.2 (15 Jul 1980) p.203-5]
[2]   J.C.Bean, J.M.Poate [Appl. Phys. Lett. (USA) vol.37 no.7 (1 Oct 1980) p.643-6]
[3]   R.T.Tung, J.C.Bean, J.M.Gibson, J.M.Poate, D.C.Jacobson [Appl. Phys. Lett. (USA) vol.40 (15 Apr 1982) p.684]
[4]   B.D.Hunt et al [Mater. Res. Soc. Symp. Proc. (USA) vol.56 (1986) p.151-6]
[5]   R.T.Tung, A.F.J.Levi, J.M.Gibson [Appl. Phys. Lett. (USA) vol.48 (10 Mar 1986) p.635]

[6]   E.Rosencher, P.A.Badoz, J.C.Pfister, F.Arnaud d'Avitaya,
        G.Vincent, S.Delage [Appl. Phys. Lett. (USA) vol.49 no.5 (4 Aug
        1986) p.271-3]

[7]   B.D.Hunt, N.Lewis, L.J.Schowalter, E.L.Hall, L.G.Turner [Mater.
        Res. Soc. Symp. Proc. (USA) vol.77 (1987) p.351]

[8]   B.M.Ditcheck, J.P.Salerno, J.V.Gormley [Appl. Phys. Lett. (USA)
        vol.47 no.11 (1 Dec 1985) p.1200-2]

[9]   B.D.Hunt, N.Lewis, E.L.Hall, C.D.Robertson [J. Vac. Sci. &
        Technol. B (USA) vol.5 no.3 (May/June 1987) p.749-50]

[10]   T.L.Lin, R.W.Fathauer, P.J.Grunthaner, C. d'Anterroches [submitted
        to Appl. Phys. Lett. (USA) (1987)]

[11]   J.Henz, M.Ospelt, H.vonKanel [Solid-State Commun. (USA) vol.63
        no.6 (1987) p.445-50]

[12]   M.L.Huberman, J.Maserjian [to be published in Superlattices and
        Microstruct. (GB) (1987)]

[13]   B.D.Hunt [EMIS Datareview RN=16176 (Oct 1987) 'Growth of
        Si/CoSi2/Si heterostructures']

27.5 PROPERTIES OF a-Si:H/a-SiN(X):H MULTILAYERS

by H.Fritzsche

May 1986
EMIS Datareview RN=15562

Up to 1 micron thick films consisting of alternating layers of amorphous
silicon (a-Si:H) and silicon nitride (a-SiN(x):H) have been prepared by
periodically switching the gas flow into a glow discharge reactor
between pure silane (SiH4) and a mixture of SiH4 and NH3 that has a
volume ratio between 1:4 and 1:6 [1,2,3]. The thickness d(S) and d(N) of
the a-Si:H and a-SiN(x):H sublayers, respectively, are chosen between
10 and several thousand Angstrom.

According to electron transmission microscopy [4] the layers are smooth
and flat. Low angle X-ray reflections up to 4th order have been observed
[1,5] with a 1st order FWHM of 0.06 degrees. The folding of the
longitudinal acoustic (LA) phonon branch along the direction
perpendicular to the layers into a Brillouin zone of dimension
2pi/(d(S)+d(N)) gives rise to new Raman peaks between wavenumbers
5-60 cm*-1 [6]. With decreasing layer thickness d(S) less than 50 A the
transverse optical (TO) phonon Raman peak in the a-Si:H layers broadens
and decreases in amplitude relative to that of the LA phonons [7,8].
Analysis of this evidence yields a width of 4.5 A for the interface
region [8]. The silicon-on-nitride and nitride-on-silicon interfaces are,
however, different. When a-Si:H is deposited on a-SiN(x):H there are
about 10*15/cm*2 extra hydrogen at that interface and in the
first 20 A of the a-Si:H layer as compared to the bulk [8]. The
hydrogen causes a low density of interface defects, about 1.4 X 10*11
/cm*2 as evidenced by optical absorption [9]. Only 10*10/cm*2 [10] or
6 X 10*10/cm*2 [11] neutral Si dangling bonds were observed at the
interfaces by paramagnetic resonance. This does not preclude the
additional presence of charged dangling bonds which are spinless.

The difference between the two types of interfaces causes internal
fields in the a-Si:H layers of order 10*5 V/cm which point in the growth
direction and originate from an interface charge density of 6 X 10*12
/cm*2 at the silicon-on-nitride interface according to measurements of
the bias dependence of the electro-absorption signal [12].

In the infrared there are absorption bands at wavenumbers 3340 cm*-1
(N-H stretching mode), 2175 cm*-1 (Si-H back-bonded to N3 in the
a-SiN(x):H layers), 2000 cm*-1 (Si-H stretching mode in a-Si:H),
2080 and 2155 cm*-1 (Si-H modes in the interface region) and at 840
cm*-1 (Si-N mode in a-SiN(x):H) [8].

For normal incidence of light the complex dielectric constant epsilon*
of the multilayer film is the thickness average of the dielectric
constants epsilon(S)* and epsilon(N)* of the two materials:
epsilon* X (d(S)+d(N)n) = epsilon(S)* X d(S) + epsilon(N)* X d(N) [13].
One finds that the optical gap Eos of a-Si:H increases from 1.7 eV to
2.1 eV and the refractive index n(S) decreases from 3.4 to 3.0 as
the a-Si:H layer thickness d(S) is decreased from 50 A to 10 A. At the
same time the conductivity activation energy Ea increases by 0.3 eV
[1,3,14]. This is tentatively attributed to quantum confinement of the
electrons in the thin a-Si:H layers. Alternatively contamination of

a-Si:H with nitrogen may produce the observed increase. The exponential part of the absorption (Urbach tail) broadens such that the Urbach tail parameter Eo increases from 0.075 eV to 0.15 eV as d(S) is decreased from 100 A to 10 A [1]. In this d(S) range the energy of the photoluminescence peak increases by 0.1 eV and broadens from 0.3 to 0.5 eV FWHM. The low temperature (T up to 77 deg K) integrated photoluminescence intensity I(PL) decreases by a factor 10 whereas at 300 deg K I(PL) increases by a factor 100 as d(S) is decreased from 100 to 10 A [15,16].

Preliminary photoemission and photoabsorption spectroscopy studies show that the 3.9 eV gap of a-SiN(x):H joins the 1.75 eV gap of a-Si:H with a 1.4 eV step at the valence band and a 0.75 eV step at the conduction band edge [17].

The electrical conductivity of these multilayers is extremely anisotropic. The transverse conductivity is governed by the thickness and quality of the insulating nitride layers and is less than 10*-14 mho/cm for d(N) greater than 20 A. The in-plane conductivity of multilayers having d(S) greater than 100 A is 10*-5 mho/cm, that is several orders of magnitude larger than bulk a-Si:H films, because charges are transferred from the nitride to the silicon layers [2,3,9,14,16]. This space charge doping effect raises the Fermi level and lowers the conductivity activation energy Ea by 0.2 eV. In this range of d(S) greater than 100 A the photoconductivity is essentially independent of d(S) [3]. For d(S) less than 100 A the dark and photoconductivity decrease with decreasing d(S) and Ea increases from 0.4 eV at d(S) = 100 A to 0.9 eV at d(S) = 12 A [2,3,9,14,16]. This has been attributed to an upward shift of the electron mobility edge resulting from the quantum confinement effect.

The electron drift mobility mu(d) of the multilayers has been measured with a surface-acoustic travelling-wave method between 300 and 450 deg K [18]. For d(S) greater than 200 A the values of mu(d) agree with those of thick unlayered a-Si:H films. Typical values are mu(d) = 0.5 cm*2/V.s at 300 deg K and mu(d) = 1.9 cm*2/V.s at 450 deg K. For d(S) less than 200 A mu(d) decreases with decreasing d(S). For d(S) = 47 A the drift mobility was found to be mu(d) = 0.05 cm*2/V.s at 300 deg K and mu(d) = 0.1 cm*2/V.s at 450 deg K. This drop in mu(d) below d(S) = 200 A is attributed to the increasing proximity of surface traps to the transport path.

One finds a linear current voltage relation for current flow parallel to the layers in a limited voltage range only. At applied fields not much larger than 100 V/cm a number of new effects appear [19,20] which become more pronounced as the applied field is increased: (i) the conductance increases by as much as a factor 100 above its low voltage ohmic value; (ii) after applying a field say of 500 V/cm the current approaches its steady value only after about 60 min; (iii) once established, this high conductance state can be probed with a small voltage and it requires about 60 min or longer to decay to the original ohmic value; (iv) this metastable bias-induced anomalous conductance can be removed by annealing above 400 deg K if no or only a small voltage is applied; (v) the time constants for reaching and for eliminating such anomalous conductances are reduced to seconds or less when the multilayer sample is illuminated. These bias-induced changes in conductance are particularly pronounced in high impedance multilayers. They are caused by inhomogeneities in the layers or at the contacts to the layers such that the equipotential surfaces are no longer planes normal to the layers. Potential differences across the insulating nitride layers require then space charge regions in the amorphous silicon layers. These in turn give rise to the changes in parallel conductance. The long time constants for establishing these space charge regions and hence the steady state result from the large transverse capacitances and longitudinal resistances. The latter are greatly reduced by illumination. Such self-induced field effect distortions readily occur whenever thin insulating layers are interleaved with resistive semiconducting layers.

REFERENCES
----

[1]   B.Abeles, T.Tiedje [Phys. Rev. Lett. (USA) vol.51 (1983) p.2003]
[2]   J.Kakalios, H.Fritzsche, N.Ibaraki, S.R.Ovshinsky [J. Non-Cryst. Solids (Netherlands) vol.66 (1984) p.339]
[3]   N.Ibaraki, H.Fritzsche [Phys. Rev. B (USA) vol.30 (1984) p.5791]
[4]   Ruguang Cheng, Shulin Wen, Jingwei Feng, H.Fritzsche [Appl. Phys. Lett. (USA) vol.46 (1985) p.592]
[5]   B.Abeles et al [J. Non-Cryst. Solids (Netherlands) vol.66 (1984) p.351]
[6]   P.Santos, M.Hundhausen, L.Ley [J. Non-Cryst. Solids (Netherlands) vol.77/78 (1985) p.1069; and Phys. Rev. B (USA) in press]
[7]   N.Maley, J.S.Lannin [Phys. Rev. B (USA) vol.31 (1985) p.5577]
[8]   B.Abeles, P.D.Persans, H.S.Stasiewski, L.Yang, W.Lanford [J. Non-Cryst. Solids (Netherlands) vol.77/78 (1985) p.995]
[9]   T.Tiedje, B.Abeles [Appl. Phys. Lett. (USA) vol.45 (1984) p.179]
[10] B.A.Wilson, Z.E.Smith, C.M.Taylor, J.P.Harbison [Solid State Commun. (USA) vol.55 (1985) p.105]
[11] C.C.Tsai, M.J.Thompson, R.A.Street, M.Stutzmann, F.Ponce [J. Non-Cryst. Solids (Netherlands) vol.77/78 (1985) p.995]
[12] C.B.Roxlo, B.Abeles, T.Tiedje [Phys. Rev. Lett. (USA) vol.52 (1984) p.1994]
[13] H.Ugur, R.Johanson, H.Fritzsche [in 'Tetrahedrally-bonded Amorphous Semiconductors', Ed. D.Adler, H.Fritzsche (Plenum, New York, 1985) p.425]
[14] T.Tiedje, B.Abeles, P.D.Persans, B.G.Brooks, G.D.Cody [J. Non-Cryst Solids (Netherlands) vol.66 (1984) p.345]
[15] S.Nishikawa, H.Kakinuma, H.Fukuda, T.Watanabe, K.Nihei [J. Non-Cryst. Solids (Netherlands) vol.77/78 (1985) p.1077]
[16] P.G.LeComber et al [J. Non-Cryst. Solids (Netherlands) vol.77/78 (1985) p.1081]
[17] B.Abeles, I.Wagner, W.Eberhardt, J.Stohr, H.Stsiewski, F.Sette [AIP Conf. Proc. (USA) vol.120 (1984) p.394]
[18] H.Ugur, H.Fritzsche [Solid State Commun. (USA) vol.52 (1984) p.649]
[19] H.Ugur, H.Fritzsche [J. Non-Cryst. Solids (Netherlands) vol.77/78 (1985) p.1085]
[20] H.Ugur [Phys. Rev. B (USA) in press]

CHAPTER 28

SILICON OXIDES: OPTICAL FUNCTIONS

28.1 OPTICAL PROPERTIES OF AMORPHOUS SiO: DISCUSSION

by H.R.Philipp

September 1987
EMIS Datareview RN=16130

Silicon monoxide (SiO) is used as the dielectric material in certain
microelectronic devices and as a protective layer and antireflecting
coating in optical applications. These layers are usually prepared
by the rapid evaporation of silicon monoxide powder under high-vacuum
conditions. If the rate of deposition is reduced and/or carried out in
the presence of a partial pressure of oxygen, the condensate acquires
excess oxygen and the O to Si atom ratio rises above unity. These
films are often labelled SiO(x) and x may have any value in the
range 1 to 2. There has been some conjecture for the formation of a
compound of stoichiometry SiO(3/2) [1,2]. However it is currently
believed that the bonding in SiO(x) (x = 1 to 2), while
tetrahedrally coordinated about each Si atom, is random with Si-O and
Si-Si bonds mixed on an atom scale [3]. Hence a continuous range of
stoichiometries exist between SiO and SiO2. The optical properties
of SiO and SiO(x) are different and hence care should be taken in
only applying the optical functions given in the table [4] to
materials of stoichiometry close to SiO. For completeness references
are given below which also describe the optical behaviour of SiO(x)
layers.

The optical properties of any material may be described by the complex,
quantum energy dependent, dielectric constant e = e1 - ie2 which, in
turn, is related to the optical constants n and k by e1 = n*2 - k*2
and e2 = 2nk where N = n - ik is the complex index of refraction. The
extinction coefficient k describes the attenuation of the
electromagnetic wave as it traverses the material and is related to
the absorption coefficient, alpha, by alpha = 4 pi k/Y where Y is the
wavelength of the light in vacuum.

Although SiO has some important applications there is not very much
information available on its optical properties. However in the visible
and near ultraviolet spectral regions where the data from several
sources can be compared, the results agree quite well with one another.
This would indicate that SiO is a reasonably well defined material
with definitive optical properties when prepared, as described above,
in the absence of ambient oxygen. A review of the optical properties
of SiO has recently been published [5] and the discussion below
reiterates its findings.

The optical properties in the infrared region between 0.0886 eV and
1.24 eV were obtained from Hass and Salzberg [6]. The values were
derived through analysis of transmission and reflectance measurements
obtained on a series of SiO layers of varying thicknesses. The overall
accuracy of these results cannot be simply specified. They fit
measured transmission and reflectance to within 2%.

In the region 1.24 eV and 25 eV, the values are taken from Philipp
[3]. They were derived by Kramers-Kronig (KK) analysis of
reflectance data augmented, in the region of low absorption, by
transmission measurements. These results agree reasonably well with

those of Hass and Salzberg [6], whose measurements extend to about 5 eV, as well as with the absorption data of Cremer et al [7]. The results of Philipp [3] are given in the table [4] because they cover a wider energy range and more explicitly evaluate the extinction coefficient in the region of low absorption compared to the other results. They are however not necessarily more accurate than those of references [6] and [7]. The differences found among these measurements, about 5% or less in n and slightly larger in k, may be due in part to sample preparation (vacuum conditions, evaporation rate, substrate temperature, etc) and, as stated, it is gratifying that they agree as well as they do. For energies above about 12 eV the values depend somewhat on the extrapolation used in the KK analysis and should be treated with caution until more definitive measurements are made at energies above 25 eV which more clearly delineate the proper behaviour of the reflectance.

It is recommended that anyone using the table [4] examine the references cited in [3,6 and 7] as well as other references [8-10] which also describe the optical behaviour of materials of stoichiometry SiO(x) (x > 1).

REFERENCES
----

[1]   G.H.Wagner, A.N.Pirres [Ind. Eng. Chem. (USA) vol.44 no.2 (Feb 1952) p.321-6]
[2]   E.Ritter [Opt. Acta (GB) vol.9 no.2 (April 1962) p.197-202]
[3]   H.R.Philipp [J. Phys. & Chem. Solids (GB) vol.32 (1971) p.1935]
[4]   H.R.Philipp [EMIS Datareview RN=16131 (Sept 1987) 'Optical functions of amorphous SiO: table (0.09-25 eV)']
[5]   H.R.Philipp [in 'Handbook of Optical Constants of Solids' Ed. E.D.Palik (Academic, New York, 1985) p.765-9]
[6]   G.Hass, C.D.Salzberg [J. Opt. Soc. Am. (USA) vol.44 no.3 (Mar 1954) p.181-7]
[7]   E.Cremer, T.Kraus, E.Ritter [Z. Elektrochem. (Germany) vol.62 no.9 (1958) p.939-41]
[8]   G.Hass [J. Am. Ceram. Soc. (USA) vol.33 no.12 (1 Dec 1950) p.353-60]
[9]   A.P.Bradford, G.Hass [J. Opt. Soc. Am. (USA) vol.53 no.9 (Sept 1963) p.1096-100]
[10]   A.Bradford, G.Hass, M.McFarland, E.Ritter [Appl. Opt.(USA) vol.4 no.8 (Aug 1965) p.971-96]

28.2 OPTICAL FUNCTIONS OF AMORPHOUS SiO: TABLE (0.09-25 eV)

by H.R.Philipp

September 1987
EMIS Datareview RN=16131

A set of optical parameters n, k and absorption coefficient, alpha, are given in TABLE 1 for the energy range 0.09-25 eV. The table entries are given to four places in most cases. The accuracy of the values does not warrant this precision [1]. This is done to show the possible presence of weak structure in the optical properties that can often be discovered by comparison of precise, but not necessarily accurate values over small energy ranges. The values apply to materials of stoichiometry SiO. The optical properties of SiO(x) (x > 1) are different from those given here [1].

TABLE 1: optical functions for amorphous SiO at room temperature for selected energy values.

n is the refractive index, k is the extinction coefficient and alpha is the absorption coefficient.

Wavelength in A = approx. 12395/(energy in eV)

Energy (eV)	n	k	1000 alpha (/cm)	Ref
0.08856	2.01	0.30	2.69	[2]
0.09537	2.04	0.20	1.93	[2]
0.1033	2.13	0.14	1.47	[2]
0.1078	2.50	0.20	2.19	[2]
0.1127	2.82	0.40	4.57	[2]
0.1153	2.86	0.58	6.78	[2]
0.1181	2.85	0.90	10.8	[2]
0.1240	2.00	1.38	17.3	[2]
0.1305	1.20	1.20	15.9	[2]
0.1378	0.91	0.75	10.5	[2]
0.1459	0.90	0.18	2.66	[2]
0.1550	1.15			[2]
0.1653	1.42			[2]
0.1771	1.60			[2]

Energy (eV)	n	k	1000 alpha (/cm)	Ref
0.2066	1.70			[2]
0.2480	1.75			[2]
0.3100	1.80			[2]
0.4133	1.82			[2]
0.6199	1.84			[2]
1.24	1.87			[2]
1.40	1.913			[3]
1.60	1.929	0.00151	0.245	[3]
1.80	1.948	0.00523	0.954	[3]
2.00	1.969	0.01175	2.38	[3]
2.20	1.994	0.02153	4.80	[3]
2.40	2.021	0.03533	8.59	[3]
2.60	2.053	0.05544	14.61	[3]
2.80	2.085	0.08374	23.77	[3]
3.00	2.116	0.1211	36.82	[3]
3.20	2.144	0.1706	55.33	[3]
3.40	2.160	0.2287	78.82	[3]
3.60	2.162	0.2872	104.8	[3]
3.80	2.157	0.3453	133.0	[3]
4.00	2.141	0.4006	162.4	[3]
4.20	2.119	0.4499	191.5	[3]
4.40	2.094	0.4948	220.7	[3]
4.60	2.066	0.5364	250.1	[3]
4.80	2.034	0.5723	278.4	[3]
5.00	2.001	0.6052	306.7	[3]
5.25	1.957	0.6383	339.7	[3]
5.50	1.914	0.6663	371.5	[3]
5.75	1.871	0.6890	401.6	[3]

Energy (eV)	n	k	1000 alpha (/cm)	Ref
6.00	1.829	0.7084	430.8	[3]
6.50	1.746	0.7348	484.1	[3]
7.00	1.667	0.7479	530.7	[3]
7.50	1.593	0.7473	568.1	[3]
8.00	1.530	0.7333	594.6	[3]
8.50	1.482	0.7153	616.3	[3]
9.00	1.445	0.7002	638.8	[3]
9.50	1.412	0.6920	666.3	[3]
10.0	1.378	0.6843	693.6	[3]
10.5	1.345	0.6701	713.2	[3]
11.0	1.320	0.6529	728.0	[3]
11.5	1.311	0.6293	733.5	[3]
12.0	1.307	0.6464	786.2	[3]
12.5	1.283	0.6523	826.5	[3]
13.0	1.259	0.6602	869.9	[3]
13.5	1.231	0.6666	912.2	[3]
14.0	1.199	0.6698	950.5	[3]
14.5	1.166	0.6692	983.5	[3]
15.0	1.132	0.6651	1011	[3]
15.5	1.098	0.6566	1032	[3]
16.0	1.066	0.6453	1047	[3]
16.5	1.036	0.6309	1055	[3]
17.0	1.008	0.6147	1059	[3]
17.5	0.9825	0.5961	1057	[3]
18.0	0.9596	0.5771	1053	[3]
18.5	0.9376	0.5578	1046	[3]
19.0	0.9178	0.5362	1033	[3]
19.5	0.9007	0.5140	1016	[3]

Energy (eV)	n	k	1000 alpha (/cm)	Ref
20.0	0.8853	0.4919	997.2	[3]
20.5	0.8721	0.4688	974.1	[3]
21.0	0.8610	0.4456	948.5	[3]
21.5	0.8519	0.4222	920.1	[3]
22.0	0.8454	0.3987	889.1	[3]
22.5	0.8391	0.3761	857.7	[3]
23.0	0.8371	0.3505	817.1	[3]
24.0	0.8444	0.3060	744.4	[3]
25.0	0.8690	0.2717	688.5	[3]

REFERENCES
----

[1] H.R.Philipp [EMIS Datareview RN=16130 (Sept 1987) 'Optical
     properties of amorphous SiO: discussion']
[2] G.Hass, C.D.Salzberg [J. Opt. Soc. Am. (USA) vol.44 no.3 (Mar
     1954) p.181-87]
[3] H.R.Philipp [J. Phys. & Chem. Solids (GB) vol.32 (1971)
     p.1935-45]

28.3 OPTICAL PROPERTIES OF NON-CRYSTALLINE SiO2: DISCUSSION

by H.R.Philipp

August 1987
EMIS Datareview RN=16132

Non-crystalline silicon dioxide, SiO2, is an important material in
integrated circuit technology. In this application it is generally
used in thin-film form and can be prepared by a variety of deposition
techniques including a) chemical vapour deposition using a mixture of
gases containing silicon and oxygen (for example, silane and
oxygen or nitrous oxide), b) sputtering, c) RF glow-discharge
techniques and d) thermal oxidation of Si using wet or dry oxygen.
Bulk silicon dioxide glass (quartz) is made by entirely different
techniques. The optical properties of these materials depend on the
process temperature [1,2], the silicon to oxygen atom ratio [3] and the
presence of impurities such as water or nitrogen [4]. The discussion
below and optical functions given in the table [5] apply, to the
extent possible, to strain-free, pure, non-crystalline silicon dioxide
and care should be taken in applying these results to more arbitrary
silicon-oxide materials.

The optical properties of any material may be described by the complex
quantum energy dependent dielectric constant e = e1 - ie2 which, in
turn, is related to the optical constants n and k by e1 = n*2 - k*2 and
e2 = 2nk where N = n-ik is the complex index of refraction. The
extinction coefficient, k, describes the attenuation of the
electromagnetic wave as it traverses the material and is related to the
absorption coefficient, alpha, by alpha = 4 pik/Y where Y is the
wavelength of the light in vacuum.

The room temperature optical properties of non-crystalline SiO2 have
been the subject of numerous studies; however few attempts have been
made to analyse this information to obtain a set of optical functions,
n and k, for this material especially in the regions of strong
absorption in the infrared and vacuum ultraviolet. This is somewhat
surprising considering the technological importance of this material.
In the region of low absorption, the index of refraction, n, can be
evaluated from prism data and this has been accomplished with great
precision for silicon dioxide. However the measured k values can be
strongly affected by impurity and defect absorption. The presence of
water or OH adsorption in the samples makes the determination of the
intrinsic k values extremely difficult (if not impossible) in certain
parts of the infrared and vacuum ultraviolet spectral regions. The
importance of SiO2 in fibre optic applications may provide a strong
incentive to clean up this material as much as possible. A review of
the optical properties of non-crystalline silicon dioxide has recently
been published [6] and the discussion below reiterates its main
findings.

In the far infrared, 0.0025 to 0.0124 eV, the table [5] values were
obtained from Randall and Rawcliffe [7] who used an interferometric
technique. Their n values appear to be reasonably precise while their
k values may be in error by as much as 30%.

In the infrared, 0.031 to 0.322 eV, the values were taken from Philipp [8]. They were obtained by Kramers-Kronig (K-K) analysis of reflectance data augmented by absorption measurements. His reflectance values are similar to those of Miller [9] but generally slightly lower in absolute magnitude. While Miller [9] also performed a K-K analysis of his data, his computing technique leads to non-physical results in certain parts of the spectrum. Miller [9] correctly points out that on the high energy side of the absorption peak near 0.135 eV, the reflectance shows a deep minimum whose absolute value is difficult to evaluate. Philipp used absorption measurements and a K-K analysis connecting n and alpha to evaluate the reflectance in this region.

In the regions of strong infrared absorption (k > 0.5), the values for n, k and alpha given in the table [5] should be reasonably good (+ or - 15 percent) although there is certainly room for improvement and further optical studies should be undertaken to improve their accuracy. In the region of weak infrared absorption, the k and alpha values determined by K-K analysis are not reliable. A second set of alpha values from absorption data [8] are given in the table [5] for certain energy ranges. They are listed under the heading alpha (abs). While they are perhaps better than the K-K derived values, they should also be treated as unreliable. This is mainly due to the possible presence of impurities and defects in the samples used. In fact no k values are given in the table [5] for infrared energies above 0.16 eV where impurity absorption, especially that associated with OH bonding, can totally mask the intrinsic SiO2 absorption. It should be pointed out however that Drummond [10] has examined the absorption spectrum for silicon dioxide in this energy range and his results show a variety of features which have been duplicated for the most part by Galeener and Lucovsky [11] with alpha roughly 20 percent larger. While these results are not considered further here for the above reasons, they should be examined by anyone interested in the absorption of SiO2 in this region. Some values obtained from [10] are given in [6].

For energies in the range 0.33 to 5.8 eV, the index of refraction of SiO2 has been determined with great precision by Malitson [12] using prism data. His results have been fitted to three and four term Sellmeier equations and the values generated by Brixner [13] are given in the table [5]. It should be pointed out that the very precise Sellmeier-equation fits to index of refraction data are not very reliable for energies outside of the fitted range. This occurs because these fits tend to be purely mathematical and do not necessarily relate to the actual physical absorption processes which determine the spectral dependence of the index of refraction.

No k values are given in the above energy range. For energies below 2 eV the measured absorption is dominated by the presence of OH impurity bands. For energies above 2 eV, the ultimate lower limit to the intrinsic absorption has been estimated by Keck et al [14]; however the reliability of their values appears questionable especially at the higher energies.

For energies above 6 eV, the optical functions given in the table [5] were obtained from a variety of sources. Both Philipp [3] and Lamy [15] have evaluated n and k up to energies of 25 eV from K-K analysis of reflectance data. Lamy's results [15] make use of the reflectance data of Platzoder and Steinmann [16] for energies above 12 eV. For energies less than 11 eV, the n and k values of Philipp and Lamy differ somewhat while for energies above 11 eV, they are in reasonably good agreement.

The decision to use the n values of Philipp [3] rather than those of Lamy [15] for energies below 11 eV was based on the close agreement of Philipp's k values in the region 10 to 11 eV with the results of Weinberg et al [17] who made very careful transmission measurements on thin layers of SiO2 prepared by thermal oxidation of silicon. This agreement was accomplished with a slight modification of Philipp's results as they describe in their paper [17]. The k values given in the table for energies between 9.5 and 11 eV are taken from this work [17]. For energies below 9.5 eV, where absorption in SiO2 is weak, the k values were obtained from a smooth curve drawn through the transmission results of [17], their thickest sample, those of Appleton et al [18] and those of Kaminow et al [19], sample B2. Care should be taken in using the k values below 9.5 eV because they may have a component of impurity absorption, especially OH, and defect absorption in them. For this reason, no k values are given for energies < 7.6 eV.

For energies above 11 eV, where the results of Philipp [3] and Lamy [15] are in reasonably good agreement, the n and k values given in the table [5] were obtained from a smooth curve drawn through both sets of data. The deviations of each set of data points from the smooth curve are generally less than 5 percent. This does not attest to their absolute accuracy however because in both studies the K-K evaluations were based on somewhat arbitrary extrapolations of the reflectance for energies above 25 eV as required by the analysis. In addition the reflectance values themselves in the measured range may be in some error.

REFERENCES
----

[1]   E.A.Taft [J. Electrochem. Soc. (USA) vol.125 no.6 (Jun 1978) p.968-71]
[2]   E.A.Taft [J. Electrochem. Soc. (USA) vol.127 no.4 (Apr 1980) p.993-4]
[3]   H.R.Philipp [J. Phys. & Chem. Solids (GB) vol.32 (1971) p.1935]
[4]   D.M.Brown, P.V.Gray, F.K.Heumann, H.R.Philipp, E.A.Taft [J. Electrochem. Soc. (USA) vol.115 no.3 (Mar 1968) p.311-17]
[5]   H.R.Philipp [EMIS Datareview RN=16133 (Sept 1987) 'Optical functions of non-crystalline SiO2: table (0.0025-25 eV)']
[6]   H.R.Philipp [in 'Handbook of Optical Constants of Solids', Ed. E.D.Palik (Academic Press, New York, 1985) p.749-63]
[7]   C.M.Randall, R.D.Rawcliffe [Appl. Opt. (USA) vol.6 no.11 (Nov 1967) p.1889-95]
[8]   H.R.Philipp [J. Appl. Phys. (USA) vol.50 no.2 (Feb 1979) p.1053-7]
[9]   M.Miller [Czech. J. Phys. Sect. B (Czechoslovakia) vol.18 (1968) p.354]
[10] D.G.Drummond [Proc. R. Soc. London (GB) vol.153 (1935) p.328]
[11] F.L.Daleener, G.Lucovsky [Proc. Int. Conf. Light Scatt. Solids 3rd, Campenas, 1975, Eds M.Balkanski, R.C.C.Leite, S.P.S.Ports (Flammarion Sciences, Paris, France, 1976) p.641]
[12] I.H.Malitson [J. Opt. Soc. Am (USA) vol.55 no.10 (Oct 1965) p.1205-9]
[13] B.Brixner [J. Opt. Soc. Am. (USA) vol.57 no.5 (May 1967) p.674-6]

[14]   D.B.Keck, R.D.Maurer, P.C.Schulty [Appl. Phys. Lett. (USA)
       vol.22 no.7 (Apr 1973) p.307-9]
[15]   P.L.Lamy [Appl. Opt. (USA) vol.16 no.8 (Aug 1977) p.2212-14]
[16]   K.Platzoder, W.Steinmann [J. Opt. Soc. Am. (USA) vol.58 (Apr
       1968) p.588]
[17]   Z.A.Weinberg, G.W.Rubloff, E.Bassous [Phys. Rev. B (USA) vol.19
       no.6 (15 Mar 1979) p.3107-17]
[18]   A.Appleton, T.Chiranjivi, M.Jafripour-Dhayvini [in 'The Physics
       of Silicon Dioxide and its Interfaces', Ed. S.T.Pantelides
       (Pergamon Press, New York, USA, 1978) p.94]
[19]   I.P.Kaminow, B.G.Bagley, C.G.Olson [Appl. Phys. Lett. (USA)
       vol.32 no.2 (15 Jan 1978) p.98-9]

28.4 OPTICAL FUNCTIONS OF NON-CRYSTALLINE SiO2: TABLE (0.0025-25 eV)

by H.R.Philipp

August 1987
EMIS Datareview RN=16133

A set of optical parameters n, k and absorption coefficient, alpha, are
given in TABLE 1 for the energy range 0.0025-25 eV. The energy and
index of refraction n values are given to four places and the
extinction coefficient, k, and alpha values are given to three places
in most of the table entries. The accuracy of these values does not
warrant this precision [1]. This is done to show the possible presence
of weak structure in the optical properties that can often be discerned
by comparison of precise, but not necessarily accurate, values over
small energy ranges. On the other hand, the very accurate energy and n
values of Brixner [2] are given to five and six places respectively in
the table. For certain energies in the infrared, a second set of
absorption coefficient values obtained from transmission measurements
are also given and are listed under the heading alpha (abs). They are
not very accurate but probably better than the other alpha values
given [1]. It should also be pointed out that the table entries apply
to pure, strain-free, non-crystalline silicon dioxide. Care should be
taken in applying them to more arbitrary silicon-oxide materials.

TABLE 1: optical functions for noncrystalline silicon dioxide at room
temperature for selected energy values

n is the refractive index, k is the extinction coefficient and alpha
is the absorption coefficient

Wavelength in A = approx. 12395/(energy in eV).

Energy (eV)	n	k	1000 alpha (/cm)	1000 alpha (abs) (/cm)	Ref
0.002480	1.955	0.00796	0.00200		[3]
0.004959	1.957	0.00696	0.00350		[3]
0.007439	1.959	0.00862	0.00650		[3]
0.009919	1.962	0.0119	0.0120		[3]
0.01240	1.967	0.0159	0.0120		[3]
0.03100	2.100			0.14	[4]
0.03410	2.147			0.19	[4]
0.03720	2.210			0.25	[4]
0.04030	2.284			0.38	[4]
0.04339	2.388			0.57	[4]

Energy (eV)	n	k	1000 alpha (/cm)	1000 alpha (abs) (/cm)	Ref
0.04649	2.537	0.199	0.938	0.94	[4]
0.04959	2.739	0.397	1.20		[4]
0.05207	2.912	0.738	3.90		[4]
0.05455	2.936	1.29	7.13		[4]
0.05703	2.308	2.29	13.2		[4]
0.05827	1.484	2.39	14.1		[4]
0.05889	1.002	2.22	13.3		[4]
0.05951	0.7517	1.86	11.2		[4]
0.06075	0.5777	1.28	7.88		[4]
0.06199	0.6616	0.822	5.16		[4]
0.06323	0.8857	0.524	3.36		[4]
0.06447	1.050	0.415	2.71	2.1	[4]
0.06534	1.161	0.377	2.50	1.7	[4]
0.06695	1.235	0.341	2.31	1.4	[4]
0.06943	1.337	0.298	2.10	1.3	[4]
0.07191	1.401	0.264	1.92	1.2	[4]
0.07439	1.450	0.235	1.77	1.1	[4]
0.07749	1.502	0.202	1.59	0.86	[4]
0.08059	1.555	0.182	1.49	0.72	[4]
0.08369	1.598	0.168	1.43	0.65	[4]
0.08679	1.643	0.157	1.38	0.63	[4]
0.08989	1.698	0.157	1.43	0.68	[4]
0.09299	1.756	0.177	1.67	0.75	[4]
0.09423	1.779	0.192	1.83	0.81	[4]
0.09547	1.810	0.227	2.20	1.1	[4]
0.09671	1.811	0.275	2.70	1.6	[4]
0.09795	1.789	0.314	3.12		[4]
0.09919	1.753	0.343	3.45		[4]

Energy (eV)	n	k	1000 alpha (/cm)	1000 alpha (abs) (/cm)	Ref
0.1004	1.701	0.341	3.47		[4]
0.1017	1.658	0.323	3.33		[4]
0.1029	1.615	0.267	2.78		[4]
0.1041	1.619	0.204	2.15	1.8	[4]
0.1054	1.652	0.152	1.62	1.3	[4]
0.1066	1.690	0.116	1.25	0.83	[4]
0.1091	1.784	0.0775	0.857	0.29	[4]
0.1116	1.869	0.0506	0.572	0.25	[4]
0.1147	2.038	0.0460	0.535	0.27	[4]
0.1178	2.224	0.102	1.22	0.37	[4]
0.1209	2.448	0.231	2.83	0.83	[4]
0.1240	2.694	0.509	6.40		[4]
0.1271	2.839	0.962	12.4		[4]
0.1302	2.760	1.65	21.8		[4]
0.1333	2.250	2.26	30.5		[4]
0.1351	1.616	2.63	36.0		[4]
0.1364	1.043	2.55	35.3		[4]
0.1376	0.5846	2.27	31.7		[4]
0.1389	0.3705	1.85	25.8		[4]
0.1401	0.3563	1.53	21.7		[4]
0.1413	0.3915	1.32	18.9		[4]
0.1426	0.4309	1.17	16.9		[4]
0.1438	0.4563	1.07	15.6		[4]
0.1451	0.4656	0.978	14.4		[4]
0.1463	0.4746	0.903	13.4		[4]
0.1475	0.4730	0.840	12.6		[4]
0.1488	0.4600	0.771	11.6		[4]
0.1500	0.4530	0.704	10.7		[4]

Energy (eV)	n	k	1000 alpha (/cm)	1000 alpha (abs) (/cm)	Ref
0.1513	0.4329	0.635	9.74		[4]
0.1525	0.4020	0.553	8.55		[4]
0.1537	0.3931	0.446	6.95		[4]
0.1550	0.4113	0.323	5.07		[4]
0.1562	0.4677	0.216	3.42		[4]
0.1575	0.5456	0.132	2.11		[4]
0.1587	0.6232	0.0768	1.24	1.2	[4]
0.1599	0.7037	0.0474	0.768	0.84	[4]
0.1612	0.7719	0.0372	0.608	0.64	[4]
0.1624	0.8213				[4]
0.1637	0.8600				[4]
0.1649	0.8897				[4]
0.1661	0.9175				[4]
0.1674	0.9488				[4]
0.1686	0.9702				[4]
0.1711	1.014				[4]
0.1736	1.053				[4]
0.1761	1.084				[4]
0.1785	1.107				[4]
0.1810	1.135				[4]
0.1835	1.158				[4]
0.1860	1.175				[4]
0.1922	1.212				[4]
0.1984	1.239				[4]
0.2108	1.278				[4]
0.2232	1.306				[4]
0.2480	1.342				[4]
0.2728	1.365				[4]

Energy (eV)	n	k	1000 alpha (/cm)	1000 alpha (abs) (/cm)	Ref
0.2976	1.383				[4]
0.3224	1.395				[4]
0.33449	1.39936				[2]
0.34863	1.40418				[2]
0.35354	1.40568				[2]
0.37542	1.41155				[2]
0.38221	1.41314				[2]
0.53317	1.43292				[2]
0.60243	1.43722				[2]
0.68384	1.44069				[2]
0.73225	1.44226				[2]
0.74663	1.44267				[2]
0.84372	1.44497				[2]
0.91018	1.44621				[2]
1.1449	1.44941				[2]
1.3863	1.45185				[2]
1.7549	1.45515				[2]
1.8892	1.45637				[2]
2.1041	1.45841				[2]
2.1411	1.45877				[2]
2.2705	1.46008				[2]
2.5504	1.46313				[2]
2.8448	1.46669				[2]
3.3967	1.47453				[2]
3.5770	1.47746				[2]
3.7105	1.47976				[2]
4.1034	1.48719				[2]
4.2848	1.49099				[2]

Energy (eV)	n	k	1000 alpha (/cm)	1000 alpha (abs) (/cm)	Ref
4.5040	1.49592				[2]
4.6751	1.50004				[2]
5.1674	1.51338				[2]
5.3858	1.52009				[2]
5.4680	1.52276				[2]
5.7976	1.53429				[2]
6.000	1.543				[5]
6.250	1.554				[5]
6.500	1.567				[5]
6.750	1.582				[5]
7.000	1.600				[5]
7.200	1.616				[5]
7.400	1.633				[5]
7.600	1.653				[5]
7.800	1.676	0.0000047	0.00372		[5-8]
8.000	1.702	0.000032	0.0259		[5-8]
8.100	1.716	0.000122	0.100		[5-8]
8.200	1.730	0.000463	0.524		[5-8]
8.300	1.747	0.00140	1.18		[5-8]
8.400	1.764	0.00317	2.70		[5-8]
8.500	1.783	0.00557	4.80		[5-8]
8.600	1.803	0.00838	7.30		[5-8]
8.700	1.825	0.0109	9.61		[5-8]
8.800	1.850	0.0132	11.8		[5-8]
8.900	1.876	0.0156	14.1		[5-8]
9.000	1.904	0.0189	17.2		[5-8]
9.100	1.935	0.0228	21.0		[5-8]
9.200	1.969	0.0271	25.3		[5-8]

Energy (eV)	n	k	1000 alpha (/cm)	1000 alpha (abs) (/cm)	Ref
9.300	2.006	0.0339	32.0		[5-8]
9.400	2.047	0.0430	41.0		[5-8]
9.500	2.092	0.0561	54.1		[5,6]
9.600	2.140	0.0770	74.9		[5,6]
9.700	2.190	0.119	117		[5,6]
9.800	2.243	0.168	167		[5,6]
9.900	2.292	0.236	237		[5,6]
10.00	2.330	0.323	327		[5,6]
10.10	2.332	0.460	471		[5,6]
10.20	2.240	0.715	775		[5,6]
10.25	2.152	0.810	842		[5,6]
10.30	2.048	0.925	966		[5,6]
10.35	1.919	1.045	1100		[5,6]
10.40	1.772	1.13	1190		[5,6]
10.45	1.645	1.136	1200		[5,6]
10.50	1.567	1.11	1180		[5,6]
10.60	1.492	0.914	982		[5,6]
10.70	1.513	0.725	786		[5,6]
10.80	1.587	0.618	677		[5,6]
10.90	1.687	0.565	624		[5,6]
11.00	1.739	0.569	634		[5,6]
11.20	1.766	0.718	815		[5,9]
11.40	1.716	0.810	936		[5,9]
11.60	1.635	0.859	1010		[5,9]
11.80	1.554	0.874	1050		[5,9]
12.00	1.475	0.861	1050		[5,9]
12.25	1.410	0.824	1020		[5,9]
12.50	1.383	0.793	1000		[5,9]

Energy (eV)	n	k	1000 alpha (/cm)	1000 alpha (abs) (/cm)	Ref
12.75	1.372	0.766	990		[5,9]
13.00	1.368	0.747	984		[5,9]
13.25	1.371	0.755	1010		[5,9]
13.50	1.363	0.775	1060		[5,9]
13.75	1.320	0.795	1110		[5,9]
14.00	1.265	0.808	1150		[5,9]
14.25	1.225	0.799	1150		[5,9]
14.50	1.195	0.771	1130		[5,9]
14.75	1.175	0.739	1100		[5,9]
15.00	1.168	0.711	1080		[5,9]
15.25	1.167	0.699	1080		[5,9]
15.50	1.172	0.696	1090		[5,9]
15.75	1.178	0.703	1120		[5,9]
16.00	1.172	0.717	1160		[5,9]
16.25	1.156	0.737	1210		[5,9]
16.50	1.137	0.755	1260		[5,9]
16.75	1.124	0.765	1300		[5,9]
17.00	1.072	0.768	1320		[5,9]
17.25	1.030	0.763	1330		[5,9]
17.50	0.999	0.750	1330		[5,9]
17.75	0.975	0.731	1320		[5,9]
18.00	0.957	0.712	1300		[5,9]
18.50	0.927	0.677	1270		[5,9]
19.00	0.902	0.645	1240		[5,9]
19.50	0.879	0.613	1210		[5,9]
20.00	0.859	0.585	1190		[5,9]
21.00	0.827	0.530	1130		[5,9]

Energy (eV)	n	k	1000 alpha (/cm)	1000 alpha (abs) (/cm)	Ref
22.00	0.797	0.480	1070		[5,9]
23.00	0.774	0.434	1010		[5,9]
24.00	0.753	0.375	912		[5,9]
25.00	0.733	0.325	824		[5,9]

REFERENCES
----

[1] H.R.Philipp [EMIS Datareview RN=16132 (Sept 1987) 'Optical properties of non-crystalline SiO2: discussion']
[2] B.Brixner [J. Opt. Soc. Am. (USA) vol.57 no.5 (May 1967) p.674-6]
[3] C.M.Randall, R.D.Rawcliffe [Appl. Opt. (USA) vol.6 no.11 (Nov 1967) p.1889-95]
[4] H.R.Philipp [J. Appl. Phys. (USA) vol.50 no.2 (Feb 1979) p.1053-7]
[5] H.R.Philipp [J. Phys. & Chem. Solids (GB) vol.32 (1971) p.1935]
[6] Z.A.Weinberg, G.W.Rubloff, E.Bassous [Phys. Rev. B (USA) vol.19 (15 Mar 1979) p.3107]
[7] A.Appleton, T.Chiranjivi, M.Jafaripour-Dhayvini [in 'The Physics of Silicon Dioxide and its Interfaces', Ed. S.T.Pantelides (Pergamon Press, New York, USA, 1978) p.94]
[8] I.P.Kaminow, B.G.Bagley, C.G.Olson [Appl. Phys. Lett. (USA) vol.32 no.2 (15 Jan 1978) p.98-9]
[9] P.L.Lamy [Appl. Opt. (USA) vol.16 no.8 (Aug 1977) p.2212-14]

CHAPTER 29

SILICON NITRIDE: OPTICAL FUNCTIONS

29.1 OPTICAL PROPERTIES OF AMORPHOUS SiN4: DISCUSSION

by H.R.Philipp

September 1987
EMIS Datareview RN=16134

Non-crystalline silicon nitride, Si3N4, is an important material in
integrated circuit technology. In this application it is used in
thin-film form and can be prepared by a variety of deposition
techniques, including (a) chemical vapour deposition from a mixture of
gases containing silicon and nitrogen (for example, SiH4 and NH3),
(b) sputtering, and (c) RF glow-discharge methods. Silicon oxynitride,
SiO(x)N(y), films can also be readily formed by adding small amounts of
NO or O2 to the reactive gases. The optical properties of these films
are strongly dependent on the deposition temperature and the Si-to-N
and Si-to-O atom ratios. Other studies also indicate the presence of
chemically bound hydrogen in the films. Thus, the stoichiometry of an
arbitrary film can be more accurately indicated by SiO(x)N(y)H(z), and
the refractive index and other optical properties will be a function of
x, y and z. Samples prepared by high-temperature pyrolysis are generally
considered to have, or come close to, the ideal stoichiometry, Si3N4.
The term 'ideal', as used here, refers only to stoichiometry and not to
the actual properties of the films nor to their technological use. The
optical functions given in the table [1] pertain to this ideal
stoichiometry and care should be taken in applying them to films
containing oxygen or hydrogen or to films having reduced density or
index of refraction. For completeness, references are included below
which also describe the optical behaviour of non-ideal silicon nitride
materials.

The optical properties of any material may be described by the complex,
quantum energy dependent dielectric constant e = e1 - ie2 which, in
turn, is related to the optical constants n and k by e1 = n*2 - k*2 and
e2 = 2nk where N = n - ik is the complex index of refraction. The
extinction coefficient k describes the attenuation of the
electromagnetic wave as it traverses the material and is related to the
absorption coefficient, alpha, by alpha = 4 pi k/Y where Y is the
wavelength of the light in vacuum.

While there have been a number of studies on the optical properties of
silicon nitride materials, the data have generally been limited to the
evaluation of the index of refraction in the visible region of the
spectrum [2-17] and to infrared measurements of the lattice absorption
bands near 0.11 eV and other bands, particularly those which show up
N-H and Si-H bonding [2-8,10,13,16,18-25]. The infrared measurements
have been used mainly for structural and chemical evaluation purposes.
Taft [13] measured the absorption coefficient of silicon nitride films
prepared by several different techniques in the region 0.05 to 0.18 eV
where the infrared absorption is strongest. Eriksson and Granqvist [25]
evaluated the complex dielectric constant, from which n, k and and alpha
can be determined as described above, for a number of silicon oxynitride
films of different O-to-N atom ratios for the energy range 0.025 to 0.5
eV. While these results are not included in the table [1] because of the
inclusion of a measurable amount of oxygen in their films, the data for
their film of stoichiometry SiO(0.25)N(1.52) when combined with Taft's
[13] alpha values for pyrolytic films can probably be used to obtain

useful, although approximate, values for the optical functions in the infrared. Unfortunately no attempt has been made to determine a set of optical parameters in the infrared for the stoichiometric material, Si3N4.

A review of the optical properties of silicon nitride in the region of strong electronic absorption has recently been published [26] and the discussion below reiterates its findings. Both Philipp [27] and Bauer [28] evaluated n and k from the visible to the vacuum ultraviolet regon. The results of Philipp [27], which extend to 24 eV, were determined by Kramers-Kronig analysis of reflectance and absorption data. The films he used were prepared by pyrolytic decomposition at 1000 deg C of a mixture of SiH4 and NH3 in the ratio 1 to 40,000. Bauer's [28] measurements extend to only 7.5 eV but were obtained on samples prepared by a variety of deposition techniques including pyrolysis of a mixture of SiCl4 and NH3 gases at substrate temperatures in the range 720 to 1000 deg C. His n and k values were obtained by error-function analyses of reflectance and transmission data. For pyrolytic silicon nitride films, the results of Philipp [27] and Bauer [28] show reasonable but not precise agreement.

The set of optical parameters n and k for silicon nitride given in the table [1] are taken entirely from the work of Philipp [27]. These results are not necessarily more accurate than those of Bauer [28] but were primarily chosen because they cover the widest energy range. In addition, Theeten et al [29] in a study of silicon nitride in the region from 1.5 to 5.8 eV using spectroscopic ellipsometry found good agreement with the values of Philipp [27] and it is in this energy range that the largest differences occur between the results of Philipp [27] and Bauer [28].

REFERENCES
----

[1]   H.R.Philipp [EMIS Datareview RN=16135 (Sept 1987) 'Optical functions of amorphous Si3N4: table (1-24 eV)']
[2]   V.Doo, D.Nichols, G.Silvey [J. Electrochem. Soc. (USA) vol.113 no.12 (Dec 1966) p.1279-81]
[3]   K.Bean, P.Gleim, R.Yeakley, W.Runyan [J. Electrochem. Soc. (USA) vol.114 no.7 (July 1967) p.733-7]
[4]   T.Chu, C.Lee, G.Gruber [J. Electrochem. Soc. (USA) vol.114 no.7 (July 1967) p.717-22]
[5]   V.Doo, D.Kern, D.Nichols [J. Electrochem. Soc. (USA) vol.115 (1968) p.61]
[6]   D.Brown, P.Gray, F.Heumann, H.Philipp, E.Taft [J. Electrochem. Soc. (USA) vol.115 no.3 (Mar 1968) p.311-17]
[7]   M.Grieco, F.Worthing, B.Schwartz [J. Electrochem. Soc. (USA) vol.115 (1968) p.525]
[8]   R.Levitt, W.Zwicker [J. Electrochem. Soc. (USA) vol.114 no.11 (Nov 1967) p.1192-3]
[9]   B.Deal, P.Flemming, P.Castro [J. Electrochem. Soc. (USA) vol.115 (1968) p.300]
[10]   G.Brown, W.Robinette,Jr, H.Carlson [J. Electrochem. Soc. (USA) vol.115 (1968) p.948]
[11]   E.V.Shitova, I.A.Yasnova, N.A.Genkina [Opt. & Spectrosc. (USA) vol.43 (1977) p.140]

[12] Y.N.Volgin, O.P.Borisov, Y.I.Ukhanov, N.I.Sukhanova [J. Appl. Spectrosc. (USA) vol.24 (1976) p.115]

[13] E.A.Taft [J. Electrochem. Soc. (USA) vol.118 no.8 (Aug 1971) p.1341-6]

[14] M.J.Rand, D.R.Wonsidler [J. Electrochem. Soc. (USA) vol.125 no.1 (Jan 1978) p.99-101]

[15] T.Wittberg, J.Hoenigman, W.Moddeman, C.Cothern, M.Gulett [J. Vac. Sci. & Technol. (USA) vol.15 no.2 (Mar/Apr 1978) p.348-52]

[16] D.Schalch, A.Scharmann, R.Wolfrat [Thin Solid Films (Switzerland) vol.124 no.3/4 (22 Feb 1985) p.301-8]

[17] H.Watanabe, K.Katoh, S.Imagi [Thin Solid Films (Switzerland) vol.136 no.1 (1 Feb 1986) p.77-83]

[18] S.Yoshioka, S.Takayanagi [J. Electrochem. Soc. (USA) vol.114 no.9 (Sep 1967) p.962-4]

[19] H.J.Stein [Appl. Phys. Lett. (USA) vol.32 no.6 (15 Mar 1978) p.379-82]

[20] S.M.Hu [J. Electrochem. Soc. (USA) vol.113 no.7 (Jul 1966) p.693-8]

[21] H.J.Stein, H.A.R.Wagener [J. Electrochem. Soc. (USA) vol.124 no.6 (June 1977) p.908-12]

[22] P.S.Peercy, H.J.Stein, B.L.Doyle, S.T.Picraux [J. Electron. Mater. (USA) vol.8 no.1 (Jan 1979) p.11-24]

[23] L.F.Cordes [Appl. Phys. Lett. (USA) vol.11 no.12 (15 Dec 1967) p.383-5]

[24] W.R.Knolle, J.W.Osenbach [J. Appl. Phys. (USA) vol.58 no.3 (1 Aug 1985) p.1248-54]

[25] T.S.Eriksson, C.G.Granqvist [J. Appl. Phys. (USA) vol.60 no.6 (15 Sep 1986) p.2081-91]

[26] H.R.Philipp [in 'Handbook of Optical Constants of Solids' Ed. E.D.Palik (Academic Press, New York, USA, 1985) p.771-4]

[27] H.R.Philipp [J. Electrochem. Soc. (USA) vol.120 no.2 (Feb 1973) p.295-300]

[28] J.Bauer [Phys. Status Solidi a (Germany) vol.39 (1977) p.411]

[29] J.B.Theeten, D.E.Aspnes, F.Simondet, M.Errman, P.C.Murau [J. Appl. Phys. (USA) vol.52 no.11 (Nov 1981) p.6788-97]

29.2 OPTICAL FUNCTIONS OF AMORPHOUS Si3N4: TABLE (1-24 eV)

by H.R.Philipp

September 1987
EMIS Datareview RN=16135

A set of optical parameters n, k and absorption coefficient, alpha, are
given in TABLE 1 for the energy range 1-24 eV. The table entries are
given to four places in some cases. The accuracy of the values does not
warrant this precision [1]. This is done to show the possible presence
of weak structure in the optical properties that can often be
discussed by comparison of precise, but not necessarily accurate,
values over small energy ranges. The values apply to materials of
stoichiometry Si3N4. Care should be taken in applying them to more
arbitrary silicon-nitride materials and to those containing oxygen.

TABLE 1: optical functions for amorphous silicon nitride at room
temperature for selected energy values.

n is the refractive index, k is the extinction coefficient and alpha
is the absorption coefficient.

Wavelength in A = approx. 12395/(energy in eV)

Energy (eV)	n	k	1000 alpha (/cm)	Ref
1.00	1.998			[2]
1.50	2.008			[2]
2.00	2.022			[2]
2.50	2.041			[2]
3.00	2.066			[2]
3.50	2.099			[2]
4.00	2.141			[2]
4.25	2.167			[2]
4.50	2.198	0.00022	0.100	[2]
4.75	2.234	0.0012	0.547	[2]
5.00	2.278	0.0049	2.48	[2]
5.25	2.331	0.011	5.85	[2]
5.50	2.393	0.029	16.2	[2]
5.75	2.464	0.057	31.8	[2]
6.00	2.541	0.102	62.0	[2]

Energy (eV)	n	k	1000 alpha (/cm)	Ref
6.25	2.620	0.174	110	[2]
6.50	2.682	0.273	180	[2]
6.75	2.724	0.380	260	[2]
7.00	2.752	0.493	350	[2]
7.25	2.766	0.612	450	[2]
7.50	2.753	0.750	570	[2]
7.75	2.711	0.866	680	[2]
8.00	2.651	0.962	780	[2]
8.50	2.492	1.16	999	[2]
9.00	2.326	1.32	1200	[2]
9.50	2.162	1.44	1390	[2]
10.00	2.000	1.49	1510	[2]
10.50	1.827	1.53	1630	[2]
11.00	1.657	1.52	1690	[2]
12.00	1.417	1.43	1740	[2]
13.00	1.247	1.35	1780	[2]
14.00	1.111	1.26	1790	[2]
15.00	1.001	1.18	1790	[2]
16.00	0.902	1.11	1800	[2]
17.00	0.810	1.03	1770	[2]
18.00	0.735	0.936	1710	[2]
19.00	0.676	0.841	1620	[2]
20.00	0.635	0.743	1510	[2]
21.00	0.617	0.647	1380	[2]
22.00	0.611	0.560	1250	[2]
23.00	0.625	0.481	1120	[2]
24.00	0.655	0.420	1020	[2]

REFERENCES
----

[1] H.R.Philipp [EMIS Datareview RN=16134 (Sept 1987) 'Optical
     properties of amorphous Si3N4: discussion']
[2] H.R.Philipp [J. Electrochem. Soc. (USA) vol.120 no.2 (Feb 1973)
     p.295-300]

CHAPTER 30

METAL SILICIDES: MISCELLANEOUS ASPECTS

30.1 GROWTH OF Si/CoSi2/Si HETEROSTRUCTURES

by B.D.Hunt

October 1987
EMIS Datareview RN=16176

A. INTRODUCTION

In recent years it has been found that both CoSi2 [1-5] and NiSi2 [6,7]
can be grown epitaxially on Si. High quality epitaxy is possible
because both silicides have a cubic, CaF2-type structure with close
lattice matches to Si (-0.4% for NiSi2 and -1.2% for CoSi2 at room
temperature). It has also been shown that growth techniques exist
which allow the production of high quality epitaxial Si/CoSi2/Si [4,8,9]
and Si/NiSi2/Si [10,11] heterostructures. Because both CoSi2 and NiSi2
are metals, such heterostructures are relevant to the formation of
buried interconnects for high speed and VLSI applications. Si/silicide/Si
structures are also important for fabrication of novel devices such as
the metal base transistor (MBT) or the permeable base transistor (PBT).
Although much work has been devoted to NiSi2/Si heterostructures, this
Datareview focuses on epitaxial CoSi2, because of its greater thermal
stability and lower resistivity (approximately 15 micro-ohm cm at room
temperature).

B. CoSi2/Si(111) HETEROSTRUCTURES

Most growth studies of CoSi2 have utilized the Si(111) surface rather
than the (100) surface, because surface free energy considerations
indicate that islanding is less likely with the (111) orientation [12].
There is one recent report by Cheng et al of single-crystal CoSi2 films
produced using a non-ultrahigh vacuum (UHV) technique [13], but high
quality single crystal growth generally requires UHV processing.
Typically Si substrates are chemically precleaned followed by an in-situ
cleaning procedure. Epitaxial CoSi2 films can be grown by a variety of
techniques, including solid phase epitaxy (SPE), reactive deposition
epitaxy (RDE), and molecular beam epitaxy (MBE). The SPE process
consists of a room temperature deposition of a thin Co film followed by
a high temperature reaction step (550-700 deg C). Films grown by RDE
are produced by the deposition of pure Co at elevated temperatures
(600-700 deg C). Both SPE and RDE are characterized by reaction into
the substrate with a final silicide thickness 3.5 times the total Co
thickness. The MBE growth method involves the codeposition of Co and Si
at elevated temperatures, generally at close to a stoichiometric ratio.

Hunt et al [4] have investigated the surface morphologies of very thin
(70-175 A) CoSi2 films produced using these techniques. They find
that in all cases the silicide films form with a high density of
pinholes (> 10*8/cm*2). This result is consistent with the findings
of Tung et al [8], although lower-pinhole-density CoSi2 layers grown
using similar techniques have been reported by Rosencher and coworkers
[9]. Pinhole formation is believed to be driven by the high (111)
surface-free-energy (SFE) of CoSi2 relative to the Si(111) SFE [12].
Very recently, Hunt et al [5], and Tung et al [8] have found the pinhole
density in thin, epitaxial CoSi2 films can be controlled using a

modified SPE growth technique. This method utilizes the room temperature deposition of a thin Co layer followed by a Si cap layer. The bilayer sandwich is then annealed to 600-650 deg C to form an epitaxial CoSi2 film. It is found that the silicide pinhole density and size distribution are strong functions of the Si overlayer thickness. Using this technique, the pinhole density can be decreased to < 10*6/cm*2 or increased to > 10*9/cm*2. This finding has important applications in MBT and PBT device fabrication. Other workers have recently discovered that the CoSi2 pinhole density can be decreased still further by using SPE of room-temperature codeposited Co/Si layers with [14] or without [18] a Si cap layer.

C. Si/CoSi2/Si HETEROSTRUCTURES

A number of groups have investigated Si overgrowth of thin, epitaxial CoSi2 layers [4,8,9,15,18]. The growth process involves the deposition of pure Si at elevated substrate temperatures immediately following CoSi2 epitaxy. Channelling minimum yield measurements indicate that the Si epitaxial quality improves as the Si growth temperature is raised from 550 to 800 deg C. However, the improvement in epitaxial quality is accompanied by an increase in Si surface roughness, as well as an increase in the silicide average pinhole size and density by a factor of approximately four times [4]. For most practical applications it appears that an intermediate growth temperature of 700 deg C produces sufficiently high quality Si/CoSi2/Si heterostructures. At this growth temperature, there are few defects in the Si overlayer and the silicide/Si interfaces are flat and uniform. The Si epilayer surface is still somewhat nonuniform, but this is not a serious problem for relatively thick Si overlayers (> 1000 A). For thinner Si epilayers, however, particular applications may demand more uniform Si growth, so it is important to understand the roughening process.

The evidence indicates that the roughening of the Si surface is caused by islanding of the Si epilayer during growth [4,15,16]. Dark-field, cross-sectional TEM studies show that the Si islands generally have either the type-A or type-B orientations [4]. Here the type-A orientation refers to Si overlayers having the same orientation as the underlying CoSi2, while for the type-B orientation the Si epilayer shares the (111) surface normal with the CoSi2, but is rotated 180 deg about this axis relative to the silicide. The CoSi2 layers are usually of the type-B orientation with respect to the Si substrate, although some fraction of the films can be type-A for different processing conditions [7]. Recently it has been found that a 20-50 A Si layer deposited over the CoSi2 near room temperature prevents Si islanding during subsequent Si growth at higher temperatures (590-640 deg C) [15]. A similar technique has also been reported to control the orientation of Si on CoSi2 [7,8]. This predeposition growth process allows the production of very thin, uniform Si epilayers on CoSi2. Such epilayers are essential for the production of CoSi2/Si multilayers (see Datareview [19]) and are also relevant to the optimization of MBT emitter junctions.

Work has also been reported on high dose ion implantation of Co and annealing to form buried CoSi2 layers [17]. Using this technique, White and coworkers have been able to produce single-crystal Si/CoSi2/Si structures with approximately 600 A Si overlayers and 1000 A CoSi2 layers on both (111) and (100)-oriented Si wafers. The best results are obtained with a 350 deg C implant temperature and a 1000 deg C anneal for 1/2-1 hr. Channelling minimum yield measurements show that the silicide layers exhibit reasonable crystal quality, and electrical transport results indicate that these films have resistivity ratios better than those of CoSi2 films grown by UHV deposition techniques. At this time it is not known whether this technique can produce the very thin (< 200 A) CoSi2 layers required for some applications.

REFERENCES
----

[1]   S.Saitoh, I.Ishiwara, S.Furukawa [Appl. Phys. Lett. (USA) vol.37 no.2 (15 July 1980) p.203-5]
[2]   J.C.Bean, J.M.Poate [Appl. Phys. Lett. (USA) vol.37 no.7 (1 Oct 1980) p.643-6]
[3]   R.T.Tung, J.C.Bean, J.M.Gibson, J.M.Poate, D.C.Jacobson [Appl. Phys. Lett. (USA) vol.40 no.8 (15 Apr 1982) p.684-6]
[4]   B.D.Hunt et al [Mater. Res. Soc. Symp. Proc. (USA) vol.56 (1986) p.151-6]
[5]   B.D.Hunt, N.Lewis, E.L.Hall, C.D.Robertson [J. Vac. Sci. & Technol. B (USA) vol.5 no.3 (May/June 1987) p.749-50]
[6]   R.T.Tung, J.M.Gibson, J.M.Poate [Phys. Rev. Lett. (USA) vol.50 no.6 (7 Feb 1983) p.429-32]
[7]   R.T.Tung, A.F.J.Levi, J.M.Gibson [J. Vac. Sci. & Technol. B (USA) vol.4 no.6 (Nov/Dec 1986) p.1435-43]
[8]   R.T.Tung, A.F.J.Levi, J.M.Gibson [Appl. Phys. Lett. (USA) vol.48 no.10 (10 Mar 1986) p.635-7]
[9]   E.Rosencher, P.A.Badoz, J.C.Pfister, F.Arnaud d'Avitaya, G.Vincent, S.Delage [Appl. Phys. Lett. (USA) vol.49 no.5 (4 Aug 1986) p.271-3]
[10]   A.Ishizaka, P.A.Cullen, Y.Shiraki [Extended Abstracts 16th Int. Conf. Solid State Devices and Materials, (Kobe, Japan, 1984) p.39]
[11]   R.T.Tung, J.M.Gibson, A.F.J.Levi [Appl. Phys. Lett. (USA) vol.48 no.19 (12 May 1986) p.1264-6]
[12]   K.Ishibashi, S.Furukawa [Jpn. J. Appl. Phys. (Japan) vol.24 no.8 (Aug 1985) p.912-17]
[13]   H.C.Cheng, I.C.Wu, L.J.Chen [Appl. Phys. Lett. (USA) vol.50 no.4 (26 Jan 1987) p.174-6]
[14]   T.L.Lin, R.W.Fathauer, P.J.Grunthaner, C.d'Anterroches [submitted to Appl. Phys. Lett. (USA) (1987)]
[15]   B.D.Hunt, N.Lewis, L.J.Schowalter, E.L.Hall, L.G.Turner [Mater. Res. Soc. Symp. Proc. (USA) vol.77 (1987) p.351]
[16]   B.M.Ditcheck, J.P.Salerno, J.V.Gormley [Appl. Phys. Lett. (USA) vol.47 no.11 (1 Dec 1985) p.1200-2]
[17]   A.E.White, K.T.Short, R.C.Dynes, J.P.Garno, J.M.Gibson [Appl. Phys. Lett. (USA) vol.50 no.2 (12 Jan 1987) p.95-7]
[18]   J.Henz, M.Ospelt, H.vonKanel [Solid State Commun. (USA) vol.63 no.6 (1987) p.445-50]
[19]   B.D.Hunt [EMIS Datareview RN=16175 (Oct 1987) 'Growth of epitaxial CoSi2/Si multilayers']

30.2 FORMATION TEMPERATURE OF TiSi2

by P.J.Rosser

November 1986
EMIS Datareview RN=12632

TABLE 1 gives values of the formation temperature of titanium disilicide formed using a variety of deposition and anneal systems.

TABLE 1

Method	Formation Temperature deg C	Activation Energy eV	Ref
annealing COSPUTTERED Ti+Si -			
in a furnace for 30 minutes (in hydrogen)	600	1.8	[1]
in a furnace (in hydrogen and vacuum)	600	1.8	[2]
in a furnace (in nitrogen)	700		
in a rapid thermal anneal system (10 seconds in nitrogen)	900		[3]
annealing EVAPORATED Ti/Si -			
in a furnace	525		[4]
	550	1.8	[5]
Ti and Si deposited in same system	450	1.8	[4]
Rapid thermal annealing (10 sec vac.)	750		[6]
Ar	700-750	2.6	[7]
Ar	750-800	1.5	[7]
annealing SPUTTERED Ti/Si -			
in a furnace	600		[4]
vac.	600	2.17	[8]
in a RTA system	700		[9]
Sputtered at an elevated substrate temperature	450		[10]
Alloy sputtering of Ti+Si at an elevated substrate temperature	200		[10]
annealing PECVD TiSi2 (> 5 min)	650		[11]
LPCVD at 600 deg C	(600)		[12]
Laser CVD at 400 deg C	(400)		[13]
+ an anneal	800		[13]

This table can only serve as an indication of the minimum temperatures at which the disilicide will be formed. It is particularly noteworthy that the minimum formation temperature decreases with improved film quality, and that many groups emphasise the importance of minimising oxygen contamination. It is also worth mentioning that the definition of disilicide formation varies from group to group. For many groups the chosen criterion is that the resistivity of the film be low and stable. This is relevant for many process applications, but very often the disilicide is formed (as shown by RBS) significantly before this. The discrepancy is a result of increased resistivity caused by grain boundary scattering and the presence of both the titanium disilicide structures, i.e. equilibrium (C54, orthorhombic, ABCDA stacking of distorted hexagonal planes [18]) structure [18] and the intermediate (C49, orthorhombic ZrSi2) structure [14,18].

The lowest formation temperatures have been reported by those groups best able to provide a clean interface between the titanium and silicon. If a native oxide is present on the silicon then the observed formation temperature is strongly dependent on the thickness of this oxide and is unlikely to be less than 600 deg C.

Finally the effect of dopant in the silicon should be considered. This can of course result in a thicker native oxide in which case the above comments apply, but several groups have reported that high doping levels significantly retard silicide formation [15-17].

REFERENCES
----

[1] S.P.Murarka, D.B.Fraser [J. Appl. Phys. (USA) vol.51 (1980) p.350]

[2] S.P.Murarka, D.B.Fraser [J. Appl. Phys. (USA) vol.51 (1980) p.342]

[3] P.J.Rosser, G.Tomkins [J. Phys. (France) vol.44 (1983) p.C5-445]

[4] L.S.Hung, J.Gyulai, J.W.Mayer, S.S.Lau, M-A.Nicolet [J. Appl. Phys. (USA) vol.54 (1983) p.5076]

[5] G.G.Bentini, R.Nipoti, A.Armigliato, M.Berti, A.V.Drigo, C.Cohen [J. Appl. Phys. (USA) vol.57 (1985) p.270]

[6] C.S.Wei, J.Van der Spiegel, J.J.Santiago [Appl. Phys. Lett. (USA) vol.45 (1984) p.527]

[7] D.Levy, J.P.Ponpon, A.Grob, J.J.Grob, R.Stuck [Appl. Phys. A (Germany) vol.38 (1985) p.23]

[8] H.Kato, Y.Nakamura [Thin Solid Films (Switzerland) vol.34 (1976) p.135]

[9] T.Okamoto, K.Tsukamoto, M.Shimizu, T.Matsukawa [J. Appl. Phys. (USA) vol.57 (1985) p.5251]

[10] M.Tanielian, S.Blackstone [Appl. Phys. Lett. (USA) vol.45 (1984) p.673]

[11] R.S.Rosler, G.M.Engle [J. Vac. Sci. & Technol. B (USA) vol.2 no.4 (1984) p.733-7]

[12] P.K.Tedrow, V.Ilderem, R.Reif [Mater. Res. Soc. Symp. Proc. (USA) vol.37 (1985) p.619]

[13] L.J.Brillson, M.L.Slade, H.W.Richter, H.Vander Plas, R.T.Fulks [Appl. Phys. Lett. (USA) vol.47 (1985) p.476]

[14] R.Beyers, R.Sinclair [J. Appl. Phys. (USA) vol.57 (1985) p. 5240]

[15] H.K.Park, J.Sachitano, M.McPherson, T.Yamaguchi, G.Lehman [J. Vac. Sci. & Technol. A (USA) vol.2 (1984) p.264]

[16] T.P.Chow, W.Katz, G.Smith [Appl. Phys. Lett. (USA) vol.46 (1985) p.41]

[17] C.Y.Wong, F.S.Lai, P.A.McFarland, F.M.d'Heure, C.Y.Ting [J. Appl. Phys. (USA) vol.59 (1986) p.2773]

[18] F.M.d'Heurle [Proc. Spring ECS meeting (1982) p.194]

30.3 RESISTIVITY OF TiSi2

by P.J.Rosser

November 1986
EMIS Datareview RN=12631

TABLE 1 gives values for the thin film resistivity of titanium disilicide films deposited and annealed using a wide range of techniques. It is of interest that the accepted bulk value of 16-18 micro-ohm cm [1-4] has been matched if not bettered by many of these thin film values. This is probably indicative of the improvements in film quality that have occurred since the bulk values were determined.

TABLE 1

Deposition	Anneal conditions Temp. (deg C)	Time		Resistivity (micro-ohm cm)	Ref
EVAPORATED Ti/X	750	10-15 sec	vac.	21	[5]
		(improved vacuum)		14	[6]
	800	> 1 sec	Ar	20.6	[7]
SPUTTERED Ti/X	800	60 min	vac.	16	[8]
	700	60 sec	Ar + N2	15	[9]
at 450 deg C	900	30 min	Ar	15	[10]
SPUTTERED Ti/POLY				13-16	[11]
ION BEAM Ti/POLY	800	30 min	Ar	14	[12]
		(residual resistivity		3)	[12]
		(room temp. intrinsic		11)	[12]
COEVAPORATION				15-25	[13]
				21	[14]
	700	> 30 min		21	[15]
	800-1000	> 5 min		21	[15]
COSPUTTERED					
[B] TiSi2/oxide	900	30 min	H2	55	[16]
[B] TiSi3/oxide	900	30 min	H2	33	[16]
[B] TiSi4/oxide	900	30 min	H2	55	[16]
TiSi2/oxide	700	30 min	N2	23	[17]
TiSi2/oxide	900	10 sec	N2	23	[17]
[B] TiSi2/poly	900	30 min	H2	25	[16]

TABLE 1(continued)

Deposition	Anneal conditions Temp. (deg C)	Time		Resistivity (micro-ohm cm)	Ref
ALLOY SPUTTERED				40-70	[13]
	800-900	60 min	vac.	25	[8]
[A] (over oxide)	700	10-70 min		230	[15]
	800	"		173	[15]
	900	"		130	[15]
	1000	"		90	[15]
[A] (over poly)	800-1000	"		40	[15]
	700	30-70 min		156	[15]
at elevated	200-450			60	[18]
temperature of					
	450-600			16	[18]
REACTIVELY SPUTTERED	800			20	[19]
APCVD at 800-1000 deg C				20	[20]
LPCVD at 600 deg C				22	[21]
PECVD	750	60 min		20	[20]
	650	> 5 min		15	[22]
LASER CVD	800	5 min		20	[23]
at 400 deg C				300	[24]

NOTES:

[A] The hot pressed target used in this study was found to contain
15% oxygen.

[B] For these film compositions only the disilicide phase was observed
by X-ray diffraction.

The wide range of values listed in this table is indicative principally
of the impact of impurities on the resistivity. For the cosputtered
films it is probable that the resistivity is also dependent on film
stress which has been observed to be lowest in films of composition
$TiSi_3$ [16]. Where the silicide has been formed at low temperature it is
also likely that silicide phases other than the equilibrium high
temperature $TiSi_2$ (C54, orthorhombic, ABCDA stacking of distorted
hexagonal planes [25]) phase are present. The intermediate $TiSi_2$
(C49, orthorhombic $ZrSi_2$ structure), Ti_5Si_3 and TiSi phases have all
been observed [9,11,16].

REFERENCES
----

[1]   D.Robins [Philos. Mag. (GB) vol.3 (1958) p.313]
[2]   V.S.Neshpor, G.V.Samsonov [Sov. Phys.-Solid State (USA) vol.2
      (1970) p.1966]
[3]   R.Keiffer, F.Benesovsky [in 'Hartstoffe' (Springer-Verlag,
      Vienna, 1963) p.455]
[4]   R.Wehrmann [in 'High temperature materials and technology' Eds
      I.E.Campbell, E.M.Sherwood (Wiley, New York, USA, 1967) p.399]
[5]   C.S.Wei, J.Van der Spiegel, J.J.Santiago, L.E.Seiberling [Appl.
      Phys. Lett. (USA) vol.45 (1984) p.527]
[6]   J.J.Santiago, C.S.Wei, J.Van der Spiegel [Mater. Lett.
      (Netherlands) vol.2 (1984) p.477]
[7]   D.Levy, J.P.Ponpon, A.Grob, J.J.Grob, R.Stuck [Appl. Phys. A
      (Germany) vol.38 no.1 (Sep 1985) p.23-30]
[8]   F.Runovc, H.Norstrom, R.Buchta, P.Wiklund, S.Petersson [Phys.
      Scr. (Sweden) vol.26 (1982) p.108]
[9]   R.Beyers, R.Sinclair [J. Appl. Phys. (USA) vol.57 no.12
      (15 Jun 1985) p.5240-5]
[10] M.Tanielian, S.Blackstone [Appl. Phys. Lett. (USA) vol.45 (1984)
      p.673]
[11] S.P.Murarka, D.B.Fraser [J. Appl. Phys. (USA) vol.51 no.1
      (Jan 1980) p.342-9]
[12] V.Malhotra, T.L.Martin, J.E.Mahan [J. Vac. Sci. & Technol. B
      (USA) vol.2 (Jan-Mar 1984) p.10]
[13] K.L.Wang et al [IEEE Trans. Electron Devices (USA) vol.ED-29
      (1982) p.547]
[14] M.J.H.Kemper, P.H.Oosting [J. Appl. Phys. (USA) vol.53 no.9
      (Sep 1982) p.6214-19]
[15] R.F.Pinizzotto, K.L.Wang, S.Matteson [Proc. Spring ECS meeting
      (1982) p.562]
[16] S.P.Murarka, D.B.Fraser [J. Appl. Phys. (USA) vol.51 no.1
      (Jan 1980) p.350-6]
[17] P.J.Rosser, G.Tomkins [J. Phys. Colloq. (France) vol.44 no.C-5
      (Oct 1983) p.C5445-8]
[18] M.Iwami, A.Hiraki [Jpn. J. Appl. Phys. Part 1 (Japan) vol.24
      no.5 (May 1985) p.530-6]
[19] H.-O.Blom, S.Berg, M.Ostling, C.S.Petersson, V.Deline,
      F.M.d'Heurle [J. Vac. Sci. & Technol. B (USA) vol.3 no.4
      (Jul-Aug 1985) p.997-1003]
[20] M.J.H.Kemper, S.W.Koo, F.Huizinga [Proc. Fall ECS meeting,
      abst 337, (1984)]
[21] P.K.Tedrow, V.Ildrem, R.Reif [Mater. Res. Soc. Symp. Proc. (USA)
      vol.37 (1985) p.619-22]
[22] R.S.Rosler, G.M.Engle [J. Vac. Sci. & Technol. B (USA) vol.2
      no.4 (Oct-Dec 1984) p.733-7]
[23] G.A.West, A.Gupta, K.W.Beeson [Mater. Res. Soc. Symp. Proc. (USA)
      vol.29 (1984) p.124]
[24] G.A.West, A.Gupta, K.W.Beeson [Appl. Phys. Lett. (USA) vol.47
      no.5 (1 Sep 1985) p.476-8]
[25] F.M.d'Heurle [Proc. Spring ECS meeting (1982) p.194]

30.4 FORMATION TEMPERATURE OF NiSi2

by R.E.Harper

July 1987
EMIS Datareview RN=15785

Several techniques have been employed to produce thin films of NiSi2 on
silicon substrates. The most common method is the thermal reaction in a
furnace or rapid thermal anneal system of nickel films deposited on
single-crystal (s-c), polycrystalline or amorphous (evaporated)
silicon substrates [1-8,12,13]. Films thus produced tend to grow
epitaxially on single-crystal substrates but are not necessarily
monocrystalline. Single crystal NiSi2 films have been grown using MBE
[9] and by a technique where ultra-thin (< 30 A) Ni films are
deposited in UHV and annealed in-situ at a low temperature to form a
NiSi2 'template' [10,11]. Thicker layers are then obtained by depositing
Ni or Ni and Si onto this template at a temperature of 550 deg C or
more [10].

TABLE 1 below gives values for the formation temperature of NiSi2
produced using the above methods. The results for conventional thermal
reaction techniques are in the range 600-900 deg C: this variation
largely reflects differences in cleanliness of the initial Ni/Si
interface and in the impurity concentration of the deposited Ni film.
It is interesting to note that RTA yields lower values than furnace
annealing. This difference could be genuine, but it is possible that
the RTA results are erroneous because of difficulties in thermometry
in these systems. It should also be noted that there is a large
difference between the formation temperatures on s-c and amorphous
substrates. Lien et al [6] have suggested several reasons for this,
including the fact that amorphous silicon has a positive formation
energy, that the Ni/Si interface is cleaner in the amorphous case
(because the amorphous Si and the Ni can be deposited sequentially
without breaking vacuum), and that the resultant silicides have
different microstructures.

The reductions in formation temperature obtained using the 'template'
methods are remarkable, and must be due to the fact that the
substrate cleaning, metal deposition and annealing are all performed
in-situ in UHV, thus ensuring minimal contamination. However it must
be pointed out that only very thin films can be obtained at these low
temperatures. The most surprising result is that of Lu et al [12] who
observed the formation of a thin interfacial layer of NiSi2 after
annealing of 300 A Ni films on BF2+ and Si+ pre-amorphised B+ implanted
s-c substrates. The remaining Ni was unreacted. This is in complete
contrast to the expected behaviour at this temperature of complete
conversion to the low temperature phase Ni2Si [14]. Lu et al [12]
conjectured that the presence of the dopant atoms substantially lowers
the activation energy for NiSi2 growth. However, they also found that
complete conversion to NiSi2 only occurred after annealing at 750 deg C
which is in good agreement with the other furnace annealing results
presented here.

TABLE 1

Preparation Method	Formation Temp (deg C)	Reference
2-stage anneal (30 min each) in vacuum of evaporated Ni on s-c Si	300,800	[1]
Anneal in vacuum or H2 of Ni on polysilicon	900	[2]
Anneal in vacuum or H2 of co-sputtered alloy Ni + Si	900	[2]
Anneal in He (0.5-1 hr) of evaporated Ni on s-c Si	800	[3]
Anneal in UHV of evaporated (UHV) Ni on s-c Si	750	[4]
Anneal in vacuum (30 min) of evaporated Ni on s-c Si	800	[5]
Anneal in vacuum of Ni on amorphous (evaporated) Si	350-425	[6]
Anneal in vacuum of Ni on s-c Si	750	[6]
Rapid thermal anneal in N2 (xenon lamp) or evaporated Ni on s-c Si	675	[7]
Rapid thermal anneal in vacuum (electron beam of quartz halogen lamp) of evaporated Ni on s-c Si	650	[8]
Co-deposition of Ni and Si (MBE) on s-c Si at elevated temperature	> 550	[9]
In-situ anneal of ultra-thin (< 30 A) Ni on s-c Si to form 'template'	450	[10]
Deposition of Ni or Ni + Si on 'template' at elevated temperature	> 550	[10]
Alternate deposition of 4 A Ni in UHV, then anneal in-situ, to max 70 A	320-350	[11]
Anneal in N2 (3 min-7.5 hr) of evaporated Ni on BF2+ implanted or B+ implanted Si+ pre-amorphised s-c Si	250-280	[12]
Anneal in N2 (1 hr) of evaporated Ni on 0.6-10 micron windows in SiO2 on s-c Si	550-900	[13]

Very little data on the activation energy for NiSi2 growth has been published. Lien et al [6] found a value of 1.65 +- 0.2 eV for growth by furnace annealing of Ni on amorphous Si. Maenpaa et al [15] found a value of 1.2 +- 0.2 eV for epitaxial regrowth on NiSi2 following amorphisation by Si+ ion irradiation. No value could be found for growth on single crystal Si.

REFERENCES
----

[1]   S.Saitoh, H.Ishiwara, T.Asano, S.Furukawa [Jpn. J. Appl. Phys. (Japan) vol.20 (1981) p.1949]

[2]   S.P.Murarka, M.H.Read, C.J.Doherty, D.B.Fraser [J. Electrochem. Soc. (USA) vol.129 (1982) p.293]

[3]   L.J.Chen, J.W.Mayer, K.N.Tu, T.T.Sheng [Thin Solid Films (Switzerland) vol.93 (1982) p.91]

[4]   R.J.Tung, J.M.Poate, J.C.Bean, J.M.Gibson, D.C.Jacoson [Thin Solid Films (Switzerland) vol.93 (1982) p.77]

[5]   Y.Sorimachi, H.Ishiwara, H.Yamamoto, S.Furukawa [Jpn. J. Appl. Phys. Part 1 (Japan) vol.21 (1982) p.752]

[6]   C.D.Lien, M.A.Nicolet, S.S.Lau [Phys. Status. Solidi a (Germany) vol.81 (1984) p.123]

[7]   A.N.Larsen, J.Chevallier, G.Sorensen [Mater. Res. Soc. Symp. Proc. (USA) vol.23 (1984) p.727]

[8]   R.E.Harper, C.J.Sofield, I.H.Wilson, K.G.Stephens [Mater. Res. Soc. Symp. Proc. (USA) vol.37 (1985) p.573]

[9]   J.C.Bean, J.M.Poate [Appl. Phys. Lett. (USA) vol.37 (1980) p.643]

[10]   R.T.Tung, J.M.Gibson, J.M.Poate [Phys. Rev. Lett. (USA) vol.50 (1983) p.429]

[11]   H.von Kanel, T.Graf, J.Henz, M.Ospelt, P.Wachter [Superlattices & Microstruct. (GB) vol.2 (1986) p.363]

[12]   S.W.Lu, C.W.Nieh, L.J.Chen [Appl. Phys. Lett. (USA) vol.49 (1986) p.1770]

[13]   C.S.Chang, C.W.Nieh, L.J.Chen [Appl. Phys. Lett. (USA) vol.50 (1987) p.259]

[14]   J.O.Olowalafe, M.A.Nicolet, J.W.Mayer [Thin Solid Films (Switzerland) vol.38 (1976) p.143]

[15]   M.Maenpaa, L.S.Hung, M.A.Nicolet, D.K.Sadana, S.S.Lau [Thin Solid Films (Switzerland) vol.87 (1982) p.77]

30.5 RESISTIVITY OF NiSi2

by R.E.Harper

May 1987
EMIS Datareview RN=15714

In TABLE 1, values are given of the room temperature resistivity
of thin films of NiSi2 on single crystal (s-c) or polycrystalline
silicon substrates, produced using various deposition methods and
thermal treatments.

TABLE 1

Preparation Method	Resistivity (micro-ohm cm)	Reference
Ni evaporation on s-c Si + 2-stage anneal 300 deg C 30 min + 800 deg C 30 min in vacuum	35	[1]
UHV Ni evaporation on s-c Si + 750-800 deg C anneal in UHV	35	[2]
Non-UHV Ni evaporation on s-c Si + 750-800 deg C anneal in vacuum (non-UHV)	35	[2]
Ni deposition on polysilicon + 900 deg C anneal in vacuum or H2	50	[3]
Cosputtered Ni and Si + 900 deg C anneal in vacuum or H2	50-60	[3]
UHV Ni evaporation + 800 deg C 30 min anneal in UHV	47	[4]
Ni evaporation on s-c Si + RTA (xenon lamp) 715 deg C 15 sec in N2	51-57	[5]
Ni evaporation on s-c Si + RTA (electron beam or quartz halogen lamp) 650-700 deg C 60 sec in vacuum	34-37	[6]

The values above are much lower than that published for bulk NiSi2 of
118 micro-ohm cm [7]. This difference arises because the methods
employed to produce bulk NiSi2 allow the incorporation of high
concentrations of impurities which adversely affect the resistivity.
Similarly the spread in tabulated values above can largely be
attributed to differences in the cleanliness of the deposition and
anneal systems and of the initial Ni-Si interface. Harper et al [6] have
found that an interfacial contamination of 2.7×10^6 oxygen atoms/cm*2
increases the NiSi2 resistivity to 70 micro-ohm cm, compared to 35
micro-ohm cm for a contamination level below 10*15 atoms/cm*2.

The temperature dependence of NiSi2 resistivity has been studied by Hensel et al [8] in the range of 1 to 300 deg K. They found that the resistivity varied approximately linearly with temperature down to around 50 deg K where it levelled off at 20-27 micro-ohm cm. They determined that the lowest attainable room temperature resistivity as determined by the lattice contribution is 14 micro-ohm cm, well below any experimental value. They therefore concluded that NiSi2 films must contain a high concentration of intrinsic defects.

REFERENCES
----

[1]   S.Saitoh, H.Ishawara, T.Asano, S.Furukawa [Jpn. J. Appl. Phys. (Japan) vol.20 (1981) p.1949]

[2]   R.T.Tung, J.M.Poate, J.C.Bean, J.M.Gibson, D.C.Jacobson [Thin Solid Films (Switzerland) vol.93 (1982) p.77]

[3]   S.P.Murarka, M.H.Read, C.J.Doherty, D.B.Fraser [J. Electrochem. Soc. (USA) vol.129 (1982) p.293]

[4]   Y.Sorimachi, H.Ishawara, H.Yamamoto, S.Furukawa [Jpn. J. Appl. Phys. Part 1 (Japan) vol.21 (1982) p.752]

[5]   A.N.Larsen, J.Chevallier, G.Sorensen [Mater. Res. Soc. Symp. Proc. (USA) vol.23 (1984) p.727]

[6]   R.E.Harper, C.J.Sofield, I.H.Wilson, K.G.Stephens [Mater. Res. Soc. Symp. Proc. (USA) vol.37 (1985) p.573]

[7]   V.S.Neshpor, G.V.Samsonov [Sov. Phys.-Solid State (USA) vol.2 (1960) p.66]

[8]   J.C.Hensel, R.T.Tung, J.M.Poate, F.C.Unterwald, D.C.Jacobson [Mater. Res. Soc. Symp. Proc. (USA) vol.25 (1984) p.575]

30.6 FORMATION TEMPERATURE OF CoSi2

by R.E.Harper

July 1987
EMIS Datareview RN=15784

Thin films of CoSi2 have been produced using several techniques, the
most common being thermal reaction of thin Co films on single-crystal
(s-c), polycrystalline or amorphous Si substrate using furnace
annealing [1-7] or rapid thermal annealing [8-12]. The CoSi2 films thus
produced on s-c Si can be fine-grained polycrystalline with no lattice
match to the substrate, epitaxial but multicrystalline, or epitaxial
and single crystal. Monocrystalline films have also been obtained using
MBE [4]. TABLE 1 below lists values of the formation temperature of
CoSi2 obtained in a variety of systems, and indicates where epitaxial
and/or single crystalline growth has been observed.

The formation temperatures for CoSi2 on s-c substrates can be split
into two groups: those in the range 500-850 deg C where polycrystalline
non-epitaxial CoSi2 is obtained, and those above 850 deg C where the
growth is epitaxial. The spread in the former values is to a large
extent due to differences in cleanliness of the initial Co-Si interface
and in the purity of the deposited Co film. For example, Collins et al
[10] found that if the native oxide was not removed prior to Co
deposition, there was no reaction between the Co and Si until the
temperature exceeded 800 deg C. They also found that where the native
oxide was removed, ion-induced CoSi2 formation could occur at only
450 deg C. It is also interesting to note that Lien et al [6] obtained
a lower formation temperature on amorphous Si than any of the s-c
results. They suggested several reasons for this, including a cleaner
Co-Si interface (since the amorphous Si and Co are deposited in the
same system without breaking vacuum), the fact that the amorphous Si
has a positive energy formation, and that the resultant silicides have
different microstructures.

There is little published data on the activation energy of CoSi2
formation. Lien et al [6] obtained a value of 2.3 +- 0.1 eV for growth
on amorphous Si, while d'Heurle and Petersson [7] and Van der Hove et
al [11] obtained a value of 2.6 eV for growth on s-c Si using furnace
annealing and rapid thermal annealing respectively.

TABLE 1

Preparation Method	Formation Temp. (deg C)	Reference
Anneal in vacuum (30 min) of evaporated Co on s-c Si	900 (epitaxial)	[1]
Anneal in vacuum or H2 of Co on polysilicon	900	[2]

TABLE 1 (continued)

Preparation Method	Formation Temp. (deg C)	Reference
Anneal in vacuum or H2 of co-sputtered alloy Co+Si	900	[2]
Anneal in He (upto 4 hrs) of Co on s-c Si	850	[3]
Anneal in He (upto 1 hr) of Co on s-c Si	900 (epitaxial)	[3]
Anneal in UHV (30 min) of UHV evaporated Co on s-c Si	850 (epitaxial, s-c)	[4]
MBE (deposition of Co or Co and Si in ratio 1:2) on substrate at elevated temperature	600 (epitaxial, s-c)	[4]
Anneal in UHV (30 min) of evaporated Co on s-c Si	900 (epitaxial)	[5]
Anneal in vacuum of evaporated Co on s-c Si	500	[6]
Anneal in vacuum of evaporated Co on evaporated (amorphous Si)	405-500	[6]
Anneal in He (upto 126 hr) of evaporated Co on s-c Si	500	[7]
Rapid thermal anneal (electron beam or quartz halogen lamp) of evaporated Co on s-c Si	600	[8]
Rapid thermal anneal (electron beam) in vacuum of evaporated Co on B+ implanted s-c Si	600	[9]
Rapid thermal anneal (quartz halogen lamp) of evaporated Co on s-c Si	600	[10]
Ar+ Ion bombardment at elevated temperature of evaporated Co on s-c Si	450	[10]
Rapid thermal anneal (quartz halogen lamp) in N2 of sputtered Co on unimplanted and B+ or As+ implanted s-c Si	> 500	[11]
Rapid thermal anneal (quartz halogen lamp) in N2 of evaporated Co on unimplanted and B+, As+ or P+ implanted s-c Si	700	[12]

REFERENCES
----

[1]   S.Saitoh, H.Ishiwara, T.Asano, S.Furukawa [Jpn. J. Appl. Phys.
        (Japan) vol.20 (1981) p.1949]
[2]   S.P.Murarka, M.H.Read, C.J.Doherty, D.B.Fraser [J. Electrochem.
        Soc. (USA) vol.129 (1982) p.293]
[3]   L.J.Chen, J.W.Mayer, K.N.Tu, T.T.Sheng [Thin Solid Films
        (Switzerland) vol.93 (1982) p.91]
[4]   R.T.Tung, J.C.Bean, J.M.Gibson, J.M.Poate, D.C.Jacobson [Appl.
        Phys. Lett. (USA) vol.40 (1982) p.684]
[5]   Y.Sorimachi, H.Ishiwara, H.Yamamoto, S.Furukawa [Jpn. J. Appl.
        Phys. Part 1 (Japan) vol.21 (1982) p.752]
[6]   C-D.Lien, M.-A.Nicolet, S.S.Lau [Appl. Phys. Lett. (USA) vol.41
        (1984) p.249]
[7]   F.M.d'Heurle, C.S.Petersson [Thin Solid Films (Switzerland)
        vol.128 (1985) p.283]
[8]   R.E.Harper, C.J.Sofield, I.H.Wilson, K.G.Stephens [Mater. Res.
        Soc. Symp. Proc. (USA) vol.37 (1985) p.573]
[9]   C.J.Sofield, R.E.Harper, P.J.Rosser [Mater. Res. Soc. Symp.
        Proc. (USA) vol.35 (1985) p.445]
[10]   R.A.Collins, S.C.Edwards, G.Dearnaley [Vacuum (GB) vol.36
        no.11/12 (1986) p.821]
[11]   L.Van den Hove, R.Wolters, K.Maex, R.de Keersmaecker, G.Declerk
        [J. Vac. Sci. & Technol. B (USA) vol.4 (1986) p.1358]
[12]   M.Tabasky, E.S.Bulat, B.M.Ditchek, M.A.Sullivan, S.S.Latas
        [Mater. Res. Soc. Symp. Proc. (USA) vol.52 (1986) p.271]

30.7 RESISTIVITY OF CoSi2

by R.E.Harper

May 1987
EMIS Datareview RN=15715

In TABLE 1, values are given for the resistivity at room temperature
of thin films of CoSi2 on single crystal (s-c) or polycrystalline
silicon substrates, produced using a variety of deposition methods
and thermal treatments.

TABLE 1

Preparation Method	Resistivity (micro-ohm cm)	Reference
Co evaporation on s-c Si + 900 deg C 30 min anneal in vacuum	15	[1]
UHV Co evaporation on s-c Si + 850-950 deg C anneal in UHV	15	[2]
Non-UHV Co evaporation on s-c Si + 850-950 deg C anneal in vacuum (non-UHV)	18-20	[2]
Co and Si co-deposition (MBE) on s-c Si at substrate temp. of 550-650 deg C	10	[2]
Co deposition on polysilicon + 900 deg C anneal in vacuum or H2	18-20	[3]
Cosputtered alloy Co+Si + 900 deg C anneal in vacuum or H2	25	[3]
UHV Co evaporation on s-c Si + 900 deg C anneal in UHV	17	[4]
Co evaporation on s-c Si + 600 deg C RTA in vacuum (e-beam or halogen lamp)	15	[5]
Co sputtered on s-c Si + > 500 deg C RTA in N2	16	[6]

It is interesting to note that the values above are much lower than the
published value for bulk CoSi2 of 68 micro-ohm cm [7]. This is almost
certainly because the powder metallurgical techniques used to produce
bulk CoSi2 allow the incorporation of high levels of impurities which
adversely affect the resistivity. Similarly the spread in the tabulated
values above is largely attributable to variations in the film purity.

The temperature dependence of CoSi2 resistivity has been investigated in
the 1-300 deg K range by Hensel et al [8]. They found that the
resistivity varied approximately linearly with temperature down to
around 50 deg K, falling from 15 micro-ohm cm at room temperature to
about 2.6 micro-ohm cm at 50 deg K and below. They concluded that the
lowest attainable room temperature resistivity is around 12 micro-
ohm cm, which is determined by the lattice contribution.

REFERENCES
----

[1]   S.Saitoh, H.Ishawara, T.Asano, S.Furukawa [Jpn. J. Appl. Phys.
      (Japan) vol.20 (1981) p.1949]
[2]   R.T.Tung, J.M.Poate, J.C.Bean, J.M.Gibson, D.C.Jacobson [Thin
      Solid Films (Switzerland) vol.93 (1982) p.77]
[3]   S.P.Murarka, M.H.Read, C.J.Doherty, D.B.Fraser [J. Electrochem.
      Soc. (USA) vol.129 (1982) p.293]
[4]   Y.Sorimachi, H.Ishawara, H.Yamamoto, S.Furukawa [Jpn. J. Appl.
      Phys. Part 1 (Japan) vol.21 (1982) p.752]
[5]   R.E.Harper, C.J.Sofield, I.H.Wilson, K.G.Stephens [Mater. Res.
      Soc. Symp. Proc. (USA) vol.37 (1985) p.573]
[6]   L.Van den Hove, R.Wolters, K.Maex, R.de Keersmaecker, G.Declerk
      [J. Vac. Sci. & Technol. B (USA) vol.4 (1986) p.1358]
[7]   V.S.Neshpor, G.V.Samsonov [Sov. Phys.-Solid State (USA) vol.2
      (1960) p.66]
[8]   J.C.Hensel, R.T.Tung, J.M.Poate, F.C.Unterwald, D.C.Jacobson
      [Mater. Res. Soc. Symp. Proc. (USA) vol.25 (1984) p.575]

CHAPTER 31

DEFECT RELATED FAILURE IN DEVICE STRUCTURES

31.1 DEFECT RELATED FAILURE MODES IN Si p-n JUNCTIONS

by K.V.Ravi

June 1987
EMIS Datareview RN=15723

A. INTRODUCTION

The electrical properties of p-n junctions that are strongly influenced
by the presence of imperfections and impurities are junction leakage
and junction breakdown. Generation and recombination in the space charge
region of p-n junctions are assisted by intermediate centres in the
band gap, and the presence of these centres affects the current-voltage
characteristics of p-n junctions. The influence of defects and
impurities in the device is to distort the ideal I-V characteristics of
the diode.

Imperfections and impurities such as dislocations, stacking faults and
second phase inclusions such as oxide and carbide precipitates and
metallic impurities have distinct effects on p-n junctions. Their
particular influence is based upon several factors including: the
location of the defect with respect to the depletion field of the p-n
junction; the degree and type of impurity decorating the defect; the
point in the fabrication sequence at which the defect is introduced into
the crystal; the presence of other defects in the vicinity and perhaps
most importantly, the process sequence to which the wafer is subjected
in the course of device or circuit fabrication.

B. DISLOCATIONS

A direct correlation between the density of dislocations in the active
volume of a diode and the diode reverse current has been established
indicating an unambiguous relationship between these defects and the
electrical characteristics of the device [1]. The electrical influence
of dislocations in p-n junctions has been a subject of long standing
controversy. Dislocations can introduce mid gap states into the
semiconductor as a result of the disruption in the periodicity of the
lattice their presence causes, as well as a result of the strain they
generate in the lattice. These gap states can cause excess reverse
leakage currents. It has been suggested that bulk leakage currents can
be explained in terms of a single donor type defect 0.055 eV below the
centre of the band gap [2]. On the other hand dislocations have been
suggested as being electrically active only when decorated by a metallic
impurity with 'clean' undecorated dislocations having negligible effects
on p-n junctions [3]. The major effects due to dislocations are due to
the collective influence of dislocations and impurities. The presence
of dislocations can influence the behaviour of impurities in the
crystal. Impurities such as the fast diffusing metals, Cu and Fe and
the slow diffusers such as O can be caused to precipitate at the
dislocations.

C. MICRODEFECTS

Microdefects (swirls) and attendant resistivity inhomogeneities
(striations) cause pronounced electrical effects in p-n junctions [4].
However it is generally difficult to distinguish between the effects
due to microdefects and process induced modifications of these defects.
Swirl defects have been demonstrated to function as local
recombination centres in p-n junctions [5]. The inhomogeneous nature
of the distribution of microdefects in wafers presents particular
problems with regard to the distribution of electrical properties of
devices fabricated in wafers containing swirl defects. A spiral
distribution of leakage paths has been demonstrated in vidicon targets
fabricated in wafers containing microdefects [6]. By employing electron
beam techniques it has been demonstrated that variations in the local
leakage currents in shallow (< 0.5 microns) p-n junctions of as
much as six orders of magnitude can be observed in adjacent diodes
fabricated in the same wafer with a one-for-one positional
correspondence between the presence of a swirl defect within the
perimeter of the diode and excess reverse leakage currents [5].

D. OXIDATION INDUCED STACKING FAULTS

The large majority of defects giving rise to generation recombination
effects within swirl bands are found to be oxidation induced stacking
faults. Oxidation is a common accompaniment to the fabrication of p-n
junctions, either prior to the diffusion of the dopant in the
fabrication of planar devices or accompanying diffusion. In addition
the diffusion of phosphorus into silicon has been shown to inject excess
interstitials into the silicon, much as oxidation, with the excess
interstitials reacting with nucleation centres such as swirl defects
within the crystal to form stacking faults. Consequently the
electrical influence of stacking faults on the characteristics of p-n
junctions has received wide attention.

An extrinsic stacking fault has no dangling bonds associated with it,
since a stacking fault retains the four fold coordination of all the
near neighbor atoms with atomic rearrangement occurring only between
third and higher nearest neighbors [7]. As a result stacking faults are
not expected to be electrically active in the same sense as
dislocations. However the presence of bounding partial dislocations and
the effects of impurity decoration cause stacking faults to be
electrically active. Several interdependent parameters determine the
electrical activity of stacking faults. These include the size of the
fault, the position of the fault with respect to the depletion field of
the p-n junction, and the degree and the type of impurity decorating the
fault. Stacking faults have been shown to function as local regions of
low minority carrier lifetime as well as functioning as
generation-recombination regions in the depletion field of p-n junctions
[8]. The effects due to stacking faults can be characterised in several
ways. A direct interaction between the fault and the depletion field of
the p-n junction can be a result of the strain field associated with
the fault. Alternatively, the stacking can locally perturb the
planarity of the p-n junction. The fault can have a retarding effect
on the diffusion front that causes a local sharp curvature in the
otherwise planar junction. The curvature introduced in the junction
can lead to excess reverse currents. The effects of impurities at or
around the fault are probably paramount in determining the electrical
activity of these defects.

E. ELECTRICAL BREAKDOWN OF p-n JUNCTIONS

The breakdown voltage, i.e the voltage in the reverse direction above which the diode will no longer block the passage of current is also a function of the presence of defects within the device. The breakdown voltage of a defect free diode is a function of the resistivity of the substrate and the abruptness and the curvature of the p-n junction. Defects and precipitates in the junction region are known to reduce the breakdown voltage below the bulk value [9]. Non-uniform junction breakdown, generally as a result of impurities and imperfections results in the generation of excess junction reverse currents, large amplitude fluctuations of the current near the region of breakdown and the observation of light emission at discrete spots within the perimeter of the device. The observation of light emission has been termed microplasma emission. This effect has generally been attributed to the generation of local high electrical fields at an imperfection, generally a metallic or oxide precipitate, whose magnitude exceeds that of the average value of the field. Microplasma breakdown has also been observed at stair rod dislocations bounding epitaxial stacking faults [10]. Local resistivity variations or striations in the crystal can also lead to localised breakdown in high power devices. Breakdown patterns consisting of concentric rings or segments of rings are observed, indicating localised bulk breakdown at striations [11]. In general the same defects that give rise to excess leakage currents in p-n junctions also contribute the premature breakdown.

REFERENCES

[1]   K.V.Ravi [Imperfections and Impurities in Semiconductor Silicon (J. Wiley & Sons, New York 1981) p.234]
[2]   M.V.Whelan [Solid State Electron. (GB) vol.12 (1969) p.963]
[3]   W.Rinder, I.Braun [J. Appl. Phys. (USA) vol.34 (1963) p.1958]
[4]   K.V.Ravi, C.J.Varker [Appl. Phys. Lett. (USA) vol.25 (1974) p.69]
[5]   C.J.Varker, K.V.Ravi [Semiconductor Silicon, 1973 Eds. H.R.Huff, R.R.Burgess (The Electrochemical Society Princeton, NJ,1973) p.670]
[6]   A.J.R.Dekock [Philips Res. Rep. (Netherlands) suppl. no.1 (1973)]
[7]   J.Hornstra [J. Phys. & Chem. Solids (GB) vol.5 (1958) p.129]
[8]   K.V.Ravi, C.J.Varker, C.E.Volk [J. Electrochem. Soc. (USA) vol.120 (1973) p.533]
[9]   H.Kressel [RCA Rev. (USA) vol.28 (1967) p.175]
[10] H.J.Queisser, A.Gotzberger [Philos. Mag. (GB) vol.8 (1963) p.1063]
[11] K.Roy [J. Cryst. Growth (Netherlands) vol.9 (1971) p.139]

31.2 DEFECT RELATED FAILURE MODES IN Si BIPOLAR TRANSISTORS

by K.V.Ravi

June 1987
EMIS Datareview RN=15721

A. INTRODUCTION

The bipolar or the junction transistor is perhaps the most widely
employed active semiconductor device. It is typically a three layer
device containing two junctions in either the npn or the complementary
pnp configuration. The presence of two junctions in close proximity to
each other further complicates the influence of imperfections and
impurities on the performance characteristics of transistors as compared
to diodes.

The three most important effects that imperfections can have on
transistor characteristics are: (1) the introduction of excess leakage
currents in the two p-n junctions; (2) reduction of the current gain of
the beta transistor; and (3) emitter collector shorts which are commonly
termed 'spikes' or 'pipes'. Of these, emitter collector pipes are the
most detrimental and have a large impact on the yield of integrated
circuits.

B. JUNCTION LEAKAGE AND TRANSISTOR GAIN

Junction leakage is caused by the introduction of generation -
recombination (G-R) centres in the depletion region of the p-n
junctions. Defects can also affect the geometric characteristics of
the junctions through their influence on the local diffusion of dopants
employed to create the junctions [1]. Collector base leakage currents
often indicate corresponding emitter base leakage although the converse
is not always the case since emitter diffusion, typically involving high
concentration phosphorus diffusion can create defects in the emitter
region without the defects necessarily penetrating the collector base
junction. It has been demonstrated that emitter leakage currents due to
diffusion induced dislocations can be significantly higher than the
collector base leakage currents. It has been suggested that the higher
density of intersecting dislocations passing through the emitter base
depletion region gives rise to an observed current dependence I is
proportional to $V*n$ (n = 2 to 4) at high voltages, whereas at the
collector base junction, the high power dependence of the I-V
characteristics is not observed due to the much lower density of
intersecting dislocations. In addition the higher doping level at the
emitter base junction can promote internal field emission at
dislocations, which is also suggested as the mechanism for the observed
voltage dependence of the leakage current [2,3]. In addition to the
influence of defects in the emitter on the leakage current the current
gain of the transistor is reduced by the presence of defects. The higher
the leakage currents in the emitter and hence the higher the density of
defects in the emitter base depletion region, the lower the current gain
of the transistor. Gain is a function of the carrier lifetime in the
emitter. The presence of lifetime reducing defects and impurities in
the emitter reduces current gain.

C. ELECTRICAL PIPING

Electrical piping in bipolar transistors is a phenomenon which is
generally observed as an excess component of current between the
emitter and the collector in a variety of three terminal measurements.
In order to achieve high switching speeds in transistors the base width
is maintained as narrow as possible and the dopant level in the emitter
is maintained at as high a level as possible. Both these factors result
in very shallow junctions and narrow base widths which make maintaining
electrical and physical separation between the emitter base and
collector base junctions a difficult proposition. Localized electrical
or physical connection between the two junctions are commonly referred
to as emitter pipes.

Physically, pipes in npn transistors are visualised as local regions of
n-type material extending through a p-type base to make contact with
the n-type collector. Electrically a pipe can be detected by measuring
the collector base junction characteristics. In a piped transistor the
first increment of current in the I-Vcbo configuration occurs at a
voltage corresponding to the emitter base breakdown voltage, BVebo.
With further increases in the voltage an approximately linear region in
the current-voltage characteristics occurs until the collector base
breakdown voltage, BVcbo, is reached. The linear portion of the
current-voltage curve between BVebo and BVcbo is characterised by the
resistance of the pipe.

The nature of the defects causing pipes in transistors has been the
subject of much study and speculation [4,5]. Dislocations have most
often been implicated as the sources of pipes. A dislocation cutting
across both the emitter base and the collector base junctions can
provide a low resistance path for conduction between the emitter and the
collector if the defect is decorated by an impurity and a cylindrical
G-R zone is established around the defect. A dislocation can also
promote locally accelerated diffusion of one of the dopants, such as
phosphorus, thus establishing a physical connection between the two
junctions.

The most common defect leading to pipe formation is the oxidation
induced stacking fault and process induced modifications of this
defect. Stacking faults formed during oxidation can be modified during
subsequent base and emitter diffusion processes in such a manner as to
form the emitter pipe which establishes a connection between the
emitter base and the collector base junctions. An exact correlation
between the stacking fault and the eventual pipe has been obtained by
the use of a nondestructive imaging technique employing a scanning
electron microscope operated in the electron beam induced current mode
of operation of the instrument [6,7]. It has been demonstrated that
the defect nucleus which eventually leads to the pipe is established
very early in the fabrication process, often extending back to the
crystal growth step.

The establishment of a physical link between the emitter base and the
collector base junctions is generally sufficient to cause piping.
However impurities and in particular the fast diffusers have a strong
influence in determining the electrical activity of the pipe. The
presence of gold, which is frequently diffused into transistors to
increase their switching speed has been shown to increase the incidence
of piping. The clustering of gold atoms at stair rod dislocations
bounding epitaxial stacking faults has been postulated to cause pipes.

REFERENCES
----

[1] J.E.Lawrence [Semiconductor Silicon, 1973 Eds. H.R.Huff,
     R.R.Burgess (The Electrochemical Society, Princeton, NJ 1973)
     p.17]
[2] P.Ashburn, C.Bull, K.H.Nicholas, G.R.Booker [Solid-State
     Electron. (GB) vol.20 (1977) p.731]
[3] A.G.Chynoweth, G.L.Pearson [J. Appl. Phys. (USA) vol.29 (1958)
     p.1103]
[4] G.H.Platinga [IEEE Trans. Electron. Devices (USA) vol.ED-16 (1969)
     p.394]
[5] F.Barson, M.S.Hess, M.M.Roy [J. Electrochem. Soc. (USA) vol.116
     (1969) p.304]
[6] C.J.Varker, K.V.Ravi [Semiconductor Silicon, 1977 Eds. H.R.Huff,
     E.Sirtl (The Electrochemical Society, Princeton, NJ 1973) p.785]
[7] P.C.Parekh [Solid-State Electron. (GB) vol.14 (1971) p.273]

31.3 DEFECT RELATED FAILURE MODES IN Si MOS DEVICES

by K.V.Ravi

June 1987
EMIS Datareview RN=15722

MOS devices are majority carrier devices and as such are not subject
to defect related malfunctions to the same degree as minority carrier
devices. Nevertheless imperfections and impurities play an important
part in the performance characteristics of these devices. Several
parameters are of interest in the operation of an MOS device, either
a metal oxide silicon capacitor or a transistor. These include the
leakage currents of p-n junctions formed between the source and the
drain with the substrate, the flat band voltage, the density of
surface states at the oxide silicon interface and the holding and
refresh times in capacitors. The investigations of the influence of
crystallographic defects on MOS device performance have been based on
studies of generation lifetime in MOS capacitors, junction leakage
currents and phenomena associated with defects at the oxide silicon
interface.

A linear relationship between the carrier generation rate G and the
width of depletion region is theoretically predicted. In practice
however it is found that such a linear relationship is seldom observed
and a nonlinearity is introduced by the presence of defects or
impurities at the oxide-silicon interface. A correlation of the observed
nonlinearity between the generate rate and the depletion layer width
with oxidation induced stacking faults has been demonstrated [1]. The
total carrier generation rate is dominated by the normal bulk
generation rate when the defect concentration is small. However when a
high density of defects is present, the total generation rate is
dominated by generation in the vicinity of the defects. One mechanism
proposed for the effect is field enhanced emission at bulk defect
centres such as stacking faults as a result of the presence of large
local electric fields [2].

The turn-on voltage, that is, the gate voltage corresponding to the
onset of strong inversion in the capacitor is a function of the charge
at the Si-SiO2 interface. This interface charge has been demonstrated
to be a function of the dislocation density in the capacitor when it is
in excess of 10*6/cm*2 [3]. The presence of dislocation at the interface
can cause localised bond disruption resulting in an increase in the
fixed charge density and the attendant shift in the flat band voltage
of the MOS device. Each dislocation can have the same effect on the C-V
curve as approximately 10*4 charges/cm*2. Dislocations generated by
phosphorus diffusion can also introduce a high density of surface
states.

A direct correlation between oxide breakdown sites and the presence
of oxidation induced stacking faults has been demonstrated by the use
of electron beam techniques [4]. Whether the presence of the stacking
faults at the interface by themselves is sufficient to cause the
observed electrical malfunction continues to be a subject of debate. It
is strongly suggested that the presence of impurities, either metallic
impurities or oxides of silicon is required to electrically activate the

defects. A model for oxide breakdown that seems to have wide acceptance is that of electron injection from the cathode into the oxide by Fowler-Nordheim tunneling which is accelerated by the oxide field [5]. It has been proposed that precipitates at stacking faults can cause locally enhanced electron tunnelling from the cathode. Metal precipitates at the silicon-oxide interface have been shown to reduce the dielectric breakdown strength of oxides. It is not clear whether the metal precipitates require a local heterogeneous nucleating agent for precipitation. From the observation of other impurity related phenomena in silicon the presence of a heterogeneous nucleating agent for precipitation seems to be necessary.

An important parameter in MOS structures such as dynamic random access memories and charge coupled devices is the storage time for electrical charge. Storage time is the time required for the pulsed MOS element to reach the steady state condition. It is a function of the flat band voltage, the pulse voltage, the density of interface states and generation-recombination (G-R) centres in the adjacent semiconductor bulk. In an ideal capacitor with no bulk G-R centres, storage times as high as 100 sec. can be realized. In practical devices such long storage times are seldom observed. Imperfections intersecting the surface have a pronounced effect on the storage time. A relationship between the stacking fault density and the relaxation time (the inverse of the storage time) has been demonstrated [6].

Information in a dynamic RAM cell is retained in the form of a charge in an MOS capacitor with a capacitance typically less then one picofarad. This charge is volatile, that is it tends to decay due to leakage in the device. To maintain the charge in the capacitor and hence the information it represents the capacitor is periodically refreshed to bring it back to its original state of charge. The circuit refresh time is the maximum allowable time to sense the charge on the capacitor and to restore it to its original level. Refresh loss occurs when one can no longer reliably sense the charge on the capacitor and as a result the information is irretrievably lost. Refresh time is short in the presence of lifetime reducing imperfections and impurities in the capacitors. By gettering impurities to internally precipitated oxide-defect complexes refresh times can be significantly increased. A direct correlation between refresh loss and the presence of precipitates functioning as internal gettering agents has been demonstrated [7].

In general defect densities required to markedly influence MOS transistor parameters are significantly higher than the defect densities required to cause electrical aberrations in bipolar devices.

REFERENCES
----

[1] D.W.Small, R.F.Pierret [Appl. Phys. Lett. (USA) vol.27 (1975) p.148]
[2] G.H.Schuttke, K.Brack, E.W.Hearn [Microelectron. and Reliab. (GB) vol.10 (1977) p.467]
[3] D.V.McCaughan, B.C.Wonsiewicz [J. Appl. Phys. (USA) vol.45 (1974) p.4982]
[4] P.S.D.Lin, R.B.Marcus, T.T.Sheng [J. Electrochem. Soc. (USA) vol.130 no.9 (1983) p.1878]
[5] T.H.Di Stefano, M.Shatzkes [J. Vac. Sci. & Technol. (USA) vol.12 no.1 (1975) p.37-46]

[6] E.L.Jansses, G.L.Declerck [in 'Semiconductor Characterisation
    Techniques' Eds. P.A.Barnes, G.A.Rozgonyi (The Electrochemical
    Society, 1978) p.376]

[7] H.Futagami, K.Hoshi, N.Isawa, T.Suzuki, Y.Okubu, Y.Kato, Y.Okamoto
    [Semiconductor Silicon, 1986 Eds. H.R.Huff, T.Abe, B.Kolbesen, (The
    Electrochemical Society, 1986) p.939]

eutectic temperature
   Ag in Si 14.11
   As in Si 13.15
   Au in Si 14.12
   B in Si   13.9
   Co in Si 14.13
   Cr in Si 14.14
   Cu in Si 14.15
   Fe in Si 14.16
   Mn in Si 14.17
   Ni in Si 14.18
   P in Si   13.14
   Sb in Si 13.16
   Ti in Si 14.20
extended defects 11.1
extinction coefficient 2.1, 2.4, 2.5, 2.6
   amorphous Si3N4 29.1, 29.2
   amorphous SiO 28.1, 28.2
   liquid Si 2.7
   non-crystalline SiO2 28.3, 28.4
extrinsic gettering 25.1

-F-

Fermi level 3.2B, 17.1D
   heavily doped n-type Si 9.1
Fick's law 13.1A, 13.6A, 13.7A, 13.8A
Fowler-Nordheim tunnelling 17.4E, 17.5A, 17.11, 17.15B
fractional interstitialcy component of diffusion 13.1B, 13.6B,
   13.7B, 13.8B
fracture stress
   ion-implanted Si 1.11B
fracture toughness 1.11
   ion-implanted Si 1.11B
   poly-Si 1.11A
Frank-van der Merwe growth 19.1C, 19.3C, 19.7C
Frohlich's theory 17.11

-G-

gauge factor 3.4A
   poly-Si 3.4A, 3.4D
generation-recombination centres 31.2B, 31.3
gettering 25.1, 25.2
   extrinsic 25.1
   high temperature oxidation 25.1B
   intrinsic 25.2
   nitridation 25.1B
   stacking fault annihilation 25.1B
grain boundary segregation coefficients for dopants in poly-Si 26.4D
   As 26.4D
   P 26.4D
Gruneisen relation 1.12

-I-

IR absorption
    as-grown   12.10B, 12.11B
    buried nitride layers   18.11
    buried oxide layers   18.10
    calibration constants   12.7, 12.8, 12.12A
    carbon impurities   12.11
    heat-treated Si   12.10D, 12.11D, 12.12A
    hydrogen-acceptor pairs   17.16E
    irradiation effects   12.10C, 12.11C
    nitrogen impurities   12.12
    oxygen impurities   12.10, 12.12C
    superlattices   27.5
    temperature dependence   12.10B
    ultra short pulses   24.1B2

                              -K-

Knoop hardness   1.9H
    ion implantation effects   1.10B
Knoop indentation   1.9D
    ion implantation   1.10B
Kronig-Penny potential   17.1C

                              -L-

lamp annealing: see incoherent light annealing
laser annealing   2.7, 24.1, 24.2
laser-assisted etching   23.5
    anisotropy   23.5
    photoablation   23.5
    photon assisted chemical etching   23.5
    photon enhanced plasma etching   23.5
laser based oxidation   16.2
    oxidation rates   16.2
laser melting   24.1B, 24.2B
lattice parameter   1.2, 1.12
    GdSi2   19.11C
    high pressure phases   1.3
    impurity effects   1.2
lifetime degradation factor   8.5
liquid Si
    density   1.19
    electrical conductivity   2.7
    extinction coefficient   2.7
    molar volume   1.19
    optical functions   2.7
    refractive index   2.7
liquid solubility
    C   12.5
    N   12.6

                              -M-

magnetic Czochralski growth   25.2
magnetoresistance coefficients
    SOS   18.7H
MBE silicon
    deep levels   10.3

mechanical strength
    oxygen precipitation effects 11.2F
melting point 1.18, 1.21, 2.7, 12.7, 14.13-14.18
    impurity effects 1.18A
    pressure dependence 1.19
melting properties 24.1
metal base transistors (MBT) 30.1
molar volume 1.19
    liquid 1.19
    solid 1.19
MOSC structures 17.16, 17.17, 25.2
    latch up 25.2
MOS devices 31.3
    defect related failure modes, 31.3
    dielectric breakdown 17.11
    interface structure 17.6
    interface traps 17.3, 17.4
    oxide traps 17.5
    refresh time 31.3
    stacking faults 31.3
    storage time for electrical charge 31.3
    thermal dehydrogenation 17.16F
MOSFET 5.2, 5.3, 6.2, 17.1C, 17.3D, 17.5A, 17.20, 17.21,
    18.2A, 18.3, 18.4, 25.3
multilayers 27.1-27.5, 30.1C

-N-

neutron irradiation 11.6
NiSi2 30.4, 30.5
    activation energy of formation 30.4
    crystal structure 30.4
    formation temperature 30.4
    resistivity 30.5
nitridation
    SiO2 films 17.25
nitridation, internal 18.11
    dose dependence 18.11
    ion implantation 18.11
nitrogen in Si 12.3, 12.6, 12.9, 12.12
non-bridging oxygen hole centre (NBOHC) 17.2, 17.5A
non-equilibrium crystal growth 24.2E, 24.2H

-O-

Onsager's theory 17.8
optical absorption coefficient 2.1, 2.3, 2.4, 2.5, 2.6
    amorphous Si 26.3D
    amorphous Si3N4 29.1, 29.2
    amorphous SiO 28.1, 28.2
    non-crystalline SiO2 28.3, 28.4
    poly-Si 26.3D
    temperature dependence 2.4
optical functions 2.1, 2.2, 2.3, 2.4, 2.5, 2.6, 2.7, 28.2, 28.4, 29.2
out-diffusion 12.2, 25.2
oxidation enhanced diffusion 11.3C, 13.1, 13.2, 13.3, 13.4, 13.6, 13.7
oxidation, internal 18.10
    dose dependence 18.10
    ion implantation 18.10

VLSI
    acceptor impurities 13.4
    Al alloy metallization 20.2
    breakdown behaviour of SiO2 films 17.22
    heterostructures 30.1A
    high pressure technology 16.3
    hydrogenation rates 17.17
    interface trap density 17.2, 17.3, 17.4
    oxide trap density 17.5
    poly-Si 26.4
    silicon oxynitrides 17.25, 17.26
    SOS 18.5
    TiN stability 20.11, 20.13
    W stability 20.14

-W-

wave functions 17.1B
work function 19.2B, 19.10B, 19.10C, 19.10D, 19.12B,
    19.18D, 19.21B

-Y-

Young's modulus 3.4, 17.24F
    SiO2 17.24F

NOTES

NOTES

NOTES

NOTES

NOTES
